The Biology of Nitric Oxide Part 5

PORTLAND PRESS PROCEEDINGS

The Biology of Nitric Oxide Part 5

Proceedings of the 4th International Meeting on the Biology of Nitric Oxide, Amelia Island, Florida, U.S.A.

Editors

S. Moncada
J. Stamler
S. Gross
E.A. Higgs

PORTLAND PRESS

Published by Portland Press, 59 Portland Place,
London W1N 3AJ, U.K.

ISBN 1 85578 102 6 ISSN 0966-4068

British Library Cataloguing in Publication Data
A catalogue record for this book is available from the British Library.

Printed in Great Britain by Whitstable Litho Printers Ltd

Contents

Papers from Poster Communications

Biochemistry

Physiology

Preface

The Fourth International Meeting on the Biology of Nitric Oxide was held at Amelia Island, Florida, U.S.A. in September 1995. Approximately 400 posters were presented at this meeting, along with about 60 oral communications and a number of hot topics. This book provides an up-to-date overview on the current status of the field, with contributions from over 1350 specialists on a variety of areas, ranging from the role played by nitric oxide in cancer and tissue rejection through to methods of detecting nitric oxide, including its visualization.

This is the sixth book on the biology of nitric oxide to be published by Portland Press since 1992. During this period, interest in the area has escalated at a rapid rate and the importance of nitric oxide in many basic physiological processes as well as in some forms of pathology is now widely accepted. Indeed, our current knowledge of the biological chemistry of nitric oxide has transformed our understanding of cellular signalling and toxicity.

Details of the *Biology of Nitric Oxide Parts 1–4* (including Physiological and Clinical Aspects, and Enzymology, Biochemistry and Immunology) as well as *Nitric Oxide: Brain and Immune System*, all of which are published by Portland Press, are listed on the back cover.

Structure-function studies of cloned, rat neuronal nitric oxide synthase in its intact and modular forms

BETTIE SUE MASTERS, LINDA J. ROMAN, PAVEL MARTASEK, JONATHAN NISHIMURA, [*]KIRK McMILLAN, [#]ESSAM A. SHETA, [+]STEVEN S. GROSS, TIMOTHY J. MCCABE, [Y]WEN-JINN CHANG and [Y]JAMES T. STULL

The Univ. of Texas Health Sci. Ctr. at San Antonio, TX, USA 78284-7760, Pharmacopeia, Inc., Princeton, NJ 08540, USA[*], Alexandria University, Alexandria, Egypt[#], Cornell Univ. Medical Col., New York, NY 10021, USA[+], and The Univ. of Texas Southwestern Medical Ctr. at Dallas, TX, 75235-9040, USA[Y]

Studies of the structural and spectroscopic characteristics of rat neuronal nitric oxide synthase (nNOS[1], NOS-I[1] , constitutive) in its intact and modular forms have been facilitated by the availability of milligram quantities of protein. In attempts to increase the availability of this enzyme for biophysical studies, we have cloned and expressed a number of cDNA constructs of nNOS in *Escherichia coli*. The following will describe the results obtained with such constructs and the appropriate references for more details. The determination of the prosthetic group constituency of this enzyme, FAD[1] and FMN[1] in the C-terminal domain [1,2], iron protoporphyrin IX (heme) in the N-terminal domain [3-6], and BH_4[1] [7,8], each of which is required for maximal enzymatic activity, has been a *tour de force* by a number of laboratories. Several mechanistic schemes have been proposed [9-11], none of which has been proven. In an attempt to understand the structural features which uniquely suit this enzyme to perform the monooxygenation of L-arginine to produce L-citrulline and NO• and to distinguish the isoforms of nitric oxide synthase from one another, we have expressed and purified the wild type and a number of site-directed and deletion mutations of nNOS.

Table 1 shows the constructs we are currently examining in order to determine the sites of binding of the various prosthetic groups and the biophysical and spectroscopic properties of these expressed proteins.

TABLE 1. nNOS Constructs Expressed in *E. coli*

Intact neuronal nitric oxide synthase
$nNOS_{1\text{-}1429}$
Fusion with glutathione-S-transferase
$nNOS_{1\text{-}220}$ (GLGF)
$nNOS_{220\text{-}720}$ (Module III[12])
$nNOS_{220\text{-}558}$ (Module II[12])
$nNOS_{558\text{-}720}$ (Module I, DHFR[12])

The successful expression of intact neuronal NOS in a prokaryotic system provides a straight-forward, low-cost procedure for producing large quantities of this protein. Gerber and Ortiz de Montellano [13] and Roman, *et al.* [14] have succeeded in producing >25 mg nNOS protein/liter in different strains of *E. coli*. The long-term stability of the purified protein upon storage will permit its use in biophysical studies such as electron paramagnetic resonance, circular dichroism, magnetic circular dichroism, and x-ray absorption fine structure analyses. However, intact isoforms of NOS range in molecular mass from 130-160 kDa, sizes which do not lend themselves easily to three-dimensional structure analysis by either nuclear magnetic resonance spectroscopy or x-ray crystallography. Therefore, our laboratory has taken the approach of expressing modules of nNOS in *E. coli* expression systems with and without the inclusion of a fusion protein, glutathione-S-transferase [12]. This has facilitated the purification of these modules whether or not the prosthetic group(s) is(are) bound. Nishimura *et al.* [12] and Martasek *et al.* (this volume; unpublished observations) have now demonstrated that the residues between 220 and 558 contain the necessary binding site for BH_4. The surprising finding that module I contained substrate binding determinants has led to the production of additional constructs to determine the limiting peptide for BH_4/substrate interactions, since this module was found to bind NNA[1] in the presence or absence of BH_4 [12]. Binding of NNA by the oxygenase domain (residues 1-714) isolated by McMillan and Masters [15] was stimulated by BH_4 (unpublished observations, S. Gross), similar to that observed with the intact nNOS, expressed in *E. coli* [14].

Finally, the examination of residues 1-220, which are unique to the neuronal NOS isoform, is in progress. This module contains a tetrapeptide sequence (GLGF) often occurring, in up to three repeats, in proteins that associate with cytoskeletal or microtubular structures [16]. Recently, in examining skeletal muscle tissue from Duchenne Muscular Dystrophy patients (DMD) and the DMD mouse model, *mdx*, we (W-J Chang, P. Martasek, B. S. Masters, and J. Stull, unpublished observations) have shown that nNOS is diminished to negligible levels, using activity measurements and immunoblotting techniques. Association of nNOS with the dystrophin-glycoprotein complex is weakened as determined by binding and elution from a wheat germ-agglutinin column. The specific relationship between residues 1-220 of nNOS containing the GLGF motif and binding to skeletal muscle proteins has yet to be demonstrated. Recently, Bredt's laboratory (17 and this volume) also reported that the dystrophin complex interacts with an N-terminal GLGF-containing domain of nNOS.

[1] Abbreviations used: nNOS and NOS-I, neuronal nitric oxide synthase; FAD, flavin adenine dinucleotide; FMN, flavin mononucleotide; BH_4, tetrahydrobiopterin; NNA, N^{ω}-nitro-L-arginine; DHFR, dihydrofolate reductase.

REFERENCES

1. Bredt, D.S., Hwang, P.M., Glatt, C.E., Lowenstein, C. Reed, R.R. and Snyder, S.H. (1991) Nature **351**, 714-718
2. Mayer, B., John, M., Heinzel, B., Werner, E.R., Wachter, H., Schultz, G. and Böhme, E. (1991) FEBS Lett. **288**, 187-191
3. White, K.A. and Marletta, M.A. (1992) Biochemistry **31**, 6627-6631
4. Stuehr, D.J. and Ikeda-Saito, M. (1992) J. Biol. Chem. **267**, 20547-20550
5. McMillan, K., Bredt, D.S., Hirsch, D.J., Snyder, S.H., Clark, J.E. and Masters, B.S.S. (1992) Proc. Natl. Acad. Sci. USA **89**, 11141-11145
6. Klatt, P., Schmidt, K. and Mayer, B. (1992) Biochem J. **288**, 15-17
7. Tayeh, M.A. and Marletta, M.A. (1989) J. Biol. Chem. **264**, 19654-19658
8. Kwon, N.S., Nathan, C.F., and Stuehr, D.J. (1989) J. Biol. Chem. **264**, 20496-20501
9. Marletta, M.A. (1993) J. Biol. Chem. **268**, 12231-12234
10. Griffith, O.W. and Stuehr, D.J. (1995) Annu. Rev. Physiol. **57**, 707-736
11. Masters, B.S.S. (1994) Annu. Rev. Nutrition **14**, 131-145
12. Nishimura, J.S., Martasek, P., McMillan, K., Salerno, J.C., Liu, Q., Gross, S.S. and Masters, B.S.S. (1995) Biochem. Biophys. Res. Commun. **210**, 288-294
13. Gerber, N.C. and Ortiz de Montellano, P.R. (1995) J. Biol. Chem. **270**, 17791-17796
14. Roman, L.J., Sheta, E.A., Martasek, P., Gross, S.S., Liu, Q. and Masters, B.S.S. (1995) Proc. Natl. Acad. Sci. USA **92**, 8428-8432
15. McMillan, K. and Masters, B.S.S. (1995) Biochemistry **34**, 3686-3693
16. Ponting, C.P. and Phillips, C. (1995) Trends Biochem. Sci. **20**, 102-103.
17. Brenman, J.E., Chao, D.S., Xia, H., Aldape, K. and Bredt, D.S. (1995) Cell **82**, 743-752

Structural analysis of neuronal nitric oxide synthase

PETER KLATT, KURT SCHMIDT, DIETER LEHNER*, OTTO GLATTER*, HANS P. BÄCHINGER# and BERND MAYER

Institut für Pharmakologie und Toxikologie, Karl-Franzens-Universität Graz, Universitätsplatz 2, A-8010 Graz, Austria, *Institut für Physikalische Chemie, Karl-Franzens-Universität Graz, Heinrichstraße 28, A-8010 Graz, Austria and #Research Department of the Shriners Hospital for Crippled Children, 3101 S.W. Sam Jackson Park Road, Portland, Oregon 97201, USA.

Neuronal nitric oxide synthase (NOS) catalyzes the conversion of L-arginine to L-citrulline and NO [1]. The enzyme appears to function as a self-sufficient cytochrome P450, combining an FAD- and FMN-containing reductase and a P450-like monooxygenase domain within a single polypeptide. However, NOS represents a unique cytochrome P450 as it contains variable amounts of tetrahydrobiopterin (H_4biopterin) as tightly bound prosthetic group. Exogenous H_4biopterin was shown to stimulate enzyme activity several fold, to trigger the coupling of oxygen activation to L-arginine oxidation and to allosterically interact with the substrate domain of NOS [2]. To investigate the allosteric role of the pteridine, we have analyzed its effect on the conformation of partially H_4biopterin-deficient NOS purified from porcine brain (≤ 0.3 mol H_4biopterin per mol NOS) by means of circular dichroism (CD), velocity sedimentation, dynamic light scattering and low-temperature SDS-polyacrylamide gel electrophoresis (LT-PAGE).

CD spectra of purified NOS were characterized by minima at 208 and 220 nm, a maximum at 192 nm and a baseline cross-over at 200 nm (Figure I). The secondary structure of NOS calculated by means of the singular value decomposition and variable-selection methods [3] yields 30 % α-helix, 19 % turns, 14 % antiparallel β-sheet, 7 % parallel β-sheet and 31 % other structures. Virtually identical spectra were obtained in the presence of exogenous H_4biopterin (0.1 mM) and/or L-arginine (1 mM), demonstrating that binding of these ligands did not induce any detectable repacking of the secondary structure. Thermal unfolding of NOS monitored as temperature-dependent decrease of the CD signal at 220 nm revealed a very sharp and monophasic transition at 58 °C indicative of a high degree of cooperativity in the unfolding process (Figure I, inset, solid line). Exogenous H_4biopterin (0.1 mM; Figure I, inset, broken line) and/or L-arginine (1 mM, not shown) had no effect on the thermal stability of NOS structure.

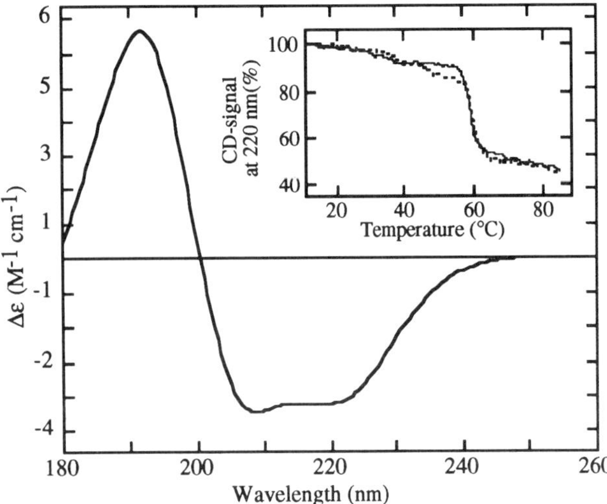

Figure I. CD-spectrum and thermal unfolding of neuronal NOS

We have analyzed the quaternary structure of NOS by means of velocity sedimentation in 5 - 20 % sucrose gradients and dynamic light scattering. The sedimentation coefficient, diffusion coefficient and hydrodynamic radius of native NOS were 7.9 ± 0.2 S, 2.0 ± 0.3 x 10^{-11} m^2 s^{-1} and 7.3 ± 0.9 nm (n = 4), respectively. From the independently determined sedimentation and diffusion coefficients, the molecular mass of NOS was calculated as 332 ± 51 kDa. The obtained frictional coefficient (1.8 x 10^{-7} g s^{-1}) and frictional ratio of NOS (1.91) were consistent with an elongated shape of the protein (axial ratio ~20/1). Virtually identical results were obtained when these experiments were performed in the presence of exogenous H_4biopterin (0.1 mM) and/or L-arginine (1 mM), demonstrating that native NOS represents an elongated homodimer irrespective of pteridine or substrate binding.

LT-PAGE [4] analysis of NOS revealed two clearly separated protein species (Figure 2, upper panel, Lane A). The calculated molecular masses of 153 ± 9 kDa and 296 ± 14 kDa (n = 5) were consistent with the occurence of NOS monomers and dimers, respectively. Incubation of NOS with increasing concentrations of H_4biopterin (0.1 - 100 µM) prior to LT-PAGE increased the fraction of SDS-resistant NOS dimers from ~20 % up to >90 % (Lane B - E). The half maximally active concentration of the pteridine (~1 µM) was in good accordance with the affinity of the cofactor determined with other methods [5]. Figure 2 (lower panel) shows that saturating concentrations of L-arginine (0.1 mM) caused partial formation (~60 %) of SDS-resistant dimers (Lane A) and decreased the concentration of exogenous H_4biopterin required for SDS-resistant dimerization (Lanes B - E). The potency of other pteridines to induce the formation of SDS-resistant NOS dimers was found to correlate with their affinities determined in radioligand binding studies [5]. More than 90 % of NOS were converted to SDS-resistant dimers by H_4biopterin(K_D ~0.2 µM), whereas 7,8-dihydro-L-biopterin (K_D ~2 µM) induced only ~60 % SDS-stable dimerization, and other pteridines with considerably lower affinities (K_D >20 µM) were ineffective (not shown). We also tested the other NOS cofactors for ligand-induced dimerization but found that neither Ca^{2+}/calmodulin nor pure NADPH stabilized NOS dimers.

In conclusion, our data demonstrate that H_4biopterin and L-arginine - allthough not affecting the secondary and quaternary structure of NOS - synergistically convert NOS into an exceptionally stable, SDS-resistant dimer, suggesting a novel role of H_4biopterin in the allosteric regulation of protein subunit interactions.

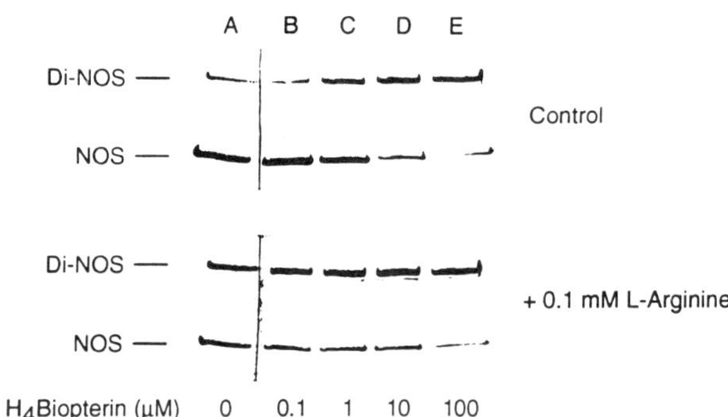

Figure 2. LT-PAGE of neuronal NOS
NOS (4 µg) was incubated at 37 °C with increasing concentrations of H_4biopterin in 40 µl of 50 mM triethanolamine/HCl (pH 7.0) in the absence (upper panel) or presence (lower panel) of L-arginine (0.1 mM). After 10 min, 40 µl of Laemmli-buffer were added, and aliquots (2 µg protein) were analyzed by LT-PAGE. The 6 % SDS gels were stained for protein with Coomassie blue. NOS and Di-NOS refer to NOS monomers and dimers, respectively.

Abbreviations used: NOS, nitric oxide synthase; H_4biopterin, (6R)-5,6,7,8-tetrahydro-L-biopterin; CD, circular dichroism; LT-PAGE, low-temperature SDS-polyacrylamide gel electrophoresis

Supported by the Fonds zur Förderung der Wissenschaftlichen Forschung in Österreich (P 10098, P 10573 and P 10682) and a grant from the Shriners Hospital (H.P.B.).

1. Griffith, O.W. & Stuehr, D.J. (1995) Annu. Rev. Physiol. **57,** 707-736
2. Mayer, B. & Werner, E.R. (1995) Naunyn-Schmiedeberg´s Arch. Pharmacol. **351,** 453-463
3. Compton, L.A., Mathews, C.K. & Johnson, W.C. (1987) J. Biol. Chem. **262,** 13039-13043
4. Takasuka, T., Sakurai, T., Goto, K., Furuichi, Y. & Watanabe, T. (1994) J. Biol. Chem. **269,** 7509-7513
5. Klatt, P., Schmid, M., Leopold, E., Schmidt, K., Werner, E.R. & Mayer, B. (1994) J. Biol. Chem. **269,** 13861-13866

Nitric oxide synthase: characterization of substrate and inhibitor binding interactions by electron paramagnetic resonance spectroscopy and steady-state kinetics

JOHN C. SALERNO[*], CHRISTOPHER A. FREY[†], KIRK MCMILLAN[‡], BETTIE SUE SILER MASTERS[‡] and OWEN W. GRIFFITH[†]

[*]Department of Biology and Center for Biochemistry and Biophysics, Rensselaer Polytech Institute, Troy, NY 12180 USA, [†]Department of Biochemistry, Medical College of Wisconsin, Milwaukee, WI 53226 USA, [‡]Department of Biochemistry, The University of Texas Health Science Center at San Antonio, San Antonio, TX 78284 USA

Nitric oxide synthase (NOS) catalyzes sequential NADPH- and O_2-dependent monooxygenase reactions converting L-arginine to N^{ω}-hydroxy-L-arginine and N^{ω}-hydroxy-L-arginine to citrulline and nitric oxide. The homodimeric enzyme contains one heme/monomer which mediates both partial reactions (1). In previous studies using optical difference spectroscopy, we showed that neuronal NOS (nNOS) as isolated contains mainly high spin, penta-coordinate heme and 15-20% low spin, hexa-coordinate heme; binding of the substrate L-arginine fully converts the low spin fraction to high spin (2). In contrast, binding of L-thiocitrulline [δ-thioureido-L-norvaline (3)] modestly increases the fraction of low spin heme (4). In the present studies we have examined these transitions more directly using electron paramagnetic (EPR) spectroscopy. We have also examined the steric constraints on substrate and inhibitor binding.

Materials and Methods: Rat nNOS was purified from stably transfected kidney 293 cells as described (2,5); nNOS activity was monitored by following the conversion of L-[^{14}C]arginine to [^{14}C]citrulline (2,4). For EPR studies nNOS samples were flash-frozen in quartz EPR tubes. Enzyme concentrations are expressed on the basis of heme concentration. Spectra were obtained at 9.445 GHz using a Bruker 300 EPR spectrometer and a liquid helium cryostat (microwave power, 20 mW; modulation frequency, 100 KHz; modulation amplitude, 14.3 G.) (6).

Results and Discussion: The X-band EPR spectrum of 20 μM nNOS as isolated at 11° K shows high spin features at g = 7.65, 4.04 and 1.89 and low spin features at g = 2.43, 2.28 and 1.89 (Table 1); a flavin radical, probably FMNH·, gives a sharp signal at g = 2. Addition of 100 μM L-arginine to NOS eliminated the low spin signals and increased the high spin signals ~ 15%. Most interestingly, the high spin signal is shifted (g = 7.56, 4.09, 1.8) showing that L-arginine distorts the heme binding pocket from its resting state even though it does not ligate as a sixth (axial) ligand to heme iron (i.e., the iron is all high spin).

Table 1. High and Low Spin Ferriheme Species of nNOS

Species Detected	Spin	g_x	g_y	g_z
nNOS alone (HS); majority	5/2	7.65	4.04	1.8
nNOS alone (LS)	1/2	1.89	2.28	2.43
nNOS-L-ARG (HS); majority	5/2	7.56	4.09	1.8
nNOS-L-TC (HSa); major	5/2	(7.68)	3.98	1.8
nNOS-L-TC (HSb); major	5/2	(7.76)	3.83	1.8
nNOS-L-TC (LSa)	1/2	1.87	2.27	2.47
nNOS-L-TC (LSb)	1/2	1.9	2.27	2.39

The spin state and principal g values of the most important species detected in these studies are shown. Abbreviations used: HS, high spin, LS, low spin; a and b are used to distinguish multiple HS or LS species; L-TC, L-thiocitrulline. Majority and major species represent > 50% and > 20% of the enzyme, respectively. The minority nNOS-L-TC HS species is not listed.

Addition of L-thiocitrulline also increases the fraction of high spin heme but, in this case, two new and distinct low spin signals remain (Table 1); g values for one species are consistent with the sulfur of thiocitrulline acting as a sixth (axial) heme ligand as was previously inferred from optical data (4). In the other low spin species, the sixth ligand is a nitrogen or an oxygen atom of unknown origin.

L-Thiocitrulline binding also induced two majority and one minority high spin states and all are distinct from the high spin states of either nNOS as isolated or the nNOS-L-arginine complex. Because the low field signals are at g = 7.68, we conclude that L-thiocitrulline distorts the heme binding pocket oppositely from the distortion caused by L-arginine. The difference between g_x and g_y (7.65 and 4.04 in the enzyme as isolated) decreases upon L-arginine addition and increases upon L-thiocitrulline addition. This reflects the relative energies of the d_{xz} and d_{yz} iron orbitals; a more complete description in terms of spin hamiltonian and zero field splitting parameters is presented elsewhere (6). The single binding mode detected for L-arginine stands in contrast to the multiple modes seen for L-thiocitrulline and suggests that the binding site has evolved to hold substrates uniquely.

The changes in the EPR spectrum of the high spin ferriheme caused by the binding of L-arginine and arginine analogs are indicative of specific interactions within the heme pocket. In all of the high spin complexes there is no sixth iron ligand adjacent to the arginine binding site; the effects we observe are therefore probably mediated by interactions between the ligand and the nNOS polypeptides forming both the arginine binding site and the adjacent heme pocket. Steric effects, van der Waals interactions, coulomb interactions and hydrogen bond formation, all leading to small local conformational readjustments, are potential causes of ligand-specific changes in the coordination of the heme iron. These effects would be transmitted to iron primarily through the enzyme's thiolate axial ligand, and result in well-defined spectroscopic species which reflect the stabilization of distinct states, each having characteristic interactions between ligand and binding site. The specific changes in the EPR spectra thus provide a probe of the arginine site, with the closely associated heme acting as an endogenous reporter group.

To characterize the interactions of nNOS with its substrates and inhibitors further, we have also carried out steady state kinetic studies with a variety of N^{ω}-alkyl-L-arginines and related derivatives. These studies establish that the binding region surrounding the L-arginine guanidino nitrogen nearest to the heme cofactor and therefore subject to oxidation can accommodate simple alkyl substituents of 3-4 carbons; in contrast, the binding region surrounding the non-reactive guanidino N is much less tolerant of substituents.

Conclusion: One guanidino nitrogen of L-arginine and the sulfur of L-thiocitrulline are bound in close proximity to the heme iron of nNOS. Binding of these species affects the d orbitals of even high spin penta-coordinate iron indicating that substrate or inhibitor binding perturbs the heme binding pocket, and affects heme iron through that perturbation.

Work was supported in part by National Institutes of Health grants DK48423 (to OWG) and HL30050 (to BSSM) and by the Robert A. Welch Foundation grant AQ-1192 (to BSSM).

1. Griffith, O.W. and Stuehr, D.J. (1995) *Ann. Rev. Physiol.* 57, 707-736.
2. McMillan, K. and Masters, B.S.S. (1993) *Biochemistry* 32, 9875-9880.
3. Narayanan, K. and Griffith, O.W. (1994) *J. Med. Chem.* 37, 885-887.
4. Frey, C., Narayanan, K., McMillan, K., Spack, L., Gross, S.S., Masters, B.S. and Griffith, O.W. (1994) *J. Biol. Chem.* 269, 26083-91.
5. McMillan, K., Bredt, D.S., Hirsch, D.J., Snyder, S.H., Clark, J.E. and Masters, B.S.S. (1992) *Proc. natl. Acad. Sci. USA* 89, 11141-11145.
6. Salerno, J.C., Frey, C., McMillan, K., Williams, R.F., Masters, B.S.S. and Griffith, O.W. (1995) *J. Biol. Chem.* In Press.

A novel mechanism of action of aspirin-like drugs: effect on inducible nitric oxide synthase

ASHOK R. AMIN[*†‡¶], PRANAV VYAS[*], MUKUNDAN ATTUR[*], JOANNA LESZCZYNSKA-PIZIAK[*], INDRAVADAN R. PATEL[§], GERALD WEISSMANN[‡], and STEVEN B. ABRAMSON[*‡]

[*]Department of Rheumatology, Hospital for Joint Diseases, New York, NY 10003; Departments of [†]Pathology & [‡]Medicine, New York University Medical Center, New York, NY 10016; and [§]Department of Biochemistry, Glaxo, Inc., Research Triangle Park, NC 27709

Summary. Nitric oxide (NO) synthesized by inducible nitric oxide synthase (iNOS) has been implicated as a mediator of inflammation in rheumatic and autoimmune diseases. We report that, in murine macrophages, therapeutic concentrations of aspirin (IC_{50} = 3 mM) and hydrocortisone (IC_{50} = 5 μM) inhibited the expression of iNOS as assess by nitrite accumulation, Western blotting and enzyme activity. There is no effect of NSAIDs on iNOS mRNA. In contrast, sodium salicylate (1-3 mM), indomethacin (5-20 μM) and acetaminophen (60-120 μM) had no significant effect at pharmacological concentrations. In addition, aspirin inhibits the catalytic activity of iNOS, an effect mimicked by N-acetylimidazole (NAI), another acetylating agent. We therefore conclude that the aspirin-like drugs differ in their mode of action, and that acetylation may be a critical difference.

Introduction. Inducible NOS (iNOS) plays an important role in inflammation, host-defense responses, and tissue repair ([1-3]. NO formation is increased during inflammation (rheumatoid arthritis, ulcerative colitis, Crohn's disease), and several classic inflammatory symptoms (erythema, vascular leakiness) are reversed by NOS inhibitors [1-3]. While NSAIDs clearly inhibit the synthesis and release of prostaglandins [4,5], these actions are by no means sufficient to explain all the anti-inflammatory effects of NSAIDs. NSAIDs also inhibit activation of neutrophils [6], which provoke inflammation by releasing products other than prostaglandins [7]. In these studies we examined the effect of NSAIDs on NO production. Among the agents studied in an effort to elucidate the effect of NSAIDs on iNOS expression and function, we have selected three: an acetylated salicylate (aspirin, an effective inhibitor of COX); a non-acetylated salicylate (sodium salicylate, an ineffective inhibitor of COX); and a non-acetylated nonsteroidal compound (indomethacin, a potent inhibitor of COX).

Materials and methods. Western blots, enzyme assay and Northern blots of iNOS and COX-2 were carried out as described [8].

Results and discussion. Murine macrophage cells (RAW 264.7) stimulated with LPS in the presence of NSAIDs (at pharmacological concentrations) showed a concentration-dependent (1-3 mM aspirin) inhibition of nitrite accumulation. Suprapharmacological concentrations of aspirin (5 and 10 mM) further inhibited nitrite accumulation [by 50% $\pm$ 6 and 80% $\pm$ 5 (p <0.005), respectively]. Sodium salicylate (3 mM) and indomethacin (5 μM) at pharmacological concentrations had no significant effect. Suprapharmacolog-ical concentrations of sodium salicylate (IC_{50} 20 mM) inhibited nitrite accumulation. Acetaminophen (60-120 μM) failed to block nitrite production, whereas hydrocortisone (5 μM) inhibited endotoxin-induced NO production by >60%.

Aspirin (3 mM) and indomethacin (20 μM) inhibited the specific activity of COX-2 by 79 $\pm$ 6.7% (p <0.001) and 84 $\pm$ 4.0% (p <0.002), respectively, while sodium salicylate (3 mM) had no effect [16 $\pm$ 11% (p <0.28)]. These data indicate that aspirin does not inhibit nitrite production by inhibiting COX, since aspirin shares this latter effect with indomethacin. The specific enzyme activity of iNOS from cells exposed to aspirin in cell-free extracts showed a significant inhibition in activity in a dose-dependent fashion (IC_{50} = 3 mM). Western blot analysis showed a significant effect of 3 mM of aspirin on iNOS expression. Aspirin at 10 mM concentration further decreased the expression of iNOS by ~70%, as determined by Western blot analysis, whereas sodium salicylate (2 mM) caused ~15% inhibition of iNOS expression. However, increased concentration of sodium salicylate (5 and 20 mM) did not cause increased inhibition of iNOS expression, unlike the increasing effects seen with 10 mM aspirin (~70%).

There was no significant difference in the expression of iNOS mRNA (at 16 h) in cells treated with LPS in the presence or absence of NSAIDs at pharmacological concentrations. Kopp and Ghosh [9] showed that aspirin (3 mM) or sodium salicylate (5 mM) inhibit NF-κB-dependent transcription, using sensitive assays based on plasmids containing two IgK-κB sites driving a luciferase reporter gene. It should be noted that, in the same studies, the same concentrations of aspirin and sodium salicylate had no significant effect on NF-κB activation, judged by gel shift assays. Our studies indicate that 3 mM aspirin is probably not sufficient to block the transcription of the iNOS gene, as observed with 30 μM of pyrrolidine dithiocarbamate, which blocked >90% of nitrite accumulation in our studies.

The potency of aspirin in inhibiting iNOS activity compared to the other NSAIDs may be attributable to the acetylation by aspirin of proteins such as COX-1 and COX-2 [10,11]. We therefore examined the effects of aspirin, sodium salicylate and indomethacin in *in vitro* iNOS enzyme assays. Aspirin at 0.1 and 1 mM concentration inhibited the conversion of L-[^{3}H]-arginine to L-[^{3}H]-citrulline in cell-free extracts by 10-12% and 45-68%, respectively, whereas no significant differences (7%) were observed in extracts treated with 1 mM of sodium salicylate. Similarly, 5 μM indomethacin or equivalent volume of absolute alcohol had no effect. These studies demonstrated that aspirin, but not sodium salicylate or indomethacin, directly interfered with the catalytic activity of iNOS, possibly by acetylating an important functional component of the enzyme or its co-factors.

Unlike aspirin, which acetylates Ser^{530} of COX and inactivates the cyclooxygenase and not the peroxidase activity [11], NAI acetylates and inhibits both the cyclooxygenase and the peroxidase activity of COX [12]. In contrast to aspirin, which does not seem to inhibit the iNOS activity significantly (10-12%) at 0.1 mM, NAI at similar concentrations inhibited ~45% of iNOS activity, whereas imidazole at similar concentrations had no significant effect (<5% inhibition). These observations may explain the differential potency of aspirin and sodium salicylate. Aspirin inhibits iNOS by effects on both synthesis of the iNOS protein and on the catalytic activity of the enzyme. In summary, we conclude that the inhibition of iNOS expression/function represents a novel mechanism of action of aspirin-like drugs and may explain individual differences in response to NSAIDs in patients with inflammatory diseases. In addition, a search for agents which can acetylate iNOS or its co-factors may be an important pharmacological strategy for developing newer aspirin-like drugs.

REFERENCES

1. Nathan C, Xie Q (1994) *Cell* **78,** 915-918.
2. Schmidt HHHW, Walter U (1994) *Cell* **78,** 919-925.
3. Marletta MA (1994) *Cell* **78,** 927-930.
4. Vane J (1994) *Nature* **367,** 215-216.
5. Furst DE (1994) *Arthr. Rheum.* **37,** 1-9.
6. Abramson SB, Leszczynska-Piziak J, Clancy RM, Philips M, Weissmann G (1994) *Biochem. Pharmacol.* **47,** 593-572.
7. Abramson S, Korchak H, Ludewig R, *et al.* (1985) *Proc. Natl. Acad. Sci. USA* **82,** 7227-7231.
8. Amin AR, Vyas P, Attur M, Leszczynska-Piziak J, Patel IR, Weissmann G, Abramson SB (1995) *Proc. Natl. Acad. Sci. USA* **92,** 1926-1930.
9. Kopp E, Ghosh S (1994) *Science* **265,** 956-959.
10. Shimokawa T, Smith WL (1992) *J. Biol. Chem.* **267,** 12387-12392.
11. Lecomte M, Laneuville O, Ji C, DeWitt DL, Smith WL (1994) *J. Biol. Chem.* **269,** 13207-13215.
12. Wells I, Marnett LJ (1992) *Biochemistry* **31,** 9520-9525.
13. Mitchell JA, Akarasereenont P, Thiemermann C, Flower RJ, Vane JR (1993) *Proc. Natl. Acad. Sci. USA* **90,** 11693-11697.
14. Salvemini D, Misko TP, Masferrer JL, Seibert K, Currie MG, Needleman P (1993) *Proc. Natl. Acad. Sci. USA* **90,** 7240-7244.

TABLE: Summary of the action of NSAIDs on expression of iNOS and COX-2.

Modulating agent	Percent inhibition of iNOS at 16 h					
	nitrite release	**specific activity in cell-free extracts**	**protein expression**	**mRNA**	**specific activity in *in vitro* assay**	**% inhibition of COX-2**
Aspirin (3 mM)	32.0*	47†	~53	NS	~45 (1 mM)	>75†
Sodium salicylate (3 mM)	7.0‡	2§	~15	NS	~1 (1 mM)	NS
Indomethacin (5 μM)	7.0‡	2§	0	NS	0 (5 μM)	>75†
Hydrocortisone (5 μM)	63.0¶	ND	ND	ND	ND	ND
N-acetylimidazole (1 mM)	ND	ND	ND	ND	~74 (1 mM)	ND

The data (expressed as percent inhibition) are compiled from the present study. RAW 264.7 cells were induced with 100 ng/ml of LPS to stimulate iNOS and COX-2 activity. After 16-18 h of incubation, COX-1/COX-2 activity was assayed as described by Mitchell *et al.* [13]. COX-1 activity was not detected in these cells, as previously described [14]. "*In vitro* assay" indicates the effect of NSAIDs (at concentrations noted in parentheses) on the catalytic activity of iNOS in cell-free extracts. The protein expression data are represented as approximate percent inhibition based on the densitometry data from one of the two representative experiments. ND = not done. NS = not significant. Note p values: §p <0.45, ‡p <0.36, *p <0.05, ¶p <0.01, †p <0.006.

Nonsteroidal anti-inflammatory drugs prevent transcriptional expression of the inducible NO synthase gene in macrophages by interference with NF-κB activation

L.J. IGNARRO, E.E. AEBERHARD and S.A. HENDERSON

UCLA School of Medicine, Center for the Health Sciences, Los Angeles, California, USA 90095

SUMMARY: Inflammation is closely associated with induction of inflammatory NO synthase (iNOS) and high-output production of NO. NO causes symptoms of acute and chronic inflammation and inhibitors of iNOS possess anti-inflammatory activity. The objective of this study was to determine whether NSAIDs inhibit high-output NO production by activated rat macrophages. Six NSAIDs (salicylate, aspirin, indomethacin, ibuprofen, diclofenac, naproxen) each added at the time of cell activation by LPS + IFN-γ inhibited or abolished NO_2^- + NO_3^- (NO_X^-) accumulation in cell-free media 24 hr later. NSAIDs did not directly inhibit iNOS enzymatic activity. No inhibitory effects were observed when NSAIDs were added 6 hr after cell activation, thus indicating that NSAIDs affected early steps in iNOS expression. NSAIDs inhibited both the appearance of iNOS enzyme recoverable from intact cells and iNOS mRNA (Northern blots). The mechanism of action of NSAIDs in preventing iNOS mRNA transcription was attributed to interference with NF-kB activation, as determined by electrophoretic mobility shift assays. In vivo experiments revealed that administration (ip) of NSAIDs to rats prevented LPS (2 mg/kg, ip)-induced NO_X^- accumulation in plasma. Thus, NSAIDs inhibit high-output NO production in vitro and in vivo by interfering with NF-kB-mediated transcription of the iNOS gene.

INTRODUCTION: High doses of aspirin and other NSAIDs are effective in the treatment of inflammatory disease. One important mechanism of action of NSAIDs is inhibition of cyclooxygenase activity and, therefore, prostaglandin production. In addition to prostaglandins, NO appears to be involved in promoting acute and chronic inflammation (1,2). Moreover, NO may activate cyclooxygenase and, thereby, enhance the formation of prostaglandins (3). In view of these findings, we tested the hypothesis that the anti-inflammatory action of NSAIDs is attributed at least in part to inhibition of NO production. Rat alveolar macrophages were activated with lipopolysaccharide (LPS) plus IFN-γ, resulting in induction of iNOS and marked NO production. NSAIDs were tested in cell culture for effects on NO production and iNOS induction.

METHODS: Culture of rat alveolar macrophages (10^6/ml) and activation with LPS (2 μg/ml) + IFN-γ (100 U/ml) were described previously (4). NO production was monitored by determining accumulation of nitrite + nitrate in the medium by chemiluminescence detection (5), and iNOS was recovered from intact cells as described (6). Expression of iNOS mRNA was determined by standard Northern blot analysis (6). Cell viability was assessed by monitoring LDH release from cells into the culture medium. The electrophoretic mobility shift assay (EMSA) will be described elsewhere. Male Sprague-Dawley rats (250 g) were given LPS (2 mg/kg; i.v.) and plasma levels of NO_X^- were determined 5-hr later. NSAIDs were given (i.p.) 1-hr prior to LPS.

RESULTS and DISCUSSION: All NSAIDs inhibited NO production by activated macrophages in a concentration-dependent manner. The approximate ED_{50} values (mM) were salicylate (5), aspirin (2.5), indomethacin (0.1), ibuprofen (1), diclofenac (0.05) and naproxen (0.05). The NSAIDs were about 10-times more potent when LPS alone was used to activate macrophages, presumably because LPS but not IFN-γ activates NF-κB and NSAIDs appear to inhibit NF-κB activation, as described below. The NSAIDs also markedly and concentration-dependently reduced the quantity of iNOS catalytic protein recoverable from activated macrophages. None of the NSAIDs directly inhibited iNOS enzymatic activity. The NSAIDs inhibited NO production only when added to macrophage cultures within 6-hr of addition of LPS or LPS + IFN-γ, thus suggesting that the drugs interfere with either transcription of DNA into mRNA or ribosomal translation of mRNA into iNOS protein. Northern blot analysis revealed that all NSAIDs interfere with iNOS mRNA formation, and there was a close correlation between inhibition of NO production and inhibition of appearance of iNOS mRNA.

These data indicate that NSAIDs inhibit NO production by activated macrophages by interfering with transcriptional expression of the iNOS gene. Further experiments using the EMSA revealed that the NSAIDs prevent iNOS mRNA expression by interfering with NF-κB activation. Thus, the action of NSAIDs was similar to that of pyrrolidine dithiocarbamate, an NF-κB inhibitor that we previously showed to inhibit NO production by activated macrophages (7).

In vivo experiments were conducted to determine whether NSAIDs could inhibit LPS-induced production of NO in rats. At doses used to treat chronic inflammation (adjuvant polyarthritis) in rats, the NSAIDs markedly reduced plasma NO_X^- levels, measured at 5-hr after LPS administration. Approximate ED_{50} values (mg/kg) were salicylate (150), aspirin (50), indomethacin (0.5), ibuprofen (75), diclofenac (10) and naproxen (20). Preliminary experiments using reverse transcriptase-PCR to analyze the quantity of iNOS mRNA formed in vivo indicate that LPS induces iNOS mRNA in lung tissue, and that the NSAIDs prevent such induction of iNOS.

CONCLUSION: These observations indicate that NSAIDs possess the properties of inhibiting NO production by LPS-activated rat alveolar macrophages by mechanisms related to interference with NF-κB activation and consequent prevention of the inducible expression of the iNOS gene. Therefore, the mechanism of anti-inflammatory action of the NSAIDs may involve not only inhibition of prostaglandin production but also inhibition of NO production.

REFERENCES:

1. Vane, J.R., Mitchell, J.A., Appleton, I., Tomlinson, A., Bishop-Bailey, D., Croxtall, J. and Willoughby, D.A. (1994) *Proc. Natl. Acad. Sci. USA* 91: 2046-2050.
2. Stefanovic-Racic, M., Meyers, K., Meschter, C., Coffey, J.W., Hoffman, R.A. and Evans, C.H. (1994) *Arthritis and Rheumatism* 37: 1062-1069.
3. Salvemini, D., Misko, T.P., Masferrer, J.L., Seibert, K., Currie, M.G. and Needleman, P. (1993) *Proc. Natl. Acad. Sci. USA* 90: 7240-7244.
4. Griscavage, J.M., Rogers, N.E., Sherman, M.P. and Ignarro, L.J. (1993) *J. Immunol.* 151: 6329-6337.
5. Bush, P.A., Gonzalez, N.E., Griscavage, J.M. and Ignarro, L.J. (1992) *Biochem. Biophys. Res. Commun.* 185: 960-966.
6. Aeberhard, E.E., Henderson, S.A., Arabolos, N.S., Griscavage, J.M., Castro, F.E., Barrett, C.T. and Ignarro, L.J. (1995) *Biochem. Biophys. Res. Commun.* 208: 1053-1059.
7. Sherman, M.P., Aeberhard, E.E., Wong, V.Z., Griscavage, J.M. and Ignarro, L.J. (1993) *Biochem. Biophys. Res. Commun.* 191: 1301-1308.

Glucocorticoids inhibit transcription of inducible NO synthase II (iNOS) by attenuating cytokine-induced activity of transcription factor NF-κB

HARTMUT KLEINERT, CHRISTIAN EUCHENHOFER, IRMGARD IHRIG-BIEDERT, AND ULRICH FÖRSTERMANN

Department of Pharmacology, Johannes Gutenberg University, Obere Zahlbacher Str. 67, D-55101 Mainz, Germany

The inducible isoform of NO synthase (iNOS or NOS II) is usually expressed after cytokine induction of cells [1]. The promoters of the murine [2, 3] and human NOS II gene [4, 5] contain consensus sequences for the binding of transcription factors associated with stimuli that induce NOS II expression [2, 3] Functional analysis of the *murine* NOS promoter has revealed the importance of the NF-κB site [2, 3, 6]. Glucocorticoids are efficacious inhibitors the NOS II induction [7, 8]. However, consensus glucocorticoid responsive elements (GRE) have not been found on either promoter. On the other hand, the activated glucocorticoid receptor has been shown to repress gene transcription by interacting negatively with stimulating transcription factors (e.g. NF-κB) [9, 10].

Inhibitors of the NOS II mRNA expression. In A549/8 cells, maximum expression of NOS II mRNA was with interleukin-1β (50 U/ml), interferon-γ (100 U/ml), and tumor necrosis factor-α (10 ng/ml) (*cytomix*). Dexamethasone markedly reduced the *cytomix*-induced NOS II mRNA levels. This inhibition was reversed by the glucocorticoid receptor antagonist RU 38486 (30 μM) (Fig. 1).

Activity of the *murine* NOS II promoter in *human* A549/8 cells. The activity of the *murine* NOS II promoters was analyzed in A549/8 cells transiently transfected with a plasmid containing the *murine* NOS II promoter inserted before a luciferase reporter gene. An up to 20-fold induction of the promoter activity was obtained in A549/8 cells incubated with *cytomix*.

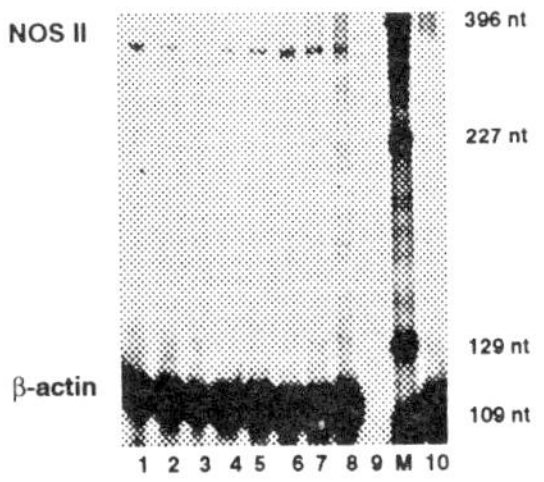

Fig. 1 S1-Nuclease-protection analysis of RNAs from A549/8 cells using human NOS II and β-actin cDNA probes. Cells received the following treatments before RNA was prepared: lane **1:** cells induced with IL-1β (50 U/ml), INF-γ (100 U/ml) and TNFα (10 ng/ml) (*cytomix*); lanes **2-4:** cells incubated with *cytomix* in the presence of dexamethasone (0.1, 1.0, and 10 μM, respectively); lane **5:** cells incubated with the glucocorticoid receptor antagonist RU 38486 (30 μM); lanes **6-8:** cells incubated with *cytomix* in the presence of dexamethasone (0.1, 1.0, and 10 μM, respectively) and RU 38486 (30 μM); lane **9:** tRNA control; lane **10:** untreated control cells; **M:** molecular weight markers

This induction was inhibited by co-incubation with dexamethasone or the NF-κB inhibitor pyrrolidine dithiocarbamate (Fig. 2).

Cytokine-induced NF-κB binding activity in A549/8 cells, inhibition by dexamethasone. Electrophoretic mobility shift assays showed a low level of NF-κB binding activity in untreated A549/8 cells. Incubation of A549/8 cells with *cytomix* for 3 h markedly increased this activity. The band shift was prevented by a double stranded oligonucleotide containing the putative NF-κB binding motif of the *human* NOS II promoter. This stimulation of specific NF-κB binding by cytokines was markedly inhibited by co-incubation with dexamethasone (1 to 5 μM) (Fig. 3). Incubation of A549/8 cells with *cytomix* for 3 h increased the nuclear content of NF-κB p65- and -p50-immunoreactivities as determined by Western blot, but dexamethasone (1 μM) had no effect on the protein amount of either NF-κB monomer (Fig. 4). Also incubation of A549/8 cells with *cytomix* for 14 h increased the expression of NF-kB p65 mRNA and I-κB mRNA as measured by S1-nuclease protection analyses. Addition of dexamethasone (0.1; 1.0 or 10 μM) to *cytomix*-treated A549/8 cells did not change the level of expression of either mRNA (data not shown).

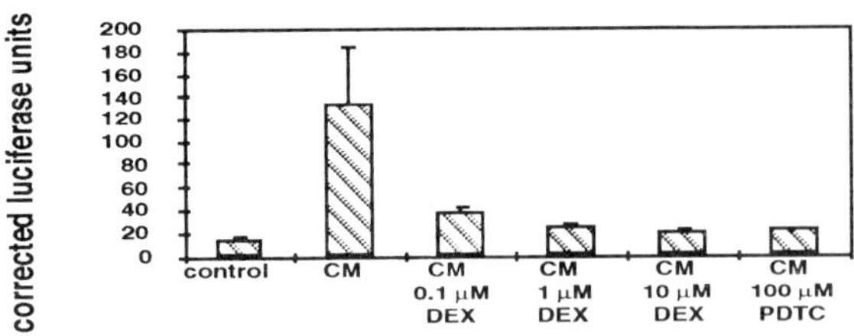

Fig. 2. Luciferase assay of transfected A549/8 cells. The ordinate shows the corrected luciferase activity of A549/8 cells transfected with both a murine NOS II promoter/luciferase reporter gene construct and an eucaryotic β-galactosidase expression plasmid (pCH110, for normalization). Transfected cells were incubated with the following agents: culture medium alone (**control**); IL-1β (50 U/ml), INF-γ, (100 U/ml), and TNF-α (10 ng/ml) (*cytomix*, **CM**); *cytomix* in the presence of dexamethasone (**DEX**, 0.1, 1.0, and 10 μM, respectively); and *cytomix* in the presence of pyrrolidine dithiocarbamate (**PDTC**, 100μM).

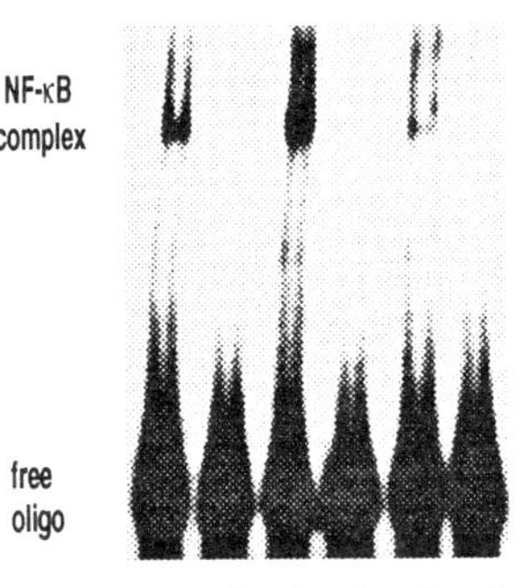

Fig. 3. Electromobility shift assay using a 5'-end labeled consensus oligonucleotide for NF-κB binding and nuclear extracts from A549/8 cells. In lanes **1** and **2**, untreated A549/8 cells were used. In lanes **3** and **4**, A549/8 cells induced for 3 h with IL-1β (50 U/ml), INF-γ (100 U/ml) and TNF-α (10 ng/ml) (*cytomix*) were used. In lanes **5** and **6**, A549/8 cells treated for 3 h with *cytomix* in the presence of dexamethasone (5 μM) were used. In all three cases, complex formation was completely prevented with an unlabeled oligonucleotide containing the putative NF-κB binding site of the human NOS II promoter (100-fold excess, lanes **2**, **4** and **6**).

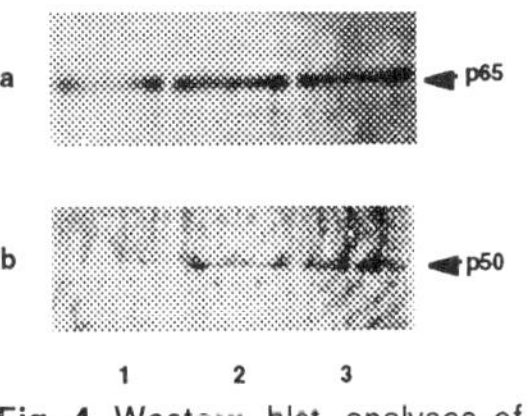

Fig. 4. Western blot analyses of A549/8 cell nuclear extracts using a polyclonal anti-NF-κB p65 antibody (a) or a polyclonal anti-NF-κB p50 antibody (b). In lane **1**, untreated A549/8 cells were used. In lane **2**, A549/8 cells induced for 3 h with IL-1β (50 U/ml), INF-γ (100 U/ml), and TNF-α (10 ng/ml) (*cytomix*) were used. In lane **3**, A549/8 cells treated for 3 h with *cytomix* in the presence of dexamethasone (1 μM) were used.

Therefore, the reduced NF-κB-binding activity in nuclear extracts of dexamethasone-treated A549/8 cells is unlikely to result from reduced expression or impaired nuclear translocation of NF-κB protein. Thus, these data suggest that, in A549/8 cells, the dexamethasone-activated glucocorticoid receptor interacts with the cytokine activated NF-κB complex, thereby repressing the binding of this complex to the NF-κB-RE in the 5'-flanking sequence of the NOS II gene.

References

1. Förstermann, U., *et al. Hypertension* **23**, 1121-1131 (1994).
2. Xie, Q.W., Whisnant, R. & Nathan, C. *J Exp Med* **177**, 1779-1784 (1993).
3. Lowenstein, C.J., *et al. Proc Natl Acad Sci USA* **90**, 9730-9734 (1993).
4. Nunokawa, Y., Ishida, N. & Tanaka, S. *Biochem Biophys Res Commun* **200**, 802-807 (1994).
5. Chartrain, N.A., *et al. J Biol Chem* **269**, 6765-6772 (1994).
6. Xie, Q.W., Kashiwabara, Y. & Nathan, C. *J Biol Chem* **269**, 4705-4708 (1994).
7 Radomski, M.W., Palmer, R.M. & Moncada, S. *Proc Natl Acad Sci USA* **87**, 10043-10047 (1990).
8. Di Rosa, M., Radomski, M., Carnuccio, R. & Moncada, S. *Biochem Biophys Res Commun* **172**, 1246-1252 (1990).
9. Mukaida, N., *et al. J Biol Chem* **269**, 13289-13295 (1994).
10. Ray, A. & Prefontaine, K.E. *Proc Natl Acad Sci U S A* **91**, 752-756 (1994).

Regulation of human inducible NOS2 expression by the tumor suppressor gene p53

STEFAN AMBS, KATHLEEN FORRESTER, SHAWN E. LUPOLD, WILLIAM G. MERRIAM, DAVID A. GELLER[2], TIMOTHY R. BILLIAR[2], and CURTIS C. HARRIS

LABORATORY OF HUMAN CARCINOGENESIS, NCI, BETHESDA MD 20892-4255, and [2]DEPARTMENT OF SURGERY, UNIVERSITY OF PITTSBURGH, PA 15261.

Nitric oxide (NO) is mutagenic in bacteria and mammalian cells, inducing DNA strand breaks as well as deamination of purine and pyrimidine residues [1-3]. NO is endogenously produced by the cytochrome P-450 enzyme family of NO synthases (NOS). There are three isoforms [4]. Two are Ca^{2+}-dependent and constitutively expressed (NOS1 and NOS3). The third isoform is inducible (NOS2), and transient changes of the intracellular Ca^{2+}-level do not regulate the enzyme activity. NOS2 can produce high, persistent levels of NO upon induction with cytokines, potentially resulting in tissue or DNA damage [5,6].

NO-related DNA damage which abrogates the function of the tumor suppressor gene p53 may contribute to the pathogenesis of human cancer [7]. We hypothesized that NO causes the high frequency of C to T transitions at CpG dinucleotides in the p53 gene in colon cancer. Excessive NO production by human NOS2 in ulcerative colitis (UC) may also contribute to increased DNA damage which subsequently causes the high frequency of UC cancers. Since p53 is part of the cellular response to DNA damage from exogenous chemical and physical mutagens [8,9], we supposed that p53 performs a similar role in response to putative endogenous mutagens. Therefore, we studied the regulatory function of wild-type (WT) versus mutant p53 on human NOS2 expression. Fragments of the human NOS2 5′ untranslated region starting at +33 and extending 0.4, 1.3 and 7 kb were prepared and subcloned into a promoterless luciferase reporter gene vector [10]. The p53 expression vectors contained human wild-type or mutant p53 under control of the cytomegalovirus promoter [11].

Expression of WT p53 resulted in a 2-fold decrease in NOS2 protein expression and enzymatic activity in cytokine-induced DLD-1 cells. The decrease in NOS2 protein expression corresponded to a decrease in human NOS2 promoter activity using the 1.3 kb NOS2 promoter construct, indicating that inhibition occurred at the level of expression. Furthermore, WT p53 repressed dose-dependently the activity of the same luciferase reporter construct in the human tumor cell lines Calu-6 (lung), SK-OV-3 (ovarian) and PC-3 (prostate) which all lack endogenous p53 expression. Cotransfection of the 175^{his} and 273^{his} dominant negative p53 mutants into Calu-6 partially abolished WT p53-mediated repression of the 1.3 kb human NOS2 promoter construct. In addition, transfections with p53 mutants alone did not affect NOS2 promoter activity suggesting that down-regulation of NOS2 expression is WT p53 specific.

In order to examine the effects of p53 on cytokine induced human NOS2 promoter activity, we cotransfected WT p53 and the human NOS2 luciferase reporter constructs pNOS2(1.3)Luc and pNOS2(7.0)Luc into AKN-1 liver cells and into WT p53 null Li-Fraumeni-syndrome human fibroblasts (LFS-041). WT p53 repressed basal and cytokine-induced NOS2 promoter activity in both cell lines. Cytokines induced 6-fold the pNOS2(7.0)Luc promoter activity in AKN-1 liver cells, and WT p53 down-regulated this activity by 90%. In LFS-041 cells, WT p53 repressed basal activity of the 1.3 kb and the 7.0 kb NOS2 promoter fragments 3-fold and 7.5-fold, respectively. Cytokine-treatment of LFS-041 doubled the pNOS2(1.3)Luc promoter activity in these cells but had only a small effect on the pNOS2(7.0)Luc promoter activity. Both were repressed by WT p53. We further tested whether WT p53 mediated repression of NOS2 promoter activity is conserved in the murine NOS2 promoter sequence which is less than 50% homologous to the human NOS2 promoter. WT p53 repressed basal and cytokine-induced murine NOS2 promoter activity in mouse p53 knockout fibroblasts using a 1.7 kb murine NOS2 promoter construct fused to a luciferase reporter gene [12]. Additional experiments with a pNOS(0.4)Luc reporter construct, which lacks the putative p53 binding consensus sequence but was inhibited by WT p53 suggest that the NOS2 promoter region required for WT p53 regulation is localized within a 400 bp region upstream of the transcription start site which contains a TATAA box.

DNA damage triggers p53 accumulation. We explored whether putative NO-induced DNA damage could induce p53 accu-mulation. NO released by 1 mM S-nitroso-glutathione induced p53 accumulation in normal human fibroblasts with a peak after 16 hr. Endogenously produced NO induced p53 accumulation in human NOS2 transfected THLE-5b immortalized human liver epithelial cells. Addition of the NOS inhibitor L-NMA diminished this effect.

The data, showing that a) WT p53 represses basal and cytokine-induced human and murine NOS2 promoter activity and b) NO induces p53, are consistent with the hypothesis of a negative feedback loop in which NO-induced DNA damage results in WT p53 accumulation and p53-mediated transrepression of NOS2 gene expression, and therefore in a reduction in NO production. In addition, our results suggest a common mechanism of p53-mediated repression of NOS2 expression, perhaps through the common TATAA sequence located in both human and murine promoter. In addition, the data may explain the already described anti-proliferative effect of NO [6] since expression of WT p53 induces G1 growth arrest by transcriptional activation of cell cycle inhibitors [8].

References

1. Wink, D.A., Kasprzak, K.S., Maragos, C.M., et al. (1991) DNA deaminating ability and genotoxicity of nitric oxide and its progenitors. Science 254, 1001-1003.

2. Nguyen, T., Brunson, D., Crespi, C.L., Penman, B.W., Wishnok, J.S., and Tannenbaum, S.R. (1992) DNA damage and mutations in human cells exposed to nitric oxide in vitro. Proc. Natl. Acad. Sci. U.S.A. 89, 3030-3034.

3. Fehsel, K., Jalowy, S., Qi, S., Burkart, V., Hartmann, B., and Kolb, H. (1993) Islet cell DNA is a target of in-flammatory attack by nitric oxide. Diabetes 42, 496-500.

4. Nathan, C., and Xie, Q.-W. (1994) Nitric oxide synthases: roles, tolls, and controls. Cell 78, 915-918.

5. Geller, D.A., Lowenstein, C.J., Shapiro, R.A., et al. (1993) Molecular cloning and expression of inducible nitric oxide synthase. Proc. Natl. Acad. Sci. U.S.A. 90, 3491-3495.

6. Gross, S.S., and Wolin, M.S. (1995) Nitric oxide: pathophysiological mechanisms. Annu. Rev. Physiol. 57, 737-769.

7. Harris, C.C. (1993) p53: at the crossroads of molecular carcinogenesis and risk assessment. Science 262, 1980-1981.

8. Lane, D.P.(1992) Cancer. p53, guardian of the genome. Nature 358, 15.

9. Kastan, M.B., Onyekwere, O., Sidransky, D., Vogelstein, B., and Craig, W.R. (1991) Participation of p53 protein in the cellular response to DNA damage. Cancer Res. 51, 6304-6311.

10. de Vera, M.E., Shapiro, R.A., Nussler, A.K., et al. Transcriptional regulation of human inducible nitric oxide synthase (NOS2) gene by cytokines: initial analysis of human NOS2 promoter. Manuscript submitted.

11. Baker, S.J., Markovitz, S., Fearon, E.R., Wilson, J.K., and Vogelstein, B. (1990) Suppression of human colorectal carcinoma cell growth by wild-type p53. Science 249, 912-915.

12. Lowenstein, C.J., Alley, E.W., Raval, P., et al. (1993) Macrophages nitric oxide synthase gene: two upstream regions mediate induction by interferon gamma and lipopolysaccharide. Proc. Natl. Acad. Sci. 90, 9730-9734.

CLONING AND REGULATION OF RAT GTP CYCLOHYDROLASE GENE: CO-ORDINATE TRANSCRIPTIONAL INDUCTION WITH iNOS BY IMMUNOSTIMULANTS

Joseph SMITH and Steven S. GROSS

Department of Pharmacology, Cornell University Medical College, 1300 York Ave, New York, N.Y. 10021, USA

It is well established that tetrahydrobiopterin (BH4) is a required cofactor for activity of all isoforms of NO synthase. Importantly, cytokines which induce expression of iNOS are also known to trigger expression of BH4 in many cell types. Indeed, we have previously shown that *de novo* synthesis of BH4 is essential for the induction by cytokines of NO synthesis in vascular endothelial and smooth muscle cells and that BH4 availability limits NO production in these cells[1,2]. Immunostimulant-evoked BH4 synthesis by vascular cells is preceded by an increase in expression of mRNA encoding GTP Cyclohydrolase I (GTPCH), the first and rate-limiting enzyme of the de novo pathway of BH4 synthesis.[3] To explore the basis for co-ordinate induction of iNOS and GTPCH mRNA's, we have now cloned the GTPCH gene to examine its promoter region for regulatory elements.

Results and Discussion: A rat genomic P1 library containing 75-100 kb inserts was screened for GTPCH clones using a PCR-based strategy. Two GTPCH clones were identified and termed 1913 and 1914. Southern blot analysis showed that the restriction patterns were essentially identical. However, while clone 1913 contains 3' and 5' gene ends, 1914 is devoid of the 3'-end. Restriction enzyme digests of rat genomic DNA, probed with a 115 bp sequence found in the 5'-end of GTPCH cDNA, reveal single bands that are identical in size to that observed with clone 1913. This suggests that there is only one GTPCH gene in the entire rat genome, and further confirms the identity of clone 1913 as GTPCH.

The transcription start site of GTPCH was identified by primer extension analysis. The precise transcription start site of the GTPCH gene in vitro corresponded to positions -1 and -2 of the GTPCH cDNA. Moreover, we found that the promoter of GTPCH was functional in an *in vitro* transcription assay. An 8.5 kB fragment containing the 5'-end of the GTPCH gene and its upstream flanking sequence was studied for *in vitro* transcription. In the absence of nuclear extract, transcription did not occur. Addition of nuclear extract from vascular smooth muscle cells which were either untreated or treated with LPS/IFN for 4 hours, elicited increasing transcription with increasing protein concentration. A single dominant transcription start site was observed which corresponded to that found by primer extension using mRNA from immunostimulant-activated vascular smooth muscle cells. An identical start site was found when *in vitro* transcription was initiated by nuclear extract from either HeLa cells or PC-12 cells.

By double-stranded dideoxynucleotide sequencing, the cytokine-induced rat GTPCH promoter was found to be TATAless. Of potential regulatory significance, the first 1 kb upstream of the GTPCH gene was found to be replete with consensus sequences for transcriptional activators that are prototypic of cytokine-inducible genes. These include an NF-kB site, an NF-IL6 site, several consensus sequences for alpha-INF.2, a g-IFN response element, an INF type 1 site, a GMCSF consensus sequence, an AP-1 site and multiple SP-1 sites. In addition, putative consensus sequences were observed for binding of CREB as well as liver and muscle specific transcriptional activators.

To study GTPCH promoter function, rat aortic smooth muscle cells were stably transfected with a construct containing a 3 kb fragment of the GTPCH promoter which was cloned in front of a reporter gene that encodes the secreted form of human placental alkaline phosphatase (SEAP). SEAP activity which is released into the cell culture medium was measured by a sensitive 96-well chemiluminescent assay. Figure 1 depicts accumulated SEAP activity, 24 hours after treatment of cells with the indicated immunostimulants. Note that the GTPCH promoter was activated by LPS, IL-1, TNF and various combinations of these immunostimulants. IFN by itself had no effect, nor did it potentiate the activation by LPS. The lack of synergism between IFN and other immunostimulants in driving the GTPCH promoter contrasts with their synergistic actions on iNOS promoter activity.

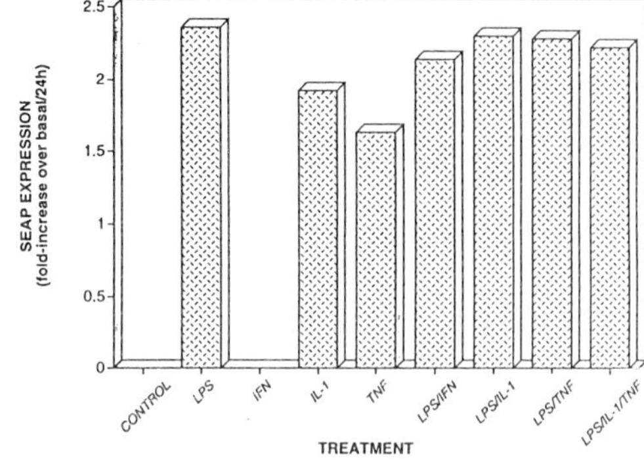

Figure 1. 3 Kb GTPCH promotar/SEAP reporter expression in stably-transfected rat VSMC After 24-Hour treatment with immunomodulators.

In further experiments, the increase in SEAP expression elicited by LPS was found to be concentration-dependent and biphasic. Interestingly, LPS induced an identical biphasic stimulation of an iNOS promoter/SEAP reporter gene while only the higher concentrations of LPS triggered enhanced NO synthesis by the transfected cells. Like LPS, IL-1 also elicited concentration-dependent expression of the GTPCH promoter/reporter gene; IL-1's effect was maximal at 100 ng/ml, a concentration which we found to trigger maximal stimulation of NO synthesis.

Sequencing of the GTPCH promoter revealed a consensus sequence for binding of CREB, a transcriptional activator that mediates the effects of cAMP. Therefore we tested whether the cell permeant cAMP analog, dibutyryl-cAMP could drive expression of the GTPCH promoter/SEAP reporter. As shown in figure 2, cAMP caused a modest but concentration-dependent increase in SEAP expression. Whether this effect requires an intact CREB site awaits testing

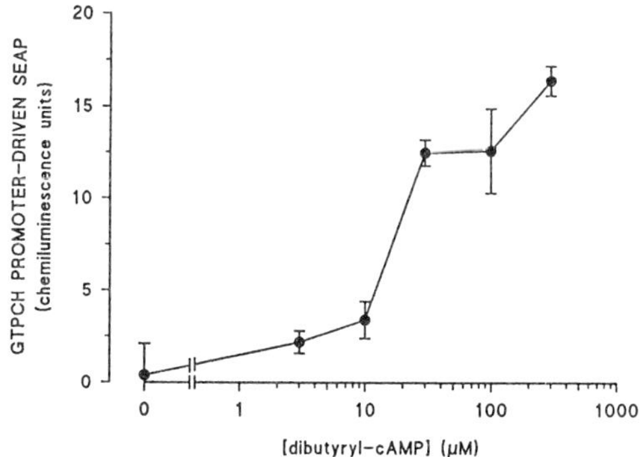

Figure 2. Induction by cAMP of GTPCH promoter/SEAP reporter in rat VSMC: inhibition by cycloheximide.

In summary we have cloned the rat GTPCH gene, established in vitro function of its promoter, and defined its transcription start site. Contained within the GTPCH promoter are numerous consensus sequences common to cytokine inducible genes and shared with the iNOS gene promoter. LPS, IL-1, TNF, cAMP and low concentrations of cycloheximide, all shown to trigger GTPCH expression, were found to enhance GTPCH gene transcription. Identification of cellular mechanisms for co-ordinating iNOS and GTPCH gene expression and assessment of the function of specific promoter sequences in mediating GTPCH transcription by immunostimulants awaits further investigations. (Support: HL50656 and 40403)

REFERENCES:

1. Gross, S.S. and Levi, R., J. Biol. Chem. 267: 25722-25729, 1992.
2. Gross, S.S., Jaffe, E.A., Levi, R. and Kilbourn R., Biochem. Biophys. Res. Commun. 178: 823-829, 991.
3. Hattori, Y. and Gross, S.S., Biochem. Biophys. Res. Commun. 195: 435-441, 1993.

Spin-trapping NO in nNOS-deficient mice: indications for stroke therapy

Mark E MULLINS[a], Neal J SONDHEIMER[b], Zhihong HUANG[c], David J SINGEL[d], Paul L HUANG[c], Mark C FISHMAN[c], Francis E JENSEN[b], Stuart A LIPTON[b], Jonathan S STAMLER[e], and Michael A MOSKOWITZ[b]

[a]Department of Chemistry, Harvard University, Cambridge, MA 02138; [b]Neuroscience Department, Children's Hospital, Boston, MA 02115; [c]Neuroscience Department, Massachusetts General Hospital, Boston, MA 02129; [d]Department of Chemistry, Montana State University, Bozeman, MT 59715; [e]Duke University Medical Center, Durham, NC 27710

Introduction

"NO-related" processes are highly dependent on the redox state of the NO molecule.[1] Many of the pharmacological applications of NO biochemistry will undoubtedly rest on the issue of NO redox specificity.

Research laboratories have long employed EPR (electron paramagnetic resonance) to identify and study species of interest in free radical biology. NO• is a relatively long-lived free radical but its half-life in biologic systems is too brief for analysis in an untrapped form. One spin-trapping protocol which has been used successfully *in vivo*, involves the formation of a paramagnetic mononitrosyl-iron complex (MNIC). Animals are injected with diethyldithiocarbamate (DETC) and Fe^{2+}. Iron complexes with DETC and binds NO and H_2O to form the paramagnetic (EPR-active) MNIC complex NO•-Fe^{2+}-$(DETC)_2$ (**Figure 1**).

The EPR-sensitive NO• complex can be detected in normal rat and mouse brains after exogenous introduction of Fe^{2+} and DETC. These results suggest that low levels of NO• exists endogenously under basal conditions. Our data can not exclude an RSNO reservoir or distinguish NO• from RSNO as DETC converts RSNO to NO• (Arnelle and Stamler, unpublished) Experimental data were quantified using (standard) signal subtraction methods and then compared to findings processed using integration methods based on a ratio of complexed NO• to copper signal. Using computer-aided subtraction of the Cu signal, we obtained spectra of the NO•-complex.

Using nNOS knockout mice (Kn), the effects of middle cerebral artery occlusion (MCAO) on cerebral infarction and cerebral ischemia have been previously documented.[2] These mice are phenotypically normal in most respects but are deficient in nNOS mRNA and protein.[3] In this investigation, a novel method of quantitative analysis is employed in which ischemic and control NO• levels in Kn and wild-type (WT) mice are directly measured by EPR spectroscopy.

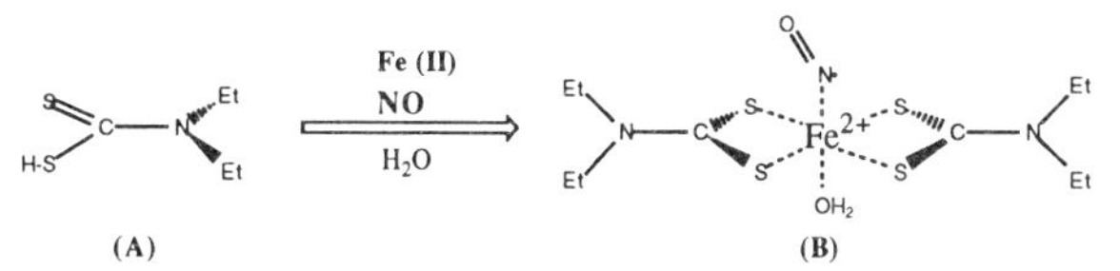

Fe (II), NO, H_2O ⟹ (A) → (B)

Figure 1. DETC and injected iron (II) citrate can cross the blood-brain barrier and form compound **B** (NO•-Fe^{2+}-$(DETC)_2$) in the presence of nitric oxide *in vivo*.

Results And Discussion

In Kn experiments, the MNIC complex produced the predicted strong EPR signal at $g_\perp = 2.035$ and $g_{||} = 2.020$ (**Figure 1**). The triplet HFS of the MNIC overlapped with the second downfield peak of the Cu-DETC spectrum. The WT mice showed a definite MNIC presence under control conditions (n = 4). The NO•/copper signal ratio was significantly elevated in samples harvested ten minutes after ischemia ($n = 4$, $p < .05$) (**Figure 2**). The NO• signal ratio was much lower after an hour of ischemia. ($n = 4$, $p < .001$ against both groups). The Kn mice, by comparison, had significantly lower levels of MNIC formation than the WT mice (WT vs. Kn control ($n = 3$) $p < .005$). There was no statistically significant difference between the Kn control, Kn ischemic ($n = 4$) and Kn hour ($n = 4$) groups (**Figure 2**).

This study demonstrates that NO-DETC MNICs are significantly increased in ischemic, wild-type SV129 mouse cortex. This effect was most pronounced after 10 min. of ischemia and decreased substantially over the course of an hour. The signal shift at one hour may be due to a depletion of oxygen or other substrate necessary for the production of NO in ischemic tissue, or to a general loss of DETC through thiol pool depletion.[4] Kn animals showed a complete lack of NO• at all times surveyed, indicating that there was no NO contribution from other isoforms during this period. The lack of NO activity as measured by NO• spin-trapping in the Kn mouse correlated positively with decreased stroke volume. This result suggests a relationship between NO-related activity and neural tissue damage. As has been previously stated, the apparent resistance of the nNOS-deficient animals to ischemic injury cannot be due to protective vascular metabolites directly because the animals have a normal cerebrovascular blood flow as measured by Doppler.[2] It is noteworthy that endothelial NOS does not produce NO as measured by spin-trapped NO• under the same conditions. In particular, recent studies indicate that most NO-associated activity in the vasculature exists as RSNO, [5] which would not be as readily detected by spin-trapping. [6]

Our data suggest that nNOS produces NO•, and that levels are increased by ischemic insult. It is unclear whether NO• or a subsequent metabolite is responsible for the increased stoke volume in mice. What remains clear is that NO and associated NO activity (as measured by NO•) is implicated directly in this neuropathology and that Kn mice are protected from ischemic injury.

Current therapeutics for cerebrovascular ischemia in humans are limited to antithrombotic drugs. Neural destruction during this period may be limited through effective current therapy but not avoided altogether. Our results give further support to involvement of NO•. The selective inhibition of nNOS combined with targeted NO delivery or promotion of eNOS might offer a specific and effective intervention.

Figure 2. NO• Levels in Ischemic Animals versus Control Animals.

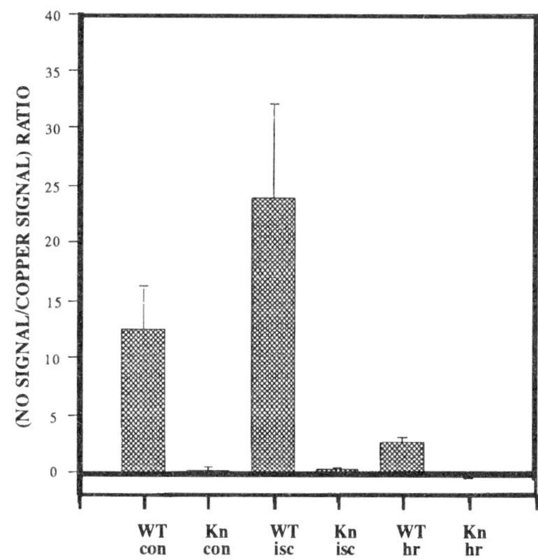

1 Stamler, J.S., Singel, D.J. and Loscalzo, J. (1992) Science **258**, 1898-1902

2 Huang, Z., Huang, P.L., Panahian, N., Dalkara, T., Fishman, M.C., Moskowitz, M.A. (1994) Science **265**, 1883-1885

3 Huang, P.L.,Dawson, T.M. , Bredt, D.S., Snyder, S.H., and Fishman, M.C. (1993) Cell **75**, 1273-1286

4 Stamler, J.S. (1994) Cell **78**, 931-936

5 Stamler, J.S., Jaraki, O., Osborne, J., Simon, D.I., Keaney, J., Vita, J., Singel, D.J., Valeri, C.R., and Loscalzo, J. (1992). Proc. Natl. Acad. Sci. U. S. A. **89**, 7674-7677

6 Gaston, B., Reilly, J., Drazen, J.M., Fackler, J., Ramdev, P., Sugarbaker, D., Mullins, M., Singel, D., Loscalzo, J., and Stamler, J. S. (1993) Proc. Natl. Acad. Sci. U. S. A. **90**, 10957-10961

Targeted disruption of the endothelial nitric oxide synthase gene

Paul L. HUANG*, Zhihong HUANG†, Michael A. MOSKOWITZ†, John A. BEVAN§, and Mark C. FISHMAN*

* Cardiovascular Research Center, Massachusetts General Hospital, Charlestown, MA 02129 USA
† Stroke Research Laboratory, Massachusetts General Hospital, Charlestown, MA 02129 USA
§ University of Vermont, Burlington, VT 05405 USA

There are three major isoforms of nitric oxide synthase, the enzymes that make nitric oxide. We have used homologous recombination to disrupt individual NOS genes in mice, in order to separate their effects and to study processes in intact animals.

Gene knock-out allows us to examine the effects of individual genes (e.g. NOS isoforms) in physiology and development. However, embryonic lethality and developmental abnormalities occur in knock-outs of some genes. In interpreting knock-out phenotypes, one must be careful to examine effects on development, and alterations in the expression of other genes that may compensate for the gene that has been knocked-out.

We have previously reported the generation of nNOS knock-out mice [1]. These mice have been extremely valuable to elucidating the role of nNOS in gastrointestinal motility [1], cerebral ischemia [2], and long term potentiation [3]. Several processes, including hypercapnic cerebral blood flow increase [4] and anesthetic requirements [5], are quantitatively normal, due to the presence of non-NO dependent compensatory mechanisms.

We disrupted the eNOS gene using homologous recombination, replacing the exons encoding NADPH ribose and adenine binding sites with a neomycin resistance gene [6]. Absence of eNOS expression was demonstrated by Western blot analysis and enzymatic assay of NOS catalytic activity.

EDRF activity is absent in eNOS mutant mice. Aortic rings from wild-type mice contract in response to norepinephrine and relax in a dose-dependent manner to acetylcholine. Acetylcholine response is blocked by L-NA. Aortic rings from eNOS mutant mice contract normally to norepinephrine, but show no response to acetylcholine. This provides, for the first time, *genetic* evidence that NO is responsible for EDRF activity.

Because NO is an endogenous vasodilator, it is thought to be important to blood pressure regulation. We measured the mean arterial blood pressure in eNOS mutant mice by femoral arterial catheterization, using several anesthetics and in the awake state. The arterial blood pressure was significantly higher in the eNOS mutant mice than in wild-type littermates (Table 1), under each of the conditions studied, indicating that absence of eNOS results in hypertension.

Blood pressure regulation is controlled by complex homeostatic mechanisms, including the renin-angiotensin system, vasodilators like NO, vasoconstrictors like the endothelins, and central baroreceptor reflexes. In the chronic absence of one of these, one might expect the other mechanisms to compensate and return the blood pressure to normal. The fact that they do not, or cannot, normalize the blood pressure in eNOS mutant mice, indicates that endothelium-derived NO plays a fundamental role in blood pressure regulation and possibly its setpoint.

The phenotype of eNOS mutant mice confirms the importance of eNOS to EDRF activity and to blood pressure regulation. These mice will hopefully be useful in studying the role of eNOS in many biologic processes.

Abbreviations used: NOS, nitric oxide synthase; eNOS, endothelial NOS; nNOS, neuronal NOS; EDRF, endothelium-derived relaxing factor

Table 1. Mean arterial blood pressure
Each value represent the mean ± S.D. for ≥14 animals.

	Mean arterial blood pressure (mmHg)	
	Wild-type	eNOS mutant
Urethane	81 ± 9	110 ± 8
Halothane	90 ± 12	109 ± 11
Awake	97 ± 8	117 ± 10

Acknowledgements
We thank Brigid Nulty, Elizabeth Joyce, and Lisa Baldwin for excellent technical assistance. This work was supported by a sponsored research agreement with Bristol-Myers Squibb (MCF), NIH grants NS33335 (PLH), NS10828 (MAM, PLH), HL32985 (JAB), and a grant from the Harcourt General Foundation (PLH).

References

1. Huang, P.L., Dawson, T.M., Bredt, D., Snyder, S. H., and Fishman, M.C. (1993) Targeted Disruption of the Neuronal Nitric Oxide Synthase Gene. Cell **75**, 1273-1286.
2. Huang, Z., Huang, P.L., Panahanian, N., Dalkara, T., Fishman, M.C., Moskowitz, M.A. (1994) Effect of Cerebral Ischemia in Mice Deficient in Neuronal Nitric Oxide Synthase. Science **265**, 1883-1885.
3. O'Dell, T.J., Huang, P.L., Dawson, T.M., Dinerman, J.L., Snyder, S.H., Kandel, E.R., and Fishman, M.C. (1994) Endothelial NOS and the Blockade of LTP by NOS Inhibitors in Mice Lacking Neuronal NOS. Science **265**, 542-546.
4. Irikura, K., Huang, P.L., Ma, J., Lee, W.S., Dalkara, T., Fishman, M.C., Dawson, T.M., Snyder, S.H., Moskowitz, M.A. (1995) Cerebrovascular alterations in mice lacking neuronal nitric oxide synthase gene expression. Proc. Natl. Acad Sci. USA **92**, 6823-2827.
5. Ichinose, F., Huang, P.L., Zapol, W.M. (1995) Effects of Targeted Neuronal Nitric Oxide Synthase Gene Disruption and NGL-arginine methylester on the threshold for isoflurane anesthesia. Anesthesiology **83**, 101-108.
6. Huang, P.L., Huang, Z., Bloch, K.D., Moskowitz, M.A., Bevan, J.A., and Fishman, M.C. (1995) Hypertension in mice lacking the gene for endothelial nitric oxide synthase. Nature, in press.

VISUALIZING THE CELLULAR RELEASE OF NITRIC OXIDE

ANNA M. LEONE, VANESSA W. FURST, NEALE FOXWELL, SELIM CELLEK, PETER N. WIKLUND* and SALVADOR MONCADA

Wellcome Research Laboratories, Langley Court, Beckenham, Kent BR3 3BS. * Dept. of Physiology and Pharmacology and Institute of Environmental Medicine, Karolinska Institute, 17176, Sweden

Introduction

The L-arginine:nitric oxide (NO) pathway is known to play a role in many physiological and pathophysiological processes[1,2]. Various techniques for the detection of NO have been developed[3-8], but none permits its direct measurement in tissues. In the absence of experimental data, researchers have computer modelled the diffusion of NO[9,10]. To gain a better understanding of the properties of NO we have developed a novel method for its visualisation. Using a high transmission microscope coupled to a high sensitivity photon counting camera, we have attempted to visualise the release and diffusion profile of NO from activated murine macrophages.

Materials and Methods

Cells:J774 16 and J774.3C3 murine monocytic cell lines, were grown to approximately 20% confluence on glass coverslips. Activation was achieved as previously described[13]. **Imaging Medium (IM) :** The IM consisted of 5mM luminol in Krebs containing, initially, 15μM hydrogen peroxide. The medium was left for two hours before use. Hardware: An Argus 50 photon counting system (kindly lent by Hamamatsu Photonics UK) coupled to an inverted Zeiss microscope was used, in a dark box environment to acquire both brighfield and photon counted images. Brightfield images were overlaid with pseudo-colourised, intensity coded, photon counted images.

Results and Discussion

Although nitric oxide does not react directly with luminol[11], NO can react with hydrogen peroxide to form peroxynitrite[8], which would generate light when it reacts with luminol[11]. In addition, NO has been reported to react with hydrogen peroxide to form singlet oxygen[12], which is also capable of generating light when reacted with luminol. Thus either or both mechanisms may be involved in our method.

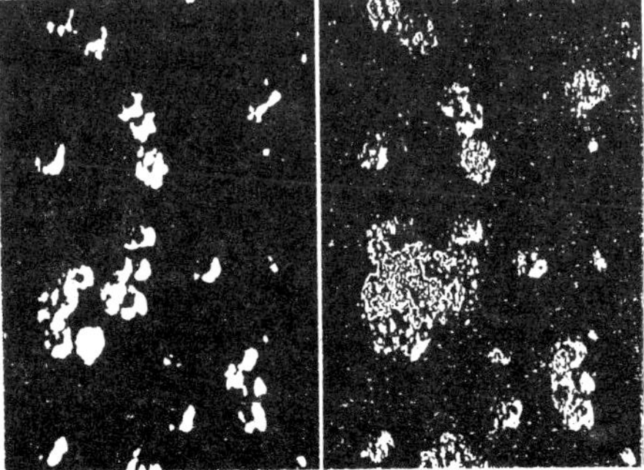

Fig. 1 shows a typical bright field image and the associated pseudo-colourised overlay obtained from activated J774.16 murine macrophages. Almost all the cells in the field showed NO being released, although the intensities varied considerably.

In contrast, cells in L-arginine-free IM showed greatly reduced light emissions Similarly, when L-NAME was added to L-arginine-containing IM, light intensities were significantly reduced This is consistent with previous studies which have shown that NO synthesis is markedly reduced in macrophages when exogenous L-arginine is absent[1,13]. To demonstrate further that the light signal was derived almost exclusively from NO rather than ROIs, we studied both J774.16 and J774.C3C cells (superoxide deficient cells, obtained as a gift from Dr. B. Bloom, Albert Einstein College of Medicine, New York). . No significant difference was observed in peak light intensities between the two cell lines (n=8).

Diffusion of NO was observed up to 175 μm radially from activated macrophages, which is at the lower end of theoretical predictions[9,10]. The method we have developed is therefore highly specific for NO and sufficiently sensitive to visualise the release of NO from individual activated macrophages.

References

1. Moncada, S., Palmer, R.M.J. & Higgs, E.A. *Biochem. Pharmacol.* **38,** 1709-1715 (1989).
2. Moncada, S. & Higgs, A. *New Engl. J. Med.*, **329**, 2002-2012 (1993).
3 Leone, A.M., Francis, P.L., Rhodes, P. & Moncada, S. *Biochem. Biophys. Res. Commun.* **200**, 951-957 (1994).
4. Kamura, E., Yoshimine, T., Tanaka, S., Hayakawa, T., Shiga, T. & Kosaka, H. *Neurosci. Lett.* **177**, 165-167 (1994).
5. Shibuki, K. & Okada, D. *Nature*, **349**, 326-328 (1991).
6. Malinski, T. & Taha, Z. *Nature*, **358**, 676-678 (1992).
7. Leone, A.M., Gustafsson, L.E., Francis, P.L., Persson, M.G., Wiklund, N.P. & Moncada, S. *Biochem. Biophys. Res. Commun.* **201**, 883-887 (1994).
8. Kikuchi, K., Nagano, T., Hayakawa, H., Hirata, Y. & Hirobe, M. *J. Biol. Chem.* **268**, 23106-23110 (1993).
9. Lancaster, J. Jr. *Proc. Natl. Acad. Sci. USA* **91**, 8137-8141 (1994).
10. Wood, J. & Garthwaite, J. *Neuropharmacology* **33**, 1235-1244 (1994).
11. Radi, R., Cosgrove, T.P., Beckman, J.S. & Freeman, B.A. *Biochem. J.* **290**, 51-57 (1993).
12. Noronha-Dutra, A.A., Epperlein, M.M. & Woolf, N. *Febs Lett.* **321**, 59-62 (1993).
13. Assreuy, J., Cunha, F.Q., Liew, F.Y. & Moncada, S.M. *Br. J. Pharmacol.* **108**, 833-837 (1993).

The influence of glucose in the bathing solution on the response of rings of rabbit aorta to peroxynitrite

ROBERT F. FURCHGOTT and DESINGARAO JOTHIANANDAN

Department of Pharmacology, State University of New York Health Science Center, Brooklyn, NY 11203, USA

Peroxynitrite ($ONOO^-$; PN), the product of the extremely rapid reaction of NO and superoxide anion (O_2^-), rapidly rearranges to inorganic nitrate at physiological pH. However, PN (or intermediate products in its decay) has been shown to be a strong oxidant for a number of biological targets [1,2]. A few recent reports indicate that PN can produce relaxation of vascular smooth muscle in organ chamber and perfusion-bioassay experiments, but its potency is much less than that of NO [3-6]. In the present study, which was begun to determine the potency of PN as a relaxant of rings of rabbit thoracic aorta, we found that certain constituents of the bathing solutions used in organ chamber experiments had significant effects on potency.

Rings of rabbit aorta were prepared and mounted in 20-ml organ chambers as previously described. The bathing solutions used were Krebs-bicarbonate solution with and without the usual 10 mM glucose (KBS+GL and KBS-GL, respectively); and Krebs-HEPES solution with and without glucose (KHS+GL and KHS-GL, respectively) in which 10 mM HEPES buffer adjusted to pH 7.4 substituted for the bicarbonate buffer. KBS was bubbled with 95% O_2/5% CO_2 and KHS with 100% O_2. Alkaline stock solutions of PN were made from H_2O_2 and $NaNO_2$ [1]. Spectrophotometric analysis showed that out of a maximal potential yield of 200 mM PN, a yield of close to 150 mM was consistently obtained, with unreacted NO_2^- making up the difference. Unreacted H_2O_2 was removed by treating the stock PN solution with MnO_2 powder.

In early experiments, it was found that glutathione (GSH, 1mM) added to the standard KBS+GL increased the relaxing potency of the PN solution about ten-fold, confirming findings of others with vascular preparations from other species [3,4]. We also found that cysteine (CYS) similarly potentiated.

In a series of experiments, it was found that the potency of PN solutions in relaxing endothelium-denuded rings was significantly greater in KBS+GL than in KBS-GL (-log ED50 4.194 vs 4.0264; n=8,P=0.02). Concentration-relaxation curves from a typical experiment are shown in Fig. 1A. Also shown is the concentration-relaxation curve for sodium nitrite in the same experiment, which was the same in both KBS+GL and KBS-GL. Since the concentration of NO_2^- in the PN solution was one-third of that of the PN itself, the relaxation by the PN solution of the ring in KBS-GL could be completely accounted for by the NO_2^- present in the solution. We concluded from experiments of this type that relaxation by PN solutions in KBS+GL is the resultant of the relaxing activities of the NO_2^- present and of some product of the reaction of the GL with the PN during its rapid decay (possibly a nitrite or nitrate ester).

Abbreviations: CYS, L-cysteine; GSH, glutathione; GL, D-glucose; HEPES, N-2-hydroxyethylpiperazine-N'-2-ethanesulfonic acid; KBS, Krebs bicarbonate solution; KHS, Krebs HEPES solution; PE, phenylephrine; PN, peroxynitrite

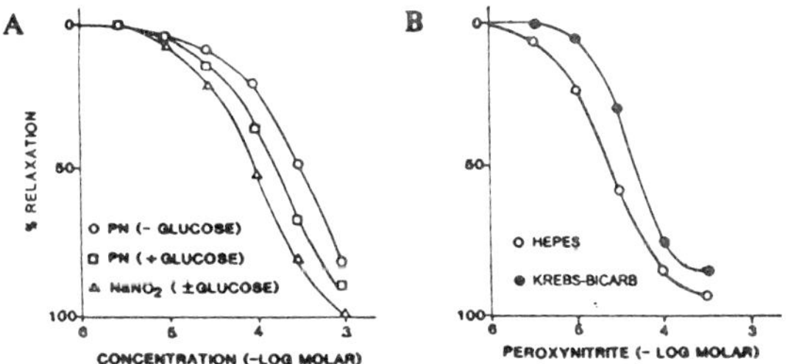

Fig. 1. A. Concentration-response curves to PN and to $NaNO_2$ in KBS+GL and KBS-GL in a typical experiment. **B.** Concentration-response curves to PN on adjacent rings from a single aorta in KBS+GL and KHS+GL. In the same experiment, the curve in KHS-GL was very close to that in KBS+GL and that in KBS-GL was about 0.3 log units to the right (not shown).

In agreement with the finding of Liu et al. [3] on the relaxation of canine coronary arteries by PN, we found that the potency of PN was greater in HEPES-buffered than in bicarbonate-buffered bathing solution (Fig. 1B). We suggest that this is due to the additional relaxing activity of a nitrite or nitrate of HEPES (on its hydroxyl group) formed during the rapid decay of added PN.

When rings in KBS were exposed to 1 mM PN for 10-20 min in the absence or presence of GL and then after thorough washing were contracted with PE and exposed to CYS (L or D) or GSH, the thiol (0.1-1.0 mM) produced a strong transient relaxation only in those rings which had been exposed to PN in the presence of GL (Fig. 2). On rings in KHS, such an exposure to PN "primed" them for marked transient relaxation by the thiols whether or not GL was present during the exposure (not shown).

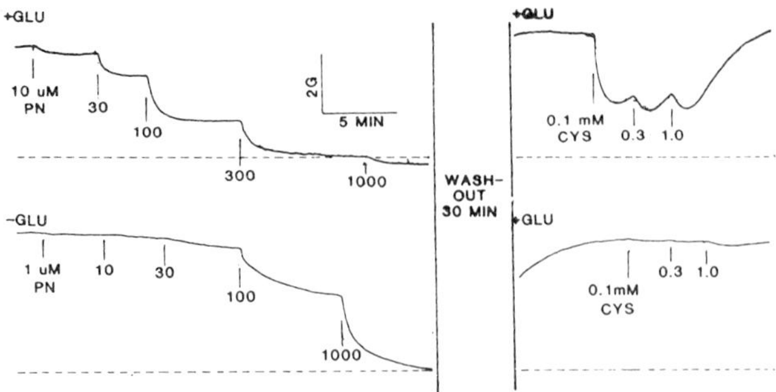

Fig. 2. Exposure to PN in KBS+GL but not in KBS-GL "primes" the aortic ring to give a marked transient relaxation to CYS. Exposure to 1000 μM PN was for 20 min prior to washout. Somewhat longer lasting transient relaxations were obtained with GSH in similar experiments (not shown).

In further experiments, PN (2 mM) was mixed in KBS or KHS with or without GL present outside of the organ chambers, and the resulting solution of decayed PN (decay being complete within seconds) was tested for its ability to prime aortic rings for transient relaxation by CYS. The decayed PN solution was substituted for 25 to 50% of the solution bathing a ring in an organ chamber for 20 min, and then after a thorough washout, the ring was tested 30-60 min later. Those rings that had been incubated with PN decayed in KBS-GL gave no relaxation in response to the thiol, whereas those that had been incubated with PN decayed in KBS+GL, KHS+GL or KHS-GL all gave strong transient relaxations (Fig. 3). The presence of GL or HEPES in the organ chamber during incubation or final testing did not influence the results.

The present findings suggest that during its rapid decay at physiological pH in KSH+GL or KHS+GL or KHS-GL, PN or an intermediate product reacts with an hydroxyl group(s) in GL and HEPES to form nitrosated or nitrated products that can be bound by the tissue and can subsequently react with added thiol to produce an active relaxant (probably a nitrosothiol) of the smooth muscle. (Shortly after completion of the present work, another laboratory reported biological and chemical findings suggesting that PN reacts with sugars and other molecules containing an alcohol functional group to form NO donors with the characteristics of nitrate/nitrites [6].)

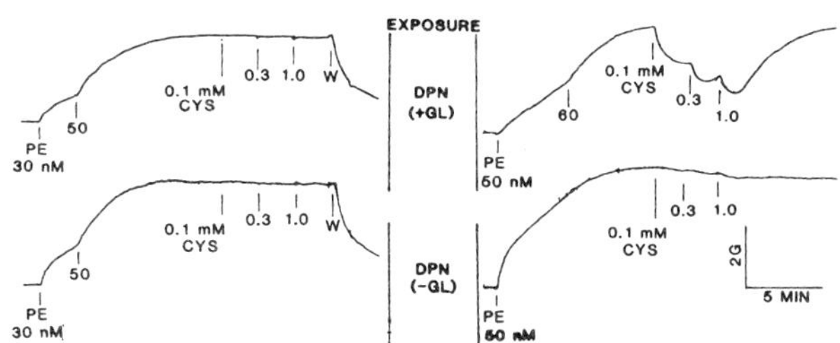

Fig. 3. Exposure of an aortic ring to a solution of PN decayed in KHS+GL but not in KHS-GL primes the ring for responding to CYS after washout with a transient relaxation. Exposure was to decayed products (0.5 mM original PN) for 20 min followed by 30-min washout. Solutions of PN decayed in KHS+GL or KHS-GL also primed for transient relaxations by thiols (not shown).

Acknowledgement: Research support by USPHS grant HL21860.

References

1. Beckman, J.S., Chen, J., Ischiropoulos, H. and Crow, J.P. (1994) in Methods in Enzymology, vol. 233 pp. 229-240, Academic Press, New York
2. Pryor, W.A. and Squadrito, J.L. (1994) Am. J. Physiol. **268**, L699-L722
3. Liu, S., Beckman, J.S. and Ku, D.D. (1994) J. Pharmacol. Exp. Therap. **268**, 1114-1121
4. Wu, M., Pritchard, K.A., Kaminski, P.M., Fayngersh, R.P., Hintze, T.H. and Wolin, M.S. (1994) Am. J. Physiol. **266**, H2108-H2113
5. Villa, L.M., Salas, E., Darley-Usmar, V.M., Radomski, M.W. and Moncada, S. (1994) Proc. Natl, Acad. Sci. U.S.A. **91**, 12383-12387
6. Moro, M.A., Darley-Usmar, V.M., Lizasoain, I., Su, Y., Knowles, R.G., Radomski, M.W. and Moncada, S. (1995) Brit. J. Pharmacol. **116**, 1999-2004

STORAGE FORMS OF NITRIC OXIDE (NO) IN VASCULAR TISSUE

Martin Feelisch, Department of Nitric Oxide Research, Schwarz Pharma AG, Alfred-Nobel-Str. 10, 40789 Monheim, FRG.

Vascular tissue is known for long to respond with reversible relaxation on illumination with UV light, a phenomenon coined photorelaxation (PR) (1). It has been shown that this response is mediated via cleavage from a vascular store of a preformed vasorelaxant (2,3) and that it is enhanced by incubation with nitrite anions (NO_2^-) as well as NO-donor compounds (3,4). Moreover, PR is inhibited by hemoglobin and methylene blue and is associated with increases in intracellular cGMP levels (5), reminescent of the characteristics of endothelium-dependent vasorelaxation. Contrary to earlier findings showing PR to be independent of the presence of the endothelium, it was shown recently that repriming of a depleted vascular store is absolutely dependent on the production of endothelium-derived nitric oxide (NO) (6).

In the present study we sought to investigate the nature of the endogenous vascular storage pool by using isolated endothelium-intact rat aortic rings in organ baths challenged by illumination with either polychromatic (400-800 nm) or monochromatic light. We further investigated whether NO and related N-oxides generated from chemically distinct classes of NO-donors are similarly trapped in the vasculature and to which extent the store(s) produced could serve as source of NO.

Repeated exposure for 30 s of tissues with monochromatic light (range: 300-600 nm) resulted in reproducible transient PR responses with maximum sensitivity in the range of 340 to 360 nm. PR was accompanied by a maximally 5-fold increase in the level of cGMP with maximal sensitivity in the same wavelength range (see Fig.1). Previous exposure of tissues to acetylcholine (1 µM) followed by a wash-out and contraction with phenylephrine was found to increase PR to UV but not visible light when compared to control rings exposed to either saline or an equieffective concentration of papaverin instead. Similarly, tissue exposure with glyceryl trinitrate (GTN) or S-nitrosoglutathione (GSNO) (both at their respective EC_{95}) followed by a wash-out also potentiated PR to UV light. The extent of potentiation of PR differed largely between individual NO-donor compounds, authentic NO and its reaction products NO_2^-, nitrate (NO_3^-) and peroxynitrite ($ONOO^-$), even when compared at equipotent concentrations. Potentiation was more pronounced in endothelium-intact compared to endothelium-denuded rings.

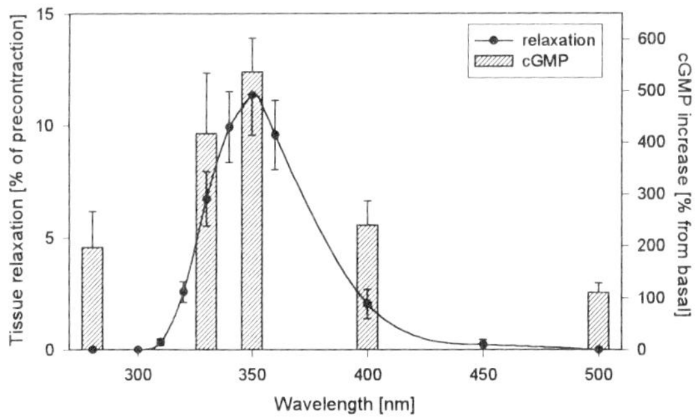

Fig. 1: Comparison of photorelaxation and cGMP increase on tissue exposure to monochromatic light (means ± SEM of 5 individual tissue preparations

On the other hand, prior illumination of tissues with polychromatic visible light or monochromatic UV light (350 nm) for 30 min was shown to completely abolish PR in response to exposure with UV light (see Fig. 2). The response to with either light source was biphasic with a transient component superimposed on a sustained component. Methylene blue, oxyhemoglobin, nitronyl nitroxides and potassium depolarisation also inhibited PR. Blockade of thiol groups by pretreatment of tissues with ethacrynic acid (30 µM) completely abrogated PR. Interestingly, in the presence of L-nitroarginine (100 µM) only PR responses to wavelengths > 400 nm were inhibited but not those elicited by exposure to lower wavelengths. Moreover, L-cysteine concentration-dependently abolished PR to both, UV and visible light. This effect was found to be due to the depletion by cysteine of the vascular store accounting for PR rather than by inhibition of NO-mediated vasorelaxation.

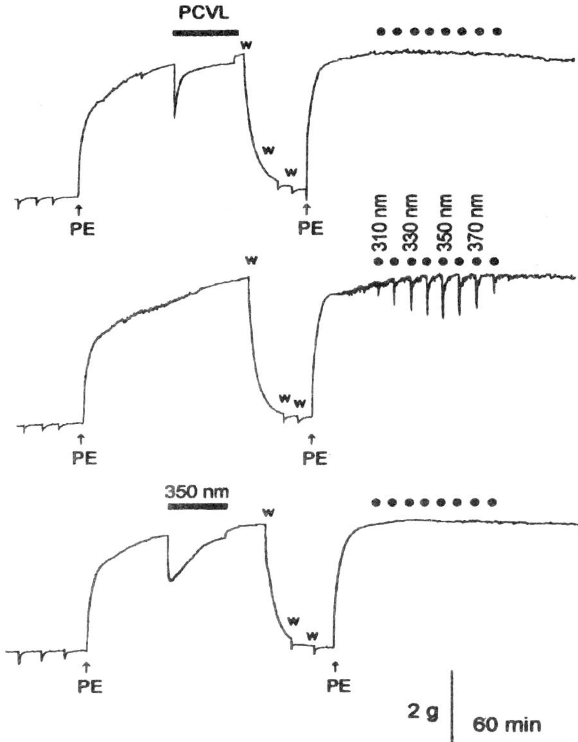

Fig. 2: Preexposure of tissues to polychromatic visible light (PCVL) or monochromatic UV light (350 nm) abolishes photorelaxant responses.
PE = phenylephrine 2 x 10^{-7} M, w = wash

In conclusion, the endothelium appears to play a critical role in modulating PR. The pharmacological characteristics of PR support the notion that these responses are largely mediated by the release of NO. Functional results with inhibitors of PR, different NO-donor compounds and illumination with light of discrete wavelengths suggest the existence of at least two distinct NO stores in the vasculature. The action spectrum of PR and the susceptibility to thiol modification furthermore suggest that one of the stores may be a nitrosothiol. Whether these photolabile vascular stores are part of an inactivation pathway for endogenously produced NO or represent redox-activatable sources of NO which contribute to the regulation of vascular tone will require further investigation.

References:

1) Furchgott, R.F. (1955) *Pharmacol.Rev.* 7:183-265.
2) Furchgott, R.F.and Jothianandan, D. (1991) *Blood Vessels* 28:52-61.
3) Venturini, C.M., Palmer, R.M.J., and Moncada, S. (1991) *J.Pharmacol.Exp.Ther.* 266: 1497-1500
4) Ehrreich, S.J. and Furchgott, R.F. (1955) *Nature* 218:682-4.
5) Karlsson, J.O.G., Axelsson, K.L., Elwing, H. and Andersson, R.G.G. (1984) *J.Cycl.Nucl.Prot.Phosph.Res.* 11:155-66.
6) Megson, I.L., Flitney, F.W., Bates, J. and Webster, R. (1995) *Endothelium*, 3:39-46

S-Nitrosohemoglobin: A new activity of blood involved in regulation of blood pressure

LEE JIA, CELIA BONAVENTURA, JOSEPH BONAVENTURA and JONATHAN S. STAMLER

Duke University Medical Center, Durham, North Carolina 27710, USA

Simulation of the diffusion and reactivity of nitric oxide NO in biological systems dictates that virtually all free NO in the vasculature should be scavenged by hemoglobin [1]. This raises the fundamental question of how NO exerts biological activity. There are two potential answers to this paradox. First, that NO-related activity, such as EDRF is not mediated exclusively by NO•. The second is that hemoglobin within red blood cells is not a net NO sink. Here we show that S-nitrosohemoglobin is endogenously formed, is biologically active, and counteracts the NO-scavenging activity of hemoglobin's metal centers.

Hemoglobin (Hb) is a tetramer comprised of two alpha and two beta subunits. In human Hb, each subunit contains one heme while the beta subunits also contain highly reactive SH groups (cysβ93)[2]. We examined the reactions of naturally occurring N-oxides such as NO and RSNO with hemoglobin (Hb). Nitric oxide (NO) converted oxyHb to metHb and nitrate, confirming previous reports. In contrast, RSNOs were found to participate in transnitrosation reactions with sulfhydryl groups of hemoglobin (forming S-nitrosohemoglobin (SNO-Hb)) which preserves the oxygen-delivery functionality.

Since oxygenation of hemoglobin is associated with conformational changes that increase the reactivity of cysβ93 [3], we examined the influence of conformational state on rates of S-nitrosylation. We found that S-nitrosylation of Hb was markedly accelerated in the oxy conformation. Moreover, deoxygenation was found to greatly enhance the rate of SNO-Hb decomposition. These data illustrate that oxygen-metal interactions influence S-NO affinity, and suggest a new allosteric function for Hb. To determine the physiological relevance of these findings, we developed an analytical approach to assay the S-nitrosothiol content of erythrocytes. Interestingly, we found that arterial blood (i.e. taken from the left ventricle) contained significant levels of SNO-Hb, whereas levels were virtually undetectable in venous blood (i.e. right ventricular blood). Two conclusions can be deduced from these findings: 1) SNO-Hb appears to be synthesized in the lung and 2) that SNO-Hb is metabolized during arterial-venous transit.

Arterial red blood cells contain two major isoforms of hemoglobin: oxyHb (Hb[FeII]O_2) and metHb (Hb[FeIII]) [2]. Intriguingly, SNO-Hb(FeII)O_2 was found to possess modest NO activity when tested in a vascular ring bioassay. By comparison, SNO-Hb(FeIII) was a more potent vasodilator (IC_{50} of approximately 5 μM). Notably, free NO was devoid of relaxant activity in the presence of oxyHb or metHb. As red blood cells contain millimolar concentrations of glutathione, we examined the vasoactivity of SNO-Hbs in its presence. We found that glutathione markedly potentiated the vasodilator activity of SNO-Hb(FeII)O_2 and SNO-Hb(FeIII). S-nitrosoglutathione formed under these conditions appeared to fully account for this effect. Further kinetic analyses revealed that transnitrosation involving glutathione was more strongly favored in the equilibrium with SNO-Hb(FeIII) than SNO-Hb(FeII)O_2. These data suggest that hemoglobin has evolved an electronic switching mechanism to achieve NO homeostasis. Specifically, NO scavenging by the metal center of SNO-Hb(FeII)O_2 would be sensed through its conversion to met (Hb[FeIII]). This electronic event would then affect S-nitrosothiol formation.

If the generation intracellular low molecular weight RSNO through such transnitrosation reactions is to be of physiological significance, then this activity must be freely accessible to the vessel wall. To test this possibility, we loaded red blood cells with SNO-Hb and measured the accumulation of extracellular RSNO. Our data reveal that red blood cells export pharmacologically relevant levels of low molecular weight RSNOs. Thus, the red blood cell does not appear to be a limiting factor in the transduction of intracellular NO-related activity to the vessel wall.

What are the therapeutic implications of these data? There is compelling need for blood substitutes because of periodic shortages of blood and the rising incidence of blood-borne illness. The use of cell-free Hb blood substitutes, however, has been associated with a variety of cardiovascular, respiratory and gastrointestinal side effects which can be attributed to scavenging of NO. For example, infusions of Hb cause hypertensive responses that relate to NO scavenging by the metal center. In contrast, we find that SNO-Hb(FeII)O_2 has no such hypertensive response, suggesting that cell-free Hb solutions that mimic blood by containing SNO-Hb may be useful as a blood substitute.

In summary, our data highlight new allosteric and/or electronic properties of hemoglobin involved in regulation of vasomotor tone, argue against the importance of free NO in transduction of such NO-related activity, and suggest that SNO-Hbs could be used to overcome the hypertensive side effects of Hb-based blood substitutes.

References:

1. Lancaster, J.R. (1994) *Proc. Natl. Acad. Sci. USA*, **91**, 8137-8141.
2. Antonini, E. & Brunori, M. (1971) *in Hemoglobin and Myoglobin in Their Reactions with Ligands*, pp. 29-31, American Elsevier Publishing Co, Inc., New York.
3. Craescu, C.T. et al. (1986) *J. Biol. Chem.* **261**, 14710-14716.

Modulation of superoxide-dependent oxidation and hydroxylation reactions by nitric oxide

Allen M. MILES, D. Scott BOHLE*, Peter A. GLASSBRENNER*, Bernhard HANSERT*, David A. WINK†, and Matthew B. GRISHAM.

Department of Physiology and Biophysics, Louisiana State University Medical Center, Shreveport, Louisiana 71130.
*Department of Chemistry, University of Wyoming, Laramie, Wyoming 82071.
†Radiation Biology Branch, National Cancer Institute, Bethesda, MD 20892.

Introduction

It is becoming increasingly apparent that certain types of inflammatory tissue injury are mediated by reactive metabolites of oxygen and nitrogen. For example, it has been demonstrated that administration of superoxide dismutase (SOD) is effective at attenuating the tissue injury observed in experimental models of arthritis, chronic gut inflammation and immune complex-induced pulmonary injury. Furthermore, models of joint, bowel and lung inflammation have been shown to be associated with enhanced production of nitrogen oxides derived from the free radical nitric oxide (NO). Indeed, recent studies have demonstrated that inhibition of NO synthase (NOS) also provides substantial protection against the inflammatory tissue injury observed in these models of acute and chronic inflammation [1-3]. These data suggest that both superoxide (O_2^-) and NO are important mediators of inflammation-induced tissue injury and dysfunction. The mechanisms by which O_2^- and NO may either separately or in tandem mediate tissue injury during inflammation remain the subject of active debate.

Recent biochemical studies have demonstrated that O_2^- and NO rapidly interact via a radical-radical reaction at a diffusion-limited rate ($k = 6.7 \times 10^9\ M^{-1}s^{-1}$) to generate the potent oxidant peroxynitrite ($ONOO^-$) [4]. Beckman and co-workers [5] have suggested that the interaction between these two free radicals to yield $ONOO^-$ and its conjugate acid, peroxynitrous acid (ONOOH), enhances dramatically the toxicity of either O_2^- or NO alone. Indeed, it has been demonstrated *in vitro* using pre-formed (e.g. chemically synthesized) $ONOO^-$/ONOOH that these oxidants are capable of directly oxidizing carbohydrates, sulfhydryls, lipids and DNA bases as well as mediating bactericidal and endothelial cell toxicity. It has also been demonstrated that the simultaneous production of NO and O_2^- by macrophages may result in the formation of $ONOO^-$/ONOOH. However, a series of recent reports demonstrate that NO may actually inhibit O_2^--dependent, iron (or hemoprotein)-catalyzed reactions in vitro oxidant-dependent inflammation in vivo.

The reasons for these apparent discrepant results are not clear. However, recent evidence by Rubbo et al [6] suggests that the relative fluxes of the two free radicals may be an important determinant as to whether NO enhances or inhibits O_2^--dependent, iron catalyzed lipid peroxidation [6]. In these studies the effects of NO were assessed only in the presence of ferric iron (Fe^{+3}). A recent preliminary study from our laboratory suggests that the affects of NO on O_2^--dependent oxidative reactions may be quite different depending upon whether redox-active transition metals are present or absent [7]. Therefore, the objectives of the present study were to: a) systemically quantify the oxidizing and hydroxylating activity of NO and/or O_2^- in the absence or presence of redox-active Fe and b) characterize these reactions using different fluxes of each radical. The physiologic significance of our findings is discussed.

Results and Discussion

Using the hypoxanthine/xanthine oxidase system (HX/XO) to generate O_2^- and H_2O_2 and the spontaneous decomposition of the spermine/NO adduct to produce varying fluxes of NO, we found that in the absence of redox-active iron (Fe), and presence of catalase to insure elimination of O_2^--driven Fenton chemistry, neither O_2^- nor NO was capable of oxidizing substantial amounts of dihydrorhodamine 123 (DHR) to yield its fluorescent product rhodamine 123 (RH) or to hydroxylate benzoic acid (BA) to yield its fluorescent product 2-hydroxybenzoate (HB). However, simultaneous production of equimolar fluxes of O_2^- and NO increased the formation of RH from normally undetectable levels to 16 μM suggesting the formation of a potent oxidant. Addition of SOD inhibited this oxidative reaction suggesting that O_2^- interacts with NO to generate a potent oxidizing agent. Maximum oxidation was obtained only when the fluxes of both O_2^- and NO were approximately equal. Excess production of either radical virtually eliminated the oxidation of DHR. In the presence of trace amounts of EDTA-chelated ferric Fe (Fe^{+3}; 5 μM for each) and absence of catalase to insure optimum O_2^--driven Fenton chemistry, NO enhanced modestly HX/XO-induced DHR oxidation (15 μM RH in the absence of NO vs 19 μM in the presence of equimolar fluxes of NO). As expected, both SOD and catalase inhibited this Fe-catalyzed oxidation reaction. Excess NO production with respect to O_2^- flux produced only modest inhibition (33%) of DHR oxidation. In a separate series of studies, we found that neither O_2^- nor NO alone were able to hydroxylate benzoic acid (BA) in the absence of Fe. The simultaneous generation of O_2^- and NO in the absence of Fe and presence of catalase only modestly enhanced hydroxylation of benzoic acid from undetectable levels to 0.6 μM 2-hydroxybenzoate (HB). SOD inhibited this reaction suggesting that O_2^- interacts with NO to generate a product with only weak hydroxylating activity. In the presence Fe^{+3}-EDTA and absence of catalase, HX/XO-mediated hydroxylation of BA increased dramatically from undetectable levels to 4.5 μM HB. SOD and catalase were both effective at inhibiting this classic O_2^- driven Fenton reaction. Interestingly, NO inhibited this Fe-catalyzed hydroxylation reaction in a concentration-dependent manner such that fluxes of NO approximating those of O_2^- and H_2O_2 virtually abolished the hydroxylation of BA. We conclude that in the absence of iron, equimolar fluxes of NO and O_2^- interact to yield potent oxidants such as $ONOO^-$/ONOOH which oxidize organic compounds. In the presence of low molecular weight, redox-active Fe complexes, NO may enhance or inhibit O_2^- -dependent oxidation and hydroxylation reactions depending upon their relative fluxes.

Acknowledgements: This work was supported by grants from the National Institutes of Health (DK47663 and CA63641; MGB) and the American Heart Association (DSB).

References

1. McCartney-Francis, N., Allen, J. B., Mizel, D. E., Albina, J. E., Xie, Q. W., Nathan, C. F., and Wahl, S. M. (1993) J. Exp Med. **178**, 749-754.
2. Grisham, M. B., Specian, R. D. and Zimmerman, T. E. (1994) J. Pharmacol. *Exp. Therap.* **271**, 1114-1121.
3. Mulligan, M. S., Moncada, S., and Ward, P. A. (1992) Br. J. Pharmacol. **107**, 1159-1162.
4. Huie, R. E. and Padmaja, S. (1993) Free Radic. Res. Commun. **18**, 195-199.
5. Beckman, J. S., Beckman, T. W., Chen, J., Marshall, P. A. and Freeman, B. A., (1990) Proc. Natl. Acad. Sci. USA **87**, 1620-1624.
6. Rubbo, H., Raddi, R., Trujillo, M., Telleri, R., Kalyanaraman, B., Barnes, S., Kirk, M., and Freeman, B. A. (1994) J. Biol. Chem. **269**, 26066-26075.
7. Miles, A. M., Mureeba, P., Mears, J. R. and Grisham, M. B. (1995) FASEB J. **9**, A853.

Role of nitric oxide in the activation of sperm from *Arbacia punctulata.*

Diane E. Heck[1]*, Walter Troll[2]*, and Jeffrey D. Laskin[3]*,

Departments of Pharmacology and Toxicology, Rutgers University, Piscataway NJ[1],08855, USA, Environmental Medicine, New York University Medical Center, NY, NY[2], 10016, USA, Environmental and Community Medicine, UMDNJ-Robert Wood Johnson Medical School, Piscataway, NJ[3], 08854, USA, and Marine Biological Laboratory, Woods Hole, MA* 02543, USA.

Sea urchin sperm provide a simple cellular model to study the early biochemical and physiological events in fertilization. In response to environmental stresses, sea urchins release their gametes into the sea (1). Eggs and sperm are then required to locate each other and interact unprotected by the parent animals. To enhance the likelihood of survival, sea urchin eggs are shielded by jellycoat. This complex protective covering comprised primarily of high molecular weight fucose-sulfate-rich glycoconjugates contains a variety of sperm-chemoattractive molecules (2). Successful fertilization of eggs from the sea urchin *Arbacia punctulata* requires activation of the released sperm. This process is bimodal consisting of initiation of motility and a dramatic increase in respiratory activity (3). Sperm activation is followed by induction of the acrosome reaction, induced by glycoprotein components of egg jelly (4). We have found that activation of sperm results in production of nitric oxide. Nitric oxide is enzymatically produced in many organisms, ranging from Limulus to mammalian species, through the oxidation of l-arginine by the nitric oxide synthase family of oxidoreductases (5-7). Within the mammalian vasculature, nitric oxide functions to regulate the relaxation of smooth muscle. In these cells nitric oxide initiates relaxation by interacting with critical heme moieties of the enzyme guanylate cyclase (8). Interestingly, in activated sea urchin sperm, which contain one of the highest concentration of guanylate cyclase of any cell type, motility is regulated by this second messenger (3, 9). Even more intriguing is the finding that motility of sea urchin sperm is calcium and calmodulin dependent (10). This observation suggests that the calcium-dependent neuronal or endothelial nitric oxide synthase isoforms may be important in sperm motility.

STIMULUS	NITRITE (nmol/50 μg sperm)
sea water	3.59 ± 1.11
egg water	18.36 ± 1.49
resact	15.87 ± 3.59
NMMA/sea water	2.46 ± 2.11
NMMA/egg water	11.21 ± 1.21
aminoguanidine/sea water	1.45 ± 1.78
aminoguanidine/egg water	4.01 ± 1.07

Table 1. Sperm (10^9/mL), freshly retrieved from *Arbacia punctulata*, were stimulated with sea water (control), egg water or resact (1 ng/mL), in the presence and absence of NMMA (2 mM) or aminoguanidine (500 μM); after 12 hr nitrite accumulation was determined using the Greiss reaction.

Production of nitric oxide could be initiated either by treating sperm with egg water, or through incubation with resact, a sperm-activating peptide derived from *Arbacia punctulata* eggs (1-1000 nM) (11). Nitric oxide production was inhibited by incubating sperm with n-monomethyl l-arginine (NMMA) or aminoguanidine (AG), specific inhibitors of nitric oxide synthase (6, 7, 12-13). Inhibition of nitric oxide production resulted in diminished sperm motility (Table 1) (14).

A question arises as to the biological consequences of inhibiting sperm nitric oxide production. To address this we incubated sperm with nitric oxide synthase inhibitors or nitric oxide sinks prior to the addition of eggs. We observed that successful fertilization was inhibited both in the presence of nitric oxide synthase inhibitors (NMMA or aminoguanidine) and in the presence of NBT, a scavenger of nitric oxide (Table 2) (15).

Treatment	% Fertilization
none	95.6 ± 1.7
aminoguanidine	30.8 ± 2.7
NBT	12.9 ± 8.6
Hb	92.1 ± 5.8

Table 2. Sperm (10^6/mL), pretreated for 1 hr with aminoguanidine (500 μM), nitro blue tetrozolium (NBT, 0.2 μM), or reduced hemoglobin (Hb, 5 mM) were added to eggs freshly retrieved from *Arbacia punctulata*. After 3 hours successful fertilization was determined as the percentage of dividing cells.

Nitric oxide was not released by the sperm during this process as hemoglobin a nitric oxide scavenger not taken up by sperm, had no effect (16). These data indicate that nitric oxide is produced during sperm activation. We speculate that initiation of sperm motility is regulated by nitric oxide and that this process is required for fertilization of *Arbacia punctulata*. Supported by NIH grant ES05022.

References

1. Foltz, K.R. and Lennarz, W.J. (1993) Dev. Biol. **158**, 46-61
2. Ward, G.E., Brokaw, C.J., Garbers, D.L. and Vacquier, V.D. (1985) J. Cell Bio.**101**, 2324-2329
3. Shimomura, H. and Garbers, D.L. (1986) Biochemistry **25**, 3405-3410
4. Keller, S.H. and Vacquier, V.D. (1994) Dev. Biol. **162**, 304-312
5. Laskin, J.D., Heck, D.E. and Laskin, D.L. (1994) Trends Endo. Metab. **5**, 377-382
6. Moncada, S., Palmer, R.M.J. and Higgs, E.A. (1991) Pharmacol.. Rev. **43**, 109-142
7. Nathan, C.F. and Xie, Q.W. (1994) J. Biol. Chem. **269**, 13725-13728.
8. Gruetter, C.A., Barry, B.K., McNamara, D.B., Gruetter, D..Y., Kadowitz, P.J. and Ignarro, L.J. (1979) Adv. Cyclic Nucleotide Res. **5**, 211-224
9. Chodavarapu, S.R. and Garbers, D.L. (19855) J. Biol. Chem. **260**, 8390-8396
10. Cook, S.P., Brokaw, C.J., Muller, C.H. and Babcock, D.F. (1994) Dev. Biol. **165**, 10-19
11. Ward, G.E., Garbers, D.L. and Vacquier, V.D. (1984) Science **227**, 768-770
12. Griffiths, M.J.., Messeret, M., MacAllister, R.J. and Evans, T.W. (1993) Br. J. Pharmacol. **110**, 963-968
13. Heck, D.E., Gardner, C.R., and Laskin, J.D. (1992) J. Biol. Chem. **267**, 21277-21280
14. Heck, D.E., Laskin, J.D., Zigman, S. and Troll, W. (1994) Biol. Bull. **187**, 248-249.
15. Wolff, D.J. and Datto, G.A. (1992) Biochem. J. **285**, 201-206
16. Stamler, J.S., Singel, D.J. and Loscalzo, J. (1992) Science **258**, 1898-1902

Nitric oxide initiates growth arrest during differentiation of neuronal cells

G. ENIKOLOPOV and N. PEUNOVA

Cold Spring Harbor Laboratory, Cold Spring Harbor, New York, 11724, USA

Arrest of cell division is a prerequisite for cells to enter a program of terminal differentiation. In the developing nervous system, growth arrest defines roughly the size of the cellular population that is further committed to become a domain of differentiated neurons (1-3). Mitogenesis and cytostasis of neuronal cell precursors can be induced by the same or by different growth or trophic factors. Differentiation of neuronal PC12 cells induced by nerve growth factor (NGF) involves a proliferative phase that is followed by cytostasis and differentiation (4, 5). We have found that NGF induces different forms of nitric oxide synthase (NOS) in neuronal PC12 cells, that NO acts as a cytostatic agent in these cells, that inhibition of NOS leads to reversal of NGF-induced cytostasis and thereby prevents further development of the differentiated phenotype, and that capacity of a mutant cell line to differentiate can be rescued by NO (6). Our experiments suggest that the cytostatic effect of NGF is mediated by NO and that induction of NOS is an important step in the commitment of neuronal precursors during differentiation.

NGF treatment induces NOS activity

When untreated PC12 cells were tested for the diaphorase cytochemical reaction, no staining was observed. But after the cells were treated with NGF, they gradually started to acquire an intense NADPH-dependent blue color after diaphorase staining, indicating that NOS accumulates in PC12 cells in response to NGF treatment. Importantly, the NGF-treated cells that were first to undergo initial morphological changes characteristic of the differentiated phenotype were also the first to show bright staining. This suggests that elevation of NOS activity precedes development of the differentiated phenotype. Induction of NOS was confirmed using in vitro assays and immunocytochemistry.

NO can act as an antimitogenic agent in PC12 cells

NO can inhibit DNA synthesis in several systems (7-9). It acts as an antimitogenic agent in PC12 cells as well. Several NO donors (SNP, SNAP, SNOP) suppressed DNA synthesis in a concentration-dependent manner (albeit with different potencies), implying that NO is an active inhibitor of DNA replication in PC12 cells. This inhibition was caused by a decrease in cell proliferation (cytostasis) and was not due to cytotoxicity.

The results of the FACS analysis show that NO induces specific changes in cell-cycle phase distribution of PC12 cells. Cells accumulated specifically in G2 phase, while the proportion of cells in S phase decreased. Remarkably, within a range of NO donors' concentration, proportion of PC12 cells in G2 and S phases was similar to the levels reached after prolonged treatment with NGF. These results with NO donors suggest that NO molecules produced by the cell during the course of NGF action could play a similar role in promoting NGF-induced cytostasis.

Inhibition of NOS activity prevents cytostatic action of NGF

After several days of NGF treatment PC12 cells' proliferation ceases and differentiation occurs. If NO produced by the NGF-induced NOS in PC12 cells can indeed act as an antiproliferative factor, then inhibition of the enzyme should uncouple the proliferative and cytostatic components of NGF action and prolong the proliferative phase while suppressing the cytostatic effect of NGF. NOS inhibitor L-NAME indeed reversed the cytostatic action of NGF and forced the cells to continue to proliferate instead of ceasing to divide after 6-8 days of NGF treatment. In the absence of NGF, similar concentrations of L-NAME did not have a detectable effect on PC12 cell growth; in particular, L-NAME did not accelerate cell proliferation.

The most visible consequence of NGF action on PC12 cells is neurite outgrowth and this was also affected by NOS inhibition. Under normal conditions, almost every cell gradually extended neurites in response to NGF but, when PC12 cells were treated with a combination of NGF and NOS inhibitor, the number of cells with neurites decreased dramatically, in accordance with the observation that most of the cells continued to divide.

The crucial phase of NOS action is cell cycle-specific and there is a critical time window for NO action within the first days of NGF addition.

Mutant cells' capacity to differentiate can be rescued by NO

Mutant PC12-U2 cells retain the early steps of the response to NGF, but have lost the capacity to execute the later steps (5). They do not stop dividing after NGF treatment and, as a consequence, they do not develop the fully differentiated phenotype; in particular, they do not send out processes. To ask if the deficit is connected to NO production, we tested if this phenotype can be overcome by the addition of NO. While neither NGF nor NO alone had any effect on the phenotype of U2 cells, in combination these treatments restored the differentiated neuronal phenotype. The cells stopped dividing and grew extensive branched neurites. Thus it was possible to rescue the defect in U2 cells and to restore the complete differentiation program by complementing NGF action with NO.

We suggest a model for NGF action on PC12 cells in which at least three stages can be outlined. In the first, proliferative stage NGF activates a cascade of genes, eventually leading to the induction of the *NOS* gene(s). In the second stage, the accumulated NOS enzyme produces enough NO to inhibit DNA synthesis and, probably, to "alert" further checkpoints, thereby blocking further progression of the cell cycle, completing the mitogenic phase of NGF action, and switching the cells to the cytostatic phase; probably, NO also directly induces some of the later differentiation markers by promoting gene activity (10). Finally, as soon as the cell perceives and processes the cytostatic signal, it starts to implement the remaining program of differentiation traits (such as neurite outgrowth), which can only occur after cell division has ceased. Given the unconventional properties of NO, it is possible that NO diffuses from the cell that produces it to promote cessation of growth in adjacent cells and that this phenomenon contributes to the synchronization of development of a domain of neuronal precursors.

References

1. Jackobson, M. In: *Developmental Neurobiology* (Plenum, NY) (1991).
2. McKay, R.D.G. *Cell*, 58, 815-821 (1989).
3. Anderson, D.J. *Annu. Rev. Neurosci.* 16, 129-158 (1993).
4. Green, L.A., & Tishler, A.S. *Proc. Natl. Acad. Sci. USA* 73, 2424-2428 (1976).
5. Burstein, D.E. & Greene, L. *Devel. Biol.* 94, 477-482 (1982).
6. Peunova, N., & Enikolopov, G. *Nature* 375, 68-72 (1995).
7. Garg, U.C. & Hassid, A. *J. Clin. Invest.* 83, 1774-1777, (1989).
8. Lepoivre, M. et al. *BBRC*, 179, 442-448 (1991).
9. Kwon, N.S., Stuehr D.J. & Nathan, C.F. *J. Exp. Med,* 174, 761-767 (1991).
10. Peunova, N., & Enikolopov, G. *Nature* 364, 450-453 (1993).

Parasympathetic signaling mediated by activation of endothelial constitutive (NOS 3) nitric oxide synthase in cardiac myocytes.

JEAN-LUC BALLIGAND*, LESTER KOBZIK#, XINQIANG HAN*, DAVID M. KAYE*, LAURENT BELHASSEN*, DONALD S. O'HARA*, RALPH A. KELLY*, THOMAS MICHEL* and THOMAS W. SMITH*.

*Department of Medicine, Cardiovascular Division, Brigham and Women's Hospital, Harvard Medical School, Boston, MA 02115, and #Department of Pathology, Brigham and Women's Hospital, Harvard Medical School, and Physiology Program, Harvard School of Public Health, Boston, MA 02115, USA.

The three NO synthase isoforms identified to date are discovered in an increasing variety of cell types with different roles in cell signaling. The constitutive isoform originally identified in brain (or *NOS1*)[1] was recently found to be expressed in skeletal muscle cells [2], and the constitutive isoform first characterized in endothelial cells (or *NOS3*) [3] has been identified in LLC-PK1 kidney epithelial cells [4] and rat hippocampal pyramidal cells [5].

Nitric oxide regulates the contraction of cardiac myocytes and isolated cardiac muscle preparations in vitro, and of hearts studied in experimental animals [for review, see 6]. The inducible NOS (or *NOS2*) initially characterized in a macrophage cell line [7] is expressed in ventricular myocytes from adult rats in response to specific cytokines [8]. Independent of *NOS2* induction, however, receptor-dependent signaling is modulated by a constitutive NOS isoform in cardiac cells [9], but the specific cell type producing NO to alter cardiac myocyte function has not been identified, nor has the identity of the constitutive isoform expressed in these cells been determined.

We first determined that the NO synthase activity measured by the assay for the conversion of ^{3}H-L-arginine to ^{3}H-L-Citrulline in extracts of purified ventricular myocytes in the absence of cytokine treatment was calcium-sensitive. This was consistent with the expression of either *NOS 1* or *NOS 3*, which are distinguished from *NOS2* by their calcium sensitivity for the binding of activator calmodulin and enzymatic activity within the physiological range for intracellular calcium. The NO synthase isoforms are also distinguished on the basis of their subcellular distribution. *NOS1* and *NOS3* are found primarily in the particulate subcellular fraction [2, 10], but *NOS2* is cytosolic in most tissues. We resolved whole cell extracts of cardiac myocytes into cytosolic and particulate fractions by ultracentrifugation, and analyzed both fractions for the presence of a calcium-sensitive NOS activity. Seventy-five percent of the activity segregated with the particulate fraction, a result also consistent with the expression of either *NOS1* or *NOS3* in ventricular myocytes. We definitively identified the isoform constitutively expressed in these cells by molecular cloning and immunohistochemical analysis. First, we designed isoform- and species-specific oligonucleotide probes to amplify cDNA for the constitutive isoforms of NOS in cardiac myocytes by reverse transcription-PCR. No PCR product was obtained from freshly isolated ventricular myocytes when amplified with primers specific for rat *NOS1* or with our primers previously characterized for rat *NOS2* [8]. Using the amplimers specific for rat *NOS3*, a single 324-bp product was amplified from reverse-transcribed RNA from purified myocytes. Cloning and nucleotide sequence analysis of the PCR product revealed a sequence 84% identical to the bovine *NOS3* cDNA, with a deduced amino acid sequence 91% identical to the bovine peptide. Using the PCR product labeled with ^{32}P as a probe, *NOS3* transcripts were detected in total RNA extracted from purified ventricular myocytes by Northern analysis. Of note, treatment with specific cytokines down-regulated the abundance of *NOS3* transcript in ventricular myocytes. Using a monoclonal antibody directed against human *NOS3*, we also detected the corresponding protein by Western analysis in whole extracts of myocytes. The protein was mainly detected in the particulate subcellular fraction, consistent with the distribution of NOS enzymatic activity as described above. With the same antibody, we demonstrated the expression of *NOS3* in single ventricular myocytes from adult rat hearts and from human cardiac tissue by immunohistochemistry.

We further studied the regulation by *NOS3* of both L-type calcium current (I_{Ca-L}) and contractility in response to cholinergic and adrenergic agonists in single myocytes using the nystatin-perforated patch-clamp technique in the whole cell configuration to allow simultaneous recordings of I_{Ca-L} along with cell shortening, measured with a video-motion analyzer. Carbamylcholine attenuated the isoproterenol-stimulated increase in I_{Ca-L} and amplitude of shortening in paced ventricular myocytes. After extracellular perfusion of the cell with methylene blue, the antagonistic effect of carbachol on isoproterenol-stimulated I_{Ca-L} and shortening was totally abolished. Internal dialysis of cardiac myocytes with methylene blue or the NO synthase inhibitor L-N-monomethyl-Arginine using the membrane ruptured patch-clamp technique also completely abrogated carbachol's effect on I_{Ca-L}, an effect that was reversed with an excess of the normal substrate L-Arginine.

We conclude that endogenous NO is constitutively produced by *NOS3* in cardiac ventricular myocytes and mediates the regulation of voltage-dependent calcium current and contraction by parasympathetic agonists in these cells.

REFERENCES:

1. Bredt, DS, Hwang, PM, Glatt, CE, Lowenstein,C, Reed RR, and Snyder, SH (1991) Nature 351, 714-718
2. Kobzik L, Reid MB, Bredt DS, and Stamler, JS (1994) Nature 372, 546-548.
3. Lamas S, Marsden PA, Li GK, Tempst P, and Michel T.(1992) Proc Natl Acad Sci U.S.A. 88, 6348-6352.
4. Tracey WR, Pollock JS, Murad F, Nakane M and Forstermann U.(1994) Am J Physiol 266(CP 35) C22-C28.
5. Dinerman JL, Dawson TM, Schell MJ, Snowman A and Snyder S.(1994) Proc Natl Acad Sci USA 91, 4214-4218.
6. Ungureanu-Longrois D, Balligand JL, Kelly RA and Smith TW.(1995) J Mol Cell Cardiol 27, 155-167.
7. Xie QW, Cho HJ, Calaycay J et al.(1992) Science 256, 225-228.
8. Balligand JL, Ungureanu D, Simmons WW et al. (1994) J Biol Chem 269, 27580-27588.
9. Balligand JL, Kelly RA, Marsden PA, Smith TW and Michel T.(1993) Proc Natl Acad Sci USA 90, 347-351.
10. Busconi L and Michel.(1993) J Biol Chem 268, 8410-8413.

The immunohistochemical localization of nitric oxide synthase (NOS) in the human male reproductive tract and in spermatozoa, suggests a possible role for nitric oxide in spermatogenesis and sperm maturation and an association with subfertility.

Armand Zini[1,3], Moira K. O'Bryan[3], Margaret Magid[2] and Peter N. Schlegel[1,3].

[1]James Buchanan Brady Foundation, Department of Urology and [2]The Department of Pathology at The New York Hospital-Cornell Medical Center, and [3]The Population Council, Center for Biomedical Research, New York, NY 10021.

Introduction

Recent work has implicated nitric oxide (NO) in several aspects of male genital physiology including erectile function, androgen secretion, as well as *in vitro* effects on sperm function [1-5]. In addition, the presence of NOS has been demonstrated by immunohistochemical and histochemical studies, in the reproductive organs of both male and female rats [6,7].

The objectives of the present study were to determine whether male human reproductive organs and spermatozoa possess endothelial NOS (ecNOS) and if so, does the protein expression in the testis change under situations of pathology and does the expression in spermatozoa correlate with motility. ecNOS protein was localized using a monoclonal antibody against the ecNOS isoform in a streptavidin-biotin-amplified peroxidase technique and confirmed using NADPH diaphorase histochemistry.

Materials and Methods

Human tissues were obtained, under ongoing internal review board approval, from operative specimens of men undergoing orchiectomy for treatment of early stage prostate cancer and from tissue blocks of human testicular biopsy material of patients evaluated for infertility. Semen samples were obtained from patients consulting for infertility.

Fresh frozen tissues [testes (n=2), epididymides (n=2) and vas deferens (n=2) from men with early stage prostate cancer] were used for histochemistry and immunohistochemistry. The Bouin's fixed paraffin-embedded human testicular biopsy material [n=8, including the pathological diagnoses of Sertoli cell only syndrome, early and late maturation arrest, and 'normal' (obstructed) spermatogenesis] was used for immunohistochemistry. Semen samples were washed by Percoll gradient centrifugation [8], (a technique used to separate spermatozoa based on their density and motility). Each fraction was assessed for percent sperm motility and percent sperm head staining.

The presence of ecNOS in human testicular cytosol, epididymal cytosol, seminal plasma and spermatozoa membranes were assessed using an immunoblot method previously described [9]. Protein was fractionated on a 7.5% SDS-PAGE resolving gel and ecNOS was detected using enhanced chemiluminescence.

Immunohistochemistry was performed using a commercially available anti-endothelial NOS (ecNOS) monoclonal antibody (Transduction Laboratories, Lexington, KY, #N30020). This antibody has been extensively characterized to confirm its specificity to the endothelial form of NOS [10,11]. As controls, specimens were processed in the presence of mouse IgG and in the absence of primary antibody.

NADPH diaphorase staining was performed on fresh frozen sections [6]. Freshly fixed slides were incubated in PBS containing 0.2mg/ml nitro blue tetrazolium and 1.0 mg/ml β-NADPH at 37°C for 3 hours. Control sections were incubated without NADPH.

Results and Discussion

The ecNOS protein was detected in human testis and epididymis by Western blot. In addition to the ~130 kDA band, a second band of approximately 44 kDA was detected in both human testis and epididymal protein preparations. The 44 kDa band likely represents an immunoreactive degradation product of ecNOS [10,12].

Within the human testis, ecNOS protein was present in Leydig cells and in Sertoli cells at all stages of spermatogenesis. Parallel NADPH diaphorase histochemistry was performed to confirm the enzymatic activity of localized ecNOS protein (Fig. 1A). ecNOS was distributed throughout the cytoplasm of Leydig cells and Sertoli cells in both histologically 'normal' testes and in those from men with subfertility. Although, germ cells were for the most part unstained, abnormal (prematurely shed and apoptotic) germ cells did stain for ecNOS.

NOS immunostaining and NADPH diaphorase staining in the epididymis and vas was almost exclusively confined to the epithelium. The greatest concentration of staining was seen in the adluminal region of the epithelium (with some extension into microvilli) and in a narrow region at the base of basal epithelial cells (Fig. 1B). ecNOS was localized primarily to the head region of abnormal spermatozoa. A significant inverse correlation was found between the percentage of stained cells and the percent motile spermatozoa in a given fraction (Fig 2).

The results presented here, and those published by other groups, suggest that germ cells and spermatozoa encounter NO or its secondary reaction products *in vivo*. The presence of ecNOS protein and NOS activity in Sertoli, Leydig and abnormal germ cells, suggests a role for NO in spermatogenesis, spermatogenic arrest and androgen production. Its presence within the epididymis and vas deferens are indicative of a possible role in sperm maturation in the male reproductive tract [6], and its presence within the prostate and seminal vesicle suggest a possible role for NOS in sperm maturation within the female genital tract [6]. The inverse correlation between sperm head staining and sperm motility suggests that NOS may either play a causative role in, or a selective process against, abnormal sperm function. Such a broad distribution illustrates that NO may play an important role in the function of the human male reproductive system, and highlights the need for further investigation into this complex system in an effort to understand both normal male reproductive function and the etiology of subfertility.

References:

1. Welch C, Watson ME, Poth M, Hong T and Francis GL. (1995) Metabolism, 44: 234-238.
2. Hellstrom WJG, Bell M, Wang R, Sikka SC. (1994) Fertil Steril, 61: 1117-1122.
3. Rajfer J, Aronson WJ, Bush P, Dorey FJ, Ignarro LJ. (1992) New Engl J Med. 326: 90-94.
4. Tomlinson MJ, East SJ, Barratt CL, Bolton AE, Cooke ID. (1992) Am J Reprod Immunol, 27: 89-92.
5. Zini A, de Lamirande E, Gagnon C. Journal of Andrology in press (1995).
6. Burnett AL, Ricker DD, Chamness SL, Maguire MP, Crone JK, Bredt DS, Snyder SH, Chang TSK. (1995) Biol Reprod 52: 1-7.
7. Suburo AM, Chaud M, Franchi A, Polak J, Gimeno MAF. (1995) Biol Reprod 52: 631-637.
8. de Lamirande E, Gagnon C. (1991) Int J Androl, 14: 11-22.
9. Cheng CY and Bardin CW. J Biol Chem 1987; 262: 12768-12779.
10. Lloyd RV, Long J Qian X, Zhang S and Scheithauer BW. (1995) Am J Path, 146: 86-94.
11. Cobb, CS, Brenman JE, Aldape, KE, Bredt, DS, Israel, MA. (1995) Cancer Res 55: 727-730.
12. Pollock JS, NakaneM, Buttery LDK, Martinez A, Springall D, Polak JM, Forstermann U and Murad F. Am J Physiol 1993; 265: C1379-1387.

Fig. 1. Photomicrographs of uncounterstained frozen sections of human testis and epididymis. **Panel A:** NADPH diaphorase staining of human testis section showing granular staining in Sertoli cell (SC) and Leydig cell (LC) cytoplasm. **Panel B:** NADPH diaphorase staining of human epididymis (caput) section showing granular staining in the epithelium.

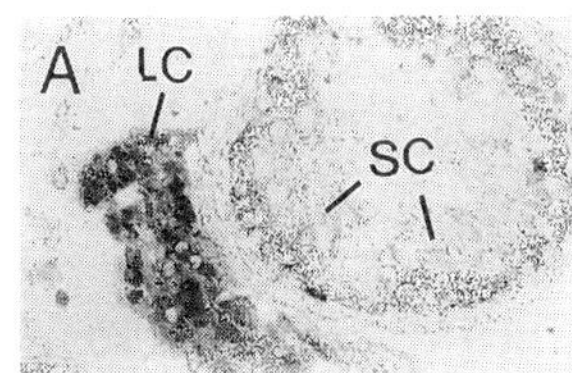

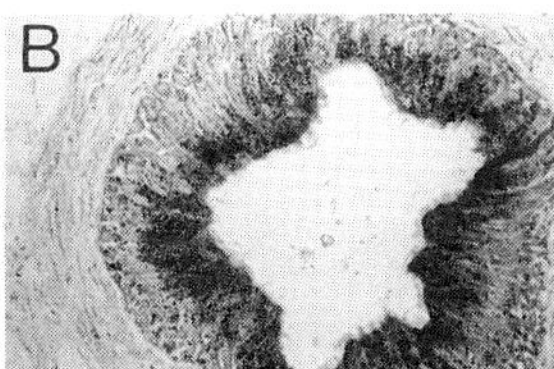

Fig. 2. Correlation between NOS staining of sperm heads and sperm motility in fractionated semen samples (n=8).

Percoll layer	% motility	% head staining
20/40 interface	10 %	23.5 %
40/65 interface	42 %	18.7 %
65/95 interface	60 %	9.5 %

Rapid flow-induced production of NO in osteoblasts

John A. Frangos*, Dameron L. Johnson and Todd N. McAllister

University of California, San Diego
Department of Bioengineering 0412
9500 Gilman Drive
La Jolla, CA. 92093

Introduction

Recent studies have indicated that nitric oxide (NO) is a potent mediator of bone remodeling by affecting both proliferative and physiological characteristics of osteoblasts and osteoclasts [1-3]. The cellular mechanisms for mechanochemical signal transduction and the physiological role of NO and other humoral/autocrine/endocrine factors, however, are still poorly understood. Previously, we hypothesized that fluid flow-induced shear stresses were an important mediator in bone remodeling. In this study we have investigated and characterized the role of fluid flow on NO production in osteoblasts, and have proposed a mechanochemical basis for bone remodeling.

Bone is a porous material, perfused by interstitial fluid which leaks from venous sinusoids in the intramedullary space. This flow is driven radially outward from the endosteum to the periosteal surface by a relatively steady transcortical pressure gradient between the sinusoids and the low resistance lymphatic drainage system on the periosteal surface. Large pulsatile flow components can be superimposed onto the steady flow by mechanically loading the bone within physiological ranges (250-3000 μstrain). High frequency low amplitude mechanical loads associated with posture maintenance may also induce fluid shear stresses above osteoblast threshold values.

Methods

New born rat calvarial cells were seeded and cultured utilizing the selective migration of osteoblasts onto glass chips. Osteoblasts explanted onto the glass were harvested and plated onto fibronectin coated slides and grown to confluence. Cells were characterized as osteoblasts morphologically as well as by their ability to respond to parathyroid hormone, produce osteocalcin, and form mineralized nodules. The cultures were also tested for endothelial cell contamination by monitoring acetylated low density lipoprotein uptake and were found to be endothelial cell-free.

Prior to shearing, the media was replaced with a phenol red and ATP free media so as not to affect the NO_x assay. Cells were exposed to a well defined shear stress of 6 dynes/cm^2 for 12 hours utilizing a parallel plate flow chamber. Samples were taken from a recirculating media system and regular intevals. NO content was then quantified using a Griess reagent and nitrate reductase. NO production rates were normalized by quantifying the total protein content using the Lowry method.

In order to isolate the source of the NO production as well as to show that the measured NO_x was induced and not endogenous to the media or sampling technique, a series of experiments were run using the non-specific NOS inhibitor N^G-amino-L-arginine (NAA) as well as the iNOS inhibitor dexamethasone. mRNA activity for endothelial type constitutive NOS was also analyzed using immunoflourescence and Western blotting techniques.

NO production rates for the osteoblasts were then compared to production rates for similarly sheared human umbilical vein endothelial cells (HUVECs) as well as cytokine (IFN-γ, TNF-α, LPS) stimulated osteoblasts.

Results

Cytokine induced production of NO had a characteristic 12 hour lag time followed by a relatively steady production rate of .6 nmoles/mg protein/hr throughout 72 hours of sampling. These results closely matched results reported by Lowik et al [4]. The addition of the iNOS inhibitor dexamethasone (100nM) largely blocked the NO production which suggests that cytokine induced NO production is mediated by the inducible isoform of nitric oxide synthase.

Fluid flow-induced NO production in osteoblasts, however, was dramatically different. Not only was the response immediate, but the production rate was over 16 fold larger (9.8 nmoles/mg protein/hr) than that observed with cytokine treatment. Cumulative production over 12 hours of exposure to shear was 127 nmoles NO/mg protein. NAA treatment largely blocked NO production which indicates that the NO is stimulated rather than endogenously present in the media. Dexamethasone pretreatment did not block the NO production, which suggests that shear activated NO production is not mediated by iNOS. The rapid response would imply there is a constitutive isoform of NOS present, however, it was not recognized by antibodies against endothelial cNOS by immunoflourescent labeling or western blotting. This data suggests that either neuronal NOS or a new osteoblast specific constitutive isoform of nitric oxide synthase may mediate this shear-induced response. This response, both in terms of overall production rate and stimulation of a constitutive isoform of NOS is remarkably similar to HUVEC behavior. HUVECs demonstrate a higher initial release rate (0-30 min) followed by a lower sustained (2-12 hours) rate. Overall production (0-12 hours), however, is nearly identical (9.8 vs 9.5 nmoles/mg protein/hr).

Discussion

To our knowledge, fluid flow represents the most potent stimulus of NO production in osteoblasts. Previously we have also shown that fluid flow is a potent stimulus for other humoral agents (PGE_2, PGI_2) [5,6] that can affect the dynamic balance between osteoclasts and osteoblasts and therefore mediate bone remodeling. These results do not alone, however, distinguish between fluid shear stress as the primary mechanical stimulus as opposed to other fluid flow related phenomena such as streaming potentials or increased mass transport. Several experiments have been reported that support a shear related mechanism [7-8].

Clearly both osteocytes and osteoblasts exhibit a sensitivity to fluid flow-induced shear stresses. Currently our objective is to further characterize this sensitivity in an effort to fully elucidate the relative importance of the frequency and amplitude of the mechanical signals as well as the roles and relative importance of osteoblasts versus osteocytes, and in doing so, develop a theory that is consistent with both physiologic and pathologic observations.

We conclude, therefore, that fluid flow-induced shear stress provides a localized mechanical signal that both osteblasts and osteocytes can detect and respond to by releasing NO and other humoral agents which mediate bone remodeling by altering proliferative and metabolic properties of osteblasts/osteocytes. We also note that the flow-induced release of NO in primary rat calvaria is constitutively present, but does not appear to the same as that present in endothelial cells and that there may be a new isoform of NOS present in osteoblasts.

References

1. Kasten, T. P., P. Collin-Osdoby, N. Patel, P. Osdoby, M. Krukowski, T.P. Musko, S.L. Settle, M.G. Currie, and G.A. Nickols. (1994) Proc. Natl. Acad. Sci. USA. 91:3569-3573.
2. Riancho, J.A., E.Salas, M.T. Zarrabeitia, J.M. Olmos, J.A. Almado, J.L. Fernandez-Luna, and J. Gonzalez-Macias. (1995) J. Bone Min. Res. 10(3):439-446.
3. MacIntyre, I., M. Zaidi, A.S. Towhidul Alam, H.K. Datta. (1991) Proc. Natl. Acad. Sci. USA. 88:2936-2940.
4. Lowik, C.W.G.M., P.H. Nibbering, M. Van de Ruit, and S.E. Papapoulos. (1994) J. Clin. Invest. 93:1465-1472.
5. Frangos, J.A., S.G. Eskin, L.V. McIntire, and C.L. Ives. (1985) Science. 227:1477-1479.
6. Reich, K.M. and J.A. Frangos. (1991) Am. J. Physiol. 261:C428-C432.
7. Kelly, P.J. and J.T. Bronk. (1990) Microvascular Research. 39:364-375.
8. Reich, K.M., C.V. Gay, J.A. Frangos. (1990). J. Cell Phys. 143:100-104.

NO as a plant growth regulator: Interaction with ethylene metabolism

YA'ACOV Y. LESHEM and E. HARAMATY

Department of Life Sciences, Bar-Ilan University, Ramat Gan 52900, Israel

ABSTRACT

Evidence obtained from stressed and senescing pea plants suggests that NO may be a plant growth factor functioning by down-regulating ethylene production: emission of both gases is markedly enhanced by the ethylene precursor aminocyclopropane-1-carboxylic acid (ACC). Levels of both NO and C_2H_4 increase upon onset of senescence or application of stress. A basic question addressed is whether stress NO emission by the plant is stress-inducing or stress-coping. Further experimentation in which the NO releasing compound S-nitroso-N-acetylpenicillamine (SNAP) was applied to senescing plants, indicated a significant inhibition of C_2H_4; moreover exogenously applied NO enhanced leaf expansion. It is concluded that in the present system NO in low concentrations acts in a stress coping capacity as well as an overall plant growth promotor.

INTRODUCTION

Apart from the cognizance of its role in N_2 fixation, virtually nothing is known about the presence and possible role of the NO free radical gas in higher plants. By employing newly developed NO probing techniques and NO producing compounds such as SNAP [3] the present research endeavored to ascertain if indeed plants produce NO endogenously and if so, to shed light upon its mode of action especially in terms of possible interaction with ethylene which invariably is associated with plant stress and senescence [6].

METHODS

Plant material was 21-day-old pea (*Pisum sativum*) explants; NO and ethylene were determined respectively by employing an Iso-NO Sensor (WPI-Florida) and a Varian FID Gas Chromotograph model 3400 utilizing experimental procedure detailed elsewhere [5]. For effect of NO on leaf growth the Pea Leaf Disc Expansion Bioassay [2] was employed. NO release into ambient air was achieved by a method outlined by Leshem and Haramaty [5]. The buffer medium throughout was 2 mM EPPS pH 8.5 containing 10^{-5} M $CaCl_2$ and when added, 2 mM ACC. Experimental results are 4 replicate means.

RESULTS AND DISCUSSION

Fig. 1-A indicates that both NO and C_2H_4, which were monitored simultaneously, are produced in the normal course of pea foliage growth and that emission of both gases is markedly increased by ACC. It is also noteworthy that NO emission exceeded that of C_2H_4. When plants were wilted before placing in the buffer medium, depending on wilting duration, plants responded by increase of NO emission [Fig. 1-B (I)]. This effect could be interpreted either as a stress-inducing response since C_2H_4 too induces stress [6], or alternatively, as a possible stress-coping one.

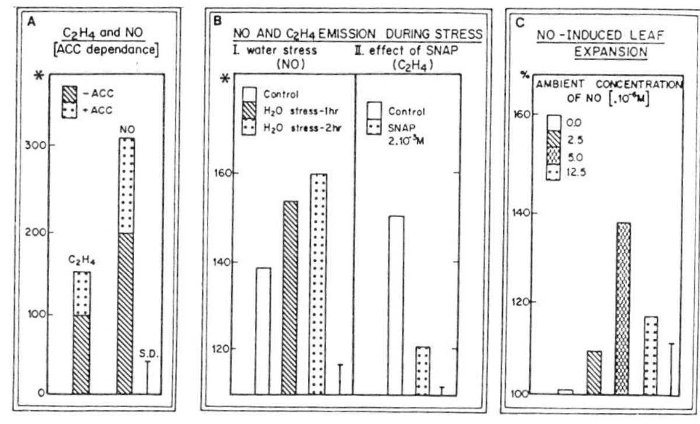

Fig. 1: ***Diagrammatic triptych indicating growth regulating effect of NO on pea foliage.*** *A and B: results after a 50 min treatment period. C. after an 18 hr incubation period.*

* *emission - nM g^{-1} fresh weight. See text for details.*

The latter possibly appears to apply since augmenting endogenous NO by application of the NO releasing SNAP cuts down C_2H_4 by two thirds [Fig. 1-B (II)]. The rationale of this experiment was that if NO is stress-coping it should inhibit ethylene (which did occur) while if it is stress-inducing it should increase C_2H_4 (which was <u>not</u> observed). Further support to this assumption as well as to the overall function of NO as plant growth regulating factor is lent by the significant enhancement of pea foliage growth after 1 hr pre-exposure to low concentration of NO in the ambient atmosphere (Fig. 1-C).

Concerning mode of interaction between NO and C_2H_4, one possible interpretation is oxidation caused by NO (or by one of its peroxynitrite derivatives) on either the 'ethylene forming enzyme', ACC oxidase, or of one of its cofactors such as ascorbate or Fe^{2+} [1]. This key enzyme in ethylene formation may well be the target for NO attack, since as a gas, NO transportation in plants may be apoplastic, the plant apoplast being directly proximal to the plant cellwall where, by immuno-cytological fluorescence staining techniques ACC oxidase has been found to be mainly located [4]: however, a similar incapacitating effect could be on cytosolic ACC oxidase as well.

REFERENCES

1. Christoffersen, R.E., McGarvey, P.J. and Savarese, P. (1993) in Cellular and Molecular Aspects of the Plant Hormone Ethylene (Peche, J.C., *et al.*, eds.), pp. 65-70, Kluwer Academic Publishers, Dordrecht
2. Elzenga, J.T.M. and Van Volkenburg, E. (1994) Membr. Biol. **137**, 227-235
3. Hery, P.J., Horowitz, J.D. and Louis, W.J. (1989) J. Pharm. Exp. **248**, 762-768
4. Laché, A., Dupille, E., Rombaldi, C., Cleyet-Marel, J.C., Lelièvre, J.M. and Pech, J.C. (1993) in Cellular and Molecular Aspects of the Plant Hormone Ethylene (Peche, J.C. *et al.*, eds.), pp. 39-45, Kluwer Academic Publishers, Dordrecht
5. Leshem, Y.Y. and Haramaty, E. *(1996)* J. Plant Phys. (in press)
6. Leshem, Y.Y., Sridhara, S. and Thompson, J.E. (1984) Plant Physiol. **75**, 329-335

Redox congeners of nitric oxide produce presynaptic inhibition of evoked synaptic currents but enhancement of spontaneous miniature synaptic currents in hippocampal/cortical neurons

Zhuo-Hua PAN, Michael M. SEGAL, and Stuart A. LIPTON

Department of Neurology, Children's Hospital, and Program in Neuroscience, Harvard Medical School, Boston, MA 02115 USA

Unlike nitric oxide (NO•), alternative redox states of the NO group such as nitrosonium equivalents (NO^+), react rapidly with thiol and thus can regulate protein function [1]. Here, using the whole-cell variant of the patch-clamp technique, we report that donors of nitrosonium equivalents, such as nitroglycerin (NTG) and *S*-nitrosocysteine (75-1000 μM each), decreased the efficacy of *evoked* neurotransmission 20-80% in a dose-dependent manner by a presynaptic mechanism (n = 26) (Figs. 1 and 2). The effect appeared to be thiol dependent, as it was occluded by pretreatment with the irreversible alkylating agent, *N*-ethylmaleimide (NEM).

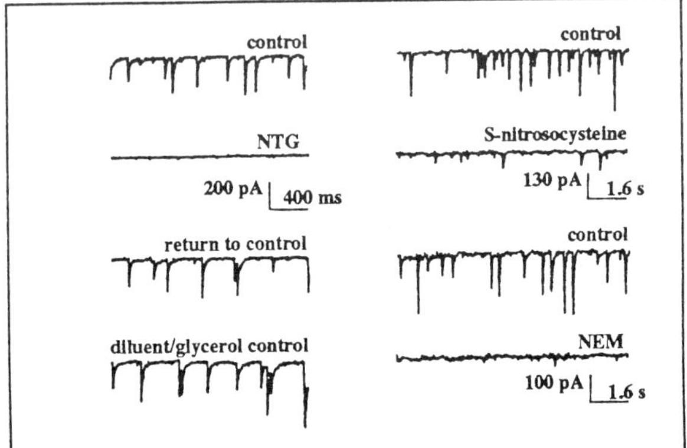

Fig. 1. NTG (1 mM) and S-nitrosocysteine (500 μM) greatly decrease synaptic activity in cortical neurons in culture within 1-2 min of addition. The effect was reversible within minutes and was not reproduced by control solutions (e.g., the glycerol backbone of NTG plus the ethanol/polyethylene glycol diluent or 24-hr-old *S*-nitrosocysteine). NEM (1 mM) completely and irreversibly blocked synaptic activity.

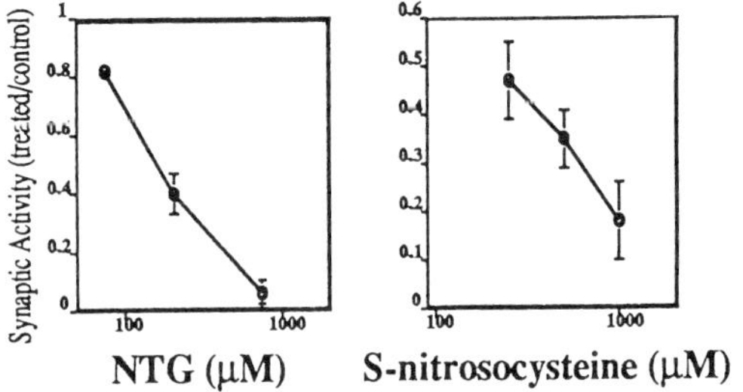

Fig. 2. Concentration-dependent inhibition of synaptic activity in cortical neurons by NTG and *S*-nitrosocysteine. The ratio of the number of synaptic events for treatment vs. control conditions is plotted for 26 neurons.

In these cortical cultures in the absence of TTX, nitrosonium donors inhibited both excitatory and inhibitory synaptic activity. The NO moiety may exert its effects by stimulating guanylate cyclase; however, the addition of 1 mM 8-bromo-cGMP to the extracellular solution did not decrease synaptic activity in this preparation.

Similar to the results in cortical neurons, we found inhibitory effects of nitrosonium donors by recording evoked autaptic excitatory postsynaptic currents from isolated hippocampal neurons in the presence of TTX (n = 6) (Fig. 3) [2]. (An autapse is a single neuron synapsing on itself). In contrast, NO• donors, such as DEA/NO (≤1000 μM), did not inhibit synaptic activity.

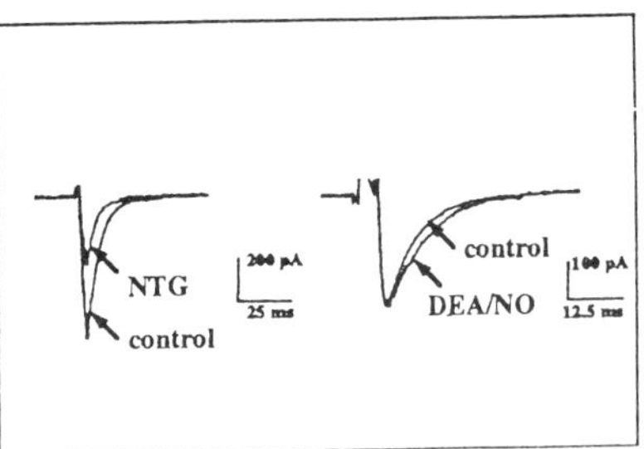

Fig. 3. Effect of NTG (a nitrosonium-like donor) and DEA/NO (an NO• donor) on evoked synaptic activity. NTG (500 μM, left) decreased evoked synaptic currents within seconds of application compared to control solution. DEA/NO (1 mM, right,) if anything slightly increased evoked synaptic current or had no effect.

In contrast to evoked synaptic activity, *spontaneous* miniature excitatory postsynaptic currents (minis) under these conditions were enhanced 42% by NTG in perforated patch recordings ($P < 0.001$, n = 4) (Figs. 4 and 5).

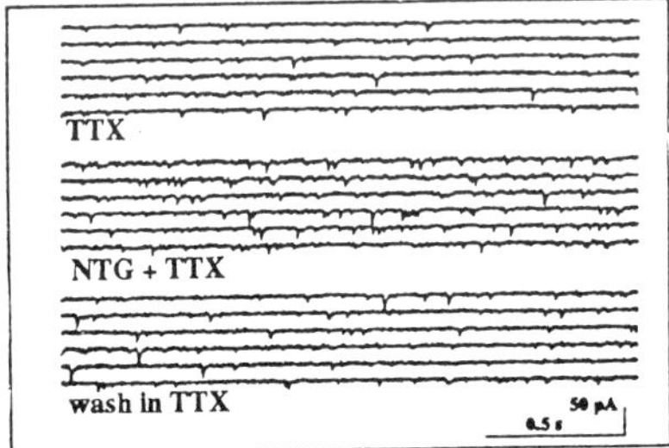

Fig. 4. Spontaneous miniature synaptic currents increase in frequency in response to NTG (1 mM). TTX (1 μM) was used to suppress sodium currents and hence action potentials which could evoke synaptic activity. As a control, the diluent for NTG was present in all solutions.

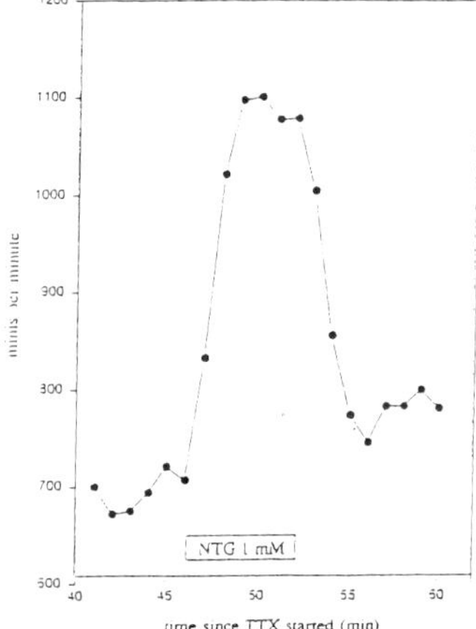

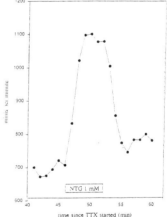

Fig. 5. Quantification of the increase in miniature synaptic current frequency in response to NTG. All solutions contained TTX (1 μM) and the diluent used for NTG.

Conclusions. NO^+ equivalents (and not NO•) decrease evoked synaptic activity but increase spontaneous synaptic activity, apparently by a presynaptic mechanism. These findings may help explain the otherwise puzzling observations that the NO moiety can either increase, decrease, or have no net effect on synaptic activity.

1. Lipton, S.A., Choi, Y.-B., Pan, Z.-H., et al. (1993) A redox-based mechanism for the neuroprotective and neurodestructive effects of nitric oxide and related nitroso-compounds. Nature **364**, 626-632

2. Segal, M.M. (1991) Epileptiform activity in microcultures containing one excitatory hippocampal neuron. J. Neurophysiol. **65**, 761-770

Nitric oxide mediates HIV-infected monocytes' hyperresponsiveness to activation.

Barbara Sherry*, Larisa Dubrovsky*, Grigorii Enikolopov[#], and Michael Bukrinsky*.

*The Picower Institute for Medical Research, Manhasset, NY 11030, USA; and [#]Cold Spring Harbor Laboratory, Cold Spring Harbor, NY 11724, USA.

HIV-1 infection leads to a progressive depletion of CD4+ T lymphocytes, associated with a general exhaustion of the immune system (reviewed in [1]). Prior to this widespread decline of immune functions, there is an evident hyperactivation of some branches of the immune system, in particular the monocyte/macrophage arm [2,3]. HIV-1-infected macrophages produce increased levels of cytokines and other biologically active molecules, which may contribute to the pathogenesis of HIV disease both by activating expression of HIV-1 provirus [4], as well as by direct effects on cytokine-sensitive tissues, such as lung or brain [5,6]. Most of these abnormally exaggerated cytokine responses require triggering by a secondary stimulus (e.g. LPS), but some can be observed in otherwise unstimulated HIV-infected monocytes [7]. In addition, as we recently demonstrated, these cells have elevated levels of iNOS and produce low but detectable amounts of NO [8]. In this paper, we provide evidence that NO is the critical factor that mediates exaggerated cytokine production by *in vitro*-infected monocytes.

To determine the role of NO in the abnormal hyperactivation phenotype of HIV-1-infected monocytes, we measured cytokine production by HIV-infected monocyte cultures in the presence of various NOS inhibitors. TNF was used as a readout for responses dependent on a secondary stimulus (LPS), while MIP-1α, which is induced by HIV-1 itself [7], characterized LPS-independent responses. Both LPS-induced (Fig. 1A) and HIV-induced (Fig. 1B) responses were reduced by NOS inhibitors. These inhibitors did not significantly change cytokine production by LPS-stimulated uninfected monocytes (Fig. 1A and 1B), in agreement with previously demonstrated lack of NO production after LPS stimulation of human monocytes. Interestingly, hemoglobin, an effective chelator of extracellular NO, did not significantly affect TNF response of HIV-infected monocyte cultures to LPS stimulation, suggesting that NO exerts its effects mostly in an autocrine fashion that does not depend on an extracellular phase.

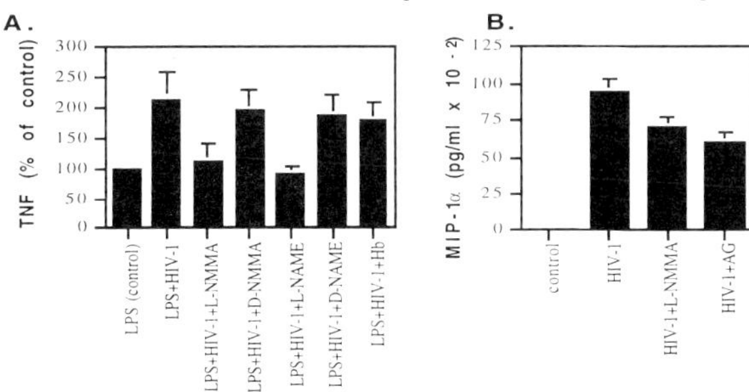

Figure 1. NOS inhibitors reduce cytokine production by HIV-infected monocytes.

HIV-1-infected monocyte cultures, stimulated with 0.5 ng/ml of LPS (A) or unstimulated (B), were treated with L-NMMA (2 mM), D-NMMA (2 mM), L-NAME (1 mM), D-NAME (1 mM), hemoglobin (Hb, 100 μg/ml), or aminoguanidine (AG, 0.5 mM). 18 hours after stimulation, TNF and MIP-1α in the culture supernatants were assayed by ELISA. Values in Fig. 1A are scaled relative to LPS-stimulated uninfected cultures taken as 100%. Error bars represent s.e.m. from three independent wells in a representative experiment.

To prove the role of NO in monocyte hyperactivation, we used several chemical donors of NO. SNAP, SIN, SNP, and spermine bis(nitric oxide) adduct upregulated production of TNF by uninfected monocyte cultures stimulated with LPS by 75-150%, resembling the effects of HIV-1 infection on monocyte responsiveness. Without LPS stimulation, NO donors did not induce any significant expression of cytokines.

Presented results demonstrate a critical role of NO in establishing an abnormal hyperactivated phenotype of HIV-1-infected monocyte cultures. Although levels of NO production by HIV-infected human monocytes [8] are about 10 times lower than those typically achieved by murine macrophages stimulated with interferon γ and LPS, intracellular concentrations of NO in HIV-1-infected monocytes seem to be sufficient to amplify the cytokine production by these cells. Interestingly, this effect is observed both with cytokines induced by a secondary stimulus (e.g. TNF) and by HIV-1 itself (e.g. MIP-1α). Since there is likely a significant difference between these two pathways, our results demonstrate a widespread activating capacity of NO. Further studies should reveal affected by NO signal transduction mechanisms leading to cytokine production in HIV-infected cells. However, identification of NO as a critical component of HIV-1-induced hyperactivation introduces a set of novel targets for therapeutic intervention designed at normalizing immune reactions of HIV-1-infected patients.

References.

1. Rosenberg, Z.F. and Fauci, A.S. (1993) Fundamental immunology (Paul, W. E., ed.) pp. 1375-1397. New York. Raven Press.
2. Edelman, A.S. and Zolla-Pazner, S. (1989) AIDS: a syndrome of immune dysregulation, dysfunction, and deficiency. FASEB J. **3**,22-30.
3. Ehrenreich, H., Rieckmann, P., Sinowatz, F., et al. (1993) Potent stimulation of monocytic endothelin-1 production by HIV-1 glycoprotein 120. J. Immunol. **150**, 4601-4609.
4. Kinter, A.L., Poli, G., Fox, L., Hardy, E., and Fauci, A.S. (1995) HIV replication in IL-2-stimulated peripheral blood mononuclear cells is driven in an autocrine/paracrine manner by endogenous cytokines. J. Immunol. **154**, 2448-2459.
5. Biglino, A., Forno, B., Pollono, A.M., Ghio, P., and Albera, C. (1995) Chest **103**, 439-443.
6. Merrill, J. and Chen, I. S. (1992) HIV-1, macrophages, glial cells, and cytokines in AIDS nervous system disease. FASEB **5**, 2391-2397.
7. Schmidtmayerova, H., Nottet, H. S. L. M., Nuovo, G., et al. (1995) HIV-1 infection alters chemokine β peptide expression in human monocytes: implications for recruitment of leukocytes into brain and lymph nodes. Proc. Natl. Acad. Sci. USA (in press).
8. Bukrinsky, M. I., Nottet, H. S., Schmidtmayerova, H., et al. (1995) Regulation of nitric oxide synthase activity in human immunodeficiency virus type 1 (HIV-1)-infected monocytes: implications for HIV-associated neurological disease. J.Exp.Med. **181**, 735-745.

The expression and regulation of nitric oxide synthase in human osteoarthritis-affected chondrocytes

ASHOK R. AMIN,[*‡§] PAUL E. DI CESARE,[‖] PRANAV VYAS,[*] MUKUNDAN ATTUR,[*] EDITH TZENG,[¶] TIMOTHY R. BILLIAR,[¶] STEVEN A. STUCHIN,[‖] and STEVEN B. ABRAMSON[*§]

Departments of [*]Rheumatology and [‖]Orthopaedic Surgery, Hospital for Joint Diseases Orthopaedic Institute, 301 East 17th Street, New York, New York 10003; the Departments of [‡]Pathology and [§]Medicine, New York University Medical Center, 550 First Avenue, New York, New York 10016; and [¶]the Department of Surgery, University of Pittsburgh, Pittsburgh, PA 15261.

Introduction

Osteoarthritis or "osteoarthrosis": which term is more appropriate? Classically, osteoarthritis (OA), unlike rheumatoid arthritis (RA), is defined as an inherently noninflammatory disorder of movable joints characterized by deterioration of articular cartilage and the formation of new bone at the joint surfaces and margins [1]. In contrast to RA, the synovial fluid in OA typically contains few neutrophils (<3,000/mm^3) and, except for advanced disease, the synovium does not exhibit significant cellular proliferation nor infiltration by inflammatory leukocytes. Studies involving animal arthritis models and analysis of human synovial fluids have implicated nitric oxide (NO) in the pathogenesis of arthritis [2,3]. Induction of arthritis in rodent models resulted in increased production of nitrite prior to the onset of clinical symptoms. The appearance of articular symptoms and joint degeneration in these animal models could be inhibited by administration of a NOS inhibitor [2]. ncNOS has been shown to be present in several cell types [4]. Its function outside the vicinity of the neuronal system has recently been recognized, where ncNOS-knockout mice show abnormalities in the gastrointestinal tract similar to pyloric stenosis [5].

Results and Discussion

We initiated our studies by standardizing conditions to extract NOS directly from OA-affected human articular cartilage prior to cell isolation. Approximately 90% of NOS from OA-affected cartilage could be isolated by freeze-milling. Western blot analysis of NOS directly from OA-affected cartilage was carried out using various α-NOS antibodies. These results show that the 150 kD human OA-NOS and ncNOS are similar in size, and are distinct from iNOS from human hepatocytes (native and transfected), human B cells, bovine and rodent chondrocytes, and murine macrophages [6,7], which are 133 kD (see Table 1 for summary of data).

These observations pose an obvious question: what is the role of human chondrocyte iNOS in OA? It should be noted that two groups [8,9] have cloned the human chondrocyte iNOS from cDNA libraries prepared from *(in vitro)* IL-1-stimulated human chondrocytes. These libraries were screened with a degenerative oligonucleotide or murine iNOS cDNA. The OA-NOS seems to be distinct from the human chondrocyte iNOS. Apparently, human chondrocytes express at least two isoforms of NOS which are differentially regulated, namely iNOS [8,9] and OA-NOS. This is not surprising, because recent studies by Peunova and Elikopolov [10] have shown that PC-12 cells (neuronal) which proliferate in the presence of nerve growth factor (NGF) show growth arrest and differentiation, partially due to differential expression of all three NOS isoforms at different periods of growth and differentiation in the same cells. The possibility that human chondrocyte 133 kD iNOS undergoes "modification" (*e.g.*, glycosylation, covalent binding to cartilage matrix components, etc.) to increase its M_r by 17 kD cannot be ruled out. Such a modified form of iNOS **(a)** would retard the mobility of the "modified" NOS on the gel and **(b)** may cause a conformational change in the enzyme to expose or conceal certain cross-reactive epitopes which then react or fail to react with regions that share ~50% homology to peptides from which the antibodies are raised.

To evaluate the biological activity of OA-NOS, we standardized parameters to set up organ culture of OA-affected cartilage in *ex vivo* conditions. We observed that ~50-100 mg of OA cartilage from various patients released ~10-80 μM of nitrite after 48 h in serum-free medium, which could be inhibited by >90% with 500 μM L-NMMA or aminoguanidine (not shown). The spontaneous accumulation of nitrite by OA-affected cartilage was evaluated for its sensitivity to cycloheximide and NF-κB inhibitor PDTC. Normal human cartilage, which showed no detectable NOS in Western blotting, did show spontaneous release of NO ($\geq$1 μM) with equivalent amount of cartilage. These experiments suggest that OA-affected chondrocytes generate micromolar concentrations of NO that have been associated with degradation of articular cartilage [3]. Exposure of the OA cartilage *in vitro* to pharmacological concentrations of IL-1β, TNF-α and endotoxin resulted in a consistent augmentation of NO production at 48 h (Table 1).

Nitrite accumulation was significantly blocked by >90% in both the cytokine + endotoxin-induced cartilage and control cartilage in the presence of cycloheximide or PDTC, thus indicating that the OA-affected chondrocyte NOS activity is sensitive to cycloheximide and PDTC with respect to production of nitrite in *ex vivo* cultures (Table 1). As expected, addition of cytokines + endotoxin to normal cartilage induced NO production. Incubation of OA-affected articular cartilage, either in basal medium alone or supplemented with BSA had insignificant impact on the release of nitrite (data not shown). In similar and parallel experiments using normal adult bovine articular cartilage or murine RAW 264.7 cells, we did not observe any detectable amounts (>1 μM) of nitrite after 48 h (data not shown), indicating that the medium used in these experiments was devoid of any stimulating agent that may have contributed to the upregulation of NOS in OA cartilage on day 2 and 3 in the *ex vivo* experiments. As expected, addition of LPS (100 ng/ml and 100 μg/ml), showed an accumulation of nitrite in the medium after 20 or 48 h in both murine RAW 264.7 cells and bovine articular cartilage, respectively (data not shown).

Potential upregulator(s) of cartilage NOS may be one or a combination of the following manifestations: ***a)*** autocrine cytokines/growth factors produced by chondrocytes in OA cartilage; ***b)*** interaction with cell surface receptor (glutamate receptors/CD53) or matrix components (fibronectin and collagen) that can modulate NOS [11,12]; ***c)*** diffusion of soluble (paracrine) factors into the cartilage *in vivo* from other cellular sources of the intra-articular region (*e.g.*, endothelial cells, lining of the synovial capillaries, local inflammatory cells, and/or synovial fibroblasts); ***d)*** abnormal mechanical forces seen by the chondrocytes in the 3-dimensional architecture, because shear stress is reported to release nitrite from articular chondrocytes [13].

In summary, articular chondrocytes are a source of increased levels of intra-articular NO which have been speculated in OA patients [3]. Our data further support the notion that NO, a known inflammatory mediator, is released in sufficient quantities [3,14] that may cause chondrocyte dysfunction and damage cartilage integrity, and thus may be one of the key mediators in the pathogenesis of OA. Further study of the mechanisms governing NOS expression, and of the consequences of increased NO production by OA-affected cartilage, will help to define potential targets for pharmacological intervention and gene therapy.

REFERENCES

1. Hough AJ (1993) *in* Arthritis and Allied Conditions. DJ McCarty, WJ Koopman, editors. Lea & Febiger, Philadelphia and London. 1699-1723.
2. McCartney-Francis N, Allen JB, Mizel DE, Albina JE, Xie QW, Nathan CF, Wahl SM (1993) *J. Exp. Med.* 178:749-754.
3. Farrell AJ, Blake DR, Palmer RMJ, Moncada S (1992) *Ann. Rheum. Dis.* 51:1219-1222.
4. Nathan C, Xie QW (1994) *Cell* 78:915-918.
5. Huang PL, Dawson TM, Bredt D, Snyder SH, Fishman MC (1993) *Cell* 75:1273-1286.
6. Xie QW, Cho HJ, Calaycay J, *et al.* (1992) *Science* 256:225-228.
7. Mannick JB, Asano K, Izumi K, Kieff E, Stamler JS (1994) *Cell* 79:1137-1146.
8. Maier R, Bilbe G, Rediske J, Lotz M (1994) *Biochim. Biophys. Acta.* 1208:145-150.
9. Charles IG, Palmer RMJ, Hickery MS, *et al.* (1993) *Proc. Natl. Acad. Sci. USA* 90:11419-11423.
10. Peunova N, Enikolopov G (1995) *Nature* **375**:68.
11. Culcasi M, Lafon-Cazal M, Pietri S, Bockaert J (1994) *J. Biol. Chem.* 269:12589.
12. Pérez-Mediavilla LA, Lopez-Zabalza MJ, Calonge M, *et al.* (1995) *FEBS Letters.* 357:121-124.
13. Das P, Schurman DJ, Smith RL (1995) *41st Ann.Mtg.Orth.Res.Soc.* 20:354.
14. Palmer RMJ, Hickery MS, Charles IG, Moncada S, Bayliss MT (1993) *Biochem. Biophys. Res. Commun.* 193:398-405.

Table 1. Properties of iNOS, OA-NOS and ncNOS.

	iNOS	OA-NOS	ncNOS
STRUCTURAL PROPERTIES:			
Size (kD)	133	150	150
Reactivity to α-ncNOS Ab	−	+	+
Reactivity to α-iNOS Ab	+	−/+	−
BIOCHEMICAL PROPERTIES:[*]			
Sensitivity to:			
cycloheximide (1-2 μg/ml)	+	+	−
TGF-β (2.5ng/ml)	+	−	?
hydrocortisone (10 μg/ml)	+/−[‡]	−	?
PDTC (30 μM)	+/−[§]	+	?
aminoguanidine (200 μM)	+	+	−
L-NMMA (500 μM)	+	+	+
Inducible by cytokines and endotoxin[‖]	+	+	−
Nitrite released in μM quantity	+	+	−[¶]
Spontaneous release of NO	−	+	+
Upregulated under pathological conditions	+	+	?

Data are compiled from experiments conducted in this study and from Amin *et al., JEM,* 1995. (+) indicates positive as compared to the respective control; data showing p values <0.01, or values representing ≥50% inhibition/ modulation, have been represented. (−) indicates undetectable or no effect; these values were ≤10%, or were statistically insignificant. (?) indicates data not available.

* Assessment of biochemical properties based on nitrite accumulation.
‡ (+) murine macrophage iNOS; (−) human chondrocyte iNOS.
§ (+) murine macrophage iNOS; (−) human hepatocyte iNOS (unpublished data).
‖ IL-1 (1 ng/ml) + TNF (1 ng/ml) + LPS (100 μg/ml)
¶ Released in the range of pM-nM.

Pathogenesis of influenza virus-induced pneumonia: Involvement of both nitric oxide and superoxide anion

Akaike, T., and Maeda, H.
Department of Microbiology, Kumamoto University School of Medicine, Kumamoto 860, Japan.

INTRODUCTION

Much attention has been paid to the multiple functions of nitric oxide (NO) in diverse biological phenomena [1, 2]. Overproduction of NO involves in the pathogenesis of endotoxin shock, inflammatory disorders, and in the neuropathogenesis of some neurotropic virus infections [3-5]. In this paper, we demonstrate the pathogenic role of NO in influenza virus-induced pneumonia.

METHODS

Production of influenza virus pneumonia. Experimental influenza virus pneumonia was produced in the mice (SPF grade male ddY mice, 5-6 wk old) with a mouse-adapted strain of influenza virus A/Kumamoto/Y5 /67(H2N2) as described in our previous report [6].

Direct proof of NO production in the lung by using electron spin resonance (ESR) spectroscopy. In vivo spin trapping for NO was performed as described [7, 8] with slight modification. Briefly, 3.6 µmol of Fe^{2+} in sodium citrate solution and 44 µmol of *N,N*-diethyldithiocarbamate (DETC) were given to the mice s.c. and i.p., respectively. Similarly, a water soluble *N*-methyl-D-glucamine dithiocarbamate (MGD)-Fe^{2+} complex was injected s.c. to the mice at a dose of 120 µmol MGD and 12 µmol Fe^{2+} per mouse. Thirty minutes after injection of dithiocarbamate-iron complexes, the lung was immediately transferred to quartz sample tubes and frozen in liquid nitrogen followed by ESR measurement at 110 K.

Treatment of virus-infected mice with L-NMMA. N^{ω}-Monomethyl-L-arginine (L-NMMA) was given i.p. once daily to mice at the doses of 2 mg/mouse and 3 mg/mouse, respectively, from 2 days to 7 days after virus inoculation (1.5 × LD_{50}), and the survival rate of the mice was recorded after infection.

RESULTS & DISCUSSION

To identify excessive production of NO in the virus-infected lung, ESR spectroscopy was used on mouse lungs obtained on day 7 after virus infection. After treatment with either DETC-iron complex or MGD-iron complex, strong ESR signal of each nitrosyl adduct was obtained in the virus-infected lung (Fig. 1).

Generation of nitrosyl dithiocarbamate-iron signals was almost completely inhibited by treatment with L-NMMA (Fig. 1), when 2 mg of L-NMMA was given i.p. 2 h before injection of dithiocarbamate-iron complexes. In contrast, there was no significant ESR signal in the normal mouse lung after administration of both DETC-iron and MGD-iron complexes (data not shown). When we tested the time profile of formation of NO-dithiocarbamate-iron complexes in the mouse lung after influenza virus infection, the generation of ESR signals of NO adduct was obtained clearly from day 6 to 8 after virus infection: maximal production was observed on day 7. No significant signal of NO adducts was found later than 8 days after infection.

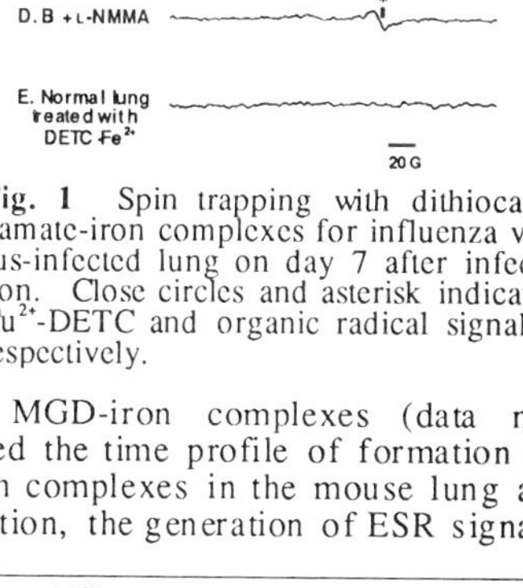

Fig. 1 Spin trapping with dithiocarbamate-iron complexes for influenza virus-infected lung on day 7 after infection. Close circles and asterisk indicate Cu^{2+}-DETC and organic radical signals, respectively.

We further investigated the role of NO production in the influenza virus-induced pneumonia. The NOS inhibitors L-NMMA were administered to mice at the doses of 2 mg/day/mouse and 3 mg/day/mouse, respectively, after virus infection, and the effect on mouse survival rate was examined. The survival rate of mice was significantly improved by L-NMMA treatment (Fig. 2).

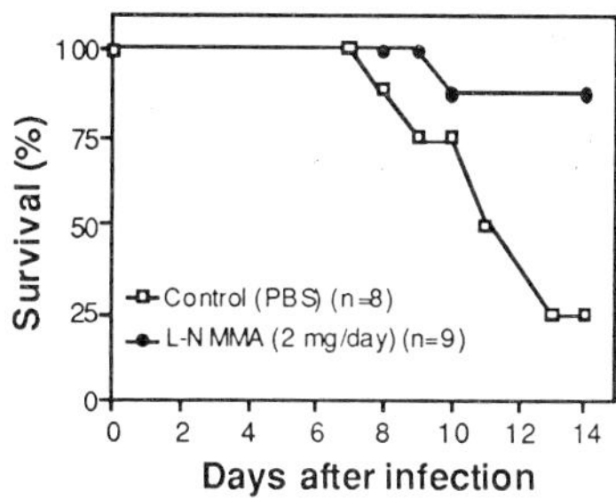

Fig. 2 The effect of NOS inhibitors on survival rate of mice after influenza virus infection. A significant difference in survival rate on day 15 was found ($p < 0.05$) between the control group and the L-NMMA-treated group.

We previously reported that neutrophils and macrophages are important cellular sources of O_2^- generation in the mouse lung of influenza virus pneumonia [9, 10]. Xanthine oxidase is highly up-regulated and is released into alveolar spaces in influenza virus infected lungs, resulting in overproduction of O_2^- in the mouse lung [9]. It should be noted that positive staining with nitrotyrosine antibody of neutrophils and macrophages as well as intra-alveolar exudate was observed in the virus-infected lung (unpublished observation). These results strongly indicate the formation of $ONOO^-$ in the influenza-virus infected lungs.

The survival rate of mice was much improved by L-NMMA treatment, indicating the pathogenic role of NO in influenza virus-induced pneumonia in mice. Our previous result showed that removal of O_2^- by superoxide dismutase improved the survival of mice [9, 10]. These findings may be interpreted as evidence that suppression of either NO or O_2^-, which will reduce the formation of $ONOO^-$ and prevent its toxic effect [11, 12], resulted in improved survival of the virus-infected mice.

In conclusion, it may be reasonable that interaction of NO with O_2^- and the formation of $ONOO^-$ may be a cardinal pathogenic principle in the process of tissue injury in some viral diseases.

Abbreviations used: L-NMMA, N^{ω}-monomethyl-L-arginine; L-NAMA, N^{ω}-nitro-L-arginine; ESR, electron spin resonance; DETC, *N,N*-diethyldithiocarbamate; MGD, *N*-methyl-D-glucamine dithiocarbamate.

REFERENCES

1. Moncada, S., Higgs, A. (1993) N. Engl. J. Med. **329**, 2002-2012
2. Maeda, H., Akaike, T., Yoshida, M., Suga, M. (1994) J. Leukoc. Biol. **56**, 588-592
3. Yoshida, M., Akaike, T., Wada, Y., Sato, K., Ikeda, K., Ueda, S., Maeda, H. (1994) Biochem. Biophys. Res. Commun. **202**, 923-930
4. Mulligan, M.S., Hevel, J.M., Marletta, M.A., Ward, P.A. (1991) Proc. Natl. Acad. Sci. U.S.A. **88**, 6338-6342
5. Koprowski, H., Zheng, Y.M., Heber-Katz, E., et al. (1993) *Proc. Natl. Acad. Sci. U.S.A.* **90**, 3024-3027
6. Akaike, T., Molla, A., Ando, M., Araki, S., Maeda, H. (1989) J. Virol. **63**, 2252-2259
7. Obolenskaya, M.Yu., Vanin, A.F., Mordvintcev, P.I., Mülsch, A., Decker, K. (1994) Biochem. Biophys. Res. Commun. **202**, 571-576
8. Komarov, A., Mattson, D., Jones, M.M., Singh, P.K., Lai, C.-S. (1993) Biochem. Biophys. Res. Commun., **195**, 1191-1198
9. Akaike, T., Ando, M., Oda, T., et al. (1990) J. Clin. Invest. **85**, 739-745
10. Oda, T., Akaike, T., Hamamoto, T., Suzuki, F., Hirano, T., Maeda, H. (1989) Science **244**, 974-976
11. Radi, R., Beckman, J.S., Bush, K.M., Freeman, B.A. (1991) J. Biol. Chem. **266**, 4244-4250
12. Lipton, S.A., Choi, Y.-B., Pan, Z.-H., et al. (1993) Nature **364**, 626-632

The role of nitric oxide and peroxynitrite in the pathogenesis of spontaneous murine autoimmune disease

Christopher T. Privalle*, Teresa Keng*, Gary S. Gilkeson‡, and J. Brice Weinberg‡

*Apex Bioscience, Inc., Post Office Box 12847, Research Triangle Park, NC 27709, and ‡VA and Duke University Medical Centers, 508 Fulton Street, Durham, NC 27705

The diseases rheumatoid arthritis (RA) and systemic lupus erythematosus (SLE) are autoimmune disorders of unknown etiology. Patients with RA and SLE have autoantibodies to various antigens including immunoglobulin, DNA, and erythrocyte membrane proteins. Arthritis, vasculitis, nephritis, and dermatitis can be seen in each disease. Tissue inflammation and destruction in these autoimmune diseases have been postulated to be mediated by products elaborated by "activated" leukocytes. Mice of the strain MRL-*lpr/lpr* have been useful in the study of human autoimmune diseases since they have disorders quite comparable to those seen in human RA and SLE [1].

We demonstrated earlier that peritoneal macrophages from MRL-*lpr/lpr* mice were "activated" as determined by a variety of parameters including increased levels of spontaneous and phorbol myristate acetate-induced reactive oxygen species generation [2]. Recently, we have also found that MRL-*lpr/lpr* mice spontaneously overproduce nitric oxide (NO), and that this overproduction is related to the development of certain aspects of autoimmune disease [3]. MRL-*lpr/lpr* mice have increased urinary excretion of nitrite/nitrate (stable products of NO oxidation) and increased expression of mRNA for inducible NO synthase (iNOS) in peritoneal spleen and kidney tissues. Macrophages from these mice have increased ability to produce nitrite/nitrate *in vitro*. Furthermore, when the mice are administered the NOS inhibitor NGmonomethyl-L-arginine (NMMA) orally, urinary nitrite/nitrate levels fall to normal, and the development of arthritis and glomerulonephritis is prevented [3]. However, NMMA treatment does not alter levels of anti-DNA antibodies in the mice [3], and it does not reduce the level of immune complex deposition in the glomeruli as determined by indirect immunofluorescence (JB Weinberg and GS Gilkeson, unpublished data).

Although reactive species such as NO, superoxide (O_2^-), and hydrogen peroxide (H_2O_2) have inflammatory and tissue destructive capabilities, peroxynitrite (formed by the reaction of O_2^- with NO) is thought to be more proinflammatory and destructive [4, 5]. Because we had noted that MRL-*lpr/lpr* mice overproduce both reactive oxygen species and NO, we hypothesized that peroxynitrite would be formed *in vivo* in these mice. Peroxynitrite can react with proteins to form nitrotyrosine which is stable and can be detected biochemically and immunologically [4]. Nitrotyrosine has been detected in tissues in humans and other animals with inflammatory disease, including lung tissue from rats exposed to hyperoxia, lung tissue from humans with acute respiratory distress syndrome or acute lung injury, and atherosclerotic lesions in human coronary vessels [6-8]. The purpose of our current study was to determine if MRL-*lpr/lpr* mice with autoimmune disease-associated glomerulonephritis had increased levels of peroxynitrite, and therefore increased nitration of renal protein tyrosines.

Protein extracts from kidneys of diseased MRL-*lpr/lpr* mice or normal BALB/c mice (age 17 to 20 weeks) were studied by immunoblot analysis with a highly specific rabbit polyclonal anti-nitrotyrosine antibody [7] and a horseradish peroxidase-linked goat anti-rabbit immunoglobulin secondary antibody using the ECL method (Amersham). At age 17 to 20 weeks, virtually all MRL-*lpr/lpr* mice have renal disease [3]. Tissues from BALB mouse kidneys contained few or no immunoreactive proteins, while those from MRL-*lpr/lpr* mice had two major bands of immunoreactivity (Mr ≈ 60,000 and 48,000) and three minor bands. The reactivity was eliminated by omitting the primary antibody, or by co-incubating the primary antibody reaction mixture with 10 mM nitrotyrosine. The identity of the nitrated proteins in the tissues from the diseased kidneys is presently unknown.

NO and peroxynitrite can react with numerous different proteins, and these reactions can alter the functions of some of the proteins [4, 9]. In an attempt to identify one of the target proteins for NO and peroxynitrite, we measured catalase activity in the mouse kidneys. Renal extracts from diseased and control mice were examined in an *in vitro* assay measuring disappearance of H_2O_2, and in an *in situ* gel catalase assay [10, 11]. These assays demonstrated that catalase activity in extracts of MRL-*lpr/lpr* mouse kidneys was decreased to 25% of the level found in BALB mouse kidneys. Treatment of diseased mice with NMMA restored the level of catalase activity in kidneys to the level seen in BALB mice. This finding suggested that catalase is one of the target proteins inactivated by NO and/or peroxynitrite.

We next examined the effect of peroxynitrite on purified bovine catalase *in vitro*. Incubation of bovine catalase with 0.1 to 1.0 mM peroxynitrite at pH 7.4 resulted in modification of the protein such that the migration of the enzyme in an 8% native polyacrylamide slab gel was altered. This altered migration was accompanied by a 25 to 35% decrease in catalase activity, as well as an increase in the level of nitrotyrosine detected by immunoblot analysis. These results indicate that catalase can serve as a target for peroxynitrite-mediated modification, and that the modified protein has decreased activity.

In conclusion, our studies indicate that MRL-*lpr/lpr* autoimmune mice with a disease comparable to human RA and SLE overproduce reactive oxygen species and NO. Increased levels of these oxidants lead to the formation of peroxynitrite *in vivo*, as evidenced by the increased levels of nitrotyrosine detected in protein extracts prepared from kidneys of diseased mice. MRL-*lpr/lpr* kidneys have decreased levels of catalase, suggesting that catalase may be one of the targets of peroxynitrite *in vivo*. Increased production of oxidants such as NO, peroxynitrite, O_2^-, and H_2O_2, coupled with decreased levels of antioxidants such as catalase, may render these mice especially susceptible to oxidant tissue damage. Preventing the development of, or blocking the actions of reactive oxygen species and/or reactive nitrogen species may be useful in the treatment of autoimmune inflammatory disorders in humans.

Supported in part by the VA Medical Research Service, the James R. Swiger Hematology Research Fund, and NIH award AR-39162

1. Cohen, P. L. and Eisenberg, R. A. (1991) Ann. Rev. Immunol. **9**, 243-269
2. Dang-Vu, A. P., Pisetsky, D. S. and Weinberg, J. B. (1987) J. Immunol. **138**, 1757-61
3. Weinberg, J. B., Granger, D. L., Pisetsky, D. S., Seldin, M. F., Misukonis, M. A., Mason, S. N., Pippen, A. M., Ruiz, P., Wood, E. R. and Gilkeson, G. S. (1994) J. Exp. Med. **179**, 651-60
4. Beckman, J. S. and Crow, J. P. (1993) Biochem. Soc. Trans. **21**, 330-4
5. Beckman, J. S., Beckman, T. W., Chen, J., Marshall, P. A. and Freeman, B. A. (1990) Proc. Natl. Acad. Sci. USA **87**, 1620-1624
6. Haddad, I. Y., Pataki, G., Hu, P., Galliani, C., Beckman, J. S. and Matalon, S. (1994) J. Clin. Invest. **94**, 2407-2413
7. Beckman, J. S., Ye, Y. Z., Anderson, P. G., Chen, J., Accavitti, M. A., Tarpey, M. M. and White, C. R. (1994) Biol. Chem. Hoppe-Seyler **375**, 81-88
8. Kooy, N. W., Royall, J. A., Ye, Y. Z., Kelly, D. R. and Beckman, J. S. (1995) Am. J. Respir. Crit. Care Med. **151**, 1250-4
9. Nathan, C. (1992) FASEB J. **6**, 3051-3064
10. Beers, R. F., Jr and Sizer, I. W. (1952) J. Biol. Chem. **195**, 133-140
11. Clare, D. A., Duong, M. N., Darr, D., Archibald, F. and Fridovich, I. (1984) Anal. Biochem. **140**, 532-7

Abbreviations used: RA, rheumatoid arthritis; SLE, systemic lupus erythematosus; NO, nitric oxide; iNOS, inducible nitric oxide synthase; NMMA, NGmonomethyl-L-arginine; O_2^-, superoxide; H_2O_2, hydrogen peroxide.

Synergistic Interactions Between Nitric Oxide and α–Tocopherol in Membrane and Lipoprotein Antioxidant Reactions

Homero Rubbo[1], Andrés Paler-Martínez[1], and Bruce A. Freeman[1,2,3]

Departments of Anesthesiology[1], Biochemistry and Molecular Genetics[2] and Pediatrics[3], University of Alabama at Birmingham, Birmingham, AL 35233

α–Tocopherol is a principal antioxidant for cell membranes and lipoproteins, acting by reducing chain-propagating peroxyl radical species (LOO·) to their corresponding hydroperoxide (LOOH) and yielding the more stable tocopheroxyl radical. The rate constant for this reaction is ~10^5 M^{-1} sec^{-1}, depending on the nature of the lipid undergoing peroxidation (1). Tocopheroxyl radical has only limited capability to further propagate radical chain reactions and can be biologically reduced by ascorbate, reduced thiols (eg. glutathione and dihydrolipoic acid) or via enzyme-dependent mechanisms to regenerate α-tocopherol.

The diffusible signal transduction and inflammatory mediator ·NO crosses cell membranes and can concentrate in lipophilic milieu by virtue of its high lipid partition coefficient. Nitric oxide, when introduced into lipid oxidation systems in the presence of α–tocopherol, preferentially reacts with lipid-derived radical species and prevents oxidation of α–tocopherol. The synergistic and potent inhibitory actions of ·NO and α–tocopherol towards lipid oxidation are revealed by the ability of ·NO to avidly terminate lipid radical-mediated chain propagation reactions and to rereduce oxidized α–tocopherol.
Liquid chromatography-mass spectroscopic (LC-MS) analysis of oxidation products of linolenic acid induced by an organic peroxyl radical initiator of lipid oxidation reactions (AMVN), showed that nitric oxide and α–tocopherol exert both independent and synergistic lipid antioxidant activity. α–Tocopherol (5 μM) inhibited AMVN-induced 18:3 oxidation by 50%. Similar inhibition of 18:3 oxidation was also afforded by the ·NO donor spermine-NONOate (20 μM, corresponding to a ·NO production of 0.5 μM/min). When 5 μM α–tocopherol and 20 μM spermine-NONOate were added in concert to AMVN-initiated 18:3 oxidation reactions, synergistic inhibition of 18:3 oxidation was afforded, with only 0.7% of 18:3 being oxidized. During this cooperative antioxidant action of α–tocopherol and ·NO, LC-MS analysis showed that α–tocopherol oxidation was prevented, if not reversed by ·NO, as indicated by decreased yields of both the $[M\text{-}H]^-$ ions of α–tocopherol oxidation products and the net extent of α–tocopherol oxidation. Spectroscopic analysis of AMVN-initiated linoleic acid (18:2) conjugated diene formation showed similar protection of 18:2 from oxidation by the independent addition of 3 μM α–tocopherol or 50 μM spermine-NONOate. There was an increase in the induction period before autocatalytic lipid oxidation when both α–tocopherol and spermine-NONOate were added. α–Tocopherol oxidation, determined fluorometrically in the organic phase after addition of 1 ml of ethanol and 5 ml hexane to 1 ml reaction systems, did not occur until diminishing rates of ·NO generation from spermine-NONOate decomposition were unable to inhibit 18:2 oxidation.

The ability of ·NO to protect α–tocopherol from oxidation by oxidizing lipids, until ·NO falls to a critical limiting concentration, is principally due to the ability of of ·NO to preferentially react with LO· and LOO· at a significantly greater rate than α–tocopherol (2-3 x 10^9 M^{-1} sec^{-1} vs. 1 to 4 x 10^5 M^{-1} sec^{1}), while at the same time yielding novel nitrogen-containing radical-radical termination products (2-4). Finally, simultaneously direct reduction of α-tocopheroxyl radical (and possibly further oxidation states of α–tocopherol) by ·NO is feasible by thermodynamic calculation and is supported by electron spin resonance (ESR) analyses, thus regenerating α–tocopherol and limiting the net extent of apparent α–tocopherol oxidation.

Nitric oxide has multiple physical and chemical qualities which make it an effective lipid antioxidant. First, ·NO readily intercepts lipid epoxyallylic and peroxyl radicals before they can react with other lipids to initiate propagation reactions or react with proteins to form Schiff's base derivatives. Second, the reaction of lipid-derived radicals with ·NO yields potentially less reactive and more innocuous non-radical products, whose secondary tissue reactivities deserve further investigation. Third, ·NO, which has a partition coefficient of 6.5 ± 0.1 for n-octanol-water, will concentrate in lipophilic milieu and cross membranes without drastically affecting the physical properties of membranes or lipoproteins because of its small molecular radius. Thus, ·NO will terminate lipid and possibly protein radical species with little or no regard to the spatial orientation of the radical intermediate within membrane or lipoprotein microenvironments. Finally, by virtue of its high reactivity with lipid radical species, ·NO will protect other lipophilic antioxidants from oxidation, and in the case of α–tocopherol, may reduce the tocopheroxyl radical to restore and better maintain tissue antioxidant defenses during periods of oxidant stress.

REFERENCES

1) Liebler, D.C. Crit. Rev. Toxicol. 23, 147-169, 1993.
2) Padmaja, S. and Huie, R.E. Biochem. Biophys. Res. Commun. 195, 539-544, 1993.
3) Rubbo, H., Radi, R., Trujillo, M., Telleri, R., Kalyanaraman, B., Barnes, S., Kirk, M. and Freeman, B.A. J. Biol. Chem. 269, 26066-26075, 1994.
4) Rubbo, H., Parthasarathy, S., Barnes, S., Kirk, M., Kalyanaraman, B. and Freeman, B.A. Arch. Biochem. Biophys, 323, 000-000, 1995.

Direct activation of the prokaryotic transcription factor oxyR by S-nitrosylation

ALFRED HAUSLADEN[1], CHRISTOPHER T. PRIVALLE[2], AND JONATHAN S. STAMLER[2]

[1]Duke University Medical Center, Division of Pulmonary Medicine, Durham, NC 27710, and [2]Apex Bioscience Inc., Research Triangle Park, NC 27709

RSNOs[1] are naturally occurring compounds which possess the unique capability of nitrosyl transfer to both metal and thiol centers. Their NO-related activities include the activation of guanylate cyclase and of the G-protein p21^{ras}, as well as the modulation of enzymes and ion channels [1]. These activities establish S-nitrosothiols as important cellular signaling molecules which act through activation/inhibition of enzymes and related signaling proteins, but it is conceivable that their role extends to the regulation of transcription factors. A candidate DNA binding protein that could be modulated by S-nitrosylation is the prokaryotic transcriptional regulator oxyR. OxyR is the redox-sensitive transcriptional activator of antioxidant enzymes in *E.coli* and *S.typhymurium*, such as the catalase/peroxidase HP1. We selected oxyR, because the oxidation of a single thiol in oxyR to a sulfenic acid (SH→S-OH) acts as a regulatory switch to activate transcription [2], making this thiol a potential target for regulation by S-nitrosylation. Furthermore, the activation of antioxidant enzymes by RSNOs would be of adaptive advantage, as RSNOs have been shown to pose a redox-threat to prokaryotes [3].

As shown in Figure 1, treatment of *E.coli* with SNO-Cys led to an increase in the HP1 activity that was comparable to the activation exerted by H_2O_2. By using an oxyR deficient mutant strain, lacking the ability to induce HP1 after H_2O_2 treatment, it was confirmed that the induction of HP1 by SNO-Cys is indeed dependent on oxyR. The induction was more pronounced in a GSH-deficient mutant. Millimolar concentrations of SNO-Cys led to an approximately 10-fold induction of HP1 in a wild type strain. In contrast, a GSH-deficient mutant showed increased HP1 activities after

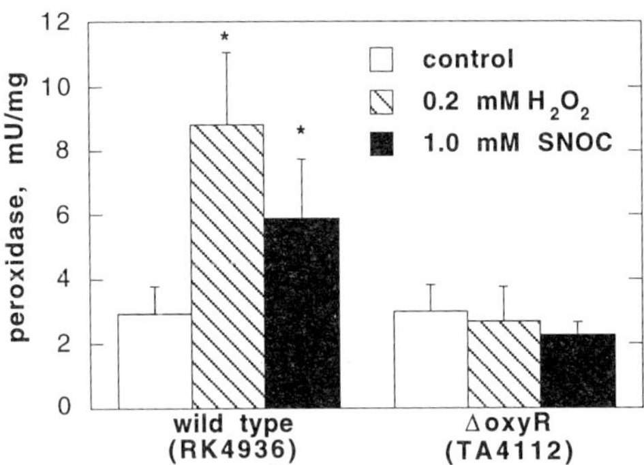

Figure 1. OxyR dependent induction of HP1 in E.coli. Cells were grown in minimal glucose medium to mid-exponential phase and then treated with the indicated amounts of hydrogen peroxide or S-nitrosocysteine.
*, different from control ($p<0.05$)

treatment with as little as 10 µM SNO-Cys, and as much as a 20-fold increase using 200 µM SNO-Cys. Western blot analysis of extracts from the GSH-deficient strain confirmed that the increase in activity was due to *de novo* protein synthesis. GSH likely acts as a sink for the NO group from SNO-Cys, thereby attenuating the S-nitrosylation of oxyR and reducing the induction of HP1. Several lines of evidence suggest that S-nitrosylation is indeed the mechanism of oxyR activation by SNO-Cys. First, we conducted the experiment under anaerobic conditions and found the activation of HP1 by SNO-Cys to be independent of oxygen, thereby excluding the S-OH modification of oxyR as the mechanism of activation. Second, authentic nitric oxide had no appreciable effect on HP1 activity. SNO-Cys readily transfers its NO-group to other thiols or metal centers [4], while authentic NO primarily targets metal centers. Therefore, the S-NO modification and concomitant activation of oxyR is a unique mechanism exerted by RSNO, and not by other N-oxides. Third, the parent compounds of SNO-Cys, nitrite and cysteine, also had no effect on HP1 activity. These results are summarized in Figure 2. Furthermore, extracts from cells

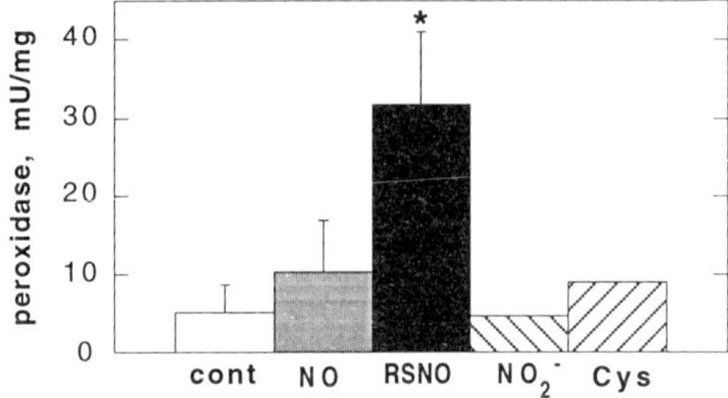

Figure 2. Anaerobic induction of HP1 by SNO-Cys. Cells were grown in an anearobic glove box to mid-exponential phase and then treated with 0.2 mM of the indicated compound. A saturated solution of nitric oxide in water was added from a sealed vial via a syringe. The asterisk indicates significant difference from the control ($p<0.05$).

treated anaerobically with SNO-Cys contained significant amounts of RSNO, confirming intracellular RSNO formation. On the other hand, under aerobic conditions RSNO was rapidly metabolized. While the half-life of SNO-Cys and GSNO in the growth medium are 30 and 70 minutes, respectively, exponentially growing *E.coli* accelerated the RSNO breakdown dramatically ($t_{1/2}\approx 10$ min.). Strikingly, wild type cells did not accumulate any nitrite after aerobic treatment with up to 100 µM SNO-Cys, whereas the oxyR mutant strain accumulated large amounts of nitrite. The oxyR mutant also showed greater growth inhibition by SNO-Cys. These results suggest the presence of an oxygen-dependent mechanism of RSNO decomposition and resistance in *E.coli*, that is at least in part under the control of oxyR.

In summary, our results demonstrate an oxyR dependent and oxygen-independent transcriptional activation of antioxidant enzymes by RSNO, which is accompanied by intracellular RSNO formation. The mechanism appears to involve the S-nitrosylation of a thiol in oxyR. The induction of antioxidant genes confers greater resistance to growth inhibition by SNO-Cys and involves an alternate mechanism of RSNO metabolism. These studies are the first to identify the presence of RSNO responsive genes and show that S-nitrosylation may serve as a functional switch in the control of gene expression.

Acknowledgments: The oxyR mutant TA4112 and its isogenic parental strain RK4936 were from Dr. B. Ames [5], the glutathione deficient mutant JTG10 was from Dr. B. Demple [6], its parental strain AB1157 was from Dr. J. Imlay. Polyclonal antibodies against HP1 were provided by Dr. P. Loewen. We thank Dr. I. Fridovich for helpful discussions and the supply of bacterial strains.

1. Stamler, J.S. (1994). Cell **78**, 931-936.
2. Toledano, M.B., Kullik, I., Trinh, F. Baird, P.T., Schneider, T.D., and Storz, G. (1994). Cell **78**, 897-909.
3. De Groote, M.A., Granger, D., Xu, Y., Campbell, G., Prince, R., and Fang, F.C. (1995). Proc.Natl.Acad.Sci. USA, **92**, 6399-6403.
4. Arnelle, D.R., and Stamler, J.S. (1995). Arch.Biochem. Biophys. **318**, 279-285.
5. Christman, M.F., Morgan, R.W., Jacobson, F.S., and Ames, B.N. (1985). Cell, **41**, 753-762.
6. Greenberg, J.T., Demple, B. (1986). J.Bacteriol. **168**, 1026-1029.

[1] Abbreviations used: GSH, glutathione; GSNO, S-nitrosoglutathione; HP1, hydroperoxidase 1, RSNO, S-nitrosothiol; SNO-Cys, S-nitrosoglutathione

The *metL* gene of *Salmonella typhimurium* is required for S-nitrosoglutathione resistance, macrophage survival, and virulence in mice

Mary Ann De Groote and Ferric C. Fang

Division of Infectious Diseases B168
University of Colorado Health Sciences Center
4200 E. Ninth Avenue
Denver, CO USA 80262

Reactive nitrogen intermediates possess antimicrobial activity against a broad range of microbial pathogens [1], but the biologically relevant intermediates and their mechanisms of action remain uncertain. GSNO (S-nitrosoglutathione) is an attractive candidate endogenous antimicrobial mediator since it exhibits broad-spectrum microbiostatic activity [2] and has been detected *in vivo* during inflammatory states [3].

A MudJ transposon library [4] in the Gram-negative bacterium *Salmonella typhimurium* was enriched for GSNO-susceptible mutants by simultaneous exposure to cycloserine and subinhibitory concentrations of GSNO. Since cycloserine is bactericidal for growing cells, this protocol provides a positive selection for mutants possessing enhanced GSNO susceptibility. A highly GSNO-susceptible clone was found to harbor a proximal transposon insertion in the *metL* gene, encoding aspartokinase II-homoserine dehydrogenase II [5]. This enzyme catalyzes two independent steps in the prokaryotic biosynthetic pathways for lysine, threonine, and methionine. Synthesis of the metabolic intermediates diaminopimelate, homoserine, and homocysteine also requires aspartokinase-homoserine dehydrogenase activity.

metL mutant *S. typhimurium* was found to possess significant hypersusceptibility to S-nitrosothiols (GSNO, S-nitroso-N-acetyl-penicillamine, and S-nitroso-N-acetyl-cysteine), but little or no alteration in susceptibility to hydrogen peroxide, paraquat, the peroxynitrite-donor SIN-1 (3-morpholinosydnonimine hydrochloride), or the NO•-generator DETA/NO [6], when assayed by a disk diffusion method [2]. Re-transduction of the *metL* transposon insertion confers S-nitrosothiol hypersusceptibility to wild-type *Salmonella*. *metL* mutant *S. typhimurium* is not auxotrophic for methionine because of the presence of a related enzyme (aspartokinase I-homoserine dehydrogenase I) encoded by the *thrA* gene [7]. Notably, *thrA* mutant *S. typhimurium* possesses hypersusceptibility to S-nitrosothiols similar to that of the *metL* mutant strain.

metL mutant *S. typhimurium* has an impaired ability to survive in thioglycollate-elicited murine peritoneal macrophages and is highly attenuated for virulence in C3H/HeN mice (*Ity*^{r} *Lps*^{n}). However, these abnormal phenotypes are only partially restored by the NO synthase inhibitors N^G-monomethyl-L-arginine and aminoguanidine, suggesting that NO-independent host defenses such as the respiratory burst may also play a role. Nevertheless, NO synthase inhibition appears to significantly impair the antimicrobial host response to wild-type *S. typhimurium* in C3H/HeN mice.

Specific metabolic intermediates or terminal products in pathways affected by *metL* were examined for their effects on S-nitrosothiol-mediated cytostasis. GSNO-induced microbiostasis is abrogated *in vitro* by homocysteine, homoserine, or methionine, but not by diaminopimelate or threonine.

Together, these observations suggest that S-nitrosothiols may have an important role as antimicrobial mediators in murine salmonellosis. Moreover, the homocysteine/methionine biosynthetic pathway appears to be important in both *Salmonella* pathogenesis and resistance to S-nitrosothiols. Homocysteine, the only thiol-containing intermediate in the methionine biosynthetic pathway, may function as an endogenous S-nitrosothiol antagonist. Transfer of NO from S-nitrosothiols to the sulfhydryl center of homocysteine may redirect NO away from other cellular protein targets. The vasodilatory, antiplatelet, antioxidant, antiproliferative, and neuroregulatory actions of S-nitrosothiols are directly opposed to actions attributed to homocysteine [8-10], and it is intriguing to speculate that a balance between S-nitrosothiols and homocysteine may be central to a diverse range of biological processes including infection, neoplasia, and atherosclerosis.

ACKNOWLEDGEMENTS

The authors are grateful to K. Sanderson and A. Hessel for the generous gift of bacterial strains, and to G. Stauffer and J. Stamler for helpful discussions. This work was supported in part by grants from the National Institutes of Health (AI32463), the U.S. Department of Agriculture (9401954), and the Thorkildsen Research Fellowship Program.

REFERENCES

1. De Groote, M.A. and Fang, F.C. (in press) NO inhibitions: Antimicrobial properties of nitric oxide, Clin. Infect. Dis.
2. De Groote, M.A., Granger, D., Xu, Y., Campbell, G., Prince, R. and Fang, F.C. (1995) Genetic and redox determinants of nitric oxide cytotoxicity in a *Salmonella typhimurium* model, Proc. Natl. Acad. Sci. U.S.A. **92**:6399-6403
3. Gaston, B., Reilly, J., Drazen, J.M., et al. (1993) Endogenous nitrogen oxides and bronchodilator S-nitrosothiols in human airways, Proc. Natl. Acad. Sci. U.S.A. **90**:10957-10961
4. Hughes, K.T. and Roth, J.R. (1988) Transitory cis-complementation: A general method for providing transposase to defective transposons, Genetics **119**:9-12
5. Zakin, M.M., Duchange, N., Ferrara, P. and Cohen, G.N. (1983) Nucleotide sequence of the *metL* gene of *Escherichia coli*, J. Biol. Chem. **258**:3028-3031
6. Wink, D.A., Kasprzak, K.S., Maragos, C.M., et al. (1991) DNA deaminating ability and genotoxicity of nitric oxide and its progenitors, Science **254**:1001-1003
7. Katinka, M., Cossart, P., Sibilli, L., et al. (1980) Nucleotide sequence of the *thrA* gene of *Escherichia coli*, Proc. Natl. Acad. Sci. U.S.A. **77**:5730-5733
8. Stamler, J.S., Osborne, J.A., Jaraki, O., et al. (1993) Adverse vascular effects of homocysteine are modulated by endothelium-derived relaxing factor and related oxides of nitrogen, J. Clin. Invest. **91**:308-318
9. Tsai, J.C., Perrella, M.A., Yoshizumi, M., et al. (1994) Promotion of vascular smooth muscle growth by homocysteine: A link to atherosclerosis, Proc. Natl. Acad. Sci. U.S.A. **91**:6369-6373
10. Stampfer, M.J. and Malinow, M.R. (1995) Can lowering homocysteine levels reduce cardiovascular risk?, N. Engl. J. Med. **332**:328-329

Nitric oxide induction of protection against rat hepatocyte toxicity from nitric oxide and from hydrogen peroxide

Y-M. KIM and J. R. LANCASTER, JR.*

Dept. Surgery, University of Pittsburgh School of Medicine, Pittsburgh PA 15261, *Depts. Physiology and Medicine, LSU Medical Center, New Orleans, LA, 70112, USA

Although much is known about the cellular actions of NO, little attention has been directed to potential cellular defensive mechanisms against the damaging effects of this radical. Compared to other cell types such as the activated macrophage, the isolated rat hepatocyte (HC) stimulated *in vitro* by inflammatory mediators to produce NO is relatively resistant to cell injury such as loss of mitochondrial electron transfer (Stadler *et al.* 1991a; Stadler *et al.* 1991b). *In vivo*, hepatic NO production can in fact serve a protective function against LPS-induced injury (Harbrecht *et al.* 1992). We have examined low level NO induction of protection against subsequent high-level NO and H_2O_2 toxicity in HC's.

Figure 1 presents the results of assaying for toxicity to hepatocytes exposed to two sequential treatments. The first treatment was either control (20 hr.), mixture of inflammatory mediators (TNF-α, IL-1β, IFN-γ, LPS, "Cyt. Mix.") without or with NMMA (20 hr.) to induce NO synthesis (Kim *et al.* 1993), or 100μM SNAP (14 hr.), a nitrogen oxide donor. The second treatment was 12 hr. treatment with NMMA alone or NMMA plus 2 mM SNAP. In control HC's the second SNAP treatment (2 mM) results in substantial (approx. 50%) toxicity. However, HC's which are first treated with either low (100 μM) SNAP or with the cytokine mix are resistant to killing.

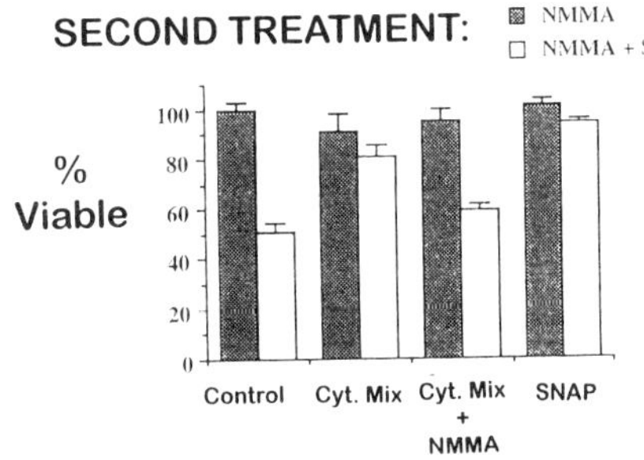

Figure 1

The effect of the cytokine mix is due to endogenous NO, since protection is largely prevented by NMMA. With SNAP, protection is imparted with only 50-100 μM concentrations and requires 8-12 hr. to develop (not shown). The parent compound of SNAP does not induce protection, and significant protection is also afforded against SNAP-induced decrease in aconitase and mitochondrial electron transfer activities (not shown).

The first four rows of Table I shows the protective effects of low (100 μM) SNAP on high (2 mM) SNAP killing. The protection is prevented by cycloheximide (CHX), demonstrating that the response involves synthesis of new protein(s). Protection is also prevented by tin- protoporphyrin-IX (SnPP), an inhibitor of heme oxygenase (Yoshinaga *et al.* 1982). In addition, SNAP induces significant cross-protection against hydrogen peroxide toxicity, which is also prevented by SnPP.

First Treatment	Second Treatment	Viability (%)
--	--	-100-
--	SNAP	51.8 ± 4.9
SNAP	--	102.1± 6.3
SNAP	SNAP	87.4 ± 5.2
SNAP + CHX	SNAP	49.0±7.4
SNAP + SnPP	SNAP	51.3 ± 5.8
--	H_2O_2	44.5 ± 4.7
SNAP	H_2O_2	80.1 ± 6.4
SNAP + SnPP	H_2O_2	40.6 ± 4.0

Table I

We have previously shown that either endogenous or exogenous NO induces disturbances in HC heme metabolism, which is attributable to NO-induced liberation of intracellular heme and consequent changes in heme-metabolizing enzymes including upregulation of heme oxygenase (Kim *et al.* 1995). Heme oxygenase is a stress response protein, and is identical to hsp-32 (Keyse *et al.* 1989). NO also upregulates heme oxygenase in rat islets (Welsh *et al.* 1994). Results here indicate that heme oxygenase upregulation may be protective against oxidative injury, although we cannot distinguish between increased resistance or repair of damaged targets, which has been demonstrated previously (Stadler *et al.* 1991a; Corbett *et al.* 1994). In any event, these results suggest that in addition to direct reaction between reactive nitrogen and oxygen intermediates, another factor which may determine the damaging *vs.* protective effects of NO in oxidative injury is the existence of cellular defensive mechanisms induced by NO exposure.

This work was supported by grants to JRL from the American Cancer Society (BE-128) and the National Institute of Diabetes, Digestive and Kidney Diseases (DK46935)

Corbett, J.A. & McDaniel, M.L. (1994) Biochem. J. 299, 719-724

Harbrecht, B.G., Billiar, T.R., Stadler, J., Demetris, A.J., Ochoa, J.B., Curran, R.D. & Simmons, R.L. (1992) Crit. Care Med. 20, 1568-1574

Keyse, S.M. & Tyrrell, R.M. (1989) Proc. Natl. Acad. Sci. USA 86, 99-103

Kim, Y.M. & Lancaster, J.R., Jr. (1993) FEBS Lett. 332, 255-259

Kim, Y.M., Bergonia, H.A., Muller, C., Pitt, B.R., Watkins, W.D. & Lancaster, J.R., Jr. (1995) J. Biol. Chem. 270, 5710-5713

Stadler, J., Billiar, T.R., Curran, R.D., Stuehr, D.J., Ochoa, J.B. & Simmons, R.L. (1991a) Am. J. Physiol. 260, C910-C916

Stadler, J., Curran, R.D., Ochoa, J.B., Harbrecht, B.G., Hoffman, R.A., Simmons, R.L. & Billiar, T.R. (1991b) Arch. Surgery 126, 186-191

Welsh, N. & Sandler, S. (1994) Mol. Cell Endocrinol. 103, 109-114

Yoshinaga, T., Sassa, S. & Kappas, A. (1982) J. Biol. Chem. 257, 7778-7785

Abbreviations used: CHX, cycloheximide; HC, hepatocyte cells; IFN-γ, interferon-γ; IL-1β, interleukin-1β; LPS, lipopolysaccharide; NMMA, N^G-monomethyl-L-arginine; SNAP, S-nitrosoacetylpenicillamine; SnPP, tin-protoporphyrin-IX

Thiols protect against nitric oxide-mediated cytotoxicity

ZAMORA RUBEN, MATTHYS KATELIJNE and HERMAN ARNOLD

Div.of Pharmacol., Dept. of Medicine, University of Antwerp - UIA, Universiteitsplein 1, B-2610 Wilrijk - Antwerpen, BELGIUM

INTRODUCTION

Nitric oxide (NO) is an effector molecule implicated in the cytotoxic activity of macrophages towards tumor cells and microbial pathogens [1]. The conditions under which NO (or NO-related species) exerts cytotoxicity have not been completely characterized yet. NO is a highly reactive free radical with a broad aqueous chemistry involving other interrelated redox forms that may also contribute to the cytotoxic effects of activated macrophages. Thus, the cytotoxicity of NO must also be viewed within the context of the intracellular redox milieu.

Previous studies showed that tumor target cells co-cultivated with cytotoxic activated macrophages stopped dividing and rapidly died unless susbtrate for glycolysis was present in their environment [2]. Furthermore, the macrophages themselves are subject to the cytotoxic effects of NO production by the L-arginine/NO pathway. Mammalian cells have evolved protective mechanisms to minimize injury by toxic radicals. In this respect, glutathione (GSH) has a major role as free radical scavenger and its depletion results in a higher sensitivity of the cells to the toxic effects of NO [3].

We investigated whether alteration of intracellular thiol levels modulates the cytotoxic effect of different NO donors in the murine macrophage cell line J774A.1.

METHODS

Experimental protocol. Cells (10^6/ml) were incubated for 24 h in DMEM (glucose: 1000 mg/l) or DMEM containing BSO or NAC (1 mM). After 24 h, the medium was replaced by low-glucose DMEM (250 mg/l) in the presence or absence of various concentrations of the NO donors. Following another 24 h incubation supernatants were collected for nitrite determination. Fresh medium containing neutral red (0.01%) was then added and after 1.5 h the dye was extracted for measuring the optical density.

Assays. Cell viability was measured using the neutral red assay and nitrite production in the culture medium was determined using the Griess reaction.

RESULTS

In DMEM (glucose: 1000 mg/l) incubation with SNAP did not result in significant cytolysis and even at 1 mM SNAP cells maintained about 80% viability. Assuming that NO-exposed cells may survive because of the sufficient supply of glucose for ATP production, the medium glucose content was lowered to 250 mg/l. Indeed, limiting glucose affected the viability in a dose-dependent manner with only ± 20% cell viability at 1 mM SNAP (Fig. 1). Addition of NAP or the SNAP metabolites NAP disulfide and NO_2^- alone did not affect the viability of control cells. The cytotoxic effect of SNAP was inhibited by the NO scavenger carboxy-PTIO. Addition of 100 µM carboxy-PTIO together with SNAP (100 and 1000 µM) increased the cell viability from 50 % to 93.1 ± 4.7 % and from 20 % to 41.2 ± 2.4 % respectively.

Thiol depletion was achieved by pretreating the cells with BSO, a potent inhibitor of γ-glutamylcysteine synthetase and hence GSH biosynthesis. BSO at 1 mM reduced intracellular glutathione to nondetectable levels. Thiol depletion increased the SNAP-mediated injury (Fig. 1). Whereas 100 µM SNAP caused only 50% cell death in control cells, viability was further reduced to 20% in BSO-pretreated cells. Incubation with the cysteine precursor NAC clearly protected the cells and almost completely restored cell viability at a concentration of 100 µM SNAP (Fig. 1). However, NAC dose-dependently decreased cellular GSH, mimicking at higher concentrations the effect of BSO (not shown).

Incubation of the cells with equimolar concentrations of other NO donors differentially affected cell viability: while SIN-1 and CysNO were not toxic, GSNO and SNAP induced cell death already at the lower concentration (Fig. 2). We compared the stability of the NO donors with the potency for cell killing. The stability was estimated by measuring the NO_2^- production in solutions of the test compounds (100 µM) incubated in culture medium at 37°C after different time intervals (not shown). The half-lives were found to be 10 min (SIN-1), 15 min (CysNO), 300 min (SNAP) and >360 min (GSNO). Both the order of stability of the test compounds and the order of potency for cell killing was GSNO > SNAP > CysNO = SIN-1.

Abbreviations used: SNAP, S-nitroso-N-acetyl-D,L-penicillamine; GSNO, S-nitrosoglutathione; CysNO, S-nitrosocysteine; SIN-1, 3-morpholino-sydnonimine.HCl; BSO, L-buthionine-[S,R]-sulfoximine; NAC, N-acetyl-L-cysteine; GSH, glutathione.

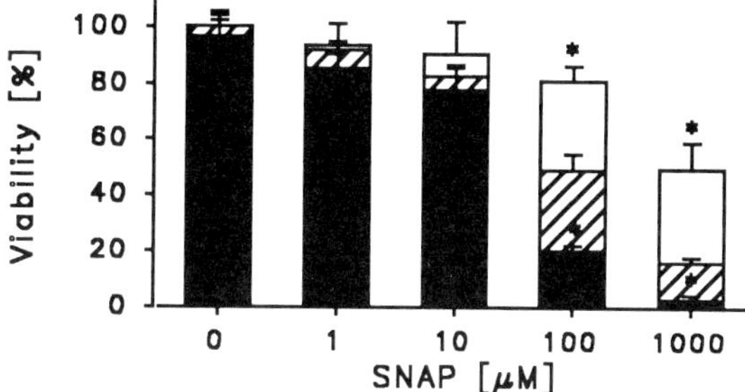

Figure 1. Effect of BSO- (solid bars) and NAC (open bars) pretreatment (both 1 mM) on SNAP-induced injury of J774 macrophages. Control cells (hatched bars) were preincubated in culture medium alone. Data are the mean ± SEM of 3 experiments. *P < 0.05.

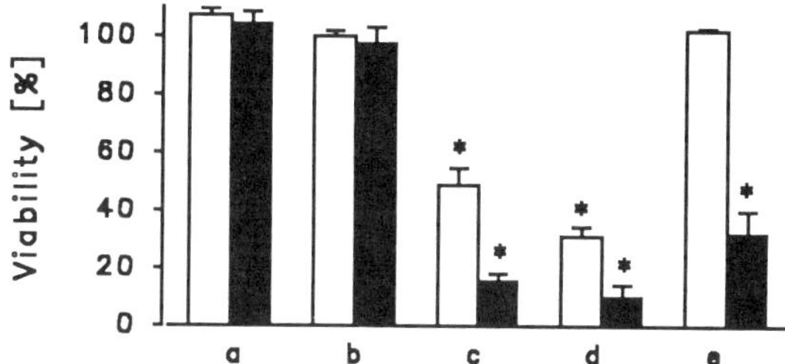

Figure 2. Effect of SIN-1 (a), CysNO (b), SNAP (c), GSNO (d) and NH_2OH (e) on viability of J774 macrophages. Cells were incubated with the test compounds at 100 µM (open bars) or 1,000 µM (solid bars) for 24 h. Data are expressed as % of viability compared to untreated cells and are the mean ± SEM of 3 experiments. *P < 0.05.

DISCUSSION

Little information is available about the conditions under which NO toxicity leads to cell death. Our observations demonstrated that exposure of J774.A1 macrophages to SNAP in low-glucose medium resulted in a significant loss of viability depending on both the glucose and the SNAP concentration. Carboxy-PTIO, a known inhibitor of NO-mediated biological responses, prevented cell injury. This points to NO or an NO-derived species as mediator of the SNAP-induced cell death.

Recent evidence shows the involvement of GSH in the protection against NO-induced cytotoxicity [3]. However, the role of other non-protein thiols has not been investigated yet. In the present study, GSH depletion of J774 cells with BSO further increased the SNAP-mediated cytotoxicity. Pretreatment with the cysteine precursor NAC prevented the toxic effects of NO. Contrary to what was expected, GSH levels were decreased by NAC pretreatment. Thus, the protective effect of NAC was clearly not mediated by GSH. NAC can react with superoxide preventing the formation of the more toxic $ONOO^-$. Alternatively, assuming that NAC is deacetylated to cysteine, a subsequent transfer of the nitroso group from SNAP to cysteine may have occurred. In support of this, equimolar concentrations of CysNO were less cytotoxic than SNAP and GSNO under the same culture conditions, suggesting a possible correlation between the cytotoxic potency of the spontaneously NO-generating compounds and the rate at which NO is liberated. Moreover, a reaction between the thiol (NAC or cysteine) and SNAP to give hydroxylamine may also occur. Indeed, we found that NH_2OH was less cytotoxic than SNAP at the same concentrations. In summary, our results demonstrate that 1) low glucose and intracellular GSH levels enhance the cytotoxic effects of NO donors, 2) long-lived NO donors are more cytotoxic than short-lived NO donors, and 3) other non-protein thiols like NAC may substitute for GSH as a key component of the cellular detoxification system.

1. Moncada, S., Palmer, R.M.J. and Higgs, E.A. (1991). *Pharmacological Reviews* **43**, 109-142.

2. Hibbs, J.B., Jr., Taintor, R.R., Vavrin, Z., Granger, D.L., Drapier, J-C., Amber, J.J. and Lancaster, J.R., Jr. (1990). ***In*** Nitric oxide from L-arginine: a bioregulatory system (Moncada, S. and Higgs, E.A., ed.), pp. 189-223, Elsevier, Amsterdam.

3. Whit Walker, M., Kinter, M.T., Roberts, R.J. and Spitz, D.R. (1995). *Pediatr. Res.* **37**, 41-49.

Nitric oxide reversibly inhibits cytochrome oxidase, catalase and cell energy metabolism, and cytochrome oxidase rapidly reduces NO

GUY C. BROWN

Department of Biochemistry, University of Cambridge, Tennis Court Road, Cambridge CB2 1QW, UK.

INTRODUCTION AND METHODS

The target proteins by which nitric oxide (NO) exerts its physiological and cytotoxic effects are unclear. NO is known to bind with high affinity to a number of haem proteins including cytochrome oxidase and catalase. We examined the effect of NO on the rates of these enzymes by incubating the isolated proteins in a gas tight vessel with an oxygen electrode (to measure oxygen production from hydrogen peroxide by catalase, and oxygen reduction to water by cytochrome oxidase) [1,2]. An aliquot of NO-saturated water was then added and the level of NO measured simultaneously with an NO electrode also in the vessel. The same vessel was used to simultaneously measure O_2 and NO with isolated mitochondria, synaptosomes, platelets, and cultured cells.

CATALASE

NO caused a rapid and reversible inhibition of catalase, with a Ki of 0.2 µM NO [1]. Since cells expressing the inducible form of NO synthase (iNOS) produce at least this level of NO [3,4] it seems likely that catalase is partially inhibited by NO in vivo, and the consequent build up of H_2O_2 might contribute to cytotoxicity.

CYTOCHROME OXIDASE

NO rapidly and reversibly inhibited cytochrome oxidase at submicromolar levels of NO in competition with oxygen [2]. NO also inhibited respiration in isolated nerve terminals from brain with a Ki of 270 nM NO at about 150 µM O_2 (roughly the arterial level of O_2) and a Ki of 60 nM NO at 30 µM O_2 (roughly the tissue level of O_2) [2]. These levels of NO did not cause any irreversible inhibition of respiration, and there was no detectable damage to EPR-visible mitochondrial iron-sulphur centres; although there was a large increase in the g=2.04 iron-sulphur dinitrosyl EPR-signal, which has in the past been thought to be indicative of cell damage [5]. Respiration in isolated mitochondria, platelets and cultured cells was also inhibited by similar levels of NO in competition with oxygen. Cytochrome oxidase is the terminal component of the mitochondrial respiratory chain, and is central to cellular energy transduction. Because the inhibition by NO is competitive with oxygen, NO raises the apparent Km of respiration for oxygen into the physiological range, and inhibits cellular ATP production and those processes dependent on ATP [3]. But do cells ever produce sufficient NO to inhibit there own respiration?

CULTURED ASTROCYTES

We induced iNOS in cultured astrocytes with interferon-γ and endotoxin, and examined whether the cellular respiration was inhibited [4]. The cells produced up to 1 µM NO and oxygen consumption was substantially inhibited compared to non-induced cells. NO levels were rapidly decreased by inhibiting NO synthase or adding haemoglobin, and these treatments rapidly reversed the inhibition of respiration. The oxygen, NO and substrate dependence of the inhibition indicated that the inhibition was at cytochrome oxidase [4]. Thus cells expressing iNOS strongly inhibit their own respiration, and presumably that of surrounding cells. In the astrocytes the decreased mitochondrial ATP production was partially compensated by an activation of glycolysis, but in other cells (such as neurons) this is less likely to occur, potentially leading to changes in cell function and cytotoxicity [3]. I have suggested that several of the physiological and pathological effects of NO may be mediated by its inhibition of cytochrome oxidase, and that NO may be an important physiological regulator of energy metabolism and its dependence on oxygen [3]. Other inhibitors of cytochrome oxidase are known to greatly increase H_2O_2 production by mitochondria, and this might also contribute to the cytotoxicity of NO.

NITRIC OXIDE BREAKDOWN BY MITOCHONDRIA

Isolated mitochondria were found to greatly increase the rate of NO breakdown. This rate was increased in the absence of oxygen, and was partly inhibited by cyanide and azide, indicating that at least part of the NO breakdown was due to reduction of NO by cytochrome oxidase. The rate of NO breakdown by mitochondria was large compared to the known rates of NO breakdown in tissue, suggesting that this mechanism may be important in vivo. Isolated cytochrome oxidase is known to reduce NO to N_2O. If N_2O is produced from NO in vivo this might be important for experimental and clinical monitoring of NO production.

This work was supported by the Royal Society (London) and Wellcome Trust.

[1] Brown, G. C. (1995) Reversible binding and inhibition of catalase by nitric oxide. Eur. J. Biochem. 232, 188-191.

[2] Brown, G. C. & Cooper, C. E. (1994) Nanomolar concentrations of nitric oxide reversibly inhibit synaptosomal respiration by competing with oxygen at cytochrome oxidase. FEBS Lett. 356, 295-298.

[3] Brown, G. C. (1995) Nitric oxide regulates mitochondrial respiration and cell functions by inhibiting cytochrome oxidase. FEBS Lett. 369, 136-139.

[4] Brown, G. C., Bolanos, J. P., Heales, S. J. R. & Clark, J. B. (1995) Nitric oxide produced by activated astrocytes rapidly and reversibly inhibits cellular respiration. Neuroscience Lett. 193, 201-204.

[5] Cooper, C. E. & Brown, G. C. (1995) The interaction between nitric oxide and brain nerve terminals as studied by electron paramagnetic resonance. Biochem. Biophys. Res. Commun. 212, 404-412.

Nitrogen oxide mediated release of iron from ferritin

Neil R. Bastian*, Meredith J.P. Foster*, Clair L. Bello III*, Daniel V. Kinikini*, and John B. Hibbs Jr*#.

*Division of Infectious Diseases, University of Utah School of Medicine, Salt Lake City, Utah 84132 USA and #VA Medical Center, Salt Lake City, Utah 84148 USA

It has been observed that tumor cells cocultivated with activated macrophages release a significant fraction of their intracellular iron in parallel with development of inhibition of DNA synthesis and mitochondrial respiration [1-2]. Reif and Simmons [3] demonstrated that the NO donor, sodium nitroprusside, mediated release of iron from ferritin *in vitro*. In this paper we present data that show that induction of NOS mediates degradation of intracellular ferritin *in vivo* and that reductive NO_x compounds mediate release of iron from ferritin *in vitro*.

Materials and Methods

In vitro iron release from ferritin was monitored colorimetrically by formation of the Fe(II)-ferrozine complex [4]. Horse spleen ferritin (Sigma) was extensively dialyzed against 50 mM Tris·HCl pH 7.5 to remove adventitious iron. Assay solutions contained, in 1 ml final volume, ferritin (200 μM iron), ferrozine (500 μM) and an NO_x donor (10 mM). Assays were carried out at pH 7.5 in 250 mM Tris·HCl buffer. Iron release is reported as the mean ± SEM of at least 4 independent measurements. Statistical analyses were done using a one tailed Student's t test. S-nitrosoglutathione (GSNO) [5], sodium α-oxyhyponitrite ($Na_2N_2O_3$) [6] and peroxynitrite ($ONOO^-$) [7] were prepared by previously described methods.

Results and Discussion

We first studied the effects of NO on iron and ferritin in cultures of human K562 cells which had been pre-incubated for 24 hours in medium containing amino acid supplements and 3 mci/ml of ^{55}Fe-citrate. The ^{55}Fe-labelled K562 cells were washed twice with phosphate buffered saline (PBS) before being co-cultivated with cytokine activated macrophages, again in the presence of the amino acid supplement. Nitrite and nitrate were assayed as described [8]. Release of iron from the K562 cells was monitored by counting the radioactivity due to ^{55}Fe in aliquots of both the medium and the cells after separation. Intracellular ferritin was quantitated by radioimmunoassay with I-125 labelled anti-human-ferritin antibodies obtained in kit form from Ramco Laboratories [9].

We found in these experiments that cytotoxic activated macrophage/K562 cocultures in medium containing L-arginine showed a significant increase in nitrite+nitrate, an increase in ^{55}Fe release, and a decrease in K562 cell intracellular ferritin (Table 1). These effects were further enhanced by the presence in the medium of 0.2 mM L-cysteine (Table 1), the rate limiting amino acid in GSH synthesis. The amount of NO produced, as evidenced by nitrite+nitrate production, correlated well with iron release, ferritin degradation, and inhibition of mitochondrial respiration. These results provide substantial evidence that biologically produced NO_x from L-arginine is responsible for the release of iron from cells and the accompanying degradation of ferritin.

Table 1. Effect of Amino Acid Supplements on Ferritin Iron

Additive	NO_2^-+NO_3^- (μM)	% Fe Release	Ferritin (ng/10^{6}cells)
Control	14	2	42
1.2 mM L-Arg	122	21	24
0.2 mM L-Cysteine	27	8	35
1.2 mM L-Arg + 0.2 mM L-Cysteine	209	42	17
Complete amino acids	158	49	17

We next designed experiments to test the effect of various nitrogen oxides on release of iron from ferritin *in vitro*. Release of iron from ferritin requires the reduction of insoluble Fe(III) in the interior of the protein to soluble Fe(II). The release of iron from ferritin by several different nitrogen oxides is shown in Figure 1. We found that very little iron was detected by the ferrozine assay in a solution of ferritin that was saturated with NO(g). However, EPR analysis of anaerobic ferritin-NO(g) solutions in the presence of reduced glutathione (GSH) showed significant formation (42 μM) of EPR active $Fe(GS)_2(NO)_2$ complexes. We also found that both $Na_2N_2O_3$, which releases nitroxyl anions (NO^-), and S-nitrosoglutathione (GSNO) induced a significant loss of ferrozine detectable iron from ferritin (Figure 1). The presence of 4 mM GSH significantly enhanced the amount of ferrozine detectable iron released by GSNO but not by $Na_2N_2O_3$ (Figure 1). In the absence of ferrozine but in the presence of glutathione, addition of either $Na_2N_2O_3$ or GSNO resulted in the formation of EPR active $Fe(GS)_2(NO)_2$ complexes. The integrated intensity of the EPR signal (42 μM) formed by anaerobic ferritin, 4 mM GSH, and saturated NO(g) was approximately the same as that formed from ferritin, 4 mM GSH and either 10 mM GSNO (41 μM) or 10 mM $Na_2N_2O_3$ (39 μM). The epr active $Fe(GS)_2(NO)_2$ complexes could be separated from the ferritin protein by Sephadex G25 chromatography (data not shown).

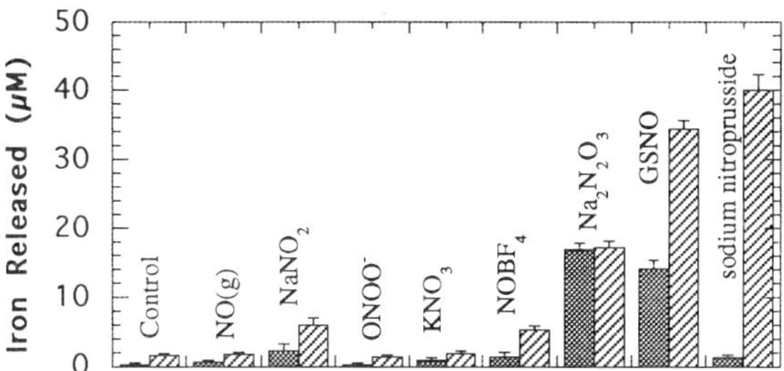

Figure 1. Release of iron from ferritin by NO_x compounds in the absence (solid bars) or presence (lined bars) of GSH.

Nitrite, peroxynitrite, and nitrate were also tested and found to release only small quantities of iron by the ferrozine assay. Addition of 4 mM GSH to the assay solution resulted in slightly enhanced release of iron by nitrite but not by nitrate or peroxynitrite. A small EPR signal was observed in solutions containing ferritin with GSH plus nitrite (integrated intensity 3μM) but not in those containing GSH plus nitrate or peroxynitrite. These results can be explained by the formation of S-nitrosoglutathione from nitrite plus GSH. The amount of iron released by nitrate and peroxynitrite was not statistically different than control values by student's t test (P=0.05). Nitrosonium tetrafluoroborate, which releases nitrosonium (NO^+) ions also failed to solubilize a statistically significant amount of ferritin iron when compared to control solutions or to ferritin reacted with tetrafluoroboric acid. A small EPR signal due to $Fe(GS)_2(NO)_2$ complex formation (integrated intensity 2 μM) was seen when GSH was added to anaerobic ferritin-$NOBF_4$ solutions, presumably due to the facile formation of S-nitrosoglutathione from GSH and NO^+.

The data presented here show that iron is released from human K562 cells cocultivated with NO_x producing macrophages and that this release is accompanied by degradation of intracellular ferritin. It also is demonstrated that iron is released from ferritin *in vitro* by reductive nitrogen oxides and that this release is accompanied by formation of epr active $Fe(RS)_2(NO)_2$ complexes.

Acknowledgements

This work was supported by Grant IRG-178B from the American Cancer Society, Grant HL51963 from NIH, by the Department of Veterans Affairs, and by the University of Utah Research Fund.

References

1. Drapier, J.-C. & Hibbs, J.B., Jr. (1988) *J. Immunol.* **140**, 2829-2838
2. Hibbs, J.B., Jr., Taintor, R.R. and Vavrin, Z. (1984) *Biochem. Biophys. Res. Commun.* **123**, 716-723
3. Reif, D.W. and Simmons, R.D. (1990) *Arch. Biochem. Biophys.* **283**, 537-541
4. Carter, P. (1971) *Analyt. Biochem.* **40**, 450-458
5. Park, J.-W. (1988) *Biochem. Biophys. Res. Commun.* **152**, 916-920
6. Addison, C.C., Gamlen, G.A., and Thompson, R. (1952) *J. Chem. Soc.* 338-345
7. Wink, D.A., Darbyshire, J.F., Nims, R.W., Saavedra, J.E., and Ford, P.C. (1993) *Chem. Res. Toxicol.* **6**, 23-27
8. Hibbs, J.B., Jr., Taintor, R.R., Vavrin, Z. and Rachlin, E.M. (1988) *Biochem. Biophys. Res. Commun.* **157**, 87-94. (erratum published in (1989) *Biochem. Biophys. Res. Commun.* **158**, 624
9. Li, P.K., Humbert, J.R., and Cheng, C. (1978) *Clin. Chem.* **24**, 1650

DNA damage induced by peroxynitrite: Formation of 8-nitroguanine and base propenals

Julieta RUBIO*, Vladimir YERMILOV and Hiroshi OHSHIMA

International Agency for Research on Cancer, 150 cours Albert-Thomas, 69372 Lyon, Cedex 08, France

INTRODUCTION

Chronic infection and inflammation are recognized risk factors for a variety of human cancers [1]. Nitric oxide (NO) produced in inflamed tissues may cause damage in DNA by several different mechanisms. NO is easily oxidized with oxygen to form N_2O_3, which nitrosates secondary amines to form carcinogenic *N*-nitrosamines and also deaminates DNA bases, leading to mutations [2,3]. NO reacts rapidly with superoxide anion ($O_2^{-}\cdot$) to form peroxynitrite anion ($ONOO^-$). Peroxynitrite (pKa=6.8) can decompose under physiological conditions to produce strong oxidants which initiate reactions characteristic of hydroxyl radical, nitronium ion and nitrogen dioxide [4,5]. Since peroxynitrite is relatively stable at physiological pH (the half life is ~1 s at pH 7.4 [4]), it could penetrate the nucleus, where it might induce damage in DNA. In this study we have studied *in vitro* the reaction of peroxynitrite with plasmid DNA, nucleosides and bases.

MATERIALS AND METHODS

Peroxynitrite was synthesized in a quenched flow reactor [6]. Base propenals and 8-nitroguanine were synthesized as previously reported [7,8]. Reactions were carried out in 0.1 M phosphate buffer (pH 4-10) containing 0.1 mM diethylenetriamine pentaacetic acid at room temperature. Reaction products from treatment of various deoxynucleosides with peroxynitrite were separated by thin layer chromatography (TLC) with Silicagel G using a solvent system of ethyl acetate: isopropyl alcohol: water (74:17:9). Base-propenals were detected by spraying the TLC plate with 0.6% 2-thiobarbituric acid (TBA) and heating for 10 min at 100°C [9]. Reaction products of guanine with peroxynitrite were separated using high performance liquid chromatography (HPLC) and analysed by mass spectrometry and ^1H NMR [8].

RESULTS AND DISCUSSION

Figure 1 shows that treatment of supercoiled pBR322 DNA (Form I) at pH 7.2 with low concentrations of peroxynitrite (1 - 3000 μM) resulted in dose-dependent formation of nicked open circular form DNA (form II). Higher concentrations of peroxynitrite also induced formation of linear DNA (form III). These results indicate that peroxynitrite can induce strand breaks in DNA.

Reaction of various deoxynucleosides with peroxynitrite was found to yield dose-dependently TBA-reactive substances, which showed the same R_F values on TLC as authentic base-propenals. Reactions of deoxynucleosides with decomposed peroxynitrite did not form TBA-reactive substances. Thus, we concluded that DNA base-propenals (base-CH=CH-CHO) were formed by the reaction between deoxynucleosides and peroxynitrite. The yield of base-propenals was higher under acidic conditions (pH <6) than at neutral or alkaline pH (pH >7.5). A similar pH dependence has been reported for malondialdehyde formation from deoxyribose and peroxynitrite [4]. Fe^{3+}/EDTA increased the yield of propenals at neutral and alkaline pH. Base propenals have been reported to be highly cytotoxic and easily formed by cleavage of the deoxyribose ring with hydroxyl radical generated during treatment of DNA with bleomycin (Fe^{2+} and oxygen) or γ-irradiation [9,10]. Therefore, hydroxyl-radical-like compound(s) generated from peroxynitrite could be involved in the base-propenal formation [4].

Guanine was found to react with peroxynitrite at neutral pH to yield several compounds, two of which were yellow. The major yellow compound (about 80 % of the all products formed) was isolated and identified as 8-nitroguanine by chemical and physical analyses (mass spectrometry, NMR etc). [8]. Kinetic studies showed that the formation of 8-nitroguanine was maximal at pH 8. Fe^{3+}/EDTA, which has been shown to catalyse nitration of tyrosine, did not affect nitration of guanine.

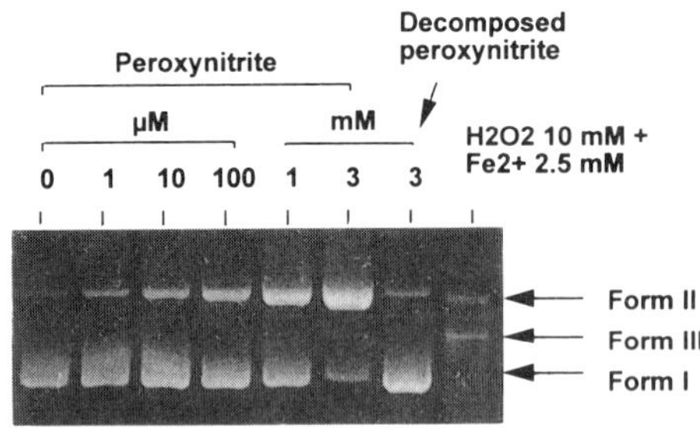

Figure 1 Induction of DNA strand breaks by peroxynitrite Plasmid pBR 322 DNA (10 μg/ml) was incubated with peroxynitrite (1-3000 μM), decomposed peroxynitrite (3000 μM) or 10 mM H_2O_2 plus 2.5 mM Fe^{2+} at pH 7.2 and 20°C.

In order to study further the biological significance of 8-nitroguanine, we have developed a sensitive and specific method to analyze this adduct in DNA. 8-Nitroguanine in acid-hydrolyzed DNA was chemically reduced into 8-aminoguanine, which was analyzed using HPLC with electrochemical detection. It was found that 8-nitroguanine was formed dose-dependently in calf thymus DNA incubated with low concentrations of peroxynitrite *in vitro*. Only peroxynitrite, but not nitrite, tetranitromethane nor NO-releasing compounds, formed 8-nitroguanine. Antioxidants and desferrioxamine inhibited the reaction. 8-Nitroguanine was rapidly depurinated from DNA incubated at pH 7.4, 37°C ($t_{1/2}$ = 4 h), yielding apurinic sites which are potentially mutagenic. Peroxynitrite, however, did not increase 8-oxoguanine levels in DNA (Yermilov, Rubio & Ohshima, submitted).

In conclusion, we have demonstrated that peroxynitrite can damage DNA by various mechanisms, possibly through generation of hydroxyl and nitronium radicals. We are currently studying whether these DNA modifications are formed *in vivo* in cultured cells and in inflamed tissues of animals and human subjects.

We thank Drs B. Pignatelli, C. Malaveille, M.D. Friesen and M. Becchi for analyses of reaction products by mass spectrometer and NMR, and also for helpful discussions and suggestions. We also thank Dr J. Cheney for editing the manuscript. JR and VY are recipients of a Universidad Nacional Autonoma de México Scholarship and a European Science Foundation fellowship respectively.

REFERENCES

1 Ohshima, H. and Bartsch, H. (1994) Mutat. Res. **305**, 253-264
2 Tannenbaum, S.R., Tamir, S., Rojas-Walker, T.D. and Wishnok, J.S. (1994) ACS Symp. Ser. **553**, 120-135
3 Liu, R.H. and Hotchkiss, J.H. (1995) Mutat. Res. **339**, 73-89
4 Beckman, J.S., Beckman, T.W., Chen, J., Marshall, P.A. and Freeman, B.A. (1990) Proc. Natl. Acad. Sci. U. S. A. **87**, 1620-1624
5 Pryor, W.A. and Squadrito, G.L. (1995) Am. J. Physiol. **268**, L699-L722
6 Reed, J.W., Ho, H.H. and Jolly, W.L. (1974) J. Am. Chem. Soc. **96**, 1248-1249
7 Johnson, F., Pillai, K.M.R., Grollman, A.P., Tseng, L. and Takeshita, M. (1984) J. Med. Chem. **27**, 954-958
8 Yermilov, V., Rubio, J., Becchi, M., Friesen, M.D., Pignatelli, B. and Ohshima, H. (1995) Carcinogenesis **16**, 2045-2050
9 Giloni, L., Takeshita, M., Johnson, F., Iden, C. and Grollman, A.P. (1981) J. Biol. Chem. **256**, 8608-8615
10 Janicek, M.F., Haseltine, W.A. and Henner, W.D. (1985) Nucleic Acid. Res. **13**, 9011-9029

*Present address: Instituto de Investigaciones Biomédicas, Departamento de Genética y Toxicología Ambiental, Apartado Postal 70228, CD Universitaria, 04510 México, DF México.

Toward design of a liver-specific nitric oxide (NO) donor

TIMOTHY R. BILLIAR*, JOSEPH E. SAAVEDRA,# DEBRA L. WILLIAMS*, and LARRY K. KEEFER+

*University of Pittsburgh Medical Center, Pittsburgh, PA 15261, USA; #Biological Carcinogenesis and Development Program, SAIC Frederick, NCI-FCRDC, Frederick, MD 21702, USA; and +Chemistry Section, Laboratory of Comparative Carcinogenesis, National Cancer Institute, Frederick Cancer Research and Development Center, Frederick, MD 21702, USA.

NO has important vasculoprotective actions and is thought to protect the microcirculation in shock states in organs such as the liver. Furthermore, recent reports have shown that NO donor drugs reduce tissue damage in endotoxemic, hemorrhagic, and traumatic shock [1-3]. With the aim of developing an NO prodrug whose effect on the vasculature is limited to the liver, we have synthesized compound **1** as a possible liver-specific NO donor. As outlined in the chapter by Keefer *et al.* elsewhere in this volume, we postulated that **1** would be stable in the circulatory system until it reached the liver, where drug-metabolizing enzymes capable of oxidatively removing the vinyl protecting group from such an enol ether are concentrated. The ion (**2**) produced in this hypothetical transformation was known from solution chemistry studies to generate NO with a very short half-life (3 seconds at pH 7.4 and 37°C), consistent with our aim of maximizing NO release in the metabolizing organ and minimizing the escape of NO into the systemic circulation. The evidence to date, reported herein, indicates that **1** might indeed be metabolized to NO liver-selectively, as summarized in Figure 1.

In vitro studies. We began by synthesizing a variety of O^2-alkylated derivatives of **2**, including the methyl, ethyl, methoxymethyl, and benzyl compounds, then screening them for susceptibility to metabolism by cultured hepatocytes. Only the vinyl

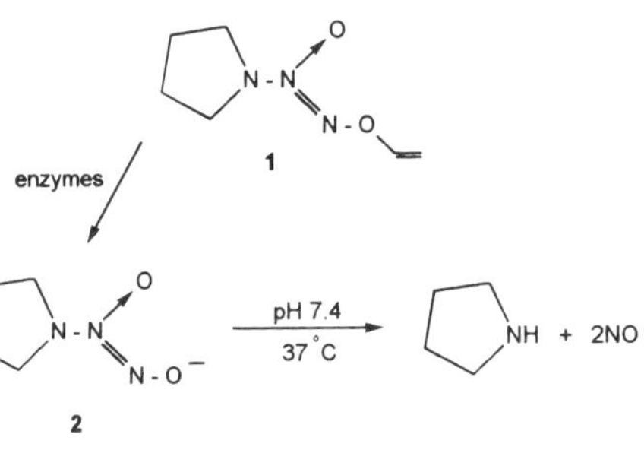

enzymes
1
2
pH 7.4
37°C
NH + 2NO

Figure 1. Predicted metabolic conversion of prodrug **1** to NO by liver enzymes.

derivative, **1**, produced significant levels of NO_X^- (i.e., nitrite plus nitrate) under the conditions employed. That this reflected generation of NO that was subsequently oxidized, rather than direct conversion of **1** to NO_X^-, was reflected in the dose-dependent increases in cGMP concentration seen in the same media as the incubation progressed at concentrations of **1** ranging from 5μM to 5 mM. The metabolism was not inhibited by N^G-methyl-L-arginine, indicating that production of NO_X^- and cGMP was not the result of NO synthase activity. The rate of NO_X^- production from **1** at a concentration of 1 mM by 400,000 hepatocytes remained fairly constant at 20-25 fmol/cell/h for 12 h before beginning to slow. Conversion of **1** to NO_X^- reached half the theoretical yield in 6 h at the lowest substrate concentration studied (5 μM). No evidence of toxicity on the part of **1** was detected, as levels of the liver injury enzymes, ALT and lactate dehydrogenase, in the supernatant at concentrations of **1** up to 5 mM were the same as those measured in the absence of **1**.

In marked contrast to the results with hepatocytes, nonparenchymal cells cultured from the livers of the same rats showed no tendency to convert **1** to NO_X^-, despite the presence of such metabolically active cell types as Kupffer cells. RAW 264.7 mouse macrophages also failed to metabolize **1**, although they produced considerable NO_X^- when treated with LPS and interferon-γ instead of **1**. The uniqueness of hepatocytes as metabolizers of **1** among all the cell types studied to date is further illustrated in Figure 2, which compares them with endothelial cells and vascular smooth muscle cells, both of which were inactive.

Abbreviations used: NO, nitric oxide; NO_X^-, nitrite plus nitrate; cGMP, guanosine 3′,5′-monophosphate; ALT, alanine aminotransferase; MAP, mean arterial pressure; SNP, sodium nitroprusside; LPS, lipopolysaccharide.

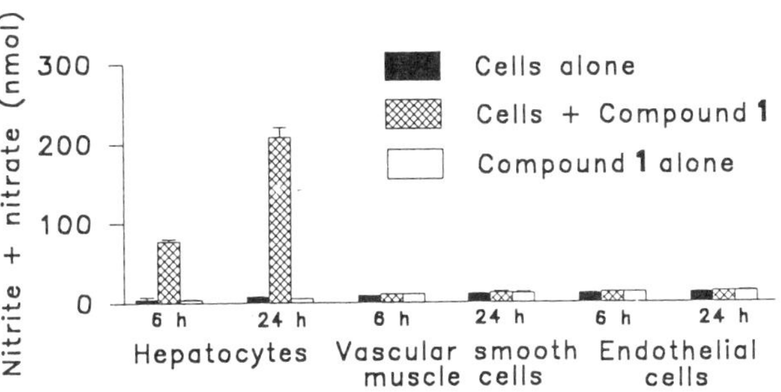

Figure 2. Metabolism of 1 mM **1** to nitrite and nitrate in 6 versus 24 h by: hepatocytes (left); vascular smooth muscle cells (center); and endothelial cell cultures (right). Values are means ± standard deviations in 1-ml incubation mixtures.

In vivo studies. Experiments with **1** in intact rodents conducted thus far support the promising in vitro indications of hepatocyte-specific metabolism described above. To address the critical question of liver-selectivity of NO release, we administered intravenous boluses of **1** at 30 nmol/kg and examined the effect on systemic blood pressure. Compound **1** induced no change in MAP, while the potent nitrovasodilator, SNP, lowered the MAP by 50% at the same molar dose. At 60 nmol/kg, however, **1** also induced a small but apparently significant drop in MAP.

Despite its minimal effect on the systemic vasculature, **1** at 60 nmol/kg substantially increased the levels of cGMP in the liver 10 min after bolus administration. This was in contrast to SNP, which at 60 nmol/kg did not increase liver cGMP over the levels seen in saline-injected controls.

To gain preliminary insight into the ability of **1** to protect the liver during endotoxemia, a model was established in which mice were injected with 100 μg of LPS each to produce circulating levels of ALT and aspartate aminotransferase essentially twice as great at 6-7 h after injection as those in control animals given saline alone. Mice given 200 μg of both **1** and LPS showed little or no increase in circulating levels of the two liver injury enzymes, however.

Summary. The evidence to date indicates that: a) **1** is selectively metabolized by hepatocytes, but not by the other cell types investigated, to a product with the bioactivity of NO; b) **1** substantially increases cGMP levels in the liver but has little or no effect on systemic blood pressure at doses up to 60 nmol/kg in rats; and c) **1** reduces LPS-induced liver injury in mice. If additional work currently in progress confirms the liver-selectivity of NO release by this compound on systemic administration, it could prove useful as a tool for studying (and possibly treating) a variety of liver disorders in which increasing local NO concentration has beneficial effects, such as infectious diseases and cancer as well as the various types of shock.

REFERENCES

1. Boughton-Smith, N.K., Hutcheson, I.R., Deakin, A.M., Whittle, B.J.R., and Moncada, S. (1990) Eur. J. Pharmacol. **191**, 485-488
2. Symington, P.A., Ma, X.-l., and Lefer, A.M. (1992) Meth. Find. Exp. Clin. Pharmacol. **14**, 789-797
3. Christopher, T.A., Ma, X.-l., and Lefer, A.M. (1994) Shock **1**, 19-24

Treatment of cutaneous mycosis with the nitric oxide donor S-nitroso-N-acetylpenicillamine (SNAP)

PATRICIO LOPEZ-JARAMILLO, CESAR RUANO, JOSE RIVERA, ENRIQUE TERAN and SALVADOR MONCADA*

Unidad de Metabolismo Mineral, Escuela de Medicina, Universidad Central e IIDES-MSP, P.O.BOX: 17-21-1060, Quito, Ecuador; and *Wellcome Research Laboratories, Beckenham, Kent, U.K.

Nitric oxide (NO) formed from L-arginine by the inducible NO synthase plays a crucial role in the immune response to various pathogens [1]. Macrophages activated with gamma-interferon (IFN-τ) and lipopolysaccharide (LPS) have a powerful cytostatic effect *in vitro* for the fungal pathogenes *cryptococus neoformans* by a process which is dependent on the presence of L-arginine, is inhibited by the NO synthase inhibitor N^G-monomethyl-L-arginine (L-NMMA) and correlates with the synthesis of L-citrulline and the NO metabolites, nitrite and nitrate [2]. This crucial role for NO in the inhibition or killing of pathogens is now know to extend to a wide variety of intra and extracellular microorganisms [3]. It is werefore possible that NO production may represent a basic antimicrobial defense reaction against pathogenic agents sequestered in all sorts of cells throughout the body. Thus, the aim of the present study was to investigate the effect of S-nitroso-N-acetylpenicillamine (SNAP), a spontaneous NO donor, on the treatment of subjects with clinical manifestation of cutaneous fungal infection.

Twenty-three patients (18-37 years, 9 female and 14 males) from the outpatient clinic of the "Ministerio de Agricultura y Ganadería", Quito, Ecuador, were asked to participate in the study, which was approved by the ethics committee of the "Instituto de Investigaciones para el Desarrollo de la Salud, Ministerio de Salud Pública de Ecuador". These subjects presented clinical manifestations of cutaneous fungal infections (skin desquamation, pruritus, polyhydrosis, odour) of the foot sole, including subungual and interdigital fungal infections. In order to be included in the study the patients must be free of any antifungal treatment for at least the 2 previous weeks. The diagnosis of fungal infection was confirmed by a positive KOH fungal investigation and by a positive culture (growth of at least one class of fungi) from samples obtained by skin scraping at the site of the lesion. The samples were cultivated in Saboreau medium and incubated at 25°C for 21 days. After this time the different classes of fungi were identified by the germinative tube technique (Table 1).

A topical cream of SNAP, with a final concentration of 200 μM was prepared using cetilic alcohol, quaternary ammonium and distilled water. The NO donor activity of the cream was verified by testing its ability to relax isolated rings of rabbit aorta in an organ bath system and by inhibiting platelet aggregation in human platelet rich plasma (Figure 1 and 2).

The SNAP cream was administered locally over the lesions every 4 hours (except while sleeping) for a total of 6 days. During the period of SNAP cream administration the blood pressure was taken daily, and a continuous monitoring system for side effects was established. After 2 days of treatment, an improvement in the clinical symptoms was observed in all patients and after 6 days all of the symptoms had disappeared. At the end of the treatment new samples were taken by skin scraping. All cultures obtained after completion of the treatment were negative (Table 1). No patients presented any side effects or changes in blood pressure. The only symptoms reported were an increased local temperature and skin blushing.

These results suggest the interesting possibility of the use of topical NO donors in the treatment of fungal and other infectious cutaneous diseases. However more studies are necessary to explore this proposal.

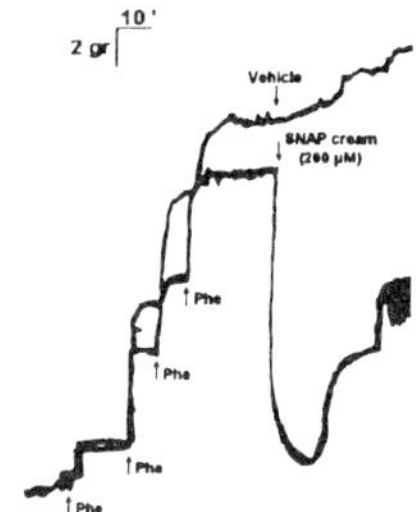

Figure 1. Relaxing effect of the SNAP cream (200 μM) on the vascular tone of rabbit aorta ring precontracted with phenylephrine (10^{-9} to 10^{-6} M).

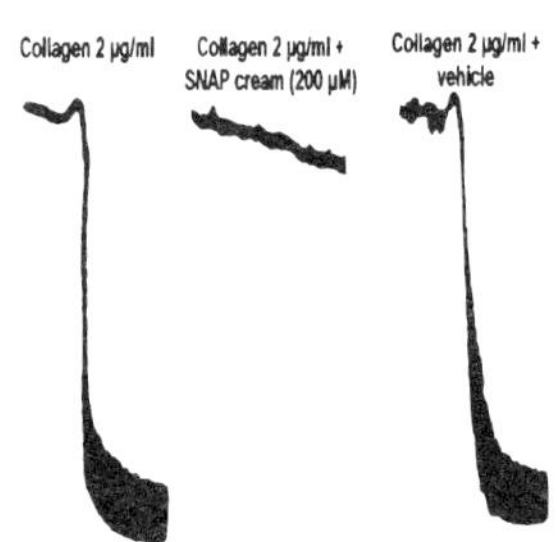

Figure 2. Inhibitory effect of the SNAP cream (200 μM) on collagen-induced platelet aggregation in human platelet rich plasma.

Table 1. Culture identification of fungus in samples obtained by skin scraping before and after treatment with SNAP cream (200 μM).

Fungal class	Before	After
Tricophyton metagraphites	16	-
Tricophyton tonsurans	7	-
Epidermophyton glocosum	4	-
Geotrichum candidum	2	-
Candida albicans	5	-

Values are the number of positive cultures

References

1. Moncada, S. and Higgs, E.A. (1993) The L-arginine nitric oxide pathway. New. Engl. J. Med. **329**,2002-2012
2. Granger, D.L., Hibbs, Jr., J.B., Perfect, J.R. and Durack, D.T. (1990) Metabolic fate of L-arginine in relation to microbiostatic capability of murine macrophages. J. Clin. Invest. **85**,264-273
3. Green, S.J. and Nacy, C.A. (1993) Antimicrobial and immunopathologic effects of cytokine-induced nitric oxide synthesis. Curr. Opin. Infect. Dis. **6**,384-396

The internal anal sphincter in children with severe idiopathic constipation: a possible therapeutic role for nitric oxide

SHAILESH B PATEL, GRAHAM S CLAYDEN*, DAVID E BURLEIGH†, HARRY C WARD and JEREMY PT WARD ‡

Depts of Paediatric Surgery, *Paediatrics and ‡Medicine, St. Thomas' Hospital, London,U.K. SE1 7EH.

†Dept of Pharmacology, Queen Mary and Westfield College, University of London.London U.K.

Introduction

Childhood constipation is a common problem, accounting for 3% of paediatric outpatient referralls and often resulting in considerable stress and anxiety in families. In some cases it can become a more severe chronic problem. Over 300 such patients are annually referred to the Children's Intestinal Motility Clinic at St. Thomas' Hospital. Symptoms in many of the patients suggest an element of distal outflow "obstruction" and treatment is directed towards overcoming the obstructive nature of the IAS. This includes vigorous anal dilatation and in approximately 10% of the patients, surgery in the form of an internal anal sphincter myectomy .

We have assessed some of the *in vivo* and the *in vitro* properties of the IAS from children with severe idiopathic chronic constipation who were undergoing myectomy as part of their treatment.

Patients and Methods

Patients

18 children undergoing myectomy were studied. There were 5 girls and 13 boys. The mean age at operation was was 7.23 yrs (age range 1.92 yrs to 17.33 years). The mean duration of symptoms was 5.71 yrs (range 1.75 yrs to 17.25 years). All children had previously failed to respond to prolonged medical treatment. Fully informed consent was obtained from the parents of all children studied, and the study was approved by the West Lambeth Health Authority Ethics Committee.

Methods *Anorectal Physiology* Anorectal manometry under ketamine anaesthesia was performed on all patients prior to surgery using a method previously described[1]. The parameters measured were:

1) Presence or absence of the normal rectoanal inhibitory reflex to rectal distension with an air filled balloon.

2) Critical rectal volume for maximal IAS relaxation.

Tissue studied The internal anal sphincter myectomy is designed to divide the upper part of the IAS using a longitudinal incision extended up from the dentate line to the lower circular smooth muscle fibres of the rectum. Full thickness longitudinal biopsies of first the mucosa, and then the muscle, 5cm long x 0.5cm wide, were taken from the edge of the incision for diagnostic histology and in vitro experiments

Once obtained, the muscle was immediately placed into Krebs-Henseleit solution previously aerated with 5% CO_2 in O_2 for transport to the laboratory.

In vitro experiments Using a dissecting microscope, the submucosal tissue was gently removed and the transverse bands of circular smooth muscle from the proximal IAS were identified. Between 2-4 muscle strips (approximately 3mm x 1mm) were then prepared from the tissue obtained from each child. The strips (total n=58) were suspended under a loading of 0.5gm in 2 ml organ baths superperfused with Krebs-Henseleit solution maintained at 37°C and aerated with 5% CO_2 in O_2 . IAS from 4 adults undergoing abdomino-perineal resections for local tumours was similarly prepared to provide control strips (n=15). All strips were allowed to equilibrate for at least 1 hour before isotonic measurement of relaxation and contraction responses. The strips were exposed to noradrenaline (NOR, 3.0 x10^{-6}M, n=58); to acetylcholine (ACH, 3.0 x 10^{-6}M, n=52); to sodium nitroprusside (SNP, 3.8 x 10^{-7}M, n=58) and to EFS (1 msec, pulse width, 10V for 30 seconds, 8 Hz, n=52)) under conditions of parasympathetic and sympathetic blockade with hyoscine and guanethidine, in the presence of NOR, 3.0 x 10^{-7}M, to maintain the strips at a higher degree of resting tone during EFS [2].

Results

Anorectal Physiology The RAR was present in 17 out of the 18 patients. It could not be elicited in one patient with massive megarectum. A megarectum (rectal capacity > 150 mls) was detected in 15 patients using manometry. It was clinically evident in another 2 patients who were too loaded for full testing .

In vitro experiments 48 study strips (83%) contracted to NOR and 4 (7%) relaxed. 21 study strips (40%) relaxed to ACH but 23 (44%) contracted and 8 (15%) showed no response. All control strips contracted to NOR and relaxed to ACH. 14 of 15 control strips (93%) responded to EFS - all showed a relaxation. However only 28 (54%) study strips relaxed to EFS. 15 (29%) showed a biphasic response i.e. relaxation followed by a contraction. 9 strips (17%) did not respond. When treated with SNP, all study and control strips showed a sustained relaxation response (p=1, χ^2 test).

Discussion

IAS relaxation in response to rectal distension is mediated via inhibitory NANC neurotransmission. There is now increasing evidence that NO is the neurotransmitter responsible for this relaxation[3,4]

Our results show that unlike the IAS from patients with Hirschsprung's disease in which a resistance to NO donor-mediated relaxation has been reported [5,6], all the IAS strips from children with non-Hirschsprung constipation in our study relaxed to exogenously delivered NO in the same way as did controls.

The treatment with topical nitroglycerin of painful conditions such as fissure-in-ano where increased anal tone may be part of the aetiology has recently been described [7,8]. Concern has also been expressed as to the long term effect on continence in adults of procedures such as anal dilatation [9].

We believe that the development of therapeutic agents that can deliver NO locally to the IAS as required may have a useful role to play in the management of severe constipation of childhood, and thus reduce the risks involved with more direct physical manipulation of the IAS.

Bibliography

1. Burleigh DE, D'Mello A, Parks AG. Responses of isolated human internal anal sphincter to drugs and electrical field stimulation (1979). Gastroenterology **77**:484-490

2. Paskins JR, Lawson JO, Clayden GS. The effect of ketamine anesthesia on anorectal manometry (1984). J Ped Surg **19**:289-291

3. Burleigh DE. Non-adrenergic, non-cholinergic inhibitory nerves of human internal anal sphincter are antagonised by L-N^g-nitro-arginine (1991). Br J Pharmacol **102**:330P.

4. Speakman CT, Burnett SJ, Kamm MA, Bartram CI. Sphincter injury after anal dilatation demonstrated by anal endosonography (1991). Br J Surg **78**:1429-1430

5. Loder PB, Kamm MA, Nicholls RJ, Phillips RK. Topical application of a nitric oxide donor reduces internal anal sphincter tone: therapeutic implications (1993). Gastroenterology **104**, No 4, Pt 2:A544(Abstract)

6. O'Kelly TJ, Brading AF, Mortensen NJM. Nerve mediated relaxation of the human internal anal sphincter: the role of nitric oxide (1993). Gut **34**:689-693

7. Bealer JF, Natuzzi ES, Flake AW, Adzick NS, Harrison MR. Effect of nitric oxide on the colonic smooth muscle of patients with Hirschsprung's Disease (1994). J Ped Surg **29**:1025-1029

8. Gorfine SR. Treatment of benign anal disease with topical nitroglycerin (1995). Dis Colon Rectum **38**:453-457

9. VanderWall KJ, Bealer JF, Adzick NS, Harrison MR. Cyclic GMP relaxes the internal anal sphincter in Hirschsprung's Disease (1995). J Ped Surg **30**:1013-1016

Abbreviations used: IAS, internal anal sphincter; RAR, rectoanal reflex; NOR, noradrenaline; ACH, acetylcholine; SNP, sodium nitroprusside; EFS, electrical field stimulation; NO, Nitric oxide.

SB Patel, current address: Dept of Paediatric Surgery, The Children's Hospital, Lewisham, Lewisham High Street, London, UK. SE13 6LH.

Nitric oxide inhibits human immunodeficiency virus replication

Joan B. Mannick*, Jonathan S. Stamler+, Edna Tang#, and Robert W. Finberg#

*Brigham and Women's Hospital
75 Francis St.
Boston, MA 02115 USA

+Duke University Medical Center
P.O. Box 2612
Durham, NC 27710 USA

#Dana Farber Cancer Institute
44 Binney St.
Boston, MA 02215 USA

Recent studies have shown that nitric oxide (NO) inhibits the replication of multiple viruses including herpes simplex, vaccinia, ectromelia and Epstein-Barr virus (EBV) (1-3). We therefore investigated the effect of NO on human immunodeficiency virus 1 (HIV-1) replication. NO generating compounds inhibit HIV-1 replication in acutely infected human peripheral blood mononuclear cells (PBMCs) at doses which do not detectably inhibit the cellular proliferation. Moreover, the NO- generating compound *S*-nitroso-*N*-acetylpenicillamine (SNAP) and azidothymidine (AZT) have an additive inhibitory effect on HIV-1 replication in PBMCs. NO inhibits HIV-1 protein expression but not reverse transcription. The mechanism by which NO inhibits HIV-1 appears to be independent of cGMP. These studies are the first description of a potential therapeutic role for NO generating compounds in HIV-1 infection.

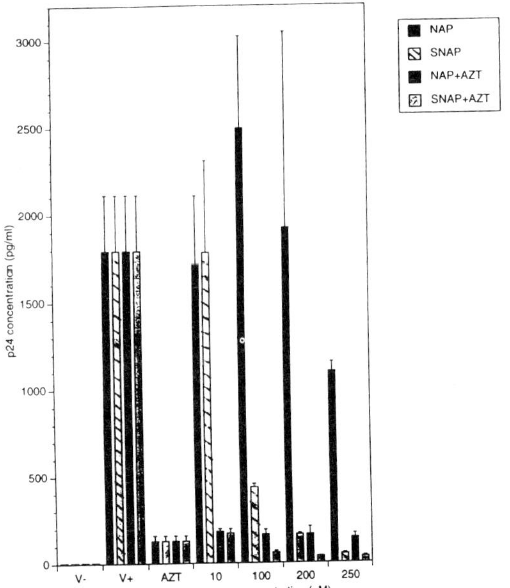

Figure I. SNAP and AZT have additive inhibitory effects on HIV-1 replication. PBMCs from healthy human donors infected with HIV-1 at an MOI of 0.1 were grown in the presence or absence of AZT (0.025 μM) and/or various concentrations of SNAP or *N*-acetylpenicillamine (NAP). HIV-1 replication was assayed 6 days after infection by measuring p24 antigen levels in cell free supernatants using an ELISA assay (NEN).

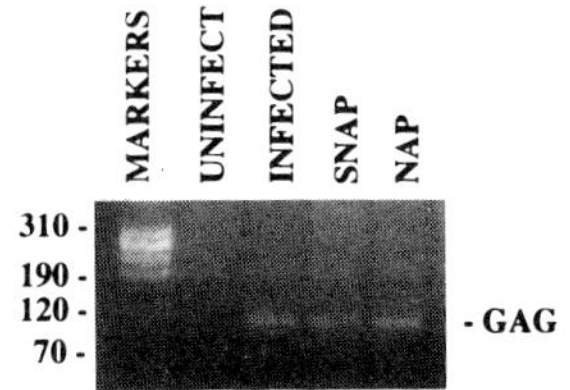

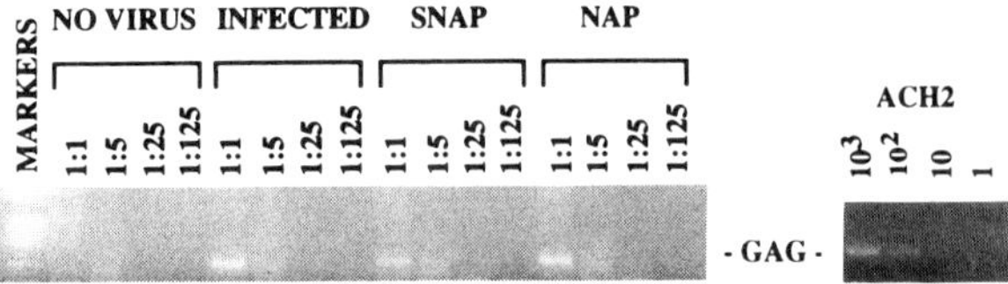

Figure II. NO donors do not inhibit HIV-1 reverse transcription.
(A) Levels of HIV-1 proviral DNA were analyzed in PBMCs infected with HIV-1 and grown for 24 hours in the presence or absence of SNAP or NAP. Proviral DNA was amplified using*gag*-specific primers which generated the expected 115 bp PCR product. (B) The relative abundance of the *gag*-specific PCR-product in each cell population was assessed by serially diluting the PCR products and comparing the relative intensities of the *gag*-specific bands on ethidium stained agarose gels. As a positive control, *gag* was amplified from DNA obtained from 10 fold serial dilutions of ACH-2 T cells which contain one proviral copy per cell. The *gag*-specific PCR product is indicated. The markers on the left indicate the base pair size of ϕX DNA fragments generated from a HaeIII digest.

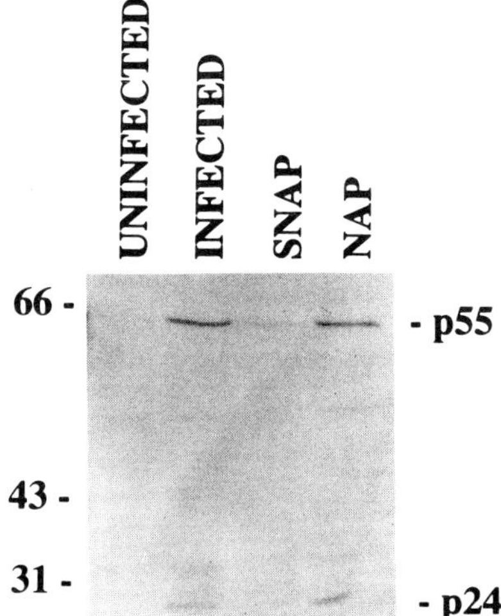

Figure III. NO donors inhibit HIV-1 protein expression. Whole cell lysates obtained from PBMCs grown in the presence or absence of SNAP or NAP for 24 hours after HIV-1 infection were blotted with pooled antisera from HIV-1-infected patients. The HIV-1 proteins p55 and p24 are indicated. Markers of molecular mass in kilodaltons are indicated on the left.

REFERENCES

1.Croen KD. Evidence for antiviral effect of nitric oxide. Inhibition of herpes simplex virus type 1 replication. J Clin Invest 1993;91(6):2446-52.

2.Karupiah G, Harris N. Inhibition of viral replication by nitric oxide and its reversal by ferrous sulfate and tricarboxylic acid cycle metabolites. J Exp Med 1995;181(6):2171-9.

3.Mannick JB, Asano K, Izumi K, Kieff E, Stamler JS. Nitric oxide produced by human B lymphocytes inhibits apoptosis and EBV reactivation. Cell 1994;79:1137-46.

Possible Roles for NO donors in Cancer Treatment

David A. Wink, James Liebmann, Roberto Pacelli, John A. Cook, Anne Marie DeLuca, Deborah Coffin, William DeGraff, Janet Gamson, Murali C. Krishna, and James B. Mitchell.
Tumor Biology Section, Radiation Biology Branch, NCI, Bethesda, MD.

The discovery that NO plays a role in the tumoricidal activity of the immune system [1] suggests that NO and NO donor complexes could serve an important function in the attenuation of various modalities used in cancer treatment. For an agent to have adequate tumoricidal activity, it must possess toxicity such that a vast majority cells die upon exposure. Are NO donor complexes sufficiently toxic to be good antitumor agents? Using a clonogenic assay with Chinese hamster V79 cells, we examined the toxicity of various classes of NO donor complexes. In the presence of 1 mM of each of the NO donor complexes, toxicity was minimal never giving more than 50% cell kill (data not shown). In fact, it has been shown that NO delivered from some of these agents can protect against peroxide or xanthine oxidase toxicity making NO an ideal antioxidant [2,3]. This lack of toxicity at mM NO concentrations is surprising in lieu of the formation of reactive nitrogen oxide species (RNOS) from reactions such as the NO/O_2 reaction which can modify bioorganic molecules. So how can NO be used in cancer treatment?

The chemical biology of NO which is defined as the chemical reactions which are pertinent to biological systems [4] provides a road map as to how NO might be used to kill tumor cells. The chemical biology of NO can be divided into two types of effects, direct and indirect. Direct effects are rapid reactions where NO directly interacts with biological molecules like the interaction between NO and heme proteins such as guanylate cyclase and hemoglobin. Indirect effects

Chemical Biology of NO

Indirect Effects
Direct Effects
H_2O
NO_2^-
GSH
O_2
GSNO
NO_x
NO
Guanylate Cyclase
Oxyhemoglobin
Enzyme Inhibition
NO_3^-
Mitochondrial Consumption
Enhanced Toxicity of Alkylating Agents
Hypoxic Radiosensitizer

involve the chemistry of RNOS formed from reactions such as the NO/O_2 reaction. One advantage of putting the effects of NO into two different categories is that different concentrations of NO are required to obtain each one. For example, low concentrations of NO are required to activate guanylate cyclase which is considered a direct effect. While at much higher concentrations, NO can irreversibly inactivate enzymes (indirect effect). This description can be a guide to strategies for improving cancer treatment (for further discussion see reference[4]). We have investigated several effects of NO donor complexes on two modalities of cancer treatment in cell culture, radiation and alkylating agents.

One of the factors which limits the effectiveness of radiotherapy in cancer treatment is the hypoxic cell population which is radioresistant. For 50 years, there has been a search, with limited success, for agents which will radiosensitize hypoxic cells. In 1957, Howard-Flanders showed that hypoxic bacteria were sensitized to radiation by NO [5]. The mechanism involves the rapid reaction between NO and DNA radicals formed from radiation exposure. Under hypoxia, these radio-induced DNA radicals can abstract neighboring protein hydrogen atoms thus repairing the DNA lesion. If NO or oxygen are present, DNA radicals react with either diatomic molecule to form adducts which cannot abstract nearby hydrogen atoms thus fixing the radiation induced DNA lesion. We have described the use of NO donor complexes on the radiosensitization of hypoxic mammalian cells. We showed that NO and the NONOate, DEA/NO ((C2H5)2N[N(O)NO]-Na+), could sensitize hypoxic cells [6]. We further examined other NO donor complexes and found that the S-nitrosothiols, S-nitrosoglutathione (GSNO) and S-nitroso-N-acetylpenicillamine (SNAP), were as effective as the NONOates, while other putative NO donor complexes, sodium nitroprusside (SNP) and 3-morpholinosydnonium (SIN-1) were ineffective. The difference between the NO donors is that measurable NO can be detected from S-nitrosothiols and NONOates, whereas NO was not detected from SIN-1 and SNP. From these results, NO and NO donor complexes are the first agents to radiosensitize hypoxic cells as effectively as oxygen representing a direct effect of NO.

Another modality of cancer treatment is the use of chemotherapeutic compounds such as alkylating agents which are toxic in part due to their ability to alkylate DNA. One example of a clinically used compound is melphalan. When MCF7 cells are exposed to 1 µM melphalan, toxicity was observed resulting in 60% cell kill (Figure 1). Though DEA/NO alone was not toxic under these conditions, the combination of DEA/NO and melphalan led to increased cytotoxicity over melphalan alone. Similar results showing that DEA/NO enhanced the toxicity of other alkylating agents in rat hepatoma H4 cells have also been reported [7]. In this paper, it was shown that the critical repair protein O^6-methyl-guanine-DNA-methyltransferase was inhibited by RNOS formed from the NO/O_2 reaction. The inhibition of repair led to increased lesions from the alkylating agent. The modification of the repair protein by RNOS is an example of an indirect effect of NO and how these effects could be used to enhance the toxicity of chemotherapeutic compounds. In conclusion, NO or

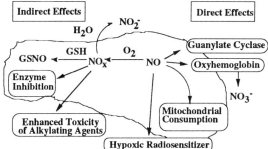

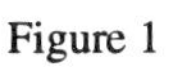
Figure 1

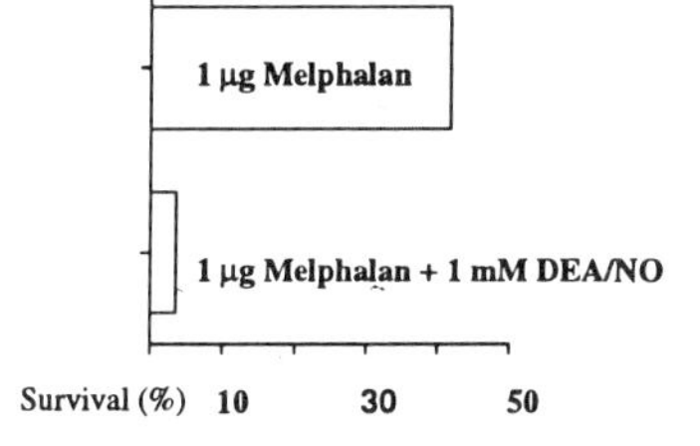

The effects of DEA/NO on melphalan (1 hr) toxicity in MCF7 cells.

RNOS delivered from NO complexes under the right circumstance may play a role in benefiting cancer treatment.

References

1. Hibbs, J.B., Z. Vavrin, and R.R. Taintor. (1987) J. Immunol. **138**, 550-565.
2. Wink, D.A., I. Hanbauer, M.C. Krishna, W. DeGraff, J. Gamson, and J.B. Mitchell. (1993) Proc. Natl. Acad. Sci. USA **90**, 9813-9817.
3. Wink, D.A., J.A. Cook, M.C. Krishna, I. Hanbauer, W. DeGraff, J. Gamson, and J.B. Mitchell. (1995) Arch. Biochem. Biophys. **319**, 402-407.
4. Wink, D.A., I. Hanbauer, M.B. Grisham, F. Laval, R.W. Nims, J. Laval, J.C. Cook, R. Pacelli, J. Liebmann, M.C. Krishna, M.C. Ford, and Mitchell JB. (1995) Current Topics in Cellular Regulation in press.
5. Howard-Flanders, P. (1957) Nature 180, 1191-1192.
6. Mitchell, J.B., D.A. Wink, W. DeGraff, J. Gamson, L.K. Keefer, and M.C. Krishna. (1993) Cancer Res. **53**, 5845-5848.
7. Laval, F., and D.A. Wink. (1994) Carcinogenesis **15**, 443-447.

Effect of Selective iNOS Inhibitors in Inflammation

Linda F. Branson, Mark G. Currie, Jane R. Connor, Daniela Salvemini, Pamela S. Wyatt and Pamela T. Manning; Inflammatory Diseases Research, G. D. Searle Co., 800 N. Lindbergh Blvd., St. Louis, MO 63167, USA.

Nitric oxide (NO) is synthesized enzymatically from L-arginine by nitric oxide synthase (NOS). Both constitutive and inducible isoforms have been characterized and are distinguished by their Ca^{++}/calmodulin dependence and by whether the enzyme is expressed constitutively (cNOS) or is induced following exposure to cytokines or endotoxin (iNOS). The constitutive isoforms are found primarily in the endothelium of the vasculature (eNOS) and in the nervous system (nNOS) and produce small amounts of NO which regulate a number of important physiological processes [1]. The expression of iNOS occurs in many cell types, including macrophages, epithelial cells and chondrocytes, in response to inflammatory and immunological stimuli. Accumulating evidence suggests that the high level production of NO generated by iNOS plays a role in the pathogenesis of tissue destruction during chronic inflammatory diseases, including diabetes, arthritis, transplant rejection and inflammatory bowel disease (IBD).

In order to assess the role of iNOS in inflammation *in vivo*, we have examined the contribution of iNOS/NO in producing inflammatory changes as well as the effect of selective iNOS inhibitors in two well-characterized animal models of inflammation. These inhibitors decrease inflammatory changes that occur in both the carrageenan-inflamed rat air pouch and dextran sodium sulfate (DSS)-induced colitis models. The results obtained in the DSS model are described in detail below.

Colitis is induced by the administration of 5% DSS dissolved in the drinking water to either C3H or Balb/c mice. Each cycle consists of the administration of DSS for seven days followed by a ten day recovery period during which the mice receive distilled water (dH_2O). An acute form of colitis intially develops and progresses to a more chronic inflammatory state with successive DSS cycles. This model is thus cyclic in nature, exhibiting similarities to human IBD, specifically ulcerative colitis. We have examined the contribution of NO generated by iNOS in producing the pathological changes associated with this model of colitis. Disease progression was monitored during the administration of two cycles of DSS and water. Okayasu et. al.[2] have previously reported that the overall length of the colon shortens during the inflammatory episodes in DSS-induced colitis. In our studies, colon length and body weight were monitored during both cycles. Colon length decreased during each DSS cycle. It recovered completely during the first recovery period but only partially during the second recovery period, suggesting the development of a more chronic inflammatory state. The DSS-treated mice also lost weight during each DSS exposure and partially recovered this weight loss during the subsequent ten day dH_2O administration. The most dramatic weight loss occurred during the first cycle decreasing to 79% of control body weight after seven days of DSS administration. Three clinical indices of disease, diarrhea, positive occult blood, and rectal bleeding, were monitored during the DSS/water cycles to assess disease severity. Only animals that displayed these symptoms during the first cycle were given the second DSS cycle to induce chronic colitis. At the end of each DSS and water cycle colon tissue was also assessed microscopically. The changes that we observed acutely and after the induction of chronic colitis included hyperemia, edema, crypt abscesses, and a prominent mucosal inflammatory infiltrate (data not shown). Many of these pathological changes also occur in human ulcerative colitis.

To demonstrate the presence of iNOS in DSS-induced colitis, RNA was isolated from colons of control mice and mice following 2 cycles of DSS. iNOS mRNA was detected by RT-PCR in all DSS-treated mice and in one control animal (Figure 1). In addition, using a rabbit polyclonal antibody to iNOS described previously [3] iNOS immunoreactive protein was also detected in apical epithelial cells and in crypt abscesses in the colons obtained from DSS-treated mice. No specific iNOS staining was found in colons obtained from control mice (photo not shown). Taken together, these results suggest that iNOS expression is induced in the inflamed colon tissue following DSS treatment.

To determine the potential role of iNOS/NO in producing tissue damage, mice received one DSS cycle (without additional treatment) and then were treated therapeutically with aminoguanidine (AG, a selective iNOS inhibitor having 26-fold selectivity for iNOS vs nNOS [3]) during the second DSS cycle beginning at the time of

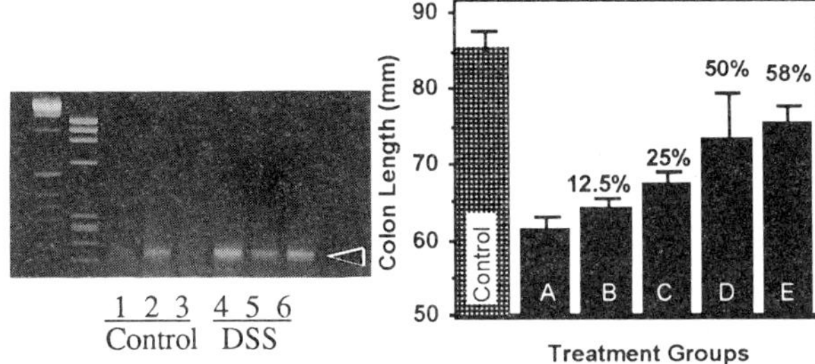

Figure 1. iNOS mRNA in colon extracts by RT-PCR.

Figure 2. Effect of AG on colon length following DSS.

DSS administration. The DSS treatment groups included, **A**. no AG, **B**. 200 mg/kg AG, s.c., b.i.d., C.400 mg/kg AG, s.c., b.i.d), **D**. 100 µg/ml AG in drinking water or **E**. 400 mg/kg AG, s.c., b.i.d. and 100 µg/ml AG in the drinking water. An untreated control group was also included. Animals were killed immediately following the seven days of DSS treatment. AG treatment partially prevented the shortening of the colon produced by the chronic DSS treatment (Figure 2) as well as the weight loss (% inhibition of weight loss: **A**= 0%, **B** = 30%, **C** = 35%, **D** = 52% and **E** = 46%). In addition, AG treatment attenuated the histopathological changes seen in this chronic model of colitis; the most dramatic protection appeared in group **E** which received AG both subcutaneously as well as in the drinking water. In addition, the inflammation and epithelial cell loss was markedly reduced in the distal colons of these animals (photo not shown). However, no significant changes in the clinical symptoms, including incidence of diarrhea, rectal bleeding, or positive occult blood occurred in any of the DSS treatment groups.

In a slightly different treatment paradigm, N-iminoethyl-L- lysine (L-NIL), another selective inhibitor of iNOS [3], was administered to mice in their drinking water (100 µg/ml) for ten days beginning after the second seven day DSS treatment to determine if the therapeutic administration of an iNOS inhibitor would increase the degree of recovery of colon length. Given in this manner, L-NIL was also partially effective both in increasing the recovery of colon length and in regaining the weight loss during the chronic phase of colitis (Table 1) .

Table 1. Effect of L-NIL Treatment During the Recovery Phase of DSS-Induced Colitis

	Colon Length (mm)	% Control (length)	%Control (body weight)
Control	79.6 ±1.7	100	100
DSS / dH_2O	66.5 ± 2.3	84 ± 3	92 ± 4
DSS / L-NIL	72.3 ± 3.2	91 ± 4	96 ± 5

Numbers are the means (N=4-5/group) ± S.E.M.

Recent clinical evidence suggests that NO produced by iNOS is involved in human ulcerative colitis [4]. In addition, NOS inhibitors including AG have been shown to attenuate pathological changes in several animal models of IBD including peptidoglycan/ polysaccharide-induced colitis [5] and TNBS-induced ileitis [6]. The data summarized here demonstrates that two selective iNOS inhibitors, AG and L-NIL, also provide protection in a chronic form of colitis in mice and it provides further evidence for the therapeutic potential of selective iNOS inhibitors in the treatment of IBD.

1. Bredt, D. S., Snyder, S. H. (1994) Annu. Rev. Biochem. 63, 175-195
2. Okayasu, I., Hatakeyama, S., Yamada, M., Ohkusa, T., Inagaki, Y., Nakaya, R., (1990) Gastroenterology 98, 694-702
3. Connor, J. R., Manning, P. T., Settle, S. L., Moore, W. M., Jerome, G. M., Webber, R. K., Tjoeng, F. S., Currie, M. G. (1995) Eur. J. Pharmacol. 273, 15-24
4. Boughton-Smith, N. K., Evans, S. M., Hawkey, C. J., Cole, A. T., Balsitis, M., Whittle, B. J. R., Moncada, S. (1993) Lancet 342, 338-340
5. Grisham, M. B., Specian, R. D., Zimmerman, T. E. (1994) J. Pharmacol. Exp. Ther. 271 (2), 1114-1121
6. Miller, M. J. S., Sadowska-Krowicka, H., Chotinaruemol, S., Kakkis, J. L., Clark, D. A. (1993) J.Pharmacol. Exp. Ther. 264, (1), 11-16

The ncNOS inhibitor 7-nitro indazole influences focal cerebral ischemia in dose dependent manner in rats.

RUDOLF URBANICS, KRISZTIAN KAPINYA, GABOR HUTAS, IMRE JANSZKY, LASZLO DEZSI and ARISZTID G.B. KOVACH*.

Experimental Research Department-2nd Institute of Physiology, Semmelweis University of Medicine, 1082 Budapest, Hungary and *Cerebrovascular Research Center, University of Pennsylvania, PA 19104-6063 USA

Nitric oxide (NO) plays a multifunctional role in cerebrovascular and neural regulation in both physiological and in pathological conditions. NO contributes to the regulation of the normal tone of different vascular beds including the cerebrocortical region, prevents platelet adhesion and activation on the endothelial surface, acts as neurotransmitter, and may regulate the activity dependent flow increase. On the other hand it can induce neurotoxicity via the NMDA pathway [1], producing cytotoxicity by free radical formation. In focal cerebral ischemia inhibition of nitric oxide synthesis can cause an increase or decrease of the infarct volume [2]. These conflicting data underline the opposing roles of NO in cerebral ischemic damage. Most of these data emerged from studies utilizing N^{ω}-substituted-L-arginine analogues for nitric oxide synthase (NOS) inhibition. Since these competitive analogues are not selective, they inhibit all forms of NOS. NOS in the nervous system is present in three isoforms. In the early phase of focal cerebral ischemia the amount of NO significantly increases and due to NMDA activation could cause neurotoxic effects and delayed irreversible ischemic damage [1]. The source of this early NO elevation is presumably the constitutive neural enzyme (ncNOS). Since both NO mediated neurotoxic processes and endothelium dependent adjustment of cerebrovascular resistance may play a role in the final outcome of focal ischemia, to determine the dominant mechanism, selective blockade of NOS is necessary. The potent antinociceptive drug 7-nitro indazole (7-NI) has no influence on blood pressure [3] and has no vasopressor activity [4], i.e. it does not appear to inhibit endothelial NOS (ecNOS).

This study was focused on the dose dependent effects 7-nitro-indazole in permanent focal cerebral ischemia in rats. Morphological changes and cardiovascular parameters were recorded in control and in 7-NI pretreated rats.

METHODS: Male Sprague-Dawley rats (300-350 gr) were anaesthetized by sodium pentobarbital (40 mg/kg ip.). Focal cerebral ischemia was produced by distal electrocoagulation of the left middle cerebral artery (MCA) and a temporary two-hour-occlusion of both common carotid arteries (CCA). Forty-five minutes before the occlusion the treated groups received 10, 20, 30 or 40 mg/kg 7-NI i.p., dissolved in arachis oil. The volume of infarction was visualized 24 hours after the MCA occlusion on coronal brain sections by 2 % triphenyltetrazolium chloride staining and measured by computerized image analysis. In separate group of animals the mean arterial blood pressure (MABP), the cardiac output (CO, measured by thermodilution) were monitored and the cerebrocortical blood flow was measured by generalized H_2 -clearance method in the left cerebral cortex supplied by the MCA. The measured parameters are expressed as mean ± SEM, and the data were analyzed by ANOVA and Student's unpaired *t* tests.

RESULTS: The single dose 7-NI pretreatment with 10 and 30 mg/kg i.p. had no effect on infarct size (Fig. 1). The pretreatment with 20 mg/kg 7-NI resulted in a significant infarct reduction (6.27 ± 1.82 % of the total cerebral volume) compared to the control group (8.58 ± 1.36 %) and in contrast the 40 mg/kg treatment caused augmented damage (10.83 ± 1.26 %).

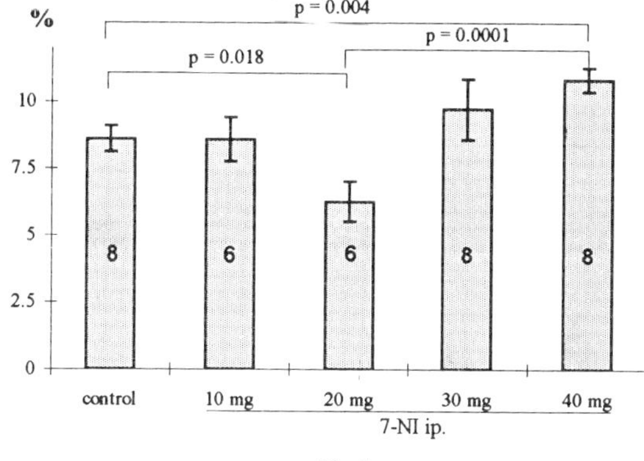

Fig. 1

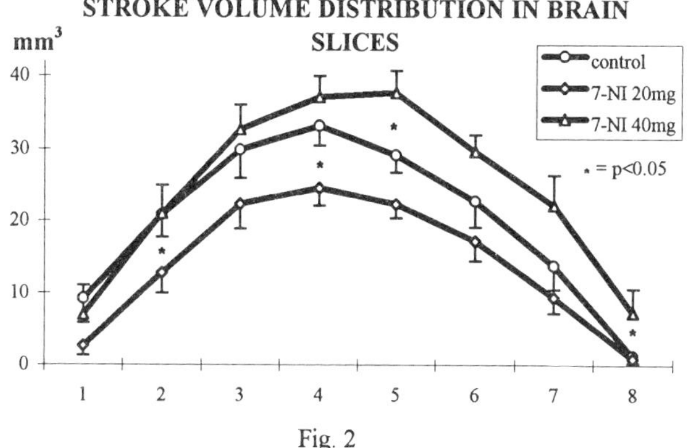

Fig. 2

Fig. 2 presents the infarct volume (mm³) distribution in the coronal brain sections in a rostro-caudal direction at 2 mm intervals. For clarity the averages of the non significantly different 10 and 30 mg/kg treatment groups are omitted. The 7-NI pretreatment with 10 and 20 mg/kg i.p. has no effect on MABP, cardiac output, and cerebrocortical blood flow, but the 40 mg/kg dose elevated the MABP (by 22 ± 9 %) and diminished the cortical blood flow (to 75 ± 8 %) 45 min after the administration. The blood pressure elevation following the CCA occlusion in this group was more pronounced and longer lasting than in the non-treated animals.

DISCUSSION: Pretreatment with 20 mg/kg 7-NI decreased the infarct size in a model of focal cerebral ischemia in rats. These data indicate that the NO increase in the early phase of cerebral ischemia plays a neurodestructive role. The selective blockade of the ncNOS enzyme with the proper dose can prevent this effect. A lower dose was not effective and the higher dose resulted in an adverse effect.

In earlier studies [3, 4] was found that 7-NI has no vasopressor activity, but according to our results the cardio-vascular effects of 7-NI at 40mg/kg dose cannot be ruled out in this rat model. The BP elevation and the decrease of CO as well as CBF may reflect an effect on the endothelial NOS. The more pronounced and longer lasting baroreceptor reflex response can be the consequence of the peripheral ecNOS effect or the inhibitory action on ncNOS in the brain stem cardiovascular center.

Supported by OTKA T17779, Hungary; NIH, NS RO1 31429-03

1. Dawson VL, Dawson TM, London ED, Bredt DS, Snyder SH. (1991) Proc.Natl.Acad.Sci.USA. 88, 6387-71
2. Dawson DA. (1994) Cerebrovasc.Brain.Metab.Rev. 6, 299-324
3. Moore PK, Wallace P, Gaffen Z, Hart SL, Babbedge RC. (1993) Br.J.Pharmacol. 110, 219-224
4. Kovach AGB, Lohinai Z, Marczis J, Balla I, Dawson TM, Snyder SH. (1994) Annals New York Acad.Sci. 738, 348-368

Normal and abnormal rates of nitrate excretion in humans.

OLUF C BØCKMAN*, ROGER DAHL*, TRULS E B JOHANSEN#, ØYSTEIN A STRAND§, CAROL O TACKET+ and TOM GRANLI*.

*Norsk Hydro Research Centre, N-3901 Porsgrunn, Norway; #Telemark Central Hospital, N-3900 Porsgrunn, Norway; § Ullevål Hospital, 0450 Oslo, Norway; +Univ. Maryland, School of Medicine, Baltimore MD 21201, USA.

Background

The concentration of NO_3^- in body fluids should in principle reflect body NO synthesis. However, in practice, various confounding factors diminish the usefulness of NO_3^- analyses for studies of NO generation in healthy and diseased states:

- analytical problems, e.g. from interfering substances, mainly in colorimetric methods
- large variations in NO_3^- concentration due to variable NO_3^- intake, notably with vegetables
- about 1/3 of NO_3^- intake seems to be metabolized, this proportion is probably variable.

Hence, we have investigated whether conditions can be found where urinary NO_3^- excretion might offer promise as a measure of NO synthesis in healthy persons and various groups of patients.

Experimental method

- sampling of 24 hrs urine preserved with Thimerosal
- nitrate analysis by Dionex Ion Chromatography with UV- and conductivity detection of NO_3^-
- samples were spiked with NO_3^- for quality assurance.

Results

Daily NO_3^- excretion was measured in 6 healthy adults on: a free-choice diet for 2 days, then a low NO_3^- diet (< 0.1 mmol/day) for 5 days, then finally on a vegetable-free diet for 3 days.
The results are summarized in fig. 1.
We found that adults on a free-choice diet can have a very variable urinary NO_3^- excretion. Hence, we studied mainly patients where NO_3^- intake was known to be low. These results are also given in fig. 1.

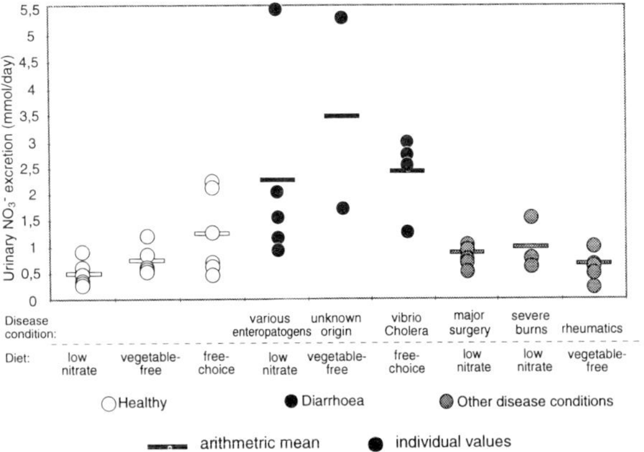

Fig. 1 Urinary NO_3^- excretion in healthy adults and various groups of patients.

Conclusion

1. It is difficult to define what are normal and abnormal rates of NO synthesis from measurements of 24 hrs urinary NO_3^- excretions, unless:
   - NO_3^- intake is known to be low
   - NO synthesis rate is so high that the urinary NO_3^- excretion is > 1.5-2.5 mmol/day.
2. Our data combined with studies by others [1,2] indicates that it is only intestinal diseases that can give so massive increases in NO synthesis that it can be clearly seen as enhanced rates of urinary NO_3^- excretion.
   This is in accordance with the experience that infant methemoglobinaemia occurs as a very rare complication in cases of enteritis and milk allergy, but has not been reported where other organs are insulted [3].
3. The role of NO in intestinal function invites further studies.

References

1. Schulz, K.R., Scheibe, J., Diener, W., Fischer, G., Namaschk, A., Thu, M.P. and Wettig, K. (1990). Endogene nitratsynthese bei ausgewölten Infektionskrankheiten. Z. Klin. Med. **45**, 925-927.
2. Forman, D., Leach, S., Packer, P., Davey, G. and Heptonstall, J. (1992). Significant endogenous syntheses of nitrate does not appear to be a feature of influenza A virus infection. Cancer Epidemiol. Biomarkers Prev. **1**, 369-373.
3. Bøckman, O.C. and Bryson, D.D. (1989). Well-water methaemoglobinaemia: The bacterial factor. In Watershed 89. The future for water quality in Europe. Proc. IAWPRC Conference, 17-20.4.89. Vol. II (Wheeler, D., Richardson, M.L. and Bridges, J., eds.), pp. 239-244. Pergamon Press, Oxford.

Measurement of basal nitric oxide generation in man using ^{15}N arginine

NIGEL BENJAMIN, LORNA SMITH, ERIC MILNE* and MARGARET MCNURLAN*

Department of Medicine and Therapeutics,
University of Aberdeen, Foresterhill, Aberdeen, U.K. AB9 2ZD
*Rowett Research Institute, Aberdeen, U.K.

Introduction

Urinary excretion of nitrate is increasingly used as a measure of NO synthesis in man and laboratory animals. There are two main problems in using nitrate excretion as a measure of endogenous NO production. Firstly, dietary nitrate intake (mainly from green vegetables) is very variable and will have a large influence on urinary nitrate excretion. Secondly, although it is clear that nitrate is endogenously synthesised from oxidation of NO and that nitrate excretion increases in inflammatory conditions, it is not clear whether all endogenous nitrate synthesis is derived from the guanidino nitrogen of L-arginine.

This study describes a novel method for measuring [15N] nitrate derived from [15N]2 L-arginine. By limiting dietary nitrate intake, studies were performed in healthy volunteers to determine what proportion of urinary nitrate originates from oxidation of NO derived from plasma arginine in man.

Methods

[15N]2 L-arginine was infused (5.175-11.5μmol/kg/hour) over 24 hours in 3 healthy male subjects who had been maintained on a low-nitrate diet for 24 hours before and during the study. Following extraction of nitrate from the urine using anion exchange resin (Imac HP-555, Merck, U.K.) and recovery with saturated NaCl, the nitrate was converted to ammonia and trapped on acidic filter paper using a modification of a diffusion method [1]. [15N] enrichment of nitrate was determined by gas isotope ratio mass spectrometry using combustion in tin foil capsules. This method has a limit of detection of 0.0004 atom% excess and requires only 100 μg nitrogen per sample. The nitrate concentration of urine and plasma were determined[2] before extraction and plasma [15N]2 L-arginine enrichment was measured by a standard GCMS method.

Results

Using the diffusion method, urinary [15N] nitrate measurements were reproducible (coeff variation inter assay= 1.1%, intra assay = 3.94%).

Plasma [15N]2 arginine enrichment reached steady state in approximately 15 hours and urinary 15N nitrate at 20 hours. At steady state urinary [15N] nitrate enrichment was 45% of plasma [15N]2 arginine enrichment. Despite reaching steady state at 20 hours, urinary [15N] nitrate enrichment failed to achieve that of plasma [15N]2 L-arginine (see fig).

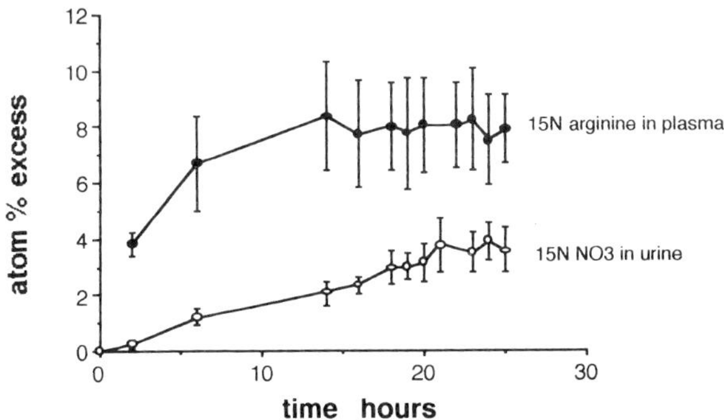

Discussion

Previous studies have used the conversion of [15N]2 arginine to [15N] nitrate as a measure of endogenous NO synthesis [3,4]. Standard GCMS methods for measuring [15N] nitrate are relatively insensitive (requiring 1 atom% excess) and use benzene which is very toxic. The method described can measure very small enrichments of [15N] nitrate and therefore less [15N]2 arginine (which is very expensive) is required. Although the diffusion method takes a longer (6 days) it does not use toxic materials.

The failure of urinary [15N] nitrate enrichment to reach that of plasma [15N]2 arginine at steady state suggests either that approximately half of urinary nitrate is synthesised from another source than oxidation of NO or that enrichment of intracellular arginine (from which NO is derived) only reaches approximately a half that seen in plasma at steady state.

The results suggest that either nitrate is generated in man by another mechanism than NO oxidation or that plasma [15N]2 L-arginine concentrations do not reflect intracellular enrichment of this amino acid.

References

1. Brooks, P.D., Stark, J.M., McInteer, B.B. and Preston, T. Diffusion method to prepare soil extracts for automated nitrogen-15 analysis (1989) Soil Sci. Soc. Am. J. **53**, 1707-1711
2. Green LC, Wagner DA, Glogowski J, Skipper PL, Wishnok JS, Tannenbaum SR. Analysis of nitrate, nitrite and [15N]nitrate in biological fluids. Analytical Chemistry 1982; **126**: 131-138.
3. Leaf CD, Wishnok JS, Tannenbaum SR. L-arginine is a precursor for nitrate biosynthesis in humans. Biochemical and Biophysical Research Communications 1989; **163**: 1032-1037.
4. Hibbs JB, Wetenfelder C, Taintor R, et al. Evidence for cytokine-inducible nitric oxide synthesis from L-arginine in patients receiving interleukin-2 therapy. Journal of Clinical Investigations 1992; **89**: 867- 877.

Thin layer chromatography: an effective method to monitor citrulline synthesis by nitric oxide synthase activity

PRANAV VYAS,[1*] MUKUNDAN ATTUR,[1*] GUO-MING OU,[1] KATHLEEN A. HAINES,[1] STEVEN B. ABRAMSON[1,2] and ASHOK R. AMIN[1,2,3]

[1]The Department of Rheumatology, Hospital for Joint Diseases Orthopaedic Institute, 301 East 17th Street, New York, New York 10003; and the Departments of [2]Medicine and [3]Pathology, New York University Medical Center, 550 First Avenue, New York, New York 10016.

Nitric oxide (NO) has been identified recently as a multifunctional mediator, produced by and acting on various cells of the body. Nitric oxide synthase (NOS) catalyzes the formation of NO from the terminal guanidino nitrogen atom of L-arginine [1]. Three major forms of NOS have been identified to date. There are two forms of constitutive NOS (cNOS), one present in endothelial cells and the other in neuronal cells [1]. Inducible NOS (iNOS) is upregulated in various cells by lipopolysaccharide (LPS), IL-1β, IFN-γ or TNF-α, and other immunological or inflammatory stimuli [1]. Production of NO from constitutive NOS is a key regulator of homeostasis, whereas the generation of NO by inducible NOS plays an important role in the host-defense response [1]. NO participates in inflammatory- and autoimmune-mediated tissue destruction.

Several methods of measuring NOS activity in cells have been reported [2]. The most frequently used include: ***a)*** monitoring the conversion of radiolabelled arginine to citrulline [3] in cell-free extracts or whole cell assays, followed by separation of radiolabelled end-product citrulline by ion-exchange chromatography [4], HPLC [2] or thin layer chromatography (TLC) [5]; ***b)*** monitoring the formation of nitrite, the stable product of NO [6]; and ***c)*** the detection of NO by oxyhemoglobin [7]. Each of these methods has its own advantages and disadvantages. In the present study we have standardized conditions to separate radiolabelled arginine and citrulline by TLC. The conversion and quantitation of L-[^{3}H]-citrulline was monitored by a Bioscan System 200 Imaging Scanner (Bioscan, Inc., Washington, DC). This method can be used for measuring both cNOS and iNOS activities in various cell types and tissues.

Fig. 1 shows the conversion of L-[^{3}H]-arginine to L-[^{3}H]-citrulline by rat brain NOS in cell-free extracts at different time intervals. Although the reaction was observed to be linear up to 15-20 min, we usually terminate the assay after 20 min which was found to be constant and reproducible. The quenching of radioactive citrulline on TLC plates can be overcome by increasing the time of exposure on Bioscan. Twenty picomoles ($\simeq$40 cpm) of citrulline were detected by this method. This estimation is based on the assumption that there is no significant difference in radioactivity during the conversion of the guanidino nitrogen atom of radiolabelled arginine (labelled in β and γ position of the carbon backbone) to citrulline.

The reaction mixtures for iNOS and cNOS assays are as described [8]. Cell-free extracts for RAW 264.7 and brain were prepared as previously described [3,6]. The reaction mixture was spiked with 1 μl (250 nM) of L-[^{3}H]-arginine (Dupont NEN, Boston, MA, catalog # NET-361, 1 mCi/ml = 37.0 MBq/ml). After 20 min the assays were terminated by heating the reaction mixture at 90°C for 5 min. Precipitates were removed by centrifuging at 14,000 rpm for 20 min. 10 μl ($\simeq$50,000 cpm) of the supernatant was spotted on activated Avicel TLC plates (Analtech, Newark, DE). The TLC plates were developed in a solvent system consisting of ethanol:water:ammonia (80:16:4). Quantitation of the spot for L-[^{3}H]-citrulline was performed by a Bioscan System 200 Imaging Scanner (Fig. 1). A phosphoimager can also be used for such quantitation. It should be noted that we screen the batches of TLC plates from the vendor, test them, and then purchase a batch. There are batch variations: there have been instances when there is no resolution.

Schultz *et al.* have detected NO from renal mesangial cells within 6 h [9]. Fig. 2 shows similar results, where 2.3 μM of nitrite could be detected ~6 h after stimulating RAW 264.7 cells with 100 ng/ml LPS. However, it should be noted that the TLC arginine-to-citrulline assay can detect significant activity of the enzyme at 4 h. We usually observe (by PCR and Northern blot) expression of iNOS mRNA 2-4 h after stimulation of RAW 264.7 cells with 100 ng/ml of LPS (data not shown) [6].

Similarly, whole cell iNOS assays using RAW 264.7 cells and bovine chondrocytes were carried out. Briefly, cells were induced with LPS for 24 h, scraped from the plate, and washed once with PSS buffer (consisting of NaCl 140 mM; KCl 4.6 mM; $CaCl_2$ 2.0 mM; $MgCl_2$ 1.0 mM; glucose 10.0 mM and Hepes 10.0 mM, pH 7.4). The required number of cells (1-2 x 10^6) were resuspended in 100 μl of PSS spiked with 1 μl (250 nM) of L-[^{3}H]-arginine, and incubated at 37°C for 10 min. Cells were spun down to remove excess of L-[^{3}H]-arginine, resuspended in 30 μl of PSS, and lysed by repeated freeze-thawing. Debris was removed by centrifuging the cells at 14,000 rpm for 20 min; 20 μl of the supernatant was spotted on TLC. Fig. 3 demonstrates a whole cell assay of iNOS which was sensitive to two competitive inhibitors of iNOS. It should be noted that cells stimulated with LPS (like cells stimulated with cytokines) increase total arginine uptake within the cells, as shown here and reported by other investigators [10]. L-[^{3}H]-arginine and N^G-methyl-L-arginine (L-NMA) utilize the same pathway for uptake [10]. In Fig. 3, cells were exposed to L-NMA and N^G-L-arginine-monomethyl ester (L-NAME) for only 15 min, then washed and assayed for the enzyme. The results indicate that the block in iNOS activity is not likely to be due to inhibition or competition of uptake of the arginine, but rather to competition at the level of enzyme activity. However, one can also use N^G-nitro-L-arginine (L-NNA) which utilizes a different transport system, distinct from L-NMA, in murine macrophages [10,11]. This method can determine NOS activity in as few as 10^6 cells (Fig. 3).

In conclusion, as little as 20 picomoles of L-[^{3}H]-citrulline formed can be detected in both these assays. It has been observed that arginase present in the cellular extracts can lead to overestimation, because ornithine formed as a result of arginase activity coelutes with citrulline when using ion exchange columns for the separation of L-[^{3}H]-arginine from L-[^{3}H]-citrulline. This problem can be overcome by addition of branch chain amino acids such as valine, which inhibits arginase [12]. Alternatively, the same problem is also overcome in the TLC method, as ornithine and citrulline are resolved separately (0.12 and 0.24 Rf, respectively) and can be quantitated, if necessary, by the imaging system. Of course, HPLC methods [13] can separate arginine from ornithine and citrulline, but the advantage of the TLC method is its capacity to estimate several samples (up to 10) at the same time on the same TLC plates. This method has been especially useful when comparing various experimental assays simultaneously.

REFERENCES

1. Nathan, C.F. 1992. *FASEB J.* **6**:3051-3064.
2. Archer S. 1993. *FASEB J.* **7**:349-360.
3. Bredt, D.S. and S.H. Snyder. 1990. *Proc. Natl. Acad. Sci. USA* **87**:682-685.
4. Riesco, A., C. Caramelo, G. Blum, M. Monton Gallego, S. Casado, and A. Lopez Farre. 1993. *Biochem. J.* **292**:791-796.
5. Snedden, J.M., T.M. Bearpark, S.A. Galton and J.R. Vane. 1990. *In* S. Moncada and E.A. Higgs (Eds.), *Nitric Oxide from L-Arginine: a Bioregulatory System,* Elsevier, Amsterdam, pp. 457-461..
6. Stuehr, J.D., H.J. Cho, N.S. Kwon, M.S. Weiss and C.F. Nathan. 1991. *Proc. Natl. Acad. Sci. USA* **88**:7773-7777.
7. Feelisch, M. and E.A. Noack. 1987. *Eur. J. Pharmacol.* **139**:19-30.
8. Misko, T.P., *et al.* 1993. *Eur. J. Pharmacol.* **233**:119-125.
9. Schultz, P.J., S.L. Archer and M.E. Rosenberg. 1994. *Kidney Intl.* **46**:683-689.
10. Schmidt, K., P. Klatt and B. Mayer. 1993. *Molec. Pharmacol.* **44**:615-621.
11. Baydoun, A.R. and G.E. Mann. 1994. *Biochem. Biophys. Res. Comm.* **200**:726-731.
12. Yan, L., R.W. Vandivier, A.F. Suffredini and R.L. Danner. 1994. *J. Immunol.* **153**:1825-1834.
13. Chenais, B., A. Yapo, M. LePoivre and J.P. Tenu. 1991. *J. Chromatography* **539**:433-441.

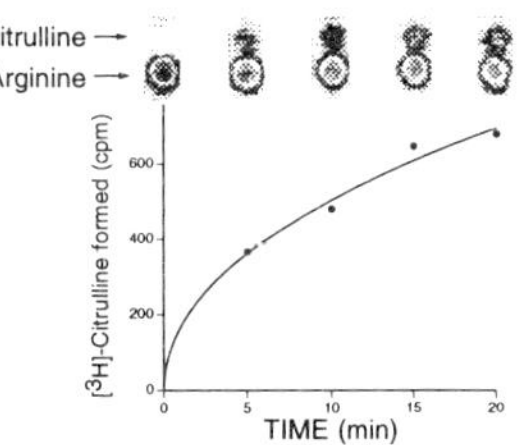

Fig. 1. Conversion of L-[^{3}H]-arginine to L-[^{3}H]-citrulline by cNOS from rat brain at different time intervals. Rat cerebellum was homogenized in 5 ml ice-cold Tris buffer 50 mM, pH 7.4, containing antipain 10 μg/ml, leupeptin 10 μg/ml, pepstatin 10 μg/ml, chymostatin 10 μg/ml and PMSF (100 μg/ml). Homogenate was centrifuged at 15,000 rpm for 30 min at 4°C, and supernatant was used for estimation of activity.

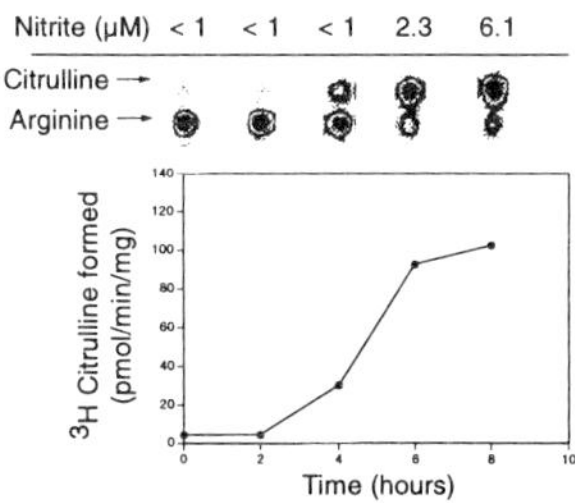

Fig. 2. Time course monitoring of iNOS activity (by TLC) and nitrite accumulation. RAW 264.7 cells were stimulated with 100 ng/ml LPS and harvested at varying time intervals to assay for iNOS in cell-free extracts; at the same time intervals, nitrite levels were monitored in the medium.

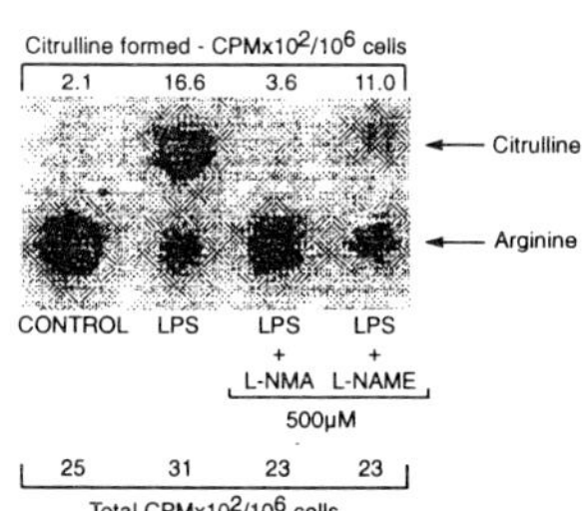

Figure 3. Whole cell assay of NOS using RAW 264.7 cells. A million cells, both unstimulated and stimulated with LPS (1 μg/ml), were resuspended in PSS buffer, and whole cell assays were carried out as described in the text. LPS-stimulated cells were pre-treated with 500 μM N^G-methyl-L-arginine (L-NMA) and 500 μM N^G-L-arginine-monomethyl ester (L-NAME) for 15 min and washed; cell-free extracts were immediately prepared in order to study the NOS-specific inhibition.

Cytochrome c reduction assay detects nitric oxide release by rat diaphragm

FRANCISCO H. ANDRADE, MELANIE R. MOODY, JONATHAN S. STAMLER*, and MICHAEL B. REID

Baylor College of Medicine, Houston TX, 77030 USA, and *Duke University Medical Center, Durham NC, 27710 USA.

Introduction: Diaphragm myocytes generate and release ROI. The release of ROI increases with contractile activity, and can be blunted by extracellular SOD [1]. While ROI may participate in the development of fatigue, they also appear to modulate contractile function in the rested muscle by enhancing excitation-contraction coupling [2]. Diaphragm myocytes also produce and release NO, which may regulate contractile function in the opposite direction as ROI [3]. To explore the roles of ROI and NO on muscle function, we have used the cytochrome c reduction assay to measure the release of redox-active species. Previously, we found that the release of redox-active species by passive diaphragm myocytes is not sensitive to SOD and catalase [4], and is only partially blocked by the inhibition of NO synthesis with 1 mM NLA. These findings suggest either incomplete inhibition of NO synthesis, or the extracellular release of other redox-active species. To address this issue, we hypothesized that inhibition of NO synthesis using 10 mM NLA would completely eliminate the release of redox-active species by rat diaphragm.

Methods: Male Sprague-Dawley rats were anesthetized, tracheotomized and ventilated with 100% O_2. The left hemidiaphragm was surgically removed with ribs and central tendon attached, and placed in Krebs-Ringer solution (in mM, 137 NaCl, 5 KCl, 2 $CaCl_2$, 1 $MgSO_4$, 1 NaH_2PO_4, 24 $NaHCO_3$). Three fiber bundles were isolated from each hemidiaphragm, tied to plastic rods, and incubated at 37° C in oxygenated Krebs-Ringer solution under the following conditions: (**1**) Control (n=18): bundles incubated in Krebs-Ringer solution only; (**2**) NLA (n=18): bundles incubated in Krebs-Ringer solution with 10 mM NLA; (**3**) NDA (n=18): bundles incubated in Krebs-RInger solution with 10 mM NDA, the inactive isomer of NLA; and (**4**) NLA/SOD+Catalase (n=18): bundles incubated in Krebs-Ringer solution with 10 mM NLA, 1000 U Cu,Zn SOD/ml, and 1000 U catalase/ml. After 60 minutes, the bundles were transferred to tubes containing the corresponding incubation medium plus 50 μM cytochrome c, and incubated for 90 minutes more. The media were removed and absorbance at 550 nm was measured. Cytochrome c reduction was calculated from the difference in absorbance between the media incubated with fiber bundles, and media co-incubated without bundles as temporal controls. Fiber bundles were weighed and the results were normalized to muscle weight and averaged for the 90-minute incubation, and analyzed as picomoles of cytochrome c reduced per mg of muscle per min.

To show that the cytochrome c reduction assay could detect NO, we measured the change in absorbance at 550 nm of cytochrome c (50 μM) when exposed to 50 μM NCys, an NO donor. NO production was monitored by measuring absorbance at 340 nm, which decreased as NCys released NO. The response of cytochrome c to other redox active molecules was tested by adding to 50 μM cytochrome c increasing amounts of GSH, GSSG (the oxidized form of GSH), ascorbate, and dehydroascorbate (the oxidized form of ascorbate). The change in absorbance at 550 nm was measured, and converted to cytochrome c reduced.

Results: As shown in Figure 1, NLA (10 mM) significantly decreased cytochrome c reduction by passive rat diaphragm to about 60% of control levels (2.31±0.23 pmoles/mg muscle/min, $p \leq 0.05$ vs. Control). NDA, the inactive isomer of NLA, had no significant effect on cytochrome c reduction rate (3.22±0.14 pmoles/mg muscle/min, not shown). The combination of NLA (10 mM) with SOD and catalase (1000 U each/ml) further decreased cytochrome c reduction to about 1/3 of control levels

Abbreviations used: ROI, reactive oxygen intermediates; NO, nitric oxide; SOD, superoxide dismutase; NLA, nitro-L-arginine; NDA, nitro-D-arginine; NCys, S-nitrosocysteine; GSH, glutathione; GSSG, glutathione disulfide.

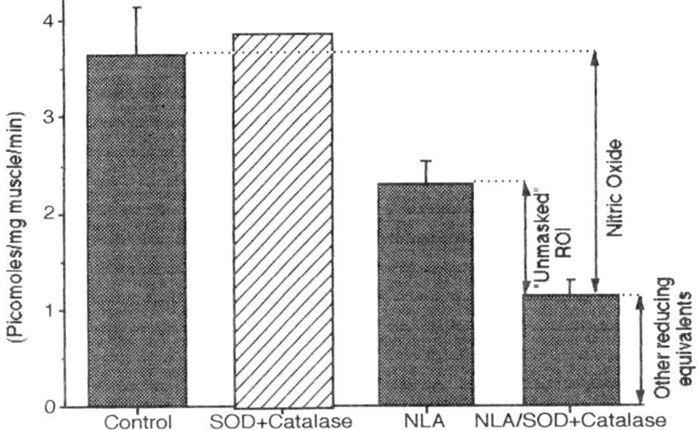

Figure 1: Cytochrome c reduction rate. SOD+Catalase data taken from [4]. See text for explanation.

(1.14±0.15 pmoles/mg muscle/min, $p \leq 0.05$ vs. Control and NLA).

In a cell-free analysis, cytochrome c was reduced by Ncys (an NO donor), in a manner that paralleled the release of NO (as measured by decreased absorbance at 340 nm) (Figure 2). Cytochrome c was also reduced by GSH, and ascorbate, but not by their oxidized forms (GSSG and dehydroascorbate).

Conclusions: NO appears to account for 2/3 of the reducing equivalents released by the diaphragm myocytes and detected by the cytochrome c reduction assay. Inhibition of NO synthesis with NLA unmasks a small amount of ROI that would be normally scavenged by NO. These "unmasked" ROI can be removed by SOD+catalase (Figure 1).

The identity of the reducing equivalents present after NLA/SOD+catalase (about 30% of the control signal) remains unknown, although they could be accounted for by release of GSH and/or ascorbate, either by damaged

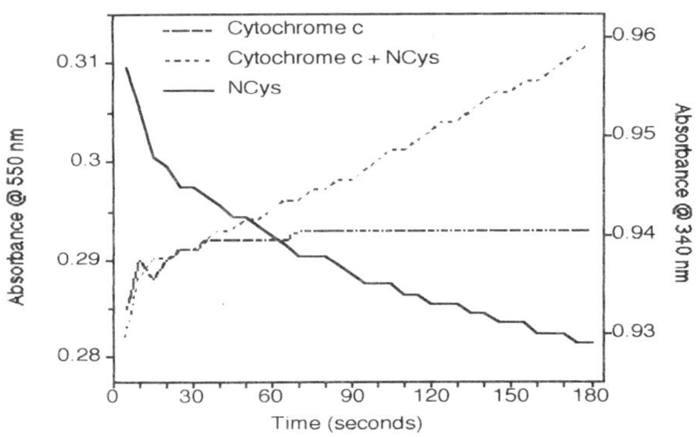

Figure 2: Cytochrome c reduction by NCys. Production of NO is shown by the change in absorbance at 340 nm (NCys dashed line).

muscle fibers, or as a physiological event.

In our system, the cytochrome c reduction assay is useful to measure release of redox-active species. While it is not specific for any given reductant or oxidant, this assay is very sensitive to a wide array of physiologically relevant species, giving an index of overall redox-active species release.

Supported by NHLBI Grant HL45721

1.- Reid, M.B., Shoji, T., Moody, M.R. and Entman, M.L. (1992) J. Appl. Physiol. **73**:1805.
2.- Reid, M.B., Khawli, F.A. and Moody, M.R. (1993) J. Appl. Physiol. **75**:1081.
3.- Kobzik, L., Reid, M.B., Bredt, D.S. and Stamler, J.S. (1994) Nature, **373**:546.
4.- Davies, A., Moody, M.R. and Reid, M.B. (1993) Am. Rev. Respir. Dis. **147**(4):A236.

Nitric oxide-sensitive guanylyl cyclase in higher plants

SILVIA PFEIFFER, GERHARD SOJA*, DORIS KOESLING**, AND BERND MAYER

Institut für Pharmakologie und Toxikologie, Karl-Franzens-Universität Graz, Universitätsplatz 2, A-8010 Graz, Austria
*Austrian Research Center Seibersdorf, A-2444-Seibersdorf, Austria, and **Institut für Pharmakologie, Freie Universität Berlin, Thielallee 67-73, D-1000 Berlin 33, Germany.

In plants, unlike animals, signal transduction studies are in their infancy. Coordination between tissues and cells is much less obvious in plants than in animals, and plants also lack the pronounced differentiation between sensory and responding cells found in mammals [1]. There is evidence that Ca^{2+} has second messenger functions, and proteins exhibiting similarly high affinity and specificity for GTP as mammalian G-proteins were found in membrane fractions of various plant cells [2]. Several studies report on the identification of cyclic nucleotides in higher plants [3-8], and recently it has been demonstrated that cyclic GMP (cGMP) may be involved in signal transduction pathways utilized by phytochromes to control expression of genes required for chloroplast development and anthocyanin biosynthesis [9]. In spite of this putative role of cGMP as cellular messenger in plants, guanylyl cyclase has not been detected in plant tissues so far. To address this issue, we have studied the effect of airborne NO on cellular cGMP levels in whole plant organisms and have partially purified an NO-sensitive guanylyl cyclase from spruce needles.

Spruce trees (*Picea abies* L.) and potato plants (*Solanum tuberosum* L.) were exposed for 84 hours to NO concentrations as occurring in heavily polluted air (i.e. 0.1 - 0.2 µl/l) using an open-top chamber. Control plants were kept under normal environmental conditions with NO concentrations ≤ 0.03 µl/l in a second chamber. Samples were taken immediatly before the experiments and 84 hrs later. Analysis of cGMP was performed by HPLC according to the method of Brown [10]. Identification of the isolated peak was based on coelution with authentic cGMP in two different solvents and time-dependent disappearance of the product upon incubation with phosphodiesterase. To confirm the HPLC data, cGMP was also determined by radioimmunoassay (RIA) as previously described [11]. Table 1 shows that spruce needles contained about one order of magnitude more cGMP than potato plants and that the intracellular levels of the cyclic nucleotide were increased about 10-fold in both tissues upon NO treatment.

sample	control	NO-treated
	(nmol cGMP/g fresh weight*)	
spruce needles	14.6±3.91	124.8±12.75
potato plants	2.76±0.47	28.8±5.4

Table 1: cGMP content in spruce needles and potatos treated for 84 hours with 0.1-0.2 µl/l NO. Control plants were kept under normal environmental conditions (NO≤ 0.03 µl/l).
* Mean ± SD, n=5.

Since these data hinted at the presence of an NO-sensitive guanylyl cyclase in the investigated plants, we have prepared native extracts from spruce needles and subjected the supernatants to fractionated precipitation with ammonium sulfate for protein concentration (0 - 20% and 20 - 75% saturation). Enzymatic cGMP formation was determined in the precipitates by RIA in the presence of GTP and $MgCl_2$ (3 mM each) with and without S-nitrosoglutathione (GSNO). In the absence of the NO donor, guanylyl cyclase activity was below the detection limit of ~0.1 pmol cGMP x mg^{-1} x min^{-1} in both fractions, but the 20-75% ammonium sulfate precipitate showed a specific activity of 0.55 pmol cGMP x mg^{-1} x min^{-1} when 0.1 mM GSNO was present during the incubations.

For further sensitive detection of guanylyl cyclase in the spruce needle extracts, we have used an anti-peptide antibody raised against a peptide corresponding to the C-terminal sequence of the α_1-subunit of bovine lung guanylyl cyclase [12]. Figure 1 (lane 1) shows that the antibody recognized a single protein band in the 20 - 75% ammonium sulfate fraction of the spruce needle extracts, which comigrated with the α_1-subunit of purified bovine lung soluble guanylyl cyclase (lane 2). The bands were not observed when the corresponding peptide (0.1 mg/ml) was present during incubation with the antibody (not shown).

Abbreviations used: cGMP, cyclic guanosine 3',5'-monophosphate; GSNO, S-nitrosoglutathione; RIA, radioimmunoassay.

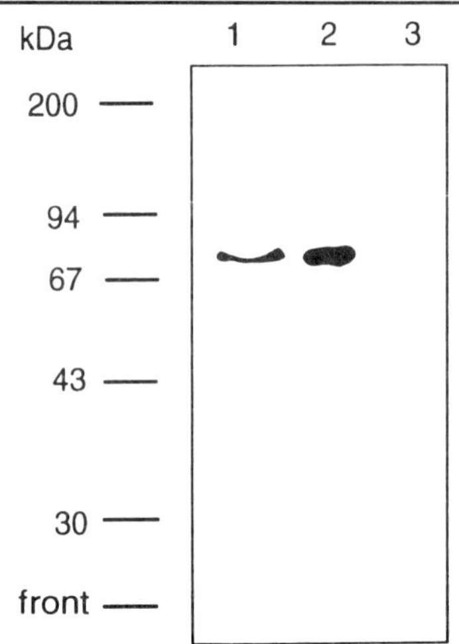

Fig. 1: Western blot of the 20-75% ammonium sulfate fraction from spruce needles (40 µg; lane 1), purified bovine lung soluble guanylyl cyclase (0.1 µg; lane 2), and the 0-20% ammonium sulfate fraction of the spruce needle extracts (40 µg; lane 3). Samples were subjected to SDS-PAGE and immunoblotting with an antibody raised against a peptide corresponding to the C-terminus of the α_1-subunit of the bovine lung enzyme.

The present study provides the first direct biochemical evidence for the occurrence of an NO-sensitive guanylyl cyclase in plants. Cross-reaction of a 82 kDa protein in the 20 - 75 % ammonium sulfate precipitate with an antibody raised against the α-subunit of bovine lung guanylyl cyclase points to structural similarities between the plant and mammalian enzymes. Our data suggest that the NO/cGMP signaling pathway described for animal tissues may also exist in higher plants.

Supported by the Fonds zur Förderung der Wissenschaftlichen Forschung in Austria.

1. Trewavas, A. & Gilroy, S. (1991) Trends in Genet. **7**, 356-361
2. Kaufman, L.S. (1994) Journal of Photochem. Photobiol. B. **22**, 3-7
3. Cyong, J. & Takahashi, M. (1982) Phytochemistry **21**, 1871-1874
4. Cyong, J., Takahashi, M., Hanabusa, K. & Otsuka, Y. (1982) Phytochemistry **21**, 777-778
5. Newton, R.P. & Brown, E.G. (1973) Phytochemistry **12**, 2683-2686
6. Newton, R.P., Kingston, E.E., Evans, D.E., Younis, L.M. & Brown, E.G. (1984) Phytochemistry **23**, 1367-1372
7. Newton, R.P., Chiatante, D., Ghosh, D., Brenton, A.G., Walton, T.J., Harris, F.M. & Brown, E.G. (1989) Phytochemistry **28**, 2243-2254
8. Janistyn, B. (1983) Planta **159**, 382-385
9. Bowler, C., Yamagata, H., Neuhaus, G. & Chua, N.H. (1994) Gene. Dev. **8**, 2188-2202
10. Brown, E.G., Newton, R.P. & Shaw, N.M. (1982) Analytical Biochemistry **123**, 378-388
11. Pfeiffer, S., Janistyn, B., Jessner, G., Pichorner, H. & Ebermann, R. (1994) Phytochemistry **36**, 259-262
12. Guthmann, F., Mayer, B., Koesling, D., Kukovetz, W.R. & Böhme, E. (1992) Naunyn-Schmiedeberg's Archives of Pharmacology **346**, 537-541

Characterization of nitric oxide synthase (NOS) in fish liver: enzyme activity and immunoblot analysis

RACHEL L. COX[#,*] and JOHN J. STEGEMAN[#]

[#]Dept. Biology, Woods Hole Oceanographic Institution, Woods Hole, MA USA 02543; Dept. Physiology, Boston University School of Medicine, Boston, MA USA 02118

ABSTRACT Nitric oxide synthase (NOS) diversity and function in non-mammalian species have received little attention. We are studying NOS in a non-mammalian vertebrate, the marine fish scup (*Stenotomus chrysops*). NOS-like activity was detected in cytosolic and microsomal fractions of scup liver. Enzyme activity decreased in the absence of Ca^{2+}, and in the presence of L-NAME or L-NMMA. Hepatic NOS was not induced in lipopolysaccharide (LPS) treated fish. In Western blots of liver homogenates from control and LPS treated fish, iNOS antibodies recognized a microsomal protein of 120 kD, and a cytosolic and microsomal doublet of 50-55 kD. Antibodies to eNOS detected a 120 kD band in microsomes of control and LPS treated fish. Anti-eNOS also recognized a 66 kD band in liver fractions of LPS treated but not control fish.

INTRODUCTION The long-term goal of these studies is to establish structure, function and phylogeny of vertebrate NOS. We have begun to characterize NOS-like activities and protein from liver and heart of a key non-mammalian vertebrate group, teleost fish. Neuronal NOS has been demonstrated in some non-mammalian species [1], including fish [2], primarily by NADPH diaphorase staining. However non-mammalian nNOS has not been characterized and there are no direct data concerning eNOS or iNOS in non-mammalian species. Reports differ on the occurrence and function of NO, and thus NOS, in fish cardiovasculature [3,4]. Expression of either inducible (Type II) or endothelial (Type III) NOS in non-mammalian vertebrates has not been documented.

METHODS Freshly dissected tissues were ground in 0.32 M sucrose, 20 mM HEPES, 0.5 mM EDTA, pH 7.2, (1 mM DTT, 1 mM PMSF, 2 ug/ml pepstatin, 5 ug/ml leupeptin, 1 ug/ml antipain, 10 ug/ml aprotinin) and centrifuged (12,000 x g, 10 minutes, 4°C). Supernatants were centrifuged (100,00 x g, 70 minutes, 2°C). Pellets were suspended in 50 mM Tris-HCl, pH 7.4, 1mM EDTA, 20% w/v glycerol, (same protease inhibitors as above).

Fish were injected intraperitoneally (i.p.) with *V. alginoliticus* ($6x10^4$ cells), and four days later, i.p. with LPS (*V. cholerae*) at 20 mg/kg. At six and twenty-four hours, tissues were dissected.

NOS activity was assayed by the conversion of radio-labelled L-arginine to L-citrulline. Incubations (20 minutes, 37°C) were stopped with the addition of 10% TCA. The mixture was applied to a cation exchange column and L-citrulline was quantified.

Liver fractions were concentrated (Centricon 10). ADP agarose was mixed with concentrated lysates (overnight, 4°C) and spun (1000 x g). Pellets were washed with 10 mM NADPH in homogenization buffer and concentrated (Centricon 30).

Scup liver homogenates were electrophoresed (6% acrylamide gels). Gels were stained (Coomassie) or transferred to Nytran. Transfers were carried out at 1 ampere for 1 hour in Tris-glycine buffer (10% methanol). All incubations (blocking, primary and secondary antibodies) were carried out for 1 hour at 22°C. Immunoreactive bands were visualized by chemiluminescence.

The two anti-peptide antibodies were: 1- made against a sequence corresponding to a conserved region of mouse eNOS; 2- made against a sequence deduced from eNOS cDNA [5].

RESULTS AND CONCLUSIONS. NOS-like activity was detected in teleost fish (scup) liver and heart. Liver cytosolic activity was 30 pmol/min/mg, microsomal activity was 10 pmol/min/mg. Heart cytosolic activity was 7.5 pmol/min/mg, microsomal activity was 4 pmol/min/mg. These rates are comparable to reported values for NOS activity in mammalian species (8.5- 74 pmol/min/mg) [5,6].

Fish liver cytosolic activity decreased in the absence of calcium and in the presence of L-NAME and L-NMMA, and when enzyme incubations were carried out at lower temperatures (Table 1).

LPS treatment of fish did not alter liver NOS activity. This result may be related to the relative resistance of fish to gram negative infection. The LD_{50} for LPS in fish is 10x - 36x that in mice [7].

Table 1. NOS Activity in fish (scup) liver cytosolic fractions.

Assay Conditions[1]	NOS Activity[2] (% of Control)[3]
L-Arginine + NADPH + Ca^{++}(control)	100
L-Arginine + NADPH	65
L-Arginine + NADPH + Ca^{++} (22°C)	84
L-Arginine + NADPH + Ca^{++} (2°C)	21
L-Arginine + NADPH + Ca^{++}+ L-NAME	74
L-Arginine + NADPH + Ca^{++}+ L-NMMA	73

1- 50 uM L-arginine, 2 mM NADPH, 3 uM H_4B, 1 mM Ca^{++}, 30 U/ml calmodulin, 1 uM [^{14}C]-L-arginine (80 Ci/mole), 200 ul reaction volume.

2- These data are means from duplicate assays, using pooled samples (4 fish).

3- Endogenous arginine content is approx. 0.5 nmoles arginine/mg protein.

In Western blots of fish liver homogenates, mammalian eNOS antibodies were cross-reactive with a microsomal protein of 120 kD in liver fractions from control and LPS-treated fish. An eNOS antibody also recognized a 66 kD in liver homogenates from LPS-treated, but not control fish. Mammalian iNOS antibodies recognized a microsomal protein of 120 kD and 50-55 kD doublet in cytosolic and microsomal liver fractions from control and LPS-treated fish. Low molecular weight proteins showing immuno-cross reactivity with NOS antibodies may be proteolytic breakdown products of parent NOS molecule(s).

Partial purification of fish liver proteins following procedures used of mammalian NOS resulted in a 110-120 kD band, visualized by SDS-PAGE. This band was not immuno-cross-reactive with mammalian NOS antibodies.

These results indicate that NOS is present in fish liver and heart. The identity and properties of fish NOS have yet to be established.

ACKNOWLEDGEMENTS We thank D. Heck and T. Michel for antibodies. This work was supported in part by NIH grant HL07501 and AFOSR grant F49620-94-1036.

REFERENCES

1- Elofsson, R., Carlberg, M., Moroz, L., Nezlin, L. and Sakharov, D. (1993) Neuroreport **4**: 279-282.
2- Li, Z.S. and Furness, J.B. (1993) Arch Hist. Cytol. **56**: 185-193.
3- Sverdrup, A., Kruger, P.G. and Helle, K.B. (1994) Acta Physiol. Scand. **152**: 219-233.
4- Miller, V.M. and Vanhoutte, P.M. (1986) Blood Vessels **23**: 225-235.
5- Evans, T., A. Carpenter and J. Cohen (1992) Proc. Natl. Acad. Sci. USA **89**: 5361-5365.
6- Knowles, R.G., Merret, M., Salter, M. and Moncada, S. (1990) Biochem J. **270**: 833-836.
7- Kodama, H., Yamada, F., Kurosawa, T, Mikami, T. and Izawa H. (1987) J. Appl. Bacteriol. **63**: 255-360.

Detection of mRNA for a nitric oxide synthase in macrophages and gill of rainbow trout challenged with an attenuated bacterial pathogen

PETER S GRABOWSKI, KERRY J LAING*, FIONA E MCGUIGAN*, LAURA J HARDIE*, STUART H RALSTON and CHRISTOPHER J SECOMBES*

DEPARTMENTS OF MEDICINE & THERAPEUTICS AND *ZOOLOGY, UNIVERSITY OF ABERDEEN, AB9 2ZD, SCOTLAND, UK.

In aquaculture, disease problems are of major economic importance to the fish farming industries, yet little is known of the mechanisms by which fish leukocytes kill pathogens. While a few vaccines exist, their development is severely hampered by this lack of fundamental knowledge. Fish phagocytes are able to kill bacterial and helminth pathogens *in vitro* [1], and the generation of reactive oxygen species is well known. However, the respiratory burst is clearly not the only method for killing some bacterial pathogens such as *Renibacterium salmoninarum* [2] and alternative mechanisms are still to be elucidated.

There is currently great interest in the biological production of nitric oxide (NO), and in its roles in physiological and pathological processes in mammals. The biological production of NO would appear to have an early phylogenetic origin and has been implicated in species such as the slime mould *Physarum polycephalum* [3]. In mammals, a family of three NO synthase enzymes generate NO from L-arginine. One isoform which is not normally expressed (iNOS), may be induced in response to pro-inflammatory cytokines and/or bacterial products such as lipopolysaccharide, in a variety of mammalian cells. In rodent macrophages, NO plays a direct role in killing intracellular parasites and extracellular pathogens, and it was recently demonstrated that mice homozygous for a disrupted iNOS gene showed reduced non-specific inflammatory responses and were susceptible to protozoan parasite infection, in contrast to wild-type and heterozygous mice which were resistant [4].

There is increasing evidence that fish produce NO by the action of an NO synthase enzyme. NO synthase activity has been detected in channel catfish head kidney leukocytes following intraperitoneal injection with live *Edwardsiella ictaluri* [5], although the cells responsible for NO synthase activity were not identified. Wang *et al.* [6] have recently shown that a long-term goldfish macrophage cell line can secrete NO, detected as nitrite accumulation, after incubation with lipopolysaccharide. In both catfish and goldfish, the dependence of NO production on metabolism of L-arginine was demonstrated using arginine analogues known to inhibit mammalian NO synthase enzymes. While there is clearly a lot of interest in the phylogeny of NO synthases, to date no non-mammalian sequences exist and the relevance of NO production to fish defences has not been established.

We have investigated the ability of rainbow trout to express NO synthase in response to challenge *in vivo* with a genetically attenuated (AroA-) fish bacterial pathogen (*Aeromonas salmonicida*) [7]. Two days post-challenge, the head kidneys were dissected and cell suspensions enriched for macrophages were isolated by Percoll density centrifugation [8], and were snap frozen in liquid nitrogen. Other tissues including gills were isolated and snap frozen immediately. Total RNA was extracted from tissues and macrophages using RNAzol, and 5 μg were reverse transcribed. We designed degenerate oligonucleotide primers (1, 2; Figure 1) against nucleotide motifs which are highly conserved both between species and between the three known isoforms of mammalian nitric oxide synthases. In a polymerase chain reaction these primers amplified products which corresponded in size to those expected from inducible NO synthase in mammalian sequences (Figure 2).

1	CCYGTBTTCCAYCAGCAGATG	2	RAAGGCRCARAASTGRGGGTA
3	GGYTGGTACATGRGCACYGAGATYGG	4	CAGASMAMCYTGGTGTTGAAG
5	AGTGACATGGAACAGGAGAGCCTT	6	GGWSGTRATGTCCAGGAAGWAGGTGAG

Figure 1. Oligonucleotide primers used in RT-PCR analysis of inducible NO synthase expression in rainbow trout (M=A/C, R=A/G, W=A/T, S=C/G, Y=C/T, B=C/G/T).

The PCR products were directly sequenced and compared against published mammalian sequences. Further degenerate primers were designed (3,4,5,6: Figure 1) taking into consideration the new sequence information, and we amplified two further products which overlapped the initial PCR sequence. The three sequences were aligned giving a product of 1,385 bases, which translated in a single open reading frame as a peptide of 461 amino acids showing ~70% homology with mammalian nitric oxide synthases. Regions corresponding to iNOS co-factor binding domains were highly conserved in the trout sequence (Figure 3), and a sequence motif shared by mammalian endothelial and neuronal NOS, and not by mammalian inducible NOS was absent from the trout sequence.

The sequence generated by reverse transcription-polymerase chain reaction corresponds to nearly half of the entire protein coding region based upon the mammalian sequences. We have produced an unrooted phylogenetic tree based upon the peptide translation which shows the relationship between the trout sequence and the known mammalian NO synthases (Figure 4). We are currently generating a cDNA library from mRNA isolated from head kidney macrophages of bacterially challenged rainbow trout, to isolate and sequence the full length iNOS cDNA.

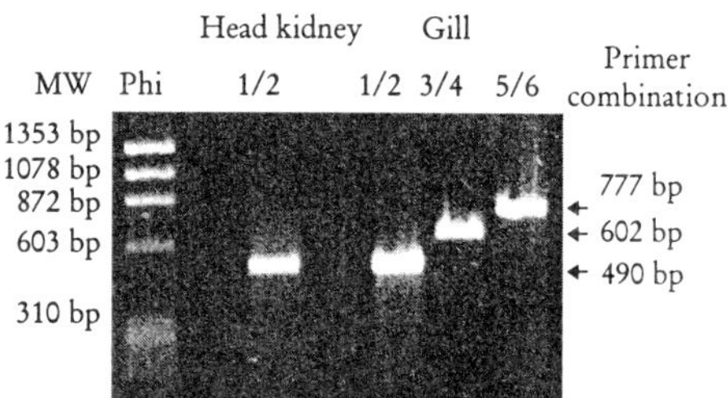

Figure 2. PCR products amplified from rainbow trout head kidney macrophages and gills using degenerate primers for nitric oxide synthase

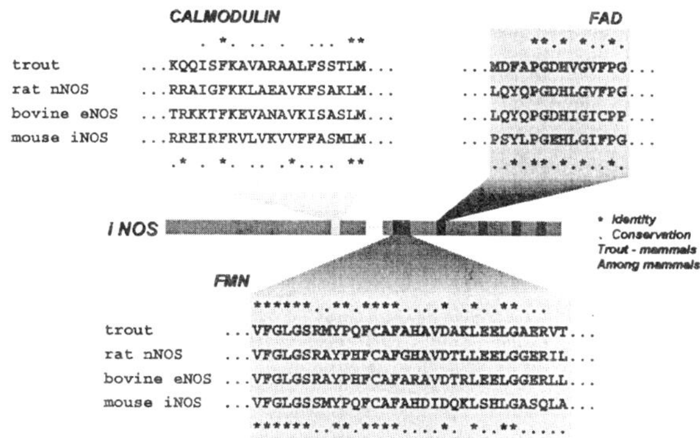

Figure 3. Conservation of co-factor binding motifs in nitric oxide synthases between rainbow trout and mammals ('*' = identity, '.' = conservation; above = trout/mammals, below = among mammals)

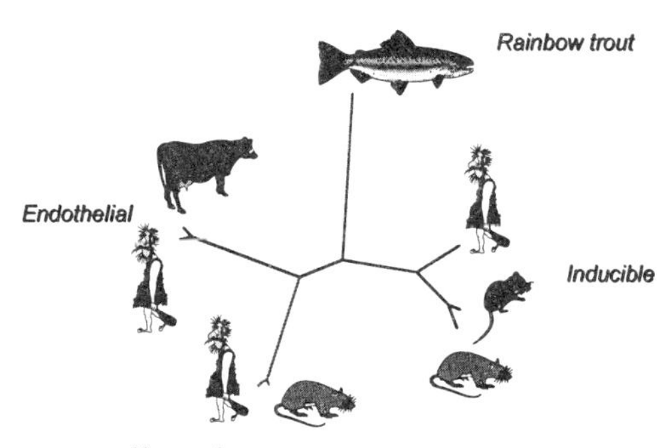

Figure 4. Unrooted phylogenetic tree of NO synthase peptide segments

References

1. Secombes, C.J. and Fletcher, T.C. (1992) The role of phagocytes in the protective mechanisms of fish. Ann. Rev. Fish Dis. **2**:53-71.
2. Bandin, I., Ellis, A.E., Barja, J.L. and Secombes, C.J. (1993) Interaction between rainbow trout macrophages and *Renibacterium salmoninarum in vitro*. Fish Shellfish Immunol. **3**:25-33.
3. Werner-Felmayer, G., Golderer, G., Werner, E.R., Grobner, P. and Wachter, H. (1994) Purification and characterization of nitric oxide synthase from the low eukaryote *Physarum polycephalum*. First Int. Cong. Biochem. Mo/ l. Biol. Nitric Oxide **73**:A30.
4. Wei, X-Q., Charles, I.G., Smith, A. *et al.* (1995) Altered immune responses in mice lacking inducible nitric oxide synthase. Nature **375**:408-411.
5. Schoor, W.P. and Plumb, J.A. (1994) Induction of nitric oxide synthase in channel catfish *Ictalurus punctatus* by *Edwardsiella ictaluri*. Dis. Aquat. Org. **19**:153-155.
6. Wang, R., Neumann, N.F., Shen, Q. and Belosevic, M.. (1995) Establishment and characterization of a macrophage cell line from the goldfish. Fish Shellfish Immunol. **5**:329-346.
7. Marsden, M.J., DeVoy, A., Vaughan, L., Foster, T.J. and Secombes, C.J. (In press) Use of a genetically attenuated strain of *Aeromonas salmonicida* to vaccinate salmonid fish. Aquaculture International.
8. Secombes C.J. (1990) in Techniques in Fish Immunology Vol. **1** (J.S. Stolen et al., eds), pp.137-154, SOS Publications, New Jersey.

Characterisation of an endogenous L-arginine/NO pathway in platelets

Reinhard Berkels, Anja Cordes, Dirk Meyer, Wolfgang Klaus and Renate Rösen

Inst. für Pharmakol., Univ. zu Köln, D - 50931 Köln, Germany

Introduction

Recently, the existence of an endogenous platelet L-arginine (L-Arg)/NO pathway has been shown [1]. During collagen induced platelet aggregation NO synthase (NOS) seems to be activated thus nitric oxide released acts as a negative feedback mechanism during aggregation.

During the L-Arg/NO metabolism a stable intermediate N^G-hydroxy-L-arginine (OHArg) is formed that by itself may serve as a substrate for NOS and/or P 450 systems [2,3].

To characterise the L-Arg/NO metabolism we examined the influence of the NO precursor L-Arg, its enantiomer D-arginine (D-Arg) and OHArg the stable intermediate of the L-Arg/NO pathway on collagen induced porcine platelet aggregation in platelet rich plasma (PRP) and in gelfiltrated platelets (GFP).

Methods: *Preparation of PRP, GFP and platelet aggregation:* PRP and PPP were prepared from blood of female pigs as described before [4]. PRP was filtrated through a Sepharose 2B column to remove plasma proteins yielding in GFP.

Platelet aggregation was induced with collagen (5 µg/ml) and online measured turbidimetrically. Data shown are related to maximal aggregation.

NADPH-Diaphorase Staining: As a histochemical marker of NOS NADPH diaphorase activity was evidenced by the conversion of nitroblue tetrazolium to the blue dye formazan [5]. Platelets in PRP and GFP were fixed with 4 % paraformaldehyde for 4 hrs, centrifugated and resuspended in PBS buffer. NADPH diaphorase staining was examined whith a light microscope.

Determination of porcine plasma L-Arg levels: A cell and protein free ultrafiltrate of PRP was obtained by ultrafiltration (5000 KD cut off ultrafilter, Satorius). Amino acids were quantified by RP-HPLC after pre column derivatisation with FMOC (9-Fluorenylmethyl N-succinimidylcarbonate). Fluorescence detection (λ_{ex} 254 nm, λ_{em} 310 nm) was used. Homoarginine was added as an internal standard with a recovery of 74 ± 6 % (n = 6).

Results: 1. Diaphorase staining was negative at control conditions (without NADPH) and in quiescent, resting platelets of PRP (with NADPH). During gelfiltration platelets become preactivated resulting in the formation of pseudopodies which corresponds to the lack of the shape change during aggregation. These platelets as well as collagen activated platelets show a prominent NADPH diaphorase staining. **2.** In porcine plasma the concentration of L-Arg was evaluated to be $9{,}31 \times 10^{-5}$ mol/l ± 10% (n=6). **3.** L-Arg and OHArg inhibited the collagen induced platelet aggregation in PRP and GFP in a dose dependent manner (Fig. 1, 2) (n = 6). OHArg was significantly less effective than L-Arg. In general in GFP concentration response curves were shifted to higher concentrations. In contrast D-Arg showed a dose dependent slight increase of platelet aggregation in porcine PRP (Fig.2), in GFP no influence on collagen induced platelet aggregation could be evaluted (Fig.1). **4.** Cytochrome P450 systems may utilize OHArg to form NO. To differentiate between metabolism of OHArg by P450 systems and/or NOS, troleandomycin (TAO) was used to inhibit P450 systems. TAO [1 µmol/l] completely blocked the OHArg induced inhibition of platelet aggregation but not that of L-Arg (n = 6) (Fig.2). Thus, OHArg is used by P450 systems rather than by NOS.

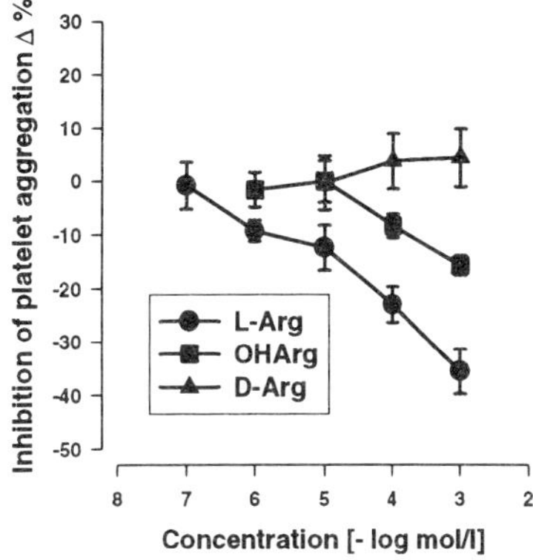

Fig 1: Influence of L-Arg, D-Arg and OHArg on collagen induced platelet aggregation in GFP

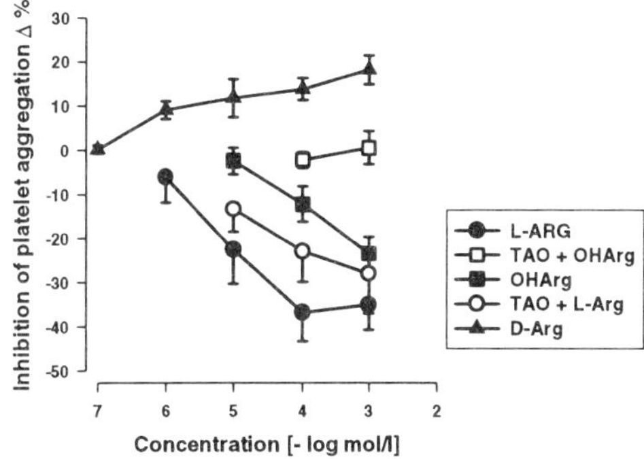

Fig 2: Inhibition of platelet aggregation (PRP) by L-Arg, D-Arg and OHArg and the influence of troleandomycin (TAO)

Discussion: Staining of NADPH diaphorase is thoght to be a histochemical marker of NOS [5]. Since NOS activity in platelets could only be evidenced after preactivation of the platelets NOS seems to be concomitantly activated. Thus, the activation of NOS might be induced by changes of its conformation or compartimentation in addition to a change in the concentration of cofactors.

Although the intracellular L-Arg level is much higher than needed for NO formation and the K_m value of the transmembrane L-Arg transporter is about ten times higher than that of NOS an increase of extracellular L-Arg reduced aggregation. This discrepancy is known as "The L-Arginine Paradoxon" [6]. OHArg inhibited platelet aggregation too, but was less potent than L-Arg. Since OHArg may be utilised by NOS and/or of cytochrome P450 dependent oxidases, cytochrome P450 was blocked with TAO. In the presence of TAO the antiaggregatory effect of OHArg but not that of L-Arg was abolished (Fig.2). Therefore it seems that not NOS but rather a cytochrome P450 oxidase utilises OHArg to form NO.

References

1. Radomski, M., Palmer, R., Moncada, S., (1990) Proc. Natl. Acad. Sci. **87**, 5193 - 5197
2. Klatt, P., Schmidt, K., Uray, G., Mayer, B., (1993) J. Biol. Chem. **268**, 14781 - 14787
3. Boucher, J.L., Genet, A., Vadon, S., Delaforge, M., Henry, Y., Mansuy, D., (1992) Biochem. Biophys. Res. Com., **187**, 880 - 886
4. Berkels, R., Klaus, W., Boller, M., Rösen, R., (1994) Thromb. Haem. **72**, 309 - 312
5. Bredt, D.S., Glatt, C.E., Hwang, P.M., Fotuhi, M., Dawson, T.M., Snyder, S.H., (1991) Neuron **7**, 615 - 624
6. Förstermann, U., Closs, E.I., Pollock, J.S, et al., (1994) Hypertension **23**, 1121 - 1131

ADP enhances NOS II gene expression in rat smooth muscle cells stimulated with interleukin-1

JUNICHI FUJII, HAN GEUK SEO, MICHIO ASAHI, and NAOYUKI TANIGUCHI

Department of Biochemistry, Osaka University School of Medicine, 2-2 Yamadaoka, Suita, Osaka 565, Japan

NO produced by vascular endothelial cells activates cytosolic guanylate cyclase of nearby VSMC by a paracrine mechanism. Produced cGMP leads to VSMC relaxation and results in vasodilatation. VSMC themselves also produces NO in response to cAMP-elevating agents and to cytokines such as IL-1β, TNF, and bFGF, however, NO production is suppressed by TGF-β. Several vasoactive factors are also known to suppress induction of NOS II, whereas plasmin can induce gene expression. It is well established that ADP causes aggregation of blood platelets through binding to the P_2 purinergic receptor. Binding of ADP to purinergic receptors on VSMC induced vasodilatation. ATP and NO, on the other hand, are considered to be transmitters in NANC inhibitory responses to VSMC. Stimulation of NANC neurons releases NO together with ATP, which work in a coordinated manner as nhibitory transmitters on VSMC. In addition, adenosine is also known to be a vasodilatatory factor through binding to adenosine receptors. We demonstrated that various nucleotides and adenosine augment production of NO in VSMC stimulated with IL-1β by enhancing NOS II gene expression.

Various nucleotides including ADP enhanced nitrite formation by VSMC stimulated with IL-1β. Since VSMC have the ability to produce NO in response to various stimuli, and have purinergic receptors, we evaluated the effects of ADP on NO formation by VSMC isolated from rat aorta by measuring levels of nitrite in the medium. The levels of nitrite were augmented by the presence of IL-1β in the medium as previously reported. Accumulation of nitrite was enhanced by addition of ADP to the medium, but ADP alone failed to stimulate production of nitrite. Levels of nitrite measured at 24 h after ADP addition, increased in a dose-dependent manner, but still the ability of the cells to produce NO was not saturated even at 1 mM ADP. The effective concentration of ADP in the medium, however, would be lower because ADP is gradually hydrolyzed to adenosine during incubation by ecto-nucleotidases in VSMC. To investigate which receptor subtype is responsible for the enhancement of NO release from VSMC with ADP, we examined effects of various nucleotides and related compounds on nitrite formation by VSMC. AMP-PNP and AMP-PCP, which have a nonhydrolizable bond between the β and γ phosphate, were effective compounds, although they can be converted to AMP and adenosine by nucleotidase. Adenine nucleotides, GTP, and adenosine seemed to have similar stimulatory effects on nitrite formation whereas AMP-CPP, CTP, and UTP were less effective. Taking these data together, the characteristics of the receptor seem to be almost consistent with the pharmacology of a P_{2y} receptor, which has been also implicated in mediating the relaxant response in NANC neurons when ATP is released.

ADP Enhanced NOS II Gene Transcription by IL-1β. Since production of NO by VSMC stimulated with IL-1β was enhanced by ADP, we examined by Northern analysis the NOS II gene expression in VSMC, stimulated with IL-1β, in the presence or absence of ADP. The NOS II mRNA, which was below the detection level, was induced by IL-1β and its expression was enhanced by administration of ADP. The induction of the mRNA was dependent on ADP concentration and roughly consistent with the levels of accumulated nitrite. The mRNA levels of NOS II in VSMC stimulated with IL-1β, in the presence of various nucleotides, were also consistent with the amount of nitrite. To confirm that nitrite accumulation following administration of ADP was due to induced NOS II, we measured the NOS activity of the VSMC both in the presence and absence of Ca^{2+}. NOS activity was negligible in an unstimulated VSMC, but a significant activity was detected in the cytosolic fraction of IL-1β-stimulated VSMC, despite the absence of Ca^{2+}, thus indicating that enhanced NO formation by ADP could be attributed to the expression of NOS II.

Involvement of purinergic and adrenergic receptors in the gene induction by ADP. Because ADP can be hydrolyzed to adenosine, we investigated the possible participation of an adenosine receptor in the induction of NOS II. 8-(*p*-sulfophenyl)theophylline, an adenosine receptor blocker, at 25 μM significantly inhibited the enhancing effect of adenosine on the nitrite accumulation and on NOS II mRNA expression, as well as those of ADP and AMP-PNP. This suggested that ADP stimulated NOS II induction through binding to the P_2 purinergic receptor, although its metabolite, adenosine, also functioned in the same way through the adenosine receptor. These data indicate that binding of extracellular nucleotides to the purinergic receptor may enhance production of NOS II by modulating signal transducing pathway of IL-1β. However, this stimulatory effect of ADP was not a direct effect on NOS II gene expression because ADP alone had no effect. Since cAMP-elevating agents are known to induce NOS II in VSMC and cAMP is one of the intracellular signals of adenosine receptor in VSMC, the enhancing effect of these compounds may be mediated by cAMP.

Roles of ADP and Adenosine-stimulated NO production. NO binds to heme in guanylate cyclase and stimulates production of cGMP, which then brings about smooth muscle relaxation, resulting in vasodilatation. Because stimulation of the purinergic receptor enhanced induction of NOS II gene expression in VSMC, a part of the vasodilatatory effect of the P_2 purinergic stimulation may be mediated by NO. Under inflammatory conditions such as endothelial injury and balloon cathetelization, activated immune response cells invade to smooth muscle layers and produce inflammatory cytokines such as IL-1β, TNF, and IFN-γ. These are also factors able to induce NOS II in VSMC and damage nearby cells to release nucleotides in the plasma. The presence of ADP enhances gene expression by these cytokines, which would result in NO production and relaxation of VSMC. This mechanism would also prolong vasodilatatory effects by NANC neurons and might occur in pathological conditions when ADP is released from damaged cells. Since the fibrinolytic enzyme, plasmin, is known to enhance NO release from VSMC, ADP in conjunction with plasmin may function in VSMC as a protective role by maintaining blood flow through induction of NOS II.

Acknowledgments. This work was supported in part by Grants-in-Aid for Scientific Research on Priority Areas and Specific Project Research on Cancer Bio-Science from the Ministry of Education, Science and Culture, Japan, and by Ono Pharmaceutical Co. Ltd.

REFERENCES

1. Seo, H. G., Tatsumi, H., Fujii, J., Nishikawa, A., Suzuki, K., Kangawa, K., and Taniguchi, N. (1994) *J. Biochem.* **115**, 602-607
2. Seo, H. G., Fujii, J., Soejima, H., Niikawa, N., and Taniguchi, N. (1995) *Biochem. Biophys. Res. Commun.* **208**,10-18

Abbreviations: VSMC, vascular smooth muscle cells; IL 1β, interleukin-1β; TNF, tumor necrosis factor; bFGF, basic fibroblast growth factor; TGF-β, transforming growth factor-β; PDGF, platelet derived growth factor; NANC, non-cholinergic, non-adrenergic.

Nitric oxide synthase from DLD-1, a human adenocarcinoma cell line: induction and sensitivity to NOS inhibitors

J. Carol Berry and Rajendra D. Ghai
Research Dept., Pharmaceuticals Div., CIBA, Summit,N.J. 07901, USA.

Introduction

There are several reports of induction of NOS in human cell lines [1,2,3,4]. Sherman et al. reported induction of NOS in a human colonic adenocarcinoma cell line, DLD-1, treated 18 to 24 hours with a combination of cytokines. Other human colon cancer cell lines have been reported to express mRNA for both inducible and constitutive NOS [3]. We have further characterized iNOS from DLD-1 cells and compared its biochemical properties against iNOS from rat liver and the constitutive HUVEC NOS enzyme.

Methods

Cell culture:

DLD-1 cells obtained from ATCC, were cultured using the conditions described by Sherman [1]. In addition the media was supplemented with 1 μM riboflavin, 5 μM nicotinic acid, 2.5 μg/ml hemin and 10 μM sepiapterin. Human umbilical vein cells (HUVEC) obtained from Clonetics Corp. were grown in endothelial medium supplemented with 3 μM tetrahydrobiopterin (BH4).

Harvest of cells, induction of NOS in DLD-1 cells and preparation of NOS from DLD-1 and HUVEC

DLD-1 cells were grown to confluence in T75 flasks and induced with cytokines according to Sherman et al.[1]. At selected times cells were washed, harvested and stored at -80°C. NOS from DLD-1 cells, rat liver and the constitutive NOS from HUVEC were prepared according to the methods of Bredt [5] and Misko [6], respectively. Each enzyme preparation was passed over a G-25 column to remove endogenous arginine and the eluate used as enzyme.

Nitrite analysis and protein determination

Nitrate/nitrite levels were measured using the Griess reagent [1,7]. Protein was determined by the method of Bradford [8].

Enzyme activity, Km determinations, inhibition of NOS activity and immunotitrations

NOS activity was determined by the conversion of ^{3}H arginine to citrulline. iNOS activity was determined according to the method of Bredt [5] and HUVEC NOS activity was measured as described by Misko [6]. Km determinations were calculated [9] using various concentrations of arginine (0.5 to 25 μM). Inhibitory constants were determined and immunotitrations performed according to Schulz [9].

Results & Discussion

We were unable to demonstrate the induction of NOS in DLD-1 cells under the conditions reported by Sherman [1]. However, NOS induction was demonstrable in DLD-1 cells treated with IFN-γ, IL-1β, IL-6 and TNF-α and cultured in media supplemented with riboflavin, hemin, nicotinic acid and sepiapterin. Under these conditions, nitrate/nitrite levels accumulated in the media and with maximal levels at 16 hrs. while NOS activity was maximal at 6-8 hrs. post induction (Figure 1). Sherman and Jenkins [1,3] reported NOS induction in DLD-1 cells, at 24 and 48 hours after treatment with the same cytokines. A possible reason for the for these differences may be the passage number of the cells used in our experiments. In agreement with Sherman and Jenkins no NOS activity was detected in unstimulated cell lysates from DLD-1 cells. The Km value for arginine was 14.9 μM for DLD-1 NOS and 11.5 and 2 μM for rat liver and HUVEC NOS. Sherman reported a Km value for arginine of 3 μM for DLD-1 NOS with assay conditions using 1000 fold purified enzyme and a BH4 concentration that was 25 times greater than that used by us. BH4 has been reported to influence both feedback inhibition of NOS by nitric oxide as well as Km for arginine [12]. The difference in assay conditions and enzyme purity are possible reasons for the differences in Km. NOS from DLD-1 and rat liver but not from HUVEC was immunoprecipitated by a monoclonal antibody to murine macrophage iNOS (Figure 2). Therefore NOS from DLD-1 and rat liver appear to have a common epitope which is different from HUVEC NOS. Three inhibitors of NOS were compared (Table 1). Nitroarginine was potent towards HUVEC NOS while aminoguanidine was most potent towards rat liver iNOS. These differences in potencies may be attributed to species differences and are in agreement with reported observations [10]. Our data suggests that DLD-1 cells can express an inducible NO synthase that is antigenically and kinetically similar to rat liver iNOS.

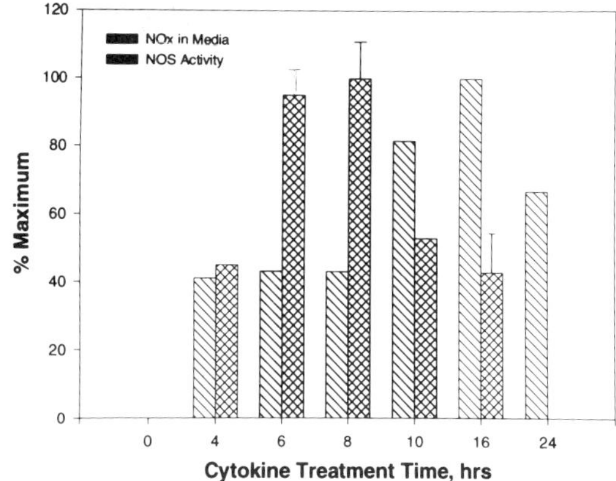

Figure 1. Time Course of NOS activity and nitrite/nitrate accumulation in media of DLD-1 cells after cytokine treatment.

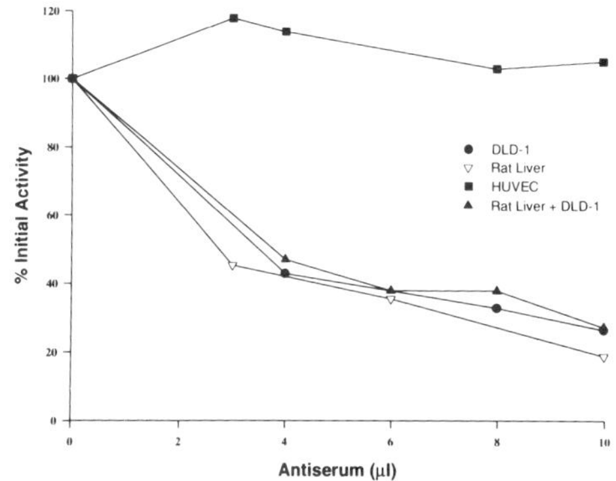

Figure 2. Immunoprecipitation of NOS from DLD-1, rat liver and HUVEC.

Table 1. IC_{50} Values for inhibition of NOS by isoform selective and non-selective inhibitors

Enzyme	N-iminoethylornithine (Non-selective)	Nitroarginine (cNOS selective)	Aminoguanidine (rodent iNOS selective)
	IC_{50} (μM)		
DLD-1	0.34	13.9	71
Rat liver	0.36	26.2	16.8
HUVEC	0.38	0.06	160

1. Sherman,P.A.,Laubach, V.E., Reep, B.R., and Wood, E.R. (1993) Biochemistry 32,11600-11605.
2. Asano,K.,Chee,C.B.E., Gaston,B., Lilly,C.M. Gerard,C., Drazen,J.M. and Stamler,J.S. (1994) Proc.Nat.Acad. Sci. 91,10089-10093.
3. Jenkins,D.C., Charles, I.G., Saylis,S.A., Lelchuck, R., Radomski,M.W. and Moncada,S. (1994) Br. J. Cancer 70, 847-849.
4. Hokari,A., Zeniya,M. and Esumi, H., (1994) J. Biochem.116,575-581.
5. Bredt,D.S., and Snyder,S.H. (1990) Proc. Natl. Acad. Sci. 87, 682-685
6. Misko, T.P., Moore, W.M., Kasten, T.P. et al.(1993) Eur. J. Pharmacol. 233,119-125.
7. Green,L.C., Wagner, D.A., Glogowski,J., Skipper, P.L., Wishnok, J.S., and Tannenbaum, S.R., (1982) Anal. Biochem. 126,131-138.
8. Bradford,M. (1976) Biochem. 72, 248-252.
9. Schulz,R., Sakane,Y., Berry,C. and Ghai,R. (1991) J. Enzyme Inhibition 4,347-358.
10. Nakane,M., Pollock,J.S.,Klinghofer,V. et al. (1995) Biochem. Biophys. Res. Commun. 206, 511-517.
11. Hyun,J., Komori Y., Chaudhuri, G., and Ignarro,L. Biochem. Biophys. Res. Commun. 206, 380-386.
12. Klatt,P., Schmid,M., Leopold,E., Schmidt,K., Werner,E.R. and Mayer,B. J. Biol Chem. 269, 13861-13866.

Interleukin-6 and TNF-α potentiate nitrite accumulation but not NO synthase induction in DLD-1 cells.

ALAN V WALLACE, IWONA HUTCHINSON, DAVID J NICHOLLS, KERRY L BODEN, JOHN F UNITT and CATHERINE HALLAM

Biochemistry Department, Astra Charnwood, Bakewell Road, Loughborough LE11 0RH, UK.

Nitric oxide (NO·) may play a key physiological role in host defence. This possibility has been supported by recent observation that NOS type II can be induced by cytokines/LPS in human lung and colon epithelial cell lines and is seen in cells shed from the urinary tract of patients with infections [1]. Epithelial tissues represent the first line of defence against infection. Induction of NOS has recently been reported from three human epithelial cell lines (derived from lung, A549 [2] and colon, DLD-1 [3], HT29 [4]) of these the colorectal carcinoma derived DLD-1 cells yield the most enzyme. The epithelial cell line, DLD-1 provides a readily available source of human type II NOS and this has already been exploited in studies of inhibitors of the enzyme [5]. Induction of NOS in human cells has been difficult to demonstrate [6] and where it has been observed a complicated combination of cytokines and LPS have usually been used [2, 3, 4, 7]. We have investigated the cytokine and LPS requirements for induction of DLD-1 cells in terms of nitrite accumulation in the culture medium and by examining enzyme activity associated with the cells.

Cells were induced in 24-well plates by addition of cytokines or LPS. The culture plates were incubated for ~20 hours before measuring the induction of NO synthase (NOS) by determining the level of nitrite in the culture medium according to the method of Griess [8].

Two methods were used to measure NOS. In the cell monolayers enzyme activity was determined by washing the wells with phosphate buffered saline and adding 0.15ml of a buffer containing 1mM arginine, 1mM NADPH & 0.1% Triton X-100. The plates were incubated at room temperature in the dark for 6 hours then excess NADPH was removed by glutamate dehydrogenase/2-oxoglutarate treatment, finally nitrite accumulation was determined. Alternatively, NOS was extracted from cultures in 225cm^2 flasks by lysing the cells in Triton X-100, endogenous arginine was removed by treatment with Dowex and enzyme activity was measured in the presence of 3μM [^{3}H]arginine and excess co-factors by the production of [^{3}H]citrulline.

Arginine uptake was assayed by incubating monolayers of control and cytokine induced DLD-1 cells with 0.5mM [^{3}H]arginine for 10 minutes.

Figure 1 shows the effect of various combinations of cytokines upon NOS induction. Induction of NOS in DLD-1 cells required both IFN-γ and IL-1β, either of these cytokines alone had no effect. TNF-α and IL-6 potentiated the effect of IFN-γ/IL-1β, and the combination of all four cytokines gave the highest accumulation of nitrite in the culture medium after overnight incubation. LPS had a negligible effect upon induction of NOS either alone or in combination with any cocktail of cytokines (not shown). When the induced monolayers were assayed for enzyme activity it was found that the enhanced nitrite accumulation seen with TNF-α and

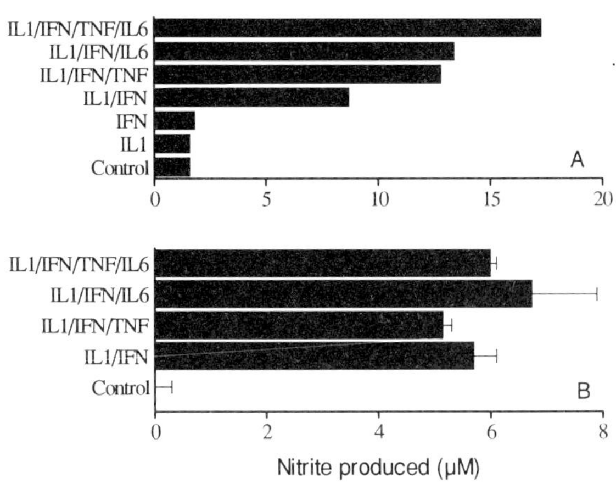

Figure 1. *Induction of NOS in DLD-1 cells.* *Panel A shows the accumulation of nitrite in the culture medium overnight. Panel B depicts the enzyme activity associated with corresponding cell monolayers.*

IL-6 did not correlate with elevated enzyme activity.

To test this observation more accurately larger scale cultures of DLD-1 cells were induced with combinations of cytokines and NOS activity was determined in lysates from the cells. As before TNF-α and IL-6 potentiated the acummulation of nitrite in the culture medium, but there was no significant effect upon the enzyme content of the cells (data not shown).

These results suggest that only IL-1β and IFN-γ are required for expression of the enzyme and that IL-6 and TNF-α provide accessory components which are important for full cellular activtiy of NOS. This may involve regulation of arginine uptake or co-factor synthesis. However, incubation of DLD-1 cells with the full cocktail of cytokines did not affect the rate of arginine uptake. Indicating that substrate deprivation was not the cause of reduced nitrite accumulation in the absence of TNF-α and IL-6.

In summary, exposure of the human colorectal carcinoma cell line, DLD-1, to IL-1β and IFN-γ induces expression of NOS type II. When monitored by the increase in nitrite accumulation in the culture medium this effect is potentiated by TNF-α and IL-6. Direct measurement of enzyme activity reveals that this potentiation is not at the level of protein production suggesting that TNF-α and IL-6 may have an indirect effect on NOS activity.

1. Smith, S.D., Wheeler, M.A. and Weiss, R.M. (1994) *Kidney International* **45**, 586-591
2. Robbins, R.A., Barnes, P.J., Springall, D.R. *et al.* (1994) *Biochem. Biophys. Res. Commun.* **203**, 209-218
3. Sherman, P.A., Laubach, V.E., Reep, B.R. and Wood, E.R. (1993) *Biochemistry* **32**, 11600-11605
4. Kolios, G., Robson, R.L., Brown, Z., Robertson, D.A.F and Westwick, J. (1995) *Br. J. Pharmacol.* **114**, 133P
5. Garvey, E.P., Oplinger, J.A., Tanoury, G.J. *et al.* (1994) *J. Biol. Chem.* **269**, 26669-26676
6. Denis, M. (1994) *J. Leukoc. Biol.* **55**, 682-684
7. Geller, D.A., Lowenstein, C.J., Shapiro, R.A. *et al.* (1993) *Proc. Natl. Acad. Sci. USA* **90**, 3491-3495
8. Griess, J.P. (1964) *Philos. Trans. R. Soc. Lond.* **154**, 667-731

Diminished Inducible Nitric Oxide Synthase Transcription, Nitric Oxide Production, and Tumoricidal Capacity in Macrophages Isolated from Tumor-Bearing Mice.

MICHAEL R. DINAPOLI and DIANA M. LOPEZ

Department of Microbiology and Immunology, University of Miami School of Medicine, Miami, FL, 33136 USA

Results and Discussion

Activated macrophages play an important role in the destruction of a variety of tumor cell, and nitric oxide (NO) has been demonstrated to be a major effector molecule for this function (1). The integrity of the pathways for macrophage activation must be preserved for NO production and cytolytic activity to be maintained. However, in many cases, developing neoplasms appear to be capable of impairing steps in this process as a means of avoiding immune destruction (2-4).

We have previously reported that following activation with LPS, PEM isolated from mammary tumor-bearing mice (TB-PEM) exhibit a diminished potential to both produce NO and lyse tumor targets when compared with PEM from normal mice (N-PEM). However, when the TB-PEM were stimulated LPS in combination with IFN-γ, they were able to recover these functions to near normal levels (4). Since tumor-associated macrophages (TAM) are in direct contact with the cells of the developing tumor, and are therefore likely to be important in controlling tumor growth, we evaluated the ability of this population of macrophages to perform these functions. Both NO production and cytolytic activity were greatly reduced in TAM compared with PEM from normal or tumor-bearing mice. These defects were apparent not only when the cells were stimulated with LPS (1 or 10 μg/ml) alone (Figure 1), but continued even when stimulated with LPS (10 μg/ml) and IFN-γ (10 or 50 U/ml) in combination (Figure 2). These alterations do not appear to be due to the enzymatic process utilized to isolate the TAM, as this process had no effects on NO production by PEM from normal or tumor-bearing mice. In addition, even though the TAM exhibited a diminished capacity to produce NO, they were physically similar to the N-PEM and TB-PEM, and retained other complex functions such as adherence to plastic, phagocytosis, mitochondrial dehydrogenase activity, and IL-6 production.

NO is produced in macrophages by the enzyme inducible nitric oxide synthase (iNOS) (5), therefore Western blots were performed to evaluate the expression of this protein in PEM and TAM. As seen in Figure 3, iNOS protein could not be detected in unstimulated PEM from either type of mice. However, when stimulated with LPS (10 μg/ml), N-PEM produced substantial quantities of this enzyme, while levels from TB-PEM remained nearly undetectable. Following stimulation with LPS (10 μg/ml) and IFN-γ (50 U/ml) in combination, both populations of PEM produced approximately equal and high levels of iNOS protein. Although N-PEM produced iNOS following activation, at similar exposure times TAM failed to produce detectable quantities of this enzyme regardless of stimulus.

Since the production of NO is known to be regulated at the transcriptional level (5), Northern hybridizations were utilized to evaluate iNOS mRNA levels in PEM and TAM. As seen in Figure 4, unstimulated PEM from both normal and tumor-bearing mice displayed no detectable message. However, upon stimulation with LPS (10 μg/ml), substantial levels of iNOS mRNA were detected in the N-PEM, while the message in TB-PEM remained almost undetectable. When stimulated with a combination of LPS (10 μg/ml) and IFN-γ (50 U/ml), N-PEM and TB-PEM produced equal and very high levels of the iNOS mRNA. At similar exposure times, almost no detectable message was found in the TAM regardless of stimulation.

The diminished levels of iNOS mRNA and protein in the tumor-bearers' macrophages appear sufficient to account for the reduced NO production and cytolytic activity of these cells. The most likely explanations for the observed alterations include a diminished rate of transcription of the iNOS gene or an increase in mRNA degradation. To distinguish between these possibilities, message stability was evaluated in PEM. The pattern and rate of iNOS mRNA decline following addition of actinomycin D appeared similar in N-PEM and TB-PEM (Figure 5), suggesting that the defect leading to the diminished levels of iNOS message may be occurring at or prior to the level of transcription.

Fig. 1: NO production and Cytotoxicity with LPS

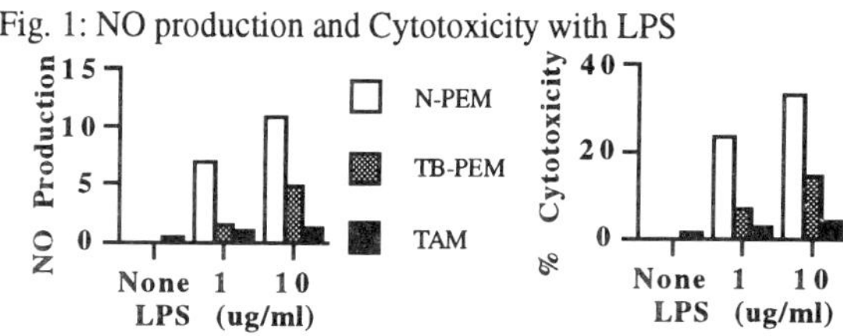

Fig. 2: NO production and Cytotoxicity with LPS and IFN

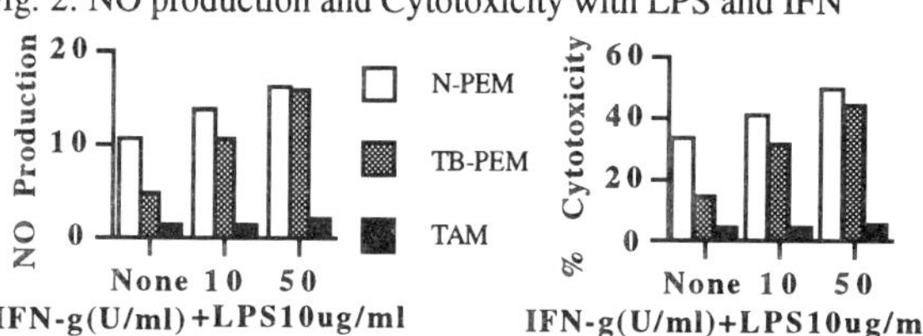

Fig. 3: Relative iNOS Protein Levels

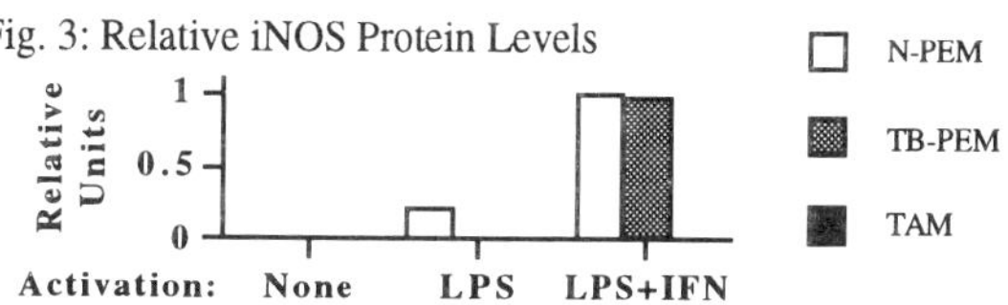

Fig. 4: Relative iNOS mRNA Levels

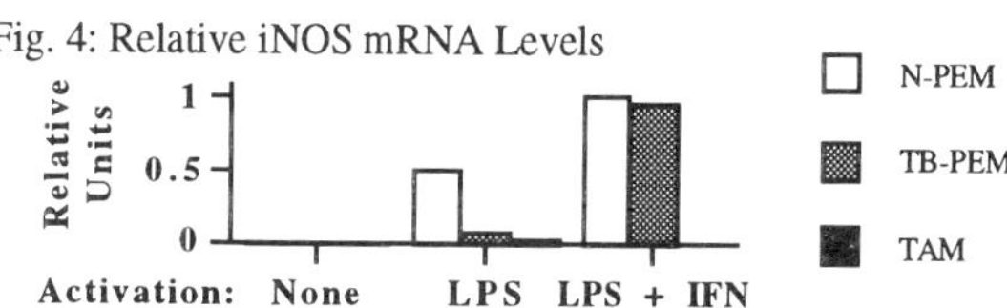

Fig. 5: iNOS mRNA Stability

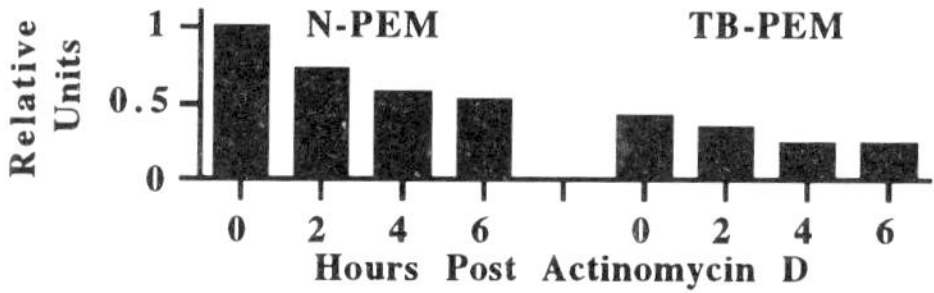

These data represent a novel examination of the molecular mechanisms by which tumor progression can alter the production of NO by host macrophages, and suggests that altered levels of NO production, through diminished iNOS transcription, is at least partly responsible for the reduced tumoricidal capacity of these cells.

Acknowledgements

This work was supported by grants CA 54226 and CA 25583.

References

1. Hibbs, J. B., R. R. Taintor, and Z. Vavrin. 1987. Macrophage cytotoxicity: Role for L-arginine deaminase and imino nitrogen oxidation to nitrite. *Science* 235:473-476.
2. Duffie, G. P., and M. R. I. Young. 1991. Tumoricidal activity of alveolar and peritoneal macrophages of C57BL/6 mice bearing metastatic or non-metastatic variants of Lewis lung carcinoma. *J. Leuk. Biol.* 49:8-14.
3. Umansky, V., M. Rocha, A. Kruger, P. Von Hoegen, and V. Schirrmacher. 1995. *In situ* activated macrophages are involved in host resistance to lymphoma metastasis by production of nitric oxide. *Int. J. Oncology.* 7:33-40.
4. Sotomayor, E. M., M. R. DiNapoli, C. Calderon, A. Colsky, Y-X. Fu, and D. M. Lopez. 1995. Decreased macrophage-mediated cytotoxicity in mammary-tumor-bearing mice is related to an alteration of nitric oxide production and/or release. *Int. J. Cancer* 60:660-667.
5. Xie, Q., H. J. Cho, J. Calaycay, R. A. Mumford, K. M. Swiderek, T. D. Lee, A. Ding, T. Troso, and C. Nathan. 1992. Cloning and characterization of inducible nitric oxide synthase from mouse macrophages. *Science* 256:225-228.

Characterization of a constitutive nitric oxide synthase from normal human keratinocytes

J.E. BAUDOUIN and P. TACHON
L'Oréal, Département de Biologie Fondamentale, Centre de recherche Charles Zviak, 90 rue du Général Roguet, 92583 CLICHY CEDEX FRANCE

INTRODUCTION

Normal human keratinocytes (NHK) after exposure to cytokines or the ligation of low affinity receptor for IgE (CD23) by the immune complex IgE/anti-IgE released nitric oxide (NO) as detected by an increase of the amount of nitrites in the culture medium involving the inducible isoform of NO synthase (iNOS) (1,2). But, little is known about the presence of a constitutive isoform of NO synthase. The present study was designed to determine whether NO synthase is constitutively present in NHK and to characterize it.

MATERIALS AND METHODS

Normal human keratinocytes (NHK) were grown in modified MCDB 153 medium. NHK were scraped off the dishes at subconfluence and homogenized in ice cold lysis buffer consisting of 50 mM Tris-HCl pH 7.5, 1 µM 5,6,7,8 tetrahydrobiopterin (BH_4), 1mM dithiotreitol (DTT), 100 µM EDTA, 1 µM leupeptin and 1 µM pepstatin A. NO synthase activity was assayed on the 14,000 xg supernatant by measuring the conversion of L[^{14}C]-arginine to [^{14}C]-citrulline. The reaction mixture containing 100 µg of homogenate protein were incubated in a buffer containing 50 mM Tris-HCl pH 7.5, 1mM NADPH, 15 µM BH_4, 5 µM FAD, 1mM EGTA, 3 mM $CaCl_2$, 50 U/ml calmodulin, 5µM L-arginine and 15 µM L-[guanido-^{14}C]arginine for 30 minutes at 37°C. The partial purification of NO synthase from 14,000 xg supernatant was performed on a 2',5'-ADP Sepharose column. Enzyme activity was eluted with lysis buffer supplemented with 10 mM NADPH.

RESULTS

The 14,000 xg supernatant, unlike pellet, contained a basal NO synthase activity with a specific activity ranging from 50 to 150 pmol/min/mg of protein. NHK NO synthase activity required the presence of calcium and was inhibited by calmodulin antagonists such as calmidazolium and trifluoperazine. NO synthase inhibition by arginine analogs : N^G-nitro-L-arginine (L-NNA) and its methyl ester (L-NAME), N^G-mono methyl-L-arginine (L-NMMA) and L-canavanine were assayed as formation of citrulline from 20 µM L-arginine by NHK 14,000 xg supernatant fraction. NO synthase activity was inhibited dose dependently with an IC_{50} of 0.3, 2, 2.5 and 100 µM for respectively L-NNA, L-NAME, L-NMMA and L-canavanine. The 14,000 xg supernatant was fractionated on a 2',5'-ADP Sepharose 4B column. NO synthase activity was eluted with buffer containing 10 mM NADPH. The NADPH eluate was examined in western blot for cross-reactivity with monoclonal antibodies raised against the three isoforms (ncNOS, ecNOS and iNOS). Figure 1 showed that a band was specifically recognized by monoclonal antibody raised against the neuronal constitutive isoform of NOS corresponding to a protein of ~ 150 kDa. The rate of citrulline formation by 0.5-2 µg of partial purified enzyme was conducted in the presence of 2-100 µM L-arginine and saturating concentrations of cofactors for 1 to 4 minutes. Km for L-arginine and Vmax were calculated from Lineweaver-Burke representation of 1/Vi versus 1/[S] and were respectively 22.3 ± 2.3 µM and 7.3 ± 0.3 nmol/min/mg of protein.

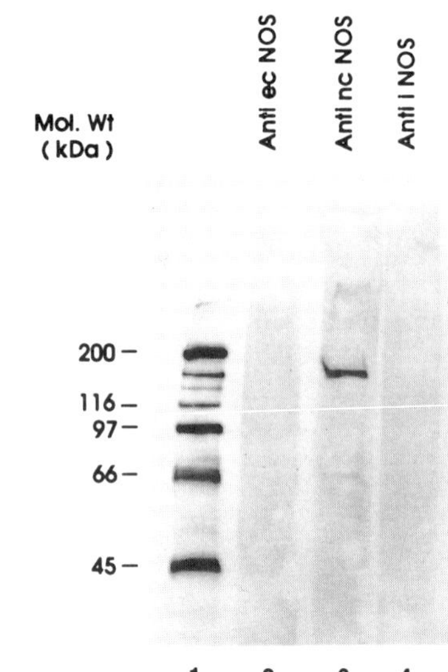

Figure 1 : Immunoblot analysis of partially purified NO synthase from normal human keratinocytes

Equal amount of protein from partially purified fraction (2.6 µg) were loaded on a 7.5 % polyacrylamide gels, transferred to nitrocellulose and immunoblotted (lanes 2-4). The immunoblotted membrane was probed with specific monoclonal antibodies raised against ecNOS, ncNOS and iNOS as specified by Transduction Laboratories. The blots were developped using the enhances chemoluminescence immunoblot system (Amersham). Lane 1 contained molecular mass markers.

CONCLUSION

Unstimulated NHK in culture contained a basal NO synthase activity which : the subcellular localization (soluble and cytosolic), the rank order in potency of arginine analogs on the inhibition of enzyme activity and the dependency on calcium/calmodulin suggest that it could be related to NOS I. The cross reactivity of a ~ 150 kDa protein with antibodies raised against neuronal isoform of NO synthase allowed us to characterize it definitely.

The authors wish to thank Isabelle Renault for editing this manuscript and Dr Marie-Madeleine CALS for her help regarding E.C.L assay.

REFERENCES

1- Heck D.E., Laskin D.L., Gardner C.R. and Laskin J.D. (1992) J. Biol. Chem **267**. 21277-21280.

2- Bécherel, P.A., Mossalayi, M.D., Ouazz, F., Le Goff, L., Dugas, B., Paul-Eugène, N., Frances, C., Chosidow, O., Kilchherr, E., Guilloson, J.J., Debré, P. and Arock, M. (1994) J. Clin. Invest. **93**. 2275-2279.

Lack of nitric oxide synthase gene induction by cytokines and lipopolysaccharides in guinea pig coronary artery smooth muscle cells

RAM V SHARMA, SHENGYUN FANG, ENQING TAN, AND RAMESH C. BHALLA

Department of Anatomy, University of Iowa, Iowa City, IA 52242.

Exposure of vascular smooth muscle cells (VSM) to interleukin 1β (IL-1β) and other inflammatory agents induce these cells to produce nitric oxide (NO) [1]. The inducible nitric oxide synthase (iNOS) gene has been cloned from cytokine-stimulated VSM cells [2]. However, immunohistochemical studies on localization of iNOS in LPS treated rat tissues have failed to demonstrate iNOS staining in VSM cells [3,4]. VSM cells undergo differentiation during subculture and change their phenotype from contractile to dividing cells [5]. Therefore, we tested the hypothesis that cytokine-induced NO production and iNOS gene expression are functions of VSM cell differentiation during subculture. This hypothesis was tested in LPS and cytokine-stimulated primary cultures and serially passaged guinea pig coronary artery (GPCA) smooth muscle cells.

Materials and Methods

VSM cells from guinea pig main coronary arteries and rat aorta were cultured as described previously [5]. The purity of VSM cell cultures was confirmed by immunocytochemical localization of smooth muscle specific α-actin. The VSM cells were serially cultured and used between passages 1-9. Inducttion of iNOS gene in VSM cells was tested by treatment of cells in phenol red free minimal essential medium with the indicated concentrations of cytokines and LPS for 20-24 hr. iNOS mRNA levels were estimated by Northern blot analysis using murine macrophage iNOS cDNA, obtained from Dr. Carl Nathan (Cornell University Medical Center, New York). iNOS activity was measured by conversion of $[^3H]$-arginine to $[^3H]$-citrulline. Nitrite levels were measured using Greiss reagent. Cells were counted using a Coulter cell counter. Data are expressed as means ± SEM.

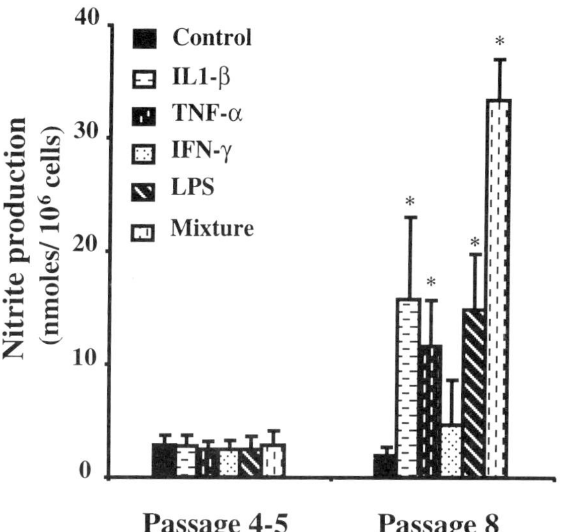

Fig. 1. Effect of serial passage of guinea pig coronary artery (GPCA) smooth muscle cells on cytokine and LPS stimulated nitrite production. Confluent cells in indicated passages were incubated in the absence (control) or presence of IL-1β (20 ng/ml), TNF-α (20 ng/ml), IFN-γ (100 ng/ml), LPS (100 μg/ml), or a mixture of all four agents for 24 hr. Nitrite production in the media was measured using Griess reagent. Data are given as the means ± SEM for 6 independent measurements; * denotes significant increase ($P<0.05$) compared to the control.

Results and Discussion

Incubation of GPCA smooth muscle cells in primary culture and early passages (up to 5th passage) with high concentrations of IL-1β (20 ng/ml), TNF-α (20 ng/ml), (IFN-γ 100 ng/ml), LPS (100 μg/ml) alone or in combination for 24 hr failed to increase nitrite production. However, an identical treatment protocol increased nitrite production in GPCA smooth muscle cells obtained after the 8th passage (Fig. 1). We also failed to detect iNOS activity in early passage GPCA smooth muscle cells treated with cytokines in combination with LPS. Treatment of GPCA smooth muscle cells in the third passage with IL-1β (10 ng/ml) for 20 hr did not increase iNOS mRNA levels. Previous attempts to induce iNOS in human aorta smooth muscle cells have also failed to demonstrate iNOS mRNA after treatment with a mixture of IL-1β (200 U/ml), TNF-α (500 U/ml) and IFN-γ (100 U/ml) [6].

We also observed that with serial passage phenotypic characteristics of GPCA smooth muscle cells change from elongated appearance to cobble shape and that coincides with the appearance of cytokine stimulated NO production. Similar to our observations, it has been recently demonstrated that phenotypic and functional changes in chondrocytes in subculture modulate iNOS induction by IL-1β and TGF-β [7]. In chondrocytes, IL-1β stimulated induction of iNOS decreases and is completely lost after 6th passage. On the other hand, while TGF-β has no effect on iNOS induction in primary cultures of chondrocytes, it starts to induce iNOS in the presence of IL-1β in 6th passage cells that have completely lost IL-1β inducibility [7]. Recently it has been demonstrated that with repeated passaging of VSM cells the levels of cGMP dependent protein kinase decreases and after several passages the enzyme levels are undetectable [8].

In conclusion, our data demonstrate lack of iNOS induction in primary cultures and early passage GPCA smooth muscle cells. As GPCA smooth muscle cells are serially cultured, their differentiation status and morphological appearance changes and is associated with iNOS induction. These observations are consistent with in vivo studies that have failed to demonstrate immunohistochemically the iNOS protein in VSM cells of rats treated with LPS [3,4].

Acknowledgments. This work was supported by USPHS grant HL-51735.

References

1. Hecker, M., Boese, M., Schini-Kerth, V.B., Mulsch, A., Busse, R. (1995) Characterization of the stable L-arginine-derived relaxing factor released from cytokine-stimulated vascular smooth muscle cells as an N^G-hydroxy-L-arginine-nitric oxide adduct. Proc. Natl. Acad. Sci. 92:4671-4675.
2. Geng, Y.J., Almqvist, M., Hansson, G.K. (1994) cDNA cloning and expression of inducible nitric oxide synthase from rat vascular smooth muscle cells. Biochim. Biophys. Acta. 1218: 421-424.
3. Buttery,L.D.K., Evans, T.J., Springall, D.R., Carpenter, A., Cohen, J., Polak, J.M. (1994) Immunochemical localization of inducible nitric oxide synthase in endotoxin-treated rats. Lab. Invest. 71: 755-764.
4. Sato, K., Miyakawa, K., Takaya, M., Hattori, R., Yui, Y., Sunamoto, M., Ichimori, Y., Ushio, Y., Takahashi, K. (1995) Immunohistochemical expression of inducible nitric oxide synthase (iNOS) in reversible endotoxin shock studied by a novel monoclonal antibody against rat iNOS. J. Leuko. Biol. 57: 36-44.
5. Bhalla, R.C., Sharma, R.V. (1993) Induction of c-fos and elastin gene in response to mechanical stretch of vascular smooth muscle cells. J. Vas. Med. Biol. 4:130-137.
6. MacNaul, K.L., Hutchinson, N.I. (1993) Differential expression of iNOS and cNOS mRNA in human vascular smooth muscle cells and endothelial cells under normal and inflammatory conditions. Biochem. Biophys. Res. Commun. 196: 1330-1334.
7. Blanco, F.J., Geng, Y., Lotz, M. (1995) Differentiation-dependent effects of IL-1 and TGF-β on human articular chondrocyte proliferation are related to inducible nitric oxide synthase expression. J. Immunol. 154: 4018-4026.
8. Cornwell, T.L., Soff, G.A., Traynor, A.E., Lincoln, T.M. (1994) Regulation of expression of cyclic GMP-dependent protein kinase by cell density in vascular smooth muscle cells. J. Vas. Res. 31: 330-337.

Immunoaffinity purification and structural characterization of recombinant human iNOS

J. Calaycay, E. McCauley, P. Griffin, K. Wong, T. Kelly, D. Geller*, K. MacNaul, N. Hutchinson, H. Williams, R. Mumford and J. Schmidt

Merck Research Laboratories, P.O. Box 2000, Rahway, NJ 07065 USA and *Dept. of Surgery, Univ. of Pitt., Pittsburgh, PA 15621 USA

Introduction

Recombinant iNOS, cloned from human hepatocytes [1] and expressed in *Sf9* cells was purified in a single step by immunoaffinity chromatography. The enzyme was isolated in the absence of NADPH and showed stability at 25°C for >4h. Specific activity ranged from 34 to 135 nmoles cit/min/mg at 25°C; a 130 kDa band was obtained by SDS-PAGE. The enzyme displayed a Soret λmax at 396 nm and a shoulder at 460 nm, attributable to flavins. A type II spectral shift was induced with 2 mM imidazole and reversed to type I by 2 mM L-arg. The ferrous-CO adduct has an absorbance max. of 446 nm indicating a P_{450}-type heme. Heme content ranged from 0.22 to 0.47 equiv./monomer. The enzyme contained CaM in a 1:1 ratio. Primary structure of CaM and iNOS were confirmed by MALDI-TOF-MS, LC-ESI-MS and Edman sequencing.

The column was constructed using a rabbit polyclonal antibody raised against a 7-residue peptide identical to the C-terminus of the enzyme. Selective elution was effected by a 25 µM solution of the cognate peptide.

Methods

The antibody was immobilized as follows: antiserum was mixed with Protein A-agarose and to the resulting complex, acetylated cognate peptide was added to occupy the antigen-binding sites of the antibody, a protective step against non-specific modification during crosslinking. The complex was then treated with dimethylpimelimidate, a homobifunctional imidoester crosslinker. The resulting column displayed high selectivity and capacity for the peptide immunogen (2.7 nmoles peptide/ml of resin).

All purification steps were done at 4°C. *Sf9* lysates in 20 mM TES, pH 7.4, 1 mM CHAPS, 100 µM BH_4, 0.1 mM DTT and 0.250 M NaCl were loaded at 3.0 ml/min and eluted by batch-mode using the cognate peptide.

Results

The immunoaffinity column totally depleted activity from the lysate in a single step as shown in Fig. 1. The peptide-eluted fractions of activity show a 130 kDa band by Coomassie Blue-stained SDS-PAGE,(inset). The yield from this procedure typically ranged from 16-32 %.

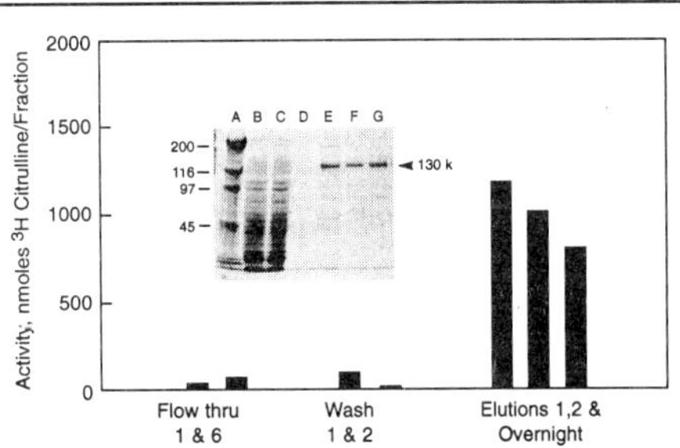

Fig. 1. Representative fractions are shown with activity measured by [^{3}H]-L-arg to [^{3}H]-L-cit conversion assay. (Inset) SDS-PAGE analysis: lane A, mol. wt. stds.; B, crude lysate; C, flow-thru 6; D, wash 2; E-G, Elutions 1, 2 and overnight.

The UV-vis and CO-ferrous spectra of the purified enzyme indicate the presence of a P_{450}-like heme as well as flavin co-factors (Fig. 2). Perturbation of the Soret absorbance with imidazole and the reversal with L-arg are shown in Fig. 3.

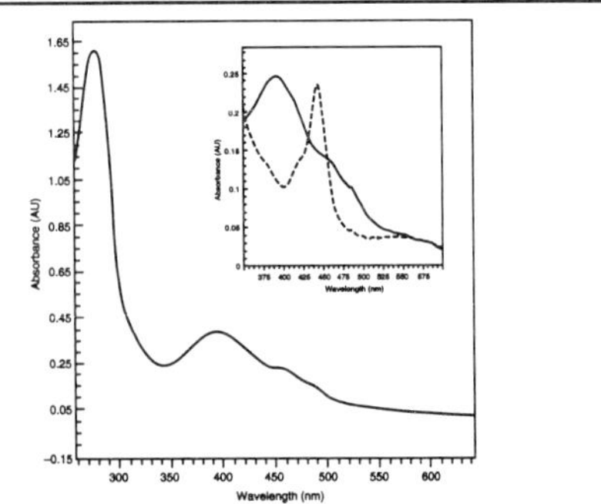

Fig. 2. UV-vis spectrum showing a peak absorbance at 396 nm and shoulders at 450 and 480. Inset, CO-ferrous spectrum with a typical peak absorbance at 466 nm.

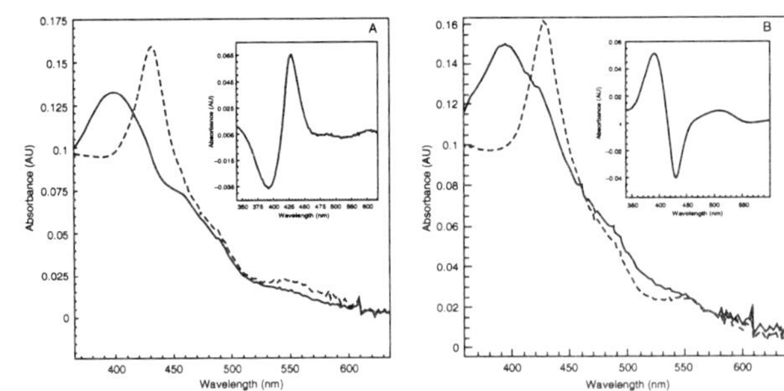

Fig. 3. A. 2.0 mM imidazole final concentration induced a shift from 396 to 430 nm. **B.** 2.0 mM L-arg added to the enzyme-imidazole complex promoted reversal to 396 nm.

Expression batches of different specific activities were measured for heme content and CaM:iNOS ratios and results are summarized in Table 1.

Table 1. Heme content and CaM:iNOS ratio from different expression batches.

Specific Activity (25°C)	Heme per Monomer	Composition	
		CaM	iNOS
nmoles cit/min/mg	*equiv.*	*equiv.*	
34	0.32	0.77	0.72
35 (61)*	0.20	1.26	1.38
135 (199)*	0.47	0.96	1.11
50 (97)*	0.25	1.00	0.78

* specific activity at 37°C

Fig. 4. shows the primary structure characterization of both CaM and iNOS by HPLC and mass spectrometry.

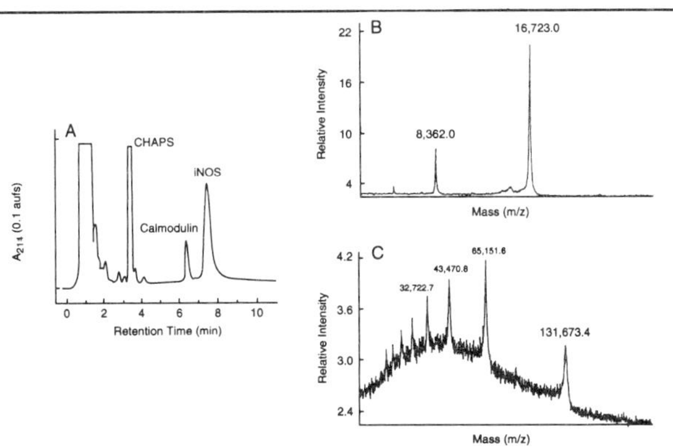

Fig. 4. Panel A, separation of the components by RP-HPLC; Panels B and C, MALDI-TOF-MS analysis of CaM and iNOS, respectively. The mass at 8362 in Panel B represents the doubly-protonated form of CaM and the group of peaks in Panel C is a series of multiply-protonated species of iNOS.

Discussion

The recombinant enzyme was isolated in a single step by immunoaffinity chromatography, a protocol that excluded NADPH, resulting in a purified enzyme with remarkable stability at 25°C for > 4h. BH_4 analysis showed that the purified enzyme was essentially saturated at ~94%. CaM which co-purified with the enzyme was measured to be in a 1:1 stoichiometry with iNOS. The heme content yielded a range of 0.20 to 0.47 equivalents per monomer. By gel filtration chromatography, the resolved dimeric species showed no enrichment in specific activity. The monomeric form was correspondingly small. These results propose that the recombinant enzyme was predominantly of the heme-deficient homodimeric form.

Mass spectrometry and microsequence analysis confirmed the fidelity of the expression system and identified CaM as a tightly-bound component of the enzyme.

We gratefully thank Dr. Steven Gross and his colleagues (Cornell University Medical College, NY, USA) for the BH_4 measurements.

Reference

1. Geller, D.A., Lowenstein, C.J., Shapiro, R.A. et al (1993) Proc. Natl. Acad. Sci. **90**, 3491-3495

Comparison of Recombinant Human Inducible Nitric Oxide Synthase Expressed in Sf9 Cells with Native Human Hepatocyte iNOS

KAREN L. MacNAUL*, M. RAJU SAYYAPARAJU *, ERMENEGILDA McCAULEY*, THERESA M. KELLY*, STEPHAN K. GRANT*, JIMMY R. CALACAY*, KENNY K. WONG*, ANDREAS NUSSLER+, JEFFREY R. WEIDNER*, RICHARD A. MUMFORD*, TIMOTHY R. BILLIAR+, MICHAEL J. TOCCI*, JACK SCHMIDT*, and NANCY I. HUTCHINSON*

*Merck Research Laboratories, P.O. 2000, Rahway, NJ, 07065 USA; +University of Pittsburgh, School of Medicine, Pittsburgh, PA, 15213 USA.

The expression of recombinant human inducible nitric oxide synthase (iNOS), which faithfully represents its native counterpart, will enable the detailed characterization of this new and potentially significant therapeutic target for inflammatory diseases. To this aim, the human hepatocyte iNOS cDNA was cloned into a baculovirus transfer vector, and recombinant virus was isolated for iNOS protein expression. Large-scale production of recombinant human hepatocyte iNOS (rH-iNOS) was optimized with respect to enzyme specific activity and yield. Native human iNOS (nH-iNOS) was obtained from isolated stimulated human hepatocytes. Cells expressing nH-iNOS and rH-iNOS were lysed in the presence of proteinase inhibitors and S-100 lysates prepared. The activity of iNOS was measured by an HPCL-based [^{3}H]-arginine to [^{3}H]-citrulline conversion assay.

Both rH-iNOS and native human iNOS migrated with an apparent M_r ~130 kDa on SDS-PAGE, and exhibitied similar Km's for arginine: 6.9±2.7 uM and 7.3±2.0 uM, respectively. Similar IC_{50}'s were obtained for N^G-Methyl-L-Arg, N^G-Nitro-L-Arg, Aminoguanidine and Diphenylene iodium.

Both enzymes were isolated to ~70-80% purity by a combination of chromatographic methods: 2',5'-ADP-Sepharose eluted with 5 mM NADPH, DEAE with a 0-500 mM NaCl gradient, Sulfopropyl sephadex, TSK G-3000SWxl size exclusion, or antibody-affinity column chromatography employing an antipeptide-antibody generated against the carboxy-terminal 7 amino acids of human iNOS. Active enzyme eluted as a dimer with a M_r=260-290 kDa, with an inactive enzyme peak eluting at a M_r=130-150 kDa. Spectral analyses of active rH-iNOS displayed heme absorbance spectra similar to that reported for the constitutively expressed NOS enzymes, with peak absorption at 280 mM and 400 nM, with a shoulder at 450-500 nm as expected for flavin containing proteins. The specific activities of the purified rH-iNOS and nH-iNOS appeared similar, and ranged between 0.1 and 0.4 umoles/min/mg at 37°C.

Unlike murine iNOS, the activities of both rH-iNOS and nH-iNOS displayed partial EGTA-sensitivity at levels as low as 30 uM EGTA. This inhibition was completely reversed by the addition of exogenous calcium. Both murine and human iNOSs were partially inhibited by high calcium concentrations.

In summary, we have optimized the expression of active human iNOS in the baculovirus system and demonstrated the fidelity of rH-iNOS with respect to nH-iNOS by both physical and kinetic criteria.

Abbreviations used: iNOS, inducible nitric oxide synthase; rH-iNOS, recombinant human hepatocyte iNOS; nH-iNOS, native human iNOS; Sf9, *Spodoptera frugiperda* .

Cytokine regulation of hepatocyte nitric oxide synthase and its effects on cytochrome P450 proteins.

Ruth E. Billings, Dawn L. Duval, and Timothy J. Carlson

Dept. of Environmental Health, Colorado State University, Fort Collins, CO 80523, U.S.A.

NO is synthesized by three NOS isoforms. The neuronal and endothelial forms are constituitive. The third form is induced in a number of cell types, including hepatocytes, by a variety of agents including cytokines, LPS and ROI (1, for review). The effects of ROI and NO are linked in a number of ways. The reaction of NO with superoxide to generate peroxynitrite may be either protective or increase cytotoxicity depending upon cellular conditions. The promoter region of NOS contains NFκB and AP-1 sites which are activated by ROI. We reported that TNFα induces NOS in cultured rat hepatocytes by mechanisms involving ROI and glutathione (2). Here we further explore the role of ROI generation by cytokines as a critical step in the induction of NOS.

In these studies, NOS activity was assessed by measuring nitrite and nitrate in the culture medium and by measuring NOS protein on immunoblots of SDS-PAGE gels. NOS gene expression was measured by reverse transcriptase-polymerase chain reaction. Untreated hepatocytes exhibit a low level of NOS activity which was attributable to low levels of inducible NOS, rather than to constitutive isoforms. This was determined by using isoform-specific antibodies. NOS was induced by TNFα in a dose-dependent manner (0.06 - 60 nM). This dose dependency correlated with TNFα receptor binding to hepatocytes. There was a close correlation between the induction of NOS activity, inducible-NOS immunoreactive protein, and mRNA. Other cytokines, IL-1β and interferon-γ IFNγ, and LPS also induced NOS whereas IL-6 was not an effective inducer (Table 1). The rank order of cytokine potency to induce NOS activity was IL-1β>TNFα = LPS >IFNγ. NOS induction by all cytokines was inhibited by antioxidant treatment which consisted of 2.5 mM Trolox, 250 μM ascorbic acid, and 10 mM sodium benzoate.

The ability of cytokines to suppress xenobiotic metabolism in general and specifically CYP activity is well established (3). More recently, NO has been implicated in the suppression of CYP content and activity (4). In the present study, we examined the role of NO in the cytokine mediated down regulation of CYP protein. CYP content and several CYP isoform levels were assessed in hepatocytes treated with a combination of cytokines (TNFα, IFNγ, and IL-1β). The combination was found to depress CYP content by 69%. Protein levels of constitutive CYP forms 1A2, 2C11, 2B1/2, and 3A2 were assessed by immunoblots. Treatment with the cytokine combination resulted in a decrease of each isoform (Table 2). The cytokine combination also induced NOS, as seen by the nitrite/nitrate values given in Table 2. Inclusion of inhibitors of NOS in the incubations significantly prevented the cytokine mediated decrease in each CYP isoform, indicating a role for NO in the down regulation of CYP. Additional support for a role of NO in the down regulation was observed by treating hepatocytes with DETA/NONOate. This NO donor caused a decrease in each CYP isoform level. CYP2B1/2 was decreased to the greatest extent by the NO donor, with 300 μM DETA/NONOate yielding CYP protein levels of 34 ± 8% of untreated cells.

The results of these studies show that NOS activity in hepatocytes is regulated primarily at the level of transcription, and expression of the NOS gene depends, in part upon the generation of oxygen radicals. An important outcome of the induction of NO in hepatocytes is the down regulation of CYP protein levels. Because CYP enzymes are responsible for the bioactivation and/or detoxification and clearance of many drugs and toxicants, the induction of NO and the NO-mediated regulation of CYP enzymes may play a critical role in the pharmacological and toxicological effects of many compounds.

Table 1. NOS induction is inhibited by antioxidants

Treatment (24 hr samples)	iNOS Immunoreactive Protein Densitometry Units/15 μg protein	+ Antioxidants
Control	1.52	0
TNFα (1μg/ml)	2.71	0
IFNγ (100 U/ml)	2.35	1.54
IL-1β (50 U/ml)	5.33	1.52
LPS (10 μg/ml)	2.61	0.86
IL-6	0.44	N.D.
TNFα + IFNγ	3.84	0.78
TNFα, IFNγ, IL-1β, LPS, IL-6	10.82	3.09

Abbreviations used: NO, nitric oxide; NOS, nitric oxide synthase; ROI, reactive oxygen intermediates; TNFα, tumor necrosis factor-α; IL-1β, interleukin-1β; IL-6, interleukin-6; LPS, lipopolysaccharide; IFNγ, interferon-γ; CYP, cytochrome P450; NMA, N^G-monomethy-L-arginine; NAME, N^{ω}-nitro-L-arginine methyl ester; DETA/NONOate, 1-hydroxy-2-oxo-3,3-bis(aminoethyl)-1-triazene).

Table 2. CYP levels in response to cytokine treatment ± NOS inhibitors.

Each CYP value represents percentage of the CYP isoform in untreated cells. The values reported are mean ±standard deviation from 3 separate experiments. Nitrite/nitrate concentration was 249±22 μM in media from cytokine-treated samples, and was reduced to 29±8 μM and 41±12 μM in samples which contained of NMA and NAME, respectively.

CYP isoform	Cyt[a]	Cyt+NMA	Cyt+NAME
CYP1A2	43±11	78±5	83±7
CYP2B1/2	33±9	96±6	99±8
CYP2C11	59±7	85±8	79±9
CYP3A2	62±4	98±11	96±10

[a]Cyt = TNFα, 0.1μg/mL; IL-1β, 100 units/mL; and IFN-γ, 100 units/mL

Acknowledgements:
This work was supported by NIH grant DK44755.

1. Kröncke,K.-D.; Fehsel, K.; and Kolb-Bachofen,V. (1995) Biol.Chem. Hoppe-Seyler, **376**:327-343.
2. Duval,D.L.; Sieg, D.J., and Billings, R.E.(1995) Arch. Biochem. Biophys. **316**:699-706.
3. Andus, T.; Bauer, J., and Gerok, W. (1991) Hepatology **13**:364-375.
4. Khatsenko, O.G., Gross, S.S., Rifkind, A.B., and Vane, J.R. (1993) Proc. Natl. Acad. Sci. USA **90**:11147-11151.

A novel system for expressing high levels of nitric oxide synthase in *Esherichia coli*

LINDA J. ROMAN, ESSAM A. SHETA, PAVEL MARTASEK, STEVEN S. GROSS*, QING LIU*, and BETTIE SUE S. MASTERS

Dept. of Biochemistry, The Univ. of Texas Health Sci. Ctr. at San Antonio, TX 78284-7760, *Dept. of Pharmacology, Cornell Univ. Medical College, New York, NY 10021.

The study of full length neuronal nitric oxide synthase (nNOS) has been limited by the small amounts of active enzyme which could be isolated from cerebellar tissue, cell culture of human kidney 293 cells, or baculovirus systems. A quick, easy, and inexpensive expression system for nNOS has been developed in *Escherichia coli* which produces 20-24 mg per liter of cells of enzyme that is replete with heme and flavins [1,2].

The expression of nNOS is under the direction of the tac promoter in the pCW vector, a vector which has been widely used to express many members of the heme-containing cytochrome P450 family [3]. This vector was chosen in an attempt to promote proper insertion of the heme moiety into nNOS. The nNOSpCW plasmid, containing the nNOS cDNA, was transformed into *E. coli* protease-minus BL21 cells along with pGroELS, which directs the expression of the *E. coli* chaperonins groEL and groES [4]. The cells were grown at 37 °C to an absorbance at 600 nm of 1.2, and protein expression was induced with 0.5 mM isopropyl β-D-thiogalactoside. Riboflavin (a flavin precursor), δ-aminolevulinic acid (a heme precursor), and ATP (a cofactor for groEL and groES activity) were also added. The cells were grown at room temperature for approximately 40 hours and nNOS purified as described [1].

Absolute spectra of purified, E. coli-expressed nNOS are shown in Figure 1. As isolated, the heme is primarily in the high-spin state, as evidenced by the peak at 400; the shoulder at 410 indicates the presence of some low-spin form. Upon addition of arginine, the enzyme is converted entirely to the high-spin state, while addition of imidazole causes conversion to the low-spin form (Figure 1). Spectrally, the nNOS enzyme produced by *E. coli* is indistinguishable from that produced by human kidney 293 cells.

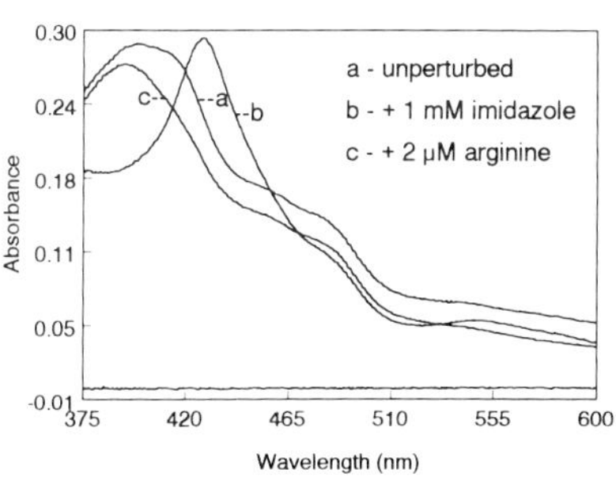

Figure 1. Absolute spectra of nNOS. Experiments contain 3.2 μM enzyme.

As isolated, the enzyme is an excellent source of pterin-deficient enzyme, as only 10% of the NOS contains tetrahydrobiopterin (BH_4), but it can be reconstituted to 70% repletion. The activities of tetrahydrobiopterin-free and reconstituted enzyme are shown in Table 1. The reconstituted enzyme is completely active, with a turnover number for L-arginine of 435 nmol/min/mg, and binds N^{ω}-nitro-L-arginine (L-NNA) with a K_d of 37 nM. The pterin-deplete enzyme binds L-NNA poorly in the absence of BH_4, and exhibits poor activity, but both are enhanced by the presence of exogenous BH_4.

Table 1

Binding of L-NNA and Enzymatic Activities of BH_4-free and BH_4-replete Preparations

	K_D (nM)	B_{max} (pmol NNA/pmol nNOS)	Activity* (nmol/min/mg)
BH_4-free			
(+) BH_4	41	0.54	202
(--) BH_4	483	0.078	75
BH_4-replete			
(+) BH_4	37	0.69	435
(--) BH_4	58	0.67	189

* as measured by conversion of [^{3}H] L-arginine to [^{3}H] L-citrulline

The ability to produce large amounts of active enzyme will greatly facilitate the study of structure/function relationships in NOS, which, in turn, will aid in drug design and development. This system will also be useful for the screening of site-directed mutants, which may give further insight into the mechanism of NO• production. Other isoforms of NOS, including those from different species, are expected to also be expressible in *E. coli*, although some modification of the system may be required.

ACKNOWLEDGEMENTS: This work was supported, in part, by Grant AQ-1192 from The Robert A. Welch Foundation and National Institutes of Health Grant HL 30050 (to B.S.S.M.) and National Institutes of Health Grant HL 50656 and HL 44603 (to S.S.G.). L.J.R. is the recipient of a Parker B. Francis Fellowship in Pulmonary Research.

REFERENCES:

1) Roman, L.J., Sheta, E.A., Martasek, P., Gross, S.S., Liu, Q., and Masters, B.S.S. (1995) Proc. Natl. Acad. Sci. USA **92**, 8428-8432.
2) Gerber, N.C. and Ortiz de Montellano, P.R. (1995) J. Biol. Chem. **270**, 17791-17796.
3) Barnes, H.J., Arlotto, M.P. and Waterman, M.R. (1991) Proc. Natl. Acad. Sci. USA **88**, 5597-5601.
4) Goloubinoff, P., Gatenby, A.A. and Lorimer, G.H. (1989) Nature (London) 337, 44-47.

COMPARISON OF THE CALCIUM AND CALMODULIN REQUIREMENTS OF THE THREE HUMAN NITRIC OXIDE SYNTHASE ISOFORMS EXPRESSED IN BACULOVIRUS-INFECTED SF9 CELLS

GLORIA C. WOLFE, KAREN L. MacNAUL, M. RAJU SAYYAPARAJU, ERMENEGILDA McCAULEY, JEFFREY R. WEIDNER, RICHARD A. MUMFORD, JOHN A. SCHMIDT and NANCY I. HUTCHINSON.

Department of Inflammation Research, Merck Research Laboratories, Rahway, NJ 07065 USA

Nitric oxide (NO) plays a variety of roles in vasodilation, neurotransmission and inflammation. NO is synthesized from L-arginine by three distinct isoforms of nitric oxide synthase (NOS). Two isoforms, ec-cNOS and n-cNOS, were initially characterized in endothelial cells and brain, respectively; and the third isoform, iNOS, can be induced in variety of cell types by inflammatory stimuli. The biochemical characterization of each NOS isozyme, derived from distinct cell sources from a number of species, have been performed. However, the only species from which full-length cDNAs of all three isoforms have been cloned is human. Expression of each of the three human NOS isoforms under identical conditions, in the same cell type, would provide a simplified system in which to compare their biochemical properties, as well as permit the screening of isozyme-specific inhibitors for potential human drug therapies.

Methods:

Complementary DNAs (cDNA) encoding human iNOS and human ec-cNOS were obtained from T. Billiar (University of Pittsburgh) and from K. Bloch (Harvard University). Complementary DNA (cDNA) for human n-cNOS was obtained by polymerase chain reaction amplification of reverse transcribed human cerebellum mRNA. The DNA sequences of the amplified fragments were confirmed by automated sequencing using an Applied Biosystems DNA sequencer. The cDNA fragments were joined by polymerase chain reaction and restriction fragment cloning to obtain the full-length (4.5 Kb) human n-cNOS cDNA.

Each human NOS cDNA was cloned into a baculovirus expression vector and co-transfected with BaculoGold DNA into Sf9.10 insect cells to obtain recombinant virus. Comparative time courses of the recombinant expression of each NOS isoform in the baculovirus system were performed at a multiplicity of infection of 5, in the presence of 8 uM δ-aminolevulinic acid. Infections were allowed to progress for 24-102 hours before harvesting. The cells were homogenized and subjected to ultracentrifugation at 100,000g for 30 minutes. The crude supernatant was passed through a Bio-Gel P-30 spin column to remove endogenous arginine, before enzyme activity was assayed. The proteins were analyzed by SDS-PAGE and immunoblotting with isoform-specific anti-NOS antibodies, as well as a common epitope antipeptide antibody which recognizes all NOS isoforms. Enzyme activity was determined in an HPLC-based [^{3}H]-L-Arginine to [^{3}H]-Citrulline conversion assay.

The three enzymes were partially purified by 2',5'-ADP-sepharose and gel filtration chromatography. The effects of exogenous Ca^{2+} [0.02-2 mM], calmodulin [CAM, 2 ug/ml], or EGTA [1 mM] on both crude and partially-purified enzymes were determined by changes in enzymatic activity.

Results

Active recombinant human iNOS, ec-cNOS, and n-cNOS were expressed in the baculovirus system. The yields of NOS activity ranged between 0.5-1.5 nmoles citrulline/min/mg of total protein.

None of the human NOS enzyme preparations required exogenous Ca^{2+} for full activity. High concentrations of Ca^{2+} had a slight inhibitory effect on iNOS and n-cNOS, but not ec-cNOS. The activity of ec-cNOS was dependent upon the addition of exogenous CAM for full activity in both the crude and partially-purified enzyme preparations, but was not required for full iNOS activity under any conditions examined. Whereas, for n-cNOS, CAM significantly enhanced enzyme activity in the partially-purified enzyme preparation, but not in the crude, spin-filtered lysate. As expected, ec-cNOS and n-cNOS were completely inhibited by EGTA; whereas iNOS was only partially inhibited. The EGTA-inhibition could be reversed by the addition of Ca^{2+}.

Conclusions

These results suggest that human ec-cNOS and n-cNOS differ in their affinity for binding calmodulin. Specifically, they suggest that n-cNOS is able to remain associated with calmodulin during spin-filtration of the crude lysates, but that that association is lost during more extensive purification procedures, or upon treatment with EGTA. In contrast, ec-cNOS requires the addition of exogenous calmodulin for full-activity in either spin-filtered crude lysate or partially-purified enzyme preparations. The association of iNOS with calmodulin was maintained throughout protein purification. These subtle distinctions in calcium/calmodulin dependence between human NOS isoforms may have implication as to the regulation of each NOS isozyme *in vivo*.

Abbreviations used: iNOS, inducible nitric oxide synthase; ec-cNOS, endothelial cell constitutive nitric oxide synthase, and n-cNOS, neuronal nitric oxide synthase: NOS, types 1, 2, and 3, respectively; cDNA, complementary DNA; CAM calmodulin; Sf9, *Spodoptera frugiperda* .

Enzyme-bound heme is required for pteridine binding to rat brain nitric oxide synthase

BARBARA M. LIST, PETER KLATT, KURT SCHMIDT, ERNST R. WERNER*, AND BERND MAYER

Institut für Pharmakologie und Toxikologie, Karl-Franzens-Universität Graz, Universitätsplatz 2, A-8010 Graz, Austria.
*Institut für Medizinische Chemie und Biochemie, Universität Innsbruck, Fritz-Pregl-Straße 3, A-6020 Innsbruck, Austria.

NO synthases (NOS) convert L-arginine to NO and L-citrulline. The reaction involves flavin-mediated shuttle of electrons from NADPH to a P450-type heme-iron which catalyzes reductive activation of molecular oxygen and substrate oxidation [1,2]. In addition to flavins and heme, NOSs contain tightly bound tetrahydrobiopterin (H_4biopterin), but the role of the pteridine in NO biosynthesis is not well understood [3]. It may act as both reactant and allosteric effector of NOS [4-7] and appears to be essential for the coupling of O_2 activation to L-arginine oxidation catalyzed by the neuronal enzyme [8,9].

Recently we have developed a method for expression of recombinant rat brain NOS in baculovirus-infected Sf9 cells [10]. This previous study showed that supplementation of the Sf9 cell culture medium with hemin is essential for expression of catalytically active NOS. In the present study, NOS was expressed in the absence of added hemin in order to investigate the role of the prosthetic heme group in enzyme structure and function. Table 1 shows the amount of heme, flavins, and H_4biopterin bound to NOS expressed in the absence and presence of 4 µg/ml hemin chloride. The hemin

Additions	heme	FAD	FMN	H_4biopterin
	(mol of bound cofactor per mol of subunit)			
none	0.37±0.08	0.39±0.02	0.40±0.03	0.22±0.04
hemin	0.95±0.09	0.40±0.09	0.46±0.11	0.47±0.06

Tab.1: Cofactor content of recombinant rat brain NOS expressed in the absence and presence of 4 µg/ml hemin chloride.

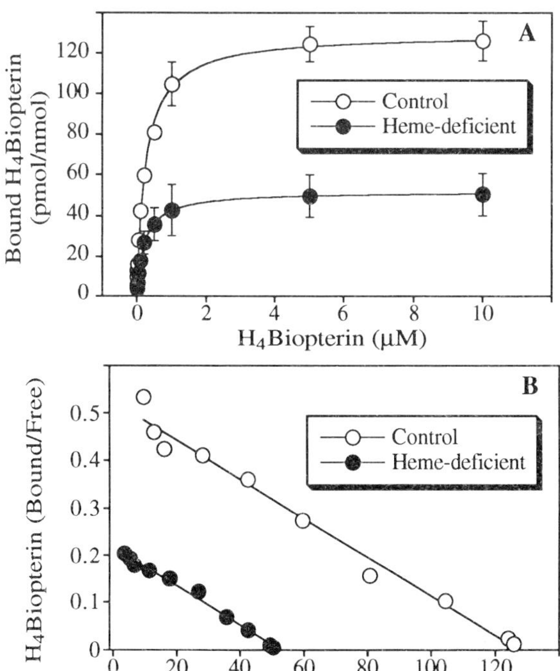

Figure 1: Binding of [^{3}H]H_4biopterin to purified rat brain NOS
A: Saturation binding (mean values ± S.E., n=3).
B: Scatchard Plot of the mean values.

Abbreviations used: NOS, nitric oxide synthase; H_4biopterin, (6R)-5,6,7,8-tetrahydro-L-biopterin; DAHP, 2,4-diamino-6-hydroxypyrimidine.

treatment of the cells increased the amount of NOS-bound heme from 0.37 to 0.95 equiv per monomer but had no effect on the flavin content. Surprisingly, however, the heme-deficient enzyme preparations appeared to contain significantly less H_4biopterin than the holoenzyme (0.22 vs 0.47 equiv per monomer). This finding indicated that the prosthetic heme group is essential for appropriate pteridine binding to NOS. To further investigate this issue, we have performed binding studies using ^{3}H-labelled H_4biopterin as radioligand [5,11]. Figure 1A shows that the heme-deficient NOS bound considerably smaller amounts of H_4biopterin than the holoenzyme, whereas both preparations exhibited similar affinities for the pteridine, as evident from the virtually identical slopes of the Scatchard plots shown in Fig. 1B. From individual curves we have calculated B_{max} values of 0.129 ± 0.004 and 0.052 ± 0.011 (mean ± S.E.; n=3) equiv of H_4biopterin bound to heme-saturated and heme-deficient NOS, respectively. The value obtained with the holoenzyme is similar to that determined previously with porcine brain NOS, and the reduction of B_{max} to 40 % of controls is in good accordance with the determined heme content of 0.37 equiv per monomer (see Tab. 1). The calculated K_D values were 0.26 ± 0.013 and 0.22 ± 0.042 µM (mean ± S.E.; n=3) for heme-saturated and heme-deficient NOS, respectively.

Considering the positive cooperativity of substrate and H_4biopterin binding [5], we have expressed pteridine-deficient NOS in the presence of hemin and 10 mM 2,4-diamino-6-hydroxypyrimidine (DAHP), an inhibitor of GTP cyclohydrolase I, the key enzyme in H_4biopterin biosynthesis [12]. If expressed under these conditions, NOS contained less than 0.1 equiv H_4biopterin per monomer, but its heme content was not significantly different from that of controls (0.91 eqiv per monomer). Moreover, the pteridine-deficient enzyme appeared to be functionally intact, since its activity was fully restored upon addition of exogenous H_4biopterin.

The present data indicate that heme-deficiency of rat brain NOS results in impaired H_4biopterin binding, whereas the pteridine is apparently not required for appropriate incorporation of the heme. Together with our previous results showing that the affinity of porcine brain NOS for H_4biopterin was markedly increased in the presence of L-arginine [5], the present study suggests that pteridine binding to neuronal NOS is tightly regulated by an allosteric interaction of the substrate site and the prosthetic heme group of the enzyme. Macrophage NOS dissociates into inactive, heme-free monomers in the absence of H_4biopterin, and reassociation to active homodimers requires the simultaneous presence of hemin, L-arginine, and H_4biopterin [4]. Thus, the synergism between L-arginine and heme to create a high affinity pteridine binding site may not be confined to the brain enzyme but represent a general mechanism to control and regulate the assembly of NOS subunits to the active dimeric proteins.

Supported by the Fonds zur Förderung der Wissenschaftlichen Forschung in Austria.

1. Mayer, B. (1995) in Nitric Oxide in the Nervous System (Vincent, S.R., ed.), pp. 21-42, Academic Press, New York
2. Griffith, O.W. & Stuehr, D.J. (1995) Annu. Rev. Physiol. **57**, 707-736
3. Mayer, B. & Werner, E.R. (1995) Naunyn-Schmiedeberg's Arch. Pharmacol. **351**, 453-463
4. Baek, K.J., Thiel, B.A., Lucas, S. & Stuehr, D.J. (1993) J. Biol. Chem. **268**, 21120-21129
5. Klatt, P., Schmid, M., Leopold, E., Schmidt, K., Werner, E.R. & Mayer, B. (1994) J. Biol. Chem. **269**, 13861-13866
6. Klatt, P., Schmidt, K., Lehner, D., Glatter, O., Bächinger, H.P. & Mayer, B. (1995) EMBO J. **14**, 3687-3695
7. Wang, J.L., Stuehr, D.J. & Rousseau, D.L. (1995) Biochemistry **34**, 7080-7087
8. Mayer, B., Heinzel, B., Klatt, P., John, M., Schmidt, K. & Böhme, E. (1992) J. Cardiovasc. Pharmacol. **20**, S54-S56
9. Klatt, P., Schmidt, K., Uray, G. & Mayer, B. (1993) J. Biol. Chem. **268**, 14781-14787
10. Harteneck, C., Klatt, P., Schmidt, K. & Mayer, B. (1994) Biochem. J. **304**, 683-686
11. Werner, E.R., Schmid, M., Werner-Felmayer, G., Mayer, B. & Wachter, H. (1994) Biochem. J. **304**, 189-193
12. Werner-Felmayer, G., Werner, E.R., Fuchs, D., Hausen, A., Reibnegger, G. & Wachter, H. (1990) J. Exp. Med. **172**, 1599-1607

Estrogen increases endothelial and neuronal nitric oxide by two mechanisms

Toshio Hayashi,# Kazuyoshi Yamada, Teiji Esaki, Emiko Mutoh and Akihisa Iguchi

Department of Geriatrics and Department of Pharmacology, Nagoya University School of Medicine, 65 Tsuruma-cho, Showa-ku, Nagoya, 466, Japan*

SUMMARY: To determine the mechanism of the antiatherosclerotic effect of estrogen, we investigated the effect of estrogen on endothelial and neuronal nitric oxide synthase (NOS-3 and NOS-1). Preincubation with a physiologic concentration of 17β-estradiol (10^{-12}-10^{-8} M) over 8 hours significantly enhanced the activity of NOS-3 in endothelial cells of cultured human umblical vein (HUVEC) and of bovine aortas (BAEC). 17β-estradiol also enhanced the release of nitric oxide (NO) as measured by an NO selective meter and $NO2^-/NO3^-$; metabolites of NO, from endothelial cells. Western blott showed a similar effect. Estrogen increases NOS-3 via a receptor-mediated system. Low concentration of 17β-estradiol (10^{-10-8}M) enhanced the activity of crude NOS-1 in cytosol fraction of rabbit cerebellums. Partially purified NOS-1 obtained from cytosol fraction by DEAE colummchromatography, showed same tendency to the activity by estrogen, and W7 or calmidazolium inhibited 90% in control as well as in any dose of estrogen. Estrogen enhanced the fluorescence of dansyl-calmodulin in low dose, augmented that in high dose.

INTRODUCTION

Estrogen retards the development of atherosclerosis. Estrogen replacement therapy suppresses the incidence of cardiovascular disease in postmenopausal women , and it reduces the plasma level of LDL-cholesterol and increases that of HDL-cholesterol. However this change in lipid profile reportedly accounts for only 25% of the profective effect. Estrogen is thought to exert a direct action on the vessel wall via an incompletely understood mechanism. NO is believed to protect against the development of atherosclerosis variously by producing vascular dilatation and inhibition of monocyte adhesion to endothelium. Estrogen may acts via an NO-mediated system. This information led us to investigate the interaction between estrogen and NOS too elucidate the mechanism of estrogen effect.

MATERIALS and METHODS

Cultured cells: Human umblical vein endothelial cells (HUVEC) were taken from the primary culture using collagenase. Endothelial cells from Bovine aorta (BAEC) were taken from the thoracic aorta of a bovine fetus .

NOS activity assay: Enzymatic reactions were conducted at 37°C for 15 min. in 50mM Tris HCl (pH7.4) containing 50 µM ^{3}H-L-arginine, 100µM NADPH, 2mM CaCl2, 100µM calmodulin, 10µM tetrahydrobiopterin, and the other test agents indicated, in a final incubation volume of 100µl. The L-[2,3,4,5-^{3}H] arginine HCl was previously purified by anionic exchange chromatography on columns of Dowex AG 1-X8, OH-form.

Measurement of nitric oxide by NO electrode: The principle of the new NO-meter (Intermedic Co.,Ltd Nagoya, Japan) is developed newly is the measurement of the pA-order redox current between a working electrode and a counter electrode.

Measurement of $NO2^-/NO3^-$: The concentration of the nitrite and nitrate ($NO2^-/NO3^-$):metabolites of NO in the culture medium were determined with an autoanalyzer (TCI-NOx 1000: Tokyo Kasei Kogyo Co., Tokyo, Japan) as recently reported.

Western blotting: Cell pellets were sonicated, following the addition of 80µl of urea solution. The cell lysate was subjected to SDS-polyacrylamide gel electrophoresis. The membrane was blocked for 1 hour at room temperature and incubated for 2 h. with mouse monoclonal NOS-3 antibody (1:1000 dilution)(Vecstatin ABC Kit Vector Laboratories, USA Co , USA) .

RESULTS

#Effect of 17β-estradiol on endothelial NOS via receptor system

Preincubation with 17β-estradiol(10^{-12} - 10^{-8}M) more than 8 hours significantly enhanced the activity of NOS-3 in HUVEC and BAEC. A high dose (10^{-7}- 10^{-6} M) of 17β-estradiol tended to reduce NOS-3 activity to the control level (Figure 1). The release of NO measured with the NO selective meter also showed the same effect . $NO2^-/NO3^-$ in the culture medium also showed an increase following incubation with physiological concentration of 17β-estradiol. Exposure to17β-estradiol for over 8 hours significantly enhanced NOS-3 activity of HUVEC and BAEC (less than 5th passage). Western blot analysis of HUVEC and BAEC using anti-monoclonal antibody to NOS-3 exhibited an effect similar to that of estrogen. The estrogen receptor antagonists, tamoxifen, and ICI182780, inhibited the effect of 17β-estradiol by 80% in NOS-3 of the endothelial cells. The effect of 17β-estradiol on the activity of NOS-3 decreased gradually when the culture passage exceeded 10. No significant effect was observed in cells that exceeded 16 passages. Immunocytochemistry showed the existence of an estrogen receptor in HUVEC and BAEC (less than 5 th passage). The marked weakness of the staining in BAEC (more than 16 th passage)(data not shown).

#Effect of 17β-estradiol nNOS through calmodulin

Low concentration of 17β-estradiol (10^{-10-8}M) enhanced the activity of crude NNOS in cytosol fraction of rabbit cerebellums, 10^{-9}M17β-estradiol increased the activity about 20%. High dose (10^{-5-4}M) 17β-estradiol augmented that about 50 % (Figure 2). W7 (100µM) or calmidazolium (0.3µM), calmodulin antagonists, inhibited those activities only about 30% in control, about 50% in low dose estrogen and less than 20% in high dose, so the activity after application of W7 or calmidazolium was almost same in any concentration of estrogen. Partially purified NNOS obtained from cytosol fraction by DEAE colummchromatography, showed same tendency to the activity by estrogen, and W7 or calmidazolium inhibited 90% in control as well as in any dose of estrogen. Estrogen enhanced the fluorescence of dansyl-calmodulin in low dose, augmented that in high dose. Tamoxifen, estrogen receptor antagonist did not affect the activity of NNOS in cerebellum, although slight decrease activity of partially purified NOS.

References

Hayashi, T, Ignarro, L. J., & Chaudhuri, G. et al.(1992). Proc Natl Acad Sci U S A, 89(23), 11259-63.

Hayashi, T, Iguchi, A et al. (1994) Biochem Biophys Res Commun, 203(2), 1013-9.

Hayashi, T, Iguchi A et al. (1995). Biochem Biophys Res Commun, in press

Hayashi T. Ignarro, L. J., & Chaudhuri, G. et al. (1995) J Cardiovascular Pharmacol. in press

Figure 1. The effect of 17β-estradiol on the activity of NOS-3 of BAEC (passage 4) *p< 0.05, **p<0.01 vs control

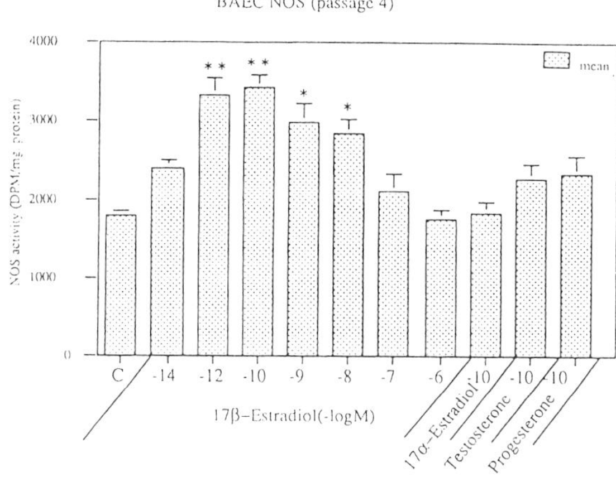

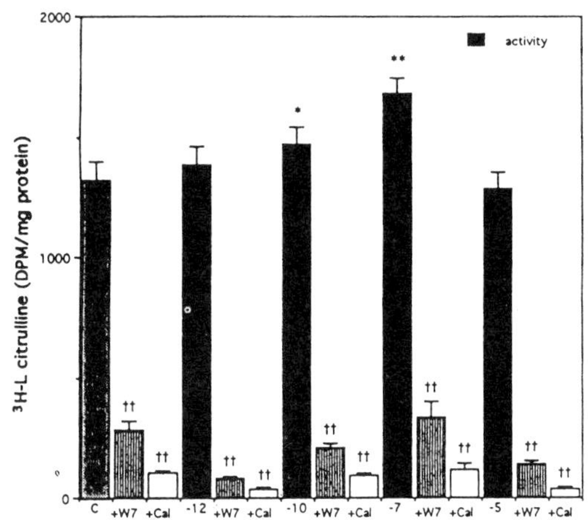

Figure 2. Inhibition of W7 (100 μM) or calmidazolium (0.3 μM) on the activity of partially purified NNOS in cytosol fraction of rabbit cerebellums with coincubation of 17β-estradiol. *p<0.05,**p<0.01 vs control(C), ++p<0.01 vs NNOS activity under same condition without calmodulin antagonist.

Regulation of endothelial nitric oxide synthase in bovine aortic endothelial cells by estradiol.

RAMESH. C. BHALLA, KAREN. F. TOTH, ROBERT. A. BHATTY, and RAM. V. SHARMA.

Department of Anatomy, University of Iowa, Iowa City, IA 52242.

Epidemiological evidence and estrogen replacement studies suggest that estrogen has a protective effect on the cardiovascular system against coronary artery diseases and also increases coronary blood flow. However, the mechanism of estrogen action on vascular cells is not understood. The presence of estrogen receptors in vascular smooth muscle cells derived from human coronary artery [1] and in endothelial cells [2] has been demonstrated. In this study, we investigated the effect of 17-β-estradiol (E_2) treatment on guinea pig coronary artery smooth muscle (VSM) cell proliferation and cytosolic calcium concentration ($[Ca^{2+}]_i$) in response to bradykinin stimulation. Further, the hypothesis that E_2 modulates NOS activity was tested in cultured bovine aortic endothelial cells (BAEC).

Materials and Methods

Endothelial cells were cultured from bovine aorta and were used between 4-8 passages. Coronary artery smooth muscle cells were cultured as described previously [3] and were used between 4-8 passages. The cultured cells were primed with E_2 (10 pg/ml) for 24-48 hr followed by treatment with E_2 (50 pg/ml) for 4-5 days. The cells were then treated with 10 pg/ml E_2 for 48 hr before $[Ca^{2+}]_i$ measurement to sensitize the E2 receptors [4]. $[Ca^{2+}]_i$ was estimated using a microscopic digital image analysis system (P T I, New Jersey) and Fura-2 as the fluorescent dye. Cells were counted using a coulter cell counter. NOS activity was measured in the membrane fraction by the conversion of $[^3H]$-arginine to $[^3H]$-citrulline. The expression of E_2 receptors in cultured BAEC and VSM cells was examined by ribonuclease protection assay using an RNA probe complementary to the hormone binding domain (obtained from Dr. Richard Karas, Tufts Univ. School of Med., Boston, Mass.) [5].

Results and Discussion

It has been recently shown that both E_2 therapy and pregnancy increase mRNA levels of NOS in heart and skeletal muscle [6]. Yet controversy exists regarding the effect of E_2 on cultured endothelial cells. We hypothesized that the culturing of endothelial and VSM cells decreases E_2 receptors that may explain the inconsistent effects of E_2 in in-vitro studies. Since steroid hormones are known to upregulate their receptors, we examined the in-vitro effects of pulsatile treatment with E_2 by culturing BAEC and VSM cells with a low concentration of E_2 (10 pg/ml, typical of nonpregnant animals) to upregulate the E_2 receptor expression in cultured cells. The expression of E_2 receptors in cultured BAEC and VSM cells was confirmed by ribonuclease protection assay. We observed that prolonged E_2-treatment of BAEC did not increase the NOS activity. Further, addition of E_2 directly in the citrulline assay had no effect on the NOS activity.

On the other hand, VSM cell proliferation 3 days after E_2 treatment was significantly lower ($P< 0.05$) compared with the control group (Fig. 1A). Similarly, E_2-treatment resulted in a significant decrease ($P< 0.05$) in the number of cells per well after six days and nine days of E_2 treatment. These data suggest that E_2 may produce the atheroprotective effect on myointimal thickening by inhibiting VSM cell proliferation in guinea pig coronary artery. In support of this contention it has recently been demonstrated that E_2 treatment of rabbits undergoing balloon injury of aorta and iliac artery significantly inhibits the myointimal thickening [7]. We also tested the hypothesis that E_2 treatment attenuates agonist-stimulated $[Ca^{2+}]_i$ in VSM cells, since E_2 treatment has been shown to produce relaxation of pre-contracted isolated rabbit coronary artery preparations [4]. In our studies E_2-treatment of guinea pig coronary artery smooth muscle cells resulted in a significant ($P < 0.05$) decrease in bradykinin (10 nM)-stimulated $[Ca^{2+}]_i$ compared with untreated cells (Fig. 1B). Our data suggest that a decrease in agonist-stimulated $[Ca^{2+}]_i$ in VSM cells may play an important role in E_2 mediated relaxation of coronary artery. Similarly, Han et.al. have shown that E_2 inhibited U46619 and K^+ -stimulated increase in $[Ca^{2+}]_i$ and contraction of isolated porcine coronary arteries [8].

In conclusion, our data suggest that estrogen may produce the atheroprotective effects by two mechanisms. First, it may acutely protect the heart from ischemia by reducing coronary artery reactivity to vasospastic hormones by decreasing agonist-stimulated $[Ca^{2+}]_i$ in VSM cells. Second, estrogen may provide a long term protection by decreasing VSM cell proliferation to atherogenic stimuli.

Acknowledgments. This work was supported by USPHS grant HL-51735.

References

1.Losordo, D.W., Kearney, M., Kim, E.A., Jekanowski, J. and Isner, J.M. (1994) Variable expression of an estrogen receptor in normal and atherosclerotic coronary arteries of premenopausal women. Circulation. 89:1501-1510.

2.Colburn, P., Buonassisi, V. (1978) Estrogen binding sites in endothelial cell cultures. Science. 201: 817-819.

3.Bhalla, R.C., Sharma, R.V. (1993) Induction of c-fos and elastin gene in response to mechanical stretch of vascular smooth muscle cells. J. Vas. Med. Biol. 4:1 30-137.

4.Collins, P., Shay, J., Jiang, C., Moss, J. (1994) Nitric oxide accounts for dose-dependent estrogen-mediated coronary relaxation after acute estrogen withdrawal. Circulation . 90: 1964-1968.

5.Karas, R.H., Patterson, B.L., Mendelsohn, M.E. (1994) Human vascular smooth muscle cells contain functional estrogen receptor. Circulation. 89: 1943-1950.

6.Weiner, C.P., Lizasoain, I., Baylis, S.A., Knowles, R.G., Charles, I.G., Moncada, S. (1994) Induction of calcium-dependent nitric oxide synthases by sex hormones. Proc. Natl. Acad. Sci. (USA) 91: 5212-5216.

7.Foegh, M.L., Asotra, S., Howell, M.H., Ramwell, P.W (1994) Estradiol inhibition of arterial neointimal hyperplasia after balloon injury. J. Vasc. Surg. 19: 722-726.

8.Han, S-H., Karaki, H., Ouchi, Y., Akishita, M., Orimo, H. (1995) 17 β-estradiol inhibits Ca^{2+} influx and Ca^{2+} release induced by thromboxane A_2 in Porcine coronary artery Circulation. 91: 2619-2626.

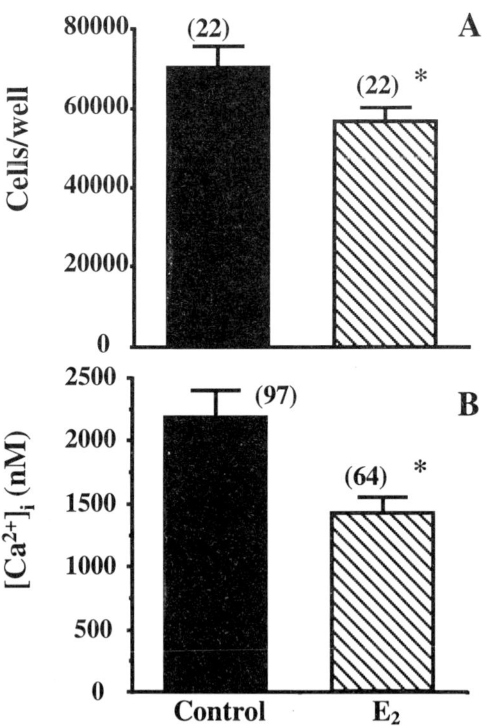

Fig.1. Effects of 17 β-estradiol (E_2) treatment on (A) proliferation and(B) Bradykinin -stimulated cytosolic calcium concentration ($[Ca^{2+}]_i$) in guinea pig coronary artery smooth muscle cells. Data given are the means ± SEM of n given in parenthesis; * denotes significant ($P<0.05$) difference between E_2 treated and control cells.

Sexual differences in the kinetics of endotoxin-stimulated increases of inducible nitric oxide synthase mRNA in both alveolar macrophages and recruited neutrophils, *in vivo*

JIANMING XIE, XINFANG ZHAO, DAVID R POWERS, THOMAS NOLAN, THOMAS D. GILES and STAN S. GREENBERG.

DEPARTMENTS OF OBSTETRICS AND GYNECOLOGY, MEDICINE and PHYSIOLOGY, LOUISIANA STATE UNIVERSITY MEDICAL CENTER, NEW ORLEANS, LOUISIANA, 70112, U.S.A.

Estrogenic hormones have been suggested to up-regulate the constitutive forms of nitric oxide synthase (cNOS) in blood vessels and other organs [1-4]. Pregnancy is also associated with both maternal and fetal up-regulation of nitric oxide (NO) synthesis [5-7]. Nitric oxide synthase (NOS) activity increases early in pregnancy and decreases prior to parturition [8,9] and may be an important modulator of placental blood flow and nutrient and oxygen exchange in the mother and fetus [5,6]. Despite the finding that estrogenic hormones may modulate the activity of constitutive NOS, a paucity of studies exist on the role of sex hormone on the cytokine and endotoxin-inducible forms of NOS (iNOS).
A preliminary communication from our laboratory demonstrated that Escherichia coli endotoxin (LPS) given to female rats up-regulated alveolar macrophage (AM) inducible nitric oxide synthase (iNOS) mRNA more than in AM from male rats, *in vivo*, two hrs. after LPS administration. This study evaluated the time course of the sexual differences in iNOS mRNA and the effect of the estrous cycle on LPS-stimulated iNOS mRNA in AM and recruited neutrophils (PMN), *in vivo*.

Methods and Materials: Male and female Sprague Dawley rats 225-350 g) were given intratracheal (i.t.) sterile saline (PBS) or LPS (1 mg/kg), 1, 2, 4, 6, 12 and 24 hr before experimentation. In addition, female rats, in which the day of estrous was determined with a vaginal smear, and an equivalent number of age matched rats, were also given i.t. LPS (1 mg/kg) four hr prior to sacrifice. The rats were anesthetized variable times after administration of LPS or sterile saline (PBS) with ketamine-xylazine. The lungs were removed and subjected to bronchoalveolar lavage (BAL). Aliquots of the BAL fluid were assayed for total nitrite and nitrate (RNI) with chemiluminescence and TNFα with an ELISA. The macrophages were isolated from the recruited PMN by FICOLL-Hypaque density gradient centrifugation. The AM or PMN (1,000,000 cells/ml) were isolated and incubated *ex vivo* for 1 hr and the ex vivo spontaneous production of RNI and TNF were measured. An aliquot of AM or PMN (1,000,000 cells) were frozen and assayed for mRNA for iNOS and TNFα with competitor equalized RT-PCR [10]. Data were analyzed with ANOVA. A value of $p<0.05$ was accepted for statistical significance of mean differences.

Results and Comments: Plasma RNI was higher in the female than in the male. Intratracheal administration of LPS into the lung increased plasma RNI at 6 and 12 hr after LPS administration. At 12 hr after i.t. LPS administration plasma RNI was increased by 61% in the male and 94% in the female rat ($P<0.05$). Intratracheal administration of sterile saline (PBS) to male and female rats did not affect either the production of RNI in BAL fluid or cellular incubate at any time period studied. Moreover, PBS did not affect LPS-induced increases in either BAL fluid or cellular incubate RNI or iNOS gene expression by AM or recruited PMN. The rate of expression of iNOS mRNA was greater in AM (Fig. 1) and recruited PMN obtained from female rats at 1, 2 and 4 hrs [12] after intratracheal administration of LPS when compared to that generated in macrophages obtained from male rats over the same time period. By 6 hr after LPS administration iNOS mRNA began to decline while reaching its peak value in the macrophages obtained from male rats (Fig. 1). Alveolar macrophage (Fig. 1) and PMN iNOS mRNA levels declined equally rapidly in macrophages (Fig. 1) and PMN obtained from male and female rats at 12 and 24 hr after the single dose of LPS (Fig. 1). No differences were observed in LPS-stimulated iNOS mRNA during the estrous cycle. Alveolar macrophage derived TNFα mRNA did not exhibit the same sexual dependence as did that of iNOS mRNA. Thus, the rate of induction of iNOS mRNA is enhanced in the phagocytic cells obtained from LPS-treated female rats when compared to that obtained from male

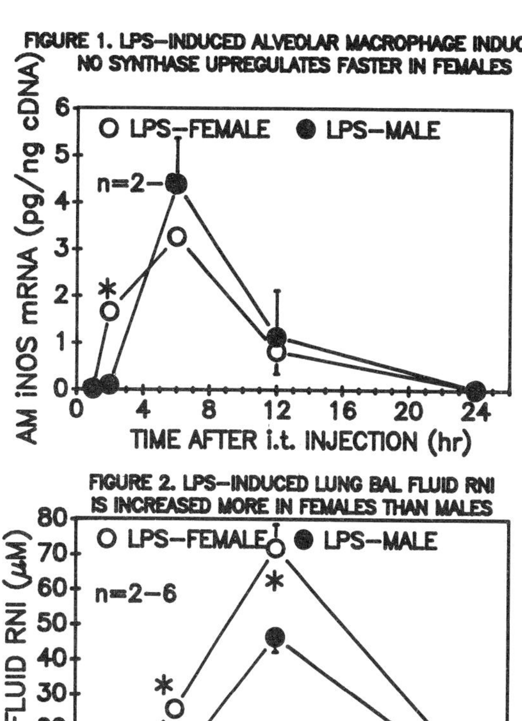

rats. The gender difference was amplified when RNI production, reflecting the enzymatic activity of iNOS, was measured. At each time period studied the BAL fluid (Fig. 2) and cellular generation of RNI was greater when obtained from the female rat than when obtained from the male rat. However, gender differences in the levels of bioactive TNFα in the BAL fluid and incubates of AM were not evident. Finally, no difference in BAL fluid RNI levels could be detected at any time period during the estrous cycle. We conclude that gender differences exist in LPS-induced gene expression for iNOS. The increased up regulation of iNOS mRNA by LPS in female macrophages and PMNs probably occurs at the level of transcription not degradation and possibly at the level of translation or enzyme activity. Moreover, the effects of LPS appear to be independent of the estrous cycle.

This research was supported by USPHS grants AA 09816 and AA09803 from the National Institute on Alcohol and Alcohol Abuse

References

1. Chaves, M.C., Ribeiro, R.A., and Rao, V.S. (1993) Braz. J. Med. Biol. Res. 26, 853-857.
2. Van Buren, G.A., Yang, D.S., and Clark, K.E. (1992) Am. J. Obstet. Gynecol. 167, 828-833
3. Sisson, J.H. (1995) Am. J. Physiol. 268, L596-L600
4. Brosnihan, K.B., Moriguchi, A., Nakamoto, H., Dean, R.H., Ganten, D., and Ferrario, C.M. (1994) Am. J. Hypertens. 7, 576-582
5. Molnar, M. and Hertelendy, F. N (1992 Am. J. Obstet. Gynecol. 166: 1560-1567
6. Chaudhuri, G. and Furuya, K. (1991) Sem. Perinatol. 15, 63-67
7. Sladek, S. M., Regenstein, A. C., Lykins, D. and Roberts, J. M. (1993) Am. J. Obstet. Gynecol. 169, 1285-1291
8. Natuzzi, E. S., Ursell, P. C., Harrison, M., Buscher, C. and Riemer, R. K. (1993) Biochem. Biophys. Res. Comm. 194, 1-8
9. Izumi, H., Garfield, R. E., Makino, Y., Shirakawa, K. and Itoh, T. (1994) Am. J. Obstet. Gynecol. 170, 236-245
10. Xie, J., Kolls, J., Bagby, G., and Greenberg, S.S. (1995) FASEB J. 9, 253-261

Cyclic AMP plays a critical role in the expression of inducible nitric oxide synthase in vascular smooth muscle cells

VALERIE B. SCHINI-KERTH, MATTHIAS BOESE, ALEXANDER MULSCH and RUDI BUSSE

Center of Physiology, J. W. Goethe University Clinic, Theodor-Stern-Kai 7, D-60590 Frankfurt/Main, Germany

Vascular smooth muscle cells and many other types of mammalian cells are able to release high amounts of nitric oxide following the expression of an inducible nitric oxide synthase (iNOS; [1]). This sequence of events is initiated by certain cytokines such as interleukin-1 ß (IL-1 ß), and may account for the endothelium-independent formation of nitric oxide by blood vessels which have been exposed *in vivo* by to endotoxin or injured by balloon catheterisation [2, 3]. In addition to cytokines, high concentrations of cyclic AMP-dependent vasodilators also cause the expression of the iNOS in vascular smooth muscle cells, and potentiate that evoked by cytokines [4]. As IL-1 ß has been shown to stimulate the synthesis of cyclic AMP in some but not all mammalian cells [5, 6], the role of this cyclic nucleotide in the signal transduction cascade initiated by activation of the IL-1 ß receptor and leading to iNOS expression was investigated using cultured rat aortic smooth muscle cells from rat aorta.

The activity of iNOS was determined indirectly by assaying the release of nitrite into the incubation medium and the expression of iNOS by Western blot analysis.

IL-1 ß-stimulated release of nitrite and expression of iNOS protein (24 h incubation) was inhibited by a selective inhibitor of cyclic AMP-dependent protein kinase, Rp-8-CPT-cAMPS (100 μM), but enhanced by a selective activator of cyclic AMP-dependent protein kinase, Sp-8-CPT-cAMPS (100 μM; Figs. 1 and 2). In the absence of IL-1 ß, Rp-8-CPT-cAMPS had no such effects but Sp-8-CPT-cAMPS increased the basal release of nitrite by 275.3 ± 31.2% (n = 9) and was associated with a weak induction of iNOS (data not shown). The release of nitrite and the expression of iNOS protein elicited by IL-1 ß was potentiated by an inhibitor of phosphodiesterase IV, rolipram (Figs. 1 and 2) and of phosphodiesterase III, trequinsin (Fig. 1 and data not shown). Exposure of cells to either rolipram (3 μM) or trequinsin (3 μM) alone increased the basal release of nitrite by 156.7 ± 16.4% (n = 16) and by 123.3 ± 25.05% (n = 9), respectively and was associated with a weak expression of iNOS (Fig. 2 and data not shown).

The present findings indicate that the intracellular level of cyclic AMP plays a decisive role in the IL-1 ß-stimulated signal transduction cascade leading to iNOS expression in vascular smooth muscle cells. This effect of cyclic AMP is most probably due to the activation of cyclic AMP-dependent protein kinases causing the phosphorylation of target proteins. As the release of nitric oxide evoked by IL-1 ß and also that evoked by cyclic AMP-dependent vasodilators either alone or in the presence of IL-1 ß are associated with an increased level of iNOS mRNA [4, 8], cyclic AMP-dependent protein kinases possibly activate proteins which are involved in the transcriptional activation of the iNOS gene such as nuclear transcription factors. Although the nuclear transcription factors implicated, remain to be identified, nuclear trancription factor-κB (NF-κB) is a likely candidate as cyclic AMP-dependent protein kinases can phosphorylate the inhibitory subunit of NF-κB, I-κB resulting in the activation of NF-κB [9] which is essential for the expression of the iNOS gene [7].

Acknowledgements

This study was supported by the Deutsche Forschungsgemeinschaft (Schi 389/1-1; Bu 436/4-3).

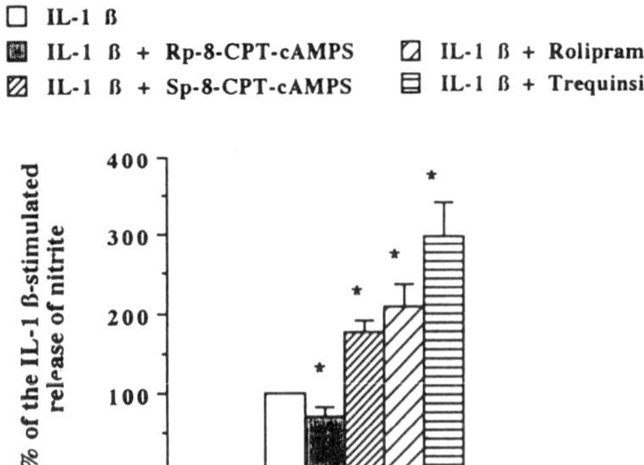

Fig. 1: Modulation of the IL-1ß (60 U/ml for 24 h)-stimulated release of nitrite by Rp-8-CPT-cAMPS (100 μM, Rp), Sp-8-CPT-cAMPS (100 μM, Sp), rolipram (3 μM) and trequinsin (3 μM). * indicates a significant inhibition or potentiation of the IL-1ß-stimulated release of nitrite.

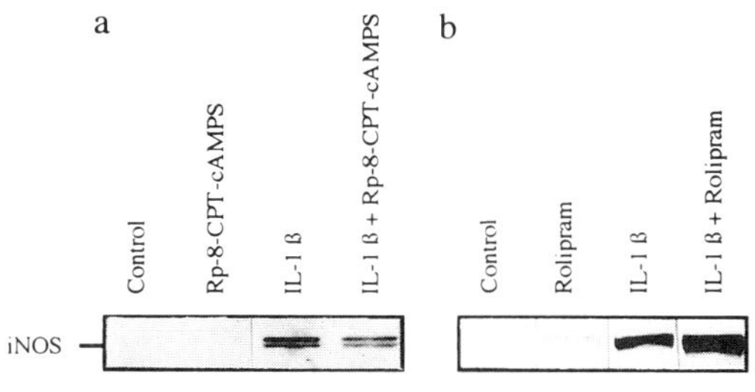

Fig. 2: Effect of Rp-8-CPT-cAMPS (100 μM) and rolipram (3 μM) on the IL-1 ß (60 U/ml for 24 h)-stimulated expression of iNOS protein.

References

1. Busse, R., Mülsch, A. (1990) FEBS Lett. **275**: 87-90
2. Julou-Schaeffer, G., Gray, G.A., Fleming, I., Schott, C., Parratt, J.R., Stoclet, J.-C. (1990) Am. J. Physiol. **259**: H1038-H1043
3. Joly, G.A., Schini, V.B., Vanhoutte, P.M. (1992) Circ. Res. **71**: 331-338
4. Imai, T., Hirata, Y., Kanno, K., Marumo, F. (1994) J. Clin. Invest. **93**:, 543-549
5. Shirakawa, F., Yamashita, U., Chedid, M., Mizel, S.B. (1988) Immunoloy **85**: 8201-8205
6. Zhang, Y., Lin, J.-X., Yip, Y.K., Vilcek, J. (1988) Proc. Natl. Acad. Sci. USA **85**: 6802-6805
7. Xie, Q.-W., Whisnant, R., Nathan, C. (1993) J. Exp. Med. **177**: 1779-1784
8. Schini-Kerth, V.B., Fisslthaler, B., Busse, R. (1994) Am. J. Physiol. **267**: H2483-H2490
9. Link, E., Kerr, L.D., Schreck, R., Verma, I., Baeuerle, P.A. (1992) J. Biol. Chem. **267**: 239-246

Differential regulation of inducible nitric oxide synthase (iNOS) by cyclic AMP in vascular smooth muscle and hepatocytes.

BRIAN G. HARBRECHT*, BRUCE JOHNSON+, BRUCE PITT+, and TIMOTHY R. BILLIAR*, Department of Surgery*, and Pharmacology+, University of Pittsburgh, Pittsburgh, PA 15261, USA.

The cytokine signals that lead to NO synthesis by iNOS in a variety of tissues have been fairly well characterized. In hepatocytes (HC) and pulmonary artery vascular smooth muscle cells (PASMC), a combination of tumor necrosis factor (TNF), interleukin-1 (IL-1), and interferon-gamma (IFN) along with lipopolysaccharide (LPS) leads to increased levels of iNOS mRNA and iNOS enzyme activity [1,2]. However, the second messenger systems that regulate cytokine-stimulated NO synthesis have not been defined. Recently, cyclic AMP (cAMP) has been shown to increase NO synthesis by renal mesangial cells [3]. Agents that increase cAMP have also been shown to improve circulatory function in shock [4]. We therefore studied whether cAMP regulates cytokine-stimulated NO synthesis in cultured HC and PASMC.

Rat HC were isolated by collagenase perfusion and purified by differential centrifugation [1]. They were cultured overnight in Williams medium E containing HEPES, L-glutamine, insulin and antibiotics with 10%fetal calf serum, washed, and then stimulated to produce NO by IL-1 alone (500 U/ml) or cytokines in combination (TNF 500 U/ml, IL-1 5 U/ml, IFN 100 U/ml, LPS 10 ug/ml). Rat PASMC were harvested from the left intrapulmonary artery and grown in 50% DMEM/50% Ham's Nutrient Mixture containing L-glutamine, antibiotics, and 10% fetal bovine serum until confluent [4]. PASMC were stimulated with the same cytokine combination to synthesize NO. All supernatants were collected at 24 hours and assessed for NO by the Greiss reaction. Data represent the mean ± SEM of triplicate cultures.

Both cytokines (Table 1) and IL-1 alone (Table 2) lead to increased NO_2^- by HC or PASMC. When added along with cytokines, cAMP analogues dibutyryl-cAMP (dbcAMP) and 8-Bromo-cAMP (8-BrcAMP) inhibited cytokine-stimulated HC NO synthesis while dbcAMP increased cytokine-stimulated PASMC NO synthesis (Table 1). cAMP analogues had no effect on NO synthesis in HC or PASMC not exposed to cytokines (not shown). In HC exposed to IL-1 alone, inhibition of adenylate cyclase with SQ 22,536 increased NO synthesis (Table 2). These data demonstrate that the second messenger cAMP differentially regulates iNOS in HC and PASMC. The mechanism for this effect in these separate tissues is unknown. Additional work will be required to determine if these effects are due to changes in iNOS mRNA levels, iNOS enzyme activity, iNOS stability, or some other mechanism. These data suggest, however, that iNOS is differentially regulated in different organs of the body and that therapeutic manipulation of iNOS may result in effects that are organ or tissue specific.

Table 1. Effect of cAMP analogues on NO_2^- produc;tion

	Hepatocytes		PASMC
	NO_2^-, nmoles/10^6HC		NO_2^-, nmoles/mg protein
M	dbcAMP	8-BrcAMP	dbcAMP
0	33.9±0.5	33.0±0.7	163.3±3.4
10^{-6}	38.4±4.9	36.5±0.8	178.8±4.4
10^{-5}	15.5±0.8	20.3±0.9	196.0±7.0
10^{-4}	14.0±1.2	9.6±0.5	269.5±11.3
10^{-3}	10.3±0.8	2.8±0.4	293.1±7.7

Table 2. Effect of adenylate cyclase inhibition on HC NO synthesis

IL-1, U/ml	SQ 22,536, 1 mM	NO_2^-, nmoles/10^6 HC
0	-	1.1±0.5
	+	0.1±0.1
50	-	1.4±0.4
	+	5.8±0.7
100	-	3.9±0.4
	+	12.0±0.7
250	-	10.4±0.4
	+	21.1±0.9
500	-	18.1±1.4
	+	32.8±1.7

1. Geller, D.A., Nussler A.K., DiSilvio, M. et al. Cytokines, endotoxin, and glucocorticoids regulate the expression of inducible nitric oxide synthase in hepatocytes. (1993) Proc. Natl. Acad. Sci. USA. **90**, 522-526.
2. Nakayama, D.K., Geller, D.A., Lowenstein, D.J., et al. Cytokines and lipopolysaccharide induce nitric oxide synthase in cultured rat pulmonary artery smooth muscle. (1992) Am. J. Respir. Cell. Mol. Biol. 7, 471-476.
3. Muhl, H., Kunz, D., Pfeilschifter, J. Expression of nitric oxide synthase in rat glomerular mesangial cells mediated by cyclic AMP. (1994) Br. J. Pharmacol. **112**, 1-8.
4. Flynn, W.J., Cryer, H.G., Garrison, R.N. Pentoxifylline but not saralasin restores hepatic blood flow after resuscitation from hemorrhagic shock. (1991) J. Surg. Res. **50**, 616-621.

Cyclic GMP-augmented but G-kinase inhibitor-insensitive induction of nitric oxide synthase in vascular smooth muscle cells

TAKESHI MARUMO[1, 2], TOSHIO NAKAKI[1], KEIICHI HISHIKAWA[1, 2], JUNICHI HIRAHASHI[1, 2], HIROMICHI SUZUKI[2], TAKAO SARUTA[2] and RYUICHI KATO[1]

Departments of Pharmacology[1] and Internal Medicine[2], Keio University School of Medicine, 35 Shinanomachi, Shinjuku-ku, Tokyo 160, Japan

The regulation of induction of nitric oxide synthase (NOS) II has not been fully understood. Involvement of cyclic AMP [1], protein kinase C [2] and tyrosine kinase [3] has been reported. Cyclic GMP (cGMP) is a signal transduction molecule of nitric oxide. It is possible that cGMP affects nitric oxide production. Therefore, effects of cGMP analog, natriuretic peptides and cGMP-elevating drugs on NOS II induction in rat aortic smooth muscle cells were examined.

Rat aortic clonal vascular smooth cells (RACS-1), obtained from a Wistar rat, were used. Cells were washed once with Dulbecco's modified essential medium without phenol red (DMEM) supplemented with sodium selenite (5 ng/ml), insulin (5 µg/ml), transferrin (5 µg/ml), penicillin (100 U/ml) and streptomycin (100 U/ml) (SIT) and were incubated in DMEM with SIT for 48 hours, with or without reagents. Aliquots of the culture medium were taken for nitrite at the end of each incubation time. Cells were then harvested for Northern blotting. Nitrite concentration was measured based on the Griess reaction. Northern blot analysis was performed using a 700 bp fragment of the 5' portion of cloned rat liver inducible NOS cDNA and human glyceraldehyde-3-phosphate dehydrogenase (GADPH) cDNA probe and the radioactivities of the membranes were measured as previously described [4]. The NOS II cDNA probe was kindly provided by Dr. Hiroyasu Esumi, National Cancer Center Research Institute, Matsudo, Japan. Cellular cGMP was determined using a commercially available radioimmunoassay kit. LPS was quantified with a commercially available kit (Endospecy ES-6 set and Toxicolor DIA set, Seikagaku Corporation, Tokyo, Japan). LPS concentrations in the culture media were below 20 pg/ml in all experiments.

The combination of IL-1α and TNF-α induces nitrite production in RACS-1 by 48 h though neither cytokine is effective when applied alone (Fig. 1), which is consistent with our previous report [5]. Nitrite production was enhanced by 8-bromo-cGMP in RACS-1 treated with IL-1α and TNF-α though 8-bromo-cGMP was ineffective when applied alone or in combination with TNF-α. The combination of 8-bromo-cGMP and IL-1α also increased nitrite production, suggesting that cGMP enhances IL-1α signal transduction pathway. Stimulatory effect of 8-bromo-cGMP was dose-dependent [6]. Natriuretic peptides stimulated intracellular cGMP accumulation in a dose-dependent manner. Natriuretic peptides

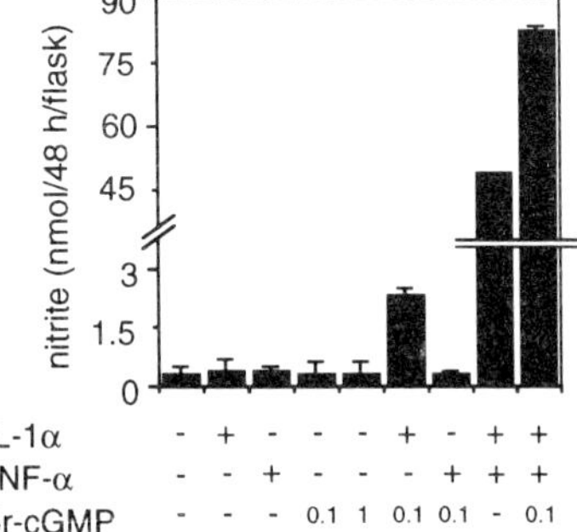

Fig. 1 Enhanced nitrite accumulation in cytokine-treated RACS-1 culture medium by 8-bromo-cGMP. RACS-1 were treated for 48 h. The following concentrations were used: 8-bromo-cGMP (0.1, 1 mM), IL-1α (200 U/ml) and TNF-α (5000 U/ml).

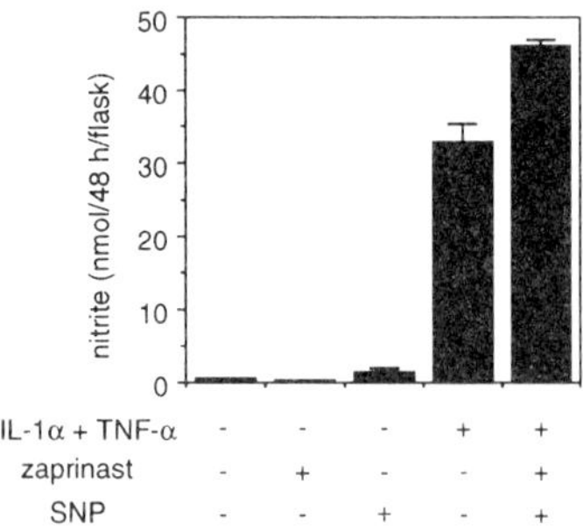

Fig. 2 Augmentation of nitrite accumulation in cytokine-treated RACS-1 culture medium by cGMP-elevating drugs. RACS-1 were treated for 48 h. The following concentrations were used: sodium nitroprusside (10 µM) (SNP), zaprinast (5 µM), IL-1α (200 U/ml) and TNF-α (5000 U/ml).

concentration-dependently enhanced nitrite production from RACS-1 in the presence of IL-1α and TNF-α but the peptides were ineffective when applied alone [6]. Sodium nitroprusside, another cGMP-elevating agent, also enhanced nitrite production from RACS-1 treated with IL-1α and TNF-α in the presence of zaprinast, a cGMP-specific phosphodiesterase inhibitor (Fig. 2). Molecular targets for cGMP include protein kinases, phosphodiesterases and ion channels [7]. In order to elucidate the role of cGMP-dependent protein kinase in NOS II induction, we examined the effect of its inhibitor, KT5823. KT5823 (up to 7 µM) alone did not induce nitrite production in the cells. Interestingly, this drug (0.7, 7 µM) did not inhibit, but rather enhanced the nitrite production induced by 8-bromo-cGMP, IL-1α and TNF-α [6]. KT5823 also enhanced nitrite production in response to the combination of IL-1α and TNF-α. Therefore, the stimulatory effect of cGMP on cytokine-induced nitrite production seems to be exerted through G-kinase inhibitor-insensitive pathway. To further evaluate whether this increased nitrite formation was due to increased NOS II mRNA expression, we performed Northern blot analysis. The combination of IL-1α and TNF-α induced NOS II mRNA at 48 h, but there was no detectable NOS II mRNA in 8-bromo-cGMP treated cells [6]. Co-treatment with 8-bromo-cGMP markedly enhanced NOS II mRNA expression stimulated by the combination of IL-1α and TNF-α. Consistent with the effects of KT5823 on nitrite production stimulated by the combination of IL-1α, TNF-α and 8-bromo-cGMP or that of IL-1α and TNF-α, co-treatment with KT5823 increased NOS II mRNA levels above those produced by the combination of these cytokines with 8-bromo-cGMP or the combination of cytokines [6].

In conclusion, cGMP augments induction of NOS II at the level of mRNA in vascular smooth muscle cells. This augmentation seems to be independent of KT5823-sensitive cGMP-dependent protein kinase. These data also suggest that natriuretic peptides may play a regulatory role in vascular remodeling via the production of large amounts of nitric oxide.

1. Koide, M., Kawahara, Y., Nakayama, I., Tsuda, T. and Yokoyama, M. (1993) J. Biol. Chem. **268**, 24959-24966
2. Hortelano, S., Genaro, A.M. and Bosca, L. (1992) J. Biol. Chem. **267**, 24937-24940
3. Marczin, N., Papapetropoulos, A. and Catravas, J.D. (1993) Am. J. Physiol. **265**, H1014-H1018
4. Marumo, T., Nakaki, T., Hishikawa, K., Suzuki, H., Kato, R. and Saruta, T. (1995) Hypertension. **25 Part 2**, 764-768
5. Marumo, T., Nakaki, T., Nagata, K., Miyata, M., Adachi, H., Esumi, H., Suzuki, H., Saruta, T. and Kato, R. (1993) Jpn. J. Pharmacol. **63**, 361-367
6. Marumo, T., T. Nakaki, K. Hishikawa, J. Hirahashi, H. Suzuki, Kato, R. and Saruta, T. (1995) Endocrinology. **136**, 2135-2142.
7. Lincoln, T.M. and Cornwell, T.L. (1993) FASEB. J. **7**, 328-338

Arginine transport in cytokine-activated bovine articular chondrocytes depends on parallel induction of nitric oxide synthase and GTP-Cyclohydrolase

Barbara Petrack, Brian Latario and Salvatore Spirito
CIBA Pharmaceutical Research, Summit, N. J. 07901

Introduction: Inflammatory cytokines induce iNOS activity in human chondrocytes [1]; iNOS inhibitors suppress arthritis in animal models; [2]; synovial fluid and serum from rheumatic patients have increased nitrite levels [3]. These findings suggest that NO over-production by chondrocytes may be causally related to cartilage damage. iNOS induction in cytokine-activated macrophages (and other cells) is coupled to co-induction of both Arg transport and BH_4 synthesis [4,5]. Arginase also is induced in activated macrophages [6]. We now report that iNOS induction in cytokine-activated BAC is accompanied by co-induction of Arg transport and GTP-CH (rate-limiting in BH_4 synthesis), but not by arginase. The lack of arginase in activated BAC could eliminate competition for substrate and enable these cells to generate high levels of NO.

Methods--Cell culture and activation: Chondrocytes, isolated according to Kuettner et al [7], were plated at 200,000 cells/well (24 well plate). Confluent, washed cells were treated for 18 hrs with a mix of 1 ug/ml LPS (E. coli 0111:b4), 50 U/ml recombinant IFN-γ and 15 ng/ml human IL-1α. Washed cells were incubated for 45 min with 500 ul Arg-deficient medium (MEM or Krebs buffer) including ^{3}H-Arg (8 uCi/ml), containing Arg, as indicated. The washed cells, on ice, were lysed in 200 uL 1% SDS in 0.1N HCL; centrifuged media and cell lysates were analyzed via HPLC.

Assays for Arg transport, iNOS, arginase and GTP-CH: ^{3}H-Cit, ^{3}H-Arg, and ^{3}H-ornithine in the medium and cell lysate were assayed via HPLC [8], using a Zorbax 300 SCX column (250 x 4.6 mm), a Waters 845 System and an INUS β-ram radioactivity detector. Cell lysate or acidified medium was analyzed at a flow rate of 1 ml/min. The mobile phase consisted of isocratic 20 mM citrate, pH 2.2 from 0-4 min; a linear gradient of 20-200 mM citrate, pH 2.2, from for 4-7 min; 200 mM citrate was maintained from 7-14 min, and returned to 20 mM citrate at 1.5 ml/min for an additional 6 min. At 45 min, >90% ^{3}H-cit remains intracellular (Fig 1):

iNOS activity = ^{3}H-Cit in cells at 45 min

Arg transport = ^{3}H-Cit in cells + ^{3}H-Arg in cells at 45 min

GTP-CH was assayed in BAC +/- LPS and cytokines. Washed cells were lysed, the homogenates were centrifuged; the cytosols were passed through G-25 and analyzed for neopterin after oxidation with iodine, as described by Viceros [10].

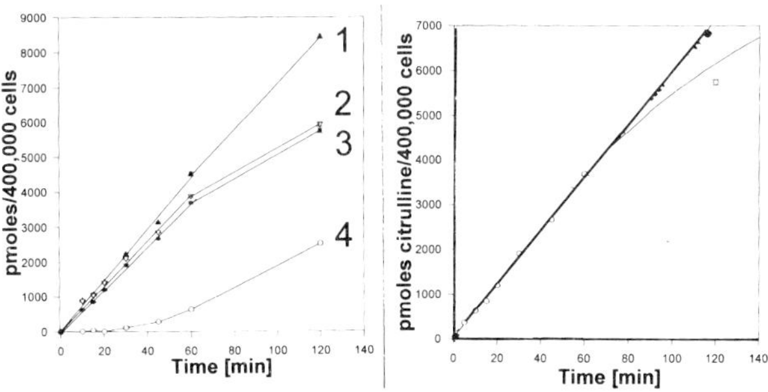

Fig.1.a. and b. Time dependence of iNOS and ^{3}H-Arg transport in activated BAC: incubations with 100 uM ^{3}H-Arg in Krebs buffer. 1=intra-cellular (^{3}H-Arg + ^{3}H-Cit) + extracellular ^{3}H-Cit; 2=intracellular(^{3}H-Arg + ^{3}H-Cit); 3=intracellular ^{3}H-Cit; 4=extracellular ^{3}H-Cit.

Abbreviations: iNOS, inducible nitric oxide synthase, NO, nitric oxide, BAC, bovine articular chondrocytes; Cit, citrulline; BH_4, tetrahydrobiopterin; IFN-γ, interferon-γ; IL-1α, interleukin-1α, LPS, lipopolysaccharide; DAHP, 2,4-diamino-6-hydroxypyrimidine; SP, sepiaptrin; GTP-CH, GTP-cyclohydrolase; L-NMA, N-methyl-L-Arg; L-NIO, N-iminoethyl-L-Arg; L-NA, nitro-L-Arg; DPI, diphenyleneiodinium-Cl.

Results and Discussion: Fig. 1 shows that cytokine-activated BAC take up ^{3}H-Arg for at least 2 hrs, converting most of it to ^{3}H-Cit, indicating potent iNOS activity. ^{3}H-Cit synthesis was linear for 1 hr and remained intracellular during this period. We did not find any ^{3}H-ornithine, either in the cell lysates or the medium, indicating absence of arginase activity in activated BAC. ^{3}H-Arg transport was saturable, with and IC_{50} of 156 uM.

Various inhibitors blocked both ^{3}H-Arg transport and iNOS activity. Inhibiting iNOS caused only a slight increase in intracellular ^{3}H-Arg (Fig. 2). Competitive iNOS inhibitors (L-NMA, L-NIO, L-NA, L-NAME) blocked ^{3}H-Arg uptake. Basic amino acids, (L-Lys, L-Orn), known inhibitors of y^+ cationic transport in other cell types, inhibited both ^{3}H-Arg transport and iNOS activity in activated BAC, whereas neutral amino acids (L-Gly, L-Gln) blocked neither transport nor iNOS activity. The noncompetitive iNOS inhibitor, DPI, also blocked ^{3}H-Arg transport. DAHP, an inhibitor of GTP-CH (rate-limiting enzyme in BH_4 synthesis) blocked both ^{3}H-Arg transport and iNOS activity. These data demonstrate that a variety of inhibitors block both ^{3}H-Arg transport and iNOS activity in parallel.

Unstimulated BAC exhibit negligible levels of both ^{3}H-Arg transport and iNOS activity and low GTP-CH, but the three activities are induced in parallel by IL-1α, LPS and IFN-γ, alone or in combination; the greatest effects were seen combining LPS, IL-1α and IFN-γ; we use this combination to activate BAC. These studies demonstrate that ^{3}H-Arg transport, iNOS activity and GTP-CH activity are induced in parallel in activated BAC.

BAC do not have arginase activity, nor is the enzyme induced by cytokines or LPS, contrary to activated macrophages, which were reported to metabolize 70% of the Arg taken up by these cells to ornithine, and only 30% to Cit + NO [11]. The absence of arginase in activated BAC may eliminate competition for Arg substrate, enabling continued production of high levels of NO, which could then contribute to cartilage pathology in arthritic patients.

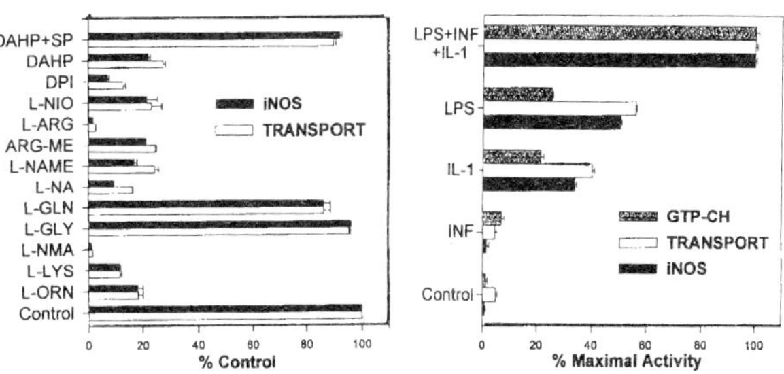

Fig. 2.a. Parallel inhibition of iNOS and ^{3}H-Arg uptake in activated BAC: 45 min; 20 uM ^{3}H-Arg in MEM; 10 mM amino acids; 0.4 uM DPI;
10 mM DAHP;100 uM SP. Curves: **(b. Parallel induction of iNOS, ^{3}H-Arg uptake and GTP-CH in cytokine-activated BAC:** 18 hrs with LPS (1 ug/ml); IFN-γ (50 U/ml); IL-1α (15 ng/ml); then, incubations with 20 uM ^{3}H-Arg in MEM.

1. Palmer, R., Hickery, M., Charles, I., Moncada, S. and Bayliss, M. (1993), Biochem. Biophys. Res. Commun., **193**, (398-405)
2. Connor, J., Manning, S., Settle, S. et al (1995) Eur. J. Pharma., **273**, 15-24
3. Farrell, A.J., Blake, D.R., Palmer, R.M.J. and Moncado, S. (1992) Ann. Rheum. Dis. **51**, 1219-1222 .
4. Bogle, R.G., Baydoun, A.R., Pearson, J.D., Moncado, S. and Mann, G.E.. (1992), Biochem. J., **284**,15-18.
5. Wang, W. W., Jenkinson, C.P., Gricavage, et al. (1995) Biochem. Biophys. Res. Commun. **210**, 1009-1016.
6 Sakai, N., Kaufman, S. and Milstein, S. (1995) J. Neurochem., **65**, 895-902.
7. Kuettner, K.E., Pauli, B.U., Gall, G., Memoli V.A. and. Shenk, R.K. (1982). J. Cell. Biol. **93**, 743-750
8. Chenais, B., Yapo, A., Lepoivre, M. and Tenu, J. P. (1991) J. Chromatog. **539**, 433-441.
9. Viveros, O.H., Lee, C-L, Abou-Donia, M.M. Nixon, J.C. and Nichol, C.A., (1981) Science **213**, 349-350
10. Granger, D.L., Hibbs, J.B., Perfect, J.R., and Durack, D.T. (1992), J. Clin. Invest, **85**, 264-273.

Regulation of the inducible L-arginine-nitric oxide pathway by endogenous polyamines in J774 cells

Anwar R. BAYDOUN & David M. MORGAN

Vascular Biology Research Centre, King's College, Campden Hill Road, London W8 7AH, U.K.

INTRODUCTION

The polyamines putrescine, spermidine and spermine are normal cellular constituents essential for cell growth and differentiation, and regulate a multitude of cellular functions [1]. Recently, exogenous polyamines have been shown to inhibit nitric oxide (NO) production both by cultured cells and in cell-free systems [2,3]. This inhibitory action required relatively high concentrations of polyamines compared to those found in plasma and may result from oxidation of the polyamines by amine oxidases present in the culture medium. In this study we have examined the effects of endogenous polyamines on the inducible L-arginine-NO pathway by examining whether difluromethylornithine (DFMO), a potent and selective inhibitor of ornithine decarboxylase and thus of polyamine biosynthesis, regulates NO production in endotoxin-activated J774 cells.

EXPERIMENTAL

Materials

Reagents for cell culture were obtained from Gibco (Paisley, U.K.). *Escherichia coli* lipopolysaccharide (LPS, serotype 0111:B4), Dowex-50W (HCR-W2) and all other chemicals were purchased from Sigma (Poole, U.K.). L-[2,3-^{3}H]arginine (53 Ci/mmol) was from Amersham International plc. N^G-monomethyl-L-arginine (L-NMMA) was a gift from Wellcome Research Lab., Kent. DFMO was a gift from Marion-Merrell-Dow Inc., (Cincinatti).

Cell culture

The murine monocyte/macrophage cell line J774 was obtained from the European Collection of Animal Cell Cultures (ECACC, Wiltshire) and maintained in continuous culture in Dulbecco's modified Eagles medium (DMEM) containing 0.4 mM L-arginine and supplemented with 4 mM glutamine, penicillin (100 units ml^{-1}), streptomycin (100 μg ml^{-1}) and 10% foetal calf serum.

Experimental protocol

J774 cells plated in 96-well culture plates (10^5 cells per well) were incubated with increasing concentrations of DFMO (1-10 mM) for either 0, 24 or 48 h proir to activation with LPS (0.1 μg ml^{-1}). Accumulated nitrite in the culture medium was determined after a further 24 h incubation with DFMO and/or LPS using the Griess reaction as described previously [4]. In other experiments DFMO (10 mM) was added at various time points up to 6 h after LPS. The direct effects of DFMO (1-10 mM) on isolated inducible nitric oxide synthase (iNOS) activity were determined by conversion of L-[^{3}H]arginine to L-[^{3}H]citrulline [5].

RESULTS

LPS-induced nitrite accumulation (0.76 $\pm$ 0.01 nmoles/μg protein/24 h) was enhanced following incubation of cells with DFMO. This effect was both time- and concentration-dependent, with 10 mM DFMO potentiating LPS-induced nitrite production by 52 $\pm$ 5.9% when pre-incubated with cells for 24 h prior to the addition of LPS. DFMO alone produced no detectable changes in basal nitrite levels and had no effect on LPS-induced nitrite production when added to J774 cells at various time points after LPS. Furthermore DFMO also failed to modify conversion of L-[^{3}H]arginine to L-[^{3}H]citrulline by iNOS isolated from LPS-activated J774 cells. Enzyme activity was unaffected by removal of calcium from the reaction buffer using 1 mM EGTA but was significantly inhibited by L-NMMA, which at 100 μM reduced L-[^{3}H]citrulline production by 69.8 $\pm$ 1.4%.

DISCUSSION

The present study demonstrates that pre-incubation of J774 macrophages with DFMO potentiates LPS-induced nitrite production. This effect is not mediated by a direct action of DFMO on iNOS since addition of DFMO to intact cells after LPS or indeed to the isolated enzyme did not alter nitrite levels or the conversion of L-[^{3}H]arginine to L-[^{3}H]citrulline. Thus the effects observed may involve regulation of iNOS expression at the molecular level. Since the only known action of DFMO to date involves inhibition of ODC [6], it is therefore reasonable to conclude that this novel effect of DFMO results from depletion of intracellular polyamine pool(s). More importantly, our data suggest that expression of iNOS, at least *in vitro*, may be critically regulated by endogenous polyamines. In this regard it is worth noting that exposure of macrophages to LPS results in enhanced activity and expression of ODC which precedes nitrite release [7]. Thus by preventing this increase in ODC activity, DFMO may have a downstream effect on the induction of NO synthase.

We conclude that DFMO significantly modulated the ability of macrophages to express iNOS and that endogenous polyamines may play an important role in regulating the production of NO under both physiological and pathophysiological conditions. Our findings could also account, at least in part, for the cytostatic and indeed antitumor properties of DFMO.

Acknowledgements

We gratefully acknowledge the support of the British Heart Foundation (PG/9309; FS/94004) and the Wellcome Trust for award of a travel grant to ARB.

REFERENCES

[1] Morgan, D.M.L. (1987). Essays Biochem. **23:** 82-115.

[2] Hu, J., Mahmoud, M.I. and El-Fakahany, E.E. (1994). Neurosci. Lett. **175**: 41-45

[3] Szabó, C., Southan, G.J., Thiemermann, C. and Vane, J.R. (1994). Br. J. Pharmacol. **113**: 757-766

[4] Baydoun, A.R., Bogle, R.G., Pearson, J.D. and Mann, G.E. (1993). Br. J. Pharmacol. **110:** 1401-1406.

[5] Brown, J.L., Tepperman, B.L., Hanson, P.J., Whittle, B.J.R. and Moncada, S.(1992). Biochem. Biophys. Res. Commun. **184**: 680-685.

[6] McCann, P.P. and Pegg, A.E. (1992). Pharmacol. Ther. **54:** 195-215.

[7] Tjandrawinata, R.R., Hawel, L. and Byus, C.V. (1994). J. Immunol. **152:** 3039-3052. 1994)

NO donors and atrial natriuretic peptide enhance the expression of inducible NO synthase in vascular smooth muscle cells: Role of cyclic GMP-and cyclic AMP-dependent protein kinases

MATTHIAS BOESE, VALERIE B. SCHINI-KERTH, ALEXANDER MÜLSCH, AND RUDI BUSSE

Center of Physiologie, J. W. Goethe University Clinic, Theodor-Stern-Kai 7, D-60590 Frankfurt/Main, Germany

Endotoxic shock is associated with an unrelenting hypotension, which is primarily attributed to an attenuated responsiveness to vasoconstrictor agents [1]. The generalized nature of this vascular hyporeactivity is due to the overproduction of nitric oxide (NO) following the induction of nitric oxide synthase in the vascular wall by endotoxin and cytokines such as interleukin 1β (IL-1β) and tumor necrosis factor-α [2-5].

As endotoxic shock is associated with increased plasma levels of atrial natriuretic peptide (ANP), experiments were designed to test wether activation of either the particulate or soluble guanylate cyclase isoforms affect the expression of iNOS in cultured rat aortic smooth muscle cells and if so to characterize the underlying mechanism.

Smooth muscle cells were obtained by elastase/collagenase digestion from endothelium-denuded thoracic aortae of male Wistar-Kyoto rats as described in detail previously [6]. Cells were cultured in Waymouth's medium supplied by non-essential amino acids, 100 U/ml penicillin and 100 mg/ml Streptomycin and 7.5 % (v/v) fetal calf serum. All experiments were performed with cultures of cells obtained from the 4th to 15th passage. When cells reached confluence the culture medium was replaced by fresh medium containing 0.1 % fetal calf serum. After 24 h the medium was exchanced again and the smooth muscle cells were used for the experiments described below. The production of NO by smooth muscle cells was estimated by the release of nitrite in the incubation medium. The expression of iNOS protein was determined by Western Blot analysis.

The incubation of confluent vascular smooth muscle cells (SMC) with IL-1β (60 U/ml for 24 h) led to the expression of iNOS as detected by immunoblotting (Fig. 1). No signal was observed in control cells (cultured in the absence of cytokine). Co-incubation of IL-1β and ANP (1μM) increased iNOS expression at any given protein concentration compared with IL-1β alone (Fig. 1a), while ANP alone had no effect on iNOS expression.

In order to evaluate wether the enhancement of the IL-1β-induced expression of iNOS protein synthesis by ANP was cyclic GMP-dependent, SMC were exposed to activators of the soluble guanylate cyclase (i.e. nitric oxide donors). As shown in Fig. 1b, IL-1β induced iNOS expression was markedly enhanced by sodium nitroprusside (SNP, 100 μM) or 3-morpholino-1-sydnonimine (SIN-1, 100 μM). Furthermore SNP induced iNOS expression also in the absence of IL-1β (Fig. 1b, 2). To demonstrate that the effects of NO on iNOS were mediated by cyclic GMP, we tested the influence of an inhibitor of cyclic GMP-dependent protein kinase (PKG), the cGMP analogue Rp-8-CPT-cGMPS (100 μM). The induction of iNOS by SNP was prevented by PKG inhibition as revealed by Western blot analysis (Fig. 2a), while Rp-8-CPT-cGMPS alone showed no effect. To assess the role of cyclic AMP on NO-enhanced iNOS expression, SMC were incubated with SNP (100 μM) and either the cAMP analogue Rp-8-CPT-cAMPS (100 μM), a specific inhibitor of protein kinase A (PKA) or the specific inhibitor of cyclic AMP phosphodiesterase type IV, rolipram (3 μ M). As shown in Fig. 2, the SNP stimulated expression of iNOS was abolished by PKA inhibition (Fig. 2b) and potentiated by rolipram (Fig 2c). Both modulators alone no visible iNOS immunoreactivity in the absence of SNP.

These observations demonstrate that activation of both isoforms of guanylate cyclase (particulate and soluble) enhance the IL-1β induced iNOS expression in cultured rat aortic smooth muscle cells. In the absence of IL-1β SNP (greater than 30 μM) but not the other vasodilators (ANP, Sin-1) caused the expression of the enzyme. The SNP-stimulated expression of iNOS protein was inhibited by selective inhibitors of cyclic GMP- and cyclic AMP-dependent protein kinases and was increased by specific inhibiton of cyclic AMP phosphodiesterase type IV indicating that activation of both types of protein kinases may play an important role in the signal transduction cascade. The data further suggest that elevated plasma levels of ANP together with local increase of NO results in the establishment of a positive feedback on the expression of inducible nitric oxide synthase.

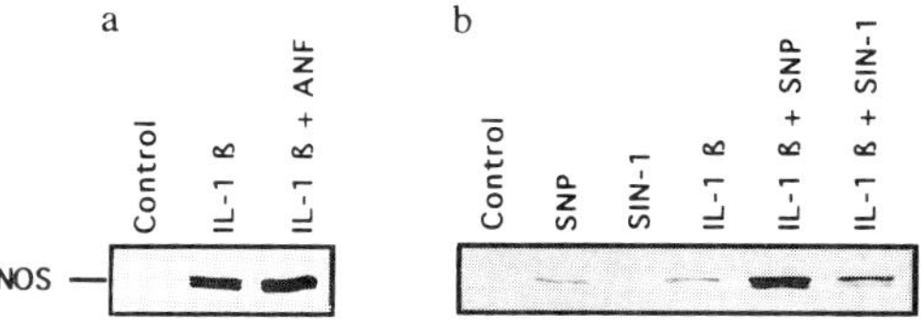

Fig. 1: Cyclic GMP-dependent vasodilators enhance the IL-1β induced iNOS expression in cultured vascular smooth muscle cells as detected by Western blot analysis. Cells were either untreated (control) or exposed to the different agonists as indicated for 24 hours. ANP: atrial natriuretic peptide (1 μM, a); SNP: sodium nitroprusside (100 μM, b) ; Sin-1: 3-morpholino sydnonimine (100 μM, b).

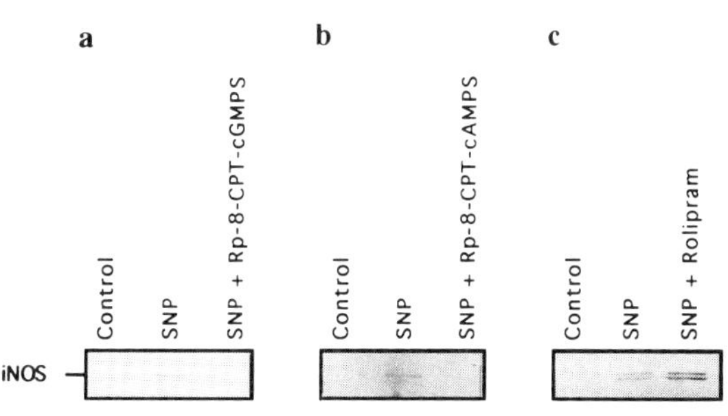

Fig. 2: Western blot analysis showing that the specific inhibitors of cyclic GMP and AMP-dependent protein kinases Rp-8-CPT-cGMPS (100 μM, a) and Rp-8-CPT-cAMPS (100 μM, b), respectively prevented and the cyclic AMP phosphodiesterase inhibitor rolipram (10 μM, c) potentiated the SNP (100 μM) stimulated expression of iNOS in cultured vascular smooth muscle cells. Cells were either untreated (control) or incubated with the different compounds for 24 hours.

References

1. Glauser, M. P., Zanetti, G., Baumgartner, J. D. and Cohen, J. (1991) Lancet 338, 732-736
2. Beasley, D., Cohen, R. A.and Levinsky, N. G. (1989) J. Clin. Invest. 83, 331-335
3. Ingrid Fleming, Géraldine Juluo-Schaeffer, Gillian A. Gray, James R. Parratt & Jean-Claude Stoclet (1991) Br. J. Pharmacol. 103, 1047-1054.
4. Busse, R., and Mülsch, A. (1990) FEBS Lett. 275, 87-90
5. Schini, V. B., Junquero, D. C., Scott-Burden, T. and Vanhoutte, P. M. (1991) Biochem. Biophys. Res. Commun. 176, 114-121

Evidence for translational control of inducible nitric oxide synthase expression in human cardiac myocytes

HARTMUT LUSS, REN-KE LI*, RICHARD A SHAPIRO, EDITH TZENG, FRANCIS X MCGOWAN+, TOSHI YONEYAMA, KAZUYUKI HATAKEYAMA, DAVID A GELLER, DONALD A G MICKLE*, RICHARD L SIMMONS, and TIMOTHY R BILLIAR

Departments of Surgery and +Anaesthesiology, University of Pittsburgh, Pittsburgh, PA 15261 USA; *Departments of Surgery and Clinical Biochemistry, University of Toronto, Toronto, Ontario, M5G 2C4 Canada

Introduction

The expression of inducible NO synthase (iNOS, or Type II NOS) upon stimulation with LPS and cytokines has been shown in a variety of cell types and tissues. GTP-cyclohydrolase I (GTP-CH), the rate-limiting enzyme during the synthesis of tetrahydrobiopterin (BH_4), can be coinduced with iNOS [1].

Cytokine-induced expression of iNOS by rodent cardiac myocytes has been well described recently [2]. Definitive proof of iNOS expression in human cardiac myocytes has been lacking. Therefore its expression was characterized in a model of nonbeating cultured human cardiac myocytes obtained from ventricular biopsies from patients with Tetralogy of Fallot.

Methods

Nonbeating human cardiac myocytes were isolated, purified, cultured, and characterized as described [3, 4]. Myocytes between passages 3 and 5 were treated for up to 96 h with LPS (10 μg/ml E.coli 0111:B4) and the cytokines IL-1β (30 U/ml), TNF-α (500 U/ml), and IFN-γ (100 U/ml), here referred to as cytokine mix (CM). Assays for NO_2^-+ NO_3^-levels [5], iNOS [5], GTP-cyclohydrolase I [6] as well as total and reduced biopterin [7] were performed as reported respectively. For Western blotting a monoclonal mouse antibody directed against iNOS (antimacNOS, Transduction Laboratories) was used. Northern blotting, cloning and expression of human iNOS cDNAs were perfomed as described [8]. Myocytes were transduced with a retroviral vector containing the coding region of the human hepatocyte iNOS cDNA (DFG-iNOS) or with a control retrovirus carrying the β-galactosidase gene (BagLacZ) [9].

Results and discussion

After treatment of human cardiac myocytes with CM mRNAs for iNOS and GTP-CH were induced. In untreated cells no transcripts were detectable.

Treatment of the cells with CM for up to 96 h did not result in any detectable NO synthesis as measured by NO_2^-+ NO_3^- accumulation and iNOS enzyme assay (Table 1). Western blot analysis confirmed the absence of iNOS protein. However GTP-CH was induced both at the mRNA and protein level as indicated by increased enzymic activity and biopterin levels in the myocytes after treatment with CM (Table 1). Sequencing and expression of a full length iNOS cDNA revealed the isolation of a functional human cardiac iNOS cDNA >99% identical to iNOS cDNAs cloned from other human tissues. When cardiac myocytes were transduced with a retroviral vector carrying only the coding region of human iNOS, iNOS was expressed in the cells as detected by Western blotting.

Table 1: NO_2^-+NO_3^-levels, iNOS and GTP-CH enzyme activities

Cell Type	RAW 264.7	Myocytes	Myocytes	Myocytes
Treatment	LPS 24h	Control	CM 24h	CM 96h
NO_2^-+ NO_3^- [μM]	+++	- [a]	-	-
iNOS [pmol/h/mg]	+++	-	-	-
GTP-CH [pmol/h/mg]	N.D. [b]	+	+++	+++
total biopterin [pmol/mg]	N.D.	0	+++	N.D.
reduced biopterin [pmol/mg]	N.D.	0	+++	N.D.

[a]not detectable, [b] not determined

Thus iNOS can be expressed in human cardiac myocytes, but its expression might be under tight and selective translational control. Reduced iNOS protein expression could explain the low or absent expression of iNOS in some human cell types.

Acknowledgements

Supported by NIH grants #GM-37753 (Richard L.Simmons), #GM-44100 (Timothy R. Billiar), Deutscher Akademischer Austauschdienst and Deutsche Forschungsgemeinschaft (Hartmut Luss). The excellent technical assistance of Paul F. Moyer is gratefully acknowledged. We thank Dr. Richard Weisel for providing us with biopsy material and Lisa Vallins for preparing and maintaining the cardiac myocyte cell cultures.

References

1. Morris, S.M. Jr., Billiar, T.R. (1994). Am. J. Physiol. 266, E829-E839
2. Balligand, J.L., D. Ungureanu-Longrois, W.W. Simmons, D. Pimental, T.A. Malinski, M. Kapturczak, Z. Taha, C.J. Lowenstein, A.J. Davidoff, R.A. Kelly, T.W. Smith, and T. Michel (1994). Cytokine-inducible nitric oxide synthase (iNOS) expression in cardiac myocytes. J. Biol. Chem. 269, 27580-27588
3. Li R-K, Weisel RD, Williams WG, Mickle DAG (1992) J. Tiss. Cult. Meth. 14, 93-100
4. Li R-K, Shaikh N, Weisel RD, Williams WG, Mickle DAG (1994) Am J. Physiol. 266, H2204-2211
5. Luss H, Watkins SC, Freeswick PD, Imro AK, Nussler AK, Billiar TR, Simmons RL, del Nido PJ, McGowan FX, J. Mol. Cell. Cardiol. (in press)
4. Hatakeyama K, Inoue Y, Harada T, Kagamiyama H (1991) J. Biol. Chem. 266, 765-769
6. Hatakeyama K, Harada T, Suzuki S, Watanabe Y, Kagamiyama H (1989) J. Biol. Chem. 264, 21660-21664
7. Fukushima T, Nixon JC (1980) Anal. Biochem. 102, 176-188
8. Geller DA, Lowenstein CJ, Shapiro RA, Nussler AK, Di Silvio M, Wang SC, Nakayama DK, Simmons RL, Snyder SH, Billiar TR (1993) Proc. Natl. Acad. Sci. USA 90, 3491 - 3495
9. Schwartz MA, Lazo JS, Yalowich JC, Allen WP, Whitmore M, Bergonia HA, Tzeng E, Billiar TR, Robbins PD, Lancaster JR, Pitt BR (1995) Proc. Natl. Acad. Sci. USA. in press
9. Charles IG, Palmer RM, Hickery MS, Bayliss MT, Chubb AP, Hall VS, Moss DW, Moncada S (1993) Proc. Natl. Acad. Sci. USA 90, 11419-11423
10. Sherman PA, Laubach VE, Reep BR, Wood ER (1993) Biochemistry 32, 11600-16005
11. Hokari A, Zeniya M, Esumi H (1994) J. Biochem. 116, 575-581

Abbreviations used: BH_4, tetrahydrobiopterin; CM, cytokine mix; GTP-CH, GTP-cyclohydrolase I; iNOS, inducible NO synthase

[1]Current address: Institut für Pharmakologie und Toxikologie der Westfälischen Wilhelms-Universität, Domagkstr. 12, D-48129 Münster, FRG

Nitric oxide synthase gene expression by endotoxin involves synergy between purine-2 receptors and protein kinase c.

STAN S. GREENBERG, XINFANG ZHAO. , DAVID R. POWERS, THOMAS D. GILES, AND JIANMING XIE.

DEPTS. OF MEDICINE AND OBSTETRICS AND GYNECOLOGY, LSU MEDICAL CENTER, NEW ORLEANS, LA, 70112.

Escherichia coli endotoxin (LPS) rapidly up-regulates gene expression for inducible nitric oxide synthase (iNOS) in alveolar macrophages (AM) in vivo [1,2]. Recent studies have clearly demonstrated that this effect of LPS did not result from LPS-induced release of tumor necrosis factor alpha (TNF) from the alveolar macrophage nor was it dependent on suppression of endogenous corticosterone [3]. The cellular mediators involved remain undefined. However, published data on peritoneal macrophages in vitro suggest that LPS-induced iNOS activity may be dependent on protein kinase C (PKC) and purine receptor antagonism or agonism [4-6]. However, the binding and cytokine cascade elicited in AM by LPS differs in vivo and in vitro. Moreover, PKC and purine receptor ligands may affect iNOS at the level of gene transcription, translation of the mRNA into iNOS enzyme or have a direct modulatory effect on iNOS enzymatic activity. This study evaluated the effect of in vivo, intratracheal (i.t.) administration of staurosporine (S) and chelerythrine (C), inhibitors of PKC, and β-methylthio-ATP, a purine-2 receptor agonist on i.t. LPS-induced upregulation of alveolar macrophage iNOS mRNA and nitrite and nitrate (RNI) levels in bronchoalveolar lavage (BAL) fluid and RNI production by alveolar macrophages obtained from LPS-treated rats.

Methods and Materials: Male Sprague Dawley rats (200-225 g) were given intratracheal (i.t.) sterile saline (PBS), chelerythrine (C, 0.1 mg/kg), staurosporine (S, 0.1 mg/kg), 2-methyl-thio-ATP (SATP, 5 mg/kg) or desferoxamine (D, 25 mg/kg), alone or in combination, 30 min before intratracheal (i.t.) administration of LPS (0.6 mg/kg). The rats were anesthetized two hr. after LPS administration. The lungs were removed and subjected to bronchoalveolar lavage (BAL). Aliquots of the BAL fluid were assayed for total nitrite and nitrate (RNI) with chemiluminescence and TNFα with an ELISA. The AM were isolated and incubated ex vivo (1 million cells/ml) for 1 or 12 hr and the ex vivo spontaneous production of RNI and TNF were measured. Alternatively, some AM were frozen and assayed for mRNA for iNOS and TNFα with competitor equalized RT-PCR [1-3]. PKC activity of AM were determined with a Pierce Spin-zyme kit and PKC content and isozymes assayed with Western blot. Data were analyzed with ANOVA. A value of $p<0.05$ was accepted for statistical significance of mean differences.

Results and Comments: Intratracheal administration of sterile saline (PBS) or desferoxamine (D) did not affect iNOS mRNA or RNI in BAL fluid or AM and did not affect LPS-induced changes in AM NOS mRNA (Fig. 1) or BAL fluid or alveolar macrophage RNI (Fig.2). Staurosporine and chelerythrine, in doses which inhibit PKC enzymatic activity, increased the influx of PMN into the lung but did not affect the non-stimulated levels of iNOS mRNA, BAL fluid RNI levels or alveolar macrophage production of RNI. In contrast to these findings, these compounds amplified LPS-induced recruitment of PMN into the alveolar space but suppressed alveolar macrophage induction of gene expression for iNOS (Fig. 1), BAL fluid concentrations of RNI and the ex vivo spontaneous generation of RNI by alveolar macrophages stimulated in vivo by LPS (Fig. 2). Pretreatment of rats with SATP given fifteen minutes before i.t. administration of LPS inhibited iNOS mRNA and depressed the RNI content of BAL fluid and AM production of RNI by approximately 50 % that of the sum of the RNI produced by SATP and LPS, when given alone (Figs 1,2). These effects were specific for iNOS and were not observed for TNFα. Since inhibition of PKC was devoid of effects on basal iNOS mRNA and RNI production it is unlikely that basal PKC activity plays a role in modulating iNOS transcription or translation. It appears likely that either PKC or a PKC-dependent mechanism is activated by LPS during the induction of iNOS gene expression. Thus, presumed inhibition of PKC antagonizes LPS-induced up-regulation of gene

FIGURE 1. LPS-INDUCED INCREASES OF BAL FLUID RNI IS BLOCKED BY PKC INHIBITORS AND A P2-RECEPTOR AGONIST

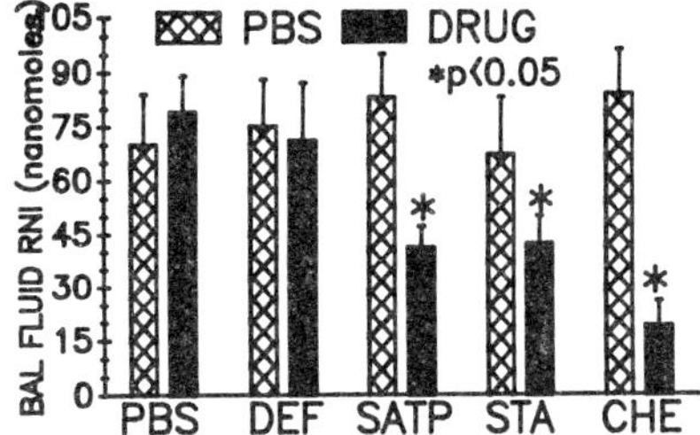

FIGURE 2. LPS-INDUCED INOS mRNA IS ATTENUATED BY PKC INHIBITORS AND METHYL-THIO-ATP

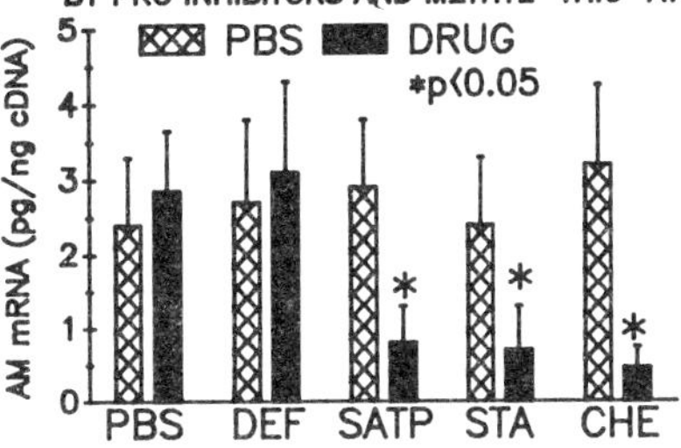

expression for iNOS. Finally, since methylthio-ATP is a purine-2 receptor agonist and inhibits both LPS-stimulated iNOS gene expression and AM production of RNI, we conclude that LPS-induced up regulation of iNOS mRNA in vivo is negatively modulated by purine-2 receptor stimulation. Thus, we suggest that LPS-induced up regulation of iNOS gene expression occurs via a PKC-dependent pathway and is inhibited by inhibition or down regulation of PKC. The down-regulation of iNOS mRNA is positively amplified at a purine-2 receptor site. Finally, our data extend the observations that PKC modulates cNOS and calcium-activated auricular iNOS activity [7-10] to cytokine and LPS-up regulated iNOS gene expression following in vivo administration of LPS.

This research was supported by USPHS grants AA 09816 and AA09803 from the National Institute on Alcohol and Alcohol Abuse

References

1. Greenberg, S.S., Xie, J., Wang, Y., Kolls, J., Malinski, T., Summer, W.R., and Nelson, S. (1994) Alcohol 11, 539-547
2. Spolarics, Z., Spitzer, J.J., Wang, J.F., Xie, J., Kolls, J., and Greenberg, S. (1993) Biochem. Biophys. Res. Comm. 197, 606-611
3. Xie, J., Kolls, J., Bagby, G., and Greenberg, S.S. (1995) FASEB J. 9, 253-261
4. Hortelano, S., Genaro, A.M. and Bosca, L. (1992) J. Biol. Chem. 267, 24937-24940
5. Severn, A., Wakelam, M.J. and Liew, F.Y. (1992 Biochem. Biophys. Res. Comm 188, 997-1002
6. Paul, A., R. H. Pendreigh, and R. Plevin. (1995) British Journal of Pharmacology 114, 482-488
7. Bredt, D. S., FerrisC.D and Snyder, S.H. (1992) J. Biol. Chem. 267, 10976-10981
8. Brune, B. and Lapetina, E.G. (1991) Biochem. Biophys. Res. Comm.. 181, 921-926
9. Davda, R. K., Chandler, L.J. and Guzman N.J. (1994) Eur. J. Pharmacol. 266, 237-244
10. Geng, Y., Maier, R. and Lotz, M. (1995) J. Cell. Physiol. 163, 545-554

Effects of tyrosine kinases inhibitors on inducible nitric oxide synthase

Joly, G., Ayres M., and Kilbourn R. G.

M. D. Anderson Cancer Center Box 13, 1515 Holcombe Bvd, Houston TX, 77030

Summary. We have examined whether specific protein tyrosine kinase (PTK) inhibitors (genistein: GE, tyrphostin: TY, or geldanamycin: GA) prevent NO· production in rat smooth muscle cells (SMC), in murine brain endothelial cells (MBEC), and in isolated rat aortas treated with endotoxin and/or cytokines. TY failed to inhibit either the release of nitrite in both MBEC and SMC or vascular hyporeactivity in rat aorta, caused by immunostimulants. GE decreased nitrite production in MBEC only at high concentration but had no effect on nitrite production in SMC and on the hypocontractility in aortic rings. In contrast, low concentrations of GA abolished the release of nitrite in MBEC and SMC treated with endotoxin and/or cytokines. GA inhibited also the hypocontractility to phenylephrine in aortic rings treated with LPS or interleukin-1. This inhibitor failed to inhibit the release of nitrite and the vascular hyporeactivity once nitric oxide synthase (NOS) was induced by immunostimulants. These data suggest that LPS- and cytokines-induced NO· production initiate a common signaling pathway involving a PTK that is inhibited by GA but not or slightly by TY or GE at a point that precedes the induction of NOS by immunostimulants.

Introduction. The intracellular signals that regulate the expression of iNOS have been only partially characterized in a small number of cell types. PTK seem to be essential in immunostimulant-induced NO· production, suggesting that PTK may be potential targets to inhibit pathogenic effects of cytokines and LPS. Furthermore, it has been shown that the phosphorylated tyrosine can control gene expression by altering the activity of several transcription factors. We have investigated whether specific and chemically different PTK inhibitors would prevent the production of NO· induced by NOS in SMC, MBEC, and isolated vessels stimulated with cytokines and/or LPS.

Materials and Methods. Cell Culture: SMC were kindly provided by Dr. Gross. MBEC were kindly provided by Dr. Nicolson. The SMC were exposed to IL-1 α (30 ng/ml), LPS (50 ng/ml) + IFN-γ (500 IU/ml), or vehicle for 24 hours in the presence or absence of PTK inhibitors added 1 hour before the NOS inducers. MBEC were exposed to TNF (10 ng/ml) + FN-γ (500 IU/ml), or vehicle in the presence or absence of PTK inhibitors added 1 hour before cytokines. Nitrite production, an indicator of NO· synthesis, was determined by colorimetric assay. Organ Chamber Studies: rings were suspended in conventional organ chambers for measurement of isometric force. To induce NOS activity, rings were incubated in 1 ml of DMEM/F12 containing IL-1 α (40 ng/ml), LPS (100 ng/ml), or vehicle for 5 hours in the presence or absence of PTK inhibitors added 1 hour before the induction.

Results. Release of Nitrite: Treatment of SMC or MBEC with cytokines and/ or LPS for 24 hours caused the accumulation of nitrite in the incubation medium. GA (0.01, 0.03, or 0.1 μM), added 1 hour before the NOS inducer, elicited a concentration-dependent inhibition of IL-1- and LPS + IFN-induced nitrite release from SMC and MBEC (Table 1).When GA (0.1 μM) was added 16 hours after the inducers, the accumulation of nitrite was not significantly different in the presence or absence of GA (data not shown). GE (1 μM or 10 μM) slightly increased the accumulation of nitrite in the incubation medium of SMC stimulated by IL-1 or LPS + IFN, while 50 μM of this inhibitor had no effect. Only high concentration (50 μM) of GE significantly reduced the release of nitrites induced by MBEC treated with cytokines. Similarly, the release of nitrite stimulated by IL-1 was greater in the incubation medium of SMC in the presence than in the absence of TY (10 μM). TY (1 or 50 μM) failed to inhibit the effects of IL-1. The pre-exposure of SMC with TY (1, 10, or 50 μM) was also associated with an increase of the accumulation of nitrite induced by LPS + IFN. Pretreatment of MBEC with TY (1, 10, or 50 μM) did not change the release of nitrites of these cells stimulated with cytokines.

Vascular reactivity: The incubation of aortic rings with endothelium for 5 hours in culture medium containing IL-1 or LPS shifted the concentration-contraction curves to phenylephrine (PE) significantly to the right. GA (0.1 μM) added 1 hour before IL-1 partially reversed the hyporeactivity to PE and higher concentrations of this inhibitor (0.5 μM and 1 μM) completely restored contractions to PE (Figure 1A). Treatment with a low concentration of GA (0.1 μM) induced a complete inhibition of the hypocontractility to PE in rings incubated with LPS and no additional inhibitory effects were obtained with higher doses of GA (0.5 μM or 1 μM) (Figure 1B). In contrast, GE or TY (1 μM) did not affect the vascular hypocontractility induced by LPS or IL-1 and 100 μM of these inhibitors significantly potentiated the vascular hyporeactivity induced by cytokines or LPS. When GA (0.5 μM) was added 4 hours after LPS or IL-1 into the culture medium, the contractions to PE were not signicantly different in the presence or the absence of this PTK inhibitor.

Table 1. Effects of increasing concentrations of GA (μM) on release of nitrite (μM) from MBEC and SMC exposed to cytokines and /or endotoxin.

MBEC		SMC			
TNF+IFN	47.7±5.8	IL-1	19.3±3	LPS+IFN	16.9±3
+0.01	40.57±10.74	+0.01	21.1±5.4	+0.1	19.2±3
+0.03	26.0±6.9	+0.1	2.3±0.1	+0.1	1.9±0.4
+0.1	1.04±0.63	+0.3	1.8±0.3	+0.3	1.8±0.4

Discussion. These findings suggest that geldanamycin may interfere with the signaling process which induces NOS, by altering the activity of a PTK at a point that precedes the induction of NOS and that PTK activation may be involved in the induction of NOS expression by cytokines and endotoxin in MBEC, SMC, and isolated arteries. However, GE and TY, two chemically distinct PTK inhibitors, do not appear to be inhibitors of NOS since they failed to inhibit the NO· production in MBEC or SMC and in isolated vessels induced with immunostimulants. The observation that the release of nitrite caused by different cytokines or endotoxin in MBEC and SMC is reduced in a dose-dependent manner by the pretreatment with GA, a potent and specific PTK inhibitor and the fact that the vascular hypocontractility of rat aortas treated with LPS or IL-1 was completely inhibited by the presence of GA, indicated the involvement of PTK in the NO· synthesis induced by NOS. Furthermore, GA does not appear to be an inhibitor of NOS since it failed to inhibit nitritc accumulation in SMC and the hypocontractility to vasoconstrictors once NOS was induced by LPS or IL-1. These results are in line with previous findings showing that the phosphorylated state of a PTK can control gene expression by altering the activity of several transcription factors. In contrast to GA, GE and TY, two chemically distinct PTK inhibitors, do not or slightly alter the signaling process leading to the induction of NOS since they do not inhibit LPS- or IL-1-induced the release of nitrite in MBEC and SMC, or the vascular hypocontractility in isolated arteries.

These results demonstrate that PTK are essential in the regulation of iNOS activity in MBEC, SMC, and in isolated vessels and that PTK inhibitors may be useful in the treatment of diseases such as septic shock.

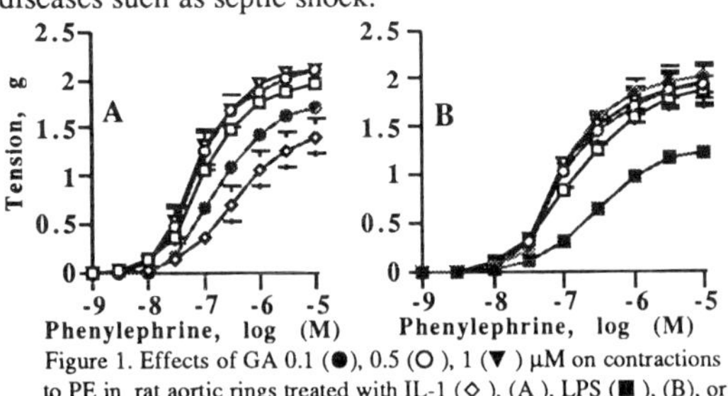

Figure 1. Effects of GA 0.1 (●), 0.5 (O), 1 (▼) μM on contractions to PE in rat aortic rings treated with IL-1 (◇), (A), LPS (■), (B), or vehicule (□)

The induction of nitric oxide synthase by lipoteichoic acid in cultured murine macrophages requires the activation of tyrosine kinase and NF-κB

MURALITHARAN KENGATHARAN, SJEF J. DE KIMPE and CHRISTOPH THIEMERMANN

The William Harvey Research Institute, St Bartholomew's Hospital Medical College, Charterhouse Square, London EC1M 6BQ, UK.

Introduction

Lipoteichoic acid (LTA; derived from the cell wall of the gram-positive bacterium *Staphylococcus aureus*) causes an enhanced formation of nitric oxide (NO) due to expression of the inducible isoform of nitric oxide synthase (iNOS) in murine macrophages [1]. The iNOS gene from murine macrophages has been cloned and expressed [2], but little is known about the intracellular signal transduction mechanisms involved in the expression of iNOS, especially when the enzyme is induced by LTA. Here, we investigate the role of (i) phosphatidylcholine-phospholipase C (PC-PLC), (ii) tyrosine kinase and (iii) the transcription factor NF-κB in the expression of iNOS induced by LTA in J774.2 macrophages.

Methods

Murine macrophages (J774.2) were cultured in 96-well plates with DMEM containing 10% foetal calf serum until confluent. The iNOS activity was induced by adding to the medium LTA (10μg/ml). Nitrite accumulation in the medium, an indicator of NOS activity, was measured at 24 h after the addition of LTA by the Griess method. Drugs were added to the medium 15 min prior to or 10 h after the addition of LTA in order to elucidate whether the drugs inhibited the induction or the activity of iNOS. Cell respiration, an indicator of cell viability, was assessed at 24 h by the mitochondrial-dependant reduction of 3-(4,5-dimethyl-thiazol-2-yl)-2,5-diphenyltetrazolium bromide (MTT) to formazan.

Results and Discussion

LTA (0.01-10μg/ml) induced within 24h a concentration-dependant increase in the concentration of nitrite in the culture medium of J774.2 macrophages. The increase in nitrite elicited by LTA (e.g. at 10μg/ml; from 2±1 to 35±2 μM) was due to an enhanced formation of NO, for (i) it was abolished by two different NOS inhibitors, N^G-methyl-L-arginine (L-NMMA) and aminoethyl-isothiourea (AE-ITU) (Table 1), and (ii) the inhibition afforded by these NOS inhibitors was restored by L-arginine, but not by D-arginine. Furthermore, the enhanced formation of NO was due to induction of iNOS, for (i) LTA caused the expression of iNOS protein in J774.2 macrophages, and (ii) both the increase in nitrite as well as the expression of iNOS protein were attenuated by dexamethasone or cycloheximide (Table 1).

Table 1. Effect of protein synthesis inhibitors and NOS inhibitors on the nitrite formation induced by LTA

Each value represents the mean ± S.E.M. from triplicate determinations (wells) from 3 separate experimental days.

\# P<0.05 vs nitrite formation at -15 min.

Inhibitor	Concentration	Nitrite (% LTA control) time of treatment after LTA (-15 min)	(10 hours)
L-NMMA	100μM	36±5	40±4
AE-ITU	10μM	34±2	38±3
Dexamethasone	1μM	34±4	94±2#
Cyclohexamide	0.3μg/ml	16±2	91±2#

The induction of iNOS caused by LTA is not due to contamination with endotoxin, for polymixin B, an agent which binds and inactivates endotoxin, abolished the increase in nitrite caused by endotoxin, but not by LTA.

Expression of proteins including iNOS by stimulation with pro-inflammatory cytokines is mediated by transcription factors such as NF-κB [3]. NF-κB is normally held in the cytoplasm in an inactivated state by the inhibitor protein IκB-α. For activation of NF-κB, degradation of IκB-α by the enzyme IκB-α protease is essential [4]. We show that PDTC which is an inhibitor of NF-κB activation [3] prevented the expression of iNOS protein and activity (nitrite) caused by LTA in J774.2 cells (Table 2). In addition, TPCK or calpain inhibitor I, which inhibit IκB protease [3,4], prevented the LTA induced NO formation. In contrast, however, chymostatin which is structurally similar to calpain inhibitor I and which does not inhibit IkB-protease did not prevent iNOS induction. This suggests that activation of IκB-α protease is important in the induction of iNOS caused by LTA. The inhibition by PDTC, TPCK or calpain inhibitor I, of the increase in nitrite stimulated by LTA was significantly less when they were added to the macrophages 10 h after LTA. These results clearly show that iNOS induction by LTA involves the activation of NF-κB.

PC-PLC controls the activation of NF-κB in response to the binding of TNF-α to the TNF-α receptor [3]. We show here (Table 2) that induction of iNOS protein caused by LTA was prevented by D609 (an inhibitor of PC-PLC) indicating that PC-PLC plays a role in the signal transduction leading to the expression of iNOS. One of the determinants of NF-κB activation is the redox status of the cell [3] and this is determined by the concentration of reactive oxygen species which are frequently produced by the mitochondrial respiratory chain [3]. The antioxidants, rotenone or BHA [3], prevented the iNOS induction by LTA suggesting that reactive oxygen species are generated during the activation of the macrophages by LTA. We also demonstrate (Table 2) that three structurally distinct tyrosine kinase inhibitors, genistein (competitive inhibitor at the ATP-binding site), erbstatin or tyrphostin AG126 (both competitive inhibitors at the substrate binding site) inhibited the expression of iNOS caused by LTA clearly showing that tyrosine phosphorylation plays an important part in the signal transduction leading to the expression of iNOS by LTA.

Table 2. Effect of inhibitors of tyrosine kinase and NF-κB activation on the nitrite formation induced by LTA

Each value represents the mean ± S.E.M. from 3-4 determinations (wells) from 3-4 separate experimental days.

Inhibitor	Concentration	Nitrite (% LTA control) time of treatment after LTA (-15 min)	(10 hours)
D609	30μg/ml	16±2	65±3
Genistein	100μM	12±1	86±4
Erbstatin	10μM	19±2	98±5
Tyrphostin	10μM	24±2	96±3
PDTC	25μM	10±1	86±2
Rotenone	30μM	14±2	92±5
BHA	30μM	37±4	95±6
TPCK	30μM	9±1	93±4
Calpain inhibitor I	30μM	14±1	96±4

In conclusion, the signal transduction events leading to the expression of iNOS by LTA in murine macrophages involve (i) the activation of PC-PLC, (ii) the phosphorylation of tyrosine kinase, and (iii) the activation of the transcription factor NF-κB.

SJDK is a recipient of a Senior Fellowship of the European Union. This project is supported by a grant from Cassella AG (Germany).

References

1. Cunha, F.Q., Moss, D.W., Leal, L.M.C.C. et al. (1993) Immunology. **78**, 563-567.

2. Lyons, C.R., Orloff, G.J. and Cunningham, J.M. (1992) J. Biol. Chem. **267**, 6370-6374.

3. Baeuerle, P.A. and Henkel, T. (1993) Ann. Rev. Immunol. **12**, 141-179.

4. Lin, Y-C., Brown, K. and Siebenlist, U. (1993) Proc. Natl. Acad. Sci. USA. **92**, 552-556.

Abbreviations used: aminoethyl-isothiourea, AE-ITU; butylated hydroxyanisole, BHA; inducible nitric oxide synthase, iNOS; nitric oxide, NO; lipoteichoic acid, LTA; pyrrolidine dithiocarbamate, PDTC; L-1-tosylamido-2-phenylethyl chloromethyl ketone, TPCK; tricyclodecan-9-yl-xanthogenate, D609; phosphatidylcholine phospholipase C, PC-PLC; N^G-metyl-L-arginine, L-NMMA; tumour necrosis facotor-α, TNFα.

In murine 3T3 fibroblasts, all second messenger pathways effectively inducing NO synthase II (iNOS) converge in the activation of transcription factor NF–κB

HARTMUT KLEINERT, CHRISTIAN EUCHENHOFER, IRMGARD IHRIG-BIEDERT, AND ULRICH FÖRSTERMANN

Department of Pharmacology, Johannes Gutenberg University, Obere Zahlbacher Str. 67, D-55101 Mainz, Germany

Transcription factor NF-κB is essential for the induction of NOS II expression in murine macrophages by bacterial lipopolysaccharide (LPS) and interferon-γ (IFN-γ) [1]. In some cell types, agents other than LPS and cytokines can induce NOS II expression. The molecular induction pathways utilized by these agents could differ from the cytokine pathway as reported for rat mesangial cells where pyrrolidine dithiocarbamate (PDTC), an inhibitor of NF-κB, suppressed interleukin-1β-, but not cAMP-stimulated NOS II mRNA expression [2].

Stimulation of Different Second Messenger Pathways Induced NOS II mRNA Expression in 3T3 cells. NOS II mRNA was markedly induced with IFN-γ (100 U/ml) or TNF-α (10 ng/ml) (Fig. 1). Similar levels of NOS II expression were inducible with TPA (50 ng/ml), or the cAMP-elevating agents forskolin (100 μM) and 8-bromo-cAMP (1 mM) (Fig. 1). 8-bromo-cGMP (1 mM) was ineffective as a stimulator of NOS II induction (Fig. 1). Thus the stimulation of the receptor tyrosine kinase pathway (by INF-γ and TNF-α), the stimulation of the protein kinase C pathway (by TPA), and the stimulation of the protein kinase A pathway (by forskolin and 8-bromo-cAMP) all induced the transcription of NOS II mRNA in 3T3 fibroblasts.

Stimulation of Three Different Second Messenger Pathways in 3T3 Cells Induced Proteins With NF–κB Binding Activity. NF-κB is a multi-subunit transcription factor that can rapidly activate the expression of genes involved in immune and acute phase responses [3]. NF–κB is composed mainly of proteins with molecular weights of 50 kDa (p50) and 65 kDa (p65). Both types of proteins share significant homology with the proto-oncogene c-rel [4-6]. These proteins can interact with each other and, following activation, bind a NF–κB response element as homo- or heterodimers [7]. In its unstimulated form, NF–κB is present in the cytosol bound to the inhibitory protein I-κB. After induction of cells by a variety of agents, NF–κB is released from I-κB and translocated to the nucleus. Agents that have been described as NF–κB activators include mitogens, cytokines and LPS, TPA and cAMP [3, 8]. Our eletrophoretic mobility shift assays (EMSA) showed that nuclei of untreated 3T3 cells contained low concentrations of proteins that bind an oligonucleotide containing a NF-κB response element. Incubation of 3T3 cells either with TPA (50 ng/ml), INF-γ (100 U/ml) or forskolin (100 μM) induced a marked increase in this binding activity (Fig. 2). The protein-DNA interaction was prevented in all cases by the addition of excess unlabeled oligonucleotide containing the NF-κB-site of the murine NOS II promoter (Fig. 2).

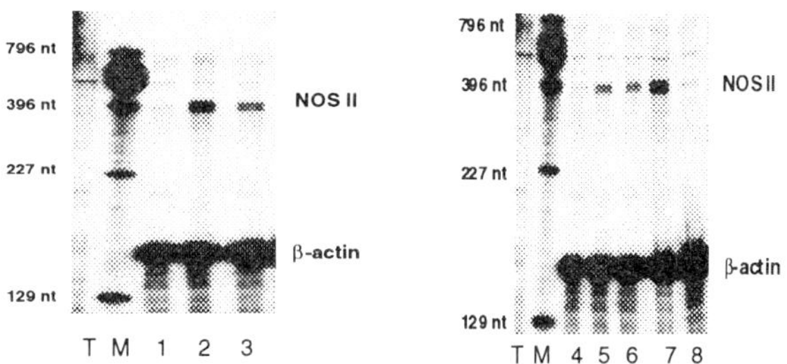

Fig. 1. S1-nuclease protection analyses using cDNA probes for murine NOS II and β-actin (for standardization). RNAs were prepared from untreated 3T3 fibroblasts (control, lane 1 and 4) and 3T3 cells incubated with human tumor necrosis factor α (TNF-α, 10 ng/ml, 4 h, lane 2), murine interferon-γ (IFN-γ, 100 U/ml, 4 h, lane 3), the protein kinase C stimulator tetradecanoyl phorbol-13-acetate (TPA, 50 ng/ml, 4 h, lane 5), the adenylyl cyclase-stimulating agent forskolin (100 μM, 4 h, lane 6), 8-bromo-cAMP (1 mM, 4 h, lane 7), or 8-bromo-cGMP (1 mM, 4 h, lane 8). T: tRNA control; M: molecular weight markers.

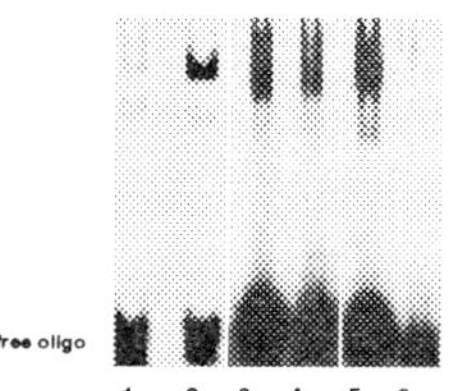

Fig. 2. Electrophoretic mobility shift assay (EMSA) using a 5'-end-labeled consensus oligonucleotide for NF-κB binding and nuclear extracts from 3T3 fibroblasts (lanes 1-4) and murine RAW 264.7 macrophages as positive controls (lanes 5 and 6). 3T3 cells were incubated for 3 h with medium alone (negative control, lane 1), tetradecanoyl phorbol-13-acetate (TPA, 50 ng/ml, lane 2), interferon-γ (IFN-γ, 100 U/ml, lane 3), or forskolin (100 μM, lane 4). RAW 264.7 macrophages were induced with bacterial lipopolysaccharide (LPS, 1 μg/ml, lane 5), and the same nuclear protein extract from RAW 264.7 cells was tested in the presence of 100-fold excess of an oligonucleotide containing the NF-κB binding site of the murine NOS I promoter (lane 6).

Inhibition of NF-κB Activation Blocks NOS II mRNA Induction. The activation of NF–κB can be blocked by thiol compounds, such as PDTC, which leave the DNA-binding activity of other transcription factors (e.g. SP1, Oct and CREB) unaffected [9]. PDTC has been shown to prevent the induction of NOS II in LPS-induced murine macrophages [1] and rat alveolar macrophages [10]. In the current series of experiments on 3T3 cells, PDTC prevented the induction of the NOS II mRNA expression in response to all inducing compounds used (Fig 3). This confirms the results of the EMSAs and indicates that in 3T3 fibroblasts NF–κB is essential for NOS II induction in response to the different second messengers. In conclusion, our data demonstrate in 3T3 cells that at least three different signal transduction pathways can stimulate NOS II mRNA expression, namely the cytokine–receptor tyrosine kinase pathway, the cAMP-protein kinase A pathway, and the protein kinase C pathway. All these pathways seem to converge in the activation of the essential transcription factor NF–κB which increases the transcription of the NOS II gene. These data suggest that the receptor tyrosine kinase pathway, the protein kinase A pathway, and the protein kinase C pathway lead to the activation of transcription factor NF-κB.

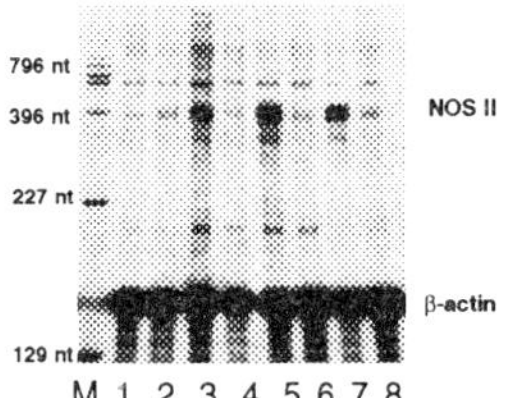

Fig. 3. S1-nuclease protection analysis using cDNA probes for murine NOS II and β-actin (for standardization). RNAs were obtained from unstimulated 3T3 fibroblasts (lane 1), 3T3 cells exposed to the inhibitor of NF-κB, pyrrolidine dithiocarbamate (PDTC, 100 μM, 4 h, lane 2), and 3T3 cells stimulated with various agents in the absence and presence of PDTC (100 μM). The following stimulating agents were used: The adenylyl cyclase-stimulating agent forskolin (100 μM, 4 h, lane 3) and forskolin in the presence of PDTC (lane 4), interferon-γ (IFN-γ, 100 U/ml, 4 h, lane 5) and IFN-γ in the presence of PDTC (lane 6), the protein kinase C stimulator tetradecanoyl phorbol-13-acetate (TPA, 50 ng/ml, 4 h, lane 7) and TPA in the presence of PDTC (lane 8). M: molecular weight markers.

References

1. Xie, Q.W., Kashiwabara, Y. & Nathan, C. *J Biol Chem* **269**, 4705-4708 (1994).
2. Eberhardt, W., Kunz, D. & Pfeilschifter, J. *Biochem Biophys Res Commun* **200**, 163-170 (1994).
3. Baeuerle, P.A. *Biochim Biophys Acta* **1072**, 63-80 (1991).
4. Bours, V., Villalobos, J., Burd, P.R., Kelly, K. & Siebenlist, U. *Nature* **348**, 76-80 (1990).
5. Ghosh, S., *et al. Cell* **62**, 1019-1029 (1990).
6. Kieran, M., *et al. Cell* **62**, 1007-1018 (1990).
7. Kunsch, C., Ruben, S.M. & Rosen, C.A. *Mol Cell Biol* **12**, 4412-4421 (1992).
8. Serkkola, E. & Hurme, M. *Febs Lett* **334**, 327-330 (1993).
9. Schreck, R., Meier, B., Mannel, D.N., Droge, W. & Baeuerle, P.A. *J Exp Med* **175**, 1181-1194 (1992)
10. Sherman, M.P., Aeberhard, E.E., Wong, V.Z., Griscavage, J.M. & Ignarro, L.J. *Biochem Biophys Res Commun* **191**, 1301-1308 (1993).

New mechanisms of action for aspirin: inhibition of NFκB activation, expression of nitric oxide synthase, and *de novo* protein synthesis.

Guim Kwon, John A. Corbett*, Jeanette R. Hill, and Michael L. McDaniel

Department of Pathology, Washington University School of Medicine, St. Louis, MO 63110
*Department of Biochemistry and Molecular Biology, St. Louis University School of Medicine, St. Louis, MO 63104

Aspirin and aspirin-like drugs (ALD) are the most commonly indicated agents for the treatment of inflammation. Aspirin was originally believed to reduce inflammation by the inhibition of cyclooxygenase (prostaglandin H synthase) activity, resulting in a reduced production of prostaglandins (1). Recently, other mechanisms have been proposed to explain the anti-inflammatory effects of aspirin, which include the interference of cellular signaling by binding key regulatory elements such as G proteins and inhibition of NFκB (2,3). NFκB, an eukaryotic transcriptional factor, plays a pivotal role in the regulation of inducible nitric oxide synthase (iNOS). In this study, we examined the effects of aspirin on NFκB activation, iNOS expression, and *de novo* protein synthesis using the insulinoma cell line, RINm5F, and the primary rat islets.

Methods

Effects of aspirin on NFκB activation and iNOS expression were studied by gelshift assays (4) and Western blot analysis (5), respectively. Inhibition of *de novo* protein synthesis by aspirin was investigated by measuring [^{35}S]methionine incorporation into rat islet proteins and in vitro translation of Brome Mosaic Virus (BMV) viral proteins (Amersham Life Science, Buckinghamshire, England).

Results

Aspirin inhibits IL-1-induced translocation of NFκB to the nucleus in a concentration-dependent manner with maximal inhibition observed at 20 mM (Fig. 1). Control cells (lane 1) show minimal NFκB binding activity. NFκB activation is required for the transcription of iNOS. As shown in Fig. 2, aspirin inhibits IL-1-induced iNOS protein expression in a concentration-dependent manner with an IC 50 ~10 mM. Aspirin was also shown to block IL-1-induced formation of nitrite, an oxidized form of nitric oxide, in the similar concentration-dependent manner as iNOS protein expression (data not shown).

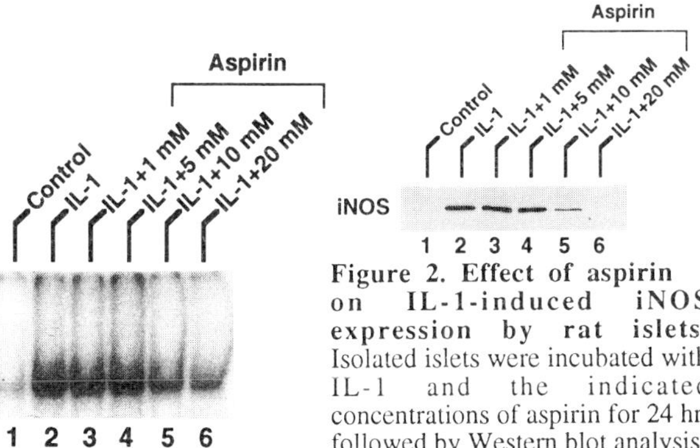

Figure 1 Effect of aspirin on DNA Binding activity of the transcriptional factor NFκB.

Aspirin
Control
IL-1
IL-1+1 mM
IL-1+5 mM
IL-1+10 mM
IL-1+20 mM
iNOS
1 2 3 4 5 6

Figure 2. Effect of aspirin on IL-1-induced iNOS expression by rat islets. Isolated islets were incubated with IL-1 and the indicated concentrations of aspirin for 24 hr, followed by Western blot analysis.

The effects of aspirin on protein synthesis were further investigated by [^{35}S]methionine incorporation into rat islets, and the translation of BMV mRNA in a rabbit reticulocyte system. As shown in Fig. 3, aspirin inhibits [^{35}S]methionine incorporation into islets with an IC 50 ~10 mM. Aspirin inhibits protein synthesis to a level comparable to the protein synthesis inhibitor cycloheximide (10 mM). As shown in Fig. 4, aspirin in a concentration-dependent manner also inhibits the translation of four BMV viral proteins with molecular weights of 109, 94, 35, and 20 kDa. The two higher molecular weight proteins, 109 and 94 kDa, are reported to often run as one band as indicated in Fig. 4. Exclusion of BMV mRNA or incubation with cycloheximide (10 mM) completely prevents protein synthesis (Fig. 4, lane 1 and 3, respectively).

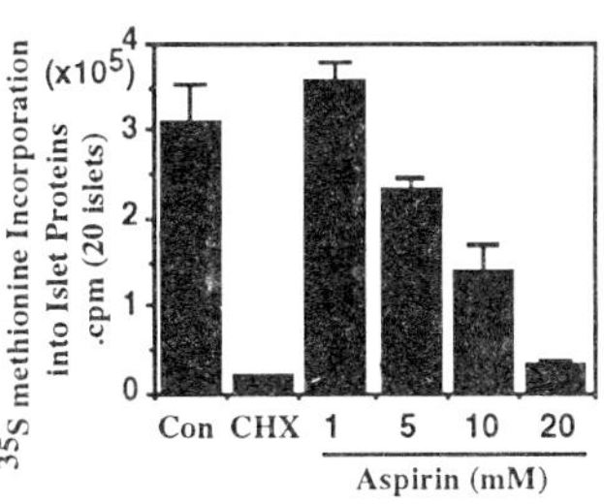

Figure 3. Effect of aspirin on [^{35}S] methionine incorporation into rat islets. Isolated rat islets (100) were cultured at 37°C for 24 hr in 1 ml of media containing 10 μM cyclo-heximide, or the indicated concentrations of aspirin. [^{35}S]methionine was included in all samples. The ^{35}S content of islet proteins was determined by liquid scintillation spectrometry.

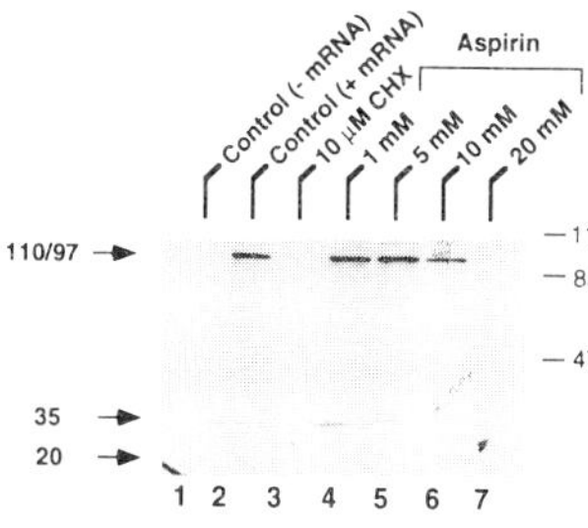

Figure 4. Effect of aspirin on *in vitro* translation of Brome Mosaic virus mRNA. Increasing concentrations of aspirin (1- 20 mM) were incubated at 30°C for 1 hr with translational mixture. Samples were run on a 10% SDS acrylamide gel, and transferred to Hybond ECL Nitrocellulose, followed by immunoblot analysis.

Discussion

In this study, we have shown that aspirin inhibits NFκB activation, and *de novo* protein synthesis determined by three different assay systems; 1) inhibition of IL-1-induced iNOS protein expression, 2) [^{35}S]methionine incorporation into cells, and 3) *in vitro* translation of BMV mRNA using a rabbit reticulocyte lysate system.

Aspirin and sodium salicylate are generally prescribed at 1 to 3 mM plasma concentration (6,7). Even higher concentrations of salicylates may be achieved at the site of inflammation such as joints due to facilitated uptake of salicylates. Our study indicates that mechanisms of action of aspirin under these conditions may be due to inhibition of the transcriptional factor NFκB and down-regulation of NFκB-dependent gene activation that are important in inflammation. The inhibition of protein synthesis stimulated by cytokines is yet another potential mechanism by which aspirin attenuates chronic inflammation and pain.

Acknowledgments

This work was supported by National Institutes of Health grants DK-06181 and a Lucille P. Markey Pathway postdoctoral fellowship.

References

1. Vane, J. R. (1971)Nature **231**, 232-235
2. Weissmann, G. (1991) Sci. Am. **January**, 84-90
3. Kopp, E. & Ghosh, S. (1994) Science **265**, 956- 958
4. Sen, R. & Baltimore, D. (1986) Cell **46**, 705-716
5. Spinger, T. A. (1991) Current Protocols in Molecular Biology **V1**, 4.2.1-4.2.6
6. Rainsford, K. D. (1984) in Aspirin and the Salicylates (Butterworths, London)
7. Famay, J. P. & Paulus, H. E. Eds., (1992) Therapeutic Applications of NSAIDS... (Dekker, New York).

Discordant responses of mRNAs encoding inducible (type II) nitric oxide synthase and other inducible enzymes in aspirin- and sodium salicylate-treated RAW 264.7 cells

Diane Kepka-Lenhart and Sidney M. Morris, Jr.

Dept. of Molecular Genetics & Biochemistry, University of Pittsburgh, Pittsburgh, PA, 15261, USA

In recent years, induction of nitric oxide (NO) synthesis has been identified as one of the major responses to inflammatory stimuli in many cell types. This reflects expression of genes coding for the inducible isoform of NO synthase (iNOS) and for enzymes involved in synthesis of substrate and cofactors for the iNOS enzyme (1). The resulting high level NO production endows cells such as macrophages with potent antimicrobial activity against a variety of pathogens. However, overproduction of NO during infection and inflammation also can have deleterious effects (e.g., in autoimmune diseases and sepsis). Although anti-inflammatory agents are often administered to prevent the action of components of the host response to inflammatory or infectious stimuli, relatively little is known regarding the impact of such agents on induced NO production.

For more than two decades, the anti-inflammatory actions of aspirin--the most commonly used anti-inflammatory drug--and its metabolite salicylate have been attributed predominantly to inhibition of prostaglandin synthesis via inhibition of cyclooxygenase activity. However, because several recent studies have identified additional effects of these agents which may alter iNOS expression (2-4), this study tested whether aspirin or sodium salicylate also would inhibit NO production and induction of iNOS mRNA in the RAW 264.7 murine macrophage cell line during stimulation by lipopolysaccharide (LPS) or interferon-γ (IFN-γ).

Results of these studies are summarized as follows: (a) Concentrations of aspirin or sodium salicylate of 10-20 mM sharply inhibited NO production by stimulated RAW cells, whereas concentrations of 1 mM or less had no effect. (b) Added prostaglandin E_2 did not overcome the inhibitory effects of aspirin or salicylate, showing that inhibition of prostaglandin synthesis was not responsible for inhibition of NO synthesis. (c) Aspirin and salicylate inhibited iNOS mRNA induction in LPS-stimulated cells but enhanced the induction in IFN-γ-stimulated cells. (d) Inhibition of NO production occurred independently of effects on iNOS mRNA induction, indicating that aspirin and salicylate act at multiple sites in the NO expression pathway. (e) Unlike induction of iNOS mRNA, induction of argininosuccinate synthetase mRNA by either LPS or IFN-γ was consistently inhibited by aspirin and salicylate.

Although two other groups have recently reported inhibition of NO production by aspirin and other nonsteroidal anti-inflammatory drugs (5-6), this study differs from previous reports in several important respects: this study is the first to examine effects of these agents on responses to IFN-γ alone; this is the first report that aspirin and salicylate can actually enhance the induction of iNOS mRNA as well as inhibiting it; this is the first study to determine that aspirin and salicylate can have different effects on multiple mRNAs induced by LPS or IFN-γ.

Several conclusions may be drawn from the present study: (1) The fact that inhibition of NO production occurred independently of effects on iNOS mRNA induction indicates that aspirin and salicylate act at multiple sites in the NO expression pathway. (2) Inhibition of LPS-dependent induction of iNOS mRNA by aspirin or salicylate is probably due to inhibition of the activation of the transcription factor NFκB. (3) The fact that aspirin and salicylate affect the inductions of iNOS and argininosuccinate synthetase mRNAs in completely divergent ways may reflect different effects of these agents on distinct IFN-γ-regulated transcription factors.

Because of the relatively high concentrations of aspirin or salicylate needed to inhibit NO production in these studies, it is unclear whether this effect contributes significantly to the anti-inflammatory effects of these agents at usual therapeutic dosages *in vivo*. Nonetheless, identification of the precise sites and mechanisms of action of these agents may be useful in designing novel strategies/drugs to modulate induced NO synthesis, as well as other inflammatory responses.

Acknowledgements. This work was supported in part by NIH grant GM50897.

References

1. Morris, S.M., Jr., and Billiar, T.R. (1994) New insights into the regulation of inducible nitric oxide synthase. Am. J. Physiol. **266**, E829-E839.
2. Wu, K.K., Sanduja, R., Tsai, A.-H., Ferhanoglu, B., and Loose-Mitchell, D.S. (1991) Aspirin inhibits interleukin 1-induced prostaglandin H synthase expression in cultured endothelial cells. Proc. Natl. Acad. Sci. USA **88**, 2384-2387.
3. Kopp, E., and Ghosh, S. (1994) Inhibition of NF-κB by sodium salicylate and aspirin. Science **265**, 956-959.
4. Takashiba, S., van Dyke, T.E., Shapira, L., and Amar, S. (1995) Lipopolysaccharide-inducible and salicylate-sensitive nuclear factor(s) on human tumor necrosis factor alpha promoter. Infect. Immun. **63**, 1529-1534.
5. Brouet, I., and Ohshima, H. (1995) Curcumin, an anti-tumor promoter and anti-inflammatory agent, inhibits induction of nitric oxide synthase in activated macrophages. Biochem. Biophys. Res. Commun. **206**, 533-540.
6. Aeberhard, E.E., Henderson, S.A., Arabolos, N.S., Griscavage, J.M., Castro, F.E., Barrett, C.T., and Ignarro, L.J. (1995) Nonsteroidal anti-inflammatory drugs inhibit expression of the inducible nitric oxide synthase gene. Biochem. Biophys. Res. Commun. **208**, 1053-1059.

Abbreviations used: LPS, lipopolysaccharide; IFN-γ, interferon-γ.

THE ROLE OF NUCLEAR FACTOR KB IN IMMUNOSTIMULANT-INDUCED GTP CYCLOHYDROLASE I GENE EXPRESSION AND TETRAHYDROBIOPTERIN SYNTHESIS IN RAT VASCULAR SMOOTH MUSCLE: RELATION TO NITRIC OXIDE SYNTHASE

Yoshiyuki Hattori[1]*, Kikuo Kasai[1], Sachiko Hattori[1], Shin-Ichi Shimoda[1], Nobuo Nakanishi[2], and Steven S. Gross[3]

[1]Department of Endocrinology, Dokkyo University School of Medicine, Tochigi, Japan; [2]Department of Biochemistry, Meikai University School of Dentistry, Saitama, Japan; [3]Department of Pharmacology, Cornell University Medical College, NY, USA

Tetrahydrobiopterin (BH4) is an essential cofactor of all isoforms of nitric oxide (NO) synthase. Bacterial lipopolysaccharide (LPS) and other immunostimulants induce an isoform of NO synthase (iNOS) in vascular smooth muscle (VSM) (1). The induction of iNOS and the overproduction of NO in VSM have been implicated in the genesis of septic and cytokine-induced circulatory shock. Whereas the induction of iNOS is necessary for immunostimulant-mediated NO overproduction, *de novo* BH4 synthesis is also elicited in VSM by immunostimulants and is essential for iNOS activity (1,2). Given that nuclear factor kB (NF-kB) mediates the induction of NOS gene expression by lipopolysaccharide (LPS), the role of NF-kB in the induction of GTPCH in LPS-stimulated rat VSM cells was assessed by examining the effects of pyrrolidine dithiocarbamate (PDTC), an inhibitor of the activation of NF-kB(3), on the abundance of GTPCH mRNA and biopterin synthesis.

MATERIALS AND METHODS

Cell culture: VSM cells were isolated from thoracic aortae of male Wistar rats as previously described (1).

* Corresponding author.

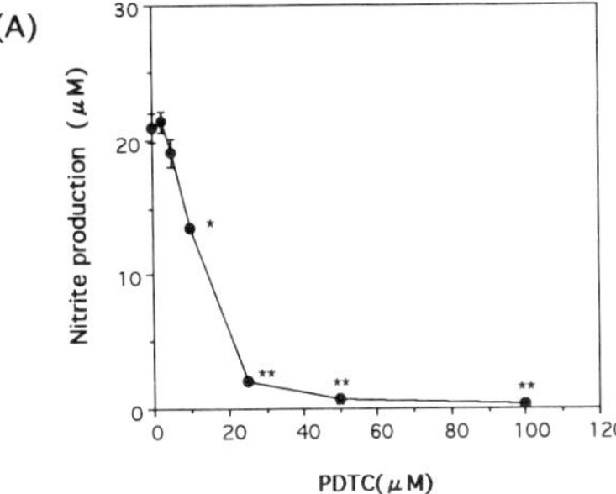

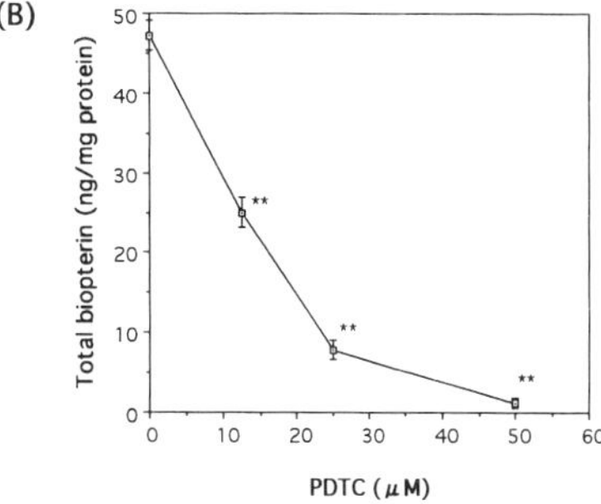

Fig.A and B Effect of PDTC on nitrite production in LPS/IFN-stimulated rat aortic VSM cells. Cells were treated with a combination of LPS (30 μg/ml) and IFNγ (50 ng/ml) in the presence of various concentrations of PDTC for 24 h, after which (A)nitrite accumulation in the culture medium and (B)total biopterin (cellular contents plus accumulation in the culture medium) were measured. Data are means ± S.E.M. (n=9). *$P < 0.05$ and **$P < 0.01$ *versus* LPS/IFN in the absence of PDTC.

Nitrite assay: Nitrite was used to indicate cellular synthesis of NO. Accumulation of nitrite in the VSM cell culture medium was quantified by colorimetric assay using the method of Griess, as previously described (1).
Biopterin assay: Cellular and extracellular biopterin (BH4 and more oxidized species) was measured by HPLC analysis after oxidation by iodine as previously described (1).
Analysis of iNOS and GTPCH mRNA abundance: Reverse transcription-polymerase chain reaction (RT-PCR) analysis was performed using gene specific primers for iNOS and GTPCH as previously described (2).

RESULTS AND DISCUSSION

NF-KB has been suggested to mediate iNOS induction in LPS- or cytokine-activated macrophages and mesangial cells, on the basis of studies with inhibitors of NF-KB. Furthermore, two NF-KB binding sites have been identified in the promoter region of the iNOS gene and have been shown to be required for inducibility by LPS in mouse macrophages (4,5). However, a role for NF-KB has not previously been demonstrated in the induction of GTPCH gene expression or synthesis of BH4, an essential cofactor of all isoforms of NO synthases.

To assess the effects of PDTC on iNOS induction, we incubated VSM cells with a combination of LPS (30 μg/ml) and interferon-γ (50 ng/ml) (LPS/IFN) in the presence of PDTC (2.5 to 100 μM). After 24 h, the accumulation of nitrite, a stable oxidation product of the unstable free radical NO, was determined as a measure of NO synthesis. Induction of NO synthesis by LPS/IFN was inhibited in a dose-dependent manner by PDTC at concentrations up to 25 μM with a half-maximal inhibitory concentration of ~12 μM; at concentrations of > 25 μM, PDTC completely inhibited NO production (Fig. A). In parallel with the nitrite assay, the cytotoxic effect of PDTC on LPS/IFN-treated VSM cells was assessed by the mitochondria-dependent reduction of MTT to formazan. Cell respiration was significantly inhibited by PDTC at concentration of 100 μM (data not shown).

Total (cellular plus extracellular) biopterin synthesis was markedly induced in VSM cells by incubation with LPS/IFN for 24 h and this increase was inhibited by PDTC in a dose-dependent manner with a half-maximal inhibitory concentration of ~17 μM (Fig. B). Up to 94% of the newly synthesized biopterin was released into the culture medium in LPS/IFN-treated VSM cells; changes in intracellular BH4 concentrations therefore appeared relatively small. This may suggest that BH4 synthesized in VSM cells serve as a cofactor for iNOS not only in VSM cells themselves but also in adjacent tissues or cells.

To evaluate the basis for the inhibition by PDTC of induction of BH4 and NO synthesis, we investigated the effect of PDTC on the LPS-induced increase in GTPCH and iNOS mRNA abundance by RT-PCR. Whereas the abundance of GTPCH and iNOS mRNA is low in untreated VSM cells, LPS/IFN markedly increased the amounts of both transcripts after 8 h. PDTC (25 μM) markedly inhibited the LPS/IFN-induced increase in both GTPCH and iNOS mRNA (data not shown).

LPS increases GTPCH mRNA abundance in the presence of a protein synthesis inhibitor, indicating no requirement for intermediary protein synthesis (2). Cloning of the 5' flanking region of the rat GTPCH gene revealed the presence of several NF-KB-binding sites (unpublished data). Together with the present data, these observations suggest that NF-KB participates in the regulation of GTPCH at the transcriptional level in LPS-treated VSM. Thus, we conclude that immunostimulants coinduce iNOS and GTPCH gene expression; both events are necessary for activation of cellular NO synthesis and are regulated, at least in part, by a common mechanism.

REFERENCES

1. Gross SS and Levi R (1992). J. Biol. Chem. 267, 25722- 25729.
2. Hattori Y and Gross SS (1993) Biochem. Biophys. Res. Commun. 195, 435-441.
3. Schreck R, Meier B, Mannel DN, Droge W and Baeuerle PA (1992) J. Exp. Med. 175, 1181-1194.
4. Xie Q, Kashiwabara Y and Nathan C (1994) J. Biol. Chem. 269, 4705-4708.
5. Xie Q, Whisnant R and Nathan C (1993) J. Exp. Med. 177, 1779-1784.

Smooth muscle cells transduced with cNOS require tetrahydrobiopterin for nitric oxide production and growth inhibition.

TIMOTHY SCOTT-BURDEN, CHRISTINE L. TOCK, DAVID A. ENGLER, JOHN J. SCHWARZ and SAMUEL W. CASSCELLS.

Vascular Cell Biology Laboratory, Texas Heart Institute, Texas Medical Center, Houston, TX 77225-0345.

Nitric oxide synthases require a number of cofactors for catalytic conversion of L-arginine to L-citrulline and NO [1]. One of these, tetrahydrobiopterin (BH4) is constitutively synthesized by a limited number of tissues, which include the endothelium [2]. Cultured endothelial cells produce BH4 at levels that support the maximal activity of their associated NO synthase (NOS III). In contrast cultured smooth muscle cells (SMC) produce BH4 only under conditions that elicit the induction of GTP cyclohydrolase expression, the rate determining enzyme for biopterin synthesis [3].

Strategies to genetically engineer cells *ex vivo* for introduction into cardiovascular devices are currently being evaluated. In this context SMC are under investigation as a robust cell type to use for *ex vivo* gene therapy. Two principle concerns related to this approach are the potential for uncontrolled proliferation by SMC and the thrombogenic surface that they present. In an attempt to counter both putative disadvantages of SMC for vascular prosthesis, cells were transfected with cDNA encoding NOS III.

METHODS:- SMC were isolated and propagated as described previously using standard procedures and reagents [4]. Transfections with $pREP_9$ vector (Invitrogen) constructs were performed using 2-8μg CsCl-purified plasmid DNA mixed with 2 x 10^6 cells, suspended in 0.27 M Tris-HCl buffered-sucrose (pH7.4) containing 10 mM $MgCl_2$ and 250 μM EDTA and square wave electroporation (400 V for 20μsec repeated 20 times). After transfection cells were grown in G418 selection medium and passaged routinely trypsinisation [4].

Expression of NOS III by transfectants was monitored by western analysis and by measurement of NO_2^- accumulation in condition medium in the presence of sepiapterin or following stimulation with an appropriate adenylyl cyclase agonist [3].

RESULTS AND DISCUSSION:- Human and bovine SMC expressed consistent levels of NOS III polypeptide as determined by western analysis following transfection with $pREP_9$ NOS III constructs (Fig. 1). Production of NO (NO_2^- accumulation in conditioned media) was evident only in the presence of 1 mM sepiapterin as a source of BH4 or following induction of GTP cyclohydrolase by elevation of intracellular cyclic AMP levels [3]. Mock transfected cells ($pREP_9$ only) neither expressed NOS III protein nor produced NO under the same conditions.

C 2° 3° 6° 9° 12° 18°

140 KDa→

Figure 1. Western analysis of passaged (2°-18°) human SMC transfected with $pREP_9$ NOS III constructs and selected with media containing G418 (250μg/ml). Material analyzed was total cell lysate from 100,000 cells for SMC and endothelial cells (c). Monoclonal NOS III antibodies were used at 1:750 dilution and second HRP coupled antibodies were used at 1:1500. Signals were developed by ECL using standard reagents from Amersham.

The growth kinetics of NOS III transfected SMC were slower as compared to their mock transfected counterparts ($pREP_9$ only) in the presence of sepiapterin (Fig. 2).

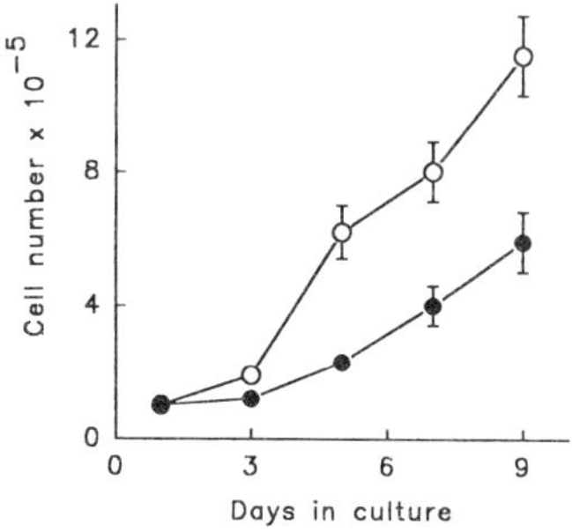

Figure 2. Growth kinetics of bovine smooth muscle cells transfected with $pREP_9$ NOS III (closed symbols) or parent vector only (open symbols) in the presence of 1 mM sepiapterin.

Vascular smooth muscle cells were successfully transfected with NOS III using Epstein-Barr virus-based vectors and electroporation. Longterm expression of NOS III was accomplished, and in the presence of a source of BH4 lead to depression of SMC growth in culture. Also transfection of SMC with $pREP_9$ GTP cyclohydrolase has been successfully carried out and in combination these constructs may facilitate the constitutive production of NO by SMC.

Successful transfection of SMC with cDNA encoding prostaglandin H synthase (PGHS) activity results in elevated production of prostacylin [5]. Stimulation of NOS III transfectants by stable analogs of prostacyclin induced GTP cyclohydrolase expression and subsequent NO production [5]. Therefore co-transfection of SMC with NOS III and PGHS constructs may allow them to be used as an alternative to endothelial cells for lining cardiovascular prosthesis.

ACKNOWLEDGEMENTS:- NIH funding grant # HL53233.

REFERENCES

1. Marletta, M.A. (1993) J. Biol. Chem. **268,** 12231-12234.
2. Schaffner, A., Blau,N., Schneemann, M., Steurer, J., Edgell, C-J.S. and Schoedon, G. (1994) Biochem. Biophys. Res. Commun. **205**, 516-523.
3. Scott-Burden, T., Elizondo, E., Ge, T., Boulanger, C.M. and Vanhoutte, P.M. (1994) Molec. Pharmacol. **46,** 274-282.
4. Scott-Burden, T., Schini, V.B., Elizondo, E., Junquero, D.C. and Vanhoutte, P.M. (1992) Circ. Res. **71,** 1088-1100.
5. Scott-Burden, T., Engler, D.A., Tock, C.L. and Casscells, S.W. (1994) in Proceedings of the ASAIO, Cardiovascular Science and Technology Conference. pp 89.

Nitric oxide synthase catalyzes the reduction of quinonoid dihydrobiopterin to tetrahydrobiopterin

Cor F.B. Witteveen, John Giovanelli and Seymour Kaufman

Laboratory of Neurochemistry, National Institute of Mental Health, Bethesda MD 20892-4096, USA.

One of the main unanswered questions concerning the mechanism of nitric oxide synthase (NOS) is the function of tetrahydrobiopterin (BH_4). The function of this cofactor has been clearly established in the reactions catalyzed by the aromatic amino acid hydroxylases, where BH_4 undergoes an oxidation-reduction cycle coupled to amino acid hydroxylation. Oxidation of BH_4 results in the formation of quinonoid dihydrobiopterin (qBH_2) which is recycled to BH_4 by dihydropteridine reductase (DHPR) in the presence of NADH. It is not known whether BH_4 undergoes similar redox reactions during NOS catalysis. NOS catalysis is not dependent on added DHPR. Therefore, if coupled oxidation of BH_4 does occur during nitric oxide synthesis, qBH_2 reduction should be an intrinsic property of NOS. As an initial step in elucidating the possible role of BH_4 recycling in NOS catalysis, we determined the capacity of NOS to reduce qBH_2 to BH_4.

The standard method for analyzing biopterins is by HPLC analysis of the sample after its oxidation under acid and basic conditions. This method does not distinguish between qBH_2 and BH_4. We therefore developed a method to determine BH_4 that shows no interference by qBH_2. This method is based on the BH_4-coupled hydroxylation of phenylalanine to tyrosine catalyzed by phenylalanine hydroxylase.

Using this method, we were able to detect the NADPH-dependent reduction of qBH_2 to BH_4 catalyzed by NOS. The reaction requires Ca^{2+} and calmodulin for optimum activity. The reduction rate with the quinonoid form of 6-methyldihydropterin is approximately twice that with qBH_2. 7,8-Dihydrobiopterin (7,8-BH_2) has negligible activity. The rate of reduction of 25 μM qBH_2 is comparable to the rate of formation of citrulline.

Reduction is inhibited by diphenyleneiodonium, indicating a role for flavins. The following findings indicate that qBH_2 reduction probably occurs at the flavin ("diaphorase") site located on the C-terminal domain [1], rather than the high affinity BH_4 binding site involved in the activation of NOS that is located on the N-terminal domain [2]: (i) Compounds with a high affinity for the BH_4-binding site (BH_4 and 7,8-BH_2) do not inhibit qBH_2 reduction. (ii) 7-Nitroindazole, which interferes with BH_4 binding to the high affinity site [3], *stimulates* qBH_2 reduction. This stimulation can tentatively be ascribed to the inhibition of reactions proceeding through heme, thereby increasing the availability of reducing equivalents for qBH_2 reduction. (iii) The concentration of qBH_2 (>25 μM) required for half maximum rate of reduction is relatively high compared to the very low concentration of BH_4 (<1 μM) required for half maximum stimulation of NOS activity.

The efficiency of added qBH_2 in stimulating NOS activity approaches that of added BH_4. qBH_2 seems to stimulate by first being converted to BH_4, which then combines with the BH_4 binding site. This postulate is supported by the observation that added qBH_2 causes little or no stimulation of NOS-catalyzed oxygenation of arginine to N^G-hydroxyarginine, a reaction that proceeds in the absence of added NADPH and consequently precludes qBH_2 reduction [4].

Whether this reduction of qBH_2 is related to the essential role of BH_4 remains to be clarified.

References

1 Bredt, D.S., Hwang, P.M., Glatt, C.E., Lowenstein, C., Reed, R.R. and Snyder, S.H. (1991) Nature **351**, 714-718.

2 Ghosh, D.K. and Stuehr, D.J. (1995) Biochemistry **34**, 801-807.

3 Klatt, P., Schmid, M., Leopold, E., Schmidt, K., Werner, E. R., and Mayer, B. (1994) J. Biol. Chem. **269**, 13861-13866

4 Campos, K., Giovanelli, J., and Kaufman, S. (1995) J. Biol. Chem. **270**, 1721-1728.

Nitric oxide inhibits the activity of GTP cyclohydrolase I

TOSHIE YONEYAMA and KAZUYUKI HATAKEYAMA

Department of Surgery, University of Pittsburgh, Pittsburgh, PA15261.

Introduction

Nitric oxide has been reported to stimulate or inhibit activity of several enzymes through covalent modifications of their amino acid residues or prosthetic group [1]. GTP cyclohydrolase I is the first and rate-limiting enzyme in the biosynthesis of tetrahydrobiopterin, one of the cofactors of nitric oxide synthases. The activity of GTP cyclohydrolase I has been suggested to be coordinately regulated with that of nitric oxide synthases. To begin to investigate the possibility that nitric oxide regulates GTP cyclohydrolase activity through direct interaction with the enzyme molecule, we examined the effect of nitric oxide on the activity of GTP cyclohydrolase I in vitro using purified preparation of the enzyme and nitric oxide solution or nitric oxide-producing chemicals, sodium nitroprusside (SNP)[1] and *S*-nitroso-*N*-acetylpenicillamine (SNAP).

Results and Discussion

The activity of GTP cyclohydrolase I was assayed after preincubation with SNP or SNAP as described previously [2]. When GTP cyclohydrolase I was preincubated with SNP, the enzyme activity was inhibited in a dose-dependent manner (Figure). SNAP also inhibited the activity of GTP cyclohydrolase I in a dose-dependent manner. The IC_{50} values

Abbreviations used: SNP, sodium nitroprusside; SNAP, *S*-nitroso-*N*-acetylpenicillamine.

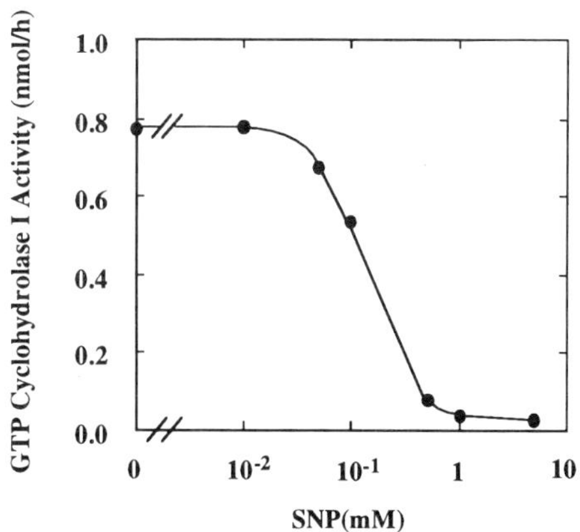

were 0.025 and 0.25 mM for SNP and SNAP, respectively. Because the enzyme activity remained to be inhibited after removal of small molecules, including SNAP or SNP, from the preincubated enzyme solution using gel filtration, the effect of nitric oxide on the enzyme activity was thought to be irreversible and hence through covalent modification. When the enzyme substrate GTP was included in the preincubation mixture, the effect of SNP and SNAP on the enzyme activity was 40% reversed at a GTP concentration of 1.0 mM. The protective effect was specific to GTP; ATP and GMP showed no protection at a concentration of 10 mM.

These results indicated that nitric oxide inhibits GTP cyclohydrolase I and the effect is through some covalent modification, partly involved in GTP binding site of the enzyme. This observation seems to be interesting in terms of the possibilities that nitric oxide may feedback regulate its production through regulating the production of a cofactor for nitric oxide synthases, tetrahydrobiopterin, and that nitric oxide produced in a cell may regulate the activity of GTP cyclohydrolase I in near-by cells. In particular, in the brain nitric oxide produced in a cell may regulate the production of dopamine, norepinephrine and serotonin in near-by cells through affecting the activity of GTP cyclohydrolase I and, consequently, affecting the activity of tyrosine hydroxylase and tryptophan hydroxylase.

References

1. Stamler, J.S. Redox signaling: Nitrosylation and related target interactions of nitric oxide. (1994) Cell **78**, 931-936.
2. Hatakeyama, K., Harada, T., Suzuki, S., Watanabe, Y., and Kagamiyama, H. Purification and characterization of rat liver GTP cyclohydrolase I. Cooperative binding of GTP to the enzyme. J. Biol. Chem. (1989) **264**, 21660-21664.

Modular structure of neuronal nitric oxide synthase: Localization of the L-arginine and pterin-binding sites

PAVEL MARTASEK, JONATHAN NISHIMURA, *KIRK McMILLAN, #QUING LIU, +JOHN SALERNO, #STEVEN S. GROSS and BETTIE SUE MASTERS

The Univ. of Texas Health Sci. Ctr. At San Antonio, TX 78284-7760 USA, Pharmacopeia. Inc., Princeton, NJ 08540, USA*, Cornell Univ. Medical Coll., New York, NY 10021, USA#, and Rensselaer Polytechnic Institute, Troy, NY 12180-3590, USA+

McMillan and Masters have reported the independent expression in *E. coli* of the heme (residues 1-714) and the flavin (residues 749-1429) domains of rat neuronal nitric oxide synthase (nNOS) [1]. The arginine and pterin (BH_4) binding sites have been localized to the heme domain. The role of BH_4 in the enzymatic conversion of L-arginine to NO has not been established; it has been suggested to serve in either a redox or allosteric role. For probing of the arginine site of nNOS, radioligand binding studies with a NOS inhibitor, [^{3}H]-N$^{\omega}$-nitro-L-arginine (NNA), have been performed [2,3]. To address more precisely the localization of the arginine-binding site and its modulation by BH_4, we expressed different modules of rat nNOS (see scheme below) as fusion proteins with glutathione S-transferase. Comparison of all three isoforms reveals that the first 220 amino acids in nNOS must play other than a catalytic role and, thus, constructs A and B start from Arg_{220}. The presence of a 160-amino acid sequence within the oxygenase domain of nNOS with apparent homology to dihydrofolate reductase (DHFR) [4], a pterin-binding enzyme, raised the possibility that this region of nNOS could contribute to the BH_4-binding subsite. This putative "DHFR module" (construct C in the scheme) was expressed in *E. coli*.

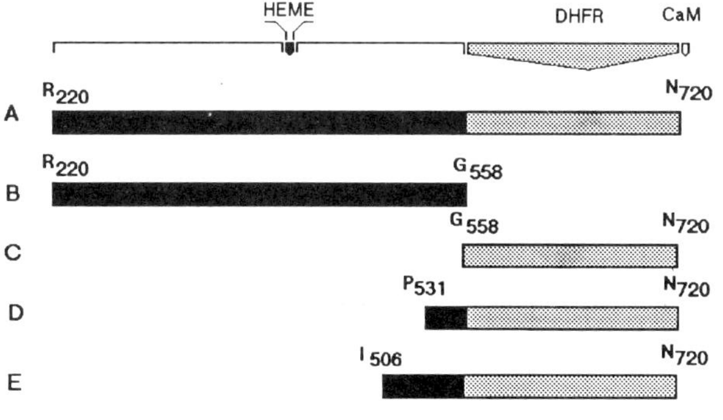

For all the constructs, plasmid pGEX-4T1 [5] (Pharmacia, Uppsala, Sweden) was digested by Bam HI and Xho I and gel purified. To amplify the DNAs coding for different motifs, a cDNA [pNOS plasmid in the Bluescript SK(-), provided by Drs. Snyder and Bredt at Johns Hopkins Medical School] was used. For transformation with resulting ligation products, *E. coli* DH5α cells were used. The cells were grown to mid-log phase at 30°C and low concentrations of isopropyl β-D-thiogalactopyranoside (0.1 - 0.3 mM) were used for induction, which was carried out at 24°C for 2-3 hours. The cells were harvested by centrifugation, resuspended in the presence of antiproteases, lysed by mild sonication and the fusion proteins were subsequently purified from the supernatant as recommended by the manufacturer.

Assays of NNA binding were performed in 96-well Millipore PVDF plates using 10 pmol of the fusion protein. The 100-μl reaction contained 50 mM TrisCl, 1 mM dithiothreitol, pH 7.6, and 300,000 cpm [^{3}H]-NNA. Simultaneous assays were conducted in the presence and absence of 10 μM BH_4 and 100 μM N$^{\omega}$-methyl-L-arginine as a competitor of NNA binding.

Construct A ($nNOS_{220\text{-}720}$) was found to bind NNA with a similar high affinity to that observed with holo-nNOS ($\approx$50 nM) and BH_4 greatly stimulated binding. While construct B ($nNOS_{220\text{-}558}$) did not bind NNA in the presence or absence of BH_4, the putative "DHFR domain" [construct C ($nNOS_{558\text{-}720}$)] was found to bind NNA, albeit with lower affinity than construct A. However, BH_4 dependence of NNA binding to construct C was not observed.

These observations are taken as a evidence that $nNOS_{558\text{-}720}$ contains sufficient sequence for arginine binding, but extension toward the N-terminus is needed for high-affinity binding and regulation by BH_4. To analyze further how far the N-terminal sequence of the "DHFR domain" ($nNOS_{558\text{-}720}$) must be extended N-terminally, we made constructs D ($nNOS_{531\text{-}720}$) and E ($nNOS_{506\text{-}720}$). In both constructs D and E, only low affinity NNA binding and no BH_4 dependence were observed. Further studies are being performed to determine more precisely the loci of L-arginine binding and BH_4 action.

Acknowledgements.
This work was supported, in part, by Grant AQ-1192 from The Robert A. Welch Foundation and National Institutes of Health Grant HL 30050 (to B.S.M.) and National Institutes of Health Grants HL 50656 and HL 44603 (to S.S.G.).

REFERENCES

1. McMillan, K. and Masters, B.S.S. (1995) Biochemistry **34**, 3686-3693.
2. Michel, A.D., Phul, R.K., Stewart, T.L. and Humphrey, P.P. (1993) Br. J. Pharmacol. **109**, 287-288.
3. Klatt, P., Schmidt, K., Brunner, F. and Mayer, B. (1994) J. Biol. Chem. **269**, 1674-1680.
4. Salerno, J.C. and Morales (1994) Biochemistry and Molecular Biology of Nitric Oxide, First International Conference (L. Ignarro and F. Murad, eds.), p. 72. UCLA, Los Angeles, CA (abstracts).
5. Smith, D.B. and Johnson, K.S. (1988) Gene **67**, 31-40.

Nitric oxide synthase: Factors concerning L-arginine mechanism

CHRISTOPHER A. FREY and OWEN W. GRIFFITH

Department of Biochemistry, Medical College of Wisconsin, Milwaukee, WI 53226, USA

Nitric oxide synthases (NOS) catalyze the five electron oxidation of L-arginine to L-citrulline and nitric oxide. The reaction proceeds in two distinct steps, with N^G-hydroxy-L-arginine (L-NOHArg) as an intermediate, shown in equation 1 (1). The enzyme functions only as a homodimer and contains one heme/monomer which mediates both partial reactions. L-Canavanine is isosteric with L-arginine, but has been reported not to be a substrate of NOS (2). The side chain of L-canavanine has a pKa of 7.0 compared with 11.0 for L-arginine (3). L-Canavanine has also been reported to inhibit NOS but the affinity of NOS for L-canavanine was not reported. In the present studies, we have examined the pH/rate profile of nNOS and binding of L-canavanine to nNOS as a function of pH.

Eqn. 1: L-Arginine + NADPH + $O_2 \rightarrow$ L-NOHArg + $NADP^+$ + H_2O
L-NOHArg + ½ NADPH + $O_2 \rightarrow$ NO + L-Citrulline + ½ $NADP^+$ + H_2O

Materials and Methods: nNOS was purified as described previously (4). NOS activity was determined by the oxyhemoglobin assay at 401 nm ($\epsilon = 0.038\ \mu M^{-1}$)(1). Difference spectroscopy was performed as previously described (5) with 1.7 μM nNOS (determined on the basis of heme concentration) at 15 °C. To force the enzyme low spin, 1mM imidazole was added first to the sample and reference cuvettes followed by L-canavanine in the sample cuvette. The volume change was < 1 %.

Results and Discussion: L-Canavanine was found to be a moderately potent inhibitor of, but was not a substrate of, nNOS at pH 5.5 or 7.5 (the level of detection was 1% the rate of the L-arginine reaction). At the pH 5.5, the side chain of L-canavanine is > 90 % protonated and therefore more closely mimics that of L-arginine; this change did not, however, result in measurable metabolism by nNOS.

The K_i of L-canavanine with nNOS was 5-fold higher than the Km of L-arginine at pH 7.5. At pH 5.5, the L-canavanine K_i to L-arginine K_m ratio was ~5:1. However, at pH 5.5, the K_i of L-canavanine and the Km of L-arginine for nNOS both increase by a factor of 4.5 relative to their values at pH 7.5. N^G-Methyl-L-arginine (L-NMA), a nNOS inhibitor which, like L-arginine, has a protonated side chain showed the same K_i trend at pH 5.5 and 7.5. From pH 5.5 to 7.5, the affinity of nNOS for all three compounds increases with the same K_i:K_m ratio. Above pH 7.5, the affinity of nNOS for L-arginine and L-NMA changes little whereas the affinity of nNOS for L-canavanine decreases substantially. The K_i and K_m changes from pH 5.5-7.5 are attributed to nNOS because the % protonation of L-arginine and L-NMA does not change throughout this range. Above pH 7.5, the % side chain protonation of L-canavanine decreases to the point where most of the side chain is unprotonated while L-arginine and L-NMA are still mostly protonated. This suggests that above pH 7.5 the increasing K_i for L-canavanine with nNOS is due to decreased side chain protonation.

A plot of V_{max} vs. pH for the L-arginine to L-citrulline nNOS reaction fits a two-proton model with an optimal V_{max} of 364 nmol/min/mg observed at pH 6.75. Titration of groups with $pK_{a1} = 5.5$ and $pK_{a2} = 8.0$ account for the ascending and descending limbs of the pH/rate curve. A plot of k_{cat}/K_m vs. pH shows a maximum at pH 7.4 with the shape controlled by changes in both k_{cat} and K_m.

Examining the K_i and K_m trends of L-canavanine and L-arginine suggests that, if the group with a pKa of 5.5 were an amino acid, then the most likely candidates would be aspartic or glutamic acid. It is our hypothesis that either aspartic or glutamic acid interacts with the positive charge on the guanidinium moiety of substrates or inhibitors, thus contributing to binding. This hypothesis is partially substantiated by the finding that % side chain protonation of L-arginine, L-NMA, and L-canavanine above pH 7.5 (>90, 99, and < 30 %) is related to their binding to nNOS.

L-Arginine binding causes a type I difference spectra with NOS, forming a high-spin, penta-coordinate heme iron (5). L-Canavanine was examined to determine if it acted like L-arginine in this respect and was found to also give a type I difference spectra. This result demonstrated that L-canavanine binding to nNOS is similar to that of L-arginine in the region near the heme cofactor.

We also considered the possibility that the changes in overall nNOS reaction rate observed on varying pH were attributable to rate changes in the reductase domain. To investigate this possibility, we examined the pH/rate profile for NADPH-dependent cytochrome c reduction. In the absence of calcium-calmodulin, the reductase domain of nNOS is effectively isolated with this assay. nNOS electron transfer activity increased markedly with increasing pH, (pH range studied was 5.5 to 9.5); a Vmax was seen at pH ≥ 9.5. The plot of V_{max} vs. pH fits a one-proton model with the titrated group having a pK_a of 8.3. The pH optimum for nNOS-catalyzed cytochrome c reduction differs markedly from nNOS-catalyzed L-arginine to NO + L-citrulline activity.

The k_{cat} vs. pH rate profile of cytochrome c reduction shows that the changes observed are due exclusively to changes in k_{cat}. The K_m for NADPH was $0.52 \pm 0.04\ \mu M^{-1}$ throughout the pH range studied.

Taken together, our pH/rate profile data on the L-arginine to NO + L-citrulline reaction and the reductase reaction suggest that K_m affects are located in the oxygenase domain; no K_m changes were observed with cytochrome c reduction. Further studies are underway to determine where in nNOS the k_{cat} effects are located.

References

1. Griffith, O. W., and Stuehr, D.J. (1995) *Ann. Rev. Physiol.* **57**, 707-36
2. Marletta, M.A. (1989) *TIBS*, **14 (12)**, 488-492
3. Boyar, A., and Marsh, R.E. (1982) *J. Amer. Chem. Soc.* **104**, 1995-1998
4. McMillan, K., Bredt, D. S., Hirsch, D. J., Snyder, S. H., Clark, J. E., and Masters, B. S. S. (1992) *Proc. Natl. Acad. Sci. U.S.A.* **89**, 11141-11145
5. McMillan, M., and Masters, B. S. S. (1993) *Biochemistry* **32**, 9875-9879

Cytokines and LPS synergize to induce NOS and arginine transporter activity in vascular smooth muscle cells.

Samantha M. WILEMAN, Anwar R. BAYDOUN and Giovanni E. MANN.

Vascular Biology Research Centre, Biomedical Sviences Division, King's College, Campden Hill Road, London W8 7AH, U.K.

INTRODUCTION

Production of nitric oxide (NO) by vascular smooth muscle cells (VSMC) has been implicated as one of the mediators of the deleterious effects associated with endotoxin shock [1]. Under these conditions, synthesis of NO from its precursor L-arginine is catalysed by the Ca^{2+}/calmodulin-insensitive nitric oxide synthase (iNOS) which is induced by pro-inflammatory cytokines and/or bacterial lipopolysaccharide (LPS). Once induced, production of NO by this enzyme appears to be critically dependent on the availability of exogenous arginine [2,3], suggesting that transport of this amino acid may be rate-limiting for iNOS activity. In this study we have therefore characterized the transport system(s) for L-arginine in rat aortic smooth muscle cells in culture (RASMC) and examined the effects of LPS and/or pro-inflammatory cytokines on transporter activity in correlation with the expression of iNOS.

EXPERIMENTAL

Materials

Reagents for cell culture were obtained from Gibco (Paisley, U.K.). *Escherichia coli* lipopolysaccharide (LPS, serotype 0111:B4) and all other chemicals were purchased from Sigma (Poole, U.K.). L-[2,3-^{3}H]arginine (37 Ci/mmol) was from NEN. , Du Pont. Recombinant murine interferon-γ (IFN-γ) was from Genzyme (Cambridge, U.K.). Human recombinant interleukin-1α (IL-1α) and tumour necrosis factor-α (TNF-α) were from British Bio-technology (Abingdon, U.K.).

Experimental protocol

VSMCs were cultured from rat aortic explants in Dulbecco's modified Eagle's medium (DMEM). Confluent monolayers in 96 well plates were activated with LPS (100 μg ml^{-1}) and/or cytokines including TNF-α (300 & 100 U ml^{-1}), IFN-γ (50 & 100 U ml^{-1}) and IL-1α (50 & 100 U ml^{-1}) for 24 h. NO production was determined by the Griess reaction [4]. Transport of L-[^{3}H]arginine (100 μM; 2 μCi ml^{-1}) was monitored in HEPES-buffered Krebs solution (50 μl; 37°C) over 30 s as described previously [4]. Arginine concentrations in lysates from control and L-arginine deprived (0.5-24 h) cells were determined by reverse phase high performance liquid chromatography (HPLC) [5].

RESULTS

Transport of L-arginine was Na^+ or pH (pH 5-8) independent. Furthermore, uptake was selectively inhibited by other cationic amino acids (L-lysine and L-ornithine) but not by neutral amino acid analogues, selective for transport systems A (2-methylaminoisobutyric acid), L (phenylalanine, leucine), N (6-diazo-5-oxo-L-norleucine, glutamine) or ASC (cysteine).

Incubation of RASMC with LPS resulted in both a concentration (0.01 - 100 μg ml^{-1}) and time- (8 - 24 h) dependent stimulation of NO synthesis, with nitrite levels elevated from 7.4 ± 1.7 (control) to 318 ± 26 pmoles μg $protein^{-1}$ by 100 μg ml^{-1} LPS over 24 h. This stimulation in NO production was accompanied by a 40% increase in L-arginine uptake, with 100 μg ml^{-1} LPS increasing transport from 6.8 ± 0.5 (control) to 9.5 ± 0.4 pmol μg $protein^{-1}$ min^{-1}. Stimulation of transport and induction of NOS were evident after a lag phase of at least 4 h and were abolished by cycloheximide (1 μM), suggesting a requirement of both processes for *de novo* protein synthesis.

Unlike LPS, activation of RASMC with either IFN-γ (100 U ml^{-1}), TNF-α (300 U ml^{-1}) or IL-1α, (100 U ml^{-1}) alone failed to stimulate L-arginine transport or increase nitrite accumulation. When applied in combination with LPS (100 μg ml^{-1}) both IFN-γ & TNF-α, but not IL-1α, potentiated the actions of LPS. IFN-γ (100 U ml^{-1}) in combination with LPS (100 μg ml^{-1}) increased nitrite levels and transport rates by 650% and 100% respectively.

HPLC analysis revealed that intracellular arginine concentrations (1.4 ± 0.3 mM in control cells) were depleted in a time-dependent manner following incubation of cells in arginine free DMEM. At 24 h arginine levels were reduced to 0.6 ± 0.2 mM. However, despite these relatively high levels, RASMC failed to produce measurable amounts of nitrite following a 24 h stimulation with LPS (100 μg ml^{-1}) and IFN-γ (50 U ml^{-1}).

DISCUSSION

The substrate specificity, Na^+ and pH independence of L-arginine transport are characteristic of the cationic amino acid transport system y^+. More importantly this study has identified that L-arginine transport and iNOS are coinduced in RASMC activated either with LPS alone or in combination with proinflammatory cytokines. These observations extend our previous study in which we demonstrated that LPS-stimulated macrophages showed an enhanced uptake of L-arginine [6]. We have now identified significant differences in the response of J774 macrophages and RASMC to LPS and cytokines. Firstly, unlike J774 cells, in which activation of L-arginine transport was detectable at doses of LPS that did not stimulate nitrite release, the increase in L-arginine transport in smooth muscle cells was always paralleled by an increase in nitrite accumulation, suggesting a closer coupling between induction of the transporter and iNOS in these cells. Secondly, incubation of RASMC with LPS and either IFN-γ or TNF-α synergistically enhanced the actions of LPS, with IFN-γ and LPS being the most potent.

Thus the ability of LPS and cytokines to enhance transport of L-arginine under conditions of increased NO production provides a unique mechanism for sustaining NO synthesis and may have important implications in the pathogenesis of endotoxin shock. Further studies of the signalling pathways mediating activation of L-arginine transport may permit the targeting of specific inhibitors to the cationic amino acid transporter(s), thereby providing a novel therapeutic approach for the management of some of the deleterious effects associated with the overproduction of NO following induction of NOS.

Acknowledgements

We gratefully acknowledge the support of the British Heart Foundation (F325, PG/91075; FS/94004) and the Wellcome Trust and Physiological Society for travel grants awarded to ARB.

REFERENCES

[1] Moncada, S., Palmer, R.M.J. & Higgs, E.A. (1991). Pharmacol. Rev., **43**, 109-142.

[2] Schott, C.A., Gray, G.A. and Stoclet, J-C. (1993). Br. J. Pharmacol., **108:** 38-43.

[3] Beasley, D., Schwartz, J.H. & Brenner, B.M. (1991). J. Clin. Invest., **87:** 602-608.

[4] Baydoun, A.R., Bogle, R.G., Pearson, J.D. & Mann, G.E. (1993). Br. J. Pharmacol., **110:** 1401-1406.

[5] Baydoun, A.R., Emery, P.W., Pearson, J.D. & Mann, G.E. (1990). Biochem. Biophys. Res. Commun., **173**: 940-948.

[6] Bogle, R.G., Baydoun, A.R., Pearson, J.D., Moncada, S. & Mann, G.E. (1992). Biochem. J., **284:** 15-18.

Diabetes and hyperglycaemia induced activation of the human endothelial cell L-arginine transporter and NO synthase: inhibitory effect of insulin in diabetes

Luis SOBREVIA, David L. YUDILEVICH and Giovanni E. MANN

Vascular Biology Research Centre, Biomedical Sviences Division, King's College, Campden Hill Road, London W8 7AH, U.K.

INTRODUCTION

Impaired endothelium-dependent relaxation occurs in diabetic patients and in vessels isolated from diabetic animals or exposed to hyperglycaemia [1]. Endothelial cells synthesize NO from L-arginine via Ca^{2+}/calmodulin dependent eNOS [2], though little is known concerning the effects of hyperglycaemia or diabetes on of the L- arginine/NO signalling pathway in human endothelium. We have reported previously that activity of the cationic amino acid transport system y^+/MCAT-1 is upregulated in human fetal cells isolated from diabetic pregnancies [3] and have now examined the mechanisms underlying modulation of the endothelial L-arginine/NO pathway by diabetes, hyperglycaemia and insulin.

EXPERIMENTAL

Endothelial cells were isolated from human umbilical cord veins from non-diabetic or gestational diabetic pregnancies and cultured in Medium 199 (5 or 25mM D-glucose), supplemented with 20% serum and 5mM L-glutamine. Transport of radiolabelled substrates was measured in Krebs. In some experiments cells were preincubated for 8 h with human insulin (0.1-10nM). cGMP accumulation and PGI_2 release were measured by radioimmunoassay. Resting membrane potential was measured by whole-cell patch voltage clamp [3]. Reagents and chemicals were from Gibco or Sigma and radiolabelled tracers from NEN.

RESULTS

Total protein, DNA, cell number and cell volume were similar in non-diabetic and diabetic cells. L-[^{3}H]leucine and [^{3}H]thymidine incorporation were decreased in actively replicating diabetic cells.

The V_{max} for saturable L-arginine transport was increased in diabetic or hyperglycaemic cells (Fig. 1A), although elevated rates of transport in diabetic cells were unaffected by 25 mM D-glucose. In contrast to L-arginine and L-lysine, transport (100 μM) of leucine, citrulline, serine, cystine or 2-deoxyglucose were unaffected by hyperglycaemia or diabetes. Basal cGMP levels were elevated by diabetes (Fig. 1B) or hyperglycaemia, whereas basal synthesis of PGI_2 was inhibited (data not shown). Histamine had no effect on L-arginine transport in either cell type, suggesting that enhanced accumulation of cGMP in response to histamine (Fig. 1B) may not necessarily be coupled directly to L-arginine transport activity.

Activation of L-arginine transport by elevated D-glucose occurred with a lag phase of 6-12 h, was unaffected by D-mannitol, abolished by cycloheximide and reversible upon re-exposure of cells to 5 mM D-glucose. Although the resting membrane potential was hyperpolarized in diabetic (-78 ± 0.3 mV) compared to non-diabetic (-70 ± 0.4 mV) cells, exposure of non-diabetic cells to 25 mM D-glucose (24 h) had no effect on membrane potential. In non-diabetic cells human insulin (1 nM) stimulated L-arginine transport and basal synthesis of NO and PGI_2. In contrast, insulin downregulated the elevated basal rates of L-arginine transport and NO synthesis in both diabetic and hyperglycaemic cells.

DISCUSSION

Our findings demonstrate that activity of the L-arginine transporter (system y^+/MCAT-1) is upregulated selectively in human endothelial cells isolated from diabetic pregnancies or exposed to hyperglycaemia or insulin. Although there are divergent reports that basal synthesis of endothelium-derived NO is increased or decreased in diabetic patients [1], our results establish that basal NO synthesis is elevated in endothelial cells exposed to high D-glucose or isolated from diabetic pregnancies. Histamine-stimulated NO production was similar in non-diabetic and diabetic cells, suggesting that levels of NO synthase were similar in these two cell types. Under the same cell culture conditions, we observed that basal and histamine-stimulated PGI_2 release were inhibited in diabetic cells.

System y^+ is activated in hepatocytes isolated from diabetic animals [4], and insulin has recently been reported to increase system y^+ activity in rat pancreas [5] and mRNA expression in rat liver cells [6]. Our findings now establish that insulin also stimulates basal NO and PGI_2 production in non-diabetic endothelial cells. Gestational diabetes induces phenotypic changes in human fetal endothelium [3,7], which are associated with a sustained membrane hyperpolarization, activation of the L-arginine/NO signalling pathway, elevation of basal NO synthesis and decreased basal and histamine-stimulated PGI_2 release.

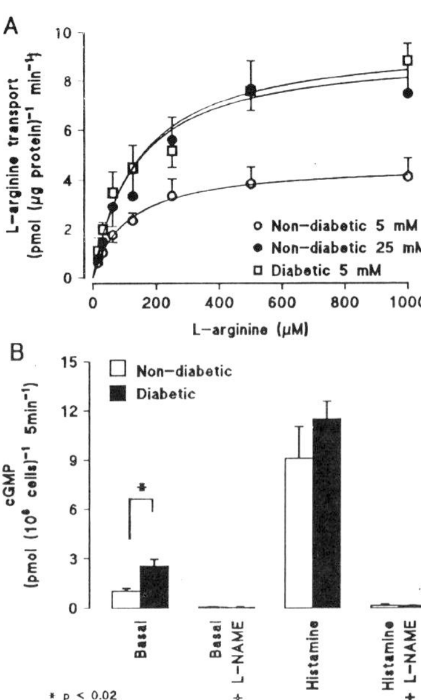

Fig. 1. Effects of diabetes and hyperglycaemia on L-arginine transport (A) and NO synthesis (B) in human endothelial cells.

Acknowledgements

We acknowledge the Wellcome Trust and British Council and thank Professor Beard and the St. Mary's Hospital labour ward midwives for the supply of umbilical cords.

REFERENCES

[1] Poston, L. and Taylor, P.D. (1995). Clin. Sci., **88**: 245-255
[2] Moncada, S., Palmer, R.M.J. and Higgs, E.A. (1991). Pharmacol. Rev. **43**:109-142.
[3] Sobrevia, L., Cesare, P., Yudilevich, D.L. and Mann, G.E. (1995). J. Physiol. (in press).
[4] White, M.F. (1985). Biochim. Biophys. Acta **822**:355-374.
[5] Muñoz, M., Sweiry, J.H. and Mann, G.E. (1995). Exp. Physiol. **80**: 745-753.
[6] Wu, J.Y., Robinson, D., Kung, H-J. and Hatzoglou, M. (1994). J. Virology **68**:1615-1623.
[7] Sobrevia, L., Jarvis, S.M. and Yudilevich, D.L. (1994). Am. J. Physiol. **267**: C39-C47.

Formation of Nitrogen Oxides upon Cytochrome P450-dependent Oxidative Cleavage of C=N(OH) Bonds: Comparison to Nitric Oxide Formation Catalyzed by NO Synthases.

Boucher, J.L., Jousserandot, A., Sennequier, N., Vadon, S., Delaforge, M. and Mansuy, D.

Laboratoire de Chimie et Biochimie Pharmacologiques et Toxicologiques, URA 400 CNRS, Université Paris V, 45 rue des Saints-Pères, 75270 Paris Cedex 06, France.

INTRODUCTION

NOSs are heme-thiolate proteins exhibiting several structural and physico-chemical properties in common with cytochromes P450 [1-3].

P450s and NOSs are also comparable for the kind of reactions they catalyze. The first step of NOS, the N-hydroxylation of L-arginine has not been found so far to be catalyzed by rat liver P450s. However, similar N-hydroxylations are classical P450 reactions already described for arylamidines and arylguanidines [4,5].

The second step of NOS, a non classical reaction involving a three-electron oxidation of the substrate, N^{ω}-hydroxy-L-arginine, is also catalyzed by hepatic P450-dependent monooxygenases [6].

The biological oxidations of N-hydroxyguanidines and of amidoximes are relatively unexplored. We have investigated the transformations of several simple compounds bearing a C=N-OH group (ketoximes, amidoximes, and N-hydroxyguanidines) in the presence of hepatic microsomal P450.

$$RR'C{=}N{-}OH \xrightarrow[NADPH+O_2]{P450} RR'C{=}O + NO_2^-(NO, NO_3^-, ...)$$

$R,R'=H$, alkyl, aryl, NHR_1, NR_1R_2

The objective of the present work was to study the mechanisms of these transformations and to suggest possible similarities between reactions catalyzed by "classical P450" or by NOSs.

RESULTS

We demonstrated that the oxidative cleavage of C=N(OH) bonds of various ketoximes, amidoximes and hydroxyguanidines is a rather general reaction, as dexamethasone-treated rat liver microsomes were found to be particularly active catalysts for the NADPH and O_2-dependent formation of the corresponding carbonyl-compounds, and of nitrogen oxides including NO [7]. Strong inhibitory effects of SOD were observed during these reactions: between 1 and 5 units SOD/ml were sufficient to give a 50% inhibition of NO_2^- formation [8].

Effect of the addition of oxygen radicals scavenging agents on the transformation of hydroxyguanidine Ia to nitrite ions and urea.

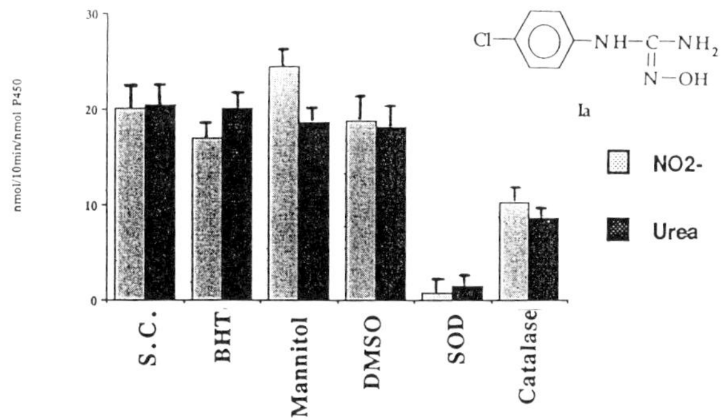

Reactions were catalyzed by dexamethasone treated rat liver microsomes in the presence of NADPH. Results are means from 3 to 5 determinations and are indicated as nmol/nmol P450/10min.

Effects of increasing amounts of SOD or of catalase on the transformation of hydroxyguanidine Ia to nitrite ions and urea.

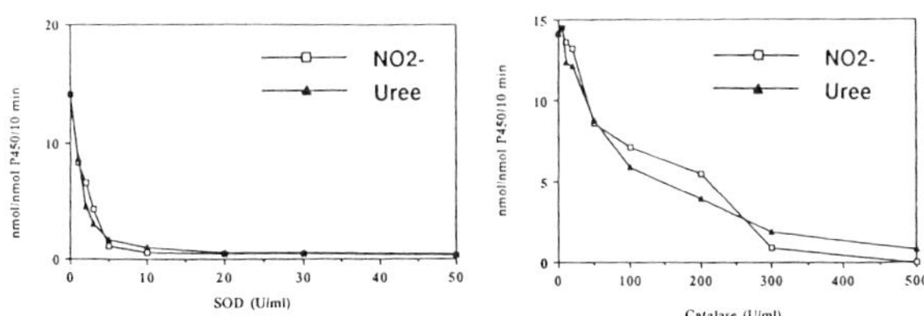

Incubations were performed in the presence of 100 μM **Ia** ; 1 mM NADPH and 1 μM P450 from dexamethasone treated rats for 10 min at 37°C. Nitrite ions were quantitated using a colorimetric method and urea after RP-HPLC analysis. Formation of 12.4 nmol 4-chlorophenyl cyanamide was also detected in the incubations.

These results suggest a key role of $O_2^{\cdot -}$ in microsomal P450-dependent oxidative cleavage of C=N(OH) bonds. They are in agreement with another series of results concerned with a study of reactions between $O_2^{\cdot -}$ and amidoximes or hydroxyguanidines. Potassium superoxide dissolved in DMSO in the presence of a crown-ether rapidly reacts with 4-chlorobenzamidoxime leading to quantitative and selective formation of 4-chlorobenzamide and NO_2^- [9]. Moreover, reaction of $O_2^{\cdot -}$ generated by xanthine and xanthine oxidase with 4-chlorophenyl-hydroxyguanidine **Ia** leads to the formation of 4-chlorophenylurea and NO_2^- [8].

DISCUSSION

Three kinds of mechanism that differ by the detailed nature of the further steps describing the reaction between NOHA and NOS-Fe(II)-O_2 have been proposed [2,3]. In all cases, the last step of the NOS reaction would be the decomposition of a peroxide adduct to NOHA by a six-center, concerted or non-concerted reaction, which regenerates NOS-Fe(III) and leads to citrulline and NO.

These mechanisms involve the dioxygen complex of NOS-Fe(II) as the key active species. As all hemeprotein-Fe(II)-dioxygen complexes, it may be viewed as a ferric complex of $O_2^{\cdot -}$ ([Fe(II) ← O_2] ↔ [Fe(III)-O-O$^{\cdot}$] ↔ [Fe^{+}(III)$O_2^{\cdot -}$]). In that sense, its ability to cleave the C=N(OH) bond of NOHA with formation of citrulline and NO could be related to the P450-dependent oxidative cleavage of the C=N(OH) bond of NOHA and many other compounds which seems to be mainly performed by $O_2^{\cdot -}$.

These data may be understood if one admits that both $O_2^{\cdot -}$ and its Fe(III) complex, Fe(III)-O-O$^{\cdot}$, are able to react with C=N(OH) bonds. In NOS, it is likely that NOHA is very well positioned in the active site in order to rapidly react with Fe(III)-OO$^{\cdot}$, leading to stoechiometric formation of NO and citrulline for 0.5 mole of NADPH consumed to reduce NOS-Fe(III) to NOS-Fe(II). On the contrary, in the active site of P450 3A, it is likely that NOHA, and exogenous amidoximes and hydroxyguanidines, are not high-affinity substrates for P450s and should not be well positioned in their active site. Therefore, their reactions with P450 Fe(III)-OO$^{\cdot}$ should be too slow to compete with P450 Fe(III)-OO$^{\cdot}$ decomposition and $O_2^{\cdot -}$ formation.

In that respect, NO-synthases may be regarded as particular P450s with a high affinity for NOHA and able to position its C=N(OH) moiety very well in the active site for a fast reaction with P450 Fe(III)-OO$^{\cdot}$, before decomposition of this complex [10].

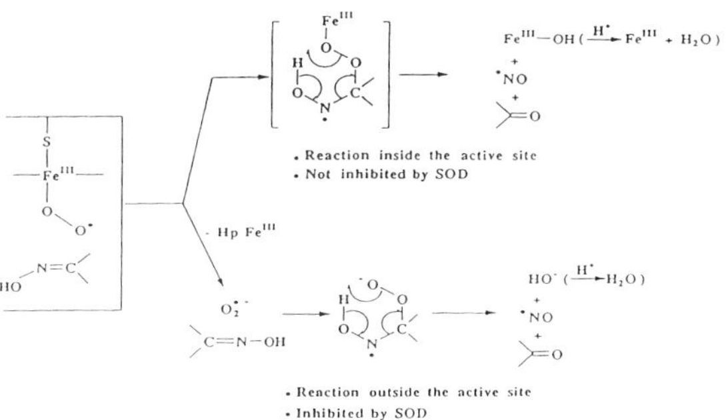

Possible mechanisms for the oxidative cleavage of C=N(OH) bonds by P450s and NOSs: contribution of $O_2^{\cdot -}$ in P450-dependent reactions. -HpFe^{III} means dissociation of the hemeprotein-Fe(III)-O-O$^{\cdot}$ complex to hemeprotein-Fe(III) and $O_2^{\cdot -}$; this reaction is especially important in P450-dependent oxidations of C=N(OH) bonds.

PERSPECTIVES

Interestingly, the P450-dependent oxidation of some amidoximes and hydroxyguanidines is not totally inhibited by a large excess of SOD, indicating that a small part of the reaction is performed by P450-Fe(III)-OO$^{\cdot}$ itself. Experiments are in progress to synthesize hydroxyguanidines that could act as high-affinity substrates for a given liver P450 isozyme. Such compounds should be oxidized by this P450 at the level of their C=N(OH) moieties without inhibition by SOD.

REFERENCES

1. Moncada, S., Palmer, R.M.J. and Higgs, E.A. (1991) *Pharmacol. Rev.* **43**, 109-142.
2. Marletta, M.A. (1993) *J. Biol. Chem.* **268**, 12231-12234.
3. Feldman, P.L., Griffith, O.W. and Stuehr, D.J. (1993) *Chem. Eng. News*, **71**, 26-38.
4. Clement, B. and Jung, F. (1994) *Drug Metab. Dispos.* **22**, 486-497.
5. Clement, B., Schultze-Mosgau, M.H. and Wohlers, H. (1993) *Biochem. Pharmacol.* **46**, 2249-2267.
6. Renaud, J.P., Boucher, J.L., Vadon, S., Delaforge, M. and Mansuy, D. (1993)*Biochem. Biophys. Res. Commun.* **192**, 53-60.
7. Jousserandot, A., Boucher, J.L., Desseaux, C., Delaforge, M. and Mansuy, D. (1995) *Bioorg. Med. Chem. Lett.* **5**, 423-426.
8. Jousserandot, A., Boucher, J.L., Delaforge, M., Mansuy, D. and Clement, B. submitted.
9. Sennequier, N., Boucher, J.L., Battioni, P. and Mansuy, D. (1995) *Tetrahedron Lett.* **36**, 6059-6062.
10. Mansuy, D., Boucher, J.L. and Clement, B. (1995) *Biochimie*, In press.

The mechanism of nitric oxide generation by one-electron oxidation of N^G-hydroxy-L-arginine

STEVEN A EVERETT, MICHAEL R L STRATFORD, KANTI B PATEL, MADELEINE F DENNIS, and PETER WARDMAN

Gray Laboratory Cancer Research Trust, PO Box 100, Mount Vernon Hospital, Northwood, Middlesex HA6 2JR, UK

Introduction

NHA has been identified as a stable product in the NOS - catalysed L-arg/NO$^•$ pathway [1]. The mechanism of oxidation of NHA by the iron-dioxygen complex in NOS remains a matter for debate [2-4]. This has been further complicated by the observation that peroxidases which exhibit one-electron oxidation activity can also induce the oxidative denitrification of NHA [5]. In this study a radiolytically generated model one-electron oxidant was used to generate putative NHA and NHA-derived radicals which were then characterized by pulse radiolysis with optical detection and EPR spectroscopy.

Materials and methods

Chemicals: NHA, carboxy-PTIO, (Cambridge Bioscience); PBN, (Sigma); NaN_3 and phosphate salts (Merck). Water was purified ('Milli-Q' system) and saturated with N_2O or N_2O/O_2 mixtures (BOC). The technique of pulse radiolysis has been described [6]. Steady-state radiolysis experiments utilized a ^{60}Co γ-source (28 Gy min^{-1}). The azide radical ($N_3^•$, E^o = 1.3V *vs.* NHE) was generated by radiolysis of N_2O-saturated solutions containing NaN_3 (20 mmol dm^{-3}) in phosphate buffer at pH 7.4. Spin-trapping experiments were performed with a Bruker EMX-band EPR spectrometer. Settings: microwave power, 20 mW; modulation amplitude, 0.2 G; time constant, 40 ms; sweep time 20 s.

Results

Oxidation of NHA by the $N_3^•$ radical (k = 8.7 x 10^8 mol^{-1} dm^3 s^{-1}) produced a NHA radical species which exhibited the reducing properties indicative of a guanidino carbon-centred radical. Figure 1a shows typical kinetic traces obtained on pulse radiolysis (2 Gy) of NHA (500 µmol dm^{-3}) in the absence and presence of 2% oxygen. In the absence of oxygen the guanidino carbon-centred radical decays via a unimolecular process (k = 1.5 x 10^3 s^{-1}) which we attribute to the oxidative elimination of NO$^•$. This was corroborated by EPR quantitation of NO$^•$ release utilizing the NO$^•$-selective conversion of carboxy-PTIO to the imino nitroxide [7].

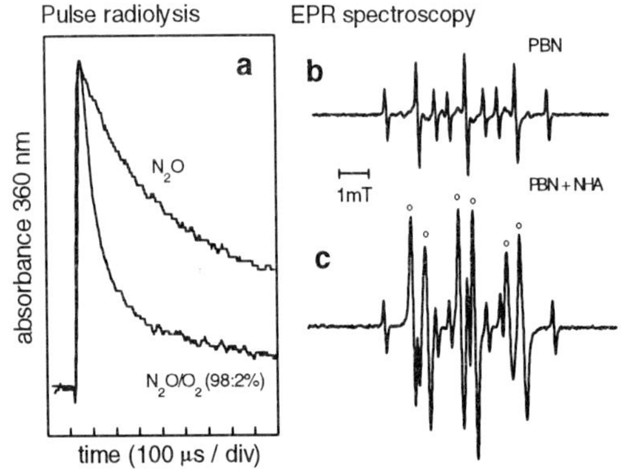

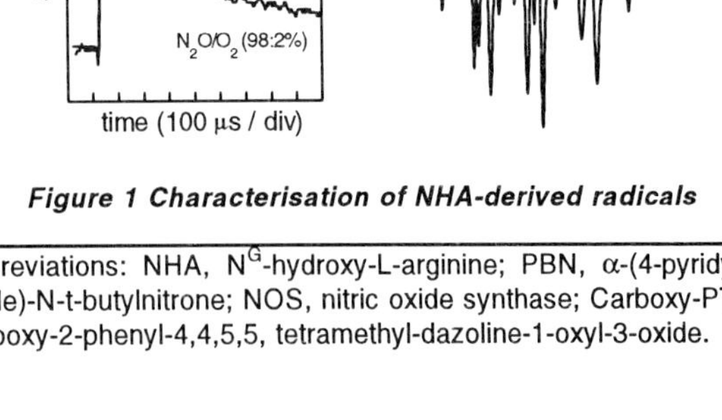

Figure 1 Characterisation of NHA-derived radicals

Abbreviations: NHA, N^G-hydroxy-L-arginine; PBN, α-(4-pyridyl-1-oxide)-N-t-butylnitrone; NOS, nitric oxide synthase; Carboxy-PTIO, carboxy-2-phenyl-4,4,5,5, tetramethyl-dazoline-1-oxyl-3-oxide.

In the presence of oxygen the NHA radical decays via a pseudo-first order process (k = 3.7 x 10^8 mol^{-1} dm^3 s^{-1}) indicative of an oxygen addition reaction. Steady-state radiolysis (*ca.* 150 Gy) of an N_2O-saturated solution containing PBN (0.5 mmol dm^{-3}), NaN_3 (20 mmol dm^{-3}) in phosphate (1mmol dm^{-3}) at pH 7.4 gave the EPR signal shown in figure 1b characteristic of the *tert*-butyl nitroxyl radical [*t*-BuN(O$^•$)H] generated by the oxidative degradation of PBN by the $N_3^•$ radical plus a minor product. However, with the inclusion of NHA (0.5 mmol dm^{-3}) a PBN-adduct radical signal appeared (figure 1b, open circles) as doublets of a triplet with hyperfine splitting constants of a_N = 1.52 mT and a_H = 0.40 mT consistent with that of a carbon-centred radical.

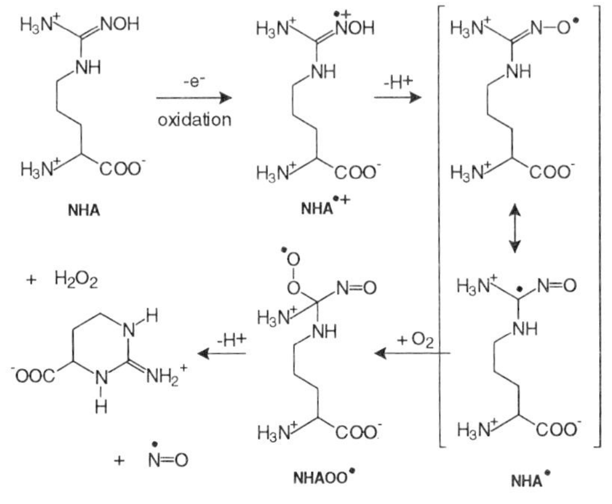

Figure 2 Mechanism for the oxidative denitrification of NHA

Conclusions

The proposed reaction scheme for the oxidative denitrification of NHA is shown in figure 2. Electron abstraction from the NHA followed by rapid deprotonation (< 1 µs) of the resultant radical cation (NHA$^{•+}$) generates a guanidino carbon-centred radical (NHA$^•$) which exists as a resonance hybrid of the nitroxyl radical. In the absence of oxygen the NHA$^•$ radical decays via a unimolecular process attributed to the elimination of NO$^•$. The NHA$^•$ rapidly adds oxygen to form a peroxyl radical (NHAOO$^•$) which also eliminates NO$^•$ and H_2O_2 without the simultaneous formation of citrulline but yielding a product which is presumed to be a cyclised carbodiimide (data not shown). No evidence was obtained for the elimination of peroxynitrite ions from the NHAOO$^•$ radical. Characterization of this alternative pathway of NO$^•$ release from NHA has important implications for NO$^•$ biosynthesis via NOS-independent enzymatic or free radical pathways.

Acknowledgements

This work is supported by the Cancer Research Campaign

References

1. Steuhr, D.J., Kwon, N.S., Nathan, C.F., Griffith, O.W., Feldman, P.L. and Wiseman, J. (1991) J. Biol. Chem. **266**, 6259-6263
2. Marletta, M.A. (1993) J. Biol. Chem. **268**, 12231-12234
3. Korth, H.-G., Sustmann, R., Thater, C., Butler, A.R and Ingold, K.U. (1994) J. Biol. Chem. **269**, 17776-17779
4. Feldman, P.L., Griffith, O.W. and Steuhr, D.J. (1993), Chem. Eng. News, **71**, 26-38
5. Boucher, J.L., Genet, A., Vadon, S., Delaforge, M. and Mansuy, D. (1992) Biochem. Biophys. Res. Commun, **184**, 1158-1164
6. Candeias, L.P., Everett, S.A. and Wardman, P. (1993) Free Rad. Biol. Med. **15**, 385-394
7. Hogg, N., Singh, R.J., Joseph, J. Neese, F. and Kalyanaraman, B. (1995) Free Rad. Res. **22**, 47-56

Effects of oxidative burst inhibitors on nitric oxide production in human neutrophils

MARIA C. CARRERAS, NATALIA RIOBO, CARLOS G DEL BOSCO and JUAN J. PODEROSO

Laboratory of Oxygen Metabolism, University Hospital, University of Buenos Aires. Córdoba 2351, 4th floor, 1120 Buenos Aires, Argentina

INTRODUCTION

Human neutrophils (PMN) generate nitric oxide (NO) simultaneously with superoxide anion (O_2^-) with formation of peroxynitrite ($ONOO^-$) in the presence of activators of NADPH oxidase like formyl peptides (fMLP) or phorbol esters (PMA) [1]. The aim of this study was to compare the response of NADPH oxidase and NO synthase signalling pathways to different inhibitors of the oxidative burst in human neutrophils.

METHODS

Preparation of human neutrophils

Human neutrophils were isolated by Ficoll-Hypaque gradient centrifugation, dextran sedimentation and hypotonic lysis of contaminant erythrocytes. The cells, resuspended in HBSS-Hepes buffer without Ca^{2+} and Mg^{2+}, were preincubated at 37°C with crescent concentrations of the corresponding inhibitors.

Nitric oxide production

NO production was measured by the oxidation of oxymyoglobin to metmyoglobin in a double beam-double wavelength 356 Perkin Elmer spectrophotometer at 581-592 nm [2].

Electrochemical detection of nitric oxide

A NO sensitive electrode from World Precision Instruments was used in some assays to measure directly NO release.

Hydrogen peroxide production

Hydrogen peroxide production was continuously measured by the horseradish-p-hydroxyphenylacetic acid assay with a Hitachi F 2000 spectrofluorometer at 315-425 nm [3].

RESULTS

Preincubation of human neutrophils with isoproterenol or adenosine (agonists of the inhibitory pathway of oxidative burst) for 5 minutes decreased H_2O_2 by 80 and 100% respect to controls (C: 0.87±0.07 nmol.min^{-1}.10^6 PMN) stimulated with 1μM fMLP alone, while NO production was decreased only by adenosine (50%) respect to controls (C: 0.23±0.02 nmol.min^{-1}.10^6 PMN).

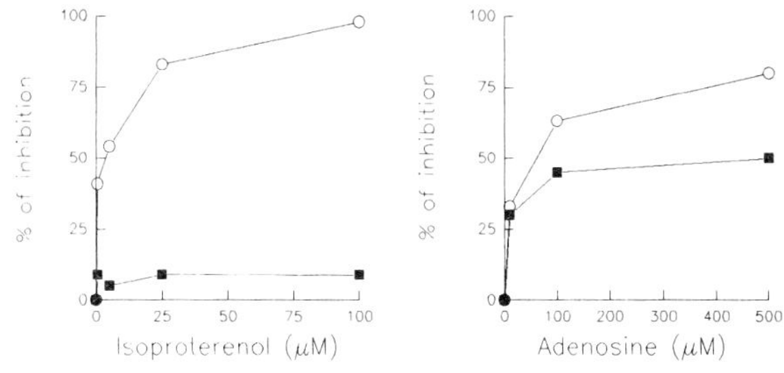

Fig. 1 **Effect of isoproterenol or adenosine on NO (squares) and H_2O_2 (circles) production by fMLP-stimulated human neutrophils.**

Abbreviations used: PMN, neutrophils; NO, nitric oxide; O_2^-, superoxide anion; $ONOO^-$, peroxynitrite; fMLP, formyl peptides; PMA, phorbol esters; H_2O_2, hydrogen peroxide.

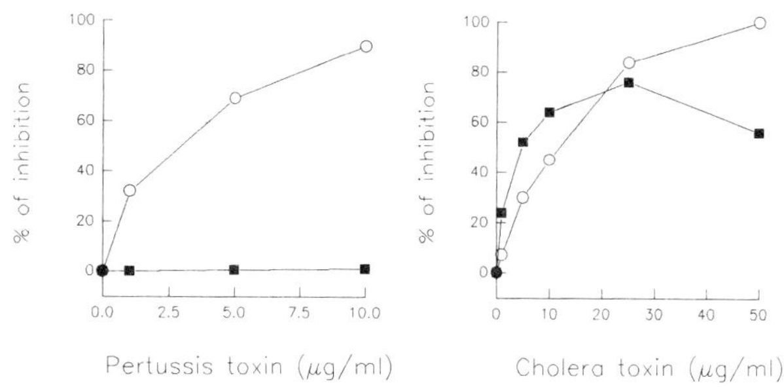

Fig. 2. **Effect of pertussis or cholera toxins on NO (squares) and H_2O_2 (circles) production by fMLP-stimulated human neutrophils.**

Preincubation with cholera or pertussis toxins , which ADP ribosylate two different GTP-binding regulatory proteins ($G\alpha_s$ and $G\alpha_i$), for 30-60 minutes abolished H_2O_2 production while NO production was decreased only by cholera toxin by 56%.

Preincubation with a tyrosine kinase inhibitor, genistein, for 30 minutes, markedly inhibited H_2O_2 production when neutrophils were stimulated with 1μM fMLP but slightly decreased it when they were activated by 0.1 μg/ml PMA. In contrast, NO production was markedly increased by genistein preincubation with both stimuli up to 160 and 300% respectively at maximal doses (fMLP: 0.38±0.03 and PMA 1.90±0.09 nmol.min^{-1}.10^6 PMN).

To detect NO electrochemically, the release of O_2^- has to be abolished or it must be removed with large amounts of superoxide dismutase, otherwise, all NO reacts with it forming peroxynitrite. We could detect NO electrochemically when PMN were preincubated with Pertussis toxin alone or with genistein.

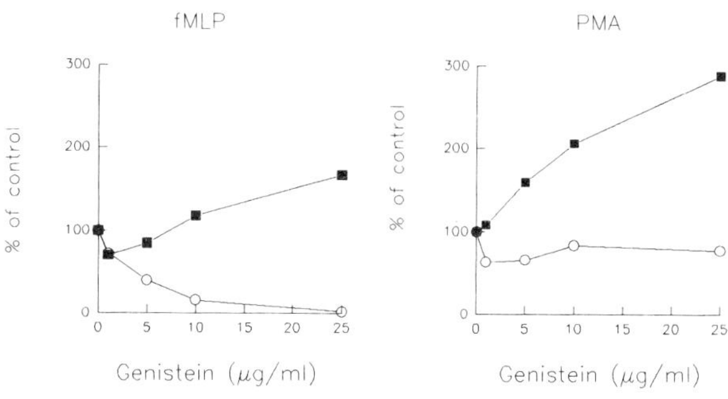

Fig. 3. **Effect of genistein on NO (squares) and H_2O_2 (circles) production by fMLP- or PMA-stimulated human neutrophils.**

CONCLUSIONS

Our data suggest that activating or inhibitory signal transduction pathways of NADPH oxidase and NO synthase are different in human neutrophils.

The activation of tyrosine kinase during the respiratory burst markedly inhibits NO production.

REFERENCES

1. Carreras, M.C., Pargament, G.A., Catz, S.D., Poderoso, J.J., Boveris, A. (1994) FEBS Lett **341**: 65-68
2. Murphy, M.E., Noack,E. (1994) Methods Enzymol **233**: 240-250
3. Wymann, M.P., von Tscharner, V., Deranleau, D.A., Baggiolini, M. (1987) Anal. Biochem. **165**: 371

Cytosol From Immumnostimulant-Activated Vascular Smooth Muscle Produces NO and Peroxynitrite in a Temporally Distinct Pattern

Michael S. SUZMAN[1], Gary A. FANTINI[1], Rachel LERNER[2] and Steven S. GROSS[2].

Departments of Surgery[1] and Pharmacology[2], Cornell University Medical College, 1300 York Avenue, New York, NY 10021, USA.

Introduction: Recent studies with purified neuronal NOS have detected peroxynitrite rather than NO as the sole nitrogenous reaction product [1]. Peroxynitrite likely arises in this setting from secondary reaction of NOS-derived NO and superoxide anion (O2•). Predominance of this reaction could be a purification artifact, such that highly purified NOS becomes largely uncoupled and generates excessive and unnatural quantities of O2•. To better appreciate the extent to which NO vs. peroxynitrite arise as intracellular products of NOS, and to identify factors affecting this balance, we have examined the products of crude iNOS-rich cytosol under a variety of conditions.

Methods: iNOS-rich cytosol (100,000 x g supernatant) was prepared from rat aortic smooth muscle cells that had been pretreated for 16h with a combination of LPS and interferon-γ, as previously described [2].

NOS activity in each sample was quantified using each of two methods. First, a kinetic spectrophotometric assay was performed based on the ability of NO and peroxynitrite to oxidize Fe^{2+}-myoglobin [2]. Incubates contained: iNOS (20-40 µg total protein), TRIS (80 µM, pH 7.6), NADPH (1mM), L-arginine (1mM), 5,6,7,8-tetrahydrobiopterin (BH_4; 10µm), dithionite-reduced myoglobin (50µM) and modifications as indicated. The rate of increase in A_{405}, refelecting ferro-heme oxidation, was continually monitored at 15 second intervals for 15 minutes. The second method of assessing NOS activity involved determination of the rate of BH_4-dependent NADPH consumption. NADPH consumption was measured at 15 sec intervals for 30 min as a decrease in A_{340}.

Direct determination of NO concentration in incubates was performed electrochemically, using an NO-specific electrode (Diamond General; Ann Arbor, MI). The electrode was pre-calibrated at 37°C against a saturated standard solution of 220 ppm NO in N_2, versus a saline blank.

Superoxide anion production by iNOS-containing cytosol was determined from chemiluminescence intensity, arising from the reaction of O_2• with lucigenin [3]. Reactions were performed in 96-well plates and were identical in composition to those used for NOS activity assays, except that 100 µM lucigenin was substituted for myoglobin. Luminescence was repeatedly measured during 10-second intervals in a 96-well scintillation counter (Wallac Microbeta Plus 1450; Gaithersburg, MD).

Results and Discussion: NOS activity, as determined by Fe^{2+}-myoglobin oxidation, was observed to be linear with protein concentration and time during a 30 min assay duration. Activity was absolutely dependent on arginine, NADPH and tetrahydrobiopterin (BH4), but unaffected by superoxide dismutase (SOD;1000 U), catalase (80 U) or DTT (1 mM). Identical results were obtained when NADPH consumption was used as a readout of NOS activity. Over a 10-fold range of protein concentration, the stoichiometry of Fe^{2+}-myoglobin oxidation/NADPH consumption was found to be 1.30 - 1.40, in accordance with prior reports of a stoichiometry ≈ 1.50 [4].

Despite linearity of NOS activity, the solution concentration of NO was observed to vary in a non-linear and unpredicted manner with time and protein concentration. Indeed, the NO-specific electrode revealed that NO levels peak within 10 min at 30 nmol/mg protein (0.25 ml rxn mix), but then rapidly decay to a plateau level which is 20% - 60% of peak, depending on [NOS] (fig. 1).

Neither the peak nor the plateau concentration of NO was found to increase linearly with iNOS protein addition. As expected, detection of any cytosol-derived NO by the electrode required arginine, NADPH and BH4. While NO levels were unaffected by catalase, SOD (1000 U) dramatically restored peak NO levels during the plateau phase (fig. 2).

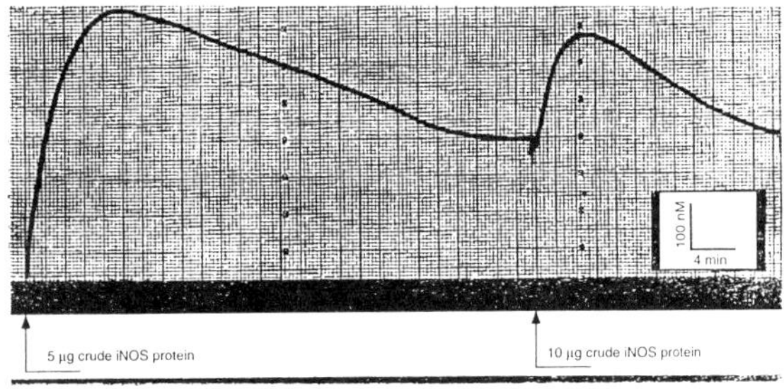

Figure 1. NO production upon cumulative addition of iNOS-containing cytosol, measured by an NO-specific electrode.

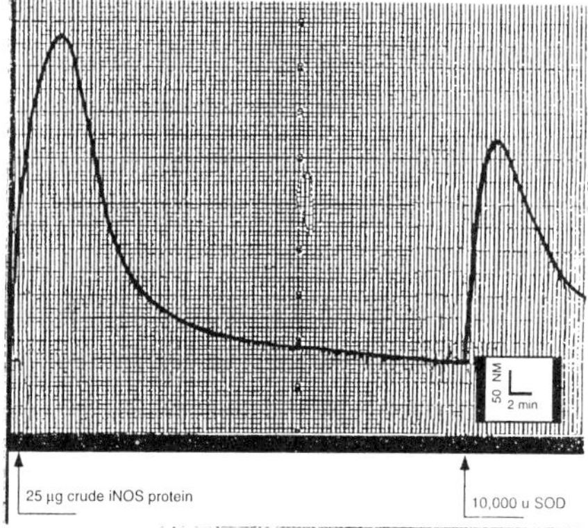

Figure 2. Restoration of NO production from iNOS-containing cytosol with addition of super oxide dismutase after 40 minutes, measured by an NO-specific electrode.

Chemiluminescence detection revealed that NOS-rich cytosol produced a burst of superoxide anion that rapidly fell to a plateau level. The rapid rise, fall and plateau of O_2• levels occured in tandem with measured changes in NO concentration. Superoxide anion production was significantly enhanced by removal of BH_4 and arginine, and it was eliminated by removal of NADPH. These findings are consistent with the view that significant quantities of superoxide anion are endogenously produced when iNOS activity is uncoupled from NADPH consumption by lack of available arginine and BH_4.

Taken together, our findings indicate that peroxynitrite, rather than NO, may become a dominant product of cytosolic iNOS over time. iNOS can potentially serve as the primary source of both the NO and superoxide which are required for peroxynitrite synthesis. The extent of NO conversion to peroxynititrite by cytosol may vary in a complex and non-linear manner with time and NOS activity. The production of peroxynitrite explains numerous biological actions which have been often ascribed to NO, but which are inconsistent with the known chemical reactivity of NO itself (e.g., thiol nitrosation).

(Supported by NIH grant HL50656)

1. Schmidt, K, Klatt, P, Mayer, B (1995) J. Biol. Chem. 270: 655-9.

2. Gross, SS and Levi, R (1992) J. Biol. Chem. 267:25722-25729.

3.Mohazzab, KM, and Wolin, MS (1994) Am. J. Physiol. 267(6 Pt 1): L815-22.

4. Stuehr, DJ, Kwon, NS, Nathan, CF, Griffith, OW, Feldman, PL et al., (1991) J. Biol. Chem. 266: 6259-6253.

Simultaneous generation of nitric oxide and superoxide by endothelial cells: ESR and ozone-mediated •NO chemiluminescence experiments

REINER F. HASELOFF, STEFAN ZÖLLNER, KATARINA MERTSCH and INGOLF E. BLASIG

Institute of Molecular Pharmacology, Alfred-Kowalke-Str. 4, D-10315 Berlin, Germany

Introduction

In recent years, very important results have been found regarding physiological functions of nitric oxide in vasoregulation [1], neurotransmission [2], and immune regulation [3]. However, like other free radicals with relatively low reactivity such as superoxide, •NO can be a precursor for the formation of very toxic species. The reaction of •NO with $•O_2^-$, with a rate constant ($k = 6.7 \cdot 10^9\ M^{-1}s^{-1}$ [4]) near to the diffusion limit is an example for such a process. Peroxynitrite formed by this reaction is a potent oxidant causing damage to biomolecules by oxidation of sulfhydryl groups [5] or other reactions. The protonated form of $ONOO^-$ decomposes yielding radical species with the reactivity of the hydroxyl radical.

The present study deals with consequences (free radical formation and viability of cells) of an influx of Ca^{2+} ions into bovine aortic endothelial cells (BAEC) produced by A23187. ESR model experiments were performed using novel 2H-imidazole-1-oxide spin traps [6] to confirm the identities of the reactive species formed under these conditions.

Methods

BAEC cultivation was made in minimum essential medium including 10% fetal calf serum, 0.2 mM glutamine without antibiotics at 37°C in 5% CO_2/95% air. Cultivation quality was verified microscopically by cobblestone appearance at confluence, factor VIII staining, contents of alkaline phosphatase and angiotensin converting enzyme. Typically, cell protein content was between 500 μg and 700 μg per flask. Experiments were carried out in PBS after washing the cells three times. *•NO detection* was accomplished by ozone-mediated chemiluminescence (CL) using a Sievers 270B analyzer. For continuous measurements, the headspace gas was forced by vacuum directly to the reaction chamber. Alternatively, nitric oxide was detected by reduction of nitronyl nitroxides to imino nitroxides (compounds synthesized by I.A. Grigor'ev, Inst. Org. Chem., Novosibirsk, Russia). *EPR experiments* were performed at room temperature using a flat quartz cell on a Bruker ECS 106 X-band spectrometer under the following standard conditions: modulation amplitude, 0.1 mT; field set, 348 mT; scan range, 10 mT; microwave power, 10 mW.

Results and Discussion

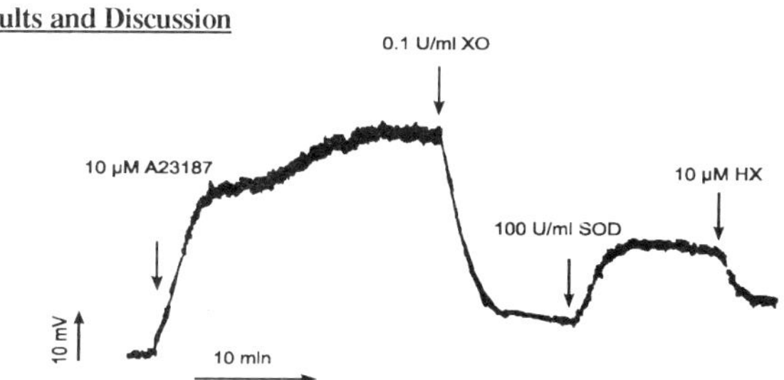

Fig. 1: Typical time course of A23187 (10 μM)-stimulated •NO liberation from BAEC detected by continuous registration of the ozone-mediated CL: effects of xanthine oxidase (XO), superoxide dismutase (SOD) and hypoxanthine (HX)

As shown in Fig. 1, there was an immediate increase of the •NO formation following the addition of A23187 reaching a maximum about 20 min after stimulation with Ca^{2+} ionophore. Addition of xanthine oxidase resulted in a rapid decrease of the signal which is assumed to be connected to the reaction of •NO (i) with the enzyme [7] and (ii) with $•O_2^-$ produced by XO from endogenous substrate (partial restoration in the presence of SOD). The latter process dominates when hypoxanthine is added (complete suppression of the signal).

Abbreviations used: BAEC, bovine aortic endothelial cells; CL, chemiluminescence; NN, nitronyl nitroxide; PBS, phosphate buffered saline; DMSO, dimethylsulfoxide; SOD, superoxide dismutase; XO, xanthine oxidase;

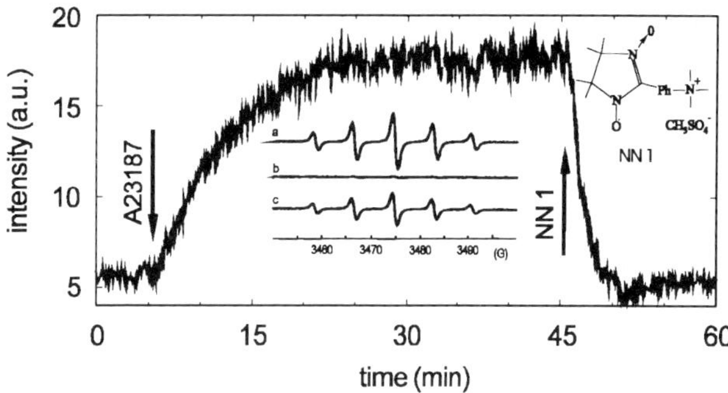

Fig. 2: •NO chemiluminescence of an endothelial cell monolayer stimulated by 10 μM A23187 and effect of 10 μM NN 1 (cf. structure). ESR spectra of NN 1 (inset): a - reference in PBS, b, c - samples taken from the supernatant after 1 min incubation of A23187 (c - presence of 100 U/ml SOD)

The addition of a nitronyl nitroxide to BAEC monolayers resulted in a rapid decrease in CL and a parallel decrease in ESR signal intensity of the nitroxide. However, the postulated corresponding imino nitroxide was not detected. Additional experiments (data not shown) indicated that nitronyl nitroxides are rapidly reduced, especially by $•O_2^-$. This is confirmed by the observed protective effect of SOD against NN reduction. Cytotoxic effects due to Ca^{2+} ionophore were verified by a neutral red assay [8]: Incubation of endothelial cells with 10 μM A23187 by 150 min resulted in a viability of 36.2 ± 1.5% of the cells. In the presence of either 100 U/ml SOD or 20 μM NN 1, viability increased to 44.6 ± 3.5 and 46.6 ± 2.4%, respectively (all values mean ± SEM).

ESR experiments were performed to confirm the identity of radical species involved. From the data obtained (Fig. 3) using the novel spin trap 4-carboxy-2,2-dimethyl-2H-imidazole-1-oxide (CIMO) it is obvious that the reaction of •NO and $•O_2^-$ results in the formation of CIMO/$•CH_3$ and CIMO/•OH. Detection of both adducts indicates that hydroxyl radicals may be formed under these conditions. Generation of •OH can contribute to the damage observed in BAEC cultures. The ability of NNs to act as

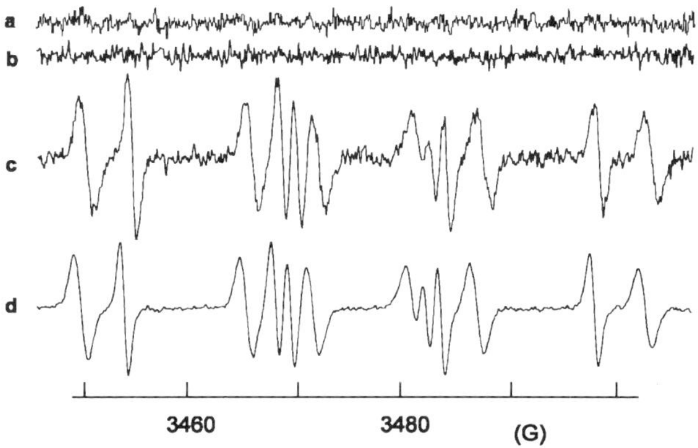

Fig. 3: CIMO spin adducts obtained by mixing with DMSO solutions of •NO and $•O_2^-$ (20% DMSO [v/v] in PBS, c_{CIMO}= 0.1 M): **a,** •NO solution (1 mM); **b,** $•O_2^-$ solution (1 mM); **c,** •NO and $•O_2^-$ solution; **d,** addition of CIMO/•OH and CIMO/$•CH_3$ spectra produced by Fenton's reagent (20% DMSO)

scavengers of both precursor species can explain the results observed in the cytotoxicity experiments. These data indicate a cytoprotective effect of nitronyl nitroxides which is worth to be investigated in more detail.

References: **1.** Palmer, R.M.J., Ferrige, A.G. and Moncada, S. (1987) Nature **327**, 524-526; **2.** Bredt, D.S. and Snyder, S.H. (1989) Proc. Natl. Acad. Sci. USA **86**, 9030-9033; **3.** Kolb, H. and Kolb-Bachofen, V. (1992) Immunol. Today **13**, 157-160; **4.** Huie, R.E. and Padmaja, S. (1993) Free Radical Res. Commun. **18**, 195-199; **5.** Radi, R., Beckman, J.S., Bush, K.M. and Freeman, B.A. (1991) J. Biol. Chem. **266**, 4244-4250; **6.** Klauschenz, E., Haseloff, R.F., Volodarskii, L.B., Blasig, I.E. (1994) Free Radical Res. **20**, 103-111; **7.** Fukahori, M., Ichimori, K., Ishidaa, H., Nakagawa, H. and Okino, H. (1994) Free Radical Res. **21**, 203-212; **8.** Mertsch, K., Grune, T., Siems, W.G., Ladhoff, A., Saupe, N. and Blasig, I.E. (1995) Cell. Mol. Biol. **41**, 243-253.

Acknowledgement: This work was supported by DFG (SFB 507), BMBF (grant BEO 21/0310015) and a grant of the Schering Forschungsgesellschaft to S.Z.

Different mechanisms are involved in the inhibition of nitric oxide synthase expression by antioxidants in cytokine-stimulated macrophages

MARKUS HECKER, CHRISTIANE PREISS, HANNELORE STOCKHAUSEN*, PETER KLEMM*, VALERIE B. SCHINI-KERTH, BEATE FISSLTHALER and RUDI BUSSE

Center of Physiology, J.W. Goethe University Clinic, Theodor-Stern-Kai 7, D-60590 Frankfurt/M. and *Pharmaceutical Research, Hoechst AG, D-65926 Frankfurt/M., Germany

The high production of nitric oxide (NO) generated by the inducible NO synthase (iNOS), e.g. in macrophages and vascular smooth muscle cells, can exert both protective and deleterious effects [1]. Inhibition of iNOS expression and/or activity therefore represents an important therapeutic goal.

Changes in NO formation in iNOS-expressing cells are usually correlated with similar changes in iNOS mRNA abundance, indicating that a major part of iNOS regulation occurs at the level of transcription. The 5'-flanking region of the iNOS gene contains several bindings sites for *cis*-regulatory elements such as nuclear factor κB (NF-κB) [2,3]. Activation of this transcription factor indeed appears to be critical for iNOS expression, e.g. in macrophages stimulated with lipopolysaccharide (LPS) [4].

Since antioxidants are thought to inhibit the cytokine-mediated activation of NF-κB [5], we investigated whether these compounds also affect iNOS expression in a macrophage cell line (RAW 264.7) activated with LPS and interferon-γ (IFNγ).

Chrysin (50 μM), 3,4-dichloroisocoumarin (DCI, 50 μM) and N-acetylserotonin (NAS, 1 mM) virtually abolished the phorbol-ester-induced generation of O_2^- by these cells. Moreover, chrysin and NAS also reduced xanthine oxidase-dependent O_2^- formation by more than 90%, thus confirming their anti-oxidative potential.

Exposure of the macrophages (2.5×10^6 cells in 2 ml medium) to LPS (140 ng/ml) and IFNγ (5 U/ml) for 6 h increased the level of nitrite in the conditioned medium from 1.7±0.3 to 16.8±1.1 μM (n=16). Stimulation with LPS or IFNγ alone resulted in a much weaker increase in nitrite production (less than 20% of the response to LPS plus IFNγ). In the presence of chrysin, DCI or NAS nitrite production was either strongly attenuated (DCI, NAS) or abolished (chrysin). This effect was paralleled by a marked reduction in iNOS protein abundance (Fig. 1a). None of the antioxidants significantly affected the conversion of [^{3}H]arginine to [^{3}H]citrulline by a cytosolic fraction prepared from cells stimulated with LPS plus IFNγ, indicating that their effect on nitrite production reflects an inhibition of iNOS expression rather than of iNOS activity.

NAS is a potent inhibitor of sepiapterin reductase, the key enzyme in the salvage pathway for tetrahydrobiopterin (BH_4) synthesis. The availability of this co-factor has been suggested not only to be important for iNOS activity, but also for expression of the iNOS gene [6]. Macrophage BH_4 synthesis, however, was not decreased after 6-h exposure to LPS plus IFNγ, even when sepiapterin reductase was additionally inhibited. Moreover, the inhibitory effect of NAS on iNOS activity or protein abundance in LPS plus IFNγ-stimulated RAW 264.7 macrophages was not affected by co-incubation with the BH_4 precursor dihydrobiopterin [7]. It is thus unlikely that NAS affects iNOS expression by limiting the availability of BH_4.

Only DCI strongly attenuated the DNA-binding activity of NF-κB in nuclear extracts from macrophages after 30-min exposure to LPS plus IFNγ, while NAS and chrysin had no effect (Fig. 1b). The inhibitory effect of DCI is presumably due to its prevention of the proteolytic degradation of the inhibitory IκB subunit [8].

The precise mechanism by which chrysin and NAS inhibit iNOS expression remains to be determined. One possibilty is an interference with the nuclear signalling of IFNγ via the Stat pathway. This notion is based on the facts: (i) the 5'-flanking region of the macrophage iNOS gene contains a binding site for the Stat1 homodimer (GAS site) in close proximity to the functionally important NF-κB site [2,3]; and (ii) the synergistic effect of LPS and IFNγ on iNOS expression in RAW 264.7 macrophages occurs at the transcriptional level [9]. In our hands, LPS and IFNγ also strongly synergized in stimulating iNOS expression in these cells, whereas LPS and IFNγ alone only caused a weak induction. Inhibition of the IFNγ/Stat1 pathway may thus exert an inhibitory effect on iNOS expression similar to that caused by blocking the LPS/NF-κB pathway. However, as yet we cannot rule out an effect of chrysin and NAS on iNOS expression at the post-transcriptional and/or posttranslational level.

Taken together, these findings suggest that antioxidants do not inhibit iNOS expression via a common pathway, but act by distinct mechanisms. They also indicate that an enhanced formation of reactive oxygen species such as O_2^- is not mandatory for the activation of NF-κB in LPS-stimulated macrophages.

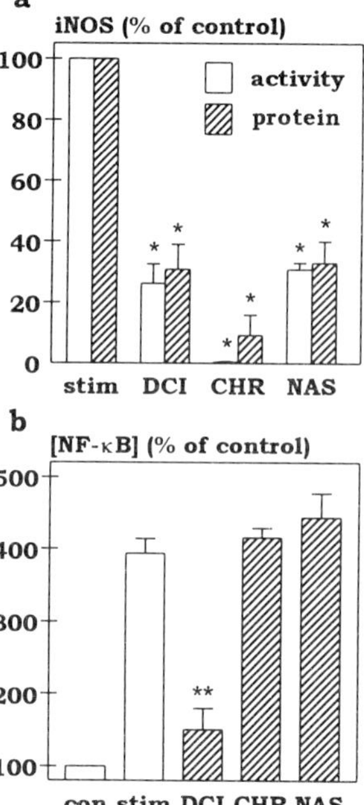

Fig. 1: Effects of DCI (50 μM), chrysin (CHR, 50 μM) and NAS (1 mM) on (*a*) iNOS protein abundance (Western blot) and activity (nitrite accumulation) and (*b*) DNA-binding of NF-κB (gel shift analysis) in RAW 264.7 macrophages stimulated with LPS plus IFNγ. Values are expressed as percentage of the level of LPS plus IFNγ-stimulated (stim) or solvent-treated control cells (con; n=5, *P<0.05).

Acknowledgements
This study was supported by the Deutsche Forschungsgemeinschaft (He 1587/5-1).

References

1. Morris, S.M. and Billiar, T.R. (1994) Am. J. Physiol. **266**, E829-E839
2. Xie, Q.-w., Wishnan, R. and Nathan, C. (1993) J. Exp. Med. **177**, 1779-1784
3. Lowenstein, C.J., Alley, E.W., Raval, P., Snowman, A.M., Snyder, S.H., Russell, S.W. and Murphy, W.J. (1993) Proc. Natl. Acad. Sci. USA **90**, 9730-9734
4. Sherman, M.P., Aeberhard, E.E., Wong, V.Z., Griscavage, J.M. and Ignarro, L.J. (1993) Biochem. Biophys. Res. Commun. **191**, 1301-1308
5. Grimm, S. and Baeuerle, P.A. (1993) Biochem. J. **290**, 297-308
6. Sakai, N., Kaufmann, S. and Milstien, S. (1993) Mol. Pharmacol. **43**, 6-11
7. Klemm, P., Hecker, M., Stockhausen, H., Wu, C.-C. and Thiemermann, C. (1995) Br. J. Pharmacol. **115**, 1175-1181
8. Henkel, T., Machleidt, T., Alkalay, I., Krönke, M., Ben-Neriah, Y. and Baeuerle, P. (1993) Nature **365**, 182-185
9. Lorsbach, R.B., Murphy, W.J., Lowenstein, C.J., Snyder, S.H. and Russell, S.W. (1993) J. Biol. Chem. **268**, 1908-1913

Is agmatine an endogenous inhibitor of inducible NO synthase ?

MICHEL AUGUET, ISABELLE VIOSSAT, JEAN-GREGOIRE MARIN and PIERRE-ETIENNE CHABRIER

INSTITUT HENRI BEAUFOUR
1 avenue des Tropiques, 91952 LES ULIS, France

Introduction

L-arginine is a nutritionally semiessential amino acid involved in a variety of physiological processes. Nitric oxide synthases (NOS) produce the potent vasodilator nitric oxide from arginine and arginine decarboxylase converts arginine to agmatine. This latter enzyme, originally observed in bacteria, has been recently found in mammalian brain. In addition, agmatine was identified as an endogenous clonidine-displacing substance (1). Thus, it was recently hypothesized that arginine may have an antihypertensive effect through nitric oxide and agmatine formation by NOS and arginine decarboxylase, respectively (2).

Methods

Functional α_2-adrenoceptor activity : Segments of ileum (1.5-2 cm long) were removed from male Hartley guinea-pigs (300-450 g) and mounted on tissue holders fitted with two parallel platinum electrodes, such that one electrode was positioned within the length of the ileal lumen. This assembly was placed in a 20 ml organ bath containing modified Krebs solution under a tension of 1 g at 37°C and gassed with 95 % O_2 / 5% CO_2. Contractile responses were measured using force displacement transducers coupled to a polygraph. The tissues were stimulated at 0.05 Hz with square wave pulses of 1ms duration and supramaximal current determined. The putative agonists were added to the organ baths after 15 min of stimulation and antagonists were introduced into the bath 30 min before agonists.

NO formation in macrophages : The murine macrophage cell line J774A1 was cultured in DMEM with 10% foetal calf serum. Cells were exposed (in presence or absence of the tested drugs) to LPS (1µg/ml) and interferon-γ (50 U/ml) during 24 h to induce iNOS. Then, nitrites in cells supernatant were determined by the Griess reaction.

Results and Discussion

Clonidine produced a potent inhibition of twitch responses (EC_{50}=13 ± 3.9 nM ; 69 ± 4.0 % reduction of the twitch responses ; n=4) which was inhibited by yohimbine (-logKb= 7.7 ± 0.05 ; n=4) attesting of an α_2-adrenoceptor mediated effect. In contrast, agmatine, up to 1 mM, only weakly affected the twitch responses (15.6 ± 5.0 % ; n=4). In addition, pretreatment with agmatine (1 mM) unaffected the response to clonidine. Thus agmatine was devoid of significant α_2-adrenoceptor activity (agonist or antagonist) in a classical model of functional pharmacology such as electrically stimulated ileum of the guinea-pig.

This result tails with the very recent observation of Pinthong et al. (3) who showed that despite recognition at α_2-adrenoceptor binding sites, agmatine failed to produce functional α_2-adrenoceptor activity not only in the guinea-pig isolated ileum but also in other functional models of α_2-adrenoceptors. In this respect, agmatine failed to mimic the effect of clonidine in rat isolated vas deferens or porcine isolated palmar lateral vein. In addition, in rat cerebral cortex, agmatine produced a concentration dependent inhibition of ^{3}H-clonidine but failed, unlike UK-14304, to inhibit forskolin-stimulated cyclic AMP. Thus collectively, the results indicate that agmatine may not possess sufficient α_2-adrenoceptor agonist properties to influence cardiovascular function (4).

LPS and interferon-γ increased nitrite concentrations in the culture medium at 24 h as a consequence of induction of iNOS in macrophages. Co-administration of LNMMA (IC_{50} = 27 ± 5 µM; n = 3) aminoguanidine (IC_{50} = 25.8 ± 2.69 µM; n = 3) or agmatine (IC_{50} = 830 ± 390 µM; n = 3) reduced the accumulation of nitrite.

This effect of agmatine was likely due to inhibition of activity, but not induction, of iNOS since agmatine inhibited NO formation from partially purified macrophage iNOS (data not shown). In addition, agmatine was about as potent and selective as aminoguanidine to inhibit the activity of the inducible, but not the constitutive, form of NOS in rat aorta (data not shown).

In conclusion, our study indicates that arginine decarboxylase pathway that produces agmatine may not play a significant role in the antihypertensive effects elicited by arginine through α_2-adrenoceptor activation. Conversely, agmatine may impair nitric oxide synthesis from arginine by inhibition of inducible NOS.

Thus, agmatine may exert a significant role as endogenous inhibitor of inducible NOS. Moreover, as arginine decarboxylase has been originally observed in bacteria, agmatine may serve as suppressant of the host immune responses associated with the systemic syndrome response. Thus, bacteria possess the armamentarium to attenuate the cytotoxic effect of NO from macrophage iNOS. Bacteria may inhibit induction of iNOS (e.g; spermine; (5)); activity of iNOS (e.g. agmatine) or may inactivate NO (e.g. pyocyanin; (6)).

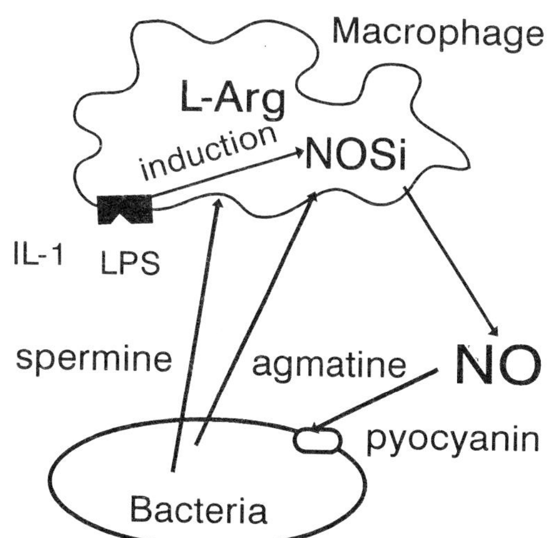

Scheme of interaction between bacteria products and NO from macrophages

References

1. Li, G., Regunathan, S., Barrow, C.J., Eshraghi, J., Cooper, R. and Reis, D.J. (1994) Science **263**, 966-969
2. Nakaki, T. and Kato, R. (1994) Jpn. J. Pharmacol. **66**, 167-171
3. Pinthong, D., Wright, I.K., Hanmer, C., Millns, P., Mason, R., Kendall, D.A. and Wilson, V.G. (1995) Arch. Pharmacol.**351**, 10-16
4. Szabo, B., Urban, R., Limberger, N. and Starke, K. (1995) Naunyn-Schmiedeberg's Arch. Pharmacol. **351**, 268-273
5. Szabo, C., Southan, G.J., Wood, E., Thiemermann, C. and Vane, J.R. (1994) Br. J. Pharmacol. **112**, 355-356
6. Warren, B., Loi, R., Rendell, N.B. and Taylor, G.W., Biochem. J. (1990) 266, 921-923

Abbreviations : LPS : lipopolysaccharide ; IL-1 : interleukin-1.

Arginine based inhibitors of induced nitric oxide synthase

E. Ann Hallinan,* Sofya Tsymbalov,* William M. Moore,# Mark G. Currie,# and Barnett S. Pitzele*

Searle
*Department of Chemistry, Skokie, Illinois 60077
#Department of Inflammatory Diseases Research, Creve Coeur, Missiouri 63167

Introduction

Nitric oxide (NO) mediates a variety of cellular processes such as regulation of vascular tone, platelet aggregation, neurotransmission, and immune activation. NO is the product of arginine metabolism by the enzyme nitric oxide synthase (NOS) yielding citrulline in addition to NO. Several isoforms of NOS both constitutive and induced have been identified and characterized.

The objective of our NOS Program is the identification of a novel inhibitor selective for macrophage induced NOS (iNOS). Rheumatoid arthritis (RA) is characterized by activation of macrophage iNOS, and it is believed that a number of the symptoms of arthritis are due to excess nitric oxide release in the joint. Thus, RA is the primary therapeutic target for the NOS program. The need for inhibitor selectivity is to avoid inhibition of key constitutive NOS such as the constitutive endothelial isoform of NOS which is crucial in the maintenance of blood pressure homeostasis.

ε-N-iminoethyl-L-lysine, a selective iNOS inhibitor,[1] has been shown to suppress the increase in plasma nitrite levels and joint inflammation associated with adjuvant-treated rats, a model for rheumatoid arthritis.[2] This poster summarizes initial synthetic explorations into the design and synthesis of novel iNOS inhibitors with enhanced selectivity based on ε-N-iminoethyl-L-lysine (NIL). The emphasis is on carboxyl replacements such as alcohols and diols.

Results and Discussion

Chemistry

Using a variety of strategies, NIL alcohol and diol analogs were synthesized. Separation of the diol diastereomers at the penultimate step was effected employing reverse phase chromatography. Attempts to achieve diastereoselectivity using Sharpless asymmetric dihydroxylation chemistry did not realize substantive enrichment of single diastereomer.

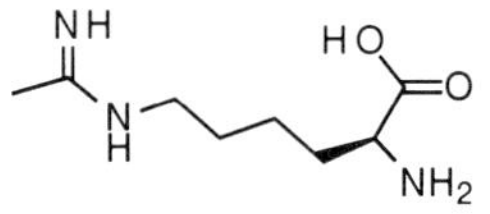

NIL

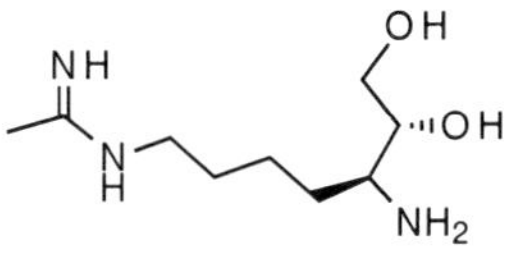

SC-60319

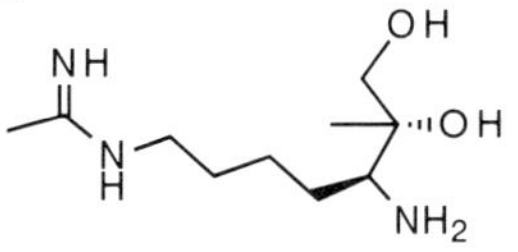

SC-63720

Pharmacology

Whereas the alcohol analogs of NIL do not show significant inhibition of human induced nitric oxide synthase (hiNOS), two vicinal diols, SC-60319 and SC-63720 have comparable inhibition of hiNOS as compared to NIL. SC-60319 has striking selectivity (685x) for hiNOS as compared to human endothelial constitutive nitric oxide synthase (hecNOS). SC-60319 shows no difference in selectivity for human neuronal constitutive nitric oxide synthase (hncNOS) versus NIL. Chirality is crucial for activity as seen with beta-hydroxy analogs of SC-60319 and SC-63720 which have significantly reduced enzyme inhibition. The necessity of free hydroxyl group is demonstrated by the minimal inhibition hiNOS by diacetyl and monoacetyl analogs of SC-60319.

Table 1 NOS Inhibition Assay Data [3, 4]

No.	hiNOS IC_{50} (μM)	hecNOS IC_{50} (μM)	hncNOS IC_{50} (μM)	hecNOS/ hiNOS	hncNOS /hiNOS
NIL	4.9	112	48	23	10
SC-60319	12	8420	150	685	12
SC-63720	9	2350	100	257	11

Conclusion

Replacing the carboxyl group of NIL with vicinal diols has yielded SC-60319 and SC-63720 which have a significant enhancement of selectivity for hiNOS versus hecNOS as compared to NIL.

References

1. Moore, W. M., Webber, R. K., Jerome, G. M., Tjeong, F. S., Misko, T. P. and Currie, M. G. (1994) J. Med Chem, **37,** 3886-8.

2. Connor, J. R., Manning, P. T., Settle, S. L., Moore, W. M., Jerome, G. M., Webber, R. K., Tjeong, F. S., and Currie, M. G. (1995) Eur. J. Pharmacol., **273**, 15-24.

3. Bredt, D. S. and Snyder, S. H. Proc. (1990) Natl. Acad. Sci. U.S.A., **87**, 682-685.

4. Misko, T. P., Moore, W. M., Kasten, T. P., Nickols, G. A., Corbett, J. A., Tilton, R. G., McDaniel, M. L., Williamson, J. R. and Currie, M. G. (1993) Eur. J. Pharmacol., **233**, 119-125.

Symmetric and asymmetric dimethylarginine inhibit arginine transport and nitric oxide synthesis in J774 macrophages.

Anwar R. BAYDOUN, Richard G. KNOWLES*, Harold F. HODSON*, Salvador MONCADA* and Giovanni E. MANN.

Vascular Biology Research Centre, King's College, Campden Hill Road, London W8 7AH, UK. *Wellcome Research Laboratories, Langley Court, Beckenham, Kent BR3 3BS, UK.

INTRODUCTION

Methylated arginine analogues have been shown to occur endogenously in man [1]. Of these compounds, N^G-monomethylarginine (L-NMMA) and N^G,N^G dimethyl-arginine (asymmetric dimethylarginine; ADMA) have been identified as potent inhibitors of nitric oxide synthase (NOS)[2,3]. By comparison symmetric N^G,$N^{G'}$ dimethyl-arginine (SDMA) has relatively little effect on isolated NOS activity [3]. In this study we have examined whether ADMA and SDMA interact with the transporter for L-arginine and whether this contributes to the regulation of NO production by the inducible pathway in bacterial lipopolysaccharide (LPS) activated J774 cells.

EXPERIMENTAL

Materials

Reagents for cell culture were obtained from Gibco (Paisley, U.K.). *Escherichia coli* lipopolysaccharide (LPS, serotype 0111:B4) and all other chemicals were purchased from Sigma (Poole, U.K.). L-[2,3-^{3}H]arginine (53 Ci/mmol) was from Amersham International plc. N^G-monomethylarginine and dimethyl-arginines (symmetric and asymmetric) were synthesised by Wellcome Research Laboratories, Kent, UK.

Cell culture

The murine monocyte/macrophage cell line J774 was obtained from the European Collection of Animal Cell Cultures (ECACC, Wiltshire) and maintained in continuous culture in Dulbecco's modified Eagles medium (DMEM) containing 0.4 mM L-arginine and supplemented with 4 mM glutamine, penicillin (100 units ml^{-1}), streptomycin (100 μg ml^{-1}) and 10% foetal calf serum.

Experimental protocol

Prior to each experiment, J774 cells were plated at a seeding density of 10^5 cells per well in 96-well plates and allowed to adhere for at least 4 h. Cells were then cultured for a further 24 h in either DMEM alone or in DMEM containing LPS (1 μg ml^{-1}). Transport of L-[^{3}H]arginine (100 μM; 1 μCi ml^{-1}) was monitored as described previously [4] in Krebs solution (50 μl; 37°C) containing increasing concentrations (0.25 to 5 mM) of either SDMA or ADMA. Cell protein was determined using the BioRad reagent and radioactivity in formic acid digest of the cells determined by liquid scintillation counting. Uptake was expressed in units of pmoles μg $protein^{-1}$ min^{-1}.

Inhibition of NO production by ADMA or SDMA in intact cells was determined by activating J774 macrophages with LPS (1μg ml^{-1}) in DMEM containing 100 μM L-arginine and either ADMA (0.01-5 mM) or SDMA (0.01-5 mM). Nitrite levels were quantified 24 h later by the Griess reaction [4]. The direct effects of ADMA or SDMA (0.03-5 mM) on isolated iNOS activity were determined by spectrophotometric analysis of the conversion of oxyhemoglobin to methemoglobin [5].

RESULTS

Incubation of J774 cells with LPS (1 μg ml^{-1}, 24 h) resulted in induction of iNOS (nitrite production 9.3 $\pm$ 1.1 nmol μg $protein^{-1}$ 24 h^{-1}), accompanied by a marked increase in uptake of L-[^{3}H]arginine with the rate of transport increasing from 3.7 $\pm$ 0.1 pmol μg $protein^{-1}$ min^{-1} (control) to 5.8 $\pm$ 0.2 pmol μg $protein^{-1}$ min^{-1} (n=3). Transport of arginine in both control and LPS activated cells was inhibited in a concentration-dependent manner by ADMA and SDMA with respective K_i values of 87 μM and 103 μM for control and 83 μM and 95 μM for activated cells. Thus these compounds were approximately equipotent, inhibiting both the stimulated and control transport rates by virtually the same extent.

ADMA (3-500 μM) caused a concentration-dependent inhibition of NO production (IC_{50}=35.5 $\pm$ 3.5 μM) by iNOS isolated from LPS-activated J774 cells. In contrast to ADMA, SDMA (0.03-5 mM) had no significant effects on enzyme activity but markedly attenuated LPS-induced nitrite production in a concentration-dependent manner when incubated with J774 cells over a 24 h period in the presence of 100 μM L-arginine. Under these conditions, accumulated nitrite levels were reduced by 22% and 54% respectively in the presence of 0.1 mM and 5 mM SDMA.

DISCUSSION

This study has identified ADMA and SDMA as inhibitors of arginine transport, with both compounds showing a similar affinity for the induced and constitutively expressed transporter protein(s). Furthermore, kinetic inhibition studies suggest that ADMA and SDMA are at least five times more potent than either N^G-monomethyl-L-arginine (K_i = 0.58 mM) or N^G-iminoethyl-L-ornithine (K_i = 0.58 mM) as inhibitors of L-arginine transport in J774 macrophages [6].

In agreement with other reports [2], we have established that isolated iNOS is selectively inhibited by ADMA, with SDMA having virtually no effect on enzyme activity at concentrations of up to 5 mM. In contrast, when applied to intact cells, SDMA inhibited NO production in a concentration-dependent manner, reducing accumulated nitrite levels by over 50% at 5 mM. Thus the ability of SDMA and ADMA to inhibit transport provides an alternative mechanism by which endogenous arginine analogues could regulate NO biosynthesis *in vivo*. This may be particularly important in disease states such as chronic renal failure where impaired excretion of these compounds results in increased circulating levels in plasma [2]. Under these conditions these compounds may cause an additive inhibition of NO production either by directly inhibiting iNOS or indirectly by limiting availability of substrate through inhibition of the L-arginine transporter(s).

Acknowledgements

We gratefully acknowledge the support of the British Heart Foundation (FS/94004) and the Wellcome Trust for the award of a travel grant to ARB. We are also grateful to miss L. Bridge for spectrophotometric analysis of NO production.

REFERENCES

[1] Kakimoto, Y. and Akazawa, S. (1970). J. Biol. Chem., **245**: 5751-5758

[2] Vallance, P., Leone, A., Calver, A., Collier, J. and Moncada, S. (1992). Lancet, **339**: 572-75.

[3] Rees, D.D., Palmer, R.M.J., Schulz, R., Hodson, H.F. and Moncada, S. (1990). Br. J. Pharmacol., **101**: 746-752.

[4] Baydoun, A.R., Bogle, R.G., Pearson, J.D. & Mann, G.E. (1993). Br. J. Pharmacol., **110**: 1401-1406.

[5] Charles, I.G., Scorer, C.A., Moro, M.A., Chubb, A., Dawson, J., Knowles, RG and Baylis, S.A. (1995). Meth. Enzymol. (in Press).

[6] Baydoun, A.R. and Mann, G.E. (1994). Biochem. Biophys. Res. Commun., **200**: 726-731.

A novel mechanism for disposal of nitric oxide in cells and tissues

RICHARD G KNOWLES, TAO LU and SALVADOR MONCADA

Wellcome Research Laboratories, Langley Court, Beckenham, Kent BR3 3BS, UK

It is now well established that nitric oxide (NO) is both an important regulator of normal cell functions and a mediator of cytotoxicity under pathological conditions. However, apart from the reactions of NO with superoxide and with haemoglobin, little is known about the disposal of NO by cells. Therefore, in the present study removal of NO by intact rat hepatocytes and murine monocytic J774 cells as well as rat liver subcellular fractions and rat brain homogenates was determined under aerobic conditions by a modification of the chemiluminescence method, with careful control of the contamination with haemoglobin.

Materials and Methods

Liver cells were prepared by collagenase digestion [1] from 24h-starved rats and incubated in a Krebs-Hepes buffer(in mM: NaCl 118, KCl 4.74, $MgSO_4$ 1.18, KH_2PO_4 1.18, $CaCl_2$ 2.5, Hepes 25, D-glucose 10, pH 7.4) at a density of 20mg wet weight/ml. Liver and brain homogenates and subcellular fractions were prepared after perfusion to remove blood, in a sucrose (320mM) Hepes (20mM) EGTA (0.1 mM) buffer pH 7.2 [2]. The murine macrophage cell line (J774) was obtained from the American Tissue Culture Collection, maintained in DMEM supplemented with 10% foetal calf serum, 2 mM L-glutamine, 100 U/ml penicillin, 100 µg/ml streptomycin and 20 mM D-glucose and incubated in Krebs-Hepes. The haemoglobin contamination of the cell suspensions and homogenates was very low (<0.1µM) as assessed spectrophotometrically at 541nm. NO solutions were prepared by dissolving NO gas (BOC Special Gases) into a sealed glass bulb filled with helium-degassed water. NO disposal was measured by a modified chemiluminescence method (Fig 1) based on reference [3]. Nitrite + nitrate production was determined by acidic vanadium conversion to NO and chemiluminescence [4].

Figure 1. Chemiluminescence method for measuring NO disposal by cells and tissues

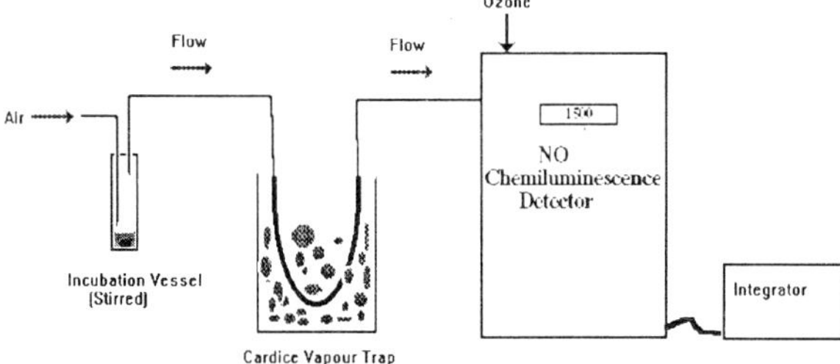

Results

Injection of NO into Krebs-Hepes buffer resulted in an overflow of NO, detected by chemiluminescence, which was proportional to the amount of NO injected and was complete by 5 minutes. This overflow was significantly reduced in the presence of hepatocytes (Fig. 2), indicating that hepatocytes removed NO effectively. This process was complete

Figure 2. Removal of NO by liver cells

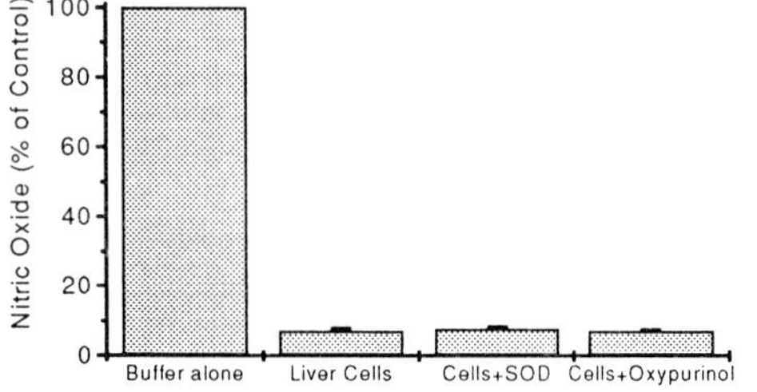

1µM NO injected into Krebs-Hepes buffer alone or with liver cells at 20mg/ml ± SOD (500U/ml) or oxypurinol (500µM)

within 3 minutes and had a high capacity since it remained undiminished after repeated NO injection. Preincubating the cells with SOD or oxypurinol before NO injection did not affect NO removal (Fig. 2) while 83±3.9% (n=3) of the NO removed was recovered as nitrite + nitrate.

Liver homogenates and subcellular fractions were also found to remove NO (Fig 3). This activity was found to be present in all the fractions, albeit with varying potency. Only a minor component of the NO removal could be mediated by superoxide produced by xanthine oxidase, as revealed by the effects of SOD and oxypurinol (Fig. 3)

Figure 3. Liver subcellular fractions remove nitric oxide

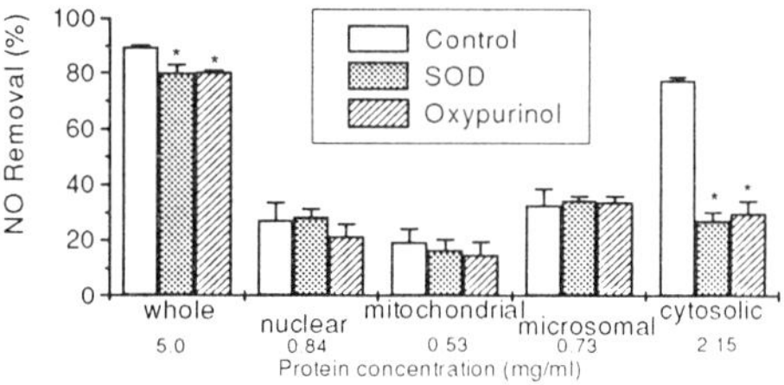

Liver fractions at 20mg of tissue per ml, ± SOD 500U/ml or oxypurinol 500µM; *, different from control P<0.05.

The NO removed by liver cells or subcellular fractions was found to be greater than 75% converted to nitrite + nitrate. J774 monocytic cells and brain homogenates were also found to remove NO (Fig. 4). The NO removal was not a non-specific reaction with protein since bovine serum albumin, ovalbumin and histone (all at 5mg/ml) did not remove NO. Liver cytosol lost its ability to remove NO after ultrafiltration through a membrane with a molecular weight cut-off of 20,000.

Discussion

The novel cellular process for NO removal found in this study catalyses the conversion of NO to nitrite + nitrate. This process does not require intact cells and is not mediated by superoxide or haemoglobin or by a non- specific reaction with protein. It is present in monocytic cells and to a lesser extent in brain as well as liver.

Figure 4. Removal of NO by cells and tissues

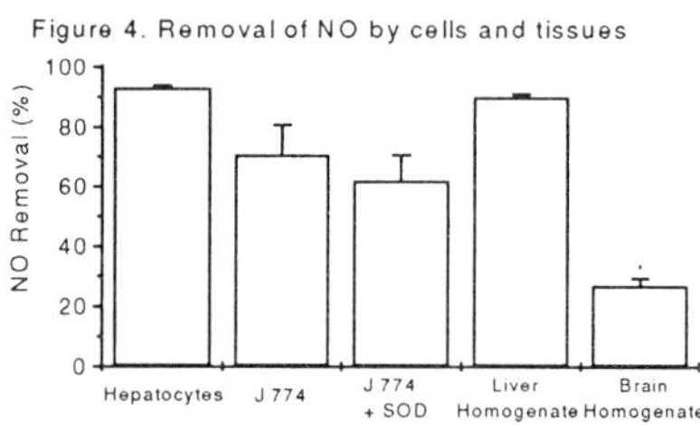

1µM NO injected into liver or J774 monocytic cell suspensions or tissue homogenates at 20mg of tissue per ml.

Within or close to blood vessels, NO is likely to be predominantly cleared by reaction with oxyhaemoglobin because of its high affinity for NO and its high concentration (≈ 5mM) in red blood cells. However in regions of tissues which are more distant from blood vessels, NO removal by this novel cellular process may predominate given a) the presence of extracellular SOD to suppress removal by reaction with superoxide and b) the high tissue concentration (>1000mg tissue/ml compared with 20mg/ml in these experiments). This process may therefore play an important role in controlling the local NO concentration in tissues.

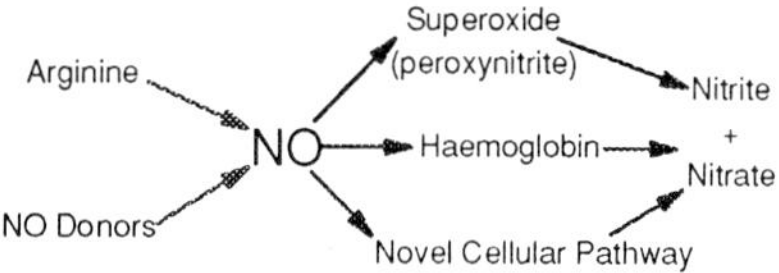

References

1. Ceppi, E.D., Knowles R.G., Carpenter, K. M. and Titheradge, M. A. (1992) Biochem. J., **284,** 761-766.
2. Reid E. & Williamson R. (1974) in Methods in Enzymology. Vol 31, (Fleischer S. and Packer L., eds.) pp. 713-747. Academic Press, New York.
3. Palmer, R.M.J., Ferrige, A.G., and Moncada, S. (1987) Nature, **327**, 524-526.
4. Bush, P.A., Gonzalez N.E., and Ignarro, L. (1992) Biochem. Biophys. Res. Commun., **186**, 308-314.

Distinct effects of nitric oxide and peroxynitrite on respiration by brain submitochondrial particles

MARIA A MORO, IGNACIO LIZASOAIN, RICHARD G KNOWLES, VICTOR DARLEY-USMAR and SALVADOR MONCADA

Wellcome Research Laboratories, Langley Court, Beckenham, Kent BR3 3BS, UK.

The brain contains a high NO synthase activity compared with other tissues [1]. During cerebral ischaemia this can result in exposure of the brain to high concentrations of NO (>1μM, [2]) and this has been implicated in the brain damage caused by ischaemia [3,4]. The mechanisms of cell damage by NO in the brain or during cell killing by cytokine-activated macrophages include inhibition of a number of cellular processes, such as DNA synthesis and mitochondrial respiration [5-7]. Some of these effects may be direct and others may arise from the reaction of NO with superoxide to form peroxynitrite ($ONOO^-$) [8]. It has been shown that both NO and $ONOO^-$ can disrupt mitochondrial function [7,9-11] and therefore either NO or $ONOO^-$ could potentially be responsible for mitochondrial damage. Because of this we have carried out a direct comparison of the effects of NO and $ONOO^-$ on respiration in which submitochondrial particles (SMP) were exposed to NO donors and to $ONOO^-$ in the presence and absence of glutathione or glucose and the effect of these treatments on the mitochondrial respiratory chain was analysed.

Materials and Methods

Rat brain mitochondria were prepared following homogenization in KCl (150mM)/potassium phosphate (20mM) buffer, pH7.6, by a modified Ficoll gradient method based on reference[12]. SMP were prepared from the mitochondria by freeze-thawing (3 cycles) followed by centrifugation. SMP respiration was measured polarographically at 0.5mg SMP protein/ml, 37°C, in potassium phosphate (50mM) / EGTA (0.1mM) buffer, pH7.2. NO concentrations were measured with a specific electrode (Diamond General, MI, USA). NO was prepared as an anaerobic solution in water. $ONOO^-$ was synthesized as described [13,14].

Results

NO produced from SIN-1 in the presence of SOD inhibited respiration from NADH, with an IC50 for NO of 2.0±0.1μM (Fig. 1). NO inhibited respiration from succinate or from TMPD/ascorbate as well as from NADH, consistent with inhibition of cytochrome oxidase (data not shown).

Figure 1: Production of NO and inhibition of respiration by SIN-1(500μM)+SOD

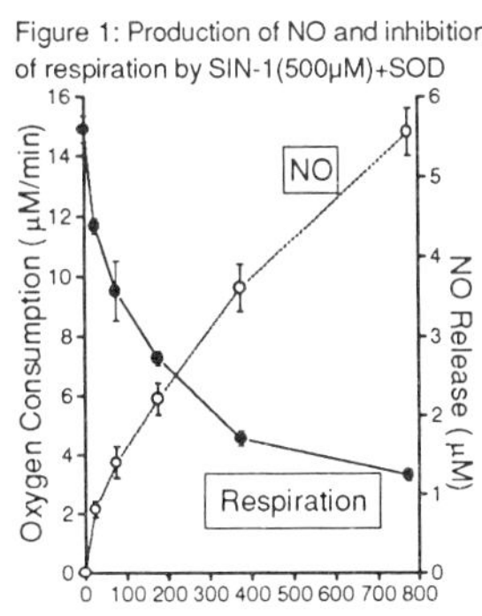

Figure 2: Inhibition of respiration by $ONOO^-$ but not decomposed $ONOO^-$

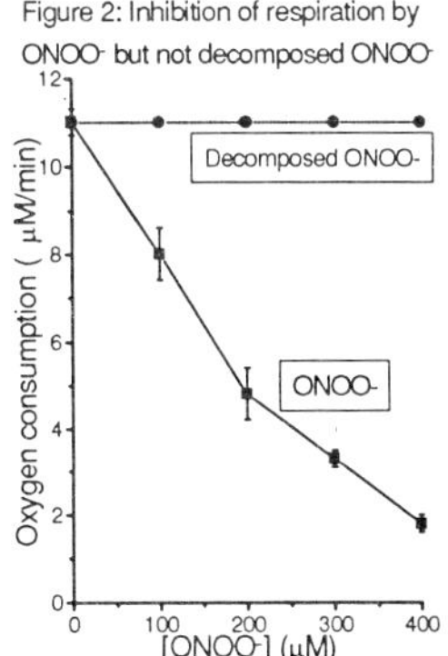

$ONOO^-$ also inhibited respiration from NADH, with an IC50 of 200±17μM (Fig. 2), and from succinate, but did not inhibit respiration from TMPD/ascorbate, implicating inhibition at sites I-III of respiration (data not shown). Inhibition of respiration by NO was reversible (by addition of oxyhaemoglobin) but the effects of $ONOO^-$ were irreversible. Glutathione(GSH) and glucose blocked the effects of $ONOO^-$ but not those of NO (Fig. 3). The EC50 values for preventing the inhibition of respiration by $ONOO^-$ were 10±2μM for glutathione and 7.9±0.9mM for glucose.

Figure 3: Glutathione and glucose prevent the inhibition of respiration by $ONOO^-$ but not that by NO

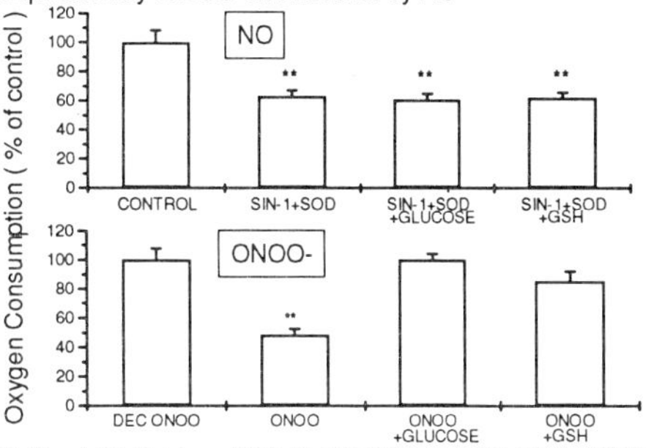

[glutathione] 100μM; [glucose] 100mM; [SIN-1] 500μM; [SOD] 400U/ml; [$ONOO^-$] 200μM

Conclusions

Nitric oxide (NO) and peroxynitrite both inhibit respiration by brain submitochondrial particles, the former reversibly at cytochrome c oxidase,

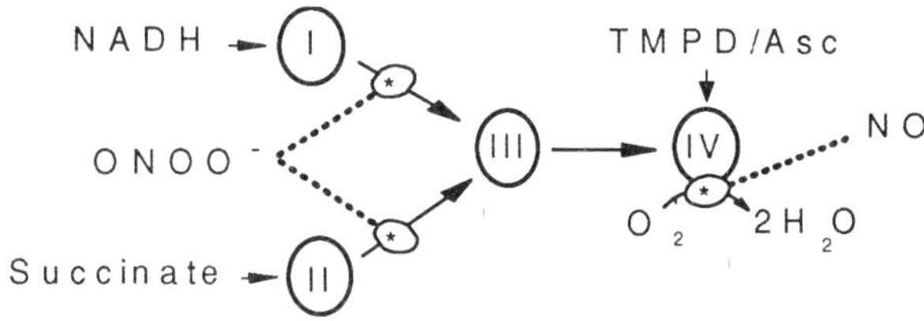

the latter irreversibly at complexes I-III. Both glutathione (IC50=10μM) and glucose (IC50=8mM) prevented inhibition of respiration by peroxynitrite, but neither glucose (100mM) nor glutathione (100μM) affected that by NO.

Thus unless peroxynitrite is formed within mitochondria it is unlikely to inhibit respiration in cells directly, because of reactions with cellular thiols and carbohydrates. However, the reversible inhibition of respiration at cytochrome c oxidase by NO is likely to occur (e.g. in the brain during ischaemia) and could be responsible for cytotoxicity.

I.L. and M.A.M. were supported by fellowships from the Commission of the European Communities.

References

1. Salter M., Knowles R.G., and Moncada S. (1991) FEBS Lett. 291, 145-149.
2. Malinski T., Bailey F., Zhang Z.G. and Chopp M. (1993) J. Cereb. Blood Flow Metab. 13, 355-358.
3. Huang Z., Huang P.L., Panahian N., Dalkara T., Fishman M.C. and Moskowitz M.A. (1994) Science 265, 1883-1885.
4. Nowicki J.P., Duval D., Poignet H. and Scatton B. (1991) Eur. J. Pharmacol. 204, 339-340.
5. Hibbs J.B.,Jr., Taintor R.R., Vavrin Z., Granger D.L., Drapier J.C., Amber I.J. and Lancaster J.R.Jr. (1990) In: Nitric Oxide from L-Arginine: A bioregulatory system. eds. Moncada S. and Higgs E.A. (Elsevier, Amsterdam, The Netherlands), pp. 189-223.
6. Bolaños J.P., Heales S.J.R., Land J.M. and Clark J.B. (1995) J. Neurochem. 64, 1965-1972.
7. Cleeter M.W.J., Cooper J.M., Darley-Usmar V.M., Moncada S. and Schapira A.H.V. (1994) FEBS Lett. 345, 50-54.
8. Beckman J.S., Beckman T.W., Chen J., Marshall P.A. and Freeman B.A. (1990) Proc. Natl. Acad. Sci.USA. 87, 1620-1624.
9. Brown G.C. and Cooper C.E. (1994) FEBS Lett. 356, 295-298.
10. Radi R., Rodriguez M., Castro L. and Telleri R. (1994) Arch. Biochem. Biophys. 308, 89-95.
11. Schweizer M. and Richter C. (1994) Biochem. Biophys. Res. Commun. 204, 169-175.
12. Partridge R.S., Monroe S.M., Parks J.K., Johnson K., Parker W.D., Eaton G.R. and Eaton S.S. (1994) Arch. Biochem. Biophys. 310, 210-217.
13. Blough N.V. and Zafiriou O.C. (1985) Inorg. Chem. 24, 3502-3504.
14. Hogg N., Darley-Usmar V.M., Wilson M.T. and Moncada S. (1992) Biochem. J. 281, 419-424.

Endothelial type nitric oxide synthase (ec-nos) in skeletal muscle fibers: mitochondrial relationships

L. Kobzik*, B. Stringer,* J.-L. Balligand#, M. B. ReidΔ, and J. S. Stamler+

*Physiology Program, Harvard School of Public Health, and #Cardiology Division, Brigham & Women's Hospital, Boston, Ma 02115; ΔPulmonary and Critical Care Medicine, Baylor College of Medicine, Houston, Texas. 77030; +Dept. of Medicine, Duke Univ. Med. Center, Durham, N. C. 27710

Introduction

Recent observations have identified nitric oxide (NO) as a physiologic modulator of skeletal muscle [1, 2]. One source of NO in skeletal muscle is neuronal NO synthase, demonstrated biochemically within muscle homogenates and localized immunohistochemically to the sarcolemma of fast-twitch muscle fibers [1]. The amount of NOS activity within different muscles correlated with their contractile properties. However, additional observations indicate that NO can directly modulate mitochondrial function [3, 4] and oxygen consumption by intact skeletal muscle [5]. We previously had noticed that nc-NOS localization did not correlate with fiber mitochondrial content (succinate dehydrogenase histochemistry). We therefore tested the hypothesis that another constitutive NOS isoform, endothelial NOS, is also expressed by skeletal muscle myocytes and regulates mitochondrial function(s).

Using a specific monoclonal antibody, we found that ec-NOS is also expressed within skeletal muscle fibres, and colocalizes strikingly with mitochondrial markers. We then identified interactions of ec-NOS and muscle mitochondria by assaying isolated mitochondria for NOS activity and by measuring inhibition of mitochondrial oxygen consumption by the NOS substrate, L-arginine.

Methods

Immunohistochemistry. Immunostaining was performed as previously described on cryostat sections fixed in 2% buffered paraformaldehyde of rat skeletal muscles (diaphragm, extensor digitorum longus, soleus, gastrocnemius) [1, 6]. Monoclonal anti-endothelial NOS was obtained from Transduction Laboratories (Knoxville, TN). Rabbit anti-neuronal NOS was generously provided by Dr. David Bredt. Histochemical stains for NADPH diaphorase, ATPase and succinate dehydrogenase were performed using standard methods.

Mitochondrial Isolation & Oxygen Consumption Assay Standard methods were used to isolate mitochondria from normal rat diaphragm and to measure oxygen consumption [7]. L- or D-arginine, or L-nitro-monomethyl-arginine (NMMA) were added to the mixture at various concentrations and mitochondrial NADH-oxygen consumption rates quantified.

NOS assay NOS activity was quantified by measuring conversion of [^{3}H]L-arginine to [^{3}H]L-citrulline, as previously described [8].

Results & Discussion

Immunolocalization showed expression of ec-NOS antigen by a subset of skeletal muscle fibres within diaphgram, extensor digitorum longus, gastrocnemius and soleus.

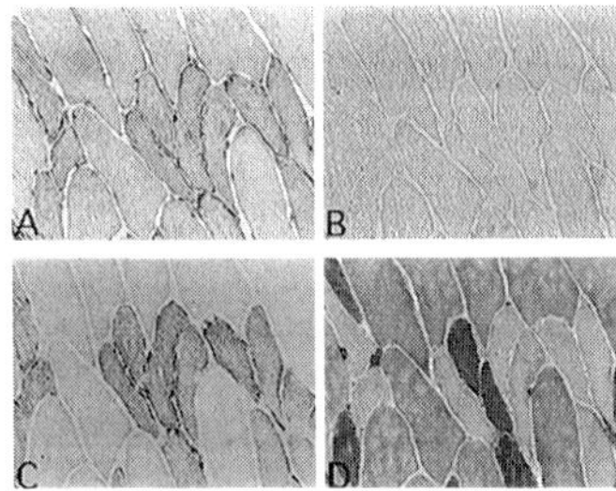

Figure 1. Immunolocalization of ec-NOS in diaphragm.

Figure 1A illustrates strong, diffuse, somewhat granular staining throughout the cytoplasm observed in a cluster of fibres within rat diaphragm. Adjacent fibers are also immunostained, albeit faintly. Vascular endothelium was strongly labeled (not shown), as previously reported [9]. Control IgG showed no significant staining (Fig. 1B). A striking correlation was observed after histochemical staining for succinate dehydrogenase, an enzymatic marker of mitochondrial content (Fig. 1C). In contrast to previous findings with nc-NOS, ec-NOS expression was not related to fiber type, as revealed by ATPase histochemical staining of the same fibres in an adjacent serial section (Fig 1D). Similar results were observed in EDL. As previously reported [1] neuronal NOS was predominantly sarcolemmal and found in type II fibres (results not shown).

The correlation of ec-NOS expression and mitochondrial content suggest a potential functional relationship. When assayed for ^{3}H-L-arginine to ^{3}H-L-citrulline conversion, rat mitochondrial preparations showed a calcium-dependent NOS activity, enriched relative to that measured in samples of total particulate fraction (1.2 $\pm$ 0.1 vs. 0.6 $\pm$ 0.1, fmol/min/mg protein, mitochondria vs. total particulate fraction respectively, N= 3, p <.01). Also, the NOS substrate L-arginine, but not D-arginine, substantially inhibited O_2 consumption of isolated mitochondria.

Table 1. Effect of L-arginine on O_2 consumption by skeletal muscle mitochondria

Treatment		O_2 Consumption (nmoles/min/mg) a	% Inhibition b
None		277 ± 28	
D-Arginine	500 uM	245 ± 5	3 ± 5
L-Arginine	10 uM	226 ± 5	14 ± 7
	100 uM	136 ± 11	52 ± 5
	500 uM	164 ± 8	44 ± 5
	1000 uM	144 ± 9	45 ± 6

[a]Means ± standard deviation, N ≥ 4; [b]relative to assay control

The results show that NO produced by NOS within the mitochondria preparations can modulate respiration, consistent with previous reports [3, 4, 10]. Our light microscopic technique cannot precisely distinguish between ec-NOS localization to the immediate vicinity of or actually within mitochondria, an issue that may be resolved by immunoelectron microscopy. Nevertheless, the data support a potential regulatory role for NO produced by the densely expressed ec-NOS present within mitochondria-rich fibers.

1. Kobzik L, Reid M, Bredt D, Stamler JS. Nitric oxide in skeletal muscle. Nature 1994;372:546-548.
2. Balon T, Nadler J. Nitric oxide release is present from incubated skeletal muscle preparations. J. Appl. Physiol. 1994;77:2519-21.
3. Cleeter M, Cooper J, Darley-Usmar V, et al. Reversible inhibition of cytochrome c oxidase by nitric oxide. FEBS Letters 1994;345:50-54.
4. Schweizer M, Richter C. Nitric oxide potently and reversibly deenergizes mitochondria at low oxygen tensions. Biochem. Biophys. Res. Commun. 1994;204:169-175.
5. King C, Melinyshyn M, Mewburn J, et al. Canine hind limb blood flow and O2 uptake after inhibition of EDRF/NO synthesis. J. Appl. Physiol. 1994;76:1166-1171.
6. Balligand J, Kobzik L, Han X, et al. Nitric oxide-dependent parasympathetic signalling is due to activation of a constitutive endothelial (type III) NO synthase in cardiac myocytes. J. Biol. Chem. 1995;270:14582-14586.
7. Stringer B, Harmon H. Inhibition of cytochrome oxidase by dibucaine. Biochem. Pharmacol. 1990;40:1077-1081.
8. Balligand J, Ungureanu-Longrois D, Simmons W, et al. Cytokine-inducible NO synthase (iNOS) expression in cardiac myocytes. J. Biol. Chem. 1994;269:27580-27588.
9. Pollock JS, Nakane M, Buttery LD, et al. Characterization and localization of endothelial nitric oxide synthase using specific monoclonal antibodies. Am. J. Physiol. 1993;265:C1379-C1387.
10.Stadler J, Billiar T, Curran R, et al. Effect of exogenous and endogenous nitric oxide on mitochondrial respiration of rat hepatocytes. Am. J. Physiol. 1991;260(5 Pt. 1):C910-C916.

Nitric oxide (nitrogen monoxide) and peroxynitrite induce Ca^{2+} release from mitochondria via separate pathways

CHRISTOPH RICHTER and MATTHIAS SCHWEIZER

Laboratory of Biochemistry I
Swiss Federal Institute of Technology (ETH)
Universitätstr. 16
CH-8092 Zürich, Switzerland

Introduction. The mitochondrial membrane potential ($\Delta\Psi$), which is built up either by respiration or ATP hydrolysis, is the driving force for the uptake of Ca^{2+} by mitochondria. Release of Ca^{2+} from mitochondria occurs either when $\Delta\Psi$ collapses (unspecific Ca^{2+} release) or when the specific Ca^{2+} release pathway operates. The latter functions with preservation of $\Delta\Psi$ (reviewed in [1]).

The specific Ca^{2+} release pathway of rat liver mitochondria has been characterized in detail by us (reviewed in [2]). It operates when a 30 kD protein located in the inner mitochondrial membrane is covalently modified by monoADPribose. This modification requires intramitochondrial Ca^{2+} and NAD^+, which is hydrolyzed enzymatically to ADPribose and nicotinamide. NAD^+ hydrolysis is controlled by the protein cyclophilin (*i.e.*, is inhibitable by cyclosporine A, CSA). Hydrolysis is only possible when some vicinal thiols are cross-linked, either by oxidation [3] or by reaction with phenylarsine oxide [4].

Objective. Nitric oxide (nitrogen monoxide, NO$^\bullet$) can bind to the iron of heme proteins [5], *e.g.*, of cytochrome oxidase [6], the terminal enzyme of the mitochondrial respiratory chain, whereas peroxynitrite ($ONOO^-$) is able to oxidize thiols [7]. Since binding of NO$^\bullet$ to cytochrome oxidase should lead to inhibition of respiration and consequently to a collapse of $\Delta\Psi$, and since $ONOO^-$ conceivably cross-links vicinal thiols, we tested whether these reactive nitrogen species induce Ca^{2+} release from mitochondria, and if so by which mechanism.

Results. We reported [8] that a submicromolar bolus of NO$^\bullet$ potently but transiently deenergizes rat liver and brain mitochondria at oxygen concentrations that prevail in cells and tissues. Deenergization is observed when mitochondria utilize respiratory substrates for the build-up of $\Delta\Psi$, but not when mitochondria are energized with ATP. The NO$^\bullet$-dependent deenergization is due to a transient inhibition of cytochrome oxidase, and is paralleled by release and re-uptake of mitochondrial Ca^{2+}. $ONOO^-$, on the other hand, activates the specific Ca^{2+} release pathway (manuscript in preparation), as concluded from the following findings: $ONOO^-$ induces Ca^{2+} release from rat liver mitochondria (i) with preservation of $\Delta\Psi$, (ii) when mitochondrial pyridine nucleotides are oxidized but not when they are reduced, (iii) in a CSA-inhibitable manner, (iv) parallel to NAD^+ hydrolysis, and (v) without entry of extramitochondrial solutes such as sucrose into mitochondria.

Discussion. We found that NO$^\bullet$ at physiologically relevant concentrations causes a transient release of Ca^{2+} from isolated mitochondria, and also from mitochondria in intact cells [8,9]. This NO$^\bullet$-dependent Ca^{2+} release from mitochondria can operate in cell signalling, as recently shown by us (Laffranchi et al., manuscript submitted) with pancreatic β-cells, where NO$^\bullet$ induces insulin secretion by mobilizing mitochondrial Ca^{2+}. It remains to be seen if also endogenously produced NO$^\bullet$ regulates this process. In fact, it has recently been suggested [10] that cytochrome oxidase activity is under the control of NO$^\bullet$.

A prolonged decrease in $\Delta\Psi$ may be dangerous since $\Delta\Psi$ is essential for many mitochondrial functions such as ion or protein transport, and oxidative phosphorylation. It is, therefore, remarkable that $ONOO^-$ induces Ca^{2+} release with preservation of $\Delta\Psi$. In mitochondria superoxide (O_2^-) formation is stimulated by Ca^{2+} uptake [11]. Thus, it is conceivable that $ONOO^-$, which is formed from O_2^- and NO$^\bullet$, acts as a physiological regulator of Ca^{2+} release from mitochondria which allows the maintenance of $\Delta\Psi$.

An important unresolved issue is the presence of nitric oxide synthase (NOS) in mitochondria. The recent report of an association of NOS with mitochondria [12] needs to be confirmed, and the exact location of the enzyme clarified. The most exciting situation would be a Ca^{2+}-dependent NOS located in the mitochondrial matrix or at the inner side of the inner mitochondrial membrane. This situation would provide a self-regulating system for mitochondrial Ca^{2+} homeostasis in which Ca^{2+} release from mitochondria, triggered by $ONOO^-$, occurs without compromising $\Delta\Psi$.

Acknowledgement. M.S. is supported by an anonymous sponsor.

[1] Richter, C. (1992) Mitochondrial calcium transport. In: *New Comprehensive Biochemistry* (Neuberger, A., and Van Deenen, L.L.M., General Editors), Volume *Molecular Mechanisms in Bioenergetics* (L. Ernster, Editor), pp. 349-358. Elsevier, Amsterdam

[2] Richter, C. and Schlegel, J. (1993) Mitochondrial Ca^{2+} release induced by prooxidants. Toxicol. Lett. **67**, 119-127

[3] Schweizer, M. and Richter, C. (1994) Gliotoxin stimulates Ca^{2+} release from intact rat liver mitochondria. Biochemistry **33**, 13401-13405

[4] Schweizer, M., Durrer, P. and Richter, C. (1994) Phenylarsine oxide stimulates the pyridine nucleotide-linked Ca^{2+} release from rat liver mitochondria. Biochem. Pharmacol. **48**, 967-973

[5] Lowenstein, C.J., Dinerman, J.L. and Snyder, S.H. (1994) Nitric oxide - A physiological messenger. Ann. Intern. Med. **120**, 227-237

[6] Gorren, A.C.F., Van Gelder, B.F. and Wever, R. (1988) Photodissociation of cytochrome c oxidase-nitric oxide complexes. Ann. NY Acad. Sci. **550**, 139-149

[7] Radi, R., Beckman, J.S., Bush, K.M. and Freeman, B.A. (1991) Peroxynitrite oxidation of sulfhydryls. The cytotoxic potential of superoxide and nitric oxide. J. Biol. Chem. **266**, 4244-4250

[8] Schweizer, M. and Richter, C. (1994) Nitric oxide potently and reversibly deenergizes mitochondria at low oxygen tension. Biochem. Biophys. Res. Commun. **204**, 169-175

[9] Richter, C., Gogvadze, V., Schlapbach, R., Schweizer, M. and Schlegel, J. (1994) Nitric oxide kills hepatocytes by mobilizing mitochondrial calcium. Biochem. Biophys. Res. Commun. **205**, 1143-1150

[10] Brown, G.C. (1995) Hypothesis. Nitric oxide regulates mitochondrial respiration and cell functions by inhibiting cytochrome oxidase. FEBS Lett. **369**, 136-139

[11] Chacon, E. and Acosta, D. (1991) Mitochondrial regulation of superoxide by Ca^{2+}: An alternate mechanism for the cardiotoxicity of doxorubicin. Toxicol. Appl. Pharmacol. **107**, 117-128

[12] Kobzik, L., Stringer, B., Balligand, J.-L., Reid, M.B. and Stamler, J.S. (1995) Endothelial type nitric oxide synthase in skeletal muscle fibers: Mitochondrial relationship. Biochem. Biophys. Res. Commun. **211**, 375-381

Redox determinants of direct and indirect modulation of L-type calcium channels in ferret ventricular myocytes : dual regulation by S-nitrosylation and cyclic GMP.

Donald L. Campbell[1], Harold C. Strauss[1,2], and Jonathan S. Stamler[2,3].

Departments of [1,2]Pharmacology, [3]Cell Biology, and [2,3]Medicine, Duke University Medical Center, Durham, North Carolina, USA 27710.

INTRODUCTION

Naturally occuring N-oxides and S-nitrosothiols could potentially serve as important, physiologically relevant modulators of working cardiac muscle through modulation of the L-type calcium current, $I_{Ca,L}$ [1]. However, it is slowly becoming recognized that the effects of such compounds on ionic channels in excitable tissues can be multiple and quite complex [2,3]. N-oxides and S-nitrosothiols could potentially modulate ventricular $I_{Ca,L}$ by two mechanisms: (i) an indirect effect via activation of guanylate cyclase, and (ii) a direct effect upon the L-type calcium channel or an associated subunit(s) or molecule. Cellular redox state could also be an important determinant of underlying mechanism [2]. While indirect, second messenger mediated modulation of cardiac $I_{Ca,L}$ has been widely studied [1], direct channel modulatory effects have yet to be demonstrated. We hypothesized that the L-type Ca^{2+} channel might be subject to discrete mechanisms of regulation by NO and S-nitrosothiols. Our data are the first to indicate that both direct and indirect channel subunit systems are involved in modulation of ventricular $I_{Ca,L}$ by NO and S-nitrosothiols.

METHODS

Right ventricular myocytes were enzymatically isolated from the hearts of 10-16 week old male ferrets as previously described [4]. Myocytes were voltage clamped using the gigaOhm seal patch clamp technique in the whole cell recording configuration and $I_{Ca,L}$ was isolated as previously described [4]. Peak $I_{Ca,L}$ was recorded (22°C) by applying a 500 msec pulse to 0 mV from a holding potential of -70 mV once every 6 seconds. Other experimental details on the properties of $I_{Ca,L}$ in these myocytes can be found in [4].

RESULTS

The compound SIN-1 generates both NO and O_2^- (superoxide), which can then combine to form $OONO^-$ (peroxynitrite). The effects of extracellular perfusion of 0.1-1 mM SIN-1 on ferret ventricular peak basal $I_{Ca,L}$ were complicated, in that it could either inhibit (6/12 myocytes) (Figure 1) or stimulate (6/12 myocytes) $I_{Ca,L}$. This variability could have been due to either the effects of NO, O_2^-, or $OONO^-$ dominating in any one given myocyte. To eliminate the effects of O_2^- and $OONO^-$ we measured the effects of 1 mM SIN-1 in the presence of superoxide dismustase (SOD; 500-3000 units/ml). Under these conditions 1 mM SIN-1 consistently produced a moderate inhibition of basal $I_{Ca,L}$. These results suggest that NO can inhibit basal $I_{Ca,L}$.

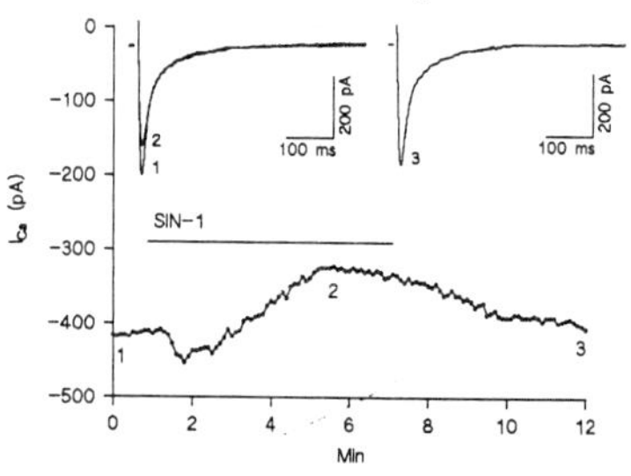

Figure 1. Effects of 1 mM SIN-1 on peak $I_{Ca,L}$ recorded at 0 mV. Representative current recordings at indicated times shown in insets.

To test the hypothesis that NO was indirectly inhibiting $I_{Ca,L}$ through activation of guanylate cyclase, the effects of 10-100 micromolar extracellular 8-Br-cGMP were determined. 8-Br-cGMP consistently inhibited peak basal $I_{Ca,L}$ (4 myocytes), suggesting that NO indirectly inhibits $I_{Ca,L}$ through activation of guanylate cyclase.

We next determined the effects of two naturally occuring S-nitrosothiols, S-nitrosoglutathione (GSNO) and S-nitrosocysteine (CySNO). In contrast to SIN-1, both 1 mM GSNO and CySNO consistently produced a significant and reversible increase in peak $I_{Ca,L}$ (4 myocytes). The stimulatory effects of both GSNO and CySNO were uneffected by 20 micromolar ryanodine (3 myocytes), indicating that their effects could occur independently of any alterations in Ca^{2+} release from the sarcoplasmic reticulum (Figure 2).

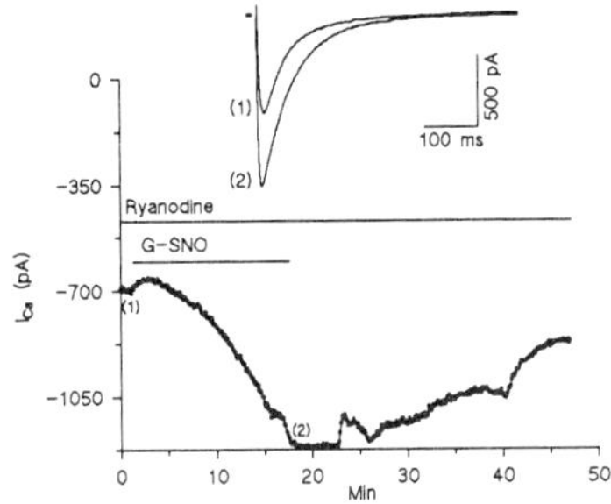

Figure 2. Stimulatory effects of 1 mM GSNO on peak $I_{Ca,L}$ recorded at 0 mV. 20 micromolar ryanodine present to block SR Ca^{2+} release.

The effects of S-nitrosothiols on $I_{Ca,L}$ were therefore consistently opposite to those produced by NO. Since it is unlikely that GSNO could obtain rapid access to intracellular sites, these data suggest that S-nitrosothiols are exerting their effects through an extracellular site that is not responsive to NO. One likely candidate for their site of action would be an extracellular modulatory thiol-containing "redox switch" associated with the channel [1].

To test for the existence of such a direct redox site we studied the effects of sulfhydryl reducing and oxidizing agents. Extracellular application of 1-5 mM dithiothreitol (DTT), a sulfhydryl reducing agent, produced a rapid inhibition of $I_{Ca,L}$. This DTT-induced inhibition could be reversed by extracellular application of 200 micromolar 5,5'-dithio-bis[2-nitrobenzoic acid] (DTNB), a thiol oxidizing agent (4 myocytes). $I_{Ca,L}$ could be repeatedly inhibited by DTT and then stimulated by DTNB (Figure 3). Furthermore, the effects of GSNO could be reversibly inhibited by 1-5 mM DTT. Taken in aggregate, our results argue for the existence of an extracellular redox site involved in direct modulation of $I_{Ca,L}$, and that S-nitrosothiols exert their effects at this site through S-nitrosylation.

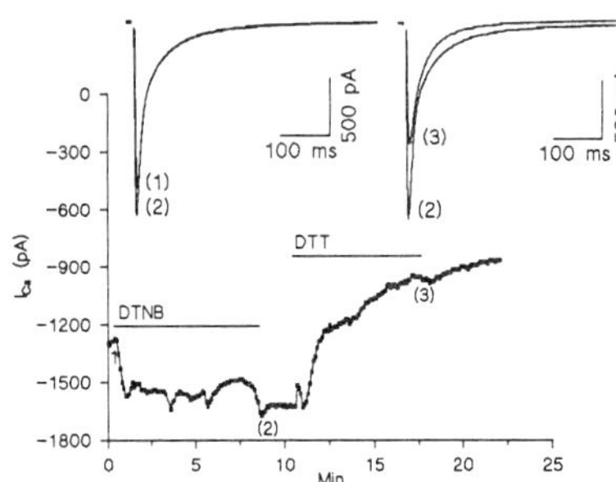

Figure 3. Direct stimulatory and inhibitory effects of DTNB (200 micromolar) and DTT (1 mM) on peak $I_{Ca,L}$ recorded at 0 mV.

SUMMARY

While most studies to date on modulation of cardiac $I_{Ca,L}$ have concentrated on receptor-mediated intracellular second-messenger cascades [1], our data strongly suggest that ventricular $I_{Ca,L}$ can also be importantly modulated by nonreceptor-mediated redox-sensitive signalling pathways directly triggered at the sarcoplasmic surface [2]. Our results indicate that: 1) both indirect (cGMP-dependent) and direct (S-nitrosylation dependent) channel subunit systems are involved in modulation of right ventricular $I_{Ca,L}$: 2) NO and S-nitrosothiols exert discrete and opposite effects on $I_{Ca,L}$; and 3) Ca^{2+} homeostasis in ventricular myocytes is directly regulated by cellular redox state.

REFERENCES

1. Campbell, D.L. and Strauss, H.C. (1995). Regulation of calcium channels in the heart. In: Advances in Second Messenger and Phosphoprotein Research (A.R. Means, ed.), pp. 25-88, Raven Press, Ltd., New York.
2. Stamler, J.S. (1994). Redox signalling: nitrosylation and related target interactions of nitric oxide. Cell. 78: 931-936.
3. Arnelle, D.R. and Stamler, J.S. (1995). NO^+, NO·, and NO^- donation by S-nitrosothiols: implications for regulation of physiological functions by S-nitrosylation and acceleration of disulfide formation. Arch. Biochem. Biophys. 318: 279-285.
4. Qu, Y., Campbell, D.L., Whorton, A.R., and Strauss, H.C. (1993). Modulation of L-type Ca^{2+} current by adenosine in ferret isolated right ventricular myocytes. J. Physiol. 471: 269-293.

Role of nitric oxide in the control of growth factor-elicited signals: phosphoinositide hydrolysis and Ca^{2+} release are inhibited via a cGMP-dependent protein kinase I pathway

*#EMILIO CLEMENTI, #CLARA SCIORATI, #JACOPO MELDOLESI and °GIUSEPPE NISTICÒ

*Dept. Pharmacology, Faculty of Pharmacy, University of Reggio Calabria, 88021 Catanzaro, Italy. #Dept. Pharmacology, C.N.R. Cytopharmacology Ctr., DIBIT-San Raffaele Scientific Institute, University of Milano, 20132 Milano, Italy. °Dept. Biology, University of Roma "Tor Vergata", 00133 Roma, Italy.

Among the molecules of the signal transduction cascades turned on by receptor agonist binding NO can play important roles not only in the intracellular activation process but also in the fine feedback regulation of signalling itself. In the cells competent for the Ca^{2+}-dependent, constitutive forms of NOS, this highly reactive radical gas, generated in response to appropriate increases of $[Ca^{2+}]_i$, works as the controller of a number of enzymes including guanylyl cyclase [1]. The ensuing increase of cGMP formation, with activation of G kinases, yields responses that may be variable from cell to cell [2]. In the case of receptors coupled to PIP_2 hydrolysis via the activation of heterotrimeric G proteins of the Gq family [3], NO has been described to exert a negative modulation with decreased generation of IP_3 and diacylglycerol, resulting in blunting of the spike of Ca^{2+} release from intracellular stores [2,4]. With these receptors the modulation was shown to depend upon G kinase activation, and the site of action was proposed to be at the G protein/PLC interface [5].

PIP_2 hydrolysis is induced not only by G protein-coupled receptors but also by growth factor receptors, working however on different PLCs and by a different activation process, *i.e.* by direct tyrosine phosphorylation of the PLCs of the γ family rather than by G protein activation of those of the β family [3,6]. For this reason their modulation by NO could not be predicted.

In the present study we have focused on PIP_2 hydrolysis and Ca^{2+} release triggered by EGF, PDGF, and FGF administration in a clone of NIH 3T3 fibroblasts overexpressing the human EGF receptor [7], and in the A431 and KB human carcinoma cell lines.

To investigate the effects of NO on PIP_2 hydrolysis and Ca^{2+} release, cell lines were routinely grown as described [7]; in the case of IP_3 measurements, their incubation medium was supplemented with tritiated *myo*-inositol [4]. Cells were detached from the Petri dish by gentle trypsinization, resuspended in KRH medium [4], loaded with fura-2AM (for Ca^{2+} release experiments; see 4) and incubated for 15 min at 37°C with various agents interfering with the L-arginine/NO pathway listed in Table I. Cell aliquots were then treated with EGF (0.1-100 nM), PDGF (0.1-10 nM) or FGF (1-30 nM) and analyzed for IP_3 generation or Ca^{2+} release as previously described [4]. Table 1 summarizes the results (± S.D.; n=6) obtained with 30 nM EGF on NIH 3T3 cells exposed to the NO generating drug, SNP, 8-Br cGMP, the NOS inhibitor, L-NAME, its less active enantiomer, D-NAME, and the G kinase inhibitor, KT5823. Consistent results were observed with all concentrations of EGF tested. A similar response pattern was observed also with PDGF and FGF, in all three cell lines investigated. These results strongly suggest that NO plays an inhibitory role on Ca^{2+} release elicited by growth factors, and that its effect is mediated somewhere upstream IP_3 generation. The fact that a cGMP analogue, 8-Br cGMP, exerts an effect similar to that of the NO donor, SNP, and that the effect of the latter is prevented by the G kinase inhibitor, KT5823, indicate in addition that NO effect is mediated by the cGMP/Gkinase signal transduction pathway.

Table 1: effects of NO on EGF signalling in NIH 3T3 cells

treatment:	IP_3 formation (cpm/mg protein)	Ca^{2+} release (% $[Ca^{2+}]_i$ increase over basal)
control	135 ±9.2	470±29
SNP (30 µM)	85.1±6.9	354±22
8-Br cGMP (200 µM)	82.0±7.4	33[illegible]±14
L-NAME (100 µM)	184 ±18	586±36
D-NAME (100 µM)	139 ±11	476±33
SNP+KT5823 (10 µM)	127 ±9.8	464±23

To elucidate the molecular level at which NO exerts its inhibition we further dissected out the growth factor-activated signal transduction pathway. The enzymes responsible for growth factor-induced IP_3 generation, the PLCs of the γ family, are known to be Ca^{2+}-dependent enzymes. In particular it is well known that they can be activated, independently of receptor coupling, by persistently high $[Ca^{2+}]_i$ levels. PLCγ isoenzymes were therefore activated by cell incubation with high (3 µM) concentrations of the Ca^{2+} ionophore, ionomycin, and IP_3 generation measured. Incubation of cell samples with SNP or 8-Br cGMP resulted in a reduction of IP_3 generation induced by the ionophore, while L-NAME increased it. When KT5823 was administered together with SNP, the inhibitory effect of the latter was prevented. These results strongly suggest that the primary site at which NO exerts its inhibitory action on Ca^{2+} release and IP_3 generation is at the level of the PLCγ isoenzymes, and that the NO effect is mediated via the activation of the cGMP/G kinase signal transduction pathway. What remains to be elucidated is the mechanism of the inhibition of the PLCγ activity by G kinase we have now described: decreased complex formation of the enzyme with growth factor receptors; of its degree of tyrosine phosphorylation; of its activation level.

NO is known to inhibit mitogenesis and proliferation [2]. In the array of intracellular signals elicited by growth factor receptor activation, impaired PIP_2 hydrolysis and $[Ca^{2+}]_i$ responses exert an inhibitory effect on growth [8]. The possibility should therefore be considered that the effects of NO and cGMP on cell growth are mediated, at least in part, by negative modulatory actions described here. Whether these actions are accompanied by others as yet unknown, remains to be investigated.

1. Bredt, D.S., and Snyder, S.H. Nitric oxide: a physiologic messenger molecule (1994) Annu. Rev. Biochem. **63**, 175-195
2. Lincoln, T.M., Komalavilas, P., and Cornwell, T.L. Pleiotropic regulation of vascular smooth muscle tone by cyclic GMP-dependent protein kinase. (1994) Hypertension **23**, 1141-1147
3. Berridge, M.J. Inositol trisphosphate and calcium signalling. (1993) Nature **361**, 315-325
4. Clementi, E., Vecchio, I., Sciorati, C., and Nisticò, G. Nitric oxide modulation of agonist-evoked intracellular Ca^{2+} release in neurosecretory PC12 cells. (1995) Mol. Pharmacol. **47**, 517-524
5. Nguyen, B.L., Saitoh, M., and Ware, J.A. Interaction of nitric oxide and cGMP with signal transduction in activated platelets (1991) Am. J. Physiol. **261**, H1043-H1052
6. Schlessinger, J., and Ullrich, A. Growth factor signaling by receptor tyrosine kinases. (1992) Neuron **9**, 383-391
7. Pandiella, A., Magni, M., Lovisolo, D., and Meldolesi, J. The effects of epidermal growth factor on membrane potential.(1989) J. Biol. Chem. **264**, 12914-12921
8. Short, A.D., Bian, J., Ghosh, T.K., *et al.* Intracellular Ca^{2+} pool content is linked to control of cell growth. (1993) Proc. Natl. Acad. Sci. USA **90**, 4986-4990

Abbreviations used: NO, nitric oxide; NOS, nitric oxide synthase; $[Ca^{2+}]_i$, intracellular Ca^{2+} concentration; PIP_2, phosphatidyl inositol 4,5,-bisphosphate; G protein, guanine nucleotide-binding protein; PLC, phospholipase C; IP_3, inositol 1,4,5-trisphosphate; EGF, PDGF and FGF, epidermal, platelet-derived and fibroblast growth factors; KRH, Krebs Ringer Hepes medium; L-NIO, L-N-(1-iminoethyl)-ornithine; SNP, sodium nitroprusside; D-NAME, N^{ω}-nitro-D-arginine methylester; L-NAME, N^{ω}-nitro-L-arginine methylester.

Growth factor-induced Ca^{2+} influx is increased by nitric oxide via activation of cGMP-dependent protein kinase I

#CLARA SCIORATI, #*EMILIO CLEMENTI, #JACOPO MELDOLESI and °GIUSEPPE NISTICÒ

#Dept. Pharmacology, C.N.R. Cytopharmacology Ctr., DIBIT-San Raffaele Scientific Institute, University of Milano, 20132 Milano, Italy. *Dept. of Pharmacology, Faculty of Pharmacy, University of Reggio Calabria, 88021 Catanzaro, Italy. °Dept. Biology, University of Roma "Tor Vergata", 00133 Roma, Italy.

NO is involved in the regulation of crucial events in intracellular signalling, including those leading to cell growth and differentiation [1]. These biological processes occur under the control of a complex network of intracellular signals originated by a variety of extracellular molecules. Among these a relevant role is played by polypeptide growth factors acting through their cognate tyrosine kinase receptors [2]. The signals delivered by these receptors include variations in $[Ca^{2+}]_i$ sustained by both release from intracellular stores and influx across the plasmalemma [2]. The latter appears to contribute critically to the overall effects of growth factors. Pharmacological blockade of Ca^{2+} influx in fact decreased substantially the mitogenic effect of EGF [3]. Moreover, virally transformed cells appear to be dependent on an intact Ca^{2+} influx system for their proliferation [4]. A detailed characterization of the relationships between NO synthesis and growth factor-induced Ca^{2+} influx could therefore be important in shedding light on the intricate network of intracellular signals responsible for the control of proliferation and transformation processes.

Until now, the effects of NO on Ca^{2+} influx have been elucidated only in the case the responses induced by activation of membrane receptors coupled to polyphosphoinositide hydrolysis and inositol 1,4,5-trisphosphate generation via heterotrimeric G proteins. With these receptors influx of the cation occurs through two independently regulated pathways, SMOC and SDC (whose current is also referred to with the acronym I_{CRAC}) Ca^{2+} channels [5]. With SDC channels the modulatory role of NO may be stimulatory or inhibitory, depending on the cell system investigated, and may be exerted either through the cGMP/G kinase or other, yet unidentified, signaling pathways [6,7]. In contrast, SMOC channels appear insensitive to the action of NO, though an effect of the gaseous messenger cannot be entirely excluded [8].

In the present study we have investigated the role of NO in the modulation of Ca^{2+} influx stimulated by the activation of growth factor receptors. As the experimental system we have focused on Ca^{2+} influx triggered by EGF and PDGF administration in a clone of NIH 3T3 cells overexpressing the human EGF receptor [9]. The cells were routinely grown as described [9], detached from the Petri dish by gentle trypsinization, resuspended in KRH medium [7], loaded with fura-2AM as described [7] and incubated for 15 min at 37°C with various agents interfering with the L-arginine/NO pathway, listed in Table I. SDC and SMOC influx pathways were analyzed separately, utilizing the Mn^{2+} quenching of fura-2 technique [7].

Table 1 summarizes the results (± S.D.; n=12) obtained with 30 nM EGF on NIH 3T3 cells exposed to the NO generating drug, SNP, 8-Br cGMP, the NOS inhibitor, L-NAME, its less active enantiomer, D-NAME, and the G kinase inhibitor, KT5823. A similar response pattern was observed also with PDGF as receptor stimulant. These results strongly suggest that NO plays a facilitatory role on both SDC and SMOC channels activated by growth factors. The observation that, when the cells were preincubated with the NOS inhibitor, L-NAME, the effects observed were opposite to those induced by the treatment with SNP, indicate that in NIH 3T3 cells NOS is physiologically active. Thus, endogenous NO may play a role in modulating the complex signalling response subsequent to growth factor receptor activation. The observation that a cGMP analogue, 8-Br cGMP, exerts an effect similar to that of the NO donor, SNP, and that the effect of the latter is prevented by the G kinase inhibitor, KT5823, indicate in addition that the NO effect is mediated by the cGMP/Gkinase signal transduction pathway.The facilitating effect of NO on Ca^{2+} influx through SDC channels confirms recent observations in pancreatic acinar cells [6] in which its effect was also mediated by cGMP-dependent protein kinase I. NO was recently reported to increase tyrosine phosphorylation in murine fibroblasts by a cGMP-dependent mechanism [10]. In addition, increasing evidence supports the notion that tyrosine kinases play a crucial role in the activation of SDCs in a variety of cell systems [5]. Taken together, these observations suggest a possible pathway by which NO activates SDCs. The role of NO in the control of Ca^{2+} processes activated by growth factor receptors might be of physiological relevance. Most studies carried out on growth and differentiation have been focused primarily on the final effects of NO rather than on the underlying mechanisms. The present data, taken together with previous reports suggesting a role for Ca^{2+} in the control of cell proliferation [3,4], suggest a possible mechanism by which the regulatory role of NO could be mediated.

Table 1: effects of NO on EGF-activated Ca^{2+} influx

treatment:	SMOC	SDCs
	(% Mn^{2+} quenching of fura-2 fluorescence)	
control	1.57±0.1	7.25±0.8
SNP (30 μM)	3.49±0.1	10.1±1.0
8-Br cGMP (200 μM)	2.98±0.9	9.72±0.8
L-NAME (200 μM)	1.09±0.1	6.11±0.6
D-NAME (200 μM)	1.50±0.2	7.23±0.7
SNP+KT5823 (10 μM)	1.39±0.1	7.02±0.3

Abbreviations used: NO, nitric oxide; $[Ca^{2+}]_i$, intracellular Ca^{2+} concentration; EGF and PDGF, epidermal and platelet-derived growth factors; SMOCs and SDCs, second messenger-operated and store-dependent Ca^{2+} channels; I_{CRAC}, Ca^{2+} release activated current; G kinase, cGMP-dependent protein kinase I; KRH, Krebs Ringer Hepes medium; SNP, sodium nitroprusside; D-NAME, N^{ω}-nitro-D-arginine methylester; L-NAME, N^{ω}-nitro-L-arginine methylester; NOS, nitric oxide synthase.

1. Forstermann, U., Closs, E.I., Pollock, J.S., *et al.*. Nitric oxide isozymes. Characterization, purification, molecular cloning, and functions. (1994) Hypertension **23**, 1121-1131
2. Schlessinger, J., and Ullrich, A. Growth factor signaling by receptor tyrosine kinases. (1992) Neuron **9**, 383-391
3. Magni, M., Meldolesi, J., and Pandiella, A. Ionic events induced by epidermal growth factor. (1991) J. Biol. Chem. **266**, 6329-6335
4. Ghosh, T.K., Bian, J., Short, A.D., Rybak, S.L., and Gill, D.L. Persistent intracellular calcium pool depletion by thapsigargin and its influence on cell growth. (1991) J. Biol. Chem. **266**, 24690-24697
5. Felder, CC, Singer-Lahat, D., and Mathes, C. Voltage-independent Ca^{2+} channels (1994) Biochem Pharmacol **48**, 1997-2004
6. Xu, X., Star, R.A., Tortorici, G., and Muallem, S. Depletion of intracellular Ca^{2+} stores activates nitric-oxide synthase to generate cGMP and regulate Ca^{2+} influx. (1994) J. Biol. Chem. **269**, 12645-12653
7. Clementi, E., Vecchio, I., Corasaniti, M.T., and Nisticò, G. Nitric oxide modulates agonist-evoked Ca^{2+} release and influx responses in PC12-64 cells. (1995) Eur. J. Pharmacol. **289**, 113-123
8. Gukovskaya, A., and Pandol, S. Nitric oxide production regulates cGMP formation and calcium influx in pancreatic acinar cells. (1994) Am. J. Physiol. **266**, G350-G356
9. Pandiella, A., Magni, M., and Meldolesi, J. Plasma membrane hyperpolarization and $[Ca^{2+}]_i$ increase induced by fibroblast growth factor in NIH-3T3 fibroblasts. (1989) Biochem. Biophys. Res. Co. **163**, 1325-1331
10. Peranovich, T.M.S., da Silva, A.M., Fries, D.M., Stern, A., and Monteiro, H.P. Nitric oxide stimulates tyrosine phosphorylation in murine fibroblasts in the absence and presence of epidermal growth factor. (1995) Biochem. J. **305**, 613-619

Nitric oxide synthase inhibition enhances protheosynthesis in rat tissues.

IVETA BERNÁTOVÁ, OĽGA PECHÁŇOVÁ and PAVEL BABÁL*

Institute of Normal and Pathological Physiology Slovak Academy of Sciences, Sienkiewiczova 1, *Department of Pathology, Medical Faculty, Comenius University, 813 71 Bratislava, Slovak Republic

NO-generating substancies have been recently identified as inhibitors of mitogenesis and proliferation [1]. The aim of this study was to determine whether L-NAME could modulate nucleic acid content and proteosynthesis in rat heart, aorta, brain, liver and kidney.

Male Wistar rats, 15 weeks old were divided into three groups. The first group (n=8) served as the control. The second (n=8) and the third group (n=8) were given L-NAME in the dose 20 and 40 mg/kg/day respectively in tap water for 4 weeks. Systolic blood pressure and heart rate were measured by tail-cuff plethysmography. After 4 weeks of L-NAME treatment heart/body weight ratio was calculated. Total RNA content and [^{14}C]leucine incorporation as a markers of proteosynthesis and DNA content as a marker of proliferation were measured.

For analysis of total RNA content, a single step method was used [2]. Total DNA content was analysed by non phenol method [3]. The proteosynthesis was determined by [^{14}C]leucine incorporation into protein [4].

L-NAME in both 20 and 40 mg/kg/day doses increased SBP by 34% and 30% respectively, while the HR decresed by 15% and 20% respectively as compared with control parameters. The heart/body weight ratio was 2.63 0.08 in control group, 2.68 0.10 in the second group and 2.86 0.21 in the third group but these changes were not significant. The total RNA content increased in the second group by 10% in myocardium, by 155% in aorta, by 47% in brain, by 8% in liver and by 13% in kidney. In the third group, the total RNA content was elevated similarly by 15% in myocardium, by 254% in aorta and by 85% in brain, by 12% in liver and by 36% in kidney. The results were significant on the level p<0.05 as compared with control parameters. The alterations of total RNA content were followed by elevation of [^{14}C]leucine incorporation into protein. These values were elevated in the second and the third group by 100% and 98% in myocardium, by 40% and 49% in aorta, by 38% and 40% in brain, by 166% and 193% in liver and by 30% and 117% in kidney (Fig.1.). The differences between the second and the third group were not significant. Total DNA content was significantly elevated in the third group by 238% in myocardium, by 200% in aorta, by 350% in brain, by 79% in liver and by 92% in kidney (p<0.05). In the second group group total DNA content was significantly elevated only in brain, liver and kidney (Fig.2.).

Abbreviations used: L-NAME, N^G-nitro-L--arginine methyl ester; SBP, systolic blood pressure; HR, heart rate

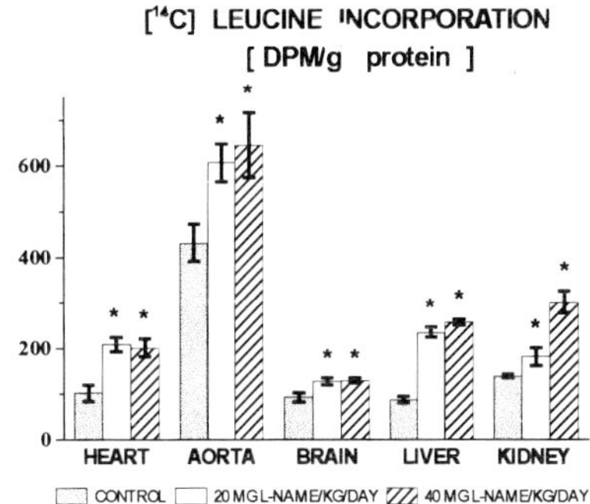

Fig. 1. [^{14}C] leucine incorporation into the protein of tissues. Results are presented as mean ± SEM, $p < 0.05$.

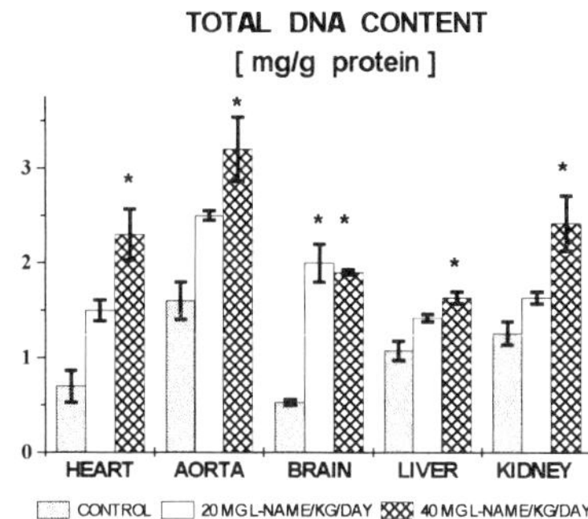

Fig. 2. Total DNA content. Results are presented as mean ± SEM, $p < 0.05$.

Taken together, the results suggesst that endogenous NO modulates total RNA and DNA content as well as proteosynthesis in the heart, aorta, brain, liver and kidney in rats. The increase in proteosynthesis may be either the result of reduced availability of nitric oxide in cells or of direct effect of blood pressure increase. However, to evaluate the mechanism of this modulation further experiments are needed.

The study was supported in part by PECO grant BMH1-CT-92-1893 and Slovak National Bank.

1.Garg,U.C. and Hassid,A.(1989)J.Clin.Invest. 83,1774
2.Chomczynski, P. and Sacchi, N. (1987) Anal. Biochem. 162,156-159
3.Sambrook, J., Fritsch, E.F. and Maniatis,T. (1989)Molecular cloning. A laboratory manual, second edition (Ford.N.,ed.), pp. 9.16-9.19, Gold Spring Harbor Laboratory Press, New York
4.Gerová.M.,Pecháňová,O.,Stoev,V.,Kittová,M., Bernátová,I. and Bárta,E.(1995) Am.J.Physiol. In press

Treatment of intact NIH 3T3 cells with nitric oxide affects proteins in the Ras signaling pathway

TARA L. BAKER and JANICE E. BUSS

Dept. of Zoology & Genetics and Biochemistry & Biophysics
Iowa State University, Ames IA 50011

Although nitric oxide and its metabolites (NO_x) are involved in several signaling systems, their importance in growth pathways, such as the Ras pathway, has been little studied. We have found that treatment of NIH 3T3 cells with S-nitroso-cysteine (SNC), a nitric oxide producer, leads to decreases in phosphorylation of MAP kinases and Ras palmitoylation, suggesting that proteins in the Ras signaling pathway are previously undetected targets for NO_x.

Despite their malignant potential, most mammalian Ras proteins normally regulate cell growth in a controlled manner, by providing the crucial link between growth factor receptors on the cell surface and cytoplasmic and nuclear target proteins. In order to form this physical connection the Ras proteins must bind to the inner side of the cell surface, a process aided through direct attachment of lipids to the carboxy terminus of the Ras protein [1]. Importantly, forms of Ras which are cytosolic are unable to transmit these growth signals. Even tumorigenic Ras proteins fail to cause oncogenic transformation unless tethered to the membrane. Membrane binding thus indirectly but potently regulates the biological activity of Ras proteins.

Important signaling proteins such as heterotrimeric G proteins and src-family tyrosine kinases have recently been discovered to be palmitoylated [2] and events that regulate palmitoylation of these proteins are just beginning to be identified. Increased [^{3}H]palmitate incorporation linked to an increase in palmitate turnover has been reported in G_s alpha subunits after exposure of cells to isoproterenol [3]. A decrease in palmitoylation of an endothelial cell nitric oxide synthetase has been reported to occur after bradykinin treatment of endothelial cells [4]. A general decrease in palmitate labeling of many proteins, including the GAP43 and SNAP-25 proteins, occurs in neurons exposed to nitric oxide, coincident with the retraction of the neuronal growth cones [5]. No agents which alter Ras palmitoylation have yet been identified.

Nitric oxide is a unique type of signaling molecule and produces its effects through direct chemical modification (e.g., nitrosylation) of target proteins [6]. The proteins which can be chemically modified by NO_x are numerous, and a current challenge of the field is to identify the proteins whose modification is relevant to particular biological responses.

When transfected NIH 3T3 cells expressing an H-Ras protein are labeled with [^{3}H]palmitate in the presence or absence of S-nitroso-cysteine (SNC), the H-Ras in the treated cells shows a 2-4 -fold stimulation of [^{3}H]palmitate incorporation (Fig. 1A). Turnover of palmitate on H-Ras is also accelerated in SNC-treated cells. Fig. 1B shows that exposure of cells to SNC increases the rate at which [^{3}H]palmitate is lost from H-Ras during a "chase" period, as detected by an ~2-fold greater rate of loss of previously attached [^{3}H]palmitate ($t_{1/2}$ decreases from ~80' to ~20'). This removal of palmitate then provides new sites for immediate re-acylation, thus explaining the observed increase in [^{3}H]palmitate incorporation. With repeated additions of SNC (4 x 30'), the rate of deacylation overtakes the rate of readdition, and overall palmitoylation of H-Ras declines.

Nitric oxide treatment also decreases the rate at which the cytosolic precursor form of H-Ras binds to membranes. In untreated cells ~50% of the newly synthesized, [^{35}S]methionine-labeled protein is found in the membrane fraction (P100) after a 30' labeling period, and ~90% after an additional 30' of chase in non-radioactive media (data not shown). If a 10' SNC treatment is given immediately after the pulse label and the chase continued for 20', the conversion of the precursor in the S100 fraction to the mature form in the P100 is inhibited. If the chase is allowed to continue for 60' or 120' after the single 10' SNC treatment, the precursor can complete processing and shift into the P100 fraction. No effect on membrane association is observed if the SNC is depleted of NO_x by allowing the solution to stand at room temperature overnight. The effect of NO_x is therefore transient as well as reversible and causes, by two methods of analysis, an effect on Ras.

A

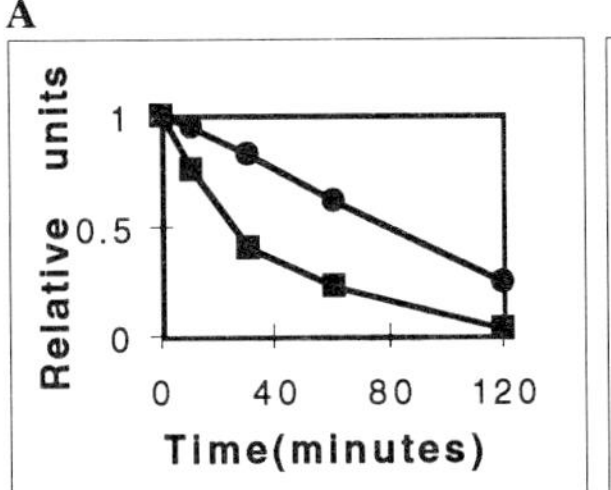

B

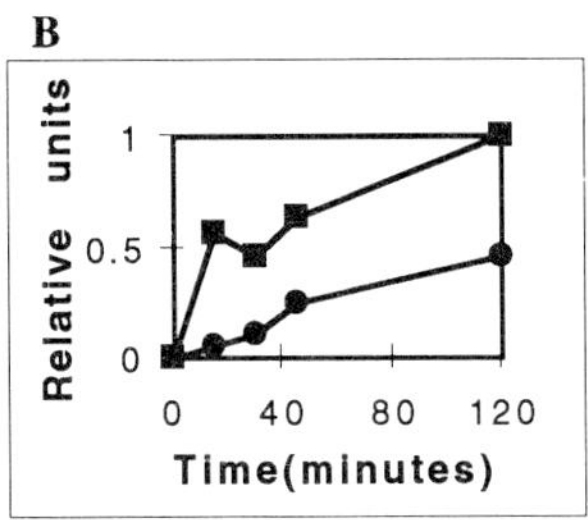

Figure 1. **Increased palmitate labeling (panel A) and turnover (panel B) on H-Ras in SNC-treated cells.** *Panel A.* Cells were labeled with 1.5 mCi/ml [^{3}H]palmitate for the indicated times in the presence of SNC or an inactive solution of SNC. Additional NO or control solutions were added at 10' intervals. *Panel B.* Cells were labeled with [^{3}H]palmitate for 2h then chased in non-radioactive medium containing 100 μM palmitate. Inactive or fresh SNC was present during the chase and was replenished every 30'. HRas was isolated by immunoprecipitation and displayed by SDS-PAGE. Scanned images of fluorograms were quantitated by densitometry with the highest value being assigned a relative value of 1. Squares (■) are SNC-treated samples; filled circles (●) are control samples.

When cells are stimulated with serum, the MAP kinases ERK1/ERK2 are phosphorylated and activated as a result of sequential interactions between Ras, Raf kinase, a MEK protein kinase and the MAPKs. Ras proteins must be lipid modified and attached to cell membranes in order to activate this pathway. Simultaneous treatment of NIH 3T3 cells with SNC and serum decreases the amount of ERK1/ERK2 phosphorylated from ~50% to <10% (data not shown). As the SNC is inactivated, phosphorylation of ERK1/ERK2 is regained. SNC thus has effects that would be predicted to inhibit the function of two proteins in the Ras signaling pathway, Ras itself and the MAP kinases.

These experiments delineate the H-Ras protein as an unprecedented target for nitric oxide, and suggest that nitric oxide may produce its effects on cell proliferation in part through destabilization of Ras protein interactions with the cell membrane. In particular we have found that nitric oxide affects palmitoylation of H-Ras. This is the first time any natural agent that affects Ras palmitoylation has been described. SNC thus provides a crucial tool for further study of the role of this modification in Ras function. Understanding how membrane binding controls Ras biological function may lead us toward novel therapeutic methods to control the unwanted activity of oncogenic Ras proteins.

References

1. Kato, K., Der, C.J. and Buss, J.E. (1992) in Seminars in Cancer Biology (Lowy, D., ed.) Vol 3, pp 179-188, W.B. Saunders, London
2. Linder, M.E., Middleton, P., Hepler, J.R., Taussing, R., Gilman, A.G., and Mumby, S.M. (1993) Proc. Natl. Acad. Sci. USA 90: 3675-3679.
3. Wedegaertner, P.B. and Bourne, H.R. (1994) Cell 77:1063-1070.
4. Robinson, L.J., Busconi, L. and Michel, T. (1995) J Biol Chem 270: 995-998.
5. Hess, D.T., Patterson, S.I., Smith, D.S. and Skene, J.H.P. (1993) Nature 366: 562-565.
6. Stamler, J.S. (1994) Cell 78:931-936.

Regulation of TNF synthesis by endogenous nitric oxide: comparison between human and murine monocytic cells.

MADDALENA FRATELLI, MIRELLA ZINETTI AND PIETRO GHEZZI

Istituto di Ricerche Farmacologiche "M. Negri", Via Eritrea 62, 20157 Milan, Italy

INTRODUCTION

While murine monocytes/macrophages produce massive amounts of NO in response to LPS, this ability is still object of debate in human cells. Although we were unable to show any nitrite/nitrate accumulation in the culture medium, we gave indirect evidence that human monocytic cells synthesize NO after LPS challenge, and this NO is implicated in the regulation of TNF synthesis and cytostatic response to LPS [1]. The purpose of this study was to explore whether the large amounts of NO produced by murine cells exert the same physiological action on TNF synthesis and cytostasis as the probably low amounts produced by human cells.

METHODS

Cell culture - RAW 264.7 cells were cultured in RPMI 1640 (Seromed, Berlin, Germany) supplemented with 10% fetal bovine serum (FBS, Hyclone Lab. Inc., Logan, UTAH). Cultures were routinely tested for mycoplasma contamination. The cells were plated for the experiments at a density of 10^6/ml in fresh medium in 24-wells plates (Falcon Becton Dickinson, Lincoln Park, New Jersey), 1ml/well and the incubations started 2 h thereafter. The number of viable cells was determined visually by counting cells that excluded trypan blue.

TNF determination - TNF secreted in the supernatant was measured by a cytotoxicity assay on L929 cells [2], using rhTNF as a standard (kind gift of BASF-Knoll, Ludwigshafen, Germany). The detection limit of the bioassay was 80 pg/ml.

Nitrites determination - Nitrites were mesaured in the supernatant 48 hours after incubation with the drugs, by spectrophotometric method using naphthylenediamine [3]

RESULTS

L-NMMA (200 μM) did not affect TNF synthesis induced by LPS (1 μg/ml) either at short (4h) or at long interval (20h) in Raw 264.7 (Fig.1, left panel), as measured by a bioassay. The same result was obtained with an ELISA measurement (not shown). In the same conditions, LPS was cytostatic to these cells (Fig. 1, right panel), and this effect was not reverted by 200 μM L-NMMA.

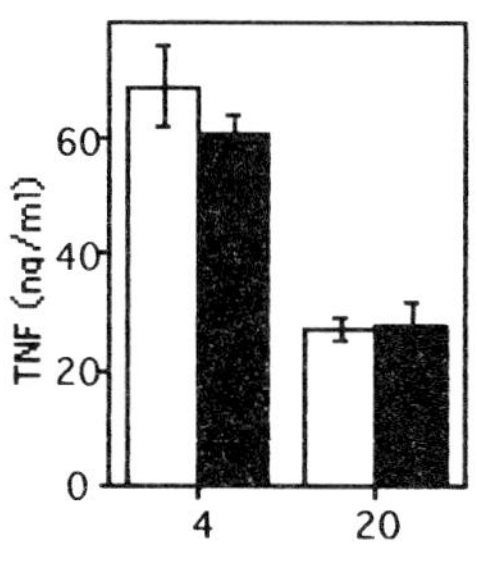

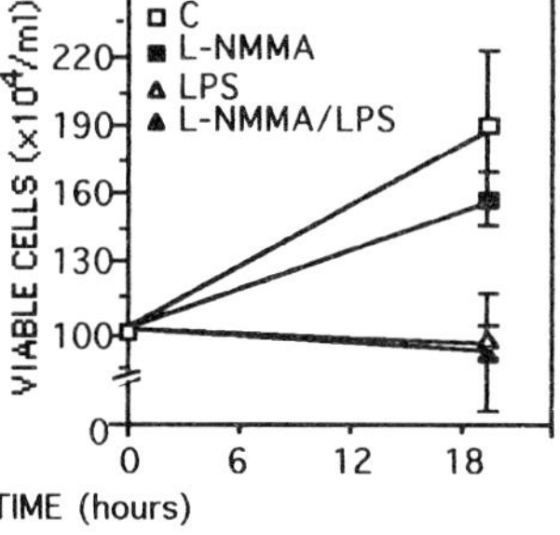

However this L-NMMA concentration, which inhibited TNF synthesis and abolished the LPS cytostatic effect in human cells, was unable to completely block nitrites accumulation in Raw 264.7 cells. This was inhibited by 93% in the presence of 1 mM of the drug (Fig.2), but still detectable nitrites levels were present.

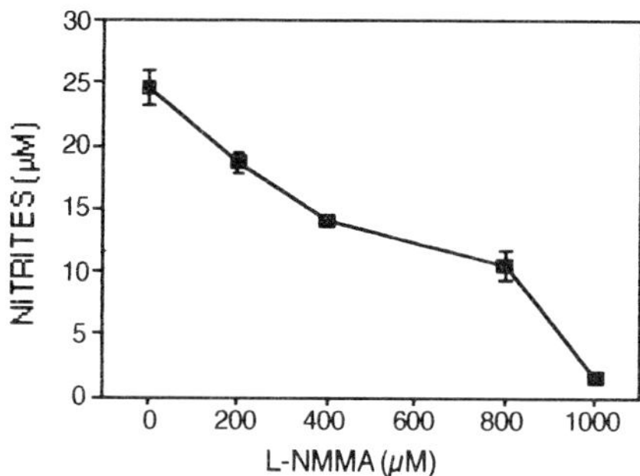

L-NMMA at 1 mM concentration was still unable to affect TNF synthesis in either direction, as measured by a bioassay, 4 and 20 h after LPS (not shown).

CONCLUSIONS

Our data do not provide evidence of a role of NO in the regulation of TNF synthesis in murine cells neither as an inducer, as we suggested for human monocytic cells, nor as an inhibitor.

However we must note that the amount of NO present even at the highest inhibitor concentration, was probably still higher than that produced by human cells, in which no detectable nitrites could be measured.

We suggest therefore the possibility that a very low amount of nitric oxide could be sufficient to potentiate TNF synthesis also in murine macrophages.

REFERENCES

1. Zinetti, M., Fantuzzi, G., Delgado, R., Di Santo, E., Ghezzi, P. & Fratelli, M. *Eur. Cytokine. Netw.* **6**:45-48, (1995).
2. Aggarwal, B. B., Khor, W. J., Hass, P. E. , et al. *J. Biol. Chem.* **260**, 2345-2354 (1985).
3. Green, L. C., Wagner, D. A., Glogowski, J., Skipper, P. L., Wishnok, J. S. & Tannenbaum, S. R. *Anal. Biochem.* **126**, 131 (1982).

Nitric oxide-dependent ADP-ribosylation of brain and liver proteins after LPS administration in mice

MIRELLA ZINETTI, FABIO BENIGNI, GRAZIA GALLI, PIETRO GHEZZI AND MADDALENA FRATELLI

Istituto di Ricerche Farmacologiche "M. Negri", Via Eritrea 62, 20157 Milan, Italy

INTRODUCTION

LPS induces a variety of central and peripheral effects, largely mediated by cytokines, including TNF, and nitric oxide. We have previously shown that peripheral (iv) administration of LPS (2.5 μg/mouse) induced TNF levels in the serum, but not in the brain, while central (intracerebroventricular, icv) LPS induced TNF production both in the brain and periphery[1]. The aim of this study was to investigate the pattern of regional production of nitric oxide after peripheral or central LPS administration. We studied circulating nitrites/nitrates, macrophage-type iNOS mRNA levels and the SNP-induced ADP-ribosylation of brain and liver proteins stimulated by SNP *ex vivo*, after *in vivo* LPS treatment. This last parameter was considered to be an indirect measurement of local NO production. In fact we have previously shown, in a different *in vivo* model, that this parameter was decreased depending on NOS activity, probably because locally produced NO causes ADP-ribosylation of tissue proteins, thereby decreasing available substrates for subsequent *ex vivo* ADP-ribosylation[2].

METHODS

Animal treatments - Male CD-1 mice (25g body weight, Charles River, Calco, Como, Italy) were used. Procedures involving animals and their care were conducted in conformity with the institutional guidelines that are in compliance with national and international laws and policies (EEC Council Directive 86/609, OJ L 358, 1, Dec. 12, 1987; NIH Guide for the Care and Use of Laboratory Animals, NIH Publication No. 85-23, 1985). LPS (from E.coli 055:B5, Sigma) was given icv or iv at the dose of 100 μg/kg (2.5 μg/mouse). In some experiments 8 mg/kg (200 μg/mouse) iv administrations were done for comparison.

Nitrites/nitrates measurements - Determinations were made in sera obtained 8 h after LPS administration, by a spectrophotometric method using naphtylethylenediamine, following nitrates reduction with cadmium[3].

ADP-ribosylation of brain and liver proteins - Livers and brains were removed 18 h after LPS administration, and the assay was performed as previously described[2]. Briefly, [^{32}P]ADP-ribosylation was routinely measured in duplicate in 0.1 ml 100 mM Tris-HCl, pH 7.5, 10 mM thymidine, 1 mM EDTA, 0.1% TRITON X-100, 0.1 mM Gpp(NH)p, 25 mM dithiothreitol (DTT), 1.7 μM [^{32}P]NAD (2.5 μCi/sample; 30 Ci/mmol, New England Nuclear, Boston, MA), 50 μl homogenate (10 mg original tissue) either in the presence or in the absence of 2 mM SNP. Samples were incubated at 30°C for 70 min. and proteins (60 μg/lane) were resolved in 10% SDS-PAGE and transferred to a nitrocellulose membrane. Blots were stained with Ponceau Red, checked for homogeneity of protein loading and subjected to autoradiography using KODAK X-OMAT AR-5 film with intensifying screens at -70°C, usually for 1-3 days. Autoradiographic bands were quantified by computerized image analysis, IBAS 2.0.

Northern blot analysis - Livers and brains were removed 4 h after LPS administration, RNA was extracted and Northern blotting analysis was performed according to standard procedures. The murine iNOS cDNA probe (a 3934-kb HindIII-SstI fragment) was a kind gift from Drs. Carl Nathan and Qiao-Wen Xie, Cornell University Medical College, New York[4].

RESULTS

Both routes of LPS administration caused an increase in plasma nitrites/nitrates levels, that were slightly, but significantly ($p<0.05$) higher in iv treated animals (23.3±7.6 vs 35.6±5.0 μM and 28.4±3.4 vs 53.4±14.7, saline vs LPS, icv and iv, respectively - means±SD, n=7-9).

We studied SNP-induced ADP-ribosylation in brain and liver proteins after both icv and iv administration. Quantification of two representative protein bands, p43 and p39 (probably corresponding to glyceraldehyde-3-phosphate dehydrogenase[5], is shown in figure (mean±sd of three independent experiments; *, ** significantly different from 100, Student's t test for a comparison to a constant value, $p<0.05$ and 0.01, respectively). A decrease in ADP-ribosylation is considered indication of local NO production.

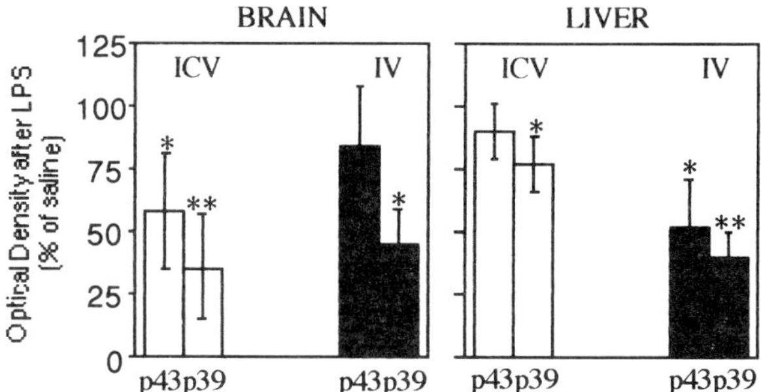

iNOS mRNA expression was detected by Northern blot analysis both in the liver and brain 4 h after iv, and only in brain after icv administration.

A summary of the results is given in table 1

	BRAIN		LIVER		SERUM	
LPS admin.	ICV	IV	ICV	IV	ICV	IV
TNF	+	-	nd	nd	+	+
NO_2^-/NO_3^-	nd	nd	nd	nd	+	++
local NO (ADPr)	++	+	+/-	++	nd	nd
iNOS mRNA	+	+/-	-	++	nd	nd

DISCUSSION

ADP-ribosylation stimulated by SNP *ex vivo* appears to be a useful tool for the study of local NO production. The advantages over mRNA expression analysis are: a higher sensitivity and the detection of actually produced NO, irrespective of its source.

Taken toghether our data suggest a complex regulation of the production of NO by central and peripheral LPS and open the way to further studies for the identification of the sources of NO during endotoxemia.

REFERENCES

1. Faggioni, R., Fantuzzi, G., Villa, P., Buurman, W., van Tits, L. J. & Ghezzi, P. *Infect. Immun.* **63**, 1473-1477 (1995).
2. Vezzani, A., Sparvoli, S., Rizzi, M., Zinetti, M. & Fratelli, M. *Neuroreport.* **5**, 1217-1220 (1994).
3. Green, L. C., Wagner, D. A., Glogowski, J., Skipper, P. L., Wishnok, J. S. & Tannenbaum, S. R. *Anal. Biochem.* **126**, 131 (1982).
4. Xie, Q., Cho, H. J., Calaycay, J. , et al. *Science* **256**, 225 (1992).
5. Zhang, J. & Snyder, S. H. *Proc. Natl. Acad. Sci. U. S. A.* **89**, 9382-9385 (1992).

S-nitrosylation of cysteine and glutathione in the anaerobic aqueous solution (pH 7.2) on contact with gaseous nitric oxide.

Anatoly F. Vanin and Irina V. Malenkova.

Institute of Chemical Physics, Russian Academy of Sciences, Kosygin Str. 4, 117977, Moscow, Russia.

It is presently shown that nitric oxide (NO) formed from L-arginine in cells and tissues can include into S-nitrosothiols (RS-NO). This inclusion provides NO stabilization which would be necessary for intra- and intercellular transport of NO. Besides, S-nitrosylation of protein thiol groups, which sharply enhances their reactivity in reactions of oxidation and addition, determines the functional role of NO in various metabolic processes [1]. A question arises on the way of S-nitrosothiol synthesis in cells and tissues at physiological pH values. In the present study it was established that S-nitrosylation of thiols can occur at neutral pH and under anaerobic conditions on contact with gaseous NO.

Experiments were designed as follows. 1-50 mM cysteine or glutathione solutions in 15 mM Hepes buffer at neutral pH were treated with NO in gas phase under the pressure of 50-700 mm Hg in the absence of oxygen at room temperature for 5 min. The treatment resulted in the appearance of S-nitrosocysteine (SNOC) or S-nitrosoglutathione (SNOG) in these solutions as detected by characteristic absorbance bands at 338 and 546 nm (Fig.1). The amount of these S-nitrosothiols increased

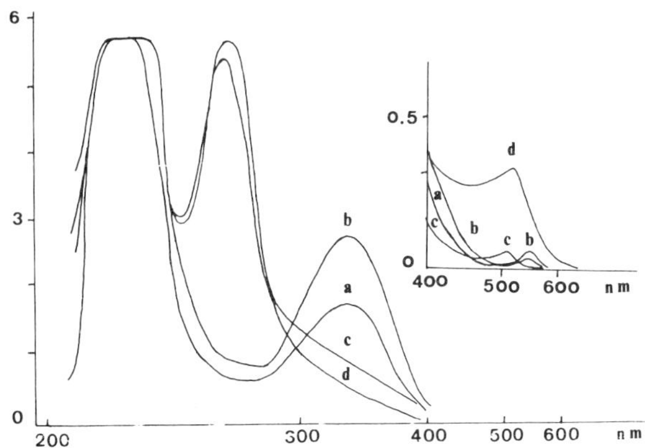

Fig.1. Absorption spectra of the solutions of 50 mM cysteine (pH 7.2) treated with NO (700 mm Hg) a - without addition of Fe^{2+} and phenanthroline; b-d - in the presence of 20 µM Fe^{2+}, 250 uM phenanthroline or 20 µM Fe^{2+} + 250 µM phenanthroline, respectively.

up to several mM with increases of thiol and NO in the gas phase above the solution. Simultaneously with the S-nitrosothiol formation, dinitrosyl iron complex (DNIC) with cysteine or glutathione, that included contaminant iron (to 5 µM), appeared in the solution. The complex formation was detected by a characteristic EPR signal from this solution with $g_{\perp}$ = 2.041, $g_{\parallel}$ = 2.014. The addition of Fe^{2+} salt in the concentration of 20 µM to the thiol solution potentiated the formation of SNOC, SNOG and DNIC while the addition of o-phenanthroline (250 µM), a selective Fe^{2+} chelator, to these solutions prior to their contact with NO completely suppressed the formation of both S-nitrosothiols and DNIC both with and without added iron (Fig.1). The addition of phenanthroline to the pre-synthesized S-nitrosothiol solution did not affect the content of the latter. Only the DNIC breakdown was observed. In contrast, the addition of Fe^{2+} salt (more than 50 µM) to this solution resulted in the breakdown of S-nitrosothiols and the increase in DNIC amount.

The data obtained decisively evidence the catalytic role of Fe^{2+} in the formation of S-nitrosothiols under the above-mentioned conditions. The catalysis is performed by only small iron concentrations in the solution: not higher then 20 µM. At this concentration, iron catalyses the formation of S-nitrosothiols in millimolar concentrations. At higher concentrations, iron not only initiates the compound formation but also results in its breakdown. What is the mechanism of Fe^{2+} -catalyzed S-nitrosothiol synthesis? We associate it with the formation of DNIC which appears in parallel with S-nitrosothiols. This complex contains NO in the form of NO^+ ion [2]. Appearance of this ion seems to provide the S-nitrosothiol formation in the process of equilibrium DNIC breakdown to its components according to the reaction 1:

$(RS^-)_2Fe^+(NO^+)_2 \longleftrightarrow Fe^{2+} + NO + RS^- + NO^+$ (1)

The presence of DNIC components, NO^+ and RS^-, in the solution may result in the formation of corresponding S-nitrosothiol. How do the NO molecules containing in DNIC transform into the NO+ form? An answer to this question is provided by the results of studies [3], which have demonstrated that DNIC synthesis is accompanied by the formation of N_2O in an equimolar amount. This means that nitroxyl ions NO^- are formed in the course of DNIC synthesis. They bind to protons with subsequent recombination to generate N_2O. In this connection one can suggest that NO+ appears in DNIC as a result of electron passage from one NO molecule to another with ionization of the latter to NO^-. Then, after the NO protonization, the HNO molecules leaves the complex with subsequent addition of the neutral NO molecule to the intermediate fragment of $(RS^-)_2Fe^{2+}(NO^+)$ complex. The NO molecule carries the unpaired electron to iron to form the final paramagnetic DNIC with the structure $(RS^-)_2Fe^+(NO^+)_2$. How did the formation of micromolar amounts of DNIC result in our experiments in the accumulation of millimolar amounts of S-nitrosothiols? Such accumulation is possible if the rate of DNIC synthesis in the reaction of Fe^{2+} with thiols and neutral NO molecules,

$Fe^{2+} + NO + RS^- \longrightarrow (RS^-)_2Fe^+(NO^+)_2 + N_2O$

(reaction 2) is considerably higher than the rate of DNIC synthesis from its components in the equilibrium reaction 1. In this instance, Fe^{2+} emerging in the equilibrium breakdown would include into the irreversible reaction of DNIC synthesis (reaction 2). This is just what result in the accumulation of S-nitrosothiols forming by reaction 1.

References.

1. Stamler, J.S. (1994) Cell, 78, 931-936.
2. Bryar, T.R. and Eaton, D.R. (1992) Can. J. Chem., 70, 1917-1926.
3. Bonner, F.T. and Pearsall, R.F. (1982) Inorg. Chem., 21, 1973-1985.

NITRIC OXIDE ACTIVATES SYNTHESIS OF HEAT SHOCK PROTEINS (HSP70)

IGOR Yu. MALYSHEV, EUGENIA B. MANUKHINA, ALEXANDER V. MALUGIN, VASAK D. MIKOYAN*, LUDMILA N. KUBRINA* and ANATOLY F. VANIN*

Inst. of General Pathology & Pathophysiology, Baltijskaya 8, Moscow 125315; *Inst. of Chemical Physics, Kosygin str. 4, Moscow 117997, Russia.

Rapid synthesis of stress proteins of the heat shock protein (HSP70) family is a characteristic common feature of cell response to actions of the environment. HSP70 play an important role in increasing the organismic resistance to stress and in the development of adaptation to environmental factors [3]. At the same time the question as what cell mechanisms are involved in the activation of HSP70 synthesis remains open in many respects. The experimental facts have drawn our attention that the same agents can activate both HSP70 and nitric oxide (NO) synthesis. Lipopolysaccharides, inductors of NO synthesis [5], activate also HSP70 synthesis [10]. Heat shock (HS), which is a conventional way of inducing HSP synthesis [1,2,4], is attended by an increase in blood NO-heme [6]. Besides, HS is known to result in a profound fall of blood pressure [9], which may also indirectly indicate an increase in generation of NO, a potent vasodilator. In adaptation of the organism to stress, a considerable activation of HSP70 synthesis in organs [3] occurs against the background of increased NO content in the same organs [7]. A similar situation was observed in inflammation: the NO content grew as the HSP70 content increased in the inflammation focus [8].

Taken together the data suggest that NO is involved in the activation of HSP70 synthesis. To verify this hypothesis we studied: 1) the effect of HS on the NO content in rat organs, 2) the effect of a NO-synthase inhibitor on the HS-induced accumulation of HSP70, 3) the effect of HS on the NO content in cultured cells, 4) the effect of a NO donor on HSP70 accumulation in cultured cells.

METHODS. Experiments were carried out on Wistar male rats. HS of rats was produced by heating of rats up to the core temperature of 42oC for 15 min. HS of cultured cells was produced by heating of human hepatocyte culture at 42oC for 2 h. NO generated in rat tissues was measured by electron paramagnetic resonance (EPR) by the NO inclusion into a NO trap Fe2+-diethyldithiocarbamate (Fe2+-DETC) with formation of paramegnetic mononitrosyl iron complexes with DETC (MNIC-DETC). Animals were decapitated 30 min following the injection of NO trap. The amount of NO trapped was evaluated by the intensity of a characteristic EPR signal from MNIC-DETC. Human hepatoblastoma Hep G-2 cells were grown on 35 mm plastic dishes. Cells were cultured in Eagle's minimum essential medium (MEM). Cells were taken to the experiment when they were at near confluency (~1.0 x 106 cells/dish). Dinitrosyl iron complex (DNIC) with glutathione was used as the NO donor. DNIC was added to cultured cells in the concentration of 180 mM and 36 mM. EPR spectra of frozen tissue and culture samples were recorded on a radiospectrometer Radiopan (Poland). Concentration of paramagnetic centers was evaluated by double integration of the EPR signal. Accumulation of HSP70 was estimated using Western blot analysis and monoclonal antibodies against HSP70.

RESULTS 1) HS increases the NO production in rat organs. Hyperthermia sharply increased the NO production in heart, liver, brain, intestine, kidney and spleen. For instance, prior to HS, the hepatic NO content was 31 ng/g wet tissue. One hour after HS, the NO content increased to 133 ng/g wet tissue. The NO increase was transient and it rapidly returned to the control level. In animals treated with L-NNA, the inhibitor of NO synthesis, the MNIC-DETC signal was detected in none of the organs studied both in control and after the thermal exposure. 2) Inhibition of NO-synthase reduces HSP70 synthesis after HS. In 24 h after HS, we observed the HSP70 accumulation in both the heart and the liver. L-NNA decreased the HS-induced HSP70 accumulation by 75% in the liver and by 45% in the heart .3) HS increases the NO production in cell culture. The NO appearance was detected by the formation of DNIC in cells. The DNIC content was 2 nmol/106 cells in heated hepatocytes. The signal was undetectable in control cells. 4) The NO donor induces HSP70 accumulation in cell culture. Incubation of hepatocytes with DNIC in the concentration of 180 μM resulted in accumulation of DNIC in the cells. The intracellular concentration of DNIC was 5nmol/106 cells after 15 min of incubation. The addition of DNIC to cell culture resulted in a pronounced time- and concetration-dependent HSP70 accumulation with emergence of new HSP70 isoforms in hepatocytes

DISCUSSION. The results obtained support our hypothesis that NO is really involved in the activation of HSP70 synthesis.

Probably, NO activates HSF, the HSP70 transcription factor. This activation may occur by at least two ways. First, NO may activate HSF due to the accelerated trimerization of HSF molecules. NO elicits S-nitrosylation of thiol groups in HSF to sharply accelerate the formation of disulfide linkage and, thereby, trimerization of the HSF molecule. The second hypothetic mechanism of HSF activation by NO is based on the suggestion that this HSF contains heme or non-heme iron. If this is the case, NO binding to iron can provide conformational changes and, thereby, activation of HSF. The data obtained open a prospect of using NO donors or pharmacological agents modulating NO sythesis to correct stress-induced damage and to increase the organismic adaptability.

REFERENCES

1. Schlesinger, M.J., Ashburner, M. and Tissieres, A. (1982) Heat Shock: From Bacteria to Man, Cold Spring Harbor Laboratory, New York.
2. Welch, W.J. and Suhan, J.P. (1986) J. Cell. Biol. **103**, 2035- 2052.
3. Meerson, F.Z. and Malyshev, I.Yu.. (1993) Phenomenon of Adaptive Stabilization of Structures and Protection of the Heart, Nauka, Moscow (in Russian).
4. Tomasovic, S.P. (1989) Life Chem. Rep. **1**, 33-63.
5. Bredt, D.S., Snyder, S.H. (1994) Annu. Rev. Biochem. **63**, 175-195
6. Hall, D.M., Buettner, G.R., Matthes, R.D. and Gisolfi, C.V. (1994) J. Appl. Physiol. 77, 548-553.
7. Meerson, F.Z., Lapshin, A.V., Mordvincev, P.I., Mikoyan, V.D., Manukhina, E.B., Kubrina, L.N. and Vanin, A.F. (1994) Byull. eksper. biol. i med. **117** (3), 242-244 (in Russian).
8. Jacquiersarlin, M.R., Fuller, K., Dinhxuan, A.T., Richard, M.J. and Polloa, B.S. (1994) Experientia **50**, 11-12.
9. Shepelev, A.P. (1976) Voprosy med. khimii **22** (1), 47-51
10. Zhang, Y., Takahashi, K., Jiang, G.-Z., Zhang, X.-M., Kawai, M., Fukada, M. and Yokochi, T. (1994) Infection and Immunity **62**, 4140-4144.

Oxidized lipoprotein(a) stimulates production of superoxide radical and inhibits vasodilation

JAN GALLE and CHRISTOPH WANNER

Dept. of Medicine, Division of Nephrology, University Hospital of Würzburg, Joseph-Schneider-Str. 2, D-97080 Würzburg, Germany

Introduction In hypercholesterolemia and atherosclerosis, endothelium-dependent dilations are impaired [1], probably due to the effects of accumulating LDL. *In vitro*, oxidized LDL has been shown to interfere with formation of endothelium-derived nitric oxide [2] and to directly inactivate endothelium-derived nitric oxide [3]. Lp(a) is another atherogenic lipoprotein which could influence vascular tone in a fashion similar to that of LDL. To evaluate this hypothesis, we studied the influence of native and oxidized Lp(a) on endothelium-dependent dilation of isolated rabbit arteries and compared its effect with the potency of LDL. Furthermore, since Lp(a) has been shown to induce free radical generation [4] which may play a role in the defective vascular relaxation in atherosclerotic arteries [5,6], we evaluated whether Lp(a)-effects on endothelial function are mediated by oxygen-derived radicals.

Methods

- ***Isolation of LDL and Lp(a):*** by ultracentrifugation and gel chromatography [7].
- ***Oxidation of Lp(a) and LDL:*** by incubation with $CuSO_4$ (5µM) for 24 hours at 23°C as described recently [8].
- ***Vessel preparation and diameter determination:*** segments of the aorta and the renal artery were obtained from rabbits and placed in organ chambers as described recently [8]. The segments were incubated for 150 min with native or oxidized LDL (100 µg/mL), native or oxidized Lp(a) (30 or 100 µg/mL), or their respective buffers as time matched control. Endothelium-dependent dilations were elicited by adding cumulative doses of acetylcholine.
- ***Detection of chemiluminescence as a parameter for O_2^- generation:*** in lipoprotein-treated aortic rabbit segments, placed in a scintillation counter with lucigenin as O_2^- detector [5].

Results

Effects of native and oxidized LDL and of native and oxidized Lp(a) on endothelium-dependent dilations. Acetylcholine-induced dilations were not impaired in arteries treated with 100 µg/mL native or oxidized LDL or native Lp(a). However, incubation with 30 or 100 µg/mL oxidized Lp(a) dose-dependently attenuated dilations (Fig. 1).

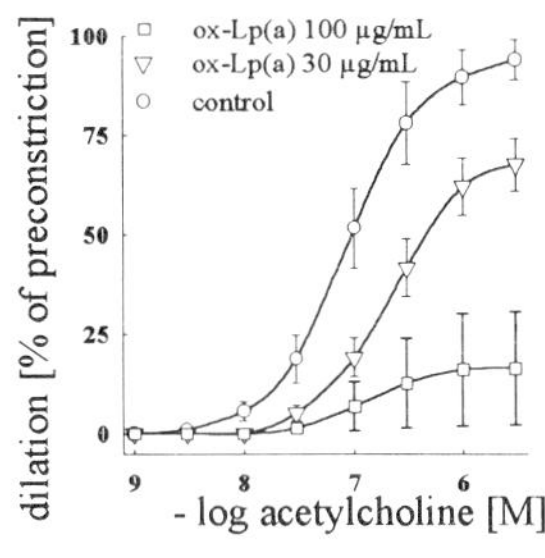

Figure 1

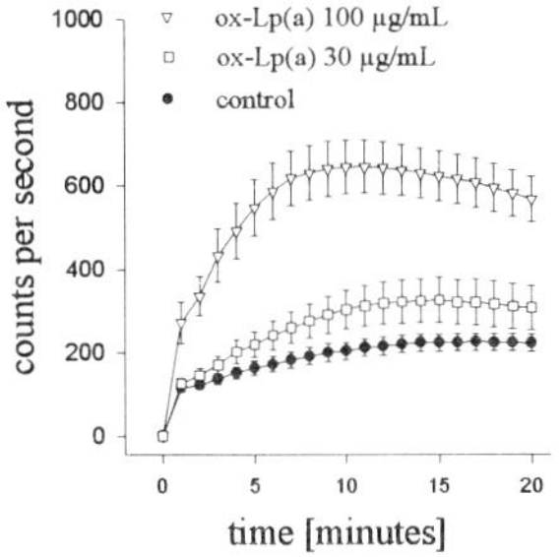

Figure 2

Detection of O_2^- generation. Reactive oxygen species may play a crucial role in the impairment of endothelium-dependent dilations in atherosclerotic arteries. We investigated whether oxidized Lp(a) stimulated O_2^- generation. Treatment of the arteries with 30 µg/mL oxidized Lp(a) induced a significant increase in the chemiluminescence signal (Fig 2).

Effect of SOD, of catalase, and of deferoxamine on endothelium-dependent dilations in the absence and presence of Lp(a). SOD and catalase, enzymes catabolizing O_2^- and H_2O_2, and deferoxamine, which inhibits the conversion of O_2^- and H_2O_2 to 0H• [9], had no influence on dilator responses in the absence of Lp(a). However, in arteries treated with oxidized Lp(a), the presence of simultaneously given SOD significantly enhanced the attenuation of endothelium-dependent dilations. The latter enhancement of attenuation by SOD was completely prevented by additional treatment with catalase or deferoxamine.

Discussion The present study demonstrates that oxidized, but not native Lp(a), attenuates endothelium-dependent dilation in isolated rabbit renal arteries. Increased O_2^- production of arteries pretreated with oxidized Lp(a) could directly be detected using a chemiluminescence assay. Hypercholesterolemia has frequently been described as a cause for disturbance of endothelial function in human and animal studies [10,11]. In hypercholesterolemic vessels, enhanced generation of O_2^- or related reaction products [12] may be responsible for impairment of endothelial function through augmented inactivation of EDNO and/or damage of endothelial cells. Our data provide direct evidence for O_2^- generation in oxidized Lp(a)-treated arteries, which might lead to attenuation of endothelium-dependent dilations through inactivation of nitric oxide. However, simultaneous treatment of the arteries with SOD and oxidized Lp(a) significantly increased its inhibitory effect on dilator responses. This unexpected effect of SOD could be explained by the deleterious effects of too much SOD activity in relation to H_2O_2 -removing enzymes. Indeed, it has been shown that SOD may favour formation of the biologically most active 0H• in the absence of catalase [13,14]. 0H• might either attack the endothelial cells, destruct nitric oxide, and/or induce further lipid peroxidation [14]. Consistent with the interpretation that 0H• contributed to attenuation of endothelium-dependent dilations after coincubation of the arteries with SOD and oxidized Lp(a), treatment with catalase completely blunted the effect of SOD. Furthermore, additional treatment with the iron chelator deferoxamine, which inhibits the formation of 0H•, also prevented attenuation of dilations in arteries treated with oxidized Lp(a) and SOD.

In conclusion, in this report we demonstrate that oxidized Lp(a) impairs endothelium-dependent dilations in rabbit renal arteries and provide evidence that generation of reactive oxygen species is the mechanism of this effect. We hypothesize that impairment of endothelium-dependent dilations by oxidized Lp(a) may contribute to attenuation of endothelial function in humans with high Lp(a) levels.

Abbreviations used: LDL: low density lipoproteins; Lp(a): lipoprotein(a); SOD: superoxide dismutase; EDNO: endothelium-derived nitric oxide. *Grant support*: By grants of the Deutsche Forschungsgemeinschaft Ga 431/1-2 and Wa 836/1-1.

References

1. Moncada, S. & Higgs, A. (1993) N. Engl. J. Med. **329**, 2002-2012

2. Kugiyama, K., Kerns, S.A., Morrisett, J.D., Roberts, R. & Henry, P.D. (1990) Nature **344**, 160-162

3. Galle, J., Mülsch, A., Busse, R. & Bassenge, E. (1991) Arterioscler Thromb **11**, 198-203

4. Hansen, P.R., Kharazmi, A.K., Jauhiainen, M. & Ehnholm, C. (1994) Eur. J. Clin. Invest. **24**, 497-499

5. Ohara, Y., Peterson, T.E. & Harrison, D.G. (1993) J Clin Invest **91**, 2546-2551

6. Mügge, A., Elwell, J.H., Peterson, T.E., Hofmeyer, T.G., Heistad, D.D. & Harrison, D.G. (1991) Circ Res **69**, 1293-1300

7. Galle, J., Stunz, P., Schollmeyer, P. & Wanner, C. (1995) Kidney International **47**, 45-52

8. Galle, J., Bengen, J., Schollmeyer, P. & Wanner, C. (1994) European Journal of Pharmacology **265**, 111-115

9. De Bruyn, V.H., Nuno, D.W., Cappelli-Bigazzi, M., Dole, W.P. & Lamping, K.G. (1994) Journal of Hypertension **12**, 163-172

10. Casino, P.R., Kilcoyne, C.M., Quyyumi, A.A., Hoeg, J.M. & Panza, J.A. (1993) Circulation **88**, 2541-2547

11. Verbeuren, T., Coene, M., Jordaens, F., Van Hove, C., Zonnekeyn, L. & Herman, A. (1986) Circ Res **59**, 496-504

12. Rubanyi, G.M. & Vanhoutte, P.M. (1986) Am. J. Physiol. **250**, H822-H827

13. Ichikawa, I., Kiyama, S. & Yoshioka, T. (1994) Kidney Int. **45**, 1-9

14. Halliwell, B. (1993) Haemostasis **23 Suppl. 1**, 118-126

Endogenous nitric oxide (NO) suppresses phospholipase A_2 - lipoxygenase pathway in rat alveolar macrophages.

KURT RACKE*#, GERNOT BRUNN#, CLAUDIA HEY*# AND IGNATZ WESSLER+

*Department of Pharmacology University of Bonn, Reuterstr. 2b, D-53113 Bonn, F.R.G.; #Department of Pharmacology University of Frankfurt, D-60590 Frankfurt, F.R.G.; +Department of Pharmacology University of Mainz, D-55131 Mainz, F.R.G.

Alveolar macrophages (AMs) are part of the non-specific defense mechanisms of the respiratory tract and also play an important role in the local control of specific immune responses and inflammatory reactions [1]. AMs can release a wide variety of mediators including LTB_4 which is a potent chemotactic factor and may act as pro-inflammatory mediator. In addition, an inducible form of NO synthase (iNOS) can by induced by various stimuli including lipopolysaccharides (LPS) and several cytokines (INFγ, TNFα and IL-1) resulting in the release of large amounts of NO [2,3].

In the present study, AMs were obtained by broncho-alveolar lavage of isolated rat lungs and cultured ($2*10^6$ cells/culture dish) as described in Fig. 1, for further details see also [2]. The cells were then incubated in 1 ml Krebs medium which contained ^{3}H-arachidonic acid (^{3}H-AA, 37 kBq). After a 2 h labelling period the release of ^{3}H-compounds was determined in 2 subsequent 50 min periods (see Fig. 1). It should be emphasized that the culture medium contained L-arginine (0.7 mM), whereas the Krebs medium is an amino acid-free physiological salt solution (see [2]). Therefore, in some experiments L-arginine the substrate for iNOS, was added to the Krebs medium. ^{3}H-AA metabolites in the incubation media were separated by reverse phase HPLC as described [4], a technique allowing the identification of all classes of eicosanoides including LTB_4. ^{3}H-LTB_4 release is expressed as DPM/sample, means$\pm$SEM of $n \geq 3$.

The spontaneous outflow of ^{3}H-AA and ^{3}H-AA metabolites was very low, that of ^{3}H-LTB_4 amounted to 143 ± 21 DPM/50 min ($n=83$) and was not significantly affected by the different culture conditions. Under control conditions, A 23187

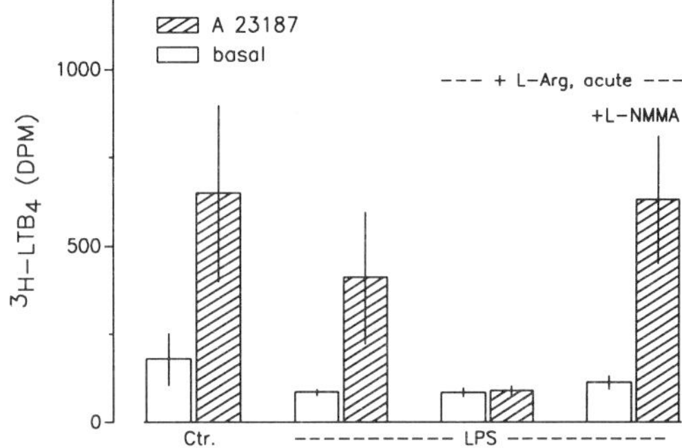

Fig. 1: Effects of A 23187 (10 μM) on the release of ^{3}H-LTB_4 from isolated rat AMs. AMs were cultured for 18 h in DMEM-F12 medium in the absence (Ctr) or presence of LPS (10 μg/ml). AMs were then incubated for 2 h in L-arginine-free Krebs medium containing ^{3}H-AA (37 kBq). Thereafter, the release of ^{3}H-compounds was determined during 2 subsequent 50 min periods. A 23187 was present during the 2nd incubation period. L-Arginine (300 μM) alone or in combination with N^G-monomethyl-L-arginine (L-NMMA, 100 μM) was added to the Krebs medium as indicated. ^{3}H-compounds in the incubation media were separated by reverse phase HPLC (see text). *Height of columns:* ^{3}H-LTB_4, expressed as DPM/sample, means$\pm$SEM of $n \geq 3$.

caused about a 5fold increase in ^{3}H-LTB_4 release, without major effects on the outflow of ^{3}H-AA or any other ^{3}H-compound. In confirmation of previous observations [5], A 23187 caused a more than 100fold increase in ^{3}H-LTB_4 release, when freshly prepared AMs were used. Thus, during the 18 h culture period the AMs lost most of their capacitiy to generate ^{3}H-LTB_4. When LPS was present during the culture period, a condition known to provoke a large induction of iNOS [2], no significant effects on ^{3}H-LTB_4 release were observed (Fig. 1). When L-arginine (300 μM), the substrate for iNOS, was added to the Krebs medium A 23187 evoked release of ^{3}H-LTB_4 was completely suppressed.

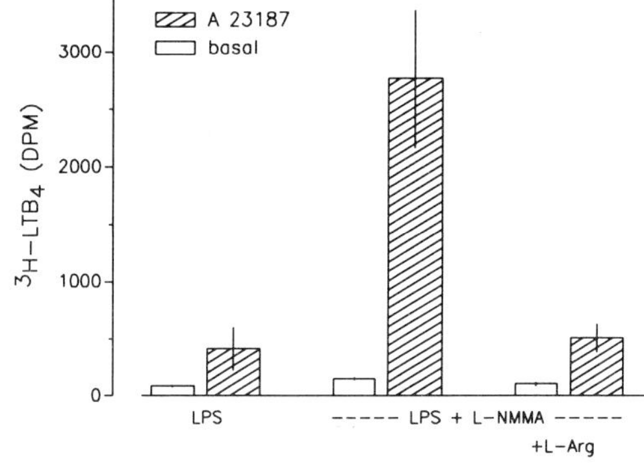

Fig. 2: Effects of A 23187 (10 μM) on the release of ^{3}H-LTB_4 from isolated rat AMs cultured for 18 h in the presence of LPS (10 μg/ml) alone or in combination with N^G-monomethyl-L-arginine (L-NMMA, 300 μM) or additionally enhanced L-arginine (+L-Arg, from normally 0.7 to 3 mM) and labelled with ^{3}H-AA (for details see Fig. 1 and text) *Height of columns:* ^{3}H-LTB_4, expressed as DPM/sample, means$\pm$SEM of $n \geq 3$.

This effect of L-arginine was prevented, when L-NMMA was additionally present, indicating the involvement of NO. 5-Lipoxygenase (5-LO) contains iron at its active site and there is evidence that NO could inhibit 5-LO by interaction with that iron [6].

Rat AMs cultured in the presence of LPS produce large amounts of NO as indicated by the large amounts of nitrate accumulating in the medium [2,7]. In order to see whether NO generated during culture period might affect the subsequent release of ^{3}H-LTB_4 L-NMMA (300 μM) was added to the culture medium. This treatment caused an about 5fold increase of the A 23187-evoked ^{3}H-LTB_4 release (Fig. 2). Elevating the concentration of L-arginine in the culture medium from 0.7 to 3 mM, opposed the effect of L-NMMA (Fig. 2), whereas the addition of 2.3 mM D-arginine had no effect (not shown) indicating the specific effects on NO synthesis. When the protein synthesis during the culture period was inhibited by cycloheximide (10 μM), the A 23187-evoked ^{3}H-LTB_4 release was abolished, independent of whether LPS and L-NMMA were present or not (data not shown). These observations indicate that at least one of the enzymes required for the synthesis of ^{3}H-LTB_4 (phospholipase A_2, 5-LO or 5-LO activating protein) have a high rate of turnover. Moreover, an intact protein synthesis is required for the facilitatory effect of L-NMMA supporting the idea that the *de novo*-synthesis of the enzymes of the phospholipase A_2 - 5-LO pathway in AMs can be inhibited by NO.

In conclusion, endogenous NO generated in AMs by suppressing the release of the potent pro-inflammatory mediator LTB_4 could have an important anti-inflammatory potential. Therefore, the use of iNOS inhibitors in the treatment of asthma as recently suggested [7] may be a two sided sword.

Supported by the Deutsche Forschungsgemeinschaft

References

1. Thepen, T. and Havenith, C.E.G (1994) In: The Handbook of Immunopharmacology (Page, C., series Ed.), Immunopharmacology of Macrophages and other Antigen-presenting cells (Bruijnzeel-Koomen, C.A.F.M. and Hoefsmit, E.C.M. Eds.) pp. 35-43, Academic Press, London
2. Hey, C., Wessler, I. and Racké, K. (1995) Naunyn-Schmiedeberg's Arch. Pharmacol., **351,** 651-659
3. Xie, Q.W. and Nathan, C. (1994) J. Leukocyte Biol. **56,** 576-582.
4. Brunn, G., Wessler, I. and Racké, K (1995) Br. J. Pharmacol. **116,** in press
5. Reimann, A., Brunn, G., Hey, C., Wessler, I. and Racké, K. (1995) In Advances in Prostaglandin, Thromboxane and Leukotriene Research, **Vol. 23**, (Samuelsson, B., Paoletti, R. and Ramwell, P.W., eds.) pp. 357-359, Raven Press, New York
6. Kanner, J., Harel, S. and Granit, R. (1992) Lipids, **27**, 46-49
7. Barnes, P.J. and Liew, F.Y. (1995) ImmunologyToday **16**, 128-130

Peroxynitrite-dependent oxidative modification of proteins in human plasma

DONATELLA PIETRAFORTE, and MAURIZIO MINETTI.

Dept. Cell Biology, Istituto Superiore di Sanità, Viale Regina Elena 299, 00161 Rome, Italy.

Peroxynitrite ($ONOO^-$) is becoming increasingly recognized as important mediator of damage in biological systems. $ONOO^-$ can induce the oxidation of several relevant targets. In the reaction with proteins, $ONOO^-$ has been shown to attack preferentially cysteine [1], tyrosine [2], tryptophan [3] and methionine residues [4]. In human plasma, $ONOO^-$ induced the depletion of low molecular weight antioxidants, the oxidation of protein -SH groups, and lipid peroxidation [5]. We studied the reaction of $ONOO^-$ with human plasma to identify radical intermediates by spin-trapping techniques and to determine the formation of nitrotyrosines by spectrophotometric methods .

Materials and Methods. $ONOO^-$ was synthesized as described by Radi et al. [6]. Thiol groups were chemically modified by incubation of plasma with 10 mM NEM in phosphate buffer (pH 7.4) for 15 min at 37 °C. Aliquots of plasma were buffered with an equal volume of phosphate buffer, pH 7.4 (150 mM final concentration of phosphate) and $ONOO^-$ (0 - 5 mM) was added as a drop (1-2 µL) from the stock solution. Plasma pH was measured after each incubation and found to change only slightly (pH 7.5 - 8.0), even at high $ONOO^-$ concentrations. Nitrotyrosine formation was evaluated by the increase in the absorption at 425 nm in 0.1 M NaOH. Thiyl radicals were spin trapped with DMPO. Samples were drawn in a gas permeable teflon tube and exposed to air at 37°C in the EPR cavity.

Results and Discussion. The addition of $ONOO^-$ (0.1-1 mM) to plasma in the presence of DMPO induced the formation of a strongly immobilized adduct. As shown in Figure 1, the adduct was completely inhibited by NEM, thus suggesting the trapping of protein thiyl radicals (DMPO-S-protein).

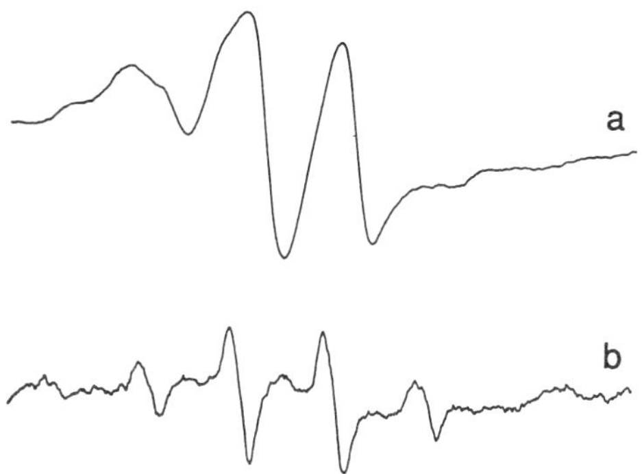

Figure 1. E.p.r. spectra of DMPO adduct formed in plasma treated with 1 mM $ONOO^-$. a) native plasma; b) plasma pretreated with 10 mM NEM. DMPO was 0.1 M. Spectra were recorded 2 min after the addition of $ONOO^-$. Modulation amplitude 2.5 G; Gain 1.6×10^5 (a) and 6.4×10^5 (b).

The DMPO-S-protein adduct is largely due to the Cys 34 of serum albumin, which represent about 80% of total SH group of plasma. Purified human albumin treated with $ONOO^-$ showed a spectrum of DMPO adduct superimposable to that observed in plasma. In NEM-treated plasma, the major adduct formed was the DMPO-OH (a_H=a_N=14.9 G) and was due to the interaction of $ONOO^-$ with the spin trap [7].

Abbreviations: DMPO, 5,5-dimethyl-1-pyrroline-N-oxide; NEM, N-ethylmaleimide; DTPA, diethylenetriamine-pentaacetic acid.

The removal of low molecular weight compounds from plasma by dialysis (cut off ~10.000) decreased by 30% the intensity of DMPO-S-protein adduct. Reconstitution of plasma with ascorbate (0.1 mM), urate (0.4 mM) and glucose (5 mM) did not significantly affect the DMPO-S-protein intensity. Notably, the addition of bicarbonate, which is present in plasma at high concentration (25 mM), increased by 146% the intensity of DMPO-S-protein adduct produced by dialyzed plasma.

The formation of nitrotyrosine in plasma was observed at high $ONOO^-$ concentrations (0.5-5 mM). As for protein thiyl radicals the levels of nitrotyrosine were decreased by dialysis (-35%), and increased in dialyzed plasma reconstituted with 25 mM bicarbonate (+84%). At variance of thiyl radicals, urate was effective in protecting protein tyrosines against $ONOO^-$ oxidation. In dialyzed plasma, the nitrotyrosine levels were reduced by 50% after reconstitution with 0.4 mM urate.

Alkylation of plasma SH groups by NEM consistently increased the nitrotyrosine formation (+27%), thus suggesting a protective role of plasma thiols. As control experiment, decomposed $ONOO^-$ (1-5 mM) did not induce the formation of both DMPO-S-protein adduct and nitrotyrosine formation. The oxidative reactions of $ONOO^-$ were not mediated by metal catalysis, since metal chelators such as EDTA and DTPA did not affect the formation of both thiyl radicals and nitrotyrosines. Our results suggest that plasma proteins represent a relevant target of $ONOO^-$ toxicity. Moreover, secondary oxidants produced by the reaction of $ONOO^-$ with bicarbonate [8] can increase the oxidative damage of plasma proteins. The SH groups of proteins (particularly the Cys 34 residue of albumin) likely represent a prominent target of $ONOO^-$ in plasma. Free radical-mediated oxidation of protein thiols may have several biological consequences since thiyl radicals are reactive toxic species that can (i) initiate lipid peroxidation [9], (ii) consume antioxidant vitamins [10], (iii) reduce molecular oxygen to the superoxide radical.[11] and (iv) induce the oxidation of protein sulfhydryls to mixed disulfides [12]. Free 3-Nitrotyrosine was found in serum and synovial fluid from rheumatoid patients [13], thus suggesting that nitrotyrosine formation of plasma proteins may be a selective marker of $ONOO^-$ production in human diseases.

References

1. Radi, R., Beckman, J. S., Bush, K. M. and Freeman, B. A. (1991) Arch. Biochem. Biophys. **288**, 481-487
2. Ischiropoulos, H., Zhu, L. Tsai, M.. Martin, J. C., Smith, C. D. and Beckman, J. S. (1992) Arch. Biochem. Biophys. **298**, 431-437
3. Ischiropoulos, H.and Al-Mehdi, A. B. (1995) FEBS Lett. **364**, 279-282
4. Moreno, J. and Pryor, W. (1992) Chem. Res. Toxicol. **5**, 425-431.
5. Van Der Vliet, A., Smith, D., O'Neill, C.A., Kaur, H., Darley-Husmar V., Cross, C. E. and Halliwell, B. (1994) Biochem. J. **303**, 295-301
6. Radi, R., Beckman, J. S., Bush, K. and Freeman, B. A. (1991) J. Biol. Chem. **266**, 4244-4250
7. Augusto, O., Gatti, R.M. and Radi, R. (1994) Arch. Biochem. Biophys. **310**, 118-125
8. Beckman, J..S., Ischiropoulos, H., Zhu, L. et al. (1992) Arch. Biochem. Biophys. **298**, 438-445
9. Schönreich, C., Asmus, K.-D., Dillinger, U. and v. Bruchhausen, F. (1989) Biochem. Biophys. Res. Commun. **161**, 113-120
10. Willson, R.L. (1983) in Radioprotectors and anticarcinogenesis. (Nygaard, O.F. and Simic, M. G., eds), pp.1-22, Academic Press, New York
11. Winterbourn, C.C. (1993) Free Rad. Biol. Med. **14**, 85-90
12. Ravichandran, V., Seres, T., Moriguchi, T., Thomas, J. A. and Johnston Jr., R. B. (1994) J. Biol. Chem. **269**, 25010-25015
13. Kaur, H. and Halliwell, B. (1994) FEBS Lett. **350**, 9-12

Peroxynitrite-induced cyclic GMP accumulation: Dependence on GSH and possible role of S-nitrosation

KURT SCHMIDT, ASTRID SCHRAMMEL, PETER KLATT, DORIS KOESLING*, AND BERND MAYER

Institut für Pharmakologie und Toxikologie, Karl-Franzens-Universität Graz, Universitätsplatz 2, A-8010 Graz, Austria.
*Institut für Pharmakologie, Freie Universität Berlin Thielallee 67-73, D-14195 Berlin, Germany

Peroxynitrite ($ONOO^-$) is widely recognized as mediator of NO toxicity, but recent studies indicate that $ONOO^-$ may have also beneficial effects. Solutions of $ONOO^-$ were shown to induce relaxation of vascular smooth muscle [1, 2], to inhibit platelet aggregation [3] and to increase cGMP accumulation in smooth muscle cells [4]. Liu et al. [1] speculated that relaxation may be due to small amounts of NO spontaneously released during $ONOO^-$ decomposition, and Wu et al. [2] reported on pronounced increases in NO release when $ONOO^-$ was incubated in the presence of GSH or tissue homogenates. Similarly, Moro et al. [3] found that $ONOO^-$ inhibited aggregation of blood platelets only when serum albumin or GSH were present. It is conceivable, therefore, that the NO-like properties of $ONOO^-$ are due to S-nitrosation of cellular proteins or GSH.

We, therefore, investigated the role of GSH in $ONOO^-$-induced cGMP accumulation. As shown in Fig. 1, $ONOO^-$ increased cGMP levels in cultured porcine aortic endothelial cells up to ~50 pmol/10^6 cells (EC_{50} ~0.2 mM). Maximal effects were comparable to those elicited by the NO donor DEA/NO (EC_{50} of 0.2 µM, 55 pmol/10^6 cells). The effect of $ONOO^-$ was potentiated in the presence of SOD (1,000 U/ml) and was inhibited by superoxide (generated by 1 mM hypoxanthine and 0.01 U/ml xanthine oxidase) suggesting NO as active intermediate. Inactivated $ONOO^-$ or $NaNO_2$ (1 mM) induced only marginal increases in cGMP accumulation (~2-fold). To investigate whether intracellular GSH mediates $ONOO^-$-induced cGMP accumulation, we pretreated cells for 30 min with 1-chloro-2,4-dinitrobenzene (100 µM) and found

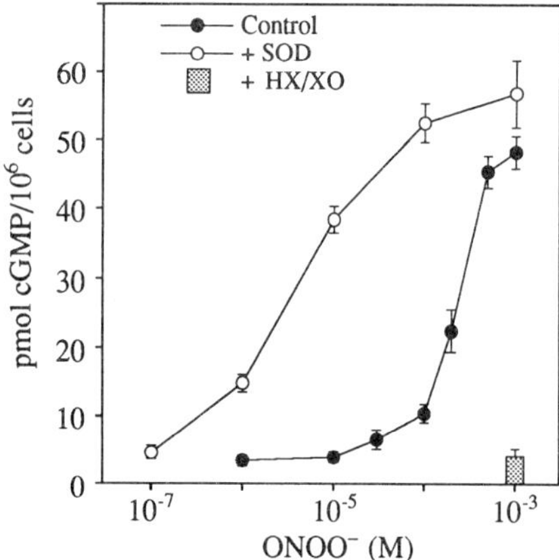

Figure 1: Effect of $ONOO^-$ on endothelial cGMP levels

that $ONOO^-$-induced cGMP accumulation was diminished by ~40% under these conditions, whereas the effect of DEA/NO remained unchanged. Similar results were obtained when 3 mM diethyl maleate was used for GSH-depletion. The essential role of GSH was further demonstrated in experiments with purified soluble guanylyl cyclase. $ONOO^-$ showed no effect on enzyme activity at concentrations up to 1 mM in the absence of added thiol, but markedly stimulated the enzyme in the presence of 2 mM GSH (Fig. 2). The effect of $ONOO^-$ was virtually abolished when it had been inactivated by incubation for 5 min at pH 7.5 prior to experiments. In contrast to $ONOO^-$, DEA/NO activated the enzyme in a thiol-independent manner. These data demonstrate that $ONOO^-$- but not NO-induced stimulation of soluble guanylyl cyclase requires presence of GSH, suggesting that the effect of the thiol was related to bioactivation of $ONOO^-$.

Abbreviations used: $ONOO^-$, peroxynitrite; DEA/NO, 2,2-diethyl-1-nitroso-oxyhydrazine; GSNO, S-nitrosoglutathione; SOD, superoxide dismutase.

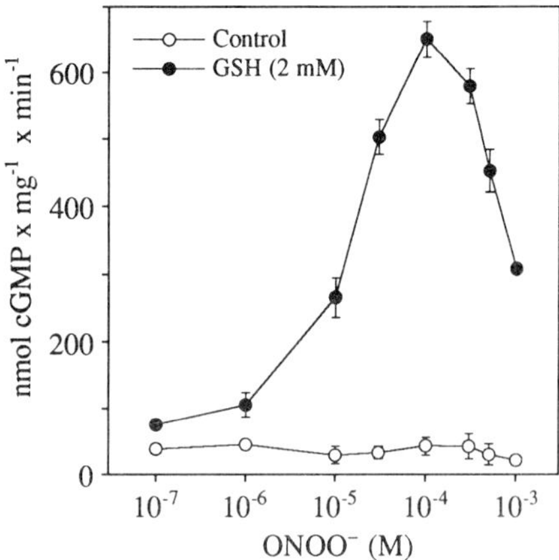

Figure 2: Effect of GSH on $ONOO^-$-induced activation of purified soluble guanylyl cyclase

Studies on the reaction between $ONOO^-$ and GSH revealed that ~1% of $ONOO^-$ was non-enzymatically converted to GSNO, a stimulator of soluble guanylyl cyclase. Formation of GSNO was increased at increasing pH; half-maximal efficiency was observed at pH 7.0, maximal yields required pH values ≥ 8.5. The reaction was not inhibited in the presence of EDTA. Authentic GSNO activated soluble guanylyl cyclase with an EC_{50} of ~10 µM up to 1 µmol cGMP x mg^{-1} x min^{-1}. Addition of 2 mM GSH increased the potency of GSNO ~100-fold suggesting an enhanced release of NO from GSNO under these conditions.

Spectrophotometric analysis of authentic GSNO revealed that in thiol-free buffer, GSNO was stable for at least 5 h at pH 2.0-9.0, but presence of 1 mM GSH induced a time-dependent, EDTA-sensitive decomposition of the nitrosothiol ($t_{1/2}$ ~3 h at pH 7.5). As shown in Fig. 3, decomposition of GSNO was accompanied by release of NO. While GSNO alone induced only a slight response of a Clark-type NO electrode, corresponding to apparent NO concentrations of less than 20 nM, a pronounced release of NO was

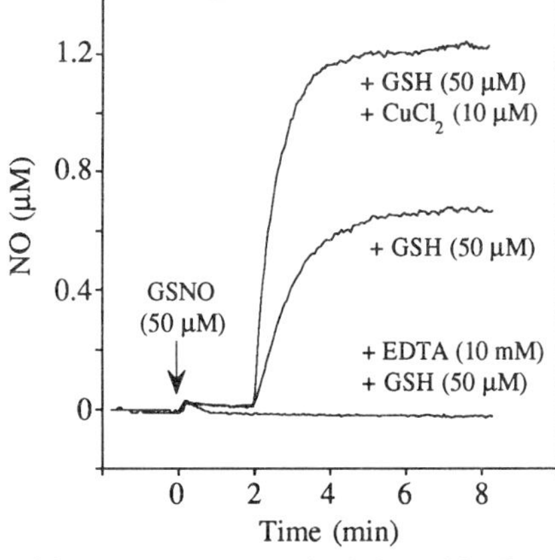

Figure 3: GSH-induced release of NO from GSNO

observed upon addition of GSH. GSH-induced NO release was potentiated by $CuCl_2$ and blocked by EDTA.

We, therefore, conclude that the NO-like properties of $ONOO^-$ are due to S-nitrosation of cellular thiols resulting in NO-mediated cGMP accumulation.

Supported by the Fonds zur Förderung der Wissenschaftlichen Forschung in Österreich.

1. Liu, S., Beckman, J.S. & Ku, D.D. (1994) J. Pharmacol. Exp. Ther. **268,** 1114-1121
2. Wu, M., Pritchard, K.A., Kaminski, P.M., Fayngersh, R.P., Hintze, T.H. & Wolin, M.S. (1994) Am. J. Physiol. **266,** H2108-H2113
3. Moro, M.A., Darley-Usmar, V.M., Goodwin, D.A., Read, N.G., Zamorapino, R., Feelisch, M., Radomski, M.W. & Moncada, S. (1994) Proc. Natl. Acad. Sci. U.S.A. **91,** 6702-6706
4. Tarpey, M.M., Beckman, J.S., Ischiropoulos, H., Gore, J.Z. & Brock, T.A. (1995) FEBS Lett. **364,** 314-318

Variants of superoxide dismutase with altered nitration activity

TERESA KENG, SANJAY JAIN, and CHRISTOPHER T. PRIVALLE

Apex Bioscience, Inc., P. O. Box 12847, Research Triangle Park, NC 27709-2847, USA

Superoxide dismutases (SODs) catalyze the dismutation of superoxide radical (O_2^-) to dioxygen and hydrogen peroxide. SODs have been utilized to prevent or reduce oxidative injury in the treatment of myocardial ischemia, stroke, head trauma, vascular occlusion, and a variety of inflammatory disorders [1]. However, to date, the results in humans treated with native human Cu, Zn SOD have been disappointing. Several groups have described an anomalous bell-shaped dose-response curve for the prevention of reperfusion-induced damage, as determined by a variety of end-points, in the isolated rat or rabbit heart. In these models, SOD was shown to have a protective effect at low dose (5-20 mg/L), followed by a dramatic decline of protection at higher doses (40 mg/L). At doses of 50 mg/L, SOD not only lost its ability to protect but in some cases exacerbated the extent of injury [1]. Explanations for these effects have included : (i) a SOD-catalase imbalance, a condition proposed to result in excess H_2O_2 levels, and (ii) a beneficial role for O_2^-, in particular the proposal that O_2^- terminates lipid peroxidation [2]. Recently, it has been demonstrated that SOD can react with peroxynitrite, resulting in generation of nitronium ion (NO_2^+), a potentially toxic species capable of nitrating protein tyrosines and subsequently altering protein function [3].

A deleterious activity of SOD was also suggested by the finding that defects in the gene for Cu, Zn SOD are associated with familial amyotrophic lateral sclerosis (FALS) [4]. These defects resulted in a decreased level of SOD activity in red blood cells of patients [5]. However, transgenic mice that express certain FALS-SOD mutant proteins did not have levels of SOD activity significantly different from that of mice expressing the wild-type enzyme [6]. Yet the transgenic mice developed symptoms of a neurodegenerative disease similar to those observed with FALS patients. These findings suggest that this neurodegenerative disorder results not from a simple diminution of SOD activity but may rather be due to a gain of a deleterious function by the mutant SODs. The deleterious function may be related to the nitration activity of SOD.

In this study, we have chemically modified bovine Cu, Zn SOD in an attempt to decrease its ability to catalyze protein nitration. In addition, we have also expressed various FALS mutant SODs in a yeast expression system and have characterized the SOD and nitration activities of these mutants.

Bovine Cu, Zn SOD was chemically modified by treatment with Traut's reagent (2-iminothiolane·HCl). Incubation of the enzyme with increasing Traut's reagent:SOD molar ratios resulted in an increase in lysine modification. At a ratio of Traut's reagent:SOD of 500:1, ~50% of the lysine residues were thiolated. Enzyme thiolation resulted in the generation of electrophoretically distinct species with altered mobilities on polyacrylamide gels. Activity gel analysis indicated that the introduction of sulfhydryl residues had only modest effects on SOD activity of these electromorphs. Spectrophotometric determination of SOD activity of Traut's-modified SOD revealed that modification of 10% of the lysine residues resulted in a 20% loss of SOD activity (Fig. 1). When lysine modification was increased to ~30% by varying the Traut's reagent:SOD ratio, SOD activity was decreased by ~35%. Additional increases in the extent of lysine modification, to a maximal level of 50% of total lysines, had no further effect on SOD activity.

In contrast to the relatively minor effects on SOD activity, thiolation of the enzyme caused a dramatic decrease in the peroxynitrite-mediated self-nitration activity of SOD (Fig. 1). In particular, modification of 10% of the lysine residues decreased SOD nitration by ~50%. Modification of greater than 30% of the lysines decreased the nitration activity to less than 10% of the level found in the unmodified enzyme. Similar results were observed when the tyrosine analog, 4-hydroxyphenylacetic acid, was utilized as the substrate for SOD-catalyzed nitration.

Abbreviations used: SOD, superoxide dismutase; H_2O_2, hydrogen peroxide; O_2^-, superoxide anion; NO_2^+, nitronium ion; FALS, familial amyotrophic lateral sclerosis.

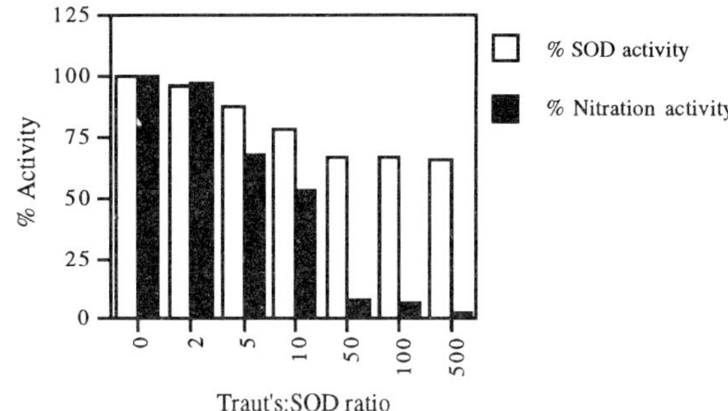

Fig. 1. Effect of Traut's modification of SOD on dismutation and nitration activities. All values are relative to the level of activity observed with the unmodified enzyme.

The nitration activity of various FALS SOD mutants was also examined. cDNAs encoding human wild-type or FALS Cu, Zn SOD mutant proteins were placed under the control of the hybrid *GAL10-TDH3* promoter and introduced into a *sod1* null mutant strain. Extracts were prepared from cells grown in the presence of galactose, and the expression of recombinant wild-type and mutant SODs was assessed by Coomassie Blue staining of a SDS-polyacrylamide gel. Several FALS SOD mutants had SOD activity at a level comparable to that found with the wild-type enzyme. However, in a nitration assay where yeast extracts were incubated with peroxynitrite, a higher level of nitrated proteins was detected with extracts containing the FALS-SOD mutants than the wild-type SOD.

The nitration activity of Cu, Zn SOD is potentially deleterious and may be responsible for the toxicity associated with the use of high doses of SOD in animal models to reduce ischemia reperfusion injury. Modification of SOD with an agent which introduces sulfhydryl groups into the enzyme dramatically decreases its nitration activity while having only slight effects on the ability of the protein to dismutate superoxide. These thiol groups may function to trap the peroxynitrite or the nitronium ion derived from reaction of SOD with peroxynitrite. Chemical modification or site-directed mutagenesis of SOD may result in a protein with enhanced therapeutic potential.

The protein nitration activity of Cu, Zn SOD may play a role in the development of FALS. The *in vitro* nitration activity of Cu, Zn SOD is enhanced in certain FALS SOD mutants, suggesting that these proteins may have a greater intrinsic ability to catalyze peroxynitrite-dependent nitration of tyrosine residues on a susceptible target in motor neurons.

ACKNOWLEDGMENTS

The authors wish to thank Drs. Joe Beckman and Yinxin Zhuang (University of Alabama, Birmingham), and Dr. Harry Ischiropoulos (University of Pennsylvania School of Medicine) for helpful discussions, antibodies to nitrotyrosine, and for providing plasmids containing FALS-SOD cDNAs.

REFERENCES

1. Omar, B. A., Flores, S. C., and McCord, J. M. (1992) Adv. Pharmacol. **23**, 109-161.
2. Nelson, S. K., Bose, S. K., and McCord, J. M. (1994) Free Rad. Biol. Med. **16**, 195-200.
3. Ischiropoulos, H., Zhu, L., Chen, J., Tsai, M., Martin, J. C., Smith, C. D., and Beckman, J. S. (1992) Arch. Biochem. Biophys. **298**, 431-437.
4. Rosen, D. R., Siddique, T., Patterson, D., *et al.* (1993) Nature **362**, 59-62.
5. Deng, H.-X., Hentati, A., Tainer, J. A., *et al.* (1993) Science **261**, 1047-1051.
6. Gurney, M. E., Pu, H., Chiu, A. Y., *et al.* (1994) Science **264**, 1772-1775.

Inhibition of γ-glutamylcysteine synthetase by S-nitrosylation

JIHONG HAN*, JONATHAN S. STAMLER', HUILING LI' and OWEN W. GRIFFITH*

*Department of Biochemistry, Medical College of Wisconsin, Milwaukee, WI 53226, USA 'Department of Medicine, Duke University Medical Center, Durham, NC USA 02117

γ-Glutamylcysteine synthetase (γ-GCS) catalyzes the reaction shown in equation 1, which is the initial and rate-limiting step in glutathione synthesis. Modification of a single cysteine residue in or near the glutamate binding site by cystamine, certain chloroketones, or related compounds has previously been shown to cause loss of enzyme activity (1,2). Active site cysteine modification is reduced or prevented by some substrates (e.g. glutamate) and by L-buthionine-S-sulfoximine (L,S-BSO), an inhibitor which, when phosphorylated by ATP, binds tightly but non-covalently to the γ-GCS substrate binding sites (3). In the present studies, we have examined the reactivity of the γ-GCS active site thiol with S-nitrosothiols (RSNOs) and with nitric oxide (NO). We also show and discuss the physiological relevance of this interaction in cells and tissue.

Eqn. 1: L-GLU + L-CYS + ATP → L-γ-GLU-CYS + ADP+Pi

Materials and Methods: γ-GCS was purified to homogeneity from rat kidney and assayed on the basis of ADP formation as described (3). The mouse peritoneal macrophage cell line RAW 264.7 was grown in DMEM media. L-Buthionine-S-sulfoximine (L,S-BSO) and L-buthionine-R-sulfoximine (L,R-BSO) were synthesized as described (4). S-Nitrosothiols were prepared by treatment of thiols with acidified nitrite (5). Other reagents were of the highest purity commercially available.

Results: Diethylamine NONOate and spermine NONOate decompose spontaneously at pH 7.4 to release 2 equivalents of NO; $t_{1/2}$ values are 2.1 and 39 min, respectively (6). In phosphate buffered saline at pH 7.4, diethylamine NONOate (5 mM) caused ~ 50% inactivation of γ-GCS in 10 min and ~ 70% inactivation in 40 min. Spermine NONOate (5 mM) caused ~ 50% inactivation in 180 min. Anaerobic exposure to NO gas also caused significant (~ 25%) inactivation of γ-GCS.

S-nitrosothiols were even more potent γ-GCS inactivators. At 5 mM, both S-nitrosocysteine (CySNO) and S-nitroso-N-acetyl-penicillamine (SNAP) caused > 80% inactivation in 10 min. Interestingly, S-nitrosoglutathione (GSNO), the RSNO most likely to be formed *in vivo*, did not inactivate γ-GCS. However, treatment of GSNO with γ-glutamyltranspeptidase released S-nitrosocysteinyl-glycine, which inhibited strongly.

Inactivation of γ-GCS by RSNOs or NO donors was substantially reduced or prevented by treatments previously shown to protect the active site thiol (1-3). A mixture of substrates (L-glutamate + ATP + Mg^{++}) offered ~ 75% protection from inactivation by SNAP. Pretreatment with cystamine, which forms a mixed disulfide with the active site thiol (1), also afforded substantial protection. Thus, if cystamine was added to γ-GCS followed by spermine NONOate, the resulting solution showed no γ-GCS activity, but ~ 75% of the initial activity could be restored with dithiothreitol, which cleaves the γ-GCS-cysteamine mixed disulfide. In contrast, ~ 60% inactivation was seen when spermine NONOate alone was added to γ-GCS, and no reversal was achieved with dithiotreitol. Finally, pretreatment of γ-GCS with L-S-BSO + ATP afforded complete protection from inactivation by SNAP; such treatment has been shown previously to protect against inactivation by cystamine or chloroketones (3,4). In contrast, treatment of γ-GCS with L-R-BSO, which does not form stable, non-covalent complex in the active site (4), did not protect against inactivation by spermine NONOate.

Treatment of RAW 264.7 macrophages with interferon-γ and lipopolysaccharide caused expression of nitric oxide synthase (iNOS), vigorous NO synthesis, and consequent 50% inhibition of macrophage γ-GCS as determined by diminished incorporation of L-[^{35}S]cysteine into glutathione (GSH); intracellular GSH levels were substantially (~ 50%) decreased. Inclusion of the iNOS inhibitors N^{ω}-methyl-L-arginine or N^{ω}-nitro-L-arginine methyl ester in the culture medium reduced NO synthesis by 83% and 77%, respectively. In the presence of iNOS inhibitors, inhibition of γ-GCS was reduced about 50% and GSH depletion was decreased about 20%.

Administration of GSNO (2 mmol/kg) to mice caused modest but statistically significant GSH depletion in several tissues. At 2 hrs, for example, liver and lung GSH levels were decreased 38% and 13%, respectively, relative to saline-treated controls; at 4 hr the kidney GSH level was decreased 11%.

Discussion: Previous studies have established that cellular GSH levels are controlled by L-cysteine availability, feedback inhibition of γ-GCS by GSH, and regulation of GSH transport out of cells (7). The present studies show that γ-GCS is also subject to regulation by RSNOs and NO. Inhibition by NO or the NO-releasing NONOates is presumably due to *in situ* formation of species with nitrosonium ion-like reactivity (e.g. metal-NO complexes or RSNOs) because direct reaction of NO with thiols is a low yield process (8). Because agents which protect the previously identified active site thiol diminish or prevent inactivation of γ-GCS, our studies strongly implicate that enzymatic thiol as the target for inactivation mediated by both RSNOs and NO. It is likely that the enzymatic thiol is initially nitrosylated (9), but our results do not exclude the possibility that further oxidation of S occurs. To date, we have not found conditions that reactivate the enzyme.

Our studies with activated macrophages and with mice given GSNO clearly establish that γ-GCS can be inactivated by NO or by NO-related products in intact cells or animals. In the examples studied, NO was generated or released at rates that are likely to occur *in vivo* at sites of inflammation where iNOS-containing activated macrophages and neutrophils have been recruited. In systemic disorders such as septic shock, iNOS is expressed and NO is formed at high rates in many tissues including vascular endothelium and smooth muscle. Our results indicate that inactivation of γ-GCS and consequent depletion of intracellular GSH levels is to be anticipated in immune and other cells expressing iNOS and producing NO and NO-related species at high rates. Because GSH is a major component of cellular antioxidant defenses (7), GSH depletion will increase the oxidant stress and tissue damage that attend *in vivo* formation of NO and other oxidants (e.g. superoxide, peroxynitrite) at inflammatory sites. Whether S-nitrosylation of γ-GCS can also serve a protective function (e.g., by preventing or delaying irreversible inactivation) or confers some other advantage to the host cell by promoting oxidant stress remains to be determined.

Acknowledgement: Studies reported here were supported in part by NIH grants DK48423 (O.W.G.) and HL02582 and HL52529 (J.S.S.). J.S.S. is a Pew Scholar in the biomedical sciences.

1. Seelig, G.F. and Meister, A. (1994) J. Biol. Chem. **259**, 3534-38.
2. Huang, C.S., Moore, W.R. and Meister, A. (1988) Proc. Natl. Acad. Sci. USA **85**, 2464-68.
3. Griffith, O.W. (1982) J. Biol. Chem. **257**, 13704-12.
4. Campbell, E.B., Hayward, M.L. and Griffith,O.W. (1991) **194**, 268-77.
5. Stamler, J.S. and Feelish, M. (1996) in Methods in Nitric Oxide Biology Research (Stamler, J.S and Feelisch, M., eds.), Chapter 36, John Wiley & Sons, New York. In Press.
6. Maragos, C.M., Morley, D., Wink, D.A. etal. (1991) J. Med. Chem. **34**, 3242-47.
7. Griffith, O.W. and Friedman, H.S. (1991) in Syngerism and Antagonism in Chemotherapy (Chou, T.C. and Rideout, D.C., eds.), pp. 245-84, Academic Press, New York.
8. Stamler, J.S. (1994) Cell **78**, 931-936.
9. Stamler, J.S., Simon, D.I. Osborne, J.A. et al (1992) Proc. Natl. Acad. Sci. USA **89**, 444-448.

Inactivation of glutathione peroxidase by nitric oxide: IMPLICATION FOR CYTOTOXICITY

MICHIO ASAHI, JUNICHI FUJII, KEIICHIRO SUZUKI, HAN GEUK SEO, and NAOYUKI TANIGUCHI

Department of Biochemistry, Osaka University Medical School, 2-2 Yamadaoka, Suita, Osaka 565, Japan

Introduction

NO is a messenger molecule with multiple biological functions including smooth muscle relaxation, neurotransmission, and macrophage-mediated cytotoxicity. NO is highly reactive with molecular oxygen, superoxide anion, and heme as well as non-heme iron. NO or its derivatives also interacts with the thiol groups of proteins and glutathione to form nitrosothiols [1]. By this mechanism, NO can be stabilized and its function prolonged. Nitrosylation of enzymes such as GAPDH [2] and protein kinase C [3] blocks their catalytic activity.

GPx, an antioxidative enzyme that scavenges various peroxides, contains a seleno-cysteine in its catalytic center. Because NO modifies the activities of several enzymes in which thiol groups are essential for catalytic function, it was of interest to determine whether NO could also affect the activity of GPx. The previous study also suggested that GPx can prevent apoptosis as well as bcl2, a proto-oncogene which blocks apoptotic death in multiple contexts [4]. Although reactive oxygen species participate in many cellular events including signal transduction and antibacterial defense, the unbalance between oxidants and antioxidants is involved in apoptosis of the cell.

We investigated whether NO inactivated the purified and intracellular GPx and discussed its physiological relevance [5].

Fig. 1. Dose-dependency [A] and time course [B] of inactivation of bovine GPx and bovine GAPDH by SNAP.

[A]

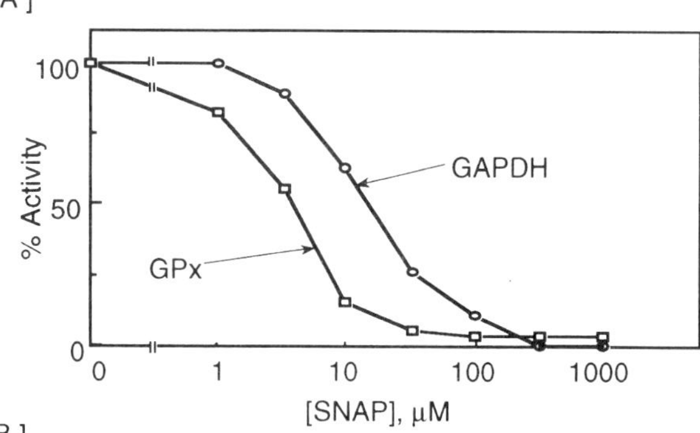

[B]

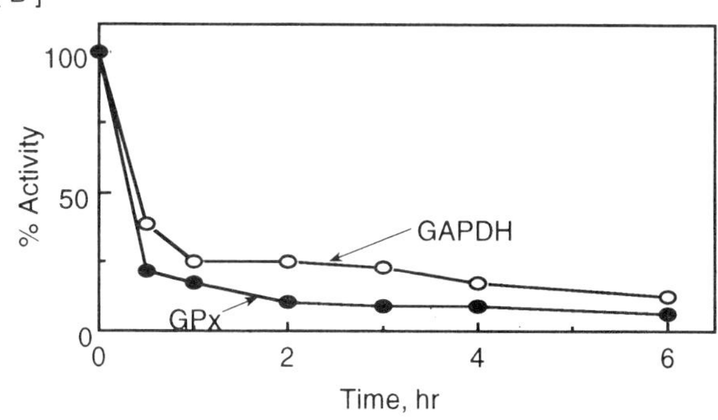

Abbreviations used are: NO, nitric oxide; GPx, glutathione peroxidase; SNAP, S-nitroso-N-acetyl-D,L-penicillamine; GAPDH, glyceraldehyde-3-phosphate dehydrogenase; SOD, superoxide dismutase.

Results & Discussion

SNAP, an NO donor, inactivated bovine GPx in a dose- and time-dependent manner (Fig. 1) . The IC_{50} of SNAP for GPx was 2 μM at 1 h of incubation and was 20% of the IC_{50} for another thiol enzyme, GAPDH, in which a specific cysteine residue is known to be nitrosylated. Incubation of the inactivated GPx with 5 mM dithiothreitol within 1 h restored about 50% of activity of the start of the SNAP incubation. A longer exposure to NO donors, however, irreversibly inactivated the enzyme. The similarity of the inactivation with SNAP and reactivation with dithiothreitol of GPx to that of GAPDH, suggested that NO released from SNAP modified a cysteine-like essential residue on GPx. When U937 cells, a human histiocytic lymphoma cell line, were incubated with 100 μM SNAP for 1 h, a significant decrease in GPx activity was observed though the change was less dramatic than that with the purified enzyme, and intracellular peroxide levels increased as judged by flow cytometric analysis using a peroxide-sensitive dye.

Table I. Alteration of enzyme activity in vitro by SNAP

	SNAP 1 hr (U/mg) (-)	SNAP 1 hr (U/mg) (+)	% Activity
GPx	36.8 ± 0.9	1.54 ± 0.13	4.2
GAPDH	4.71 ± 0.07	0.36 ± 0.02	7.6
Mn-SOD	10600 ± 50	11300 ± 990	107
Cu-Zn-SOD	13100 ± 270	13400 ± 1450	102
Catalase	2.69 ± 0.15	3.01 ± 0.11	112

Other major antioxidative enzymes, Cu,Zn-SOD, Mn-SOD, and catalase, were not affected by SNAP (Table 1), which suggested that the increased accumulation of peroxides in SNAP-treated cells was due to inhibition of GPx activity by NO. Moreover, stimulation with lipopolysaccharide significantly decreased intracellular GPx activity in RAW 264.7 cells, a mouse macrophage cell line, and this effect was blocked by NO synthase inhibitor N^{ω}-methyl-L-arginine. This indicated that GPx was also inactivated by endogenous NO.

Recent reports showed that, in addition to having a cytostatic effect, NO induced apoptotic cell death in several types of cells. Because redox regulation of cells and GPx activity are closely tied to apoptosis, inactivation of GPx by NO may be one of the causes of apoptotic cell death in these cells.

Achnowledgments

We thank Ube Industries, LTD., and Toyobo Co. LTD. for the kind gifts of recombinant human Cu,Zn-SOD, and some of the purified GPx used in this work, respectively.

References

1. Stamler, J. S., Simon, D. I., Osborne, J. A., Mullins, M. E., Jaraki, O., Michel, T., Singel, D. J., and Loscalzo, J. (1992) *Proc. Natl. Acad. Sci. USA.* **89**, 444-448.
2. Dimmeler, S., Lottspeich, F., and Brune, B. (1992) *J. Biol. Chem.* **267**, 16771-16774.
3. Gopalakrishna, R., Chen, Z. H., and Gundimeda, U. (1993) *J. Biol. Chem.* **268**, 27180-27185.
4. Hockenbery, D. M., Oltvai, Z. N., Yin, X.-M., Milliman, C. L., and Korsmeyer, S. J. (1993) *Cell* **75**, 241-251.
5. Asahi, M., Fujii, J., Suzuki, K., Seo, H. G., Kuzuya, T., Hori, M.,Tada, M.,Fujii, S., and Taniguchi, N. (1995) *J. Biol..Chem.* **270**, 21035-21039.

Nitric oxide production in human T-cell subsets and its effect on cationic amino acid transport

Sarah Chen, Dorothy H Crawford* and Richard Boyd

Dept of Human Anatomy, University of Oxford, South Parks Road, Oxford, OX1 3QX, U.K. *Dept of Clinical Sciences, London School of Hygiene and Tropical Medicine, Keppel Street, London, WC1E 7HT, U.K.

Introduction

The cationic amino acid L-arginine is an essential precursor for the production of nitric oxide (NO), an intercellular signalling molecule [1]. Cationic amino acid transport in human lymphocytes is increased following activation with the mitogen phytohaemagglutinin (PHA) [1] and maximal cationic amino acid transport occurs in activated CD8$^+$ CD45RA$^+$ subset of T-lymphocytes [3]. This population contains cytotoxic naive T-cells. Unpublished work carried out in our laboratory has shown that human peripheral blood lymphocytes produce NO. Our aims were firstly to quantitate NO production in T-cell subsets and secondly to determine the effect of NO on cationic amino acid transport using a NO donor.

Materials and methods

Lymphocyte preparation

Peripheral blood mononuclear cells (PBMC) were isolated from whole blood by centrifugation over Ficoll-Hypaque (Flow Laboratories, Irvine, U.K.). PBMC were rosetted with 2-aminoethylisothiouroniun bromide hydrobromide (AET)-treated sheep red blood cells (E), and separated into E rosette positive and negative fractions by Ficoll-Hypaque fractionation. E rosette positive (T cells) were recovered by lysis in sterile distilled water.

Cell separation procedure

T cells were pelleted and separated into CD45RO$^+$, and CD45RO$^-$ populations by resuspending the cells at 10^7/ml in culture supernatant containing CD45RO monoclonal antibody (mab) and incubated on ice for 30min. Following two washes with ice-cold RPMI-1640 containing 2% foetal calf serum (FCS), the cells were resuspended in 1ml of medium containing 10μl washed magnetic beads coated with antibody to mouse immunoglobulin (Dynal, Oslo, Norway) per 10^6 cells. This mixture was incubated on ice with gentle agitation for 10min. The positive and negative cell fractions were separated using a magnet applied to the outside of the tube. Cells were further separated into CD4$^+$ and CD4$^-$ populations in the same way using the appropriate mab.

Negatively separated cell populations were checked for purity using the reciprocal antibody and indirect immunoflurescence staining.

Cell culture

Cells were cultured at 10^6 cells/ml in RPMI-1640 medium containing 10% FCS, 100U/ml penicillin, 100μg/ml streptomycin (Flow Laboratories, Irvine, U.K.) and (10μg/ml; Sigma, Poole, U.K.) in tissue culture flasks at 37°C in a humidified atmosphere of 5% CO_2 for 24 hours.

NO measurement

This was carried out amperometrically using a commercially available NO electrode (World Precision Instruments, Hertfordshire, U.K.). Before use, the electrode was calibrated using a chemical calibration method recommended by the manufacturer. Cells were washed three time in 10ml Kreb's Ringer solution before NO production in response to the addition of L-arginine was measured in glass vials.

Measurement of cationic amino acid uptake

Lysine influx was measured as described previously [2]. Lysine was used as the probe cationic amino acid because it is metabolically much more stable than arginine. Self and cross inhibition experiments with L-arginine showed that transport was through two transport systems (y$^+$ and y$^+$L) and that both amino acids were substrates for these two transporters and had similar affinities.

NO donor

S-nitroso-*N*-acetylpenicillamine (SNAP) was used as a source of exogenous NO [4]. The parent compound of SNAP, *N*-acetyl-D,L-penicillamine (NAP, Sigma, Poole, U.K.) was used in parallel control experiments.

Results

NO production in response to the addition of L-arginine was measured. Figure 1 shows the results of one representative experiment where maximal NO production was observed in CD45R0$^-$ CD4$^-$ (CD45RA$^+$ CD8$^+$) population. Negatively separated populations checked for purity by indirect immunoflurescence staining with the reciprocal mab were found to be at least 85% pure by FACS analysis.

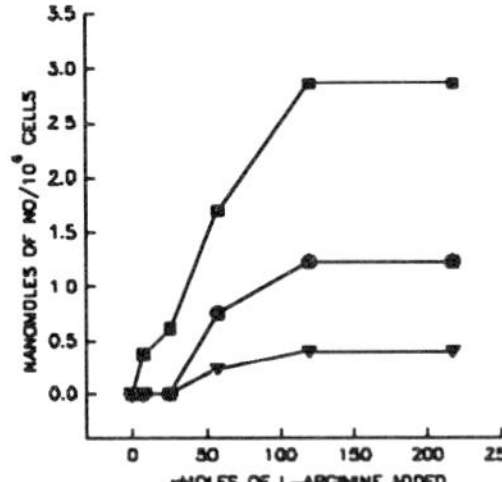

Figure 1. NO production in T-lymphocyte subsets (■) CD45RO$^-$ CD4$^-$, (●) CD45RO$^+$, (▼) CD45RO$^-$ CD4$^+$. Cells were prepared as described and NO measured. These are the results of one representative experiment (n=3).

When 1mM SNAP was added 4 hours after the addition of PHA, uptake of 2μM lysine through system y$^+$ was reduced to 58±4% control, mean ±SEM (n=11). That this effect was at least in part due to NO production was shown by experiments indicating that NAP, the parent compound of SNAP unable to generate NO, only partially inhibited transport (92±7% n=10 p=0.01) as compared to SNAP. The effects of SNAP were reversed by the presence of a NO ligand (haemoglobin) added coincidently. That the effects of NO were not non-specific is shown by concomitant experiments on lysine transport through system y$^+$L. There was no difference (p=0.7) in uptake between SNAP (78±5% control, n=11) and NAP (80±9% control, n=10) treated cells.

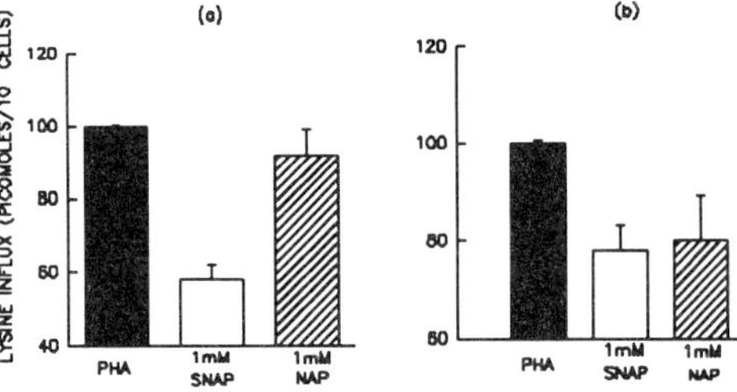

Figure 2. Lysine influx through system y$^+$ (a) and y$^+$L (b) into human PBMC cultured for 24 hrs in the presence of 10μg/ml PHA with 1mM SNAP or NAP added 4hrs after PHA. The cells were washed as described before measurement of lysine influx. The influx into control cells (PHA stimulated, without SNAP or NAP) has been normalised to represent 100% in order to account for variations in transport activity measured in different donors. The bars represent the mean ± SEM from 11 experiments. Each experiment was carried out on a separate donor.

Discussion

These results show that activated human T lymphocytes produce NO; that the rate of production is highest in the same population of lymphocytes, the cytotoxic naive T cells in which the rate of cationic amino acid transport is most rapid; that the rate of NO production by these cells is dependent upon extracellular arginine in a concentration dependent manner; and intriguingly that the rate of cationic amino acid transport is inhibited by NO. This strongly suggests that the presence of a hitherto unpredicted feedback loop whereby there is end product NO inhibition of a rate limiting step (arginine entry through system y$^+$) in the pathway of NO synthesis. The mechanism(s) responsible need investigation.

The authors thank Dr F. Ala (Regional Blood Transfusion Service, Birmingham, U.K.), Professor P.C.L. Beverley (ICRF, London, U.K.) for providing the mabs and Dr H Hodson (Wellcome Research Laboratories, Beckenhan, U.K.) for providing the SNAP used in this study. SC is in receipt of a Medical Research Council Studentship.

References

1. Schmidt, H.H. and Walter, U. (1994) Cell **78:** 919-925.
2. Boyd, C.A.R. and Crawford, D.H. (1992) Eur. J. Physiol. 422: 87-89.
3. Crawford, D.H., Chen, S. and Boyd C.A.R. (1994) Immunol. **82:** 357-360.
4. Cunningham, J.M., Mabley, J.G., Delaney, I.C., and Green I.C. (1994) Mol. Cell. Endocrinol. **102:** 23-29.

Inhibition of soluble guanylyl cyclase by copper ions

ASTRID SCHRAMMEL, PETER KLATT, DORIS KOESLING*, KURT SCHMIDT, AND BERND MAYER

Institut für Pharmakologie und Toxikologie, Karl-Franzens-Universität Graz, Universitätsplatz 2, A-8010 Graz, Austria.
*Institut für Pharmakologie, Freie Universität Berlin, Thielallee 67-73, D-14195 Berlin, Germany.

Soluble guanylyl cylcase (sGC), which catalyzes the conversion of Mg-GTP to cGMP, is stimulated several hundred-fold by submicromolar concentrations of NO [1]. Enzyme stimulation is due to binding of NO to the prosthetic heme group of sGC resulting in the formation of a nitrosyl-heme complex [2]. sGC represents the major physiological target of NO in a number of biological tissues, including blood vessels and brain, in which NO-induced accumulation of cGMP results in smooth muscle relaxation and modulation of neurotransmitter release, respectively [3,4]. Impaired NO/cGMP signalling has been implicated in various cardiovascular and neurological disease states, e.g. ischemia-reperfusion injury of the heart, stroke, and artherosclerosis [5-7].

In the present study we found that recombinant bovine lung sGC purified from a baculovirus overexpression system [8] was inhibited by low concentrations of $CuSO_4$. As shown in Fig. 1, $CuSO_4$ inhibited cGMP formation by the enzyme stimulated with 1 μM DEA/NO with an IC_{50} of ~2 μM; complete inhibition was observed with 10 μM $CuSO_4$. GSH protected sGC from copper-induced inhibition, as revealed by a more than 10-fold reduced potency of $CuSO_4$ in the presence of 1 mM of the thiol (see Fig. 1).

Hemoglobin exhibits a high-affinity binding site for Cu(I) ions, which may be involved in copper-induced, thiol-sensitive methemoglobin formation [9,10]. Since sGC was reported to contain stoichiometric amounts of copper [11], it is conceivable that copper-induced enzyme inhibition may be due to oxidation of heme-iron, resulting in reduced affinity for NO and thus deactivation of the NO-stimulated cyclase. However, our data indicate that the inhibitory effect of $CuSO_4$ is not due to interference with

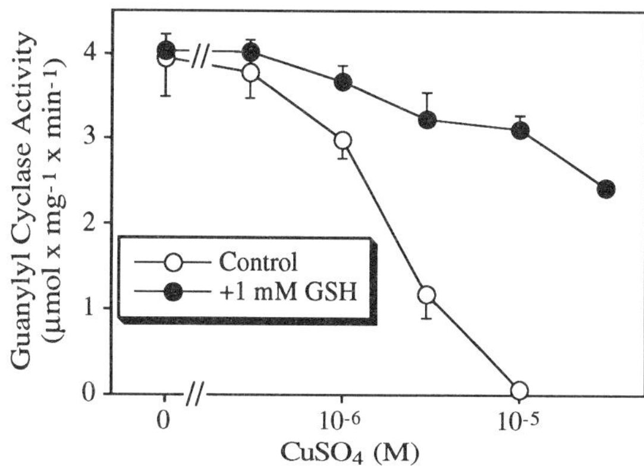

Figure 1: Inhibition of sGC by $CuSO_4$ and protection by GSH.

stimulation of the enzyme by NO, because (1) $CuSO_4$ inhibited basal and NO-stimulated enzyme activities with similar potency, (2) higher concentrations of DEA/NO did not prevent inhibition, and (3) reaction of $CuSO_4$ with NO was negligible under our experimental conditions (data not shown). Inhibition of sGC due to the oxidative properties of copper was further excluded in experiments performed with selective Cu(I) and Cu(II) chelators [12]. Fig. 2 shows that the inhibitory effect of 3 μM $CuSO_4$ was almost completely antagonized by the Cu(I)-specific antagonist neocuproine, whereas cuprizone, a compound with Cu(II) selectivity, was much less effective. GSH is known to form highly stable complexes with Cu(I) ions [13], but the protective role of the thiol was probably not due to chelation of copper, as several other thiols were also effective (not shown).

The present data show that copper ions induce a pronounced inhibition of cGMP formation by sGC. The inhibitory effect of $CuSO_4$

Abbreviations used: sGC, soluble guanylyl cyclase; DEA/NO, 2,2-diethyl-1-nitroso-oxyhydrazine.

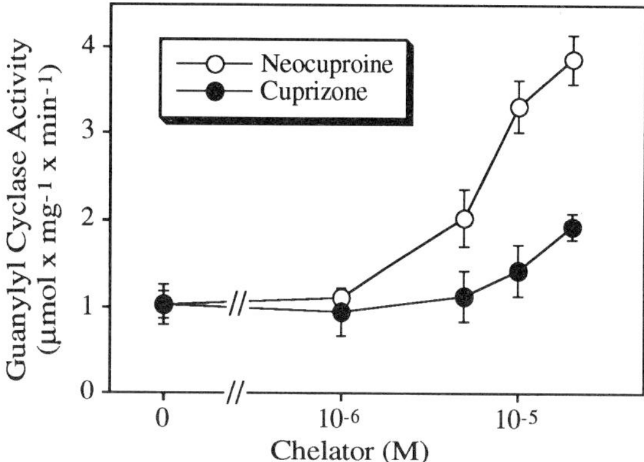

Figure 2: Effect of the Cu(I)-specific chelator, neocuproine on sGC activity in the presence of 3 μM $CuSO_4$.

may be due to binding of Cu(I) ions to one ore more sulfhydryl groups critically involved in the catalytic function of sGC, since the enzyme was protected by a Cu(I)-selective chelator and several thiols. Intracellular GSH levels are in the millimolar range and copper occurs primarily in chelated, redox-inactive forms, suggesting that Cu(I)-induced inhibition of sGC is insignificant under physiological conditions. However, oxidative stress may cause depletion of tissue GSH [14], and redox-active copper was reported to be mobilized in the course of myocardial ischemia [15] or may be released from caeruloplasmin by peroxynitrite [16]. Thus, impaired accumulation of cGMP induced by free copper may contribute to the pathophysiology of certain diseases.

Supported by the Fonds zur Förderung der Wissenschaftlichen Forschung in Austria.

1. Mayer, B., Koesling, D. & Böhme, E. (1993) Adv. Second Messenger Phosphoprotein Res. **28**, 111-119
2. Traylor, T.G. & Sharma, V.S. (1992) Biochemistry **31**, 2847-2849
3. Moncada, S., Palmer, R.M.J. & Higgs, E.A. (1991) Pharmacol. Rev. **43**, 109-142
4. Garthwaite, J. & Boulton, C.L. (1995) Annu. Rev. Physiol. **57**, 683-706
5. Bruckdorfer, K.R., Jacobs, M. & Rice-Evans, C. (1990) Biochem. Soc. Transactions **18**, 1061-1063
6. Coyle, J.T. & Puttfarcken, P. (1993) Science **262**, 689-695
7. Lefer, A.M. & Lefer, D.J. (1993) Annu. Rev. Pharmacol. Toxicol. **33**, 71-90
8. Wedel, B., Humbert, P., Harteneck, C., Foerster, J., Malkewitz, J., Böhme, E., Schultz, G. & Koesling, D. (1994) Proc. Natl. Acad. Sci. USA **91**, 2592-2596
9. Rifkind, J.M. (1981) in Metal Ions in Biological Systems, Vol. 12 (Sigel, H., ed.), pp. 191-232, Marcel Dekker, New York
10. Smith, R.C., Reed, V.D. & Webb, T.R. (1993) J. Inorg. Biochem. **52**, 173-182
11. Gerzer, R., Böhme, E., Hofmann, F. & Schultz, G. (1981) FEBS Lett. **132**, 71-74
12. Gordge, M.P., Meyer, D.J., Hothersall, J., Neild, G.H., Payne, N.N. & Noron-Hadutra, A. (1995) Br. J. Pharmacol. **114**, 1083-1089
13. Miller, M.D., Buettner, G.R. & Aust, S.D. (1990) Free Radical Biol. Med. **8**, 95-108
14. Bray, T.M. & Taylor, C.G. (1993) Can. J. Physiol. Pharmacol. **71**, 746-751
15. Chevion, M., Jiang, Y.D., Harel, R., Berenshtein, E., Uretzky, G. & Kitrossky, N. (1993) Proc. Natl. Acad. Sci. USA **90**, 1102-1106
16. Swain, J.A., Darley-Usmar, V. & Gutteridge, J.M.C. (1994) FEBS Lett. **342**, 49-52

Inhibiton of brain NO synthase by a synthetic peptide derived from the putative pteridine binding site

BERND MAYER, EVA LEOPOLD, PETER KLATT, SILVIA PFEIFFER, WALTER R. KUKOVETZ, AND KURT SCHMIDT

Institut für Pharmakologie und Toxikologie, Karl-Franzens-Universität Graz, Universitätsplatz 2, A-8010 Graz, Austria.

NO synthases (NOS) convert L-arginine to NO and L-citrulline. The reaction involves flavin-mediated shuttle of electrons from NADPH to a P450-type heme-iron which catalyzes reductive activation of molecular oxygen and substrate oxidation [1]. In addition to flavins and heme, NOSs contain tightly bound tetrahydrobiopterin (H_4biopterin) but the role of the pteridine in NO biosynthesis is not well understood [2]. It may act both as reactant and allosteric effector of NOS [3-6] and appears to be essential for the coupling of O_2 activation to L-arginine oxidation catalyzed by the neuronal enzyme [7]. Based on sequence similarities to dihydrofolate reductase, it has been proposed that the binding site of NOS for H_4biopterin is located between the heme and calmodulin sites, starting about 150 amino acids downstream from Cys_{415}, the axial ligand of the heme (see [8] and references therein).

Synthetic peptides and the corresponding anti-peptide antibodies have been used previously to study the involvement of particular sequences in the function of different enzymes, including cytochrome P450s [9-13]. To investigate the role the putative pteridine binding site in the NOS reaction, we have tested synthetic peptides derived from this sequence for their effects on the catalytic functions of recombinant neuronal NOS purified from a baculovirus overexpression system [14]. The synthetic peptides we have investigated correspond to amino acid residues 564-582 (A), 596-610 (B), 656-667 (C), and 684-698 (D) of the rat brain enzyme [15]. The peptides B, C, and D did not affect citrulline formation at up to 100 µM, but peptide A inhibited the enzyme with an IC_{50} of 10 - 30 µM. Figure 1 shows that the potency of the peptide decreased at higher NOS concentrations; an approximately 500-fold excess of the peptide was required for half-maximal inhibition of the enzyme. However, this inhibition was not overcome by exogenous H_4biopterin added in concentrations of up to 1 mM, and the peptide neither interfered with binding of radiolabelled H_4biopterin, indicating that enzyme inhibition was not due to an interaction of the synthetic peptide with the pteridine site of NOS. Both, enzyme kinetic experiments and binding studies with ^3H-labelled nitroarginine showed that inhibition of NOS by peptide A did not result from competition with the substrate (not shown). Furthermore, the peptide did not block NOS-catalyzed reduction of cytochrome c, suggesting that it had no effect on redox-cycling of the flavins. An antibody raised in rabbits against peptide A reacted with rat brain NOS in immunoblots but did not inhibit formation of L-citrulline, suggesting that it doesn't react with the native protein.

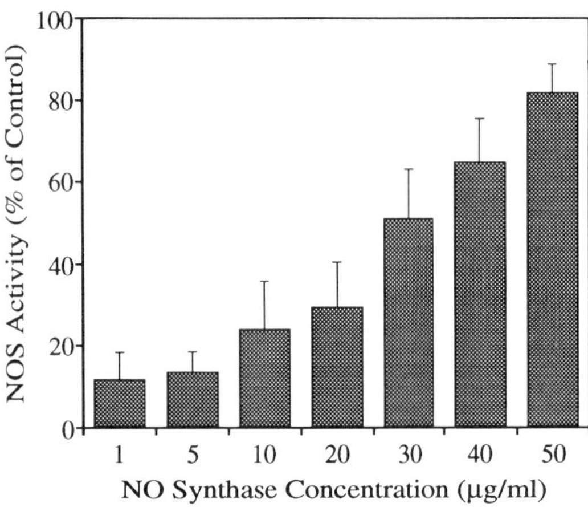

Figure 1: Inhibition of rat brain NOS by 0.1 mM peptide A

Abbreviations used: NOS, nitric oxide synthase; H_4biopterin, (6R)-5,6,7,8-tetrahydro-L-biopterin.

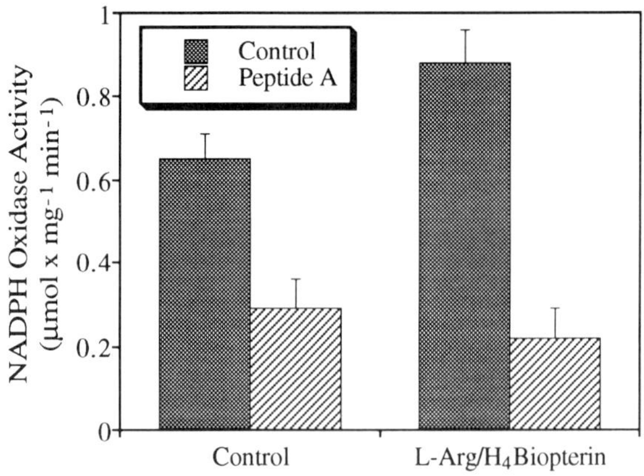

Figure 2: Inhibition of rat brain NOS-catalyzed NADPH oxidation by 0.1 mM peptide A. L-Arg, L-arginine.

Ca^{2+}/calmodulin-activated neuronal NOS reduces molecular O_2 to superoxide and hydrogen peroxide when the concentrations of L-arginine or H_4biopterin are limiting [16-18]. Figure 2 shows that peptide A (0.1 mM) markedly inhibited the NADPH oxidase activity of rat brain NOS down to approximately 45 % of controls in the absence of L-arginine/H_4biopterin. The inhibitory effect was even more pronounced in the presence of L-arginine/H_4biopterin (~25 % of controls), further confirming that the peptide-induced NOS inhibition was unrelated to substrate or pteridine binding.

Our data show that the brain NOS sequence 564-582 is critically involved in reductive activation of molecular oxgen, presumably in the electron transfer from the flavins to the heme, but is not part of the pteridine or arginine binding sites of the enzyme.

Supported by the Fonds zur Förderung der Wissenschaftlichen Forschung in Austria.

1. Klatt, P., Schmidt, K., Uray, G. & Mayer, B. (1993) J. Biol. Chem. **268**, 14781-14787
2. Mayer, B. & Werner, E.R. (1995) Naunyn-Schmiedeberg's Arch. Pharmacol. **351**, 453-463
3. Baek, K.J., Thiel, B.A., Lucas, S. & Stuehr, D.J. (1993) J. Biol. Chem. **268**, 21120-21129
4. Klatt, P., Schmid, M., Leopold, E., Schmidt, K., Werner, E.R. & Mayer, B. (1994) J. Biol. Chem. **269**, 13861-13866
5. Klatt, P., Schmidt, K., Lehner, D., Glatter, O., Bächinger, H.P. & Mayer, B. (1995) EMBO J. **14**, 3687-3695
6. Wang, J.L., Stuehr, D.J. & Rousseau, D.L. (1995) Biochemistry **34**, 7080-7087
7. Mayer, B., Heinzel, B., Klatt, P., John, M., Schmidt, K. & Böhme, E. (1992) J. Cardiovasc. Pharmacol. **20**, S54-S56
8. Nishimura, J.S., Martasek, P., McMillan, K., Salerno, J.C., Liu, Q., Gross, S.S. & Masters, B.S.S. (1995) Biochem. Biophys. Res. Commun. **210**, 288-294
9. Edwards, R.J., Singleton, A.M., Murray, B.P., Sesardic, D., Rich, K.J., Davies, D.S. & Boobis, A.R. (1990) Biochem. J. **266**, 497-504
10. Fong, Y.L. & Soderling, T.R. (1990) J. Biol. Chem. **265**, 11091-11097
11. Foster, C.J., Johnston, S.A., Sunday, B. & Gaet, F.C. (1990) Arch. Biochem. Biophys. **280**, 397-404
12. Omata, Y., Sakamoto, H., Robinson, R.C., Pincus, M.R. & Friedman, F.K. (1994) Biochem. Biophys. Res. Commun. **201**, 1090-1095
13. Shen, S. & Strobel, H.W. (1994) Biochemistry **33**, 8807-8812
14. Harteneck, C., Klatt, P., Schmidt, K. & Mayer, B. (1994) Biochem. J. **304**, 683-686
15. Bredt, D.S., Hwang, P.M., Glatt, C.E., Lowenstein, C., Reed, R.R. & Snyder, S.H. (1991) Nature **351**, 714-718
16. Mayer, B., John, M., Heinzel, B., Werner, E.R., Wachter, H., Schultz, G. & Böhme, E. (1991) FEBS Lett. **288**, 187-191
17. Heinzel, B., John, M., Klatt, P., Böhme, E. & Mayer, B. (1992) Biochem. J. **281**, 627-630
18. Pou, S., Pou, W.S., Bredt, D.S., Snyder, S.H. & Rosen, G.M. (1992) J. Biol. Chem. **267**, 24173-24176

Role of a nitric oxide donor in hydrogen peroxide-induced endothelial cell barrier dysfunction

KAREN E. MCQUAID, VALERIE C. CULLEN and ALAN K. KEENAN

Pharmacology Dept., University College, Dublin 4, Ireland.

Introduction

The structural and functional integrity of the vascular endothelium is critical for the maintenance of its barrier function. The first target of reactive oxygen species (ROS) released from circulating inflammatory cells is often the vascular endothelium whose barrier function then becomes compromised [1].

The role of NO in ROS-induced endothelial cell (EC) injury is unclear as investigations have indicated both a cytoprotective [2] and a cytodestructive [3] role. In this study the role of NO in oxidant-induced EC barrier dysfunction was investigated using hydrogen peroxide (H_2O_2) as the ROS and sodium nitroprusside (SNP) as the source of NO. Bovine pulmonary artery endothelial cells (BPAEC) grown on polycarbonate membrane assemblies were used as an *in vitro* model of EC monolayer permeability, to assess altered barrier function.

Methods

For permeability studies EC (passage numbers 9-18) seeded at a density of ~2.3×10^5 cells/cm^2 were grown to confluence on Transwell polycarbonate membrane assemblies (6.5 mm diameter, 0.4 µm pore size, Costar), precoated with 25 µg/ml fibronectin for 3 h at 37^{0}C. 72 h after seeding monolayers were incubated with test reagent(s) in Locke's buffer containing 2 mM L-glutamine, pH 7.4, for 20 min at 37^{0}C under sterile conditions.

Following removal of test reagents, monolayer permeability was assessed by applying 100 µl 4% trypan blue-labelled bovine serum albumin (TB-BSA) to the upper chamber, 600 µl HBSS containing 0.02 M HEPES to the lower chamber and incubating the system in a shaking water bath at 37^{0}C for 1 h. TB-BSA transfer across the EC monolayer (permeability) was quantified by measuring the absorbance of the solution in the lower chamber at 590 nm.

Nitrite release from the medium overlying treated cells was quantified by the Griess reaction. SOD activity in EC lysates was determined using the SOD-525 method (R + D Systems, UK). Statistical analysis was carried out using ANOVA, followed by a Bonferroni multiple comparisons test or an unpaired two-tailed Student's t test. $p< 0.05$ was taken to be significant.

Results

Treatment of EC monolayers with 0.2 mM H_2O_2 for 20 min caused a significant increase in permeability to TB-BSA, $p<0.0001$ (mean ± S.E.M., n = 9). A 20 min incubation with the NO donor SNP (0.1 mM) did not affect TB-BSA transfer, however SNP in combination with H_2O_2 significantly increased transfer above that seen with H_2O_2 alone, $p<0.0001$ (mean ± S.E.M., n = 9), (Table 1).

Table 1. The effect of SNP on H_2O_2-induced increases in BPAEC monolayer permeability to TB-BSA.

Each value represents the mean ± S.E.M. of 9 observations.

Treatment	fold increase over basal
Control	1 ± 0
H_2O_2 (0.2 mM)	2.071 ± 0.189
SNP (0.1mM)	1.007 ± 0.033
H_2O_2 + SNP	3.951 ± 0.585

This enhancement appeared to be due to NO production, since 0.1 mM SNP released nitrite into the medium overlying cells (38.86 ± 1.79 nmol/10^6 cells). Nitrite released from cells treated with 0.1 mM GTN was significantly less (5.06 ± 0.24 nmol/10^6 cells) and was undetectable in untreated cells. Potassium ferricyanide (0.083 mM) did not alter the response to H_2O_2, suggesting that cyanide release from SNP was not responsible for the enhancement of the H_2O_2 effect. Endogenous SOD activity in BPAEC treated with 0.2 mM H_2O_2 for 20 min significantly decreased (77.8% reduction), compared to untreated cells ($p<0.05$, n = 6). The thiol donor thiosalicylic acid (TSA, 1 mM) when added in combination with 0.2 mM H_2O_2, abolished the H_2O_2-induced increase in TB-BSA transfer.

Discussion

It can be concluded from this study that there is a role for NO in modulation of the EC barrier dysfunction caused by short term ROS exposure. The differences in the levels of nitrite produced by SNP and GTN may be due to the fact that SNP's liberation of NO occurs spontaneously whereas GTN's enzymic bioconversion to NO is tightly regulated [4].This would result in different quantities of NO (and consequently nitrite) being released by the two NO donors.

We propose that high levels of NO released from SNP enhance H_2O_2-induced EC barrier dysfunction directly and/or indirectly. Direct interaction between NO and H_2O_2 has previously been suggested by a study showing a catalase-inhibitable killing effect of macrophages (mediated by NO) on *Leishmania* [5]. Alternatively, superoxide could (i) be generated indirectly from H_2O_2 which can activate PKC and increase $[Ca^{2+}]_i$, two known stimuli for superoxide production [6] and (ii) be stabilised by the fall in endogenous SOD activity observed in this study, thus allowing formation of the cytotoxic species peroxynitrite [7].

H_2O_2 and NO may interact indirectly, at the level of the glutathione (GSH) redox system which protects the endothelium against damage due to H_2O_2 [8] (supported by the observed protective effect of TSA treatment above) or large amounts of NO [9]. Simultaneous exposure of cells to large amounts of H_2O_2 and NO, may overwhelm the defensive capacity of the GSH redox system, resulting in enhanced H_2O_2 and/or NO cytotoxicity.

Supported by the Wellcome Trust (grant no. 36702/Z/92/Z)

References

(1) Sinclair, D.G., Braude, S., Haslam, P.L. and Evans, T.W. (1994). *Chest*, 106: 535-539.

(2) Kim, H., Chen, X. and Gillis, C.N. (1992). *Biochem. Biophys. Res. Commun.*, 189: 670-676.

(3) Palmer, R.M.J., Bridge, L., Foxwell, N.A. and Moncada, S. (1992). *Br. J. Pharmacol.*, 105: 11-12.

(4) Lopez-Belmonte, J., Whittle, B.J.R. and Moncada, S. (1993). *Br. J. Pharmacol.*, 108: 73-78.

(5) Li, Y., Severn, A., Rogers, M.V., Wilson, M.T. and Moncada, S. (1992). *Eur. J. Immunol.*, 22: 441-446.

(6) Matsubara, T. and Ziff, M. (1986). *J. Cell. Physiol.*, 127: 207-210.

(7) Beckman, J.S., Beckman, T.W., Chen, J., Marshall, P.A. and Freedman, B.A. (1990). *Proc. Natl. Acad. Sci. USA*, 87: 1620-1624.

(8) Suttorp, N., Toepfer, W. and Roka, L. (1986). *Am. J. Physiol.*, 251: C671-C680.

(9) Walker, M.W., Kinter, M.T., Roberts, R.J. and Spitz, D.R. (1995). *Pediatr. Res.*, 37: 41-49.

EFFECTS OF NITROGLYCERINE ON ENERGY METABOLISM OF ERYTROCYTES AND RETICULOCYTES

M. M. Kostić, M. R. Petronijević, Lj. M. Djoković and S. D. Maletić

Institute of Physiology, Faculty of Medicine, 34000 Kragujevac, P.O.B.124, Yugoslavia

Glycolysis is the only energy producing prosess in mammalian erythrocytes, while in reticulocytes energy is provided by glycolysis (10 %), as well as, by oxidative phosphorylation (OxP) (90 %) /1/. Nitric oxide (NO) in many cells inactivates aconitase and mitochondrial respiratory chain, while glycolysis remains unaltered by NO /2,3/. However, there are no data concerning the effects of NO on energy metabolism of red blood cells, especially regarding their high content of hemoglobin - an effective scavenger of NO /4/. Therefore, the aim of this study was to investigate the role of nitroglycerine (NTG), a widely used donor of NO, on red blood cell energy metabolism.

MATERIALS AND METHODS

Erythrocytes and reticulocytes of rats were used in this study. Reticulocytosis was induced by phenylhydrazine /5/. From washed cells suspensions of mature erythrocytes either reticulocyte-rich suspensions (containing 60-80 % of reticulocytes), were prepared /5/. Aerobic incubation was performed without (control) or with different concentrations of NTG (0.1, 0.25, 0.5, 1, 1.5 mM).

Oxygen consumption in reticulocytes was measured by Warburg technique /6/. On the basis of coupled oxygen consumption and P/O ratio 2.5 amount of ATP produced by OxP was calculated /7/. Total energy production in reticulocytes was calculated by addition of glycolytic to mitochondrial ATP production. Glycolytic energy production was calculated on the basis of lactate/ATP ratio of 1 /7/. Glucose, lactate, adenine nucleotides (ATP, ADP, AMP) and glycolytic intermediates were determined enzymatically /8/. Cyclic GMP was determined by radioimmunoassay /9/.

RESULTS AND DISCUSSION

Reticulocyte respiration was significantly altered by NTG in dose- and time-dependent manner. After 2 h of incubation coupled oxygen consumption was significantly lower, while uncoupled was significantly higher in the presence of NTG. They were accompanied by stimulation of glycolysis, as measured by lactate accumulation and glucose consumption. However, even 3-fold stimulation of reticulocyte glycolysis, providing 40-50 % of whole energy production, was not sufficient to compensate decreased energy production due to inhibition of OxP (Table 1). Besides lower ATP production, a dose-dependent decrease of ATP level was also found. However, increase of ADP and particularly AMP content prevented the loss of adenine nucleotides, which occurred when ATP production by OxP is blocked with oligomycin as inhibitor of ATP synthase /10/. Calculated mean ATP-turnover time was prolonged even for 50 % in he presence of 1.5 mM NTG, which indicates an inhibition of ATP-consuming processes in NTG-treated reticulocytes.

What is the reason for the stimulation of glylolysis in NTG-treated reticulocytes? Levels of all glycolytic intermediates and application of "cross-over" theoreme /11/ indicate stimulation at the HK-PFK, GA3PDH and PK level. Stimulation of glycolysis accompanied with inhibition of the

Table 1. Influence of nitroglycerine (NTG) on the energy production in rat reticulocytes and erythrocytes
Each value represents the mean±S.E.M. for 5-6 expts.

NTG (mM)	ATP production (μmol/ml cells/2h)		ATP level (μmol/ml cells)	
	Rtcs	Ercs	Rtcs	Ercs
0.00	87.40±4.71	6.76±0.19	1.53±0.14	0.53±0.07
0.10	94.29±3.12	6.99±0.92	1.39±0.14	0.46±0.09
0.25	74.62±6.25	7.32±0.44	1.50±0.13	0.57±0.06
0.50	59.43±4.67	7.53±0.62	1.42±0.14	0.47±0.10
1.00	26.76±5.19	8.79±0.50	1.07±0.07	0.46±0.09
1.50	35.60±7.29	9.08±0.98	0.93±0.09	0.38±0.04

OxP, activation of HK-PFK, decrease of ATP and simultaneous rise of ADP and AMP levels (not shown), all together represent an example of Pasteur effect /1/ occuring in NTG-treated reticulocytes. However, in mature erythrocytes NTG dose-dependently increased glycolytic rate and decreased ATP level (Table 1). These data indicate that NTG itself stimulates glycolysis, but not only through the Pasteur effect which could not appear in mature erythrocytes. Therefore our data show that NO released by NTG affects glycolytic sequence as well, although the earlier data have indicated that NO leaves it unaltered /3/. Furthermore, newer data, contrary to our resultes, indicate inhibition of GA3PDH /12/.

Basal level of cGMP in rat reticulocytes amounted to 15±3 pmol/ml packed cells and increased in the presence of 0.1, 0.25, 0.5, 1 and 1.5 mM NTG for 57, 65, 89, 171 and 100 %, respectively. However, metabolic effects of NTG were not mimick by exogenous 8-Br-cGMP, except the slight, dose independent inhibition of OxP accompanied by no changes of glycolytic rate. These data indicate that NTG-induced (a) inhibition of coupled respiration in reticulocytes and (b) stimulation of glycolysis in reticulocytes and erythrocytes are mediated by NO as an effector molecule, but not by cGMP.

REFERENCES

1. Rapoport,S.M. (1986) The Reticulocyte, CRC Press, Inc., Boca Raton, Florida
2. Stadler,J., Billiar,T.J., Curran,R.D., Stuehr,D.J., Ochoa,J.B. and Simmons,R.L. (1991) Am.J.Physiol.260, C910-C916
3. Granger,D.L., Taintor,R.R., Cook,J.L. and Hibbs, J.B.Jr.(1980) J.Clin.Invest.65, 357-370
4. Feelisch,M. and Noak,E.A. (1987) Eur.J.Pharmacol. 139, 19-30
5. Kostić,M., Maretzki,D., Živković,R., Krause, E.G. and Rapoport,S.M.(1987) FEBS Lett.217, 163-168
6. Umbreit,W.W., Burris,R.H. and Stauffer,P.(1964) Manometric Techniques, Burgess Publishing Co., Minneapolis
7. Siems,W., Mtller,M., Dumdey,R., Holztter,H.G., Rathmann,J. and Rapoport,S.M.(1982) Eur.J. Biochem.139, 101-107
8. Bergmeyer,H.U.(1974) Methods in Enzymatic Analysis, Academic Press, New York
9. Brown,B.L., Albano,J.D., Barnes,G.D. and Ekins, R.P.(1974) Biochem.Soc.Trans.2, 10-12
10. Gautheron,D.C.(1984) J.Inher.Metab.Dis.7,57-61
11. Chance,B., Higgins,J., Holmes,W. and Connely,C.M. (1958) Nature 182, 1190
12. Mallozzi,C., DiStasi,A.M.M. and Minetti,M.(1995) Endothelium, 3 (Suppl), 44

Nitric oxide promotes oxygen free radicals release in heart mitochondria and submitochondrial particles

JUAN J. PODEROSO, MARIA C. CARRERAS, CONSTANZA LISDERO, FRANCISCO SCHOPFER, NATALIA RIOBO and ALBERTO BOVERIS.

Laboratory of Oxygen Metabolism. University Hospital, University of Buenos Aires. Cordoba 2351, 4th floor, 1120 Buenos Aires, Argentina.

Introduction

NO interferes with oxidative metabolism.The NO-derived peroxynitrite anion ($ONOO^-$) induces membrane lipoperoxidation, oxidation of sulfhydryls and final damage to membrane mitochondrial electron chain components [1]. However, NO reversible inhibition of cytochrome oxidase has been reported as well [2]. The main purpose of this study is to describe the effects of NO, released by an exogenous donor, on mitochondria and SMP, particularly, on O_2^- and H_2O_2 release.

Methods

Release of NO by GSNO-DTT system.

The NO released by the GSNO-DTT system was detected electrochemically with a NO sensitive electrode; NO released builds up a ramp concentration and effective NO concentration can be calculated.

Effects of NO on the enzymatic activities of SMP.

Rat and beef heart mitochondria and SMP were isolated by standard methods [3]. The activities of cytochrome oxidase, NADH-and succinate-cytochrome c reductases were determined spectrophotometrically after 4-8 min incubation of SMP with GSNO-DTT. Cytochrome spectral changes induced by NO were recorded in the 500-650 nm wavelength range.

Production of superoxide anion and hydrogen peroxide by SMP and mitochondria.

The release of O_2^- was measured by the superoxide-sensitive oxidation of 1mM adrenaline to adrenochrome monitored in a double beam-double wavelength spectrophotometer at 485-575 nm and H_2O_2 by the horseradish peroxidase/p-hydroxyphenyl acetic acid assay at 315(ex) and 425(em) nm in an spectrofluorometer with 8mM succinate and 2 μM antimycin.

Results

NO decreased enzymatic activities of SMP. Cytochrome oxidase and NADH- and succinate c reductase activities were decreased by 100, 63 and 74% respectively (Table 1); cytochrome oxidase was highly sensitive to NO (Table 2).

Table 1. Enzymatic activities and production of O_2^- and H_2O_2 in SMP exposed to GSNO

Each value represents mean ± SEM from 3-5 samples.

		Control	1.2μM NO
		nmol . min^{-1}. mg prot^{-1}	
Enzymatic Activity	Cyt. oxidase	19±1	1±0
	NADH-cyt.c-reductase	112±6	41±9
	Succinate-cyt c-reductase	199±11	36±8
O_2^- Production	+ 8 mM Succinate	undetected	0.58±0.01
	id + 2 μM antimycin	0.46±0.07	0.55±0.05
H_2O_2 Production	+ 8 mM succinate	undetected	0.20±0.03
	id + 2 μM antimycin	0.52±0.06	0.45±0.05

Abbreviations: NO, nitric oxide; GSNO, S-nitrosoglutathione; DTT, dithiothreitol; SMP, Submitochondrial particles.

Table 2. Cytochrome oxidase activity of SMP exposed to GSNO Each value represents mean ± SEM of 3-6 samples.

Treatment	Cytochrome Oxidase Activity (k'(min^{-1}).mg prot^{-1})
None	20±1
50 nM NO	12±1
50 nM NO + 8 mM succinate	15±1
id + antimycin	18±2
id + 1 μM SOD	8±1
50 μM peroxynitrite	16±3

NO promoted a significant O_2^- and H_2O_2 release by mitochondria (Fig.1) and SMP (Fig.2), even in absence of antimycin.

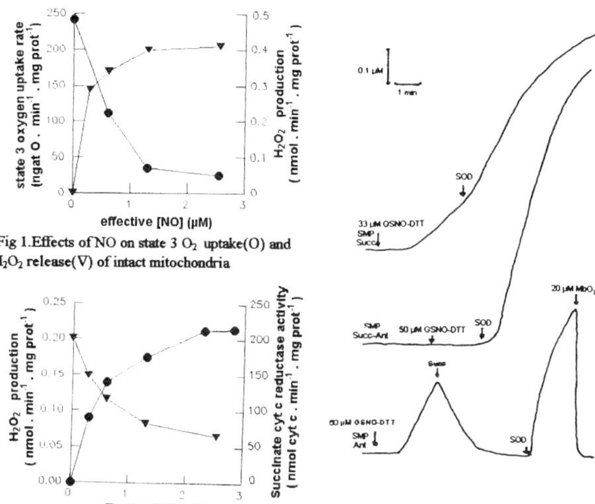

Fig 1.Effects of NO on state 3 O_2 uptake(O) and H_2O_2 release(V) of intact mitochondria

Fig 2.Effects of NO on H_2O_2 release(O) and succinate-cytochrome c reductase activity(V) by SMP.

Fig 3. Detection of NO and O_2^- production by SMP.

Also, NO partially reduced cyt b while cyt a and c remained oxidized, an effect similar to that of antimycin.

When SMP were supplemented with succinate alone or plus antimycin, electrochemical detection of NO ramp accumulation was abolished but restored by superoxide dismutase, presumably due to O_2^-/NO reaction to form peroxynitrite (Fig.3). In accord, inhibition of cytochrome oxidase by NO was reverted by supplementation in the incubation media with succinate and antimycin (Table 2).

Concluding remarks

1. NO is a multisite mitochondrial inhibitor being cytochrome-oxidase the most sensitive step.
2. NO increases the release of superoxide anion by mitochondria in an antimycin-like effect.
3. Released O_2^- can react directly with NO. In accord, NO did not block cytochrome oxidase when O_2^- was present. Probably, removing NO restores electron transfer chain activity as a part of a self-regulatory mechanism.

Acknowledgements

This work was supported by grants from Fundacion Perez Companc (Buenos Aires, Argentina), Laboratorios Roche and the University of Buenos Aires.

References

1. Radi, R. , Rodriguez, M., Catro, L, and Telleri R. (1994) Arch. Biochem. Biophys. **308**, 89-95
2. Cleeter, M.W.J, Cooper, J.M., Darley-Usmar, V.M., Moncada, S., and Shapira, A.H.V. (1994) FEBS Lett. **345**, 50-54.
3. Turrens, J.F., and Boveris ,A. (1980) Biochem. J. **191**, 421-427

Nitric oxide induces a transmembrane signal in human erythrocytes involving glyceraldehyde 3-phosphate dehydrogenase and Band 3

CINZIA MALLOZZI, A. M. MICHELA DI STASI and MAURIZIO MINETTI

Cell Biology Department, Istituto Superiore di Sanità
V. Regina Elena, 299 - 00161 Rome (Italy)

Nitric oxide (•NO) is a novel biological messenger molecule involved in multiple biological functions including vessel smooth muscle relaxation, platelet aggregation, neurotransmission modulation and phagocyte cytotoxicity [1]. The intracellular targets of •NO include heme proteins, proteins with iron-sulfur centers, and thiols. Protein thiols react readily with •NO to form stable nitrosothiols [2] which among their different functions might serve as bioactive reservoirs of •NO. Proteins that are S-nitrosylated include serum albumin, tissue-type plasminogen activator, glyceraldehyde-3-phosphate dehydrogenase (GPDH), catepsin B, and the N-methyl-D-aspartate receptor-channel complex. GPDH has received particular attention in view of the evidence that •NO induces a NAD-dependent covalent modification of a cysteine in the active site in association with a loss of the enzymatic activity [3]. This •NO-dependent reaction has been observed in a variety of different cellular extracts and intact cells, and was not limited to exogenously added •NO, but could be demonstrated also after activation of endogenous •NO-synthase [4, 5]. Moreover, several reports have recently shown that ADP-ribosylation is a mechanism of post-translational modification of cytoplasmic and membrane proteins that may play a physiological role in modulating cell function. •NO has been suggested to act as a regulator of endogenous intracellular ADP-ribosylation [6].

In this preliminary study, we found that •NO stimulated the endogenous NAD-dependent modification of the membrane-bound glycolytic enzyme GPDH in intact human erythrocytes. These data were obtained by using the "back ADP-ribosylation" method: immediately after the treatment of whole erythrocytes with a physiologic •NO donor (nitrosylated form of serum albumin, BSANO [2]), present at μM concentration in plasma of healthy subjects, the labeling reaction was performed with added •NO donors (BSANO again or sodium nitroprusside, SNP) and 1-2 μCi [^{32}P] NAD on membrane proteins. As shown in Table 1, a treatment of only 5 min of intact cells with 30 μM BSANO induced an inhibition of 40% and 80% in the labeling of membrane-bound GPDH stimulated by 30 μM BSANO and 100 μM SNP, respectively. Such decrease is taken to indicate that some of the sites had been modified by •NO in the intact cells.

In the erythrocytes, GPDH is normally bound to the N-terminus of the cytoplasmic domain of the anion transporter Band 3. Previous studies [7] have shown that the binding of GPDH to the membrane is modulated by tyrosine phosphorylation of Band 3. Band 3 tyrosine phosphorylation may prevent the GPDH binding and allow its translocation to the cytoplasm. Since the enzyme is inhibited in its bound state, the functional consequence of Band 3 tyrosine phosphorylation could be the GPDH activation and the enhance of the glycolytic rate [8]. We investigated the phosphorylation status of Band 3 when intact erythrocytes were exposed to 30 μM BSANO or 50 μM SNP as radical generators in the presence of sodium orthovanadate as a protein phosphatases inhibitor. Using anti-phosphotyrosine monoclonal antibodies for the immunoblotting analysis, we found that •NO induced a time-dependent increase in the tyrosine phosphorylation of Band 3, which increased after 15 min of incubation and peaked after 60 min. We also found that, after the removal of •NO by washing the cells with phosphate saline buffer, the erythrocytes were able to recover their low basal tyrosine phosphorylation status by incubating the cells in autologous plasma or 1 mM glucose (after 120 min of incubation at 37°C the recovery was total).

In our previous work [9], we found that methemoglobin, produced in intact cell by exposure to BSANO, was efficiently reduced to hemoglobin by autologous fresh plasma, glucose or NADH, suggesting a possible involvement of glycolysis in the recovering of functional cell.

Although conflicting evidence has been presented regarding the role of •NO in the regulation of cellular glucose metabolism [10], these results suggest that erythrocytes may represent the last step in •NO biotransformation or inactivation through the activation of a membrane signal transduction. However, the correlation between GPDH modification and Band 3 tyrosine phosphorylation remain to be clarified.

Table 1. Inibition of GPDH [^{32}P]NAD labeling

BSANO (30 μM) pretreatment (min):	0	5	15	30	60
	[^{32}P]NAD GPDH (c.p.m.)				
none	2.0	2.2	1.9	1.5	1.5
+BSANO (30 μM)	17.0	10.1	5.3	6.2	7.0
+SNP (100 μM)	58.2	9.7	8.5	6.3	5.4

'Abbreviation used': •NO, nitric oxide; GPDH, glyceraldehyde 3-phosphate dehydrogenase; BSANO, nitrosylated form of serum albumin; SNP, sodium nitroprusside.

REFERENCES

1. Moncada, S. (1992) Acta Physiol. Scand. **145**, 201-227
2. Stamler, J. S., Simon, D. I., Osborne, J. A. et al. (1992) Proc. Natl. Acad. Sci. U.S.A. **89**, 444-448
3. Brüne, B., Dimmeler, S., Molina y Vedia, L. and Lapetina, E. G. (1994) Life Sci. **54**, 61-70
4. Hauschildt, S., Scheipers, P., Bessler, W. G. and Mülsch, A. (1992) Biochem. J. **288**, 255-260
5. Dimmeler, S., Ankarcrona, M., Nicotera, P. and Brüne, B. (1993) J. Immunol. **150**, 2964-2971
6. McDonald, L. J. and Moss, J. (1994) Mol. Cell. Biochem. **138**, 201-206
7. Low, P. S., Allen, D. P., Zioncheck, T. F. et al. (1987) J.Biol. Chem. **262**, 4592-4596
8. Harrison, M. L., Rathinavelu, P., Arese, P.,Geahlen, R. L., and Low, P. S. (1991) J. Biol. Chem. **266**, 4106-4111
9. Pietraforte, D., Mallozzi, C., Scorza, G. and Minetti, M. (1995) Biochemistry **34**, 7177-7185
10. Mateo, R. B., Reichner, J. S., Mastrofrancesco, B., Kraft-Stolar, D. and Albina, J. E. (1995) Am. J. Physiol. (Cell Physiol.) **268**, C669-C675

Role of thiols in and around circulating erythrocytes in the metabolism of nitrosyl-hemoglobin

Y. MINAMIYAMA, S. TAKEMURA and M. INOUE

Department of Biochemistry, Osaka City University Medical School, 1-4-54 Asahimachi, Abeno-ku, Osaka 545, Japan.

Although physiological importance of NO and the mechanism of its synthesis have been studied extensively, only a limited information is available for the dynamic aspects of its metabolism *in vivo*. Because of extremely high affinity of NO to oxyhemoglobin and thiol compounds, significant fraction of *de novo* synthesized NO forms nitrosyl-hemoglobin (NO-Hb) and nitrosothiols. To know factors affecting the generation and degradation of NO-Hb in the circulation, the adduct was generated in RBC by incubating with [1-hydroxy-2-oxo-3-(N-methyl-3-aminopropyl)-3-methyl-3-aminopropyl]-3-methyl-1-triazene (NOC7), an NO donor, and the change in its cellular levels was studied by electron spin resonance (ESR) method. The present work describes the effect of thiol status in and around the circulating RBC on the formation and degradation of NO-Hb.

[Materials and Methods]

Reduced glutathione (GSH) was purchased from Wako Pure Chemical Co. (Osaka). Diamide and L-buthionine-[S, R]-sulfoximine (BSO) were from Sigma (St. Louis, MO). NOC7, an NO donor which spontaneously releases NO, was obtained from Dojin Co. (Kumamoto). Other reagents used were of analytical grade.

Male Wistar rats (200-220g) and Nagase analbuminemic rats were obtained from SLC, Co. (Shizuoka) and from Dr. Takahashi, respectively, and used for experiments without prior fasting. Fresh venous blood samples were collected from rats into heparinized tubes. NO-Hb was prepared by incubating blood samples with varying concentrations of NOC7 at 37°C. To know the effect of intra- and extracellular SH, blood samples and RBC suspension (Ht = 50%) were incubated with 2 mM of NOC7 at 37°C in the presence or absence of 10 mM diamide or GSH. At indicated times, 400 µl of the samples were quickly frozen in liquid nitrogen and subjected to ESR analysis at 110°K using a JES-RE1X spectrometer (JEOL, Tokyo) with 100 kHz field modulation. Analytical conditions were as previously reported [1].

Under urethane anesthesia (1 g/kg, i.p.), NOC7 (10 µmol/kg) was intravenously injected to rats and 0.4 ml of blood samples were collected from the femoral vein. The change in blood levels of NO-Hb was analyzed by ESR. To know the effect of thiol status on NO metabolism, changes in NO-Hb levels were also determined in animals which were administered either GSH (200 µmol/kg, i.p.) or BSO (0.45 mmol/day, p.o.), 30 min or 7 days before experiments or injection of NOC7, respectively.

[Results & Discussion]

Upon incubation with normal blood, NOC7 generated NO-Hb dose-dependently (Fig. 1). The amount of NO-Hb formed with the blood from Nagase analbuminemic rat was significantly larger than those from control rats (Fig. 2a). Treatment of RBC with 10 mM diamide markedly increased the amounts of NOC7-induced NO-Hb. In contrast, addition of either GSH or mercapt albumin was decreased the amounts of NO-Hb (Fig. 2b).

Intravenous administration of NOC7 also generated NO-Hb in the circulating RBC; levels of NO-Hb increased maximally at 35 min

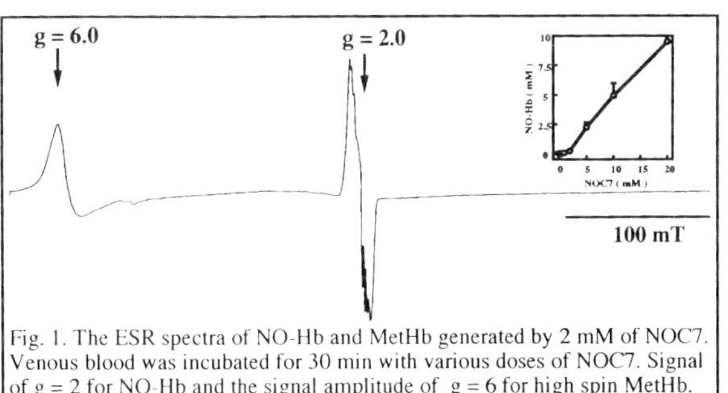

Fig. 1. The ESR spectra of NO-Hb and MetHb generated by 2 mM of NOC7. Venous blood was incubated for 30 min with various doses of NOC7. Signal of g = 2 for NO-Hb and the signal amplitude of g = 6 for high spin MetHb. Reciver gain = 50. NO-Hb levels generated by NOC7 were shown in the inserted figure.

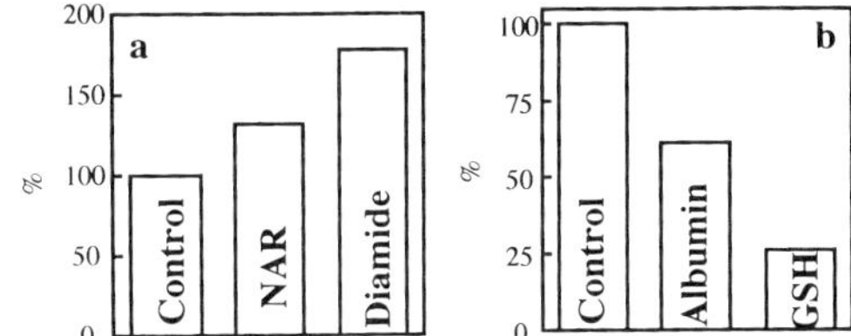

Fig. 2. Effects of thiols on NO-Hb formation by NOC7 *in vitro*. The formation of NO-Hb was induced at 37°C with 2 (b) or 5 mM (a) NOC7 for 30 min in venous blood with or without albumin (0.5 mM), GSH (10 mM), diamide (10 mM), or collected blood from analbuminemic rat (NAR).

after administration of NOC7 and decreased thereafter with an apparently half-life of 100 min (Fig. 3). To know the effect of thiol status on the formation and degradation of NO-Hb, changes in NO-Hb levels were also studied with animals which were pretreated with either GSH or BSO. In GSH-treated animals, the amount of NO-Hb generated in RBC was significantly lower than control group but retained its maximal levels for longer than 1 h ($T_{1/2}$ =10 hr). In contrast, the rate of NO-Hb formation was significantly larger with BSO-treated rats than with control group particularly during the first 10 min. NO-Hb formed in the circulating RBC of BSO-treated animals disappeared with a half life of 90 min. To know the role of thiol status in the formation of NO-Hb, changes in SH levels in blood, plasma, liver and kidney were determined with animals which were treated with either GSH or BSO. Loading of GSH increased the SH levels of blood, plasma and kidney by about 100, 300 and 30%, respectively. In contrast, BSO markedly decreased the SH levels of blood, plasma, liver, and kidney by about 40, 35, 90 and 75%, respectively.

These results suggested that thiol status in and around RBC including Cys^{34} of albumin might play important role in the formation and degradation of NO-Hb (Fig. 4). Furthermore, these findings suggest that NO and/or its metabolites also reacts with various thiols *in vivo* thereby forming fairly stable S-nitrosothiols which might release bioactive NO depending on the redox state of animals (Fig. 4).

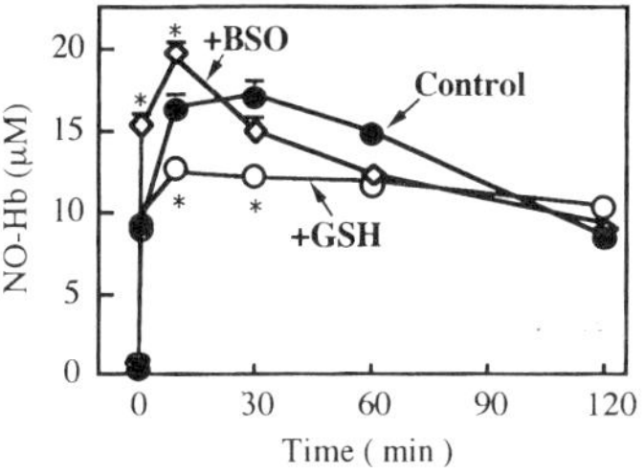

Fig. 3. Effects of GSH on the blood levels of NO-Hb in NOC7 injected rat. NOC7 was administered as described in Fig. 3. Similar experiments were carried out with animals which were administered with 0.2 mmol/kg of GSH (i.p.) 30 min before injected NOC7 and which were orally administered BSO (0.45 mmol/day for 7 days). *: P<0.05 as compared with control group.

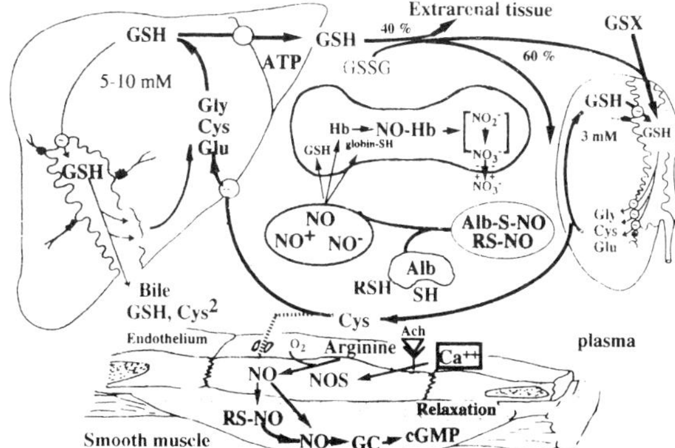

Fig. 4 Dynamic aspects of biothiols and NO metaboism

[References]

[1] Inoue, M., Minamiyama, Y., Takemura, S. (1995) in Methods in Enzymology (Packer, L., ed), (in press) Academic Press Inc. New York.

Effects of nitric oxide production on hemoglobin expression in the erythroleukemic cell line K562

Steven Rafferty, Joseph Domachowske, and Harry Malech

Laboratory of Host Defenses,
National Institute of Allergy & Infectious Diseases,
National Institutes of Health, Bethesda, Md. 20892

Here we report that heterologous expression of inducible murine macrophage nitric oxide synthase (NOS-2) in the human erythroleukemic K562 cell line results in constitutive production of nitric oxide associated with inhibition of hemoglobin synthesis. Our observations show that hemoglobin expression is inhibited indirectly by post-transcriptional repression of erythroid aminolevulinic acid synthase (eALAS), the first enzyme of the heme biosynthetic path.

The coding region of NOS-2 was cloned into the vector pCEP4 (Invitrogen). Expression is constitutive and is driven from the CMV promoter; the vector replicates with the host genome but does not integrate, and is maintained by hygromycin selection. K562 cells were transfected with the construct by electroporation; control transfections were run in parallel with plasmid pCEP4-CAT (chloramphenicol acetyl transferase).

Constitutive NOS expression was detected by Western blotting of cell lysates and by the accumulation of nitrite in the culture media of NOS-transfected but not control cells. NOS-transfected cells grew more slowly than control cells but were able to maintain constitutive production of nitric oxide for more than one month following transfection. Nitrite accumulation has three phases: a lag period of two hours; a 12 to 15 hour period in which nitrite accumulates at a rate of about 1 nmol/10^6 cells/hour; and a final plateau, with no further significant increase in nitrite. However, if the cells were resuspended in fresh media, nitrite accumulation resumed.

We noted that the cell pellets of NOS-transfected K562 cells were white while those of control cells were red-pink, which suggested that NOS activity was interfering with the expression of fetal hemoglobin in this cell line. The difference spectrum cytosols from control transfected and NOS-2 transfected cells was that of oxyhemoglobin, thus confirming our suspicions. The hemoglobin content of the NOS-transfected cells was less than one fifth that of control cells, as determined by a colourimetric diaminofluorene assay.[1] Western blots for fetal hemoglobin showed a sharp decrease in the globin content of NOS-transfected cells, which could be reversed if 1 mM N-methyl-arginine, a NOS inhibitor, was included in the culture media. There was no difference in the level of globin mRNA between NOS and control transfected cells, thus hemoglobin was being post-transcriptionally repressed.

Hemoglobin translation is tightly controlled by the availability of the heme cofactor.[2] To determine if the constitutive expression of active NOS in K562 cells was repressing hemoglobin biosynthesis by limiting heme availability, we supplemented the culture media of NOS-transfected cells with hemin and measured globin expression by immunoblotting. Supplementing the culture of NOS-transfected cells with 20 μM heme derepressed fetal globin expression. Globin expression was also restored by including 0.5 mM δ-aminolevulinate, the first intermediate of the multistep heme biosynthetic path, in the culture medium of NOS-transfected K562 cells. Sodium butyrate (1 mM), a general inducer of erythroid differentiation in K562 cells, could not restore globin expression in NOS-transfected cells.

Our observations of lowered globin expression in NOS-transfected K562 cells, and the pattern of its restoration by inducers of globin synthesis (Figure 1) suggests that nitric oxide impairs hemoglobin biosynthesis by blocking the synthesis of ALA, the first intermediate of heme biosynthesis. The level of mRNA for erythroid ALA synthase is the same in both NOS-transfected and control cells, thus its repression is post-transcriptional.

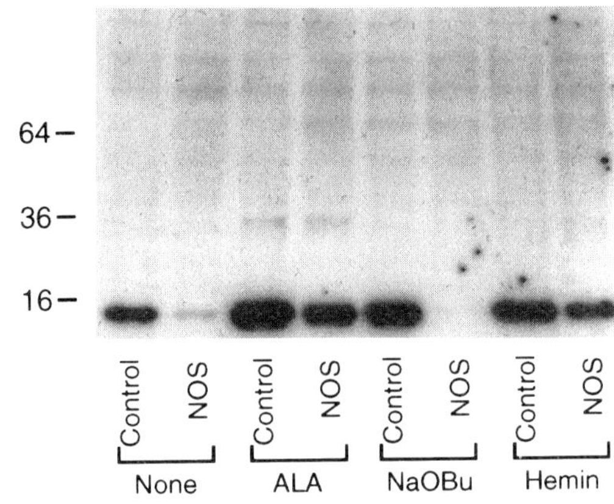

Figure 1. Western blot of fetal hemoglobin in cytosols of NOS-transfected ("NOS") or CAT-transfected ("Control") K562 cells. Cells were transfected and were grown for five days in selective media (RPMI 1640, 10% fetal calf serum, 275 units/mL hygromycin). Cells were grown for an additional four days in the presence of no further additives ("None") or with one of the following: 0.5 mM δ-aminolevulinic acid ("ALA"); 1 mM sodium butyrate ("NaOBu"); or 20 μM heme ("Hemin"). Each lane contains 100 μg of cytosolic protein. The position of size calibration markers in kilodaltons are indicated to the left of the blot.

The 5'-untranslated regions of the mRNA of erythroid ALA synthase and ferritin possess a metastable stem-loop structure (iron-responsive element, or IRE) that is the binding site for an iron-sensing protein (iron regulatory protein, or IRP).[3] Binding of IRP to an IRE in the 5' untranslated region of mRNA inhibits translation initiation. IRP binds iron reversibly as an iron-sulphur cluster; iron and mRNA binding by IRP are mutually exclusive. Recent experiments show that endogenous or exogenous nitric oxide can disrupt the iron-sulphur complex of IRP and convert it to its mRNA-binding form, which translationally represses ferritin in tissue culture.[4,5] Presumably the same mechanism causes the post-transcriptional repression of erythroid ALA synthase that we see in NOS-transfected K562 cells, with further detrimental effects on hemoglobin biosynthesis. These observations further illustrate the potential for endogenously produced nitric oxide to regulate cellular post-transcriptional events. In particular, our observations may be relevant to the role of nitric oxide in anemia and lowered blood hemoglobin concentrations that are associated with chronic infections such as tuberculosis or parasitic disease.[6]

1. Kaiho, S. & Mizuno, K. (1985) *Anal. Biochem.* **149**, 117-120.

2. Traugh, J.H. (1989) *Seminars in Hematology* **26**, 54-62.

3. Klausner, R.D., Rouault, T.A. & Harford, J.B. (1993) *Cell* **72**, 19-28.

4. Pantopoulos, K. & Hentze, M.W. (1995) *Proc. Natl., Acad. Sci. USA* **92**, 1267-71.

5. Oria, R., Sanchez, L., Houston, T., Hentze, M.W., Liew, F.I., & Brock, J.H. (1995) *Blood* **85**, 2962-66.

6. Mabbott, N. & Sternberg, J. (1995) *Infect. & Immun.* **63**, 1563-66.

Singlet and triplet nitroxyl anion (NO^-) lead to *N*-methyl-D-aspartate (NMDA) receptor downregulation and neuroprotection

Stuart A. LIPTON*, Won-Ki KIM*, Posina V. RAYUDU*, Wael ASAAD*, Derrick R. ARNELLE†, and Jonathan S. STAMLER†

*Department of Neurology, Children's Hospital, and Program in Neuroscience, Harvard Medical School, Boston, MA 02115 USA; †Department of Medicine, Duke University Medical Center, Durham, NC 27710 USA

Summary

Nitric oxide (NO·) reacts only very slowly with thiol groups under physiological conditions, while other redox-related species (NO^+, NO^-) react rapidly with thiols. Previously, our laboratories had shown that nitrosonium ion equivalents (NO^+) downregulate the activity of the *N*-methyl-D-aspartate (NMDA) subtype of glutamate receptor and prevent NMDA-induced neurotoxicity [1]. Here we studied the effects of nitroxyl anion (NO^-) in the singlet and triplet states. Exogenous donors of NO^- in the singlet state were applied to cultured rat cortical neurons while monitoring their NMDA-evoked responses during patch-clamp recording or digital Ca^{2+} imaging with fura-2. These NO^- donors inhibited NMDA-evoked responses. The effects were readily reversed by chemical reduction with dithiothreitol (DTT). The inhibitory effects of the NO^- donors were completely blocked by prior incubation in the thiol oxidizing agent 5-5'-dithio(bis-2-nitrobenzoic acid) (DTNB) or the alkylating agent *N*-ethylmaleimide. These results are best explained by reaction of singlet NO^- with critical thiol groups of the redox modulatory site of NMDA receptor, leading to disulfide bond formation. The reaction of singlet NO^- with thiol should liberate hydroxylamine, which was detected experimentally. NO^-, apparently in the triplet state, can be produced from NO· via incubation with SOD[Cu(I)]. NO^- generated in this manner also inhibited NMDA-evoked responses to some degree, possibly via reaction with O_2 to form $ONOO^-$. Consistent with these findings, NO^- donors significantly ameliorated NMDA receptor-mediated neurotoxicity, suggesting potential therapeutic uses of these redox-related congeners of nitric oxide in neurologic diseases.

Methods

Preparation of cultures of cerebral cortex. Cerebrocortical cultures were prepared as previously described from 16-17 day fetuses of Sprague-Dawley rats [2].

Single cell calcium imaging. Cortical neurons were loaded with fura-2/AM and digitally imaged for NMDA-evoked $[Ca^{2+}]_i$ responses in Mg^{2+}-free Hanks' saline supplemented with 0.5 µM tetrodotoxin (TTX) + 1 µM glycine, as previously described [2].

Electrophysiological studies with whole-cell patch recording. NMDA-evoked whole-cell currents were recorded at -60 mV using standard patch-clamp techniques in Mg^{2+}-free Hanks' saline plus TTX, as previously described [1].

Results

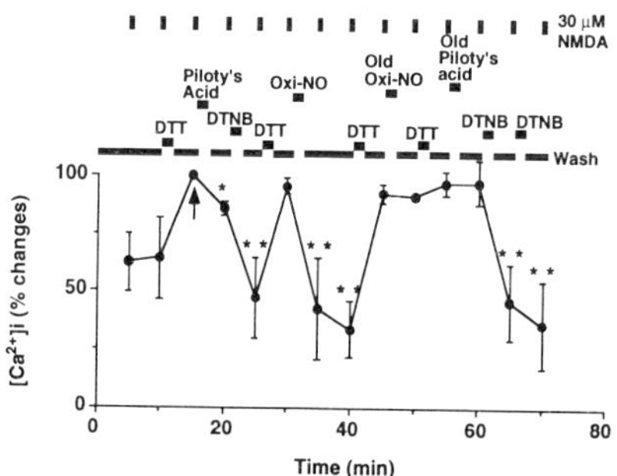

Fig. 1. NO^--generating drugs inhibit NMDA-evoked increases in $[Ca^{2+}]_i$. DTT (2 mM), DTNB (0.5 mM), and NO^--generating drugs (Piloty's acid and Oxi-NO, 5 mM each, thought to generate singlet NO^-). Statistics: an analysis of variance followed by a Scheffé multiple comparison of means [* $P<0.05$, ** $P<0.01$ compared to the response to NMDA obtained directly after DTT exposure (marked with an arrow)].

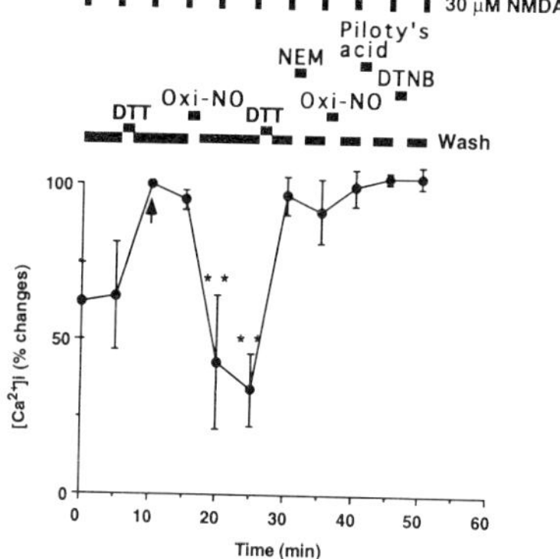

Fig. 2. Irreversible alkylation of thiol groups with *N*-ethylmaleimide (NEM) prevents the effect of NO^- on NMDA-evoked neuronal $[Ca^{2+}]_i$ responses. Alkylation of thiol (-SH) groups with NEM (1 mM) prevented the subsequent inhibition of NMDA responses by the various redox reagents.

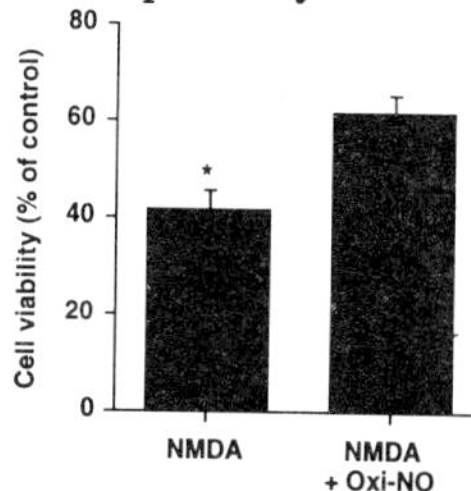

Fig. 3. Oxi-NO attenuates NMDA receptor-mediated neurotoxicity. Neuronal viability was assessed by measuring the amount of lactate dehydrogenase released from dying neurons 24 h after exposure to NMDA (300 µM), Oxi-NO (1 mM), or the combination. Data are expressed as mean ± SEM.

Conclusions

- NO^- donors can downregulate NMDA receptor activity.
- NO^- donors can prevent NMDA receptor-mediated neurotoxicity.
- Since overstimulation of NMDA receptors mediates a variety of neurologic disorders (e.g., focal stroke, epilepsy, Hungtington's disease, AIDS dementia, neuropathic pain, and others), NO^- donors may prove to be of therapeutic potential.

1. Lipton, S.A., Choi, Y.-B., Pan, Z.-H., et al. (1993) A redox-based mechanism for the neuroprotective and neurodestructive effects of nitric oxide and related nitroso-compounds. Nature **364**, 626-632

2. Lei, S.Z., Pan, Z.-H., Aggarwal, S.K., et al. (1992) Effect of nitric oxide production on the redox modulatory site of the NMDA receptor-channel complex. Neuron **8**, 1087-1099

3. Murphy, M.E., Sies, H. (1991) Reversible conversion of nitroxyl anion to nitric oxide by superoxide dismutase. Proc. Natl. Acad. Sci. U.S.A. **88**, 10860-10864

Down regulation of NMDA receptor activity in cortical neurons by peroxynitrite

Won-Ki KIM*, Posina V. RAYUDU*, Mark E. MULLINS†, Jonathan S. STAMLER¶, and Stuart A. LIPTON*

*Department of Neurology, Children's Hospital, and Program in Neuroscience, Harvard Medical School, Boston, MA 02115 USA; †Department of Chemistry, Harvard University, Cambridge, MA 02138 USA; ¶Department of Medicine, Duke University Medical Center, Durham, NC 27710 USA

Summary

Peroxynitrite (ONOO⁻), formed by reaction of nitric oxide (NO˙) with superoxide anion ($O_2^{\cdot-}$), is a potent oxidant and neurotoxin [1, 2]. The role of peroxynitrite as a potential second messenger or regulatory agent, however, remains controversial. Peroxynitrite has been reported to oxidize sulfhydryl groups [3], and NMDA-evoked responses can be modulated by sulfhydryl redox agents [2]. Therefore, the present study was undertaken to determine if peroxynitrite, at low concentrations, could modulate the activity of the NMDA receptor. Peroxynitrite, synthesized by stopped flow reaction kinetics, was rapidly applied to rat cortical neurons in culture while monitoring their NMDA-evoked calcium responses with digital calcium imaging of fura-2 or whole-cell recording with patch electrodes. Peroxynitrite, at micromolar concentrations or less ($t1/2 \approx 1$ s), decreased NMDA responses. The effect was reversed by subsequent exposure to the reducing agent dithiothreitol (DTT). Peroxynitrite did not further decrease NMDA responses of neurons previously treated with another sulfhydryl oxidizing agent 5,5'-dithio-bis-(2-nitrobenzoic acid) (DTNB). Moreover, peroxynitrite had no effect on NMDA receptors that had been irreversibly alkylated with *N*-ethylmaleimide (NEM). These findings suggest that peroxynitrite can reversibly downregulate NMDA receptor activity by oxidizing critical sulfhydryl groups compromising the redox-modulatory site(s) of the NMDA receptor.

Methods

Preparation of cultures of cerebral cortex. Cerebrocortical cultures were prepared as previously described from 16-17 day fetuses of Sprague-Dawley rats [2].

Peroxynitrite and other redox reagents

Peroxynitrite was synthesized by stopped flow reaction kinetics [2]. Prior to each experiment the concentration of ONOO⁻ was monitored at 302 nm using a spectrophotometer. A solution of ONOO⁻ (nominally 50 μM) was made by adding an appropriate aliquot of stock solution to Mg^{2+}-free Hanks' saline immediately prior to use. DTT was used at 2 mM, DTNB at 0.5 mM, and NEM at 1 mM.

Single cell calcium imaging. Cortical neurons were loaded with fura-2/AM and digitally imaged for NMDA-evoked $[Ca^{2+}]_i$ responses in Mg^{2+}-free Hanks' saline supplemented with 0.5 μM tetrodotoxin (TTX) + 1 μM glycine, as previously described [2].

Electrophysiological studies with whole-cell patch recording. NMDA-evoked whole-cell currents were recorded at -60 mV using standard patch-clamp techniques in Mg^{2+}-free Hanks' saline plus TTX, as previously described [2].

Results

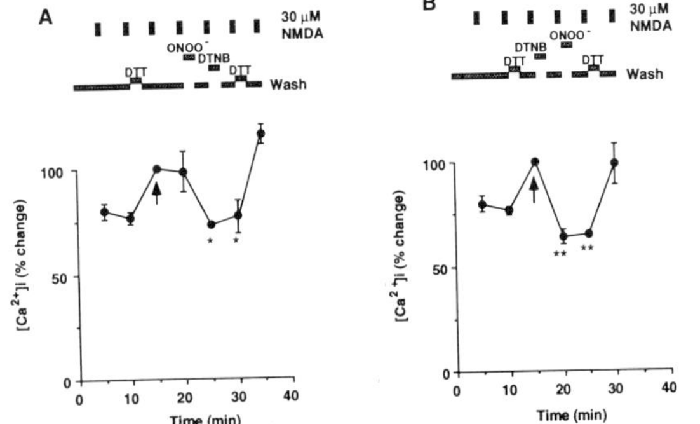

Fig. 1. Peroxynitrite decreases NMDA-evoked $[Ca^{2+}]i$ responses. Redox agents were applied for 2 min and then washed off prior to 30 μM NMDA application. ONOO⁻ and DTNB acted similarly and occluded the effects of one another (**A** and **B**). DTT reversed the effects of ONOO⁻ or DTNB. Values are mean ± SEM, normalized to the maximal NMDA response indicated by the arrow. Statistics: an analysis of variance followed by a Scheffé multiple comparison of means [* $P<0.05$, ** $P<0.01$ compared to the response to NMDA obtained directly after DTT exposure (marked by the arrow)].

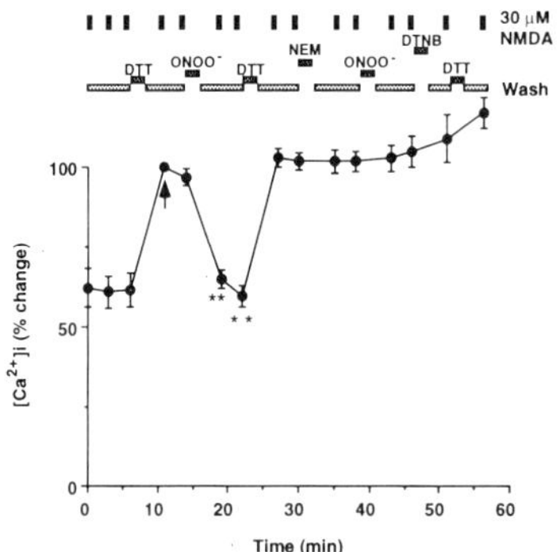

Fig. 2. Irreversible alkylation of thiol groups with *N*-ethylmaleimide prevents the effect of ONOO⁻ on NMDA-evoked calcium responses. Alkylation of thiol (-SH) groups with NEM prevented the subsequent inhibition of NMDA responses by the various redox agents including ONOO⁻.

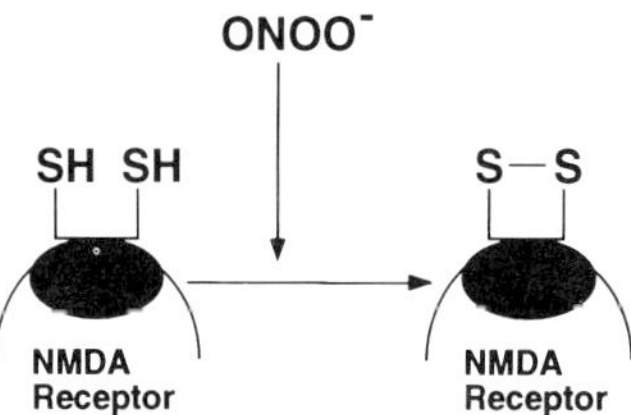

Fig. 3. Proposed mechanism for the action of ONOO⁻ on thiol (-SH) groups.

Conclusions

- ONOO⁻ can decrease NMDA receptor activity by reacting with a redox modulatory site on the receptor.
- These results raise the possibility that, at low concentrations, ONOO⁻ modulates physiological processes.

1. Beckman, J.S., Beckman, T.W., Chen, J., Marshall, P.A., Freeman, B.A. (1990) Apparent hydroxyl radical production by peroxynitrite: implications for endothelial injury from nitric oxide and superoxide. Proc. Natl. Acad. Sci. U.S.A. **87**, 1620-1624

2. Lipton, S.A., Choi, Y.-B., Pan, Z.-H., et al. (1993) A redox-based mechanism for the neuroprotective and neurodestructive effects of nitric oxide and related nitroso-compounds. Nature **364**, 626-632

3. Radi, R., Beckman, J.S., Bush, K.M., Freeman, B.A.(1991) Peroxynitrite oxidation of sulfhydryls. The cytotoxic potential of superoxide and nitric oxide. J. Biol. Chem. **266**, 4244-4250

Nitric oxide modulation of human bone marrow colony growth and differentiation

Paul J. Shami and J. Brice Weinberg

Duke and VA Medical Centers, Hematology/Oncology Division, 508 Fulton Street, Durham, NC 27705, USA.

Nitric oxide (NO) is a molecule with important functions affecting the vascular, nervous and immunologic systems [1]. In previous work, we have studied the effect of NO on malignant human hematopoiesis. We have demonstrated that NO inhibits the growth of cells of the HL-60 human myeloid leukemia cell line [2]. Along with growth inhibition, NO induces the monocytic differentiation of these cells and modulates the gene expression of cytokines and oncogenes in them [2]. We have also demonstrated that NO inhibits the growth of acute nonlymphocytic leukemia cells freshly isolated from untreated patients [3]. NO induces the expression of monocytic differentiation antigens in some of these cells [3]. The purpose of the present work was to determine the effect of NO on normal human hematopoiesis.

Bone marrow cells were obtained from patients undergoing bone marrow harvest in preparation for bone marrow transplantation. The mononuclear fraction of the bone marrow was separated using a fycoll-Hypaque density gradient and was used for culture in semi-solid media. In some experiments the CD34+ fraction of the bone marrow was enriched from the mononuclear cells using panning techniques (AIS MicroCELLector, Santa Clara, CA). Using this separation method, we obtained a cell fraction with 85-90% CD34+ cells. Bone marrow mononuclear or CD34+ cells were cultured in semi-solid media as previously described [4]. Sodium nitroprusside (SNP) (Elkins-Sinn, Cherry Hill, NJ) or S-nitroso-acetyl penicillamine (SNAP) (Chem Biochem Research Inc., Salt-Lake City, UT) were used as NO donors and were added to the cultures at the time of plating.

When added to mononuclear bone marrow cells at concentrations of 0.25 to 1 mM, SNP inhibited the growth of both CFU-E and CFU-GM (Figure 1A). At concentrations of 0.25 to 1 mM, SNAP had a more prominent growth inhibitory effect on CFU-E and CFU-GM than SNP (Figure 1B).

CD34+ cells treated with SNP showed different effects between the myeloid and erythroid colonies. At a concentration of 0.25 to 1 mM, SNP inhibited the growth of CFU-E but enhanced the growth of CFU-GM (Figure 1C). To confirm these results, we repeated the experiments using a population of 100% CD34+ cells obtained by cell sorting. When such cells were used, SNP (0.25 - 1 mM) inhibited the growth of CFU-E but did not affect the growth of CFU-GM (data not shown). When added to CD34+ cells at concentrations of 0.25 and 0.5 mM, SNAP had a more prominent growth inhibitory effect on CFU-E than CFU-GM, although at higher concentrations both lineages were equally inhibited (Figure 1D). To determine whether NO affected the growth of bone marrow cells by activating guanylate cyclase, we measured intracellular levels of cGMP. Mononuclear cells from the bone marrow were isolated by density centrifugation, cultured in RPMI-1640, and treated with 1 mM SNP. Intracellular levels of cGMP were measured at 15, 30, 45, and 60 min. SNP increased intracellular cGMP levels in a time dependent manner with almost a 6 fold increase at 60 minutes as compared to controls. We finally treated mononuclear or CD34+ cells obtained from the bone marrow with the membrane permeable cGMP analog 8-Br-cGMP (Sigma Chemical Company, St. Louis, MO). When added to bone marrow mononuclear cells at a concentration of 3 mM, 8-Br-cGMP had no effect on CFU-E but it inhibited CFU-GM by 53%. When added to CD34+ enriched cells at the same concentrations, 8-Br-cGMP had no effect on CFU-E but it inhibited CFU-GM by 34%.

We show in the present work that NO affects the growth and differentiation of normal human bone marrow colonies *in vitro*. When added to the mononuclear fraction of the bone marrow, it inhibited both CFU-E and CFU-GM growth. When added to CD34+ cells, it had a more potent growth inhibitory effect on CFU-E than on CFU-GM. SNP treatment actually potentiated the growth of CFU-GM while inhibiting the growth of CFU-E. We therefore conclude that NO can affect human bone marrow growth both directly and indirectly. It can act on accessory cells to stimulate the production of growth inhibitory cytokines. Indeed, we have previously shown that NO upregulates the gene expression of TNF-α and IL-1β [2].

Abbreviations used: NO, nitric oxide. SNP, sodium nitroprusside. SNAP, S-nitrosoacetyl penicillamine. CFU-E, colony forming unit - erythroid. CFU-GM, colony forming unit - granulocyte macrophage.

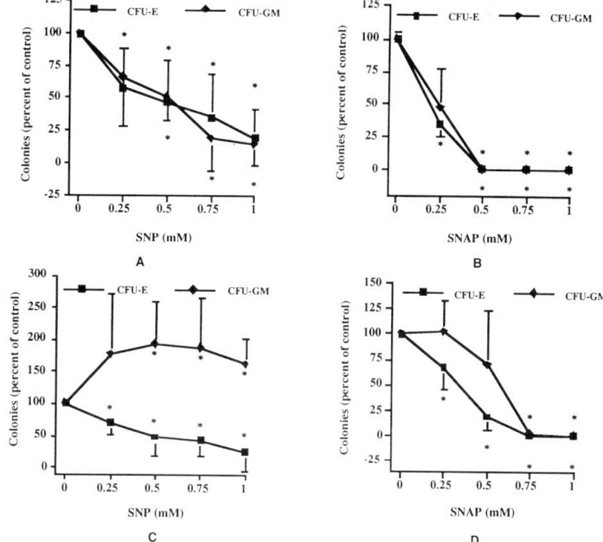

Figure 1: Effect of SNP and SNAP on human mononuclear (A and B) and CD34+ (C and D) bone marrow cells. Asterisks denote statistically significant differences between treatments and controls. (Data reproduced from reference 4).

NO also acts directly on bone marrow progenitors where it has a selective growth inhibitory effect on CFU-E while stimulating the growth of CFU-GM. These effects are cGMP independent since they were not reproduced by the addition of 8-Br-cGMP. Punjabi et al have shown that IFN-γ inhibits the growth of murine bone marrow through the induction of NO production [5] while Maciejewski et al observed that NO induces apoptosis in human bone marrow cells [6]. NO can therefore play a major role affecting the growth and differentiation of bone marrow cells *in vivo*. It could be the final mediator of the cytopenias observed in acute and chronic inflammatory states (such as septic shock and rheumatoid arthritis). It could also play a normal physiologic role modulating the production of bone marrow cells along certain lineages. Therapeutic strategies being developed to modulate NO levels *in vivo* could therefore have important effects on hematopoiesis.

References:

1. Hibbs, J.B., Taintor, R.R., Vavrin, Z., et al. (1990) in Nitric oxide from L-arginine: a bioregulatory system (Moncada, S., Higgs, E.A., ed.), p.189-195, Elsevier Science Publishers B.V., New York.
2. Magrinat, G., Mason, S.N., Shami, P.J., Weinberg, J.B. (1992) Blood **80**,1880-1884.
3. Shami, P.J., Moore, J.O., Gockerman, J.P., Hathorn, J.W., Misukonis, M.A., Weinberg, J.B. (1995). Leuk. Res. **19**, 527-533.
4. Shami, P.J., Weinberg, J.B. (1995). Blood. In press.
5. Punjabi, C.J., Laskin, D.L., Heck, D.E., Laskin, J.D. (1992). J. Immunol. **149**, 2179-2184.
6. Maciejewski, J.P., Selleri, C., Sato, T., Cho, H.J., Keefer, L.K., Nathan, C.F., and Young, N.S. (1995) J. Clin. Invest. **96**,1085-1092.

Nitric Oxide Inhibits Endothelin-Stimulated Protein Synthesis in Cardiac Myocytes and DNA Synthesis in Cardiac Fibroblasts

ANGELINO CALDERONE, CYNTHIA M. THAIK, NOBUYUKI TAKAHASHI, DONNY L. F. CHANG and WILSON S. COLUCCI

Cardiomyopathy Center and Cardiovascular Division, Departments of Medicine, Boston University School of Medicine, and the Boston Veterans Affairs Hospital, Boston, MA, 02118.

Introduction: ET is a potent vasoconstrictor which can stimulate hypertrophy of cardiac myocytes [1, 2]. NO has been shown to inhibit angiotensin and serum-stimulated proliferation in a variety of cell types, including vascular smooth muscle cells [3] and cardiac fibroblasts [4]. It is not known whether NO can inhibit ET-stimulated growth responses in cardiac myocytes or fibroblasts. We therefore tested the hypothesis that NO can inhibit ET-stimulated protein synthesis in cardiac myocytes and DNA synthesis in cardiac fibroblasts, and examined the role of cGMP in mediating these effects.

Methods: Neonatal rat ventricular myocytes were prepared as described by Kasten [5], plated in DMEM with 7% fetal bovine serum at a density of 50-100 cells/mm^2 for 24 hr, and changed to serum-free medium for 24 hr prior to experiments. Neonatal rat ventricular fibroblasts were prepared as described by Sadoshima and Izumo [6]. First and second passage fibroblasts were plated at 50-100 cells/mm^2 in DMEM with 7% fetal bovine serum for 24 hr, and changed to serum-free medium 24-48 hr prior to experiments. Protein and DNA synthesis were assessed by measuring the incorporation of [^{3}H]-leucine (2 μCi/ml) and [^{3}H]-thymidine (2 μCi/ml), respectively. ET (1 nM), SNAP (100 uM), 8-Br-cGMP (1 mM) and/or ANP (100 nM) were added to cultures for 24 hr, concurrent with either [^{3}H]-leucine or [^{3}H]-thymidine. cGMP levels were measured in the presence of isobutyl-methylxanthine (0.2 mM) using a standard radioimmunoassay kit (Biomedical Technologies, Inc., MA).

Results: In cardiac myocytes (Figure 1), the addition of ET for 24 caused a 68 +/- 4 % increase in [^{3}H]-leucine incorporation (n=17). The addition of SNAP for 24 hr had no effect on basal [^{3}H]-leucine incorporation, but caused a 56 +/- 9 % decrease (p< 0.004; n=6) in the response to ET.

In cardiac fibroblasts (Figure 2), the addition of ET for 24 hr caused a 51 +/- 3 % increase in [^{3}H]-thymidine incorporation (n= 13). The addition of SNAP for 24 hr caused a 42 +/- 7 % decrease (p<0.002; n=5) in basal [^{3}H]-thymidine incorporation, and a further 84 +/- 9 % decrease (p<0.005; n=4) in the response to ET.

8-Br-cGMP and ANP had no effect on basal [^{3}H]-leucine or [^{3}H]-thymidine in myocytes or fibroblasts, respectively. 8-Br-cGMP and ANP did not inhibit ET-stimulated[^{3}H]-leucine incorporation in myocytes (Figure 1), but markedly inhibited ET-stimulated [^{3}H]-thymidine incorporation in fibroblasts (Figure 2). SNAP and ANP each increased cellular cGMP content in myocytes (3.6 and 2.4-fold, respectively) and fibroblasts (50 and 4.6-fold, respectively).

Abbreviations used: ANP, atrial natriuretic peptide; 8-Br-cGMP, 8-bromo-cGMP; ET, endothelin; NO, nitric oxide; NOS, nitric oxide synthesis; SNAP, S-nitroso-N-acetyl-D,L-penicillamine.

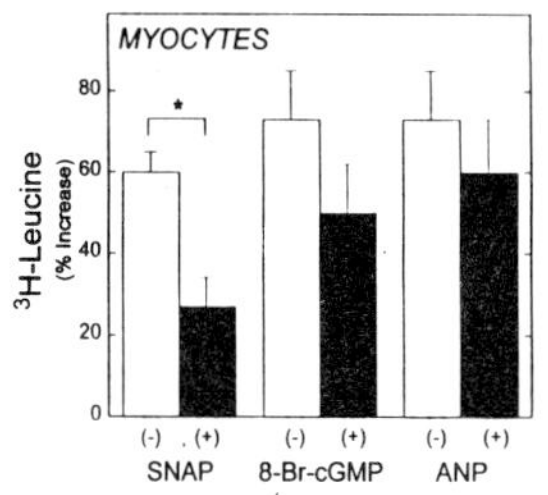

Figure 1.

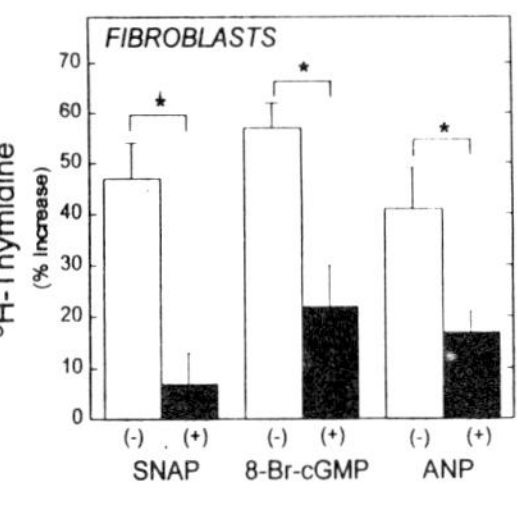

Figure 2.

Discussion: These data show that NO can inhibit ET-stimulated protein synthesis in cardiac myocytes and DNA synthesis in cardiac fibroblasts. Although SNAP increased cGMP levels in both myocytes and fibroblasts, cGMP mimicked the growth-inhibiting effect of SNAP only in fibroblasts. Thus, the inhibition of DNA synthesis in fibroblasts appears to be mediated, at least in part, by cGMP [4]. The failure of cGMP and ANP to mimmick SNAP in myocytes suggests that inhibition of ET-stimulated protein synthesis in myocytes by NO involves primarily a cGMP-independent mechanism [7-9].

ET is expressed by several cell types in the myocardium, including myocytes [10]. Blockade of ET receptors inhibits pressure overload-induced myocardial hypertrophy, suggesting that it may play a role in myocardial remodelling caused by hemodynamic factors [11]. Several cell types in the myocardium can express NOS and have the ability to produce NO under both physiologic and pathologic conditions [12]. The demonstration that NO can inhibit ET-stimulated growth responses in both myocytes and fibroblasts from the heart raises the possibility that NO, acting in an autocrine or paracrine manner, exerts a counter-regulatory effect on myocardial hypertrophy caused by ET, and possibly other hypertrophic stimuli that induce myocardial ET expression [12].

References:

1. Shubeita, D.H., McDonough, P.M., Harris, A.N., et al. (1990) J. Biol. Chem. 265: 20555-20562.
2. Suzuki, T., Hoshi, H., and Mitsui, Y. (1990) FEBS Lett. 268: 149-151.
3. Cao, L., and Gardner, D.G. (1995) Hypertension. 25: 227-234.
4. Garg, U.C., and Hassid, A. J. Clin. Invest. (1989) 83; 1774-1777.
5. Kasten, F.H. (1973) In Tissue Culture Methods and Applications. (Kruse, P.F., and Patterson, M.K., Jr.,eds.) pp. 72-86, Academic Press, New York.
6. Sadoshima, J., and Izumo, S. (1993) Circ. Res. 84:413-423.
7. Curran, R.D., Ferrari, F.K., Kispert, P.H. et al. (1991) FASEB J. 5: 2085-2092.
8. Bolotina, V., Najibi, S., Palacino, J.J., Pagano, P.J., and Cohen, R.A. (1994) Nature 368: 850-853.
9. Brune, B., Dimmeler, S., Molina, Y., Vedia, L., and Lapetia, E.G. (1994) Life Sci. 54: 61-70.
10. Ito, H., Hirata, Y. Adachi, S., et al. (1993) J. Clin. Invest. 92: 398-403.
11. Ito, H., Hiroe, M., Hirata, Y., et al. (1994) Circ. 89: 2198-2203.
12. Balligand, J-L., Unureanu, D., Kelly, R.A., et al. (1993) J. Clin. Invest. 91: 2314-2319.

Antiproliferative and Cytotoxic Effects of Nitric Oxideon Endothelial Cells

Amlan RayChaudhury, Henri Frischer and Asrar B. Malik, Dept. of Pharmacology, Rush Medical College, Chicago, IL 60612.

Nitric oxide (NO), initially identified as the endothelium-derived relaxing factor (EDRF), is produced by endothelial cells (EC), macrophages/monocytes, neutrophils, smooth muscle cells, retinal pigmented epithelial cells and neurons. Besides vasodilation, other functions including microbial and tumor cell lysis and neurotransmission have also been attributed to NO. Investigations into the effects of NO on angiogenesis, the growth of new blood vessels, have yielded conflicting results, with reports suggesting NO can stimulate [1] or inhibit [2] angiogenesis. In addition to its own effects, NO combines with superoxide, endogenously produced by EC and inflammatory cells, with high affinity to form peroxynitrite anion ($ONOO^-$), a strong oxidant implicated in vascular tissue injury [3]. In this study the growth-regulatory effect of NO and $ONOO^-$ on EC was investigated. Proliferation of cultured human pulmonary arterial, bovine pulmonary arterial and microvascular, and rat brain microvascular EC were significantly inhibited in a dose-dependent manner by NO generated by exogenous addition of an NO donor, S-nitroso-N-acetyl-D,L-penicillamine (SNAP). Optimal inhibition was achieved at 200 μM SNAP, which released 60-70 μM NO. This level is comparable to NO levels produced by cytokine or lipopolysaccharide-activated cultured SMC and RPE as reported by other investigators. Addition of nitric oxide synthase inhibitors L-NAME and L-NIO had no effect on baseline EC proliferation, indicating that NO was not produced in sufficient quantities by unstimulated EC to act as an autocrine growth regulator. These results are consistent with those of other investigators who found sodium nitroprusside to inhibit EC proliferation [4,5]. NO also inhibited the mitogenic effect of basic fibroblast growth factor, a potent EC mitogen and angiogenic agent, on bovine pulmonary arterial EC (BPAEC).

Although NO release from SNAP was complete in 24 hr, cells exhibited reduced growth rates for at least 72 hr but then recovered. In addition, the antiproliferative effect of NO was reversed when SNAP was washed off. These observations indicated that the growth inhibitory effect of NO was nontoxic. NO-treated EC also did not internalize trypan blue, an indicator of dead cells. To examine the possibility that NO may have a gradual or partial cytotoxic effect that accounts for its antiproliferative effect and that the reversibility of this effect may be due to compensatory proliferation by unaffected cells, BPAEC were treated with 5 ng/ml transforming growth factor-ß1 (TGF-ß1) for 48 hr to growth arrest the cells, following which TGF-ß1 with and without SNAP were added to the culture. Cells were counted after 4 days. There was no reduction in cell numbers between those samples that received SNAP and those that did not. Since in these cultures proliferation has been suppressed, the absence of cell number reduction indicated that NO does not exert a cytotoxic effect. Control experiments showed (i) SNAP reduced cell numbers in rapidly growing cells (ii) addition of SNAP and TGF-ß1 to rapidly growing cells had an additive antiproliferative effect, showing that TGF-ß1 did not protect cells from the antiproliferative effect of NO in the previous experiment and (iii) SIN-1, a peroxynitrite donor, had a strong cytotoxic effect on TGF-ß-arrested cells, indicating growth arrest per se did not abolish cytotoxic effects on these cells.

Peroxynitrite was highly toxic to sparse EC. 1 mM SIN-1 killed >95% cells in 24-48 hr. In contrast, confluent EC were resistant to this toxic effect. This resistance was density-dependent - and unaffected by cell number or growth arrest. Cells plated at equal cell numbers on 6-well and 24-well plates so that they were sparse in the former and confluent in the latter showed a differential response to the cytotoxic effect of 1 mM SIN-1 (added in equal volume of media to wells of both plates to remove differences in molecules of SIN-1 per cells) - sparse cells were completely killed whereas confluent cells suffered a nontoxic growth-inhibitory effect. Peroxynitrite cytotoxicity was completely abolished by the drug Euk-8 (manufactured by Eukarion, Bedford, MA), which possesses superoxide dismutase and catalase activities. Cells treated with the peroxynitrite donor SIN-1 and Euk-8 showed a growth inhibition comparable to NO-treated cultures. These enzymes are involved in neutralizing superoxide and peroxide in biological systems, indicating that one mechanism through which confluent cells acquire resistance to peroxynitrite toxicity is by upregulating these enzymes or other antioxidants. *In vivo,* production of antioxidative agents by the confluent endothelium may prevent NO from getting converted into $ONOO^-$, enabling NO to carry out its vasodilatory and tumoricidal and microbicidal functions without damaging the intact endothelium. Our recent studies have shown that NO inhibits *in vitro* microvessel formation by BPAEC. Tumorigenesis, inflammation, diabetic retinopathy, and other angiogenesis-dependent pathologic conditions cause activation of inflammatory cells, resulting in the secretion of cytokines such as IL-1 and TNF-α, which can increase NO concentrations in the endothelial microenviroment by upregulating inducible nitric oxide synthase in EC, SMC, and macrophages. The antiproliferative and antiangiogenic actions of NO may therefore serve a protective role in these situations.

REFERENCES

1. Ziche, M., Morbidelli, L., Masini, E., et al. (1994). Nitric oxide mediates angiogenesis in vivo and endothelial cell growth and migration in vitro promoted by Substance P. J. Clin. Invest. **94,** 2036-2044.

2. Pipili-Synetos, E., Sakkoula, E., Haralabopoulos, G., Andriopoulou, P., Peristeris, P. and Maragoudakis, M. E. (1994). Evidence that nitric oxide is an endogenous antiangiogenic mediator. Br. J. Pharmacol. **111,** 892-900.

3. Beckman, J. S., Beckman, T. W., Chen, J., Marshall, P. A. and Freeman, B. A. (1990). Apparent hydroxyl radical formation by peroxynitrite: implications for endothelial injury from nitric oxide and superoxide. PNAS **87,** 1620-1624.

4. Fukuo, K., Inoue, T., Morimoto, S., et al. (1995). Nitric oxide mediates cytotoxicity and basic fibroblast growth factor release in cultured vascular smooth muscle cells. A possible mechanism of neovascularization in atherosclerotic plaques. J. Clin. Invest. **95,** 669-676.

5. Yang, W., Ando, J., Korenaga, R., Toyo-oka, T. and Kamiya, A. (1994). Exogenous nitric oxide inhibits proliferation of cultured vascular endothelial cells. Biochem. Biophys. Res. Commun. **203,** 1160-1167.

The antiproliferative effect of nitric oxide in cultured fibroblasts may be mediated by its direct action on the epidermal growth factor receptor tyrosine kinase activity

ANTONIO VILLALOBO*, CARMEN GÓMEZ# and CARMEN ESTRADA#

*Instituto de Investigaciones Biomédicas, Consejo Superior de Investigaciones Científicas, Arturo Duperier 4, 28029 Madrid, and #Departamento de Fisiología, Facultad de Medicina, Universidad Autónoma de Madrid, Arzobispo Morcillo 4, 28029 Madrid, Spain.

Nitric oxide is able to arrest cell proliferation, and this property is exploited by macrophages to combat tumor cells, intracellular parasites, and other invading microorganisms [1,2]. In addition to the action of nitric oxide against extraneous and abnormal cells, this compound also inhibits the proliferation of normal cells in multicellular organisms. Such effect can be exerted by both cGMP-dependent [3-6] and cGMP-independent [7] pathways. However, nitric oxide also has a mitogenic action at low concentrations. The mitogenic effect coincide with a near-maximal stimulation of the guanylate cyclase activity [8].

Cell proliferation is a complex process that is controlled by multiple systems acting on different pathways and/or stages of the cellular cycle. Nevertheless, a convenient manner to control this process is by acting on growth factor receptors, since the signaling pathways of mitogenic factors could be shut down from their initiation point. Therefore, the goal of this study was to investigate whether the epidermal growth factor receptor (EGFR), a receptor with tyrosine kinase activity that is involved in growth and differentiation processes, is directly modulated by nitric oxide.

A mouse fibroblast cell line (EGFR-T17 fibroblasts), stably transfected with a cDNA encoding the sequence for the human EGFR, and that overexpress this receptor, was used in this study.

Two nitric oxide donors, S-nitroso-N-acetylpenicillamine (SNAP) and 1,1-diethyl-2-hydroxy-2-nitroso-hydrazine (DEA-NO), produce a significant inhibition of cell proliferation as assessed by [methyl-^{3}H]thymidine incorporation assays in fibroblasts grown in the presence of either 10 % (v/v) fetal calf serum or 10 nM EGF plus 1.7 μM insulin. Table I presents the results obtained in a typical experiment with 10 mM DEA-NO.

Table I. Antiproliferative effect of DEA-NO.

	[methyl-^{3}H]thymidine incorporation (cpm)	
Addition	Fetal calf serum	EGF + Insulin
None	56,835 ± 3,660	51,629 ± 1,970
DEA-NO	1,069 ± 284	1,099 ± 37

Each value represents the mean ± S.E.M. of four different determinations.

This inhibition was dependent on the concentration of the nitric oxide donors, and does not appear to be mediated by a rise in the concentration of cGMP, since the antiproliferative effect was not observed upon addition of 8-Br-cGMP, a permeant cGMP analog, or zaprinast, a phosphodiesterase inhibitor (results not shown).

Thereafter, the fibroblasts were permeabilized with 0.1 % (w/v) Triton X-100 to determine *in situ* the effect of SNAP and DEA-NO on the phosphorylation of both the EGFR and the exogenous substrate poly-L-(Glu:Tyr), using [γ-^{32}P]ATP as phosphate donor in the absence and presence of EGF. It was observed that both nitric oxide donors inhibit the tyrosine kinase activity of the EGFR in a concentration dependent manner (results not shown).

Table II shows the inhibitory action of 5 mM DEA-NO on the phosphorylation of the EGFR and poly-L-(Glu:Tyr). As readily observed, this inhibition was more noticeable on the EGF-dependent activity.

Table II. DEA-NO inhibits the phosphorylation of the EGFR and poly-L-(Glu:Tyr).

	Phosphorylation (%)			
	EGFR		poly-L-(Glu:Tyr)	
Addition	-EGF	+EGF	-EGF	+EGF
None	100	214 ± 18 (7)	100	206 ± 32 (5)
DEA-NO	41 ± 6 (10)	66 ± 16 (8)	62 ± 10 (8)	65 ± 11 (8)

Each value represents the mean ± S.E.M. The number of experiments performed is indicated in parenthesis.

Reduced hemoglobin, a nitric oxide scavenger, efficiently prevents the inhibitory action of DEA-NO (results not shown).

Furthermore, we excluded that nitrite peroxide, which could be formed upon reaction of nitric oxide with superoxide anion, were the inhibitory species, since this inhibition was also observed in the presence of superoxide dismutase (results not shown).

All together, these results demonstrate that nitric oxide was indeed involved in the inhibitory process, and that its antiproliferative action was due, at least in part, to its direct action on the EGFR, inhibiting both its transphosphorylation and its tyrosine kinase activity toward exogenous substrates.

Acknowledgments: This work was supported by grants to A.V. from the *Comisión Interministerial de Ciencia y Tecnología* (SAF392/93), *Consejería de Educación y Cultura de la Comunidad Autónoma de Madrid* (AE16-94), and *Dirección General de Investigación Científica y Técnica* (PR94-343), and grant to C.E. from the *Fondo de Investigaciones Sanitarias* (94/0388).

References

1. Schmidt, H.H.H.W. and Walter, U. (1994) Cell *78*, 919-925.
2. Cui, S., Reichner, J.S., Mateo, R.B. and Albina, J.E. (1994) Cancer Res. *54*, 2462-2467.
3. Garg, U. and Hassid, A. (1989) J. Clin. Invest. *83*, 1774-1777.
4. Garg, U.C., Devi, L., Turndorf, M., Goldfrank, L.R. and Bansinath, M. (1992) Brain Res. *592*, 208-212.
5. Yang, W., Ando, J., Korenaga, R., Toyo-oka, T. and Kamiya, A. (1994) Biochem. Biophys. Res. Commun. *203*, 1160-1167.
6. Cornwell, T.L., Arnold, E., Boerth, N.J. and Lincoln, T.M. (1994) Am. J. Physiol. *267*, C1405-C1413.
7. Garg, U. and Hassid, A.(1990) Biochem. Biophys. Res. Commun. *171*, 474-479.
8. O'Connor, K.J., Knowles, R.G. and Patel, K.D. (1991) J. Cardiovasc. Pharmacol. *17 (Suppl. 3)*, S100-S103.

Peroxynitrite, but not nitric oxide or superoxide, causes DNA strand breakage, activates poly-ADP ribosyl synthetase, and depletes cellular energy stores in J774 macrophages and rat aortic smooth muscle cells *in vitro*.

CSABA SZABÓ, BASILIA ZINGARELLI, MICHAEL O'CONNOR and ANDREW L. SALZMAN

Division of Critical Care, Children's Hospital Medical Center, 3333 Burnet Avenue, Cincinnati, OH, 45229

Nitric oxide (NO) and superoxide anion react to form peroxynitrite, a highly toxic oxidant species produced in inflammation and shock [1-2]. Poly (ADP) ribosyltransferase (PARS) is a protein-modifying and nucleotide polymerizing enzyme which is present in the nucleus of eukaryotic cells [3]. PARS covalently attaches ADP-ribose to various proteins, and activation of PARS may rapidly deplete the intracellular concentration of its substrate, NAD^+, slowing the rate of glycolysis, electron transport, and, therefore, ATP formation [3, 4]. Activation of PARS by DNA strand breakage, and the resultant depletion in cellular ATP, appears to be an archetypal response, observed in multiple cell types after a variety of cellular stresses, including exposure to hydrogen peroxide, a generator of hydroxyl radical [4]. Based on the hydroxyl radical-like chemical activities of the degradation products of peroxynitrite (peroxynitrous acid, for instance) [5], we hypothetized that peroxynitrite is able to initiate DNA strand breaks and activate PARS in macrophages and smooth muscle cells.

J774 macrophages and rat aortic smooth muscle (RASM) cells were cultured as described [6]. The formation of peroxynitrite by NO and superoxide donor and peroxynitrite generator compounds was measured by the peroxynitrite-dependent oxidation of dihydrorhodamine 123 to rhodamine [7]. Mitochondrial respiration was measured by the conversion of MTT to formazan [6]. DNA single strand breaks were determined by the alkaline unwinding method [4]. PARS activity was measured using radiolabeled NAD^+ [4]. Cellular NAD^+ levels were measured by a colorimetric assay. ATP levels were measured using HPLC. Cellular glutathione was measured as described [4].

Exposure to peroxynitrite (100 µM - 1 mM) caused a dose-dependent increase in DNA strand-breakage in cultured J774.2 macrophages and in rat aortic smooth muscle cells, which was not affected by PARS inhibition (Fig. 1a). We observed a dose-dependent (up to 5-fold) increase in the activity of PARS in cells exposed to peroxynitrite (Fig. 1b). Interestingly, there was also a basal activity of PARS in J774 cells, but not in the RASM cells (Fig. 1b). ADP-ribosylation was inhibited by the PARS inhibitor 3-aminobenzamide (Fig. 1b). Activation of PARS by peroxynitrite-mediated DNA strand breakage resulted in a significant decrease in intracellular energy stores, as reflected by a decline of intracellular NAD (Fig. 1c) and ATP content (Fig. 1d). 3-aminobenzamide (1 mM) inhibited the loss of NAD and ATP in cells exposed to peroxynitrite (Fig. 1c-d). Peroxynitrite also inhibited mitochondrial respiration (Fig. 1e). The loss of cellular respiration was rapid, peaking 1-3 h after peroxynitrite exposure, and reversible, with substantial recovery after a period of 6-24 h. Inhibitors of PARS, such as 3-aminobenzamide (Fig. 1e), or nicotinamide (not shown), prevented the inhibition of mitochondrial respiration in cells exposed to peroxynitrite. The loss of cellular glutathione, however (Fig. 1f) was unaffected by 3-aminobenzamide.

In contrast, impairment of cellular respiration by the NO donor diethylamine NONOate (up to 1 mM), or by the superoxide generator pyrogallol did not cause DNA strand breakage, and the depression of cell respiration in response to these agents was largely refractory to PARS inhibition (Fig. 2b). These agents did not increase the oxidation of dihydrorhodamine 123, indicating the lack of peroxynitrite production (Fig. 2a). On the other hand, SIN-1, a compound that releases NO as well as superoxide, increased dihydrorhodamine 123 oxidation (Fig. 2a), and the decrease in cell respiration in response to SIN-1 was largely prevented by inhibition of PARS (Fig. 2b).

Thus, our data demonstrate that peroxynitrite cytotoxicity is mediated by DNA strand-breakage and activation of the DNA repair enzyme PARS. Therefore, peroxynitrite, and not NO *per se* appears to be the major mediator of DNA damage and activation of PARS in the cell types studied. Activation of PARS, an energy-consuming futile repair circle, may play a central role in the cellular injury following exposure to exogenous or endogenous peroxynitrite.

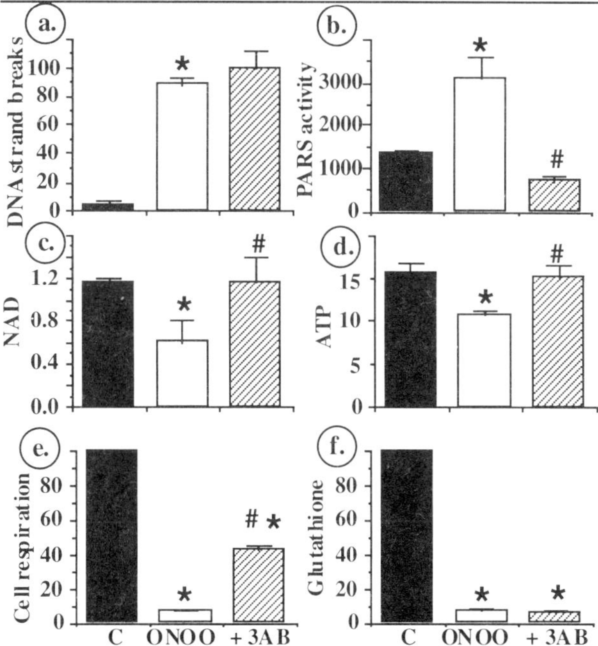

Fig. 1. Effect of peroxynitrite (ONOO, 1 mM) on J774 cells and the effect of 3-aminobenzamide (1 mM). Depicted are changes in (a): DNA strand breaks (% of maximal at 1h); (b): PARS activity (dpm, at 10 min); (c-d): NAD and ATP content (pmoles/million cells at 3h); (e-f): mitochondrial respiration and cellular glutathione (% of unstimulated control at 24h). *,# $p<0.05$ represent significant effect of ONOO or significant protection by 3-AB, respectively.

Abbreviations used: 3-AB, 3-aminobenzamide, HPLC, high pressure liquid chromatography; NO, nitric oxide; ONOO, peroxynitrite; PARS, poly (ADP) ribosyltransferase

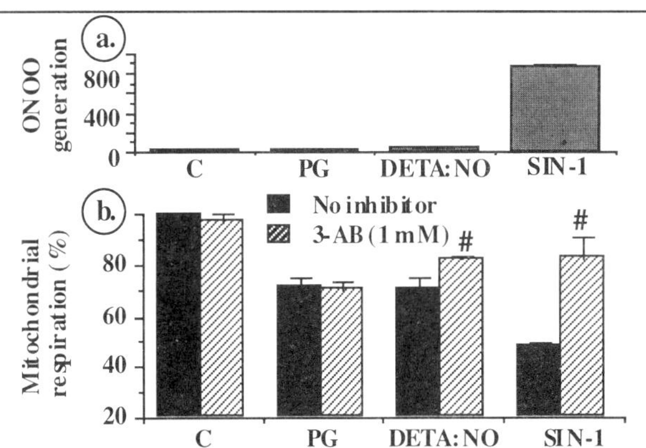

Fig. 2. (a): Peroxynitrite generation in response to pyrogallol (PG); DETA:NO and SIN-1 (1 mM each, for 60 min; expressed as pmoles/min). (b): Effect of 3-aminobenzamide (3-AB, 1 mM) on the decrease in mitochondrial respiration (percentage of control) in RASM cells exposed to pyrogallol, DETA:NO or SIN-1 (1 mM each, for 48h). #$p<0.05$: protection by 3-AB.

References

1. Szabó, C., Salzman, A.L., Ischiropoulos, H. (1995) FEBS Lett. **363**, 235-238
2. Szabó C. (1995) New Horizons, **3**, 3-32
3. Ueda, K. Hayaishi, O. (1985) Ann. Rev. Biochem. **54**, 73 -100
4. Cochrane, C.G. (1991) Molec. Aspects Med. **12**, 137-147
5. Pryor, W., Squadrito, G. (1995) Am. J. Physiol. **268**, L 699-722
6. Szabó C., Mitchell J.A., Gross, S.S., Thiemermann, C, Vane, J.R. (1993) J. Pharmacol. Exp. Ther. **256**, 674-680
7. Kooy, N., Royall, J., Ischiropoulos, H., Beckman, J. (1994) Free Rad. Biol. Med. **16**, 149-155

Apoptosis Induced by Nitric Oxide in Pancreatic β-cells

Hideaki Kaneto, Junichi Fujii, Han Geuk Seo, Keiichiro Suzuki, Masahiro Nakamura, Haruyuki Tatsumi, and Naoyuki Taniguchi

Department of Biochemistry, Osaka University School of Medicine, 2-2 Yamadaoka, Suita, Osaka 565, Japan

Introduction

Nitric oxide (NO) is responsible in part for the cytotoxity of the inflammatory process. NO has been reported to cause DNA damage in several cell types, and NO was found to induce apoptosis in macrophage cells. Insulin-dependent diabetes mellitus (IDDM) is mediated by an autoimmune mechanism or inflammatory process that is characterized by destruction of pancreatic β-cells. Interleukin-1β (IL-1β) has been proposed to play an important role in mediating both destruction and dysfunction of β-cells. The deleterious effects of IL-1β have been proposed to involve generation of reactive oxygen species including NO and inhibition of mitochondrial function. The aim of the present study was to explore DNA damage in β-cell after exposure to exogenous NO or IL-1β-induced intracellular NO generation.

Results

Apoptosis induced by exogenously produced NO in HIT cells

To investigate the mechanism by which DNA is damaged in HIT cell nuclei, two types of NO donor, S-nitroso-N-acetylpenicillamine (SNAP) and sodium nitroprusside (SNP), were employed to generate NO. After incubation of HIT cells with more than 10 μM SNAP or SNP for 6 h, internucleosomal DNA cleavage was observed.

Apoptosis induced by endogeously produced NO in HIT cells

We treated HIT cells with IL-1β to induce inducible nitric oxide synthase (iNOS) mRNA expression. The induction of iNOS mRNA expression was observed after 4 h of incubation with 5 ng/mL IL-1β. NO production increased remarkably after 12 h of incubation and the cell viability decreased concomitantly. Furthermore, internucleosomal DNA cleavage was detected after 12 h of incubation. Addition of 1 mM N^G-monomethyl-L-arginine (NMMA) or 1 mM aminoguanidine suppressed the DNA cleavage. Addition of 5 μg/mL actinomycin D or 50 μM cycloheximide also inhibited the DNA cleavage. These results were correlated with the amount of NO produced in each condition. All these results were consistent with the idea that NO produced in HIT cells by iNOS induced by IL-1β was responsible for DNA cleavage.

Morphological evidence for apoptosis

Scanning electron microscopic examination showed untreated HIT cells were covered with microvilli and ruffles. After 24 h of treatment with 5 ng/mL IL-1β, numbers of microvilli and ruffles were reduced. Instead, extensive surface blebbings and granular protrusions were often observed. Apoptotic bodies with smooth surface were sometimes found to be attached to the cells. Furhtermore, some vacuoles, lots of small vesicles, and nucleus fragmentation with condensed nuclear chromatin were observed by transmission electron microscopy. These morphologicl changes were suppressed by an addition of 1 mM NMMA.

Apoptosis induced by exogenously or endogenously produced NO in isolated rat pancreatic islet cells

We performed the experiments with freshly isolated rat pancreatic islet cells in order to validate the significance of NO in the development of diabetes. After incubation of isolated islets with more than 100 μM SNAP for 6 h, internucleosomal DNA cleavage was observed as seen in β-cell line HIT. DNA cleavage was also observed after 24 h of incubation with 5 ng/mL IL-1β. Addition of 1 mM NMMA or 1 mM aminoguanidine suppressed the DNA cleavage. Thus results obtained by using HIT cells seemed to be applicable to the freshly isolated islet cells.

Discussion

Reactive oxygen species have been demonstrated to cause apoptotic cell death. Recent reports have shown that NO also caused apoptosis in macrophage cells. Activation of pancreatic β-cells by cytokines causes a response similar to that seen with activated macrophages; the cells generate reactive oxygen species including NO. Here NO caused internucleosomal DNA cleavage leading to apoptosis in both HIT cells and islet cells, as was found in macrophages. During inflammatory processes such as pancreatitis, macrophages invade the pancreatic islets and produce NO. The macrophages also produce inflammatory cytokines such as IL-1β, which induce iNOS in β-cells. In the case of inflammation, NO produced by β-cells and macrophages could induce apoptosis of β-cells by causing internucleosomal DNA cleavage in addition to inhibiting mitochondrial function and GAPDH activity, leading to the deterioration of β-cell function and finally diabetes mellitus. NO is known to react synergistically with superoxide anion to form peroxynitrite anion and far-more-reactive hydroxyl radical, which might also play a role in the apoptotic cell death. Although NO-induced internucleosomal DNA cleavage seems to be an important initial step in the destruction of β-cells, it is not clear whether the DNA cleavage itself directly destroys the β-cells.

The effect of nitric oxide and the nitrosonium cation (donated by sodium nitroprusside) on apoptosis in Swiss 3T3 fibroblasts

SHAZIA KHAN*, MARTIN N. HUGHES*, A. DAVID EDWARDS† and HUSEYIN MEHMET†.

*Department of Chemistry, King' s College London, Strand, London WC2R 2LS, UK, and † Department of Paediatrics and Neonatal Medicine, Royal Postgraduate Medical School, Hammersmith Hospital, Du Cane Road, W12 0NN, UK.

INTRODUCTION

Nitric oxide (NO) has a variety of physiological actions: it can behave as a second messenger, a cytoprotective molecule or a cytotoxin [1]. These apparently contradictory actions of NO may reflect its complex cellular chemistry [2]. For example, NO may be converted to its redox-related species, such as the nitrosonium cation (NO^+) and the nitroxyl anion (NO^-).

We hypothesised that these species of NO can act as messenger molecules in apoptotic cell death. Apoptosis is an active physiological process characterised by defined morphological features including membrane blebbing, cell shrinkage, nuclear pyknosis and DNA fragmentation [3]. Although the signal transduction pathways that regulate apoptosis are largely undefined, NO may have a role, since it has previously been shown to induce apoptosis in activated macrophages [4].

Here, we have studied the effects of NO and NO^+ on apoptosis in Swiss 3T3 fibroblasts. These cells are well characterised at the signalling level [5] and can be reversibly arrested in the G_0 phase of the cell cycle, thus removing the heterogeneity of response in an asynchronous population. We also investigated the effects of the NO scavenger HbO_2 and the antioxidants ascorbate and NAC on the apoptotic actions of NO and NO^+.

METHODS

Quiescent Swiss 3T3 fibroblasts were washed with serum-free DMEM at 37 °C and incubated with NO (either in aqueous solution or supplied by the donor SNOG) or NO^+ (supplied by SNP) in 10 % CM (medium removed from quiescent cultures). We ensured that SNP behaved as a source of NO^+ by performing all SNP experiments in the dark. Cells were incubated for 2 or 24 hours in a 10 % CO_2 humidified atmosphere at 37 °C. The 2 hour cultures were washed and incubated for a further 22 hours in 10 % CM. At the end of the experiment cell monolayers were fixed with methanol:glacial acetic acid (3:1 v/v), washed with PBS, and stained with the nuclear stain, PI (4 μg/ml) in the presence of DNAse-free RNAse (100 μg/ml). PI-labelled nuclei were visualised by fluorescence microscopy. For quantitative analysis, a minimum of 200 nuclei were counted in six or more fields for each experimental point. For the scavenging experiments, NO / NO^+ donors were added to the cultures in 10% CM containing the scavengers.

RESULTS

Time course experiments showed that SNP could trigger apoptosis after a 2 hour exposure while NO and SNOG required a 24 hour exposure (Fig. 1).

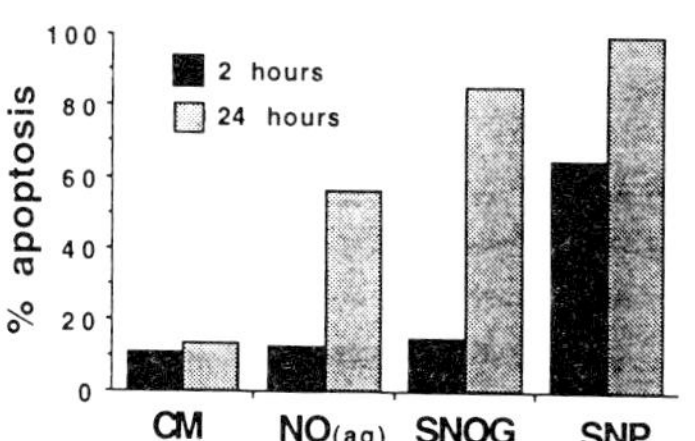

Figure 1 Time dependent effect of NO, SNOG and SNP on apoptosis in Swiss 3T3 fibroblasts. Quiescent cells were treated with 10% CM alone (CM), or with NO (aq), SNOG, or SNP (all at 1 mM) for 2 or 24 hours. At the end of the experiment, apoptosis was assayed by PI staining.

Abbreviations used: SNOG, *S*-nitrosoglutathione; SNP, sodium nitroprusside ($Na_2[Fe(CN)_5NO]$); HbO_2, oxyhaemoglobin; NAC, *N*-acetylcysteine; Dulbecco's modified Eagle's medium, DMEM; PI, propidium iodide; PBS, phosphate-buffered saline; CM, conditioned medium.

Aqueous NO and the NO-donor SNOG both induced apoptosis in Swiss 3T3 fibroblasts in a similar dose-dependent manner after 24 hours (not shown) suggesting that they induce cell death by similar mechanisms. The apoptotic effect of SNP was also dose-dependent after 2 and 24 hours (not shown).

To investigate the apoptotic actions of NO and NO^+ in more detail the NO scavenger HbO_2 and the antioxidants NAC and ascorbate were used. In the presence of HbO_2, SNOG-induced apoptosis was virtually abolished. Similarly, NAC and ascorbate reduced SNOG-induced apoptosis to almost control levels (Fig. 2).In contrast, the apoptotic effect of SNP was potentiated by these scavengers (Fig 2). This effect was dose-dependent (not shown).

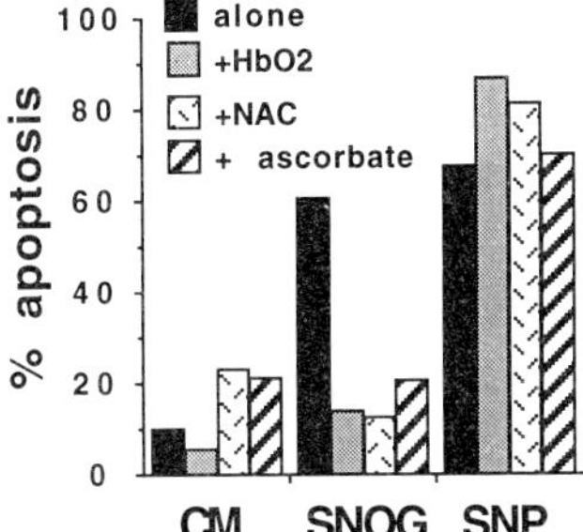

Figure 2 The effect of scavengers on NO and NO^+-induced apoptosis. Quiescent 3T3 cells were incubated with 10% CM (CM), SNOG (0.6 mM) or SNP (1 mM) either alone or with the scavengers HbO_2 (0.1 mM), NAC (10 mM) or ascorbate (0.3 mM). At the end of the experiment, apoptosis was assayed by PI staining.

DISCUSSION

These results demonstrate a clear difference between the cytotoxic effects of NO and NO^+ on quiescent Swiss 3T3 fibroblasts. The results with the NO scavengers reinforce this possibility: Irrespective of the scavenger used, NO-induced apoptosis was inhibited. Although the mechanism of NO-induced killing is unclear, it may be that NO directly induces apoptotic cell death by a free radical mechanism. However, we cannot exclude the possibility that NO competes with intracellular reactive oxygen species for scavengers, thus leading to a net increase in other free radicals.

In contrast the induction of apoptosis by NO^+ was potentiated by all three scavengers used, suggesting that the reaction between SNP and the scavengers leads to the formation of more cytotoxic species.

The contrasting kinetics also suggest that NO and NO^+ induce apoptosis via different mechanisms. The slower induction of apoptosis by NO may be a consequence of its autoxidation and subsequent reaction with a cellular target such as glutathione [5], effectively extending the life time of NO.

On the other hand, the rapid induction of apoptosis by NO^+ is consistent with the direct nitrosation of key thiol-containing proteins in the cell. This damage could then act as a direct trigger of an apoptotic signalling pathway.

We therefore suggest that nitrosation may be an important reaction in the induction of apoptosis by NO and NO^+. Under normal physiological conditions NO is unlikely to induce apoptosis in Swiss 3T3 fibroblasts. However, at high concentrations apoptosis could be induced by nitrosothiols formed by the oxidation of NO to a nitrosating agent such as nitrogen dioxide or dinitrogen trioxide.

In contrast, NO^+ donors can trigger apoptosis by directly nitrosating key thiol groups. Possible targets may include the large number of tyrosine kinase receptors known to contain reactive thiols. Since apoptosis has been suggested as the default pathway for all cells [6] NO and its redox-related species may trigger apoptotic cell death by disrupting the basal signals from such receptors.

ACKNOWLEDGEMENTS

We would like to thank the Engineering and Physical Sciences Research Council (EPSRC) for financial support for S. Khan, and Mary Kozma for technical assistance with tissue culture

REFERENCES

1. Feldman, P.L., Griffith, O.W. and Stuehr, D.J. Chem. Eng. News (1993) **71**, 26-38
2. Stamler, J.S. (1994) Cell **78**, 931-936
3. Martin, S.J., Green, D.R. and Cotter, T.G. (1994) Trends Biochem. Sci. (1994) **19**, 26-30
4. Rozengurt, E. (1986) Science **234** 161-166
5. Wink. D.A., Nims, R.W., Darbyshire, J.F. et al. (1994) Chem. Res. Toxicol. **7**, 519-525
6. Raff, M.C. Nature (1992) **356**, 397-400

Nitric oxide mediated apoptosis is time, dose and cell line dependent.

JOSE F. PONTE[1] and ANNE E. HUOT[1,2]

[1]Departments of Cell and Molecular Biology and [2]Biomedical Technologies, University of Vermont, Burlington, Vermont 05405, U.S.A.

Most animal cells have the ability to self-destruct by activation of an intrinsic cell suicide program when they have become seriously damaged. This death program is associated with characteristic morphological and biochemical changes, and has been termed apoptosis [1]. Cell shrinkage and loss of normal contacts, dense chromatin condensation, cellular budding and fragmentation, and rapid phagocytosis by phagocytes or adjacent cells are the typical events of apoptosis. DNA digestion at internucleosomal sites was recognized in 1980 as a characteristic feature of apoptosis, in which, activation of nucleases degrade the chromosomal DNA first into large (50 to 300 kilobases) and then into very small (180 to 200 base) oligonucleosomal fragments [2].

Apoptosis is a major factor in the development and homeostasis of metazoan animals including: sculpting the developing organism [3], as a mechanism for regulation of cell numbers [4,5], and as a defense mechanism to remove unwanted and potentially dangerous cells, such as self-reactive lymphocytes [6], cells that have been infected by viruses [7,8], and tumor cells [9]. The initiation of apoptosis is tightly regulated.

NO is a lipophilic, diatomic gas that acts as a diffusible messenger and is involved in a wide spectrum of biological activities. We investigated the ability of NO to induce apoptosis in several cell lines, as well as, the kinetics of this induction.

Methods

Cell Culture: p815, L1210, L-M, L929, WEHI-164, EMT-6 and NIH 3T3 cells were maintained in culture medium supplemented with serum.

Gel Electrophoresis: Cells were plated 50,000 and 200,000 per ml in 60 mm dishes and treated with SNP (0-3000 μM) for 1-3 days. Cells were suspended in Tris/glycerol buffer with RNase A and loaded to wells of a 2% agarose gel in which the top fifth was replaced with 1% agarose, 2% SDS and 53 μg/ml proteinase K. Gels ran for 3 hrs, were labeled with ethidium bromide for 1 hr, washed and visualized on an UV light box.

Cellular DNA Fragmentation ELISA: Kits were obtained from Boehringer Mannheim Cat # 1585 045. Cells were labelled with 10 μM BrdU overnight at 37°C, centrifuged at 250g for 10 min and resuspended in medium. The cells were plated at 10,000 cells/100 μl/well in 96 well plates and SNP (0, 100, 500, 1000, or 3000 μM SNP) was added. The plates were incubated for 12-96 hrs at 37°C and then centrifuged at 250g for 10 min. The supernate was assayed for DNA fragments by ELISA. Briefly, samples were transferred to the wells of microtiter plates precoated with anti-DNA antibody, incubated for 90 min at RT, washed, denatured and fixed by microwave irradiation for 5 min. The plate was cooled for 10 min at -20°C, anti-BrdU peroxidase conjugate was added and incubated for an additional 90 min at RT. After washing, immuno-complexed anti-BrdU peroxidase was detected with TMB substrate.

Results/Discussion

Agarose gel electrophoresis demonstrated DNA laddering in p815 cells when exposed to 500 or 1000 μM SNP for 24 hrs. DNA laddering was also observed in L1210 cells exposed to 1000 or 3000 μM SNP for 24 hrs. DNA damage was observed as smears when L-M, L929, WEHI-164 or NIH 3T3 cells were exposed to varying amounts of SNP for 24 to 96 hrs. No evidence of DNA damage was observed in EMT-6 cells under the conditions tested (Data not shown). To further delineate the ability of NO to induce apoptosis, DNA fragmentation ELISA assays were performed. Exposure of each cell line to 1000 μM SNP for 12, 18, 24, 36, 48, 72 or 96 hrs showed a time dependent induction of apoptosis by NO in NIH 3T3, L1210, p815, WEHI-164, L-M and L929 cells (Table 1). DNA fragmentation was not observed in the EMT-6 cells.

Abbreviations used: nitric oxide (NO); sodium nitroprusside (SNP)

Table 1. Time dependence of NO mediated apoptosis. Cells were exposed to 1000 micromolar SNP for the times indicated. The amount of fragmented DNA was quantitated using by ELISA. The amount of fragmented DNA released is expressed as a percent of the control (untreated cells), which was taken to be 100%.

Hours	3T3	L929	WEHI-164	EMT-6	p815	L-M	L1210
12	74	110	106	97	86	103	85
24	95	87	69	79	100	85	180
36	245	71	66	92	96	106	113
48	360	88	112	78	103	76	169
60	230	85	159	93	212	74	136
72	129	77	114	45	88	193	230
84	108	120	105	32	86	135	248
96	74	110	106	97	86	103	85

Dose dependent NO induced apoptosis was demonstrated by exposing cells to varying SNP doses. The time at which the cells are undergoing apoptosis is summarized as follows: NIH 3T3 cells at 24 hrs; L1210 and p815 cells at 36 hrs; WEHI-164 cells at 48 hrs; L-M and L929 cells at 72 hrs; EMT-6 cells at no time point tested (Figure 1). Figure 1 shows that many of the cell lines display peak DNA fragmentation at a particular time followed by a decrease in DNA fragmentation. It is possible that high doses of SNP may lead to necrotic cell death rather than apoptosis.

The data support the conclusion that NO induced apoptosis is time, dose and cell line dependent. Differences between cell lines may, in part, be explained by cell signalling. Further, potential NO targets within the cell may dictate when and if the apoptosis pathway is activated.

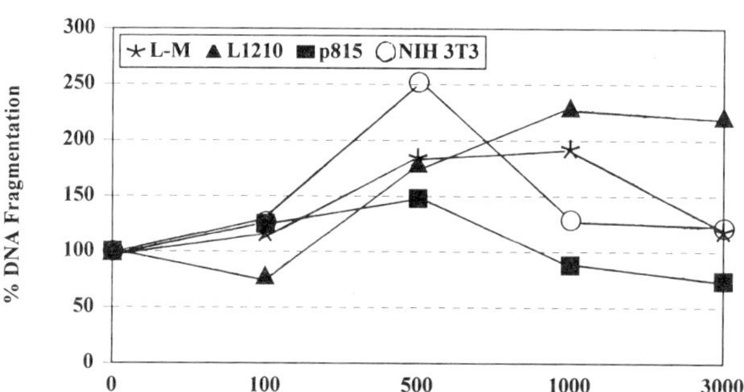

Figure 1. Apoptosis vs. NO concentration. Cells were exposed to varying concentrations of SNP for 72 hours. DNA fragmentation was quantitated by the ELISA assay. Results are expressed as the percent of the control (untreated cells). L929 cells only showed apoptosis at 3000 uM SNP (119%). EMT-6 cells did not show apoptosis at any concentration tested. Although WEHI 164 cells were induced to undergo apoptosis, there was no dose dependent relationship. The data for these three cells lines have been omitted for clarity.

Acknowledgements: This work was funded in part by a grant from the National Science Foundation (Vermont EPScOR).

References

1. Wylie, A.H., Kerr, J.F.R. and Currie, A.R. (1980) Int. Rev. Cytol. **68**, 251-306
2. Roy C., Brown, D., Little, J. et al. (1992) Exp. Cell Res. **200**, 416-424
3. Hammer, S.P. and Mottet, N.K. (1971) J. Cell Sci. **8**, 229-251
4. Raff, M.C. (1992) Nature **356**, 397-400
5. Raff, M.C., Barres, B.A., Burne, J. et al. (1993) Science **262**, 695-700
6. Goldstein, P., Ojcius, D.M. and Young, J.D.E. (1991) Immunol. Rev. **121**, 29-65
7. Vaux, D.L., Haecker, G. and Strasser, A. (1994) Cell **76**, 777-779
8. Debbas, M. and White, E. (1993) Genes Dev. **7**, 546-554
9. Williams, G.T. (1991) Cell **65**, 1097-1098

Exposure of endothelial cells to peroxynitrite inhibits tyrosine phosphorylation and induces apoptosis.

HARRY ISCHIROPOULOS

Institute for Environmental Medicine, and Dept. of Biochemistry and Biophysics, University of Pennsylvania, School of Medicine, Philadelphia, PA. USA. 19104-6068.

Vascular endothelium is a critical target of oxidant-mediated injury in disease processes such as ischemia-reperfusion, organ transplantation, sepsis, inflammation and atherosclerosis. In addition to oxidants derived from the partial reduction of oxygen, recent evidence has implicated nitric oxide-derived reactive species in cellular injury. The nitric oxide-mediated injury to endothelium can be derived by the reaction of nitric oxide with superoxide to form peroxynitrite ($ONOO^-$). Peroxynitrite is a highly reactive species capable of oxidizing tissue lipids, proteins and DNA [1].

Exposure of vascular endothelial cells to peroxynitrite induces delayed death indicative of apoptosis. Cultured pulmonary artery endothelial cells (BPAEC) were exposed to 2 μM/min $ONOO^-$ generated by SIN-1 for two hours. After exposure to peroxynitrite the cells were extensively washed and returned to normal media. The following indices of cellular function were examined: cellular redox activity, mitochondrial membrane potential, cell permeability and DNA fragmentation. The redox activity measured by the mitochondrial dehydrogenase reduction of 3-[4,5-dimethyl-thiazol-2yl]-2,5-diphenyltetrazolium bromide (MTT) declined over time in $ONOO^-$ treated cells reaching nearly undetectable levels approximately 16 hours following treatment. At the same time the mitochondrial membrane of $ONOO^-$ treated cells was depolarized. Mitochondrial membrane depolarization was evident by the loss of mitochondrial trapped rhodamine 123 that was pre-loaded 1 hour prior to $ONOO^-$ exposure. Increased cell permeability detected by the uptake of ethidium homodimer-1 was not apparent until 18 hours following exposure. Lysed control cells contained 1.3$\pm$0.3 arbitrary fluorescence units (AFU)/μg cell protein and NO exposed cells contained 1.5$\pm$0.5 AFU/mg cell protein whereas $ONOO^-$ exposed cells contained 5.6$\pm$0.8 AFU/μg cell protein, indicating a significant increase in membrane permeability. Eighteen hours after exposure staining of with propidium iodide revealed apoptotic nuclei in peroxynitrite treated cells and *in situ* DNA fragmentation was visualized by the TUNEL assay (unmodified T7 DNA polymerase-mediated dATP-biotin nick end labeling). Phase-contrast microscopical examination of $ONOO^-$ treated cells after 24 hours revealed morphological changes indicative of apoptosis; membrane blebbing, cellular shrinkage and degeneration. Similar time course and apoptotic death has been reported in cortical neurons exposed to same or higher concentrations of peroxynitrite [2]. However, a bolus addition of chemically synthesized peroxynitrite failed to induce apoptosis in cultured human umbilical endothelial cells although the same bolus addition of chemically synthesized $ONOO^-$ induced apoptosis in other cell lines [3,4]. The differences may be due to the duration of exposure and the threshold of peroxynitrite cytotoxicity in different cell types.

Peroxynitrite is a relatively long-lived, highly reactive species, capable of oxidizing biomolecules similar to other strong oxidants, (i.e. $\cdot$OH). However, due to the relative slower reaction rates of $ONOO^-$ with target molecules ($\sim 10^5$ M^{-1} s^{-1}), peroxynitrite can diffuse before it reacts with selective cellular targets, such as iron-thiolate and zinc-thiolate-containing proteins [5]. Previous data showed that peroxynitrite can diffuse through the cell and mitochondrial membranes of type II pneumocytes and react with components of the respiratory chain [6]. Peroxynitrite also exhibits selective reactivity with proteins to form nitrotyrosine by adding a nitro group at the ortho position of tyrosine residues [7]. Nitrotyrosine can be considered as a 'footprint' of *in vivo* $ONOO^-$ reactivity and extensive protein tyrosine nitration has been detected in atherosclerosis, sepsis, ischemia-reperfusion and inflammation [8-14].

Tyrosine phosphorylation is compromised in endothelial cells exposed to peroxynitrite. Tyrosine phosphorylation, an important regulator of signal transduction in endothelial cells has been implicated in cellular responses to growth factors, cytokines and calcium ionophores. Since nitrotyrosine is a major product formed by the spontaneous reaction of peroxynitrite with proteins, we examined if tyrosine nitration interferes with tyrosine phosphorylation. This hypothesis was tested: 1) *In vitro* using a peptide that is a modified gastrin analog with one tyrosyl residue at the carboxyl terminal. This peptide is a substrate for the tyrosine kinases *v-abl* ($p43^{v\text{-}abl}$) and *c-src* ($p60^{c\text{-}src}$). These purified commercially available tyrosine kinases are known to phosphorylate a number of cellular proteins after stimulation with growth factors and cytokines. The phosphorylation of the 2.2 kd synthetic peptide by these purified tyrosine kinases ($p43^{v\text{-}abl}$ and $p60^{c\text{-}src}$) was inhibited by 51% (2,037$\pm$190 vs. 1,007$\pm$152 cpm, n=3) when the peptide was first reacted with $ONOO^-$ to form nitrotyrosine. 2) In endothelial cell lysates after exposure of cells to peroxynitrite. BPAEC were exposed to 10 μM/min $ONOO^-$ for 2 hours. After exposure, cell lysates were prepared in the presence of tyrosine phosphatase and protease inhibitors, and 20 μg of cell lysate protein was separated on a 12% SDS-PAGE gels under reduced conditions. Western blot analysis of endothelial cell lysates using two different commercially available monoclonal anti-phosphotyrosine antibodies revealed approximately 7 phosphorylated bands ranging from 50 to 150 kd. The protein bands located approximately at 143, 100, 90, 80 and 70 kd showed the greatest binding to anti-phosphotyrosine antibodies. The antibody binding of these bands was markedly reduced following exposure to $ONOO^-$. Moreover, analysis of endothelial cell lysates using an affinity-purified, polyclonal, anti-nitrotyrosine antibody showed increased nitration of the protein bands which showed reduced binding with the phosphotyrosine antibody after $ONOO^-$ exposure. These results suggest that $ONOO^-$ reduced tyrosine phosphorylation by either nitrating tyrosine residue or by increasing the activity of tyrosine phosphatases. Published data indicated that reactive species by oxidizing cysteine residues of tyrosine kinases and tyrosine phosphatases can directly modulate their activity. The possibility that $ONOO^-$ may directly modulate the activity of tyrosine kinases and phosphatases was supported by the finding that the number and intensity of tyrosine phosphorylated bands was increased in lysates of $ONOO^-$ treated endothelial cells that were pretreated with orthovanadate. These data clearly indicate that peroxynitrite may play a key role in altering signal transduction mediated by tyrosine phosphorylation.

Overall, peroxynitrite is a unique and versatile oxidant capable of altering signal transduction and initiating events that lead to apoptosis.

References

1. Beckman, J.S., Chen, J., Ischiropoulos, H., and J.P. Crow. (1994) Methods in Enzymology **233:** 229-240.
2. Bonfoco, E., Krainc, D., Ankarcrona, M., Nicotera, P. and Lipton, S.A. (1995) Proc. Natl. Acad. Sci. USA **92:** 7162-7166.
3. Estévez, A.G., Radi, R., Barbeito, L., Shin, J.T., Thompson, J.A. and Beckman, J.S. (1995) In Press, J. Neuroscience,
4. Lin, K-T., Xue, J-Y., Nomen, M., Spur, B. and Wong P.Y-K. (1995) J Biol. Chem. **270:** 16487-16490.
5. Castro, L., Rodriguez, M. and Radi, R. (1994) J. Biol. Chem. **269:** 29409-29415.
6. Hu, P., Ischiropoulos, H., Beckman, J.S. and Matalon, S. (1994) Am. J. Physiol. **266:** L628-L634.
7. Ischiropoulos, H., Zhu, L., Chen, J., Tsai, J-H.M., Martin, J.C., Smith, C.D. and J.S. Beckman. (1992) Arch. Biochem. Biophys. **298:** 431-437.
8. Beckman, J.S., Ye, Y-Z., Anderson, P.G., Chen, J, Accavitti, M.A., Tarpey, M.M. and White, C.R. (1994) Biol. Chem. Hoppe-Seyler, **375:** 81-88.
9. Haddad, I.Y., Pataki, G., Hu, P., Galliani, C., Beckman, J.S. and Matalon, S. (1994) J. Clin. Invest. **94:** 2407-2413.
10. Kaur H. and Halliwell, B. (1994) FEBS Lett. **350:** 9-12.
11. Kooy, N.W., Royall, J.A., Ye, Y-Z., Kelly, D.R. and Beckman, J.S. (1995) Am. J. Resp. Crit. Care Med. **151:** 1250-1254.
12. Ma T.T., Ischiropoulos, H. and Brass, C.A. (1995) Gastroenterology **108:** 463-469.
13. Ischiropoulos, H., Al-Mehdi, A.B. and Fisher, A.B. (1995) Am. J. Physiol: Lung Cell.Mole. Physiol. **13:** 158-164.
14. Szabo, C., Salzman, A. L. and Ischiropoulos, H. (1995) FEBS Lett. **363:** 235-238.

Apoptotic cell death and p53 accumulation in response to nitric oxide

BERNHARD BRÜNE, UDO K. MESSMER and
KATRIN SANDAU

University of Erlangen-Nürnberg, Faculty of Medicine,
Department of Medicine IV-Experimental Division,
Loschgestraße 8, 91054 Erlangen, Germany

In several systems the progressive intra- or extracellular generation of nitric oxide (NO) causes apoptosis. Apoptosis is a controlled form of cell selection throughout physiology and pathology. Morphological criteria (chromatin condensation) and biochemical alterations (DNA fragmentation) are routine confirmatory markers. Apoptosis usually occurs in isolated, single cells and does not stimulate persisting tissue changes such as inflammation and scarring that characterize necrosis. NO causes apoptotic cell death in systems like RAW 264.7 macrophages (MQ), a pancreatic ß-cell line (RINm5F), rat thymocytes (TM), and rat mesangial cells (MC). The cellular self-destructive program becomes active in response to induction of the lipopolysaccharide/cytokine responsive NO-synthase or is initiated following cellular exposure to a chemically heterogeneous group of NO-releasing compounds. NO-donors allow the investigation on NO's role in apoptotic signaling pathways without interfering with transducing signals involved in NOS induction.

NO-induced apoptosis

With the use of selected nucleophile/nitric oxide adducts known as NONOates we demonstrate a correlation between the rate of NO-donor decomposition and initiation of typical apoptotic markers like chromatin condensation and DNA fragmentation. NOS-induction causes similar apoptotic alterations in RAW 264.7 macrophages with the demonstration that N^G-monomethyl-L-arginine completely blocks the process. Apoptosis is further characterized by a preserved membrane structure, i.e. no LDH release, retention of intracellular NAD^+ and ATP. Continuation of NAD^+ is compatible with the inability of 3-aminobenzamide, a nonselective inhibitor of poly ADP-ribosylpolymerase, to inhibit NO-related apoptosis in RAW macrophages.

Accumulation of the tumor suppressor p53

In RAW 264.7 and RINm5F cells, we observed accumulation of the tumor suppressor p53 in response to endogenously generated or exogenously supplied nitric oxide. Studies using N^G-monomethyl-L-arginine suggestively trace back NO-intoxication to increased p53 expression. Following DNA damage, an increased demand for DNA repair is most likely associated with p53 accumulation. The latter may have the dual role of causing cell cycle arrest via downstream genes and to stimulate DNA repair directly. However, intranuclear p53 accumulation following DNA damage can also be part of the signaling leading to apoptosis either directly acting on the DNA or again by blocking the cell cycle. Considering the role of p53, our results are consistent with a mechanism whereby NO-induced DNA damage results in p53 expression and apoptosis. Accumulation of p53 with its potential role in regulating the cell cycle clock and initiating apoptosis, may have the dual role of signaling either growth arrest or death. Negative and positive modulators of apoptosis may then determine cellular susceptibility to the potentially damaging molecule.

Redox signaling during NO-induced apoptosis

In rat TH GSNO, depending on the concentration, causes negative and positive regulation of apoptosis. GSNO below 0.6 mM produces DNA-laddering. Higher doses of the NO-donor (1-2 mM) suppress TH apoptosis initiated by the classical agonist dexamethasone. The inhibitory potency is in analogy to the action of the thiol modulating compounds N-ethylmaleimide or 1-chloro-2,4-dinitrobenzene. Likely, inhibition of apoptosis by GSNO refers to thiol modification of critical proteins in response to NO-treatment.

With MC we especially probed for a potentially damaging role of peroxynitrite. The redox cycling quinone, 2,3-dimethoxy-1,4-naphthoquinone (DMNQ) induces apoptosis comparable to GSNO. We assayed MC apoptotic cell death after 24 h under conditions of single and costimulatory exposures to DMNQ or GSNO.

Table 1. DNA fragmentation in response to DMNQ and GSNO
Each values represents the mean ± S.D. of for 4 determinations.

	control	DMNQ (10 μM)	GSNO (2 mM)	DNMQ plus GSNO
% DNA Fragmentation	4 ± 2 %	22 ± 3 %	34 ± 3 %	8 ± 2 %

Under conditions of enforced superoxide and NO production, probably resulting in enlarged $ONOO^-$ generation, we observed decreased rather than increased apoptotic DNA fragmentation. Suggestively peroxynitrite is not a major cause for MC apoptosis. The cellular redox environment, apoptosis regulating proteins, and defense mechanisms may determine how cells respond towards potentially toxic molecules like nitric oxide. Cellular self-destruction known as apoptosis together with p53 accumulation, at least in part, may be considered a patho-physiological NO-signaling system.

Acknowledgements to the Deutsche Forschungsgemeinschaft and the European Community.

Bac-to-Bac baculovirus system: An alternate method to express complex enzymes such as inducible nitric oxide synthase that require multiple co-factors

INDRAVADAN R. PATEL[1], PRANAV VYAS[2], MUKUNDAN ATTUR[2], STEVEN B. ABRAMSON[2,3], and ASHOK R. AMIN[2,3,4,*]

[1]Department of Biochemistry, Glaxo, Inc., Research Triangle Park, NC 27709 USA; [2]Department of Rheumatology, Hospital for Joint Diseases, New York, NY 10003 USA; Departments of [3]Medicine and [4]Pathology, New York University Medical Center, New York, NY 10016 USA.

Murine inducible nitric oxide synthase (miNOS) is a complex enzyme and requires various co-factors including Ca^{++}, calmodulin (CaM), NADPH and FAD for its catalytic activity [1]. The rapid development of our understanding of the biological role/action of NO has to a large degree been paralleled by our understanding of the enzyme responsible for the synthesis of NO. The relatively rapid advancement of NOS enzymology is primarily due to its multidisciplinary importance in various biological sciences, including neurology, immunology, host-defense and medicine. Furthermore, the structure/function question has crossed over several well-established enzymatic problems. In the present study we report the cloning and expression of a functional murine iNOS in baculovirus-infected insect cells using the Bac-to-Bac system, which can be scaled up to purify sufficient quantities of recombinant protein for functional enzymology studies of this complex enzyme in the future.

A full-length miNOS cDNA was cloned and expressed using a novel method of expression called Bac-to-Bac baculovirus system. The cDNA of miNOS was first cloned in the pFASTBAC-1 donor plasmid (as shown in Fig. 1), and subsequently recombinant viral DNA (composite bacmid DNA) was generated in *E. coli* by Tn7 transposition of the target DNA into the specific site in baculovirus genome constituted in a plasmid. *Spodoptera frugiperda* (Sf-9) cells were transfected with this composite bacmid DNA containing the iNOS cDNA. This resulted in a functional miNOS in 2-3 weeks. Western blot analysis revealed the presence of a 133 kD iNOS band at 24, 48 and 72 h after transfection of Sf-9 cells. Enzyme activity was also shown by the presence of nitrite accumulated in the medium for up to 72 h after transfection, after which the cells appeared granular and necrotic. Monitoring the specific activity of the enzyme (24-72 h) by L-arginine-to-L-citrulline conversion assay revealed that the activity of the enzyme (in cell-free extracts) could be detected within 24 h post transfection, although nitrite accumulation ($\leq 0.1\ \mu M$) could not be detected at this time period. The enzyme activity in crude enzyme could be inhibited by competitive iNOS inhibitors in cell-free extracts. The expression of the enzyme activity was congruent with the presence of iNOS as observed by Western blot analysis.

The strategy described in this study permits rapid and efficient generation of recombinant virus by site-specific transposition in *E. coli,* and offers several advantages over more traditional methods for generation of recombinant viruses and the expression of recombinant protein [2-5]: **a)** The time required to generate the recombinant virus is ~7-9 days, as compared to 3-4 weeks using the conventional method. **b)** The frequency of generating recombinant virus is almost 100%, as compared to 1-30% with standard methods. **c)** There is almost no background of the parent virus in the Bac-to-Bac system, since a single bacterial colony can represent a recombinant virus and can be used to transfect susceptible insect cells. **d)** This strategy eliminates the labor-intensive process of picking plaques, and eliminates 2-3 rounds of plaque purification. **e)** The initial titer of the virus is as high as 5×10^7 pfu/ml, saving ~4 days, as compared to other methods which require an additional transfection in the T-25 flask to generate an equivalent amount of virus. Thus, this method generally saves 3-4 weeks of additional work required by the conventional method. Although there are a couple of methodologies described in the literature which are similar to Bac-to-Bac [6, 7], the ease and frequency of getting recombinant viruses are far greater in the Bac-to-Bac system. The most important step in the Bac-to-Bac method is the selection of white colonies from among a large number of blue and light-blue colonies. To facilitate this, we usually incubate our plates at 4°C for 4-6 h, after the original incubation at 37°C for 36-48 h [7], which results in efficient blue-white color development.

This strategy is convenient and reliable for expressing complex and multi-factor requiring enzymes such as iNOS for enzymology studies.

*To whom correspondence should be addressed at Rheumatology Research Lab, Room 1600, Hospital for Joint Diseases, 301 East 17th Street, New York, NY 10003. Tel: (212) 598-6537. Fax: (212) 598-6168.

The authors would like to thank Dr. James Cunningham (Harvard Medical School) for providing the iNOS cDNA, and Ann Rupel for preparing and editing the manuscript.

REFERENCES

1. Bredt DS, Hwang PM, Glatt CE, Lowenstein C, Reed RR, Snyder SH (1991) *Nature* **351,** 714-718.
2. O'Reilly DR, Miller LK, Luckow V (1992) *Baculovirus Expression Vectors: A Laboratory Manual.* W.H. Freeman and Company, New York.
3. Kitts PA, Ayres MD, Possee RD (1990) *Nucleic Acids Res.* **18,** 5667-5672.
4. Sewall A, Srivastava N (1991) *The Digest* **4,** 1.
5. Hartig PC, Cardon MC (1992) *J. Virol. Methods* **38,** 61-70.
6. Peakman TC, Harris RA, Gewert DR (1992) *Nucleic Acids Res.* **20,** 495-500.
7. Patel G, Nasmyth K, Jones N (1992) *Nucleic Acids Res.* **20,** 97-104.

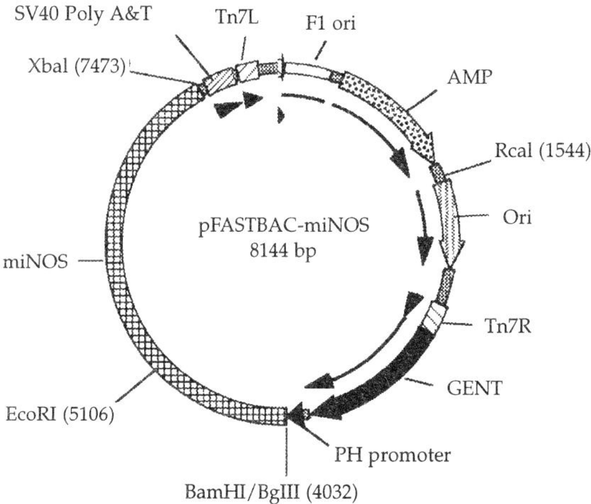

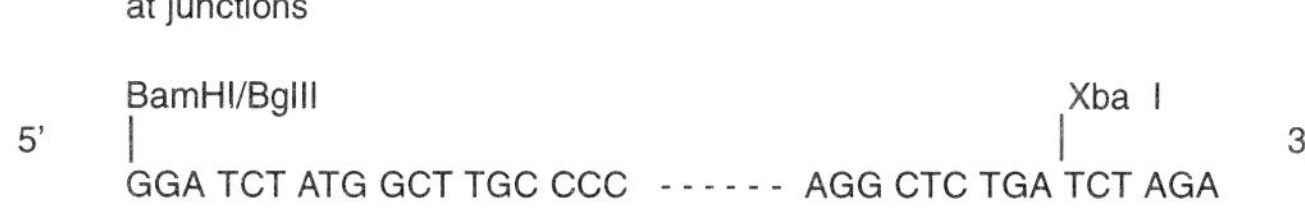

Figure 1. Schematic representation of a functional murine iNOS cDNA cloned in pFASTBAC-1.

Identification of cAMP responsiveness of the promoter region of GTP cyclohydrolase I

KAZUYUKI HATAKEYAMA, TOSHIE YONEYAMA, SEIJI ISHII#, and DAVID A. GELLER

Department of Surgery, University of Pittsburgh, Pittsburgh, PA 15261, #Department of Biochemistry, Osaka Medical College, Takatsuki, Osaka 569, Japan.

Introduction

GTP cyclohydrolase I is the first and rate-limiting enzyme for the biosynthesis of tetrahydrobiopterin[1], an essential cofactor of nitric oxide synthases. We have studied the enzymology and molecular biology of GTP cyclohydrolase I [1,2], revealing that the enzyme is an allosteric enzyme composed of decameric subunits. We have recently found that GTP cyclohydrolase I activity is feedback-regulated by the endproduct BH_4 and also by phenylalanine, the substrate for phenylalanine hydroxylase that requires BH_4 as an essential cofactor as nitric oxide synthases do [3]. The m RNA specific for GTP cyclohydrolase I is constitutively expressed in some types of cells such as the liver cells and neuronal cells and also known to be coinduced along with inducible nitric oxide synthase by cytokines in macrophages, hepatocytes, fibroblasts and smooth muscle cells. To begin to understand the transcriptional regulation of the GTP cyclohydrolase I gene, we have cloned and sequenced the promoter region of the GTP cylohydrolase I gene.

Results and Discussion

Analysis of the nucleotide sequence of 1.2 kb of the rat GTP cyclohydrolase I 5'-flanking region revealed that the promoter contained a TATA-like box, AATA, 10-base pairs upstream from a putative transcription start site. We have identified several potential regulatory elements such as cAMP response element, nuclear factor kappa beta, interferon-gamma response element, SP-1, AP-2, Myb and C/EBP. To examine the promoter activity the 5'-flanking fragment of GTP cyclohydrolase I gene was placed upstream of the luciferase reporter gene and transfected by lipofectin into the rat primary hepatocytes. The construct exhibited promoter activity achieving a level of expression 10% that of the Rous sarcoma virus promoter in primary hepatocytes. In rat PC-12 pheochromocytoma cells the construct also exhibited promoter activity achieving a level of 1% that of the virus promoter and the activity was further stimulated in the presence of dibutyl cAMP at a concentration of 1 mM. However, the construct exhibited little promoter activity in NIH 3T3 cells, which do not contain the GTP cyclohydrolase I activity. Thus the 1.2 kb 5'-flanking region of GTP cyclohydrolase I gene confers the promoter activity in a cell-specific manner and contain a cAMP responsive element. The deletion experiments further showed that the cAMP responsive element is located between 0.4 and 0.2 kb upsteam from the translational start site, which region actually contained the consensus nucleotide sequence motif for cAMP resposiveness identified (the lengths of the deleted clones were: the clone *del*3, 0.7 kb; the clone *del*12, 0.4 kb; the clone *del*8, 0.2 kb) (Figure).

We showed that the 5' promoter region of GTP cyclohydrolase I is responsible to cAMP stimulation using a reporter vector system. This result is consistent with our unpublished observation that dibutyl cAMP increased the level of mRNA specific for GTP cyclohydrolase I and its enzyme activity several-fold in PC-12 cells. In smooth muscle cells, it has been

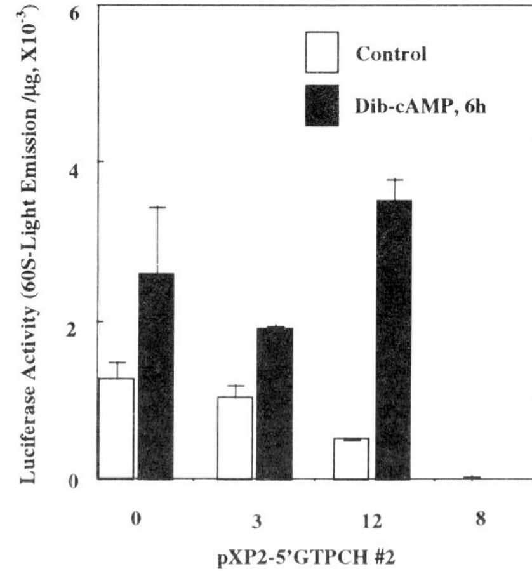

shown that interleukin-1ß induce the elevation of mRNA specific for GTP cyclohydrolase I through increasing the levels of cAMP. Thus cAMP plays an important role in GTP cyclohydrolase I induction and thus in BH_4 biosynthesis.

References

1. Hatakeyama, K., Harada, T., Suzuki, S., Watanabe, Y., and Kagamiyama, H. Purification and characterization of rat liver GTP cyclohydrolase I. Cooperative binding of GTP to the enzyme. J. Biol. Chem. (1989) **264**, 21660-21664.

2. Hatakeyama, K, Inoue, Y, Harada, T., and Kagamiyama,H. Cloning and sequencing of cDNA encoding rat GTP cyclohydrolase I. The first enzyme of the tetrahydrobiopterin biosynthetic pathway. J. Biol. Chem. (1991) **266**, 765-769.

3. Harada, T., Kagamiyama, H. and Hatakeyama, K. Feedback regulation mechanisms for the control of GTP cyclohydrolase I activity. (1993) Science **260**,1507-1510.

4. Scott-Burden, T., Elizondo, E., Ge, T., Boulanger, C.M. and Vanhoutte, P.M. Simultaneous activation of adenylyl cyclase and protein kinase C induces production of nitric oxide by vascular smooth muscle cells. (1994) J. Pharm. Exp. Therap. **46**, 274-282.

Functional analysis of the 5'-flanking region of the chicken inducible nitric oxide synthase gene.

ARTHUR W. LIN and CHARLES C. MCCORMICK

Division of Nutritional Sciences, Cornell University, Ithaca, NY 14853.

Nitric Oxide (NO) is a simple molecule which functions as a neurotransmitter, endothelial relaxing factor, and a cytotoxic and tumorcidal agent. It is synthesized by the enzyme nitric oxide synthase (NOS). There are, at least, three genetically distinct types of NOS: type 1 (bNOS), a constitutive form which was initially identified in neurons, type 2 (iNOS), an inducible form from macrophages, and type 3 (eNOS), a constitutive form which was initially identified in endothelium. All isoforms utilize the amino acid arginine and molecular oxygen as substrates and require NADPH, BH_4, FAD, and FMN as cofactors (Marletta, 1993). The two constitutive forms are activated by and dependent on changes in intracellular calcium (Nathan, 1992), whereas the inducible isoform is calcium independent apparently because calmodulin is a tightly bound subunit of the iNOS. (Cho *et al.* 1992).

Although there has been intense interest in the biological functions of NO and the regulation of NOS in humans and rodents (Nathan *et al.* 1994), little is known of NOS in any non-mammalian system. In an effort to study NOS regulation and to evaluate the evolution of NOS, we have cloned the first nonmammalian NOS cDNA from a chicken macrophage cell line. We found that the production of NO and the accumulation of iNOS mRNA in chicken macrophages is induced by LPS. This suggested that the regulation of chicken iNOS activity was at the transcriptional level. In the present study, we report the cloning of a portion of the chicken iNOS 5' flanking region and its functional analysis. The transcriptional initiation site of chicken iNOS was identified to be a ***G*** nucleotide by primer extension. There is a putative TATA box, 32 bp upstream of the transcriptional initiation site which is similar to mouse (30 bp) and human (30 bp) iNOS gene (Xie *et al.* 1993; Chartrain *et al.* 1994; Lowenstein *et al.* 1993). The 5' flanking region, 3,145 bp, of chicken iNOS gene contains full LPS inducibility. To analyze the important cis-regulatory elements which are regulated by LPS, we constructed deletion mutants of the upstream sequence of chick iNOS gene and ligated them to a luciferase reporter gene. Mutant clones were transiently transfected into a chicken macrophage cell line, HD11. The major finding of our study was that a 267 bp upstream region contained full LPS inducibility. In this LPS responsive region, several transcriptional factors consensus binding elements were identified; such as, NF-κB (Sen and Balitmore, 1986), PEA1, PEA3, and C/EBP (Williams *et al.* 1991). By employing an NF-κB inhibitor, PDTC, we demonstrated that NF-κB is involved in the induction of chicken iNOS gene. Further studies, such as footprinting and mutagenesis, are needed to further characterize the promoter of the chicken iNOS gene, and to understand how chicken iNOS gene is regulated.

References

Chartrain, N.A., Geller, D.A., Koty, P.P., Sitrin, N.F., Nussler, A.K., Hoffman, E.P., Billiar, T.R., Hutchinson, N.I. & Mudgett, J.S. (1994) J. Biol. Chem. **269**, 6765-6772

Cho, H.J., Xie, Q.W., Calaycay, J., Mumford, R.A., Swiderek, K.M., Lee, T.D. & Nathan, C. (1992) J. Exp. Med. **176**, 599-604

Lowenstein, C.J., Alley, E.W., Raval, P., Snowman, A.M., Snyder, S.H., Russell, S.W. & Murphy, W.J. (1993) Proc. Natl. Acad. Sci. USA **90**, 9730-9734

Marletta, M.A. (1993) J. Biol. Chem. **268**, 12231-12234

Nathan, C. (1992) FASEB. J. **6**, 3051-3064

Nathan, C. & Xie, Q. (1994) J. Biol. Chem. **269**, 13725-13728

Sen, R., and Baltimore, D. (1986) Cell **46**, 705-716

Williams, S.C., Cantwell, C.A., and Johnson, P.F. (1991) Gene & Development **5**, 1553-1567

Xie, Q., Whisnant, R. & Nathan, C. (1993) J. Exp. Med. **177**, 1779-1784

Effects of nitric oxide on gene expression in *Escherichia coli* and *Saccharomyces cerevisiae*

CHRISTOPHER T. PRIVALLE, and TERESA KENG

Apex Bioscience, Inc., P. O. Box 12847, Research Triangle Park, NC 27709-2847, USA

Nitric oxide (NO), a biologically generated free radical, has cytotoxic and genotoxic effects as a consequence of its ability to damage proteins and DNA. It has been shown to destroy iron sulfur centers of key enzymes such as aconitase and to inhibit the activity of heme-containing proteins. In addition, nitrosative deamination of DNA bases results in the production of mutagenic products. NO and superoxide can also react in a diffusion limited manner to form peroxynitrite which can directly oxidize tissue sulfhydryls and, via a metal catalyzed reaction, form a nitrosonium-like species capable of nitrating proteins. These destructive properties of NO and derivatives of NO are utilized by activated macrophages to kill invading microorganisms. We have investigated the effects of NO on gene expression in microorganisms. In particular, we studied the *soxRS* transcriptional regulatory system in the bacterium *Escherichia coli* and the expression of HAP1-regulated genes in the yeast *Saccharomyces cerevisiae*. NO induces expression of genes in the *soxRS* regulon and inhibits expression of genes under the control of the HAP1 activator.

The *soxRS* locus in *E. coli* has been shown to encode a two-stage control system that regulates , in a positive manner, the global response to superoxide. An intracellular redox signal, such as superoxide, converts the Fe/S-containing SoxR protein into a transcriptional activator of the *soxS* gene [1]. The SoxS protein then activates transcription of approximately ten genes in the *soxRS* regulon, including *sodA* which encodes MnSOD, and genes encoding glucose-6-phosphate dehydrogenase, endonuclease IV, and fumarase C.

The effects of NO on expression of genes in the *soxRS* regulon was investigated using *E. coli* strains bearing either a lysogenized *ΔsoxRsoxS'::lacZ* λ phage (TN530) or a *sodA::lacZ* fusion (QC1709). NO inhibited growth of both strains, with maximal inhibition observed at 100 µM NO. Exposure of strain TN530 to 50 µM NO resulted in a four-fold induction of expression of the *soxS'::lacZ* fusion (Fig. 1). Exposure of strain QC1709 to the same concentration of NO caused a seven-fold induction of expression of *sodA::lacZ*. These results indicate that NO is able to induce expression of *soxS* which in turn activates expression of genes in the *soxRS* regulon, including *sodA*. The NO mediated induction of MnSOD expression was confirmed by immunoblot analysis of bacterial extracts using anti-MnSOD antibodies. This induction by NO was dependent upon an intact *soxRS* locus as no MnSOD protein could be detected in a *ΔsoxRS* strain.

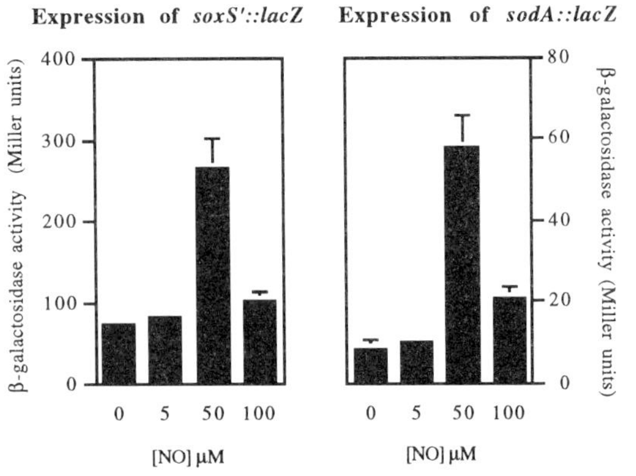

Fig. 1. Induction of *soxS'::lacZ* and *sodA::lacZ* by nitric oxide. *E. coli* strains TN530 (*soxS'::lacZ*) or QC1709 (*sodA::lacZ*) were grown anaerobically for 1.25 h at 37°C in TSY medium. NO was added from an anaerobic NO-saturated aqueous solution and cells were grown an additional 2 h. β-galactosidase activity was determined in cells permeabilized with SDS/chloroform.

In addition to effects on transcription of *sodA*, NO appears to have post-transcriptional effects on MnSOD biosynthesis. The level of MnSOD activity is increased in cells exposed to NO. Native immunoblot analysis also indicated that following anaerobic NO exposure, Mn_2-MnSOD was present. We have previously demonstrated that anaerobic electron sinks such as nitrate, a variety of oxidants, and diamide, a thiol oxidant, elicit anaerobic biosynthesis of a largely inactive, incorrectly substituted MnSOD [2]. Thus, these results suggest that NO not only acts as a *soxRS*-dependent transcriptional inducer of MnSOD polypeptide biosynthesis, it also facilitates incorporation of Mn into the active site of nascent MnSOD polypeptide, resulting in an active form of the enzyme.

In addition to examining the effects of NO on gene expression in *E. coli*, we also studied the effects of NO on the activity of the HAP1 activator protein in *S. cerevisiae*. HAP1 is required for the expression of many genes encoding proteins involved in electron transport and respiration, including *CYC1* which encodes iso-1-cytochrome *c*. Expression of these genes is heme-dependent and the DNA binding activity as well as the transcriptional activation activity of HAP1 is enhanced by heme [3]. A heme-independent allele of *HAP1*, *hap1-43*, encodes a protein that is able to activate transcription of *CYC1* even in the absence of heme [4]. These strains of *S. cerevisiae* appear to be more sensitive to NO than the bacterial strains; growth of both the *HAP1* and the *hap1-43* strains is inhibited by 50 µM NO. The presence of 10 µM NO in the medium reduced expression of the *CYC1-lacZ* fusion protein 3.5-fold in the *HAP1* strain and 2.5-fold in the *hap1-43* strain (Fig. 2). Interestingly, 20 µM NO decreased expression of the fusion protein 18-fold in the wild-type strain but only 4.6-fold in the *hap1-43* strain. In the absence of heme, expression of *CYC1-lacZ* in the *hap1-43* strain is not affected by NO and remains at approximately 10 units. These observations suggest that NO may affect expression of *CYC1* by interfering with the ability of the HAP1 activator to interact with heme. Under conditions when heme is not produced, expression of *CYC1* is not significantly affected.

The induction of the *soxRS* regulon in *E. coli* by NO has potentially important implications in infectious diseases. The stress response provided by the *soxRS* locus may play a role in bacterial defense against host-mediated cytotoxic protective mechanisms and may be an important component of bacterial virulence [5]. These results, coupled with the demonstrated transcriptional effects in yeast, indicate that NO, and possibly products derived from NO, may serve as biological signaling molecules.

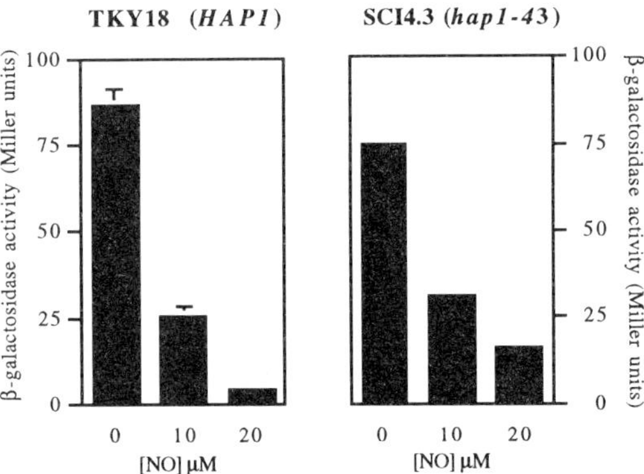

Fig. 2. Effect of NO on expression of *CYC1-lacZ* in *HAP1* and *hap1-43* yeast strains. Yeast strains TKY18 (*HAP1*) and SCI4.3 (*hap1-43*) transformed with *CYC1-lacZ*-bearing plasmids were grown in synthetic glucose medium in the presence of the indicated concentrations of NO. The levels of β-galactosidase activity were determined following permeabilization with SDS/chloroform.

REFERENCES

1. Hidalgo, E., and Demple, B. (1994) EMBO J. **13**, 138-146.
2. Privalle, C. T., and Fridovich, I. (1992) J. Biol. Chem. **267**, 9140-9145.
3. Zhang, L., and Guarente, L. (1994) J. Biol. Chem. **269**, 14643-14647.
4. Ushinsky, S., and Keng, T. (1994) Genetics **136**, 819-831.
5. Nunoshiba, T., DeRojas-Walker, T., Tannenbaum, S. R., Demple B. (1995) Infect. Immun. **63**, 794-798.

Regulation of nitric oxide synthase isoform gene expression in human neuronal cells

T. OGURA, Y. KURASHIMA AND H. ESUMI

Investigative Treatment Division, National Cancer Center Research Institute, East, 5-1, Kashiwanoha 6-chome, Kashiwa, Chiba, 277, Japan

Introduction

In the nervous system, nitric oxide (NO) is believed to be involved in many neuronal functions (1,2). It has been reported that only neuronal NO synthase (n-NOS) isoform is expressed constitutively and its enzymatic activity is strictly regulated by intra-cellular calcium concentration in neuronal cells (2). However, little is known about the regulation of n-NOS gene expression. In this paper, we described the regulation of n-NOS gene and induction of inducible NOS (iNOS) gene in neuronal cells.

Materials and Methods

Cells: Human neuroblastoma cell lines, TGW-I-nu and NB-39-nu, were cultured in alpha-MEM and RPMI-1640 containing 10% fetal calf serum at 37°C under 5% CO_2 in air. TGW-I-nu was treated with 5 µM *trans*-retinoic acid (RA). NB-39-nu was treated with 10 µg/ml LPS, 100U/ml IFN-γ, 500 U/ml TNF-α, and/or 10 ng/ml IL-1β.

Northern blot analysis: Total RNA was isolated from cells using guanidine thiocyanate extraction method (3). Twenty µg of the RNA was separated on an 0.8% or 1% agarose gel containing 6% formaldehyde, which was then blotted onto a membrane filter. To detect n-NOS, neurofilament-M (NF-M) and iNOS mRNAs, a 0.7 kb fragment of human n-NOS cDNA (4), a 1.1 kb Eco RI fragment of NF-M cDNA (5), and a 0.5 kb fragment of human iNOS cDNA (6) were used as probes, respectively. Each mRNA was detected by hybridization as described previously (4).

Result and Discussion

n-NOS expression in neuronal cell differentiation

Human neuroblastoma cell line, TGW-I-nu, was able to differentiate with growth arrest and morphological change following treatment with RA. Upon RA-induced neuronal differentiation, expression of n-NOS increased from day 1 and the level of mRNA expression 7 days after treatment with RA was about 50 times higher than the level in untreated cells (Fig. 1). However iNOS mRNA could not be detected in TGW-I-nu following RA treatment. A time-dependent accumulation of NF-M mRNA, a marker of neuronal cell differentiation, was also observed, and a striking similarity between the patterns of accumulation of n-NOS and NF-M mRNA was observed (Fig. 1). Moreover the accumulation of n-NOS mRNA was also observed in the

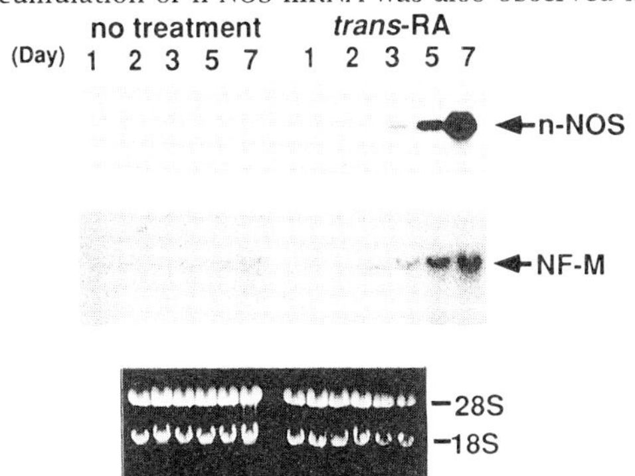

Fig. 1. **Expression of n-NOS mRNA in TGW-I-nu treated with retinoic acid.** Cells were treated with or without 5 µM retinoic acid. At the time indicated, 20 µg of total RNA was analyzed by northen blotting with n-NOS (upper) and NF-M (middle) cDNA as probes. The amount of ribosomal RNA is shown at the bottom.

Abbreviations used: NO, nitric oxide; n-NOS, neuronal nitric oxide synthase; iNOS, inducible nitric oxide synthase; RA, retinoic acid; NF-M, neurofilament-M.

development of cultured neural precursor cells. Thus it suggests that, under physiological condition, n-NOS mRNA accumulates during neuronal cell differentiation. However, induction of n-NOS may not be essential for neuronal cell differentiation, because inhibition of growth arrest and morphological change of TGW-I-nu cells by RA were not observed after treatment with 1-10 mM N^G-monomethyl-Larginine or N^G-nitro-L-arginine which are specific inhibitors of NOS. In this paper, we clearly show that n-NOS mRNA accumulate during neuronal cell differentiation and suggest that the signal transduction mechanism which is involved in the induction of neuronal cell differentiation by RA is distinct from the NO-mediated pathway.

iNOS expression in neuronal cells upon stimulation with cytokines: Expression of iNOS mRNA was examined in NB-39-nu by treatment with LPS, IFN-γ, TNF-α and IL-1β. A 4.5-kb iNOS mRNA was found to be induced in a time-dependent manner. No n-NOS mRNA could be detected in NB-39-nu upon stimulation. Expression of the iNOS mRNA

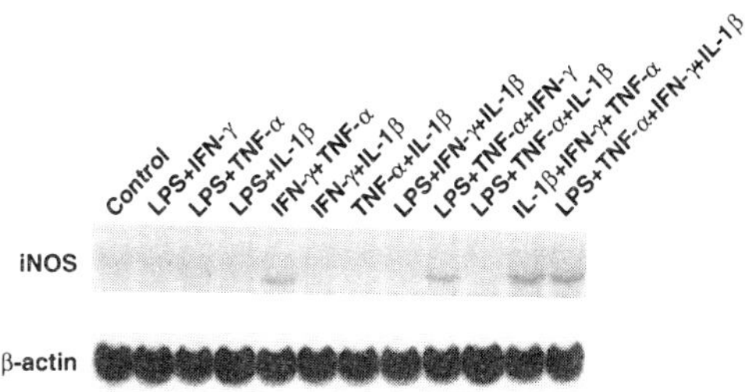

Fig. 2. **Synergistic effect of LPS, IFN-γ, TNF-α and/or IL-1β on expression of iNOS mRNA in NB-39-nu.** Expression of iNOS mRNA in 20 µg of total RNA after treatment with LPS, IFN-γ, TNF-α and/or IL-1β was analyzed by northern blotting. The amount of total RNA analyzed was evaluated by hybridization with human β-actin probe.

was first detected after 12 h of stimulation with LPS and cytokines, and was peaked at 24 h. The trace amount of iNOS mRNA was expressed in NB-39-nu by treatment with IFN-γ alone, and its level was synergistically enhanced by the simultaneous addition of TNF-α. On the other hand, LPS and IL-1β had no effect on expression of iNOS mRNA in NB-39-nu (Fig. 2). This finding indicates that iNOS can be expressed in neuronal cells like in microglia, astrocyte and inflammatory cells, and IFN-γ and TNF-α are key cytokines for regulation of iNOS gene in neuronal cells. Interestingly, iNOS mRNA was rapidly expressed in microglia, astrocyte and inflammatory cells, however, delayed response of iNOS mRNA induction was observed in neuronal cells. This finding may indicate that expression of iNOS in neuronal cells is involved in mechanisms such as ischemic delayed neuronal cell death in the nervous system.

Acknowledgements: This study was partly supported by a Grant-in-Aid from the Ministry of Health and Welfare for the Second-term of a Comprehensive 10-Year Strategy for Cancer Control, Japan, the Foundation of Toyota Riken, Japan, SRF Grant for Biomedical Research, Japan, and Ichiro Kanehara Foundation, Japan.

References

1. Dawson, T.M. and Snyder, S.H. (1994) J. Neurosci. **14**, 5147-5159.
2. Nathan, C. and Xie, Q.-W. (1994) J. Biol. Chem. **269**, 13725-13728.
3. Chomczynski, P. and Sacchi, N. (1987) Anal. Biochem. **162**, 156-159.
4. Fujisawa, H., Ogura, T., Kurashima, Y., Yokoyama, T., Yamashita, J. and Esumi, H. (1994) J. Neurochem. **63**, 140-145.
5. Myers, M.W., Lazzarini, R.A., Lee, V.Y., Schlaepfer, W.W. and Nelson, D.L. (1987) EMBO J. **6**, 1617-1626.
6. Fujisawa, H., Ogura, T., Hokari, A., Weisz, A., Yamashita, J. and Esumi, H. (1995) J. Neurochem. **64**, 85-91

A first stable inducible nitric oxide synthase antisense macrophage cell line

GERALDINE BOSSE, HANS-GEORG FISCHER* and HELGA ROTHE

Diabetes Research Institute at the Heinrich-Heine University, *Institute for Medical Microbiology and Virology, Heinrich-Heine University, 40225 Düsseldorf, Germany

Nitric oxide (NO) is involved in diverse biological functions, such as the regulation of vascular relaxation, and the cytotoxic action of macrophages [1]. Large amounts of this short lived radical is released by the inducible nitric oxide synthase (iNOS) of macrophages after stimulation with lipopolysaccharide (LPS) or certain cytokines [2]. Besides the protection of the host from infections , it also suppresses immune cell functions and participates in destructive autoimmune processes [3]. Since activated macrophages produce a lot of immune mediators, like cytokines or oxygen radicals a synergy or antagonism between NO and other products may be expected. The aim of our work was the generation of an iNOS deficient macrophage cell line to determine the contribution of iNOS expression to various macrophage functions.

Therefore we used the antisense technology, by transfection of a plasmid expressing part of the iNOS RNA in antisense orientation into the mouse macrophage cell line J774.1A. A 666bp fragment of the mouse iNOS cDNA (kind gift of C. Nathan, New York, NY, USA) was subcloned in 3'-5'-orientation in the multicloning site of the pcDNAneoI expression vector between the CMV promotor sequence and the neomycin resistence gene. 20 μg of the *Ksp*I linearized construct was mixed with $2x10^7$ J774.1A cells. Transfection was performed by a single electric pulse (240 V, 1640 μF) generated by a CellJet electroporator. Transfected cells were selected for neomycin resistence and single clones were picked after two weeks. The cells were cultured in RPMI 1640 with 10 % fetal calf serum plus 750 μg G418 per ml. A vector control cell line was generated in the same way by transfection with *Ksp*I linearized pcDNAneoI.

Analysis for iNOS activity was done after activating macrophages with LPS (400 ng/ml). The suppression of nitrite formation (measured by the method of Griess) in the iNOS antisense line was 78% at 400 ng LPS/ml when compared to the nontransfected J774.1A cell line ($p<0.001$). When compared to the control vector transfected cell line the suppression was 75 % ($p<0.001$) (Table 1). At the time of analysis the iNOS antisense cell line had been in continous culture for 6-12 weeks suggesting the stable expression of the antisense gene.

Table 1. Nitrite production (μM) after LPS stimulation*

	0 ng LPS/ml	400 ng LPS/ml
antisense	0	8±3
vector control	0	27±8
J774.1A	0	29±9

*modified after [4]

An alternative approach to the transfection of the cells with a plasmid expressing part of the iNOS RNA in antisense orientation would be the use of antisense oligonucleotides. But the use of an antisense vector has the advantage over the use of antisense oligonucleotides that antisense RNA is continously synthesized within the target cell. Therefore a continous supplementation of exogenous oligonucleotides is not required. In smooth muscle cell cultures the iNOS expression was inhibited by antisense oligonucleotides yielding a maximum of 36 % suppression of nitrite formation [5]. Besides the better suppression of nitrite formation in our iNOS antisense cell line, the stability of such cell lines allows a direct analysis of the contribution of iNOS activity to macrophage activation and functions.

We used the iNOS antisense macrophage cells to determine the impact of iNOS deficiency on LPS induced cytokine gene expression. Earlier studies had shown that iNOS expression following LPS treatment is reduced in vitro and in vivo by administration of TNFα antibody. This indicates that TNFα promotes iNOS gene expression in an autocrine manner. In return, the iNOS product NO may act on TNFα gene expression. When testing the iNOS antisense macrophage line we observed no change in LPS induced TNFα production (measured by a L929 bioassay) when compared to the vector control line or the original J774.1A cells (Table 2).

Table 2: TNFα (U/10^6 cells) in the supernatants after LPS stimulation*

	0 ng LPS/ml	400 ng LPS/ml
antisense	0	54±13
vector control	0	48±11
J774.1A	0	51±17

*modified after [4]

Another macrophage cytokine intimately associated with iNOS expression is interleukin-1 (IL-1), which is known to be a potent inducer of iNOS in several cell types. As shown in Table 3, we found no impact of iNOS deficiency on LPS induced IL-1 release (measured by a bioassay via the ConA dependent response of clone D10.G4.1) when compared to control transfected or normal macrophage cell lines.

Table 3: IL-1 (U/10^6 cells) in the supernatants after LPS stimulation*

	0 ng LPS/ml	400 ng LPS/ml
antisense	0	174±53
vector control	0	180±31
J774.1A	0	146±57

*modified after [4]

These findings argue against a role of NO in TNFα and IL-1 gene expression and underscore the specificity of the antisense DNA approach.

References:

1. Nathan, C. (1992) FASEB J. **6**, 3051-3064
2. Stuehr, D. J. and Marletta, M. A. (1985) Proc. Natl. Acad. Sci. USA **82**, 7738-7742
3. Kolb, H. and Kolb-Bachofen, V. (1992) Immunol. Today **13**, 157-160
4. Rothe, H., Bosse, G., Fischer, H.-G. and Kolb (1995) Biological Chemistry (submitted)
5. Thomae,K.R., Geller, D.A., Billiar, T.R., Davies, P., Pitt, B.R., Simmons, R.L. and Nakayama, D.K. (1993) Surgery **114**, 272-277

Inducible Nitric Oxide Synthase in the Carrageenan-Induced Paw Model of Inflammation.

Daniela SALVEMINI, David M. BOURDON, Zhi-Qiang WANG, Pamela S. WYATT, Margaret H. MARINO†, Pamela T. MANNING and Mark G. CURRIE.

Inflammatory Diseases Research and † Protein Biochemistry Department, G.D. Searle & Co., St. Louis, MO 63017, USA

Introduction The release of nitric oxide (NO) from the inducible form of nitric oxide synthase (iNOS) is involved in acute and chronic inflammation. At least two important discoveries have been made recently that help to define potential mechanisms of action of iNOS in inflammatory diseases. The first observation is that NO stimulates cyclooxygenase (COX) activity resulting in exaggerated release of pro-inflammatory PG (Salvemini et al., 1993), while the second is that NO interacts with superoxide anions (O_2^-) to form the cytotoxic radical peroxynitrite (Beckman et al., 1990). We have examined the significance of these interactions in the carrageenan-induced paw model of inflammation.

Abbreviations: AG, aminoguanidine; carr., carrageenan; COX, cyclooxygenase; L-NIL, N-iminoethyl-L-lysine; NO, nitric oxide; NO_x, NO_2- and NO_3-; PGE_2, prostaglandin E_2

Methods

Carrageenan paw edema: Male Sprague Dawley rats (175-200 g) received a subplantar injection of carrageenan (0.1 ml of a 1% suspension in 0.85% saline) in the right hind paw. Paw volume was measured by a plethysmometer (Ugo-Basile, Varese, Italy) before carrageenan and hourly afterwards (1-10 h). Edema was expressed as the increase in paw volume (ml) after carrageenan compared to the pre-injection value for each animal.

Determination of nitrite/nitrate (NO_x) and prostaglandin (PGE_2) from carrageenan-injected rat paws: At the required time point after the intraplantar injection of carrageenan, rats were sacrificed and each paw was cut at the level of the calcaneus. Paws were gently spun (250xg for 20 min) to recover a sample of the edematous fluid. Recovered fluid volume was measured for each paw and NO_x and PGE_2 concentrations were determined in duplicate, fluorometrically (Misko et al., 1993a) or by ELISA (Cayman Chemicals, Ann Arbor, MI), respectively. Results are expressed as NO_x/paw or ng PGE_2/paw.

Results and discussion The intraplantar injection of carrageenan elicited an inflammatory response that was characterized by a time-dependent increase in paw edema that correlated with increased NO_x levels in the paw exudate (Table 1).

Table 1. Time (hour, h)-dependent increase in edema (ml), NO_x (nmoles/paw) and PGE_2 (ng/paw) following carrageenan. Each value is the mean±s.e.m of n=6 animals

Time (h)	Edema	NO_x	PGE_2
0	0±0	0.5±0.05	0.4±0.1
3	1.4±0.1	20±4	3±0.1
6	1.7±0.2	40±7	2.7±0.2
10	1.6±0.1	55±7	2.9±0.1
10+L-NIL	0.6±0.1	5±0.2	0.5±0.07
10+AG	0.8±0.1	7±0.1	0.7±0.08
10+Indo	1±0.1	50±1	0.3±0.05

Paw edema and NO_x release reached a maximum by 6 h and remained elevated for up to 10 h (Table 1). The increase in NO_x at the 6th hour and thereafter is likely due to the activity of iNOS. Indeed, iNOS mRNA was detected between 3 to 10 h after carrageenan by ribonuclease protection assays and iNOS protein was detected at 6h, which was maximal at 10 h by western blot (not shown).

Two selective iNOS inhibitors, N-iminoethyl-L-lysine (L-NIL, 3-30 mg/kg i.v., Moore et al., 1994; Connor et al., 1995) or aminoguanidine (AG, 30-300 mg/kg i.v. Misko et al., 1993b) given 1 h before carrageenan both failed to inhibit carrageenan-induced paw edema during the first 4 h of the response. They did however inhibit paw edema and NO_x release at subsequent time points (from 5-10 h). Results at the 10h period for L-NIL (30mg/kg, n=8) or AG (300mg/kg, n=8) are shown in Table 1. Injection of carrageenan also elicited a time-dependent increase in PGE_2 release (Table 1) which was not only inhibited by indomethacin (10mg/kg, orally 2 hours before carrageenan, n=6), a cyclooxygenase inhibitor but also by the iNOS inhibitors (Table 1). Indomethacin had minimal effects on NO_x release (Table 1). These findings support our previous observations that NO is an important modulator of *in vivo* COX activity (Salvemini et al., 1995).

The inflammatory response observed in this model had a O_2^- component as superoxide dismutase coupled to polyethyleneglycol (PEGrhSOD, 12 x 10^3 U kg^{-1}, n=6) inhibited the increase in paw oedema by approximately 60% (from 1.7±0.2 to 0.6±0.1 ml, n=6, $P<0.001$). The generation of peroxynitrite has been proposed in a number of pathophysiological conditions that are associated with overproduction of both NO and O_2^- (see Beckman & Crow, 1992 for review). Using the presence of tyrosine nitration as a specific marker of the presence of peroxynitrite, we demonstrated that the inflammatory response evoked by carrageenan involves the generation of $ONOO^-$ at the injured site. Thus as shown in our poster at the meeting, nitrotyrosine immunoreactivity was not detected in the non-inflamed paw tissue of control rats, nor was there any staining in either control or treated tissue using nonimmune serum (not shown). In contrast, marked nitrotyrosine immunorectivity was found in the paw tissue obtained from rats 10 h following carrageenan administration. Additionally, the nitrotyrosine immunoreactivity was largely eliminated by incubating the anti-nitrotryosine antiserum with an excess of nitrotyrosine demonstrating the specificity of staining.

In summary, inhibition of iNOS activity results in an inhibition of both NO and PG release and possibly peroxynitrite formation. These properties account for the anti-inflammatory effects of iNOS inhibitors.

References

BECKMAN, J.S., BECKMAN, T.W., CHEN, T.W., MARSHALL, P.A. & FREEMAN, B.A. (1990). Proc. Natl. Acad. Sci. U.S.A., 87, 1620-1624.

BECKMAN, J.S. & CROW, J.P. (1992). Biochem. Soc. Trans., 21, 1992-1996.

CONNOR, J., MANNING, P.T., SETTLE, S.L., MOORE, W.M., JEROME, G.M., WEBBER, R.K., TJOENG, F.S. & CURRIE, M.G. (1995). Eur. J. Pharmacol., 273, 15-24.

MISKO, T.P., SCHILLING, R.J., SALVEMINI, D., MOORE, W.M. & CURRIE, M.G. (1993a). Anal. Biochem. 214, 11-16.

MISKO, T.P., MOORE, W.M., KASTEN, T.P., NICKOLS, G.A., CORBETT, J.A., TILTON, R.G., MCDANIEL, M.L., WILLIAMSON, J.R. & CURRIE, M.G. (1993b). Eur. J. Pharmacol., 233, 119-225.

MOORE, W.M., WEBBER, R.K., JEROME, G.M., TJOENG, F.S., MISKO, T.P. & CURRIE, M.G. (1994). J. Med. Chem., 37, 3886-3888.

SALVEMINI, D., MISKO, T.P., SEIBERT, K., MASFERRER, J.L., CURRIE, M.G. & NEEDLEMAN, P. (1993). Proc. Natl. Acad. Sci. U.S.A., 90, 7240-7244.

SALVEMINI, D., MANNING, P.T., ZWEIFEL, B.S., SEIBERT, K., CONNOR, J., CURRIE, M.G., NEEDLEMAN, P. & MASFERRER, J.L. (1995). J. Clin. Invest., 96, 301-308.

Inhibition of inducible nitric oxide synthase reverses endotoxin tolerance in cultured macrophages.

Elizabeth E. Mannick, Peter D. Oliver, Halina Sadowska-Krowicka, and Mark J.S. Miller

LSUMC Department of Pediatrics, New Orleans, Louisiana 70112

INTRODUCTION

On repeated exposure to endotoxin (lipopolysaccharide - LPS), macrophages produce decreased amounts of pro-inflammatory cytokines, a phenomenon known as endotoxin tolerance (1, 2). Nitric oxide (NO) synthesis by the enzyme, inducible nitric oxide synthase (iNOS), is down-regulated in vitro and in vivo by repeated doses of endotoxin (3). Since NO may act in an autocrine manner to suppress its own synthesis by inhibiting iNOS, we hypothesized that NO induced by LPS might regulate its own suppression following rechallenge with LPS (4,5). We also sought to ascertain if NO governs two other aspects of endotoxin tolerance; cell viability and TNF-α production. These hypotheses were tested using RAW 264.7 cells from a murine peritoneal macrophage cell line.

MATERIALS AND METHODS

Cell Culture: RAW 264.7 cells (Sigma, St. Louis, MO) were incubated in 5% CO2 at 37 C and grown to confluence in Dulbecco's Modified Eagle's Medium (Sigma) containing 10% heat-inactivated calf serum.

Induction of Endotoxin Tolerance: Cells were pretreated with either LPS (E. coli serotype 026:B6, Sigma) at a dose of 10 ng/ml, 100 ng/ml or 2 ug/ml; LPS and NIL (Searle/MCR, St. Louis, MO) at a dose of 2 ug/ml and 0.5 mM, respectively; NIL alone at a dose of 0.5 mM, SNAP (Sigma) alone at a dose of 500 uM or culture medium alone. After 18 hours, culture medium was removed from all wells and fresh culture medium containing 2 ug/ml LPS was added to each well. Fresh culture medium alone was added to control wells. Twenty-four hours after LPS challenge, cells were prepared for subsequent analysis.

Cell Viability Analysis: The number of cells excluding trypan blue per 100 cells was assessed by light microscopy.

Nitrite: 100 ul of supernatant was collected from each of 6 wells per treatment group and 100 ul of Griess reagent (1% sulfanilamide: 0.1% N- (1-naphthyl)-ethylenediamine dihydrochloride and 2.5% H3PO4) added. The absorbance at 595 nm was determined spectrophotometrically on a microplate reader (Bio-Rad, Burlingame, CA) by using sodium nitrite as standard.

Cytokine ELISA: Production of TNF-α and IL-1 in culture supernatant was assessed by commercially available enzyme-linked immunoabsorbance assay (CYTImmune Sciences, College Park, MD).

RESULTS

Nitrite Production: As shown in Figure 1, pretreatment with 2 ug/ml of LPS for 18 hours, resulted in a significant ($p < 0.001$; ANOVA, Tukey-Kramer Multiple Comparisons Test) reduction in nitrite production 24 hours after challenge with 2 ug/ml. A similar protective effect was achieved using pretreatment LPS doses of 10 ng/ml and 100 ng/ml (data not shown). Pretreatment with NIL alone completely prevented nitrite formation with subsequent LPS challenge ($p < 0.001$). The combination of LPS and NIL significantly reversed the protective effect of endotoxin ($p < 0.001$). Pretreatment with SNAP also resulted in a significant reduction in nitrite ($p < 0.001$).

Cell Viability: LPS pretreatment (2ug/ml) prevented the reduction in cell viability associated with subsequent LPS challenge ($p < 0.05$; Kruskal-Wallis ANOVA, Dunn's Multiple Comparison Test). LPS and NIL together failed to prevent the reduction in cell viability caused by rechallenge with LPS ($p < 0.05$). Pretreatment with NIL alone was partially protective (NS). Pretreatment with SNAP completely protected against rechallenge with LPS ($p < 0.01$).

TNF-α Production: LPS pretreatment (2 ug/ml), LPS and NIL pretreatment, NIL pretreatment and SNAP pretreatment all significantly reduced TNF-α production 24 hours following rechallenge with LPS ($p < 0.001$; ANOVA, Tukey-Kramer Multiple Comparisons Test). Although the combination of LPS and NIL as a pretreatment resulted in a greater production of TNF-α than LPS alone (10.42 vs. 4.73 ng/ml), this difference did not achieve statistical significance.

DISCUSSION

This report shows that in cultured murine macrophages, tolerance to endotoxin as measured by a decrease in nitrite production and an improvement in cell viability can be reversed by using an inhibitor of inducible NO synthase, N-iminoethyl lysine. Moreover, S-nitroso-N-acetylpenicillamine, a NO donor, mimics the effects of LPS pretreatment. Together, these results suggest that the tolerigenic effects of LPS on nitrite production and viability may be mediated by NO.

Nitric oxide has been shown to inhibit iNOS activity (4). The sustained production of NO by iNOS is known to decrease viability in murine macrophages and iNOS inhibitors have been shown to restore viability (6). The intracellular pathways involved in the negative feedback of NO on iNOS are not completely established but may to involve a reversible inhibition of protein kinase C via S-nitrosylation of the enzyme (5,7).

Addition of NIL to LPS pretreatment partially reversed tolerance as defined by an increase in TNF-α production. SNAP and NIL pretreatments also significantly inhibited TNF-α production with subsequent LPS challenge, suggesting that NO or iNOS exerts a negative feedback on TNF.

Our results are consistent with the recently reported data of Fahmi et al who demonstrated that an 18-hour pre-exposure of murine fibroblasts to LPS, SNAP or sodium nitroprusside, significantly reduced TNF-α production in response to endotoxin rechallenge and that addition of N-monomethyl-L-arginine, partially reversed this effect (9). As noted by these authors, an inhibitory role of NO in the regulation of TNF-α production is at variance with others' reports of an upregulation of TNF-α by NO (10). The variable effects of NO on TNF-α may represent species differences (human vs. murine) or differences among cell types (neutrophil vs. macrophage). Two alternative explanations are possible, however. First, since the stimulation of TNF-α synthesis by the nuclear transcription factor, NF-kB, is dependent on the oxidant status of the cell (11), the effect of NO on TNF-α synthesis may depend on the ratio of NO to superoxide (12); this may govern whether NO is an oxidizing agent (by contributing to the formation of peroxynitrite) or a free radical scavenger (by binding metals and preventing Fenton chemistry). Second, in our experiments, pretreatment with either SNAP or NIL resulted in a reduction in subsequent TNF-α production. Since SNAP and NIL both inhibit iNOS formation, it may be that the iNOS enzyme, not NO itself, is involved in upregulating TNF-α.

In summary, our results indicate that NO plays a role in endotoxin tolerance via a negative feedback on its own production.

REFERENCES

1. Greisman S.E. (1983) in Beneficial Effects of Endotoxin (Nowotny, A, ed.), pp.149-178, Plenum Press, New York.
2. Virca G.D., Kim S.Y., Glaser K.B., Ulevitch R.J. (1989) J Biol. Chem. 264, 21951-21956.
3. Szabo C., Thiemermann C., Wu C.C., Perretti M., Vane J.R. (1994) Proc. Natl. Acad. Sci. USA 91, 271-275.
4. Assreuy J., Cunha F.Q., Liew F.Y., Moncada S. (1993) Br. J. Pharmacol. 108, 833-837.
5. Jun C.D., Choi B.M., Lee S.Y., Kang S.S., Kim H.M., Chung H.T. (1994) Biochem Biophys Res Comm 204, 105-111.
6. Albina J.E., Cui S., Mateo R.B., Reichner J.S. (1993) J. Immunol. 150, 5080-5085.
7. Gopalakrishna R., Chen Z.H., Gundimeda U. (1993) J. Biol. Chem. 268, 27180-27185.
8. Erroi A., Fantuzzi G., Mengozzi M., Sironi M., Ovencole S.F., Clark B.D., Dinarello C.A., Isetta A., Gnocchi P., Giovanelli M., Ghezzi P. (1993) Infect. Immun. 61, 4356-4359.
9. Fahmi H., Charon D., Mondange M., Chaby R. (1995) Infect. Immun. 63, 1863-1869.
10. VanDervort A.L., Yan I., Madara P.J., Cobb J.P., Wesley R.A., Corriveau C.C., Tropea M.M., Danner R.L. (1994) J. Immunol. 152, 4102-4109.
11. Schreck R., Baeuerle P.A. (1991) Trend. Cell Biol. 1, 39-42.
12. Pryor W.A., Squadrito G.L. (1995) Am. J. Physiol. 268 (5 Pt 1), L699-722.

Figure 1 Nitrite Production N = 4

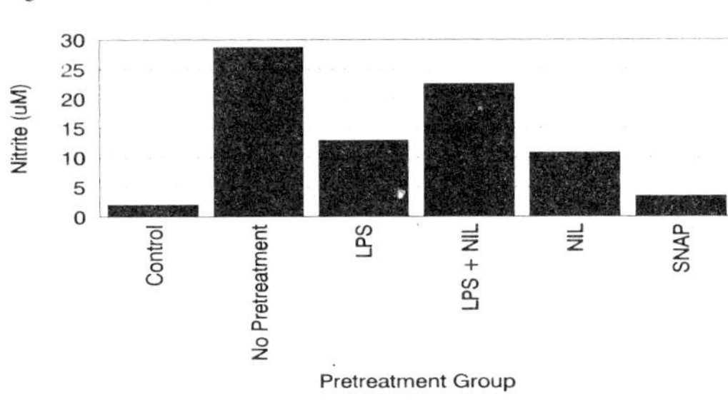

Figure 2 Cell Viability N = 8

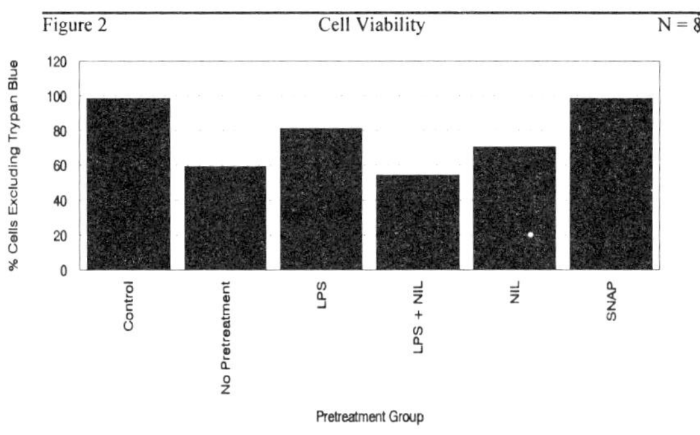

Figure 3 TNF-α Production N = 3

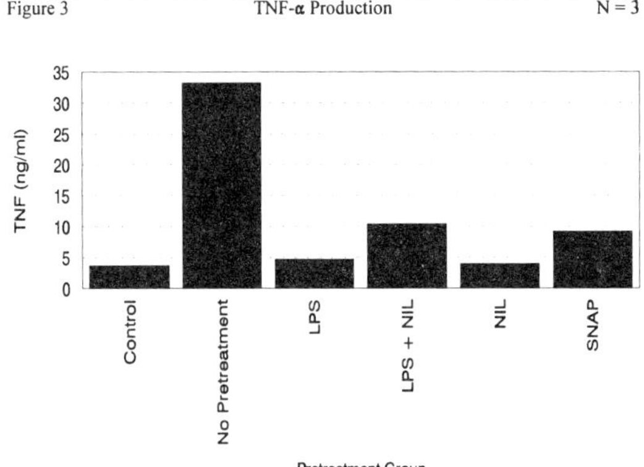

Disassembly of microfilaments inhibits induction and activity of macrophage nitric oxide synthase.

Patricia FERNANDES[#], Helena ARAÚJO[*] and Jamil ASSREUY[#].

Dept. of Pharmacology, UFRJ, PO Box 68013, 21944-970 and * Lab. Molecular Pharmacology, Instituto de Biofísica Carlos Chagas Filho, UFRJ, 21944-970, Rio de Janeiro, RJ, Brazil.

Nitric Oxide (NO) is a liposoluble gas formed from L-arginine by NO synthases (NOS). In macrophages, NO is produced by an inducible NOS following activation of cells by several stimuli, usually bacterial endotoxin (LPS) and interferon gamma (IFN). This induction process occurs for 4-6 h after the initial stimuli. Following its synthesis, macrophage NOS continues to produce NO for several hours. Released NO has an important role in the destruction of intracellular parasites and tumour cells [1]. Cell shape is maintained by an array of filaments that composes the cytoskeleton. Filaments include microtubules (made of tubulin subunits), intermediary filaments (made of several proteins) and microfilaments (made of actin monomers) [2]. Besides its structural function, the cytoskeleton plays a role in several cell functions such as secretion, organelle traslocation, signal transduction and gene activation [3]. Several compounds have been used to characterise the involvement of the cytoskeleton in these functions. Among them, cytochalasin B (CytB) and colchicine are inhibitors of subunit addition to the microfilaments and microtubules, respectively. As cells turnover the cytoskeleton by polymerization/depolymerization of the filaments through addition/ removal of its subunits, inhibition of further polymerization can lead to disassembly of the filament. Marczin et al, 1993 [4], reported that addition of colchicine to cerebral endothelial cells decreased the staining of constitutive NOS assayed by NADPH diaphorase method. In the course of experiments designed to study the relationship between phagocytosis and cytotoxic activity of macrophages, we have found that CytB decreased nitrite levels in the supernatant of activated cells. Therefore, we decided to study the relationship between cytoskeleton (mainly microfilaments) and NO production.Thioglycollate-elicited C57/Black10 macrophages were activated with LPS (100 ng/ml) + IFN (20 U/ml). Nitrite levels in the supernatant were assayed 48 h after the activation. Addition of increasing concentrations of CytB (10-300 μM) together with the stimulus decreased dose-dependently nitrite accumulation (80% maximal inhibition), whereas colchicine inhibited it by 20% at most and was not dose-dependent. Addition of CytB (50 μM) 0, 6 and 12 h after the stimulus inhibited nitrite accumulation, but the inhibitory effect decreased with time, the highest inhibition seen when CytB was added together with stimulus and smaller when it was added 12 h after. This result indicates that disassembly of microfilaments was affecting the induction and the activity of NOS. To confirm that microfilaments were important for NOS induction, CytB (50 μM) was added together with the stimulus, and the cells were washed and fresh medium was added 6 h later. Nitrite accumulation was measured 48 h after the stimulus. Nitrite levels were identical in washed compared with non-washed cells. Inhibition of nitrite accumulation by CytB was similar in both groups. The results were further confirmed by assaying NOS activity in intact macrophages, using citrulline production. Macrophages were activated and CytB was added at 0, 6 or 16 h later. Two h after CytB addition cells were washed, the medium was changed to Hanks solution containing ^{14}C-L-arginine and CytB where needed. Two hundred μl of the supernatant were taken 1, 2 and 3 h after label addition. ^{14}C-arginine was separated from ^{14}C-L-citrulline by adsorption to Dowex resin. After centrifugation, resin supernatant was assayed by liquid scintillation. Activation of macrophages led to a high citrulline production when compared to non-activated controls. To confirm that citrulline production was due to NOS activity, L-NIO (a specific inhibitor of the enzyme) was added at 0 h and a strong inhibition of citrulline production was seen. When CytB was added together with stimulus (at 0 h) an almost complete inhibition in citrulline accumulation was observed. As in nitrite accumulation smaller, although significant, inhibitory effect was seen when CytB was added at later times. The results indicate that: a) the integrity of the microfilaments is essential for NO production by activated macrophages; b) the integrity of microfilament network is required for the NOS induction process to proceed normally and c) The activity of synthesised NOS is also dependent on the integrity of the microfilament network, maybe through and anchorage-dependent enzymatic activity. Working on the visualisation of NOS in macrophages with or without the microfilament network preserved is now being carried out in our laboratory.

We are grateful to Mr. Antônio V. Conrado for his tecnical assistance. This work was supported by grants from Conselho Nacional de Desenvolvimento Científico e Tecnológico (CNPq) and Fundação Universitária José Bonifácio (FUJB).

REFERENCES

1. Moncada, S., Palmer, R.M.J. and Higgs, E.A. (1991) Pharmacol. Rev. **43**, 109-142.

2. Bershadsky, A.D. and Vasiliev, J.M. (1989) in Cytoskeleton (Siekevitz, P., ed.), Plenum Press, New York

3. Alberts, B., Bray, D., Lewis, J., Raff, M., Roberts, K. and Watson, J.D. (1993) in The Molecular Biology of the Cell (Alberts, B. et al), Garland Publishing, Inc., New York & London.

4. Marczin, N., Papapetropoulos, A., Jilling, T., Catravas, J.D. (1993) Br. J. Pharmacol. **109**, 603-605.

Induction of nitric oxide synthase by Gram-positive organisms in murine macrophages and rat aortic smooth muscle cells

SJEF J. DE KIMPE, LISA BRYAN*, CHRISTOPH THIEMERMANN and JOHN R. VANE

The William Harvey Research Institute, Medical College and *Dept. of Microbiology of St. Bartholomew's Hospital, Charterhouse Square, London EC1M 6BQ, United Kingdom

There is little information regarding the mechanism by which Gram-positive organisms induce septic shock. Lipoteichoic acid (LTA), a component of the cell wall of *Staphylococcus aureus* induces the release of cytokines and nitric oxide *in vitro* [1] and *in vivo* [2]. However, the cell wall of Gram-positive organisms is very heterogeneous and contains more components able to initiate an inflammatory response [3]. Moreover, the chemical structure of these cell wall components can vary considerably among different species. Here, the ability of different gram-positive organisms (*Staphylococcus aureus*, *Staphylococcus epidermis*, Group A Streptococcus and *Streptococcus pneumonia*) to induce nitric oxide synthase was investigated in cultured macrophages (J774.2 cells) and in rat aortic smooth muscle (RASM) cells.

The Gram-positive organisms (isolates from septic patients) were grown to stationary phase, washed in saline and killed by boiling for 15 min. In some experiments, the cell wall of the bacteria was further disrupted by sonication for 1 min. J774 macrophages or RASM cells were cultured to confluency in 96-well plates containing 200μl of culture medium (DMEM or RPMI 1640, respectively) supplemented with 10% foetal calf serum. Cells were stimulated with the different micro-organisms in the absence or presence of interferon (IFN)-γ (10 U/ml). After 24 h (J774.2) or 40 h (RASM), nitrite accumulation, an indicator of nitric oxide synthesis, was measured using the Griess reaction. Statistical evaluation was performed using ANOVA and Bonferroni's test for multiple comparison.

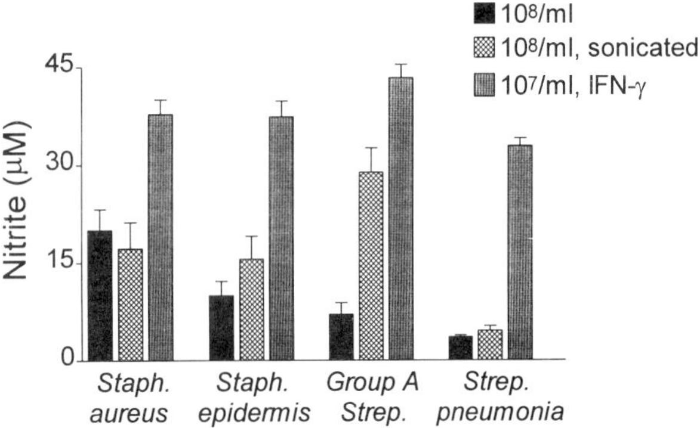

Figure 1. Induction of nitrite release by J774.2 macrophages activated for 24 h by heat-killed Gram-positive organisms. Values are mean ± S.E.M. of 4 independent experiments performed in triplicate.

Gram-positive organisms (10^8/ml) induced the release of nitrite by J774.2 cells with the following rank order of potency: *S. aureus* > *S. epidermis* = Group A streptococcus > *S. pneumonia* (Figure 1). *S. pneumonia* elicited only a minor increase in nitrite formation from the macrophages. Further disruption of the bacterial cell wall by sonication caused a four fold increase in the nitrite accumulation caused by Group A Streptococcus (P<0.01). As sonication did not influence the potency of the other bacteria, boiled and sonicated Group A Streptococcus was the most potent preparation (Figure 1). Co-stimulation with IFN-γ, greatly enhanced the release of nitrite caused by these micro-organisms (Fig. 1). In the presence of IFN-γ, the nitrite concentration in the incubation medium was similar for the four bacteria preparations. Addition of polymyxin B to the culture medium blunted the induction of nitrite release elicited by lipopolysaccharide (LPS; from Gram-negative bacteria), but not the response to the Gram-positive organisms (Table 1). Also in RASM cells, Gram-positive organisms (10^7/ml, in the presence of IFN-γ and polymyxin B) induced a significant release of nitrite with the following order of potency: *S. aureus* (31±2μM) > *S. Epidermis* (24±1μM) > *S. pneumonia* (14±2μM) = Group A Streptococcus (13±2μM).

Abbreviations used: IFN-γ: interferon-γ; LPS: lipopolysaccharide; LTA: lipoteichoic acid; NO: nitric oxide; PepG: peptidoglycan; RASM: rat aortic smooth muscle.

Table 1. Nitrite (μM) accumulation in the presence of polymyxin B Mean ± S.E.M. are given of 3 experiments in triplicate. # P<0.05.

Microorganism + IFN-γ (10^7/ml)	control	polymyxin B (1μg/ml)
	μM	μM
Lipopolysaccharide (1μg/ml)	26 ± 3	2 ± 1 #
S. aureus	35 ± 3	29 ± 3
S. epidermis	37 ± 3	38 ± 2
Group A Streptococcus	42 ± 3	38 ± 3
S. pneumonia	31 ± 3	27 ± 3

The present study demonstrates that various species of Gram-positive organisms can induce the release of nitric oxide (measured as nitrite in the incubation medium) from cultured murine macrophages and RASM cells. LPS is not involved in this response, as the induction of nitrite release elicited by the Gram-positive preparations is not influenced by polymyxin B, a compound which binds and effectively inactivates LPS. Streptococcal bacteria seem to be less potent than Staphyloccocal bacteria. After sonication the Group A Streptoccocus became the most potent inducer of nitric oxide release, suggesting the release of a potent component by further disruption of the cell wall. In contrast, the *S. pneumonia* preparation remained relatively inactive. Major components of the cell wall are peptidoglycan (40-60% of the cell wall dry weight) and teichoic acids (20-40% of the cell wall dry weight). Teichoic acid exists in two structurally unrelated forms: (i) associated with lipid, called lipoteichoic acid (LTA); (ii) covalently linked to peptidoglycan. LTA from *S. aureus* induces nitric oxide synthase in macrophages and vascular smooth muscle [1,4]. LTA is amphiphilic and generally consists of a hydrophylic poly(glycerophosphate) chain attached to a glycolipid. Interestingly, such structure is not present in *S. pneumonia* strains [5]. Thus, the low potency of *S. pneumonia* suggests that LTA is a major determinant for the induction of nitric oxide release in Gram-positive bacteria. In the presence of IFN-γ, *S. pneumonia* as well as Group A streptococcus (without prior sonication) induce a maximal response in macrophages. Interestingly, Kengatharan et al. report in this book, that isolated peptidoglycan induces nitric oxide release in macrophages only in the presence of IFN-γ. Therefore, apart from LTA, peptidoglycan may contribute to the induction of nitric oxide release elicited by the various Gram-positive strains.

This study is partially supported by Cassella AG (Germany). SJDK is a recipient of a fellowship of the European Union.

1. Cunha, F.Q., Moss, D.W., Leal,L.M.C.C., Moncada, S. and Liew, F.Y. (1993) Immunology **78**, 563-567.
2. De Kimpe, S. J., Hunter, M. L., Bryant, C. E., Thiemermann, C. and Vane, J. R. (1995) Br. J. Pharmacol. **114**, 1317-1323.
3. Bone, R. C. (1993) J. Crit. Care **8**, 51-59.
4. Lonchampt, M.O., Auguet, M., Delaflotte, S., Goulin-Schulz, J., Chabrier, P.E. and Braquet, P. (1992) J. Cardiovasc. Res. **20**(Suppl. 12), S145-S147.
5. Fischer, W. (1988) Adv. Microbial. Physiol. **29**, 233-302.

Peptidoglycan induces nitric oxide synthase activity only in combination with lipoteichoic acid or interferon-γ in cultured J774.2 macrophages

MURALITHARAN KENGATHARAN, SJEF J. DE KIMPE
CHRISTOPH THIEMERMANN and JOHN R. VANE

The William Harvey Research Institute, St Bartholomew's Hospital Medical College, Charterhouse Square, London EC1M 6BQ, UK.

Introduction

Lipoteichoic acid (LTA) and peptidoglycan (PepG) are cell wall components of Gram-positive organisms such as *Staphylococcus aureus* which do not contain endotoxin. LTA induces an enhanced formation of nitric oxide (NO) due to induction of NO synthase (iNOS) in macrophages [1]. However, very little is known about the effect of PepG on NO formation in macrophages. Here, we investigated the effect of PepG on the nitrite release (an indicator of NO formation) by murine J774.2 macrophages when added to the cells alone or in combination with LTA or interferon-γ (IFNγ).

Methods

Murine macrophages (J774.2) were cultured in 96-well plates with Dulbecco's modified eagle's medium containing 10% foetal calf serum until confluent. Nitrite accumulation in the medium, an indicator of NOS activity, was measured at 24 h after the addition of killed *S.aureus* or PepG (alone or in combination with LTA or IFNγ), by the Griess method. Cell respiration, an indicator of cell viability, was assessed at 24 h by the mitochondrial-dependant reduction of 3-(4,5-dimethyl-thiazol-2-yl)-2,5-diphenyltetrazolium bromide (MTT) to formazan.

Results

Incubation with PepG (1-100 μg/ml for 24 h) alone did not induce nitrite formation by J774.2 macrophages (Table 1). However, in the presence of IFN-γ (10U/ml), PepG caused a concentration-dependent accumulation of nitrite (Table 1). Interestingly, PepG also strongly potentiated the formation of nitrite caused by LTA. The increase in nitrite was reduced by treatment of the cells with the non-selective NOS inhibitor N^G-methyl-L-arginine (L-NMMA) or by the iNOS-selective inhibitor aminoethyl-isothiourea (AE-ITU) (Figure 1). In addition, pre-treatment of cells with cycloheximide (0.3-1.0μg/ml) prevented the rise in nitrite. Incubation of cells with polymixin B (0.05μg/ml) did not attenuate the rise in nitrite induced by PepG+IFNγ although the response to endotoxin was prevented.

Table 1. PepG synergises with LTA or IFNγ to induce iNOS activity in J774.2 macrophages
Each value represents the mean ± S.E.M. from triplicate determinations (wells) from 3 separate experimental days.

Concentration of PepG (μg/ml)	Concentration of nitrite at 24h (μM) plus vehicle	plus LTA (1μg/ml)	plus IFNγ (10U/ml)
0	2±2	9±1	7±2
1	2±1	9±2	20±3
10	3±1	15±1	35±2
30	2±1	25±3	41±2
100	3±2	30±2	44±4

Furthermore, polyclonal antibodies to PepG attenuated the rise in nitrite caused by killed *S. aureus* but not the nitrite formation induced by LTA (Figure 2).

Discussion

Our results show that PepG, a major cell wall component of Gram-positive bacterium *S.aureus*, synergises with LTA (another cell wall component of *S.aureus* or with IFNγ) in causing an enhanced formation of NO by J774.2 macrophages.

Abbreviations used: aminoethyl-isothiourea, AE-ITU; inducible nitric oxide synthase, iNOS; interferon γ, IFNγ; lipoteichoic acid, LTA; nitric oxide, NO; N^G-methyl-L-arginine, L-NMMA; peptidoglycan, PepG; tumor necrosis factor-α, TNFα

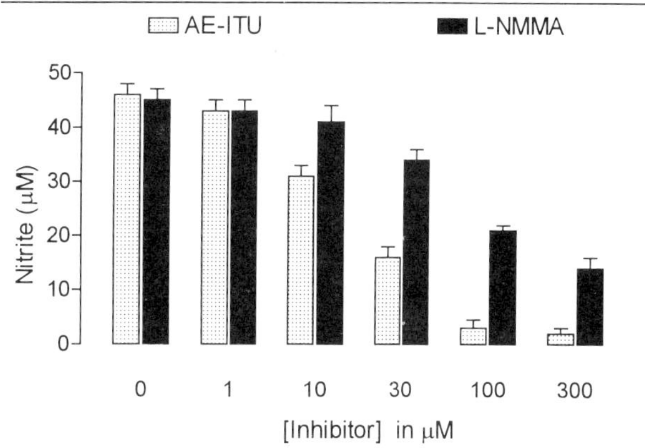

Figure 1. The NOS inhibitors, L-NMMA and AE-ITU concentration-dependantly inhibit the nitrite formation over 24 h by J774.2 macrophages activated with PepG (10μg/ml) +IFNγ (10U/ml). Results are expressed as mean±S.E.M. of four independent experiment performed in triplicate.

PepG is not found in the cell wall of Gram-negative bacteria and is not structurally similar to endotoxin since polymixin B, an agent which binds and inactivates endotoxin, did not affect the nitrite formation induced by PepG+IFNγ. Unlike endotoxin, PepG did not induce nitrite on its own although PepG induces the release of TNFα in macrophages [2]. However, PepG synergised with LTA or IFNγ in inducing iNOS in macrophages and this response was prevented by (i) the NOS inhibitors L-NMMA or AE-ITU, and (ii) the protein synthesis inhibitor, cycloheximide.

The fact that PepG antibodies attenuated the NO formation by the *S.aureus* suggests that synergy between PepG and LTA contributes to the mechanism by which *S.aureus* induces iNOS in macrophages.

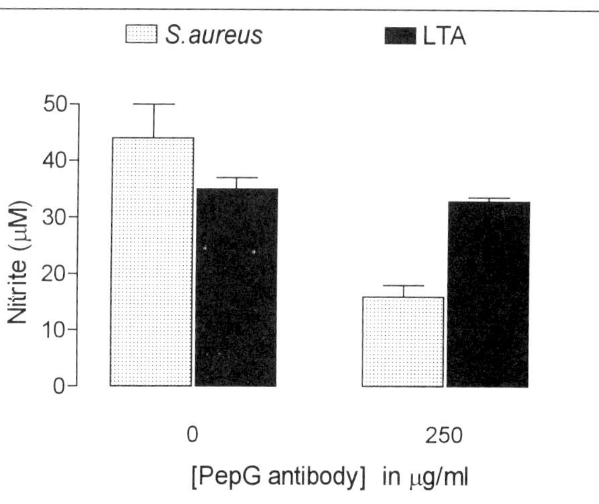

Figure 2. Enhanced formation of nitrite over 24 h by J774.2 macrophages activated with *S.aureus* (10^8 cells/ml), but not with LTA (10μg/ml), is attenuated by polyclonal antibodies to PepG. Results are expressed as mean±S.E.M. of three independent experiment performed in triplicate.

In conclusion, PepG synergises with LTA or IFNγ in inducing an enhanced formation of NO in J774.2 macrophages.

SJDK is a recipient of a Senior Fellowship of the European Union. This project is supported by a grant from Cassella AG (Germany).

References

1. Cunha, F.Q., Moss, D.W., Leal, L.M.C.C. et al. (1993) Immunology. **78**, 563-567.
2. Timmerman, C.P., Mattsson, E., Martinez-Martinez, L. et al. (1993) Infection and Immunity. **61**, 4167-4172.

The role of L-arginine-Nitric Oxide pathway and Reactive Oxygen Species in infection of resistant and susceptible mice by *Salmonella typhimurium*.

YVES, P. GAUTHIER and DOMINIQUE, R. VIDAL

Unité de microbiologie. Centre de Recherche du Service de Santé des Armées Emile Pardé, BP 87, 38702 La Tronche, Cedex, France.

Nitric oxide, which is involved in mechanisms of non specific defense against various intracellular pathogens, inhibits the growth of *Salmonella typhimurium*, the agent responsible for murine Typhoïd. Furthermore, L-arginine which is the only physiologic substrate for NO-synthase, is known to have immunomodulatory effects [1]. The question whether or not putative differences in the L-arginine/Nitric oxide pathway between ity^S and ity^r strains of mice can account for the differences in their resistance to salmonellosis remains controversial [2].

The current work examines the influence of the NO donnor L-ARG and the NO inhibitor aminoguanidine *in vivo* on the course of salmonellosis, and *in vitro* on both the bactericidal activity of macrophages and the production of Reactive Oxygen Species and NO, in a sensitive (ity^S) mouse strain BALB/c and in a resistant (ity^r) strain CBA.

A single injection of 10mM L-ARG into mice shortly before *i.p.* infection with a LD_{50} of *S. typhimurium*, led to a significant protection of BALB/c mice, as shown by a 100 % survival rate in the treated group, compared with 50 % only in the control group 12 days after challenge, whereas such a protective effect was not observed with CBA mice, yielding similar survival rates, 62.5 % in both groups. In a previous work we demonstrated that the NO inhibitor AG also induced a protective effect in BALB/c mice, 60 % survival *v-s* 30 % in the control group, but was deleterious to CBA mice, 20 % survival *v.s.* 60 % in the control group, 10 days after infection [3].

Correlatively, as early as 2 hours after infection, the number of bacteria collected from the peritoneal cavity of BALB/c mice was 3 times lower in animals injected L-ARG before infection, $7.7\text{-}10^1$ *v.s.* $2.3\text{-}10^2$ CFU, and 2 times lower in animals given AG, $1.1\ 10^2$ *v.s.* $2.3.10^2$, compared with the number in control mice. Conversely the number of bacteria in the peritoneal cavity of CBA mice was increased 2 times by L-ARG, from $2.8.10^4$ to $6.5.10^4$ and AG induced a slight increase from $2.8.10^4$ to $3.6.10^4$. Similar observations were made at the spleen level as well, the number of CFU per gram of spleen of BALB/c mice, 2 hours after challenge, decreasing from $4.9.10^2$ in the control mice to 16 or 7 only when given L-ARG or AG respectively. Conversely AG led to a 2.5 times increase in the CFU content of CBA spleen, from $1.6.10^4$ to $3.9.10^4$, and L-ARG inhibited the clearance of the bacteria during the first ten hours post infection $1.38.10^4$(10 h) *v.s* $1.3.10^4$ (2 h) (Figure 1).

In vitro experiments showed that in the presence of added 4mM L-ARG, non activated BALB/c macrophages displayed a more significantly enhanced Salmonella killing activity (+ 44 %) than CBA macrophages (+ 28 %), and that both strains of macrophages had an enhanced secretion of

---

Abbreviations used : NO, Nitric oxide ; L-ARG, L-arginine ; AG, aminoguanidine ; R.O.S., Reactive Oxygen Species ; LPS, lipopolysaccharide ; IFN, interferon ; CFU Colony Forming Unit ; *i.p.*, intraperitoneal.

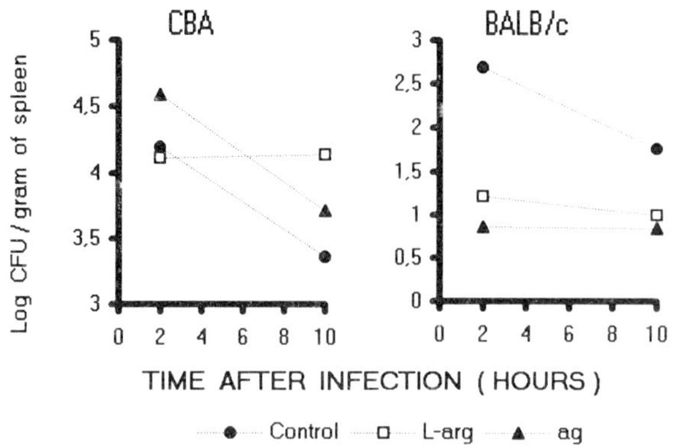

FIGURE 1 : Effect of L-Arginine and Aminoguanidine pretreatment on the number of viable bacteria in the spleens of CBA and BALB/c mice infected with a LD_{50} of *S.typhimurium*.

superoxide anion $O_2^{\dot{-}}$, from 6 to 13 nmole/10^6 cells/90min and from 3 to 7 nmole/10^6 cells/90min in BALB/c and CBA strains respectively, whereas hydrogen peroxide production (H_2O_2) was not modified. Activation of the macrophages with IFN-γ (200 U/ml) and LPS (1μg/ml) led to a significant increase in bacterial killing by CBA macrophages (+ 34 %) whereas salmonella killing activity of BALB/c macrophages was not significantly modified. Furthermore addition of AG induced an increase in Salmonella killing by activated BALB/c macrophages (+ 20 %) concomitant with an increase in $O_2^{\dot{-}}$ production (+ 94 %), and a drastic decrease in nitrite accumulation from 67 to 0.6 nmole/10^6 cells / 24 hours, whereas AG was inhibitory for Salmonella killing activity in activated CBA macrophages.

These results show that BALB/c (ity^S) and CBA (ity^r) mice differ in the respective roles of the L-ARG/NO pathway and R.O.S. production in the course of Salmonella infection, and suggest that : (a) R.O.S. production is a more potent factor of Salmonella killing in BALB/c mice than in CBA mice, and correlatively the L-ARG/NO pathway is more efficient in this respect in CBA mice than in BALB/c mice, (b) The protective effects of L-ARG or AG on the sensitive mouse BALB/c infected with *S.typhimurium* may, at least partly, result from an induced overproduction of superoxide anion by macrophages at an early stage of infection.

ACKNOWLEDGMENTS

This work was supported by a grant from Direction de la Recherche et de la Technologie. We thank E. Letourneur for typing the manuscript and M.F. Burckhart for drawing the figure.

REFERENCES

[1]Kirk S.J. and Barbul A., (1990) Role of Arginine in trauma, Sepsis, and Immunity. JPEN **14 (5)**, 226-229

[2]Zwilling B.S and Hilburger M.E., (1994) Macrophage Resistance Genes : *Bcg/ity/lsh* in Macrophage-Pathogen Interactions, (B.S. Zwilling and T.K. Eisenstein Eds), pp 133-245, Marcel Dekker, New York.

[3]Gauthier Y.P., Burckhart M.F. and Vidal D.R., (1994) Role of Iron and Nitric oxide in murine salmonellosis. Proceedings of the 3rd International Congress on the immune consequences of Trauma, shock and Sepsis, Munich (Faist, E., ed.), Pabst Science Publishers, Eichengrun, *in press*.

Modulation of TNF-α and NO responses in *S.typhimurium* infected mice after *in vivo* immunomodulatory therapy

CHANTAL BOTTEX, SYLVIE FERLAT, FLORENCE CONDEMINE, NICOLAS BURDIN, DOMINIQUE VIDAL, FRANCOISE PICOT* and PIERRE POTIER*

CRSSA Emile Pardé, 38702 La Tronche Cédex, France
*ICSN-CNRS, 91198 Gif-sur-Yvette, France

Nitric oxide (NO) participates as a significant molecule in the repertoire of host non-specific immune mechanisms [1]. This reaction requires the specific cooperation of distinct humoral signals : IFN-γ, TNF-α or LPS [2] leading to the full activation of macrophage anti-microbial functions. Consequently, major efforts are under way to develop drugs that can modulate the host's natural defenses and restore impaired immune functions. Here, we investigated the immunomodulating properties of a new compound, the synthetic fatty acid 10-hydroxy, 2-decen transoic acid, first isolated from royal jelly, and bound to glycerol as a lipophilic carrier, subsequently referred to as di DHA-glycerol. When administered in mice, we have previously shown that the drug protected mice against a lethal challenge with virulent *S.typhimurium*. But no direct correlation was seen between the level of non specific resistance and NO production. In the present study, we investigated the relationship between bacterial clearance, TNF-α and NO production both in splenocytes and peritoneal macrophages of BALB/c mice, first immunostimulated or not with the di DHA-glycerol before a virulent challenge with *S.typhimurium* (Batches of 33 mice : 4 daily injections ip of either 5 µg or 15 µg/mouse of di DHA-glycerol ; infectious challenge was performed ip the day after the last injection : 70 CFU *S.typhimurium* C5/mouse, approximately 1 LD_{50}).

Our results showed that in the first 6 days post challenge, Salmonella growth rate was identical in the 3 groups. Likewise, no TNF-α or NO_2^- production could be evidenced either in splenocytes or macrophage cultures (upon in vitro restimulation with heat killed bacteria). By day 8 post challenge, bacterial clearance began in the 2 treated groups (mean : 4.0 Log_{10} CFU in the 5 µg treated group ; 5.2 Log_{10} CFU in the 15 µg treated group ; 6.2 Log_{10} CFU in the infected controls). At that time, NO_2^- production was higher in the group treated with 5 µg di DHA-glycerol than in the 2 other groups (up to 58 µM NO_2^- *vs* 42 µM in controls) (figure 1).On the contrary, in peritoneal macrophage cultures, the highest rates of NO_2^- were seen in the infected controls, the lowest ones in the 5 µg treated group. In addition, the most abundant production of TNF-α was observed in splenocytes cultures from 5 µg treated mice (twice the EC_{50} of mr TNF-α used (0,03 ng/ml) on L929 target cells) (figure 2). At that time, no TNF-α secretion was evidenced in peritoneal macrophage cultures. By day 13, bacterial clearance continued (mean : 3.3 Log_{10} CFU in the 5 µg treated group, 5.4 Log_{10} CFU in the 15 µg treated group, 6.7 Log_{10} CFU in the infected control). NO_2^- levels were still increasing up in the 2 treated groups although not tightly bound to the bacterial counts. Likewise, peritoneal macrophage cultures showed twice higher NO_2^- levels in the

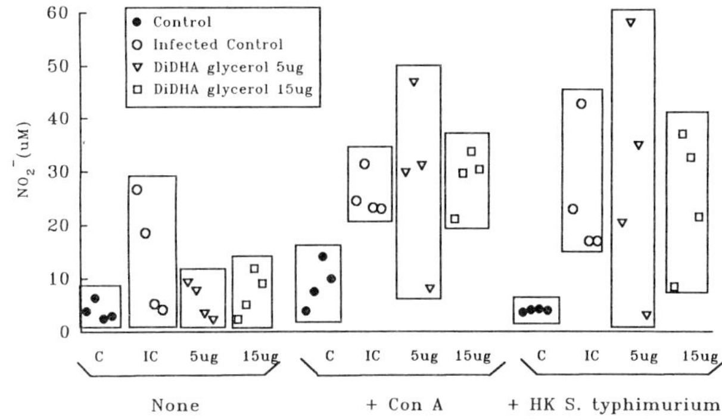

FIGURE 1 : NO_2^- production in splenocyte cultures from mice immunostimulated or not with di DHA-glycerol, 8 days after a virulent challenge with *S.typhimurium.*

5 µg treated group than in the 15 µg one. On the contrary, TNF-α secretion decreased in both treated groups below the EC_{50} level as defined above with mr TNF-α (figure 2).

In conclusion, the immune potentiation of host's natural defenses through appropriate immunomodulatory therapy can lead to enhanced non specific resistance against pathogens. Among the numerous mediators probably involved in bacterial clearance and enhanced survival rate, adequate levels of TNF-α and NO responses during the time course of infection are of major importance. The imbalance of one of these mediators with others involved in autoregulation process can lead to the full evolution of septic shock.

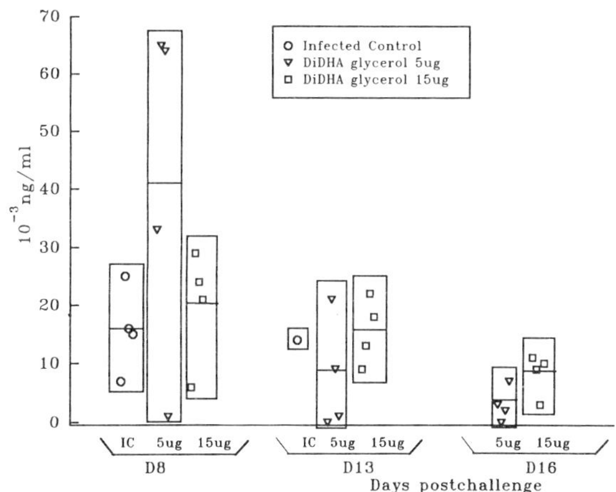

FIGURE 2 : TNF-α secretion in splenocyte cultures from mice immunostimulated or not with di DHA-glycerol 8 days, 13 days or 16 days after a virulent challenge with *S.typhimurium.*

Abbreviations used : TNF-α : Tumour Necrosis Factor ; IFN-γ : Interferon γ ; DiDHA-glycerol : Di 1,2 (10-hydroxy 2-decenoyl) glycerol ; mr TNF-α : murine recombinant TNF-α ; ip : intraperitoneal ; FCS : fetal calf serum.

ACKNOWLEDGEMENTS : This work was supported by a grant from Direction de la Recherche et de la Technologie.

REFERENCES

[1]Moncada, S., Palmer, R.M.J., and Higgs, E.A. (1991). Nitric oxide : physiology, pathophysiology, and pharmacology. Pharmac. Rev. 43, 109-142.

[2]Drapier, J.C., Wietzerbin, J., and Hibbs, J.B., Jr. (1988). Interferon-gamma and tumour necrosis factor induce the L-arginine-dependent cytotoxic effector mechanism in murine macrophages. Eur. J. Immun. 18, 1587-1592.

Nitric Oxide Appears Protective in Tanzanian Children with Malaria: Evidence for Increased NO Production in Subclinical Infection and Suppressed Production in Clinical and Cerebral Malaria.

NM Anstey[1,3], MY Hassanali[2], ED Mwaikambo[2], D Manyenga[1], J Mlalasi[1], JB Weinberg[4], D Arnelle[5], MI McDonald[1,3] and DL Granger[3]

[1]Duke-Muhimbili Clinical Research Lab and [2]Dept of Paediatrics, Muhimbili Medical Centre, Dar es Salaam, Tanzania; Divisions of [3]Infectious Diseases, [4]Hematology and [5]Pulmonary Medicine, Box 3824, Duke University Medical Center, Durham, NC, 27710 USA.

BACKGROUND AND AIMS: Parasite sequestration within cerebral blood vessels and cytokine-mediated immunopathology are both thought to be important in the pathogenesis of cerebral malaria (CM), a major cause of death in African children. Nitric oxide has been shown to be protective against *Plasmodium falciparum in vitro* [1,2]. However it has been hypothesised that high levels of tumour necrosis factor (TNF) in CM contribute to the pathogenesis of CM *in vivo* through excessive local synthesis of nitric oxide and altered neurotransmission [3]. We thus compared markers of NO production (plasma and urinary nitrate+nitrite [NOx]) in Tanzanian children with and without cerebral malaria.

METHODS: Dietary history, urine and plasma were collected at Muhimbili Medical Centre, Dar es Salaam, May 1994-Jan 1995, from five groups of children: diet controlled healthy controls (HC) [n=40], diet controlled asymptomatic parasitaemia (AP) [n=10], clinical malaria with no cerebral symptoms or signs (NCM) [n=53], cerebral malaria with unrousable coma but subsequent complete recovery (CMCR) [n=50] and cerebral malaria complicated by death or neurological sequelae (CMDS) [n=36]. Urine nitrate was quantitated by bacterial nitrate reductase conversion of nitrate to nitrite, with measurement of nitrite using the Griess Reaction. Spot urine nitrate:creatinine ratios were computed. Plasma NOx was measured by capillary electrophoresis of ultrafiltrated plasma, with subsequent correction for trace amounts of nitrate in blood collection tubes and ultrafilters. Plasma creatinine and urea levels were measured with an autoanalyzer. Differences between study groups were determined by ANOVA with post hoc analysis using Fisher's PLSD.

RESULTS: There were no significant differences in age, gender or weight among study groups. Mean duration of fasting in each group was: HC 13.0 hrs, AP 13.0 hrs, NCM 5.6 hrs, CMCR 16.2 hrs and CMDS 17.7 hrs. Urine nitrate:creatinine ratios were *inversely* correlated with disease severity, with levels lowest in fatal CM: HC 0.39, AP 0.55, NCM 0.23, CMCR 0.16 and CMDS 0.12 (CM and NCM both differ from HC and AP, and CM differs from NCM, all with $p<0.05$). Plasma NOx levels were also decreased in clinical and cerebral malaria but increased in the AP group: HC 39.6uM, AP 57.8uM, NCM 31.3uM, CMCR 28.5uM and CMDS 36.5uM (AP differs from HC, NCM, CMCR and CMDS; HC differs from NCM and CMCR, all with $p<0.05$).

Renal impairment was found only in those with CM, with 31% having an elevated plasma urea or creatinine for age. Because nitrate is excreted renally, plasma nitrate levels are increased in renal failure [4]. Plasma NOx levels were therefore corrected for renal function and expressed as plasma NOx:creatinine ratios: AP 1.84, HC 1.18, NCM 0.88, CMCR 0.62 and CMDS 0.60. The AP and HC groups differ from each of the NCM, CMCR and CMDS groups, all with $p<0.05$. Following this correction for renal function, the differences among groups become even more apparent. Corrected plasma NOx in CMDS was the lowest of all groups, and that in the AP group remained the highest.

CONCLUSIONS: Plasma NOx and urinary NOx excretion are decreased in both clinical and cerebral malaria, with levels inversely proportional to disease severity. This suggests that there is *suppression* of NO synthesis rather than excessive production in clinical malaria in children. Our results cannot be explained by differences in dietary nitrate ingestion. Because NO has been shown to inhibit Plasmodium *in vitro*, decreased levels of NOx in clinical disease imply that suppression of NO-mediated protection may be responsible for greater parasite replication and may contribute to the development of clinical and cerebral malaria in children. Because the NOx values in clinical and cerebral malaria are so low (below the level observed in the HC group) the low levels could result from diminished activity of both the constitutive and inducible isoforms of nitric oxide synthase (NOS).

In this control group of children from a holoendemic region with high rates of malaria transmission, NOx levels may not reflect constitutive synthesis alone. It is possible that control children have additional low level iNOS activity from repeated exposures to malaria-infected mosquitoes, and that this protects them from developing parasitaemia.

Plasma NOx is increased in malaria infection *without* disease (asymptomatic parasitaemia), indicating maximal NO production in this group. Enhanced NO synthesis may protect these children from developing overt disease. Nitric oxide appears to have a protective rather than pathological role in paediatric malaria in Africa.

ACKNOWLEDGEMENTS: Supported in part by an American Society of Tropical Medicine and Hygiene Becton Dickinson Fellowship (NA).

REFERENCES:

1. Gyan, B., Troye-Blomberg, M., Perlmann, P. & Björkman, A. (1994). Human monocytes cultured with and without interferon-gamma inhibit *Plasmodium falciparum* parasite growth *in vitro* via secretion of reactive nitrogen intermediates. Parasite Immunology.**1 6**, 371-375.

2. Mellouk, S., Hoffman, S. L., Liu, Z. Z., de la Vega, P., Billiar, T. R. & Nussler, A. K. (1994). Nitric oxide-mediated antiplasmodial activity in human and murine hepatocytes induced by gamma interferon and the parasite itself: enhancement by exogenous tetrahydrobiopterin. Infection and Immunity.**6 2**, 4043-4046.

3. Clark, I. A., Rockett, K. A. & Cowden, W. B. (1991). Proposed link between cytokines, nitric oxide, and human cerebral malaria. Parasitology Today.**7**, 205-207.

4. Strand OA, Granli T, Kirkeboen KA, Dahl R, Bockman OC. (1995) Serum and 24 hour urine excretion of nitrate in human septic shock. Endothelium. **3**, s98.

Nitric oxide and human cerebral malaria

KIRK A. ROCKETT†, DOMINIC K. KWIATKOWSKI*, FADWA AL YAMAN, CLIVE A. BATE*, PETER W. GAGE, MELISSA M. AWBURN, and IAN A. CLARK

Australian National University, Canberra, ACT 0200, Australia, and *Institute of Molecular Medicine, John Radcliffe Hospital, Oxford, OX3 9DU, U. K.

Induction of nitric oxide by malaria parasites

Malarial illness and pathology are now generally accepted to be caused by cytokines induced by material released when the infected red cells burst, releasing the new wave of merozoites[1]. The material released from *Plasmodium falciparum* has been partially purified and shown to stimulate macrophages to make TNF[2] . We have extended this work to show that these same preparations, isolated from parasitized erythrocytes, stimulate the mouse macrophage cell line RAW 264.7 to produce inducible nitric oxide synthase and release nitric oxide. iNOS was demonstrated using a specific antibody kindly provided by Dr Tom Evans, (Hammersmith), and nitric oxide production was monitored via the Griess reaction and by HPLC (by the generation of ^{14}C.citrulline from ^{14}C.arginine). The malarial extract, prepared by water lysis and solvent extraction of infected and control red cells, was active only in the presence of interferon-gamma, which had only low activity itself. Activity was not caused by contaminant LPS. By using cytokine-specific antisera we have found that this induction of nitric oxide is independent of TNF and IL-1α and partly independent of IL-1β. These results are consistent with arguments that various aspects of malaria pathology, including immunosuppression, cerebral malaria, and cardiovascular changes, are mediated by nitric oxide.

The effect of macrophage nitric oxide on neuronal NMDA channels

The traditional concept of human cerebral malaria, based on obstruction of cerebral blood flow by sequestered parasitised red cells, does not in our view adequately explain the changes in mental status seen in this condition, since patients recovered from cerebral malaria have a low incidence of the type of residual neurological deficit that accompanies cerebral ischaemia when it is severe enough to induce even a short episode of coma. We have proposed that the altered mental states observed in human cerebral malaria may be caused, in part, by nitric oxide, induced in cerebral vascular walls by malaria toxin, diffusing through the blood/brain barrier and, in parallel with the action of general anaesthetics and ethanol, modulating NMDA channels in post-synaptic cells[3].

To test this, NMDA receptor-mediated current in rat hippocampal neurons, were measured in whole cells using a patch clamp technique. Neurons were exposed to a buffer solution containing both NMDA and cells of mouse macrophage lineage, RAW 264.7, which had been stimulated to produce nitric oxide. The NMDA current was potentiated by the stimulated cells. The potentiation was abolished in the presence of L-NMMA or haemoglobin, but was not affected when the medium contained D-NMMA, or L-NMMA plus excess arginine. In other experiments the frequency of NMDA receptor-mediated currents in inside-out patches excised from rat hippocampal neurons was either enhanced or decreased depending on the distance between the patch and the macrophages producing nitric oxide. These results suggest that nitric oxide produced by the stimulated cells is responsible for the change in NMDA currents we observed, and are consistent with the idea that extra-neuronal nitric oxide induced by products of schizogony contributes to the seizures and unconsciousness seen in cerebral malaria.

The correlation of serum nitrate with coma in falciparum malaria.

In order to establish whether nitric oxide generation in malaria patients is consistent with these concepts, serum levels of RNI (nitrate plus nitrite) were measured in 92 coastal Papua New Guinean children with cerebral malaria. RNI levels were compared to disease severity and clinical outcome, and correlated with both the depth of coma on admission and its duration. Median levels were higher among children with the deepest coma than among those with lighter coma ($p = 0.008$) and also among children with the longest duration of coma (> 72 hours) ($p = 0.004$). RNI levels also correlated with clinical outcome, with fatal cases having higher RNI levels ($p = 0.014$). Thus high RNI levels are associated with indices of disease severity and may predict outcome in children with cerebral malaria.

Nitrate consumption in food or water will inevitably lead to some individual variation in serum RNI levels, and must be taken into account in studies such as this. Malarial patients who are more deeply comatose on admission have a significantly longer duration of coma before presentation[4], during which time they would have been unable to eat or drink, and thus ingest nitrate. Thus the proportion of admission serum RNI levels attributable to diet in the deeply comatose children, the patients in whom RNI levels were in fact highest in our study, would be lower than in any other group. Therefore very little of the serum nitrate we measured in our deeply comatose patients is likely to have been of dietary origin. These field data, taken in conjunction with the *in vitro* experiments also outlined here, are consistent with the idea that nitric oxide could contribute to the pathogenesis of the coma in human cerebral malaria. More closely focussed experimental approaches that will provide direct evidence for our hypothesis are in progress.

1. Clark I.A., Virelizier J.-L., Carswell E.A., Wood P.R. (1981) Infect. Immun. **32**, 1058-1066
2. Bate C.A.W., Taverne J., Bootsma H.J., Mason R.C.S., Skalko N., Gregoriadis G., Playfair J.H.L. (1992) Immunol. **76**, 35-41
3. Clark I.A., Rockett K.A., Cowden W.B. (1991) Parasitol. Today **7**, 205-207
4. Mabeza G.F., Moyo V.M., Thuma P.E., Biemba G., Parry D., Khumalo H., Nyarugwe P., Zulu S., Gordeuk V.R. (1995) Ann. Trop. Med. Parasitol. **89**, 221-228

Abbreviations used: NMDA (N-methyl-D-aspartate), RNI (reactive nitrogen intermediates), TNF (tumour necrosis factor), LPS (lipopolysaccharide, IL-1 (interleukin-1).

†Present Address: Department of Paediatrics, John Radcliffe Hospital, University of Oxford, Oxford, OX3 9DU, U. K.

Role of activation of poly-ADP ribosyl synthetase in the cellular energy depletion in immunostimulated macrophages *in vitro* and *ex vivo.*

BASILIA ZINGARELLI, ANDREW L. SALZMAN and CSABA SZABÓ

Division of Critical Care, Children's Hospital Medical Center, 3333 Burnet Avenue, Cincinnati, OH, 45229

Peroxynitrite, a highly toxic oxidant species is produced during circulatory shock from the reaction of nitric oxide (NO) and superoxide anion [1-2]. We have demonstrated that peroxynitrite cytotoxicity is mediated by DNA strand-breakage and activation of the DNA repair enzyme poly (ADP) ribosyltransferase (PARS), whereas NO cytotoxicity does not involve the PARS pathway [3].

Immunostimulation of macrophages and other cell types is associated with inhibition of mitochondrial respiration, which is prevented by inhibition of the inducible isoform of NO synthase (iNOS). Immunostimulation also results in the production of superoxide anion. Therefore, we hypothesized that immunostimulated macrophages produce peroxynitrite in amounts which influences cellular functions. In support of this proposal, we have shown in macrophages that the suppression of mitochondrial respiration, in response to the peroxynitrite donor compound SIN-1, authentic peroxynitrite or immunostimulation, can be prevented by uric acid, which may act as a scavenger of peroxynitrite [4]. In the present study we investigate whether activation of PARS, an energy-consuming futile repair circle, may play a role in the cellular injury in response to endogenously generated peroxynitrite.

J774 macrophages were cultured as described [4]. The formation of peroxynitrite by NO and superoxide donor and peroxynitrite generator compounds was measured by the oxidation of dihydrorhodamine 123 [5]. Mitochondrial respiration was measured by the MTT assay [4]. DNA single strand breaks were determined by the alkaline unwinding method [6]. PARS activity was measured using radiolabeled NAD^+ [6]. Cellular NAD^+ levels were measured by a colorimetric assay. ATP levels were measured using HPLC. Cellular glutathione was measured as described [6].

Activation of J774 macrophages with bacterial lipopolysaccharide (LPS, 10 μg/ml), produced superoxide (Fig. 1a), NO (Fig. 1b) and peroxynitrite (Fig. 1c) 24h after stimulation. This was associated with DNA strand breakage (Fig. 1d), NAD^+ depletion (Fig. 1e), glutathione depletion (Fig. 1f) and inhibition of mitochondrial respiration (Fig. 1g). Depletion of NAD^+ and ATP and inhibition of respiration were prevented by inhibition of NOS by N^G-methyl-L-arginine (L-NMA, 3 mM), and by inhibitors of PARS, such as 3-aminobenzamide (1 mM) (Fig 2a-c) or nicotinamide (1 mM) (not shown). The loss of cellular glutathione, however, was unaffected by 3-aminobenzamide or L-NMA (Fig. 2d). 3-aminobenzamide caused a slight (21±2 %) inhibition of nitrite production (Fig. 2c). This inhibition is unlikely to be responsible for the protective effect, as similar slight inhibition of NO production (by a lower concentration of L-NMA) was not protective.

Similarly, in peritoneal macrophages obtained from rats treated with LPS (15 mg/kg i.p.), enhanced nitrite/nitrate (from 5±1 to 102±24 μM), superoxide (from 19±13 to 261±38 pmoles/min/million cells) and peroxynitrite production (from 4±2 to 97±8 pmoles/min/million cells), DNA single strand breakage (from 38±9 to 83±15%), NAD^+ depletion (to 45±9% of control at 5h), ATP depletion (to 54±3% of control) and inhibition of mitochondrial respiration (to 65±6% of control) was found at 5h after LPS. The maximum of DNA strand breakage occurred at 1-2h after LPS, whereas the maximal NAD^+ depletion and inhibition of mitochondrial respiration occurred at 3-5h. Pretreatment of the rats with 3-aminobenzamide (10 mg/kg), largely prevented the ATP and NAD^+ depletion and the depression of mitochondrial respiration, but did not diminish the production of nitrite/nitrate, superoxide or peroxynitrite by these cells.

Thus, activation of PARS, and the subsequent energy-consuming DNA repair process, plays an important role in the energy depletion in immunostimulated macrophages. DNA strand breakage, the trigger of PARS activation, is probably caused by endogenously produced peroxynitrite and not NO *per se,* since we have found that exogenous peroxynitrite, but not NO (or superoxide) is a potent inducer of DNA strand breakage [3].

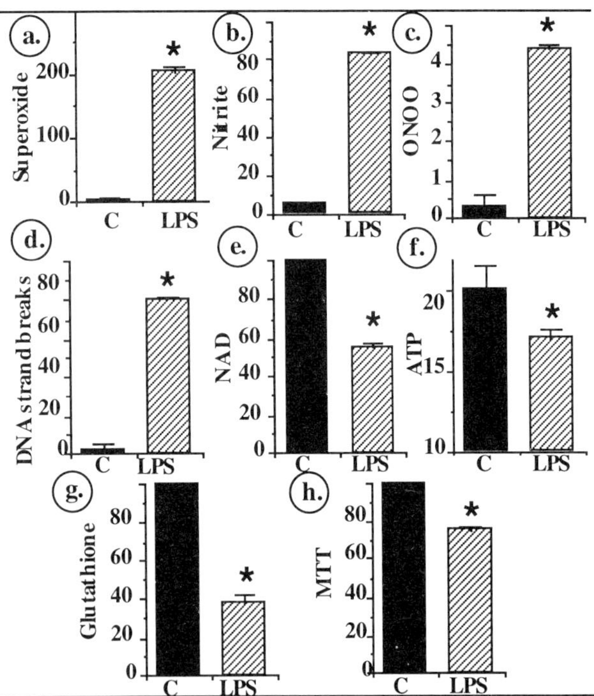

Fig. 1. Effect of LPS on the production of (a): superoxide (pmoles/ million cells/ min); (b): nitrite (μM); (c): peroxynitrite (pmoles/ million cells); (d): DNA strand breaks (% of maximal). (e-h): Suppression by LPS of NAD (% of control), ATP (pmoles/ million cells), cell respiration and glutathione content (% of control).

Abbreviations used: 3-AB, 3-aminobenzamide, L-NMA, N^G-methyl-L-arginine; LPS, lipopolysaccharide; NO, nitric oxide; PARS, poly (ADP) ribosyltransferase

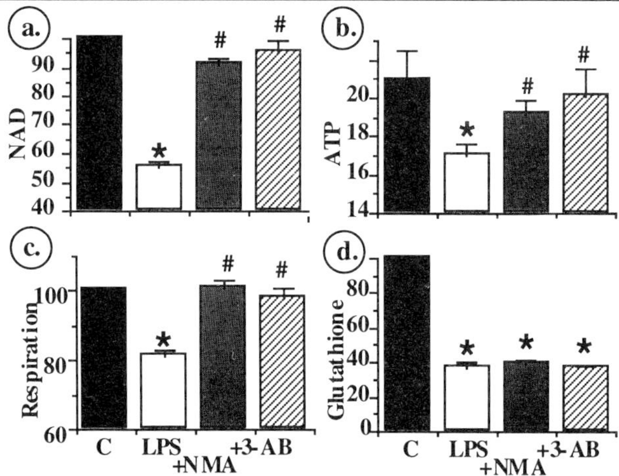

Fig. 2. (a) NAD, (b) ATP and (d) glutathione content and (c) cellular respiration (a, c, d expressed as % of control, b expressed as pmoles/million cells) in J774 cells stimulated with LPS for 24h. *represents significant effect of LPS and #represents significant protection by L-NMA or 3-AB ($p<0.05$).

References

1. Szabó, C., Salzman, A.L, Ischiropoulos, H. (1995) FEBS Lett. **363**, 235-238
2. Szabó C. (1995) New Horizons, **3**, 3-32
3. Szabó, C., Zingarelli, B., O'Connor, M., Salzman, A.L (1995) Endothelium **3** (Suppl.), 46 (abstr.)
4. Szabó, C., Salzman, A.L (1995) Biochem. Biophys. Res. Comm. **209**,739-743
5. Kooy, N., Royall, J., Ischiropoulos, H., Beckman, J. (1994) Free Rad. Biol. Med. **16**, 149-155
6. Cochrane, C.G. (1991) Molec. Aspects Med. **12**, 137-147

HIV-1 gp120 reduces the expression of nNOS and increases NGF in the hippocampus of rat

GIACINTO BAGETTA*, MARIA T. CORASANITI°, LUIGI ALOE§, LAURA BERLIOCCHI**, NICOLA COSTA°, ALESSANDRO FINAZZI-AGRO'^ and GIUSEPPE NISTICO'**

*Department of Neuroscience, University of Cagliari, Cagliari (Italy); °Faculty of Pharmacy and IBAF-CNR, 88100 Catanzaro (Italy); §CNR Institute for Neurobiology, 00137 Rome (Italy); **"Mondino-Tor Vergata" Neurobiology Center, Department of Biology and of ^Experimental Medicine and Biochemical Sciences, University of Rome "Tor Vergata", 00133 Rome

Introduction

The HIV-1 coat glycoprotein gp120 may be the aetiologic agent of human brain neuronal loss often observed at post-mortem in AIDS patients because it causes death of several types of neurones maintained in culture. In fact, exposure to picomolar concentrations of gp120 produces death of rodent hippocampal neurones, retinal ganglion cells [see 6] and cerebellar granule cells [7]. The mechanism through which gp120 produces its detrimental effects involves excessive Ca^{2+} entry into neurones via NMDA receptor associated cation channel and through voltage operated Ca^{2+} channels since NMDA antagonists and Ca^{2+} channel blockers prevent neuronal cell death [see 6,7]. *In vitro* experiments have recently shown that exposure of cortical neurones to gp120 increases nitric oxide (NO) and this seems to be involved in the mechanism of gp120-induced death because inhibition of NO synthesis abolished the cytotoxic effects of the HIV-1 coat protein [4]. Here we now report that, in rats, intracerebroventricular (i.c.v.) microinfusion of gp120 for 14 days *reduces* the expression of the constitutive, neuronal type [see 5] of NOS (nNOS) and *increases* NGF in the hippocampus and these effects are seen in the absence of neuronal cell death.

Methods

Adult male Wistar rats (250-280 g) were housed in a temperature (22 °C)- and humidity (65%)-controlled colony room. Under chloral hydrate (400 mg/kg i.p.) anaesthesia, rats were implanted with a cannula in one lateral cerebral ventricle under stereotaxic guidance. HIV-1 gp120 (from baculovirus expression system, >90% pure) was injected i.c.v. once daily via an injector connected to a 5 μl Hamilton syringe (1 μl volume of infusate; 1μl/min rate). Hippocampal RNA isolation and electrophoresis were performed as previously described [see 1] and the filters generated were hybridized to a 4.8 kb rat brain nNOS cDNA probe spanning the whole gene and labelled with ^{32}P[dATP] by random priming [see 1]. The blots were then washed and exposed for 72 h to Kodak X-Omat AR autoradiography film. NOS mRNA signal intensities were evaluated by laser densitometry [see 1]. NOS activity was measured in individual rat brain hippocampal homogenates as previously reported [see 1] and the results expressed as pmol/min/mg protein of [^{3}H]citrulline formed. NGF was evaluated by a two-site immunoassay method [see 3]. The data represent the mean±s.e. mean of 4-12 experiments per group.

Results

Densitometric analysis of nNOS mRNA signal (10.5 Kb) from the homogenate of the hippocampus of four individual rats receiving gp120 (100 ng/rat/day, given i.c.v. for up to 14 days) yielded a 2-fold decrease at 14 days, but not at 1 day and 7 days, as compared to the effect of similar treatments with BSA (300 ng/rat/day). This effect was accompanied by a significant ($P<0.01$) decrease of the Ca^{2+}- calmodulin-dependent [^{3}H]citrulline formation in the rat hippocampus (control= 115±8, gp120= 76±5, pmol/min/mg of protein); no significant changes in NOS activity were observed at 1 day and 7 days of gp120 treatment.

In parallel experiments it has been observed that the concentration of NGF in the hippocampus were doubled by i.c.v. gp120 (100 ng/rat/day) given for 7 (control= 1834±518, gp120=3100±51, pg/g tissue) and 14 (control= 1420±189, gp120= 3538±453, pg/g tissue) consecutive days but not in the cerebral cortex (control= 1006±151 and 889±31 pg/g tissue at 7 and 14 days of BSA treatment, respectively; gp120= 1081±91 and 804±88 pg/g tissue at 7 and 14 days, respectively). Histological examination of brain coronal tissue sections (10 μm; n=6 per rat) from rats (n=6) receiving i.c.v. gp120 for 14 days in no instance showed gross hippocampal damage nor did it reveal *in situ* DNA fragmentation (data not shown); the latter is in contrast with the DNA fragmentation seen in the brain cortex [2].

Discussion

The present data show that in rats i.c.v. microinfusion for 14 days of gp120 reduces the expression of nNOS and increases NGF level in the hippocampus where no signs of neuronal death have been observed. The reduction of nNOS mRNA produced by gp120 is likely to be accompanied by a parallel decrease of the gene product. This is reflected by a significant decrease of Ca^{2+}- calmodulin-dependent NOS activity detected in the same region of the rat brain. The effects of gp120 appear to be specific because a treatment with 3-fold excess of BSA for up to 14 days did not affect significantly nNOS mRNA signal and nNOS activity.

In the mammalian brain NO plays several important roles, ranging from intercellular messenger to transducing mechanism for discrete sensory stimuli; NO has also been suggested to couple cerebral blood flow to neuronal cell activity thus subserving an important trophic function in the central nervous system [see 5]. Consistent with the latter function, reduced nNOS mRNA level with consequent decreased production of citrulline, the co-product of NO synthesis, by i.c.v. gp120 may lead brain neurones to death though this may require more than 14 days of treatment to be disclosed.

On the other hand, the observed increase in NGF may prevent or delay the onset of hippocampal cell death. The latter hypothesis is supported by the occurrence of apoptotic cell death in the brain cortex of rats receiving i.c.v. injection of gp120 for up to 14 days but in which no changes in NGF were seen [2].

Acknowledgements

Supported by the VIII 1995 AIDS Project, Istituto Superiore di Sanità, Rome (Italy). The MRC AIDS Directed Programme Reagent Project and Dr H.C. Holmes (NIBSAC, South Mimms, UK) are gratefully acknowledged for generous supply of gp120.Our thanks to Ms E. Baboro for excellent typing of the manuscript and to Mr G. Politi for technical assistance.

References

1. Bagetta, G., Corasaniti, M.T., Berliocchi, L., Navarra, M., Finazzi-Agrò, A. and Nisticò, G. (1995) Biochem. Biophys. Res. Commun. **211**, 130-136.
2. Bagetta, G., Corasaniti, M.T., Aloe, L., Berliocchi, L., Costa, N., Finazzi-Agrò, A. and Nisticò, G. (1995) Proc. Natl. Acad. Sci. USA. (in press).
3. Bracci-Laudiero, L., Aloe, L., Levi-Montalcini, R., Buttinelli, C., Schilter, D., Gillessen, S. & Otten, U. (1992) Neurosci. Lett., **147**, 9-12.
4. Dawson, V.L., Dawson, T.M., Uhl, G.R. and Snyder, S.H. (1993) Proc. Natl. Acad. Sci. USA **90**, 3256-3259.
5. Knowles, R.G. and Moncada, S. (1994) Biochem. J. **298**, 249-258.
6. Lipton, S.A. (1992) Trends Neurosci. **15**, 75-79.
7. Savio, T. and Levi G. (1993) J. Neurosci. Res. **34**, 265-272.

Spontaneously-occurred inducible nitric oxide synthase (iNOS) in adenoma and adenocarcinoma in human colon.

Toru Kono, Naotoshi Ando, Akitoshi Kakisaka, Shin-ichi Kasai, and Jun Iwamoto[1]

Department of Surgery and Department Physiology[1], Asahikawa Medical College, Asahikawa, 078 Japan.

Nitric oxide synthesized by iNOS mediates cytotoxicity against parasites and tumor cells. Interestingly, human adenocarcinoma cell line would also occur iNOS *in vitro* in response to cytokines and LPS (1). In pursuit of iNOS in the actual human tumor of the colon, we surveyed both a trace of NO in the colonal gas by chemiluminescence and the endoscopically-resected tissues employing the iNOS immuno-histology. We also surveyed the surgically-resected human adenocarcinoma for iNOS-like immunoreactivity. During colonoscope, each segment of the colon (ascending, transverse, descending, rectum) was carefully inspected in 9 patients with abdominal symptoms or histories of colonal disorders. The luminal gas from each colon segment was separately collected in an air-tight syringe for further measurements of NO by chemiluminescence (Model 270B, Sievers). Calibration for NO analyzer was carried out with a 10 to 100 μl of NO calibration gas at 5 ppm, which is detailed elsewhere (2). Colonal gas collection was performed as follows. The gas inlet for colonal distension of the conventional colon fiberscope (Fuji film) was utilized to inject compressed air to wash out. Suction and injection of air was repeated several times. Immediatelly after last suction, a 100 ml of nitrogen gas was injected and kept staying for 20 seconds and then a 10 ml of sample gas was collected.

Among 36 sites inspected (Table 1), polyps were found in 13 segments among 36 inspected segments. The polyps were resected and subjected to the iNOS immunohistology and pathological examination. The iNOS-like immunoreactivity was found in all resected adenomas and adenocarcinomas. The luminal NO above 2 ppb was taken as NO-positive, since there were no values higher than 2 ppb in 12 control subjects. NO was varied from 2.3 to 99.8 ppb for the 17 NO-positive segments with 10 true positivities (TP, matchings with endoscopic positives). However, there were substantial numbers of false positives (FP) in 7 segments among 23 normally-visible segments (Table 2).

Table 1. Colon tumors and luminal NO

	Ascending colon		Transverse colon		Discending and sigmoidal colon		Rectum	
sex age	tumor size	NO (ppb)	tumor size	NO (ppb)	tumor size	NO (ppb)	tumor size	NO (ppb)
Adenomas								
M 54	N	0.0	5mm	10.1	N	0.0	N	0.0
F 54	3mm	68.0	N	69.0	N	5.2	N	0.0
M 69	N	35.4	N	99.8	5-10 mm	0.7	N	0.0
M 52	8mm	65.6	N	0.0	15mm	13.3	N	0.0
F 61	4mm	0.0	N	0.0	7mm	0.0	12mm	6.8
Malignant tumors								
M 57	25mm	46.3	N	1.0	N	2.3	N	0.0
F 57	N	8.6	N	5.2	4mm	17.2	50mm	0.0
M 40	N	0.0	large *1	21.1	N	0.3	large *2	19.6
M 50	N	0.0	N	0.0	40mm	13.9	N	0.0

*1 &*2 lymphoid hyperplasia

Table 2. Matchings for NO- and Endoscope positive segments.

	segments	matchings
True positive	10	10/17
True negative	15	15/19
False positive	7	7/17
False negative	4	4/19

The 4 FP segments out of 8FP were detected in the segments rostral to the larger tumor in caudal region and related to severe constipation. These FP might be related to non-specific dysfunction of bowel movement. The other 2 FP segments are adjacent to the higher TP region, so that they may be resulted from technical error. The rest of 19 segments had no luminal NO with only 4 false negatives (FN). These 4 FN involve 2 bleeding segments, suggesting that blood hemoglobin cleared colonal NO. In addition, we also detected colonal NO from diverticles and mucosal aphta.

A good parallelism seems to exist in the relationship between the colonal NO and the colonoscopic diagnosis for colon tumors. The colon tumors, either adenoma adenocarcinoma, or malignant lymphoma, could occur iNOS which catalyzes detectable NO in the lumen of the colon. However, the concentration of NO from these polyps and tumors are much lower than that from ulcerative colitis. Unlike abundant NO detected in the colon of ulcerative colitis (3), NO in the human tumors is less than 100 ppb. It is likely that low NO level is merely related to the low grade inflammation occured in those neoplasms. Alternatively, NO from colon tumors may be related to the specific immunological roles such as immunosuppression by the experimental colon cancer againt host immune cells (4). Further investigation is needed to clarify significance of NO in the human colon tumors.

Acknowledgement: The authors thank to Yoko Ikoma for her expert immunohistological technique. This work is supported by research grants 07557185 and 07670077 from the Japanese Ministry of Art and Education.

References

1. Sherman PA, Laubach VE, Reep BR, Wood ER. Purification and cDNA sequence of an inducible nitric oxide synthase from a human tumor cell line. (1993) Biochemistry 32, 11600-11605.
2. Iwamoto J, Pendergast DR, Suzuki H, Krasney JA. Effect of graded exercise on nitric oxide in expired air in humans. (1994) Respir. Physiol. 97, 333-345.
3. Lundberg JON, Hellstrom PM, Lundberg JM, Alving K. Greatly increased luminal nitric oxide in ulcerative colitis. (1994) Lancet 344, 1673-1674.
4. Lejeune P, Lagadec P, Onier N, Pinard D, Ohshima H, Jeannin J-F. Nitric oxide involvement in tumor-induced immunosuppression. (1994) J. Immunol. 152, 5077-5083.

Estrogen and cytokines inhibit NO mediated cGMP production in human endothelial cells.

Dan L. Wood and Jai Pal Singh

Cardiovascular Research, Lilly Research Laboratories, Indianapolis, IN, 46285

Introduction:

The protective role of estrogen in coronary artery disease is well accepted [1]. However, the mechanism for the beneficial actions of estrogen is not yet understood. Some of the estrogen benefit in vascular disease may be mediated by its effect on serum lipids, or on factors affecting thrombosis. Estrogen also appears to exert a direct effect on arteries. Recent studies have shown that estrogen replacement therapy improves endothelial dependent vasodilatation in hypercholesterolemic, post menopausal women [2]. Intravenous administration of estrogen has been shown to improve the coronary artery response to acetylcholine [3]. In non-human primates, estrogen treatment restores the coronary response to acetylcholine [4]. The mechanism of estrogen action on vascular reactivity has not been defined. Estrogen treatment has been shown to enhance the expression of NO synthase in the heart, kidney, skeletal muscle and cerebellum in non-pregnant guinea pigs [5]. A preliminary study using cultured human aortic endothelial cells has also shown that 48 hour incubation with estrogen leads to an increased expression on NO synthase protein [6]. These studies suggest that estrogen may effect vascular responses by modulating NO synthesis. In the present study we have investigated the effect of estrogen on endothelial NO generation under inflammatory conditions. Our results show that 24 hour treatment of endothelial cells with inflammatory cytokines reduced the production of NO dependent cGMP in human endothelial cells. Treatment with cytokines plus physiological concentrations of estrogen resulted in a further attenuation of cGMP synthesis. These results show that estrogen modulated cGMP synthesis in endothelial cells treated with inflammatory cytokines.

Methods:

Cell Culture: Human umbilical vein endothelial cells (HUVEC) were obtained from Clonetics Corporation. The cells were grown in gelatin coated plates in phenol red free Medium 199 containing 10% FBS, 2mM L-glutamine, 100 U/ml penicillin, 100 µg/ml streptomycin, 50 µg/ml ECGS, and 100 µg/ml heparin.

Determination of NO Dependent cGMP Synthesis: Endothelial cells in 12 well plates were grown to confluence. Cells were then washed, and transferred to HBSS supplemented with 0.5 mM isobutyl methyl xanthine, 1.2 mM calcium chloride, 0.6 mM magnesium sulfate, and 10 µM arginine for 30 minutes. Cells were then stimulated with A23187 (1 µM), histamine (10 µM), or sodium nitroprusside (1 mM). After 15 minutes the buffer was removed and cGMP was extracted in 250 µl of 0.01 N HCl. The extract was neutralized 1N NaOH, and cGMP was determined using an enzyme immuno-assay (Amersham). The protein concentration of cell extracts was determined by the BCA method (Pierce kit).

Determination of the Effect of Cytokines and Estrogen on cGMP Synthesis: Endothelial cultures were first pre-incubated in the above medium for 24 hours, with or without cytokines, {IL-1β (5U/ml), TNF-α (200 U/ml), INF-γ (200 U/ml)}, or with cytokine treatment plus the indicated concentrations of 17β-estradiol. The cGMP response to agonists was then determined.

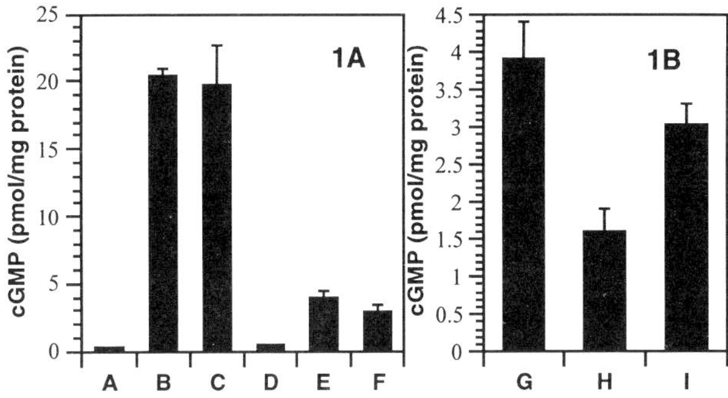

Figure 1. Effect of treatment with Cytokine or Cytokine plus 17β-estradiol on NO dependent cGMP synthesis in HUVEC. (1A) Cells were cultured for 24 hours with (D,E,F), or without (A,B,C) cytokines, and cGMP was determined in response to: control buffer (A,D), A23187 (B,E), histamine (C,F). **(1B)** Cells were cultured for 24 hours with cytokines (G), cytokine+17β-estradiol (10nM) (H), or cytokine+17β-estradiol (10nM) + ICI164384 (1µM) (I), and cGMP was determined in response to A23187.

Results and Discussion:

Figure 1A shows that treatment of HUVEC with A23187 or histamine leads to a 70-73 fold stimulation of cGMP synthesis as compared to control. A23187 stimulation of cGMP synthesis was blocked by the NO synthase inhibitor L-NAME, suggesting that cGMP synthesis was mediated via NO. Pre-incubation of HUVEC with inflammatory cytokines IL-1β, TNF-α and IFN-γ for 24 hours resulted in a significant reduction in cGMP synthesis in response to A23187 (20.38± 0.62 pmol/mg protein as compared to 3.95±0.50 pmol/mg protein, p> 0.05). These results show that treatment of HUVEC cultures with cytokines inhibited cGMP synthesis. Figure 1B shows that incubation of HUVEC simultaneously with 10 nM estrogen and cytokines produced a further attenuation of cGMP synthesis compared to cytokine treatment alone. cGMP synthesis was reduced from 3.95±0.50(G) to 1.6±0.31(H) pmol/mg protein, $p<0.05$.

The effect of estrogen was antagonized by anti-estrogen compound ICI164384, suggesting that the activity was mediated via the estrogen receptor. These results demonstrate that exposure of endothelial cells to inflammatory cytokines in-vitro may lead to inhibition of endothelial activities mediated through NO or cGMP. Furthermore these results demonstrate that in the presence of cytokines, 17β-estradiol treatment significantly reduced cGMP synthesis in HUVEC.

REFERENCES:

1. Gura, T. (1995) Science 269, 771-773.
2. Lieberman, E.H., Gerhard, M., Uehata, A., Walsh, B.W., Selwyn, A.P., Ganz,P., Yeung, A.C., and Creager, M.A. (1994) Ann. Intrn. Med. 121, 936-941.
3. Reis, S.E., Gloth, S.T., Blumenthal, R.S., Resar, J.R., Zacur, H.A., Gerstenblith, G., and Brinker, J.A. (1994) Circulation 89, 52-60.
4. Williams, J.K., Adams, M.r., Herrington, D.M., and Clarkson, T.B. (1992) J. Am. Coll. Cardiol. 20, 452-457.
5. Weiner, C.P., Lizasoain, I., Baylis, S.A., Knowles, R.G., Charles, I.G., and Moncada, S. (1994) Proc. Natl. Acad. Sci. (U.S.A) 91, 5212-5216.
6. Hishikawa, K., Nakaki, T., Marumo,T., Suzuki, H., Kato, R. and Saruta, T. (1995) FEBS Lett. 360, 291-293.

INDUCTION OF NITRIC OXIDE SYNTHASE IN HUMAN THYROCYTES

Kikuo Kasai[1]*, Yoshiyuki Hattori[1], Nobuyuki Banba[1], Satoshi Motohashi[1], Nobuo Nakanishi[2], and Shin-Ichi Shimoda[1]

[1]Department of Endocrinology, Dokkyo University School of Medicine, Tochigi, Japan; [2]Department of Biochemistry, Meikai University School of Dentistry, Saitama, Japan

Recent data support important roles for cytokines in the inflammatory and/or immune response in the thyroid. Under these conditions, NO synthase (NOS) is likely to be induced in the thyroid. In the present study, we investigated the possible induction of NOS in human thyroid epithelial cells (thyrocytes) upon stimulation with cytokines.

MATERIALS AND METHODS

Cell culture: We obtained specimens of seven resected human thyroid glands. Diagnoses were: adenomatous goiter, paraneoplastic (possibly normal) thyroid tissues excised for localized medullary thyroid cancer, and follicular adenoma. Each specimen was digested with collagenase; the thyrocytes (more than 90%) were suspended in Ham F-12 medium supplemented with five hormones (insulin, hydrocortisone, glycyl-histidyl-lysine, transferrin, and somatostatin) and 10% fetal bovine serum (FBS), and seeded into 96-well plates, 6-well plates or 100-mm petri dishes (1). The medium was replaced every three days with fresh medium. Cells were used for experiment after they reached confluence.

Nitrite assay: Nitrite was used to indicate cellular synthesis of NO. Acuumulation of nitrite in the cell culture medium was quantified by colorimetric assay using the method of Griess, as previously described (2).

* Corresponding author.

Biopterin assay: Total biopterin (BH4 and more oxidized species) was measured by HPLC analysis after oxidation by iodine as previously described (3).

Analysis of NOS and GTP cyclohydrolase I (GTPCH) mRNA: Reverse transcription-polymerase chain reaction (RT-PCR) was performed using gene specific primers as previously described (2). Primers used were: NOS forward 21-mer, 5'-TGCCCTGGCAATGGAGAGAAA-3'; NOS reverse 21-mer, 5'-GAGCTGATGGAGTAGAACCTG-3'; GTPCH forward 21-mer, 5'-CCGCCTACTCGTCCATCCTGA-3';GTPCH reverse 21-mer, 5'-ACCTCGCATTACCATACACAT-3'. All PCR reactions resulted in the amplification of a single product of the predicted size for inducible NOS (944 bp) and GTPCH (434bp). The direct analysis of sequence of PCR product for NOS revealed that human thyrocyte inducible NOS mRNA showed 98.1% identity with that of human hepatocytes (4).

RESULTS AND DISCUSSION

Although unstimulated thyrocytes produced little NO (measured as nitrite), interleukin-1α or β (IL-1α/β) substantially increased NO formation. Interferon-γ (IFNγ) by itself failed to stimulate NO formation but markedly increased the IL-1-stimulated NO production. Tumor necrosis factor-α (TNF-α) alone did not induce NO production, but did slightly in the presence of IFNγ (Fig.1).

Fig. 2 shows nitrite production by primary cultured human thyrocytes over 48 hr evoked by 20 ng/ml IL-1α and by 20 ng/ml IL-1α in the presence of 100 U/ml IFNγ. Human thyrocytes obtained from 7 resected thyroid specimens [2 paraneoplastic (possibly normal) tissues, 3 adenomatous goiters and 2 follicular adenomas] were well respond to IL-1α or IL-1α/IFNγ, although the extent of nitrite production and synergistic action of IFNγ with IL-1α differed individually.

Induction of NO formation by thyrocytes upon stimulation with IL-1α plus IFNγ was accompanied by the synthesis of tetrahydrobiopterin (BH4), an obligatory cofactor of NOS. Coinduction of NO and BH4 synthesis in thyrocytes was preceded by coexpression of mRNAs for NOS and GTPCH, the rate-limiting enzyme for de novo synthesis of BH4 (not shown). NO synthesis was prevented by an inhibitor of GTPCH, 2,4-diamino-6-hydroxypyrimidine (DAHP), and this inhibition was completely reversed by administration of sepiapterin, a substrate for BH4 synthesis via pterin salvage pathway (not shown).

In contrast to IFNγ, some cytokines such as interferon-α, interleukin-4 and transforming growth factor-β1 inhibited the IL-1-induced NO production. Thus, cytokines such as IL-1, IL-1/IFNγ and TNF-α/IFNγ stimulate human thyrocytes to produce NO: this process can be modulated by other cytokines and coregulated with a cofactor BH4 biosynthesis, and resulting NO may affect cell function including thyroid hormone synthesis.

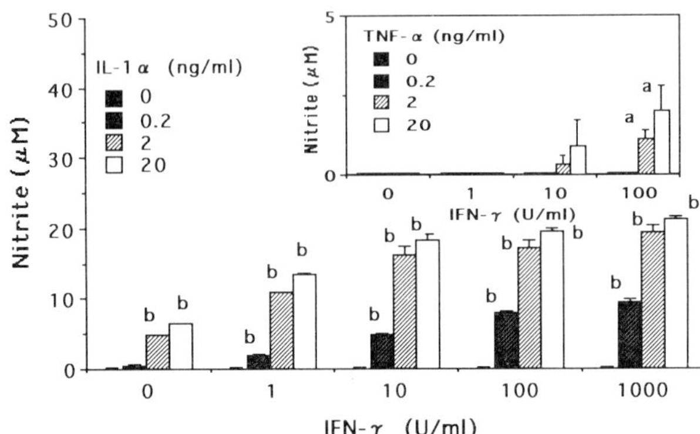

Fig. 1. Effects of various doses of IL-1α and TNF-α on nitrite production in the presence or absence of IFNγ in human thyrocytes. Thyrocytes were incubated with various doses of IL-1α or TNF-α in the presence or absence of graded doses of IFNγ over 48 hr in Ham F-12 containing 6H and 10 % FBS. Values are expressed as a mean of three wells ± SD. a; p<0.05, b; p<0.01 (*vs.* the value in the absence of any cytokine).

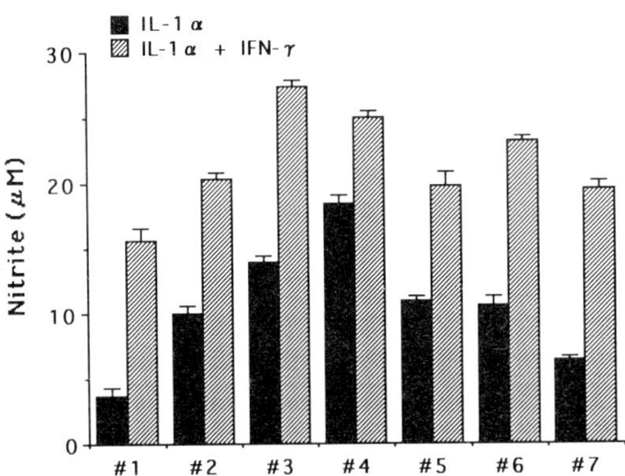

Fig. 2. Nitrite production by primary cultured human thyrocytes prepared from seven resected thyroid tissues. Thyrocytes were prepared from seven resected specimens: #1 and #2 from paraneoplastic (possibly normal) tissue; #3 - # 5 from adenomatous goiter; #6 and #7 from follicular adenoma and after confluency, they were incubated with IL-1α (20 ng/ml) or IL-1α (20 ng/ml) plus IFNγ (100 U/ml) in Ham F-12 containing 6H and 10 % FBS over 48 hr. Values are expressed as a mean of three wells ± SD.

REFERENCES

1. Kasai K, Ohmori T, Koizumi N, Hosoya T, Hiraiwa M, Emoto T, Hattori Y, Shimoda SI (1989) Life Sci 45:1451-1459
2. Hattori Y, Gross SS (1993) Biochem Biophys Res Commun 195:435-441
3. Suzuki H, Nakanishi N, Yamada S (1988) Biochem Biophys Res Commun 153:382-387
4. Geller, DA, Lowenstein CJ, Shapiro RA, Nussler AK, DiSilvio M, Wang SC, Nakayama DK, Simmons RL, Snyder SH, Billiar TR (1993) Proc Natl Acad Sci USA 90:3491-3495

Human inflammatory cells express mRNA for inducible nitric oxide synthase but lack detectable translated product or enzyme activity

ASHOK R. AMIN[1,2,3*], MUKUNDAN ATTUR[1], PRANAV VYAS[1], JOANNA M. LESZCZYNSKA-PIZIAK[1], JOHN REDISKE[4], KALPIT VORA[5] and STEVEN B. ABRAMSON[1,2]

[1]Department of Rheumatology, Hospital for Joint Diseases, New York, NY 10003; Departments of [2]Medicine and [3]Pathology, NYU Medical Center, New York, NY 10016; [4]Ciba-Geigy Corp., Summit, NJ 07901; [5]Thomas Jefferson University, Pittsburgh, PA.

Generation of nitric oxide (NO) by inducible nitric oxide synthase (iNOS) plays an important role in inflammation, host-defense responses, immunity and tissue repair. Biochemical evidence that human leukocytes cells express NOS protein or its product, NO, has been inconclusive and a source of controversy. We therefore examined the expression and function of NOS in various human and rodent cell types, including neutrophils, monocytes, HL-60 cells and T cells. As expected, human articular chondrocytes and umbilical vein endothelial cells (HUVEC), and murine macrophages/rat brain cells, expressed iNOS and cNOS following exposure to cytokines + endotoxin or other relevant stimuli. In contrast, human peripheral monocytes/macrophages, neutrophils, HL-60 and T cell lines (Jurkat), treated in the same fashion as above, did not exhibit detectable amounts of NOS expression or activity as assayed by Western blot (sensitivity ≥10 pg), Griess reaction (sensitivity ≥0.1 μM), and radiolabelled L-arginine-to-L-citrulline conversion (sensitivity ≥20 pmol L-citrulline), despite ample evidence in the literature that addition of NO donors or competitive iNOS inhibitors to these cells modulates their function. These studies also show that human peripheral monocytes/macrophages, neutrophils, HL-60 and Jurkat T cells did express a full-length NOS mRNA as seen by RT-PCR analysis and Northern blot (4.4 kD). In spite of the constitutive expression of mRNA in neutrophils and the lack of detectable NOS activity (based on Western blotting and L-arginine-to-L-citrulline conversion assay), stimulation of human neutrophils with FMLP or PMA *in vitro* induced the ADP-ribosylation of an intracellular NO target, glyceraldehyde-3-PO_4 dehydrogenase (GAPDH), in a NO-dependent manner. Therefore, the method of detection of NO activity is an important issue. Table 1 summarizes some of the published studies [1-6].

These studies indicate that, unlike the rodent system, although human inflammatory leukocytes and T cells express mRNA for NOS, the translated product is tightly regulated, is expressed at low levels *in vitro,* and can be detected with sensitive methods. We therefore speculate that human inflammatory cells may express iNOS mRNA in a dual fashion that ***(a)*** can be translated optimally *in vivo* in the presence of soluble factors together with ligands on the surface of the cell in the microenvironment [7] in a close cell-cell interaction [8] that may contribute to phagocytic and tumoricidal activity as described in the rodent system, and ***(b)*** acts as an autacoid mediator involved in signal transduction.

*To whom correspondence should be addressed at Rheumatology Research Lab, Room 1600, Hospital for Joint Diseases, 301 E. 17th St., New York, NY 10003. Tel: (212) 598-6537. Fax: (212) 598-6168.

We thank Ms. Ann Rupel for editing and preparing the manuscript.

REFERENCES

1. Mehta JL, Lawson DL, Nicolini FA, *et al.* (1990) *Biochem. Biophys. Res. Comm.* **173,** 438-442.
2. Wright CD, Mülsch A, Busse R, *et al.* (1989) *Biochem. Biophys. Res. Comm.* **160,** 813-819.
3. Riesco A, Caramelo C, Blum G, Monton M, Gallego MJ, Casado S, Lopez Farré A (1993) *Biochem. J.* **292,** 791.
4. Lärfärs G, Gyllenhammar H (1995) *J. Immunol. Meth.* **184,** 53-62.
5. Clancy RM, Leszczynska-Piziak J, Levartovsky D, Amin AR, Abramson SB (1995) *J. Leukocyte Biol.* **58,** 196-202.
6. Salvemini D, De Nucci G, Gryglewski RJ, *et al.* (1989) *Proc. Natl. Acad. Sci. USA* **86,** 6328-6332.
7. Pérez-Mediavilla LA, Lopez-Zabalza MJ, Calonge M, Montuenga L, Lopez-Moratalla N, Santiago E (1995) *FEBS Lett.* **357,** 121-124.
8. Zembala M, Siedlar M, Marcinkiewicz J, Pryjma J (1994) *Eur. J. Immunol.* **24,** 435-439.

Table 1. Detection of nitric oxide activity in human neutrophils.

Cells/conditions	Method of Detection	Reference
Resting/activated PMN	RT-PCR and Northern blotting (mRNA)	This study
Resting PMN	Aortic ring relaxation	[1]
Resting PMN	Chemiluminescence	[2]
Resting/activated PMN	L-arginine-to-L-citrulline conversion assay cAMP formation	[3]
Resting PMN	Oxyhemoglobin to methemoglobin conversion	[4]
Resting/activated PMN	ADP-ribosylation of GAPDH	This study
Resting/activated PMN	ADP-ribosylation of actin	[5]
Resting PMN	Inhibition of platelet aggregation	[6]

The examples described above represent some of the different methods used to detect nitric oxide activity in human neutrophils, assayed by direct and indirect functional assays using activating agents (*e.g.,* LTB4, DTI, FMLP, PAF, etc.) ± superoxide dismutase and/or attenuation of hemoglobin/methylene blue.

Nitric oxide synthase in circulating vs. extravasated polymorphonuclear leukocytes

Allen M. MILES*, Michael W. OWENS+,Shawn MILLIGAN+,Glenda G. JOHNSON*, Jeremy Z. FIELDS#@, Todd S. ING@, Venkata KOTTAPALLI#, Ali KESHAVARZIAN#@ and Matthew GRISHAM*.

*Department of Physiology and Biophysics, Louisiana State University Medical Center, Shreveport, Louisiana 71130.
+Department of Medicine, Louisiana State University Medical Center, Shreveport, Louisiana 71130.
#GI Division, Department of Medicine, Loyola University Medical School, Maywood,IL 60153.
@Medical and Research Services, Hines V. A. Hospital, Hines, IL.

Introduction

There is a growing body of both experimental and clinical data that demonstrates that certain forms of acute and/or chronic inflammation are associated with enhanced production of NO. For example, it has been shown in experimental animals and humans that arthritis, glomerulonephritis, sepsis, endotoxemia, hepatitis, colitis, graft vs. host disease, and pulmonary inflammation are associated with the upregulation of the inducible NO synthase (iNOS) and an increased production of NO as measured by increases in plasma and/or urinary levels of nitrate and nitrite. It has been suggested that some of the pathophysiology observed in these models of inflammation may arise as a result of enhanced release of NO since large amounts of NO are known to injure cells and tissue by promoting intracellular iron release, inhibiting mitochondrial function and inhibiting DNA synthesis. Although the sources of NO in these pathophysiological situations are not known it is thought that phagocytic leukocytes (e.g., polymorphonuclear leukocytes, monocytes, macrophages) represent major sources.

A substantial amount of information has been reported on the molecular mechanisms by which inflammatory cytokines and/or bacterial products upregulate iNOS in rodent macrophages. However, relatively little is known regarding the regulation of iNOS in polymorphonuclear leukocytes (PMNs) or whether human PMNs even contain an active iNOS. It has been demonstrated that elicited (extravasated) rat peritoneal PMNs produce substantial amounts of NO_2^- *in vitro* [1,2]. Also, previous studies from our laboratory [1] as well as others [2] have demonstrated that unlike macrophages, elicited rat PMNs produce NO ex vivo in the absence of any exogenous cytokines and/or bacterial products. These data suggest that rat PMN-associated iNOS expression may be regulated by very different mechanisms compared to macrophages. Although several laboratories have reported that human PMNs do release a factor with properties similar to NO [3,4], there has been no molecular nor biochemical confirmation of the expression of iNOS in human PMNs. Indeed, several other laboratories have reported that human PMNs do not synthesize NO. Therefore, we quantified and compared the levels of iNOS mRNA, protein, enzyme activity and cellular production of NO in circulating versus extravasated rat and human PMNs.

Results and Discussion

Circulating rat and human PMNs were purified from peripheral blood and extravasated PMNs were elicited in rats by intraperitoneal injection of 1% oyster glycogen or in humans by peritoneal dialysis of patients with peritonitis. Inducible NOS mRNA from circulating and elicited PMNs was quantified using slot blot hybridization analysis with a cDNA probe specific for iNOS. iNOS protein was identified using Western immunoblot analysis and NOS activity was quantified by measuring the N^G-monomethyl-L-arginine (L-NMMA)-inhibitable conversion of ^{14}C-labeled L-arginine to ^{14}C-L-citrulline. In a separate series of experiments, circulating or extravasated PMNs were cultured for 4 hrs and the accumulation of L-NMMA- inhibitable nitrite (NO_2^-) in the supernatant was determined and used as a measure of NO production in vitro. We found that circulating PMNs (rat or human) contained no iNOS mRNA, protein nor enzymatic activity. Furthermore, circulating rat or human PMNs ($2x10^6$ cells/well) were unable to generate significant amounts of NO_2^- when cultured for 4 hrs in vitro. In contrast, iNOS mRNA levels in 4- and 6-hr elicited rat PMNs increased 21- and 42-fold, respectively when compared to circulating cells. Western blot analysis revealed the presence of iNOS protein in the elicited rat PMNs and iNOS enzymatic activity increased from normally undetectable levels in circulating rat PMNs to 81 and 285 pmol/min/mg for the 4- and 6-hr elicited rat PMNs, respectively. Approximately 20 to 30% of the total iNOS activity was Ca^{+2}-dependent. Nitrite formation by elicited rat PMNs in the absence of any exogenous stimuli increased from normally undetectable amounts for circulating PMNs to approximately 8 and 11 $\mu M/10^6$ cells for the 4- and 6-hr elicited PMNs, respectively. Highly enriched preparations of extravasated human PMNs contained neither message, protein nor iNOS enzymatic activity. Taken together, our data demonstrate that inflammation-induced extravasation of rat PMNs upregulates the transcription and translation of iNOS in a time-dependent fashion and that 20-30% of the total inducible NOS was Ca^{+2}-dependent. In contrast, neither circulating nor extravasated human PMNs contained iNOS message, protein nor enzymatic activity. These data suggest that the human PMN iNOS gene is under very different regulation than is the rat gene.

Acknowledgements: Some of the work reported in this manuscript was supported by grants from the National Institutes of Health (CA63641 and DK47663).

References

1. Grisham, M. B., K. Ware, H. E. Gilleland Jr., L. B. Gilleland, C. L. Abell, and T. Yamada. (1992) Gastroenterology. **103**:1260-1266.
2. McCall, T. B., N. K. Boughton-Smith, R. M. J. Palmer, B. J. R. Whittle and S. Moncada. (1989) Biochem. J. **261**:293-296.
3. Wright, C. D., A. Mulsch, R. Busse, and H. Osswald. (1989) Biochem. Biophys. Res. Comm. **160**:813-819.
4. Salvemini, D. G. DeNucci, R. J. Gryglewski, and J. R. Vane. (1989) Proc. Natl. Acad. Sci. USA, **86**:6328-6332.
5. Keller, R., R. Keist, P. Erb, T. Aebisher, G. De Libero, M. Balzer, P. Groscurth and H. U. Keller. (1990) Cell. Immunol. **131**:398-403.
6. Klebanoff, S. J., and C. F. Nathan. (1993) Biochem. Biophys. Res. Commun. **197**:192-196.

Human B and T lymphocytes express the constitutive endothelial isoform of nitric oxide synthase

NORBERT REILING, ROLF KRÖNCKE, ARTUR J. ULMER, JOHANNES GERDES, HANS-DIETER FLAD and SUNNA HAUSCHILDT*.

Department of Immunology and Cell Biology, Forschungsinstitut Borstel, Parkallee 22, D-23845 Borstel and * Department of Immunobiology, University of Leipzig, Talstr. 33, D- 04103 Leipzig, Germany.

Nitric oxide (NO) is a pleiotropic mediator of a variety of cellular processes such as vasorelaxation, neurotransmission, and cytotoxicity [1]. NO is synthesized from L-arginine by the enzyme NO synthase (NOS). There are at least three distinct isoforms of NOS present in human cells, which have been isolated and cloned. Two enzymes, the endothelial and the neural isoform, are reported to be constantly present and termed constitutive NOS (ecNOS, ncNOS) [2,3]. The third isoform is inducible (iNOS) and expressed in response to immunostimulants, e.g. in hepatocytes [4]. Little is known about the formation of NO and the expression of NO synthases in cells of the human immune system. It has recently been shown by independent groups that human monocytes/macrophages express iNOS upon stimulation [5-7]. Apart from iNOS, ecNOS expression is also present in resting monocytes/macrophages [7]. To investigate whether ecNOS is expressed in human immunocompetent cells other than monocytes/macrophages, we focussed our studies on lymphocytes. We studied the expression of the human constitutive endothelial NOS isoform in highly purified human T cells from peripheral blood mononuclear cells (PBMC) and in frozen sections of tonsillar tissues.

By intron-spannning reverse transcription polymerase chain reaction (RT-PCR) using gene specific oligonucleotide primers [7] ecNOS mRNA expression was analyzed. T cells were isolated from PBMC by counterflow elutriation[8] and subsequent nylon wool filtration. RT-PCR of these highly purified lymphocytes (FACS analysis: >96% CD3+) yielded a 422 bp long ecNOS-specific fragment. Culture of the cells (RPMI 1640 supplemented with 10% (v/v) fetal calf serum (FCS) and antibiotics) for 6 and 15 h with phorbol-12-myristate-13-acetate (PMA) (10 ng/ml) / concanavalin A (10 μg/ml) led to a decrease of the mRNA signal. To analyze the ecNOS mRNA expression in T and B lymphocytes from tonsillar tissue various subpopulations, namely germinal center B cells ($CD19^+/IgD^-$), mantle zone B cells ($CD19^+/IgD^+$), germinal center T cells (GCTC ; $CD4^+/CD57^+$), and tonsillar T cells (TTC; $CD4^+/CD57^-$) were isolated by magnetic- and fluorescence-activated cell sorting. Freshly isolated germinal center B cells as well as mantle zone B cells could be shown to express ecNOS mRNA by amplification of an ecNOS-specific cDNA fragment by RT-PCR. However when the B cells were cultured in the presence of PMA (10 ng/ml) / IL-4 (50 ng/ml) prior to mRNA extraction, the ecNOS mRNA diminished completely. Stimulation of the TTC with PMA/ConA for 12 also led to a decrease of the ecNOS mRNA. In contrast to TTC, GCTC neither express ecNOS mRNA in the unstimulated nor in the stimulated state. To further substantiate the existence of ecNOS in human lymphocytes, we examined mRNA expression in human lymphoid T and B cell lines. The B cell lines Raji and Daudi as well as the T cell line Jurkat could be shown to express ecNOS mRNA.

To investigate the expression of ecNOS on protein level immunohisto-chemical staining using a mab against ecNOS (Transduction Laboratories, Lexington, KY; USA) was performed applying the alkaline phosphatase anti-alkaline phosphatase (APAAP) method [9]. Staining of frozen tonsillar sections showed a strong positivity in the endothelial cells of the blood vessels indicating the existence of ecNOS as previously described. In addition a fine granular staining was found in many cells of the lymphoid tissue. This staining was also seen in human T cells from PBMC. To further specify the ecNOS expression in B cells double immunofluorescence stainings of frozen tonsillar sections with a mab against ecNOS and an antiserum against IgD was performed. These experiments clearly demonstrate the existence of the enzyme in B cells. The ecNOS mRNA expression of the B and T lymphocyte subpopulations are summarized in Table 1.

Abbreviations used: NOS, nitric oxide synthase; GCTC, germinal center T cells; TTC, tonsillar T cells; PMA, phorbol-12-myristate-13-acetate; PBMC, peripheral blood mononuclear cells

Table I. EcNOS mRNA and protein expression by human lymphocytes

cell	source (phenotype)	EcNOS		
		mRNA		protein
		unstimulated	stimulated	
T cell	peripheral blood ($CD3^+$)	++	—	+
	tonsillar ($CD4^+/CD57^-$)	++	—	n.d.
	germinal center ($CD4^+/CD57^+$)	—	—	n.d.
B cell	mantle zone ($CD19^+/IgD^+$)	+	—	+
	germinal center ($CD19^+/IgD^-$)	+	—	n.d.

The ability of NOS to convert colorless tetrazolium salts and NADPH to darkblue formazan under L-arginine free conditions has been used to localize constitutive NOS in neuronal tissue [10]. We applied the same staining method to frozen tonsillar sections and T cell preparations. Beside a strong dark blue reactivity in the endothelial cells, a fine granular, blue staining inside many cells of the lymphoid tissue was observed. Although enzymes other than ecNOS are known to act as NADPH diaphorases, this results might suggest the presence of functional ecNOS in human lymphocytes.

These data are the first to demonstrate the expression of the endothelial cNOS on mRNA and protein level in human lymphocytes. A role of the product NO in human T lymphocytes has recently been suggested by Dong et al. [11], who showed that NO activates cystic fibrosis transmembrane conductance regulator Cl^- currents in human cloned T cells.

This study was supported in part by a grant from BMBF, AIDS Verbund Hamburg, project A02.

1. Moncada, S., Palmer, R.M.J. and Higgs, E.-A.(1991) Pharmacol. Rev. **43**,109-142
2. Marsden, P.A., Schappert, K.T., Chen, H.S., et al.(1992) FEBS Lett. **307**,287-293
3. Nakane, M., Schmidt, H.H.H.W., Pollock, J.S., Förstermann, U. and Murad, F.(1993) FEBS Lett. **316**,175-180
4. Geller, D.A., Lowenstein, C.J., Shapiro, R.A., et al.(1993) Proc. Natl. Acad. Sci. USA **90**,3491-3495
5. De Maria, R., Cifone, M.G., Trotta, R., et al.(1994) J. Exp. Med. **180**,1999-2004
6. Bukrinsky, M.I., Nottet, H.S.L.M., Schmidtmayerova, H., et al.(1995) J. Exp. Med. **181**,735-745
7. Reiling, N., Ulmer, A.J., Duchrow, M., Ernst, M., Flad, H.-D. and Hauschildt, S.(1994) Eur. J. Immunol. **24**,1941-1944
8. Grage-Griebenow, E., Lorenzen, D., Fetting, R., Flad, H.-D. and Ernst, M.(1993) Eur. J. Immunol. **23**,3126-3135
9. Cordell, J.L., Falini, B., Erber, W.N., et al.(1984) J. Histochem. Cytochem. **32**,219-229
10. Hope, B.T., Michael, G.J., Knigge, K.M. and Vincent, S.R.(1991) Proc. Natl. Acad. Sci. USA **88**,2811-2814
11. Dong, Y., Chao, A.C., Kouyama, K., et al.(1995) EMBO J. **14**,2700-2707

Characterisation of inducible nitric oxide synthase in human monocytes/ macrophages after ligation of the CD_{23} receptor.

VALENTINA RIVEROS-MORENO, MIRIAM PALACIOS, JOSE RODRIGO*, RICARDO MARTINEZ-MURILLO*, and SALVADOR MONCADA .

Wellcome Research Laboratories, Langley Court, Beckenham, Kent BR3 3BS, U.K. * Instituto Cajal Avda. Dr Arce 37, Madrid, Spain.

We have studied the induction of nitric oxide (NO) synthase by a novel mechanism operating in human monocytes/ macrophages. This mechanism involves the induction of the FcεRII/CD_{23} receptor on the macrophage membrane and its subsequent cross-linking (ligation). The receptor is induced by cytokines such as interleukin 4 (IL_4) and GM-CSF and the ligation may be either by IgE, the natural low affinity ligand for the receptor, or by a monoclonal antibody to CD_{23} [1]. We have analysed the correlation between the amount of CD_{23} receptor expression and the extent of NO synthase induction, as well as the sub-cellular localisation of the induced enzyme.

Methods: *Human monocytes/ macrophages:* AB human buffy coats were supplied by the National Blood Service, London and South East Zone. Monocytes/macrophages were isolated by a centrifugation step on a Ficoll gradient, the subsequent elimination of platelets by centrifugation at 800 rpm for 3x and binding to Dynabeads coated with mouse anti-human platelets (clone HPL-1, Harlan Sera-lab Ltd.) and finally, by adherence to plastic. The resulting cell population was characterised by flow cytometry using suitable fluorescent antibodies. *Induction of NO synthase:* The cells were incubated with either IL_4 (10 ng/ml, Boehringer Mannheim Biochimica) or GM-CSF (Genzyme) for 48h. Expression of CD23 was assessed by flow cytometry using a FITC-anti CD_{23} antibody (Serotec). Monocytes/macrophages were subsequently incubated with either anti-CD_{23} monoclonal (10 µg/ml, clone 135, supplied by Dr. M.D. Mossalayi) or human IgE (10 µg/ml, Aalto Bioreagents Ltd.) and anti-IgE (10µg/ml, Nordic Immunology) for 48 h, in order to induce NO synthase. *Enzyme analysis:* Cells were homogenised in a previously described buffer [2], with 3 cycles of freeze/thawing. The NO synthase activity was measured in the total homogenate by the oxyhaemoglobin assay or by the citrulline assay. Production of nitrite was measured in the cell-free culture supernatants by the chemiluminescence method. *Immuno-electron microscopy:* Cell pellets were fixed in 4% paraformaldehyde/0.1% glutaraldehyde in 0.1M phosphate buffer pH 7.4, for 30 min, then washed in 30% sucrose/phosphate buffer followed by the immuno-cytochemical procedure described by Shu et al [3]. The antibody used was an anti-peptide described previously [4], specific for the mouse macrophage iNOS at a dilution of 1:5 000.

Results and Discussion : To assess the expression of CD_{23}, the population of monocytes/macrophages was treated with different inducers alone or in combination. CD_{23} expression occurred over 75% of the monocyte population. Contamination by lymphocytes was always less than 5%. Platelets were undetectable as determined by FITC-anti-CD_{41}. Fig.1 shows a direct correlation between the

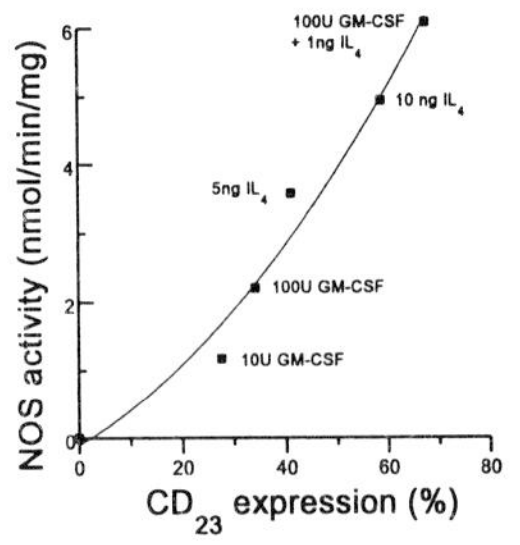

Figure 1.

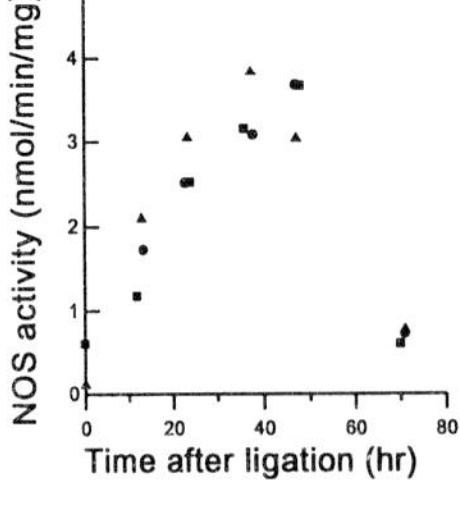

Figure 2.

quantity of CD_{23} receptor expressed on the surface of the monocytes/macrophages and the amount of NO synthase induced after ligation of the receptor. To study the time course of NO synthase induction, 10 ng/ml of IL_4 and 10µg/ml of anti-CD_{23} were used. Fig. 2 shows that maximum enzyme activity is obtained after 40 h of ligation. Triangles, dots and squares correspond to different experiments, i.e. different blood donors.

Centrifugation of the cell homogenate at 10 000g produced loss of the enzyme activity. This is in sharp contrast to the murine NO synthase which is found in the 100 000g supernatant. Detergents such as CHAPS (5mM), sarkosyl (0.25%) and n-octyl -β-D-glucopyranoside (0.25%) solubilised the homogenate and maintained the enzyme activity. We have studied the subcellular distribution of the enzyme by immunoelectron microscopy and compared it to that in the macrophage cell line J774. Panel A in Fig. 3 shows activated human monocytes containing a diversity of clearly circumscribed immunopositive cytoplasmic granules. Some were large and were surrounded by a clear membrane structure,

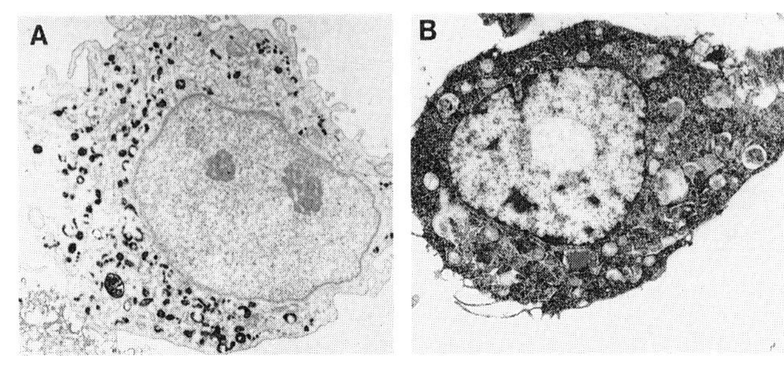

Figure 3

others showed an inner granulated appearance. Panel B shows activated murine J774 cells. The NO synthase immunoreactivity is very intense and found throughout the cytoplasm, including the cytoplasmic protrusions, confering a dark, finely granulated background. In both cell types the non-activated controls showed no immunoreactivity.

The inducible NO synthase present in human macrophages requires a very different induction process to its murine counterpart. The induction of the CD_{23} membrane receptor prior to the actual enzyme induction is a unique feature of the process. The cross-linking or ligation of this receptor, leading to the onset of the synthesis of NO synthase, must involve membrane signalling pathways. The involvement of cytokines in the overall process of NO synthase induction has not been ruled out but, as many have shown previously, they cannot operate on their own in human macrophages.

The fact that the enzyme is induced into discrete membranous vesicles suggests its usage possibly within such vesicles. This is a challenging concept, particularly considering that NO can diffuse freely through membranes.

References:

1] Mossalayi, M.D., Paul-Eugène, N., Ouaaz, F. et al. (1994) Int. Immunol. **6**, 931-934.

2] Charles, I., Chubb, A., Gill, R. et al. (1993) Biochem.Biophys. Res. Commun. **196**, 1481-1489.

3] Shu, S., Ju, G. and Fan, L. (1988) Neurosci. Lett. **85**, 169-171.

4] Hamid, Q., Springall, D.R., Riveros-Moreno, V. et al. (1993) Lancet **342**, 1510-1513.

Soluble CD23 (sCD23, 25 kDa) triggers different nitric oxide synthases in human monocytic cells

JEAN-PIERRE KOLB*, NATHALIE DUGAS#, and BERNARD DUGAS+

*U365 INSERM, Institut Curie, 75231 Paris, #Laboratoire de Neuro-Immunologie, Hôpital de Bicêtre, 94010 Kremlin-Bicêtre, +CNRS URA 625, Hôpital Pitié-Salpêtrière, 75013 Paris, France.

Introduction. The CD23 antigen, or low affinity IgE receptor is a type II transmembrane protein (1) that can be expressed on various hemopoietic cells. Two isoforms differing only by 6/7 amino acids in their intracytoplasmic N-terminal tail result from alternative splicing, CD23a (B cells) and CD23b (other cells). Through proteolyic events that are in part autocatalytic, CD23 yields soluble fragments displaying various biological properties, the 25 kDa fragment being the more stable (2). CD23 is a ligand for CD21 (3) and for CD11b and CD11c (4). The functions ascribed to membrane and soluble CD23 (sCD23) are multiple and depend on the cell type (5). This antigen is implicated in the control of parasitic and allergic manifestations (6), in antigen presentation and in the regulation of IgE synthesis. sCD23 also display IgE-independent activities, such as early differentiation of T cells (7) and induction of secretion of pro-inflammatory cytokines by monocytes. CD23 is an important NO-regulatory receptor in human monocytes (8) and we have shown recently that part of the IL-4-induced NO production by human monocytes could be inhibited with a neutralizing anti-CD23 mAb Fab fragment (9). Recombinant 25 kDa sCD23 was therefore tested for its capacity to elicit signalling pathways, notably nitric oxide synthase (NOS) activation, in human monocytes and monocytic cells, such as the U937 cell line.

Results and Discussion. sCD23 was found to trigger the generation of cGMP in unstimulated monocytic cells (peak about 5-10 min), the effect being specific of sCD23 since it was inhibited by monovalent anti-CD23 mAb Fab fragment. The cGMP increase was suppressed with N^G monomethyl L-arginine (L-NMMA), a competitive inhibitor of L-arginine for the NOS pathway and with inhibitors of NO-driven stimulation of soluble guanylate cyclase. It was also suppressed with EGTA and with W-7 and calmidazolium, two inhibitors of the calcium/calmodulin (Ca/CaM) complex. By FACS analysis and microspectrofluorometry, sCD23 was found to elicit an increase in the $[Ca^{2+}]_i$ of monocytes and U937 cells that resulted from a calcium influx in these cells. These data suggested that the sCD23-driven cGMP increase resulted from the activation of soluble guanylate cyclase via NO generated through the stimulation of a calcium-sensitive NOS. This hypothesis was strengthened by the detection by RT-PCR of type III (EC: endothelial cells) NOS mRNA in monocytic cells, and of EC-NOS protein by immunofluorescence. These events were mimicked by mAbs against CD11b (clone 44; Serotec) and anti-CD11c (clone 3.9; Serotec), but not anti-CD11a (clone B-B15; Serotec). In addition, these reagents were found to stimulate TNF-α secretion by monocytes, that was reduced in the presence of L-NMMA.

sCD23 was also found to trigger NO accumulation in human monocytic cells preincubated with IFN-γ, or IL-4, or both, as detected by L-NMMA-inhibited cGMP increase. However, this increase was only partially inhibited (about 50 %) in the presence of calcium chelator and of Ca/CaM inhibitors. In IL-4-preincubated U937 cells, about 50% of the NOS catalytic activity stimulated by sCD23 was sensitive to EGTA, suggesting that sCD23 stimulated a different NOS isoform than in control resting cells. Whether EC-NOS protein expression was little affected by IL-4 pretreatment, a weak expression of iNOS could be detected in these cells by immunochemistry and immunofluorescence. In addition to cNOS mRNA, iNOS mRNA was detected by RT-PCR in activated monocytic cells, as the corresponding protein (10, 11). It is thus conceivable that sCD23 might stimulate different NOS in human monocytic cells, depending on their activation state .

The similitude of action on Ca^{2+} mobilisation and cGMP production of both sCD23 and anti-CD11b, -CD11c antibodies are in agreement with previous finding demonstrating that CD11b and CD11c are receptors for CD23 on monocytes (4). Engagement of CD11c by an anti-CD11b/c mAb or sCD23 induced the production of the pro-inflammatory cytokine, TNF-α, and this production was NO-dependent. These data indicate that the activation of a cNOS pathway by sCD23 could be at the origin of the development of pro-inflammatory responses.

References

1. Kikutani, H., Inui, S, Sato, R., et al. (1986). Molecular structure of human lymphocyte receptor for immunoglobulin E. Cell **47**, 657-65.

2. Delespesse, G., Sarfati, M., and Hofstetter, H. (1989). Human IgE-binding factors. Immunol Today **10**, 159-164.

3. Aubry, J.P., Pochon, S., Graber, P., Jansen, K.U., and Bonnefoy, J.Y. (1992). CD21 is a ligand for CD23 and regulates IgE production. Nature **358**, 505-508.

4. Lecoanet-Henchoz S., Gauchat, J-F., Aubry, J-P., et al. (1995). CD23 regulates monocyte activation through a novel interaction with adhesion molecules CD11b-CD18 and CD11c-CD18. Immunity **3**, 119-125.

5. Delespesse, G., Suter, U., Mossalayi, D.M., et al. (1991). Expression, structure and function of the CD23 antigen. Adv. Immunol. **49**, 149-191.

6. Capron, A., Dessaint, J.P., Capron, M., et al. (1986). From parasites to allergy: a second receptor for IgE. Immunology Today **7**, 15-18.

7. Mossalayi, M.D., Lecron, J.C., Dalloul, A.H., et al. (1990). Soluble CD23 (FcεRII) and IL-1 synergistically induce early human thymocytes maturation. J. Exp. Med. **171**, 959-964.

8. Dugas, B., Mossalayi, D.M., Damais, C., and Kolb, J.P. (1995). Nitric oxide production by human monocytes/macrophages. Evidence for a role of CD23. Immunology Today. In press.

9. Paul-Eugène, N., Mossalayi, M.D., Sarfati, M., et al. (1995). Evidence for a role of FcεRII/CD23 in the IL-4-induced nitric oxide production by normal human mononuclear phagocytes. Cell. Immunol. **163**, 314-318.

10. Reiling, N., Ulmer, A.J., Duchrow, M., Ernst, M., Flad, H.D., and Hauschildt, S. (1994). Nitric oxide synthase: mRNA expression of different isoforms in human monocytes/macrophages. Eur. J. Immunol. **24**, 1941-1944.

11. Weinberg, J.B., Misukonis, M.A., Shami, P.J., et al. (1995). Human mononuclear phagocytes inducible nitric oxide synthase iNOS): analysis of iNOS mRNA, iNOS protein, biopterin, and nitric oxide production by blood monocytes and peritoneal macrophages. Blood **86**, 1184-1195.

Relationship of nitric oxide production to interleukin-2 induced antitumor killer cell activation *in vivo* and *in vitro* in splenocytes of healthy and tumor bearing mice

AMILA ORUCEVIC and PEEYUSH K. LALA

Department of Anatomy, The University of Western Ontario, London, Ontario N6A 5C1, Canada

Nitric oxide (NO[1]), a short lived molecule, plays an active role in many physiological as well as pathological processes [1]. Since IL-2 therapy of human cancer patients is known to induce high NO production [2], we have recently begun to examine the contributory role of NO on toxic side effect of IL-2 therapy known as "capillary leak syndrome". We found that treatment with N^G-Nitro-L-Arginine methyl ester (L-NAME), a potent inhibitor of NO synthesis, ameliorated IL-2 induced capillary leakage in normal [3] and tumor bearing mice [4]. Furthermore, L-NAME had antitumor and antimetastatic effects, and in combination with IL-2, caused an augmentation of early antitumor effects induced by IL-2 [4]. The objectives of the present study were to examine the possible role of NO on IL-2 induced generation of antitumor killer cells *in vivo* and *in vitro*.

NO has been reported to have opposing effects on the immune system: suppression of lymphocyte proliferation [5], and promotion of antitumor cytotoxicity of macrophages [6]. Experiments were therefore designed to test the effects of inhibition of NO production with L-NAME therapy (0.1, 0.5 and 1 mg/ml in drinking water) on IL-2 (10 inj., 15,000 Cetus U/inj., i.p., every 8 h) induced generation of antitumor cytotoxicity *in vivo* in splenocytes of healthy and C3-L5 mammary carcinoma (Ca) bearing C3H/HeJ mice. Parallel experiments were conducted *in vitro* to measure antitumor cytotoxicity of unfractionated or macrophage (Mϕ)-depleted splenocytes incubated for 4 d with IL-2±L-NAME (IL-2: 1000 Cetus U/ml/4x10^6 cells; L-NAME: 0.1, 0.5 or 1 mg/ml/4x10^6 cells; 20x10^6 cells/well/treatment). YAC-1 lymphoma (NK-sensitive) and C3-L5 Ca (NK resistant) cells were used as killer cell targets in a ^{51}Cr release assay.

[1]Abbreviations used: NO, nitric oxide; IL-2, interleukin-2; L-NAME, N^G-Nitro-L-Arginine methyl ester; Mϕ, macrophage; Ca, carcinoma.

Results: ***In vivo:*** Splenocyte killer activity against both targets was markedly stimulated by IL-2 therapy in healthy and tumor bearing mice; this was significantly enhanced with additional L-NAME therapy (Table 1), which also blocked IL-2 induced rise in NO levels in the serum (data not shown). Enhancement of IL-2 induced killer activity with L-NAME was dose dependent (data not shown).

Table 1.
In vivo killer cell generation (^{51}Cr release assay)

Percent specific lysis at 50:1 effector:target ratio[a]

	Healthy mice		Tumor-bearing mice	
Groups	YAC-1	C3-L5	YAC-1	C3-L5
Control	13.8 ± 0.3	-2.9 ± 0.5	16.5 ± 0.5	-2.6 ± 0.3
IL-2	64.9 ± 1.2	4.47 ± 0.5	52.6 ± 2.9	1.5 ± 0.7
IL-2 + L-NAME (1 mg/ml)	73.3 ±1.6*	11.6 ±0.5*	65.3 ± 0.4*	4.5 ± 0.4*

[a] mean ± SE of triplicate measurements with effector cells pooled from 5 mice / group; similar effects were noted at other (25:1 and 100:1) effector : target ratios.
* Significantly different from IL-2 ($p < 0.05$).

In vitro: Addition of L-NAME to IL-2 *in vitro* also caused a significant increase in cytotoxicity of unfractionated or Mϕ depleted splenocytes of healthy, but not tumor-bearing mice, although NO production measured in the culture media was significantly blocked in both cases (data not presented).

Summary and conclusions: We have shown that IL-2 induced increase in NO production *in vivo* interferes with LAK cell activation, which can be overcome with L-NAME therapy. This provides at least one reason for the beneficial effect of adding L-NAME to IL-2 therapy in reducing the tumor burden, reported by us elsewhere [4]. Stimulatory effects of L-NAME on IL-2 induced splenocyte killer activity observed only *in vivo* (but not *in vitro*) in the case of tumor bearing mice may be explained by the presence of additional cells (eg. endothelial cells of the tumor vasculature, and Mϕ within the tumor) which can express iNOS [7], and thus produce additional NO after IL-2 therapy. Present findings, combined with our observation that L-NAME can also mitigate IL-2 induced capillary leakage in healthy [3], as well as in tumor bearing mice [4], suggest that L-NAME could be a valuable adjunct to IL-2 therapy of cancer and infectious diseases.

References

1. Moncada, S. and Higgs, A. (1993) The L-arginine-nitric oxide pathway. N. Engl. J. Med. **329**, 2002-2012

2. Hibbs, J.B.,Jr., Westenfelder, C., Taintor, R., et al (1992) Evidence for cytokine-inducible nitric oxide synthesis from l-arginine in patients receiving interleukin-2 therapy. J. Clin. Invest. **89**, 867-877

3. Orucevic, A. and Lala, P.K. (1995) Prevention of IL-2 induced capillary leak syndrome in healthy and tumor bearing mice with N^G-Nitro-L-Arginine. Proceedings of the American Association for Cancer Research **36**, 496 (Abstract)

4. Orucevic, A. and Lala, P.K. (1995) N^G-Nitro-L-Arginine methyl ester, an inhibitor of nitric oxide synthesis, ameliorates interleukin-2 induced capillary leakage and reduces tumor growth in adenocarcinoma bearing mice. Br. J. Cancer **(in press)**

5. Albina, J.E., Abate, J.A., and Henry, W.L.,Jr. (1991) Nitric oxide production is required for murine resident peritoneal macrophages to suppress mitogen-stimulated T cell proliferation: role of IFNγ in the induction of nitric oxide-synthesizing pathway. J. Immunol. **147**, 144-148

6. Stuehr, D.J. and Nathan, C.F. (1989) Nitric oxide: a macrophage product responsible for cytostasis and respiratory inhibition in tumor target cells. J. Exp. Med. **469**, 1543-1555

7. Thomsen, L.L., Miles, D.W., Happerfiels, L., Bobrow, L.G., Knowles, R.G., and Moncada, S. (1995) Nitric oxide synthase activity in human breast cancer. Proceedings of the American Association for Cancer Research **36**, 503 (Abstract)

Induction of nitric oxide biosynthesis and its role in solid tumor

Doi, K., Akaike, T., Noguchi, Y., Horie, H., Ogawa, M. and Maeda, H.
Departments of Microbiology and Surgery II, Kumamoto University School of Medicine, Kumamoto 860, Japan

Introduction

Rapid tumor growth is sustained by a number of factors such as angiogenesis factors, growth factors, but permeability enhancement has received less focus in this regard. We have previously reported a possible involvement of NO in enhanced vascular permeability in solid tumors [1]. In various tumor cell lines and solid tumors, the expression of different isoforms of NO synthase (NOS) has been documented recently [1-4]. In this experiment, role of NO in solid tumor pathology was investigated.

Methods

Implantation of AH136B tumor. AH136B tumor cells, cultured serially in ascitic fluid in rats, were implanted subcutaneously (s.c.) in the dorsal site in the foot of Donryu rats.

ESR measurement for NO generation in solid tumors. NO adduct of *N*-(dithiocarboxy)sarcosine (DTCS)-Fe^{2+} complex was formed de novo in the tissues after administration of iron complexes to the rats. DTCS-Fe^{2+} complex (180 mg/kg and 40 mg/kg, respectively, was injected s.c., and 30 min after the administration of the iron complex, ESR measurement was carried out by using X-band electron spin resonance (ESR) spectrometer (Bruker 380E) at 110K.

The effect of inhibition of NOS on the formation of the NO adducts was also tested in vivo. Briefly, NOS inhibitors such as N^{ω}-monomethyl-L-arginine (L-NMMA) (100 mg/kg) or *S*-methylisothiourea (SMT) (50 mg/kg) was injected i.p. 90 min before administration of DTCS-Fe^{2+} complex to rats, and ESR measurement for the solid tumors was performed.

Identification of inducible NOS (iNOS) mRNA expression in solid tumor. The induction of iNOS mRNA was tested by reverse transcriptase-polymerase chain reaction (RT-PCR) with Southern blot analysis as described previously [5].

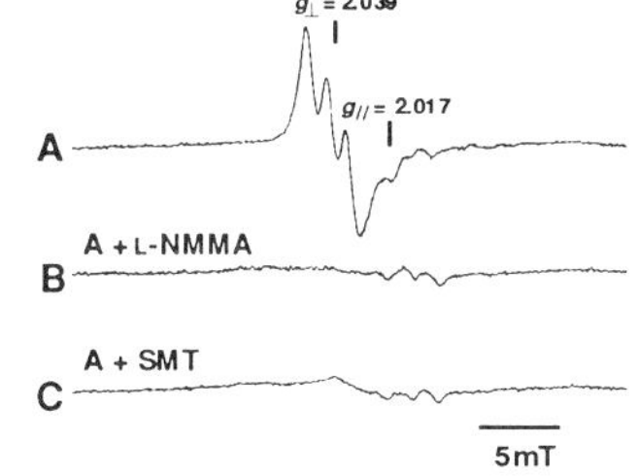

Fig. 1 ESR spectrum of NO-DTCS-iron complex generated in the solid tumors (AH136B) (A), and the effects of L-NMMA (B) and SMT (C).

Assessment of Extravasation in solid tumor. Both SMT and L-NMMA were administered to the rats i.p. 4 times every 2 hr at a total dose of 10 mg/kg SMT and L-NMMA, respectively, and ^{51}Cr-bovine serum albumin (BSA) was injected at 1 hr after the initial administration of SMT or L-NMMA via the femoral vein of tumor bearing rats. Six hours after injection of ^{51}Cr-BSA, the radioactivity of ^{51}Cr-BSA extravasated in the solid tumor was quantitated by a gamma counter as described previously [6].

Results & Discussion

The strong ESR signals of NO adduct of DTCS-Fe^{2+} complex were generated in the solid tumor (Fig. 1A). Administration of either L-NMMA or SMT to the tumor bearing rats resulted in complete suppression of the generation of the NO-DTCS-Fe^{2+} ESR signal as shown in Fig. 1A and B.

Abbreviation used: L-NMMA, N^{ω}-monomethyl-L-arginine; SMT, *S*-methylisothiourea; iNOS, inducible isoform of NO synthase; DTCS, *N*-(dithiocarboxy)sarcosine; RT-PCR, reverse transcriptase-polymerase chain reaction; ESR, electron spin resonance spectroscopy.

Small signals originated from NO-hemoglobin were observed even after treatment with these NOS inhibitors. This may be reflecting the trace amount of NO-hemoglobin that had been formed before initiating the treatment with NOS inhibitors. The signal height of NO-DTCS-Fe^{2+} observed in AH136B solid tumors was increased as the tumor gained weight up to 1.75 g. Induction of iNOS mRNA expression was confirmed by RT-PCR analysis (data not shown).

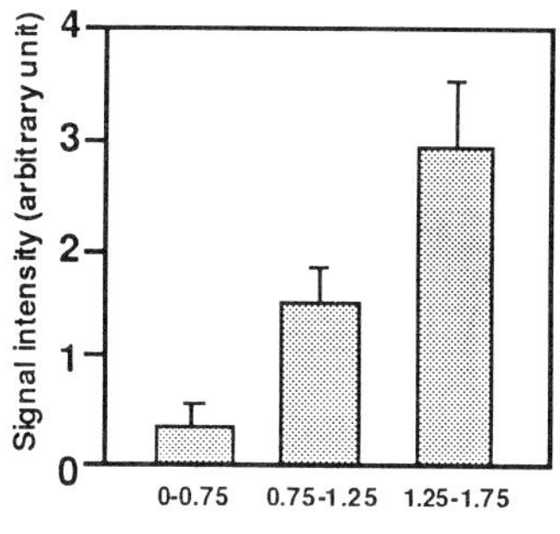

Fig. 2 Correlation of concentration of NO-(DTCS)2-Fe^{2+} formed in the solid tumors with their tumor size (weight).

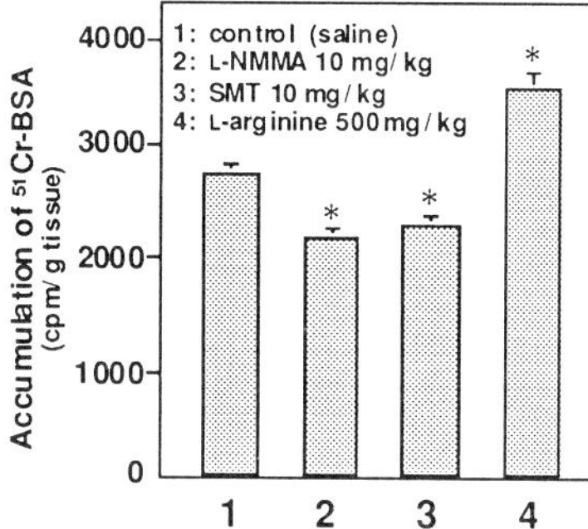

Fig. 3 Effects of L-NMMA, SMT, and L-arginine on vascular permeability in the solid tumor. *, p<0.01.

The enhanced vascular permeability was suppressed by NOS inhibitors L-NMMA and SMT and augmented with administration of L-arginine.

A number of structural evidences indicate that the vasculature in solid tumors possesses quite unique characteristics, e.g., extensive angiogenesis leading to hypervasculature, irregular morphology with defective architecture, enhanced vascular permeability, and poorly developed lymphatic system. It is reported that various vascular permeability factors including bradykinin [6-7], tumor necrosis factor α, interleukin-2 [8] and a tumor vascular permeability factor [9] are involved in the universal phenomena of enhancement of vascular permeability in the solid tumors. Our preliminary experiment showed that the enhanced vascular permeability of tumor tissues was mediated by NO [1].

Recent investigation revealed that NO synthesis pathways are remarkably upregulated in the brain and gynecological tumors in humans and in the experimental solid tumor in mice [1-4]. It is also reported that L-arginine-dependent NO pathway mediates angiogenic activity, which leads to hypervasculature in solid tumors [10]. Consequently, it is conceivable that the angiogenic potential as well as vascular permeability enhancing effect of NO may facilitate the rapid growth of solid tumors for which are various nutrients in great demand.

References

1. Maeda, H., Noguchi, Y., Sato, K., Akaike, T. (1994) Jpn. J. Cancer Res. **85**, 331-334
2. Bastian, N.R., Yim, C-Y., Hibbs, Jr.J,. Samlowski, W.E. (1994) *J Biol Chem*; 269: 5127-31.
3. Cobbs C.S., Brenman J.E., Aldape K.D., Bredt D.S., Israel M.A. (1995) Cancer Res. **55**, 727-730
4. Thomsen, L.L., Lawton, F.G., Knowles, R.G., Beesley, J.E., Riveros-Moreno, V., Moncada, S. (1994) Cancer Res. **54**, 1352-1354
5. Zheng, Y.M., Schäfer, M.K.H., Weihe, B., et al. (1990) J. Virol. **67**, 5786-5791
6. Matsumura, Y., Kimura, M., Yamamoto, T., Maeda, H. (1988) Jpn. J. Cancer Res. **79**, 1327-1334
7. Maeda, H., Matsumura, Y., Kato, H. (1988) J. Biol. Chem. **263**, 16051-16054
8. Ettinghausen, S.E., Puri, R.J., Rosenberg, S.A. (1988) J. Natl. Cancer Inst. **80**, 177-187
9. Senger, D.R., Galli, S.J., Dvorak, A.M. et al. (1983) Science **219**, 983-985
10. Leibovich, S.J., Polverini, P.J., Fong, T.W., Harlow, L.A., Koch, A.E. (1994) Proc. Natl. Acad. Sci. U.S.A. **91**, 4190-4194.

Tumor cell apoptosis may represent a novel cytotoxic mechanism resulting from IL-2 induced nitric oxide (NO·) synthesis.

Wolfram E. Samlowski*, John R. McGregor*, Neil R. Bastian*, Oh-Deog Kwon†, Chang-Yeol Yim†;

*University of Utah, Salt Lake City, UT 84132 U.S.A. and †Chonbuk National University, Chonbuk 560-182 Korea.

Introduction and methods

IL-2 therapy strongly induces nitric oxide (NO·) synthesis in mice and humans [1]. In vitro, NO· has potent antitumor activity, including release of intracellular iron, inhibition of mitochondrial respiration and DNA synthesis. Very little is known about the potential of NO· to act as an antitumor mechanism in vivo. We used a murine model to evaluate macrophage activation and NO· synthesis during IL-2 treatment. A lymphokine-activated killer (LAK) cell resistant tumor was used, to diminish the effects of lymphocyte cytotoxicity.

BALB/c mice were implanted with 2 x 10^6 syngeneic Meth A tumor cells intraperitoneally. This ascites variant is lethal to mice 12-18 days following injection, and is 50 fold more resistant to LAK cell killing (at a 100:1 effector to target cell ratio) than sensitive murine tumor cell lines.

Beginning on day 7 following tumor implantation, groups of 4-8 mice were treated with a 5 day course of IL-2 (180,000 IU subcutaneously every 12h for 5 days). Some mice were implanted with osmotic minipumps (Model 2001, Alza Corporation, Palo Alto, CA) containing 0.2 ml of 3.38 M N^{ω}-monomethyl-L-arginine (MLA)[2]. Following the last dose of IL-2, ascites was harvested, by irrigating the peritoneal cavity with sterile PBS.

In some experiments, in vivo nitric oxide synthesis was analyzed, by placing mice on a nitrite/nitrate free diet in metabolic cages and analyzing total urine nitrite/nitrate excretion [2].

Results and Discussion

Evidence for intraperitoneal NO· synthesis

Ascites cells were placed into cell culture for 24h and nitrite accumulation measured in culture supernatants [2]. Cells from IL-2 treated mice demonstrated increased nitrite production compared to untreated mice (63±14 µM versus 3.2±1.5 µM). N^{ω}-monomethyl-L-arginine (MLA), a NO· synthase inhibitor, decreased nitrate production in a dose-dependent manner, confirming the origin of the nitrite from the L-arginine:NO· pathway. Electron paramagnetic resonance (EPR) spectroscopy demonstrated evidence for nitrosyl-heme formation in ascites from IL-2 treated mice, which were not present in untreated animals. Furthermore, when parallel ascites cell cultures were pulsed with [^{3}H]-thymidine, NO· production correlated in an inverse fashion with tumor cell proliferation. These findings established that IL-2 treatment induced nitric oxide synthesis within the tumor microenvironment and inhibited tumor cell proliferation

Cellular composition of ascites following IL-2 therapy

Differential cell counts (200 cells) were performed on Wright stained cytocentrifuge slides of ascites. Results were expressed as absolute cell recovery per mouse (mean ±SD of 2 animal groups at each timepoint). The result of this experiment revealed that Meth A tumor grew progressively in control mice, with an increase of 0.618 x 10^8 cells/day. Tumor cell recovery was significantly decreased following IL-2 treatment (0.179 x 10^8 cells/day (p=.0011). Cells from IL-2 treated mice more frequently demonstrated lymphocytes and macrophages adherent to tumor cells. Ascites cells from IL-2 treated mice also exhibited a greater frequency of nuclear condensation and cell membrane blebbing, suggestive of tumor cell apoptosis.

Macrophage recruitment into ascites was enhanced in the IL-2 treated group compared to controls. The IL-2 treated group had an average increase of 0.145 x 10^8 macrophages/day in comparison to -0.003 cells/day in the control group (p=.0007).

Effect of IL-2 induced NO· synthesis on survival

To test the role of NO· in mediating antitumor responses produced by IL-2 treatment, mice were treated with IL-2 with or without MLA co-administration [3]. IL-2 therapy of Meth A ascites bearing mice prolonged median survival to 23 days, compared to 16 days in control mice (p=0.0001). In contrast, chronic MLA administration diminished median survival both of untreated control (12 days), as well as that of IL-2 treated mice (14 days, p=0.04 versus IL-2 alone). Evaluation of urinary nitrite/nitrate excretion demonstrated that IL-2 treatment resulted in an 8-fold increase, which was completely inhibited by MLA [3].

In a further experiment, we evaluated whether NO· could directly induce programmed cell death in Meth A cells (Figure 1). Meth A tumor cells were exposed to pure NO· gas, to avoid other agonists that could be present in malignant ascites. NO· exposed Meth A demonstrated DNA fragmentation characteristic of apoptosis, in contrast to unexposed control tumor cells.

Previous investigations have shown that NO· induces anticancer activity by inhibiting mitochondrial respiration and DNA synthesis in vitro. The results of experiments reported in this manuscript extend these observations by demonstrating that subcutaneous IL-2 treatment elicits significant macrophage activation and NO· synthesis within malignant ascites in mice. NO· production correlated with reduced tumor cell proliferation, a decrease in tumor cells in ascites, and increased survival. These studies suggest that NO· synthesis in vivo can contribute to IL-2 induced anticancer responses. Our studies further suggest that cellular apoptosis may represent an anticancer effector mechanism induced by NO·.

Figure 1: Induction of Meth A apoptosis by direct NO· exposure

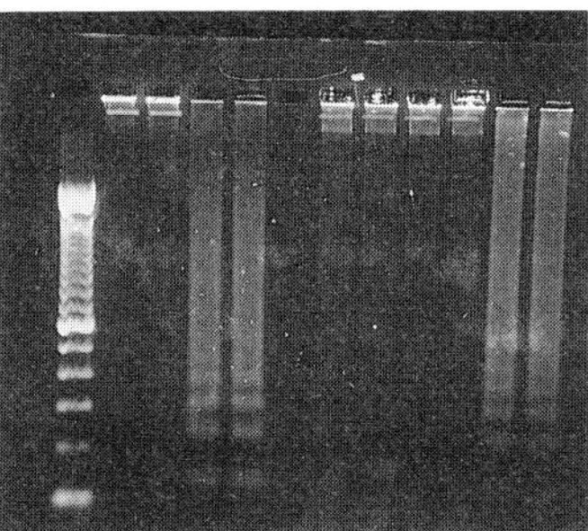

A B C D E F G H I J K L

Meth A tumor cells were depleted of oxygen in a hypobaric nitrogen gas atmosphere for 15 minutes, the exposed to NO· gas for 30 minutes. Following 18h in cell culture, DNA was extracted and electrophoresed (1 µg/lane) on a 1% agarose gel (Lanes K, L). Controls included: DNA from untreated Meth A tumor cells (lanes G, H), or cells exposed to anaerobic environment only (lanes I, J). Murine thymocytes activated with CD3 MAb (lanes D, E), and untreated thymocytes (lanes B, C) also served as positive and negative controls. Lane A contained size markers (100 bp).

References:

1. Hibbs, J.B.,Jr., Westenfelder, C., Taintor, R., et al (1992) J. Clin. Invest. **89**, 867-877

2. Yim, C.-Y., Bastian, N.R., Smith, J.C., Hibbs, J.B.,Jr. and Samlowski, W.E. (1993) Cancer Res. **55**, 5507-5511

3. Yim, C.-Y., McGregor, J.R., Kwon, O.-D., et al (1995) J. Immunol. (In Press).

Mechanism of action of immunosuppressive drugs on inducible nitric oxide synthase expression

MUKUNDAN ATTUR, PRANAV VYAS, DAVID LEVARTOVSKY, SYED NAQVI, RAFI RAZA, GEETA THAKKER, STEVEN B. ABRAMSON and ASHOK R. AMIN*

Department of Rheumatology, Hospital for Joint Diseases, New York University Medical Center, New York, NY, 10003 USA.

Cyclosporin, FK-506 and rapamycin are microbial products with potent immunosuppressive properties that result primarily from inhibition of T lymphocyte activation. Although these compounds are structurally unrelated, they share several mechanistic similarities at the cellular and molecular levels, such as their ability to inhibit peptidyl-prolyl *cis-trans* isomerase (PPIase) activity of their respective binding proteins/adaptors, which include FK-BP, calcineurins and transcription factors. While many compounds such as glucocorticoids and antimetabolites can inhibit a lymphocytic proliferative response *in vitro* and some are potent immunosuppressives *in vivo*, the selectivity of these three compounds for certain types of activation events clearly sets them apart from other agents.

In the present study we report that cyclosporin (at 1-25 μg/ml) and rapamycin (at 0.1-10 nM), but not FK506 (at 5-10 nM), inhibit the expression (as measured by nitrite production) of inducible nitric oxide synthase (iNOS) by 0-50% and 5-55%, respectively, in murine macrophages activated with LPS (see Table 1). The mode of action of cyclosporin is at the level of transcription of iNOS mRNA, which leads to a decrease in iNOS protein expression and, consequently, decreased production of nitric oxide. Although there was an effect on the nitrite accumulation and enzyme activity of iNOS in the presence of rapamycin, the mRNA inhibition was not impressive like that observed with cyclosporin, thus indicating that the actions of these two compounds on iNOS may be distinct. There was no direct effect of cyclosporin, rapamycin or FK506 on the activity of iNOS when added in cell-free extracts in a radiolabelled L-arginine-to-L-citrulline conversion assay. Cyclosporin was also found to inhibit nitrite accumulation in human chondrocytes expressing iNOS.

These observations are intriguing from the standpoint that, while neuronal NOS is known to be sensitive to FK506, the same concentration of the drug has no significant effect on iNOS. These studies indicate that **(a)** cyclosporin and, to a lesser extent, rapamycin are effective drugs that modulate iNOS expression in murine and human cells, and **(b)** the action of each of these immunosuppressive drugs can be differentiated based on their ability to modulate iNOS expression.

*To whom correspondence should be addressed at Rheumatology Research Lab, Room 1600, Hospital for Joint Diseases, 301 E. 17th St., New York, NY 10003. Tel: (212) 598-6537. Fax: (212) 598-6168.

The authors thank Ms. Ann Rupel for preparing and editing the manuscript.

Table 1. Action of immunosuppressants on iNOS expression in murine macrophage cells (RAW 264.7) activated with LPS.

Immunosuppressant	Percent inhibition of iNOS expression as assessed by: Nitrite accumulation	Western blotting	L-arginine-to L-citrulline assay	Northern blotting
Cyclosporin (1 - 25 μg/ml)	0 - 50	20 - 80	10 - 50	20 - 80
FK506 (5 - 10 nM)	<1	<20	<15	<10
Rapamycin (0.1 - 10 nM)	5 - 55	10 - 50	5 - 30	<20

The table summarizes the effect of immunosuppressants on iNOS expression at various levels. The data represent percent modulation of iNOS expression in LPS-stimulated cells pretreated with immunosuppressants (0 h), as compared to control cells stimulated with LPS only (taken as 100%). The values indicate the range of inhibition of the immunosuppressants at various concentrations.

Cyclosporin A suppresses expression of two isoforms of nitric oxide synthase in cultured macrophages and vascular smooth muscle

GREGORY J. DUSTING, HARUYO HICKEY, KAZUO AKITA, KIM JACHNO, DAVID MUTCH and IRENE NG

Department of Physiology, The University of Melbourne, Parkville, Victoria 3052, Australia.

The immunosuppressant drugs CsA and FK 506 suppress the production of NO in the acute phase of allograft rejection. We have previously shown that therapeutic concentrations of CsA, and to a lesser extent FK 506, block the production of nitrite induced in a murine macrophage cell line (J774.2 cells) and rat VSMC in culture [1,2]. We have now investigated the mechanism of this inhibition, determining the influence of CsA and FK 506 on NO production, NOS gene expression, and the isoforms of NOS affected.

Methods

The cell line J774.2 and rat VSMC were incubated (for 24 and 48h, respectively) with inducing agents (LPS, interferon gamma or interleukin 1ß) and the influence of CsA or FK 506 on NO production was determined in both cell lines by nitrite accumulation (Griess reaction). Cell viability at the end of incubations was assessed by the Trypan Blue exclusion test, and was higher than 90% for all except some cultures with the highest concentration of CsA, and these were excluded from analysis. NOS activity was determined by conversion of ^{3}H-L-arginine to ^{3}H-L-citrulline in cell homogenates obtained after 24h treatment in culture. Homogenates were incubated in the presence of cofactors in each of three buffers containing calcium (2 mM), EGTA or EGTA plus NMMA (1 mM). Calcium-dependent NOS activity was calculated from the difference in L-citrulline counts recovered between incubations in calcium and EGTA buffers, and calcium-independent NOS activity was obtained from the difference between incubations in EGTA and EGTA/NMMA buffers. The transcription of mRNA for the inducible iNOS was examined after extraction from J774 cells, followed by RT-PCR using a iNOS specific primer with appropriate standards, and visualized on agarose gels.

Results and Discussion

In both cell lines, the accumulation of nitrite was reduced concentration-dependently by the immunosuppressants, for CsA (0.1-10 μM) by up to 90% and for FK 506 by a lesser amount, without compromising cell viability. Neither compound inhibited nitrite accumulation when added 12 or 24 h (to J774 and VSMC, respectively) after the inducing agents, but NMMA (1 mM) virtually abolished nitrite accumulation. This suggests that CsA and FK 506 suppress the induction of NOS by LPS or specific cytokines, but do not affect the activity of NOS enzymes once expressed. This contrasts with the recently reported inhibition of neuronal NOS activity by FK506-induced phosphorylation in renal cortical cultures and a kidney cell line [3].

In control J774 cells, the levels of both calcium-dependent and calcium-independent NOS activity were low, and both activities were increased markedly (15- and 13-fold, respectively) by LPS, 24% of total NOS activity being of calcium-dependent isoforms (Fig.1). Co-incubation with CsA (1 μM) reduced the LPS-induced, calcium-independent activity (from 100 to 27±8 %, n=5), and also reduced the calcium-dependent activity (from 32±9 to 10±3%, n=5). NMMA eliminated all NOS activity. Similar results were obtained in VSMC, except that LPS-induced, calcium-dependent NOS represented only 5 % of total NOS activity, and calcium-dependent and -independent

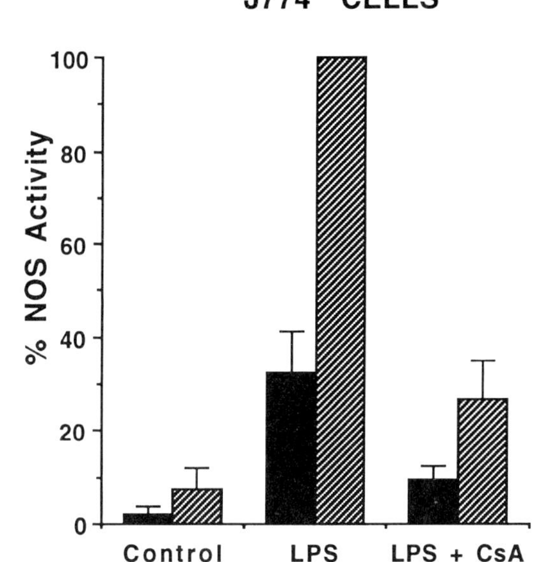

Figure 1. NOS activity in J774 cells obtained 24h after incubation with LPS or LPS plus CsA (1 μM). Values are in terms of L-citrulline generated in pmol/mg protein (expressed as % of LPS value in each culture) for calcium-dependent (filled columns) and calcium-independent (hatched columns) activities (see text). n=5 in each case.

activities were inhibited by CsA to 35 and 48 % (n=1) of LPS-induced activity, respectively. Finally, control J774 cells did not express mRNA for iNOS, but LPS induction produced a band corresponding to the iNOS fragment on agarose gels. CsA (0.1-10 μM) suppressed the expression of mRNA for iNOS but not that for ß-actin. Taken together, these data indicate that CsA blocks the transcription of iNOS in J774 cells, and probably does the same in vascular smooth muscle. In addition, CsA appears to block the expression of a calcium-dependent NOS that is also induced by LPS in both cell types, but the mechanism of this inhibition remains to be studied.

Thus CsA exerts a suppressant effect on NO production at therapeutic concentrations used to prevent interleukin-2 expression and transplant rejection. FK 506, like CsA, is thought to act in T-cells as a suppressor of gene transcription promoters through interaction of its specific binding protein with a calcineurin/calmodulin complex that, in turn dephosphorylates relevant nuclear factors initiating transcription. FK 506 is 30-100 times more potent than CsA as an immume suppressor, but it had minimal effects on NO production in J774 cells. Clearly, CsA has intracellular actions additional to those of FK 506, and these might be related to its side effects of nephrotoxicity and hypertension. Perhaps after transplantation the kidney is affected by circulating cytokines, and the renal circulation is maintained by newly expressed NOS isoforms. Suppression of NOS may also have a role in graft protection by CsA.

Abbreviations used: CsA, cyclosporin A; LPS, lipopolysaccharide; NMMA, N-monomethyl L-arginine NOS, NO synthase; VSMC, vascular smooth muscle cells

1. Akita, K., Dusting, G.J. and Hickey, H. (1994) Clin. exp. Pharmacol. Physiol. **21**, 231-233.
2. Dusting, G.J. and Macdonald, P.S. (1995) Annals Med **27**, 395-406.
3. Dawson, T.M., Steiner, J.P., Dawson, V.L. et al. (1993) Proc. Natl. Acad. Sci. USA **90**, 9808-12.

Nitric oxide, Oxyradicals, TGF-β1 and FK506 increased Suppression by Dexamethasone in Histamine and Ischemic Paw Edema of Mice.

Yoshihiko OYANAGUI (Drug Develop. Laboratories, Pharmacology 2, Fujisawa Pharmacerut. Co., 2-1-6 Kashima, Yodogawa-ku, 532, Osaka, Japan).

Dexamethasone (Dex) needed 1 hr lag time to suppress histamine and ischemic models, because Dex works via synthesis of anti-inflammatory protein(s) such as vasoregulin (1). Histamine (3 μg/paw, 10μl) was injected and increase of paw thickness was measured with a gauge. Average increase of control was 0.68 to 0.78 mm (n=5, 14 expts.). Ischemic paw edema was induced by 10 times binding of right legs just above articulation with commercial rubber ring (1$\times$ 1 mm, d=42 mm). The rubber was scissored off after 20 min and the swollen paw was measured after another 20 min (2). Average increase of paw thickness was 0.81 to 0.88 mm (n=5, 14 expts).

Results. Both edemata were never suppressed by Dex (even 10 mg/kg) before 1 hr. Dex (0.3 mg/kg, s.c.) alone showed maximum suppression after 3 hr. However, a new immunosuppresant, FK506 (0.01 - 10mg/kg, oral) with Dex showed suppressions as early as 30 min. TGF-β1 (0.01 - 1 μg/kg, i.p., human recombinant, 2×10^7 U/mg) accelerated the edema suppressions by Dex among many agents tested. TGF-β1 alone had no effect. TGF-β1 was most effective at 0.3 μg/kg. This suppression was nullified by TGF-β1,2,3 antibodies. Increase of edemata by 3 μg/kg TNF-α (i.p.) was blocked by TGF-β1, but the suppressions by 10 mg/kg

Fig. 1

Drug (i.p.)	Dose (mg/kg)	Suppression by 0.3 μg/kg TGF-β1 + 0.3 mg/kg Dex: Histamine Edema (0 20 40 o/o)	Ischemic Edema (0 20 40 o/o)
Control	-		
L-NMMA	300	**	**
-	100		*
L-NAME	100	**	**
-	30		*
D-NAME	100		
L-Arg	1,000		
D-Arg	1,000		
L-NAME + L-Arg (1,000)	100		
L-NAME + D-Arg (1,000)	100	**	**
SOD	300	**	**
-	30	**	**
Catalase	100	**	**
-	30	**	**
3-AT	30	**	*
Catalase + 3-AT (30)	100		
Mannitol	300	**	**
-	100	**	**
Cycloheximide	10	**	**

(oral) FK506 was not influenced. As redox control of gene expression is known (3), we tested NO inhibitors and oxyradical scavengers whether they modify the edema suppressive capacity by TGF-β1 and Dex (Fig. 1). Drugs (i.p.) were injected 30 min before TGF-β1 and Dex (3 tests with 3 mice, n=9 in total). L-NNMA and L-NAME impaired the suppressions of both edema models suggesting that endogenous NO is required for TGF-β1-enhanced suppressions by Dex. D-NAME was without effect and L-arginine blocked the effects by L-NAME. D-arginine showed no influence. SOD and catalse also impaired the suppression by TGF-β1 plus Dex suggesting that O_2^- and H_2O_2 were involved to

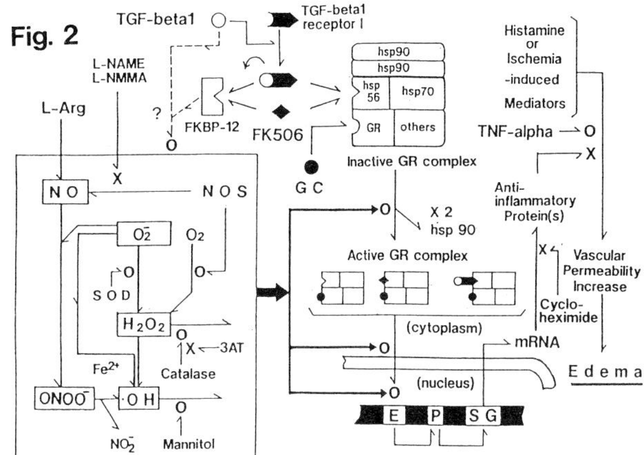

sustain this suppressive action. Blockade of suppression by 3-AT (catalase inhibitor) supported the role of H_2O_2. A hydroxy radical ($\cdot$OH) scavenger, D-mannitol impaired the suppressions, so that the directly acting oxyradical seemed to be $\cdot$OH. NO forms $\cdot$OH via generation of peroxynitrite ($ONOO^-$) and O_2^- plus H_2O_2 results in the generation of $\cdot$OH under a trace amount of Fe^{2+} ion. Edema formation in TGF-β1-treated mouse under a protein synthesis inhibitor, suggested that TGF-β1 works via stimulating GR signalling pathway. Cycloheximide impaired also the suppressions at 3 hr after the injection of Dex alone.

Discussion. TGF-β1 is a multifunctional cytokine presenting immunosuppressive and anti-inflammatory actions. TGF-β1 null mutation in mice causes excessive inflammation response and early death (4). However, its vascular permeability inhibitory action *in vivo* was proven for the first time in this work. Edema compresses the vessels and reduced blood flow. This causes necrosis and graft rejection together with the invaded inflammatory cells. We already reported that FK506 accelerated histamine edema suppression(5). FK506 binds hsp56, a component of glucocorticoid (GC) receptor (GR) and enhanced CAT expression *in vitro* (6). As FK506, TGF-β1 receptor I (TR-I) interacts with FK binding protein-12 (FKBP-12), it is very plausible that TR-I binds hsp56 as FK506 does. TGF-β1 (ligand) attachment to TR-I may enhance the binding of TR-I to hsp56 of a GR complex which in turn accelerate GR signalling to express rapidly the anti-inflammatory proteins (Fig. 2). Exact site of NO or oxyradicals to sustain GR signalling was impossible to be specified, but perhaps hydroxyl radical ($\cdot$OH) enhances the dissociation of hsp90, the penetration of active GR through nuclear membrane and/or the attachment on glucocorticoid response element (GRE, shown as E). This leads to work on promotor (P) and structual gene (SG) of anti-inflammatory proteins. Our results may give a theoritical support for clinicians to use always a little amount of GC with lowered dose of FK506 to avoid their each toxicities and showed a glimpse of TGF-β1 action in *in vivo*. In contrast to the proinflammatory action of a large amount of NO or oxyradicals, a normal level of these radicals was demonstrated to be essential for anti-inflammatory action of GC.

References

1)Y.Oyanagui and S.Suzuki: Agents and Actions 17, 270-277 (1985). **2**)Y.Oyanagui and S.Sato: Free Rad. Res. Comms. 9, 87-99 (1990). **3**)R.Schreck et al.: J. Exp. Med. 175, 1181-1194 (1992). **4**)A.B.Kuruvilla et al.: Proc. Natl. Acad. Sci. USA. 90, 770-774 (1993). **5**)Y.Oyanagui: Life Sci. 55, PL-177-185 (1994). **6**)A.M.Yem et al.: J. Biol. Chem. 267, 2868-2871 (1992). **7**) T.Wang et al.: Science 265, 674-676 (1994).

Participation of Superoxide and Nitric Oxide in Antimicrobial Host Defense at Low Mediator Concentrations

Sandra S. Kaplan*, J.R. Lancaster, Jr.+, R.E. Basford*, and Richard L. Simmons*

*University of Pittsburgh School of Medicine, Pittsburgh, PA 15261 and +Louisiana State University, New Orleans, LA

Phagocyte-derived RNI[a] such as ·NO[b] are cytotoxic to pathogenic microorganisms [1] and cell-derived ROI[c] such as O_2^-[d] also play an essential role in anti-microbial host defense [2]. Bacteria and their metabolites, small formylated peptides and LPS[e], and cytokines induce the generation of both ROI and RNI thereby bringing to bear the individual and combined cytotoxic effects of these mediators [3]. The interactive effects, however, have been unclear. Many studies have shown that the simultaneous presence of ·NO and O_2^- produces $ONOO^-$[f], a long lived powerful oxidant with great potential for cytotoxicity [4]. Other studies have shown a cytoprotective effect because ·NO scavenges O_2^- [5]. Our previous work supported the cytoprotective role by showing decreased killing of staphylococci when both mediators were present, and further showing that these radicals killed staphylococci differently: early onset killing by O_2^- and many hours delayed killing by ·NO[6]. (Manuscript submitted).

Since stimuli for ROI and RNI would result in different patterns of release (immediate for ROI and many hours delayed for RNI), it seemed reasonable that these cytotoxic agents would act sequentially rather than simultaneously. Now, conditions for scavenging would be reduced and the possibility of cooperative interaction enhanced. To test this hypothesis, S. aureus was incubated with the O_2^- donor HX[g] and XO[h] using concentrations of HX of 0.25, 0.50 and 1.0 mM together with corresponding concentrations of XO of 25, 50 and 100 mU/ml respectively for one hour at 37°C. These bacteria were washed and resuspended in enriched buffer containing the ·NO donor SNAP[i] at concentrations ranging from 0.01 to 1.0 mM. Samples were removed for bacterial counts at the outset and at 3, 5, and 24 hours. When HX/XO concentrations were 1.0 mM and 100 mU no interactive enhancement of staphylococcal killing was observed when 0.1 mM SNAP was added. Similarly, the concentrations of HX/XO of 0.50 mM and 50 mU or 0.25 mM and 25 mU, followed by 0.1 mM SNAP did not alter the pattern or degree of bacterial killing of the dominant killing mechanism at any time point. The effect of decreasing the concentration of SNAP to 0.01 mM is shown in Table 1.

TABLE 1. Effect of Sequential O_2^- and NO on S. aureus
Each value represents the mean + SEM

HX/XO @ 50	% Surviving Staphylococci		
SNAP @ 0.01	O_2^- Alone	·NO Alone	Sequ. O_2/·NO
3 hr	6±4	92±15	6±5
5 hr	7±3	102±12	9±5
24 hr	20±4	116±17	25±7
HX/XO @ 25			
SNAP @ 0.01			
3 hr	82±35	95± 9	67± 1
5 hr	114±23	111±11	65±20
24 hr	59± 7	102± 5	39±13[a]

[a]$p<0.05$ compared to O_2^- alone and SNAP (0.01) alone

Abbreviations Used:

a. Reactive nitrogen intermediates
b. Nitric Oxide
c. Reactive oxygen intermediates
d. Superoxide
e. Lipopolysaccharide
f. Peroxynitrite
g. Hypoxanthine
h. Xanthine oxidase
i. S-Nitroso-N-acetyl-Penicillamine

Table 1 shows that when the concentrations of O_2^- and ·NO are low such that staphylococcal killing is poor by either mediator, the sequential presence of O_2^- followed by ·NO resulted in greater killing than with either mediator alone.

In other studies S. aureus was incubated with 1 mM SNAP for 2 or 4 hours before centrifugation and addition of the O_2^- generating system. Viability was tested at the outset, when the SNAP was removed, 2 hours later, and at 24 hours.

TABLE 2. Effect of Sequential NO and O_2^- on S. aureus
Each value represents the mean + SEM

SNAP for 2 hrs	% Surviving Staphylococci		
HX/XO @25	SNAP Alone	O_2^- Alone	Sequ. NO/O_2
2 hr	97± 7	32±4	97± 9
4 hr	120±35	38±3	33± 8
24 hr	26±11	57±9	41+16
SNAP for 4 hrs			
HX/XO @25			
4 hr	57±7	38±3	57±2
6 hr	43±9	43±6	20±2[a]
24 hr	17±6	57±9	23±2

[a]$p<0.05$ compared to SNAP alone and O_2^- alone

This sequential study also was performed using human neutrophils by incubating S. aureus with 50% authentic ·NO for one hour before adding these bacteria to neutrophils. Native neutrophils and neutrophils heated to 46°C for seven minutes to create a model of impaired production of ROI were used. Heated neutrophils failed to produce ROI with a peptide mediator or when placed in an uncoated plastic culture dish and staphylococcal killing was decreased to 43±9% of the inoculum compared to 94±3% killing by native neutrophils.. When the ·NO treated bacteria were added to the heated neutrophils, the neutrophils killed 57±7% of the inoculum ($p\leq0.001$).

Neutrophils stimulated in a physiologic manner exhibit a transient oxidative burst. ·NO, if present would act to neutralize the O_2^- and diminish cytotoxicity. Extensive studies of ·NO producing cells, however, have shown that when a stimulus acts to induce ·NO formation, there is a many hour delay before ·NO is detectable, and it is unlikely that O_2^- and ·NO would be co-produced. Instead, production would be sequential.

This study showing a sequential cooperative effect, and our previous work showing impairment of killing by simultaneous O_2^- and ·NO support a hypothesis whereby neutrophils kill bacteria by overlapping mechanisms.

References:

(1) Hibbs, J.B.Jr., Taintor, R.R., Vavrin Z., Rechlin, E.M. (1988) Nitric oxide: A cytotoxic activated macrophage effector molecule. Biochem Biophys Res Commun 157:87-94.

(2) Babior, R.M. Kipnes R.S., Curnurtte J.T. (1973) Biological defense mechanisms. The production by leukocytes of superoxide, a potential bactericidal agent. J Clin Invest 51:741-744.

(3) Bastian, N.R., Hibbs J.B.,Jr., (1994) Assembly and regulation of NADPH oxidase and nitric oxide synthase. Curr Op Immunol 6:131-139.

(4) Beckman, J.S., Crow J.P. (1993) Pathological implications of nitric oxide, superoxide and peroxynitrite formation. Biochem Soc Transact 21:330-334.

(5) Rubanyi, G.M., HO, E.H, Cantor E.H., Lumma, W.C., Botelho L.H.P. Cytoprotective function of nitric oxide: inactivation of superoxide radicals produced by human leukocytes.

Quantification and characterisation of the effect of Reactive Nitrogen Intermediates on *Leishmania mexicana*.

William H.H. Reece, Lorna Proudfoot, Graham H. Coombs*, Foo-Y. Liew.

Dept Immunology, Western Infirmary, Glasgow, G11 6NT, Scotland. * Division of Infection and Immunity, IBLS, Glasgow University, Glasgow, G12 8QQ, Scotland.

Reactive Nitrogen Intermediates (RNI), which include nitric oxide, play a fundamental role in protective immunity against *Leishmania* [1-3]. *Leishmania* is a genus of intracellular protozoan parasite transmitted by sandflies. The parasites cause a range of pathologies from local cutaneous lesions to diffuse visceral infections in mammals. Many animals used experimentally are resistant to this parasite, and this resistance is probably due to macrophages (the cellular host of *Leishmania*) producing RNI. As yet, however, we do not know how RNI kill the parasite.

We are studying the effect of RNI on axenically grown amastigotes of *Leishmania mexicana* [4-6]. Amastigotes are the life cycle stage that is found in mammalian macrophages, whereas promastigotes are the stage in the sandfly gut. Hence amastigotes are the stage that is exposed to RNI *in vivo*.

We show that RNI are toxic to *Leishmania mexicana*, and that RNI appear to be cytotoxic rather than cytostatic, causing the parasite to lose structural integrity. We also show that the toxicity is unlikely to be due to cleavage of the sugar moiety from glycoinositolphospholipids (GIPLs), which are the major surface molecules of amastigotes.

Materials and methods. Amastigotes from *L. mexicana* (MNYC/BZ/62/M379) were cultured according to the method of Bates *et al* (1992) in Schneider's Drosophila medium with 20% FCS and 25 µg/ml gentamycin, pH 5.4-5.6. RNI were produced by adding sodium nitrite to this acidified medium, and calculating the concentration of nitrous acid (HONO) (See Fig. 1). [^{3}H]-Thymidine uptake was measured by adding 1 µCi [^{3}H]-thymidine to 200 µl of amastigotes (2 x 10^6 ml^{-1}) in 96 well plates, and incubating for 24 h at 32 °C in the presence of sodium nitrite (Fig 1). Structural integrity was assayed by microscopic observation of washed and fixed amastigotes after a 24 h treatment with sodium nitrite (Fig. 2). Viability of these amastigotes was assayed by placing them in HOMEM medium with 10% FCS, and allowing them to transform to promastigotes at 28°C (Fig. 2). GIPLs labelled on their oligosaccharide moiety were obtained by incubating live parasites with [^{3}H]-glucosamine, and extracting the GIPL as described by McConville and Bacic [9]. After treating the GIPLs with sodium nitrite in sodium acetate buffer pH 4.0 for 24 hr 32°C, their integrity was assayed by water/butanol phase partitioning of the label. All figures are representative of at least 3 separate experiments.

Results and discussion. Using this acidified nitrite method, we have managed to produce RNI at a constant concentration throughout these experiments. We can quantify the concentration of at least one of the RNI - nitrous acid (HONO), which correlates with the toxicity of the solution.

Fig. 1 shows that parasites proliferated at different rates in different pH's, but that RNI could kill whatever the pH. Nitrite itself was not toxic since different concentrations were toxic at different pH's (Fig. 1a).

The results in Fig. 2 are consistent with RNI being considered cytotoxic rather than cytostatic. If they were cytostatic, one would expect a large range of concentrations where the numbers of parasites after 24 h was the same as the start number. This was not the case.

The mechanism of toxicity of acidified nitrite is unlikely to be deglycosylation of the surface molecule GIPL. The labelled oligosaccharide moiety was not cleaved from the lipid moiety below millimolar concentrations of HONO (Fig. 3), whereas the parasite was killed in the micromolar range (Figs 1 and 2).

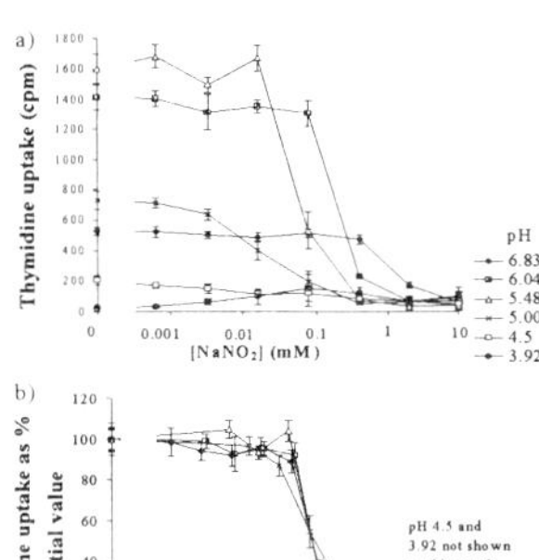

Figure 1 a) Thymidine uptake by amastigotes at varying pH's, and different concentrations of sodium nitrite.
b) The same data as in a), but thymidine uptake normalised for the initial uptake at the different pH's, and plotted against the calculated concentration of nitrous acid.
Results are the mean ± SEM (n=4)

$$[HONO] = \left(\frac{[NO_2^-]}{10^{(pH - pKa(HONO))}}\right)$$

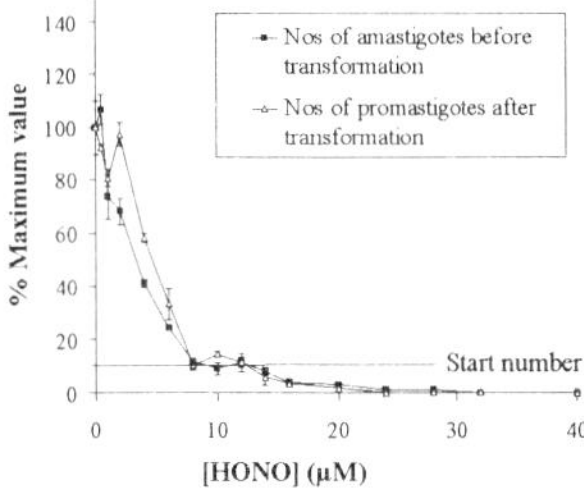

Figure 2 The effect of RNI on parasite integrity and viability. Amastigotes were treated for 24 h with HONO, washed, fixed and counted. Viability was assayed by efficiency of transformation to promastigotes after treatment with RNI. Results are the mean ± SEM (n=4).

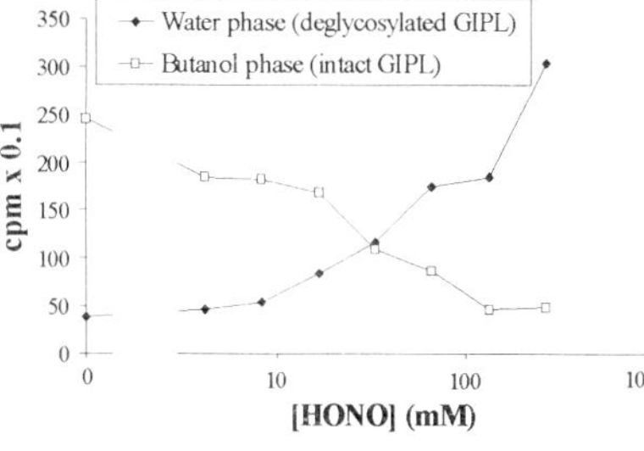

Figure 3 The effect of RNI on GIPL structure. Labelled and purified GIPLs were treated with acidified nitrite for 24 h at 32°C. Integrity of the GIPL was assayed by butanol/water phase partitioning.

This work was funded by The Wellcome Trust.

References

1. Assreuy, J., Cunha, F.Q., Epperlein, M., *et al* (1994) Eur. J. Immunol. **24,** 672-676
2. Green, S.J., Meltzer, M.S., Hibbs, J.B., Jr., and Nacy, C.A. (1990) J. Immunol. **144,** 278-283
3. Wei, X.-Q., Charles, I.G., Smith, A., *et al.* (1995) Nature **375,** 408-411
4. Bates, P.A. (1993) Parasitol. Today **9,** 143-146
5. Bates, P.A., Robertson, C.D., Tetley, L., and Coombs, G.H. (1992) Parasitology **105**, 193-202
6. Bates, P.A. (1994) Parasitology **108,** 1-9
7. Murray, H.W. (1981) J. Exp. Med. **153,** 1690-1695
8. Loewe, S. (1957) Pharmacol. Rev. **22**, 237-242
9. McConville, M.J. and Bacic, A. (1990) Mol. Biochem. Parasitol. **38**, 57-68

Abbreviations used in this paper: RNI - Reactive Nitrogen Intermediates, HONO - nitrous acid, SEM - Standard error of the mean, GIPL - glycoinositolphospholipid.

A study of the antibacterial activity of nitric oxide donor compounds

MICHAEL G. BURDON,[1] ANTHONY R. BUTLER[2] and LOUISE M. RENTON[2]

[1]School of Biological and Medical Sciences and [2]School of Chemistry, University of St Andrews, St Andrews, Fife KY16 9ST, UK.

Introduction

Nitric oxide (NO) is formed by many types of cell in the body as part of the immune or inflammatory response system. NO is produced in neutrophils, lymphocytes and macrophages as part of the cytotoxic function of these cells[1].

The antiparasitic activity of NO has been demonstrated *in vivo* with the parasitic diseases malaria[2] and leishmaniasis[3]. The World Health Organization tested the NO donor compounds SNAP, SNP and RBS (Roussin's black salt) *in vitro* on *P. falciparum* D6 (Sierra Leone) and W2 (Indochina) clones. They established that the drugs had a very slight effect on the parasites with SNP being most active[4]. However, in comparison with standard antimalarials such as chloroquine and qinghaosu their level of activity was not significant.

We have developed a wide range of NO donor compounds and present evidence for their antibacterial activity with *Escherichia coli* cultures.

Methods

In order to investigate the toxicity of test compounds towards bacteria, *Escherichia coli* (DH5 strain) was used. To obtain a standard stock culture a loopful of *E. coli* from a pure slope of agar was transferred to sterile nutrient broth or a minimal defined media and incubated at 37°C.

The test compounds SNAP, GSNO, RBS and SNP were prepared in PBS. Dibenzylhydroxyguanidine (DBHG) was prepared in DMSO and peroxynitrite prepared and used in situ. *E. coli* was added to each test solution then transferred to an agar plate and incubated at 37°C for 24hours. After this time the viable colonies were counted.

Results and discussion

Although the arginine-NO pathway occurs in activated macrophages, NO-donor drugs (SNAP, GSNO and DBHG) were not potent antibacterial agents. Cu(II) acetate, which is known to catalyse NO release enhanced the toxicity of SNAP (graph **(a)**); however at higher concentrations it was itself toxic. Species containing NO^+ (i.e. RBS and SNP) were *substantially,* more toxic (graph **(b)**) than those releasing NO. Similar results with *Clostridium sporogenes* have been reported[5]. We suggest that this is because they nitrosate important thiol groups involved in bacterial metabolism. Peroxynitrite is more toxic than SNP but less than RBS. Its toxicity could be because of HO· production but was not diminished by a HO· trap (ascorbic acid, graph **(c)**) and NO^+ formation is a possible explanation for toxicity.

$$\mathrm{HOONO} \longrightarrow \mathrm{HOO^{\cdot}} + \mathrm{NO^+}$$

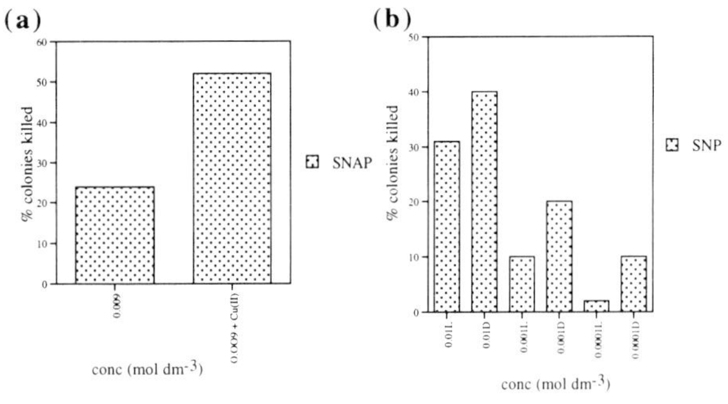

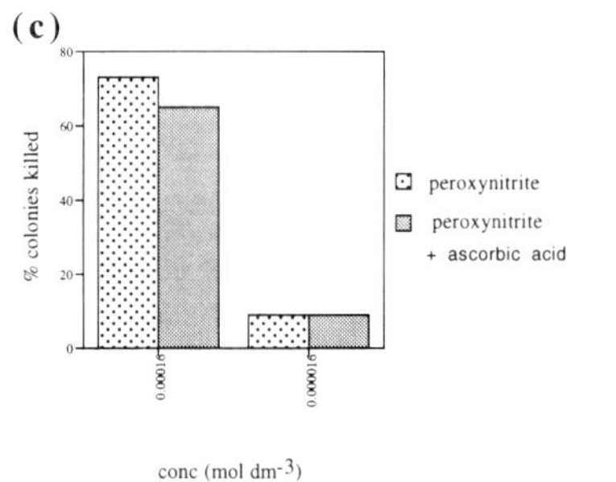

(a) effect of Cu(II) on SNAP toxicity

(b) effect of light on SNP toxicity

(c) effect of ascorbic acid on peroxynitrite toxicity

References

1. Bjorkman, A. and Phillips-Howard, P.A. (1990) *Trans. R. Soc. Trop. Med. Hyg.*, **84**, 17780.
2. Taylor-Robinson, A.W., Phillips, R.S., Severn, A., Moncada, S. and Liew, F.Y. (1993), *Science*, **260**, 1931-1934.
3. Liew, F.Y., Millott, S., Parkinson, C., Palmer, R.M.J. and Moncada, S. (1990), *J. Immunol.*, **144**, 4794-4797.
4. Davidson, D.E. (1992) Steering Committee on the Chemotherapy of Malaria, World Health Organization, Geneva.
5. Cui, X., Joannou, C.L., Hughes, M.N. and Cammack, R. (1992) *FEMS Microbiology letters*, **98**, 67-70.

Direct quantitation of nitric oxide released from cells using liposome-encapsulated PTIO

Akaike, T. and Maeda, H.
Department of Microbiology, Kumamoto University School of Medicine, Kumamoto 860, Japan.

Introduction

Stable nitronylnitroxide derivatives, 2-phenyl-4,4,5,5-tetramethyl-imidazoline-1-oxyl 3-oxide derivatives (PTIOs), show a specific scavenging action against NO released in ex vivo and in vivo systems [1-3]. The reaction of PTIOs with NO gives a clear change in electron spin resonance (ESR) signals, yielding 2-phenyl-4,4,5,5-tetramethyl-imidazoline-1-oxyl derivative (PTIs) [1, 4]. Here, we describe quantitative assay procedure using liposome PTIO for NO produced in the cells.

Methods

ESR spectroscopy for measurement of PTIOs and PTIs: ESR spectrum was obtained at room temperature with X-band JES-RE1X ESR spectrometer (JEOL) by using quartz flat cell with an effective volume of 180 µl.

Preparation of liposome encapsulated PTIO: PTIOs are susceptible to the nonspecific reduction by various reducing agents such as ascorbate, thiol compounds, and superoxide anion. It is reported that encapsulation in a liposome vesicle made nitronylnitroxide remarkably stable [5].

We have prepared successfully much more stable liposome PTIO than that reported earlier. Liposome PTIO was prepared by reverse phase evaporation method [6] by using L-α-phosphatidylcholine and dimylstoylamido-1,2-deoxyphosphatidyl choline and trimethylammonio-PTIO (TMA-PTIO).

Quantitation of NO released from cells by using liposome PTIO: Liposome PTIO was incubated with a murine macrophage cell line RAW264 cell in culture with or without stimulation by both 10 µg/ml of lipopolysaccharide (LPS) and 200 U/ml of interferon-γ (IFN-γ). After various stimulation-periods, liposome PTIO was added to each culture with or without either Cu,Zn-superoxide dismutase (SOD) or L-arginine, followed by the incubation. After 30 min-incubation period, ESR measurement was performed to quantitate the amount of liposome PTI produced.

Results & Discussion

The amount of nitroxide radicals was measured by double integration of experimental signals after normalization with manganese (MnO) signal recorded simultaneously.

Stoichiometric conversion of liposome PTIO to liposome PTI via the reaction with NO is clearly shown in Fig. 1. In this experiment, propylamine NONOate (1-hydroxy-2-oxo-3-(*N*-methyl-3-aminopropyl)-3-methyl-1-triazene) was used as a source of NO generation. When various concentrations of liposome PTIO were given to the reaction system containing NO-NOate in PBS (pH 7.4), a linear generation of liposome PTI was observed (the slope is 2.0), which is strictly dependent on the concentration of NONOate added (Fig. 1).

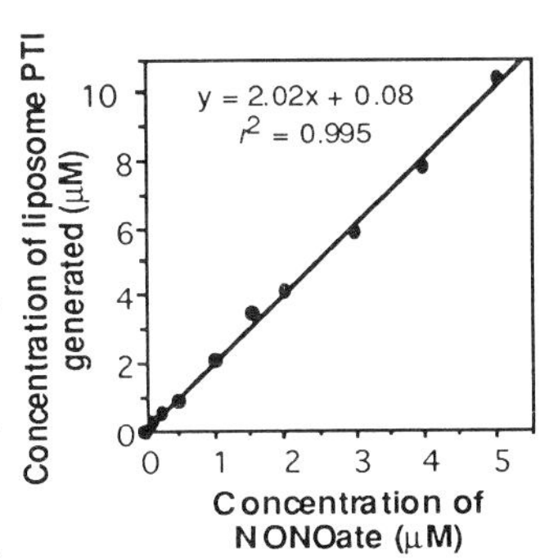

Fig. 1 Formation of liposome PTI in the reaction system of NONOate and liposome PTIO.

These results indicate that two moles of NO is released from one mole of the NONOate and that liposome PTIO reacts with NO in a completely stoichiometric manner. The minimal detectable limit of NO is above 0.05 µM (>9 pmol).

NO released extracellularly was examined by the reaction of liposome PTIO with RAW264 cells stimulated with LPS and IFN-γ. As demonstrated in Fig. 2A, the generation of liposome PTI was readily identified and quantitated. The ESR signals are very stable for several hours of incubation periods with RAW264 cells even in the presence of serum (blood) and reducing agents, e.g., ascorbate and thiol compounds. Therefore, NO production by the cells can be measured accurately based on the quantitative generation of liposome PTI in the reaction system.

Addition of Cu,Zn-SOD to the reaction mixture resulted in increment of the generation of liposome PTI (not shown) due to the removal of superoxide generated by the cells. Deletion of L-arginine or addition of an NOS inhibitor such as N^{ω}-monomethyl-L-arginine (L-NMMA) in the mixture gives no appreciable production of liposome PTI.

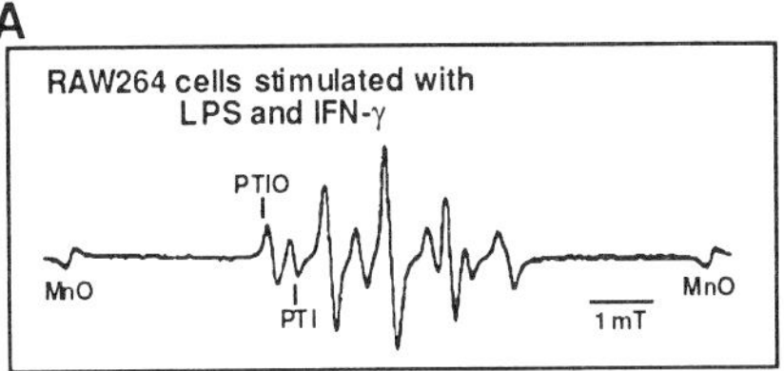

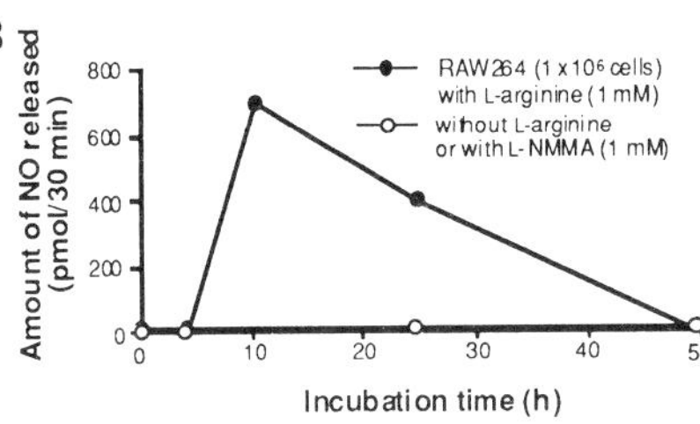

Fig. 2 Quantitation of NO produced by RAW 264 cells by using liposome PTIO with ESR spectroscopy. A. ESR spectra observed with liposome PTIO incubated with RAW264. B. Time profile of NO production by RAW264 cells stimulated with LPS and IFN-γ both added at time zero.

The time profile of NO production determined by liposome PTIO/ESR method (Fig. 2B) correlates well to that of mRNA expression of inducible NO synthase examined by Northern blot analysis (data not shown).

Liposomal membrane functions as a barrier against various polar or water soluble reducing compounds to PTIO, and it will permit only diffusion of hydrophobic small molecules, especially NO, into the PTIO comportment of liposome. The chemical reaction forming PTIs from PTIOs is essentially specific for NO because no other substance that reacts with PTIOs forming PTIs has been found so far. In addition, PTIOs directly react with NO but not with other NO-related species such as nitrosothiols, iron-nitrosyl complex, and nitrite.

Therefore, a specific, sensitive and quantitative measurement for NO produced in biological systems now becomes possible by application of PTIOs and liposome PTIO by using ESR spectroscopy.

Abbreviation used: PTIOs, 2-phenyl-4,4,5,5-tetramethyl-imidazoline-1-oxyl 3-oxide derivatives; L-NMMA, N^{ω}-monomethyl-L-arginine; ESR, electron spin resonance; IFN-γ, interferon-γ; SOD, superoxide dismutase; LPS, lipopolysaccharide.

References

1. Akaike, T., Yoshida, M., Miyamoto, Y., et al. (1993) Biochemistry **32**, 827-832
2. Yoshida, M., Akaike, T., Wada, Y., Sato, K., Ikeda, K., Ueda S., Maeda, H. (1994) Biochem. Biophys. Res. Commun. **202**, 923-930
3. Maeda, H., Akaike, T., Yoshida, M., Suga, M. (1994) J. Leuk. Biol., **56**, 588-592
4. Az-ma, T., Fujii, K., Yuge, O. (1994) Life Science **54**, PL185-190
5. Woldman, Y.Y., Khramtsov, V.V., Grigor'ev, I.A., Kiriljuk, I.A., Utepbergenov, D.I. (1994) Biochem. Biophys. Res. Commun. **202**, 195-203
6. Szoka, F., Papahadjopoulos, D. (1978) Proc. Natl. Acad. Sci., USA **75**, 4194-4198

Techniques for the measurement of nitric oxide and its reaction products

Richard S. Hutte
Sievers Instruments, Inc., Boulder, CO USA

Since the discovery that nitric oxide is biologically produced, a wide range of analytical techniques have been used to measure NO and its reaction products [1]. These include spectrophotometry (Griess reaction and methemoglobin), fluorometric techniques, mass spectrometry, electrochemical detection, electron spin paramagnetic resonance, and chemiluminescence (ozone and luminol). Other researchers monitor NO production indirectly including assays for citrulline, cGMP, physiological assays and inhibition of nitric oxide synthase. Each of these techniques has its advantages and disadvantages in terms of sensitivity, selectivity, cost of equipment, sample size requirements, ease of use and the capabilities of a laboratory and it staff.Ozone-based chemiluminescence is particularly well suited for NO research due to the high sensitivity and selectivity of this technique and the wide range of applications that can be performed using chemiluminescence.

Measurement of nitric oxide in biological systems is complicated by low concentration of NO produced and the reactivity of this molecule. In solution, NO is oxidized by molecular oxygen to form nitrite (NO_2^-), oxidized by superoxide anion and oxyhemoglobin to form peroxynitrite and nitrate (NO_3^-). Nitric oxide also reacts with thiols to form S-nitroso compounds, amines to produce nitrosamines and NO reacts with a variety of metals to form metal nitrosyl complexes. Other species such as NO^+ and NO^- have been postulated to exist in biological system.Since NO is a gas it will also rapidly diffuse from solution into the gas headspace above the liquid. In the gas phase at low concentration, NO reacts slowly with oxygen to produce nitrogen dioxide and other oxides of nitrogen. At higher concentrations, such as those used for inhalation therapy, significant amount of NO_2 can be produced.

At Sievers Instruments, we have been producing nitric oxide analyzers based on ozone-induced chemiluminescence since 1988. Unlike air pollution monitors, our NO/O_3 detectors are designed for measurement of NO in biological systems. Specifically, our detectors have always employed small reaction cells to permit sensitive measurement of NO at low flow rates (100-300 mL/min versus 500 to 1000 mL/min for air pollution monitors). By using sensitive photomultiplier tubes, PMT cooler and more efficient vacuum pump we can achieve sensitivity better than or equal to air pollution equipment for gas-phase measurement of NO. Our special purge systems permit measurement of NO in solution and combined with chemical reducing agents (KI/Acetic Acid for NO_2^- measurement, VCl_3/HCl for nitrate and nitrite measurement) provide sensitive measurement of NO and its reaction products in liquid samples with sensitivity 250 times better that air pollution monitors. The low flow rates and fast response time of our electronics permit real time, breath-by breath measurement of NO in exhaled breath.

In response to the changing needs of NO researchers and requirements for monitoring NO inhalation therapy, Sievers Instrument has developed a new detector, the Model 280 Nitric Oxide Analyzer (NOA™). At the heart of the new detector are the same components that have provided high sensitivity and fast response for hundreds of researchers using our Model 270B NOA. But the Model 280 NOA features new electronics that include microprocessor control, analog, digital and printer outputs, onboard data storage using PCMCIA RAM cards, an easy to use menu-based operation system, all packaged in a smaller instrument. A schematic of the analyzer is shown in Figure 1. The detector consists of a sample inlet system, an ozone generator, a chemiluminescent reaction cell, an optical filter to reduce interference for other species that chemiluminesce with ozone, a vacuum pump and chemical trap to remove ozone and protect the vacuum pump and the new electronics including a back-lit LCD display and four button panel. The sample inlet systems including equipment for direct gas measurement, a complete line of mouth-breathing face masks for monitoring NO in exhaled breath, an NO_x converter and our special purge system for measurement of NO, NO_2^- and NO_3^- in liquid samples. The detector can also be used for measurement of NO using headspace sampling or measurement of exhaled NO using Tedlar gas bags. Gas flow into the analyzer is controlled either with a ceramic restrictor to provide a constant flow rate or a needle valve can be used for researcher who want to vary the flow into the analyzer.

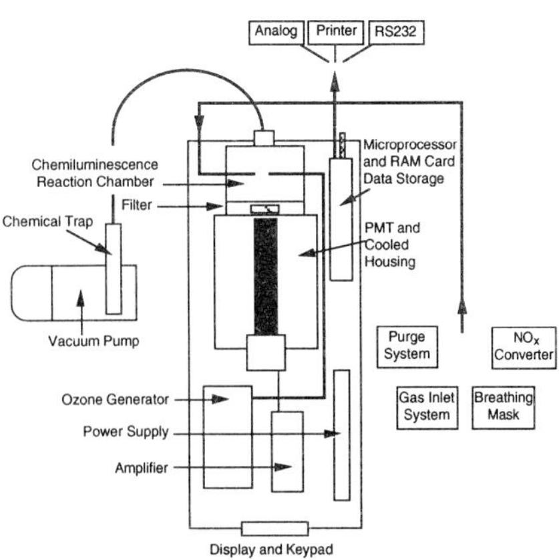

Figure 1. Schematic of Model 280 Nitric Oxide Analyzer (NOA™).

A new feature of the Model 280 NOA is gas calibration software to permits direct output of gas concentration in ppbV or ppmV. A mV output can also be selected for liquid samples, headspace and researchers who want to measure total amount of NO in a gas sample. In order to provide high sensitivity and a wide dynamic range, the 280 NOA uses a dual gain amplifier. The high gain mode permits measurement of gas phase NO over a range from <5 ppb to 1 ppm. for liquid samples the range is from 1 picomole NO, NO_2^- or NO_3^- to ~300 picomoles. Switching to the low gain permits measurement up to 500 ppmV for gases and ~100 nanomoles for liquids. Range switching for gas measurements can be automatic or you can select the desired range in software.

The 280 NOA has five outputs; the front panel display, an analog output, RS232 output for direct downloading of data to a PC or Apple computer, a printer port, an analog output and the RAM card. The RS232 port supports baud rates from 1200 to 32.5K with sampling rates ranging from 20 samples per second to 0.002 samples per second (1 hour per sample). The RS232 output consists of the date and time of the measurement, the measured value and the units (mV, ppbV or ppmV) in a comma separated format that can be imported into spreadsheet and graphing programs for easy display. The analog output (0-1V or 0-10V) can be set to output mV's or gas concentration (ppbV or ppmV). The output can also be scaled in software to permit output of any range of mV or gas concentration. This permits you to set to 0-1 V full-scale output to correspond to 0-5 ppbV, 5-10 ppmV, 10-50 mV or any desired range. The print interval is selected in software from 20 samples per second to 1 sample per hour. The printer output shows the minimum, maximum and average value (mV, ppbV or ppmV) for the selected interval. The front panel display shows the instantaneous value and is updated every second.

The internal RAM Card can be used to store data at intervals from 20 samples per second to 10 samples per hour. Either average values or instantaneous values at the end of the sampling interval can be stored. Data can be retrieved from the RAM card in three ways. The data can be viewed from the front panel display, the data can be downloaded to a computer via the RS232 port or the card can be removed from the analyzer and read using software from Sievers and either the internal PCMCIA card slot on the computer or an external card reader.

The microprocessor monitors the instrument and consumables and reports possible problems and alerts the user when maintenance is required using "error" messages. These error messages are stored in battery-backed RAM and can be view from the front panel display, output to a printer or downloaded via the RS232 to a computer. Operation of the detector is performed using a four buttons (up down, enter and clear) and an easy to use menu-based software. The software permits the status of the consumables (vacuum pump oil, chemical trap, reaction cell cleaning and PMT cooler servicing) to be monitored and reset when service is performed. The software also allows all of the outputs to be configured, provides for gas calibration, and allows the user control of the outputs and power to the vacuum pump, ozone generator, photomultiplier tube.

Applications

A wide range of applications can be performed using the new NOA, but to illustrate the performance of the new analyzer, two applications will be presented; measurement of NO in exhaled breath and measurement of nitrite in liquid samples.

NO in exhaled breath - Nitric Oxide is present in low concentration in exhaled breath and the levels may correlate with diseases states such as asthma. Many different sources of NO have been identified including nasal sites, and sites in the airway (epithelial, macrophage, and mast cells) [2].Several different approaches to

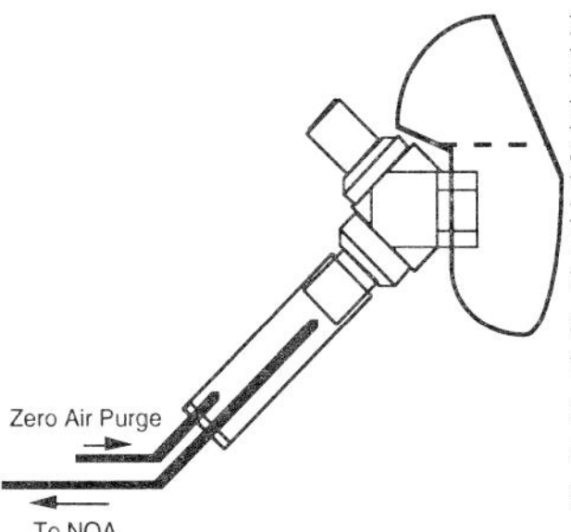

Figure 2. Apparatus for continuous measurement of exhaled NO

measuring exhaled NO in humans have been reported ranging from placing subjects in a sealed room and monitoring the increase in NO in the air, having subjects breath into Tedlar bags and continuous breath-by-breath measurements. The slow response time and large gas flow rates of air pollution monitors have generally prevented continuous measurement of exhaled NO.

Figure 2 shows the experimental set-up for continuous measurement of exhaled NO using the new analyzer. The subject is fitted with a mouth-breathing face mask that has a partition that seals the nose and minimizes nasal contributions to exhaled air. The outlet of the mask is a Y-shaped two way non-rebreathing valve.with tubing on the outlet of the valve to isolate the exhaled air from ambient. A short length of 1/8" OD Teflon tubing is positioned inside the tubing, near the outlet of the valve and the other end of the Teflon tubing connect to the inlet of the NOA. A ceramic frit restrictor is used to control the flow rate of gas into the analyzer at 200 mL/min.

One of the difficulties in measuring exhaled NO in this continuous manner is the presence of NO in ambient air. Since the analyzer is always drawing gas into the reaction chamber, the sample alternates between exhaled air and ambient air. If the levels of NO in exhaled breath are greater than ambient, a positive peak is observed. If the levels of less them ambient, a negative peak is observed and if the concentration of NO in exhaled air is the same as ambient, more peak will be observed. One approach to this problem is shown in Figure 2. The hose connected to the Y valve is purged with zero air (gas containing < 1 ppb NO). The flow rate of zero gas is greater than the flow rate being drawn into the analyzer. Once ambient air has been purged from the hose, the gas inlet is moved near the outlet of the hose to prevent ambient air from being drawn into the hose between breaths. To completely eliminate any effects from ambient NO, zero air can also be connected to the inlet side of the y piece, allow the subject to breath gas containing low levels of NO.

Figure 3 shows an example of continuous monitoring of exhaled NO in an adult smoker using this zero gas purge technique. The NO concentration for the six breaths range from 2 to 4 ppb. The volume of the tube connected to the Y-value was sufficiently large so that not all of the exhaled air was drawn into the analyzer between breath. As a result, the NO concentrations between breaths does not drop to zero and instead represents a mixture of the zero gas purge and remaining exhaled breath.

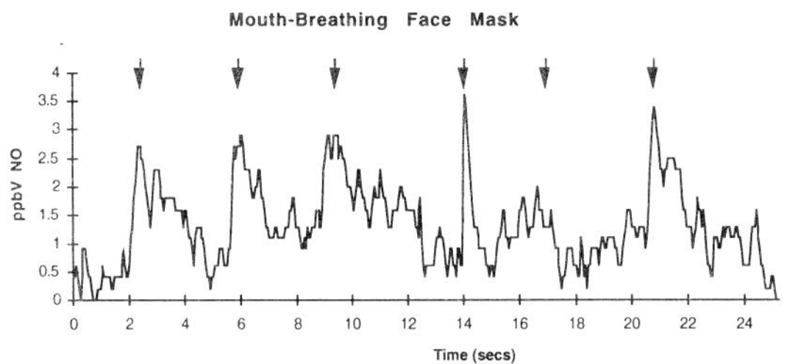

Figure 3. Continuous Measurement of Exhaled NO from Male Smoker. Arrows indicate exhalation

Figure 4 shows continuous monitoring of the NO concentration of air exhaled through the nose. For this experiment, a mask was not used, rather the Teflon line from the NOA was simply held under the nose. NO concentrations range from 8 to 17 ppb for 6 exhalations. In this sampling mode, ambient air is drawn into the analyzer between exhaled breaths.

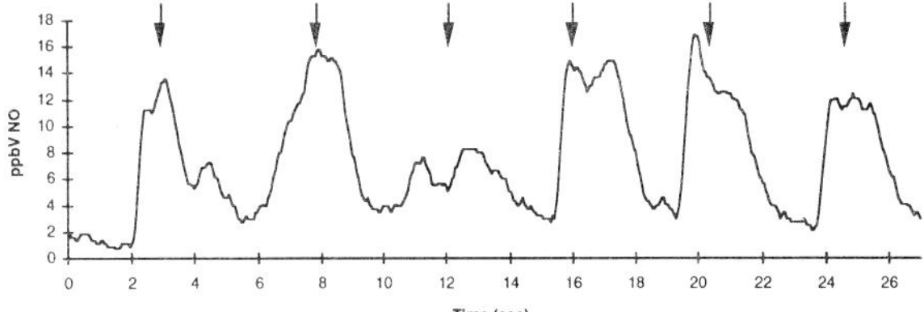

Figure 4. Continuous Measurement of Exhaled NO through Nose.

Liquid Samples - One of the major applications in NO research is measurement of NO and its reaction products in liquid samples ranging from cell culture media, plasma, urine and other biological fluids. For the analysis of these samples, two techniques are used; headspace analysis and gas purging. By using chemical reducing agents in conjunction with these analytical techniques, reaction products including nitrite, nitrate S-nitroso and N-nitroso compounds can be measured.

In headspace analysis, a liquid sample is placed in a sealed container with a gas headspace above the liquid sample. To prevent oxidation of NO in the gas phase, oxygen is usually purged from the gas headspace prior to sealing the container. NO in solution will partition into the gas and after a short period of time, an equilibrium will be established between the concentration of NO in the gas and NO in solution. The gas headspace is then sampled, either using a gas tight syringe and injection directly into the NOA or the entire gas headspace can be withdrawn by the vacuum pump for improved sensitivity. Headspace analysis can also be combined with chemical reducing agents to permit measurement of solution nitrite/nitrate. .

An alternative to headspace analysis is to use a gas to purge all of the NO from solution. Figure 5 shows the purge system used with the new NOA. The system consists of a glass purge vessel equipped with a heating jacket, cold water condenser, a glass frit, a septum for injection of the samples a gas inlet and needle valve to control the flow rate of an inert purge gas (helium, nitrogen or argon). For nitrate analysis, a gas bubbler containing aqueous NaOH is connected to the outlet of the purge vessel to remove acid vapors from the gas stream. Stopcocks are posited at the gas inlet, outlet of the purge vessel, and the outlet of the gas bubbler to permit isolation of the purge vessel and gas bubbler from the vacuum of the NOA.

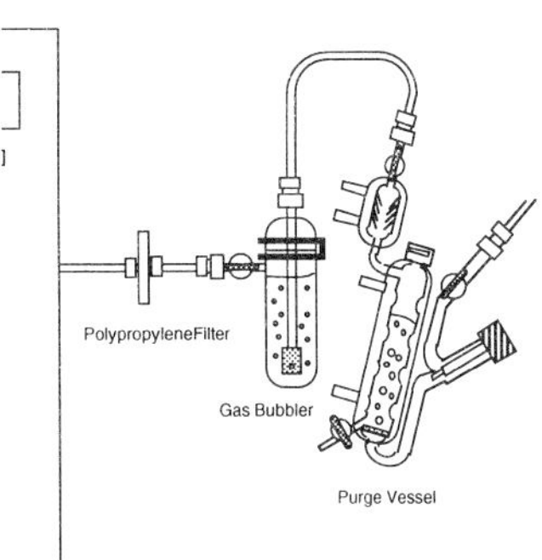

Figure 5. Purge System for Liquid Samples

A new component for the purge system is a polypropylene filter. The hydrophobic membrane in this disposable element will prevent water from entering the analyzer, but passes NO. This prevents liquids from entering the detector's reaction cell and reduces the need for cleaning the reaction cell.

Most of the applications using the purge system use chemical reducing agents for converting nitrite and/or nitrate and other reaction products to NO. For nitrite measurement, the recommended reducing agent is 1% (by weight) KI or NaI in acetic acid. In a typical analysis, 4 mL of glacial acetic acid (or concentrated acetic acid) is added to the purge vessel and the acid purged for ~5 minutes to remove dissolved oxygen. A 50 milligram sample of iodide is dissolved in deionized water and added to the purged acetic acid. Finally ~100 µL of diluted anti-foaming agent is added to the reagent to minimize foaming due to proteins. Typically this amount of reagent is sufficient for several hours of measurements

In operation, the flow rate of purge gas is adjusted to bring the level of the liquid reagent near the top of the purge vessel so that injections through the septum are made into the reagent. The flow rates of the purge gas entering the purge vessel is less than the pumping speed of the vacuum pump resulting in a slight vacuum in the purge vessel. This vacuum speeds the removal of NO from solution and results in sharp NO peaks in the detector.

Figure 6 shows the detector response obtained from injection of 1.5, 6, and 12 picomoles of nitrite. The detection limit is <1 picomole and analysis time is ~30 seconds. In this sample mode, the NOA is a mass sensitive detector, thus the concentration detection limit corresponds to <1 µM for a 1 µL injection, <10 nanomolar for a 100 µL injection. A calibration curve is shown in Figure 7. The new NOA has a linear response from the detection limit up to the highest amount tested (130 picomoles) and for the high gain setting on the amplifier, the linear respond extends to ~300 picomoles.

The purge vessel can also be used for measurement of nitrate and nitrite in plasma, urine and other fluids. For these samples the reducing agent is VCl_3 in 1M or 2 M HCl. A 0.1 M solution of VCl_3 is prepared, filtered and 5 mL of the filtrate added to the purge vessel. For nitrate reduction, the heating jacket is connected to a circulated water bath set to deliver water in excess of 90 °C. Antifoaming agent

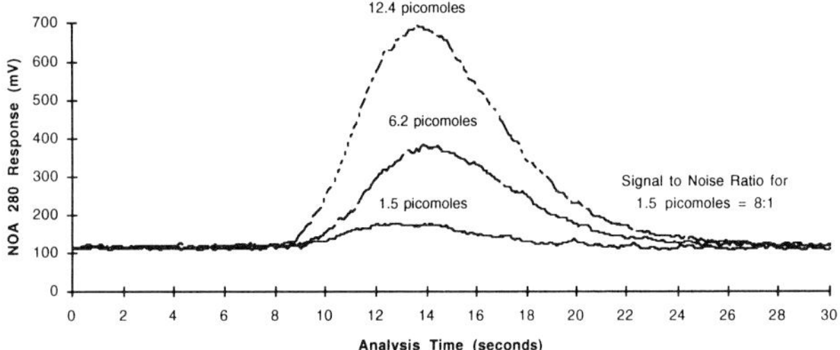

Figure 6. Analysis of Nitrite using the Purge Vessel with KI/Acetic Acid. Injection of 1μL Nitrite Standards.

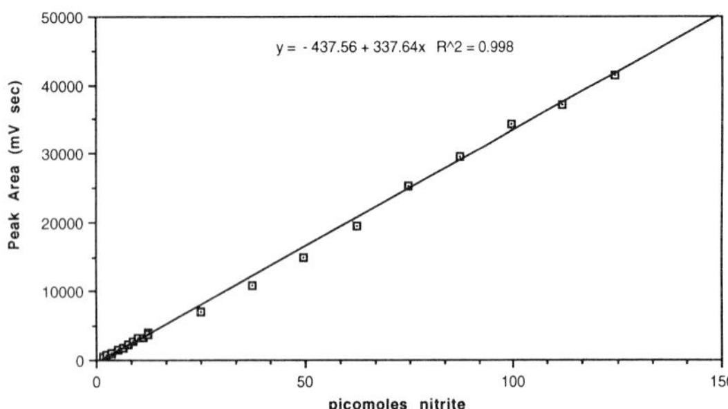

Figure 7. Calibration Curve for Analysis of Nitrite using Purge Vessel with KI/Acetic Acid.

must be added to the reagent. The flow rate of purge gas is adjusted to increase the pressure in the purge vessel to prevent vacuum distillation of the reagent, but low enough to maintain a vacuum in the purge vessel.

The antifoaming agent reduces the foaming from samples containing high levels of protein and measurement of nitrate/nitrite in plasma, without deproteinization, can be easily performed. Depending on the sample size injected, 10-30 plasma samples can be analyzed before changing the VCl_3 reagent.

The purge system offers many advantages over similar apparatus constructed from organic synthesis glassware. The low dead-volume and presence of the vacuum in the purge vessel results in a sharp NO peak for nitrate samples, reducing the volume of sample required. Analysis time is less than one minute per sample and the small volume of vanadium reagent means that only a brief purge of 5-10 minutes is required when fresh reagent is added to the purge vessel. In contrast, most researchers need to purge the VCl_3 reagent for hours when using the larger volumes required with the organic synthesis glassware.

The purge system can be used for measurement of any liquid sample with sample volumes from < 1 microliter up to several milliliters and can also be used for continuous, on-line measurement of NO and its reaction products in perfusates and other flowing streams.

Conclusions - A new chemiluminescence nitric oxide analyzer has been developed that provides a valuable new tool for researchers. Improved data output capabilities, gas calibration and a wide array of sample inlet systems permit measurement of NO and its reaction products for all aspects of NO researcher.

References

1. Archer, S. (1993) FASEB J. **7**, 349-360.

2. Gaston, B., Drazen, J. M., Loscalzo, J., Stamler, J.S. (1994) Am. J. Respir. Crit. Care Med.**149**, 538-551.

Measurement of nitric oxide in biological materials by high performance ion chromatography and EPR spectroscopy: a comparative study

MICHAEL R L STRATFORD, MADELEINE F DENNIS, GILL M TOZER, VIVIEN E PRISE and STEVEN A EVERETT

Gray Laboratory Cancer Research Trust, Mount Vernon Hospital, Northwood, Middlesex HA6 2JR, UK.

Introduction

The lifetime of NO *in vivo* is very short, and its involvement in many physiological processes can readily be assessed by the analysis of nitrite and nitrate which are the end-products of NO oxidation [1]. We describe here an improvement to our previously published HPIC method [2], using a silica based ion exchange resin, for the determination of nitrite and nitrate in biological fluids using electrochemical and spectrophotometric detection respectively. Comparisons are drawn with the direct detection of NO release in rat plasma by EPR spectroscopy.

Methods

Chemicals: carboxy-PTIO (Cambridge Bioscience UK); NOC-7 (Alexis Biochemicals UK); LPS (Sigma); all other chemicals were from Merck. HPIC of nitrite and nitrate was carried out using a Millennium system (Waters) as follows: column Exsil SAX (150 x 4.6mm) (Hichrom) with a reversed phase (Hypersil 5ODS) and anion exchange (Exsil SAX) guard column (10 x 2mm); eluent 25mM KH_2PO4, 5mM H_3PO_4, 11% acetonitrile (perfusates), 15mM KH_2PO4, 30mM H_3PO_4, 6% acetonitrile (plasma), flow rate 1.6ml/min. Electrochemical detection, Coulochem 5100A, 1st electrode +0.35V, 2nd electrode +0.65V; spectrophotometric detection, Waters 996 at 214nm. Direct detection of NO was by EPR spectroscopy utilising the NO selective conversion of carboxy-PTIO to the iminyl nitroxide carboxy-PTI [3]. Tumours were perfused ex vivo as previously described [4]. For in vivo studies with LPS, rat plasma samples were taken from a cannulated artery and deproteinised with acetonitrile (HPIC) as previously reported [2] or treated with ferricyanide (EPR) to reconstitute the radical forms of the imidazolineoxyls [5].

Abbreviations: HPIC, high performance ion chromatography; LPS, lipopolysaccharide; NO, nitric oxide; Carboxy-PTIO, carboxy-2-phenyl-4,4,5,5, tetramethyl-dazoline-1-oxyl-3-oxide.

Results

Figure 1 shows a chromatogram illustrating the increase in plasma nitrate which occurs after administration of LPS. The levels are increased by ~ 10 fold over control after 7 h. Nitrite levels are not measured in plasma because of its nearly quantitative oxidation to nitrate by in vivo oxidants such as haemoglobin [1].

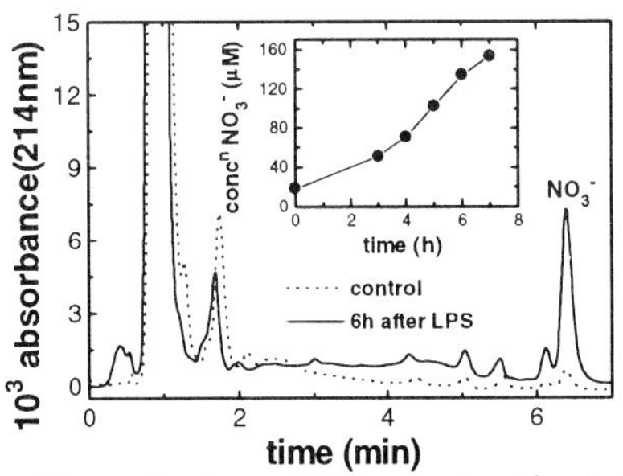

Figure 1: Chromatograms of rat plasma before and after stimulation with LPS (20mg/kg). **Inset** shows the time course of the increase in plasma nitrate

Figure 2 shows EPR spectra of rat plasma containing the NO-specific probe carboxy-PTIO (ca. 100μM) obtained (a) before and (b) 6h after the administration of LPS. Basal levels of NO are difficult to resolve, but the carboxy-PTI signal (solid circles) indicative of NO release was clearly observable following stimulation by LPS (6h).

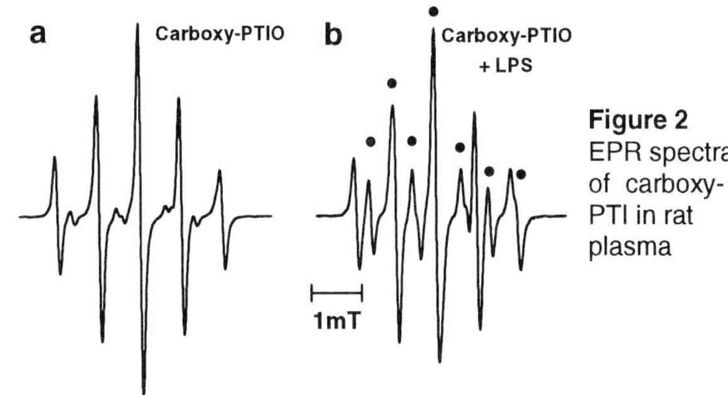

Figure 2 EPR spectra of carboxy-PTI in rat plasma

Figure 3 is a chromatogram from the ex vivo venous perfusate of a BD9 rat P22 carcinosarcoma showing the increase in nitrite concentration which is observed after including the NO donor NOC-7 in the perfusate. In this case, the nitrite is not further oxidised due to the absence of significant amounts of oxidants in the perfusion medium, and basal nitrate levels are unchanged (data not shown).

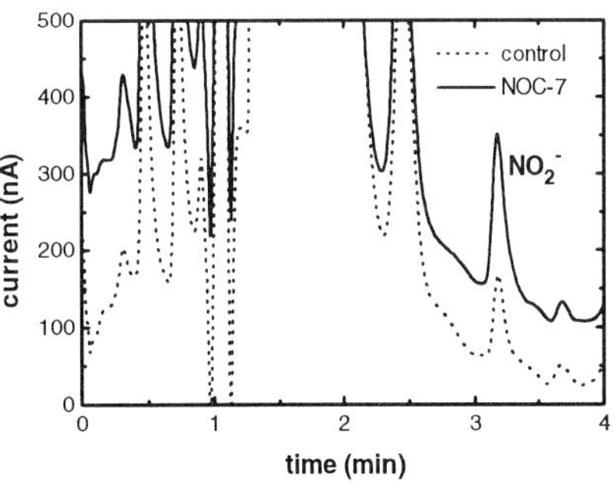

Figure 3 Release of NO from NOC-7 in a tumour perfusate

Discussion

HPIC is a valuable means of assessing the role of NO in tumour physiology; the use of electrochemical detection facilitates quantitation of NO_2^- at micromolar concentrations where endogenous materials frequently interfere with absorbance detection. We previously used a polymeric ion exchange column [2], which suffers from a short life-time with biological samples, combined with high cost. The use of an additional reversed-phase guard column with the silica-based ion exchange column has resulted in a system which is robust. In common with other ion exchange methods, there is a gradual shortening of retention times which is countered by reducing the ionic strength of the eluent.

In vivo, basal concentrations of nitrate are easily detected but these may reflect dietary and other sources of nitrate. Quantitation of basal NO by EPR spectroscopy has proved difficult in plasma and tumour perfusates. Nevertheless, the combination of chromatographic and EPR methods clearly distinguishes between NO and other nitrogen oxides eg NO^-, NO^+ in physiological processes.

Acknowledgements

This work is supported by the Cancer Research Campaign. Financial support from Dionex UK is gratefully acknowledged.

References

1. Ford, P.C., D.A. Wink, and D.M. Stanbury. (1993) FEBS Letts. **326**, 1-3.
2. Everett S.A., Dennis M.F., Tozer G.M., Prise V.E., Wardman P. and Stratford M.R.L. (1995) J. Chromatogr. **706**, 437-442
3 Hogg N., Singh R.J., Joseph J., Neese F. and Kalyanaraman B. (1995) Free Rad. Res. **22**, 47-56
4. Tozer G.M., Prise V.E., Bell K.M. (1995) Acta Oncol. **34**, 373-377
5 Yoshida M., Akaike T., Wada Y., Sato K., Ikeda K., Ueda S., Maeda H. (1994) Biochem Biophys Res Comm. **202**, 923-930

Reactions of nitric oxide with organic and biological molecules.

MOHAMMED AFZAL, JONATHAN S.B.PARK and JOHN C.WALTON.

School of Chemistry, University of St.Andrews, St.Andrews, Fife, Scotland. KY16 9ST

Nitric oxide, formed by the action of NO synthases on L-arginine, reacts rapidly with oxygen to give NO_2 and with haem containing species. It is likely that NO will react with other organic molecules present in the cellular environment and this paper outlines our study with model cell compounds, particularly polyenes and enones.

In the complete absence of oxygen, NO did not react with dienes such as 1,4-diphenylbutadiene, 2,5-dimethylhexa-2,4-diene or anthracene, or even extensively conjugated polyenes such as retinyl acetate. Direct addition to give a nitroxide appears to occur only with specially designed dienes like 1,2-bis(*exo*-isopropylidene)cyclohexa-3,5-diene [1]. However, in the presence of traces of oxygen, NO_2 adds to a double bond to give a carbon-centred radical **1** which combines with NO to give a nitroso compound. The latter species picks up a second radical **1** to give a long-lived nitroxide which can be detected by EPR spectroscopy.

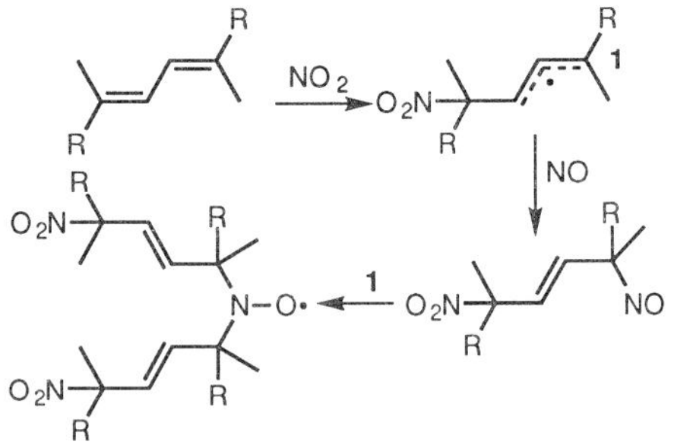

Pure NO also fails to react with unsaturated carbonyl compounds but, in the presence of oxygen, addition of NO_2 to an enone initiates a similar sequence. For example, with retinyl acetate series of carbonyl compounds and alkylaromatics were produced.

In conclusion, nitric oxide does not react with polyunsaturated compounds when no oxygen is present. It follows that biological membranes and other lipid-rich structures will be unaffected by NO. When NO synthases produce NO under aerobic conditions, the more reactive NO_2 radical is then formed. This rapidly attacks unsaturated lipids and initiates formation of carbon-centred radicals which couple with NO to give active nitroso compounds. Decomposition of intermediates affords further radicals which cause extensive oxidative degradation of unsaturated biomolecules.

We thank the Wellcome Foundation and the EPSRC for generous financial support of this research.

Reference

1. Korth, H-G., Ingold, K.U., Sustmann, R., de Groot, H., Sies, H. (1992) Angew. Chem. Int. Edn. Engl., **31**, 891-893.

The interaction of nitric oxide with cobalamins

JOHN REGLINSKI and JUDITH NAISMITH

Department of Pure & Applied Chemistry
Strathclyde University Glasgow G1 1XL UK

Nitric oxide is now recognised as one of the most important cell signaling species in-vivo. Its versatility in biological systems probably results from the broad range of chemistry in which it can participate. For example, it reacts with superoxide to form peroxynitrite (equ 1), as an electrophile with thiolates forming S-nitrosothiols (equ 2) and as a π-acid in reactions at metal centres (equ 3).

$$^{\bullet}NO + O_2^{\bullet-} \longrightarrow [O_2NO]^- \quad -(1)$$
$$^{\bullet}NO + RS^- \longrightarrow [RSNO] \quad -(2)$$
$$^{\bullet}NO + Fe^{III}L_n \longrightarrow L_{(n-1)}Fe^{III}N{=}O \longrightarrow L_{(n-1)}Fe^{II}N{\equiv}O^+ \quad -(3)$$

Although the biology of nitric oxide is already rich, its chemistry predicts that it can react via yet another route. Nitric oxide can be induced into coupling reactions at sulphite [1] (equ 4) or cobalt(II) (equ 5) [2].

$$[SO_3]^{2-} + 2\,^{\bullet}NO \longrightarrow [O_3SNONO]^{2-} \quad -(4)$$
$$Co^{II} + NH_3\ ^{\bullet}NO \longrightarrow ([Co(NH_3)_5]_2N_2O_2)^{4+} \quad -(5)$$

Although the biological significance of the reaction at sulphite (equ 4) is as yet unknown, parallels can be drawn between the penta-aminecobalt(III) motif and the environment of cobalt(III) in vitamin B_{12}.

Vitamin B_{12} contains as d^6 low-spin cobalt (III) centre in a slightly distorted octahedral environment. This arrangement is responsible for the relatively slow rate of exchange of the axial ligand with solvent and solute species. Thus, any significant interaction of nitric oxide with vitamin B_{12} in-vivo will occur through an interaction with one of its reduced forms (i.e. co(II)enzymeB_{12} or co(I)balamin) both of which contain labile axial ligands. VitaminB_{12} can be reduced to co(II)enzymeB_{12} via dithionite or co(I)balamin under more severe conditions using $NaBH_4$.

Prolonged exposure of cobalamin to BH_4^- can lead to irreversible reduction of the corrin ring. As such, it is best to prepare the co(I)balamin just prior to use and furthermore follow the BH_4^- reduction by spectrophotometry to establish the endpoint of the reaction (figure).

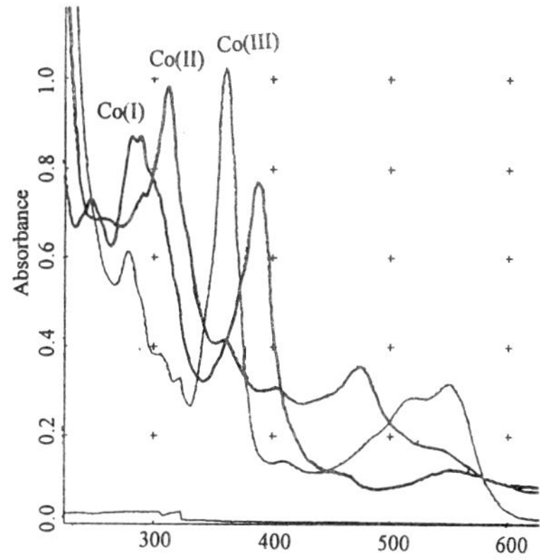

Consistent with the early report by Firth [3] nitric oxide has no effect on co(II)enzyme. However, treatment of the co(I)balamin with nitric oxide leads to a rapid re-oxidation of the metal centre to cobalt(III). This process can be probed further by examining the products of the ligand exchange reaction permitted by the reduced vitamin B_{12}. 5-deoxyadenosylCo(III)B_{12} has a characteristic band in the u.v.-visible spectrum at 350nm which on reduction by BH_4^- collapses. Re-oxidation of the co(I)balamin by nitric oxide in the presence of excess cyanide leads to the preferential formation of cyanoCo(III)B_{12} (λ_{max}= 370nm). Cyanide coordination can be confirmed by ^{13}C NMR, which shows resonances consistent with cyanoCo(III)B_{12} (125ppm) in the presence of cyanide (115ppm).

Surface enhanced resonance Raman scattering (SERRS) provides vibrational information from the Co-corrin ring and can distinguish between co(II)enzyme and vitamin B_{12} in our mixtures. Thus using SERRS the integrity of the corrin ring during these reactions and the lack of B_{12} degradation can be demonstrated. Furthermore, the presence a trace amount of co(II)enzyme, which was deliberately left in the BH_4^- reduced solution and which is incapable of oxidation by nitric oxide can be seen. This is clear evidence that the mechanism of the reaction is occurring via a concomitant two electron transfer Co(I) ⇒ Co(III) and the co(II)enzyme does not form as an intermediate.

An inspection of the atmosphere above the solution during nitric oxide mediated oxidation by gas phase infra red reveals that there are two major infra-red active products stable N_2O (ν 2230 cm^{-1}, ν 1305 cm^{-1}) and transient N_2O_3 (ν 1800 cm^{-1}). N_2O is the major product of the acid digestion (equ 6) of the coupled nitric oxide species described above (equ 1,2)

$$[O_3SNONO]^{2-} + H^+ \longrightarrow [HSO_4]^- + N_2O \quad -(6)$$

Thus, equation 6 and the infra red evidence suggests that nitric oxide coupling occurs at co(I)balamin prior to acid catalysed disproportionation to a cobalt(III) corrin product (CNCo(III)B_{12}) and N_2O. The steps in this reaction have been modeled using ab-initio calculations on the sulphite reactions (equ 1,6) with the information derived transferred to the vitamin motif.

Co(I) —NO→ Co(I) —NO→ Co(I) → Co(III) —H^+→ Co(III) + N_2O OH^-

The mechanism presented provides the rational for the coupling reaction observed and explains how the co(II)enzyme can be omitted in the process by allowing the direct transfer of 2 electrons from Co(I) to the N_2O_2 ligand (equ 7)

$$2NO + 2e \longrightarrow [N_2O_2]^{2-} \longrightarrow N_2O + OH^- \quad -(7)$$

Co(I)balamin will further reduce N_2O to N_2 and O_2 [4]. The trace oxygen generated will be consumed by nitric oxide to form NO_2. It is thought that in turn co(I)balamin will be able to couple NO to NO_2 through an N-N interaction to form N_2O_3. However, this reaction arises as a consequence of the experimental conditions. In-vivo, where the concentration of vitamin B_{12} and NO are low, we should not expect this reaction to occur.

Concluding remarks

Although nitric oxide does not react with the co(II)enzyme, it is possible that it could interfere in its reactions. For example, during mutase reactions nitric oxide may be capable of trapping radical intermediates leading to the formation of nitroso- compounds. The metabolism of certain metals (e.g. arsenic) is postulated to occur via carbo-cation (Me^+) transfer from methylco(III)balamin leading to the formation of co(I)balamin. Thus, under certain stress conditions, the key species will form and the reaction discussed above may become viable. An overall reaction similar to that discussed above has been shown to occur at bacterial nitrate reductase. It would not be surprising if some isoform of this enzyme (e.g. P450's) did not exist in the body and that the reaction discussed was not more prevalent. Under any of these circumstances derivative assays for nitric oxide may be highly misleading.

References

1. Fremy, H. (1845) Ann. Clim. Phys. **15**, 408
2. Nast, R. and Rohmer, M (1956) Z. Anorg. Allg. Chem. **285**, 271
3. Firth, R.A. Hill, H.A.O., Pratt, J.M., et al. (1969) J. Chem. Soc. A. 381
4. Banks, R.G.S., Henderson, R.J. and Pratt, J.M. (1968) J. Chem. Soc. A. 2886

Reactivity of Peroxynitrite as Studied by Pulse Radiolysis

K. KOBAYASHI, M. MIKI, and S. TAGAWA

The Institute of Scientific and Industrial Research, Osaka University, Mihogaoka 8-1, Ibaraki, Osaka 567, Japan

Peroxynitrite has been increasingly recognized as a reactive oxidizing species that may participate in many injury processes associated with oxidative biological damage [1, 2]. In most studies, however, peroxynitrite was synthesized by the reaction of nitrite and hydrogen peroxide with acid followed by a base quench [1]. With this method, the possibility of contamination by other reactive specie is considered. On the other hand, there is a conflict report that NO functions as a protecting role [3], though the chemical mechanism was not presented in the study. To clarify the effect of peroxynitrite, it is necessary to investigate the reactivity of peroxynitrite toward various biological molecules in a simple system.

The spectral and kinetic behavior of NO and its redox states are conveniently studied by pulse radiolysis [4, 5]. By pulse radiolysis of aqueous solutions of NO_2^-, it is possible to produce NO to follow its reaction spectrophotometrically [5]. With this technique, we have succeeded in a direct observation of the reaction of NO and O_2^- with a second order rate constant of 3.8 x 10^9 M^{-1} s^{-1}. The present paper describes the kinetic behavior of peroxynitrite by the use of pulse radiolysis techniques.

Result and Discussion

We have previously reported that NO is produced via Reaction 1 by pulse radiolysis of aqueous solution of 1-5 mM $NaNO_2$ under the deaerated condition [5]. This was confirmed by the formation of the NO complex of ferrous myoglobin (Mb) after pulse radiolysis in the presence of ferric Mb [5]. Thus, peroxynitrite is produced via Reactions 1, 2, and 3 by pulse radiolysis of oxygen saturated solutions in the presence of $NaNO_2$.

$$NO_2^- + e_{aq}^- \longrightarrow NO_2^{2-} \xrightarrow{H^+} NO + OH^- \quad (1)$$

$$O_2 + e_{aq}^- \longrightarrow O_2^- \quad (2)$$

$$NO + O_2^- \longrightarrow ONOO^- \quad (3)$$

When an solute is also present in the solution, the reaction of peroxynitrite with the solute would be observed.

Oxidation by Peroxynitrite

On pulse radiolysis of oxygen-saturated solutions containing 5 mM $NaNO_2$, 100-400 μM NADH, and 0.1 M formate, at pH 7.4, the resulting absorbance at 340 nm decreased after the decay of peroxynitrite. The spectrum obtained after the pulse shows that NADH is oxidized by peroxynitrite in Reaction 4.

$$ONOO^- + NADH \xrightarrow{H^+} NO_2^- + NAD^+ + H_2O \quad (4)$$

This reaction proceeded nearly quantitatively, and was not affected by the addition of OH radical scavengers such as formate and ethanol (50 - 500 mM). From the data, it can be said that peroxynitrite does not react with these OH radical scavengers, and that OH radical is not released during the decomposition of peroxynitrite.

Peroxynitrite has a pK_a of 6.8, as shown in Reaction 5.

$$ONOO^- + H^+ \rightleftharpoons ONOOH \quad (5)$$

It has been reported that peroxynitrous acid (ONOOH), rather than $ONOO^-$, is potent oxidant with the reactivity of hydroxyl radical [6]. To examine the reactivity of ONOOH, the effect of pH in Reaction 4 was examined. The rate constant was increased with an decrease of pH with pK_a value of 6.8. However, pH-effect on the reaction was very small.

The reactions of cysteine, GSH, and methionine with peroxynitrite were performed by pulse radiolysis. The rate of the decay of peroxynitrite followed at 300 nm increased with the increase of the concentration of these compounds. The rate constants of the reactions of peroxynitrite with cysteine, GSH, and methionine with peroxynitrite are 2.6 x 10^3 M^{-1} s^{-1}, 6.5 x 10^2 M^{-1} s^{-1}, 4.2 x 10^2 M^{-1} s^{-1} at pH 7.4, respectively. The values are similar to those reported before [2, 7].

The reaction of ascorbate or trolax with peroxynitrite was performed by both pulse radiolysis and stopped-flow methods. However, we could not observe the reactions of ascorbate or trolax. The rates in the decay of peroxynitrite were not affected upon the addition of 1 mM ascorbate or trolax at various pH (pH 5-8).

Nitration of Phenol Derivatives by Peroxynitrite

It has been reported that superoxide dismutase or Fe^{3+}-EDTA catalyzes the nitration by peroxynitrite of a wide range of phenolics including tyrosine in proteins [8]. Beckman et al. [8] have proposed that metal ions catalyze the heterolytic cleavage of peroxynitrite to form a nitronium ion (NO_2^+). In the present study, the reactions of phenol derivatives with peroxynitrite were performed in the absence or the presence of Fe^{3+}-EDTA. Of various phenol derivatives, effective nitration of *p*-hydroxy benzoate by peroxynitrite was observed. Above pH 8, an absorption increase at 400 nm was observed with the decay of peroxynitrite. From the spectrum obtained after the pulse shows that the nitration occurs nearly quantitatively.

$$HO\text{-}C_6H_4\text{-}COO^- + ONOO^- \xrightarrow{H^+} HO\text{-}C_6H_3(NO_2)\text{-}COO^- + OH^- \quad (6)$$

Similar nitrations of phenolic compounds such as tyrosine, phenol, and *p*-hydroxy phenylacetic acid were observed. Upon the addition of these compounds, the rates in the decay of peroxynitrite increased, and concomitantly the absorption at 400 nm increased above pH 8. In contrast, a similar process was not observed in the presence of benzoate.

To test the contribution of transition metals to the nitration of phenolics, the effects of metal chelator EDTA and ferric-EDTA were examined. EDTA and ferric-EDTA (100 μM) had no effect on the nitration.

In the present experiment, we have no data showing that peroxynitrite is a strongly oxidizing species. This is in conflict with those obtained by using peroxynitrite prepared in the quenched flow reactor. We do not yet known reason for these discrepancies. It may simply be due to technical problems relating to the purity of peroxynitrite. In contrast, NO and O_2^- can be produced concomitantly by the present method, and therefore reactive chemical species other than peroxynitrite can be excluded. Rather present findings are in accordance with the proposal that NO can function as a chemical barrier to cytotoxic free radical [3].

References

1. Beckman, J. S., Beckman, T. W., Chen, J., Masshall, P. A., and Freeman, B. A. (1990) Proc. Natl. Acad. Sci. U. S. A. **87**, 1620-1624.
2. Radi, R., Beckman, J. S., Bush, K. M., and Freeman, B. A. (1991) J. Biol. Chem. **266**, 4224-4250.
3. Wink, I. A., Hanbauer, I., Krishna, M. C., DeGraff, W., Gamson, J., Mitchele, J. B. (1993) Proc. Natl. Acad. Sci. U. S. A. **90**, 9813-9817.
4. Czapski, G., Holcman, J., Bielski, B. H. J. (1994) J. Am. Chem. Soc. **116**, 11465-11469.
5. Kobayashi, K., Miki, M., and Tagawa, S. J. Chem. Soc. Dalton Transactions in press.
6. Koppenol, W. H., Moreno, J. J. Pryor, W. A., Ischiropoulos, H., and Beckman, J. S. (1992) Chem. Res. Toxicol. **5**, 834-842.
7. Pryor, W. A., Jin, X., and Squadrito, G. L. (1994) Proc. Natl. Acad. Sci. U. S. A. **91**, 11173-11177.
8. Beckman, J. S., Ischiropoulos, H., Zhu, L., van der Woerd, M. Smith, C., Chen, J., Harrison, J., Martin J. C., and Tsai, M. (1992) Arch. Biochem. Biophys. **298**, 438-445.

Studies towards the preparation of substituted hydroxyguanidines with tumour specific properties.

ANTHONY R. BUTLER[1], PETER DAVIS[2] and GARRY J. SMITH[1].

[1]School of Chemistry, University of St. Andrews, Fife, KY16 9ST, Scotland, U.K. and [2]Celltech Ltd, 216 Bath Rd, Slough, Berkshire, SL1 4EN, U.K.

Introduction.

In 1987 Hibbs *et. al.* [1] found that the cytotoxicity of macrophages against tumour cells was dependent upon the presence of L-arginine. This activity is a function of the nitric oxide (NO) released and results in the formation of nitrite and citrulline. The production of NO by macrophages was confirmed by three independent groups in 1988/89 [2]. It seems likely that this process parallels that occurring in endothelial cells, where N-hydroxy-L-arginine is an intermediate, enzymatic oxidation of which results in the release of NO [3].

The role of NO in the growth of tumours is still poorly understood, but it may be possible for NO to both increase and reduce tumour growth. The growth of tumours is regulated, to some extent, by tumour vasculature. Release of NO in the tumour would improve tumour vasculature, thus increasing the growth of the tumour. It has also been shown that NO is cytotoxic towards tumour cells, having the effect of reducing tumour growth. To this end it has been suggested that NO may promote or deminish tumours depending upon local concentration [4]. In spite of the dual role of NO, tumour-specific NO-donor drugs could be of therapeutic value.

Results and Discussion.

Hydroxyguanidines and hydroxyamidines were chosen for study as a result of the structural relationship they have with N-hydroxy-L-arginine. The hypothesis is that these compounds will release NO on enzymatic oxidation and are likely to be NO-synthase independent.

Scheme 1 shows the method for the preparation of the target compounds. Other synthesis were attempted, but were less successful. The method chosen was a modification of a number of methods and allowed the preparation of hydroxyguanidines in both reasonable yield and a pure state. The preparation of hydroxyamidines, scheme 2, was simpler with no problems encountered. All compounds prepared have a maleimido function, the purpose of which is to allow the compounds to be chelated to tissue specific antibodies.

All the hydroxyguanidines and hydroxyamidines were tested both chemically and physiologically, to prove the release of NO. Testing via the Griess test [5] and a NO electrode showed that NO was released on chemical oxidation. All compounds were active to the Griess Test at 10^{-5}M and NO was detected at 10^{-7}M with the NO electrode. All compounds were also found to be vaso-active [6] at 10^{-6}M, also showing that NO is released on enzymatic oxidation.

Studies to determine the cytotoxicity of these compounds towards tumour cells is in progress and the results will be published at a later date.

Conclusions

Difficulties with the synthetic procedures highlighted the instability of some hydroxyguanidines, but it was still possible to prepare them in reasonable yields and to a high degree of purity. Chemical testing led us to the conclusion that all the compounds that were prepared released NO on oxidation. All compounds tested were physiologically active and when linked to tumour-specific antibodies, could give a tumour-specific NO-donor drug. NO release will be effected by enzymatic oxidation of the guanidine *within the tumour.*

Scheme 1 : Preparation of Hydroxyguanidines [7].

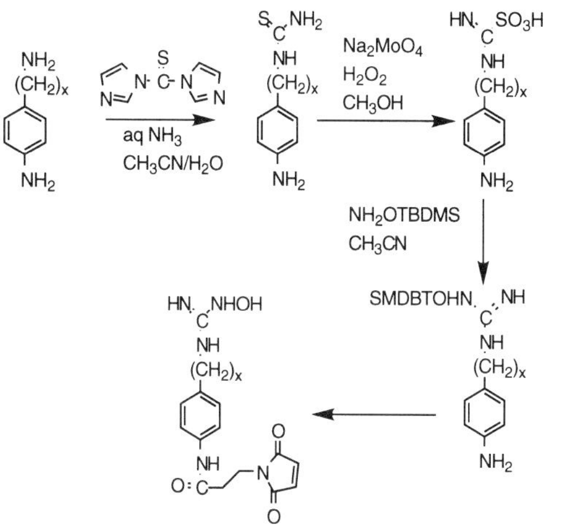

Scheme 2 : Preparation of Hydroxyamidines [8]

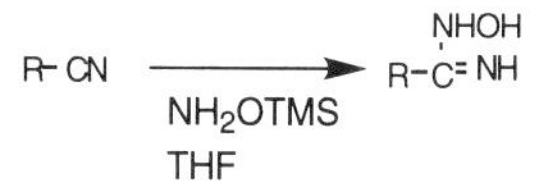

References

1) Hibbs, J.B. Jr., Vavrin, Z., Taintor, R.R. (1987) *J. Immunol.*, **138**, 550-565.

2) Marletta, M.A., Yoon, P.S., Iyengar, R., Leaf, C.D., Wishnok, J.S. (1988) *Biochemistry*, **27**, 8706-8711. Hibbs, J.B. Jr., Taintor, R.R., Vavrin, Z., Rachlin, E.M. (1988) *Biochem. Biophys. Res. Commun.*, **157**, 87-94. Stuehr, D.J., Gross, S.S., Sakuma, I., Levi, R., Nathan, C.F. (1989)*J. Exp. Med.*, **169**, 1011-1020.

3) Marletta, M.A., Yoon, P.S., Iyengar, R., Leaf, C.D., Wishnok, J.S. (1988) *Biochemistry*, **27**, 8706-8711. Wallace, G.C., Fukuto, J.M. (1991) *J. Med. Chem.*, **34**, 1746-1748.

4) Jenkins, D.C., Charles, I.G., Thomsen, L.L., Moss, D.W., Holmes, L.S., Baylis, S.A., Rhodes, P., Westmore, K., Emson, P.C., Moncada, S. (1995) *Proc. Natl. Acad. Sci. USA*, **92**, 4392-4396.

5) Fukuto, J.M., Wallace, G.C., Hszieh, R., Chaudhuri, G. (1992) *Biochem. Pharmacol.*, **43**, 3, 607-613, 1992.

6) Flitney, F.W., Megson, I.L., Flitney, D.E., Butler, A.R. (1992) *Br. J. Pharmacol.*, **107**, 842-848, 1992.

7) Clement, B. (1986) *Arch. Pharm. (Weinheim)*, **319**, 968-972. Miller, A.E., Bischoff, J.J., Pae, K. (1988) *Chem. Res. Toxicol.*, **1**, 169-174. Miller, A.E., Feeney, D.J., Ma, Y., Zarcone, L., Aziz, M.A., Magnuson, E. (1990), *Synth. Comm.* **20**, 2, 217-226. Muller, G.W., Walters, D.E., DuBois, G.E. (1992) *J. Med. Chem.*, **35**, 4, 740-743.

8) Gautier, J., Miocque, M., Farnoux, C.C.(1975) Patai : The chemistry of amidines and imidates. Wiley, New York, 283-348.

Substituted *N*-methyl-*N*-nitrosoanilines are a novel class of protein *S*-nitrosation agents

Peng G. WANG, Zhengmao GUO, Andrea MCGILL, Libing YU, Jun LI, and Johnny RAMIREZ

Department of Chemistry, University of Miami, Coral Gables, P. O. Box 249118, FL 33124, U.S.A.

Under physiological conditions, NO is synthesized from L-arginine and reacts with O_2, O_2^- and transition metal centers to form oxidized adducts such as NO_x, $OONO^-$ or metal NO adducts.[1] Biologically prevalent thiols react with these adducts to form *S*-nitrosothiols. *S*-nitroso proteins have been detected in human airway, plasma, platelets and neutrophiles, and have been shown to exhibit endothelium-derived relaxing factor-like effects.[2] The *S*-nitrosation of proteins, therefore, is believed to be a general way to store, transport and finally release NO.[3] In addition, *S*-nitrosation of cysteine dependent enzymes is an established mechanism by which NO exhibits its inhibiting or regulatory activities.[2] Here we report our latest finding that substituted *N*-methyl-*N*-nitrosoanilines serve as stable NO^+ donors for *S*-nitrosation of proteins.

This work was based on our previous synthesis and bioassay of a novel protein tyrosine phosphatase (PTPase) inhibitor, dephostatin (2,5-dihydroxyl-*N*-methyl-*N*-nitrosoaniline), a natural product isolated from the culture broth of *streptomyces sp.*[4] We found that not only dephostatin inactivated both the recombinant *Yersinia* and mammalian PTPases but also the unsubstituted compound, *N*-methyl-*N*-nitrosoaniline (**3**), exhibited weak inhibition. Subsequently, we synthesized a series of substituted *N*-methyl-*N*-nitrosoanilines and investigated their inhibition ability against thiol-dependent enzymes.[5] We first tested the representative member of thiol proteases, papain. The enzymatic assay revealed that the substituted *N*-methyl-*N*-nitrosoanilines exhibited different inhibiting activities against papain. Dephostatin is the most active inhibitor with a pseudo first order rate constant of 0.328 min^{-1}, which is followed by compounds **1** and **2** with pseudo first order rate constants of 0.243 and 0.009 min^{-1}, respectively. Detailed kinetic measurements showed that K_I and k_i were 2.813 mM and 0.102 min^{-1}, respectively, for the inhibition of papain with nitrosamine **2**.[6] Chlorine as an electron withdrawing group at the *para* position of aniline makes **4** less active than the unsubstituted **3**. Substitution at either the *meta* (**5**) or the *ortho* position (**6** and **7**) results in poor inhibitors. Psedo-first order rate constants (1/min) are: dephostatin (0.328), **1** (0.243), **2** (0.090), **3** (0.057), **4** (0.017), **5** (0.036), **6** (0.011) and **7** (0.026). The inactivated papain (0.5 mg/mL, 100 μL) fully regained its original activity when cysteine (1 mM, 100 μL) was added. Furthermore, the inactivation was partially prevented in the presence of Gly-Gly-Tyr-Arg, a competitive inhibitor of papain. This suggests that the inactivation with nitrosamines takes place within the active site of the enzyme. Based on these results, we believe that the inactivation process is *S*-nitrosation of papain in which the *N*-methyl-*N*-nitrosoanilines function as a NO^+ donor.

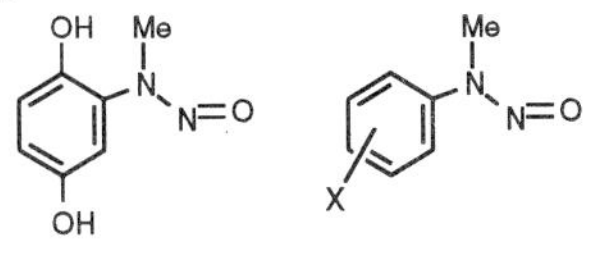

Dephostatin

#	posi.	X
1	*p*	NMe_2
2	*p*	OH
3	*p*	H
4	*p*	Cl
5	*m*	OH
6	*o*	OH
7	*o*	NO_2

To support the S-nitrosation mechanism, papain (10 mg) and compound **2** (10 mg) in 2 mL of 50 mM sodium phosphate buffer (pH 7.0, 10% acetonitrile, 1 mM EDTA) were incubated at room temperature until papain was completely inactivated. From the concentrated reaction mixture, *p*-hydroxyl-*N*-methylaniline **8**, the product of the *S*-nitrosation (**eq.1**), was isolated by preparative thin layer chromatography and confirmed by mass spectroscopy. The UV-Vis spectrum of the inactivated papain showed absorption maxima in the 330-370 nm wavelength range which are characteristic for S-nitrosothiols.[7] Finally, the formation of the S-NO bond was further verified by FTIR spectroscopy, running on cast films obtained by evaporation of the inactivated papain solution on a CaF_2 crystal. The spectra clearly showed the S-NO bond absorption bands at 1166 cm^{-1} and 1153 cm^{-1}.

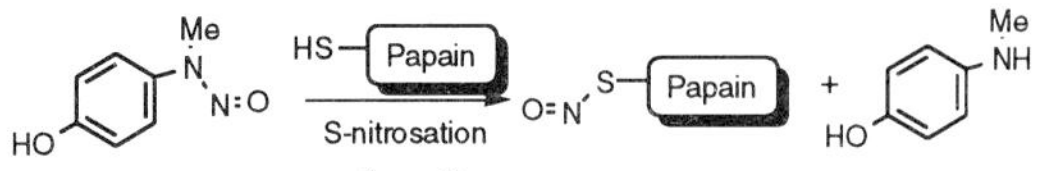

2 + HS–Papain —(S-nitrosation)→ O=N–S–Papain + **8**

(eq. 1)

We also tested another thiol protease, bromelain. It was found that nitrosamines (**1** and **2**) exhibited strong inhibition activities as in the case of papain. In contrast, no inhibition was detected when serine proteases such as α-chymotrypsin and substilisin BPN' solutions were incubated with these nitrosamines. Besides thiol proteases, other sulfhydryl-dependent enzymes were also inactivated by substituted *N*-methyl-*N*-nitrosoanilines. For example, aldehyde dehydrogenase has a critical cysteine at its active site and it was inactivated by nitrosamine **2**.

In summary, we observed that properly substituted *N*-alkyl-*N*-nitrosoanilines underwent *S*-transnitrosation with sulfhydryl-dependent enzymes *under physiological conditions*. An electron-donating substitution at the *para*-position of the phenyl ring apparently assists the formation of the protonated reaction intermediate. The higher nucleophilicity of the critical cysteine of the enzyme than simple thiols makes the transnitrosation possible. Our finding points out that *N*-alkyl-*N*-nitrosoaniline structure can function as a stable, functional moiety in the design of nitric oxide releasing agents for biomedical and pharmaceutical applications.

Acknowledgment

This work was partially supported by a research grant (F95UM-2) from The American Cancer Society, Florida Division, Inc. Special thanks to Dr. Zhong-Yin Zhang at Albert Einstein College of Medicine for collaborating research on the enzymatic assay of protein tyrosine phosphatases. We thank Professor Carl Hoff for running FTIR spectroscopy and Professor Luis Echegoyen for helpful discussions.

References and notes

1. Stamler, J. S., Singel, D. J. and Loscalzo, J. (1992) Science, **258**, 1898-1902.
2. Stamler, J. S. (1994) Cell **78**, 931-936.
3. Girard, P., Potier, P. (1993) FEBS Lett. **320**, 7-8.
4. Yu, L.-B., McGill, A., Ramirez, J., Wang, P. G. and Zhang, Z. - Y. (1995) Bioorg. Med. Chem. Lett. **5**, 1003-1006.
5. Compounds **3**, **4** and **7** were synthesized by *N*-nitrosation of substituted *N*-alkylanilines with $NaNO_2/HOAc$; **2** was synthesized by de-*O*-methylation of *p*-hydroxyl-*N*-methylaniline with BBr_3 and subsequent *N*-nitrosation; **1** was synthesized from *p-N',N'*-dimethylamino-aniline; **5** and **6** were synthesized from *N*-methylation of substituted anilines with formaldehyde and $NaCNBH_3$ prior *N*-nitrosation.
6. Kitz, R. and Wilson, I. B. (1962) J. Biol. Chem. **237**, 3245-3249. K_I and k_i represent the dissociation constant for breakdown of enzyme-inactivator complex (E-I) and the rate of inactivation respectively.

$$E + I \underset{k_{-1}}{\overset{k_1}{\rightleftharpoons}} E:I \xrightarrow{k_i} E\text{-}I,\ K_I = k_{-1}/k_1,\ k_i\text{: } k_{inact}$$

7. Williams, D. L. H. (1985) Chem. Soc. Rev. **14**, 171-195.

pH profiles of S-nitrosothiol derivatives of amino acids, peptides, and proteins

Derrick R ARNELLE and Jonathan S. STAMLER

Divisions of Respiratory and Cardiovascular Medicine, Duke University Medical Center, Durham, NC 27710, U.S.A.

Summary

The pH profiles of several amino acid, peptide and protein nitrosothiols were examined in physiological buffers. In general, RSNO were found to be quite stable between pH 1 and 5; however, stability declines markedly from pH 7 to 10. Large variations in the lifetime of different RSNO were seen over the pH range of 4-8 and these differences were influenced by buffer composition. Addition of 0.5 mM DTPA to buffers obviated pH and buffer composition effects on S-nitrosothiol stability. Under these conditions, S-nitrosothiols were stable, at least for several hours, in the pH range of 1-10. Conversely, addition of nanomolar concentrations of copper(II) in Chelex treated buffers markedly decreased S-nitrosothiol stability at alkaline pH, mirroring the results using untreated buffers. From these data, we conclude: (1) RSNO are considerably more resistant to changes in pH than previously assumed. (2) The pH dependent effects on S-nitrosothiol stability are largely due to *in vitro* copper and other trace metal contaminants. (3) S-Nitrosothiol stability in untreated buffers appears to correlate with pH dependent metal ligating properties of the amino acid, peptide or protein.

Methods

S-nitrosothiols were synthesized by adding equal volumes of 200 mM thiol in 0.5 M HCl, 0.5 mM Na_2DTPA to 200 mM $NaNO_2$ in 0.5 mM Na_2DTPA unless otherwise noted. A partially reduced solution of bovine serum albumin (BSA) was nitrosated in 85 mM HCl , 0.1 mM EDTA with one equivalent of nitrite. After one hour, the concentration of SNO-BSA was determined by the Saville assay; 1.1± 0.2 RSNO per mole BSA was found.

These studies were conducted using solutions containing 100 mM of the following buffers neutralized with 5M KOH: citric acid (pH 3,4, and 5), acetic acid (pH 5), MES (pH 6, 6.5, and 7), potassium phosphate monobasic (pH 1, 7 and 7.4), HEPES (pH 7.4 and 8), and sodium borate (pH 9 and 10). Changes in the absorbance at 340 nm (A_{340}) of a 100 µM solution of RSNO in 100 mM buffer were recorded by a Perkin-Elmer Lambda 2S spectrophotometer over the course of 4-10 hours.

Results and Discussion

The stability of S-nitrosothiols (RSNO) is believed to be strongly influenced by pH; it is generally believed that RSNO are stable in acid and that increasing pH leads to decreasing RSNO stability. This notion probably arose from two lines of evidence: 1) the most common method of RSNO synthesis involves treatment of a thiol with an equivalent of nitrite in acid, and 2) observations that amino acid RSNO, such as SNO-cysteine, are unstable at pH 7.4 or higher. Interestingly, protein RSNO are thought to be much less susceptible to changes in pH. Given recent studies by a number of groups suggesting that metal ion contaminants are involved in RSNO decomposition at physiological pH [1-4], we examined the stability of RSNOs synthesized from cysteine, homocysteine, glutathione and bovine serum albumin between pH 1-10.

If no measures are taken to control trace metal ions, decomposition of SNO-cysteine (SNO-Cys) decreases more rapidly as pH increases. SNO-BSA, SNO-homocysteine (SNO-Hcy) and SNO-glutathione (GSNO) are less susceptible to these pH dependent changes. Decreases in A_{340} for SNO-Cys and SNO-Hcy are not reproducible under these conditions; biphasic decomposition is often observed. Although Chelex-100 treatment of buffers to remove trace metals increases the stability of RSNO, S-nitrosothiols are significantly more stable (≥6 hours) in buffers containing 0.5 mM Na_2DTPA in the acidic and physiological ranges. Decomposition of GSNO does occur slowly in the alkaline range; however SNO-Cys and SNO-Hcy appear to be stable at high pH in the presence of DTPA. SNO-BSA appears to be stable at all pH values except for pH 4 where A_{340} increases with time. Our data show that S-nitrosothiols of amino acids, peptides and proteins decompose very slowly between pH 1-10 in the absence of trace metal ion contamination. This suggests that the presence of metal ions, either as hydrated ions or coordination complexes, constitute the major factor in determining S-nitrosothiol stability over this pH range. GSNO was less stable at pH 9 and 10 than either SNO-Cys or SNO-Hcy. This may be explained by an intramolecular reaction involving the N-terminal amino group of the γ-glutamyl residue, which has a pK_a of about 9.5. Likewise, the instability of SNO-BSA at pH 4 may be due to changes in the electrostatic environment near the nitrosothiol thereby altering its susceptibility to attack by proximal nucleophiles, such as carboxylates, and to related rearrangements such as those of the Fischer-Hepp type.

Abbreviations used: DTPA, diethylenetriamine pentaacetic acid; SNO-Cys, SNO-cysteine; DTT, dithiothreitol; SNO-BSA, S-nitroso bovine serum albumin, SNO-Hcy, SNO-homocysteine; GSNO, SNO-glutathione; RSNO, S-nitrosothiol(s).

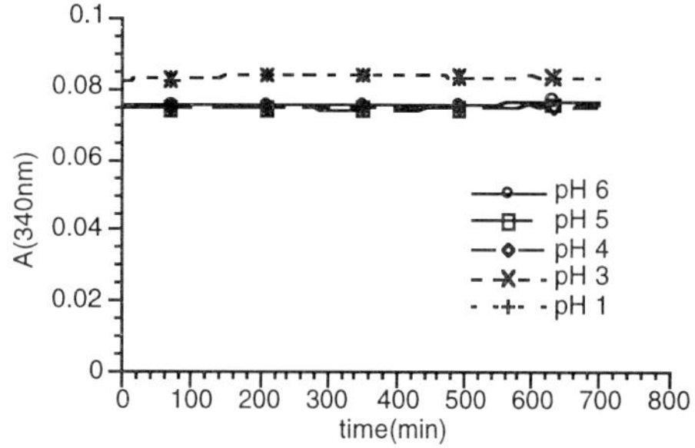

Figure 1: Decomposition of 100 µM SNO-Cys in the presence of 100 mM buffer, 0.5 mM Na_2DTPA between pH 1 - 6. Similar data was obtained for SNO-Hcy and SNO-BSA between pH 1-10

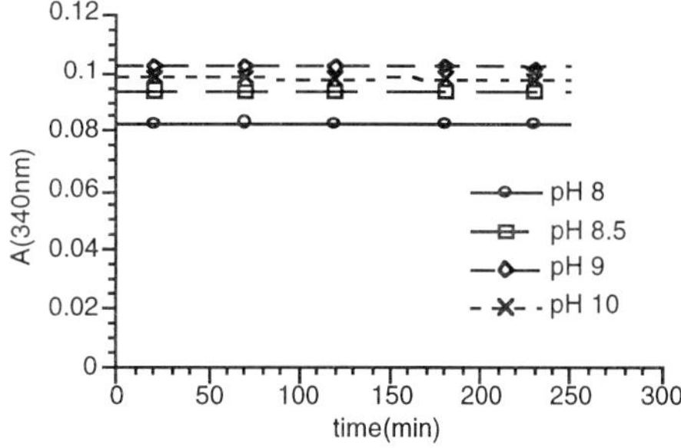

Figure 2: Decomposition of 100 µM SNO-Cys in the presence of 100 mM buffer, 0.5 mM Na_2DTPA between pH 8-10. Similar data was obtained for SNO-Hcy and SNO-BSA between pH 1-10

References

1. McAninly, J., D.L.H. Williams, S.C. Askew, A.R. Butler, and C. Russell, *Metal Ion Catalysis in Nitrosothiol (RSNO) Decomposition.* J. Chem. Soc., Chem. Comm., 1993. **23**: p. 1758-1759.

2. Ramdev, P., J. Loscalzo, M. Feelisch, and J.S. Stamler, *Biochemical properties and bioactivity of a physiologic NO reservoir.* Circulation, 1993. **88**: p. I-522.

3. Gaston, B., J. Reilly, J.M. Drazen, J. Fackler, P. Ramdev, D. Arnelle, M.E. Mullins, D.J. Sugarbaker, C. Chee, D.J. Singel, J. Loscalzo and J.S. Stamler, *Endogenous nitric oxides and bronchodilator S-nitrosothiols in human airways.* Proc. Natl. Acad. Sci. USA, 1993. **90**: p. 10957-10961.

4. Feelisch, M., M. te Poel, R. Zamora, A. Deussen, and S. Moncada, *Understanding the controversy over the identity of EDRF.* Nature, 1994. **368**: p. 62-65.

Stability of S-nitrosothiols related to glutathione

JAYNE M TULLETT*, HAROLD F HODSON#, ANDREAS GESCHER*, DAVID E G SHUKER* and SALVADOR MONCADA#

*MRC Toxicology Unit, University of Leicester, Leicester, LE1 9HN, U.K. and #Glaxo Wellcome Research Laboratories, Langley Court, Beckenham, Kent, BR3 3BS, U.K.

Introduction

RSNOs are potential pro-drugs of NO. RSNOs are generally unstable decomposing to give the disulphide and NO [1]. In previous studies CysNO was found to be short lived but GSNO was stable [2]. The reason for this difference was studied by examining the stability of the S-nitroso analogues of the glutathione related dipeptides CysglyNO and GlucysNO.

The decomposition of RSNO seems to be catalysed by trace metals such as copper [3]. The hypothesis was tested that the removal of Cu^{2+} would prolong the half-lives of the RSNOs related to GSNO.

Materials and Methods

Thiols were nitrosated according to Hart [4]. Half-lives were determined in phosphate buffer pH 7.4, at 37°C, by monitoring the u.v. absorbance at approximately 330nm as a function of time. Data were collected for 24h or 3 half-lives for the less stable compounds. For CysNO and CysglyNO half-lives were also determined in the presence of DTPA, a copper chelator. All half-life values were calculated assuming pseudo first-order kinetics.

Results

Table 1 and Fig.1 show that CysNO and CysglyNO were short-lived, GlucysNO and GSNO were stable. The presence of DTPA increased the stability of CysNO and CysglyNO (Table 1, Fig. 2).

Table 1. Half-lives of GSNO and related compounds

Compound	Half-Life Without DTPA	Half-Life With DTPA
CysNO	32.4±8.9 min*	11.0±1.1 h
CysglyNO	18.9±6.4 min	4.3±0.3 h
GlucysNO	> 24 h	> 24 h
GSNO	> 24 h	> 24 h

* Mean ± SD for 3 experiments

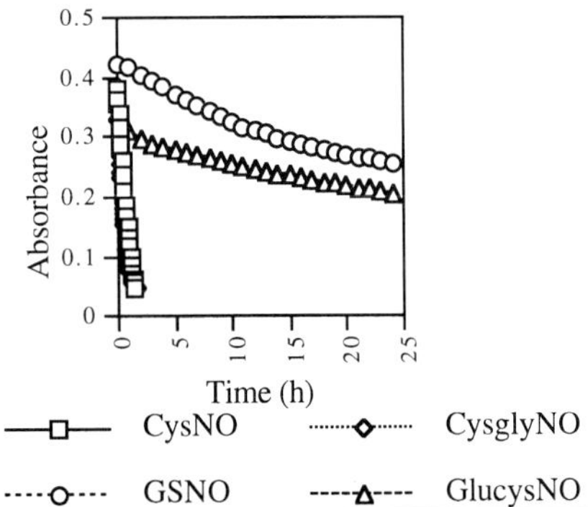

Figure 1. Difference in stability between the four RSNO

Abbreviations used: NO, Nitric Oxide; RSNO, S-nitrosothiol; CysNO, S-nitroso-L-cysteine; CysglyNO, S-nitroso-L-cysteinylglycine; GlucysNO, S-nitroso-L-γ-glutamyl-L-cysteine; GSNO, S-nitroso-L-glutathione; DTPA, Diethylenetriaminepentaacetic acid.

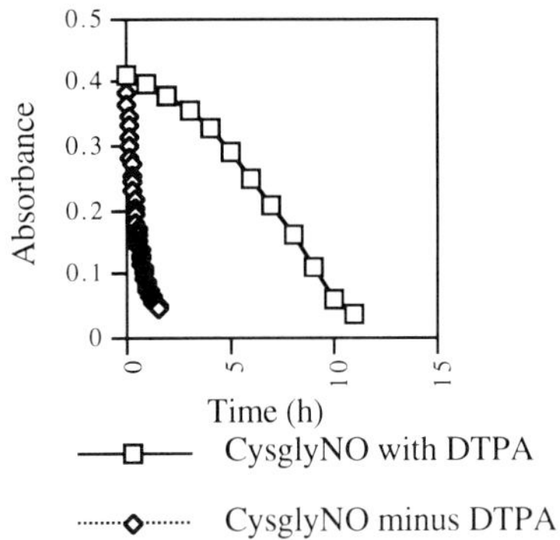

Figure 2. Effect of copper on the decomposition of CysglyNO

Discussion

Of the four RSNO studied CysNO and CysglyNO were the least stable irrespective of the presence of DTPA. DTPA increased the half-lives of CysNO and CysglyNO significantly clearly showing the effect of the presence of copper on the decomposition of these RSNO. The metal ion catalysis is thought to take place by co-ordinating with the RSNO between the N of the NO and the N of the NH_2, forming a ring structure (Fig. 3). This facilitates expulsion of NO from the molecule [5]. The results presented suggest that molecules for which an energetically favoured ring structure involving Cu^{2+} can be formed are less stable (i.e. CysNO and CysglyNO) than those in which a stable ring cannot be formed (i.e. GlucysNO and GSNO).

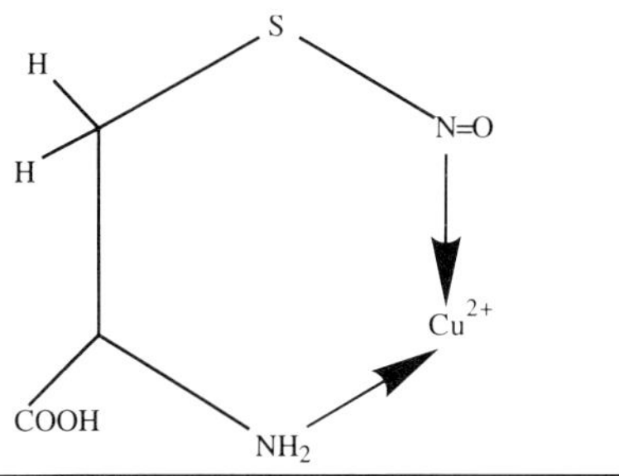

Figure 3. Expected structure of ring formed in the Cu^{2+} decomposition of CysNO

This work was funded by a collaborative studentship from the Medical Research Council with Glaxo Wellcome Ltd.

References

1. Mathews, W.R. and Kerr, S.W.(1993) J. Pharmacol. Exp. Therapeutics **267**(3), 1529-1537
2. Oae, S. and Shinhama, K. (1983) Org. Prep. Poc. Int. **15**(3), 165-198
3. McAninly,J., Williams, D.L.H., Askew, S.C., Butler, A.R. and Russell, C. (1993) J. Chem. Soc., Chem. Commun. **23**, 1758-1759
4. Hart, T.W., (1985) Tetrahedron Letters, **25**(16), 2013-2016
5. Askew, S.C., Barnett, D.J., McAninly, J. and Williams, D.L.H. (1995) J. Chem. Soc., Perkin Trans. 2,**4**, 741-745

Diethyldithiocarbamate induced decomposition of S-Nitrosothiols

Derrick R ARNELLE, Brian DAY, James D. CRAPO and Jonathan S. STAMLER

Divisions of Respiratory and Cardiovascular Medicine, Duke University Medical Center, Durham, NC 27710, U.S.A.

Summary

Diethyl dithiocarbamate (DETC) is used extensively as an inhibitor of CuZn superoxide dismutase (SOD) in the study of superoxide and nitric oxide interactions, and as a nitric oxide trap in EPR studies, in the form of the bis-DETC-iron nitrosyl complex. Addition of DETC to solutions of S-nitrosoglutathione caused a rapid decrease in S-nitrosothiol (RSNO) concentration with concomitant formation of nitrite, nitrate, disulfuram, GSSG, and mixed disulfide. Similarly, S-nitrosocysteine and S-nitroso-N-acetyl-penicillamine (SNAP) are unstable in the presence of DETC. Using both a NO electrode and an oxymyoglobin trap, we verified that NO was released during this reaction. Moreover, NO liberation was markedly potentiated by addition of SOD. Hydroxylamine and N_2O were not detectable in these reactions. Taken in aggregate, our results show that 1) DETC converts RSNO to NO and that 2) O_2^- is produced in the process. These findings complicate interpretation of experiments in which DETC is used to alter NO mediated responses. Some biological actions of DETC may result from RSNO elimination rather than SOD inactivation. Moreover, apparent DETC induced potentiation of superoxide effects may derive from superoxide produced in the conversion of RSNO to NO.

Results and Discussion

Attempts to synthesize S-nitroso-DETC by nitrosation of DETC with sodium nitrite under acidic conditions (pH 2.5 - 5) or with tert-butyl nitrite at pH 7.4 were unsuccessful as determined by UV/VIS spectroscopy. In contrast, nitrosation of glutathione occurs readily under the same conditions. Addition of DETC to solutions containing GSNO results in nitrosothiol decomposition (Fig 1) with rapid nitric oxide release (Fig 2). The increase in the NO signal in the presence of SOD is indicative of superoxide production during GSNO decomposition. The majority of GSNO ($\geq$ 95%) is converted to nitrite at low DETC/GSNO ratios; however at high DETC/GSNO ratios, 10-25% of the nitrosothiol decomposes to nitrate (Fig 3). Disulfiram and a DETC-GSH mixed disulfide are formed during the reaction as confirmed by cellulose TLC; however, disulfiram concentrations exceed GSNO equivalents by several fold.

We have examined the effect of DETC on several low molecular weight S-nitrosothiols. In all cases, addition of DETC caused rapid decomposition of the nitrosothiol to disulfides, nitrite, and nitrate via nitric oxide and superoxide. In theory, these results might be explained by homolytic or heterolytic breakdown of S-nitroso-DETC. However, homolytic decomposition is favored over heterolytic pathways for several reasons. First, a free radical mechanism is most consistent with the effect of SOD on NO production. Second, a homolytic mechanism best rationalizes the presence of non-stoichiometric amounts of disulfiram. Third, we have been unable to detect products of the heterolytic mechanism, such as hydroxylamine or nitrous oxide. A good fit for the second order rate constant , consistent with a rate determining bimolecular reaction between DETC and GSNO, was obtained from the data in Figure 1; however, we have no direct evidence for the formation S-nitroso-DETC.

Complete inhibition of SOD by DETC in tissues requires millimolar concentrations of DETC and lengthy incubations (15-30 min). Initial concentrations of RSNOs in such systems is probably in the nanomolar-micromolar range, thus DETC concentrations exceed RSNO concentrations by several orders of magnitude. Under these conditions, essentially all low molecular weight nitrosothiols are rapidly converted to nitric oxide (nitrite and nitrate) with production of superoxide before complete inhibition of SOD. This rapid reaction between DETC and S-nitrosothiols thus complicates interpretation of experiments in which DETC is used to alter NO and O_2^- mediated responses. Some biological actions of DETC may result from RSNO elimination rather than SOD inactivation. Moreover, apparent DETC induced potentiation of superoxide effects may derive from conversion of RSNOs to NO and O_2^-.

Abbreviations used: DETC, diethyl dithiocarbamate; RSNO, S-nitrosothiol(s); SOD, superoxide dismutase; DTPA, diethylenetriamine pentaacetic acid; TLC, thin-layer chromatography; GSNO, S-nitroso-glutathione; GSH, reduced glutathione; GSSG, oxidized glutathione.

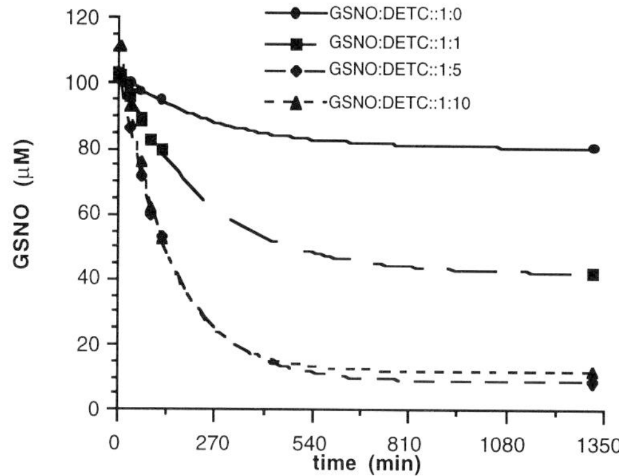

Figure 1: Time course of GSNO decomposition: 100µM GSNO in 100 mM KH_2PO_4, 0.5 mM DTPA pH 7.4 with 0, 0.1, 0.5, and 1.0 mM DETC. Nitrosothiol and nitrite concentrations were measured by a modified Saville assay.

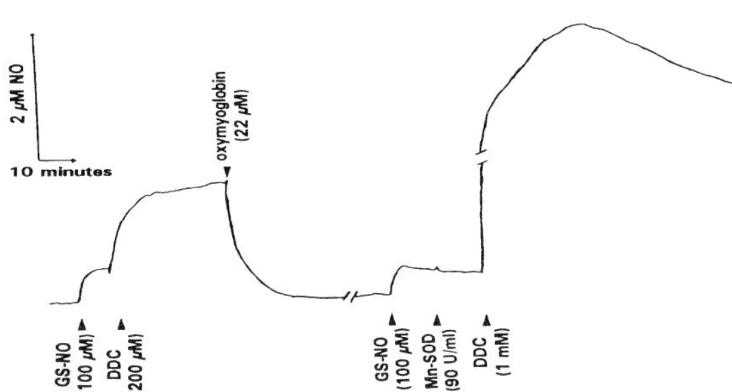

Figure 2: NO release from GSNO in the presence of DETC. Addition of 2 µmoles of DETC to a solution containing 1 µmole GSNO in Krebs buffer results in release of NO. The NO signal was abolished by the addition of 0.2 µmoles of oxymyoglobin. Inclusion of 90 u/ml of SOD increased the NO signal.

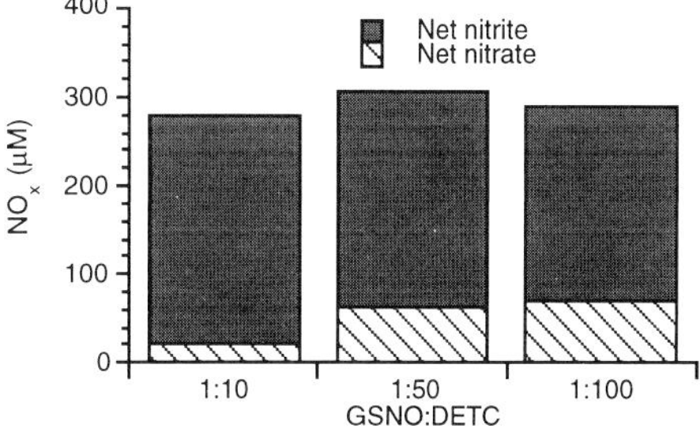

Figure 3: Nitrite and nitrate formation: 300 µM GSNO in 100 mM potassium phosphate, 0.5 mM DTPA pH 7.4 with 3, 15 and 30 mM DETC (10, 50 and 100 equivalents, respectively) after 180 minutes. Nitrate and nitrite concentrations were determined by capillary zone electrophoresis.

THE TRANSDERMAL DELIVERY OF AN NO DONOR DRUG: A NEW APPROACH TO RAYNAUD'S SYNDROME.

ANTHONY.R.BUTLER[1], IAIN.R.GREIG[1] AND FAISEL KHAN[2].

[1]School of Chemistry, University of St.Andrews, Fife KY16 9ST, Scotland, U.K. and [2]Deparment of Medecine, Ninewells Hospital, Dundee, Scotland U.K.

Introduction.

There are currently many NO donor drugs available for the treatment of various physiological defects. Perhaps the most effective of these are nitroglycerine and isosorbide dinitrate. These however need to be metabolised in order to produce nitric oxide. This occurs in the vasculature of microvessels larger than 100μm in diameter. In parts of the microcirculation too small to carry out the metabolic conversion, e.g. the skin , a similar drug which does not require conversion is required.

For this purpose we have been investigating the synthesis and uses of S-nitrosothiosugars. These are ideal for study for a number of reasons:

-the hydroxyl groups allow a wide variety of derivatives to be produced to give a range of stabilities and solubilities,

-the size and composition is similar to that of nitroglycerine,

-the breakdown product is an inert disulphide,

-none of the compounds have been adequately investigated prior to this study.

Raynaud's Syndrome [1,2,3,4]

This is a condition which afflicts as many as 5% of the population, of which 80% are female. The condition is caused by a paroxysmal or intermittent spasm of the digital arteries. To the sufferer this means very painful fingers and toes when exposed to lower temperatures, sometimes as high as 17°C. In more advanced cases this can lead to cyanosis and the loss of digits through superficial gangrene. The condition can be caused either by a disease or by vibrational means e.g. piano players. The reasons for these microcirculational defects are not known and modern treatment is limited in effectiveness. With these compounds we hope to investigate and perhaps treat this condition.

Transdermal Delivery [5]

Transdermal Delivery is a method of adminstering a drug directly thorugh the skin. This avoids the difficulties encountered with oral administration, such as breakdown in the stomach and also allows the drug to be delivered at a constant rate. The drug is designed so that it can pass through the skin membrane and the dosage controlled in a number of ways.

We have been making use of a system which makes use of a laser emitter and detector to record any change in the dermal blood flow. The principle is that the frequency of the reflected laser beam differs from that of the incident beam if it strikes a moving object -i.e. blood cells. With this method any increase in blood flow and/or any increase in blood cell count can be observed for the compound in question.

An example of a transdermal delivery system is that of the nicotine patch. These supply nicotine to the body in a manner currently thought to be less harmful than smoking. The drug is absorbed onto an inert matrix. This reservoir is covered by a microporous membrane which will keep the delivery rate constant.

Results and Discussion

A number of S-nitrosothiosugars have been made. These have been based on mono- and disaccharides [scheme 1], on other sugars such as dianhydroalditols [scheme 2] and on glycerol derivatives.

Scheme 1 [6,7,8]: Preparation of SNAG

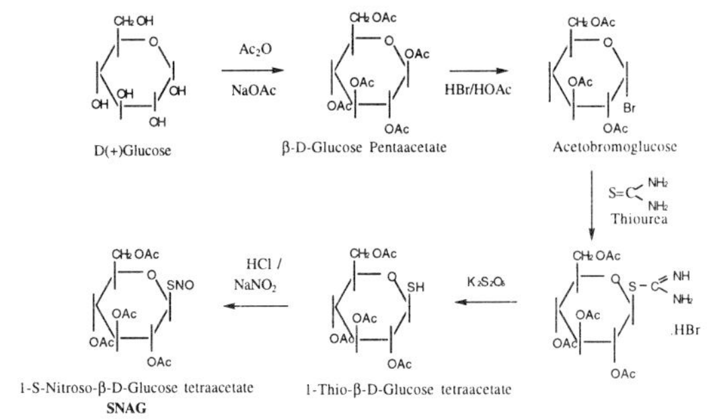

D(+)Glucose → (Ac_2O, NaOAc) → β-D-Glucose Pentaacetate → (HBr/HOAc) → Acetobromoglucose → (Thiourea) → (HBr salt) → ($K_2S_2O_5$) → 1-Thio-β-D-Glucose tetraacetate → (HCl / $NaNO_2$) → 1-S-Nitroso-β-D-Glucose tetraacetate **SNAG**

Scheme 2 [9,10]:

Isomannide → (Tf_2O) → Isomannide ditriflate → ($K^+CH_3COS^-$) → (NaOMe) → 2,5-dithioisomannide ⇌ (N_2O_3)

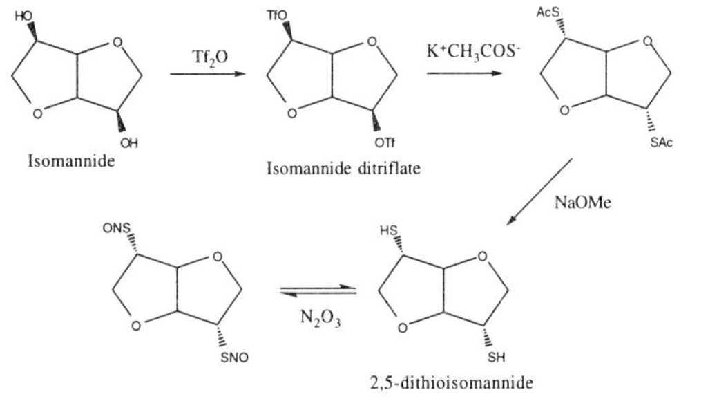

Physiological Testing

SNAG has been tested using the system described above. Initial test results have been very positive. A 1% SNAG solution in ethanol gave an eight fold increase blood flow over a 30 minute period. Whilst a 0.1% solution gave a three fold increase under the same conditions.

Conclusions

We have made a number of nitroso derivatives of thiosugars which are stable. One of these, SNAG, has been tested as a means of transdermal delivery for the skin microcirculation. It proved highly effective in a series of preliminary tests and will be used for a study of the pathophysiology of Raynaud's syndrome.

References

1) Stelwagon *Diseases of the Skin*
2) Mackey, H.O. *A Handbook of Diseases of the Skin*
3) Roxburgh, A.C. *Common Skin Diseases*
4) Rook, Wilson and Ebling *Textbook of Dermatology*
5) Parikh, N.K., Babar, A., Plakogiannis, F.M. (1985) *Pharm. Acta. Helv.*, **60**, 34-40.
6) Stanek, J., Sindlerova, M., Cerny, M., (1965) *Collection Czechoslov. Chem. Comm.*, **30**, 297-301.
7) Cerny, M., Pacak, J., (1963) *Montash*, **94**, 290-95.
8) Wolfrom, M.L., Thompson, A., *Methods in Carbohydrate Chemistry Vol II.*
9) Field, R.A., (1989) *PhD Thesis*, University of East Anglia.
10) Thiem, J., Luders, H., (1986) *Makromol. Chem.*, **187**, 2775-90.

Formation of iron-sulfur-nitrosyl complexes: pH dependence

DEBRA F. BLYTH and CHRISTOPHER GLIDEWELL

School of Chemistry, University of St Andrews, St Andrews, Fife, KY16 9ST, Scotland, UK

Introduction

In the presence of iron(II) salts, both nitrite and nitric oxide react with the sulfur containing amino acids, cysteine [1] and methionine [2] under a range of experimental conditions relevant to food processing to yield the iron-sulfur-nitrosyl complexes, $Na[Fe_4S_3(NO)_7]$ and $[Fe_2(SMe)_2(NO)_4]$ respectively; quantitative conversion of nitrite to $[Fe_4S_3(NO)_7]^-$ can be achieved. Both compounds are potentially able to transport large quantities of nitric oxide and indeed, the tetranuclear complex, $Na[Fe_4S_3(NO)_7]$ has been shown to exhibit vasodilatory properties [3]. The complexes contain coordinated NO ligands formally present as NO^+ and under certain experimental conditions, transfer of the nitrosyl groups can occur. The ready nitrosation of a range of secondary amines by $[Fe_2(SMe)_2(NO)_4]$ has been demonstrated [4-7]. However the dinuclear cluster, $[Fe_2(SMe)_2(NO)_4]$ has been shown to promote the tumorigenic activity of some environmental carcinogens, including N-nitrosamines and some polycyclic aromatic hydrocarbons such as those found in cigarette smoke and therefore is a cause for concern [8]. Due to the acidic nature of many food products and the acidic conditions found in the human stomach, the formation of these iron-sulfur-nitrosyl complexes from the reactions of cysteine and methionine respectively with iron(II) salts and nitrite, have been studied with decreasing levels of pH under conditions relevant to food processing.

Experimental

pH measurements were recorded using a Piccolo ATC pH-meter on buffered solutions of KH_2PO_4 & NaOH at pH 6 & 7 and H_2SO_4 solutions at pH 6 and below. The autoclave employed had a capacity of 14.3 dm^3.

Cysteine reactions; L-Cysteine (16.5 mmol), iron(II) sulfate heptahydate (1.80 mmol) and sodium nitrite (1.45 mmol) were dissolved with sodium ascorbate (10.1 mmol) in a aqueous solution of known pH (500 cm^3) and the mixture autoclaved at 118°C for 30 min. After cooling, the reaction mixture was filtered and exhaustively extracted with diethyl ether. The combined ether extracts were washed with water and dried over magnesium sulfate before removal of the solvent. Duplicate runs were made omitting the sodium ascorbate. Control reactions were carried out in a similar manner using $Na[Fe_4S_3(NO)_7]$ (0.18 mmol).

Methionine reactions; DL-Methionine (6.7 mmol), iron(II) sulfate heptahydrate (6.1 mmol) and sodium nitrite (14.5 mmol) were dissolved with sodium ascorbate (13.6 mmol) in an aqueous solution of known pH (100 cm^3) and the mixture refluxed under nitrogen for 2 hours. After cooling, the reaction mixture was filtered and exhaustively extracted with diethyl ether. The combined ether extracts were washed with water and dried over magnesium sulfate. After evaporating to dryness the extracts were dissolved in the minimum volume of methylene chloride and eluted through a silica chromatography column. Control reactions were carried out in a similar manner using $[Fe_2(SMe)_2(NO)_4]$ (0.31 mmol).

Results and discussion

When cysteine was autoclaved at 118°C with sodium nirite and iron(II) sulfate in the presence of sodium ascorbate at pH 7, $Na[Fe_4S_3(NO)_7]$ was isolated in 87% yield (based on nitrite). The yield of $Na[Fe_4S_3(NO)_7]$ isolated fell slightly as the pH was decreased, but at very acidic conditions no $Na[Fe_4S_3(NO)_7]$ was detected. In the absence of sodium ascorbate the yield of $Na[Fe_4S_3(NO)_7]$ isolated remained fairly constant on decreasing pH, but again at very acidic conditions no $Na[Fe_4S_3(NO)_7]$ was detected. Control reactions demonstrated quantitative recovery of $Na[Fe_4S_3(NO)_7]$ at moderate pH levels, with a dramatic loss in yield at very low pH, indicating that $Na[Fe_4S_3(NO)_7]$ itself decomposes at very acidic conditions.

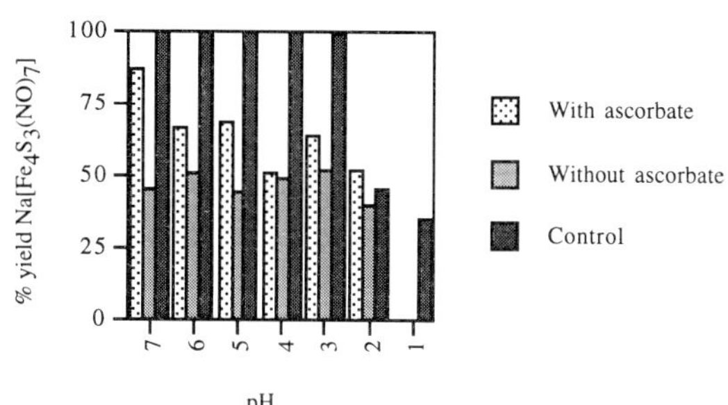

Graph 1; Percentage yield of $Na[Fe_4S_3(NO)_7]$ obtained from the reaction of cysteine with iron(II) salts and nitrite at different values of pH

Methionine, when heated to 100°C with sodium nitrite, iron(II) sulfate and sodium ascorbate under anaerobic conditions at pH 7 gave the dinuclear cluster $[Fe_2(SMe)_2(NO)_4]$ in ca. 13% yield (based on iron(II)). On lowering pH, the yield of $[Fe_2(SMe)_2(NO)_4]$ isolated remained fairly constant but at very acidic conditions, less than 1% $[Fe_2(SMe)_2(NO)_4]$ was detected. Control reactions showed almost quantitative recovery during the pH range studied, demonstrating that $[Fe_2(SMe)_2(NO)_4]$ is relatively stable in acidic conditions, but its formation from methionine, iron(II) salts and nitrite is severely hindered at low pH.

However at the acidity levels typically found in foodstuffs, the yields of the tetranuclear cluster, $Na[Fe_4S_3(NO)_7]$ and the neutral dinuclear complex, $[Fe_2(SMe)_2(NO)_4]$ from the reactions of cysteine or methionine respectively with iron(II) salts and nitrite with sodium ascorbate can approach quantitative capture of the nitrite.

This work was funded by the Engineering & Physical Sciences Research Council (U.K.)

References

1. Butler, A.R., Glidewell, C. and Glidewell, S.M. (1990) *Polyhedron*, **9**, 2349
2. Butler, A.M., Glidewell, C. and Glidewell, S.M. (1992) *Polyhedron*, **11**, 591
3. Flitney, F.W., Megson, I.L., Flitney, D.E. and Butler, A.R. (1992) *Br. J. Pharmac.*, **107**, 842
4. Wang, G.H., Zang, W.X. and Chai, W.G. (1980) *Acta Chim. Sin.*, **38**, 95
5. Wang, M., Li, M.H., Jiang,Y.Z., Sun, Y.H., Li, G.Y., Zhang, W.X., Chai, W.G. and Wang, G.H. (1983) *Cancer Res. Prevent. Treat.*, **10**, 145
6. Croisy, A., Ohshima, H. and Bartsch, H. (1984) *I.A.R.C. Sci. Publ.*, **57**, 327
7. Lees, A. (1992) PhD Thesis, University of St. Andrews
8. Li, M.H., Lu, S.H., Ji, C., Wang, Y., Yang, M., Cheng, S. and Tiam, G. (1979) *Proc. Int. Symp. Princess Takamatsu Cancer Res. Fund*, 139; Chem. Abstr. (1982) **97**, 180397

Dinitrosyl Iron complexes: formation, stability, and physiological implications

Mark E MULLINS[a], Jonathan S STAMLER[b], David J SINGEL[c] and Andrew R BARRON[a]

[a]Department of Chemistry, Harvard University, Cambridge, MA 02138; [b]Duke University Medical Center, Durham, NC 27710; [c]Department of Chemistry, Montana State University, Bozeman, MT 59715

Dinitrosyl iron complexes (DNICs) have been identified in biological systems [1,2] and may be involved in NO transport and function. The goal of this study was to understand their behavior under various physiologies.

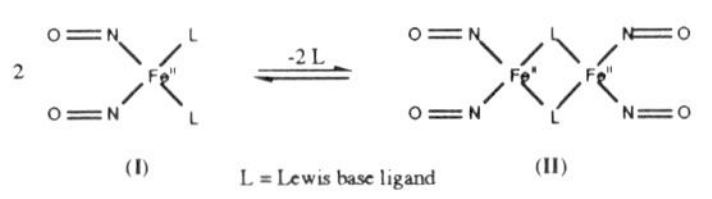

(I) L = Lewis base ligand (II)

Figure 1. DNIC monomer-dimer equilibrium.

DNICs exist in equilibrium between the paramagnetic monomeric (I) and diamagnetic dimeric (II) species (**Figure 1**). In the complex biological milieu, there are a number of Lewis bases (L) that could potentially act as donor ligands. One of the most relevant, considering its prevalence and facile complexion with iron, is thiol.

The effects of thiol concentration (relative to NO and Fe^{2+}) on the formation and structure of DNICs has been determined by thiol titration experiments monitored by EPR (electron paramagnetic resonance) spectroscopy. In the absence of any suitable donor ligand, the reaction of iron(II) sulfate ($FeSO_4$) with $NaNO_2$ in the presence of a reducing agent results in the formation of a high spin d^7 octahedral complex, $[Fe^{I}(NO)(H_2O)_5]^{2+}$. The EPR spectrum of DNICs prepared in the absence of cysteine shows features of this complex. As the cysteine concentration is increased from 0.3 x to 3 x $[Fe^{2+}]$ the signal due to the aqueous complex decreases in intensity, until, at cysteine concentration in excess of 3 x $[Fe^{2+}]$, it is lost. The signal due to the cysteine-DNIC is observed at cysteine concentrations as low as 0.3 x $[Fe^{2+}]$. Here the EPR signal overlaps with that of the aqueous complex. At cysteine concentrations in excess of 15 x $[Fe^{2+}]$, the cysteine-DNIC complex is the only paramagnetic species observed in solution. Increasing the cysteine concentration increases the intensity of the cysteine-DNIC EPR spectral signal; however, a maximal value could not be determined because a precipitate formed at thiol concentrations greater than 30 x $[Fe^{2+}]$. Furthermore, we observed a non-linear increase in signal intensity with cysteine concentration suggesting oxidation of thiol to thioethers. In summary, we have demonstrated that only three molar equivalents of thiol per iron are required to ensure complete thiol-DNIC formation in exclusion of all other paramagnetic species. Issues of applicability to the endogenous biologic formation of DNICs may be further enhanced by using a relatively low thiol : iron ratio.

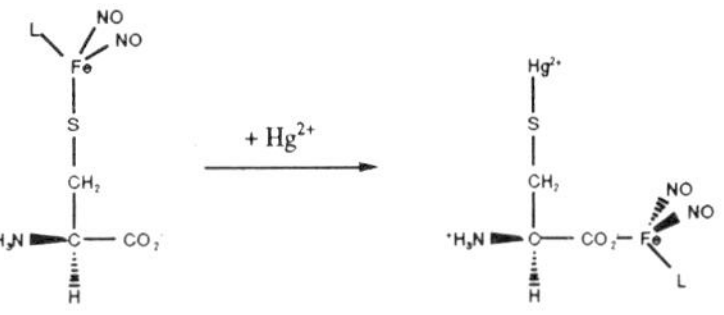

Figure 2. Proposed Effects of Mercury(II) on L-Cysteine-DNICs.

To address the issue of DNIC stability relative to pH, a representative small molecular weight thiol (cysteine)-DNIC was formed under standard conditions and diluted with buffered solutions with pH values from 1 to 10. The standard method of DNIC formation results in a pH = 6 which was considered the reference value. Our results indicate that DNICs are more stable at high pH (no upper limit reached through pH = 10). In addition, DNIC production in the physiologic pH range (6 to 8) is near optimal and further rationalizing their presence *in vivo*.

The EPR spectra of standard solutions of either cysteine-DNIC or BSA-DNIC were recorded with various molar equivalents of $HgCl_2$ (**Figure 3**). Above 0.167 equivalents of $HgCl_2$ the EPR signal due to the cysteine-DNIC was observed to decrease in intensity. However, the effect of Hg^{2+} is more complex that simply attenuating the EPR signal. In fact upon addition of Hg^{2+}, two EPR signals are observed. The larger signal at g = 2.033 shows no discernible hyperfine splitting, and a half-width of about 8 G. This signal is similar to the aqueous-type complex which is observed for DNIC solutions in the absence of thiol. The second, smaller, signal at g = 2.020 has a distinctive 13-line hyperfine structure. This signal is similar to the non-sulfur binding spectrum previously discussed for thiaproline-DNIC. Thus, the thiol-DNIC is not completely destroyed but has been rearranged. While the thiol residue of cysteine is clearly no longer bound to iron, the exact fate of the thiol group is unknown. Clearly complexation to Hg^{2+} is most likely, but, we cannot rule out the possibility that is has been oxidized to a thioether.

In the case of Hg^{2+} addition to the BSA-DNIC there is an apparent change in the symmetry of the complex, from axial (**Figure 3a**) to rhombic (**Figure 3b**). The EPR spectrum observed upon addition of 1 - 2 molar equivalents of Hg^{2+} has an appearance similar to that observed for imidazole-DNICs. Thus, it would appear that the mercury may effect the mode of binding of BSA to the iron center due to possible conformational changes as a result of its complexation with the thiol termini. At a three fold excess, or greater, of Hg^{2+} per cysteine-DNIC no EPR signal is observed. Similarly, at a five-fold excess or greater of Hg^{2+}, all signals disappeared from a BSA-DNIC solution. Thus, at large excesses of Hg^{2+} all the iron-NO centers have been removed or converted into diamagnetic dimers (*c.f.*, **Figure 2**). In summary, mercury may alter the potential bioactivity of the DNIC due to changes in ligation about the iron center.

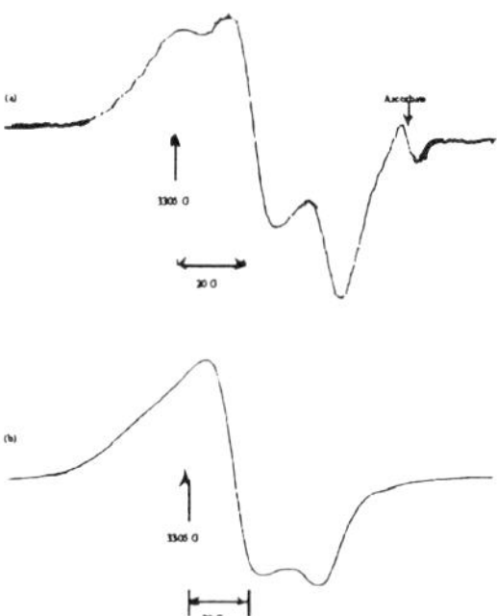

Figure 3. EPR spectra of bovine serum albumin (BSA)-DNIC in the absence (a) and presence (b) of Hg^{2+}.

1 Vanin, A.F. (1991) FEBS Lett. **289**, 1-3

2 Vedernikov, Y.P., Mordvintcev, P.I., Malenkova, A.F. and Vanin, A. (1992) Eur. J. Pharmacol. **211**, 313-317

Release of nitric oxide (NO) from SIN-1 (3-Morpholino-sydnonimine) in the presence of model compounds for cytochrome P-450

SABINE RIEDERER, CHRISTIN SCHELLENBERGER and HANS-JÜRGEN DUCHSTEIN

Institute of Pharmacy, Freie Universität Berlin, Königin-Luise-Str. 2+4, 14195 Berlin, Germany

Idea

In the literature we find evidence for the participation of cytochrome P-450 in the release of nitric oxide from glycerol trinitrate [1, 2]. We imitate this and similar reactions with metal-complexes as model compounds in a reductive pathway [3].

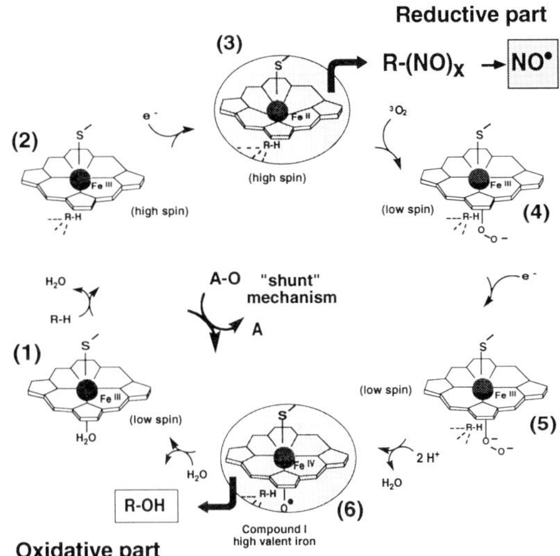

Fig. 1

Tools

Because of the different oxidation states of the NO-containing drugs, we try to activate SIN-1 with tetraphenylporphyrins in the presence of oxygen donors and with Co-Salen in the presence of oxygen. Under these conditions we imitate the oxidative part of the cytochrome P-450 cycle.

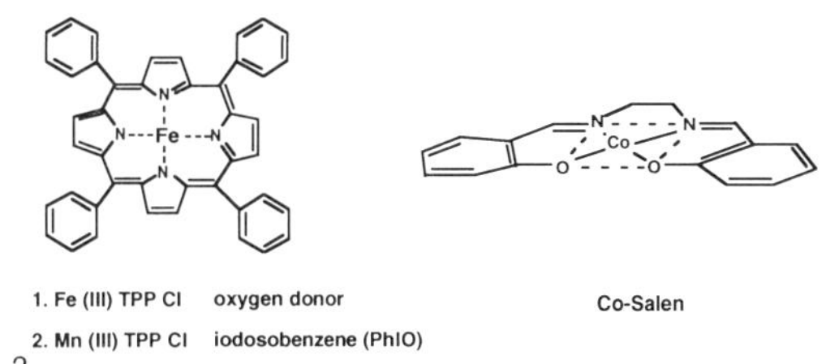

Fig. 2

Results before our investigation [4, 5]:

1. Spontaneous release of nitric oxide.
2. NO is oxidized to nitrite and nitrate.
3. Cysteine enhances the NO - release.
4. The reaction needs O_2, which reacts to $O_2^{\bullet -}$.
5. Molsidomine is metabolized quantitatively.
6. The amount of NO correlates with the increase of the cGMP-level
7. The conversion of SIN-1 does not correlate with the increase of the cGMP-level.

Analytics

The analytical assay for detecting NO is a chemiluminescence method, based on the measurement of intensity of the fluorescent radiation emitted after chemical oxidation of nitric oxide by ozone. The detection limit is approximately 1 nmole and a signal concentration plot shows a linear region up to 100 nmole.

Results

1. The reaction needs O_2.
2. $O_2^{\bullet -}$ catalyses the reaction.
3. The NO-release is increased by Co-Salen/O_2 (a singlet oxygen imitating system [3]) with simultaneous decrease of SIN-1A. The reaction is quenched by DABCO (1,4-diazabicyclooctane).
4. The reaction with chemically produced $^1O_2$ show the same characteristics as the reaction with Co-Salen / O_2.
5. The reaction with photochemically produced $^1O_2$ has a much lower reaction rate.
6. The NO-release is temperature dependent in contradiction to the conversion of SIN-1.

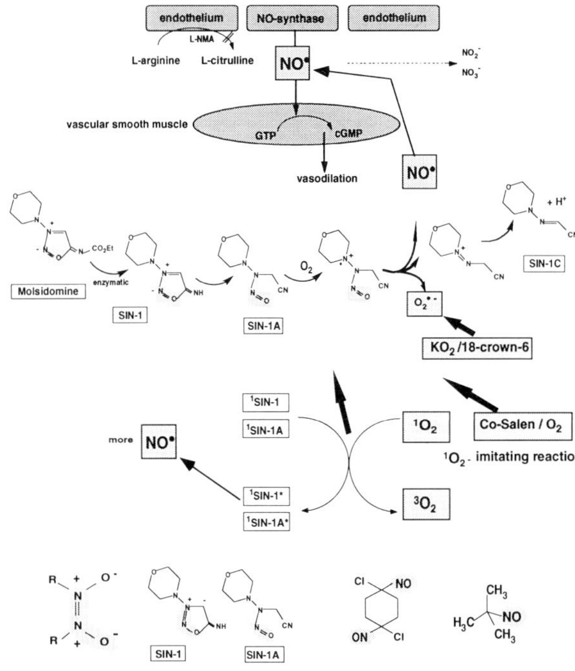

Fig. 3

A possible explanation of this obvious contradiction is the behaviour of SIN-1A as an efficient singlet oxygen quencher (proofed by the inhibition of the $^1O_2$-dependent synthesis of juglone), like other nitroso-compounds found in the literature [6]. Therefore we investigated methyl-nitrosopropane under comparable activation conditions, and this compound releases nitric oxide with simultaneous quenching properties.

Table 1. Degradation of molsidomine under different reaction conditions

	NO•	SIN-1A	SIN-1C
1. OH^-	↑↑	↑↑	↑
2. Argon	↓	↓	↓
3. O_2 / buffer	↓	↑	↑↑
4. $O_2^{\bullet -}$	↑↑	↑↑↑	↑↑↑
5. Co-Salen / O_2	↑↑↑	↓↓	↑↑↑
6. $^1O_2$ / chemical	↑↑	↓↓↓	↑↑↑
7. $^1O_2$/ photochem. (22 °C)	—	↓↓↓	↓↓↓
8. $OH^\bullet$	↓↓↓	—	—
9. cysteine	↓	↓	↓

Reactions 1 to 6, 8 and 9 were performed at 37°C

Arrows mean the enhanced or deminished release of NO and the formation of SIN-1 and SIN-1C, in comparison to the spontaneous degradation of molsidomine at 37°C in the presence of air.

Literature

1. Servent,D., Delaforge,M., Ducrocq,C., Mansuy,D. and Lenfant,M. Nitric oxide formation during microsomal hepatic denitration of gylceryl trinitrate: Involvement of cytochrome P-450 (1989) Biochem. Biophys. Res. Commun. **163**, 1210-1216
2. Schröder,H. Cytochrome P-450 mediates bioactivation of organic nitrates (1992) J. Pharmacol. Exp. Ther. **262**, 298-302
3. Duchstein,H.-J. and Riederer,S. Mechanistische Vorstellungen zur Freisetzung von Stickstoffmonoxid aus NO-Pharmaka. Modellreaktionen in Gegenwart von Licht und Übergangsmetallkomplexen 1.Mitt.(1995) Arch. Pharm. (Weinheim) **328**, 317-324
4. Feelisch, M., Ostrowski,J.and Noack,E. On the mechanism of he NO release of sydnonimines (1989) J. Cardiovas. Pharmacol. **14** (Suppl. 11), S 13 -S 22
5. Bohn,H. and Schönafinger, K. Oxygen and oxidation promote the release of nitric oxide from sydnonimines (1989) J. Cardiovas. Pharmacol. **14** (Suppl. 11), S 6- S 12
6. Bellus,D. Physical quenchers of singlet molecular oxygen (1979) Adv. Photochem. **11**, 105-205

SIN-1A/cyclodextrin complexes. Novel, stable and biologically active NO releasing agents

MARIA VIKMON[1], LAJOS SZENTE[1], JOSEPH M. GÉCZY[2], HIDDE BULT[3], TADEUSZ MALINSKI[4] and MARTIN FEELISCH[5]

[1]Cyclolab Cyclodextrin R & D Ltd., Budapest, Hungary; [2]Therabel Research, Bruxelles, Belgium; [3]Universitaire Instelling Antwerp, Belgium; [4]Oakland University, Rochester, USA; [5]Schwarz Pharma AG, D-40789 Monheim, Germany.

SIN-1 (linsidomine) experiences widespread use as a nitric oxide (NO) donor. It undergoes, in a pH- and temperature-dependent manner, hydrolytic ring-opening to form SIN-1A (N-morpholino-N-nitrosamino-acetonitril; Fig. 1), which can release NO upon oxidation. The potential utility of SIN-1A as a research tool is limited by its chemical instability, requiring storage at -80 °C under nitrogen. We therefore sought to examine the usefulness of cyclodextrins (CDs) for improving stability of SIN-1A in the solid state and, furthermore, to test the influence of complexation on the compound's decomposition kinetics and biological activity.

SIN-1A/CD inclusion complexes were prepared by hydrolysis of SIN-1 in the presence of β- or γ-cyclodextrins under an anaerobic atmosphere in the dark. Solid material was obtained by subsequent freeze-drying of the aqueous solution [1]. The concentration of SIN-1, SIN-1A and SIN-1C in the cyclodextrin preparations were determined by reversed-phase HPLC. The conversion of SIN-1A to SIN-1C was also monitored using conventional UV-spectrophotometry. The kinetics of NO release from SIN-1, SIN-1A and SIN-1A/CD complexes was measured using three different techniques; i) the oxyhemoglobin assay [2], ii) the gas-phase chemiluminescent reaction with ozone [3] and iii) electrochemical detection using a porphyrinic microsensor [4]. Nitrite (NO_2^-) and nitrate (NO_3^-) were determined as stable end-products of SIN-1 and SIN-1A decomposition using anion-exchange HPLC [2]. The biological activity of SIN-1 and SIN-1A preparations was assessed by measurement of the inhibition of platelet aggregation and vascular relaxation, respectively.

Fig. 1: Structure of SIN-1A

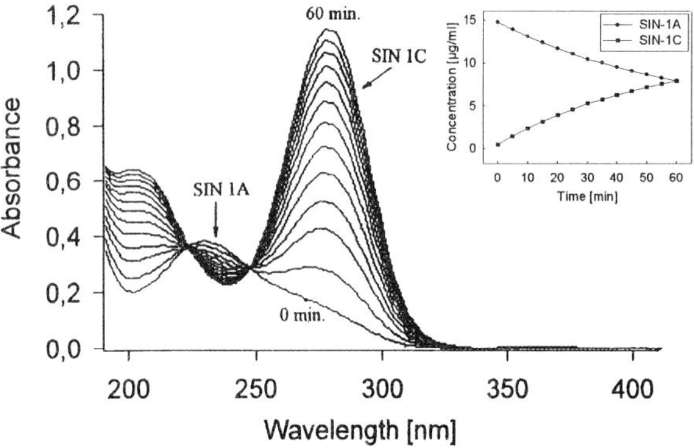

Fig. 2: UV-Spectral changes associated with the decomposition of SIN-1A/γ-CD and formation of SIN-1C (inset) in aqueous solution (100 mM phosphate buffer pH 7.4; 37 °C).

In the presence of inorganic salts, such as ammonium acetate, β- and γ-cyclodextrins were found to catalyze the quantitative conversion of SIN-1 to SIN-1A with simultaneous stabilization of the latter by formation of a non-covalent inclusion complex. Freeze-dried SIN-1A/CD complexes (the content of SIN-1A, SIN-1 and SIN-1C typically is 10, <0.05 and < 0.5 %, respectively) appeared as light pale yellow powder, which is highly soluble (> 10 mg/ml) in saline and aqueous buffer solutions and can be stored as solid at room temperature for several years without major decomposition.

In aerated aqueous solution, SIN-1A/CD decomposes with stoichiometric formation of SIN-1C and NO. The half-life of SIN-1A/CD ranged from 50 - 120 min at pH 7 at room temperature, depending on the type of cyclodextrin used (Fig. 2). Whereas decomposition of SIN-1A was almost independent of pH in the range of 4 - 10, rates of NO formation increased with decreasing pH. In contrast to SIN-1, there was no lag-phase for NO formation with either authentic SIN-1A or SIN-1A/γ-CD. Constant steady-state levels were obtained already after a few seconds and remained stable for at least 30 min (Fig. 3).

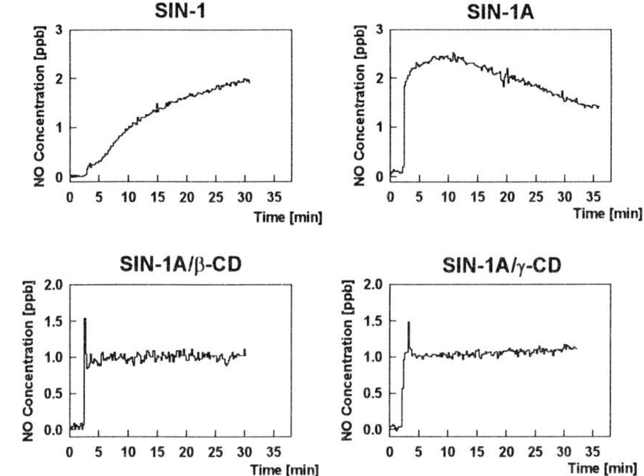

Fig. 3: Representative tracings of the kinetics of NO release from SIN-1, SIN-1A and SIN-1A/β- and γ-CD complexes as determined by chemiluminescence (1 mM each in 100 mM phosphate buffer pH 7.4; 37 °C under nitrogen).

SIN-1A decomposition and NO formation was accelerated in the presence of oxygen and transition metals, in particular copper. Under aerobic conditions, addition of superoxide dismutase (SOD) increased NO recovery using either chemiluminescence or electrochemical detection by scavenging superoxide (O_2^-), thus preventing formation of peroxynitrite ($ONOO^-$). This effect of SOD was less evident (<10 % difference vs. control) using the oxyhemoglobin assay.

The rate of NO generation was almost linearly dependent on the concentration of the SIN-1A/CD complex. In the presence of oxygen, almost equimolar concentrations of NO_2^- and NO_3^- were detected as stable endproducts of SIN-1A decomposition. The observed NO_2^-/NO_3^- ratio suggests that i) peroxynitrite formation from SIN-1A is not stoichiometric, and ii) peroxynitrite may react with excess NO to form a new N-oxide species, which decomposes to nitrite rather than nitrate.

Illumination with ambient light did not influence the apparent rate of NO release from SIN-1A/γ-CD. Thiols (even in five-fold molar excess) had an only marginal inhibitory effect on NO-formation.

SIN-1A/β- and γ-CD-complexes were found to inhibit human washed platelet aggregation (EC_{50} 3×10^{-6} and 3×10^{-5} M for SIN-1A/γ-CD and SIN-1, respectively) and to relax vascular tissue in a concentration-dependent manner. No difference in biological activity was observed between free and complexed SIN-1A in either bioassay system. SIN-1A/β-CD retained >80 % of its biological activity even after 3.5 years of storage at room temperature.

In conclusion, complexation with cyclodextrins leads to a dramatic increase in chemical stability of SIN-1A without effecting its biological activity. Under aerobic and anaerobic conditions, NO release from SIN-1A/CD complexes is instantaneous, almost linearly concentration-dependent, largely unaffected by the presence of thiols and constant over prolonged periods of time. The good predictability and reproducibility of NO generation from SIN-1A/CD complexes makes these compounds particularly attractive for the controlled, spontaneous generation of NO in aqueous solution and for application as NO-donors to biological systems.

References:

[1] Vikmon, M., Szemán, J., Szejtli, J. and Géczy, J. (1995) Proc. 1st World Meeting APGI/APV, pp. 587-588, Budapest, Hungary.

[2] Feelisch, M. and Noack, E. (1987) Eur. J. Pharmacol. **139**, 19 - 30.

[3] Archer, S. (1993) Faseb J. **7**, 349 - 360.

[4] Malinski, T. and Taha, Z. (1992) Nature **358**, 676 - 678.

Biochemical and pharmacological characterization of the novel NO-donor, SP/W-5186

KARLHEINZ KNÜTTEL, CLAUS O. MEESE, HILMAR BÖKENS, ROLF SPAHR, HUGO FRIEHE, DARYL REES*, MICHAEL J. FOLLENFANT*, BRENDAN J.R. WHITTLE* and MARTIN FEELISCH

Schwarz Pharma AG, D-40789 Monheim, Germany, *The Wellcome Foundation Ltd., Beckenham, Kent, BR3 3BS, UK.

SP/W-5186 [N-(3-Nitratopivaloyl)-S-(N′-acetylglycyl)-L-cysteine ethylester, Fig. 1] is a new NO-donor compound comprised of a nitrato fatty acid (3-nitratopivalic acid; SP/W-4744) linked to an S-protected cysteine moiety which is under development as anti-anginal drug. We here briefly describe some of the key pharmacological and metabolic characteristics of this compound.

At pH 7.4, SP/W-5186 is hydrolyzed to form compounds with free sulfhydryl group [N-(3-nitratopivaloyl)-L-cysteine ethylester, SP/W-3672; N-(3-nitratopivaloyl)-L-cysteine, SP/W-4853]. These metabolites represent the proposed active principle and are in redox equilibrium with the corresponding disulfides.

In vitro, SP/W-5186 requires reaction with cysteine to release NO. In contrast, its principal hydrolysis product, SP/W-3672, is a spontaneous NO-donor (Fig. 2). Thus, SP/W-5186 is the prodrug of a spontaneous NO-generating entity.

In accordance with this prodrug concept, SP/W-5186 is an only weak stimulator of soluble guanylyl cyclase (sGC) whereas SP/W-3672 potently activates sGC already in the absence of cysteine. In contrast to classical nitrates such as glyceryl trinitrate (GTN), sGC stimulation by SP/W-3672 is not dependent on the presence of a cytochrome P_{450}-related enzyme system.

Fig. 1: Structure of SP/W-5186

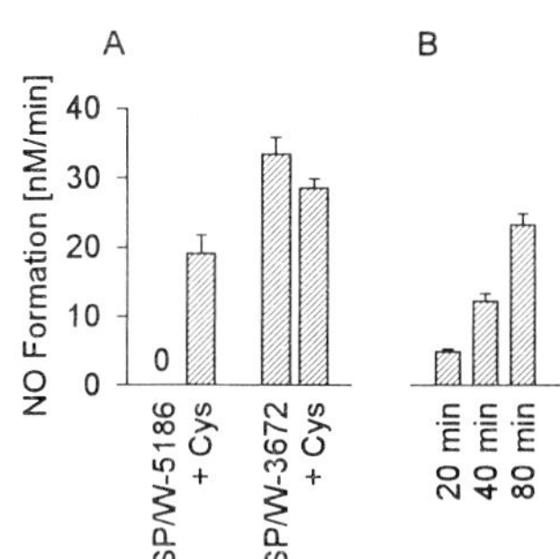

Fig. 2: A) NO release from SP/W-5186 or SP/W-3672 (1 mM each) in the absence and presence of L-cysteine (5 mM) as determined by the oxyhemoglobin technique (pH 7.4, 37 °C; means ± SEM; n = 3); **B)** NO release from SP/W-5186 (5 mM) after pre-incubation for different time intervals in buffer pH 7.4, 37 °C (means ± SEM; n = 3-4)

SP/W-5186 and SP/W-3672 elicited, in a methylene blue- and oxyhemoglobin-sensitive manner, concentration-dependent relaxation of phenylephrine-precontracted rat aortic rings in organ baths. SP/W-3672 is only marginally more potent than SP/W-5186 (Tab. 1), suggesting rapid tissue conversion of SP/W-5186 to the active metabolite, SP/W-3672. Vascular relaxation by either compound was accompanied by concentration-dependent increases in cGMP levels. SP/W-6373, the nitrate-free analogue of SP/W-5186, was without any effect on cGMP content, indicating that the nitrate group is a prerequisite for both NO formation and sGC activation to occur.

Classical organic nitrates are known to be only weak inhibitors of collagen-induced human platelet aggregation *in vitro*. In the same system, SP/W-5186 is about 10-fold more potent than isosorbide-5-nitrate (IS-5-N). Despite containing a single nitrate group only, SP/W-3672 was found to be considerably more potent in inhibiting platelet aggregation than GTN (Tab. 1) suggesting that SP/W-5186 may be endowed with more potent *in vivo* antiplatelet activity than classical nitrates.

In conscious Beagle dogs, *i.v.* application of SP/W-5186 lead to a dose-dependent (1.6-26.8 µmol/kg) decrease in systolic arterial (SAP), pulmonary arterial (PAPs) and central venous pressure (CVP) with predominant action on capacitance vessels (GTN>SP/W-5186>SP/W-4744>IS-5-N). After oral application of SP/W-5186 (Fig. 3), hemodynamic effects were slower in onset and markedly longer lasting compared to IS-5-N. Arterial vessel compliance remained elevated even 6 h post application.

Table 1: Vasorelaxing and anti-aggregatory effects of SP/W-5186 in comparison to other nitrates.

EC_{20} and EC_{50} values, respectively, were determined after construction of concentration response-curves for either compound in endothelium-denuded phenylephrine-precontracted isolated rat aortic rings (n = 6-10) and on collagen-induced human washed platelet aggregation (n = 3).

Compound	Vasorelaxation EC_{50} [M]	Inhibition of platelet aggregation EC_{20} [M]
SP/W-5186	7×10^{-7}	5×10^{-5}
SP/W-3672	4×10^{-7}	1×10^{-6}
SP/W-4744	1×10^{-4}	3×10^{-4}
GTN	3×10^{-8}	9×10^{-6}
IS-5-N	5×10^{-4}	3×10^{-4}

After *i.v.* and *p.o.* application to Beagle dogs plasma concentrations of SP/W-5186 were below the detection limit (<25 pg/ml; GC-MS) indicating rapid metabolism. Disulfides of SP/W-3672 and SP/W-4853 were found to be the major metabolites in dog plasma, which is consistent with the proposed metabolic pathway. In all cases, plasma levels of the two metabolites preceded the hemodynamic effects.

An *in vivo* tolerance model was developed which allows to investigate changes in hemodynamic responses after a 3-day oral pretreatment phase. In this model, SP/W-5186 was found not to develop tolerance with respect to changes in CVP, PAPs and arterial compliance. The effects on SAP were slightly reduced after a high-dose homologuous pretreatment. SP/W-5186, but not SP/W-4744, thus displays a tolerance profile distinct from that of classical nitrates which reveal reduced hemodynamic responses in the arterial and venous circulation under these conditions.

General pharmacology investigations with SP/W-5186 revealed an inconspicuous action profile. Approx. LD_{50} values were determined to be > 4640 and 3463 mg/kg after oral application to rats and mice, respectively. Four weeks subacute toxicity studies in rats and dogs showed an inconspicuous behaviour and body weight gain of the animals. Histopathological examinations revealed no substance-related changes. Ames and Micronucleus tests showed no mutagenic potential for SP/W-5186. Although a spontaneous NO-donor, no mutagenic effects were observed with SP/W-3672.

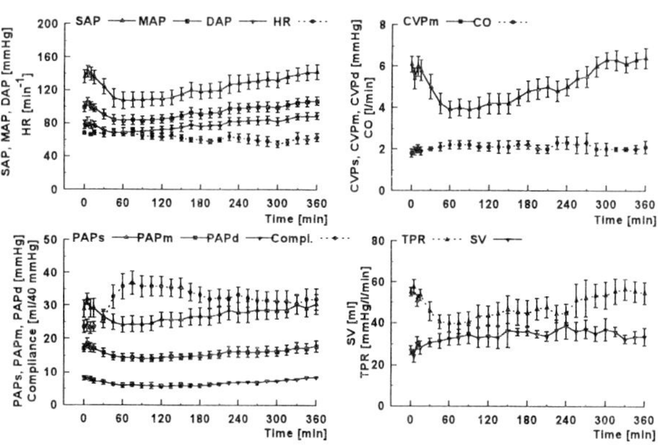

Fig. 3: Hemodynamic effects of SP/W-5186 (10.5 mg/kg) after oral application to conscious Beagle dogs (means ± SEM, n = 6).
SAP, MAP, DAP = systolic, mean and diastolic arterial pressure, HR = heart rate, CVPm = mean central venous pressure, CO = cardiac output, PAPs, PAPm, PAPd = systolic, mean and diastolic pulmonary arterial pressure, Compl = arterial compliance, SV = stroke volume, TPR = total peripheral resistance.

In conclusion, SP/W-5186 is a new, orally available NO donor with a favourable safety profile. It is a prodrug, which is rapidly metabolized to the active principle, SP/W-3672, which spontaneously generates NO. Both, SP/W-5186 and SP/W-3672 have considerably more potent vasorelaxant and anti-platelet activity than classical organic nitrates. SP/W-5186 elicits long-acting hemodynamic effects and is associated with a low tendency of tolerance development *in vivo*.

N-Hydroxyl-N-nitrosamines, redox-sensitive nitric oxide donors

Peng G. WANG, Luis ECHEGOYEN, Andrea MCGILL, Yifang YANG, Jun LI and Libing YU

Department of Chemistry, University of Miami, Coral Gables, P. O. Box 249118, FL 33124, U.S.A.

While the endogenous NO is generated by the catalysis of nitric oxide synthase (NOS) from arginine, exogenous production of NO from NO releasing agents offers a valuable tool in biological research on the function of nitric oxide. Moreover, many pharmaceutical applications of NO releasing agents have been suggested. Besides organic nitrates (such as nitroglycerin) which release NO through metabolic pathways, sodium nitroprusside (SNP), S-nitrosothiols and NONOates are three classes of NO donor compounds currently in use. SNP is an ion-nitrosyl with strong NO character. S-nitrosothiols decompose in aqueous solution into nitric oxide and thiiyl radicals. Electrophilic attack by NO on various nucleophiles (such as primary and secondary amines, polyamines, oxides and sulfites) produces isolatable adducts (NONOates). These types of compounds decompose spontaneously in aqueous solution and generate NO.

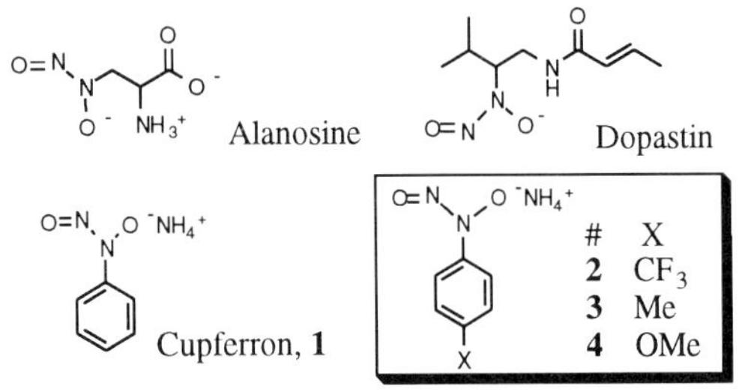

Alanosine Dopastin

Cupferron, 1

#	X
2	CF_3
3	Me
4	OMe

N-hydroxyl-N-nitrosamine compounds have gained more attention because of their NO releasing ability [1]. Natural product alanosine is a potential anticancer drug[2-3] and dopastin is a dopamine β-hydroxylase inhibitor [4]. Synthetic compound cupferron (**1**) is commercially available. Keefer and coworker have demonstrated that certain N-hydroxyl-N-nitroso compounds decompose under physiological conditions to release NO and suggested that this class of compounds could be used as antihypertensive agents [5]. Here, we report the substitution effect on the generation of NO from the N-hydroxyl-N-nitrosamines.

Scheme 1. Generation of NO by Electro or Enzymatic Oxidation

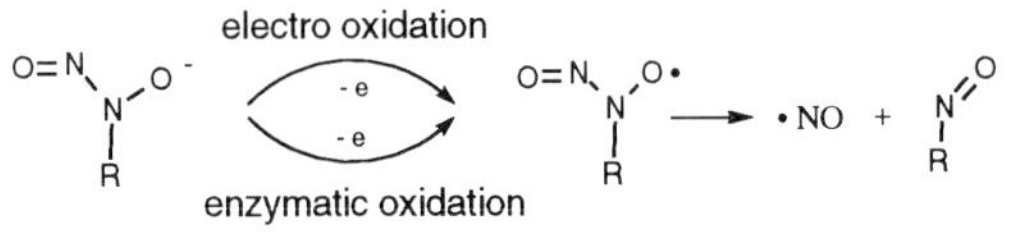

electro oxidation

enzymatic oxidation

We adopted a general route for the synthesis of compounds **2** - **4**. The N-hydroxylamines were readily synthesized from reduction of the corresponding nitro compounds and **2** - **4** were obtained by nitrosation of the N-hydroxylamines. We found that **1** - **3** were stable in 50 mM sodium phosphate buffer (pH 7.0). Compound **4** exhibted considerable extent of decomposition over a long period of time.

Scheme 2. Synthesis of Substituted N-Hydroxyl-N-nitrosamines

Zn EtONO NH_3

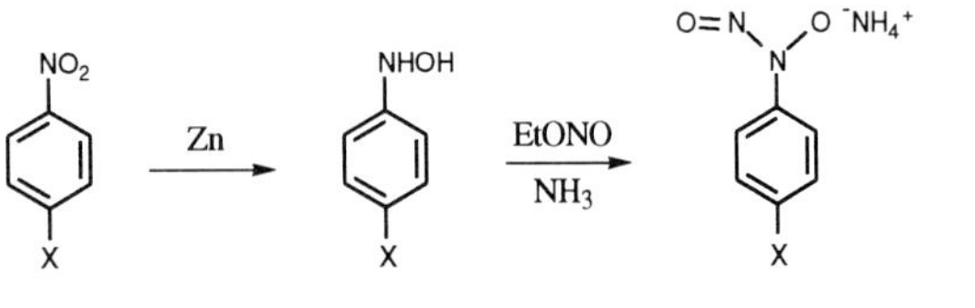

We studied the enzymatic oxidation of these N-hydroxyl-N-nitrosamines with horseradish peroxidase. After a long period of time, compound **4** was completely oxidized into 4-hydroxyl-nitrosobenzne, while **1** and **3** were only partically oxidized and **2** was slightly oxidized. At the same substrate concentration, electron-donating substitution made the nitrosamine easier to be oxidased to give NO. Compound **4** has about 10 times more activity than **2**.

Table I. Initial oxidation rate of **1** - **4** with peroxidase[a]

	conc. of **1** - **4**	90 m	175 mM
#	substitition	υ_0 ($M^{-1}s^{-1}x10^6$)	υ_0
2	CF_3	0.2	0.3
1	H	0.8	0.032[b]
3	Me	1.1	2.2
4	OMe	1.9	3.3

[a]The assay was done in 50 mM potassium phosphate buffer containing 10 mM H_2O_2, 114 mM of peroxidase at pH 7.0 and 25 °C. [b]Too high absorption at this concentration affected the accuracy of the measurement.

The substitution effect was further verified by electrochemical studies on **1** - **4** [6]. Generation of nitric oxide from the eletro-oxidation was detected by methemoglobin spectrophotometry assay [7]. Compound **4** had the lowest peak potential (**Table II**). Electrochemical oxidation / reduction of **2** showed a small extent of reversibility. It is clear that electron-donating substitution on the phenyl ring decreases the peak potential.

Since the stability, oxidation potentials and enzymatic kinetics can be readily modified by substitution on the phenyl ring, this class of compounds is the first electrochemical-NO-generating system with variable oxidation potentials. A micro-NO-generating electrode can be assembled for a variety of biomedical applications. These redox-variable compounds can also be used to produce NO by oxidation *in vivo* according to different "oxidative stress" of cellular milieu.

Table II. Half-wave potentail, peak-to-peak potential, and peak potentials for **1** - **4**.

-X	$-CF_3$[b]	-H[a]	-Me[a]	-OMe[a]
E	+0.54 (0.49)[a]	+0.62	+0.60	+0.49
$E_{(cathodic)}$	-0.15	-0.32	-0.35	-0.35
$E_{(anodic)}$	+0.05	+0.09	+0.16	+0.01

[a]Peak potential. [b]Half-wave potential.

Acknowledgment

This work was partially supported by a research grant (F95UM-2) from The American Cancer Society, Florida Division, Inc.

References

1. Alston, T. A., Porter, D. J. T. and Bright, H. J. (1985) J. Biol. Chem. **260**, 4069-4074.
2. Murthy, Y. K. S., Thiemann, J. E., Coronelli, C. and Sensi, P. (1966) Nature **211**, 1198-1199.
3. Fumarola, D. (1970) Pharmacology **3**, 215.
4. Iinuma, H., Takeuchi, T., Kondo, S., Matsuzaki, H., Umezawa, H. and Ohno, M. (1972) J. Antibiot. **25**, 497-499.
5. Keefer, L. U.S. Pat. Appl. (US 423279 A0 900301).
6. Lawless, J. G. and Hawley, M. D. (1968) Anal. Chem. **40**, 948-951.
7. Archer, S. (1993) FASEB J. **7**, 349-360.

Synthesis of a novel series of biologically active *S*-nitrosothiols

ANTHONY R. BUTLER and HAITHAM H. AL-SA'DONI

School of Chemistry, University of St. Andrews, St. Andrews, Fife KY16 9ST, Scotland, UK

Introduction.

S-Nitrosothiols (RSNO) are an important class of NO-donor drugs. They decompose giving NO and a disulfide:

$$2\ \mathrm{RSNO} \longrightarrow \mathrm{RSSR} + 2\ \mathrm{NO}$$

They have been used clinically and occur naturally where they may have a role in smooth muscle relaxation and in the prevention of platelet aggregation. The most extensively examined nitrosothiols are *S*-nitroso-*N*-acetyl-D,L-β,β-dimethylcysteine(SNAP) and *S*-nitroso-L-glutathione(GSNO). We set out to extend the range of compounds of this type and to look for a correlation between chemical structure and biological activity. The situation has been complicated by the recent discovery that the main route for the release of NO from a nitrosothiol is a copper(I)-catalysed process(A. R. Butler, D. L. H. Williams, H. H. Al-Sa'doni, S. C. Askew, A. P. Dick and H. R. Swift; unpublished results). Therefore we examined the effect of copper ions on the stability of the compounds we synthesised.

Materials and Methods.

SNAP(**1**) and GSNO(**13**) were prepared by literature methods[1, 2]. The dipeptides were prepared by coupling *N*-acetyl-D,L-β,β-dimethylcysteine with the methyl esters of a number of amino acids. The coupling agent was *N*-cyclohexyl-*N'*-2-[*N*-morpholino]ethylcarbodiimide metho-*p*-toluenesulphonate and we found that using a *N*-acetyl-D,L-β,β-dimethylcysteine derivative, it was unnecessary to protect the thiol group during the reaction. The thiol group was nitrosated by the use of *t*-butyl nitrite. For the kinetic studies NO release was monitored by observing the decrease in absorbance at the λ_{max} (normally 339 nm). The vasodilator effect of nitrosothiols was examined using a length of pre-contracted isolated rat tail artery. The drug was delivered either by bolus injection or as an internal prefusate. The prevention of platelet aggregation was studied using a Platelet Aggregation Profiler. For the detection of NO a WP NO sensor was used.

Results and Discussion.

The *S*-nitrosothiols were prepared according to the following scheme:

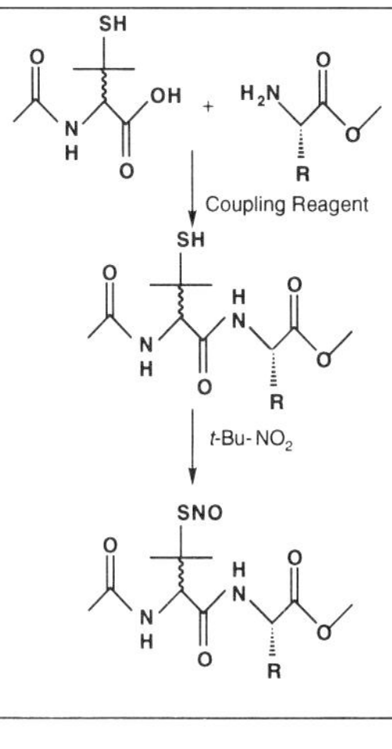

SH ... OH + H_2N ... R

↓ Coupling Reagent

SH ... R

↓ *t*-Bu-NO_2

SNO ... R

The new compounds (**2**-**12**) prepared are shown in the figure below:

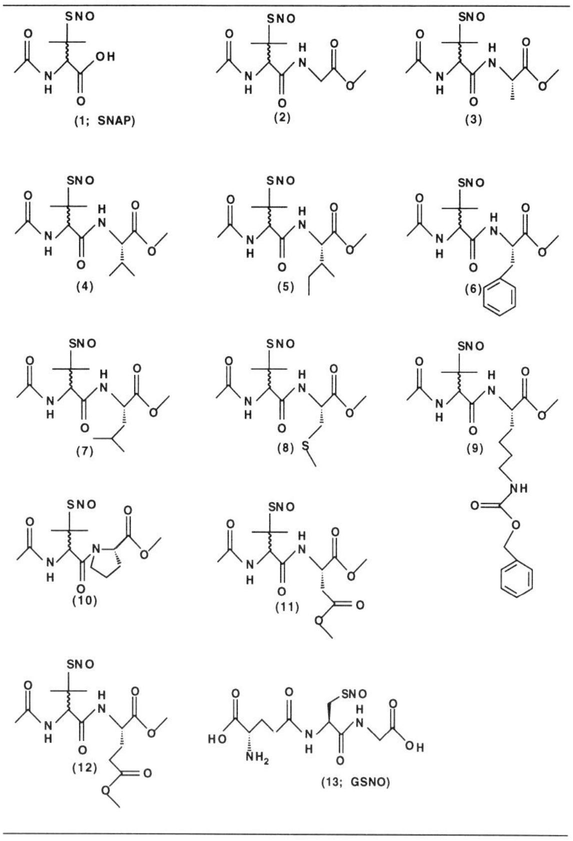

(1; SNAP) (2) (3) (4) (5) (6) (7) (8) (9) (10) (11) (12) (13; GSNO)

The copper catalysed release of NO from nitrosothiols is known((A. R. Butler, D. L. H. Williams, H. H. Al-Sa'doni, S. C. Askew, A. P. Dick and H. R. Swift; unpublished results) to occur by the following mechanism:

$$RS^- + Cu^{++} \longrightarrow RS\cdot + Cu^+ \quad \text{(i)}$$

$$RS\cdot \longrightarrow 1/2\ RSSR$$

$$RS^- + Cu^+ \longrightarrow RS^- + Cu^{++} \quad \text{(ii)}$$

$$RS^- + Cu^{++} \longrightarrow RS\cdot + Cu^+ \quad \text{(iii)}$$

$$RS\cdot \longrightarrow 1/2\ RSSR$$

All the new compound examined show less susceptibility to copper(I)-catalysed release of NO than SNAP but are more reactive than GSNO. However, they are all active as vasodilator[3] and in the prevention of platelet aggregation(A. R. Butler, H. H. Al-Sa'doni, A. J. Bancroft and M. McLaren ; unpublished results).

As a class they combine the favoured property of chemical stability with a high level of biological activity.

References

1. Field, L., Dilts, R. V., Ravichandran, R. , Lenhert, P. G. and Carnahan, G. E. (1978) J. C. S. Chem. Comm. , **6**, 249-250.
2. Hart, T. W. (1985) Tetrahedron letters, **26**, 2023-2016.
3. Butler, A. R., Al-Sa'doni, H. H., Megson, I. L. and Flitney, F. W.(this volume).

Oxygen consumption and reduction of nitro blue tetrazolium by NO-donors SIN-1 and GEA 3162 in aqueous solutions

PÄIVI HOLM, HANNU KANKAANRANTA, EEVA MOILANEN and TIMO METSÄ-KETELÄ*

University of Tampere, Medical School, Department of Pharmacology, Clinical Pharmacology and Toxicology, P.O.Box 607, FIN-33101 Tampere, Finland.

*Deceased

SIN-1 is a NO-releasing compound belonging to sydnonimines and is known to release NO spontaneously in aqueous solutions by an oxygen-dependent mechanism. In this process both superoxide anion and NO are released [1]. Superoxide reacts rapidly with NO leading to formation of peroxynitrite which is a strong oxidant and highly toxic [2]. This reaction serves also as an inactivation mechanism for both of these radicals [3].

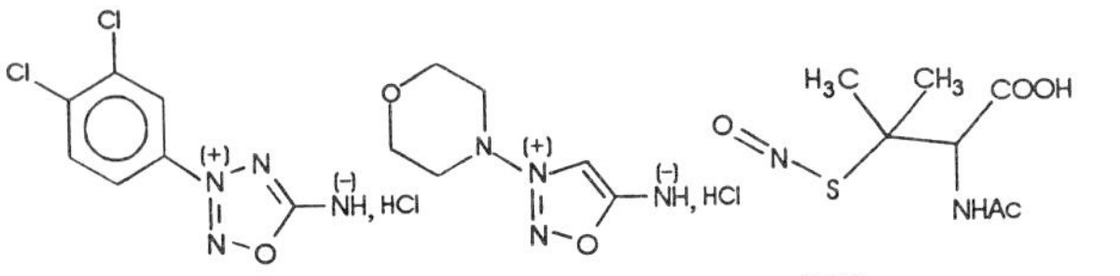

Figure 1. The chemical structures of GEA 3162, SIN-1 and SNAP.

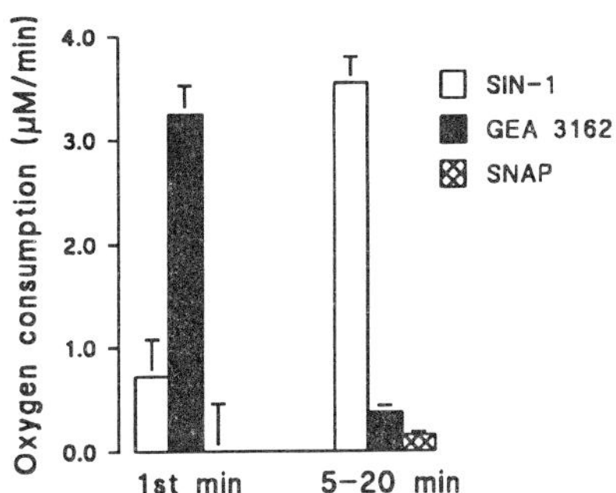

Figure 2. The effects of NO-donors (500 μM) on oxygen consumption in 10 mM phosphate buffer (pH 7.4, 25°C). Mean ± SE, n= 6.

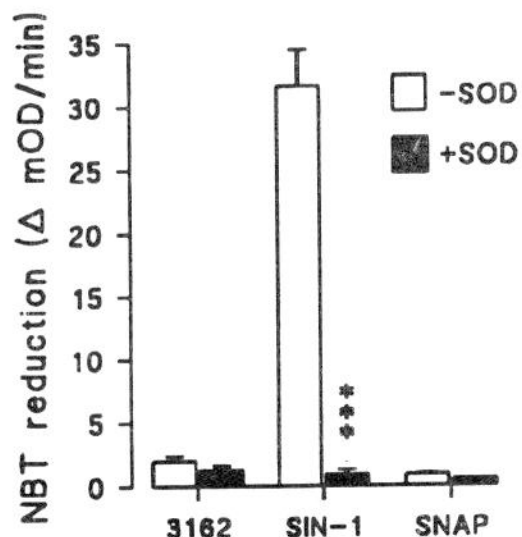

Figure 3.The effects of NO-donors (500 μM) on NBT reduction in 50mM phosphate buffer (pH 7.7, 37°C). Mean ± SE, n= 3-4, *** indicates $p < 0.001$.

Recently, a new group of NO-donors was synthesized by GEA Ltd, (Copenhagen, Denmark) (Figure 1). The mechanism of NO release from these 3-aryl substituted oxatriazole-5-imines is not known in detail [4,5]. In the present study, we measured oxygen consumption and reduction of nitro blue tetrazolium (NBT) as a marker of superoxide production in solutions of GEA 3162, SIN-1 and SNAP.

The oxygen consumption was measured by a Clark electrode at 37 °C. The test compounds were first dissolved in DMSO and diluted 1:10 in 10 mM phosphate buffer, pH 7.4. Results are shown in figure 2. GEA 3162 and SIN-1 reduced the oxygen content of the buffer but showed different kinetics. The effect of GEA 3162 was most pronounced during the first minute of incubation while SIN-1 consumed oxygen at highest rate during the later phase (5-20 min). SNAP did not significantly consume oxygen during the 20 min incubation time. The difference in the kinetics of NO release from GEA 3162 and SIN-1 [4] fits with the present data of oxygen consumption.

The generation of superoxide by NO-donors was measured by the superoxide dismutase-inhibitable reduction of NBT [1]. NBT reduction was monitored at 560 nm for 20 min. As expected, SIN-1 was able to reduce NBT and the effect was inhibited by SOD (100 U/ml). GEA 3162 and SNAP were very weak reductants and SOD did not significantly affect the response (Figure 3).

Abbrevations used: NBT (nitro blue tetrazolium), NO (nitric oxide), SIN-1 (3-morpholino-sydnonimine), SNAP (S-nitroso-*N*-acetylpenicillamine), SOD (superoxide dismutase)

Formation of peroxynitrite in solutions of SIN-1 and GEA 3162 was estimated indirectly by measuring production of nitrotyrosine [6]. Nitrotyrosine was produced in the presence of SIN-1 but not in solutions of GEA 3162. This suggests that peroxynitrite is not formed as a break-down product of GEA 3162 (data not shown).

In conclusion, the different rate of oxygen consumption by GEA 3162 and SIN-1 in aqueous solutions is comparable with the different kinetics of NO release from these compounds. The weak ability of GEA 3162 to reduce NBT and induce formation of nitrotyrosine as compared with SIN-1 suggest that GEA 3162 does not release O_2^- simultaneously with NO.

References:

[1] Feelisch,M., Ostrowski,J. and Noack,E. (1989). On the mechanism of NO release from sydnonimines. J.Cardiovasc.Pharmacol.14(Suppl 11),S13-S22.

[2] Pryor, W.A. and Squadrito G.L. (1995) The chemistry of peroxynitrite: a product from the reaction of nitric oxide with superoxide. Am.J.Physiol. **268**,L699-L722.

[3] Gryglewski,R.J., Palmer,R.M.J. and Moncada, S. (1986) Superoxide anion is involved in the breakdown of endothelium-derived vascular relaxing factor. Nature **320**,454-455.

[4] Kankaanranta,H., Rydell,E., Petersson,A.-S. et al. (1996). Nitric oxide-donating properties of mesoionic 3-aryl substituted oxatriazole-5-imine derivatives. Br.J.Pharmacol.(in press).

[5] Karup,G., Preikschat,H., Wilhelmsen,E.S. et al. (1994). Mesoionic oxatriazole derivatives - a new group of NO-donors. Pol.J.Pharmacol. **46**,541-552.

[6] Kaur,H. and Halliwell,B. (1994) Evidence for nitric oxide-mediated oxidative damage in chronic inflammation. Nitrotyrosine in serum and synovial fluid from rheumatoid patients. FEBS Lett. **350**,9-12.

Non-intermittent long-term administration of pentaerithrityl-tetranitrate results in unexpected, tolerance-devoid coronary- and venodilation

EBERHARD BASSENGE*, DIRK STALLEICKEN+ and BRUNO FINK*

*Institute of Applied Physiology, 79104 Freiburg, Germany; +ISIS-Pharma, 08002 Zwickau, Germany

Tolerance to nitrates is a multifactorial phenomenon which is difficult to analyze and to quantitate. Nitrovasodilators/NO-donors of various structures and compositions are associated with different degrees of tolerance which affect specific vessel sections in a different mode. Most nitrovasodilators induce tolerance within 2-3 days (e.g. nitroglycerin = NTG) during long-term therapy.

Aim of the study

To analyze pentaerithrityl-tetranitrate (PETN) induced dilator responses in different vascular sections such as the coronary and the peripheral vascular bed. We tested the NO-donor PETN (which releases 3 vasoactive metabolites: PE-trinitrate, PE-dinitrate, PE-mononitrate with different durations af action) in a chronically instrumented canine tolerance model, in which non-intermittent administration of NTG (1,5 μg/kg/min) and similar nitrovasodilators induced complete tolerance after 2 days.

Methods

PETN was administered orally 3mg/kg, 4 x daily for 5 days to dogs (24 to 30 kg, n = 5) instrumented to measure chronically coronary flow and -diameters as well as end-diastolic pressures and other hemodynamic parameters. Perivascular piezoelectric crystals were used for continuous coronary diameter recordings and an implantable pressure transducer (Konigsberg) to monitor left ventricular end-diastolic pressures (LVEDP). The technique is described elsewhere [1, 2]. In all dogs left circumflex coronary diameter (CD_{LC}), LVEDP, mean arterial pressure (MAP), and heart rate (HR) were evaluated on day 0 under control conditions and during 5 following days of non-intermittent, long-term administration of PETN as well as 4h, and 24h after the administration had been discontinued (i. e. day 5 and 6).

Results

As shown in fig. 1 the diameters of large coronary arteries increased significantly ($p<0.01$) from 1.82 ± 0.16 to 2.02 ± 0.17 mm 90 min after the first administration of PETN and remained in the dilated state throughout the whole 5-day period.

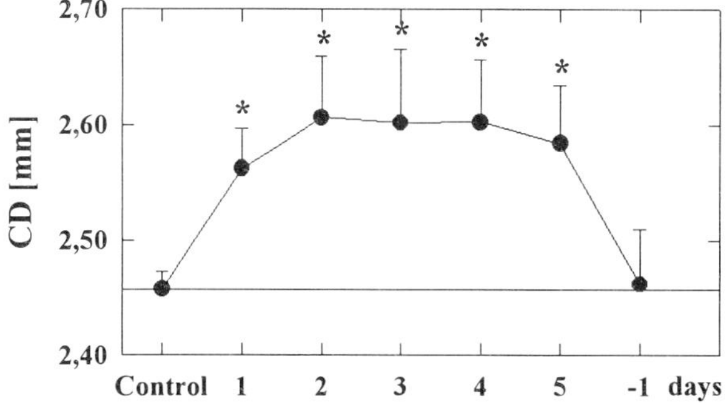

Fig. 1: Large coronary artery dilator effects induced by 5 days chronic administration of PETN (4 x 3 mg/kg daily per os). Data are mean ± SEM (n=5) . *-p < 0.05 control vs. data of chronic administration.

Hemodynamic measurements were carried out twice daily immediately before PETN-administration. LVEDP fell significantly ($p<0.01$) from 6.2 ± 0.6 to 3.4 ± 0.2 mm Hg 90 min after the first administration of PETN and remained at this reduced level for 5 following days of PETN administration (fig. 2).

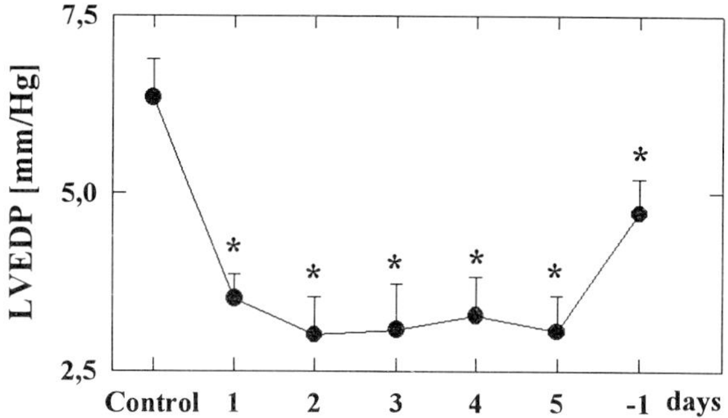

Fig. 2: Long-term reductions in LVEDP as an index of venodilation induced by non-intermittent administration of PETN (4 x 3 mg/kg daily per os). Data are mean ± SEM (n=5). *-p < 0.05 control vs. data of chronic administration.

Heart rates increased from 81 ± 4 to 92 ± 7 min^{-1} and mean arterial blood pressure fell from 98 ± 3 to 93 ± 4 mm Hg (no statistical significance) throughout the 5-day period. Dose-response curves of CD, LVEDP and MAP (short term i.v. infusions of PETN from 0.03 μg/kg/min to 30.0 μg/kg/min) obtained before and at the end of PETN administration were not shifted to higher concentrations (to the right) significantly.

Discussion and conclusions

In contrast to various continuously infused nitrates into the same animal model, PETN remained effective as a long-term vasodilator releasing 3 vasoactive metabolites [3]. Coronary arteries and the venous bed were dilated for the whole observation period of 5 days without any evidence of tolerance. This remarkable difference compared to other organic nitrates studied with the same experimental approach so far is unexpected and difficult to explain, since the mode of action of PETN or NTG to bring about cGMP-mediated large artery dilation and venodilation is very similar [4].

The explanation is probably based on the differences in redox-potentials triggering NO and NO_2 release from PETN and other organic nitrates, as well as differences in the rates of NO and NO_2 release which can be responsible for different rates of peroxynitrite production associated with administration of organic nitrates. Peroxynitrite can provoke the oxidation of sulfhydryl groups of specific enzymes involved in cGMP formation in the vasculature. Inactivation of these enzymes due to changes in protein conformation of enzymes induced by disulfide bridges are likely to occur. This effect can be particularly destructive in case of soluble guanylyl cyclase (sGC), since the heterodimeric structure is necessary for the activation by NO [5]. Nitrosation of tyrosine residues associated with complete inactivation of various kinases may also become relevant when peroxynitrite is formed in the vasculature. Thus, a substantial deficit in cGMP formation can be observed even when sGC and NO-mediated activation of sGC is not affected.

Thus in a subsequent series of experiments more detailed data on the concomitant peroxynitrite formation in the vasculature induced by various nitrovasodilators (exogenous NO-donors) are necessary.

1 Holtz, J., Giesler, M. and Bassenge, E. (1983) Z.Kardiol. 72, 98-106
2 Stewart, D.J., Elsner, D. Sommer, O., Holtz, J. and Bassenge, E. (1986) Circulation 74,573-582
3 Bassenge, E. (1995) in Pentaerithrityltetranitrat (Schneider, H.T. and Stalleicken, D., ed.), pp.54-65, Steinkopff, Darmstadt
4 Bassenge, E. (1994) Cardiovasc. Drugs and Therapy 8, 601-610
5 Skatchkov, M., Sommer, O., Larina, L., Vanin, A. and Bassenge, E. (1995) J.Mol.Cell Cardiol. 27, A93

A copper-mediated mechanism in the anti-platelet action of S-nitrosothiols.

MICHAEL P GORDGE, JOHN S HOTHERSALL, GUY H NEILD and ALBERTO A NORONHA-DUTRA

Institute of Urology and Nephrology, UCL, London W1N 8AA, UK

Introduction

RSNOs are adducts of NO with sulphydryl (-SH) groups on carrier molecules. RSNOs are potentially useful as tissue-specific NO donors, for example GSNO has been used as a platelet-selective agent (1). The explanation for this target-selectivity is unknown.

RSNO formation occurs at neutral pH under aerobic conditions (2) and in human blood and tissues, RSNOs may represent a storage or transport pool of biologically active NO (3). The mechanism by which RSNOs donate NO to target cells is poorly understood, but RSNO breakdown may be catalysed by membrane-associated enzyme(s) (4). Cleavage of GSNO by γGT has been proposed (5).

We have previously demonstrated (6) that the anti-platelet actions of GSNO and S-nitrosocysteine were reduced by the copper (I) specific chelator BCS, and that this inhibition was not explained by prevention of copper-catalysed RSNO breakdown in solution. We have now extended our investigations into the mechanism of action of RSNOs

Materials and Methods

Measurement of inhibition of platelet aggregation by NO donors and the effect of copper chelation with BCS

NO donors were added 15 seconds prior to induction of thrombin-induced platelet aggregation. Percentage inhibition of aggregation was calculated by comparison with responses in the absence of NO donor. Experiments were also performed in the presence of BCS (10μM).

Effect of BCS on the actions of different NO donors

The effect of BCS on the platelet inhibitory activity of (a) non-RSNO compounds: NO, $ONOO^-$ and NaNP, and (b) RSNO compounds: SNAP, GSNO and TASNO, was assessed as described above.

Comparison of the potency of GSNO and TASNO.

Aggregation inhibition by GSNO and the membrane permeable compound TASNO (both 10^{-9} -10^{-4}M) was compared.

Effect of platelet pre-treatment with BCS on the action of GSNO

Platelets were treated with BCS (100μM), washed and resuspended in BCS-free buffer. The action of GSNO (10^{-9} - 10^{-5} M) was then compared in pre-treated and untreated platelets.

Effect of the γGT inhibitor, acivicin, on the action of GSNO

Platelets were incubated with acivicin (1mM) for 20 minutes prior to measurement of action of GSNO (10^{-9} - 10^{-5} M).

Results

Effect of BCS on platelet aggregation inhibition by NO donors.

BCS significantly reduced the action of the RSNO compounds SNAP, GSNO and TASNO (all 1 μM). The actions of NO (1 μM), $ONOO^-$ (100μM) and NaNP (10 μM) were unchanged (Fig 1).

Comparison of the potency of GSNO and TASNO

GSNO was a more potent inhibitor of platelet aggregation than TASNO ($p<0.001$)(data not shown).

Abbreviations used: S-nitrosothiols-RSNOs; S-nitrosoglutathione-GSNO; γ-glutamyl transpeptidase-γGT; bathocuproine disulphonic acid-BCS; S-nitroso-N-acetyl-penicillamine-SNAP; S-nitrosothioctic acid-TASNO.

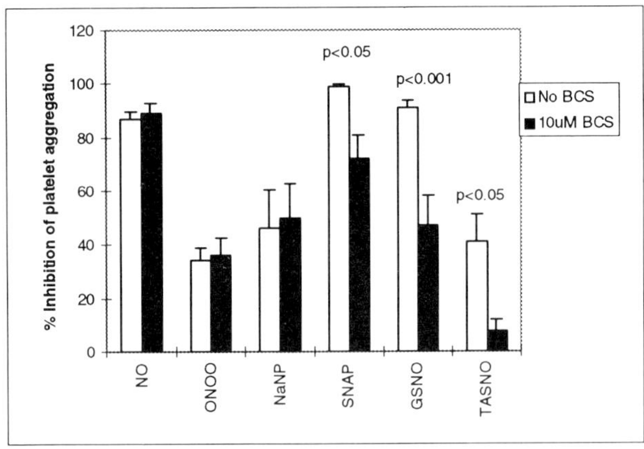

Fig 1. Effect of BCS (10μM) on platelet aggregation inhibition by different NO donors. Values are mean (SEM) from 3-9 experiments.

Effect of pre-treatment of platelets with BCS on the action of GSNO

The anti-aggregatory action of GSNO was significantly inhibited using BCS-pre-treated platelets, compared with untreated platelets ($p<0.05$) (data not shown).

Effect of the γGT inhibitor, acivicin, on the action of GSNO

Platelets treated with acivicin (1mM) showed no significant change in response to the anti-aggregatory action of GSNO.

Discussion

The copper (I) chelator BCS reduced the platelet inhibitory actions of the RSNO compounds tested but showed no effect on non-RSNO-type NO donors. The low potency of TASNO, indicates that RSNO action is independent of the membrane permeability of the carrier molecule. Resistance of BCS-treated platelets to the action of GSNO, even when resuspended in BCS-free buffer, suggests that BCS must act by binding to a platelet-related structure. Breakdown of GSNO by γGT cannot be an important pathway for the transfer of NO to platelets, since treatment of target platelets with acivicin failed to block the action of GSNO.

A mechanism specific for RSNO-type NO donors, and requiring platelet associated copper, is involved in mediating their anti-platelet activity. This may be a membrane receptor/enzyme responsible for processing and targeting of NO from RSNO donors. The enzyme is not γGT. Differential expression of such an enzyme/receptor would confer tissue-selectivity on the actions of RSNOs. Understanding this system will be important in allowing development of novel therapeutic agents for conditions such as hypertension, thrombosis and atherosclerosis.

Acknowledgements

We thank the St Peter's Trust for supporting this project.

References

1. Langford E.J., Brown, A.S., Wainwright, R.J., et al. (1994) *Lancet* **344**:1458-1460.

2. Wink D.A., Nims R.W., Darbyshire J.F., et al. (1994) *Chem Res Toxicol* **7**: 519-525.

3. Stamler J.S., Jaraki O., Osborne J. et al. (1992) *Proc Natl Acad Sci USA* **89:** 7674-7677.

4. Kowaluk, E. A. and Fung H. (1990) *J Pharmacol Exp Ther* **256**:1256-1264.

5.Askew,S.C., Butler,A.R., Flitney,F.W., Kemp,G.D. and Megson,I.L. (1995) *Bioorganic and Medicinal Chemistry* **3**:1-9.

6. Gordge,M.P., Meyer,D., Hothersall,J.S., Neild,G.H., Payne,N., and Noronha-Dutra, A. (1995). *Br J Pharmacol.* **114**:1083-1089.

Vasodilator properties of some novel *S*-nitrosated dipeptides: comparative chemical and pharmacological studies.

HAITHAM H. AL-SA'DONI*, ANTHONY R. BUTLER*, IAN L. MEGSON# and FREDERICK W. FLITNEY#

Schools of Chemistry* and Biological & Medical Sciences#, University of St. Andrews, St. Andrews, Fife, KY16 9ST, Scotland, UK.

Introduction.

S-Nitrosothiols (RS-NO) are relatively unstable compounds and break down thermally, photochemically [1] and in a metal ion catalysed process [2] to give a disulphide and nitric oxide (NO):

$$2\ RSNO \longrightarrow RSSR + 2\ NO$$

These compounds are of interest because of their potent pharmacological properties and possible physiological role in smooth muscle relaxation and inhibition of platelet aggregation [3]. Endothelium-derived relaxing factor, or EDRF, has pharmacological properties identical to NO or a closely related compound, such as an *S*-nitrosothiol [4]. We have synthesised *S*-nitroso-*N*-acetyl-D,L-β,β-dimethylcysteine (**1**; SNAP) and three novel *S*-nitrosated dipeptides, *S*-nitroso-*N*-acetyl-D,L-β,β-dimethylcysteinylglycine methyl ester (**2**), *S*-nitroso-*N*-acetyl-D,L-β,β-dimethylcysteinyl-L-alanine methyl ester (**3**) and *S*-nitroso-*N*-acetyl-D,L-β,β-dimethylcysteinyl-L-valine methyl ester (**4**). The vasodilator properties of compounds **2-4** were compared with that of SNAP to investigate the relationship of structure to chemical stability and biological activity.

Materials and Methods.

S-Nitrosothiols: SNAP was prepared according to the method of Field *et al* [5]. The synthesis of the *S*-nitrosated dipeptides (**2-4**) is described by Butler & Al-Sa'doni [6]:

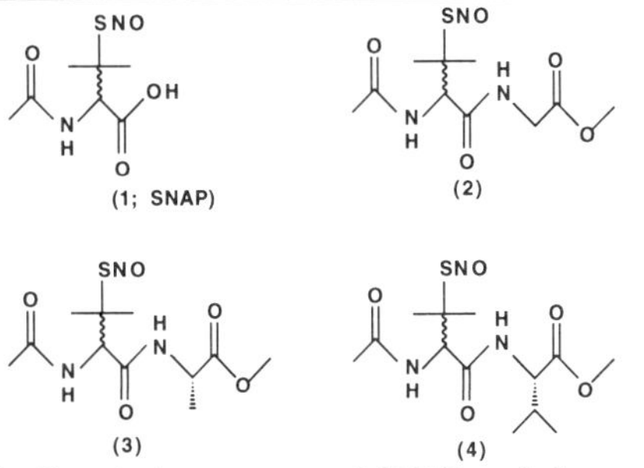

(1; SNAP) (2) (3) (4)

Figure 1: Chemical structures of SNAP and three novel *S*-nitrosated dipeptides

Rat Tail Artery Bioassays: Male Wistar rats (300-400g) were killed by cervical dislocation and their tails removed. The tail artery was dissected free, cannulated, and perfused internally with Krebs solution (37°C) at a constant flow rate (2ml min^{-1}). Vessels were precontracted with phenylephrine hydrochloride (PE; 1-7μM), generating perfusion pressures of 100-120mmHg.

Results and Discussion.

The biological activity of *S*-nitrosothiols depends upon the structure of R [7]. Thus, changing R results in a new compound with different chemical and physiological properties.

The chemical stability of the *S*-nitrosated dipeptides was monitored spectrophotometrically (at λ_{max} for -SNO group; 30°C and pH 7.4). All four compounds decomposed to the corresponding disulphide and NO, but the *S*-nitrosated dipeptides were found to be more stable than SNAP. The addition of cysteine, a transnitrosating agent, accelerated the rate of decomposition of all compounds.

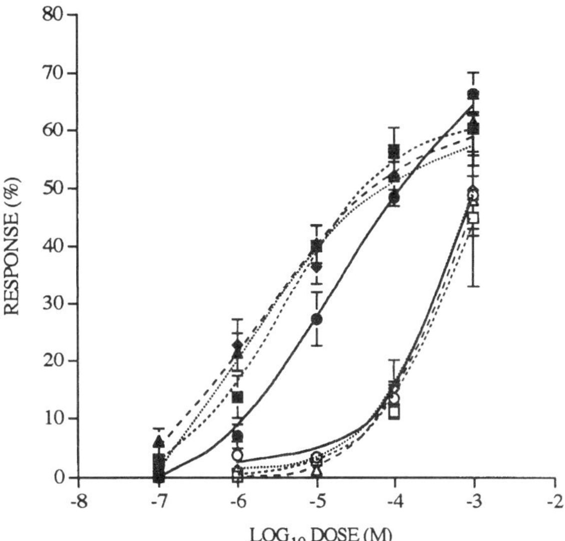

Figure 2: Log dose-response curves comparing the vasodilator effects of SNAP (filled circles), **2** (filled squares), **3** (filled triangles), and **4** (filled diamonds). Open symbols shows responses in presence of oxyhaemoglobin (15μM).

Figure 2 shows log-dose response curves resulting from bolus injections of compounds **1-4** into pre-contracted tail arteries. Oxyhaemoglobin (Hb; 15μM) significantly attenuated responses to all four compounds ($P<0.01$ for all doses of compounds **1-4** when compared to control values using an unpaired Student's t-test). Hb is a well recognised NO scavenger and these results indicate that NO is the effector agent in vasodilation to all four compounds.

S-Nitrosated dipeptides are more effective vasodilators than SNAP (ED_{50}s of 4.1, 2.8, 3.7μM, respectively compared with 17.8μM for SNAP). SNAP responses were statistically significantly different for 10^{-5}M (**3**) and 10^{-6}M (**2**, **4**) doses even though the local chemical environment of the -SNO groups are almost identical. This finding is contrary to what might be expected on the basis of their chemical stabilities *in vitro* and may indicate that the *S*-nitrosated dipeptides interact with tissue components or show greater permeability than SNAP.

References

1. Williams, D. L. H. (1988) Nitrosation, 1st Ed. Cambridge University Press, 171-196.
2. McAninley, J. , Williams, D. L. H., Askew, S. C., Butler, A. R. and Russell, C. (1993) J. Chem. Soc., Chem. Commun., **23**, 1758-1759.
3. Ignarro, L. J., Lippton, H., Edwards, J. C., Baricos, W. H., Hyman, A. L., Kadowitz, P. J. and Gruetter C. A., (1981) J. Pharmacol. Exper. Ther., **218**, 739-749.
4. Myers, P. R., Minor, R. L. JR., Guerra, R. JR., Bates, J. N. and Harrison D. G.(1990) Nature **345**, 161-163. Rubanyi, G. M., Johns, A., Wilcox, D., Bates, F. N. and Harrison, D. (1991) Cardiovasc. Pharmacol., **17**, S41-S45.
5. Field, L., Dilts, R. V., Ravichandran, R., Lenhert, P. G. and Carnahan, G. E. (1978) J. C. S. Chem. Comm., 249-250.
6. Butler, A. R. and Al-Sa'doni, H. H. (this volume).
7. Mathews, W. R. and Kerr, S. W. (1993) J. Pharmacol. Exper. Ther., **3**, 1529-1537.

Role of reactive oxygen species in DNA damage and loss of viability induced by the nitric oxide donors SIN-1 and GSNO in an insulin-secreting cell line

Maria Di Matteo[1], Carol A Delaney[1*], Michael HL Green[2], Jillian E Lowe [2], and Irene C Green[1]

[1] Biochemistry Laboratory, School of Biological Sciences, [2] MRC Cell Mutation Unit, Sussex University, Brighton, BN1 9QG, UK

Introduction

SIN-1 (3-morpholinosydnonimine) is a useful compound to model the cytotoxic immune response because it combines with molecular oxygen to release nitric oxide, superoxide anion and H_2O_2 [1]. By removing superoxide with SOD (superoxide dismutase), and H_2O_2 with catalase, we can test interactions between reactive species, and assess the importance of peroxynitrite in DNA damage and cell killing [2]. All experiments have been performed in the insulin-secreting SV40-transformed hamster cell line HIT-T15.

Fig. 1 Nitric oxide and reactive oxygen species from **SIN-1** might interact to cause DNA damage in at least four ways

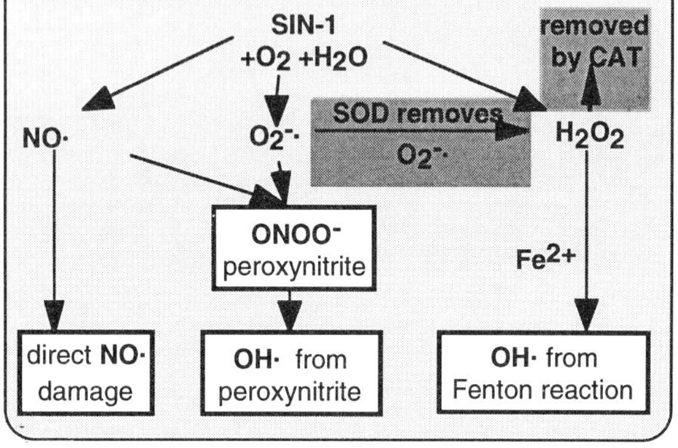

SIN-1
+O2 +H2O
removed by CAT
NO·
O2⁻·
SOD removes O2⁻·
H2O2
ONOO⁻ peroxynitrite
Fe2+
direct **NO·** damage
OH· from peroxynitrite
OH· from Fenton reaction

Results

Fig. 2. Effect of **SOD** and **catalase** on **SIN-1** and **GSNO** DNA strand breakage as measured by the comet assay

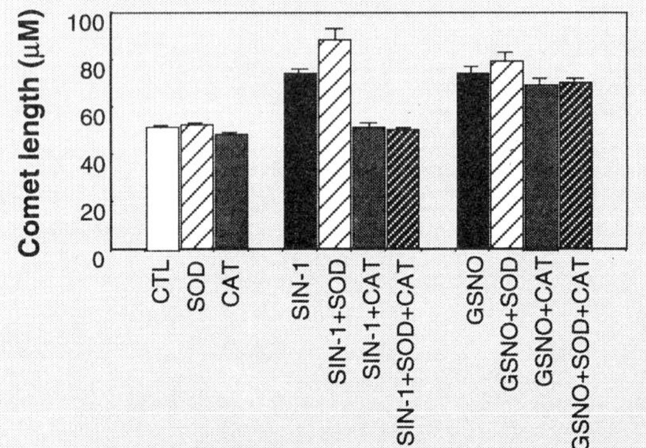

Comet assay performed as described previously [3]. Cells were embedded in agar, lysed, placed in alkali and an electric current applied. DNA with strand breaks migrated from the nucleus in a comet tail but undamaged DNA remained trapped. 60 min treatment; SIN-1, 100µM; GSNO, 300 µM; SOD, 200 U/ml; catalase, 100 U/ml; nose to tail comet length measured; values means of 3-6 independent experiments.

- **Superoxide dismutase** does not protect against **SIN-1**-induced DNA strand breakage by either donor.
- **Catalase** protects *completely* against **SIN-1** strand breakage.
- **SOD + catalase** give *no greater protection* against **SIN-1**-induced DNA strand breakage than **catalase** alone.
- **Catalase** does not protect against **GSNO** strand breakage.
- **Therefore SIN-1-induced DNA strand breakage is accounted for by H_2O_2, not peroxynitrite**

Fig. 3. Loss of viability as determined by the MTT assay

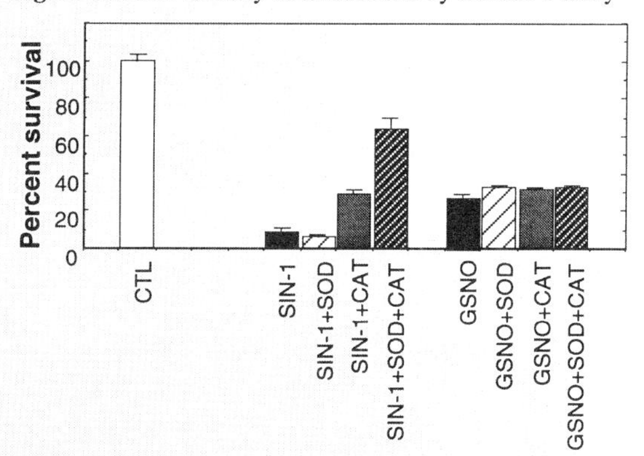

The cells were treated for 24 h with the nitric oxide donor in the presence or absence of SOD and catalase, then incubated 4 h with MTT [4]. SIN-1 1.5 mM; GSNO 0.5 mM; mean of 4 independent experiments.

- A different result was obtained when viability was determined by the MTT assay which measures loss of cellular reducing ability.
- Although **SOD** *alone* did not protect against **SIN-1**, it *did protect* when **catalase** was also present.
- **This suggests that *both* peroxynitrite and H_2O_2 are involved in cell killing by SIN-1, but peroxynitrite is *not* involved in DNA strand breakage.**
- **SIN-1 inhibition of DNA synthesis (24 h treatment) and of glucose-stimulated insulin secretion (60 min treatment) showed no protective effect of superoxide dismutase + catalase, thus resembling DNA strand breakage rather than cell killing.**

Conclusions

- **SIN-1** induces DNA damage in HIT-T15 cells mainly through hydrogen peroxide, not through peroxynitrite or nitric oxide.
  - *The evidence for this is that the effect of **SIN-1** in strand breakage is almost completely reversible by **catalase***
- Peroxynitrite is a less effective DNA damaging agent in HIT-T15 cells than OH· from H_2O_2.
  - *The evidence for this is that **SOD + catalase** has no greater protective effect than catalase alone against **SIN-1**-induced*
    - *DNA strand breakage*
    - *inhibition of DNA synthesis*
    - *inhibition of glucose-stimulated insulin secretion*
- There may be a role for peroxynitrite in decreasing cell 'viability' as determined by the MTT assay
  - *The evidence for this is that **SOD + catalase** give greater protection against **SIN-1** in the MTT assay than **catalase** alone*
- The action of **GSNO** is not affected directly by **SOD** or **catalase** and another mechanism appears to be involved

Acknowledgements

Work supported in part by British Diabetic Association, BBSRC and Commission of European Communities (EV5V-CT91-0004)

References

1. Feelisch, M., Ostrowski, J. and Noack, E. (1989) J. Cardiovasc. Pharmacol., **14 (suppl. 11),** S13-S22.
2. Beckman, J.S., Beckman, T.W., Chen, J., Marshall, P.A. and Freeman, B.A. (1990) Proc. Natl. Acad. Sci. USA, **87,** 1620-1624.
3. Delaney, C.A., Green, M.H.L., Lowe, J.E. and Green, I.C. (1993) FEBS Lett., **333,** 291-295.
4. Di Matteo, M., MacArthur, D., Delaney, C.A., Cunningham, J. and Green, I.C. (1994) Diabetologia, **37,** A52 abstract 200.

* *Current address: Department of Medical Cell Biology, Biomedicum, University of Uppsala, Box 571, S-751 23, Uppsala, Sweden*
Abbreviations: GSNO, S-nitrosoglutathione; SIN-1, 3-morpholinosydnonimine; SOD, superoxide dismutase

Effects of dietary antioxidants in rats on tolerance to glyceryl trinitrate *in vitro*.

DAVID W. LAIGHT and ERIK E. ÄNGGÅRD

William Harvey Research Institute, St. Bartholomews Hospital Medical College, Charterhouse Square, London, EC1M 6BQ, England

Oxygen free radicals may conceivably mediate cellular tolerance to organic nitrates by disabling the enzyme(s) important in bioactivation and in the case of superoxide anion, scavenging and thereby inactivating NO [1]. Antioxidants have previously been shown to diminish pharmacological tolerance to organic nitrates in the rabbit aorta *in vitro* [2]. Most recently, the endothelial production of superoxide anion has been proposed to largely account for tolerance to GTN in the aorta isolated from rabbits exposed to GTN *in vivo* [3]. The present study investigates the effects of dietary antioxidants probucol and vitamin E on cellular tolerance to GTN in the rat aorta *in vitro*.

Male Wistar rats were maintained on one of three diets: standard chow (control); chow supplemented with 0.5% w/w probucol; or chow supplemented with 0.5% w/w vitamin E (α-tocopherol acetate). After approximately 4 months, rats were anaesthetised with pentobarbitone sodium (60 mg/kg i.p.) and sacrificed. The thoracic aorta was removed after a thorectomy and aortic rings prepared and mounted in 10 ml organ baths in PSS warmed to 37°C and gassed with 95% O_2/ 5% CO_2. The composition of the PSS was (in mM): NaCl 133; NaH_2PO_4 1.35; KCl 4.7; $MgSO_4$ 0.61; $NaHCO_3$ 16.3; $CaCl_2$ 2.52; glucose 7.8. Rings were allowed to equilibrate for 1h. The tolerance protocol consisted of incubating rings with GTN (30μM or 100μM) for 30 min followed by washing for 30 min. Non-tolerant rings were incubated with GTN vehicle (0.9% w/v saline) in a similar manner. A concentration-response curve to GTN (1nM - 300μM) was then constructed in tolerant and non-tolerant rings submaximally precontracted with noradrenaline (100nM). Tolerance was assessed by inspecting pD_2 values and evaluating AUC.

The pD_2 values for GTN relaxation in non-tolerant rings were: control group 6.41±0.40 (n=4); probucol group 6.98±0.22 (n=5); vitamin E group 6.39±0.10 (n=5) (N.S.) (see Fig.1). The pD_2 and AUC values for GTN relaxation were significantly depressed in rings previously exposed to GTN (30μM or 100μM) compared to non-tolerant rings in all dietary groups. However, the pD_2 values for GTN relaxation were significantly elevated from 5.25±0.23 (n=4) in the control group to 6.44±0.16 (n=5, p<0.01) in the probucol group in 30μM GTN tolerant rings and from 4.92±0.36 (n=4) to 6.29±0.23 (n=5, p<0.01) in 100μM GTN tolerant rings. pD_2 values in the vitamin E group in 30μM and 100μM GTN tolerant rings were 5.33±0.08 (n=5) and 4.88±0.15 (n=5), respectively and not different from control group values. In addition, the AUC in 100μM GTN tolerant rings as a percentage of non-tolerant AUC was significantly increased in the probucol group (81.5±5.5%, n=5) compared to the control group (56.0±7.3%, n=4) and vitamin E group (56.9±6.0%, n=5) (P<0.05). The AUC in 30μM GTN tolerant rings as a percentage of non-tolerant AUC was apparently increased in the probucol group (82.4±2.7%, n=5) compared to the control group (72.0±6.1%, n=4) and vitamin E group (73.0±6.0%, n=5).

Abbreviations used: GTN=glyceryl trinitrate; PSS=physiological salt solution; AUC =area under curve; N.S.=not significant; s.e.m.=standard error of the mean

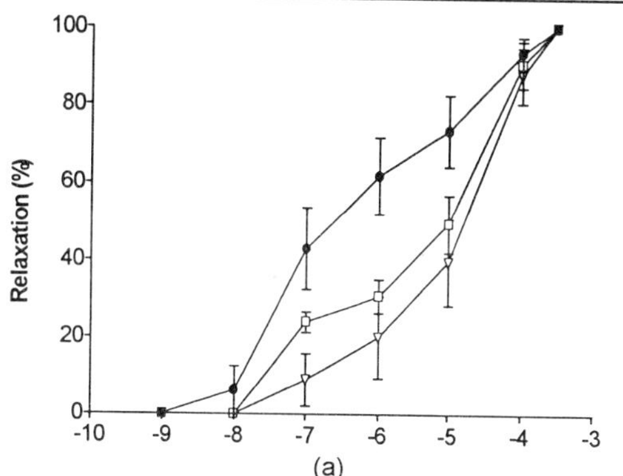

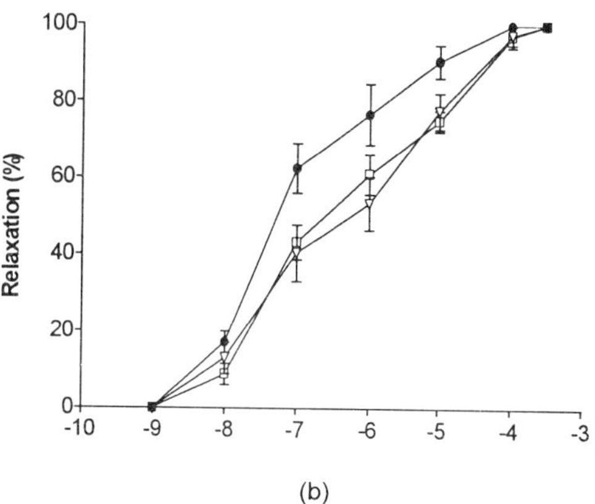

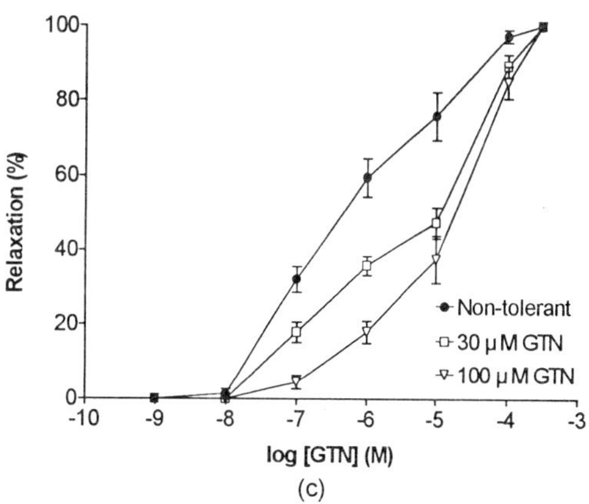

Figure 1. Relaxation of the rat aorta to GTN with or without previous exposure to GTN after (a) standard diet (n=4), (b) dietary probucol (n=5) and (c) dietary vitamin E (n=5). Values are mean±s.e.m.

In conclusion, dietary probucol partially prevents the development of pharmacological tolerance to GTN in rat aortic smooth muscle *in vitro*. This would support the notion that oxidative stress is involved in some aspects of tolerance to organic nitrates. However, the inefficacy of another dietary lipophylic antioxidant, vitamin E, may suggest an alternative action of probucol.

1. Gryglewski, R.J. (1987) Agents & Actions **22**, 351-352

2. Yeates, R.A. and Schmid, M. (1992) Arzneim.-Forsch./Drug Res. **42**, 297-302

3. Münzel, T., Sayegh, H. and Freeman, B.A. (1995) J. Clin. Invest. **95**, 187-194

Suppression of nitrate induced tolerance by vitamin C and other antioxidants

EBERHARD BASSENGE and BRUNO FINK

Institute of Appl. Physiology, Univ. of Freiburg, Germany

When analyzing induction of tolerance caused by a number of nitrovasodilators we observed that antioxidants were effective in suppressing the progressive decay of coronary and veno-dilator responses. Antioxidants may protect NO from immediate inactivation prior to the stabilization of an EDRF-like compound or by augmenting NO-action which is reduced in tolerance by a stepwise inactivation of enzymes involved in the reduction of nitrates to NO due to an additional, unfavourable production of oxidizing side-products of NO and NO_2.

Methods

Six adult mongrel dogs of either sex (24 to 30 kg weight) were chronically instrumented under sodium pentobarbital anaesthesia (25 mg/kg). The left circumflex coronary artery was instrumented with perivascular piezoelectric crystals for continuous diameter recordings and with implantable pressure transducers (Konigsberg) to monitor left ventricular end-diastolic pressures (LVEDP). The technique is described elsewhere [1, 2]. A silicone catheter was implanted into the pulmonary artery for long-term infusion of nitroglycerin (NTG) or of NTG along with vitamin C (Vit-C) as a coinfusion. NTG (1.5 µg/kg/min) was infused continuously for 5 days in the presence or absence of Vit-C (55 µg/kg/min) as an effective antioxidant. Coronary flow and diameters as well as LVEDP as an index of venodilator action were continuously monitored along with changes in platelet cGMP-levels: during short-term i.v. infusions (10 min) of increasing NTG dosages (dose response relationships, 0.15, - 15.0 µg/kg/min before, after 5 day continuous NTG or NTG/Vit-C infusion and 24 h after discontinuation of infusion).

Platelet isolation and determination of cyclic GMP

Blood was drawn from the carotid artery into citric acid solution (6:1, vol./vol.) consisting of (final concentration in mMol) sodium citrate 8, citric acid 3.3, glucose-monohydrate 12, iloprost (5 nMol), acetylsalicylic acid 0.025 and apyrase 0.24 U/ml. Platelet-rich plasma was obtained by centrifugation at 200 g for 10 minutes at room temperature after preincubation for 15 min. This fraction was centrifuged at 450 g for 15 min., and cells were washed two times in buffer (pH 6.4) containing (mMol/l) NaCl 136, KCl 2, $MgCl_2$ 2.4, glucose 5, glutathion 1, HEPES -10, EGTA 0.02, fatty acid-free bovine serum albumin (BSA, 0.1%). and prostacyclin analogon iloprost (PGE_1 0.1 nMol). After the last centrifugation platelets were finally resuspended in HEPES-buffer (pH 7.4) of the same composition without PGE_1, and BSA. Platelets were counted microscopically and adjusted to a final concentration of 120,000/µl in suspension. The suspension washed platelets (350 µl) was after addition of 50 µl of SIN-1 (100 µg/ml) incubated or the buffer alone for 1 min at 37°C. The reaction was terminated by adding 100 µl $HClO_4$. After neutralization of the samples with 145 µl 1.5 M Na_2CO_3 and centrifugation at 10,000 x g for 5 min, cyclic nucleotides were determined in the supernatants by radioimmunoassay [3].

Results

Continuous infusion of NTG, 1.5 µg/kg/min or NTG with coinfusion of Vit-C 55 µg/kg/min resulted in a pronounced vasodilation (up to 70 % of the maximal dilator response observed with short-term NTG infusions) of the left circumflex artery (LCX) diameter (tab.1).

Difference %:	day 1	day 2	day 3	day 4	day 5
CD without VC	10± 0.4	8±0.3	5±0.5	0±0.5	-1±0.2
CD with VC	11± 0.9	11± 0.4	11± 0.7	10± 0.3	10± 0.4
LVEDP without VC	-34±5	-34±6	-33±4	-31±3	-35±6
LVEDP with VC	-24±5	-27±6	-24±4	-27±7	-25±5

LCX diameters during continuous administration of NTG decreased progressively to reach after 3 day control values. These changes during continuous NTG infusion are characteristic for the development of tolerance. NTG/Vit-C coinfusion did not result in tolerance, but resulted in a steady-state level of coronary dilation maintained throughout the whole experimental period of 5 days. The interpretation of this phenomenon is based upon the measurements of NO-mediated cGMP-release by SIN-1 (Fig. 1).

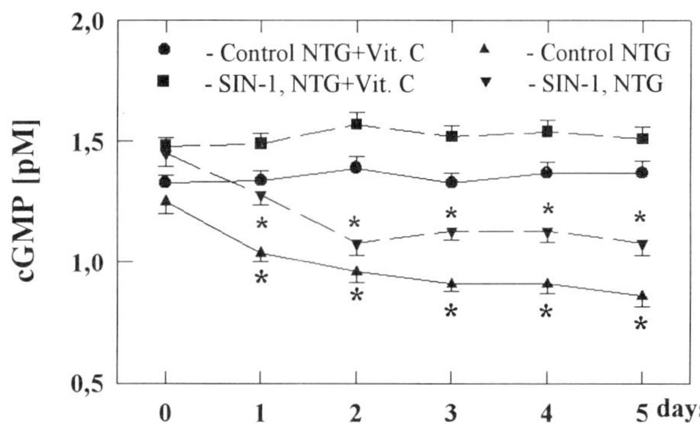

Fig. 1: cGMP-production of platelets during continuous administration of nitroglycerin with and without antioxidant protection. Administration of NTG (1.5 µg/kg/min i.v.), or NTG with Vit.C (55 µg/kg/min i.v.). Data from washed ex vivo platelets are mean ± SEM (n=4). * $p<0.05$ control day (0) vs. data after chronic administration. Stimulation of cGMP-release was obtained with SIN-1 (100 µg/ml).

It is obvious that the coinfusion of NTG and Vit-C prevents the suppression of the activity of soluble guanylyl cyclase. In addition it is important that no drop in coronary diameters was observed 24 h after the cessation of the coinfusion in contrast to the NTG infusion without antioxidants.

Discussion and Conclusions

Summarising our findings on the development of tolerance it is likely that an additional rapid transformation of NO into peroxynitrite by various mechanisms [4] is a decisive factor in the induction of tolerance to nitrates. The enhanced performation can be suppressed by additional antioxidant administration. Administration of Vi-C as antioxidant prevented the progressive loss of dilator responses to NTG throughout a 5-day infusion protocol possibly by a protection from premature inactivation of NO and in addition by inhibiting an unexpected upregulation of platelet activity.

1 Holtz, J., Giesler, M., and Bassenge, E. (1983) Z.Kardiol. **72**, 98-106
2 Stewart, D.J., Elsner, D. Sommer, O., Holtz, J. and Bassenge, E. (1986) Circulation **74**,573-582
3 Schröder,H., Strobach, H. and Schrör, K. (1992) Biochem. Pharmacol. **43**, 533-537
4 Ischiropousol H., Zhu, L. and Beckman, J. S. (1992) Arch Bioch. Bioph. **298**, 431-437

Antioxidant mediated prevention of nitrate tolerance is achieved through immediate inactivation of peroxinitrite

MIKHAIL SKATCHKOV[+], BRUNO FINK[+], SERGEY DIKALOV[*], OLAF SOMMER[+]; EBERHARD BASSENGE[+]

[+]Institute of Applied Physiology, Univ. of Freiburg, 79104 Freiburg, Germany; [*]Institute of Chemical Kinetics and Combustion, Novosibirsk, 630090 Russia

Activation of soluble guanylyl cyclase (sGC) in smooth muscle cells (SMC) stimulates formation of cGMP and relaxation of blood vessels. Nitric oxide (NO) activates sGC. Long-term administration of NO-donors *in vivo* causes nitrate-induced tolerance (Tol) resulting in impaired cGMP formation and reduced dilatation after continuous NO-donor administration. Exogenous NO-donors do not only stimulate sGC, but also enhance formation of peroxynitrite (Per) a toxic non-radical product of the diffusion-limited reaction of NO radicals with superoxide radicals (SOR). Per in protonated form underlies decomposition resulting in release of OH°-like radicals together with NO_2° radicals. Per inactivats SH groups of proteins [1] at much higher rate than pure NO_2. Subsequent formation of inter- and intra-molecular disulfide bonds can cause unfavourable changes in activity of enzymes involved in cGMP release. Development of Tol [2] can be associated with formation and decomposition of Per and subsequent inactivation of enzymes involved in cGMP formation, particularly sGC. sGC is known to possess 12 SH groups, 8 of which are easily titrated [3]. Inactivation of SH-groups and/or nitrosation of tyrosine residues may disturb the heterodimer structure of α_1/β_1 subunits of sGC. Lack of heterodimer structure renders sGC inactive [3].

Aim of the study: To analyze the action of ascorbate and dimethylsulfoxide on peroxynitrite formation in the process of activating sGC by nitroglycerin (NTG) in platelets.

Materials and Methods:

Electron spin resonance (ESR) analysis

The production of superoxide radicals and Per was verified monitoring ESR spectra of spin-adduct DMPO-OH. ESR spectrometer settings were: 3350 mT field set, 100kHz modulation frequency, 0.1 mT modulation amplitude, microwave frequency 9.43 GHz, 5 mWt microwave power, temperature 25°C, samples volume 50μl. 0.1 M dimethyl-1-pyrroline N-oxide (DMPO) was used as spin-trap.

Spin-trapping. 0.2 mM diethylentriaminepentaacetic acid (DTPA) was used as transition metal chelating agent. To avoid the formation of H_2O_2 catalase (0.1mg/ml) was added to cell suspensions. Formation of spin-adduct DMPO-OH is not the result of H_2O_2 present in the cells in our experiments, but is the product of the reaction of DMPO and superoxide radicals together forming DMPO-OOH. This adduct in turn is reacting quickly with glutathione peroxidases to form DMPO-OH detected by ESR. Life-time of Per in 15 mM HEPES buffer pH 7.3, 25°C, is about 0.2 s [4]. Per is known to react with DMPO forming spin-adduct DMPO-OH. Decreasing DMPO-OH spin adduct formation when NO donors were added to the platelets in the presence of dimethyl-sulfoxide (DMSO) was the result of the competition of DMSO for Per (reaction constant $K=10^5\ M^{-1}s^{-1}$, [4]) generated in the cell containing media. The same effect (decreasing intensity of the ESR spectra corresponding to DMPO-OH) was observed when ascorbate was added instead of DMSO. The presence of ascorbate radicals can be detected in ESR spectra. Catalase and DTPA prevents the formation of OH° radicals in presence of ascorbate (5 mM) as a potential metal reducing agent. Thus the action of ascorbate in our experiments is mainly scavenging superoxide radicals as precursors of Per, because the direct reaction between Per and ascorbate is too slow ($K_{per\text{-}asc}=235\ M^{-1}s^{-1}$, compared to $K_{per\text{-}sor}=10^5 M^{-1}s^{-1}$ in the alternative reaction between ascorbate and superoxide, [5]).

Peroxynitrite assay:

Per production was determined spectrophotometrically at λ=420 nm employing the reaction of nitration of 4-hydroxyphenylacetic acid (4-HPA) by Per catalyzed by Cu-Zn-superoxide-dismutase (0.1mg/ml) [4]. 100% of Per corresponds to the formation of 6 nMol/min of NO_2-HPA in the presence of 10^8 cells/ml/min in 1mM solution of 4-HPA. Activation of **sGC** was performed by incubation of platelets with 0.1mM **NTG** in the presence of 1mM of cysteine in 15 mM HEPES buffer at pH 7.3 and assessed by radioimmunoassay for cGMP. Ascorbate or DMSO were added to cells 20 min before **NTG**. Data SEM; n=4.

Results: Additions of ascorbate or DMSO diminished superoxide radicals [SOR] and peroxynitrite concentrations, and promoted activation of **sGC** in **NTG** stimulated platelets

	Ascorbate 5mM	DMSO 0.05%	DMSO 0.1%	DMSO 0.2%
[SOR] drop in %	75±7	10±3	10±3	12±4
[Peroxynitrite] drop in %	55±6	50±6	72±8	80±9
[cGMP]increase, in %	43±6	28±7	40±9	65±7

Basal release of cGMP in NTG stimulated platelets in the absence of ascorbate or DMSO was 4.1±0.8 pmol/min/10^8 cells. There was an additional increase in cGMP release (about 40%) due to the action of ascorbate or DMSO. Thus by preventing formation of Per (by Na-ascorbate) or scavenging of Per (by DMSO) one can promote the activation of **sGC** in platelets by NO. Dimethyl-sulfone used instead of DMSO did not elicit any increase in cGMP or any drop in Per concentration in these experiments.

Discussion: Two approaches to prevent Per-induced inactivation of **sGC** and other enzymes are possible. The first one is to scavenge O_2^- radicals as precursor of Per. Ascorbate is a potent scavenger of O_2^- and NO_2 radicals. Coadministration of ascorbate with nitroglycerin (**NTG**) prevents the developement of Tol *in vivo* (Bassenge/Fink, in this vol. and [6]). The second approach is to scavenge Per effectively. DMSO is known to scavenge Per (formaldehyde is the end-product) [1]. Coadministration of DMSO and **NTG** prevents Tol *in vivo* [6]. The action of DMSO is not yet identified. However the nitrosating agent (NO_2) released after the reaction of DMSO with Per appears to be less potent in the process of either inactivation of SH-groups or in nitrosation of tyrosyl residues of proteins [4]. Ascorbate induced scavenging of NO_2 is much faster than scavenging of Per by ascorbate [5]. In the absence of such a special reductant as dithiothreitol in cells *in vivo,* sGC is unprotected against the action of Per during long-term administration of NO-donors. Inactivation of sGC due to the formation of disulfide bonds can be reversible to some extend. But recovery of sGC may be a slow process comparable to the time required to eliminate Tol *in vivo* (days). Duration of this process in cells *in vivo* is determined by the intracellular concentration of Na-ascorbate and the intracellular ratio of thiol/thiol-disulfides as well as by the rather small uptake rate of ascorbate into the cells. If DMSO is added to NTG stimulated cells, NO_2, but not the more reactive Per will affect the enzymes. To eliminate solely NO_2 actions, intracellular concentrations of ascorbate and thiols may be then quite sufficient.

Conclusions: Ascorbate and DMSO in adequate concentrations can effectively protect cellular enzymes associated with NO-mediated activation of sGC against destructive actions of peroxynitrite in platelets. Comparing the inhibitory effects of ascorbate or DMSO on tolerance observed *in vivo* [6] and actions of ascorbate and/or DMSO on a cellular level, it is likely that enhanced formation of Per and Per-induced changes in enzymes activities are responsible for the major part of the multifactorial phenomenon of tolerance to organic nitrates.

1. Beckman J. S., Chen J., Ischiropoulos H., Crow J. P. (1994) Meth. Enzymol., **233**, 229-240
2. Bassenge E. (1995) Basic Res. Cardiol. 90, 125-141.
3. Harteneck C., Koesling, D., Söling, A., Schultz, G. Böhme, E. (1990) FEBS Lett. **272**, 221-223.
4. Ischiropoulos H., Zhu L. and Beckman J. S. (1992) Arch. Bioch. Bioph. 298, 431-437.
5. Barlett D. (1995) Free Rad. Biol. Med. 18, 85-92.
6. Bassenge E., Fink B., XXIII Meeting FEBS, 1995, p.206.

Nipradilol As NO Donor : Increase In Plasma Nitrite Concentration Following Acute Vasodilatation

T Adachi, S Hori*, K Miyazaki, E Takahashi, M Nakagawa, A Udagawa**, N Hayashi**, N Aikawa*, S Ogawa;
Cardiopulmonary Division, Department of Internal Medicine, Department of Emergency Medicine*, Keio University, School of Medicine, Nihon University, Memorial Critical Emergency Center** Tokyo, Japan

Nipradilol, a beta-adrenoceptor blocking agent with vasodilation activity, has a structure which contains a NO_2 group. To clarify if nipradilol acts as a nitric oxide (NO) donor in vivo, the time course of the systemic vasodilatation was studied following intravenous administration of nipradilol (1mg/kg) in 11 closed-chest anesthetized dogs. Plasma nitrite and nitrate (NO_2^-, NO_3^-, measured by an automated system using the Griess method with micro-dialysis membrane), cyclic guanosine monophosphate (cGMP), human atrial natriuretic polypeptide (hANP), norepinephrine, and epinephrine concentration were also measured, at control, 3, 5, 10, and 30 min. Cardiac output was measured with 8F Opti-Swan Gantz Catheter.

A fall in systemic vascular resistance was observed at 1 min (64±4%* of control), followed by a significant increase in NO_2^- at 3 min (0.74±0.24** vs 0.55±0.07μM at control level), which reached the peak level at 10 min (1.03±0.29μM**). Despite of a sustained release of NO_2^- (0.96±0.11μM**), systemic vascular resistance increased (123±10%* of control) at 30 min, which was associated with a release of norepinephrine (558±206* vs 157±35pg/ml at control) and epinephrine (684±263** vs 46±13pg/ml at control). Plasma cGMP increased at 30 min (23.1±3.1* vs 20.7±2.8 pmol/ml at control) without a change in hANP (62.9±15.4 vs 63.7±22.6pg/ml at control). NO_3^-didn't change significantly .

In conclusion, NO_2^- increased rapidly and significantly following the nipradilol-induced systemic vasodilatation. It is suggested that nipradilol exerts its vasodilator effect by acting as a NO donor. (mean±SE, *: P<.05, **: P<.01)

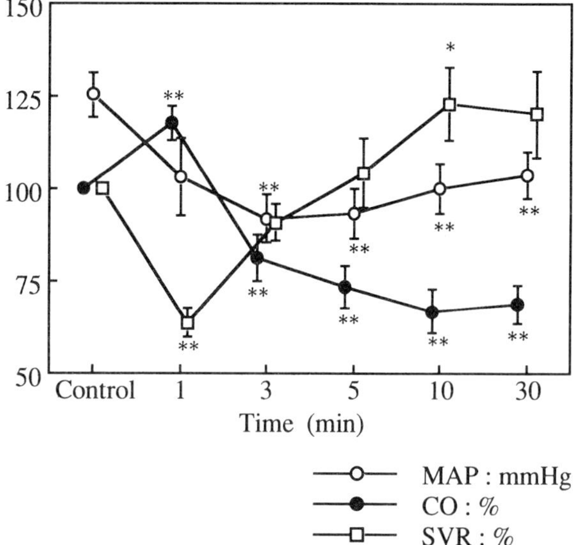

Figure 2 Hemodynamic changes following intravenous administration of nipradilol

MAP: mean aortic pressure, CO : cardiac output, SVR: systemic vascular resistance
MAP : P<0.001, CO : P<0.001, SVR : P<0.001 (two-way ANOVA)
*: P<0.05, **: P<0.01 vs. control (Dunnet's method)

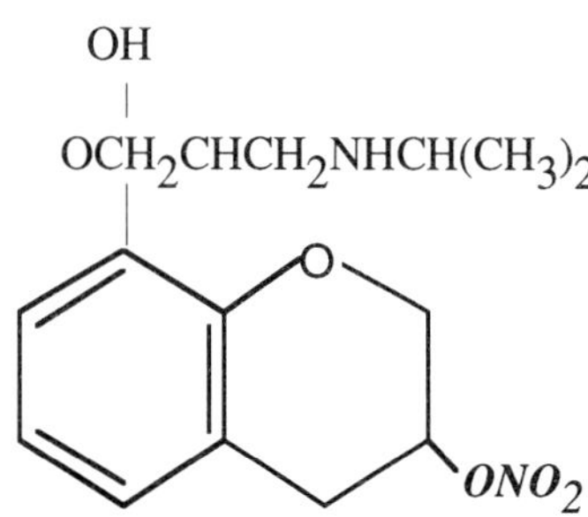

Figure 1 The Structure of Nipradilol

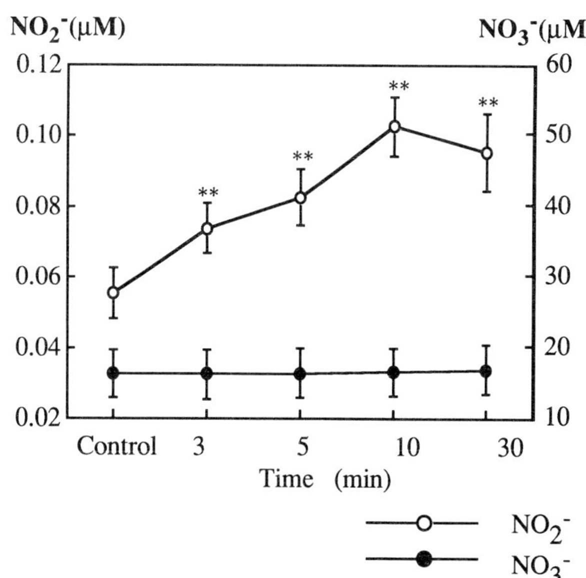

Figure 3 Plasma NO_x concentration following intravenous administration of nipradilol

NO_2^- : P<0.001, NO_3^- : n. s. (two-way ANOVA)
**: P<0.01 vs. control by Dunnet's method

Nitric oxide mediates isoflurane elevation of cGMP in cultured human endothelial cells.

IROKA J. UDEINYA, MELVILLE Q. WYCHE, LUKE LIM, and PETER JIANG.

Department of Anesthesiology, College of Medicine, Washington, DC 20059 U.S.A.

Abstract. Isoflurane, an inhalational anesthetic agent causes vascular relaxation, decrease in blood pressure, enhanced blood flow, and elevation in endothelial cell cyclic GMP. In this report, we provide evidence that elevation in cyclic GMP by isoflurane is mediated by vascular NO. Confluent umbilical vein endothelial cells in 25^{cm2} flasks were treated with 5% isoflurane in air, or air alone, and incubated 30 minutes at 37^{oc}. Some cultures were pretreated with hemoglobin (Hb) 1 mg/ml, or N^G Methyl-L-arginine (L-NMMA) 300 uM, and some with 4uM 3-isobutyl-1-Methyl-xanthine (IMX). Treated and control cultures were extracted with 6% chilled TCA, and analyzed for cyclic GMP content by RIA. Treatment of the endothelial cells with isoflurane caused cyclic GMP level to increase by 110 ± 10 (n=8): in the presence of L-NMMA the level was reduced by 23%±0.13 (n=8). Hemoglobin completely blocked the cyclic GMP elevation. In the presence of IMX (4uM) the cyclic GMP levels in the isoflurane treated cells rose by 189% ± 50 (n=6). The results confirm that isoflurane induced elevation in intracellular cyclic GMP in endothelial cells is mediated at least in part through enhanced NO activity.

Background. Isoflurane use may be accompanied by hypotension which is thought to be due to decreased vascular resistance especially in peripheral vascular beds.[1] The vasoactivity of isoflurane in humans and some animals appears to involve vascular endothelium, and to be mediated through pathways related to prostanoid and/or the endothelial derived nitric oxide (NO).[2] Previous studies in our laboratory (Anesthesiology, submitted) have shown that isoflurane causes elevation of cGMP in cultured human umbilical vein endothelial cells. It is shown in the present study that the isoflurane induced elevation in cGMP is mediated through enhanced NO activity.

Method. Cryopreserved human umbilical vein endothelial cells were thawed and cultured by standard method, in T-25 culture flasks until confluent.[3] After washing with HBSS, the cells were allowed to equilibrate in 2 ml warm medium 199 for 30 minutes at 37°C, and then incubated 30 minutes at 37^{oc} in medium containing nitric oxide modulator at appropriate concentrations. Control cultures were incubated in medium without NO modulators.

After incubation with modulators, the cultures were treated for 5 minutes with air alone, or air with 5.0 percent isoflurane, using an Ohio Anesthesia Machine (Ohio Medical Products, Madison, WI) and gas analyzer, (DATEX Instruments, Helsinki, Finland), followed by incubation at 37°C for 60 minutes.

Medium was removed and the cells were extracted in 1 ml 6% ice-cold thichloroacetic acid (TCA) and collected with a rubber policeman (Costar). The collected cells were sonicated for 5 seconds with a ultrasonic cell disrupter (Heat Systems Inc., Farmingdale, NY) boiled 5 minutes, and centrifuged at 2000 g for 10 minutes. After centrifugation, the cGMP levels in the cultures were determined by radioimmune assay (RIA).

All experiments were performed in triplicates, and results expressed as Mean ± Standard Error of the Mean (SEM) for the stipulated number of independent experiments.

Results. In 3 separate experiments, the average cGMP levels in cultures treated with 5.0 percent isoflurane were consistently higher than those treated with air alone. In control cultures incubated with medium alone and not treated with air or isoflurane the cGMP level was 30 ± 5 fmol/10^6 cells. In cultures treated with air and isoflurane, the cGMP levels were 11 ± 1 and 110 ± 10 percent higher than coltrols respectively.

Table 1. Effect of Isoflurane with NO Modulators on cGMP synthesis.

Percent change in cGMP levels.

	Hb (1 mg/ml)	L-NMMA (300 uM)	L-NAME (10uM)	IMX (4uM)
Isoflurane	-41±2	-23±0.13	-29±6	+189
Air	-25±4	-18±11	-17±2	+252

The effect of modulators on the isoflurane induced cGMP synthesis is shown in Table 1. The NO synthesis inhibitors L-NAME (10 uM) and L-NMA (300 uM) caused decrease in cGMP levels of 23 ± 0.13 and 29 ± 6 percent respectively in cultures treated with isoflurane; and 18 ± 11 and 17 ± 2 in those treated with air. Hemoglobin 1 mg/ml, caused reduction of 41 ± 2 and 25 ± 4 percent in cGMP levels in isoflurane and air treated cultures respectively. In the presence of 1-methyl-3-isobutyl xanthine (4 uM) the cGMP levels increased by 189 ± 50 and 252 ± 100 percent in isoflurane and air treated cultures respectively.

Discussion. It has been shown in this, as in our previous report, (Anesthesiology submitted) that in the presence of isoflurane, there is enhanced synthesis of cGMP by human vascular endothelial cells. It was shown in the earlier study that the mechanism of the isoflurane action on the endothelial cells may involve both an NO, and prostanoid metabolic pathways.

Earlier studies by others in a variety of animal species and tissues, had led to conflicting conclusions about the role of NO and cGMP in the vascular action of isoflurane. In studies with rat thoracic aorta for example, it was observed that isoflurane, which caused vasodilation at 1, 2 and 3 percent had no effect on the cGMP content of either endothelium intact or endothelium-denuded vessels[4]. In contrast, Greenblatt et. al showed that intravenous infusion of L-NMMA caused an increase in coronary vascular resistance in rats anesthetized with isoflurane[2]. The conflicting results in these studies may be related to species as well as tissue differences. The present study was carried out to establish evidence for the role of NO in the vascular effects of isoflurane. It has been shown definitively that isoflurane causes elevation in cGMP production in human vascular endothelial cells. The action is mediated at least in part by enhanced NO and guanylate cyclase activity. It may be concluded that isoflurane enhancement of NO and cGMP activities may contribute to its mechanism of action on the vascular system in vivo.

References.

1. Stevens W.C., Cromwell T.H., Salsey M.J., Eger E.L., Shakespeare T.F. and Bahlman S.H. (1971) The cardiovascular effects of a new inhalation anesthetic, Furane in human volunteers at constant arterial carbon dioxide tension. Anesthesiology 35, 8-16.
2. Greenblatt E.D., Loeb A.L., Longnecker D.E. (1992) Endothelium-dependent circulatory control: A mechanism for the differing peripheral vascular effects of isoflurane versus halothane. Anesthesiology 77: 1178-1185.
3. Brendel J.K., and Johns R.A. (1992) Isoflurane does not vasodilate rat thoracic aotic rings by endothelium-derived relaxing factor or other cyclic GMP-mediated mechanisms. Anesthesiology 77: 126-131.
4. Jaffe E.A., Nachman R.L., Becker C.G., Minick C.R. (1973) Culture of human endothelial cells derived from umbilical veins. Identification by morphologic and immunologic criteria. J. Clin. Invest. 52: 2745-2756.

Transduction of nitric oxide synthase activity from endothelium to vascular smooth muscle: involvement of S-nitrosothiols

Brian J DAY, Li JIA, Derrick R ARNELLE, James D CRAPO and Jonathan S STAMLER

Department of Medicine, Duke University, Durham, NC 27710, U.S.A.

Summary

High hemoglobin (Hb) concentrations in blood establish a diffusion gradient for endothelial-derived nitric oxide (NO) which results in the net movement of NO towards the vessel lumen and away from underlying smooth muscle (1). This leads to a conceptual problem if endothelial-derived relaxing factor (EDRF) responses are to be attributed (solely) to free NO. One solution to this paradox is that EDRF acts on adjacent smooth muscle as a NO adduct, for example a S-nitrosothiol (RSNO). Here we identify the presence of intracellular RSNO in endothelial cells. To differentiate between NO and RSNO effects, we employed small cell permeable probes which selectively attenuated NO or RSNO responses. We then assessed the effects of these probes on EDRF activity in rabbit thoracic aortic rings constricted with phenylephrine. Probes which largely inhibited NO-induced relaxations had little effect on RSNO or EDRF responses. Conversely, selective inhibition of RSNO relaxations resulted in attenuated EDRF activity. Moreover, conditions supporting breakdown of RSNO to nitric oxide attenuated EDRF and RSNO relaxation potency. These data suggest involvement of S-nitrosothiols in transduction of NOS activity from the endothelium to underlying smooth muscle, and help rationalize EDRF responses *in vivo*.

Results and Discussion

We found that RSNOs are formed constitutively by endothelial cells and their production was stimulated with a calcium ionophore (figure 1). We then used a oxidative stress model to create an NO sink and compared relaxation responses of NO, EDRF and S-nitrosoglutathione (GSNO) in a rabbit aortic ring bioassay. In this model we found that, unlike NO responses, EDRF was resistant to inhibition as was GSNO (figure 2). Similiarly, diethyldithiocarbamate (which releases NO from RSNO) decreased EDRF relaxations to a much greater extent than seen with NO (figure 3).

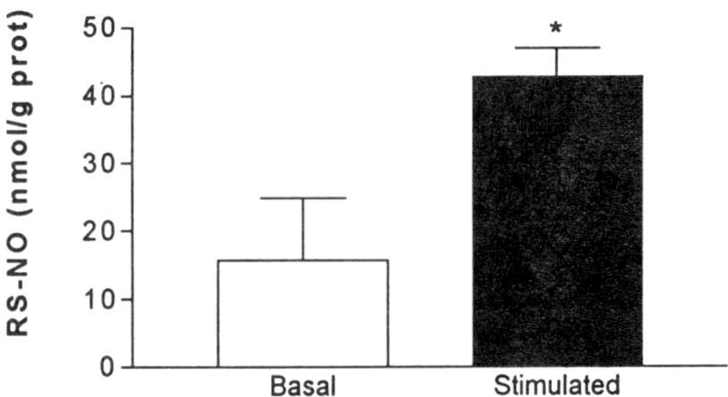

Figure 1. Stimulation of ecNOS by A23187 increases intracellular levels of S-nitrosothiols. CPA-47 cells were grown on cytodex-2 beads, packed in 1 ml columns, and treated with vehicle (basal) or 10 μM of A23187 (stimulated) for 10 minutes. RSNO content was measured on cell lysates. * p=0.037.

Our data suggest that S-nitrosothiols play an important role in the transduction of NOS activity to underlying smooth muscle, especially under conditions of oxidative stress. These finding help rationalize EDRF responses in the vasculature where heme-based and oxygen-based sinks are prevalent.

Abbreviations used: Hb, hemoglobin; NO, nitric oxide; EDRF, endothelium derived relaxing factor; RSNO, S-nitrosothiol; NOS, nitric oxide synthase; GSNO, S-nitrosoglutathione; CPA-47, calf pulmonary artery cells; H_2TBAP, tetrakis(4-benzoic acid)porphyrin; MnTBAP, manganic tetrakis(4-benzoic acid)porphyrin.

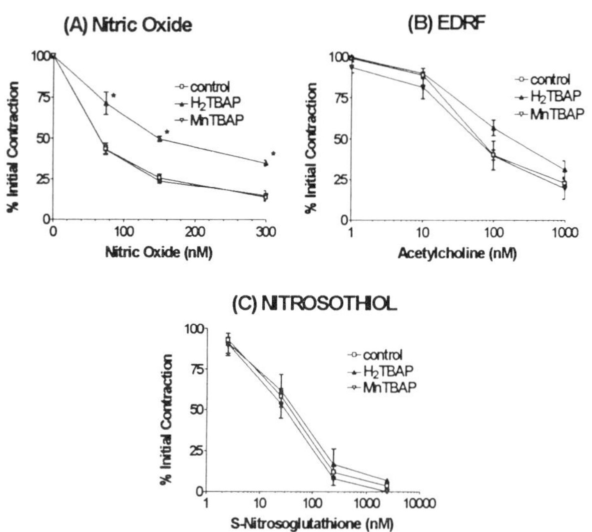

Figure 2. Superoxide attenuates NO responses without effecting either EDRF or GSNO responses. Rabbit aortic rings were pre-constricted with 0.1 μM phenylephrine and relaxed with (A) NO (B) acetylcholine (EDRF) or (C) GSNO (nitrosothiol) in the presence of 100 μM of either the superoxide generator (H_2TBAP) or the SOD mimetic (MnTBAP). * p<0.05.

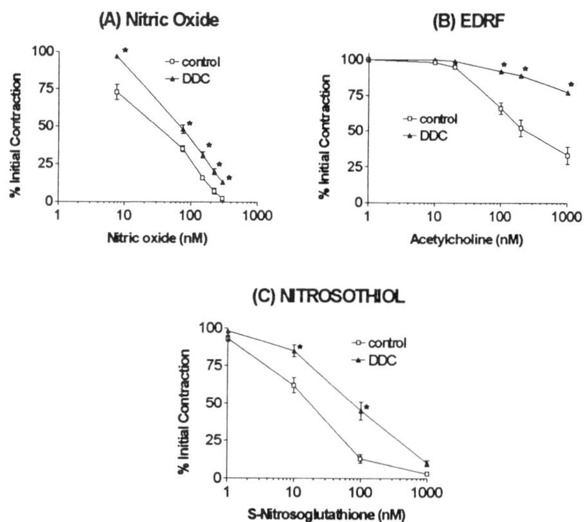

Figure 3. Release of NO from RSNOs along with increased superoxide formation with diethyldithiolcarbamate treatment inhibits NO-related relaxations. Rabbit aortic rings were pre-constricted with 0.1 μM phenylephrine and relaxed with (A) nitric oxide, (B) acetylcholine or (C) GSNO in the presence of 1 mM diethyldithiocarbamate. * p < 0.01.

References

1. Lancaster, J.R. (1994) Proc. Natl. Acad. Sci. U.S.A. **91**, 8137-8141

Differentiation between EDRF and NO in organ chamber experiments on rings of rabbit aorta

ROBERT F. FURCHGOTT, DESINGARAO JOTHIANANDAN and NASRIN ANSARI

Department of Pharmacology, State University of New York Health Science Center, Brooklyn, NY 11203, USA

We have previously reported that in organ chamber experiments on rings of rabbit thoracic aorta, superoxide (O_2^-) generated by addition of xanthine plus xanthine oxidase (X+XO) was a rapidly acting potent inhibitor of relaxation produced by added NO, whether the X+XO was added before or after the NO. However, the immediate effectiveness of X+XO as an inhibitor of endothelium-dependent relaxation by acetylcholine (ACh) was much greater when the X+XO was added several minutes after the ACh (during maintained relaxation) rather than before or immediately after (Fig. 1) [1]. This suggested that the EDRF first released from endothelial cells in response to ACh may not be simply NO, but rather some adduct of NO that is more resistant to inactivation by O_2^- than is NO itself [1]. Also, superoxide dismutase (SOD) appeared to be more effective in protecting ACh-induced relaxations than NO-induced relaxations against inhibition by X+XO. Since all of the earlier experiments were conducted with single additions of NO (75-150 nM) which produce only transient relaxations, it was decided that a more meaningful comparison of the susceptibilities of NO and ACh-released EDRF to inactivation by O_2^- could be made by using continuous infusions of NO into the organ chambers to produce sustained relaxations equivalent to those obtained with ACh.

Rings of rabbit thoracic aorta (intact, +EC; or denuded of endothelial cells, -EC) were equilibrated in 20 ml organ chambers under 2-3 g basal tension. The bathing solution was Krebs-bicarbonate at 37^0. Phenylephrine (PE), 50-200 nM was used to produce sustained contractions. Unless otherwise indicated, the concentration of added X was 0.3 mM and that of XO was 3.2 mU/ml. This combination of X+XO generates O_2^- at a rate of about 0.01 mM/min for about 25 min in the 20 ml of Krebs solution in each organ chamber at 37°. Since H_2O_2 can produce endothelium-dependent relaxation as well as some endothelium-independent relaxation of rings of rabbit aorta [1,2], catalase (CAT; 200 U/ml) was always added to the bathing solution in experiments with X+XO in order to rapidly remove the H_2O_2 generated by the dismutation of O_2^-, either spontaneous of catalyzed by added SOD. Solutions of NO in water at a concentration of 0.015 mM were prepared as previously described in gas-sampling chambers (tonometers) with a self-sealing rubber stopper at the entry port [3]. For continuous infusion of a NO solution into an organ chamber, a sample of about 5 ml of solution was withdrawn from the tonometer through the entry port into a gas-tight syringe which was then placed in position in an electronically controlled infusion pump. The solution was pumped from the syringe through a short length of small-diameter teflon tubing into an organ chamber at a rate appropriate for obtaining the desired degree of sustained relaxation.

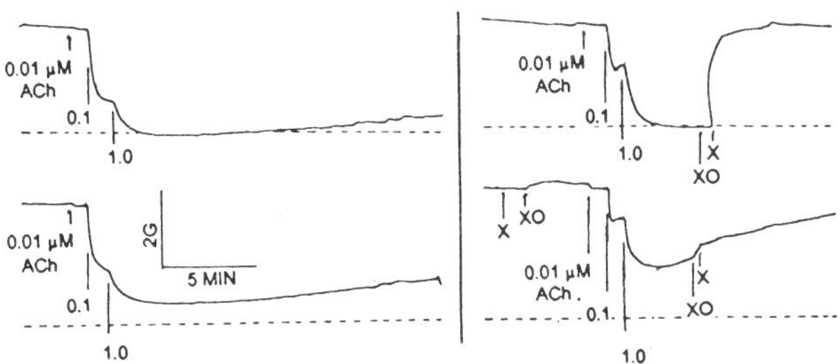

Fig. 1. Initial inhibition of ACh-induced endothelium-dependent relaxation by the superoxide generated by xanthine (X, 0.3 mM) plus xanthine oxidase (XO, 3.2 mU/ml) is much more effective when the X+XO is added several min after the ACh rather than before the ACh. The left panel shows relaxation of two intact rings from the same aorta in response to ACh. After washout and recontraction with phenylephrine, the rings were again tested with ACh (right panel), with X+XO being added before (lower tracing) and after ACh (upper tracing).

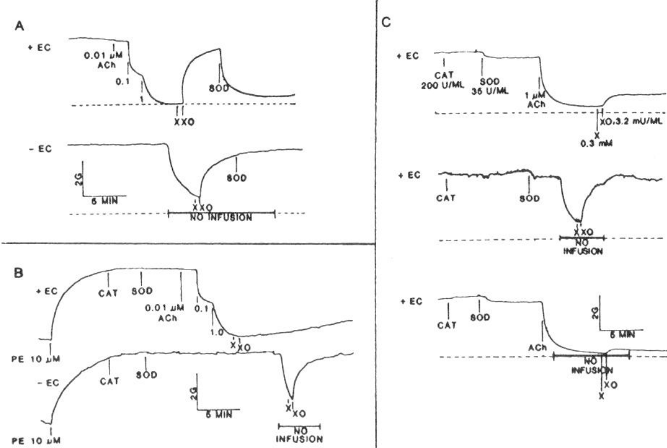

Fig. 2. A. SOD largely reverses the blockade by superoxide (generated by X+XO) of ACh-induced relaxation but not the blockade by superoxide of relaxation induced by an infusion of NO. The intact ring (+EC) and the endothelium-denuded ring were from the same aorta. In all of the experiments shown in this figure, the NO infusion (indicated by horizontal bar) was 0.15 ml/min of 15 μM NO into 20 ml of Krebs solution in the organ chamber; XO, 3.2 mU/ml; X. 0.3 mM; SOD, 35 U/ml; CAT, 200 U/ml. **B.** Pre-added SOD largely prevents the blockade by superoxide of ACh-induced relaxation but not the blockade by superoxide of relaxation induced by an infusion of NO. The two rings were from the same aorta. **C.** SOD largely prevents the blockade by superoxide of ACh-induced relaxation even in the presence of a continuous infusion of NO. The three intact rings were from the same aorta. Since SOD protects ACh-induced relaxation against blockade just as effectively in the presence of NO (lower trace) as in its absence (upper trace), the failure of SOD to protect NO-induced relaxation against blockade (middle trace) cannot be attributed to inactivation of SOD by NO.

Results of typical experiments using infusions of NO are shown in Fig. 2. Fig. 2A and 2B clearly show that SOD can largely protect ACh-induced endothelium-dependent relaxation against inhibition by superoxide generated by X+XO, but that it fails to protect NO-induced endothelium-independent relaxation against inhibition by similarly generated superoxide. The possibility that the failure of SOD to protect NO-induced relaxation against the generated superoxide is due to inhibition of SOD catalytic activity by NO was ruled out by experiments of the type shown in Fig. 2C.

The ability of SOD to largely protect relaxation induced by EDRF (released by ACh) but not relaxation induced by infused NO against inhibition by superoxide generated by X+O indicates that in the presence of added SOD, the rate of inactivation by superoxide of freshly released EDRF is less than that of the infused NO. One hypothesis to explain this difference is that freshly released EDRF is not simply NO but is some adduct or precursor of NO that delivers NO to the smooth muscle cells and is not inactivated as rapidly as NO by superoxide. A second hypothesis is that freshly released EDRF is NO (EDNO); however, as a result of the physical parameters of the testing system, SOD is more effective in protecting EDNO during its diffusion from the endothelial cells to the adjacent smooth muscle cells than it is in protecting infused NO throughout the bathing solution.

Acknowledgement: This research was supported by USPHS grant HL21860

References

1. Furchgott, R.F., Jothianandan, D. and Ansari, N. (1994) in Endothelium-Derived Factors and Vascular Functions (Masaki, T., ed.) pp. 3-11, Elsevier, Amsterdam
2. Furchgott, R.F. (1991) in Resistance Arteries, Structure and Function (Mulvaney, M., ed.) pp 216-220, Elsevier, Amsterdam
3. Furchgott, R.F., Khan, M.T. and Jothianandan, D. (1990) in Endothelium-Derived Relaxing Factors (Rubanyi, G.M. and Vanhoutte, P.M., eds.), pp. 8-21, Karger, Basel

Differentiation of endothelium-derived nitric oxide- and hyperpolarizing factor-mediated relaxations in the rat coronary vascular beds using potassium channel blockers

ICHIRO SAKUMA, HIROSHI ASAJIMA and AKIRA KITABATAKE

Department of Cardiovascular Medicine, Hokkaido University School of Medicine, N-15, W-7, Kita-ku, Sapporo 060, Japan

Bolus intracoronary injection (10 pmol) of endothelin (ET)-3 to the isolated rat heart elicits an endothelium-dependent decrease (≈30 %) in the coronary perfusion pressure (CPP)(Fig. 1) [1]. We had reported previously that this relaxation of rat coronary vascular beds induced by ET-3 is blocked by oxyhemoglobin (oxyHb)(5 μM) and high potassium (K^+)(16.9 mM), but not by indomethacin (5 μM), indicating that endothelium-derived nitric oxide (EDNO) and a mechanism(s) related to potassium channels, but not prostacyclin (PGI_2), are involved in the vasorelaxation. In the present study we further tried to elucidate the mechanisms of the ET-3-mediated coronary vasodilatation by using potassium channel blockers.

Methods: Hearts obtained from male Wistar rats, weighing 180-200 g, were perfused at a constant flow of 10 ml/min by the Langendorff technique, using Krebs-Henseleit (KH) solution continuously aerated in a reservoir with 95 % O_2 and 5 % CO_2 to yield a pH of 7.4 (37 °C). Changes in CPP were measured with a pressure transducer connected to a side arm of the aortic cannula, and were recorded continuously. ET-3 at a dose of 10 pmol, which dose had been shown to elicit a maximum decrease in CPP [1], was dissolved in KH solution and administered as a 10-μl bolus into the perfusate from a point 2-cm proximal to an aortic cannula.

The mode of decrease in CPP elicited by ET-3 was expressed in terms of magnitude of decrease in CPP defined as the maximum percent decrease from the basal CPP, or duration of decrease in CPP, half recovery time ($T_{1/2}$), defined as time from the point of maximum decrease in CPP to that of half between the maximum decrease to the basal CPP.

Results, expressed as mean± SEM, were analyzed by one-way analysis of variance and differences of means were further assessed with Student's *t* test for paired values.

Results: The control CPP and the ET-3-induced decrease in CPP were not modified by indomethacin (IND; 5 μM)(Fig. 1) or glibenclamide (GLIB; 10 μM), an ATP-sensitive potassium channel blocker (Fig. 2, $T_{1/2}$: control, 0.74±0.08, GLIB, 0.78±0.11). L-N$^{\omega}$-nitroarginine (L-NNA; 10 μM), oxyHb (5 μM) and tetraethylammonium (TEA; 5 mM), a Ca^{2+}-activated potassium channel blocker, raised the control CPP and, although did not affect the magnitude of the ET-3-induced decrease in CPP, significantly shortened its duration (Fig. 2)($T_{1/2}$: control, 0.90±0.22, TEA, 0.48 ±0.09, P<0.001). The effects of L-NNA were reversed by addition of L-arginine (100 μM) to L-NNA (Fig. 1). In contrast, tetrabutylammonium (TBA; 1 mM), a Ca^{2+}-activated potassium channel blocker raised the control CPP and abolished the ET-3-induced decrease in CPP (Fig. 2). Apamin (0.1 μM), a Ca^{2+}-activated potassium channel blocker, although did not modify the control CPP, inhibited the magnitude of the ET-3-induced decrease in CPP significantly (Fig. 2).

Discussion: In the rat coronary vascular beds ET-3 elicits a transient endothelium-dependnet relaxation through the stimulation of ET_B receptors on the endothelial cells [2]. Since the relaxation was inhibited by L-NNA, oxyHb, TBA, TEA and apamin, but not by IND and GLIB, the factors released from the endothelial cells to relax coronary vasculature are thought to be EDNO and one that is related to Ca^{2+}-activated potassium channels but not to ATP-sensitive potassium channels, possibly endothelium-derived hyperpolarizing factor (EDHF) [3]. The contribution of PGI_2 to the ET-3-mediated vasomotion was negligible. In addition, effects of agents on the control CPP revealed that EDNO and EDHF, but not PGI_2, are involved in the shear stress-mediated relaxation in the present experimental settings.

The mode of inhibition of the ET-3-induced relaxation was considerably different among inhibitors of EDNO and potassium channel blockers. The EDNO inhibitors only shortened the duration of the relaxation, whereas potassium channel blockers inhibited its magnitude. Thus, these results suggest that the later portion of the ET-3-induced dilatation of the rat coronary vascular beds is mediated by EDNO and the rapid early portion by EDHF.

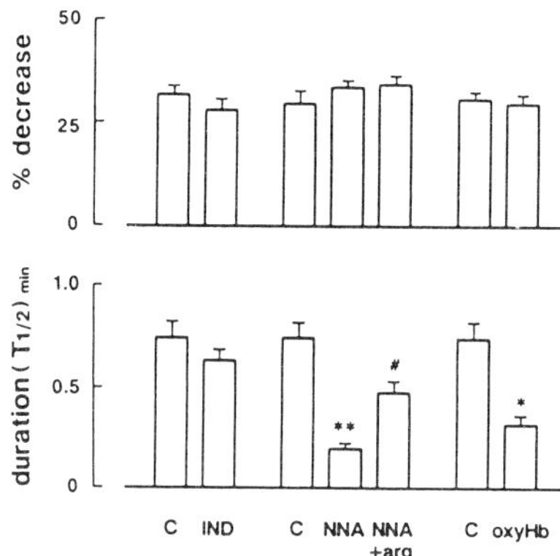

Figure 1. Effects of indomethacin (IND; 5 μM) and EDNO inhibitors, L-N$^{\omega}$-nitroarginine (NNA; 10μM) and oxyhemoglobin (oxyHb; 5 μM) on the ET-3-induced decrease in the coronary perfusion pressure in terms of % decrease (upper panel) and duration ($T_{1/2}$)(lower panel). C: control, arg: L-arginine (100 μM). *: P<0.05, **: P<0.01 vs control. #: P<0.05 vs NNA.

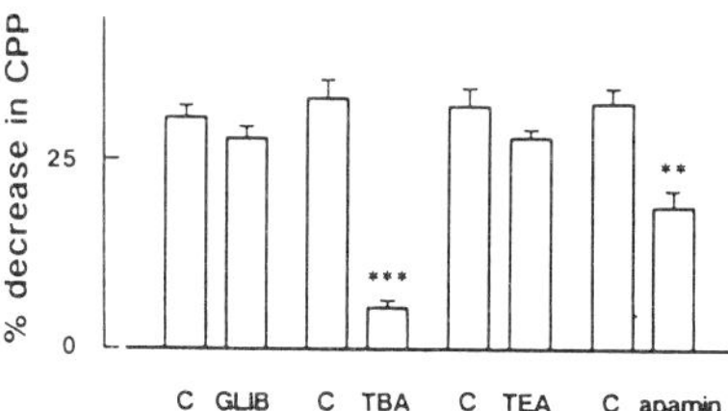

Figure 2. Effects of potassium channel blockers on the ET-3-induced decrease in the coronary perfusion pressure (CPP) in terms of % decrease. C: control, GLIB: glibenclamide 10 μM, TBA: tetrabutylammonium 1 μM, TEA: tetraethylammonium 5 μM, apamin 0.1 μM. **: P<0.01, ***: P<0.001 vs control.

Ca^{2+}-activated potassium channel blockers possess selectivity to a certain type(s) of Ca^{2+}-activated potassium channels. TEA is more selective to the large conductance channel and apamin is selective to the small conductance channel. TBA has relative selectivity to the intermediate conductance channel. In the present study TBA and apamin inhibited the magnitude of the ET-3-induced relaxation, while the action of TEA resembled that of inhibitors of EDNO. Thus, it is assumed that TEA inhibited Ca^{2+}-activated potassium channels on the endothelial cells, but not those on the vascular smooth muscle. Because Ca^{2+}-activated potassium channels on the endothelial cells are thought to function to facilitate Ca^{2+} entry upon receptor stimulation, TEA might have inhibited the EDNO release only. The effects of apamin on the relaxation may indicate that the small conductance potassium channels are involved in the action of EDHF released by ET-3.

In conclusion, EDNO and EDHF were thought to be involved in the shear stress-mediated and the ET-3-induced relaxation of the rat coronary vascular beds. The rapid early portion might be mediated by EDHF and the later portion by EDNO. Ca^{2+}-activated potassium channel blockers were useful to assess the mechanisms related to the release and action of EDNO and EDHF.

This work was supported by a Grant-in-Aid 06670683 from the Ministry of Education, Science and Culture of Japan and by the Japan Heart Foundation Research Grant for 1993.

1. Sakuma, I., Asajima, H., Fukao, M., Tohse, N., Tamura, M. and Kitabatake, A. (1993) J. Cardiovasc. Pharmacol. **22**(Suppl. 8), S232-S234
2. Sakuma, I., Asajima, H., Fukao, M., Tohse, N., Tamura, M. and Kitabatake, A. (1995) J. Cardiovasc. Pharmacol. **26**(Suppl. 3), S400-S403
3. Garland, C.J., Plane, F., Kemp, B.K. and Cocks, T.M. (1995) Trend. Pharmacol. Sci. **16**, 23-30

Light dependent relaxation of vascular smooth muscles. Experimental proof for the participation of nitric oxide under various activation conditions

HANS-JÜRGEN DUCHSTEIN and SABINE RIEDERER

Institute of Pharmacy, Freie Universität Berlin, Königin-Luise-Str. 2+4, 14195 Berlin, Germany

Idea

After studying the work of Venturini [1] about photorelaxation, we evaluated the literature (see below) and decided to repeat the experiments under the same conditions, because this group has only determined the activation of cGMP, but not a direct measurement of nitric oxide. For this aim we establish an analytic chemiluminescence assay for NO, which is sensitive enough for an amount of 1 nmole nitric oxide.

Selected literature about photorelaxation

1. Furchgott [2] described photorelaxation (1955)
2. Investigation of light dependancy (Furchgott 1961[3] and 1968 [4])
3. Photorelaxation independent of endothelium and accompanied with an increase of cGMP (Furchgott 1985 [5])
4. S-nitroso derivatives are potent vasodilators and can be photolysed to release NO (Flitney 1990 [6] and 1993 [7])
5. Photorelaxation of smooth muscles without endothelium, which can be reactivated by NO-donors (Venturni 1993 [1])

Table 1. Results of photoactivation with vascular smooth muscles

	Release of NO
1. with endothelium + carbachole (1 mM) in the dark	0.54 nmol = 52%
2. without endothelium + carbachole (1 mM) in the dark	0%
3. with endothelium + h•v	1.06 nmol = 100%

experimental conditions see below

In table 1 we show the results for the release of NO from rat aortic smooth muscles under photoactivation conditions, in which we confirm the findings of Venturini [1].

Tools

For the different chemical activations we try as model systems tetraphenylporphyrins in the presence of oxygen donors or in the presence of sodium borohydride (Fig. 1). Under these conditions we imitate the different parts of the cytochrome P-450 cycle (see abstract number 266 this symposium).

1. Fe (III) TPP Cl   oxygen donor
2. Mn (III) TPP Cl   iodosobenzene (PhIO)

Fig. 1

Table 2. Results of chemical activation with vascular smooth muscles

	Release of NO
Reductive activation	
1. without endothelium + Fe(III)TPP Cl (0.02 mM) + $NaBH_4$ (= 0.04 mM) in the dark	0%
2. without endothelium + Fe(III)TPP Cl (0.02 mM) + $NaBH_4$ (= 0.04 mM) + h•v	0.62 nmol = 59%
Oxidative activation	
1. without endothelium + Fe(III)TPP Cl (0.02 mM) in the dark	0%
2. without endothelium + Fe(III)TPP Cl (0.02 mM) + iodosobenzene in the dark	0.27 nmol = 26%
3. without endothelium + Fe(III)TPP Cl (0.02 mM) + iodosobenzene + h•v	0%

experimental conditions see below

These results are summarized in table 2. It turns out, that a reductive mechanism like in the release of glycerol trinitrate (GTN), enzymatic [10, 11] or under participation of model compounds [8], has no evidence in the activation of smooth muscles. NO is not released in the dark in the presence of Fe(II)TPP (Fig. 2).

Reductive metabolism of NO-releasing drugs

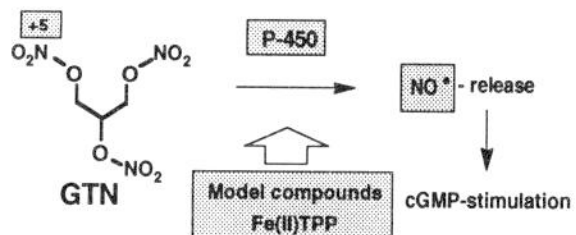

Fig. 2

In the simulation of the oxidative shunt mechanism of cytochrome P-450 [9] we observe a release of NO, which is comparable to the amount during photoactivation. This is done with aortic strips in the presence of Fe(III)TPPCl and iodosobenzene as oxygen donor. In an independent experiment NOHA (N^{ω}-hydroxy-L-arginine), the intermediate of the biosynthesis of NO, is activated with the same model reaction under oxidative conditions (Fig. 3).

Therefore it seems to be possible, that NO is derived under these conditions from L-arginine, and we mimic the reaction of the heme containing NO-synthase with model compounds of cytochrome P-450.

Biosynthesis of nitric oxide (Oxidative activation)

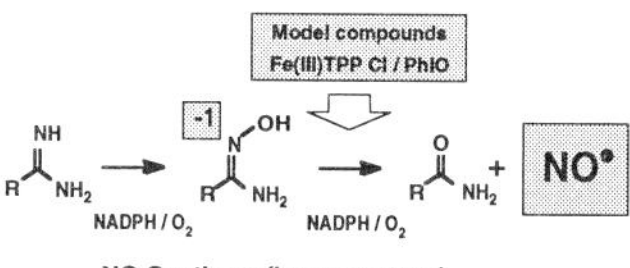

Fig. 3

Analytics

The analytical assay for detecting NO is a chemiluminescence method, based on the measurement of intensity of the fluorescent radiation emitted after chemical oxidation of nitric oxide by ozone. The detection limit is approximately 1 nmole and a signal concentration plot shows a linear region up to 100 nmole.

Experimental conditions

Male rat aortic strips denuded from endothelium, placed in Krebs-Henseleit buffer with 0.2 M vit. C, measured in 5 mL phosphate buffer (pH = 6.88).

Literature

1. Venturini,C.M., Palmer,R.M.J. and Moncada,S. Vascular smooth muscle contains a depletable store of a vasodilator which is light-activated and restored by donors of nitric oxide (1993) J. Pharmacol. Exp. Ther. **266**, 1497-1500
2. Furchgott,R.F. The pharmacology of vascular smooth muscle (1955) Pharmacol. Rev. **7**, 183-265
3. Furchgott,R.F., Ehrreich,S.J. and Greenblatt,E. The photoactivated relaxation of smooth muscle of rabbit aorta (1961)J. Gen. Physiol. **44**, 499-519
4. Ehrreich,S.J. and Furchgott,R.F. Relaxation of mammalian smooth muscle by visible and ultraviolet radiation (1968) Nature (London) **218**, 682-684
5. Furchgott,R.F., Martin,W., Cherry,P.D., Jothianandan,D., and Villani,G. (1985) Endothelium dependent relaxation photorelaxation and cyclic GMP, in Vascular Neuroeffective Mechanisms (Bevan,J.A.,Godfraind,T.,Maxwell,R.A., Stoclet,J.C. and Worcel,M. eds.) pp. 105-114, Elsevier, Amsterdam
6. Flitney,F.W., Megson,I.L., Cligh,T. and Butler,A.R. , Nitrosylated iron-sulphur clusters, a novel class of vasodilators: Studies on the rat isolated tail artery (1990) J. Physiol. **430**, 42P
7. Flitney,F.W., Megson,I.L., Thomson,J.L.M. and Kennovin,G.D. Photochemical release of nitric oxide from iron-sulfur cluster nitrosyls: laser potentiation of vasodilator actions on rat isolated tail artery (1993) J. Physiol. (London) **459**, 90P
8. Duchstein,H.-J. and Riederer,S. Mechanistische Vorstellungen zur Freisetzung von Stickstoffmonoxid aus NO-Pharmaka. Modellreaktionen in Gegenwart von Licht und Übergangsmetallkomplexen 1.Mitt. (1995) Arch.Pharm. (Weinheim) **328**, 317-324
9. Riederer,S., Schellenberger,C. and Duchstein,H.-J. Release of nitric oxide (NO) from SIN-1 (3-morpholinosydnonimine) in the presence of modelcompounds for cytochrome P-450 (1995) This symposium, Abstract No. **266**
10. Servent,D., Delaforge, M., Ducrocq,C., Mansuy,D. and Lenfant,M. Nitric oxide formation during microsomal hepatic denitration of gylceryl trinitrate: Involvement of cytochrome P-450 (1989) Biochem. Biophys. Res. Commun. **163**, 1210-1216
11. Schröder,H. Cytochrome P-450 mediates bioactivation of organic nitrates (1992) J. Pharmacol. Exp. Ther. **262**, 298-302

Direct proof of nitric oxide formation from a nitro-vasodilator in rats and metabolism by erythrocytes

HIROAKI KOSAKA, SATONORI TANAKA* and EIJI KUMURA; 1st Dept. Physiol., *1st Dept. Surg., Medical School, Osaka Univ., Suita, Osaka 565, JAPAN.

Introduction: Nitrovasodilator has been widely used as a therapeutic agent, although the mechanisms on the bioactivation are not fully clarified. Nitrovasodilator is converted by vascular smooth muscle, into an active vasodilator species. However, we detected ^{15}N-HbNO (hemoglobin-NO) after mixing human blood and ^{15}N-isosorbide dinitrate (ISDN), a long-acting nitrovasodilator.

In vivo: When ISDN solution (2 μmoles) was injected into rats, HbNO spectra with a 3-line hyperfine structure was detected with ESR in the jugular blood. When the ^{15}N-compound was injected into rats, the blood showed spectra of ^{15}N-HbNO with a 2-line hyperfine structure, proving directly that NO is derived from ISDN [1]. Arterial blood did not show the hyperfine structure. Why? The 3-line hyperfine structure must be associated with pentacoordinate Hb α-NO in the low affinity, tense state (T state) quaternary structure.

Affinity of NO to Hb α becomes stronger than that to Hb β in the low affinity state in venous phase. Because HbNO concentration was less than 1% of total Hb, HbNO is in subsaturation. Heterogeneous population of Hb with various low O_2 saturation presents the T-state pentacoordinate Hb α-NO with distinct 3-line hyperfine structure in venous phase. HbNO in the high affinity relaxed state (R state) is hexacoordinate and shows no 3-line hyperfine structure [2].

In whole blood: Deoxygenated fresh human blood was mixed with ^{15}N-ISDN solution anaerobically. A HbNO signal with the 3-line hyperfine structure was predominant at first (Fig. 1). Then ESR spectra of HbNO depicting a 2-line hyperfine structure in the pentacoordinate Hbα-^{15}NO species overwhelmed the ^{14}N-HbNO spectrum.

In Red blood cells (RBC): When RBC was washed and then mixed with ^{15}N-ISDN, no 3-line hyperfine structure was present (Fig. 1). Major metabolic activity of ISDN in blood exists in RBC [1].

The difference in the HbNO concentration between whole blood and washed RBC is due to the presence of ^{14}N-HbNO. In fact, ^{14}N-HbNO was detected in untreated deoxygenated blood of the donor (Fig. 2A). When other blood of 4 healthy adults was deoxygenated without addition of ISDN, two (Fig. 2D and 2E) of the four showed the spectra of ^{14}N-HbNO, suggesting various amount of nitrosothiols rather than NO_2^- in the whole blood.

After mixing ISDN with plasma anaerobically in the presence of Hb (1 mM), the amount of HbNO detected at 15 min was a little, suggesting plasma is not responsible for the metabolic site and the ^{14}N-NO releasing factor [1].

Hemolysate & Metabolism: Hemolysate yielded similar or slightly lower catalytic activity of ISDN than that of washed RBC [1]. When the hemolysate was treated for 3 h with N-ethylmaleimide which forms covalent thiol-adducts, N-ethylmaleimide did not inhibit the production of HbNO, suggesting thiol-independent metabolism of ISDN (Fig. 3).

It has been proposed that NO_2^- release by glutathione-*S*-transferase is an important initial step in NO production from nitroglycerin. RBC have also glutathione-*S*-transferase. Glutathione-*S*-transferase inhibitor, *S*-hexylglutathione, did not inhibit the formation of HbNO, suggesting that its activity is not responsible for the bioactivation of ISDN to NO in RBC (Fig. 3).

The present study indicates that NO is produced from ISDN in blood in the ischemic region, where the life-span of the generated NO will be prolonged because of the absence of the reaction between HbO_2 and NO.

SUMMARY: We proved the production of NO from nitrovasodilator by detecting ^{15}N-HbαNO having a 2-line hyperfine structure in the venous blood of rats after administration of ^{15}N-isosorbide dinitrate (ISDN), a long-acting nitrovasodilator. When human blood was mixed with ^{15}N-ISDN anaerobically, ESR spectra showed initially ^{14}N-HbNO, which was then overwhelmed with the increased ^{15}N-HbNO. The amount of HbNO produced in washed RBC was similar to that in whole blood, showing that major metabolic activity of ISDN in blood exists in RBC.

References

1. Kosaka, H., Tanaka, S., Yoshii, T., Kumura, E., Seiyama, A. and Shiga, T. (1992) Biochem. Biophys. Res. Commun. 189, 392-397.

2. Kosaka, H., Sawai, Y., Sakaguchi, H., Kumura, E., Harada, N., Watanabe, M., and Shiga, T. (1994) Am. J. Physiol. 266, C1400-C1405.

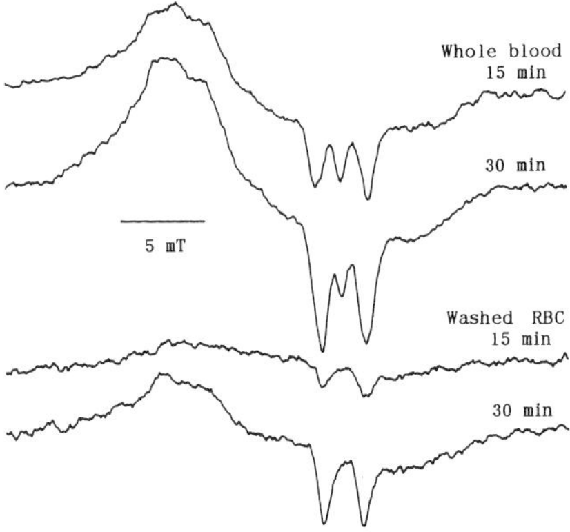

Fig. 1. ^{15}N-Nitric oxide hemoglobin production after addition of ^{15}N-ISDN solution (1 mM) to fresh human whole blood or to washed RBC, anaerobically at 37°C.

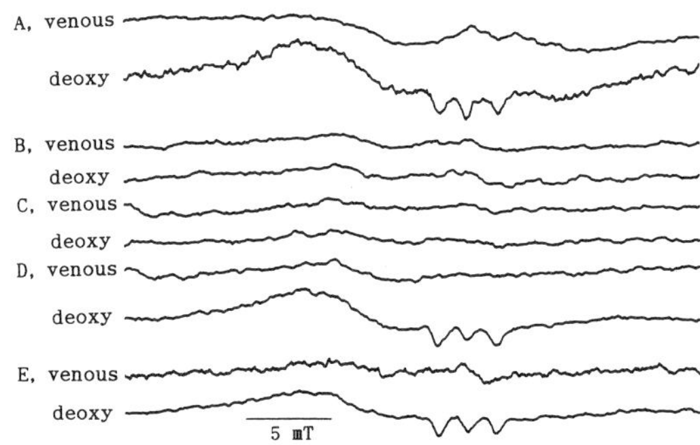

FIG. 2. Detection of HbNO after deoxygenation alone in fresh human blood (A, D, and E) of non-smokers.

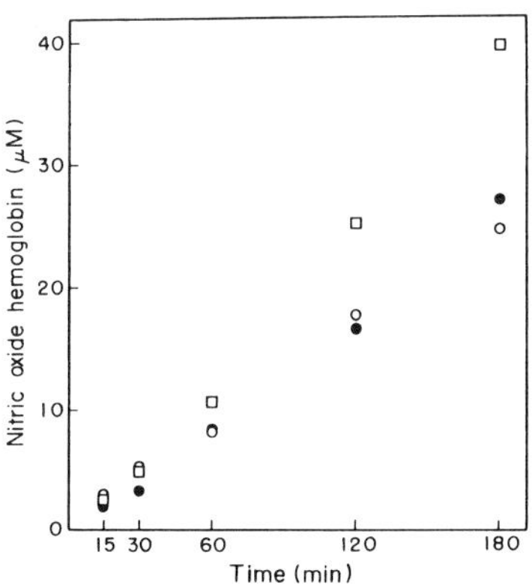

FIG. 3. Effect of *S*-hexylglutathione (1.8 mM) and N-ethylmaleimide (18 mM) on the metabolism of isosorbide dinitrate by hemolysate. ○, isosorbide dinitrate alone; ●, plus *S*-hexylglutathione; □, plus N-ethylmaleimide.

Inhibition of vasorelaxant effect of nitric oxide/cyclic GMP by chronic administration of high dose of K+ channel opener

HIDEAKI KARAKI, KAMOLCHAI TRONGVANICHNAM, MINORI MITSUI-SAITO and HIROSHI OZAKI

Department of Veterinary Pharmacology, Graduate School of Agriculture and Life Sciences, The University of Tokyo, Bunkyo-ku, Tokyo 113, Japan.

One of the mechanisms of vasorelaxant effect of NO/ cGMP system is to open K^+_{Ca} channel and to hyperpolarize the membrane. K^+ channel openers are the potent vasodilators acting on K^+_{ATP} channel. Strong activation of cellular regulation mechanisms sometimes induces desensitization. In the present experiment, we examined the effects of long-term administration of high dose of a K^+ channel opener, levcromakalim (LEM), to know if the K^+ channels are desensitized. We also examined the effects of long-term administration of high doses of verapamil and isosorbide dinitrate (ISDN).

Methods: A high dose of LEM (0.75 mg/kg x 3/day), or vehicle was orally given to normotensive or spontaneously hypertensive rats (SHR) for two weeks. Pharmacological dose of LEM that decreased blood pressure in SHR (0.15 mg/kg x 3/day x 2 weeks)[1], high dose of ISDN (30 mg/kg x 3 /day x 2 weeks) or high dose of verapamil (5 mg/kg x 3 /day x 4 weeks) was also given to the normotensive rats. Increasing the dose of verapamil to 10 mg/kg x 3 /day killed rats in 3 to 7 days.

Thoracic aorta was then isolated and cut into rings (2 - 3 mm wide) and placed in a normal physiological salt solution (PSS). PSS contained (mM): NaCl 136.9, KCl 5.4, $CaCl_2$ 1.5, $MgCl_2$ 1.0, $NaHCO_3$ 23.8, glucose 5.5 and ethylene diamine tetraacetic acid 0.01, and aerated with 95 % O_2 and 5 % CO_2 at 37°C and pH 7.4. The endothelium was removed by gently rubbing the intimal surface of aortic ring with a tip of forceps moistened with PSS. Contraction of muscle strip was measured isometrically.

Results and Discussion: In the LEM-treated aorta stimulated with 100 nM norepinephrine (NE), concentration-inhibition curve for LEM was shifted to the higher concentrations (fig. 1), suggesting that chronic LEM-treatment desensitized the site of action of LEM, possibly K^+ channels. Unexpectedly, LEM-treatment also attenuated the relaxant effects of sodium nitroprusside (SNP) (fig. 1) and 8-Br-cGMP. However, the relaxant effects of

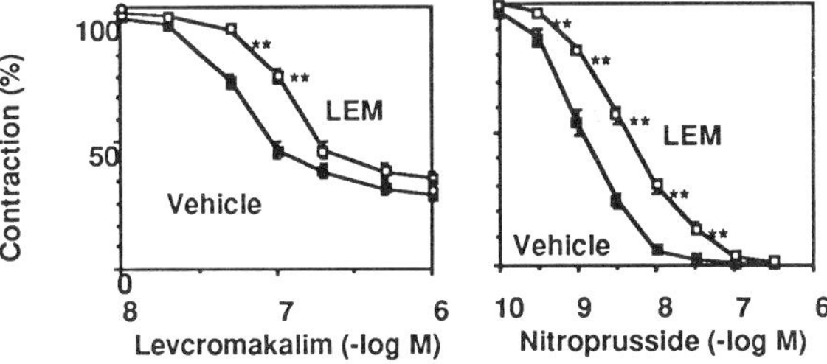

Fig. 1. Effects of chronic treatment of the rat with levcromakalim (LEM: 0.75 mg/kg x 3/day x 2 weeks) on relaxation induced by levcromakalim and sodium nitroprusside in the isolated aorta stimulated by 100 nM NE. After the NE-induced contraction reached a steady level, inhibitor was cumulatively added.

forskolin was only slightly attenuated and the relaxant effect of verapamil was not affected. Pharmacological dosage of LEM did not have such effects. These results suggest that chronic treatment with high dose of LEM desensitized also the NO/cGMP system. Since production of cGMP by SNP was not changed and since the effect of 8-Br-cGMP was attenuated by LEM-treatment, site of action of cGMP but not production of cGMP seemed to be desensitized.

It was also found that LEM-treatment augmented the isometric contractile responses of the aorta to KCl, tetraethyl ammonium (TEA) (fig. 2) and NE. In the LEM-treated SHR aorta

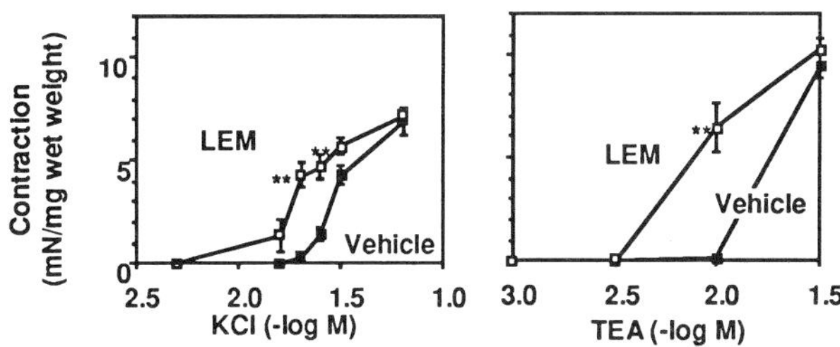

Fig. 2. Effects of chronic treatment of the rat with levcromakalim (LEM: 0.75 mg/kg x 3/day x 2 weeks) on contraction induced by cumulative addition of KCl or tetraethyl ammonium (TEA).

but not in vehicle-treated SHR aorta, spontaneous rhythmic contractions superimposed on the increase in muscle tone were observed that were inhibited by verapamil.

To examine if LEM-treatment depolarized smooth muscle membrane, we observed the effects of membrane depolarization. In the aorta isolated from untreated rat, 15.4 mM KCl augmented the NE-induced contraction. In the depolarized muscle, furthermore, the relaxant effects of LEM, SNP and 8-Br-cGMP were attenuated whereas the relaxant effects of forskolin and verapamil were not changed. Thus, the effect of membrane depolarization was quite similar to that of chronic LEM-treatment.

To desensitize NO/cGMP system, ISDN was chronically administered. In the aorta isolated from ISDN-treated rat, the relaxant effect of ISDN was attenuated. However, the relaxant effects of SNP, 8-Br-cGMP, LEM and verapamil were not affected. These results suggest that ISDN-treatment desensitized the NO-generating step but not the processes after production of cGMP.

To know the long-term effect of Ca^{2+} channel blocker, rats were chronically treated with high dose of verapamil. In the aorta isolated from verapamil-treated rat, neither the contractile effects of KCl and NE nor the relaxant effect of verapamil, SNP and LEM were changed (fig. 3). These results suggest that chronic inhibition of Ca^{2+} channel does not induce desensitization.

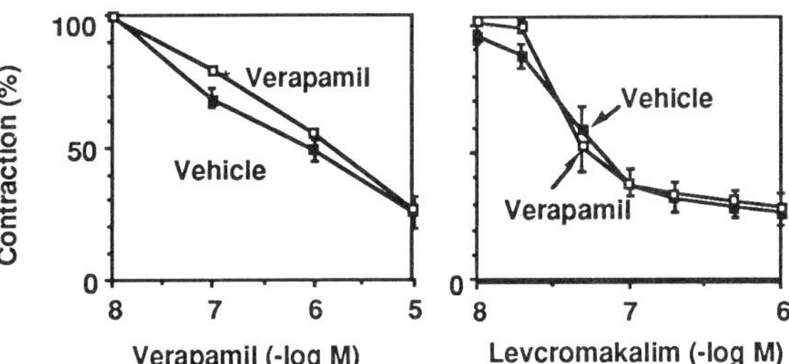

Fig. 3. Effects of chronic treatment of the rat with verapamil (5 mg/kg x 3 /day x 4 weeks) on relaxation induced by verapamil or levcromakalim in the isolated aorta stimulated by 100 nM NE. After the NE-induced contraction reached a steady level, inhibitor was cumulatively added. Rats were killed by a higher dose of verapamil (10 mg/kg x 3 /day) in 3-7 days.

These results suggest that chronic, strong activation of K^+_{ATP} channel by high dose of LEM desensitizes K^+ channel (both K^+_{ATP} and K^+_{Ca} channels), depolarizes the membrane, opens Ca^{2+} channel, and augments the effects of vasoconstrictors. Since it has been shown that the relaxant effect of cGMP is attenuated in depolarized muscle [2], inhibition of the relaxant effects of SNP and 8-Br-cGMP may also be due to membrane depolarization.

Supported by Grant-in-Aid for Scientific Research from the Ministry of Education, Culture, Sports and Science, Japan.

References

1. Clapham, J.C., Hamilton, T.C. and Longman, S.D. (1991) Arzneimmittelforsch. **41**, 385-391.

2. Karaki, H., Sato, K., Ozaki, H. and Murakami, K. (1988) Eur. J. Pharmacol. **156**, 259-266.

PEROXYNITRITE IS A VASORELAXANT WHICH ATTENUATES CATECHOLAMINE HEMODYNAMIC RESPONSES *IN VIVO*

NEIL W. KOOY, JAMES A. ROYALL, and STEPHEN J. LEWIS

The University of Iowa, Iowa City, Iowa 52242

Introduction: Vascular dysfunction, manifested as decreased vascular tone and hyporeactivity to vascular constrictors, is a characteristic and determinant pathophysiological feature of the systemic inflammatory response syndrome. Nitric oxide is a powerful vasodilator that plays an important physiological role in the normal regulation of vascular tone. In response to inflammatory cytokines, however, vascular endothelium and smooth muscle express inducible nitric oxide synthase leading to the release of nitric oxide in high concentration [1]. Whether pathological vascular relaxation is mediated directly by nitric oxide or indirectly through the formation of secondary reaction products has not been established. Nitric oxide reacts at a near diffusion-limited rate with superoxide anion to form the potent oxidant peroxynitrite [2]. Peroxynitrite has been demonstrated within the vasculature in human inflammatory disease processes [3], therefore, characterization of the *in vivo* vascular responses to peroxynitrite and the subsequent alterations in vascular reactivity may be relevant for understanding the pathophysiology of vascular dysfunction in the systemic inflammatory response syndrome and other inflammation-mediated disease states.

Methods: Male Sprague-Dawley rats (316 ± 11 g, n=8) were anesthetized with pentobarbital (50 mg/kg, i.p.) and were surgically implanted with femoral arterial and venous catheters for the measurement of pulsatile and mean arterial blood pressure and heart rate, and the administration of drugs, respectively. Immediately following catheterization, a midline laparotomy was performed and miniature pulse Doppler flow probes were placed around the lower abdominal aorta, renal and superior mesenteric arteries for the measurement of hindquarter, renal, and mesenteric blood flow velocities respectively, and for the determination of hindquarter (HQR), renal (RR), and mesenteric (MR) vascular resistances. Following stabilization of the hemodynamic parameters, the effects of norepinephrine (0.5, 1, and 2 µg/kg, i.v.) and epinephrine (0.5, 1, and 2 µg/kg, i.v.) were examined prior to and following the administration of peroxynitrite. Peroxynitrite was administered in ten consecutive bolus doses of 10 µmol/kg, i.v. (Total peroxynitrite dose = 100 µmol/kg). The effects of norepinephrine, epinephrine, and peroxynitrite are expressed as mean ± s.e.m of the percentage changes from baseline. The data were analyzed by repeated measures analysis of variance (ANOVA) followed by Student's modified *t*-test with the Bonferroni correction for multiple comparisons. The SE terms were derived from the formula $(EMS/n)^{1/2}$ where EMS is the error mean square term from the ANOVA and n is the number of rats. A value of $p < 0.05$ was taken to denote statistical significance.

Results: Peroxynitrite injection produced significant decreases in MAP, HQR, and MR, but no change in RR. These effects were subject to rapid tachyphylaxis with subsequent injections of peroxynitrite producing progressively smaller hemodynamic effects (Table 1).

Table 1. A summary of the effects of repeated injections of peroxynitrite (10 µmol/kg, i.v.) on MAP, HQR, and MR in pentobarbital-anesthetized rats (n=8). Ten total injections of peroxynitrite were given and the hemodynamic effects of the first, sixth and tenth injections are presented.

	Peroxynitrite (Injection #, 10 µmol/kg, i.v.)		
Parameter	1	6	10
Δ MAP (%)	-49 ± 3	-25 ± 3*	-16 ± 3*
Δ HQR (%)	-61 ± 4	-37 ± 3*	-21 ± 7*
Δ MR (%)	-49 ± 4	-12 ± 3*	-3 ± 2*

* $p < 0.05$ sixth and tenth injection vs. first injection.

The MAP and vascular resistances returned to baseline values after each dose of peroxynitrite, consequently, the loss of peroxynitrite-mediated responses was not due to alterations in baseline hemodynamic values.

The intravenous administration of norepinephrine and epinephrine produced dose-dependent increases in MAP, RR, and MR. Norepinephrine produced dose-dependent increases while epinephrine produced dose-independent decreases in HQR. Following peroxynitrite administration norepinephrine- and epinephrine-induced alterations in MAP, HQR, RR, and MR were significantly attenuated (Table 2).

Table 2. A summary of the effects of norepinephrine (1 µg/kg, i.v.) and epinephrine (1 µg/kg, i.v.) on MAP, HQR, RR, and MR in pentobarbital-anesthetized rats (n=6) prior to (Pre) and following (Post) administration of peroxynitrite (100 µmol/kg, i.v. total).

	Norepinephrine		Epinephrine	
Parameter	Pre	Post	Pre	Post
Δ MAP (%)	44 ± 3	18 ± 3*	26 ± 1	13 ± 2*
Δ HQR (%)	25 ± 4	4 ± 3*	-34 ± 7	-9 ± 6*
Δ RR (%)	70 ± 12	36 ± 7*	42 ± 4	31 ± 6*
Δ MR (%)	210 ± 24	63 ± 17*	198 ± 18	88 ± 14*

* $p < 0.05$ post-peroxynitrite vs. pre-peroxynitrite.

Discussion: Excessive nitric oxide production via the cytokine-induced expression of inducible nitric oxide synthase has been implicated in the pathophysiology of vascular dysfunction, however, whether pathological decreases in vascular tone are mediated directly by nitric oxide or indirectly through the formation of secondary reaction products has not been established. The presence of peroxynitrite, a secondary reaction product of nitric oxide, has been demonstrated within the vasculature of humans with inflammatory diseases suggesting that peroxynitrite may play a pivotal role in mediating vascular dysfunction. Previous reports have demonstrated that peroxynitrite exhibits nitric oxide-like biological activities *in vitro*, inducing vascular relaxation [4] and inhibiting platelet aggregation [5] apparently via the secondary formation of S-nitrosothiols [6]. The present study demonstrates that peroxynitrite is a potent *in vivo* vasorelaxant which is subject to rapid tachyphylaxis. If *in vivo* vasorelaxation to peroxynitrite is mediated through the secondary formation of S-nitrosothiols, tachyphylaxis to the vasodilatory effects of peroxynitrite may be due to peroxynitrite-mediated inhibition of the cellular and/or molecular mechanisms responsible for the vascular response to nitric oxide or related nitrosyl factors.

While decreases in vascular tone may be attributed to excess nitric oxide directly, nitric oxide does not directly inhibit vasoconstrictor function. The present study demonstrates that peroxynitrite inhibits catecholamine-mediated vascular responses. Therefore, peroxynitrite may be responsible for the refractory decrease in vascular tone characteristic of the systemic inflammatory response syndrome.

In conclusion, peroxynitrite is a potent *in vivo* vasorelaxant, the effects of which are subject to rapid tachyphylaxis, and peroxynitrite attenuates vasoconstrictor and vasodilator catecholamine responses. Endogenous peroxynitrite may therefore be important in the vascular pathology of sepsis and other inflammatory conditions.

References:

1. Moncada, S. and Higgs, A. (1993) N. Engl. J. Med. **329**, 2002-2012.
2. Huie, R.E. and Padmaja, S. (1993) Free Radic Res Commun **18**, 195-199.
3. Kooy, N.W., Royall, J.A., Ye, Y.Z., Kelly, D.R., and Beckman, J.S. (1995) Am. J. Resp. Crit. Care Med. **151**, 1250-1254.
4. Liu, S., Beckman, J.S., and Ku, D.D. (1994) J. Pharm. Exp. Ther. **268**, 1114-1121.
5. Moro, M.A., Darley-Usmar, V.M., Goodwin, D.A., Read, N.G., Zamora-Pino, R., Feelisch, M., Radomski, M.W., and Moncada, S. (1994) Proc. Natl. Acad. Sci. USA **91**, 6702-6706.
6. Mayer, B., Schrammel, A., Klatt, P., Koesling, D., and Schmidt, K. (1995) J. Biol. Chem. **270**, 17355-17360.

NO synthase inhibition, regional cerebral blood flow and vascular reactivity in the cat.

ARISZTID G.B. KOVACH, J. MARCZIS, A. KALMAR, GY. VASAS, *R. URBANICS and M. REIVICH

Cerebrovascular Res. Center, Univ. Pennsylvania, Philadelphia, PA 19104-6063, USA and *II. Inst. of Physiology, Semmelweis Univ. Med. School, 1082 Budapest, Hungary

It has been demonstrated [1] that N^G-nitro-L-arginine (NOLA), which inhibits the endothelial and the neural NO synthase, and the predominantly neural ncNOS inhibitor 7-nitro-indazole (7-NI) [2] both decrease cerebral blood flow (CBF) in a regionally heterogeneous manner. Inhibition of NOS by NOLA increases mean arterial pressure (MAP) and caused changes in vascular endothelium dependent reactivity. 7-NI treatment, in rats [3] and in our studies in cats [2], does not increase MAP, and does not affect endothelial dependent vascular reactivity. The present experiments were designed to investigate the influence of inducible NOS (iNOS) inhibitors on rCBF and cerebral vascular reactivity.

METHODS. Experiments were carried out in 31 male cats, anesthetized ip with 50 mg/kg chloralose and 200 mg/kg urethane. rCBF was measured by radiolabeled microspheres. Tissue samples were taken from the following brain areas: parietal cortex, cerebellar cortex, thalamus, hypothalamus, white and medulla oblongata and the cervical, thoracic and lumbar spinal cord and the pituitary. rCBF was measured in control state and 15 and 40 min after 40 mg/kg aminoguanidine (AG) iv administration.

In a separate group of cats, cerebral vascular reactivity was studied in vitro after reperfusion injury. MAP was reduced by bleeding to 90, 70 and 50 mmHg for 20 min at each level followed by retransfusion (HHR). Middle cerebral artery (MCA) segments were prepared 20 min after reperfusion both from control and from AG treated cats and from animals subjected to HHR with or without pretreatment with AG. Dose response curves to different endothelium dependent and independent relaxing and contracting agents were compared. Furthermore, the in vitro vascular effect of 7-NI, S-methyl-isothiourea sulfate (SMUT), and S-aminoethyl-isothiourea sulfate (AE-UT) was studied. The influence of angiotensin converting enzyme (AC) inhibitor captopryl, adrenergic blocking drug phentolamine and endothelin antagonist PD14 on the contracting responses were also studied.

RESULTS. In six brain structures (cortex, hypothalamus, medulla, cervical-, thoracic-, and lumbar spinal cord) and the pituitary gland, blood flow remained unaltered after AG administration. In contrast, in three regions (cerebellum, thalamus and medulla) blood flow decreased to 75 ($p<0.05$), 72 (p,0.05) and 76 (n.s) % at 15 min and to 65 ($p<0.01$), 61 ($p<0.01$), and 64 ($P<0.05$)% of baseline 40 min after iv 40 mg/kg AG administration, respectively see Fig. 1.

Fig. 1. EFFECT OF iv AG ON rCBF

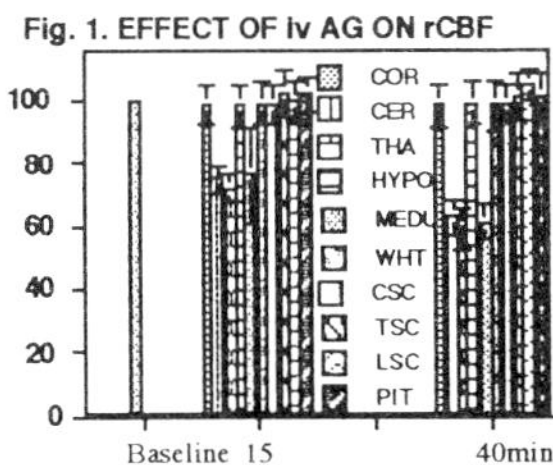

Our previous studies [2] demonstrated that HHR-induced reperfusion injury significantly enhances the contractile responses of the MCA to noradrenaline in vitro, while endothelium dependent relaxations to ACh and ATP, but not to SIN-1, are attenuated (endothelium dysfunction). According to our present study in vivo 40 mg/kg AG treatment administered 20 min before HHR prevented the impairment of the cholinergic relaxations (as seen in Fig. 2) and reduced the enhancement of the adrenergic contractions.

Fig. 2. Effect of iv AG on HHR induced impaired MCA relaxation to ACh

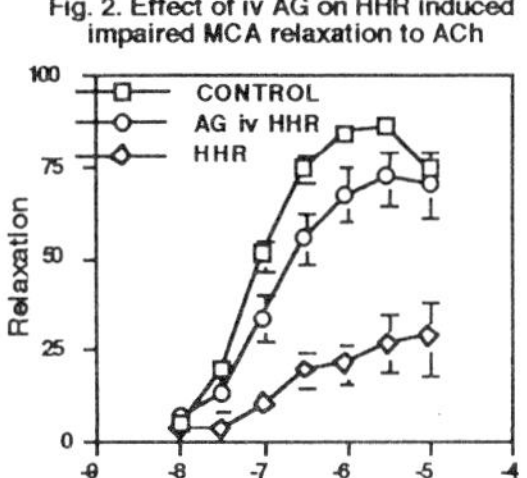

AG in vitro dose dependently contracted the MCA (Fig. 3). 7-NI and the two iNOS inhibitors (SMUT and AT-UT) studied, contracted the MCA also significantly, see Fig 3. The ACh induced endothelium dependent MCA vasodilator responses remained intact in control vessels, both by in vivo and in vitro AG, 7-NI and by in vitro SMUT or AT-UT application. The in vitro constrictor effect of iNOS inhibitors (AG, SMUT, AE-T) was significantly greater (480-700 mg) compared to the ncNOS inhibitor 7-NI (200-250 mg).

Fig 3. AG,7-NI,SMUT.AE-UT

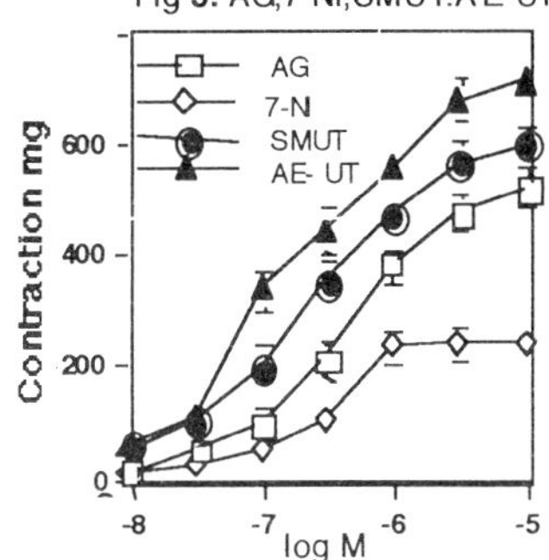

Pronounced differences can be observed by which the NO synthase inhibitors evoke vasoconstriction. The AG contractions are significantly inhibited by the endothelin antagonist PD14 (10^{-7} M), and by phentolamine (10^{-7} M) see Fig. 4. The AC inhibitor captopryl has no effect.

Fig.4 Effect of phentolamine PD14 on AG contraction

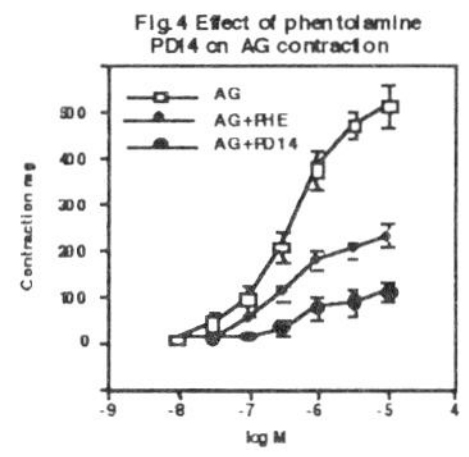

The 7-NI-induced contractions are significantly inhibited by the endothelin antagonist PD14. Captopryl and Phentolamine were ineffective.

SMUT constriction is decreased significantly by both phentolamine and the converting enzyme antagonist captopryl but not by the endothelin antagonist PD14.

Captopryl significantly inhibits the AE-UT induced MCA constriction. Phentolamine and the endothelin antagonist had no effect.

The MCA constricted in vitro by the 4 NOS inhibitors showed normal endothelium dependent vascular relaxation to ACh.

CONCLUSION. CBF decreased significantly after AG administration in three of the ten brain regions studied. These results suggest that either the NOS in these structures differs from the other brain regions or the blood flow decrease is related to a vascular effect of AG. The in vitro results demonstrate that the iNOS inhibiting compounds more powerfully constrict cerebral vessel than ncNOS inhibitors. The data demonstrate that adrenergic activation, endothelin release, and converting enzyme activation are involved differently in the complex regulatory process of vasoconstriction after NOS inhibition, providing information regarding the mechanism of cerebral vascular constriction after ncNOS and iNOS inhibition.

Supported by NIH, NS RO1 31429-03.

REFERENCES 1. Kovách et al. J. Physiol. 449 183-196, 1992; 2. Kovách et al. Ann. New York Acad. Sci. 738348-368,1994; 3. Moore et al. Brit. J. Pharmacol, 108, 296-97, 1993.

Effect of oxidized low-density lipoprotein on contraction and nitric oxide-induced relaxation of porcine coronary artery and rat aorta

DAVID A. COX AND MARLENE L. COHEN

Lilly Research Laboratories, Eli Lilly and Co., Indianapolis, IN 46285

Introduction

The coronary vasodilation that normally results from intracoronary infusion of 5-hydroxytryptamine (5-HT) is converted to a vasoconstriction in humans with atherosclerosis, most likely due to an inhibition of endothelium-dependent, nitric oxide (NO)-mediated vasorelaxation [1]. Oxidized low-density lipoprotein (oxLDL) has been implicated in the initiation and progression of atherosclerosis [2], and has been shown to inhibit endothelium-dependent relaxation [3]. Although these data suggest a role for this lipoprotein in atherosclerosis-induced vascular dysfunction, data on the effect of oxLDL on vasoconstriction is limited. Therefore, the purpose of this study was to compare the effect of human oxLDL on 5-HT-induced contraction and endothelium-dependent, NO-mediated relaxation in isolated rings of porcine coronary artery (PCA) and rat aorta (RA).

Methods

Human LDL was purchased commercially and oxidized by exposure to $CuSO_4$ (50 µM) for 8 hrs. Arterial rings were prepared from PCA and RA and mounted in tissue baths containing 10 ml Krebs'-Heinseliet buffer. Changes in force in response to additions of test compounds were detected and recorded as described previously [4].

Results and Discussion

OxLDL (100 µg/ml) significantly enhanced 5-HT-induced contraction in the PCA (Fig. 1A). In contrast, oxLDL (100 µg/ml) had no effect on 5-HT-induced contraction in the RA (Fig. 1B).

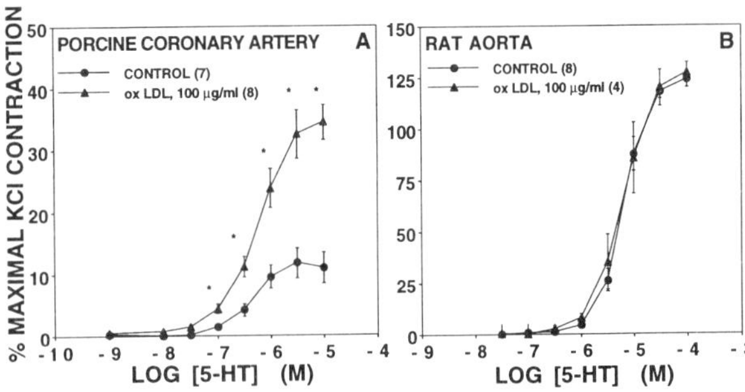

Figure 1: Effect of oxLDL (100 µg/ml) on 5-HT-induced contraction of (A) porcine coronary artery and (B) rat aorta

Thus, oxLDL enhanced 5-HT-induced contraction selectively in the PCA.

Since oxLDL was suggested to affect endothelial function, we compared the effect of 5-HT on the endothelium in the PCA and RA. Arterial rings were incubated with ketanserin (1 µM) to block the 5-HT_{2A} contractile receptors, and were contracted with the thromboxane A_2 analog U46619 (30 nM). 5-HT relaxed precontracted PCA but not the RA (Fig. 2A). The 5-HT-induced relaxation in the PCA was blocked by L-NAME (100 µM)(Fig. 2A) and endothelial removal (data not shown), confirming previous results [4] and indicating that 5-HT relaxed the PCA via release of NO from the endothelium. Although 5-HT did not relax RA, acetylcholine-induced relaxation in RA was blocked by L-NAME (100 µM)(Fig. 2B), confirming the presence of endothelium-dependent, NO-mediated relaxation in this tissue. Taken together, these data suggest that, unlike the PCA, the RA lacks endothelial 5-HT receptors coupled to NO release and vascular relaxation. Thus, while removal of the endothelium in PCA abolished both basal and 5-HT-stimulated NO release, the same manipulation in RA resulted in the loss of only basal NO release.

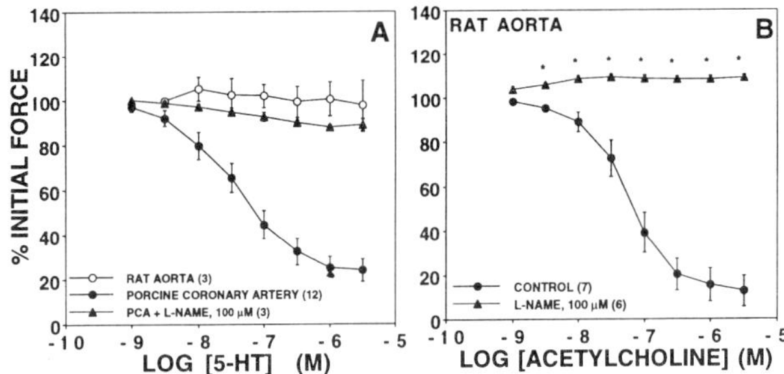

Figure 2: (A) Effect of 5-HT on PCA and RA precontracted with U46619 (30 nM). (B) Effect of acetylcholine on RA precontracted with U46619.

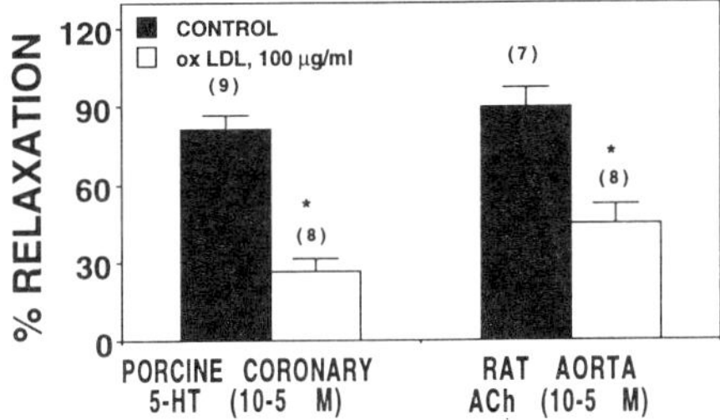

Figure 3: Effect of oxLDL (100 µg/ml) on endothelium-dependent relaxation to 5-HT (10^{-5}M) and acetylcholine (10^{-5}M) in PCA and RA, respectively.

Since 5-HT increased NO release in the PCA but not in the RA, an effect of oxLDL on agonist-induced NO release would explain the selective enhancement of 5-HT-induced contraction in PCA. Indeed, oxLDL (100 µg/ml) significantly inhibited 5-HT-induced relaxation in the PCA (Fig. 3). Therefore, oxLDL enhanced 5-HT-induced contraction via an effect on endothelial NO release or breakdown. OxLDL (100 µg/ml) also inhibited acetylcholine-induced relaxation in the RA (Fig. 3). Thus, the failure of oxLDL to enhance 5-HT-induced contraction in the rat aorta was most likely due to the inability of 5-HT to induce NO release from the endothelium in this tissue rather than an inability of oxLDL to inhibit agonist-induced NO release in RA. Furthermore, the selective enhancement of 5-HT-induced contraction in PCA relative to the RA suggests that oxLDL does not affect directly 5-HT-induced smooth muscle contraction.

In summary, these studies demonstrate that alterations in coronary arterial contraction to 5-HT can be induced by oxLDL *in vitro* and mimic the effect of atherosclerosis on human coronary arteries *in vivo*. Thus, oxLDL may be important in the genesis of vascular dysfunction produced by atherosclerosis.

References

1. Golino, P., Piscione, F., Willerson, J.T., et al. (1991) Divergent effects of serotonin on coronary artery dimensions and blood flow in patients with coronary atherosclerosis and control patients. N. Eng. J. Med. 324, 641-648.

2. Young, S.G. and Parthasarathy, S. (1994) Why are low-density lipoproteins atherogenic? West. J. Med. 160, 153-164.

3. Kugiyama, K., Kerns, S.A., Morrisett, J.D., et al. (1990) Impairment of endothelium-dependent arterial relaxation by lysolecithin in modified low-density lipoproteins. Nature 344, 160-162.

4. Cox, D.A. and Cohen, M.L. (1995) Selective endothelial dysfunction in porcine coronary artery produced by incubation in cell culture medium *in vitro*. Endothelium 3, in press.

Platelet-activating factor contributes differently to the induction of nitric oxide synthase elicited by lipoteichoic acid or lipopolysaccharide in anaesthetised rats

SJEF J. DE KIMPE, CHRISTOPH THIEMERMANN and JOHN R VANE

The William Harvey Research Institute, St Bartholomew's Hospital, Medical College, Charterhouse Square, London EC1M 6BQ, United Kingdom

Receptor antagonists for platelet-activating factor (PAF) exert beneficial effects in animal models of Gram-negative shock. The vasodilator nitric oxide (NO) contributes importantly to the circulatory failure (hypotension and vascular hyporeactivity) elicited by endotoxin (lipopolysaccharide; LPS). Interestingly, the PAF receptor antagonist WEB2086 inhibits the expression of the inducible isoform of nitric oxide synthase (iNOS) by LPS in rats [1]. Gram-positive organisms do not contain LPS and thus may elicit septic shock in a different way from Gram-negative bacteria. The cell wall component lipoteichoic acid (LTA) from *Staphylococcus aureus* induces iNOS expression in anaesthetised rats [2]. Here, we investigate the effect of two structurally different PAF antagonists, WEB2086 and BN52021, on the induction of iNOS and hypotension elicited by LTA or LPS in anaesthetised rats.

Macrophages (J774.2) were cultured to confluency in a 96 well plate containing Dulbeco's modified essential medium and activated with LTA (10 μg/ml) to induce iNOS. After 24 h, nitrite concentration as an indicator of NO production was measured in the supernatant by the Griess reaction. WEB2086 or BN52021 were added to the wells 20 min prior to LTA. Male Wistar rats (200-325g) were anaesthetised with thiopentobarbitone sodium (120mg/kg, ip). The trachea was cannulated to facilitate spontaneous respiration, the carotid artery to monitor mean arterial blood pressure (MAP) and heart rate, and the jugular vein to administer compounds. WEB2086 (5 mg/kg) or BN52021 (20 mg/kg) were administered 20 min prior and 160 min after the injection of LTA (10 mg/kg), LPS (10 mg/kg) or PAF (30 ng/kg) at time 0. Induction of iNOS activity was determined by the conversion of [3H]L-arginine to [3H]L-citrulline in lung homogenates. Statistical evaluation was performed using Bonferoni's t test for multiple comparisons.

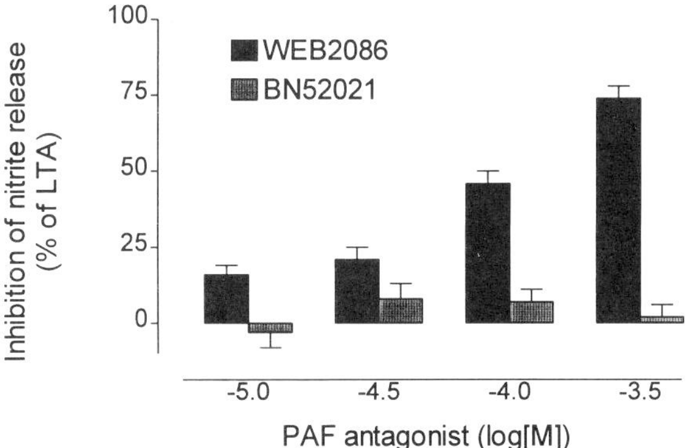

Figure 1. WEB2086, but not BN52021, inhibits the formation of nitrite by cultured macrophages activated with LTA (10 μg/ml for 24 h). WEB2086 or BN52021 were added 20 min prior to LTA. Results are expressed as mean ± S.E.M. of four independent experiments performed in triplicate.

Intravenous injection of PAF (30 ng/kg) resulted in a fall in MAP of 52 ± 3 mmHg (n=3). The decrease in MAP elicited by PAF was strongly inhibited by treatment of the rats with WEB2086 (5 mg/kg, 8 ± 2 mmHg, $P<0.01$) or BN52021 (20 mg/kg, 9 ± 2 mmHg, $P<0.01$) 20 min prior to the injection of PAF. Thus, at the doses used in this study, both WEB2086 or BN52021 inhibit the hypotension elicited by exogenous PAF to a similar extent. Activation of cultured macrophages by LTA for 24 h resulted in an increase in nitrite accumulation from 1.9 ± 0.4 mM (n=12) to 30.0 ± 3.1 mM (n=12), which was inhibited by WEB2086, but not by BN52021 (Fig. 1). These results suggest that WEB2086 acts intracellularly to interfere with PAF as a second messenger in the induction of nitrite release in macrophages activated by LTA. In contrast, BN52021 inhibits only the hypotensive effect of PAF, indicating that BN52021 acts only extracellularly. This is supported by findings that BN52021 preferentially binds to extracellular PAF receptors [3], and that WEB2086 inhibits the actions of PAF released intracellularly in leukocytes and endothelial cells [4]. Interestingly, in many different cell types, including leukocytes and endothelial cells, the majority of PAF is retained intracellularly upon stimulation [5].

Table 1. Effect of PAF antagonists *in vivo*
Values are mean ± S.E.M. of 5-12 experiments; nd: not done; # $P<0.05$ PAF antagonist vs LTA or LPS alone.

	LTA at 300 min		LPS at 180 min	
	MAP	iNOS	MAP	iNOS
	mmHg	pmol/min/mg	mmHg	pmol/min/mg
vehicle	108 ±5	0.3 ±0.1	111±6	nd
LTA or LPS	75 ±6	3.3 ±0.6	74±9	7.1 ±1.3
+ BN52021	74 ±6	3.8 ±0.7	100±5 #	2.2 ±0.5#
+ WEB2086	99 ±6#	1.3 ±0.5#	nd	nd

Injection of LTA resulted in a sustained hypotension at 300 min (Table 1). Treatment of rats with WEB2086 prevented the delayed hypotension compared to sham-operated rats receiving vehicle only. In contrast, treatment of rats with BN52021 had no significant effect on the hypotension elicited by LTA at 300 min. Similarly, the induction of iNOS activity in lungs by LTA at 300 min was significantly inhibited by WEB2086, but not by BN52021. As BN52021 acts only extracellularly, these results indicate that LTA causes the intracellular release of PAF, which contributes to the induction of iNOS activity and, subsequently, delayed hypotension.

Injection of LPS results in the induction of iNOS and hypotension in anaesthetised rats at 180 min (Table 1) In contrast to LTA, the responses to LPS were inhibited by the treatment with BN52021 (Table 1). Thus, LPS elicits the extracellular release of PAF, which contributes to the induction of iNOS activity.

Sepsis resulting from Gram-positive organisms has risen markedly in the last decade, and it is possible that cases of Gram-positive sepsis may predominate in the years to come [7]. Therefore, it is important to develop animal models without endotoxaemia in order to elucidate the pathopysiology of Gram-positive shock and to evaluate the effects of novel therapeutic agents. In a recent multicenter clinical trial, the PAF receptor antagonist BN52021 appeared to be beneficial in patients with severe Gram-negative, but not with Gram-positive sepsis [8]. This surprising difference between Gram-negative and Gram-positive organisms, is also observed in our rat models of septic shock employing LPS or LTA.

This study is partially supported by Cassella AG (Germany). SJDK is a recipient of a fellowhip of the European Union.

1. Szabó, C., Wu, C.-C., Mitchell, J.A., Gross, S.S., Thiemermann, C. and Vane, J.R. (1993) Circ. Res. **73**, 991-999.
2. De Kimpe, S. J., Hunter, M. L., Bryant, C. E., Thiemermann, C. and Vane, J. R. (1995) Br. J. Pharmacol. **114**, 1317-1323.
3. Marcheselli, V. L., Magdalena, J. R., Domingo, M.-E., Braquet, P. and Bazan, N. G. (1990) J. Biol. Chem. **265**, 9140-9145.
4. Stewart, A. G., Dubbin, P. N., Harris, T. and Dusting, G. J. (1990). Proc. Natl. Acad. Sci. USA **87**, 3215-3219.
5. Lynch, J. M. and henson, P. M. (1986) J. Immunol. **137**, 2653-2661.
6. Tenaillon, A., Dhainaut, J. F., Letulzo, Y., et al. (1993). Am. Rev. Respir. Dis. **147**, A196.
7. Bone, R. C. (1994) Arch. Intern. Med. **154**, 26-34.

Abbreviations used: BN52021: ginkgolide B; iNOS: inducible nitric oxide synthase; LPS: lipopolysaccharide; LTA: lipoteichoic acid; MAP: mean arterial pressure; NO: nitric oxide; PAF: platelet activating factor; WEB2086: apafant.

The character of NIPRADILOL:β adrenergic blocker and NO donor in Atherosclerotic Vessels

Toshio Hayashi,[#] Kazuyoshi Yamada, Teiji Esaki, Emiko Muto and Akihisa Iguchi

Department of Geriatrics and Department of Pharmacology, Nagoya University School of Medicine, 65 Tsuruma-cho, Showa-ku, Nagoya, 466, Japan*

Summary

Nipradilol is a characteristic β-adrenergic antagonist (β-blocker) containing a nitroso compound. We studied the character and vasoactive effect of this drug in atheroslcerotic vessels. Nipradilol spontaneously released NO in the presence of endothelial cells. Rabbits were divided into four groups: GpI; standard diet, GpII; standard diet plus nipradilol, GpIII; atherogenic diet, GpIV: atherogenic diet plus nipradilol; treatment lasted 9 weeks. No changes were induced in body weight, blood pressure,or plasma HDL cholesterol by either atherogenic diet or by the presence of nipradilol. The atherogenic diet significantly increased serum total cholesterol and triglycerides, and no change was observed between GpIII and IV. Nipradilol restored the acetylcholine-induced NO mediated relaxation response that was diminished in atherosclerotic aortic rings. The endotheliium-independent relaxation induced by nitroglycerin(NG) was not affected by either condition. Results indicate that the NO released by nipradilol may protect endothelial function in atherosclerosis with no deleterious effect on the metabolism of NO or lipids.

Introduction

There are many reports indicating the impairment of endothelium-dependent vasodilation in the coronary arteries of patients with atherosclerosis or risk factors for atherosclerosis (1). Animal experiments and human studies have shown that endothelial dysfunction can occur when there is a subtle, but appreciable, intimal thickening, which suggests that a loss of endothelium-dependent vasodilator function may be an early manifestration of atherosclerosis(2). Nitric oxide (NO) is believed to protect against the development of atherosclerosis variously by producing vascular dilatation, inhibiting the adhesion of monocytes adhesion to endothelium, or inhibiting of platelet aggregation(3). Nipradilol (3,4-dihydro-8-[2-hydroxy-3-isopropylamino] propoxy-3-nitroxy-2H-1-benzopyran), β-blocker that contains a nitrosyl group, has been reported to increase cGMP in vascular smooth muscle *in vitro*(4).

Data from animal experiments have shown that β blockers is antiatherogenic, particularly among individuals who are behaviourally predisposed to coronary artery atherosclerosis(5). It is possible that the well-known deleterious effect of β blockers on serum lipids is one reason for the discrepancies among reports of the effect of β blockers on atherosclerosis. Our objective was to elucidate the mechanism of action of Nipradilol on atherosclerotic blood vessels. .

METHODS

Animals 30 male New Zealand white rabbits were divided into 4 gourps with 7 to 9 animals in each; GpI: standard diet, GpII: standard diet with nipradilol 10mg/kg.day, GpIII: standard diet with 1% cholesterol, GpIV: standard diet with 1% cholesterol and nipradilol.

Cultured endothelial cells Endothelial cells from bovine aorta (BAEC) were taken from the thoracic aorta of a bovine fetus.

Blood Pressure Measurement We monitored blood pressure every 4 weeks, using a indirect cuff technique.

Preparation of isometric tension measurement Preparation of rabbit aortic rings was similar to that described by Furchgott and Zawadzki(6). Briefly, the thoracic aortas were cut into 3-mm- wide transverse rings which were mounted under 1 g of resting tension on stainless-steel hooks in 20 ml-capacity muscle chambers and were bathed in Kreb's bicarbonate solution, pH 7.4 at 37°C. Some experiments were done after indomethacin (5X10^{-6}M) was incubated for 60 min to rule out the contribution of prostanoids.

Measurement of nitric oxide by NO electrode The principle of the NO-meter (Model NO501, Intermedic Co.,Ltd Nagoya, Japan) is developed newly is the measurement of the pA-order redox current between a working electrode and a counter electrode using amodified microsensor (7).

Measurement of $NO2^-/NO3^-$ The concentration of the nitrite and nitrate ($NO2^-/NO3^-$) in the plasma were determined with an autoanalyzer (TCI-NOx 1000: Tokyo Kasei Kogyo Co.,Tokyo, Japan).

Histological evaluation of atherosclerosis Cross sections of the descending thoracic aorta adjacent to each segment used in the *in vito* experiments were stained with hematoxylin-eosin and with van Gieson's elastic stain.

RESULTS

No significant differences were observed in body weight and serum total protein between 4 groups. The application of nipradilol did not affect plasma lipid levels (data not shown). Blood pressure and heart rate (measured in morning) did not show any differences between 4 group rabbits before and after 9 weeks treatment(data not shown). ACh (Figure 1) and A23187 (data not shown) produced concentration-dependent relaxation of endothelium-intact aortic rings precontracted with prostaglandin F2α. In GpIII, the magnitude of relaxation of these aortic rings to these endothelium-dependent vasodilators was diminished. No significant differences in relaxation to ACh and A23187 between aortic rings obtained from either control (GpI or II) or hypercholesterolemic animals with nipradilol (GpIV) were observed. NG produced concentration-dependent relaxant responses of endothelium-denuded aortic rings which were of equal magnitude in rings from control animals (GpI and II)(data not shown). Histological examination showed atheromatous lesions in GpIII and IV compared to the control animals. There was not a significant difference in between Gp IV and III, as indicated by mean lesion area and mean percentage of luminal encroachment(Table 1). Nipradilol released NO spontaneously, in vitro, in the presence of the aortic endothelium, as indicated by the NO-selective electrode.

DISCUSSION

These data suggest that Nipradilol did not show any deleterious effects in lipid metabolism and did not show remarkable β blocking effects in this study. The endothelium-dependent relaxation induced by ACh was attenuated in the arteries from GpIII , although the response was preserved in GpIV. Nipradilol can reverse the capacity of NO release stimulated with various agonists which was decreased in atherosclerotic vessels. This effect could be attributed to the preservation of endothelium. Recently, probucol was to be able to reserve the endothelial function in atherosclerotic coronary artery as antioxidant (8). NO is easily speculated to act as scavenger for $O2^-$ as superoxide-dismutase.

REFERENCES

1. Ludmer PL et al. (1986) N Engl J Med, 315, 1046-105

2. Luscher TF et al. (1993) Annu Rev Med 144:395-418

3. Hayashi T et al. (1991) Atherosclerosis, 87, 23-38.

4. Radmski MW et al.(1987) Br J Pharmacol 192:639-46

5. Bondjers G et al. (1994) Eur Heart J 15 s8-s15

6. Furchgott RF and Zawadzki J(1980) Nature(Lond.), 288, 373-376.

7. Malinski T et al. (1987) Nature 358:676-8

8. Plane F et al.(1993). Atherosclerosis, 103(1), 73-9.

Figure 1. Cumulative concentration-response cures to acetylcholine during contraction evoked by prostaglandin F2α in the thoracic aortas from each group of rabbits

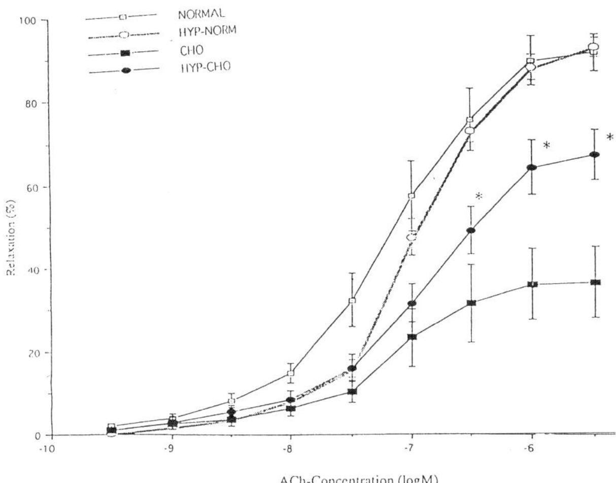

TABLE 1.

NOx

	Before	3 W	6 W	After
GpI	3.80±0.63	3.01±0.55	3.40±0.24	3.50±0.24
II	same	4.11 ±0.65	4.10±0.20	4.05±0.25
III	same	3.80±0.72	3.90±0.65	3.40±0.68
IV	same	4.05±0.59	3.00±0.67	3.36±0.52

ATHEROSCLEROSIS

	Surface Involvement	Area Occupied Lesion
GpI	0	0
II	0	0
III	9.4 ± 3.0	9.4± 3.0
IV	40.6±9.1	43.4±7.1

The Effects of Peroxynitrite on the Myocardium During Ischemia-Reperfusion is Dependent upon the Biological Environment

Bernard L Lopez, Theodore A Christopher, Xin-liang Ma

Division of Emergency Medicine, Thomas Jefferson University, Philadelphia, Pa.

Reperfusion of ischemic tissue can initiate the simultaneous production of NO and superoxide (O_2^-). The product of NO and O_2- is peroxynitrite ($ONOO^-$), a potent free radical that causes oxidative tissue damage. $ONOO^-$ appears to have significant importance in the development of myocardial ischemia/ reperfusion (IR) injury.

An important issue is the role of the environment in the pathophysiology of $ONOO^-$-induced myocardial IR injury. The vast majority of studies examining $ONOO^-$ involves biochemical or cell culture studies. To date, there is only one study examining the effects of $ONOO^-$ in a heart preparation. Villa et al examined the effects of peroxynitrite in a crystalloid-perfused, isolated rat heart preparation, finding that $ONOO^-$ induces both vasodilation and impaired vascular relaxation. It is unknown how $ONOO^-$ would affect cardiac function during IR in a blood-perfused environment.

The objective of our study was to investigate the effects of blood vs. crystalloid perfusion on the pathophysiologic effects of $ONOO^-$ in rat myocardial IR in an isolated heart preparation.

Methods

Male Sprague-Dawley rat hearts were perfused with either Krebs-Henseleit (KH) solution or blood on a Langendorff perfusion apparatus and subjected to 30 min of ischemia followed by 60 min of reperfusion. SIN-1, an $ONOO^-$ generator, was administered immediately before reperfusion. In a separate series of experiments, hearts perfused with KH were treated with SOD or SIN-1 + SOD prior to reperfusion. Cardiac function (% recovery of left ventricular developed pressure [LVDP] and coronary flow [CF] after reperfusion) as well as cellular injury (release of LDH and % necrosis by NBT staining) were measured. At 30 min of reperfusion:

Table. Cardiac function/injury 30 min after Reperfusion

	%LVDP	%CF	LDH(U/ml)	%Necr
Blood	38±4	45±4	2996±327	55±6
Blood + SIN-1	57±6*	66±7*	1648±233**	31±7**
KH	59±9	57±6	848±139	29±6
KH+SIN-1	49±5	64±5	1434±203t	48±7t
KH+SOD	71±6t	69±6	574±104t	18±2t
KH+SIN-1+SOD	79±6yy	73±5y	496±98yy	11±3yy

*$p<0.05$, **$p<0.01$ vs blood group; [t]$p<0.05$ vs KH group; [y]$p<0.05$, [yy]$P<0.01$ vs KH+SIN-1; n=10-15 hearts per group

Summary

1. SIN-1 (3-morpholinosydnonamine) is a well-described simultaneous donor of NO and O_2- that generates $ONOO^-$.

2. Blood-perfused hearts demonstrated significantly worse injury after IR than crystalloid-perfused hearts. The presence of leukocytes and platelets enhanced this injury.

3. The administration of SIN-1 to blood-perfused hearts caused significant protection against IR-induced injury.

4. The administration of SIN-1 to crystalloid-perfused hearts worsened cellular injury after IR; this effect was reversed by the co-administration of SOD. This superoxide scavenging property is a likely mechanism of protection against $ONOO^-$-induced myocardial IR injury.

5. Blood products, most notably blood thiols such as glutathione, scavenge superoxide, preventing the formation of $ONOO^-$ as well as enhancing the presence of NO. In addition, thiols regenerate NO from $ONOO^-$.

6. The net effect of SIN-1 in a blood-perfusion environment is the increased presence of NO. NO is therefore protective against IR-induced injury.

Conclusion

1. SIN-1 exerts a deleterious effect on crystalloid-perfused rat hearts during IR that is reversed by the co-administration of SOD.

2. SIN-1 exerts a protective effect on blood-perfused rat hearts during IR. It is likely due to the presence of NO.

3. The actions of $ONOO^-$ are critically dependent on the biological environment.

Effect of modification of nitric oxide levels on tumour vascular tone

GILLIAN M. TOZER, VIVIEN E. PRISE, STEVEN A. EVERETT, MICHAEL R. L. STRATFORD, MADELEINE F. DENNIS, S. WORDSWORTH and DAVID J. CHAPLIN

Tumour Microcirculation Group, Gray Laboratory Cancer Research Trust, Mount Vernon Hospital, Northwood, Middlesex, U.K., HA6 2JR

Introduction

NOS inhibitors have been shown to reduce tumour blood flow [1]. They therefore have potential for improving cancer chemotherapeutic regimes which are optimised under ischaemic conditions. We investigated the effects of the NOS inhibitor, L-NNA, and the NO· scavenger, C-PTIO, on vascular parameters of tumours and normal tissues in control rats and rats stimulated with LPS.

Materials and Methods.

Tumours: P22 carcinosarcoma (early transplants), s.c. in BD9 rats, 1.41 ± 0.06g. Chemicals: C-PTIO (Cambridge Bioscience, Cambridge, U.K.), L-NNA (Sigma Chemical Co. Ltd., Poole, U.K.), $LPS_{e.\ coli}$ (Sigma), ^{125}I-IAP (Institute of Cancer Research, Sutton, U.K.).

L-NNA (1 mg kg^{-1} i.v.) or C-PTIO (1.7 mg kg^{-1} min^{-1} i.v.) were administered at 20 or from 15 minutes prior to blood flow measurement respectively. LPS (20 mg kg^{-1} i.p.) was normally administered to conscious rats 3 or 7 hours prior to blood flow measurement. Blood samples were taken to assay plasma nitrate by anion-exchange chromatography and plasma C-PTIO/C-PTI levels by EPR spectroscopy.

Arterial blood pressure was monitored. Blood flow to tumour and normal tissues was measured in anaesthetised (Hypnorm and midazolam) rats using uptake of ^{125}I-IAP [2]. Blood flow values were used to calculate tissue vascular resistance (= perfusion pressure ÷ blood flow), where perfusion pressure was taken to be MABP.

Results

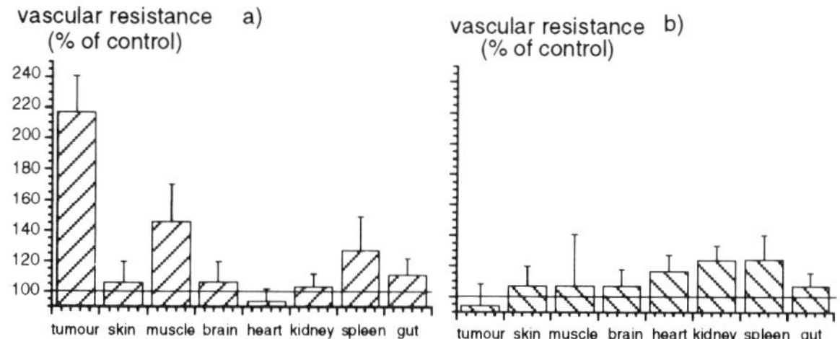

Figure 1: Effect of L-NNA (a) and C-PTIO (b) on tissue vascular resistance. Results are means ± 1 SEM for n > 5 for all groups

1 mg kg^{-1} L-NNA significantly increased tumour vascular resistance (i.e. vasoconstricted) by a factor of 2.2 ($p < 0.001$) (Figure 1a). MABP remained unchanged at this dose such that blood flow changes (results not shown) were the direct reverse of the changes shown in Figure 1. None of the other tissue changes reached statistical significance at the 5% level. Figure 1b) shows the same results for 1.7 mg kg^{-1} min^{-1} C-PTIO, which also had no effect on MABP. In this case, there was no increase in tumour vascular resistance although there was a small (x1.2) increase in resistance in the kidney ($p < 0.05$).

After stimulation by LPS, concentration of nitrate in rat plasma was 46 ± 2 μM and 146 ± 7 μM (n = 4) at 3 and 7 h respectively. EPR analysis showed that plasma levels of C-PTIO at the end of infusion were 103 ± 7 μM (n = 7). Spectra showed a secondary peak, typical of C-PTI, which indicates significant scavenging of NO· (results not shown).

Abbreviations used: ^{125} I-IAP, ^{125} I-iodoantipyrine; C-PTIO, carboxy-2-phenyl-4, 4, 5, 5,tetramethylimi-dazoline-1-oxyl-3-oxide; EPR, electron proton resonance; L-NNA, N^{ω}-nitro-L-arginine; LPS, lipopolysaccharide (e. coli); MABP, mean arterial blood pressure; NO· , nitric oxide; NOS, nitric oxide synthase

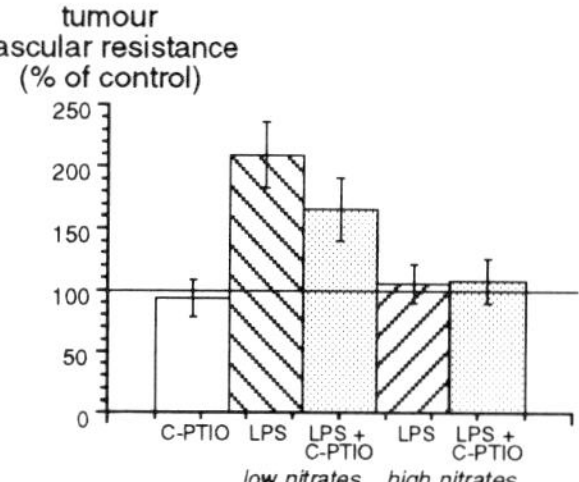

Figure 2: Effect of C-PTIO and LPS on tumour vascular resistance. "Low nitrate" and "high nitrate" correspond to 3 and 7 h post LPS administration respectively - see text. Data are means ± 1 SEM for n n > 5 for all groups

LPS significantly increased tumour vascular resistance by 100% ($p < 0.05$) at 3h post LPS (*low nitrate*) (Figure 2). Addition of C-PTIO had no further effect. At 7h (*high nitrate*), tumour vascular resistance had returned to control levels. Addition of C-PTIO had no further effect. Results for normal skin, skeletal muscle, small intestine and brain were very similar to those for tumour. Similarly, MABP increased at 3h, then decreased to normal at 7h. (results not shown). C-PTIO did not reverse the effects of LPS for any tissue. In spleen, there was no effect of LPS at 3h and resistance decreased slightly at 7h (results not shown). This may reflect high local levels of NO·. No effects were observed in kidney and heart.

Discussion

The tumour vasculature was found to be uniquely sensitive to inhibition of NOS by L-NNA. Therefore, L-NNA or a related compound has potential for optimising cancer treatments which benefit from ischaemic tumour conditions (e.g. bioreductive drugs). However, the NO· scavenger C-PTIO, did not affect tumour vascular resistance. The explanation for this remains unclear: plasma levels of C-PTIO appeared to be adequate (>100μM). Yoshida *et al.* [3] explained a lack of response to C-PTIO in normal rats by suggesting that an NO· bioadduct(s) (e.g. nitrosothiols) may play a role in normal haemodynamic control. However, this concept is controversial [4].

A small *hypertensive* effect and vasoconstriction in tumour and some normal tissues at 3 h post LPS followed by return to normal at 7 h corresponded to low and high levels of plasma nitrate respectively. This suggests that increased levels of NO· could account for the effects at 7h. C-PTIO did not reverse any of the LPS effects. C-PTIO has been found to reverse the *hypotensive* effect of i.v. LPS [4]. The explanation for the mild *hypertension*, induced in our rats by i.p. administration, is most likely different local tissue concentrations of LPS. Vasoconstrictive compounds such as endothelin, angiotensin II and neuropeptide Y are known to be produced following LPS [5] which could induce hypertension.

In conclusion, inhibition of NOS by L-NNA or a similar compound has potential for selective modification of tumour blood flow for therapeutic benefit. There is less potential for C-PTIO except for mechanistic studies.

Acknowledgements

This work was funded by the Cancer Research Campaign.

References

1. Meyer,R.E., Shan,S., DeAngelo,J., et al. (1995) Nitric oxide synthase inhibition irreversibly decreases perfusion in the R3230Ac rat mammary adenocarcinoma, Br. J. Cancer 71, 1169-1174
2. Tozer,G.M. and Shaffi,K.M. (1993) Modification of tumour blood flow using the hypertensive agent, angiotensin II, Br. J. Cancer 67, 981-988
3. Yoshida,M., Akaike,T., Wada,Y., et al. (1994) Therapeutic effects of imidazolineoxyl N-oxide against endotoxin shock through its direct oxygen-scavenging activity, Biochem. Biophys. Res. Commun. **202**, 923-930
4. Feelisch,M., te Poel,M., Zamora,R., Deussen,A. and Moncada,S. (1994) Understanding the controversy over the identity of EDRF, Nature **368**, 62-65
5. Weitzberg,E. (1993) Circulatory responses to endothelin-1 and nitric oxide with special reference to endotoxin shock and nitric oxide inhalation, Acta Physiol. Scand. **148, Suppl. 611**, 1-72

Induction of inducible nitric oxide synthase mRNA in rat aortic endothelial cells and human neuroblastoma cells following treatment with paclitaxel (an antitumor agent) and interferon-γ

Y. KURASHIMA, T. OGURA, *Y. SASAKI and H. ESUMI

Investigative Treatment Division, National Cancer Center Research Institute, East. *Division of Hematology and Oncology, National Cancer Center Hospital, East, 5-1, Kashiwanoha 6-chome, Kashiwa, Chiba, 277, JAPAN

Introduction

Paclitaxel isolated from the Western Yew, *Taxus brevifolia* is a new antimicrotubular antineoplasic agent which induces tubulin polymerization. This agent has been shown to have a wide range of antitumor activity (1). However, in some clinical cases, paclitaxel caused serious adverse effects such as hypotension, neuropathy and neutropenia. In the present work, we asked if NO is involved in the mechanisms of those adverse effects of paclitaxel. We found that paclitaxel is a inducer of inducible NO synthase (iNOS) in endothelial and neuronal cells.

Materials and Methods

Cells : Rat aortic endothelial cell, RACE,(2) which was gift from Dr. Nakaki (Keio Univ, Tokyo, Japan) and human neuroblastoma cell line, NB-39-nu, were cultured at 37°C under 5 % CO_2 in air in Earle's M199 and RPMI 1640 medium supplemented with 10% fetal calf serum, respectively. .

Treatment : Paclitaxel (gift from Bristol-Myers Squibb Company) was dissolved in dimethyl sulfoxide to a stock concentration of 10 or 50 mM. Culture medium was changed 3 h before treatment and cells were treated with or without 30 μM paclitaxel, and 1000 U/ml recombinant mouse interferon-γ (IFN-γ) or 100 U/ml recombinant human IFN-γ at 70-80% confluency.

Northern blot analysis : Total RNA was isolated from cells using guanidine thiocyanate extraction method (3). Ten μg of total RNA was separated on a 1 % agarose gel containing 6 % formaldehyde, which was then blotted onto a membrane filter. To detect rat and human iNOS mRNA, a 0.5 kb fragment of rat iNOS cDNA (4) and a 0.5 kb fragment of human iNOS cDNA (5) were used as probes, respectively. Each mRNA was detected by hybridization as described previously (5).

Results and Discussion

A trace amount of iNOS mRNA was detected in RACE and NB-39-nu upon stimulation with IFN-γ . As shown in Fig. 1., iNOS mRNA was remarkably induced at 3 h after treatment with IFN-γ alone, then rapidly decreased. The mRNA expression was synergistically enhanced by a combination of IFN-γ and paclitaxel. No iNOS mRNA could be detected in RACE after treatment with paclitaxel alone. High level of iNOS mRNA expression was observed in NB-39-nu by treatment with IFN-γ and paclitaxel simultaneously, but not either with IFN-γ alone or paclitaxel alone at 48 h (Fig. 2). The time course of induction of iNOS mRNA in endothelial and neuronal cells were completely different. In clinical cases, hypotension and peripheral neuropathy were observed within 3 h, and 48 to 120 h after start of paclitaxel infusion, respectively, and these time courses of the appearance of adverse effects are well concordant with those of iNOS mRNA expressions in endothelial cells and neuronal cells upon stimulation with IFN-γ and paclitaxel, respectively. These findings strongly suggest that induction of iNOS expression play an important role in hypotension and neurotoxicity by paclitaxel.

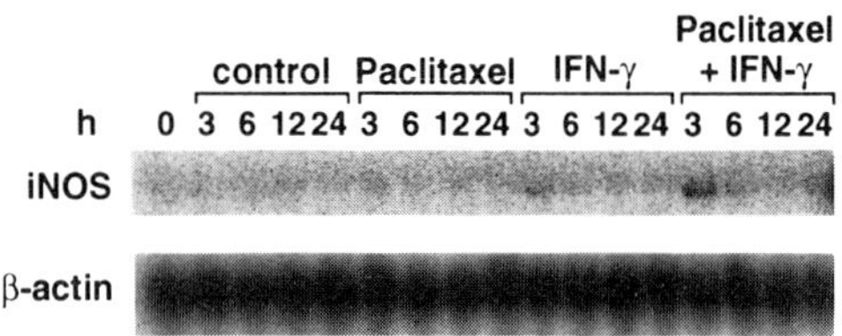

Fig.1 Expression of iNOS mRNA in RACE treated with paclitaxel and/or IFN-γ. RACE was treated with 30 μM paclitaxel and/or 1,000 U/ml recombinant mouse IFN-γ. At the time indicated, 10 μg of total RNA was analyzed by northern blotting using rat iNOS cDNA probe. The amount of total RNA analyzed was evaluated by hybridization with rat β-actin probe.

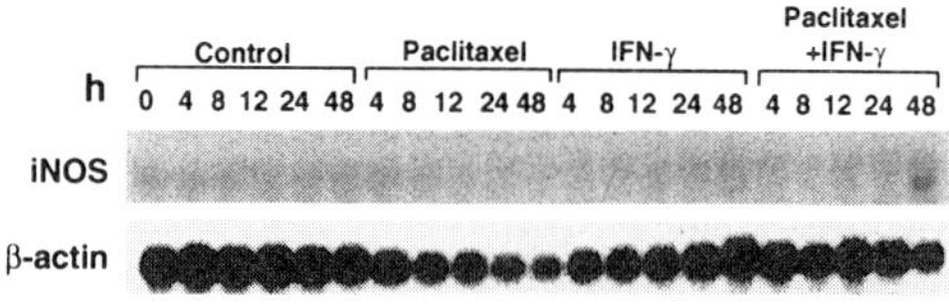

Fig.2 Expression of iNOS mRNA in NB-39-nu after treatment with paclitaxel and/or IFN-γ. NB-39-nu was treated with 30 μM paclitaxel and/or 100 U/ml recombinant human IFN-γ. At the time indicated, 10 μg of total RNA was analyzed by northern blotting using human iNOS cDNA probe. The amount of total RNA analyzed was evaluated by hybridization with human β-actin probe.

Abbreviations used : RACE, rat aortic endothelial cell; IFN, interferon; iNOS, inducible nitric oxide synthase.

Acknowledgments : This study was partly supported by a Grant-in Aid from the Ministry of Health and Welfare for the Second-term of Comprehensive 10-Year Strategy for Cancer Control, Japan, a Grant-in-Aid from the Ministry of Education, Science and Culture for the Scientific Research on Priority Areas, Japan, the Foundation of Toyota Riken, Japan, and SRF Grant for Biomedical Research, Japan.

References

1. Gelmon, K.(1994) The Lancet, **344**, 1267-1272
2. Nakaki, T., Nakayama, M., Yamamoto, S and Kato, R.(1990) Mol. Pharmacol., **37**, 30-36
3. Chomczynski, P. and Sacchi, N.(1987) Anal. Biochem. **162**, 156-159
4. Adachi, H., Iida, S., Oguchi, S. et al. (1993) Eur. J. Biochem. **217**, 37-43
5. Hokari, A., Zeniya, M. and Esumi, H. (1994) J. Biochem. **116**, 575-581

Head and neck squamous cell carcinomas express a spectrum of constitutive nitric oxide synthase

Brandon G. Bentz M.D., G. Kenneth Haines III M.D., David G. Hanson M.D., and James A. Radosevich Ph.D.

Northwestern University/ VA Lakeside Medical Center, Departments of Otolaryngology-Head & Neck Surgery and Pathology, Searle Building 12-561, 303 East Chicago Ave., Chicago, IL 60611-3008, U.S.A., (312) 503-2051, Fax (312) 503-4943

ABSTRACT:

Little has been reported of the role of NO· in the pathogenesis of human neoplasia. This study was designed to assess the expression of the cNOS in HNSCCa. One hundred fourteen paraffin-embedded specimens of HNSCCa were immunostained with an anti-cNOS monoclonal antibody and graded on a 0-4+ scale. cNOS was found to be strongly expressed in skeletal muscle, endothelium, respiratory and large salivary duct epithelium. Normal squamous mucosa demonstrated regional variation in cNOS expression, with high levels in the basal and parabasal cells, decreasing amounts in the lower and midportion of the spinosa layer, and either remaining low or significantly increasing in the upper spinosa and stratum corneum. Areas of squamous metaplasia generally demonstrated strong expression of cNOS without regional variation. Squamous carcinomas demonstrated a range of diffuse cNOS expression without apparent correlation to TNM staging, grade, regional lymph-node or distant metastasis, disease-free survival, overall survival, or final patient status. However, cNOS was found to change from a localized expression in normal squamous mucosa to diffuse expression in metaplastic and neoplastic conditions.

INTRODUCTION:

Within the past few years, mounting evidence has been reported to implicate NO· as having antimicrobial activity [1]. NO· in chronic inflammation, however, may actually have detrimental effects by inducing immunosuppression or by generating carcinogenic radicals [2,3]. To date, little has been reported of the role of NO· in carcinogenesis. In light of these previous findings, this study was designed to assess the frequency and pattern of expression of cNOS in normal squamous mucosa and HNSCCa.

MATERIALS AND METHODS:

Anti-cNOS Mab: Tissue specimens were formalin fixed and paraffin embedded, and H&E stained sections were reviewed for histopathologic diagnosis. Immunostaining was performed directly utilizing a commercially available cNOS Mab (Transduction Laboratories, Lexington, KY) and the Avidin-Biotin complex (ABC) method (Vector Laboratories, Burlington, CA) as previously described [4]. Slides were lightly counterstained with Harris's Hematoxylin, subsequently reviewed by the study pathologist (GKH), and scored for staining pattern and intensity on a scale of 0-4+.
Statistical Analysis: Student-*t* test analysis.

RESULTS:

Examples of staining characteristics for normal squamous mucosa as well as squamous metaplasia and squamous cell carcinoma are shown in Figure 1. cNOS was found to be strongly expressed in skeletal muscle, endothelium, respiratory and large salivary duct epithelium. Normal surrounding squamous mucosa demonstrated specific regional variation of cNOS expression. Intense focal staining for cNOS was found in the basal and parabasal areas of the stratum basalis, which was lost in the lower and middle layers of the stratum spinosum. The expression was also found to either decrease or significantly increase in the upper spinous and stratum corneum layers. Examination of dysplastic mucosa demonstrated intense, diffuse staining for cNOS which lost regional variation. Neoplastic conditions also demonstrated diffuse staining with a broad range of intensity. No correlation was noted between cNOS expression and TNM staging, grade, site of primary, disease free survival, total survival, or final patient status (unpublished work).

Abbreviations used: nitric oxide: NO·, constitutive nitric oxide synthase: cNOS, head and neck squamous cell carcinoma: HNSCCa, monoclonal antibody: Mab

DISCUSSION:

Our data did not demonstrate a significant correlation between the expression of cNOS and clinically relevant factors. While cNOS was found in several normal cell types, specific regional expression of cNOS was demonstrated in normal squamous mucosa. The expression of cNOS in normal mucosa and dysplastic/neoplastic conditions appears to change from regional to diffuse respectively, and dysplastic conditions demonstrate intense staining whereas neoplasias were found to have a wide range of cNOS staining intensities. This change from regional to diffuse expression of cNOS may increase the overall NO· within the microenvironment of a developing focus of squamous dysplasia or carcinoma. Therefore, NO· may be a contributing element in the initial events of squamous carcinogenesis, and through this change in expression provide a possible mechanism to explain the later manifestations of immunosuppression seen in patients with this disease.

REFERENCES:

[1] Nathan CF, Hibbs JB. (1991) Role of nitric oxide in macrophage antimicrobial activity. Current Opinion in Immunol. **3:1**, 65-70.

[2] Eisenstein TK, Huang D, Meissler JJ, Al-Ramadi B. (1994) Macrophage nitric oxide mediates immunosuppression in infectious inflammation. Immunobiol. **191**, 493-502.

[3] Ohshima H, Bartsch H. (1994) Chronic infections and inflammatory processes as cancer risk factors: possible role of nitric oxide in carcinogenesis. Mut. Res. **305**, 253-64.

[4] Hsu SM, Raine L, Fanger H. (1981) The use of avidin-biotin-peroxidase complex (ABC) in immunoperoxidase techniques: a comparison between ABC and unlabelled antibody (PAP) procedures. J. Histochem. Cytochem. **29:4**, 577-81.

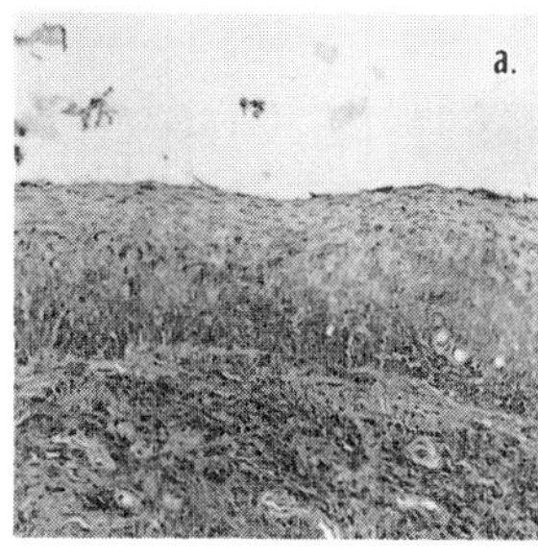

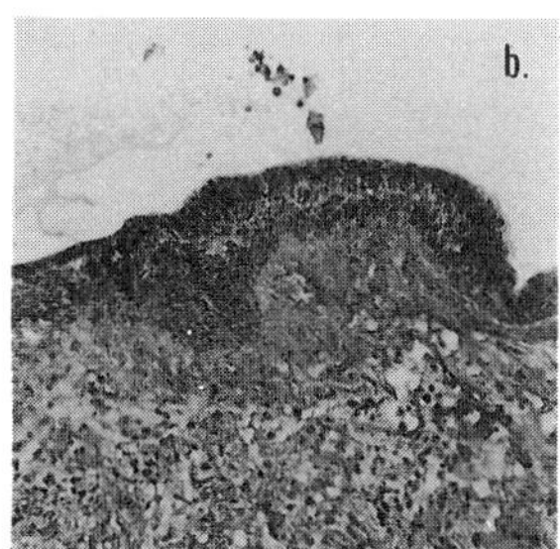

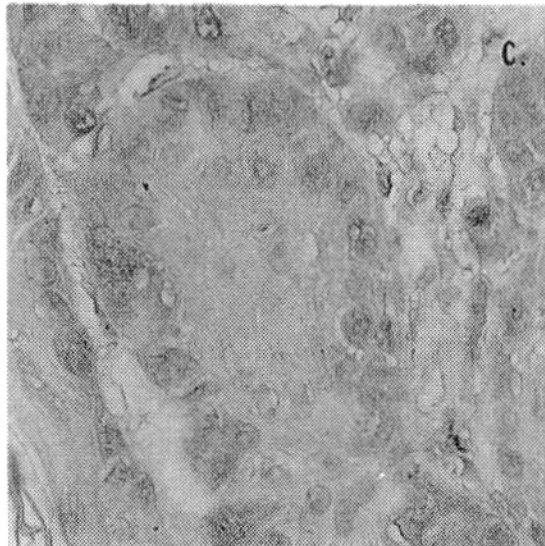

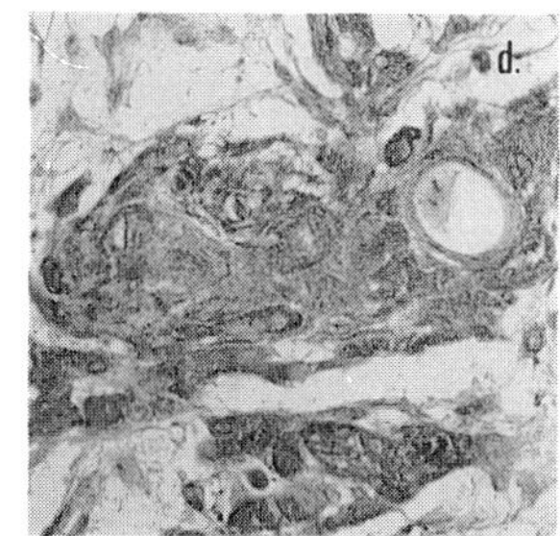

FIGURE LEGEND

a.) Normal surrounding squamous mucosa (200x) demonstrating regional variation in expression of cNOS. b.) Squamous metaplasia (200x) demonstrating intense global staining for cNOS. c.) Squamous cell carcinoma (400x) demonstrating 1+ global staining intensity for cNOS. d.) Squamous cell carcinoma (400x) demonstrating 4+ global staining intensity for cNOS.

Evidence for vasoactive intestinal peptide-induced nitric oxide release from perivascular nerves in bovine cerebral arteries

CARMEN GONZÁLEZ, CARMEN MARTIN and CARMEN ESTRADA

Departamento de Fisiología, Facultad de Medicina, Universidad Autónoma de Madrid, Arzobispo Morcillo 1, 28029, Spain.

Cerebral arteries are endowed with an adventitial network of nitric oxide synthase (NOS)- and vasoactive intestinal peptide (VIP)-containing nerve fibers [1,2]. Nitric oxide (NO) has been shown to participate in the relaxation observed in these vessels upon transmural nerve stimulation (TNS) [2,3]; however, the role of VIP remains undetermined. The purpose of this work was to analyze whether VIP is involved in the neurogenic relaxation either by directly relaxing smooth muscle cells or by releasing NO from the nitrergic nerve terminals.

Experiments were carried out in arterial rings isolated from the bovine anterior cerebral artery, with and without endothelium. Rings were suspended on two intraluminal parallel wires, introduced in an organ bath containing a physiological solution, and connected to a Piodem strain gauge for isometric tension recording. TNS (200 mA, 0.2 ms, 1-8 Hz) was applied by using two parallel platinum electrodes, one at each side of the vessel, connected to a CS-20 stimulator (Cibertec). The neurogenic nature of the stimuli was demonstrated because TNS responses were blocked by tetrodotoxin. TNS was performed in vessels previously contracted with prostaglandin $F_{2\alpha}$ and in the presence of 5 μM indomethacin, 1 μM phentolamine, 1 μM propranolol, and 0.1 μM atropine, to prevent any effect mediated by prostaglandins, noradrenaline or ACh.

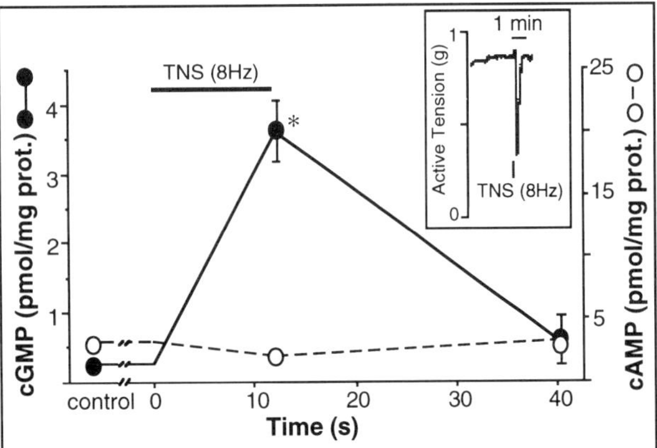

Figure 1.- Changes in intracellular cyclic nucleotide concentrations produced by transmural nerve stimulation (TNS) in bovine cerebral arteries. Data represent means±SE. n=4; *p<0.05. Inset: Recording of the relaxation induced by TNS.

VIP concentration curves were performed in a cumulative manner , in the presence or absence of L-nitroarginine methyl ester (L-NAME). For cyclic nucleotide measurements, arterial rings were mounted as for the functional experiments and frozen with liquid nitrogen embedded forceps, in control conditions and during TNS or 10^{-8} M VIP administration, in the absence or in the presence of L-NAME. cGMP and cAMP concentrations were determined by radioimmunoassay.

TNS produced a transient relaxation in arterial rings with and without endothelium, which was completely abolished by the NOS inhibitor L-NAME (100 μM) and by hemoglobin (4 μM). Intracellular cGMP concentration was significantly enhanced during TNS, returning to control levels 30 s after termination of the stimulus, with a time course parallel to that of the vessel relaxation; the cGMP increase was abolished in the presence of L-NAME. By contrast, intracellular cAMP, the second messenger coupled to VIP receptors, did not change during TNS. Figure 1 shows the effects of TNS on arterial rings with endothelium. Similar results were obtained in endothelium-denuded vessels. These data indicate that the neurogenic vasodilatation was entirely mediated by NO and that VIP contained in the perivascular nerves did not directly activate adenylate cyclase-coupled receptors in the vascular smooth muscle.

Table I. Effect of VIP on bovine cerebral arteries: intracellular cGMP concentration and arterial tone, in presence and absence of L-NAME

	cGMP	Arterial tone
	fmol / mg prot	% of control
Control	60 ± 8 (n=15)	100
VIP	113 ± 10 * (n=14)	52 ± 7.0 * (n=4)
L-NAME + VIP	72 ± 8 (n=10)	79 ± 4 (n=4)

Data represent means±SE. * p<0.05 as compared with control values and with values obtained in the presence of L-NAME.

Addition of VIP to arterial rings with and without endothelium produced a concentration-dependent relaxation that was partially reduced in the presence of L-NAME. Intracellular cGMP concentration was enhanced 1 min after addition of VIP, a time at which the relaxation rate was maximal. In the presence of L-NAME the VIP-induced cGMP increase was abolished. Table 1 summarizes the changes of cGMP concentrations and arterial tone produced by VIP (10 nM for cGMP measurements and 2 nM for relaxation studies) in endothelium-denuded vessels, in the presence and absence of L-NAME. These results indicate that part of the relaxing effect of VIP in cerebral arteries was mediated by NO production and guanylate cyclase activation.

Two sources of NO have been identified in cerebral arteries: endothelial cells and perivascular nerves. Since the NO-mediated effects of VIP on bovine cerebral arteries occurs in vessels without endothelium, the activation of NOS by VIP probably takes place in nitrergic nerve terminals. Therefore, VIP may participate in the neurogenic relaxation of bovine cerebral arteries by inducing NO release from perivascular nitrergic nerves.

Acknowledgments

This work was supported by grants from the Fondo de Investigaciones Sanitarias (92/0290, 94/0380 and 95/1422)

References

1.-Suzuki, N., Hardebo, J.E. and Owman, Ch. (1988) Origins and pathways of cerebrovascular vasoactive intestinal polypeptide-positive nerves in rat. J. Cereb. Blood Flow Metab. **8**, 697-712

2.-Estrada,C., Mengual, E. and González, C. (1993) Local NADPH-diaphorase neurons innervate pial arteries and lie close or project to intracerebral blood vessels: A possible role for nitric oxide in the regulation of cerebral blood flow. J. Cereb. Blood Flow Metab. **13**, 978-984

3.-Toda N, Okamura T (1990) Mechanism underlying the response to vasodilator nerve stimulation in isolated dog and monkey cerebral arteries. Am. J. Physiol. **259**, H1511-H1517

Effect of L-N^{G}-nitroarginine on hippocampal prostaglandin synthesis in rats

SATHASIVA B. KANDASAMY

Radiation Pathophysiology and Toxicology Department, Armed Forces Radiobiology Research Institute, Bethesda, MD 20889-5603, U.S.A.

INTRODUCTION

Exposure to ionizing radiation induces a decrease in hippocampal norepinephrine (NE) release. Nitric oxide synthase (NOS) mediates this effect because pretreatment with L-N^{G}-nitroarginine (L-NA), an NOS inhibitor before irradiation, prevented the radiation-induced decreases in NE release [1, 2]. It has been shown that prostaglandin E_2 (PGE_2) inhibits NE release from sympathetic nerves in the peripheral and central nervous systems; conversely, inhibition of PGE_2 synthesis leads to an increase in NE release [3]. The purpose of this study was to find out if the prevention of radiation-induced decreases in NE release by L-NA is due to the inhibition of PGE_2 synthesis. Therefore, the effect of L-NA on PGE_2 synthesis induced by recombinant human interleukin-1β (rhIL-1β) in the hippocampus was determined.

MATERIALS AND METHODS

Male Sprague-Dawley Crl:CD (SD) BRD rats weighing 250-300 g were euthanized by decapitation. The hippocampus was dissected and homogenized and the PGE_2 levels were measured by radioimmunoassay. Statistical analyses were performed with the Student's t-test. Multiple comparisons with control/treated values were done by analyses of variance and Dunnett's test. Data were identified as significant if $p<0.05$.

RESULTS

Treatment with 3-10 μg/kg IP of rhIL-1β increased hippocampal PGE_2 levels (Fig. 1). The rhIL-1β-induced PGE_2 level in the hippocampus was inhibited by pretreatment with 1-3 mg/kg IP of indomethacin (Fig. 2). However, pretreatment with 5-20 mg/kg of L-NA did not inhibit rhIL-1β-induced PGE_2 levels (Fig. 3).

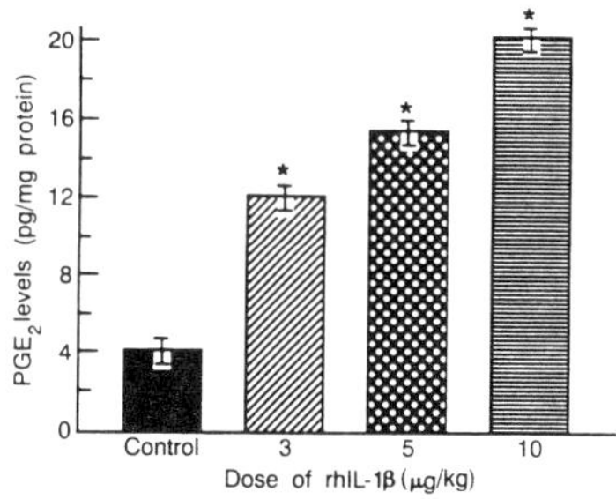

Figure 1. Hippocampal prostaglandin E_2 (PGE_2) levels after treatment with rhIL-1β. Values are expressed as the mean of levels from five rats ± SEM. *Significantly different from control values: $p<0.05$.

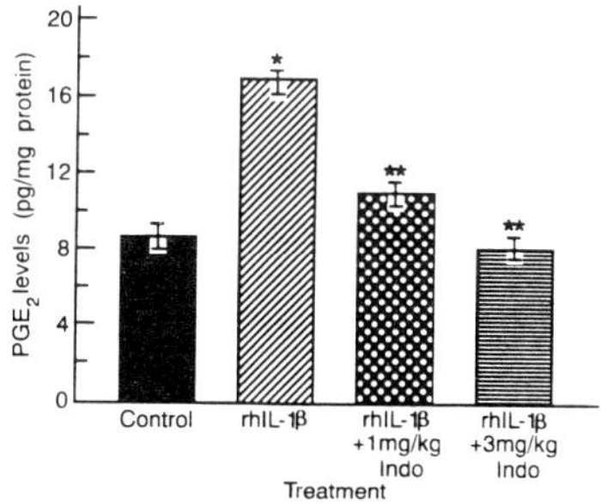

Figure 2. Effect of indomethacin (Indo) on hippocampal PGE_2 levels induced by 5 μg/kg rhIL-1β. Values are expressed as the mean of levels from five rats ± SEM. *Significantly different from control values: $p<0.05$. **Significantly different from rhIL-1β treated values: $p<0.05$.

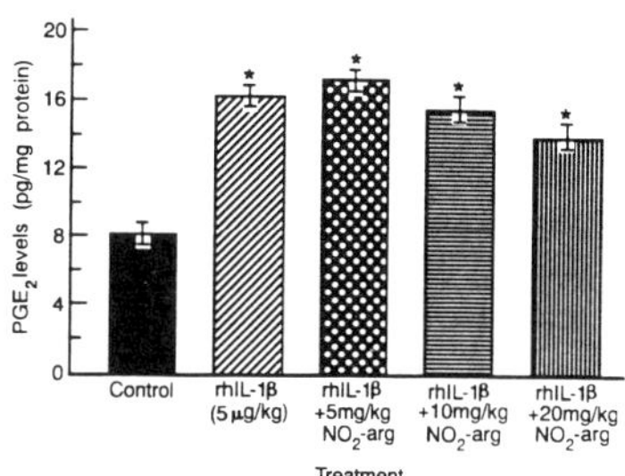

Figure 3. Effect of L-N^{G}-nitroarginine (L-NA) on hippocampal PGE_2 levels induced by 5 μg/kg rhIL-1β. Values are expressed as the mean of levels from five rats ± SEM. *Significantly different from control values: $p<0.05$.

DISCUSSION

A radio-ligand-binding study has shown that rhIL-1β receptors are widely distributed throughout the brain, especially in neuron-rich sites such as the granule cell layer of the dentate gyrus, the pyramidal cell layer of the hippocampus, the granule cell layer of the cerebellum, and the hypothalamus [4]. Although rhIL-1β cannot penetrate the blood-brain barrier, it exerts its central effects by entering the brain through the organum vasculosum of the lamina terminalis, a hypothalamic area that lacks the blood-brain barrier [5]. Systemic or central administration of rhIL-1β has been shown to stimulate PGE_2 synthesis in the hippocampus and the hypothalamus, and treatment with the cyclooxygenase inhibitor indomethacin blocked the synthesis [6, 7].

Immunochemical localization of NOS has been demonstrated in most areas of the rat brain, including the hippocampus [8]. NOS forms NO from L-arginine. L-NA is a competitive inhibitor of NOS, and it inhibits NOS activity irreversibly in the brain after systemic administration [9]. RhIL-1β increased PGE_2 levels in the hippocampus, and inhibition of PGE_2 synthesis by indomethacin confirmed it. The failure of L-NA to inhibit PGE_2 synthesis induced by rhIL-1β suggests that the prevention of the radiation-induced decreases in NE release by L-NA was not due to PGE_2 synthesis inhibition. These results suggest that L-NA has no inhibitory action on PGE_2 synthesis induced by rhIL-1β in the hippocampus.

REFERENCES

1. Kandasamy, S.B., Stevens-Blakely, S.A., Dalton, T.K. and Harris, A.H. (1992) in The Biology of Nitric Oxide, Part I (Moncada, S., Marletta, M.A., Hibbs, J.B. Jr. and Higgs, E.A., eds.), pp. 252-254, Portland Press, London
2. Kandasamy, S.B., Dalton, T.K. and Harris, A.H. (1994) in The Biology of Nitric Oxide, Part 3 (Moncada, S., Feelisch, M., Busse, R. and Higgs, E.A., eds.), pp. 353-356, Portland Press, London
3. Bergstrom, S., Farnebo, L.O. and Fuxe, K. (1973) Eur. J. Pharmacol. **21**, 362-368
4. Farrar, W.L., Hill, J.M., Harel-Bellan, A. and Vinocour, M. (1987) Immunol. Rev. **100**, 361-378
5. Katsuura, G., Arimura, A., Koves, K. and Gottschall, P.E. (1990) Am. J. Physiol. **258**, E163-E171
6. Komaki, G., Arimura, A. and Koves, K. (1992) Am. J. Physiol. **25**, E246-E251
7. Sirko, S., Bishai, I. and Coceani, F. (1989) Am. J. Physiol. **256**, R616-R624
8. Snyder, S.H. and Bredt, D.S. (1991) Trends Pharmocol. Sci. **12**, 125-128
9. Dwyer, M.A., Bredt, D.S. and Snyder, S.H. (1991) Biochem. Biophys. Res. Commun. **176**, 1136-1141

Peroxynitrite induces hyperreponsiveness of the guinea pig airways in vitro and in vivo.

GERT FOLKERTS, GUDARZ SADEGHI HASJIN, PAUL A.J. HENRICKS, HENK J. van der LINDE, INGRID van ARK, ALFONS K.C.P. VERHEYEN* and FRANS P. NIJKAMP

Utrecht Institute for Pharmaceutical Sciences, Utrecht University, P.O. Box 80082, 3508 TB Utrecht, The Netherlands. *Janssen Research Foundation, B-2340 Beerse, Belgium.

Nitric oxide (NO·) is an important messenger molecule released from a variety of cell types in the airways [1]. The concentration of nitric oxide in exhaled air of asthmatic patients is increased [2]. Moreover, the airways of asthmatic patients are often inflamed and it has been demonstrated that the production of superoxide anion ($O_2^{\cdot-}$) by alveolar macrophages of allergic asthmatics is increased after segmental antigen challenge [3]. Peroxynitrite ($ONOO^-$) is the product of the rapid reaction between nitric oxide and superoxide and can oxidize sulfhydryl groups and initiate lipid peroxidation [4]. The aim of the present study was to investigate whether peroxynitrite could alter airway responsiveness and morphology in the guinea pig.

Peroxynitrite was synthesized in a quenched flow reactor as described before [5]. *Airway responsiveness in vitro:* Specified pathogen-free male Dunkin-Hartley guinea-pigs (450 to 550 g; Harlan Olac, Ltd., Bicester, England) received a lethal dose of pentobarbital sodium (300 mg/kg), and the trachea was isolated, divided into two equal parts of approximately 15 rings and perfused in an organ bath as described before [6]. Changes in tracheal tension were measured at an optimal counterweight of 2 g. The inside of the trachea was perfused with Krebs (37°C) solution at a constant flow of 2 ml/min with a peristaltic pump. Every 15 min the Krebs buffer was refreshed on both sides until a stable tone was reached (usually within 75 min). The preparations were treated with 10 µM peroxynitrite or control solution from mucosal or serosal sides. After 15 min, cumulative concentrations of histamine or methacholine were applied to the mucosal or serosal side. In the control groups, the average maximal effect was taken as 100% response to the corresponding agonist and the response of the treated group was normalized accordingly. *Histology:* At the end of the organ bath experiments, control and peroxynitrite-treated tracheas were fixed in phosphate-buffered formaldehyde (10%) and embedded in paraffin blocks. Sections measuring 5 µm were stained with Luna's method for eosinophilic granules and were evaluated by light microscopy. *Major Basic Protein ELISA:* Each tracheal tube was divided into two equal parts of 15 rings each and suspended in 1 ml Krebs buffer at 37°C. One group was exposed to 10 µM peroxynitrite for 15 min while the control group was treated only with the buffer. The solutions were stored at -70°C prior to analysis. The presence of major basic protein in the buffer solution was measured with a specific ELISA [7]. *Airway responsiveness in vivo:* Guinea pigs were anaesthetized with ether and 100 nmol of peroxynitrite in 0.1 ml plasma expander medium or this solvent (control group) was instilled intratracheally. One, 3, 5, or 10 days after treatment, the animals were anaesthetized with urethane (2.8 g/kg i.p.) and the airway reactivity was investigated *in vivo* as described before [8]. In short, the trachea was cannulated and connected to a pneumotachograph with a flow head. Lung resistance (R_L) was determined breath by breath using a computerized respiratory analyzer. Responses are presented as increase in R_L above baseline. *Statistics:* Various parameters were averaged for each experimental group and the results were expressed as mean ± s.e. mean. Differences between groups were considered statistically significant when $p < 0.05$ with Student's unpaired *t*-test.

Histamine caused concentration-dependent contractions of tracheal tubes when applied to the mucosal side. Mucosal pretreatment with 10 µM peroxynitrite significantly increased the maximal contractions to histamine and methacholine by 30% and 42% respectively ($p < 0.05$, $n = 6$-8). Peroxynitrite (10 µM) administered either on the mucosal or on the serosal side had no significant effects on the responsiveness of tracheal tubes when exposed to histamine on serosal side. Mucosal preincubation of tracheas with decomposed peroxynitrite (initially 10 µM solution) for 15 min had no effect on their responsiveness to histamine given from the same side. In the non-treated tracheas, the mucosal epithelium was fairly intact and the mucosal and submucosal eosinophils contained closely packed, intensely red-stained granules. However, after treatment with 10 µM peroxynitrite, the epithelial cells were largely desquamated often leaving a thin layer of basal cells. Many eosinophils beneath this damaged epithelium showed loosely packed granules; in addition, eosinophilic granules seemed frequently distributed within the extracellular matrix. Total major basic protein released from control and treated preparations were 1.37 ± 0.27 µg and 3.21 ± 0.43 µg, respectively ($p < 0.01$, $n = 5$). One day after tracheal instillation of peroxynitrite, no difference in the lung resistance between control and treated groups was observed. However, three days after inoculation the increase in R_L in response to intravenous administration of histamine was significantly ($p < 0.01$, $n = 4$-5) potentiated compared with the responsiveness of vehicle-treated animals. Airway hyperresponsiveness to histamine was still present on day 5 and 10 after peroxynitrite treatment.

In the present study it is demonstrated that peroxynitrite is able to induce a long-lasting airway hyperresponsiveness *in vivo* at least up to day 10 after inoculation. In addition, a non-specific tracheal hyperresponsiveness *in vitro* is observed which was associated with eosinophil destruction and/or activation. A concentration of peroxynitrite as low as 10 µM resulted in an extensive epithelial damage of the trachea after 15 min. From this, one can propose that local epithelial damage, as seen in many asthmatic subjects, might be a consequence of peroxynitrite formation. Epithelial damage may lead to the loss of the barrier function against noxious substances and cause non-specific airway hyperresponsiveness. This may well explain the mechanism through which peroxynitrite caused an increased reactivity of the respiratory airways towards histamine and methacholine.

In conclusion, the effects of peroxynitrite were identified on the guinea pig airway responsiveness *in vitro* and *in vivo*, as well as on the morphology of tracheal tissue. Suppression of excessive production of nitric oxide, superoxide or peroxynitrite may prevent epithelial damage and the induction of airway hyperresponsiveness.

References

1. Nijkamp, F.P., and G. Folkerts. (1994) Clin. Exp. Allergy **24**, 905-914.
2. Kharitonov, S.A., D. Yates, R.A. Robins, R. Logan-Sinclair, E.A. Shinebourne, and P.J. Barnes. (1994) Lancet **343**, 133-135.
3. Calhoun, W.J., H.E. Reed, D.R. Moest, and C.A. Stevens. (1992) Am. Rev. Respir. Dis. **145**, 317-325.
4. Beckman, J.S., T.W. Beckman, J. Chen, P.A. Marshall, and B.A. Freeman. (1990) Proc. Natl. Acad. Sci. USA **87**, 1620-1624.
5. Radi, R., J.S. Beckman, K.M. Bush, and B.A. Freeman. (1991) J. Biol. Chem. **266**, 4244-4250.
6. Nijkamp, F.P., H.J. van der Linde, and G. Folkerts. (1993) Am. Rev. Respir. Dis. **148**, 727-734.
7. Oosterhout, A.J.M., Ark van I., Folkerts G., et al. (1995) Am. J. Respir. Crit. Care Med. **151**, 177-83.
8. Folkerts, G., H. van der Linde, and F.P. Nijkamp. (1995) J. Clin. Invest. **95**, 26-30.

Broncholytic effect of new nitric oxide donors in guinea-pig and rat bronchi *in vitro*

KIRSI VAALI, RIIKKA NEVALA, BEATRIX REDEMANN, LIANG LI, ILARI PAAKKARI and HEIKKI VAPAATALO

Institute of Biomedicine, Department of Pharmacology and Toxicology, P.O. Box 8, FIN-00014 University of Helsinki, Finland.

Nitric oxide (NO) released from NANC nerves is thought to be an important bronchodilating mediator [1]. High concentration of NO has been found in exhaled air in asthmatic patients [2]. Of the asthmatics, the NSAID-treated subjects had significantly higher peak expired NO concentrations than the controls, but asthmatic patients receiving inhaled corticosteroids had levels similar to controls [3]. Accordingly, we compared bronchorelaxing efficacies of new NO donors GEA 3175 and 3162 (oxatriazole imines) to SNP and SIN-1 and to the adrenergic β_2-agonist, salbutamol, in bronchi of guinea-pigs and rats *in vitro*.

Dunkin Hartley guinea-pigs of either sex (350-600 g) were anesthetized with pentobarbital (75 mg/kg), or male unanesthetized Wistar rats (250-300 g) were decapitated and 4 mm wide rings were excised from the principle bronchi. They were studied in the modified Krebs-Ringer solution after constriction by 1 μM metacholine or 40 mM KCl.

Abbreviations used: MC (metacholine), NANC (non-adrenergic, non-cholinergic), NO (nitric oxide), SNP (sodium nitroprusside), SIN-1 (3-morpholino-sydnonimine), NSAID (non-steroidal anti-inflammatory drugs), HIST (histamine).

In **guinea-pig** bronchial smooth muscle after 1μM MC preconstriction salbutamol exerted the strongest effect (EC_{50} 10 nM) compared to GEA 3175 and GEA 3162 (EC_{50} 2.5 μM for both), while EC_{50} values for SNP and SIN-1 were 1 μM and 20 μM, respectively. After 1 μM HIST preconstriction, salbutamol (EC_{50} 0.6 μM) was more effective than the NO donors. SIN-1 was weak relaxant (EC_{50} not measurable) and EC_{50} for the others were 0.6-1.3 μM. GEA 3175 relaxed maximally 95% and GEA 3162 to 90%, while salbutamol relaxed only 70%, SIN-1 35% and SNP 85%. After KCl 40 mM preconstriction the NO donors relaxed the bronchi less than 20%, whereas salbutamol induced nearly 100% relaxation (EC_{50} 33 nM) (Figure 1).

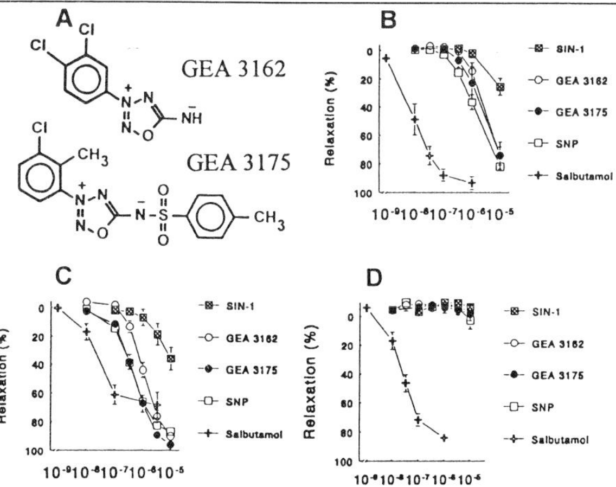

Figure 1. The chemical structure of GEA 3162 and GEA 3175 **(A)**. Relaxing effects of NO donors and salbutamol on guinea-pig bronchi *in vitro* with 1μM MC **(B)**, 1μM histamine **(C)** or 40 mM KCl **(D)** induced preconstriction.

In **rat** bronchial muscle after MC preconstriction relaxations by both GEA-compounds amounted to 90% while that by salbutamol was only 50% and those by SIN-1 and SNP were less than 60%. Both GEA 3162 and 3175 showed EC_{50} of 1-2 μM, which was of the same magnitude as that of salbutamol or SNP. EC_{50} of SIN-1 was 50 μM. In 40 mM KCl preconstriction, the maximal relaxation by salbutamol was less than 30%, while that of GEA 3175 was 70% (EC_{50} 4 μM). Both GEA compounds induced 50% relaxation (Figure 2).

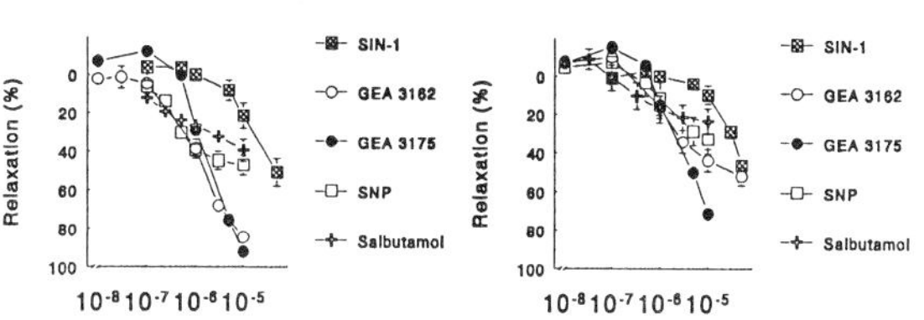

Figure 2. Relaxing effects of NO donors and salbutamol on rat bronchi *in vitro*: 1μM MC **(A)** 40 mM KCl **(B)** induced preconstriction.

In conclusion, the new NO donors GEA 3162 and GEA 3175 are more potent bronchorelaxing agents than the old ones SNP and SIN-1, but in the guinea-pig bronchi salbutamol was more potent than the NO-donors, whereas in the rat bronchi GEA 3175 induced the strongest relaxation.

This study was supported by the Gea A/S, (Hvidovre, Denmark), the Academy of Finland, University of Helsinki, and the Emil Aaltonen Foundation, (Tampere, Finland).

References:

[1] Barnes, P.J. (1992) Neural mechanisms in asthma. Br. Med. Bull. **48,** 149-168.

[2] Persson M.G., Zetterström O., Agrenius V., Ihre E. and Gustafsson LE. (1994) Single-breath nitric oxide measurements in asthmatic patients and smokers. *Lancet* **343**(8890):146-147.

[3] Kharitonov S.A., Yates D., Robbins R.A., Logan-Sinclair R., Shinebourne E.A. and Barnes P.J. (1994) Increased nitric oxide in exhaled air of asthmatic patients. *Lancet.* **343**:133-135.

Intramolecular rearrangement of aminoethyl-isothiourea (AETU) into mercaptoethylguanidine (MEG) and aminothiazoline (ATZ): MEG accounts for the selective inhibitory effect of AE-TU on the inducible isoform of nitric oxide synthase.

GARRY J. SOUTHAN, BASILIA ZINGARELLI, ANDREW L. SALZMAN and CSABA SZABÓ

Division of Critical Care, Children's Hospital Medical Center, 3333 Burnet Avenue, Cincinnati, OH, 45229

Aminoethylisothiourea (AETU), aminopropropylisothiourea (APTU) and their derivatives appear to be potent inhibitors of iNOS activity in immunostimulated J774 macrophages [1]. Furthermore, when compared to N^G-methyl-L-arginine (L-NMA) they show an apparent selectivity for iNOS over the endothelial, constitutive isoform (ecNOS). However, these compounds undergo chemical conversion in neutral or basic solution to give mercaptoalkylguanidines and a smaller amount of cyclic derivatives [2, 3]. For example, AETU gives rise to mercaptoethylguanidine (MEG) and a cyclic compound, 2-aminothiazoline (ATZ). These rearrangement products are themselves potent inhibitors of NOS and are probably responsible for the inhibition of NOS by aminoalkylisothioureas *in vitro* [2]. Here we report on some of the *in vivo* pharmacological properties of AETU, ATZ, MEG and other mercaptoalkylguanidines.

The pressor effects of known NOS inhibitors were used to assess the ability of the drug to inhibit ecNOS activity *in vivo*. Male Wistar rats were anesthetized and instrumented as described [1]. After recording baseline hemodynamic parameters, animals received NOS inhibitors (0.1-10 mg/kg i.v.) in a cumulative fashion. The separate bolus injections were performed every 5 minutes. N^G-methyl-L-arginine (L-NMA) and ATZ caused very pronounced pressor responses, whereas AETU (after initial, transient pressor responses, not shown) caused no change in mean arterial blood pressure (MAP) up to the dose of 10 mg/kg (Fig. 1). MEG caused slight depressor responses, whereas mercaptopropylguanidine (MPG) caused pronounced pressor responses (Fig. 1). The order of potency of these compounds on MAP was the similar to their potency on ecNOS *in vitro*; L-NMA>MPG=ATZ>AETU>MEG; L-NMA being the most potent and MEG being the least potent inhibitor [2].

In animals subjected to endotoxin shock, the pressor effect of a NOS inhibitor is an indicator of the ability of the agent to inhibit iNOS. The effects of infusions of selected NOS inhibitors into control animals and into rats subjected to endotoxin shock then allows the effects of the agents on iNOS and ecNOS *in vivo* to be compared. In these studies, rats were injected with *E. coli* LPS (15 mg/kg i.v.) at time 0. In the control group, rats were injected with vehicle (saline, 0.1 ml/kg i.v.) at 90 minutes and then saline was infused at a rate of 0.2 ml/kg for a further 90 min. In subsequent groups, rats were injected with LPS as above and, after 90 min, AETU, MEG or ATZ was injected (at 10 mg/kg i.v.) and then infused (30 mg/kg/h) for a further 90 min. In separate experiments, changes in MAP of normal rats (without endotoxin) in response to bolus injections and subsequent infusion of AETU, MEG or ATZ (doses as above) were monitored for 90 min. We found that AETU and MEG, when given as infusions, produced slight decreases in MAP in control rats. Infusion of AETU or MEG into endotoxin-treated rats, however, caused an increase in MAP and restored 80% of the endotoxin-induced fall in MAP (Fig. 2). In contrast, ATZ was a potent pressor agent both in control and endotoxemic rats (Fig. 2)

Thoracic aortae from rats were taken and prepared for the measurement of isometric contraction. Concentration-response curves to norepinephrine (10^{-9}-10^{-5} M) in the presence of NOS inhibitors (100 µM, 30 min treatment) were obtained in endothelium-denuded aortic rings taken from either control rats or rats treated with LPS (15 mg/kg i.v for 180 min). AETU, MEG and ATZ (at 100 µM) all caused a similar restoration of the contractile responses in rings obtained from endotoxemic rats (not shown).

In endothelium-intact rings, the effect of pretreatment with NOS inhibitors on the endothelium-dependent relaxations elicited by acetylcholine (10^{-6} M) was also investigated. ATZ (10 µM) caused a reduction in the relaxation in response to acetylcholine in line with its potent effects on ecNOS, but MEG did not inhibit the relaxant responses up to 100 µM. AETU had intermediate potency in inhibiting the relaxations (Fig. 3).

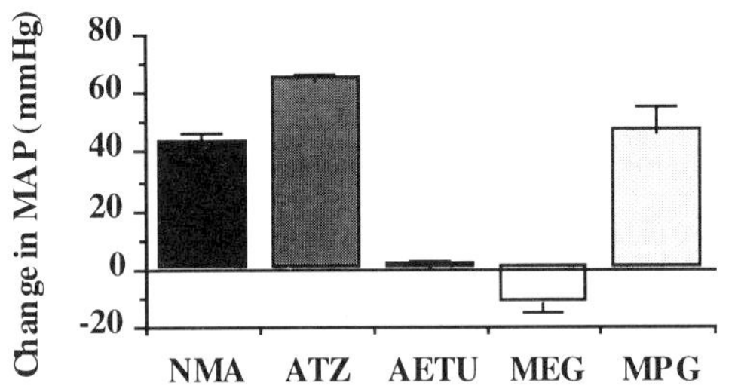

Fig. 1. Effect of L-NMA, ATZ, AETU, MEG and MPG on the blood pressure in anesthetized rats. All drugs were given at 10 mg/kg i.v.; blood pressure was measured at 5 min after the injection.

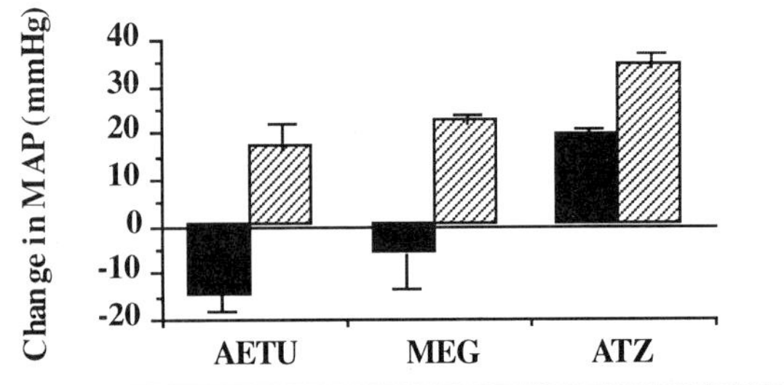

Fig. 2. Delayed effect of AETU, MEG and ATZ on the blood pressure in anesthetized control rats (closed bars) and rats injected with endotoxin (hatched bars). Drugs were given at 10 mg/kg i.v. bolus and infused at 30 mg/kg/h for 90 min.

Abbreviations used: AETU, aminoethyl-isothiourea; ATZ, aminothiazoline; L-NMA, N^G-methyl-L-arginine; MEG, merkapto-ethylguanidine; MPG, merkaptopropylguanidine; NO, nitric oxide; ecNOS, constitutive NOS isoform; iNOS, inducible NOS isoform

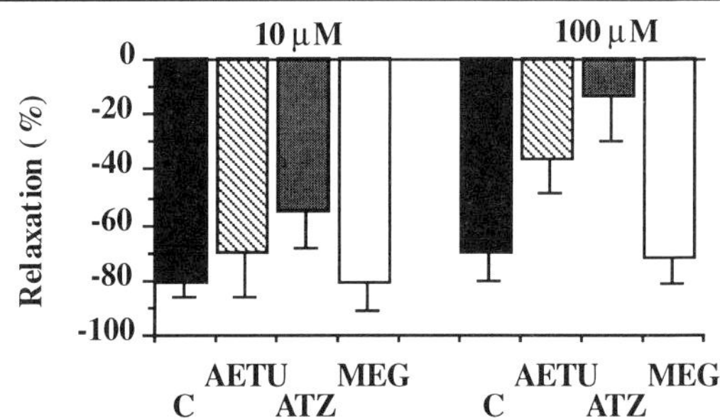

Fig. 3. Inhibition of endothelium-dependent relaxations elicited by 10 µM acetylcholine by AETU, ATZ and MEG at 10 µM (left panel) and at 100 µM (right panel).

In conclusion, the indications of potential iNOS-selectivity of MEG are reinforced by: (i) The lack of increase in MAP when MEG was injected into normal rats at doses up to 10 mg/kg, (ii) the absence of an increase in MAP when either AETU or MEG were infused into normal rats, (iii) substantial restoration (80%) of the hypotension induced in rats by treatment with LPS, (iv) the absence of a significant reduction, by MEG, in the relaxations in response to acetylcholine. AETU can give rise to small amounts of the potent ecNOS inhibitor, 2-aminothiazoline, in addition to MEG, which can account for the differences in the *in vitro* and *in vivo* effects of AETU and MEG.

References

1. Southan, G.J., Szabó, C., Thiemermann, C. (1995) Br. J. Pharmacol. **114**, 510-516
2. Southan, G.J., Salzman, A.L., Szabó, C. (1995) Endothelium **3** (Suppl), 79 (abstr.)
3. Khym, J.X., Doherty, D.G., Shapira, R. (1958) J. Am. Chem. Soc., **80**, 3342-3349.

Mercaptoalkylguanidines are competitive inhibitors of nitric oxide synthase *in vitro* and *in vivo*: preference towards the inducible isoform.

GARRY J. SOUTHAN, ANDREW L. SALZMAN and CSABA SZABÓ

Division of Critical Care, Children's Hospital Medical Center, 3333 Burnet Avenue, Cincinnati, OH, 45229

The generation of nitric oxide (NO) from L-arginine by NO synthases (NOS) can be inhibited by compounds containing the amidine function, such as guanidines and S-alkylisothioureas [1,2,3]. Unlike most L-arginine based inhibitors, however, some guanidines and S-alkylisothioureas are selective towards the inducible isoform (iNOS) over the constitutive endothelial isoform (ecNOS) and so may be of therapeutic potential. Of the S-substituted isothioureas, S-aminoethylisothiourea (AETU) shows a marked selectivity towards iNOS in rodent enzyme systems [2]. Here we investigate the effects of AETU and other aminoalkylisothioureas (AATUs) on the activities of iNOS, ecNOS and bNOS, the NOS isoform present in neurons. In view of the chemical nature of AETU, we have also investigated whether AETU itself, or, some other species, is the active mediator of its inhibitory effects.

Generation of thiols in solutions of aminoalkylisothioureas was measured using "Aldrithiol-2" (2,2'-dipyridyl disulfide). J774 macrophages and rat aortic smooth muscle (RASM) were cultured and induced to express iNOS with bacterial lipopolysaccharide as described [2] and nitrite production and mitochondrial respiration were measured at 24h and 48h, respectively [2]. NOS assays (measurement of conversion of radiolabeled L-arginine to L-citrulline) were performed, using homogenates of bovine endothelial cells for ecNOS, homogenates of lungs from septic rats for iNOS and homogenates of rat brains for bNOS activity [2]. The effect of NOS inhibitors on blood pressure was studied in anesthetized rats [2]. AATUs and merkaptoalkylguanidines (MAGs) were synthesized using standard methods.

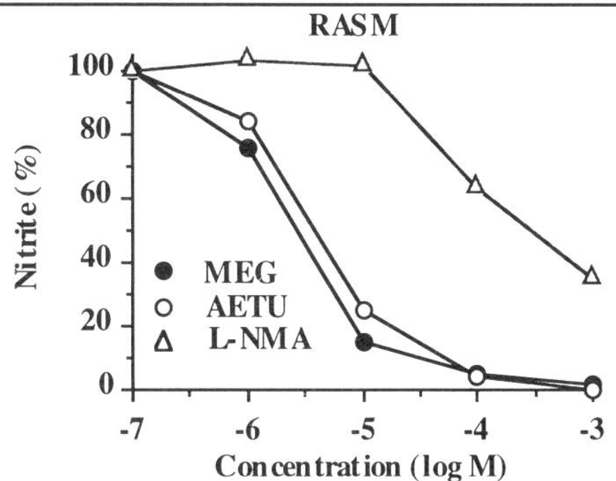

Fig. 1 Inhibitory effect of AETU, MEG and L-NMA on nitrite accumulation in RASM cells stimulated with LPS and gamma-interferon for 48h.

Aminoethylisothiourea (AETU), aminopropylisothiourea (APTU) and their derivatives containing alkyl substituents on one of the amidino nitrogens, potently inhibit nitrite formation by immunostimulated J774 macrophages and RASM cells (Fig. 1) with EC_{50} values ranging from 6 - 30 µM (EC_{50} values for N^G-methyl-L-arginine (L-NMA) and N^G-nitro-L-arginine were 159 and > 1000 µM, respectively). For instance, EC_{50} values in the J774 cells for AETU, aminopropyl-TU, aminobutyl-TU, N-methyl-AETU, N-ethyl-AETU and N-allyl-AETU were 14, 8, 51, 29, 27 and 24. A second substitution reduced the inhibitory potency: EC_{50} values for N,N'-ethylene-AETU and N,N'-dimethyl-APTU were 96 and 350 µM. The inhibitory effects of these aminoalkyl-isothioureas were attenuated by L-arginine in the medium, indicating that these agents may compete with L-arginine for its binding site on NOS.

Abbreviations used: AATU, aminoalkyl-isothiourea; AETU, aminoethyl-isothiourea; L-NMA, N^G-methyl-L-arginine; MAG, mercaptoalkylguanidine; MEG, merkaptoethylguanidine; MPG, merkaptopropyllguanidine; bNOS, brain NOS; ecNOS, constitutive NOS; iNOS, inducible NOS; RASM, rat aortic smooth muscle

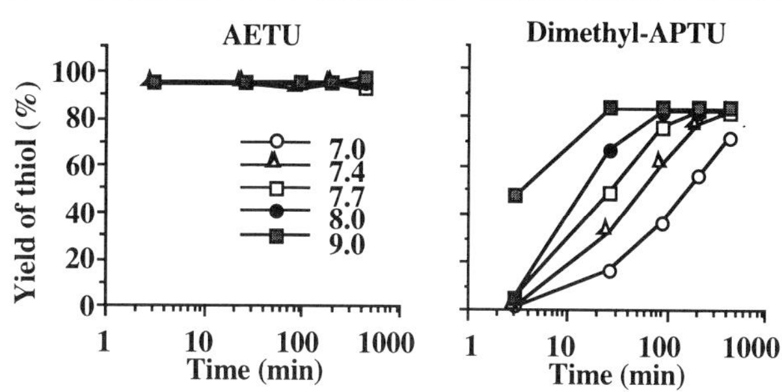

Fig. 2. Time courses for the generation of free thiol (sulphurhydryl, -SH) from AETU and N,N'-dimethyl-APTU in phosphate buffer at various pH values.

	EC_{50} (µM)		
Compound	**iNOS**	**ecNOS**	**bNOS**
N^G-Methyl-L-arginine (L-NMA)	17	5	20
N^G-Nitro-L-arginine	300	2	0.8
Aminoguanidine	80	2600	220
Aminoethyl-TU (AETU)	1.7	50	27
Aminopropyl-TU (APTU)	0.9	5.5	3
N-methyl-APTU	7	3	15
Mercaptoethylguanidine (MEG)	11.5	110	60
Mercaptoproylguanidine (MPG)	7	4	80
S-methyl-MEG	1.4	43	8
S-ethyl-MEG	30	850	460

Table 1. EC_{50} values of AATUs, MAGs and other compounds for their inhibition of the activities of iNOS, ecNOS, bNOS in tissue homogenates.

However, the above AATUs undergo chemical conversion in neutral or basic solution as indicated by (1) the disappearance of AATUs from solution as measured by HPLC, (2) the generation of free thiols not previously present (Fig. 2.) and (3) the isolation of species (as picrate and flavianate salts) from neutral or basic solutions of AATUs that are different from those obtained from acid solutions (Fig. 2). The explanation of these observations is the rearrangment of AATUs to MAGs [4]. Therefore, we prepared MAGs and found them to be potent iNOS inhibitors with EC_{50} values comparable to those of their isomeric AATUs. For instance, EC_{50} values for MEG, MPG, S-methyl-MEG and S-ethyl-MEG were 13, 15, 80 and 328 µM, respectively. Substitution of the amidine nitrogens slows the rate of rearrangement, and decreases the potency of iNOS inhibition (compare AETU and N,N'-dimethyl-APTU; Fig. 2).

As the inhibitory potency of NOS inhibitors in whole cell systems may be affected by cellular uptake, the effect of AATUs and MAGs on the NOS activity by iNOS, ecNOS and bNOS were compared in various cell/tissue homogenates in the presence of appropriate co-factors. AETU and APTU were extremely potent when compared to L-NMA, L-NA or aminoguanidine (Table 1). S-methyl-MEG and S-ethyl-MEG were potent inhibitors, despite their weaker inhibitory effects in intact cells (Table 1). The order of pressor potency of these various agents in anesthetized rats, in general, correlated with the order of inhibitory potency on ecNOS.

Thus, the *in vitro* effects of AETU and related aminoalkyl-isothioureas can be explained in terms of their intramolecular rearrangement to generate mercaptoalkylguanidines, which represent a new class of NOS inhibitors, some with selectivity towards iNOS.

References

1. Misko, T.P., Moore, W.M., Kasten, T.P. et al. (1993) Eur. J. Pharmacol. **233**, 119-125
2. Southan, G.J., Szabó, C., Thiemermann, C. (1995) Br. J. Pharmacol. **114**, 510-516
3. Garvey, E.P., Oplinger, J.A., Tanoury, G.J. et al. (1994) J. Biol. Chem. **269**, 26669-26676.

Homopiperidine amidines as selective inhibitors of inducible nitric oxide synthase

Donald W. Hansen Jr.*, Karen B. Peterson*, Mahima Trivedi*, R. Keith Webber#, Foe S. Tjoeng#, William M. Moore+, Gina M. Jerome+, Mark G. Currie+, Barnett S. Pitzele*

*4901 Searle Pkwy, G.D. Searle, Skokie, IL 60077, USA
#700 Chesterfield Village Pkwy, G.D. Searle, Chesterfield, MO 63198, USA
+800 N. Lindbergh Blvd, G.D. Searle, Creve Coeur, MO 63167, USA

The current intense nitric oxide (NO) research effort in both academic and industrial institutions is due to the accumulating data suggesting that NO is a messenger molecule with more diverse functions than any other known [1]. Imbalances in finely tuned NO synthesis can lead to serious consequences and selective control of NO synthesis has considerable therapeutic potential. Localized excess NO release has been implicated in the pathology of diseases such as osteo and rheumatoid arthritis, inflammatory bowel disease, and sepsis. An attractive approach to the treatment of these inflammatory conditions is through the selective inhibition of human inducible nitric oxide synthase (hiNOS), the isoform that is apparently triggered in these disease states.

The natural substrate for NOS, arginine, has been the obvious prototype of inhibitor design. Much pharmacology and biology has been determined utilizing some of the early relatively nonspecific inhibitor derivitives of Arg [1,2]. Since it appears that a guanidine or amidine is a necessary pharmacophore for NOS inhibition, we chose to investigate some cyclic amidines as potential selective inhibitors of hiNOS. Our efforts have led to the discovery of the amidinium salts whose general structure is illustrated in Table 1.

The synthesis of these analogs involves the initial preparation of appropriately substituted cyclohexanones. 2-substituted cyclohexanones are generated by either by direct alkylation of cyclohexanone or alkylation of 2-carboalkoxy-cyclohexanone followed by decarboxylation. Alternatively, cyclohexene oxide can be opened with an organometallic agent and the resulting alcohol oxidized to the 2-substituted cyclohexanones. These intermediates and other substituted cyclohexanones are then converted to their oximes and subjected to a variety of Beckmann Rearrangement conditions to provide separable regioisomeric mixtures of substituted caprolactams. In the case of 2-substituted cyclohexanones, the 7-substituted caprolactam predominates in the mixture. The caprolactam is subsequently treated with an alkyloxonium fluoroborate salt to generate the respective imino ethers. Treatment of these materials with ammonium chloride provides the target amidinium hydrochlorides whose preliminary biology is described below.

Table I. Enzyme Inhibition of Substituted Homoiminopiperidinium Salts

R^4, R^3, R^5, R^6, R^7, =NH, NH, .HCl

Inhibition of NO Synthase[2] IC50 (μM)

Compound	(a) hiNOS	(b) hecNOS	(c) hncNOS	Ratios b/a-c/a
1, R^3 = H	2	18	4	10-2
2, R^3 = Ethyl	1.1	31.5	2.6	29-1
3, R^3 = Allyl	3.8	342	14	90-4
4, R^5 = Methyl	5.9	33.5	13.4	6-2
5, R^6 = Methyl(R)	7	46		7-
6, R^7 = Allyl	0.14	19	1	136-7
7, R^7 = Propyl	0.08	7.8	0.5	98-6
8, R^7 = Butyl	0.4	89	1	222-2
9, R^7 = 2-Et-butyl	2.1	660	5.4	314-3
10, R^7 = Phenyl	12.4	179	7	14-0.5

The homoiminopiperidinium salts indicated in Table I were evaluated for their ability to inhibit each of the three human isoforms of NOS [2]. The parent caprolactam amidinium chloride, **1**, has micro molar potency for hiNOS and only 10-fold selectivity over hecNOS. When an ethyl function is inserted into the three position, the potency of **2** is slightly enhanced and its selectivity over hecNOS is nearly tripled. The analog **3**, containing an allyl group at position three, has comparable potency for hiNOS but is 90 fold selective over hecNOS. All these amidinium salts are near equally potent at hiNOS and hncNOS. Analogs that are substituted at positions five and six, **4** and **5**, have an enzyme inhibition profile less potent but similar to their parent, **1**.

However, relatives possessing an aliphatic substitutent at position seven, compounds **6**, **7**, **8**, and **9**, exhibit dramatic improvement in potency, selectivity, or both as compared to **1**. Analog **7** has a hiNOS IC_{50} of 80 nM and selectivity of nearly 100 while **9** is greater than 300 fold selective for hiNOS over hecNOS. It is interesting to note that the presence of a phenyl function at position seven, compound **10**, greatly reduces both potency and selectivity.

Acknowledgments:

We wish to thank Dr. Ann Hallinan for helpful discssions, Ms. Dorothy S. Honda, Mr. Randy J. Fronek and Ms. Sharon H. Kinder for chromatographic purifications, and Dr Scott Laneman for hydrogenations.

References:

1. Marletta, M. (1994) J. Med. Chem. **37**, 1899-1907 and references cited therein.
2. Moore, W.M., Webber, R.K., Jerome, G.M., Tjoeng, F.S., Misko, T.P., Currie, M.G. (1994) J. Med. Chem. **37**, 3886-3888 and references cited therein.

S-2-Amino-5-(2-nitroimidazol-1-yl) pentanoic acid: A potential bioreductively-activated inhibitor of nitric oxide synthase activity for use in cancer therapy

SARAJ ULHAQ, MATTHEW A.NAYLOR, MICHAEL D. THREADGILL*, EDWIN CHINJE and IAN. J. STRATFORD.

MRC Radiobiology Unit, Chilton, Oxon, OX11 ORD, U.K,
*School of Pharmacy & Pharmacology, University of Bath, Bath, Avon, BA2 7AY, UK.

Introduction

The participation of nitric oxide (NO) in vasodilation is now well documented. The NO formed originates from L-arginine (Fig 1). It involves the oxidation of one of the guanidino nitrogens and is catalysed by the enzyme nitric oxide synthase (NOS) [1].

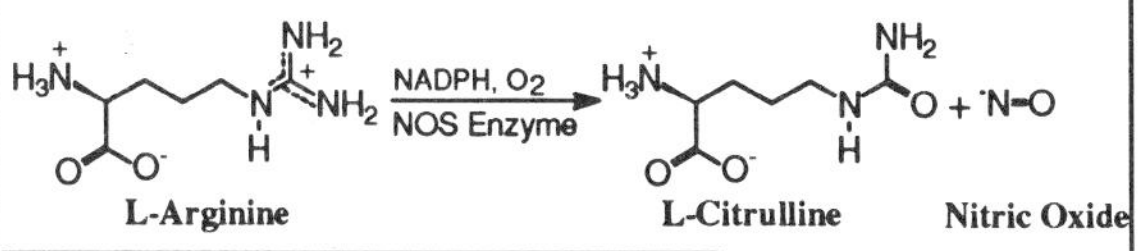

NADPH, O_2 / NOS Enzyme

L-Arginine → L-Citrulline + Nitric Oxide

Fig 1. Biosynthesis of nitric oxide (NO)

L-arginine analogues are likely to inhibit NOS and thus reduce NO production. This will produce vasoconstriction and consequent reduction in blood flow. Compounds that are selectively reduced to yield such analogues within hypoxic tissue (e.g. solid tumours) should result in selective reduction in blood flow and hence further increase hypoxia. This can then be exploited with other bioreductive therapies for the treatment of solid tumours.

Research Proposal

We have synthesised L-arginine analogues *S*-2-amino-5-(2-nitroimidazol-1-yl) pentanoic acid **5** and *S*-2-amino-5-(2-aminoimidazol-1-yl) pentanoic acid **6** (Scheme I). The 2-nitroimidazole analogue **5** should be bioreductively reduced to give the 2-aminoimidazole analogue **6**, which is a formal direct analogue of L-arginine. If **6** is able to act as an NOS inhibitor and **5** is devoid of activity, then we would have the required system where **5** could act as a prodrug of **6**. These compounds were evaluated *in vitro* for their NOS inhibition.

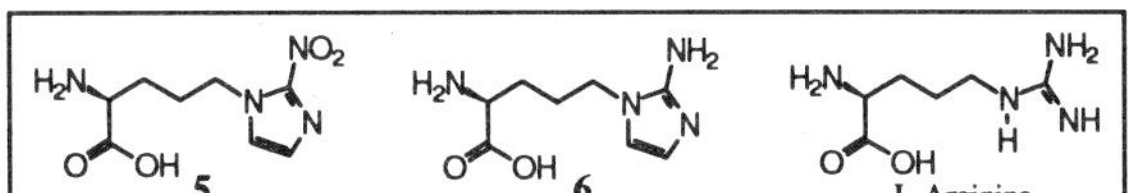

5 6 L-Arginine

Scheme I

Experimental

Synthetic Chemistry

The L-arginine analogues *S*-2-amino-5-(2-nitroimidazol-1-yl) pentanoic acid **5** and *S*-2-amino-5-(2-aminoimidazol-1-yl) pentanoic acid **6** were synthesised stereospecifically from the chiral protected amino acid α-*t*-butyl-N-BOC-L-Glutamic acid (Scheme II).

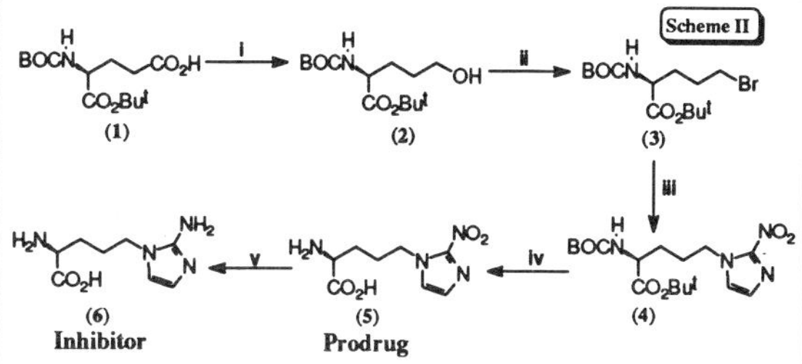

Scheme II

(1) (2) (3) (4) (5) Prodrug (6) Inhibitor

Key: i. $ClCO_2Et$, TEA, $NaBH_4$ ii. CBr_4/Ph_3P/THF iii. 1-potassio-2-nitroimidazole iv. 6N.HCl, EtOAc v. H_2, Pd/C, MeOH

Biological Evaluations

Compounds **5** and **6** were evaluated as inhibitors of nitric oxide synthase (NOS) activity, based on the conversion of [^{3}H]-L-arginine to [^{3}H]-L-citrulline [2]. Cytosolic fraction of rat brain was used as the source of Ca^{2+} dependent form of NOS.

Results

Results obtained by the enzyme assay (Fig.2) show that the 2-aminoimidazole analogue of L-arginine **6** inhibits the NOS activity (IC_{50}=1.98mM) and the inhibition is concentration dependent. The 2-nitroimidazole analogue of L-arginine **5** on the other hand shows very little inhibition (IC_{10}> 6mM).

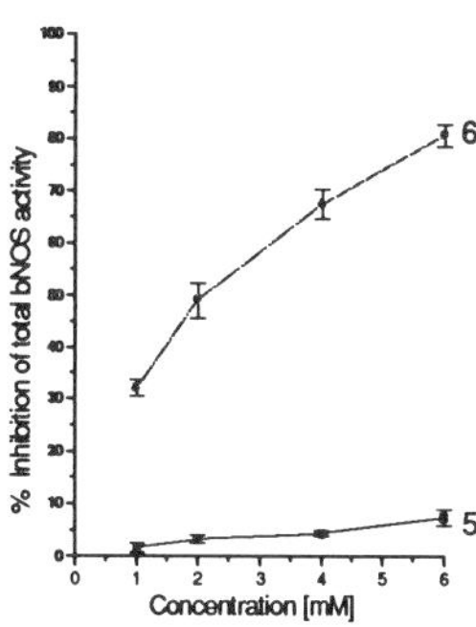

Fig **2**. Inhibition of brain NOS activity by **5** and **6**

Conclusions

Since *S*-2-amino-5-(2-aminoimidazol-1-yl) pentanoic acid **6** is an NOS inhibitor and *S*-2-amino-5-(2-nitroimidazol-1-yl) pentanoic acid **5** shows very little activity, we can regard **5** as a potentially hypoxia-selective prodrug of **6** with a possible application in the selective modulation of tumour blood flow.

Acknowledgements

We wish to thank the MRC for their financial support for this project.

References

1. D.J. Stuehr, O.W. Griffith.; Mammalian nitric oxide synthases *Adv. Enzym. Rel. Mol. Biol.*, 1993, **65**, 287-346.
2. C.Szabo, J.A. Mitchell, C. Thiemermann, J. R. Vane., Nitric oxide-mediated hyporeactivity to noradrenaline precedes the induction of nitric-oxide synthase in endotoxin shock. *Br J. Pharmcol.*, 1993, **108**, 786-792.

THE ROLE OF NITRIC OXIDE IN THE EVOLUTION OF LIVING SYSTEMS FROM PREMORDIAL PROKARYOTES TO EUKARYOTES - A HYPOTHESIS

M. Anbar
Department of Biophysical Sciences, School of Medicine and Biomedical Sciences, University at Buffalo, Buffalo, NY 14214.

As a chemical messenger nitric oxide is less specific and its concentration is less controllable than most other neurotransmitters or hormones.[1,2] Its extensive role as a primary or secondary chemical messenger in such a large variety of functions in current biology seems, therefore, surprising.[2,3] Since different enzymes evolved to secure its production in different cell lines and organ systems, and its role had not be taken over by more specific and better controllable agents, its biological function must be both ancient and essential. Like polyphosphates, amino acids or nucleotides, NO seems to have a fundamental biological role, that cannot be replaced by a different chemical species. It has been suggested that the biological function of NO evolved early in the anaerobic stage of biological evolution when exogenous NO was available.[1] Under those conditions, NO must have reacted with copper and iron binding proteins, modulating their enzymatic activity. Since those premordial enzymes had a variety of biochemical functions, and since NO diffuses very rapidly throughout the living milieu, it is very likely that NO acted as an intracellular synchronizing chemical messenger in early prokaryotes.[1] Carbon monoxide might have played a similar role at that time. Thus NO and CO were probably the first chemical modulators of biochemical redox activity in premordial cells.[1] In order to act as a biochemical modulator, NO must have been either oxidized or reduced within the biological milieu, since the concentration of a modulator must change with time. How was NO periodically removed from the system?

We know that NO is being reduced to N_2O and eventually to ammonia in denitrifying bacteria. Copper-binding proteins and iron-binding cytochromes are currently involved in the reduction process,[4-6] which seems to date back to anaerobic biology.[7] In fact, based on structural similarities in the pertinent enzymes, it was suggested that the oxygen-reducing respiratory chain developed from the anaerobic, denitrifying respiratory system.[7] It is very likely that NO bound to Cu^{2+} (formally equivalent to NO^+ bound to Cu(I)), which is prone to a nucleophilic attack by water,[4] reacts as an electrophile with sulfhydryls, e.g., cysteine,[8] to form an RS-NO transient. The facile binding of NO to Cu(II) has been confirmed also in other systems.[9] Sulfhydryls (RSH, including HSH) may then reduce NO^+ by double electron transfer to NOH (i.e., to nitrous oxide).

Since the premordial cell, in a non-oxidizing atmosphere, contained polypeptide bound Cu(II) and low molecular weight sulfhydryls (RSH, including H_2S), there seems to have been a non-enzymatic effective mechanism to reduce NO to N_2O, with RSNO as an intermediate. (The contemporary involvement of protein-bound Fe(III) as a binding site for NO which can then react as an electrophile,[10] was not feasible under those non-oxidizing conditions.) It is conceivable that the same Cu(II) entity could bind both RS^- and NO as ligands. Since RS^- binding increases the oxidation potential of Cu(II), it favors the formation of copper bound NO^+ as an intermediate, thus enhancing its eventual reduction. This makes the rate of NO removal second order in RSH. It is noteworthy that while H_2S and RSH can act similarly as ligands, only RSH can act as an electron donor to NO after RSNO was formed. Since the concentration of RSH is diminished in the presence of NO by the Cu(II) catalyzed reaction with NO, and since the rates of replenishment of RSH and NO from exogenous extracellular sources is different for NO and the polar RSH or charged RS^-, the rate of removal of NO will oscillate. The frequency of this oscillation depends on the steady state concentrations of NO and RSH, on the rates of diffusion of these two species, on the rates of transport of NO and RSH through the premordial biological membrane, and obviously on the rate constant k of the $k[NO][RSH]^2[Cu(II)][peptide]$ kinetic equation. Although there are too many unknowns to allow a reliable estimate the frequency of this periodic modulation of the intracellular level of NO, there is little doubt the under those conditions the intracellular level of NO did oscillate, which may explain its suggested role as a synchronizing pacemaker in the premordial cell.[1] In other words, these oscillations may have become the basis for a premordial biological clock that synchronizes intracellular biochemical activities.

Since the level of NO in the non-oxidizing atmosphere remained constant for hundreds of millions of years, while the extracellular concentration of RSH varied in different micro environments, the early prokaryotes probably started to produce RSH biosynthetically from H_2S, in parallel to the photooxidation of H_2S to sulfur, used as an early source of free energy. The biosynthesis of cysteine from H_2S and O-acetyl-L-serine observed in chloroplasts and prokaryotes (from which chloroplasts probably originated),[12] presents a late example of such a process. The biosynthesis of RSH from H_2S might have coupled the NO-modulated synchronization of copper and iron binding enzymes with the availability of free energy needed for the activity of those early enzymes.

Once the atmosphere became oxygenated, NO had to be biosynthesized by a premordial NOS in order to maintain its function under aerobic conditions. It may be rewarding to deconvolute that ancestral NOS, which must have evolved when the level of atmospheric O_2 reduced the extracellular NO concentration below an effective threshold. The evolution of eukaryotes by symbiosis of prokaryotes, which occurred about the same time, offered NO a new role. At that stage in biological evolution, it may have become the synchronizing messenger between the different organelles inside the early eukaryotic cell. It may be speculated that NO was absolutely necessary for the evolution of eukaryotes because it facilitated instantaneous communication between different encapsulated living systems, each containing iron binding proteins, that shared the same cytoplasm. At a later stage, several isoenzymes evolved to produce NO in different cell lines.

The mechanism of NO formation in eukaryotes under aerobic conditions involves both positive and negative feedback. Here are some examples: NO modulates NOS by binding to its active site; NO releases Fe(II) from certain iron binding proteins such as ferritin,[13] and Fe(II) competes for HO_2 radicals, the most effective scavenger of NO; Fe(II) is also used in the biosynthesis of NOS; NO can modulate mitochondrial ATP production[14,15] while ATP is needed in the biosynthesis of NOS, in the activation of iNOS[16,17] and in enhancing the Ca[11] influx needed to activate cNOS;[18] lowering the rate of respiration changes the caloric output cells and with it the rate of NO production, which is temperature sensitive;[19] finally, in certain cells lines extracellular NO can enhance NO production.[20] Any of these processes, all of which can occur inside cells, is likely to induce an oscillatory behavior in the level of NO, which is essential for the role of a synchronizing inter-organelle intracellular modulator, as well as a intercellular modulator of clusters of homologous cells produced by subdivision.

Nitric oxide probably still maintains its intracellular synergic synchronizing role in contemporary prokaryote and eukaryote cells. Some suggestions to test this hypothesis were discussed elsewhere.[1] As metazoan organisms evolved, NO took on an additional role -- that of a chemical messenger between different eukaryote cell lines. A classic example is the function of endothelial NO in the vascular system. In many cases the chemical message of NO is inhibition of certain enzymes or groups of enzymes, exemplified by the interference of NO with cell respiration.[14,15] In certain cells such inhibition may lead to lethal ATP deprivation.[21] This cytotoxic function of NO was perfected by lymphocytes, especially by macrophages, to control the proliferation of alien cells. However, a more important function in higher organisms has been the synchronizing effect of NO on target cells in different organ systems, in analogy with its intracellular role inside individual cells. This unique function of NO is discussed at length in the accompanying paper.[3]

Notwithstanding the reversible binding of NO to hemoglobin or the occasional formation of HO_2 radicals, eukaryotes seem not to have developed an enzyme to rapidly eliminate nitric oxide, in analogy with carbon monoxide dehydrogenase. Such an enzyme might have, however, interfered with the inter-organ synchronizing functions of NO, defeating this unique feature of nitric oxide. Interestingly, unlike NO, carbon monoxide which can act as a neurotransmitter, does not cause synchronous vasodilation in the brain.[22] Moreover, in spite of its similarity to NO in many respects, there is little evidence for CO as an intracellular messenger, nor are there examples of carbon monoxide acting as an inter-organ communicator. Nitric oxide is, therefore, a very unique chemical messenger indeed.

1. Anbar, M. (1995) *Experientia* 51, 545-550.
2. Anbar, M. (1995) *Thermologie Oesterreich* 5,15-27.
3. Anbar, M. (1995) These proceedings, paper #346 (poster presentation).
4. Ye, R.W., Toro-Suarez ,I., Tiedje, J.M. and Averill, B.A. (1991) *J. Biol. Chem.* 266, 12848-12851.
5. Ehrenstein, D. and Nienhaus, G.U. (1992) *Proc. Natl. Acad. Sci. USA* 89, 9681-9685.
6. Zumft, W.G., Braun, C. and Cuypers H. (1994) *Europ. J. Biochem.* 219, 481-490.
7. Saraste, M. and Castresana, J. (1994) *FEBS Lett.* 341:1-4.
8. Lipton, S.A., Stamler, J.S. (1994) *Neuropharm.* 33, 1229-1233.
9. Musci, G., Di Marco, S., Bonaccorsi di Patti, M.C. and Calabrese, L. (1991) *Biochemistry* 30, 9866-9872.
10. Lancaster, J.R. Jr. and Hibbs J.B. Jr. (1990) *Proc. Natl. Acad. Sci. USA* 87, 1223-1227.
11. Jaffrey, S.R., Cohen, N.A., Rouault, T.A. et al. (1994) *Proc. Natl. Acad. Sci. USA* 91, 12994-12998.
12. Rolland, N., Job, D. and Douce, R. (1993) *Biochem. J.* 293(Pt 3), 829-833.
13. Reif, D.W. (1992) *Free Radic. Biol. Med.* 12, 417-427.
14. Henry, Y. and Guissani, A. (1994) *Transfusion Clin. Biolog.* 1, 157-164.
15. Salzman, A.L., Menconi, M.J., Unno, N. *et al.* (1995) *Am. J. Physiol. - GI & Liver Physiol.* 31, G 361-G 373, 1995.
16. Yagi, K., Nishino, I., Eguchi, M., Kitagawa, M. *et al.* (1994) *Biochem. Biophys. Res. Commun.* 203, 1237-1243.
17. Tonetti, M., Sturla, L., Bistolfi, T., Benatti, U. and De Flora, A. (1994) *Biochem. Biophys. Res. Commun.* 203, 430-435.
18. Korenaga, R., Ando, J., Tsuboi, H. *et al.* (1994) *Biochem. Biophys. Res. Commun.* 198, 213-219.
19. Bernard, C., Merval, R., Esposito, B. and Tedgui, A. (1994) *Europ. J. Pharmacol.* 270, 115-118.
20. Garcia-Welsh, A., Laskin, D.L., Hwang, S.M. *et al.* (1994) *J. Leukocyte Biol.* 56, 488-494.
21. Virta, M., Karp, M. and Vuorinen, P. (1994) *Antimicrobial Agents & Chemotherapy* 38:2775-2779.
22. Brian, J.E., Heistad, D.D. and Faraci, F.M. (1994) *Stroke* 25, 639-643.

Visualising the release of nitric oxide from a variety of cellular sources

VANESSA W FURST, ANNA M LEONE, NEALE FOXWELL, SELIM CELLEK, PETER N WIKLUND* and SALVADOR MONCADA.

Wellcome Research Laboratories, Beckenham, Kent, UK, BR3 3BS. *Dept of Physiology and Pharmacology, and Institute of Environmental Medicine, Karolinska Institute, 17176 Stockholm, Sweden.

Introduction

An activated luminol chemiluminescent method for visualising the cellular release of nitric oxide (NO) has been developed and its specificity for NO demonstrated with activated macrophages [1]. We have applied this method to visualise the release of NO from a cytosolic preparation of inducible NO (iNOS) synthase, derived from activated J774.16 cells, in microtitre plates. We further demonstrate the application of this method to cultured endothelial cells and rat aortae stimulated with bradykinin and cultured NG 108 cells stimulated with acetylcholine.

Method

The imaging medium (IM) consisted of 5 mM luminol (sodium salt) in Krebs with 15 μM hydrogen peroxide, and was prepared approximately two hours before use to allow the background chemiluminescence and hydrogen peroxide concentrations to minimise. Bright-field images were taken and stored for reference. All cells were cultured on glass coverslips and viewed with a Zeiss Axiavert 135 TV inverted microscope (Carl Zeiss (Oberkochen) Ltd). The chemiluminescent signal was captured by a photon sensitive CCD camera (Hamamatsu Photonics UK Ltd) attached directly below the Zeiss microscope for maximum light transmission. The data was processed by the Argus 50 photon counting system (Hamamatsu Photonics UK Ltd). The bright-field images were overlaid with photon counted images in slice mode, or shown separately. Slice mode is a signal enhancement mode enabling colour assignment to a signal intensity scale. Photon counts were collated in gravity mode for quantification.

Results

The chemiluminescent signal from the iNOS preparation was inhibitable with L-NAME and signal intensity was enzyme concentration dependent (Fig. 1). In Figure 1 the top row (boxes A, B and C) shows the control, i.e. IM only with 100 μM L-arginine and iNOS co-factors: NADPH 100 μM, BH_4 5 μM. The second row (boxes D, E and F) shows a concentration dependent signal increase from D, 0.025 nM/min/mg protein, E, 0.05 nM/min/mg protein to F, 0.1 nM/min/mg protein of iNOS. The bottom row (G, H and I) shows the same concentrations of iNOS as above but with 1 mM L-NAME. The image aquisition time was 2 sec and black microtitre plates (Dynatech) were used to block light reflection between wells.

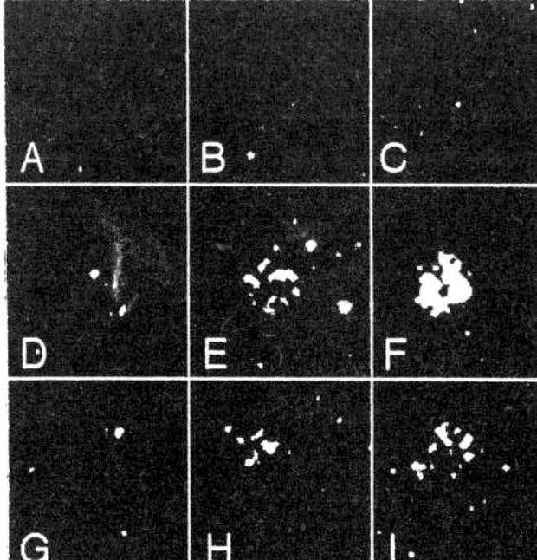

Figure 1.

The cultured endothelial cells gave an increased chemiluminescent signal with bradykinin (BK) which was inhibitable with L-NAME (Fig. 2). The endothelial cells were seeded at a low density in order to try and visualise the NO release from individual cells and cell clusters. The signal intensity from endothelial cells, particularly when seeded at low density, approached the detection level limit of the method.

Images from rat aortae, taken sequentially at 5 second intervals, showed a basal level of signal that increased when stimulated with acetylcholine (Ach). The signal originated at the site of Ach administration and spread down the aortae in sequential images.

The signal from cultured NG 108 cells was inhibitable with L-NMMA and as with the macrophages [1] and endothelial cells the signal was most intense over the cell clusters.

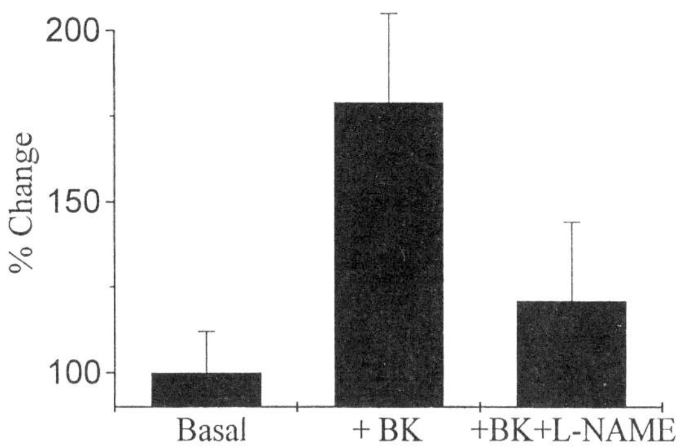

Figure 2. Graph showing the % change of photon counts (n=5) for cultured endothelial cells with bradykinin (10nM) with and without L-NAME (200uM).

Discussion

A chemiluminescent signal with activated luminol has been demonstrated before by other workers [2] but we describe the first use of this reaction for visualising the distribution of an NO signal spacially. The chemiluminescent signal was shown to be derived from NO by the fact that the signal was enhanced under conditions known to increase NO production and diminished in the presence of NO inhibitors. Further work is necessary to establish the exact chemistry of the chemiluminescent signal, but preliminary work [vide infra 1] suggests that luminol is activated by the addition of hydrogen peroxide to form a new 'activated luminol' species which gives a chemiluminescent signal directly with NO.

Some applications of this method benefit from integration times of minutes due to improved noise/signal ratios. However the IM sensitivity to NO can generate an image in seconds, giving this method the potential for generating image sequences [3] or real time 'films' [4] of NO release. This is particularly applicable were NO can be 'switched on' within the image capture time [3, 4].

In conclusion this method has demonstrated specificity for NO from a wide range of biological sources and has the potential for real time imaging of NO release.

References

1. Leone, A.M., V.W. Furst, N.A. Foxwell, S. Cellek, P.N. Wiklund and S. Moncada (1995) Biology of nitric oxide, this volume.
2. Kikuchi, K., T. Nagan, H. Hayakawa, Y. Hirata and M. Hirobe (1993) Anal. Chem. **65**, 1794-1799.
3.Cellek,S., A.M. Leone, V.W. Furst, P.N. Wiklund and S. Moncada (1995), Biology of nitric oxide, this volume.
4.Wiklund,P.N., A.M. Leone, V.W. Furst, S. Cellek and S. Moncada (1995), Biology of nitric oxide, this volume.

Real-time imaging of nerve-induced nitric oxide release

N PETER WIKLUND[1], HENRIK H IVERSEN[2], ANNA M LEONE[3], LARS E GUSTAFSSON[2], VANESSA FURST[3], SELIM CELLEK[3], ÅKE FLOCK[2] and SALVADOR MONCADA[3]

Dept Surgery section of Urology[1], Karolinska Hospital and Dept of Physiology and Pharmacology[2], Karolinska Institute, 171 77 Stockholm, Sweden. Wellcome Research Laboratories[3], Langley Court, Beckenham, Kent, BR3 3BS, U.K.

In the peripheral nervous system nitric oxide (NO) has been suggested as a mediator of non-adrenergic non-cholinergic (NANC) autonomic neurotransmission. Release of NO is likely responsible for adaptive gastric dilatation in the stomach [1] and for the relaxation of smooth muscle during the peristaltic reflex[2]. The short half-life has made it difficult to study and detect the release NO in biological tissues. In order to gain better understanding of the pattern of release and diffusion of NO, nerve-induced release of NO in the guinea pig ileum with intact myenteric plexus was visualised according to the method by Leone et al (abstract this meeting).

Methods; NO was visualised in a luminol/hydrogen peroxide-based imaging medium (IM). The light produced by the interaction of NO with IM (referred to as NO/luminol activity) was quantified using an Argus 50 photon counting system (Hamamatsu Photonics UK Ltd.) coupled to a Reichart inverted microscope. Several studies have quantified the NO/luminol-induced chemiluminescence as a measurement of NO production in various cell types although they did not visualise the release of NO [3-4]. The present method is an indirect method to study NO release since NO is not reacting directly with luminol. The mechanism for the generation of luminescence in the presence of NO and IM is not clear. Both peroxy nitrite and singlet oxygen have been suggested to be formed during similar conditions and could thus react with luminol to generate luminescence [5-6]. The longitudinal muscle layer together with the underlying myenteric plexus was spread out on to a glass cover-slip. It was placed in Krebs' kept at 37°C. Both ends of the preparations was hooked on silver electrodes. Stimulation was performed at 0.3-2ms pulse duration, 5-30 Hz.

Results and Discussion; The ganglions as well as thicker nerve trunks were clearly visible in the myenteric plexus preparation used (Fig 1A). A weak light signal was seen in the preparation after IM was added. There was no change in background light signal after application N^G-nitro-L-arginine methylester (L-NAME)100 μM (Fig 1C). Electrical stimulation induced a marked increase in NO/luminol activity in the myenteric plexus (Fig 1B). The light emission was evenly distributed over the preparation and there was similar NO/luminol activity over ganglions and thick nerve trunks as over areas with only thin nerve trunks and single neurons (Fig 1B). L-NAME (10^{-4} M, n=6) inhibited the stimulation-induced activity (Fig 1B) by 65±11%, and the addition of tetrodotoxin (10^{-6} M, n=6) further inhibited the activity to 91±6 %. The increase in NO/luminol activity was frequency dependent and increased over the duration of the simulation period (1-2 min). The increase in release seen over time was present at all stimulation frequencies studied. The NO/luminol activity was rapidly reduced after the end of the stimulation period and 20 sec after the stimulation similar NO/luminol activity as prior to stimulation was observed. Application of authentic NO (10nM-5μM) to the IM generated light in a concentration-dependent fashion. Since the NO/luminol activity was markedly inhibited by L-NAME and tetrodotoxin it is likely that NO released from neuronal structures caused the light emission in our experiments. The level of inhibition of the NO/luminol activity by L-NAME is in good agreement with our previous finding regarding inhibition of nerve-induced NO release as measured by bioassay or the NO oxidation products NO_2^- and NO_3^-[7]. Since NO will diffuse quickly in both aqueous and lipid milieu the diffusion pattern for NO is going to be of great importance for it effects. Our experiments

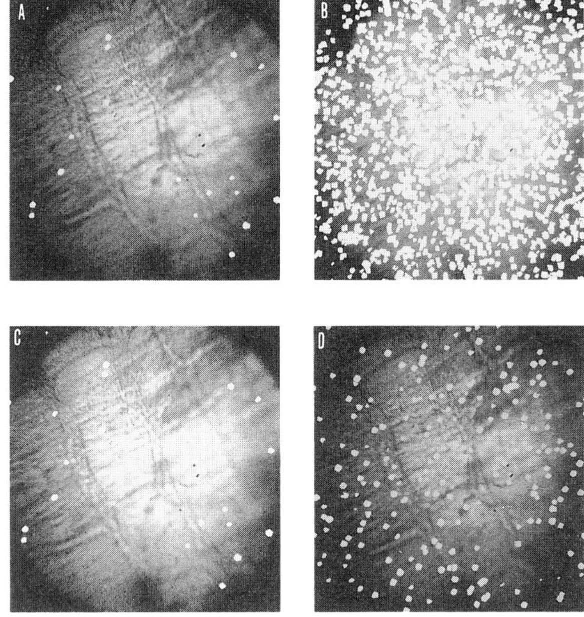

Figure 1. Bright field image of the guinea pig ileum myenteric plexus overlaid with white photons counted under a 20 sec acquisition in order to study the localisation of NO release; a) basal condition, b) during nerve stimulation (1 ms, 30 Hz, 20 sec), c) basal condition in the presence of L-NAME (100 μM), d) during nerve stimulation in the presence of L-NAME. Note the inhibition of nerve-induced NO/luminol activity during L-NAME application.

show that there was an increase in the NO/luminol activity over time during stimulation. This most likely reflects the diffusion and inactivation of NO in the tissue. In the theoretical model describing the kinetic and concentration profile for NO generated from single or multiple sources used by Wood and Garthwaite [8] it is predicted that when there are multiple simultaneously active NO releasing sources within the tissue volume, and in the absence of decay of NO or of a time-dependent reduction in NO release, the concentration of NO simply rises linearly with time. Even distant sources of NO (200 - 500 μm away) will have a significant contribution to the steady-state NO concentration in this situation even when the half-life of NO is short (0.5 - 5 sec). In conclusion, quantification of NO release by real time visualisation gives important new information regarding its role as a neurotransmitter. Thus, diffusion profiles for NO generation as well as the localisation of release can be studied.

Acknowledgements; The study was supported by Stiftelserna Maud och Birger Gustafsson, Loo och Hans Osterman and Swedish MRC (11199 and 7919)

References

1. Desai, K.M., Sessa, W.C. and Vane, J.R. (1991) *Nature* **351**, 477-479.
2. Hata, F. *et al.* (1990) *Biochem. Biophys. Res. Commun.* **172**, 1400-1406.
3. Kikuchi, K., Nagano, T., Hayakawa, H., Hirata, Y. and Hirobe, M. (1993) *J. Biol. Chem.* 268, 23106-231110.
4. Wang, J-F., Komarov, P. de Groot, H. (1993) *Arch. Biochem. Biophys.* 304, 189-196.
5. Noronha-Dutra, A.A., Epperlein, M.M. and Woolf, N. (1993) *FEBS letters* 321, 59-62.
6. Radi, R., Cosgrove, T.P., Beckman, J.S. and Freeman, B.A. (1993) *Biochem. J.* 290, 51-57.
7. Wiklund N.P., Leone, A.M., Gustafsson L.E. and Moncada, S. (1993) *Neuroscience* 53, 607-611.
8. Wood, J and Garthwaite, J.(1994) *Neuropharmacol.* 33, 1235-1244.

Nitric oxide production from human skin

RICHARD WELLER*, SIMON PATULLO, LORNA SMITH, MICHAEL GOLDEN, ANTHONY ORMEROD*, NIGEL BENJAMIN

Department of Medicine and Therapeutics, University of Aberdeen Medical School, Foresterhill, Aberdeen AB9 2ZD and *Department of Dermatology, Aberdeen Royal Hospitals, Aberdeen AB9 2ZB.

Introduction

Both constitutive and inducible NO synthesis in mammalian cells requires NO synthase enzymes using L-arginine as a substrate[1,2]. An alternative method of production is by sequential reduction of nitrate by bacterial nitrate reductase and further reduction of nitrite to NO. Reduction of nitrite will occur chemically, particularly in acidic conditions. We, and others, have recently described the chemical synthesis of NO in the mouth and stomach [3,4,5] which relies on the secretion of nitrate in saliva and conversion to nitrite on the tongue with reduction to NO in stomach acid.

The purpose of the present study was to ascertain the mechanism by which NO is generated on another epithelial surface; the skin; and to consider whether the large amounts generated may also be important in protection from microbial pathogens.

We have therefore measured NO production from human skin and, using inhibitors of NO synthase, topical application of antimicrobials, inorganic nitrite, and agents altering skin acidity, have studied the importance of these separate mechanisms for NO generation.

Methods

Studies were performed in 14 healthy male volunteers, (age 21 - 39).A further 6 subjects (age range 17-40) who had been on therapy for at least 5 weeks with oral tetracylcline or tetracycline analogue were studied.NO produced by the skin was sampled by placing the subject's hand in a glass vessel which was sealed at the wrist with a rubber membrane. NO-free air was drawn into the vessel at 81ml/min and subsequently sampled using a chemiluminescence NO analyser. NO concentration in the sampled air normally reached equilibrium after 5-10 minutes. Skin NO synthesis was expressed in moles cm-2.min-1.Sweat was collected from the back of 6 subjects after 40 minutes in a 40°C sauna, every 5min by scraping sweat from the skin using clean glass slides.To assess the effects of skin suface acidity on NO generation the subject's hand was immersed in citrate/phosphate buffer (0.1M, pH3) or Tris(hydroxymethyl) aminomethane buffer (0.2M, pH9) immediately before NO measurement. Similarly the effect of nitrite was determined by immersion into potassium nitrite solution (2-400µM). The effect of all topical applications was compared to topically applied distilled water alone, which we found has no effect on NO generation.

Seven subjects were studied on 2 different days, at least 2 days apart. On one experimental day saline then LNMMA (4 and 8µmoles/min) and again saline was infused into the left brachial artery, each for 13 minutes. Blood flow was measured in both arms by forearm venous plethysmography during the final 3 minutes of each infusion. Changes in blood flow were expressed as %change in blood flow ratio between the two arms as previously described [5]. On the control day no brachial artery infusion was given.The order of study day was determined by balanced, random allocation.

Results

NO production by the skin of the hands of healthy volunteers ranged from 4.2-19.4 femtomoles cm-2.min-1(mean 12.05±1.68). Individual subjects studied on different days tended to have similar NO production.

NO production by the skin increased significantly after the duringapplication of pH3 buffer compared to the normal saline control

(from 7.5±1.7 to 10.1±2.4 fmoles cm-2.min-1; $p < 0.05$, paired Students t-test). On the application of the pH9 buffer the skin NO production significantly decreased (8.5±2.1 to 5.4±1.4 fmoles cm-2.min-1; $p < 0.01$).

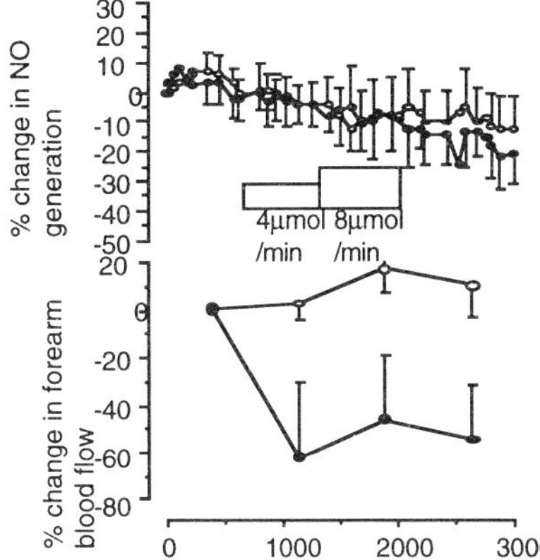

Change in hand skin NO generation (top panel) and forearm blood flow (bottom panel) during brachial artery LNMMA infusion (closed symbols) or saline (open symbols)

Potassium nitrite (2-400µM) when applied to the hand caused a dose-dependent increase in NO generation which was linear. ($P<0.000004$, $r>0.99$).

There was no significant change in hand NO production following chlorhexidine (5.2±0.3 to 7.4±0.5 fmoles cm-2.min-1). Two hours following antibiotic spray (Tribiotic) mean NO production showed no change in 3 volunteers, from 14.7±5.3 to 14.8±5.2 fmoles cm-2.min-1.

Mean hand skin NO production was reduced to 3.05±0.2 fmoles cm-2.min-1 in patients on long-term antibiotics ($P<0.001$) compared to age-matched normal controls.

Infusion of normal saline into the brachial artery was associated with a gradual fall in NO generation from the hand during the course of the experiment (approximately 1 hour) from 16.2±1.7to 14.9±1.5 fmoles cm-2.min-1on the day when no brachial artery infusion was used. When LNMMA was infused

chemiluminescence NO analyser. NO concentration in the sampled air normally reached equilibrium after 5-10 minutes. Skin NO synthesis was expressed in moles cm-2.min-1.Sweat was collected from the back of 6 subjects after 40 minutes in a 40°C sauna, every 5min by scraping sweat from the skin using clean glass slides.To assess the effects of skin suface acidity on NO generation the subject's hand was immersed in citrate/phosphate buffer (0.1M, pH3) or Tris(hydroxymethyl) aminomethane buffer (0.2M, pH9) immediately before NO measurement. Similarly the effect of nitrite was determined by immersion into potassium nitrite solution (2-400µM). The effect of all topical applications was compared to topically applied distilled water alone, which we found has no effect on NO generation.

Seven subjects were studied on 2 different days, at least 2 days apart. On one experimental day saline then LNMMA (4 and 8µmoles/min) and again saline was infused into the left brachial artery, each for 13 minutes. Blood flow was measured in both arms by forearm venous plethysmography during the final 3 minutes of each infusion. Changes in blood flow were expressed as %change in blood flow ratio between the two arms as previously described [5]. On the control day no brachial artery infusion was given.The order of study day was determined by balanced, random allocation.

Results

NO production by the skin of the hands of healthy volunteers ranged from 4.2-19.4 femtomoles cm-2.min-1(mean 12.05±1.68). Individual subjects studied on different days tended to have similar NO production.

NO production by the skin increased significantly after the duringapplication of pH3 buffer compared to the normal saline control

(from 7.5±1.7 to 10.1±2.4 fmoles cm-2.min-1; $p < 0.05$, paired Students t-test). On the application of the pH9 buffer the skin NO production significantly decreased (8.5±2.1 to 5.4±1.4 fmoles cm-2.min-1; $p < 0.01$).

NOS - MAPPING of human nasal mucosa under physiological and pathophysiological conditions.

OLAF MICHEL, WILHELM BLOCH*, JAN ROCKER, SILKE PETERS* and KLAUS ADDICKS*

*Department of Anatomy, University of Cologne, 50931 Cologne, FRG

introduction

Human nasal mucosa is rich in blood vessels and nerve fibers which innervate the vessels as well as epithelium of secretory glands and surface mucosal membranes. All these components are involved in the excessive swelling concomitant with inflammatory and allergic diseases. Nitric oxide, which is produced by several isoforms of NOS [1] is considered to be an important regulator of vascular tone, neuronal function and inflammatory processes [2]. There is growing evidence for the presence of all known NOS-isoforms in nose and nasal sinus of different species [3-7]. The aim of the present study was to investigate, under physiological and pathophysiological conditions, the distribution of various NOS-isoforms by NADPH-d as well as the distribution of the bNOS and the iNOS through specific antibodies. Inflammatory effects were furtherly investigated by induction studies with LPS conducted in an organ bath followed by morphological studies.

material and methods

Specimens were selected mostly from biopsies and probes of operations of the medial conchae. The samples were taken from tissue of the nasal mucosae under inflammatory, allergic and physiological conditions. For the organ bath treatment nasal probes without pathophysiological alterations were selected during surgical intervention and were immediately transferred in a tyrode solution prewarmed to 37°C. Next the specimens without inflammatory and allergic alterations were exposed to LPS for 30, 60, 120, 180 and 240minutes. The controls were incubated without any additions for the same time. NADPH-diaphorase staining of 20µm thick cryostat section of the human nasal mucosae was done for

Abbreviations used: inducible nitric oxide synthase (iNOS), neuronal nitric oxide synthase (bNOS), nicotinamide adenin dinucleotide phosphate (NADPH), NADPH-diaphorase (NADPH-d), lipopolysaccharide (LPS), bovine serum albumin (BSA)

one hour at 37°C in a Tris-buffer solution (pH 8.0) which contained 83 mg ß-NADPH (Biomol, Hamburg), 40 mg nitro blue tetrazolium (Biomol, Hamburg), 125 mg monosodium malate and 0.1% triton X-100 at 100 ml. The sections were thaw-mounted onto gelatine coated glas slides and studied microscopically with an axiophot (Zeiss, Oberkochen). For immunohistochemistry the cryostate sections were pretreated with 3% H2O2 in a 60% methanol solution, 0.2% Triton in 0.1 M PBS and 5% BSA solution. The incubation with the primary antibody was conducted in a solution of 0.8% BSA and 20 mM NaN3 in PBS containing the iNOS (Biomol, Hamburg) specific mouse antibody (1:500 or 1:1000 dilution) or bNOS (Afiniti, Nottingham) specific rabbit antibody for 12 hours at 4°C. After rinsing, the sections were incubated with the secondary biotinylated goat-anti-mouse or goat-anti-rabbit antibody (1:100 dilution) (Vector Laboratories) for 1 hour at room temperature. Then a streptavidin-horseradish-peroxidase complex was utilized as detection system (1:200 dilution) for another hour and finally developed for 15 minutes with 3,3 diaminobenzidine tetrahydrochloride in 0,05 M Tris HCl-buffer and 0,1% H2O2 and counterstained with methylgreen.

results

The immunohistochemical assay containing the bNOS and NADPH-d revealed a rare nitrinergic innervation of the subepithelial secretory glands, the surface epithelium and the arteries and veins in the mucosa. Under physiological conditions the arteries and arterioles displayed a distinct staining through the NADPH-d reaction whereas the endothelium of capillaries, veins and venous sinus were only weakly stained. An iNOS antibody reaction could not be detected in nasal blood vessels without inflammatory and allergic alterations. Under inflammatory conditions capillaries, veins and venous sinuses appeared with a strongly increased NADPH-d staining. This has been confirmed by a positive immunoreaction of the vessels with the iNOS-antibody in samples orginating from patients with inflammatory and allergic diseases. Positively NADPH-d stained and iNOS positive leucocytes in the subepithelium could be observed to a varying degree in specimens from patients with clinical signs of inflammatory or allergic diseases. Nasal mucosae without signs of inflammatory and allergic alterations only showed sporadic iNOS positive leucocytes. In isolated cases bNOS positive mast cells were observed. NADPH-d staining was also apparent to a varying degree in the glandular epithelium and the surface epithelium of the nasal mucosae under physiological as well as pathophysiological conditions. The expression of NADPH-d in inflammatory and allergic nasal mucosa was much more intensive and extensive when compared to physiological conditions. The distribution pattern of NADPH-d in the surface epithelium was confirmed by iNOS immunoreactivity. The glandular epithelium also showed NADPH-d reaction but the immunostaining with the iNOS antibody was negative under inflammatory and allergic conditions. The positive NADPH-d staining of the glandular epithelium could not be reproduced by immunohistochemical methods using the b-NOS antibody. These results were confirmed by the organ bath studies, where LPS treatment was used in order to induce iNOS with the exception of the glandular epithelium and the leucocytes. After the organ bath treatment the glandular epithelium was iNOS positive. Leucocytes containing iNOS were only observed in small number.

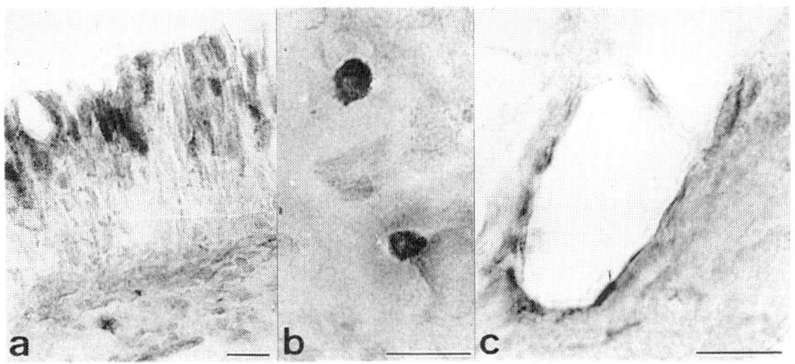

Immunhistochemical analysis of the human nasal mucosa using iNOS-antibody shows immunreactive surface epithelium (a), leucocytes (b) and endothelium (c) under inflammatory conditions. bar 50µm

discussion

The basal nasal mucosal blood flow is subject to a circadian cycle [8]. The distribution of NOS in nerve fibers and endothelium under normal conditions in unaltered human nasal mucosa suggests a minor role of NO for the circadiane cycle. The focal expression of iNOS in the unaltered surface epithelium provides evidence for the involvement of NO, released by iNOS, in the host mechanisms under physiological conditions such as the bacteriostatic effects of NO [9] and the NO-induced increase of ciliary beat frequency [10]. This theory is supported by findings of Lundberg et al. [3] which suggested a role of NO for host defence in the nasal sinus. The excessive expression and the altered distribution of iNOS observed under inflammatory conditions, e.g. in venous sinus, and capillaries leads us to the assumption that NO plays a role in the regulation of blood flow and secretory function. The increased expression of NOS in inflammatory and allergic diseases in the endothelium of sinusoidal veins which are suggested to be involved in the regulation process of nasal swell bodies [11] also point towards an involvement of NO. Furthermore subepithelial nasal blood vessels which reveal an increased NOS expression are generally accepted to be an important structure for the secretory process in the nasal mucosa [12]. Therefore we suggest a regulative function of NO for the secretory process in the nasal mucosa under pathophysiological -like inflammatory or allergic-conditions.

references

1. Förstermann, U., Schmidt, H.H.H.W., Pollock, J.S., Sheng, H., Mitchell, J.A., Warner, T.D., Nakane, M.M. and Murad, F. (1991) Biochem. Pharmacol., 42: 1849, 1991.
2. Moncada, S., Palmer, R.M.J. and Higgs, E.A. (1991) Pharmacol.Rev. 43,109-142
3. Lundberg, J.O.N., Farkas-Szallasi,T., Weitzberg,E., Rinder,J., Lidholm,J., Änggard,A., Hökfelt,T., Lundberg,J.M. and Alving,K. (1995) Nature Med. 1,370-373
4. Hanazawa, T., Konno, A., Kaneko, T., Tanaka, K., Ohshima, H., Esumi,H. and Chiba,T. (1994) Brain Res. 657,7-13
5. Kulkarni, A.P., Getchell, T.V. and Getchell, M.L. (1994) J.Comp.Neurol. 345,125-138
6. Kishimoto, J., Keverne, E.B., Hardwick, J. and Emson, P.C. (1993) Eur.J.Neurosci. 5,1684-1694
7. Bacci, S., Arbi-Riccardi, R., Mayer, B., Rumio, C. and Borghi-Cirri, M.B. (1994) Histochem. 102,89-92
8. Kenning, J. (1968) Int.Rhinology 6,99-136
9. Mancinelli, R.L. and McKay, C.P. (1983) Applied Environ.Microbiol. 46,198-202
10. Jaln, B., Rubenstein, I., Robbins, R.A., Leishe, K.L. and Sisson, J.H. (1993) Biochem.Biophys.Res.Comm. 191,83-88
11. Grevers, G. and Herrmann, U. (1987) Laryng.Rhinol.Otol. 66,152-156
12. Grevers, G. (1993) Laryngoscope 103,1255-1258

Cigarette smoke elevates plasma but not platelet cyclic guanosine monophosphate or systemic nitrate levels in healthy female smokers

Åke Wennmalm and Christina Rångemark

Department of Clinical Physiology, Göteborg University, Sahlgrenska University Hospital, S-413 45 Göteborg, Sweden

Cigarette smoke contains large amounts of nitric oxide (NO) (400-1000 ppm; 1). NO inhibits platelet activation by stimulating the soluble guanylate cyclase in the platelets and increasing the levels of cGMP (2). NO is also a vasodilator, operating by elevating cGMP in vascular smooth muscle and thereby inducing vessel wall relaxation. The gaseous phase of cigarette smoke has been shown to induce an increase in the activity of guanylate cyclase in animal tissues, suggesting that NO is responsible for the observed vasodilator effects of cigarette smoke (3, 4).

Cigarette smoking is linked to an increased incidence of cardiovascular disease, both in men and women (5, 6). Platelet activation is involved in thrombogenesis, thereby contributing to vascular occlusion and tissue infarction (7). Several studies have demonstrated that cigarette smokers have a chronic but reversible form of platelet activation, resulting in increased formation of thromboxane A_2 (TxA_2; 8-10). Hence, smoking-associated cardiovascular disease may, at least partly, be based on an increased platelet formation of TxA_2.

Since cigarette smoke contains NO with anti-platelet activity, but also elicits pro-thrombogenic platelet activation, the final outcome of smoking on platelet activity might be an integrated response from the activity of smoke-derived, inhaled NO on the one hand and the proaggregatory activity of platelet TxA_2 on the other. To address the possibility that the inhaled NO is taken up from the lungs into the blood and thereby affects platelet cGMP in the pulmonary vascular bed we studied, firstly, whether female smokers had higher plasma levels or urinary excretion of nitrate, the stable metabolite of NO (11), and secondly, if the smokers had elevated platelet, plasma, or urinary levels of cGMP, in comparison to nonsmokers.

Twenty-three healthy, habitual cigarette smokers (mean daily consumption 16±1 cigarettes/day) and 26 matched nonsmokers were studied. A 24 hour portion of urine was collected from all participants, and at the end of the urine collection period a blood sample was drawn. Plasma and urine nitrate levels were analysed with a gas chromatography/mass spectrometry method as described earlier (11), and platelet, plasma, and urine cGMP were determined with radioimmunoassay.

Data on plasma and urine nitrate are shown in Table 1. There were no differences in plasma or urine nitrate levels between smokers and nonsmokers. Table 1 also displays platelet, plasma, and urine levels of cGMP. Platelet cGMP was lower ($p<0.05$) in smokers compared to nonsmokers. In contrast, plasma cGMP was significantly ($p<0.05$) higher in smokers compared to nonsmokers. The urine excretion of cGMP was also higher ($p<0.05$) in smokers compared to nonsmokers.

Table 1. Plasma and urine nitrate, and platelet, plasma, and urine cGMP in smokers and nonsmokers. Data presented as mean±SEM.

	nitrate		cyclic GMP		
	plasma µmol/L	urine mmol/24h	platelet pmol/10^{10}	plasma nmol/L	urine µmol/24h
smokers	30±1.4	1.5±0.2	8.1±0.7	2.5±0.2	0.63±0.04
nonsmokers	26±2.5	1.2±0.1	10±0.8	1.9±0.2	0.51±0.04
level of sign.	n.s.	n.s.	0.05	0.05	0.05

The lack of difference in plasma or urine nitrate levels between smokers and nonsmokers may seem to indicate that the smoke-derived NO was not retained in the lungs. However, calculation of the amount of NO inhaled during normal smoking indicates that the lack of difference in plasma or urine nitrate between smokers and nonsmokers was not unexpected. Smoking of one cigarette, i.e. about 10 puffs, yields about 18 µmol of NO. Smoking 16 cigarettes per day (present consumption) consequently yields about 300 µmol/24 h. This figure closely resembles the numeric but insignificant difference in urine nitrate excretion between smokers and nonsmokers in the present study. It is therefore suggested that NO in cigarette smoke is quantitatively insufficient to affect the plasma level of nitrate in smoking women with normal smoking habits.

Platelet cGMP was lower in the present smokers compared to the nonsmokers. This supports the assumption that NO from the smoke was not taken up into the blood in its active form in amounts sufficient to elicit biological activity. The observed plasma level and urine excretion of cGMP were slightly higher in smokers than in nonsmokers. A probable explanation to this finding is that NO derived from the smoke, during its diffusion from the alveoli to the pulmonary capillaries, i.e. before being metabolized to nitrate, activated guanylate cyclase in the pulmonary vascular smooth muscle cells. If this was the case the increased plasma and urinary levels of cGMP observed in smokers may reflect the effect of the inhaled NO in the pulmonary vascular bed. Supporting this assumption, inhalation of filtered cigarette smoke has been shown to cause consistent pulmonary vasodilation in pigs. The major part of this vasodilator response was probably caused by NO (12).

In conclusion, direct measurements of its major metabolite and determination of its second messenger levels demonstrate that NO absorbed from the inhaled smoke does not counterbalance the proaggregatory activity of TxA_2 in healthy female smokers. Hence, the negative effect of smoking on platelet function appears to be unopposed by smoke-related factors with potential antiplatelet activity.

Supported by the Swedish Medical Research Council (project 4341), and by the Swedish Tobacco Company. The authors are grateful to Mrs. Maud Peterson for skilful technical assistance.

References

1. Norman, V., and Keith, C.H. (1965) Nitrogen oxides in tobacco smoke. Nature **205**, 915-916
2. Radomski, M.W., Palmer, R.M.J., and Moncada, S. (1990) An L-arginine/nitric oxide pathway present in human platelets regulates aggregation. Proc. Natl. Acad. Sci. U.S.A. **87**, 5193-5197
3. Arnold, W.P., Aldred, R., and Murad, F. (1977) Cigarette smoke activates guanylate cyclase and increases guanosine 3',5'-monophosphate in tissues. Science **198**, 934-936
4. Gruetter, C.A., Barry, B.K., McNamara, D.B., Kadowitz, P.J., and Ignarro, L.J. (1980) Coronary arterial relaxation and guanylate cyclase activation by cigarette smoke, N'-nitrosonornicotine and nitric oxide. J. Pharmacol. Exp. Ther. **214**, 9-15
5. Kannel, W.B. (1981) Update on the role of cigarette smoking in coronary artery disease. Am. Heart J. **101**, 319-328
6. Rosenberg, L., Kaufman, D.W., Helmrich, S.P., Miller, D.R., Stolley, P.D., and Shapiro, S. (1985) Myocardial infarction and cigarette smoking in women younger than 50 years of age. JAMA **253**, 2965-2969
7. Davies, M.J., and Thomas, A.T. (1984) Thrombosis and acute coronary-artery lesions in sudden cardiac ischemic death. N. Engl. J. Med. **310**, 1137-1140
8. Nowak, J., Murray, J.J., Oates, J.A., and FitzGerald, G.A. (1987) Biochemical evidence of a chronic abnormality in platelet and vascular function in healthy individuals who smoke cigarettes. Circulation **76**, 6-14
9. Barrow, S.E., Ward, P.S., Sleightholm, M.A., Ritter, J.M., and Dollery, C.T. (1989) Cigarette smoking: profiles of thromboxane- and prostacyclin-derived products in human urine. Biochim. Biophys. Acta **993**, 121-127
10. Rångemark, C., Ciabattoni, G., and Wennmalm, Å. (1993) Excretion of thromboxane metabolites in healthy women after cessation of smoking. Arteriosclerosis Thrombosis **13**, 777-782
11. Wennmalm, Å., Benthin, G., Edlund, A., Jungersten, L., Kieler-Jensen, N., Lundin, S., Nathorst Westfelt, U., Petersson, A.-S., and Waagstein, F. (1993) Metabolism and excretion of nitric oxide in humans; and experimental and clinical study. Circ. Res. **73**, 1121-1127
12. Alving, K., Fornhem, C., and Lundberg, J.M. (1993) Pulmonary effects of endogenous and exogenous nitric oxide in the pig: relation to cigarette smoke inhalation. Br. J. Pharmacol. **110**, 739-746

Nitric oxide-dependent cerebral arteriolar relaxation in the rat: effects of estrogen depletion and repletion.

DALE A. PELLIGRINO, QIONG WANG, and VERNA L. BAUGHMAN.

Dept. of Anesthesiol., U. of Illinois-Chicago, Chicago, IL 60612, USA.

In the periphery, correlations exist between levels of circulating estrogen, constitutive (i.e., Ca^{2+}-dependent) nitric oxide synthase (cNOS) expression, and vasodilating capacity [1,2,4,5,8,9,12]. The principal aim of this investigation was to determine whether chronic reductions in circulating estrogen levels result in concomitant reductions in ***brain*** cNOS activity to the extent that cerebral vasodilating capacity is diminished. Within that context, we endeavored to establish whether such changes favor one or both of the Ca^{2+}-dependent isoforms--i.e., the endothelial NOS (eNOS) or the neuronal NOS (nNOS), and whether the suppressed vascular reactivity, if present, could be "corrected" by 17 ß-estradiol (E_2) treatment. Three groups were evaluated: normal female, chronically ovariectomized, and E_2-treated chronically ovariectomized rats. The major experimental approach involved the use of intravital microscopy and a closed cranial window system. We monitored pial arteriolar diameter changes during cortical suffusions of an eNOS-dependent dilator, acetylcholine (ACh), a nNOS-dependent dilator, N-methyl-D-aspartate (NMDA), a direct NO donor, S-nitroso acetylpenicillamine (SNAP), and a NO-independent dilator, adenosine (ADO). In addition, we evaluated pial arteriolar responses to hypercapnia in the different groups. Hypercapnia has been shown to elicit cerebrovasodilation in rats through a partly nNOS-dependent, but eNOS-*in*dependent mechanism [10,11]. To confirm NO-dependency in all of the above experiments, arteriolar responses were measured prior to and after initiating suffusion of the non-specific NOS inhibitor, nitro-L-arginine (L-NA). As a complement to the vascular reactivity evaluations, we also examined, in separate rats from the same groups, Ca^{2+}-dependent NOS activity in samples of cerebral cortical tissue.

Methods

The study protocol was approved by the Institutional Animal Care and Use Committee. Sprague-Dawley rats, 250-350g, were used. The cranial window design and surgical implantation were described in detail in previous publications [11]. The windows were placed 24 h prior to experimentation. On the day of study, anesthesia was induced with halothane and the rat was paralyzed (curare), tracheotomized, and mechanically ventilated. Bilateral femoral arterial and venous catheters were inserted and the cranial window was exposed. Anesthesia during the study was iv fentanyl (25 $\mu g \cdot kg^{-1} \cdot h^{-1}$) plus ventilation with 70% N_2O/30% O_2. Cannulae were secured into the inflow, outflow and intracranial pressure (ICP)-monitoring ports of the window. Artificial cerebrospinal fluid (aCSF) was suffused at a rate of 1.0 $ml \cdot min^{-1}$, and maintained at a temperature of 37°, a PCO_2=40-45 mmHg, PO_2=50-60 mmHg and pH≈7.35. The ICP was controlled at 5-10 mmHg by adjustments of the height of the outflow cannula. The reactivity of 25-50 μm pial arterioles on the exposed cortical surface was assessed via measurement of diameter changes. A microscope (Nikon) and color video camera (Sony) arrangement was equipped with an epi-illumination, darkfield system (Fryer Co. Inc., Carpentersville, IL). Magnifications of ≥800X were displayed on a video monitor. Measurements of vessel diameters were made using a calibrated video microscaler (Optech).

In all experiments, initial diameter measurements were made following a 30 min period of cortical suffusion with drug-free aCSF. Next, topical applications of aCSF containing SNAP (10^{-5} M), ACh (10^{-4} M), ADO (10^{-5} M), and NMDA (10^{-6} M) were sequentially applied (3-5 min each). Hypercapnia ($PaCO_2$ ~65 mmHg) was also imposed for 3 min. Drug-free aCSF was suffused for 5-10 min between vasodilator applications and during and following hypercapnia. L-NA (1 mM) was then introduced, and after 1 hour, the above sequence was repeated (with L-NA in the aCSF).

In additional normal female, ovariectomized, and E_2-treated ovariectomized rats (n=4, in each group), brain cNOS activity was measured in cerebral cortex samples using a well-documented isotopic conversion assay [7,10]. The rats were anesthetized with halothane and quickly decapitated. The brains were rapidly removed and samples from the cerebral cortex, approximating the area under the cranial windows in the intravital microscopy studies, were dissected out, frozen on dry ice, and stored at -80°C until analysis.

Statistical comparisons of pial diameter values within groups were made using a 2-way analysis of variance (ANOVA), with a post-hoc C matrix test for multiple comparisons (Systat, Evanston, IL). For comparisons between groups, we employed a multiway ANOVA (Systat). Statistical significance was taken at the $p<0.05$ level.

Results

No significant variations in PO_2, PCO_2, pH, or MABP were observed in any of the groups over the course of each experiment. Exposure to L-NA did not significantly affect pial arteriolar diameters in any of the groups. The initial pial arteriolar diameters measured in the normal female, ovariectomized + E_2, and ovariectomized groups were 31.2 ± 4.6 μm, 32.7 ± 2.6 μm, and 33.0 ± 4.5 μm, respectively. Baseline diameters measured between exposures to the various vasodilating stimuli, both in the absence and in the presence of L-NA showed, statistically insignificant minor variations (<10%) from the initial values. The pial arteriolar responses are presented in figure 1.

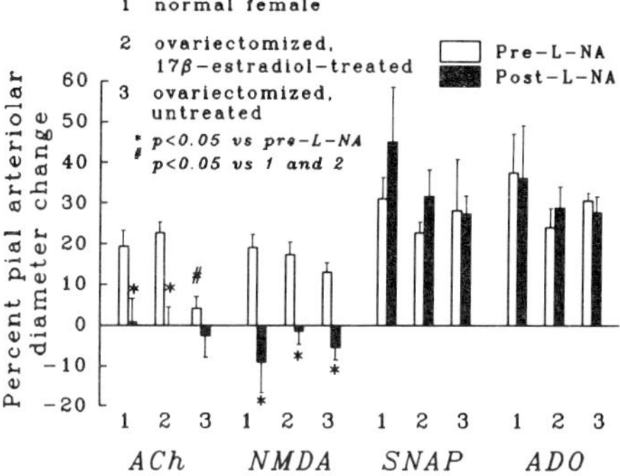

Figure 1. *Pial arteriolar responses to vasodilator suffusions prior to and after NOS inhibition. Means ± SE.*

Untreated ovariectomized rats exhibited little response to ACh. On the other hand, normal females displayed the expected ACh-induced vasodilatory response. E_2 treatment in the ovariectomized rats was accompanied by a vasodilatory response to ACh that was equivalent to that seen in normal females. The arteriolar dilation elicited by NMDA was about one-third lower in ovariectomized (untreated) versus normal females. However, this difference was not statistically significant. With E_2 treatment, numerically greater (~30%), but statistically different NMDA-induced diameter increases, compared to untreated ovariectomized females, were measured. The NO-dependence of the vasodilations accompanying ACh (E_2-treated and normal females) and NMDA (all groups) suffusions was confirmed by the finding that those responses were completely blocked after L-NA. The absence of any differences in the responses to SNAP or ADO indicates that neither L-NA nor E_2 treatment had any direct influence on vascular smooth muscle function. CO_2 reactivity in ovariectomized rats was ~50% of that seen in normal females. E_2 treatment produced only a modest increase in the CO_2 response (fig. 2). Interestingly, the magnitude of the CO_2 reactivity in the ovariectomized rats was nearly identical to that seen in normal males (fig. 2, data taken from ref. 11). Ovariectomy also was accompanied by a loss of NO-dependency in the CO_2 response. Thus, whereas ≥2h L-NA suffusion reduced the CO_2 reactivity by ≥50% in normal males and females, it had no effect on CO_2 reactivity in ovariectomized rats, whether treated with E_2 or not (fig. 2). Cerebral cNOS activity was ~40% lower in ovariectomized versus normal female rats (fig. 3). E_2-treated ovariectomized rats showed a NOS activity that was significantly higher than that seen in their untreated counterparts, but not significantly different from that measured in normal females (fig. 3).

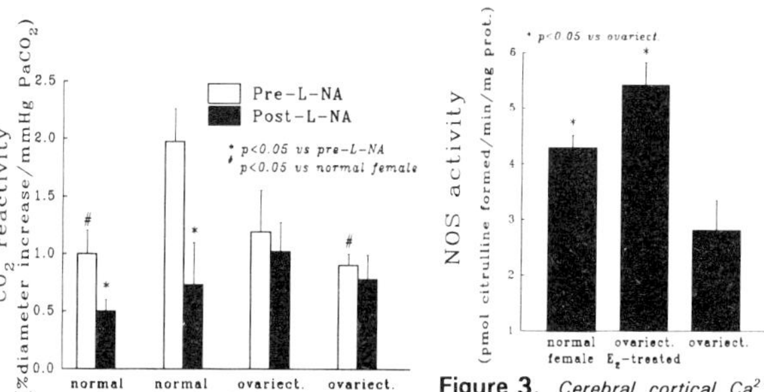

Figure 2. *Pial arteriolar responses to hypercapnia prior to and following L-NA. Means ± SE.*

Figure 3. *Cerebral cortical Ca^{2+}-dependent NOS activity. Values are means ± SE.*

Discussion

These results indicate that ovariectomy decreases, while

Abbreviations used: NOS--nitric oxide synthase; E_2--17 ß-estradiol; ACh--acetylcholine; NMDA--N-methyl-D-aspartate; SNAP--S-nitroso-acetylpenicillamine; ADO--adenosine; L-NA--nitro-L-arginine.

subsequent E_2 treatment increases brain cNOS activity and cerebral vasodilatory capacity. At first glance, those findings would appear to favor changes in eNOS over nNOS. Thus, ACh-induced vasodilation was suppressed by ovariectomy and restored by E_2 treatment, whereas the effects of ovariectomy and E_2 replacement on NMDA-induced dilation were rather modest. It is well established that, in the cerebral circulation, the vasodilatory actions of ACh and NMDA are mediated almost exclusively through eNOS-derived [11] and nNOS-derived NO [3], respectively. However, one cannot ignore possible estrogen influences on brain nNOS for two reasons. First, the magnitude of the cerebrovascular CO_2 response and the NO-dependency of that response did appear to be directly related to amount of estrogen present. We have found that the NO-dependent portion of the CO_2 response in the cerebral circulation of male rats is almost exclusively a function of nNOS-derived NO [10,11]. If one assumes a similar nNOS-dependency in normal females, *and* that the reduced CO_2 reactivity following ovariectomy is indeed related to diminished NO generating capacity, one might conclude that estrogen does have a direct influence on cerebral nNOS activity. The second, and perhaps more compelling indication of an influence of estrogen on brain nNOS, can be taken from the measurements of Ca^{2+}-dependent NOS activity. Recent findings indicated that ~95% of the total cNOS activity in a sample of cortical tissue is nNOS [6]. Thus, it is rather unlikely that the striking differences in brain tissue cNOS activity we observed among the 3 groups could have been exclusively due to differences in eNOS activity. In summary, the E_2-associated changes we observed probably included *both* nNOS and eNOS, although, based on the ACh and NMDA responses, eNOS would seem to be affected more.

Nevertheless, despite an apparently strong correlation between levels of circulating estrogen and brain nNOS activity, nNOS-dependent vasodilating stimuli were not affected in a consistent manner. One explanation for the very modest estrogen-related variations in NMDA-induced arteriolar responses is that the magnitude of the estrogen effect on nNOS activity was insufficient to elicit significant changes in NMDA-mediated NO generation. The changes in CO_2 reactivity associated with chronic manipulations in estrogen present a different and somewhat confusing picture. Thus, on the one hand, CO_2 reactivity was reduced after ovariectomy--a finding consistent with a reduced nNOS activity (see earlier). On the other hand, E_2 treatment was not able to restore CO_2 reactivity to the levels seen in normal females, despite the fact that Ca^{2+}-dependent NOS activity was "normalized" by chronic E_2 therapy. Furthermore, the vasodilatory response to hypercapnia in ovariectomized rats, whether E_2-treated or not, was rendered insensitive to NOS inhibition. Unfortunately, any attempt to explain why E_2 treatment failed to restore normal reactivity and NO-dependency in the hypercapnic response goes beyond the scope of this study and must await additional experimentation.

In conclusion, these results suggest that long-term changes in circulating estrogen levels can have a direct influence on brain Ca^{2+}-dependent NOS activity and cerebral vasodilatory capacity. The changes in vascular reactivity appear to favor regulation of eNOS over nNOS. However, the finding of significant estrogen-related differences in Ca^{2+}-dependent NOS activity in samples of cortical tissue, where nearly all of the cNOS activity can be attributed to nNOS, would suggest that both isoforms are affected by chronic changes in estrogen.

References

1. Bell, D.R., Rensberger, H.J., Koritnik, D.R. and Koshy, A. (1995) Amer.J.Physio.l **37**, H377-H383
2. Cheng, D.Y., Feng, C.J., Kadowitz, P.J. and Gruetter, C.A. (1994) Life Sci. **55**, PL187-PL191
3. Faraci, F.M. and Breese, K.R. (1993) Circ.Res. **72**, 476-480
4. Goetz, R.M., Morano, I., Calovini, T., Studer, R. and Holtz, J. (1994) Biochem.Biophys.Res.Commun. **205**, 905-910
5. Hishikawa, K., Nakaki, T., Marumo, T., Suzuki, H., Kato, R. and Saruta, T. (1995) FEBS Lett. **360**, 291-293
6. Huang, Z.H., Huang, P.L., Panahian, N., Dalkara, T., Fishman, M.C. and Moskowitz, M.A. (1994) Science **265**, 1883-1885
7. Irikura, K., Maynard, K.I. and Moskowitz, M.A. (1994) J Cereb.B-lood Flow Metab. **14**, 45-48
8. Lieberman, E.H., Gerhard, M.D., Uehata, A., Walsh, B.W., Selwyn, A.P., Ganz, P., Yeung, A.C. and Creager, M.A. (1994) Ann.Intern.Med. **121**, 936-941
9. Miller, V.M. and Vanhoutte, P.M. (1991) Am.J Physiol **261**, R1022-R1027
10. Wang, Q., Pelligrino, D.A., Baughman, V.L., Koenig, H.M. and Albrecht, R.F. (1995) J.Cereb.Blood Flow Metab. **15**, 774-778
11. Wang, Q., Pelligrino, D.A., Koenig, H.M. and Albrecht, R.F. (1994) J Cereb.Blood Flow Metab. **14**, 944-951
12. Weiner, C.P., Lizasoain, I., Baylis, S.A., Knowles, R.G., Charles, I.G., and Moncada, S. (1994) Proc.Natl.Acad.Sci. USA **91**, 5212-5216

Effect of estradiol on neuronal NO synthase (nNOS) is dose-and duration-dependent

S NELSON[1], I LIZASOAIN[2] JC LEZA[2], RG KNOWLES[3], L THOMPSON[1], S MONCADA[3], MA MORO[2] and CP WEINER[1]

Perinatal Research Laboratory[1], Univ. of Iowa College of Medicine, Iowa City, IA, Departmento de Farmacologia[2], Univ. Complutense, Madrid, and the Wellcome Research Laboratory[3], Beckenham, Kent

Introduction

Estrogen replacement therapy (ERT) has been used over the past 40 years to reduce the symptoms of menopause. During this time, it has become increasingly clear that ERT also protects women from coronary artery disease, heart attacks and stroke. [1]

We have recently shown that pregnancy and short term ERT of ovulating guinea pigs increases eNOS and nNOS activity in several tissues including skeletal muscle and brain. [2]

Evidence suggests NO is involved in the regulation of cerebral blood flow, neurotransmission, long-term potentiation and memory. Disruptions in the physiological role of NO in the brain may result in stroke, epilepsy, Huntington's chorea, Alzheimer's, edema and meningeal inflammation. [3-6]

Purpose

The purpose of this study was to investigate the effect of estrogen supplementation and replacement on NO synthase activity and NO synthase-specific mRNA in the forebrain and cerebellum of intact and long term castrate guinea pigs.

Methods

Short term supplemental estradiol.

Gonadal intact, random cycling female guinea pigs received estradiol benzoate (vehicle, 10, 50, 150, 500 or 1500 µg/kg i.p.) for 5 days and killed 24 h after the final injection.

Long term supplemental estradiol.

Gonadal intact, random cycling female guinea pigs received constant release estradiol pellets (0.5 or 15 mg) and were killed on day 2, 5, 10 and 20.

Long term replacement estradiol.

Female guinea pigs (12 weeks post castration) received constant release estradiol pellets (0.25, 0.5, 1.5 or 7.5 mg) and were killed on days 19-20. Non-pregnant and pregnant animals were included for comparison.

For all animals, the forebrain and cerebellum were quickly excised, frozen in liquid nitrogen and stored at -80°C until assayed.

Measurement of NO Synthase Activity

Forebrain and cerebellum were homogenized mechanically at 0°C in 4 volumes of buffer containing 320 mM sucrose, 50 mM Tris, 1 mM EDTA, 1 mM DL-dithiothreitol and appropriate protease inhibitors. The crude homogenate was centrifuged at 0°C at 15,000 x g for 20 minutes and the postmitochondrial supernatant place on ice.

NO synthase activity was determined by measuring in duplicate the conversion of L-[U-^{14}C]-arginine to [U-^{14}C]-citruilline. Measurements are presented as pmol of citrulline formed/min per gram wet weight.

mRNA Quantification

mRNA was quantified by ribonuclease protection assay (RPA). using guinea pig nNOS (472 bp) and G3PDH (350 bp) labelled RNA transcripts.

Results

Short term (5d) estradiol <u>supplementation</u> caused a dose-dependent increase in cerebellar Ca^{2+}-dependent NO synthase activity. The forebrain was less sensitive to estradiol than the cerebellum with the only significant increase in activity occurring at 500 µg/ml.

Long term <u>supplementation</u> with the 0.5 mg pellet significantly increased cerebellar Ca^{2+}-dependent NO synthase activity by day 2 of therapy. The increase in activity was sustained through day 20. In the forebrain, Ca^{2+}-dependent NO synthase activity appeared to rise progressively with the 0.5 mg estradiol pellet for the duration of the study, but did not reach significance with the present sample size. Supplementation with the 15 mg pellet initially stimulated Ca^{2+}-dependent NO synthase activity in the cerebellum (d2-10), but by day 20 levels were less than control. In the forebrain this dose caused a cummulative decline in activity (d2-20).

Ca^{2+}-dependent NO synthase activity in forebrains is increased by pregnancy and castration above non-pregnant controls. Long term estrogen <u>replacement</u> caused a dose-dependent decrease in activity reaching non-pregnant levels at the highest dose (7.5 mg).

nNOS-specific mRNA in forebrains was increased by pregnancy and castration above non-pregnant controls as seen with NO synthase activity. With increasing doses, there is a decrease in nNOS mRNA.

Conclusions

1) Estradiol stimulates nNOS activity in forebrain and cerebellum and mRNA in forebrain.
2) The effect of estradiol on nNOS is dependent on tissue, dose and duration of treatment.

Hypotheses

1) One mechanism by which ERT provides its beneficial effects on the central nervous system is by stimulating nNOS mRNA and NO synthase.
2) Dose and duration are important factors in ERT since it appears that excessive estradiol may be counterproductive.

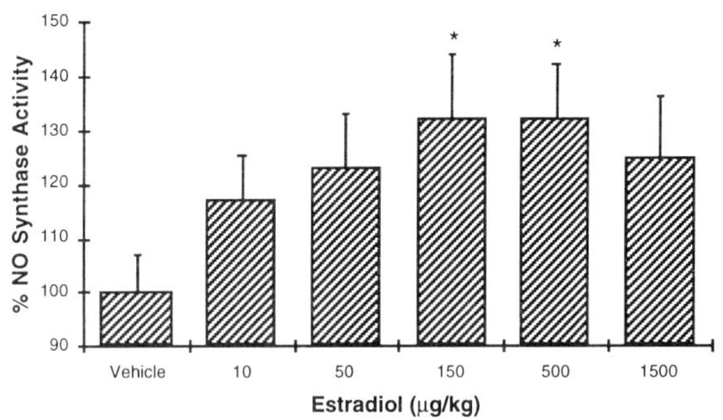

Figure 1. Effect of short term estradiol supplementation on NOS activity in cerebellums of cycling guinea pigs. Results are mean ± SEM (n=8). *=p<0.05

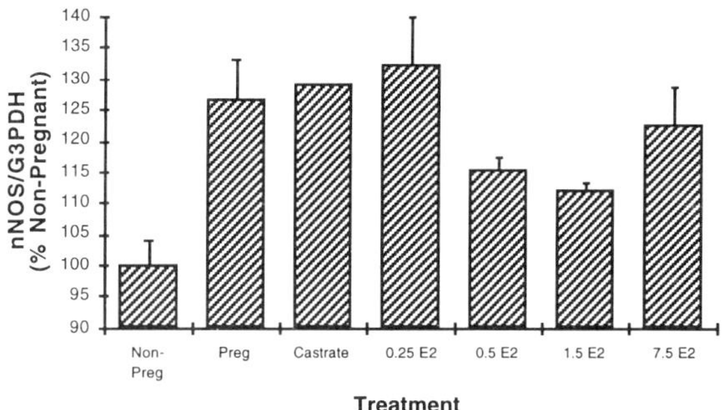

Figure 2. Effect of estrogen replacement on nNOS mRNA in long term castrate guinea pigs. Results are mean ± SEM (n = 3 animals per group).

References

1. Wren B.G. (1992) The effect of oestrogen on the female cardiovascular system. Med J Australia **157**(3):204-8.
2. Weiner C.P., Lizasoain I., Baylis S.A., Knowles R.G., Charles I.G. and Moncada S. (1994) Induction of calcium-dependent nitric oxide synthases by sex hormones. PNAS **91**(11):5212-6.
3. Lowenstein C.J., Dinerman J.L. and Snyder S.H. (1994) Nitric oxide: a physiologic messenger. Ann Int Med **120**(3):227-37.
4. Moncada S. and Higgs A. (1993) The L-arginine-nitric oxide pathway. New Eng J Med **329**(27):2002-12.
5. Koedel U., Bernatowicz A., Paul R., Frei K., Fontana A. and Pfister H.W. (1995) Experimental pneumococcal meningitis: cerebrovascular alterations, brain edema, and meningeal inflammation are linked to the production of nitric oxide. Annals of Neurology **37**(3):313-23.
6. Dorheim M.A., Tracey W.R., Pollock J.S. and Grammas P. (1994) Nitric oxide synthase activity is elevated in brain microvessels in Alzheimer's disease. Biochem Biophys Res Comm **205**(1):659-665.

Low but not high dose estradiol decreases thromboxane contraction of coronary arteries.

LOREN P. THOMPSON and CARL P. WEINER

Dept. of Obstetrics/Gynecology and Perinatal Research Laboratory, The University of Iowa, Iowa City, IA 52242

The incidence of coronary artery disease in women is one half that of age-matched men [1]. After menopause, it more than doubles and is reduced by postmenopausal estrogen therapy. Estrogen has been implicated in providing long term protection against cardiovascular disease by several mechanisms including decreasing vasoconstrictor responsiveness and coronary vasospasm. Recently, estrogen has been shown to both increase mRNA expression of endothelial nitric oxide (NO) synthase and neuronal NO synthase in skeletal muscle and increase NO synthase activity in skeletal muscle, heart and brain [2]. Thus, we tested the hypothesis that estrogen decreases coronary artery reactivity by increasing NO synthase activity and NO release. We tested this by measuring the effect of 17-β estradiol supplementation of castrate guinea pigs on the contraction to thromboxane of isolated coronary arteries.

Methods. Reproductively mature female guinea pigs were anesthetized (xylazine, 1 mg/kg; ketamine 80 mg/kg; i.p.) and the ovaries surgically removed through bilateral flank incisions. Castrate animals were allowed to recover for at least 12 weeks to reduce estradiol levels. Animals were then sedated and estradiol pellets (0.25, 0.5, 1.5, or 7.5 mg 17β-estradiol) were placed subcutaneously in the abdomen. These pellets provide constant estradiol release over 21 days. Animals were killed on day 19 or 20. Control animals were castrated but did not receive pellets. *Tissue Preparation.* Hearts were removed from the anesthetized animals through a thoracotomy and the left coronary arteries were excised and cleaned in iced buffer. Ring segments of the left descending branch were mounted onto wire myographs for measurement of isometric force. Tissues were immersed in 10 ml baths containing a physiologic bicarbonate buffer, gassed with 95% oxygen and 5% carbon dioxide and maintained at 37°C. *Experimental Protocol.* Contractile responses to the cumulative addition of the thromboxane analog, U46619 (10^{-10}-10^{-5}M), were measured in the presence and absence of nitro-L-arginine (LNA; 10^{-4}M; 30 min.) to inhibit NO synthase, and methylene blue (10^{-5}M; 30 min.) to inhibit guanylate cyclase. *Statistical Analysis.* Contractile responses (mean ± SE) to U46619 of each ring were normalized to the 120mM KCl contraction. Dose response curves were compared using ANOVA with concentration, estradiol dose and treatment as independent variables and contractile response as the dependent variable. -Log ED_{50} value is the -log concentration which produces 50% contractile response to U46619.

Results. Maximal contractile responses to U46619 were decreased in arteries from guinea pigs receiving 0.25mg estradiol compared to untreated castrates and castrates receiving estradiol doses of 0.5, 1.5, and 7.5mg. Negative log ED_{50} values for U46619 were not different between arteries from untreated and estradiol-treated castrates. LNA and methylene blue potentiated both -log ED_{50} values and maximal contraction of arteries from castrates receiving 0, 0.25, and 0.5mg estradiol. In arteries from animals receiving 1.5 and 7.5mg estradiol, contractile responses to U46619 were unaffected by LNA and methylene blue. In the presence of either LNA or methylene blue, maximal contraction and sensitivity (measured by -log ED_{50} values) to thromboxane were unaltered by estradiol.

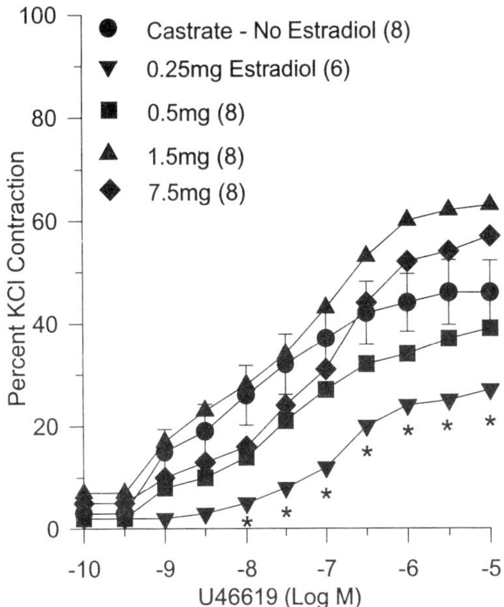

Figure 1. Effect of estradiol supplementation on contractile responses to U46619 of guinea pig coronary arteries. * = P<0.05 vs Castrate. ( ) = # of animals in each group. SE bars omitted for comparison.

Conclusions. This study demonstrates that estradiol has a variable effect on thromboxane contraction of coronary arteries. Compared to castrates receiving no estradiol, estradiol inhibited thromboxane contraction at relatively low doses and increased contraction at higher doses. After inhibition of the NO/cGMP pathway by LNA and methylene blue, thromboxane contraction did not vary with the estradiol dose. Therefore, the decrease in contraction at low estradiol during control conditions is consistent with stimulation of NO synthase activity and the increase in contraction at high estradiol to inhibition of NO synthase activity. These data suggest that low dose estradiol inhibits thromboxane contraction by stimulating basal NO release and potentiates contraction at high doses via inhibiting NO release.

1. Kannel, W.B., Jhortland, M.C., McNamara, P.M., and Gordon, T. (1976) Menopause and risk of cardiovascular disease. The Framingham study. Ann. Inter. Med. **85**, 447-452.

2. Weiner, C.P., Lizasoain, I., Baylis, S.A., Knowles, R.G., Charles, I.G., and Moncada, S. (1994) Induction of calcium-dependent nitric oxide synthases by sex hormones. Proc. Natl. Acad. Sci. **91**, 5212-5216.

Umbilical arterial S-nitrosothiol concentration as a marker for normal human perinatal circulatory transition.

EMORY A. FRY[*], STEVE SEARS[*],DERRICK ARNELLE[#], RUSSELL E BYRNS[+], LOUIS J. IGNARRO[+], JONATHAN S. STAMLER[#], and BENJAMIN GASTON[*].

[*]Department of Pediatrics, Naval Medical Center, San Diego, CA 92134; [#]Departments of Cardiovascular and Pulmonary Medicine, Duke University Medical Center, Durham, NC 27710; [+]Department of Pharmacology, Center for Health Sciences, University of California, School of Medicine, Los Angeles, CA 90024.

Background

Physiological studies in both animal [1] and human newborns [2,3] demonstrate that inhaled nitric oxide ($NO^{\cdot}$) is a selective pulmonary vasodilator and are the basis for the clinical use of $NO^{\cdot}$ gas in the treatment of neonatal diseases such as meconium aspiration and persistent pulmonary hypertension. Indirect evidence [4-6], suggests a physiological role for nitrogen oxide species and nitric oxide synthase (NOS) activation in normal human perinatal transition. However, the details are poorly understood, limiting our understanding of the variable efficacy of nitric oxide gas therapy.

S-nitrosothiols (RS-NO) are stable endogenous vasodilators that are biologically relevant plasma markers for NOS activity. Their identification in human plasma [8] and lung lining fluid [9] has provided a potential tool for the measurement of NOS activity in normal subjects including newborns.

We hypothesize that human cord RS-NO may reflect acute changes in perinatal $NO^{\cdot}$ production. Further, these RS-NO compounds may represent more than simply markers for NOS activity; their bioactivity profiles [7-9] suggest that they could be important effector end-products of the enzyme in the human newborn. Understanding the role of these molecules in neonatal transitional physiology could thus 1) provide evidence that endogenous nitrogen oxides are involved in the process for which exogenous $NO^{\cdot}$ is being used in human infants, and 2) help to tailor therapy so that the least toxic chemical species are delivered to the patients most likely to benefit.

Objectives

- To establish normal fetal/umbilical arterial and placental/umbilical venous concentrations of circulating RS-NO.
- To determine whether abnormal transition from fetal to neonatal life is associated with abnormal umbilical arterial or venous plasma levels of RS-NO.

Methods

Paired umbilical artery and vein blood samples from 31 newborns were collected in EDTA. Samples underwent centrifugation at 7000g for 5 min. and were frozen in liquid nitrogen within 20 min. of delivery. Information detailing each infant's gestational age, delivery, APGAR scores, transition course, cord gases, and diagnosis was recorded. S-nitrosothiol concentrations were determined using a post-photolysis chemiluminescence technique [8]. Nitrite (NO_2^-) was assayed by chemiluminescence after reduction in KI. Nitrate (NO_3^-) levels were calculated by subtracting nitrite concentrations from cumulative NO_2^- and NO_3^- assayed after reduction in VCl_3 [10]. Means were compared using paired and non-paired two-tailed t-testing as appropriate. $P < 0.05$ was considered significant. Data are presented as mean ± standard error.

Results

Arterial RS-NO levels (33.6 ± 4.4 ηM) were nearly twice those of matched venous samples (20 ± 3.4 ηM) in healthy term infants (n =26; $p < 0.001$). Arterial RS-NO in term infants were four times those of distressed preterm infants (8.2 ± 8.1 ηM; n = 5; $p < 0.025$) while venous levels did not differ (p =NS). There was also no difference in the nitrate concentrations between term (31 ± 2.3 μM; n = 12) and preterm (30.8 ± 4.8 μM; n = 4) venous samples (p = NS). A set of 36 wk twins illustrates these differences. Twin A had an abnormal perinatal transition with a one minute APGAR of 1. Arterial and venous RS-NO levels were 0 and 8 nM respectively. Twin B transitioned well with a one minute APGAR of 9 and had corresponding values of 41 and 29 ηM.

Conclusions

We have established fetal/umbilical arterial RS-NO levels to be 60% higher than those of the mixed placental/umbilical venous pool in term infants. In distressed preterm infants, fetal/umbilical arterial RS-NO levels are depleted while placental/umbilical venous levels are stable. Concentrations of nitrate, a stable inert endproduct of NO metabolism, are similar in term and preterm umbilical veins, suggesting that RS-NO are not simply marker by-products of NOS function and may be formed from catalyzed or redox sensitive S-nitrosolating reactions for a specific purpose. We speculate that RS-NO depletion may adversely affect human perinatal transition and that RS-NO synthesis and/or release may be regulated by the fetus at the cellular and/or chemical level.

Supported by U.S. Navy grant NMCSD S-93-L-117.

References

1. Kinsella JP, McQueston JA, Rosenberg AA, and Abman SH (1992) Am. J. Physiol. 262:H875-80.
2. Roberts JD, Polaner DM, Lang P, and Zapol WM (1992) Lancet 340:818-19.
3. Kinsella JP, Neish SR, Shaffer E, and Abman SH (1992) Lancet 340:819-20.
4. Abman SH, Chatfield BA, Hall SL, and McMurtry IF (1990) Am. J. Physiol. 259: H1921-27.
5. McQueston JA, Cornfield DN, McMurtry IF, and Abman SH (1993) Am. J. Physiol. 264:H865-71.
6. Lui SF, Hislop AA, Haworth SG, and Barnes PJ (1992) Br. J. Pharmacol. 106:324-30.
7. Mathews R and Kerr S (1993) J. Pharmacol. Exper. Ther. 267:1529-37.
8. Stamler JS, Jaraki O, Osborne J, et al. (1992) Proc. Natl. Acad. Sci. USA 89:7674-77.
9. Gaston B, Reilly J, Drazen JM, et al. (1993) Proc. Natl. Acad. Sci. USA 90:10957-61.
10. Ignarro LJ, Fukuto JM, Griscavage JM, Rogers NE, and Byrns R (1993) Proc. Natl. Acad. Sci. USA 90:8103-07.

Expression of inducible nitric oxide synthase in the developing mouse placenta.

SALLY. A. BAYLIS, CARL P. WEINER*, ALEXANDER SANDRA*, RACHEL J. RUSSELL, XIAO-QING WEI, SALVADOR MONCADA, PAUL J. L. M. STRIJBOS and IAN G. CHARLES.

Wellcome Research Laboratories, Langley Court, South Eden Park Road, Beckenham, Kent, BR3 3BS, U.K. and * University of Iowa College of Medicine, Iowa City, LA 52242, U.S.A.

We and others have previously shown that the endothelial form of nitric oxide synthase (eNOS or NOS III) is expressed in the human placenta [1, 2]. We hypothesised that the increase in cytokines present at the placental: maternal interface would stimulate the transcription of the cytokine-inducible nitric oxide synthase (iNOS or NOS II). Using a variety of approaches we now demonstrate that the iNOS gene is expressed in the murine placenta.

In initial studies the polymerase chain reaction coupled to the reverse transcription of mRNA (RT-PCR) was used to determine whether iNOS was produced in the developing mouse placenta which could be dissected from the embryo from day 12 onwards. Messenger RNA was prepared from mouse placentae at 12, 14, 16 18 and 20 days post-coitum. At each of these times examined, iNOS-specific cDNA was amplified [3] and direct DNA sequence analysis of the PCR products using an ABI 373 sequencer confirmed that the PCR products were identical to the published sequence for the murine macrophage cDNA [4] (data not shown). Northern blot analysis of these mRNAs (shown in figure 1) reveals that the 4 kb iNOS transcript was maintained at a stable level from day 12 to day 20. The placental iNOS message was present at significantly lower levels than in J774 murine macrophages treated with interferonγ and bacterial lipopolysaccharide (LPS). Probing the northern blot for β-actin confirmed the integrity and loading of mRNAs.

The spatial localisation of iNOS in the developing placenta was determined by *in situ* hybridisation. Using this technique iNOS mRNA was found as early as 8 days post-coitum where there was extensive staining for iNOS in the polar trophoectoderm. Subsequently staining was observed in the cytotrophoblast (data not shown).

Placental tissue was analysed for the expression of calcium-dependent and calcium-independent NOS activity. NOS activity was determined by measuring the conversion of L-[U-^{14}C]arginine to [U-^{14}C]citrulline (5). No significant difference was observed in the levels of calcium-dependent NOS activity with the change in gestational age of the tissue; the average activity observed was 1.79 pmol/mg protein. In contrast, the calcium-independent NOS activity was observed to decrease with time (figure 2). The level of calcium-independent NOS activity from placental tissue from 14 days post-coitum was similar to that found in the spleens of mice treated with LPS. Significantly, the calcium-independent NOS activity was absent in placentae from iNOS deficient mice (whose iNOS gene had been inactivated by homologous recombination[6]) at 16 days post-coitum. This demonstrates that the calcium-independent NOS activity observed is directly attributable to the iNOS gene product. Similarly no calcium-independent NOS activity was found in the spleens of iNOS deficient mice treated with LPS.

Fig. 1 Northern blot analysis of iNOS transcription in the murine placenta.

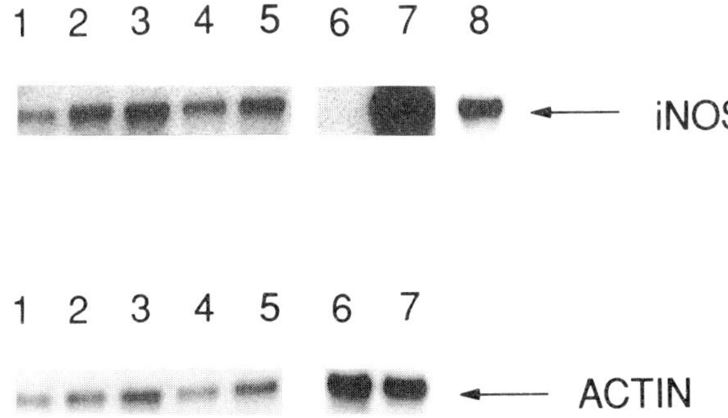

Equivalent quantities of mRNA (1.25 μg) were loaded onto a formaldehyde agarose gel, transferred to nylon membrane by northern blotting and then hybridised with labelled probes to iNOS (top panel) and β-actin (bottom panel). Lanes 1-5, mRNA extracted from placentae at days 12, 14, 16, 18 and 20 post-coitum respectively. Lane 6 and 7 mRNA from untreated and interferonγ/LPS treated J774 macrophages respectively. Lane 8 represents a shorter exposure for the sample in lane 7.

Fig. 2 Calcium-independent NOS activity in the murine placenta in wild type and iNOS deficient mice

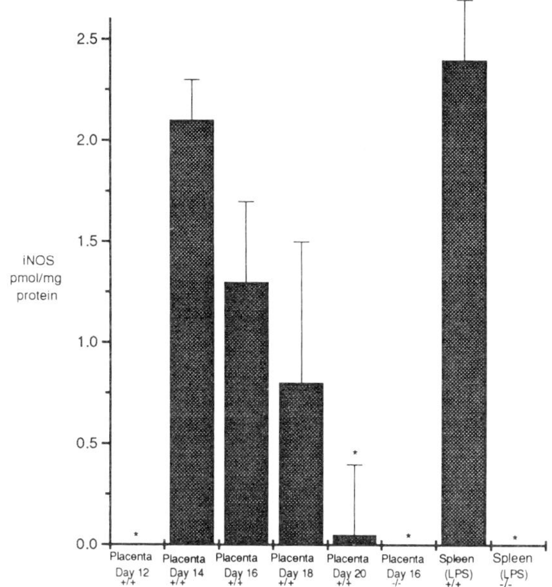

Calcium-independent NOS activity was determined for placental tissue from wild type (+/+) and iNOS deficient mice (-/-) at various days post-coitum. Control samples were determined from spleens of wild type and iNOS deficient mice following administration of LPS (*p<0.05 significantly different from day 14).

Here we demonstrate that iNOS gene is expressed in the murine placenta. The gene is transcribed as early as day 8 post-coitum, and calcium-independent NOS activity is detected from 14 days post-coitum. Such activity is absent in mice deficient in the iNOS gene, demonstrating that the calcium-independent NOS activity observed is directly attributable to the iNOS gene product. Early expression of the iNOS gene is localized by *in situ* hybridisation to the portion of the murine placenta in most intimate contact with the maternal surface. This suggests that NO released by the inducible NOS enzyme may play a role in placental function in the mouse. However there does not appear to be an absolute requirement for iNOS gene expression in murine development, since the iNOS deficient mice develop and reproduce in an apparently normal fashion [6].

We thank Hugh Spence for oligonucleotide synthesis, Marcus Oxer for DNA sequencing and Nick Davies and Barry Warburton for their assistance.

1. Myatt, L., Brockman, D. E., Langdon, G. and Pollock, J. S. (1993) Placenta **14**, 373-383.
2. Garvey, E. P., Tuttle, J. V., Covington, K., Merrill, B. M., Wood, E. R., Baylis. S. A. and Charles, I. G. (1994) Arch. Biochem. Biophys. **311**, 235-241.
3. Jenkins, D. C., Charles, I. G., Thomsen, L. L., *et al.* (1995) Proc. Natl. Acad. Sci. USA, **92**, 4392-4396.
4. Lyons, C. R., Orloff, G. J. and Cunningham, J. M. (1992) J. Biol. Chem. **267**, 6370-6374.
5. Salter, M., Knowles, R. G. & Moncada, S. (1991) FEBS Lett. **291**, 145-149.
6. Wei, X., Charles, I. G., Smith, A., *et al.* (1995) Nature **375**, 408-411.

LNAME produces fetal deformities and growth retardation by a mechanism independent of inhibition of nitric oxide.

DAVID R POWERS, THOMAS NOLAN, JIANMING XIE, JACK LANCASTER JR., THOMAS D. GILES and STAN S. GREENBERG.

DEPARTMENTS OF OBSTETRICS AND GYNECOLOGY, MEDICINE and PHYSIOLOGY, LOUISIANA STATE UNIVERSITY MEDICAL CENTER, NEW ORLEANS, LOUISIANA, 70112, U.S.A.

Pregnancy is associated with both maternal and fetal up-regulation of nitric oxide (NO) synthesis [1-3]. Nitric oxide synthase (NOS) activity increases early in pregnancy and decreases prior to parturition [3-5] and may be an important modulator of placental blood flow and nutrient and oxygen exchange in the mother and fetus [1,2]. Since NO inhibits DNA synthesis in smooth muscle and osteocytes [6] maternal and fetal production of NO may also be involved in angiogenesis and fetal development [7,8].

Ingestion of LNAME, an L-arginine derived inhibitor of both inducible and constitutive NOS by pregnant rats on gestational days 13 through 19 or 20 produced fetal intrauterine growth retardation (IUGR) and selective impairment of fetal hindlimb development [8]. It was suggested that these effects of LNAME were probably related to inhibition of fetal constitutive NOS. However, measurement of RNI levels were not performed and the cyclic GMP content of the amniotic fluid was not inhibited by doses of LNAME which produced the fetal deformities [7]. Thus, inhibition of NOS could not be established as causal to IUGR and fetal hindlimb deformities. To critically test this relationship this study compares the effects of LNAME to those of L-N^6-iminoethyl-L-ornithine (LNIO), high doses of aminoguanidine and amiloride, a guanidine-like compound devoid of inhibitory effects on NOS, on IUGR and fetal limb development.

Methods and Materials: On day 13 of pregnancy female Sprague Dawley rats (275-300 g)the animals were randomly assigned to one of five treatment groups (n=6/gp) and given LNAME (1 mg/ml), amiloride (50 ug/ml) or aminoguanidine (500 ug/ml) in their drinking water, LNIO (10 mg/kg/day, ip) or pyrogen-free drinking water. LNIO was given ip because of our uncertainty of the magnitude to which it was absorbed following its oral administration. Maternal body weight and the total drug, food and water intake and total 24 hr urine output were measured each day [7,9]. On gestational day 21 the dams were anesthetized with ether, a laparotomy was performed and the pups were gently removed from the uterus. Pup position, number and weight and placental weight were recorded along with the incidence of stillbirths and fetal resorptions and the incidence and type of fetal malformation. Approximately 50% of the normal and deformed fetuses were fixed in 10% formalin for subsequent dehydration in alcohol and paraffin embedding prior to histologic analyses after staining with hematoxylin and eosin or Boones fixative. The remainder of the fetuses were dissected and their heart, lungs, kidney and liver weighed and assessed for developmental malformations. Plasma nitrite and nitrate were assayed by chemiluminescence and with the Griess reaction [9]. Data were analyzed with ANOVA . The incidence and type of fetal anomalies as well as the number of living, dead and resorbed fetuses within each group were compared to the control group and to each other with Fishers exact probability test and Chi Squared analyses. A p value of 0.05 or less was accepted for statistical significance of mean differences.

Results and Comments: The administered dose of LNIO was 10 mg/kg/day and increased from approximately 2.5 to 3.5 mg/rat during the gestation period. The daily intake of LNAME and amiloride on the first day of drug administration was less ($p<0.05$) than the average daily dose of 160 mg/kg/day and 3.6 mg/kg/day, respectively, maintained during gestational days 14 through 20. The average daily consumption of aminoguanidine was approximately 20 mg/kg/day. The plasma concentration of RNI in the control pregnant rats was artifactually elevated during treatment with LNAME as measured by chemiluminescence but the plasma nitrite concentration was decreased as measured by the Griess reaction [9, Fig. 1]. Aminoguanidine and LNIO

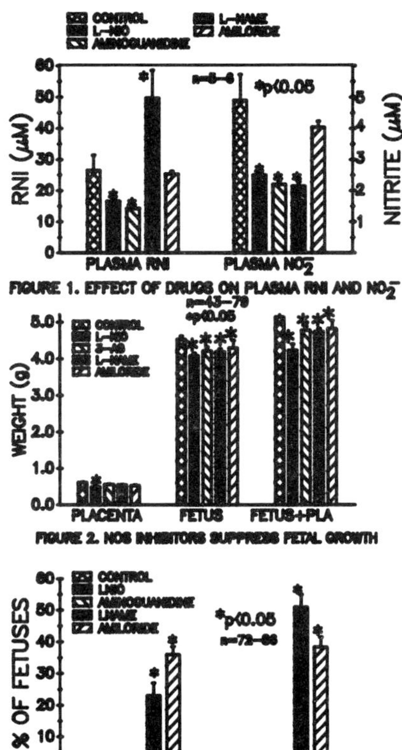

FIGURE 1. EFFECT OF DRUGS ON PLASMA RNI AND NO_2^-

FIGURE 2. NOS INHIBITORS SUPPRESS FETAL GROWTH

FIGURE 3. EFFECT OF NOS INHIBITORS ON THE FETUS

decreased total RNI. Aminoguanidine decreased and LNIO increased plasma nitrite. Amiloride was devoid of any effect on either RNI or nitrite [Fig. 1]. Nevertheless, each of the drugs tested produced IUGR. LNIO not only produced IUGR but also acted directly on the fetus since the fetus/placenta ratio was also depressed by this compound [Fig. 2]. Despite the ability of all the compounds tested to produce IUGR, only LNAME and amiloride produced fetal resorptions and fetal limb defects [Fig. 3]. These were characterized by the absence of digits and loss of ossification of the long bones. Since LNIO and aminoguanidine inhibited NOS and produced IUGR, LNIO directly suppressed fetal growth and amiloride did not affect NOS, there is no evidence supporting a causal relationship between inhibition of NOS and fetal limb deformities. This action of LNAME and amiloride appears to be independent of inhibition of NOS

References

1. Molnar, M. and Hertelendy, F. N (1992 Am. J. Obstet. Gynecol. 166: 1560-1567

2. Chaudhuri, G. and Furuya, K. (1991) Sem. Perinatol. 15, 63-67

3. Sladek, S. M., Regenstein, A. C., Lykins, D. and Roberts, J. M. (1993) Am. J. Obstet. Gynecol. 169, 1285-1291

4. Natuzzi, E. S., Ursell, P. C., Harrison, M., Buscher, C. and Riemer, R. K. (1993) Biochem. Biophys. Res. Comm. 194, 1-8

5. Izumi, H., Garfield, R. E., Makino, Y., Shirakawa, K. and Itoh, T. (1994) Am. J. Obstet. Gynecol. 170, 236-245

6, Lowik, C. W., Nibbering, P. H., Van De Ruit, M. and Papapoulos, S. E. (1994) J. Clin. Invest. 93, 1465-1472.

7. Diket, A. L., Pierce, M. R., Munshi, U. K., Voelker, C. A., Eloby-childress, S., Greenberg, S. S., Zhang, X. J., Clark, D. A. and Miller, M. J. (1994) Am. J. Obstet. Gynecol. 171, 1243-1250.

8. Lee, Q. P. and Juchau, M. R. (1994) Teratology 49, 452-464

9. Greenberg, S., Xie, J., Spitzer, J., Wang,,J., Lancaster, J., Grisham, M.B., Powers, D. and Giles, T. (1995) Life Sci In Press

Nitric oxide synthases in the rat ovary

BRADLEY J. VAN VOORHIS*, KIMBERLEY MOORE*, PAUL JLM STRIJBOS#, SCOTT NELSON*, SALLY BAYLIS# and CARL P. WEINER*

*Department of Obstetrics and Gynecology, University of Iowa College of Medicine, Iowa City, IA 52242-1080, USA and #The Wellcome Research Laboratories, Beckenham, Kent BR3 3BS, United Kingdom

Several lines of evidence suggest that NO may regulate ovarian processes. We have previously demonstrated eNOS within human granulosa-luteal cells [1]. Furthermore, NO inhibits granulosa-luteal cell steroidogenesis suggesting NO may be an autocrine regulator of steroid production [1]. NO has also been implicated in the mechanism of ovulation [2-4]. Recently, NO was shown to suppress apoptosis in cultured pre-ovulatory ovarian follicles [5]. We have recently demonstrated that both eNOS and iNOS, but not nNOS, are expressed in the rat ovary [6]. We have also demonstrated that messenger RNA levels for these enzymes change after gonadotropin stimulation. The purpose of the present investigation was to localize iNOS and eNOS in the rat ovary.

Methods

Immature rat stimulation

The mature rat ovary is heterogeneous, containing developing follicles, ovulatory follicles and corpora lutea simultaneously. Therefore, to demonstrate relative changes in NOS-specific mRNA at various time points during folliculogenesis, ovulation, and corpus luteum formation, we utilized a gonadotropin-stimulated immature rat model known to produce relatively homogeneous populations of these ovarian structures when compared to the mature rat ovary. Immature 23 day old Sprague-Dawley female rats were obtained and some of the rats were sacrificed immediately while the remaining rats received a single sc injection of 20 IU pregnant mare serum gonadotropin. Forty eight h later, some of the rats were sacrificed. The remaining rats received an ovulatory dose of 10 IU hCG and then groups of these rats were sacrificed 12, 24 and 72 hours after the hCG injection. At each time point, ovaries were rapidly excised and frozen in liquid nitrogen for further analysis.

In situ hybridization for iNOS

Cryostat sections (12μm) were prepared from ovaries which were frozen in liquid nitrogen. Sections were mounted onto gelatin-coated glass slides, fixed for 20 minutes in freshly-made paraformaldehyde (2%), washed in phosphate-buffered saline and dehydrated through a graded series of ethanol. A three probe cocktail complementary to nucleotide bases 445-480, 1257-1292 and 1842-1879 of the coding sequence of the murine iNOS cDNA was used throughout. Oligonucleotide probes were labelled with $\alpha[^{35}S]$dATP.

Ovarian sections were incubated with of hybridization solution and $\alpha[^{35}S]$dATP-labelled oligonucleotide probes (300,000 c.p.m./50μl per probe). Following washing, sections were dehydrated through ethanol and apposed to film for three to four weeks. After developing and fixing, sections were counter stained with Cresyl Violet (0.1%).

Immunohistochemistry for eNOS

Five micron frozen sections of ovaries were fixed in 4° C methanol for 15 seconds. All subsequent steps were carried out in a moist chamber at room temperature. The slides were washed in PBS before adding 0.2% BSA in PBS for 10 min. The primary antibody (diluted 1:25 in PBS) was added to the slides for 60 min. This monoclonal antibody was raised in mouse against a 20.4 kDA protein fragment corresponding to amino acids 1030-1209 of human endothelial NO synthase (Transduction Laboratories, Lexington, KY) and it has previously been shown to cross react with rat eNOS. Following PBS rinsing, FITC conjugated secondary antibody was added to slides for 30 min. After PBS washing, the slides were stained for 15 min in 4', 6-diamidino-2-phenylindole, a DNA binding stain. The slides were then mounted and examined by fluorescent microscopy under 600X magnification. Control cover slips had mouse non-immune serum substituted for the primary antibody.

Results

In situ hybridization studies were performed on ovaries excised at the described time points. iNOS mRNA localized to the granulosa cell layer of secondary, pre-antral follicles found predominantly in the ovaries from unstimulated, immature rats. It also was present in pre-antral follicles that failed to develop following gonadotropin stimulation in ovaries excised at other time points. iNOS mRNA was also localized to the granulosa cell layer of primary follicles although the signal was less intense than that seen in secondary follicles. The granulosa cell layer of some smaller antral follicles also contained iNOS mRNA but the grain density was less than in secondary follicles. No iNOS mRNA could be localized in large antral follicles, pre-ovulatory follicles, early corpora lutea, mature corpora lutea, or any other cell type other than granulosa cells.

Endothelial NOS was localized in the ovary by immunofluorescent microscopy. Specific staining could only be detected in blood vessels located in the hilum and stromal tissues of the ovary.

Discussion

A wide range of cell functions have been attributed to NO. In this report, we demonstrate that NO produced by two NOS isozymes may have important physiologic roles in the ovary. iNOS mRNA is expressed in the granulosa cells of primary, secondary and small antral follicles in the ovary. Endothelial NOS protein is localized to the blood vessels in the ovary. The fact that mRNA levels of both isozymes change after gonadotropin stimulation of the ovary suggests they may have important roles in ovarian physiology.

Macrophages produce iNOS in response to microbial products and numerous cytokines and NO mediates some of the cytotoxic actions of the activated macrophage [7]. Inducible NOS expression can also be detected in a variety of other cell types following cytokine stimulation in vitro but the expression of iNOS in normal tissues is relatively rare [8]. Reported examples of tissues normally expressing iNOS include the rat uterus, the pregnant rabbit uterus and large airways in humans [7-9]. In the present investigation, we demonstrate that the rat ovary also expresses iNOS and that its expression is predominantly in the granulosa cells of secondary, pre-antral follicles. Our findings support previous studies which have demonstrated the presence of NOS activity in whole ovarian dispersates from immature rats both before [3] and after Il-1β stimulation [2,3]. We extend these observations by identifying iNOS as an isozyme expressed in the intact, *in vivo* ovary and by localizing expression by *in situ* hybridization methods.

Inhibitors of NOS when injected either intraperitoneally or into the ovarian bursae inhibit ovulation in the rat suggesting that NO is involved in the ovulatory process [4]. Because there are no completely specific inhibitors of any particular isoform of NOS, the isoform involved in ovulation is uncertain. Our findings support a role for eNOS but not iNOS in the ovulatory process. eNOS mRNA expression peaks in ovaries containing ovulatory follicles [6]. In contrast, iNOS mRNA reaches its nadir in ovaries containing ovulatory follicles and we could not localize message in the granulosa cells of large antral or ovulatory follicles. Given the profound vasodilatory effects of NO, it is likely that eNOS participates in ovulation by mediating some of the ovarian blood flow changes documented to occur in the peri-ovulatory period.

Acknowledgments

This work was supported by the Berlex Scholar Award from the Berlex Foundation (BVV). We would also like to thank Sean Murphy, PhD, Frank Longo PhD, and Jeanne Snyder, PhD for help and advice with various aspects of this study.

References

1. Van Voorhis,B.J.,Dunn,M.S.,Snyder,G.D. and Weiner,C.P.(1994) Nitric oxide:an autocrine regulator of human granulosa-luteal cell steroidogenesis. Endocrinology 135,1799-1806.
2. Ellman,C.,Corbett,J.A.,Misko,T.P.,McDaniel,M. and Beckerman, K.P.(1993) Nitric oxide mediates interleukin-1-induced cellular cytotoxicity in the rat ovary:a potential role for nitric oxide in the ovulatory process. J.Clin.Invest. 92,3053-3056.
3. Ben-Shlomo,I.,Kokia,E.,Jackson,M.J.,Adashi,E.Y. and Payne,D.W.(1994) Interleukin-1B stimulates nitrite production in the rat ovary:evidence for heterologous cell-cell interaction and for insulin-mediated regulation of the inducible isoform of nitric oxide synthase. Biol.Reprod. 51,310-318.
4. Shukovski,L. and Tsafriri,A.(1994) The involvement of nitric oxide in the ovulatory process in the rat. Endocrinology 135,2287-2290.
5. Chun,S.Y.,Eisenhauer,K.M.,Kubo,M. and Hsueh, A.J.W.(1995) Interleukin-1B suppresses apoptosis in rat ovarian follicles by increasing nitric oxide production. Endocrinology 136,3120-3127.
6. Van Voorhis,B.J.,Moore,K.,Strijbos,P.J.L.M.,et al.(In Press) Expression and localization of inducible and endothelial nitric oxide synthase in the rat ovary:effects of gonadotropin stimulation in vivo. J.Clin.Invest.
7. Nathan,C. and Xie,Q.(1994) Nitric oxide synthases:roles, tolls, and controls. Cell 78,915-918.
8. Sladek,S.M.,Regenstein,A.C.,Lykins, D. and Roberts,J.M.(1993) Nitric oxide synthase activity in pregnant rabbit uterus decreases on the last day of pregnancy. Am.J.Obstet.Gynecol. 169,1285-1291.
9. Huang,J.,Roby,K.F.,Pace,J.L.,Russell,S.W. and Hunt,J.S.(1995) Cellular localization and hormonal regulation of inducible nitric oxide synthase in cycling mouse uterus. J.Leukoc.Biol. 57,27-35.

HYPERACTIVATION IN HAMSTER SPERM MAY INVOLVE NITRIC OXIDE AND HYDROGEN PEROXIDE.

Richard R. Yeoman and Ian H. Thorneycroft

Department of Obstetrics and Gynecology, University of South Alabama, Mobile, AL 36688

Mature hamster sperm need to be reacted to acquire fertilizing capability. Initial stimulation is followed by a 1-2 hour capacitation period involving poorly understood mechanisms which prepare sperm for subsequent hyperactivated movement, acrosomal reaction and oocyte penetration. Our previous work in the hamster epididymal sperm model indicated that expression of hyperactivated motility could be diminished by inhibition of nitric oxide synthase [1,2]. Methyl-L-arginine but not the inactive enantomere caused a 49% decrease in the motility index. A similar inhibition was observed with methylene blue and LY83583 induced blockade of guanylate cyclase. Indirect support for nitric oxide involvement with sperm capacitation is suggested by the well known stimulatory effects of albumin which has recently been proposed as a reservoir for nitric oxide [3].

Hydrogen peroxide or superoxide anions have also been implicated in capacitation of hamster and human sperm respectively [4,5]. The interaction of these reactive oxygen species with nitric oxide during stimulation of hyperactivation in hamsters was thus investigated.

MATERIALS AND METHODS

Human tubal fluid media (Irvine Scientific, Irvine,CA) supplemented with polyvinyl alcohol (1mg/ml), penicillamine (20 uM), and hypo-taurine (100 uM) was used in all experiments. All chemicals were purchased from Sigma (St.Louis, MO). Catalase was from Calbiochem (San Diego, CA).

Mature golden hamsters (17-27 wks) were sacrificed in a high CO_2 chamber, epididymi removed and cauda contents expressed under warmed oil. Sperm ($6X10^5$/500 ul) were cultured in 24 well plates (Falcon/Becton Dickinson, Lincoln Park, NJ) at 37^oC with 5% CO_2 for 6 hrs. Test treatments of 1 uM epinephrine (EPI), EPI plus 5 mM nitro-L-arginine (NLA), 100 nM sodium nitroprusside (SNP), EPI plus 100 ug/ml catalase (CAT) or SNP plus CAT were established at the start of culture and compared to control.

Hyperactivation was evaluated on a warmed microscope stage using the motility index (MI) which is the percent motile times the grade squared. The grading of motility used the standard scale of 0 = no movement to 5 = hyperactivation. Significance of comparisons were evaluated with the T-test.

RESULTS

All hyperactivation results reported here were collected after 6 hours of incubation since our previous studies indicated this was the peak time of activation and are shown in Figure 1. The MI was markedly increased with the addition of 1 uM EPI (15106±70) compared to control conditions (806±70). However, the stimulatory effects of EPI were inhibited when NLA was added to the media (558±186; P<0.01) indicating a role of nitric oxide. Further indication of nitric oxide involvement in sperm hyperactivation was established by addition of 100 nM SNP which increased the MI compared to control (1413± 77; P<0.01).

Addition of the hydrogen peroxide metabolic enzyme CAT (100 ug/ml) to media stimulated with either EPI or SNP induced a marked reduction in MI (751±115 and 537±77, respectively; P<0.01) suggesting involvement of hydrogen peroxide in addition to nitric oxide in sperm hyperactivation.

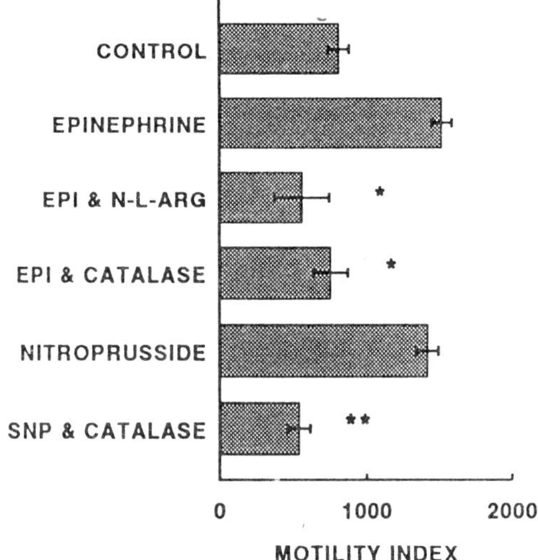

Figure 1. Motility index of hamster sperm after different incubation conditions. Labels and dosages are detailed in text. Asterisks indicate significant change.

DISCUSSION

Epinephrine has long been known to stimulate sperm, is present in oviductal tissue and stimulates nitric oxide [6-8]. Sperm generate superoxide and hydrogen peroxide which has been implicated in capacitation [4,5,9]. The present study strongly indicates participation of both nitric oxide and hydrogen peroxide in sperm capacitation with possibly the nitric oxide step occurring first. Details of this mechanism are yet to be determined.

REFERENCES

1. Yeoman,R.R., Huff,C.A., Aksel,S. (1993) Fert. Steril. 60:S76
2. Yeoman,R.R. (1994) Biol. Reprod.50,S198
3. Stamler,J.S., Jaraki,O., Osborne,J.et al. (1992) Proc. Nat. Acad. Sci. 89,7674-7677
4. Bize,I., Santander,G., Cabello,P., Driscoll,D. and Sharpe,C. (1991) Biol. Reprod. 44,398-403
5. DeLamirande,E. and Gagnon,C. (1993) Free Rad. Biol. Med. 14,157-166
6. Bavister,B.D., Chen, A.F., Fu, P.C. (1979) J. Reprod. Fert. 56,507-513
7. Bodkhe,R.R., Harper,M.J.K. (1972) Biol. Reprod. 6,288-299
8. Lin,A.M., Schaad, N.C., Schulz, P.E., Coon,S.L., Klein, D.C. (1994) Brain Res. 651,160-8
9. Holland,M.K., Storey,B.T. (1981) Biochem. J. 198,273-280

The cyclic expression of nitric oxide synthase (NOS) in the rat testis is suggestive of a role for NOS in spermatogenesis.

Moira K. O'BRYAN[1], Armand S. ZINI[1,2], Peter N. SCHLEGEL[1,2], AND C. Yan CHENG[1].

[1]The Population Council, Center for Biomedical Research, New York, NY 10021 and [2]The James Buchanan Brady Foundation, Department of Urology, The New York Hospital-Cornell Medical Center, New York, NY 10021. U.S.A.

Introduction

Work by our laboratory has recently demonstrated the presence of the endothelial isoform of NOS (ecNOS) within the human testis (presented at this meeting) in a manner suggestive of a role in spermatogenesis. However, because of difficulties in obtaining human testicular material, and particularly normal tissue, the human is not a good system in which to explore the role of ecNOS in spermatogenesis. As such, in this study we have sort to determine, using immunohistochemical and histochemical means, the distribution of ecNOS protein and activity within the testis of the rat.

Rationale and Methods

ecNOS protein localization was demonstrated using an avidin-biotin amplified immunoperoxidase technique. Testes were dissected from adult, male Sprague Dawley rats, fixed in Bouin's fixative (2-4hrs), processed into paraffin blocks and sectioned using standard procedures. Endogenous peroxidase activity and non-specific antibody (Ab) were blocked on sections using 0.45% hydrogen peroxide and 10% non-immune goat serum respectively. ecNOS protein was visualized by incubating sections sequentially in an anti-human ecNOS Ab (Transduction Laboratories, #N30020) (0.5μg/ml), biotinylated anti-mouse immunoglobulin (Ig) (Zymed Laboratories), streptavidin conjugated peroxidase (Zymed Laboratories), and 0.5mg/ml diaminobenzidine tetrahydrochloride (DAB: Sigma).

NADPH diaphorase activity was used as a means of confirming the distribution of ecNOS observed immunohistochemically. NADPH diaphorase histochemistry was achieved using the method of Burnett et al [1].

An immunoblotting analysis of reproductive tissue extracts was carried out in order to confirm the presence of ecNOS and to assess their apparent molecular weights and the existence of degradation products. Cytosolic preparations of protein from rat testis, Sertoli cells, Leydig cells, germ cells, concentrated Sertoli cell conditioned medium (SCCM) and human endothelial cells were separated onto a 7.5%T sodium dodecyl sulphate polyacrylamide gel and transferred to nitrocellulose for immunoblotting using the method of Cheng et al [2]. Immunostaining was achieved using the above mentioned antibody (0.1μg/ml), a rabbit anti-mouse conjugated peroxidase secondary Ab (Sigma) and a chemiluminescence substrate (Boehringer Manneheim).

Results and Discussion

The distribution of ecNOS protein and activity, as indicated by immunostaining and NADPH diaphorase histochemical staining respectively, were seen at all stages of spermatogenesis, however, the pattern of staining was found to vary cyclically with the stages of sperm development. Although invariably seen within the cytoplasm of Sertoli cells, at stages I-VI a relatively higher concentration of fine, particulate ecNOS positive granules were observed. Within Sertoli cells, ecNOS protein and activity was seen in association with residual bodies, towards the lumen of seminiferous tubules, and in phagolysosomes at the base of the epithelium (stages VII-IX). Such a distribution is indicative of a function for ecNOS in the degradation of excess elongating spermatid cytoplasm subsequent to it be endocytosed by Sertoli cells in the form of residual bodies.

Germ cells were also stained for ecNOS, however, only at specific stages of development. ecNOS immunostaining was seen within round spermatids at stage IX-X in association with chromatoid bodies. At present the function of the chromatoid bodies is not known. Additionally, the head region of elongated spermatids were stained for ecNOS at stages XII-II. It is during this stage of spermatogenesis that extensive rearrangement of the nuclear chromatin occurs, involving amongst other things the exchange of histone proteins for protamines [3].

Leydig cells were also found to possess ecNOS protein and activity. Other groups have speculated that NOS may be involved in the regulation of testosterone secretion into the systemic circulation [4].

The presence of ecNOS in Sertoli cell (in the form of SCCM) and Leydig cells was confirmed by immunoblotting.

Conclusions

ecNOS is present within the rat testis in a manner suggestive of a role in spermatogenesis. Its presence within residual bodies and phagolysosomes indicates a role for NO in the degradation of residual bodies and its presence within elongating spermatid nuclei a role in nuclear packing. The presence of ecNOS in Leydig cells is supportive of a function for NO in the regulation of testosterone secretion.

References

1. Burnett, A.L., Ricker, D.D., Chamness, S.L, et al. Localization of nitric oxide synthase in the reproductive organs of the male rat. Biol. Reprod. (1995) **52**, 1-7.
2. Cheng, C.Y., Musto, N.A., Gunsalus, G.L., and Bardin, C.W. The role of the carbohydrate moiety in the size heterogeneity and immunologic determinants of human testosterone-estradiol-binding globulin. J. Steroid Biochem. (1985) **22**, 127-134.
3. Clarke, H.T. Nuclear and chromatin composition of mammalian gametes and early embryos. Biochem. Cell Biol. (1992) **70**, 856-866.
4. Welch, C., Watson, M.E., Poth, M., Hong, T., and Francis, G.L. Evidence to suggest nitric oxide is an interstitial regulator of Leydig cell steroidogenesis. Metabolism (1995) **44**, 234-238.

Localization of nitric oxide synthase in the human prostate: Histochemical and immunhistochemical study using light- and electronmicroscopical techniques.

Klaus Addicks, Theodor Klotz*, Christina Loch*, Gunther Schmidt*, Udo Engelmann* and Wilhelm Bloch

Department of Anatomy, University of Cologne, Joseph Stelzmann Str.9, D-50931 Cologne and Clinic for Urology, University of Cologne, 50924 Cologne

introduction

Previous studies have suggested that NO as a regulation factor of the genitourinary tract[1,2] released from two different NOS isoforms called ecNOS and neuronal bNOS[3] which are located in endothelial cells and nitrinergic nerve fibers. Apart from the neuronal localization bNOS also exists in other cell types e.g. epithelial cells[4,5]. Evidence, brought foward recently, suggests an epithelial NO-production in the genitourinary tract [6]. There are functional implications for NO-mediated regulation of blood flow, smooth muscle tone, neurotransmitter release and secretory functions [7]. The aim of the present study is to evaluate the localization of the constitutive NO-synthase isoform in human prostatic tissue as presupposition for functional implications.

material and methods

Prostatic specimens were procured from 10 patients without prostatic obstruction. The patients underwent radical prostectomy for low-volume localized prostate or bladder cancer without pretreatment. The specimens from periurethral or posterolateral gland regions were collected by prostatectomy, cystectomy, suprapubic prostatectomy or transurethral resection. A possible infiltration through the prostate or bladder cancer was excluded as described by Burnett et al. [8]

NADPH-diaphorase staining of 20µm thick cryostat section of the prostate was done for one hour at 37°C in a Tris-buffer dilution (pH 8.0) which contained 83 mg nicotinamide adenin dinucleotide phosphate (ß-NADPH)(Biomol, Hamburg), 40 mg nitro blue tetrazolium (Biomol, Hamburg), 125 mg monosodium malate and 0.1% triton X-100 at 100 ml.

Abbreviations used: NADPH-diaphorase (NADPH-d), endothelial nitric oxide synthase (ecNOS), neuronal nitric oxide synthase (bNOS), benign prostate hyperplasia (BPH)

For lightmicroscopy, the sections were thaw-mounted onto gelatine coated glas slides and studied microscopically with an axiophot (Zeiss, Oberkochen). For electronmicroscopy, the slices were postfixed with 2% osmium tetroxide in 0.1 M PBS for 20min at 4°C. This procedure led to the formation of poorly soluble osmium-coordinated complexes with the formazan, generated from nitro blue tetrazolium through NADPH-d activity, appearing as black spots at the ultrastructural level. Ultrathin sections (30-60nm) were obtained and examined with a Zeiss EM 902A electron microscope (Zeiss, Wetzlar, FRG).

For immunohistochemistry cryostate sections were pretreated with 3% H_2O_2 in a 60% methanol solution, 0.2% Triton in 0,1 M PBS and 5% bovine serum albumin (BSA) solution. The incubation with the primary antibody was done in a solution of 0.8% BSA and 20 mM NaN_3 in PBS containing the bNOS (Affiniti, Nottingham) specific mouse antibody (1:250 or 1:500 dilution) for 12 hours at 4°C or with ecNOS (Biomol, Hamburg) specific rabbit antibody (1:1000 dilution) . After rinsing, the sections were incubated with the secondary biotinylated goat-anti-mouse or goat-anti-rabbit antibody (1:100 dilution) (Vector Laboratories) for 1 hour at room temperature. Then a streptavidin-horseradish-peroxidase complex was utilized as detection system (1:200 dilution) for another hour and finally developed for 15 minutes with 3,3 diaminobenzidine tetrahydrochloride in 0,05 M Tris HCl-buffer and 0,1% H_2O_2 and counterstained with methylgreen.

Results

1. In the prostatic tissue NADPH-d and immunohistochemistry using bNOS-antibody revealed the existence of dense nitrinergic nervous network (fig.1), which gave off single fibres characterized by distinct formation of axon varicosities: Blood vessels, smooth muscle compartment and the entire prostate glands were nitrinergically innervated without differences between the periurethral and posterolateral zone. The ultrastructural NADPH-d staining demonstrated axon bundles with directly adjacent NADPH-d positiv and negativ axons. 2. The NADPH-d positive epithelium did not display a specific immunohistochemical staining with bNOS or ecNOS without an unspecific staining of the nuclei by bNOS-antibody. 3. Expression of NADPH-d in the endothelium of the prostatic vessels varied depending on the region of the vascular bed and the diameter of the vessels. The arteries of most endothelial cells revealed a pronounced expression of NADPH-d, while veins in the endothelium were not entirely NADPH-d positiv and only a few endothelial cells displayed NADPH-d positivity. Capillaries and postcapillary vessels of the prostate only had sporadic NADPH-d in their endothelium. The NADPH-d results are confirmed by ecNOS immunohistochemistry. 4. Preliminary studies using prostatic tissue from patients with BPH suggest a decrease of nitrinergic innervation of the prostate.

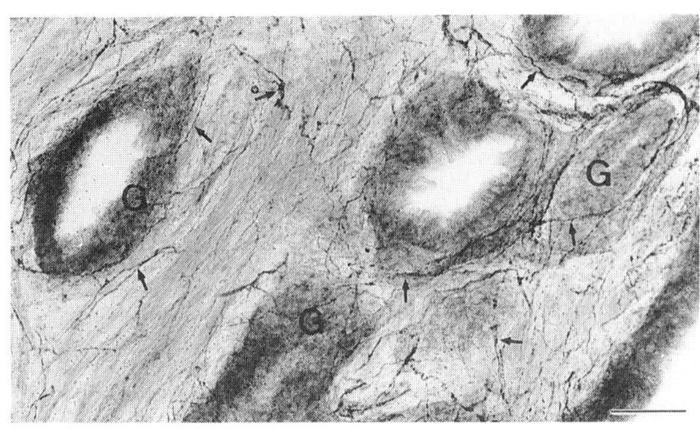

figure 1 NADPH-d staining reveals dense nitrinergic innervation (arrows) of the human prostate. The glandular epithelium (G) shows NADPH-d reaction which is not confirmed by bNOS immunoreaction. bar 150µm

discussion

The human prostate contains a very dense, differentiated nitrinergic innervation which implies that NO plays a role in the neural vegetative regulation of different compartments in the prostate. The dense nitrinergic innervation of the secretory epithel indicates an NO-mediated control of epithelial secretion as proposed by Burnett et al. [6,8]. The smooth muscular innervation of the fibromuscular stroma suggests NO to be a regulatory factor for smooth muscle tone. This theory is supported by findings of electric field stimulation experiments, which demonstrated an influence of NO on the smooth muscular tone of the prostate [2]. The pattern of nitrinergic vascular innervation suggests a NO-mediated regulation of the regional circulation. Furthermore, our ultrastructural findings showing mixed nitrinergic nervous fibres, support the theory of a direct interaxonal modulation by NO which we have suggested for the rat heart's adrenergic system [9]. The uneven expression of NOS in the endothelium provides evidence for a segmental differentiation of the NO-mediated vascular regulation which we have investigated in the heart [9]. The pattern of neural and endothelial distribution imply a control of the regional vascular regulation mediated through the endothel as well as the autonomous nervous system. The reason for the lack of co-localization of the epithelial NADPH-d with bNOS and ecNOS antibody remains unclear. Possible explanations for the lack of co-localization of NADPH-d with NOS-isoforms might be either that this is an unknown isoform of NOS or that this particular NADPH-d is not associated with NOS, as it is suspected to be the case, e.g. in the nasal mucosa, in the liver and the adrenal glands [10]. Our preliminary findings of a decreased nitrinergic innervation of the prostate in BPH and a NO-influence on smooth muscular relaxation [2] points at an involvement of NO in BPH.

references

1. Burnett, A.L., Lowenstein, C.L., Bredt, D.S., Chang, T.S.K. and Snyder, S.H.: Science, 257: 401, 1992.
2. Takeda M, Tang R, Shapiro E, Burnett AL, Lepor H Urology, 45(3): 440, 1995
3. Förstermann, U., Schmidt, H.H.H.W., Pollock, J.S., Sheng, H., Mitchell, J.A., Warner, T.D., Nakane, M.M. and Murad, F. (1991) Biochem. Pharmacol., 42: 1849, 1991.
4. Pollock, J.S., Nakane, M., Förstermann, U. and Murad, F.: J. Cardiovasc. Pharmacol., 20(Suppl 12): S50, 1993.
5. Schmidt, H.H.H.W., Gagne, G.D., Nakane, M., Pollock, J.S., Miller, M.F. and Murad, F.: J. Histochem. Cytochem., 40: 1439, 1993.
6. Burnett, A.L., Ricker, D.D., Chamness, S.L., Maguire, M.P., Crone, J.K., Bredt, D.S., Snyder, S.H. and Chang, T.S.: Biol. Reprod., 52: 1, 1995.
7. Kerwin, J.F. and Heller, M. Med. Res. Rev., 14: 23, 1994.
8. Burnett, A.L., Takeda, M., Maguire, M.P., Lepor, H., Chamness, S., Chang, T.S.K. and Ricker, D.D. Urology ,45: 435, 1995.
9. Addicks, K., Bloch, W. and Feelisch, M.: Microsc. Res. Tech., 29: 161, 1994.
10. Bredt, D.S., Glatt, C.E., Hwang, P.M., Fotuhi, M., Dawson, T.M. and Snyder, S.H. Neuron, 7: 615, 1991.

Nitric oxide synthase activity in the urogenital tract.

Ingrid Ehrén, Margareta Hammarström*, Jan Adolfsson and N. Peter Wiklund.

Department of Urology and Department of Obstetrics and Gynecology*, Karolinska Hospital, S-171 76 Stockholm, Sweden.

Nitric oxide (NO) has been shown to be a autonomic non-adrenergic non-cholinergic neurotransmitter in nerves, causing smooth muscle relaxation by activating guanylate cyclase and increasing cyclic guanosine monophosphate (cGMP) [1]. In the urogenital tract NO has been suggested as a neurotransmitter mediating erection and also the dilatation of the bladder neck and urethra during the micturition reflex [2]. NO is synthesized from L-arginine by three different NO-synthases (NOS) that has been cloned. Both endothelial NOS (e-NOS), which can be found in vascular endothelial cells and neuronal NOS (n-NOS), which is present in both the peripheral and central nervous system, are constitutive and Ca-dependent. Inducable NOS (i-NOS) which is Ca-independent, is activated as a host defense mechanism and has been identified in e.g. macrophages [3]. In an in vivo study of the urinary bladder in rat, it was shown that inhibition of nitric oxide synthase production cause bladder hyperactivity [2]. It has also been shown that estrogen induces the activity of Ca^{2+}-dependent NOS in the guinea pig uterine artery, heart, skeletal muscle, esophagus and cerebellum [4].

Our aim was to study the NOS activity in the human urogenital tract and we also investigated female guinea-pig bladders during hormone treatment. The NOS activity was measured indirectly by the conversion of L-[U-^{14}C] arginine to L-[U-^{14}C] citrulline and the amount of citrulline produced was counted by liquid scintillation and ex-pressed as picomoles per gram of tissue (wet weight) per minute.

Samples of human renal pelvis, ureter, urinary bladder, bladder nèck, proximal urethra, corpus cavernosum, prostate, seminal vesicle, vas deferens, epididymis and testis were obtained from patients undergoing various surgical procedures. In the human renal pelvis, but not in the ureter, we found a significant Ca^{2+}-dependent NOS activity. In the lower urinary tract Ca^{2+}-dependent NOS activity was comparatively low in the urinary bladder, intermediate in the bladder neck and high in the proximal urethra. No significant Ca^{2+}-dependent NOS activity was found in the testis or epididymis but there was a high activity in the vas deferens, seminal vesicle, prostate and corpus cavernosum. There was no significant Ca^{2+}-independent NOS activity obtained in any of the organs.

Table 1 .Nitric oxide synthase (NOS) activity in the human urogenital organs as measured by the formation of L-citrulline from L-arginine (pmol g^{-1} min^{-1}). Means ± SEM. * = p< 0,05, **= p<0,01, ***= p<0,001 . ND = not detectable. NS = not significant.

	Ca-dependent NOS	Ca-independent NOS
renal pelvis (n=6)	21±5**	2±2 (NS)
ureter (n=5)	10±5 (NS)	9±8 (NS)
urinary bladder (n=11)	26±7 **	5±2 (NS)
bladder neck (n=17)	69±15***	13±5 (NS)
proximal urethra (n=7)	153±54*	2±2 (NS)
testis (n=4)	3±4 (NS)	ND
epididymis (n=5)	30±34 (NS)	7±5 (NS)
vas deferens (n=10)	159±41**	13±5 (NS)
seminal vesicle (n=9)	260±53***	0,3±0,3 (NS)
prostate (n=5)	175±38***	34±24(NS)
corpus cavernosum (n=2)	114±2	ND

Female guinea-pigs were castrated whereafter six were treated with estrogen for 10 days, six with progesterone for 10 days and six were untreated. The animals were then sacrificed, their bladders were removed and the NOS activity was measured. There was significantly higher Ca^{2+}-dependent NOS activity in both fundus and trigone in the bladders from estrogen treated animals as compared to those untreated. In the bladders from progesterone treated guinea-pigs we found significantly higher Ca^{2+}-dependent NOS activity only in the trigone as compared to those untreated. No significant Ca^{2+}-independent NOS was found in any of the groups.

Table 2. Nitric oxide synthase (NOS) activity in the urinary bladder smooth muscles in female guinea-pigs as measured by the formation of L-citrulline from L-arginine (pmol g^{-1} min^{-1}). Means ± SEM. * = p< 0,05 denotes significant difference between oophorectomized and hormone treated guinea-pigs according to Student't-test for unpaired variates. NS = not significant. Prog.= progesterone.

	Ca-dependent NOS	Ca-independent NOS
oophorectomized (n=6)		
fundus	304±108	7±3 (NS)
trigone	172±26	7±3 (NS)
oophorectomized + estrogen (n=5)		
fundus	635±65*	6±5 (NS)
trigone	328±23*	10±9 (NS)
oophoorectomized + prog. (n=5)		
fundus	510±64	3±2 (NS)
trigone	336±55*	5±2 (NS)

It is well known that NO is a mediator of erection and we did find a high Ca^{2+}-dependent NOS activity in the corpus cavernosum but also in the prostate, vas deferens and seminal vesicle. The role for NO in the reproductive organs is still unknown but it is possible that NO is a modulator involved in fertility and further studies are needed to clarify this possibility.

In the urinary tract there was a significant Ca^{2+}-dependent NOS activity in the renal pelvis suggesting that NO may be a regulator of the smooth muscle activity in the upper urinary tract. In the lower urinary tract we found a high Ca^{2+}-dependent NOS activity in the proximal urethra, intermediate in the bladder neck and comparatively low in the bladder which is in good agreement with previous studies suggesting NO as a dilator of the bladder neck and urethra during the micturition reflex.

In the bladders of female guinea-pigs there were, in contrast to the findings in the human bladders, a higher NOS activity in the fundus as compared to the trigone. It is possible that NO may have an inhibitory role in guinea-pigs, keeping the bladder relaxed during the filling phase. There was a significant rise of NOS in the group with estrogen treatment as compared to the ones who were only castrated. Since it is well known that postmenopausal women have a high incidence of voiding disorders that is reduced with estrogen treatment, it is possible that a decreased NO synthase activity can be involved in the pathophysiological mechanism leading to urgency and incontinence.

References

1. Moncada, S. (1992) The L-arginine: nitric oxide pathway. The 1991 Ulf von Euler lecture. Acta Physiol Scand 145:201-227.
2. Andersson, K.-E. (1993). Pharmacology of lower urinary tract smooth muscles and penile erectile tissues. Pharmacological Reviews 45(3), 253-308.
3. Knowles, R. G. & Moncada, S. (1994). Nitric oxide synthases in mammals. Biochem J 298, 249-258.
4. Weiner CP, Lizasoain I, Baylis. S.A., Knowles, R.G., Charles, I.G. & Moncada, S. (1994). Induction of calcium-dependent nitric oxide synthases by sex hormones. Proc Natl Acad Sci USA 91(11):5212-5216.

Localization and co-expression of nitric oxide synthase and neuropeptides in ureterovesical ganglia of man

ZARKO GROZDANOVIC*, BERND MAYER# and HANS GEORG BAUMGARTEN*

*Department of Anatomy, Free University of Berlin, D-14195 Berlin, Germany and #Department of Pharmacology and Toxicology, Karl Franzens University, A-8010 Graz, Austria

Introduction

The ureterovesical ganglion complex is located in the anterior portion of the pelvic plexus. The constituent neurons provide autonomic innervation to the pelvic ureter, the ureterovesical junction and the bladder trigone, possibly acting to modulate and co-ordinate the muscular activity of these structures during micturition. The ganglia are predominantly composed of a mixed population of cholinergic and adrenergic nerve cell bodies and terminals comprising the so-called urogenital short neuron system [1-4]. Recently, NOS was histochemically localized to a subpopulation of neurons in the pelvic ganglion of the mouse, some of which were associated with the ureterovesical junction [5,6]. In this study we have examined the distribution and peptide co-expression pattern of NOS in the human ureterovesical ganglia by combining NOS immunolabeling and NADPH-d histochemistry with immunoreactivity for VIP, NPY, and CGRP.

Material and methods

Specimens of the pelvic ureter (3 cm) and the bladder cuff (diameter about 1.5 cm) were obtained from two organ donors (aged 19 and 54 years) and two patients undergoing surgery for invasive bladder cancer (aged 63 and 74 years). The indirect immunofluorescent technique was employed to demonstrate NOS and neuropeptide immunoreactivities. NOS catalytic activity was visualized using the histochemical staining for NADPHd. To investigate the possible co-existence of NOS and neuropeptides, tissue sections were first incubated for the detection of peptide immunoreactivities and, after taking photographs, were subjected to the NADPHd reaction.

Results

NOS immunostaining and NADPHd activity were strictly colocalized in about 20% of neurons in the ureterovesical ganglia, which were located close to the juxtavesical ureter or were scattered in the fatty connective tissue surrounding the terminal ureter. Single axons were seen to leave the ganglia or to travel within nerve fascicles in the adventitia of the ureter or the bladder. In specimens from elderly individuals, NADPH-d staining revealed neurons with obvious structural alterations, such as enlargement of the cell body diameter, bizzare contours and hypertrophic processes. In addition, deposits of lipofuscin pigment were evident in unstained cells. Double labeling showed that all VIP-IR ganglion cells also contained NOS activity. A very minor population of NOS neurons was devoid of VIP immunoreactivity. Less than half of NOS-containing ganglion cells also reacted for NPY immunoreactivity, although there were prominent differences between individual ganglia. CGRP immunoreactivity was only observed in varicose terminal-like nerve fibers, some of which appeared to be associated with NOS-positive perikarya.

Abbreviations used: CGRP, calcitonin gene-related peptide; DMPP, 1,1-dimethyl-4-phenylpiperazinium; IR, immunoreactive; NADPHd, NADPH diaphorase; NO, nitric oxide; NOS, nitric oxide synthase; NPY, neuropeptide Y; VIP, vasoactive intestinal peptide.

Discussion

These findings show that NOS is expressed in a subset of ureterovesical ganglion cells and is in part colocalized with VIP or NPY. Since all VIP-positive cells contain NOS, it may be assumed that VIP and NOS are costored with NPY in a subpopulation of ganglion cells. NOS-containing ganglion cells are contacted by CGRP-IR nerve endings, suggesting an input from preganglionic efferent and/or dorsal root afferent neurons [7,8].

Within the wall of the ureter, NOS-containing nerve fibers are seen to run in the smooth muscle coat, in the subepithelial connective tissue, and in close association with blood vessels [9]. NOS has been found to co-exist with VIP and/or NPY, but not with tyrosine hydroxylase, the rate-limiting enzyme of catecholamine biosynthesis, in ureteric nerves [10]. Physiological and pharmacological data support a role for NO in regulating the ureteric motility. SIN 1, an NO donating substance, was found to relax muscle strips from human renal pelvis and ureter precontracted with potassium chloride [11]. In the pig intravesical ureter, L-N^G-nitroarginine, an NOS inhibitor, reduced electrically evoked relaxations [12]. Exogenous application of NO (e.g., acidified solution of $NaNO_2$) or NO-donors (e.g., S-nitroso-N-acetyl cystein, sodium nitroprusside) had strong inhibitory effects on contractions of the pig ureter induced by electrical field stimulation, noradrenaline or the nicotinic receptor agonist DMPP [12,13]. Furthermore, S-nitroso-N-acetyl cystein increased the tissue level of cyclic GMP. These findings are consistent with the observation that 8-bromo-cyclic GMP, a phosphodiesterase-resistant analog of cyclic GMP, reduced electrically elicited contractions of the guinea-pig ureter [14].

We thus conclude that the passage of urine across the ureterovesical junction is under relaxatory control of a local NO/VIP(NPY) pathway which may be modulated by preganglionic efferent and/or primary afferent CGRP-input.

Acknowledgements

This study was supported in part by the "Fonds zur Förderung der Wissenschaftlichen Forschung in Österreich" (P-8836). We thank Dr. H.H. Knispel, Urology Clinic, Benjamin Franklin Medical Center, Berlin, for supplying us with human specimens.

References

1. Elbadawi, A. and Schenk, E.A. (1971a) J. Urol. **105**, 368-371
2. Elbadawi, A. and Schenk, E.A. (1971b) J. Urol. **105**, 372-374
3. Schulman, C.C., Duarte-Escalante, O. and Boyarski, S. (1972) Br. J. Urol. **44**, 698-712
4. Schulman, C.C., Duarte-Escalante, O., Boyarski, S. and Gregoir, W. (1973) J. Urol. **109**, 381-384
5. Grozdanovic, Z., Baumgarten, H.G. and Brüning, G. (1992) Neuroscience **48**, 225-235
6. Grozdanovic, Z., Brüning, G. and Baumgarten, H.G. (1994) Acta Anat. **150**, 16-24
7. Senba, E. and Tohyama, M. (1988) Brain Res. **449**, 386-390
8. Papka, R.E. (1990) Neuroscience **39**, 459-470
9. Goessl, C., Grozdanovic, Z., Knispel, H.H., Wegner, H.E.H. and Miller, K. (1995) Urol. Res. **23**, 189-192
10. Smet, P.J., Edyvane, K.A., Jonavicius, J. and Marshall, V.R. (1994) J. Urol. **152**, 1292-1296
11. Stief, C.G., Taher, A., Meyer, M. et al. (1993) J. Urol. **149**, 492A
12. Hernandez, M., Prieto, D., Orensanz, L.M., Barahona, M.V., Garcia-Sacristan, A. and Simonsen, U. (1995) Neurosci. Lett. **186**, 33-36
13. Chiu, A.W., Babayan, R.K., Krane, R.J. and Saenz de Tejada, I. (1994) J. Urol. **151**, 335A
14. Cho, Y.H., Biancani, P. and Weiss, R.M. (1984) Fed. Proc. **43**, 353

Androgen dependence of neuronal nitric oxide synthase content and erectile function in the rat penis

D.F. PENSON, C. NG, J. RAJFER, and N.F. GONZALEZ-CADAVID

Division of Urology, Department of Surgery, UCLA School of Medicine, Harbor-UCLA Medical Center, Torrance, CA 90509, USA

Nitric oxide (NO) is the main mediator of smooth muscle relaxation in the corpora cavernosa that triggers penile erection [1-3]. The factors affecting the NO-dependence of the erectile response can be studied in vivo in animal models, particularly the rat [3-10], by electrical field stimulation (EFS) of the cavernosal nerve. Only one isoform of nitric oxide synthase (NOS), the neuronal type (nNOS), has thus far been identified in the penis [4,7-10], and it appears to be restricted to the nerve terminals [4,11]. iNOS has been shown in cultures of penile smooth muscle cells, but only upon induction [12]. The 140 kD eNOS band is scarcely detectable by western blot in the rat penis (unpublished).

Penile erection is an androgen-dependent event in which dihydrotestosterone appears to be the primary active androgen involved in this process in the rat model [6,8-10,13]. Castration of the adult rat for one week reduces the erectile capability of the animal by 40-50%, as measured by EFS [7,8]. This effect is not enhanced when the rats are submitted to EFS up to two months after orchiectomy [13]. The erectile impairment is accompanied by an even greater decrease of penile NOS activity, as measured by the L-arginine/citrulline conversion assay [6,9,10]. However, there are conflicting reports as to whether this reduction is also accompanied by down-regulation of nNOS content, as determined by immunodetection [9,10].

The putative regulation of penile nNOS by androgens may have implications for conditions in humans where erectile dysfunction is associated with reductions in serum T, such as chronic diabetes, aging, or hypogonadism [14]. These alterations are present in the rat models, accompanied by a decrease of penile NOS activity [5-7,9,10], with or without the reduction of nNOS content [7,9,10].

The present study focuses on whether the obliteration of androgen effect at the tissue level, by androgen receptor blockade combined with castration, is able to down-regulate nNOS content in the rat penis in conjunction with the reduction in total penile NOS activity. The functional correlation of this putative penile nNOS androgen-dependence is established by determining if the erectile response to EFS can be completely blocked by this treatment. Furthermore, the possible modulation of penile NOS by estrogens, or by manipulation of the hypothalamic/pituitary axis is studied using castrated animals treated with estrogen and intact animals treated with hypophysectomy or GnRH antagonist.

For EFS determinations and western blot analysis of penile cytosol, three groups of 5-6 Fisher 344 (5-month-old) rats were injected daily as follows: vehicle only subcutaneously; the anti-androgen Flutamide 25 mg/kg intraperitoneally (Fl); GnRH antagonist 1.25 mg/kg subcutaneously (GnRHA). Three additional groups were castrated and submitted to daily injections of vehicle, or flutamide, or implanted with a subcutaneous pellet of beta-estradiol 3-benzoate (E2). An additional group underwent hypophysectomy. All treatments were terminated after 7 days.

The effects on erectile function were measured by assessing maximal intracavernosal pressure (MIP) in response to EFS [5-8]. In comparison to intact animals (84 ± 7.9 mm Hg), the MIP in castrated animals (44 ± 9.6 mm Hg) was decreased by 50% ($p<0.01$), as expected. Fl given to intact animals reduced the erectile response to levels (57 ± 8 mm Hg) slightly above those seen in castrate animals. The combination of Fl with castration led to the nearly complete blockade of erectile response (12 ± 3.3 mm Hg), significantly lower than in castrate animals ($p<0.05$). Castrate animals treated with E2 also had a significant MIP decrease (26.5 ± 3 mm Hg) when compared to castrate animals ($p<0.05$). Finally, GnRHA or hypophysectomy blocked the erectile response to levels found in complete androgen ablation (18.4 ± 2.8 and 9 ± 2 mm Hg, respectively).

Penile nNOS levels were measured by western blot in the cytosol from the penis (shaft+bulb) of rats submitted to EFS [7,8,10]. The 160 kD band was detected with an antibody against the human cerebellum nNOS and a secondary antibody followed by a luminol detection reaction. Castration alone, or administration of Fl to intact rats did not change the intensity of the band, whereas the remaining four experimental paradigms reduced it moderately. A semi-quantitative densitometric analysis revealed that these reductions were not statistically significant. When the penis was divided in sections, only the nNOS content of the glans was significantly decreased by 65% ($P<0.01$) by androgen depletion.

The group which showed the greatest decrease in erectile response, castrates given Fl, was selected for the estimation of NOS activity by the L-arginine/citrulline conversion assay in the total penile cytosol from rats not submitted to EFS [10]. As expected there was a significant 50% decrease in NOS activity when comparing castrated to intact animals ($P<0.05$). However, the addition of Fl in castrate animals did not reduce NOS activity beyond that seen in animals submitted to castration alone.

This study confirms our previous assumption [10] that the androgen-dependence of NOS activity in the rat penis occurs primarily through the modulation of the enzyme itself rather than by the control of penile NOS content. This conclusion is supported by the much higher decrease of total NOS activity than specific nNOS level induced by short-term androgen ablation. The factors responsible for this stable inhibition remain to be identified. However, the moderate down-regulation of penile nNOS content by androgen depletion suggests an additional point of control of NO synthesis, which is most evident in the glans. This is likely to result from a preferential loss of nerve terminals in the distal region, rather than from true NOS down-regulation. The latter may be responsible for the nNOS decrease found in more prolonged androgen depletion conditions such as chronic diabetes [7].

Functionally, the androgen control of penile NOS activity may be causally related to our finding that penile erection in the rat is almost entirely dependent on androgens, and not just partially as suggested by previous studies [6,14,15]. The fact that a full androgen ablation, achieved with castration and androgen receptor blockade, reduced erectile response below that seen in castration alone provides strong evidence for the complete androgen dependence of penile erection. However, since the blockade of the hypothalamic/pituitary axis was as effective as full androgen receptor ablation, pituitary factors regulated by androgens may complement the NO cascade in the erectile response. Alternative smooth muscle relaxation pathways independent from penile NOS are likely to participate. This may explain why androgen receptor blockade in castrated rats is more effective than castration alone in inhibiting the erectile response without inducing an additional decrease of penile NOS activity. The study of the factors regulating nNOS in the rat penis may clarify the interaction of these relaxation pathways in a functionally relevant in vivo model.

References

1. Ignarro, L.J., Bush, P.A., Buga, G.M., Wood, K.S., Fukuto, J.M. and Rajfer, J. (1990) Biochem. Biophys. Res. Commun. **170**, 843-850
2. Rajfer, J., Aronson, W.J., Bush, P., Dorey, F.J. and Ignarro, L.J. (1992) N. Engl. J. Med. **326**, 90-94
3. Lugg J.A, Rajfer, J. and González-Cadavid, N.F. (1995) J. Androl. **16**, 2-5
4. Burnett, A.L., Lowenstein, C.J., Bredt, D., Chang, T.S.K. and Snyder, S.H. (1992) Science **257**, 401-403
5. Garbán, H., Vernet, D., Freedman, A., Rajfer, J. and González-Cadavid, N.F. (1995) Am. J. Physiol. **268**, H467-475
6. Lugg, J., Rajfer, J. and Gonzalez-Cadavid, N.F. (1995) Endocrinology **136**, 1495-1501
7. Vernet, D., Cai, L., Rajfer, J. and González-Cadavid, N.F. (1995) Endocrinology, In press
8. Garban, H., Marquez, D., Cai, L., Rajfer, J. and Gonzalez-Cadavid, N.F. (1995) Biol. Reprod., In press
9. Chamness, S.L., Ricker, D.D., Crone, J.K. et al. Fertil. Steril. (1995) **63**, 1101-1107
10. Lugg, J., Ng, C., Rajfer, J. and González-Cadavid, N.F. (1995) J. Urol. **153**, Suppl. 509A, Abstr. 1122
11. Burnett, A.L., Tillman, S.L., Chang, T.S.K. et al. (1993) J. Urol. **150**, 73-76
12. Hung, A., Vernet, D., Xie, Y. et al. (1995) J. Androl. In press
13. Mills, T.M., Stopper, V.S. and Wiedmeyer, V.T. (1994) Biol. Reprod. **51**, 234-238
14. Murray, F.T., Geisser, M. and Murphy, T.C. (1995) Am. J. Med. Sc. **309**, 99-109

The effect of adaptation to exercise on endothelium-dependent responses and nitric oxide production in the rat.

EUGENIA B.MANUKHINA, ALEXANDER V.LAPSHIN, FELIX Z.MEERSON, VASAK D.MIKOYAN*, LUDMILA N.KUBRINA* and ANATOLY F.VANIN*

Inst. of General Pathology & Pathophysiology, Baltijskaya 8, Moscow 125315; *Inst. of Chemical Physics, Kosygin str. 4, Moscow 117997, Russia.

Adaptation to exercise (AE) is known to play an important role in treatment and prevention of cardiovascular diseases, in particular, of those associated with reduced endothelium-dependent relaxation of blood vessels [1,2,3]. Local mechanisms, by which AE exerts its beneficial effect, are not yet well understood and the available information about changes in local regulation of vascular tone, especially about the role of endothelium-dependent responses, is contradictory [4,5]. The aim of the present work was to study the effect of AE on functional state of the endothelium. To this aim we evaluated constrictor and dilator responses of isolated blood vessels and directly measured the nitric oxide (NO) production in the organism.

Experiments were carried out on Wistar male rats. AE consisted of 36 sessions of passive swimming (unloaded) for 1 h daily. A ring preparation of isolated thoracic aorta was placed in an "organ-bath" system. Contractions of intact and denuded aortas were recorded in parallel by a two-channel recorder Gemini (Ugo Basile, Italy). The endothelium was removed mechanically using a special catheter. Contractions were induced with norepinephrine (NE) (10^{-8} - 5 x 10^{-7} M). Adrenoreactivity was expressed as ED_{50}. Endothelium-dependent relaxation of NE (5 x 10^{-7} M)-precontracted aorta was induced by acetylcholine (ACh) (10^{-8} - 10^{-5} M).

The NO produced in rat tissues was assayed using the EPR method by NO inclusion in a complex with Fe^{2+}-diethyldithiocarbamate (DETC) with formation of paramagnetic mononitrosyl iron complexes with DETC (MNIC-DETC) which give a characteristic EPR signal. The amount of MNIC-DETC and, thereby, the amount of NO included was evaluated by the intensity of EPR signal [9]. To form MNIC in the organism, rats were injected with Na-DNIC (50 mg/100 g body weight, i.p.) and Fe+citrate (20 mg + 95 mg per 100 g body weight). Rats were decapitated 30 min following the injection of Na-DETC. Isolated organs were grounded and frozen in a press-form. The EPR signal was recorded on a radiospectrometer Radiopan (Poland).

It follows from Table 1 that the NE sensitivity of the intact aorta from adapted animals was decreased by approximately 30% as compared to the control while the NE sensitivity of the denuded aorta was similar in AE and in control. The net decrease in adrenoceptor sensitivity of the aorta with intact endothelium seems to be due mainly to the changed sensitivity of endothelial adrenoceptors while that of smooth-muscle adrenoceptors remains unchanged.

Table 1. The adrenergic sensitivity of isolated rat aorta from control and exercise-adapted rats (ED_{50}, nM) (M±m)

	Control (n=9)	AE (n=10)
With endothelium	2.76±0.20	3.88±0.44*
Without endothelium	1.71±0.16	1.58±0.12

The AE potentiated the endothelium-dependent relaxation to ACh of the aorta at all concentrations used. For instance, the maximum relaxation induced by ACh at the concentration of 10^{-5} M was 47+2.4% vs. 37+2.2% ($p<0.05$) of the value of contraction response to NE in exercise training and in control respectively. Therefore, AE not only potentiated the inhibitor effect of endothelium on contractions but also enhanced the endothelium-induced relaxation of the isolated aorta.

To elucidate the mechanism of this effect we have measured the tissue production of NO which plays the key role in the local control of vascular tone [7]. Table 2 demonstrates that AE increased the formation of MNIC-DETC complex more than 6 times in the liver, 3.3 times in the intestine and 1.3 times in the spleen. A considerable amount of the complex was detected in the heart and kidney from adapted animals while in control the complex concentration was below the level of detection. This signifies corresponding increases in NO content.

Table 2. The NO production in organs of control and exercise-adapted rats (M+m).

	NO production, ng/g wet tissue	
Organ	control (n=5)	adaptation (n=5)
Liver	16,6±6	100±10*
Intestine	250±50	830±160*
Heart	0	10±3*
Kidney	0	27±7*
Spleen	50±20	67±20

It is known that both transient [8] and chronic increase in blood flow [9] potentiates generation and release of the endothelium-derived relaxing factor, that is, essentially NO, as a result of enhanced shear stress. This is why regular intermittent increases in the blood flow during daily exercise training could augment the NO production and, probably, increase the capacity of NO deposition in a biologically active depot [10].

On the whole, the results obtained show that AE potentiates the NO generation in the organism and, thereby, enhances both the inhibitor effect of endothelium on vascular contractions and the endothelium-dependent relaxation. In effect this may provide the protective effect of AE in diseases associated with attenuated endothelium-dependent relaxation.

References

1. Sellier,P.(1995) J.Cardiovasc.Pharmacol. **25** (Suppl.1), S9-S14
2. Kaplan,N.M.(1991) Hypertension **18** (Suppl.1), I-153-I-160
3. Kramsch,D.M., Aspen,A.J., Abramowitz,B.M., Krelmendahi,T. and Hood, W.B. (1981) N.Engl.J. Med. **305**, 1483-1489
4. Chen,H., Li,H.-T. and Chen C.-C. (1994) Circulation **90**, 970-975
5. Green,D.J., Cable,N.T., Fox,C., Rankin,J.M. and Taylor,R.R. (1994) J. Appl. Physiol. 77, 1829-1833
6. Vanin,A.F., Mordvintcev,P.I. and Kleshchev,A.L.(1984)Studia Biophys. **107**, 135-142.
7. Bredt,D.S. and Snyder, S.H. (1994) Annu. Rev. Biochem.**63**, 175-195
8. Rubanyi,G.M., Romero,J.C. and Vanhoutte, P.M. (1986) Am.J.Physiol. **250**, H1145-H1149
9. Miller,V.M. and Burnett,J.C. (1992) Am.J.Physiol.**263**, H103-H108
10. Mulsch,A., Mordvintcev,P., Vanin,A.F. and Busse, R. (1991) FEBS Lett. **294**, 251-256

Photorelaxation of vascular smooth muscle by visible light is qualitatively different from that produced by ultraviolet irradiation.

IAN L. MEGSON, CAROL E. SCOTT & FREDERICK.W. FLITNEY

Division of Cell & Molecular Biology, School of Biological & Medical Sciences, University of St Andrews, St Andrews, Fife KY16 9TS, Scotland, UK.

Introduction. Nitric oxide (NO) can be released photochemically from a molecular 'store' contained within vascular smooth muscle (VSM) cells. [1,2]. The store is discharged rapidly by exposing a precontracted artery to visible laser light, producing a *transient* photorelaxation [2]. A 'depleted' artery will fail to respond to a second (identical) exposure when it is delivered immediately (1-5min) after the first, but its photosensitivity recovers slowly ($T_{1/2}$ = *ca*100min at 33-37°C) in the dark. The recovery process (repriming) displays an absolute requirement for endothelium-derived NO and it is inhibited by ethacrynic acid, a thiol alkylating agent.

These observations are at variance with earlier studies which (generally) have made use of endothelium-denuded arterial rings and exposure to UV light. Under these circumstances, UV exposure elicits a response which is *sustained* throughout the period of illumination. Also, a second (or subsequent) irradiation produces a similar response even when delivered immediately after the first. Thus, the first exposure does not induce a refractory state in the vessel.

We have compared UVA and visible light-induced photorelaxations, employing isolated, internally-perfused segments (1-2cm) of rat tail artery with intact endothelium. Our results suggest that at least two photolabile sources of NO are present in VSM, both of which can be degraded by UVA light, but only one by visible light.

Methods. Experiments were performed in a darkened laboratory. Isolated segments of tail artery from male Wistar rats (300-400g) were perfused internally with Krebs solution (2mls.min^{-1}, 33°C) containing phenylephrine (1-7μM, generating perfusion pressures of 100-120mmHg). Vessels were illuminated periodically with UVA (λ = 366nm; intensity = 1.2mW.cm^{-2}; duration = 6min) or visible laser light (λ = 514 5nm; intensity = 6.3mW.cm^{-2}; duration = 6min) and afterwards kept in the dark for varying time intervals (ΔT = 10, 20, 40, 72, 150 and 300min) before being irradiated again.

Results and Discussion. UVA-induced responses are composite in nature, comprising a *transient* component superimposed on a *sustained* component (Fig 1A). The latter reverses when illumination ceases. By contrast, visible light-induced photorelaxations are transient and show almost full recovery *during the period of irradiation* (Fig1B). The transient component of the UVA response is only evident during the *first* in a series of irradiations when ΔT is kept small (*ca* <5min). However, it gradually reappears as ΔT is progressively increased and so in this respect it resembles repriming of visible light-induced photorelaxations [2,3]. The sustained component of the UVA-response is largely independent of ΔT.

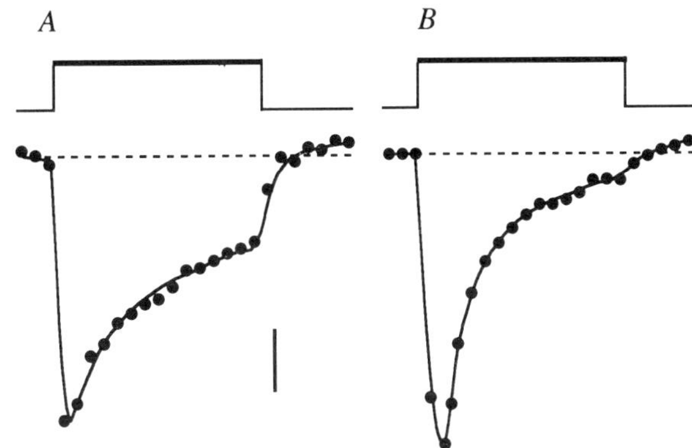

Figure 1. Computer averaged photorelaxant responses to UVA (A; n = 7) and visible (B; n = 5) light (ΔT = 72min). Upper traces indicate period of irradiation (6min). Vertical scale bar = 2.5% for A and 5% for B.

These results can be explained if we postulate that VSM contains *two* photodegradable sources of NO (*A* and *B*). Visible and UVA light are both able to release NO from store *A*, which has a relatively small capacity and is therefore exhausted rapidly. Depletion of *A* accounts for the transient nature of visible light-induced photorelaxations and also for the transient component of the UVA-induced response. Additionally, UVA (but *not* visible) light can liberate NO from another, more durable source (*B*). The photochemical release of NO from store *B* persists throughout the exposure period [4] and is responsible for the sustained component of the UVA response. We show elsewhere [3] that the relative contributions of the transient and sustained components of UVA-induced photorelaxations are altered in vessels from spontaneously hypertensive animals.

Acknowledgements. Supported by the British Heart Foundation.

References

1. Venturini, C.M., Palmer, R.M.J. and Moncada, S. (1994). J. Pharmacol. Exp. Ther. **266**, 1497-1500.
2. Megson, I.L., Flitney, F.W., Bates, Jennifer and Webster, R. (1995). Endothelium. **3**, 39-46.
3. Flitney, F.W., Megson, I.L. and Scott, Carol E. (this volume)
4. Kubaszewski, E., Peters, A., McClain, S., Bohr, D. and Malinski, T.(1994) Biochem. Biophys. Res. Commun. **200**, 213-218.

Ultra-violet and visible light-induced photorelaxation of vascular smooth muscle from normotensive and spontaneously hypertensive rats.

FREDERICK W. FLITNEY, IAN L. MEGSON and CAROL E. SCOTT.

Division of Cell & Molecular Biology, School of Biological and Medical Sciences, University of St Andrews, St Andrews, Fife KY16 9TS, Scotland, UK.

Introduction. Vascular smooth muscle (VSM) cells contain a finite molecular 'store' of NO which can be discharged by irradiating a vessel with visible laser light [1]. This causes a transient photorelaxation. We show elsewhere [2] that UVA irradiation generates a more complex response, consisting of a transient component superimposed upon a sustained component. The former resembles visible light-induced photorelaxation in that it is temporarily lost after the initial exposure but then gradually recovers with time in the dark.

Here we compare UVA and visible light-induced photorelaxations, using internally-perfused tail arteries from spontaneously hypertensive (SHRs) and normotensive rats (NTRs). The results show that recovery (repriming) following visible light exposure is *impaired* in vessels from SHRs as compared to NTR controls, whereas repriming following exposure to UVA light is *enhanced* in SHRs.

Methods. Experiments were made with isolated, perfused tail arteries from age-matched male SHRs and NTRs (n = 7 per group) using experimental procedures described previously [2,3]. Vessels were precontracted with phenylephrine and exposed periodically to UVA (λ = 366nm; 1.2mW.cm^{-2}; 6min) or visible (λ = 514.4nm; 6.3mW.cm^{-2}; 6min) light. Blood pressure recordings (tail cuff method) were made during a 3-4 week acclimatisation period prior to sacrificing the animals.

Results and Discussion. Systolic and diastolic pressures (mmHg) were significantly ($p<0.001$; Student's t-test) higher in SHRs (192 +/-2.12 and 138+/- 2.39) than in NTRs (135 +/-1.8 and 84.7+/-1.8). Repriming curves for visible-(A) and UVA-induced (B) photorelaxations are shown in Figure 1.

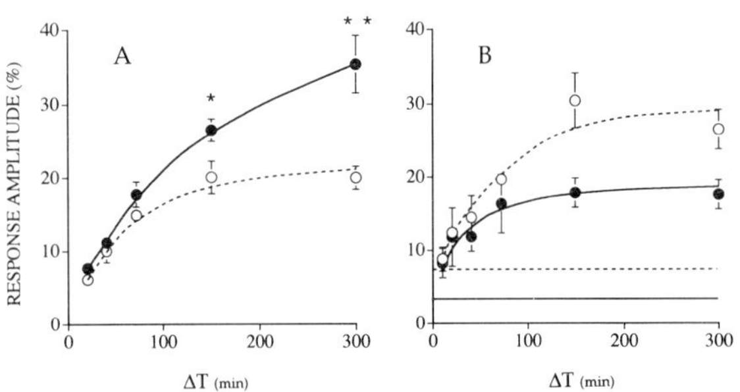

Figure 1. Repriming curves for visible (A) and UVA-induced (B) photorelaxations in vessels from NTRs (filled circles) and SHRs (empty circles). Mean +/- 1 SEM shown (n = 3-5 for A and 3-7 for B).

Recovery following exposure to visible light (Figure 1A) is impaired in vessels from SHRs (empty circles) as compared to those from NTRs (filled circles). Conversely, Figure 1B shows that UVA-induced photorelaxations are enhanced in arteries from SHRs (open circles) compared to NTRs (filled circles). This is largely accounted for by an increase (*ca* 2x) in amplitude of the sustained component of the response, represented in Figure 1B by the solid (NTRs) and dashed (SHRs) horizontal lines.

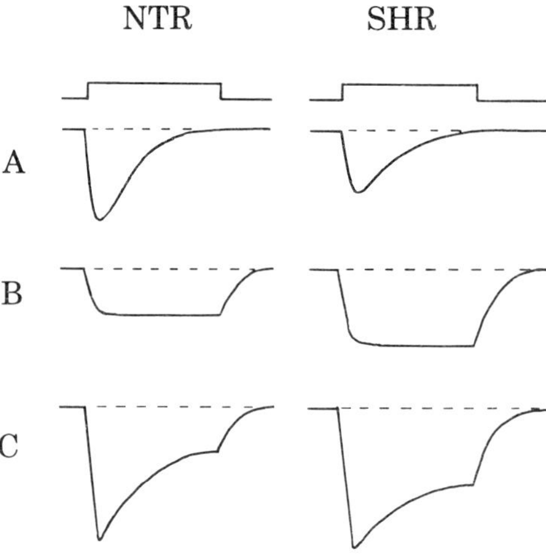

Figure 2. Transient (*A*) and sustained (*B*) components of the UVA-induced response in NTRs and SHRs. UVA response (*C*) is the sum of components *A* and *B*.

These observations provide additional support for the working hypothesis outlined previously [2] and represented diagramatically in Figure 2. We postulate that there are *two* photosensitive sources of NO in VSM cells (*A* and *B*), both of which can be discharged by UVA light (*A*+*B*) but only one by visible light (*A*). The transient component of the UVA-induced response and virtually all of the visible light-induced response is attributed to the photochemical release of NO from store *A*, whereas the sustained component of the UVA-induced response is due to NO released from store *B*. The results presented here imply that the capacity of *A* is *decreased* in SHRs, whereas *B* is *increased*.

The chemical nature of *A* and *B* remains to be established. Repriming of visible light photorelaxation (store *A*) is inhibited by oxyhaemoglobin and/or by L-NMMA and potentiated by either authentic NO or *S*-nitroso-*N*-acetylpenicillamine. It is also prevented by pre-treating arteries with ethacrynic acid, a thiol-alkylating agent [2]. On the basis of these observations, we postulate that *A* might be a photosensitive *S*-nitrosothiol and that the availability of one or more of the reactants required for its synthesis (e.g endothelium-derived NO or endogenous tissue thiols) is diminished in hypertensive animals

Acknowledgements. This work is supported by the British Heart Foundation

References

1 Megson *et al* (1995) *Endothelium:* **3**: 39-46.
2 Megson, I.L., Scott, C.E. and Flitney, F.W. (This volume).
3. Flitney, F.W., Megson, I.L., Flitney, D.E. and Butler, A.R. (1992). Br. J. Pharmacol. **107**, 842-848.

Lack of effects of nitric oxide on baroreflex inhibition of sympathetic tone in vivo

J. Zanzinger, J. Czachurski and H. Seller

Institute of Physiology, University of Heidelberg, D-69120 Heidelberg, Germany

Introduction: A number of histological studies revealed the presence of NO-synthases (NOS) within or nearby all areas which are involved in baroreflex transmission [1]. This was interpreted, equivocally, as a sign of potential functional importance of NO in baroreflex transmission. On the other hand there are conflicting results from functional studies in vivo. While NOS-inhibition seemed to increase sympathetic baroreflex responses in acute experiments [2] the opposite was observed after chronic administration of L-NAME in rats [3]. In contrast, NO had no effects on baroreflex function when injected in the NTS of rabbits though lowering baseline sympathetic tone [4]. In another recent study on rabbits, increases in carotid sinus nerve (CSN) activity evoked by application of constant pressure ramps in an vascularly isolated preparation of the carotid sinus were changed by high doses of NO or S-nitrosocysteine whereas inhibition of NO-synthesis was ineffective [5]. In the present study we investigated the neuronal effects of local administration of the NOS-inhibitor nitro-L-arginine (L-NNA) and of different NO-donors on baroreceptor function reflected by changes in CSN-activity in an intact preparation of the carotid sinus in cats. Possible effects on central baroreflex transmission were tested by studying the inhibitory effects of standardized baroreflex activation (electrical stimulation of the CSN) on pre- and postganglionic sympathetic nerve activity (SNA) upon microinjection of these substances into the nucleus tractus solitarii (NTS) and in the rostral and caudal ventrolateral medulla (RVLM and CVLM, respectively) in anaesthetized cats.

Methods: Experiments were performed on chloralose anaesthetized cats (n=18). The animals were vagotomized and arterial baroreceptors and chemoreceptors were denervated by cutting the CSN and vago-aortic nerves respectively. For recording of pre- or postganglionic SNA, the left white ramus of the 3rd thoracic segment (n=10) or the left renal nerve (n=8), respectively, were prepared as previously described [6,8], placed on bipolar platinum electrodes and kept in a pool of warm paraffin oil. Neural signals were amplified (20000 - 50000x; Tektronix AM 502), filtered (2 Hz - 3 kHz) and stored and analysed with a CED 1401 interface combined with an 80486 PC computer. To activate baroreflex inhibition of SNA, the central portion of the left CSN was placed on bipolar electrodes connected to an isolated stimulator (Digitimer). CSN stimulation was performed to produce effective but submaximal SNA-inhibition by consecutive trains of 3-8s length using 3-8 V, 50 Hz and 0.5 ms pulse duration at intervals of 30 - 60s. Measurements of baroreflex responses to CSN-stimulation were taken after an adaptation period of at least 10 min. 5 sweeps were routinely averaged prior to further evaluation to reduce influences of physiological fluctuations in reflex activities. Baroreceptor function was tested in 6 cats. In these animals the right carotid sinus and its CSN were carefully dissected free and surrounded by a pool made of cotton pads soaked with agar agar. Drug solutions were subsequently administered on the carotid sinus by filling the pool thereby reaching the whole surface of the carotid sinus. CSN-activity, was recorded from the distal end of the right CSN. For microinjections in the NTS, CVLM and RVLM, the surfaces of these areas were exposed. Injections were performed with glass micropipettes having a tip diameter of approximately 20 µm. 500 nl per injection site were injected at depths of 1.2 mm below the surfaces by pneumatic pressure.. Data were analysed by ANOVA for repeated measurements and comparison of means was carried out by the Tukey method. Values are reported as mean ± SD.

Results: Figure 1 shows effects of NO on baroreceptor function. CSN= carotis sinus nerve activity, MAP = mean arterial blood pressure. Concentrations of L-NNA and NO-donors: mM.

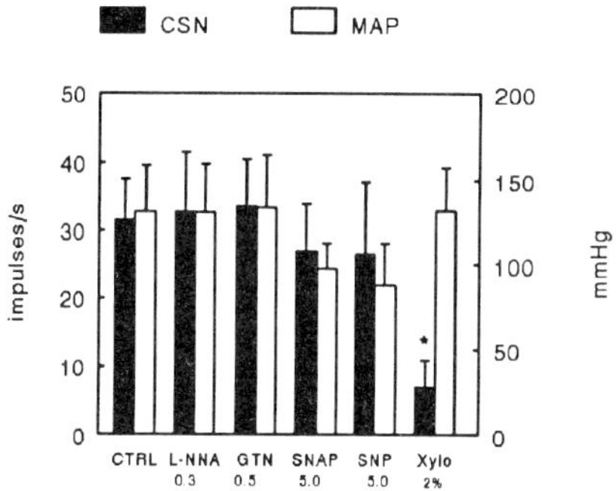

Neither NO-blockade nor NO-donors in increasing doses up to those shown in the figure, had any significant blood pressure independent effects on baroreceptor function. In contrast, chemoreceptor responses to injections of bicarbonate solution into the A.carotis were markedly attenuated by GTN and slightly enhanced by L-NNA in the same experiments. Furthermore, CSN-activity was suppressed by the local anaesthetic xylocaine (2%). Similarly, as we have shown recently [6], microinjections of the same substances in the NTS, CVLM or RVLM did not alter baroreflex transmission as measured by changes in the time of complete SNA-inhibition during electrical stimulation of the CSN. However, microinjections of glutamate in the NTS markedly prolonged, and kynurenate as well as bicuculline in the CVLM and RVLM, respectively, inhibited sympathoinhibition via the baroreflex.

Discussion: We were unable to detect any significant specific effects of endogenous or exogenous NO on mechanoreceptor function at the carotid sinus. Furthermore, NO has no significant effects on the central transmission of the baroreflex [6]. Sufficient amounts of the applied substances at the receptors within the carotid sinus can be assumed since L-NNA and NO-donors significantly modulated chemoreceptor responses in the same experiments and, in addition, caused systemic effects. We used the intact carotid sinus preparation to maintain the physiological pulsatile pressure stimuli for the baroreceptors and to preserve blood flow effects on the endothelium. Our observations are strengthened by recent studies on rabbits which came to similar conclusions [7]. In conclusion, our experiments suggest that sympathetic baroreflex mechanisms are preserved during variations in local NO-supply. Since NO acts as a direct vasodilator and as inhibitor of excitatory sympathetic mechanisms [8], this absence of an interaction of NO with the sympathoinhibitory baroreflex may be a prerequisite for the appropriate control of cardiovascular functions by NO in vivo.

References

1. Mizukawa, K. , Vincent, S. R., McGeer, P. L. & McGeer, E. G. (1988). J.Comp.Neurol. **279**, 281-311
2. Kumagai, H. , Averill, D. B., Khosla, M. C. & Ferrario, C. M. (1993). Hypertension **21**, 476-484.
3. Scrogin, K. E., Veelken, R. & Luft, F. C. (1994). Hypertension **23**, 982-986.
4. Harada, S. , Tokunaga, S. , Momohara, M. , Masaki, H. , Tagawa, T. , Imaizumi, T. & Takeshita, A. (1993). Circ.Res. **72**, 511-516.
5. Matsuda, T. , Bates, J. N., Lewis, S. J., Abboud, F. M. & Chapleau, M. W. (1995). Circ.Res. **76**, 426-433.
6. Zanzinger, J., Czachurski, J.& Seller, H. (1995) Neurosci. Lett. **197**, 199-202
7. Jimbo, M. , Suzuki, H. , Ichikawa, M. , Kumagai, K. , Nishizawa, M. & Saruta, T. (1994). J.Autonom.Nerv.Syst. **50**, 209-219.
8. Zanzinger, J. , Czachurski, J. & Seller, H. (1995) Am.J.Physiol. **268**, R958-R962.

LNIO inhibits endothelium-dependent responses but does not elevate pulmonary vascular resistance.

PHILLIP J. KADOWITZ, BRYAN J. DEWITT, THOMAS D. GILES, MICHAEL GIVEN, DAVID R. POWERS and STAN S. GREENBERG,

DEPT. OF PHARMACOLOGY, TULANE UNIVERSITY MEDICAL CENTER AND DEPT. OF OBSTETRICS AND GYNECOLOGY, MEDICINE and PHYSIOLOGY, LOUISIANA STATE UNIVERSITY MEDICAL CENTER, NEW ORLEANS, LOUISIANA, 70112, U.S.A.

Inhibition of endothelial nitric oxide synthase (cNOS) activity is associated with vasoconstriction, increased vascular resistance and inhibition of endothelium dependent vasodilation, in vivo [1,2]. Vascular cNOS spontaneously and continuously generates small amounts of NO (basal release). This is believed to be responsible for the maintenance of low vascular tone [3,4]. The endothelium also generates NO in response to agonist-induced stimulation and shear stress (stimulated NO release). This is believed to mediate the vasodilator responses to many agonists and physiologic interventions [5]. Recent studies suggest that basal and agonist-stimulated cNOS activity may be differentially regulated, in vitro and in vivo. However, the concept that the pressor response to NOS inhibitors is associated with both inhibition of basal NO release as well as agonist-stimulated release of NO has been challenged [6]. We tested this concept, in vivo, using the isolated lung of the closed chest cat, perfused with autologous blood.

Methods and Materials: Adult cats of either sex (2-4 kg) were anesthetized with ketamine-xylazine. The left lower lung lobe of the closed-chested cat was cannulated with a specifically designed 5-French triple lumen catheter under fluoroscopic guidance. The animals were heparinized and the lung was perfused at constant flow with autologous blood obtained from the femoral artery. Lung lobe perfusion pressure was 18 mm Hg. Left atrial pressure was measured by a catheter inserted trans-septally into the left atrium. Since perfusion was performed at constant flow changes in lobar perfusion pressure (LPP) were proportional to changes in lobar vascular resistance [7]. After a thirty min equilibration period U46619 was infused into the left lower lung lobe to increase pulmonary vascular tone. Mean systemic arterial pressure (MAP), LPP and the vasodepressor responses to intralobar injections of acetylcholine (Ach, 1 μg), bradykinin (BK, 0.3 μg), substance P (SP, 1 μg), pinacidil (PIN, 100 μg) or adenosine (ADO, 100 μg) were obtained prior to and 15 min after the intralobar infusion of saline (control), LNIO (1-20 mg/kg) or LNAME (1-100 mg/kg). In a separate series of animals the responses to LNAME were evaluated in the presence of 30 mg/kg of LNIO. These doses of LNIO and LNAME inhibit the endothelial synthesis of nitrite and nitrate (RNI) as assayed by chemiluminescence and the Griess reaction [8]. Data were analyzed with ANOVA . A p value of 0.05 or less was accepted for statistical significance of mean differences.

Results and Comments: Administration of LNIO (1-30 mg/kg, iv) did not significantly elevate either MAP or pulmonary vascular resistance [Fig. 1] but produced-dose dependent inhibition of the pulmonary vasodilator responses of the cat lung to intra-arterial injections of ACH, SP and BK [Fig. 2]. LNIO did not attenuate the decrease in LPP produced by adenosine or PIN [Fig. 2]. In contrast to these findings, doses of LNAME (100 mg/kg, iv) which inhibited the endothelium-dependent vasodilation to ACH, BK and SP by the same magnitude as that produced by LNIO, elevated both MAP and LPP. Thus, LNIO and LNAME qualitatively differ in their ability to elevate pulmonary vascular resistance and MAP despite identical inhibition of endothelium-dependent relaxation. These findings suggest that either 1) basal NO production is resistant to suppression by LNIO or 2) basal production of NO may not be the dominant mediator of the low vascular resistance of the pulmonary and systemic circulations and 3) LNAME-induced elevation of systemic pressure is mediated by an mechanism independent of inhibition of NOS. The data obtained with the sequential administration of LNIO and LNAME support the latter postulate. Administration of 30 mg/kg iv of LNIO to rats reduces plasma and urine RNI by 80% within 24 hr and prevents endotoxin-induced up-regulation of NOS in alveolar macrophages and neutrophils [8]. In cats pretreated with LNIO (30 mg/kg, iv) LNAME

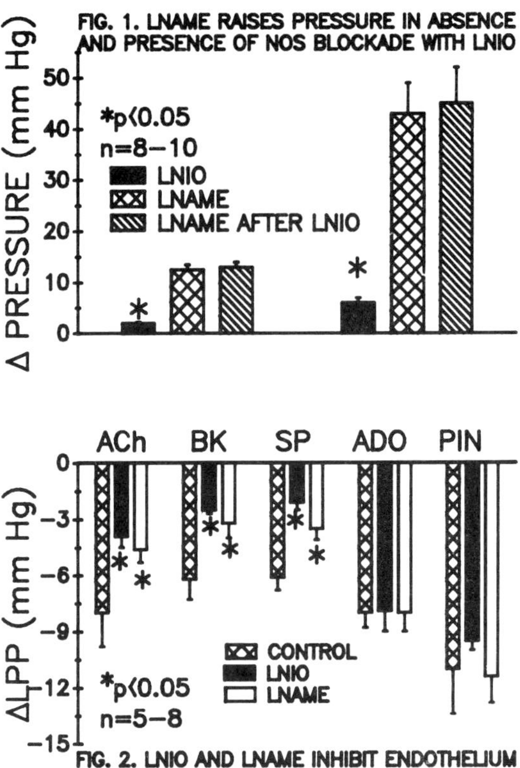

FIG. 1. LNAME RAISES PRESSURE IN ABSENCE AND PRESENCE OF NOS BLOCKADE WITH LNIO

FIG. 2. LNIO AND LNAME INHIBIT ENDOTHELIUM DEPENDENT DILATION OF THE PERFUSED LUNG

did not produce any further inhibition of the endothelium-dependent vasodilator responses to ACH, BK or SP (Fig 1). Moreover, LNAME mediated increases in MAP and pulmonary vascular resistances did not differ in control and LNIO treated cats. These data support the conclusion that LNIO and LNAME inhibit basal and stimulated release of NO differently. It is unlikely that LNIO inhibits neural NOS rather than endothelial NOS since 7-nitro-indazole, an inhibitor of neural NOS, does not elevate pressure but also does not inhibit endothelium dependent relaxation [9,10]. We also suggest that LNAME mediated increases in systemic and pulmonary vascular resistance may not result solely from inhibition of cNOS. Finally, the role of basal NO release in maintaining the pulmonary and systemic circulation in the dilated state under resting conditions requires further investigation.

References

1. Rees, D. D., Palmer, R. M., Schulz, R., Hodson, H.F. and Moncada, S. (1990) Br. J. Pharmacol. 101: 746-752.
2. Gardiner, S. M., Kemp, P.A., Bennett, T., Palmer, R.M. and Moncada, S. (1992) Eur. J. Pharmacol. 213, 449-451.
3. Hasunuma, K., Yamaguchi, T., Rodman, D.M., O'Brien, R.F. and McMurtry, I.F.(1991) Am. J. Physiol. 260, L97-L104.
4. Aisaka, K., Gross, S.S., Griffith, O.W. and Levi, R. (1989) Biochem. Biophys. Res. Comm. 160, 881-886.
5. Narayanan, K., Spack, L. McMillan, K., Kilbourn, R.G., Hayward, M.A., Masters, B.S. and Griffith, O.W. (1995) J. Biol. Chem. 270: 11103-11110
6. Lot, T. Y., Stark, G. and Wilson, V.G. (1993) Naunyn- Schmiedebergs Arch. Pharmacol. 347, 115-118.
7. DeWitt, B. J., Cheng, D.Y., McMahon, T.J. Nossaman, B.D. and Kadowitz, P.J. (1994) Am. J. Physiol. 266, H2256-H2267
8. Greenberg, S., Xie, J., Spitzer, J., Wang,,J., Lancaster, J., Grisham, M.B., Powers, D. and Giles, T. (1995) Life Sci In Press
9. Moore, P. K., Babbedge, R.C., Wallace, P., Gaffen, Z.A. and Hart, S.L. (1993) Br. J. Pharmacol. 108: 296-297, 1993.
10. Yoshida, T., Limmroth, V. Irikura, K and Moskowitz, M.A. (1994). J. Cerebral Blood Flow Metab.14: 924-929

Dynamic properties of intermediate filaments in cultured endothelial cells: the effects of controlled fluid shear stress.

FREDERICK W. FLITNEY[#], RODERT D. GOLDMAN[*], OMAR SKALLI[*], KWESI O. MERCURIUS[+] and PETER F. DAVIES[+]

[#]Division of Cell & Molecular Biology, University of St Andrews, Scotland KY16 9TS. [*]Cell & Molecular Biology, Northwestern University, Chicago Ill. 60611 USA. [+]SCOR in Atherosclerosis, Department of Pathology, University of Chicago, Chicago Ill 60637-1470 USA

Introduction. Fluid shear stress (FSS) is a potent stimulator of endothelial cell (EC) metabolism and gene expression [1]. The cellular mechanism(s) responsible for *sensing* altered shear stress remains uncertain. Previous studies have drawn attention to microfilament aggregates (actin stress fibres, or SFs) and their interactions with focal adhesion sites as possible mechanotransducer elements. The involvement of other components of the CSK, notably *intermediate filaments* (IFs), has been largely ignored. Here we show that ECs possess a dense, vimentin-containing IF network which extends throughout the cytoplasm and affords direct mechanical continuity between lumenal and ablumenal cell surfaces and the nucleus. Its 3D organisation is profoundly altered by subjecting ECs to FSS. Moreover, our observations offer new immunocytochemical evidence for a structural connection between actin SFs and IFs *at focal adhesion sites*, via a known [2] 300kDa IF-associated protein.

Methods. Cultured bovine aortic ECs were grown to confluence and subjected to fluid shear stress using a cone and plate type apparatus. After shearing (12 dynes.cm^{-2}; 30min-12hr) they were fixed (5-10 min) in 2% formaldehyde (EM grade) and extracted with Nonidet NP-40 detergent (0.1%, 10-15 min). The distribution of CSK elements was studied subsequently using single or double indirect immunofluorescence staining.

Results and Discussion. Immunofluorescent staining of *unsheared* ECs using a monoclonal antibody to bovine lens vimentin revealed a dense network of IFs, emanating from a cup-shaped, juxtanuclear body. Confluent (but *not* non-confluent) cells typically contained prominent, circumferentially-disposed bundles of IFs, forming a multi-stranded 'ring', reminiscent of keratin-containing tonofilaments which typify epithelial cells. They lie in close proximity to dense peripheral bands (DPBs) of actin microfilaments.

This arrangement is dramatically altered after subjecting cells to moderate shear stresses in steady laminar flow. Cytological changes show up initially as (a) an apparent 'unravelling' of the tonofilament-like bundles, forming extended whorls which intersect at nodes; and (b) loss of DPBs and the appearance throughout the cytoplasm of parallel arrays of actin SFs. After only two hours, some cells lack a discernible IF network and instead contain short, discontinuous strands or globular foci of vimentin. Following longer shear times, the cells elongate and the network reappears. The restructured network differs significantly from the original, in that IFs are less dense, they are disposed in a direction which parallels the long axis of the cell and tonofilament-like, circumferential bundles are absent.

Monoclonal antibody to an IF-associated protein of M_r 300kD, called IFAP300 [2], co-localised with IFs, as described for other cell types. However, a proportion of cells also exhibited strong additional IFAP300 immunoreactivity, in the form of fine spicules scattered over the basal cell surface. Spicules are especially prevalent towards the periphery, giving the cell a 'spiky' appearance. Double staining of cells with rhodamine-labelled phalloidin and IFAP300 antibody showed that the spicules *precisely* co-localise with the tips of actin SFs and also with a monoclonal antibody to vinculin, a known protein component of focal adhesion sites.

Shearing results in an increase in IFAP300 staining. The majority of cells exhibit spicules and their density is greater than in unsheared cells. The increase in IFAP300 staining and 'unravelling' of the tonofilament-like bundles of IFs both coincide with the loss of dense peripheral bands (DPBs) of actin and their replacement with parallel arrays of SFs. The association between IFAP300 immunoreactivity, vinculin and SF tips is retained after shearing: accordingly, IFAP300 spicules often appear in parallel linear arrays, marking both ends of the actin SFs.

The earliest detectable changes occur within *ca* 30-60mins Remodelling of the *actin* CSK occurs on a comparable time scale, though the temporal resolution of the techniques used thus far is insufficient to be certain that the two events occur simultaneously. The results strongly suggest that IFAP300 is a component of focal adhesion sites. Interestingly, IFAP300 was recently shown to be a component of hemidesmosomes and desmosomes in epithelial cells where it appears to be involved in linking IFs to both types of junction [3].

The possibility that IFAP300 serves as a 'linker' protein at focal adhesion sites is clearly important. One can speculate that the tips of actin SFs remain attached to IFs via IFAP300 throughout the restructuring process. The redistribution of one element could then occur passively, as a direct result of the other. For example, actin SFs could move by interacting with 'tracks' of myosin and thereby 'drag' IFs into their new position. The simultaneous reorganisation of actin DPBs and tonofilament-like rings of IFs, implied by our observations, would be one consequence of such a mechanism. An alternative hypothesis is that the 'linker' function of IFAP300 is lost early on, permitting the actin and IF components to reconfigure *independently* of one another. This might be achieved through phosphorylation of IFAP300. It is thought that IFAP300 functions as a 'cross-bridging' protein, stabilising the network by linking neighbouring IFs [2]. Phosphorylation appears to reduce lateral adhesion and destabilises IFs in preparation for their disassembly during mitosis [4]. A similar phosphorylation-dependent mechanism may be a neccessary prelude to restructuring of IF networks in response to flow-induced stresses in ECs.

In conclusion, we postulate that IFAP300 is an integral component of the focal adhesion complex in ECs which serves to link the actin and IF components of the cytoskeleton. This raises the possibility that IFAP300 may be involved in transmitting mechanical forces to the cytoskeleton via transmembrane proteins (integrins) which couple components of the extracellular matrix to intracellular proteins at focal adhesion sites.

Acknowledgements. FWF is supported by the British Heart Foundation.

References

1. Davies, P.F. Flow-mediated endothelial mechanotransduction (1995). Physiol. Rev. **75**, 519-560.
2. Yang, H-Y, Lieska, N. and Goldman, R.D. A 300,000 mol-wt intermediate filament associated protein in BHK-21 cells.(1985). J. Cell Biol. **100**, 620-631.
3. Skalli, O., Jones, J.C.R., Gagescu, R. and Goldman, R.D. IFAP300 is common to desmosomes and hemidesmosomes and is a possible linker of intermediate filaments to these junctions. .(1994). J. Cell Biol. **125**, 159-170.
4. Skalli, O., Chou, Y-H. and Goldman, R.D. Cell cycle-dependent changes in the organisation of an intermediate filament-associated protein: correlation with phosphorylation by p34^{cdc2}. (1992) Proc. Nat. Acad. Sci. (USA). **89**, 11959-11963.

Oxygen sensitivity of nitric oxide-mediated relaxation differs between fetal guinea pig arteries.

LOREN P. THOMPSON and CARL P. WEINER

Department of Obstetrics/Gynecology and Perinatal Research Laboratory, The University of Iowa, Iowa City, IA 52242 USA

The fetus alters its vascular resistance and cardiac output distribution in response to a reduction in oxygen delivery [1,2]. However, little is known of what regulates fetal vascular tone during hypoxemia. Nitric oxide (NO) is an important endothelium-derived modulator of vascular tone in both the fetal and adult circulations and its synthesis is dependent on oxygen. However, the effect of hypoxia and anoxia on endothelium-dependent relaxation varies among vessel beds. The role of NO in modulating fetal vascular tone and cardiac output redistribution during hypoxemia is not fully understood. Thus, the purpose of this study is to determine the effect of oxygen tension on NO-mediated relaxation among different fetal guinea pig arteries.

Methods. Adult pregnant guinea pigs were anesthetized with xylazine (1mg/kg) and ketamine (80 mg/kg; i.p.). Near term fetuses were removed from pregnant guinea pigs via a hysterotomy. Carotid and iliac arteries were excised and cut into ring segments. Arterial rings were mounted onto wires and placed in temperature-regulated tissue baths containing physiological buffer aerated with 95% oxygen and 5% carbon dioxide. Optimal length of each ring was determined with 40mM KCl. *Experimental Protocol.* Relaxation responses to cumulative addition of the endothelium-dependent relaxing agents, acetylcholine (ACh) and the calcium ionophore, A23187, and the endothelium-independent relaxing agent, sodium nitroprusside (SNP), were measured in $PGF_{2\alpha}$ contracted rings. ACh relaxation was also measured in endothelium intact and denuded rings and in the presence and absence of nitro-L-arginine (LNA; 10^{-4}M). To determine the effect of oxygen tension on relaxation, rings were aerated with either 95% O_2 or 0% O_2 by substituting 95% N_2/5%CO_2 with 95% O_2/5% CO_2. Bath PO_2 values were measured with a Strathkelvin oxygen electrode and meter and were 567±32 and 39±3 mmHg for each gas mixture, respectively. *Statistical Analysis.* Responses are mean ± SE. N values indicate the number of animals for each group. Dose response curves were compared using ANOVA with concentration and treatment as independent variables and relaxation as the dependent variable.

Results. Endothelium removal and LNA decreased maximal relaxation to ACh of both fetal carotid and iliac arteries. Lowering oxygen tension with 0% O_2 reduced the dose dependent relationship to both ACh and A23187 of fetal carotid arteries but had no effect on either maximal ACh relaxation or the -log ED50 value of fetal iliac arteries. Low oxygen tension increased the vascular sensitivity (measured by -log ED50 value) to sodium nitroprusside of both fetal carotid and iliac arteries compared to responses in high oxygen tension. Low oxygen tension had no effect on 8-bromo-cGMP mediated relaxation.

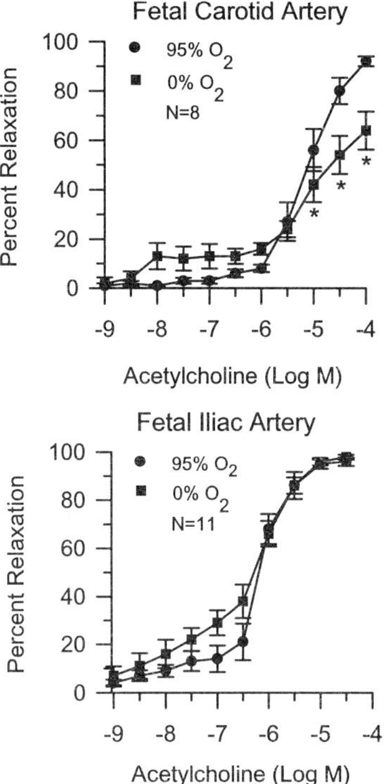

Figure 1. Relaxation to acetylcholine of fetal carotid (top) and iliac (bottom) arteries aerated with 95% O_2-5% CO_2 and 0% O_2(95% N_2)-5% CO_2. * = $P<0.05$.

Conclusions. This study demonstrates a difference in oxygen sensitivity between fetal arteries that is endothelium-dependent. Oxygen tension levels that inhibit endothelium-derived NO relaxation of fetal carotid arteries do not inhibit relaxation of iliac arteries. In contrast, low oxygen tension increased sodium nitroprusside relaxation of both fetal arteries in a similar manner. We attribute these findings to an inhibition of NO synthase activity and an increase in guanylate cyclase activity. Lowering oxygen tension had no effect on 8-bromo-cGMP relaxation. Thus, the enhanced sodium nitroprusside relaxation is unlikely due to increased cGMP sensitivity of the vascular smooth muscle. The findings of this study suggest that NO synthase activity of different fetal arteries may have different oxygen sensitivities which may provide a mechanism by which the fetal blood flow is redistributed during hypoxemia.

References

1. Calvert, S.A., Widness, J.A., OH, W., and Stonestreet, B.S. (1990) The effects of acute uterine ischemia on fetal circulation. Pediatr. Res. **27**(6), 552-556.

2. Rurak, D.W., Richardson, B.S., Patrick, J.E., Carmichael, L., and Homan, J. (1990) Blood flow and oxygen delivery to fetal organs and tissues during sustained hypoxemia. Am. J. Physiol. **258**, R116-R112.

Intrinsic basal nitric oxide production does not affect O_2 consumption or function in control or hypertrophic hearts

Harvey R. Weiss, John D. Sadoff and Peter M. Scholz

UMDNJ-Robert Wood Johnson Medical School, Piscataway, New Jersey 08854-5635 USA

Nitric oxide (NO) is produced by nitric oxide synthase (NOS) and is an endogenous stimulator of guanylate cyclase [1]. It causes increased intracellular cyclic GMP, relaxing vascular smooth muscle in heart and other organs [1,2]. Most studies report negative inotropic or metabolic effects with NO induced guanylate cyclase stimulation [2,3]. Blockade of endogenous NO may not affect myocardial O_2 consumption [4], which would imply that endogenous NO is not sufficient to affect myocardial cyclic GMP. Cyclic GMP levels have been reported to be elevated in some types of hypertrophy [5,6] and failure [7]. In aortic valve stenosis-induced left ventricular hypertrophy (LVH), there is elevated basal cyclic GMP (5). This study was designed to test the hypothesis that basal endogenous myocardial NO production is a major controller of myocardial cyclic GMP and that this is enhanced in hypertrophy.

Methods

This study was performed in 14 dogs. In 5 dogs, LVH was created using aortic valve plication [5] and the animals were studied 6 months later. Dogs were anesthetized and ventilated. An ultrasonic flow probe was placed on the left anterior descending coronary artery (LAD). A diagonal branch and an adjacent vein were cannulated. Ultrasonic dimension crystals were inserted. Miniature force gauges were placed adjacent to the length crystals. Baseline measurements were recorded. Biopsies were obtained. Acetylcholine (1 μg) was infused into the LAD. N^G-nitro-L-arginine methyl ester [L-NAME, N=5, control; N=5, LVH (6 mg/kg)] or N^G-monomethyl-L-arginine [L-NMMA, N=4, control (3 mg/kg)] were infused. Measurements were obtained and acetylcholine (to test NOS blockade) was infused. The heart was frozen in liquid nitrogen. Regional segment work was calculated by determining the area under the force-length loop. Regional myocardial O_2 consumption was calculated from flow, O_2 extraction and hemoglobin data. Samples were analyzed to determine cyclic GMP levels using a radioimmunoassay.

Results

There was a gradient between systolic left ventricular pressure (162±13 mmHg) and arterial pressure (116±6) in LVH. Left ventricular pressure was also higher than control (111±8). In both groups, intracoronary infusion of L-NAME had no significant effects on heart rate, arterial blood pressure, left ventricular pressure or left ventricular dP/dt_{max}. Before blockade, acetylcholine caused a significant increase in LAD flow in both controls and LVH. After L-NAME, LAD flow was not significantly increased with acetylcholine.

L-NAME did not affect regional myocardial mechanics, e.g., segment force, % shortening and segment work, Table 1. L-NAME caused local vasoconstriction and increased O_2 extraction. This was accompanied by a drop in the venous O_2 saturation from 56±3 to 50±3 %. Local O_2 consumption remained unchanged after NOS blockade. Cyclic GMP was not changed after L-NAME, but cyclic GMP was significantly higher in LVH. L-NMMA had no effect on hemodynamics in control dogs. L-NMMA did not change LAD work (1794±460 to 1578 g*mm*min^{-1}), peak force or % shortening, O_2 consumption (6.6±1.5 to 6.2±1.2 ml O_2/min/100g), O_2 extraction or LAD flow. Cyclic GMP levels in the LAD region were also unaltered after L-NMMA (1.2±0.1 to 1.1±0.3 pmol/g).

Discussion

The results of this study demonstrated that blockade of basal nitric oxide production with L-NAME and L-NMMA was not associated with changes in cardiac hemodynamics, myocardial force, shortening, regional work, cyclic GMP or O_2 consumption. The NOS blockade did lead to local vascular effects. Nitric oxide relaxes smooth muscle and increases cyclic GMP [1,2]. We observed significant vasoconstriction as a direct result of NO reduction. We confirmed NO blockade by showing decreased vasodilatation in response to acetylcholine. In cardiac myocytes, altering cyclic GMP also appears to affect metabolism and function.

TABLE 1. Effects of intracoronary L-NAME on Local myocardial parameters in control and LVH dogs.

	CONTROL		LVH	
	Baseline	L-NAME	Baseline	L-NAME
Peak Force (g)	11.5±1.2	10.8±1.6	10.3±0.3	10.3±0.4
% Shortening	11.6±3.1	10.9±2.3	10.3±2.3	9.4±2.2
Segment Work (g*mm*min)	1710±566	1476±477	1268±264	1250±318
LAD Flow (ml/min/ 100g)	87.6±9.3	75.7±11.1*	102.3±20.3	92.5±19.4*
O_2 Extraction (ml O_2/ 100ml)	8.9±0.9	10.4±1.1*	10.8±0.4	12.1±0.6
O_2 Consumption (ml O_2/min/ 100g)	7.7±1.0	7.7±1.3	11.3±2.7+	11.4±2.5+
Cyclic GMP-LAD (pmol/g)	1.3±0.2	1.5±0.5	2.6±0.7+	3.3±0.9+

*different from pre. +different from control group.

Negative inotropic and metabolic responses occur after either increased NO or cyclic GMP in both isolated hearts and myocytes [2,3]. From the present results, it appears that basal endogenous NO production is not sufficient to affect either myocardial function or cyclic GMP.

In our LVH model, basal cyclic GMP levels were markedly elevated. Myocardial cyclic GMP levels are elevated in pressure-overload hypertrophy [5,6] and heart failure [7]. We found normal cyclic GMP-phosphodiesterase activity in our LVH model [5]. We determined that increased NO synthase activity was not the cause of the increased cyclic GMP with LVH.

If the basal nitric oxide was sufficient to affect myocytes *in vivo*, blockade of NOS should lead to a decrease in myocardial cyclic GMP [2]. This change in cyclic GMP would change myocardial function and/or metabolism [3,5]. We found no effect of NOS blockade on cyclic GMP levels and also no metabolic or functional changes. Since we demonstrated NOS blockade, this implies that endogenous basal production of NO is not sufficient to alter myocardial function in control or hypertrophic hearts. (Supported by U.S.P.H.S. grant HL40320).

References

1. Moncada, S., Palmer, R.M.J., and Higgs, E.A. (1991) Pharmacol. Rev. **43**, 109-142
2. Lohmann, S.M., Fischmeister, R., and Walter, U. (1991) Bas. Res. Cardiol. **86**, 503-514
3. Brady, A.B., Warren, J.B., Poole-Wilson, P., Williams, T., and Harding, S.E. (1993) Am. J. Physiol. **265**, H176-H182
4. Gurevicius, J., Salem, M.R., Metwally, A.A., Silver, J.M., and Crystal, G.J. (1995) Am. J. Physiol. **268**, H39-H47
5. Roitstein, A., Kedem, J., Cheinberg, B., Weiss. H.R., Tse, J., and Scholz, P.M. (1994) J. Surg. Res. **57**, 584-590
6. Nichols, J.R., and Gonzalez, N.C. (1982) J. Molec. Cell. Cardiol. **14**, 181-183
7. Jakob, G., Mair, J., Pichler, M., and Puschendorf, B. (1995) Brit. Heart J. **73**, 145-150

Nitric Oxide Antagonizes Pressure-Flow Autoregulation in the Superior Mesenteric Artery

M. PAULA MACEDO and W. WAYNE LAUTT

Department of Pharmacology and Therapeutics, Faculty of Medicine, University of Manitoba, Winnipeg, Manitoba, Canada, R3E OW3.

INTRODUCTION: The intrinsic ability of an organ to maintain constant blood flow through a wide range of perfusion pressures is referred to as autoregulation [1,2]. Autoregulatory mechanisms lead to vasoconstriction in response to elevated blood flow. However, an increase in blood flow has also been shown to cause increased shear stress with consequent increased nitric oxide release and vasodilation in the superior mesenteric artery as well as other tissues [3,4]. The present investigation was carried out to investigate the possible role of nitric oxide to antagonize autoregulation.

METHODS: Fasted cats of either sex (n=5) were anesthetized with sodium pentobarbital (32.5 mg/kg ip). Anesthesia was maintained by i.v. infusion (6 ml pentobarbital/500 ml dextran) adjusted as required. Body temperature was maintained at 37.5°C with thermal control unit regulating a heated surgical table. Arterial pressure and central venous pressure were monitor via catheters in a carotid artery and a femoral vein. Laparotomy, splenectomy and occlusion of the inferior mesenteric artery were performed. The superior mesenteric artery was isolated and its nerve plexus ligated and cut. Both femoral arteries were cannulated and connected to the superior mesenteric artery through a pump-controlled long circuit previously described [5].

Autoregulation was determined by a decrease in pressure from 140 to 75 mmHg. This protocol was repeated in the control and in the presence of the nitric oxide synthase inhibitor, NG-nitro-L-arginine methyl ester (L-NAME) (2.5 mg/kg i.v. bolus). Autoregulation was assessed by: ARI = $1 - [SMAF_C - SMAF_R/SMAF_C] / [SMAP_C - SMAP_R/SMAP_C]$ where SMAF refers to superior mesenteric arterial flow, SMAP to superior mesenteric arterial pressure and the subscripts C and R refer to control levels and reduced levels respectively.

Data analysis was expressed as mean $\pm$ S.E. Statistical analysis was by the student t-test. Significance was accepted at $P<0.05$.

RESULTS: The decrease in flow per unit decrease in pressure was shown to be of a lesser extent after L-NAME when compared to the control state and this effect was reversed by L-arginine. Autoregulation was assessed using the autoregulatory index (ARI) expressed as the ratio of percent change in flow to percent change in pressure. L-NAME significantly increased autoregulation from an ARI of 0.07 $\pm$ 0.04 to 0.32 $\pm$ 0.08 ($P< 0.03$).

DISCUSSION: Our experiments demonstrated that autoregulatory responses in the superior mesenteric artery were augmented after treatment with the inhibitor of nitric oxide synthase, L-NAME. Recently similar findings were reported in the heart [4]. Endothelial cells release nitric oxide in response to an increase in shear stress [6]. Since shear stress is linearly related to flow rate [7], a flow-induced nitric oxide release mechanism is more prominent at higher flow rates such that, as blood flow increases, shear stress increases thus releasing nitric oxide and promoting vasodilation. This proposed mechanism is consistent with our hypothesis that nitric oxide antagonizes autoregulation in the superior mesenteric artery over an autoregulatory range.

ACKNOWLEDGEMENTS: This work was funded by the Medical Research Council of Canada and the Manitoba Heart & Stroke Foundation. Manuscript preparation was by Karen Sanders.

REFERENCES:

1. Johnson, P.C. (1986) Autoregulation of blood flow. Circ. Res. 59, 483-495.
2. Lang, D.J. and Johnson, P.C. (1988) Elevated ambient oxygen does not affect autoregulation in cat mesentery. Am. J. Physiol. 255, H131-H137.
3. Macedo, M.P. and Lautt,W.W. (1994) Nitric oxide suppression of norepinephrine release from nerves in the superior mesenteric artery. Proc. West. Pharmacol. Soc. 37, 103-104.
4. Kuo, L., Chillian, M.C., and Davis, M.J. (1991) Interaction of pressure and flow-induced responses in porcine coronary resistance vessels. Am. J. Physiol. 261, H1706-H1715.
5. Lockhart, L.K. and Lautt, W.W. (1990) Hypoxia-induced vasodilation of the feline superior mesenteric artery is not adenosine mediated. Am. J. Physiol. 259, G605-G610.
6. Smiesko, V., Lang, D.J., and Johnson, P.C. (1989) Dilator response of rat mesenteric arcading arterioles to increased blood flow velocity. Am. J. Physiol. 257, H1958-H1965.
7. Kamika, A. and Togawa, T. (1980) Adaptative regulation of wall shear stress to flow change in the canine carotid artery. Am. J. Physiol. 239, H14-H21.

Shear-Induced Modulation by Nitric Oxide of Sympathetic Nerves in the Superior Mesenteric Artery

M. PAULA MACEDO and W. WAYNE LAUTT

Department of Pharmacology & Therapeutics, Faculty of Medicine, University of Manitoba, Winnipeg, Manitoba, Canada, R3E OW3.

INTRODUCTION: Nitric oxide has been implicated in the mechanism of flow-induced dilation dependent upon shear stress on endothelial cells [1-3]. Vascular endothelial cells appear to be involved in the modulation of vascular responses to sympathetic nerve stimulation [4-7]. The purpose of this study was to test the hypothesis that endothelium derived nitric oxide (NO) release is promoted by an increase in shear stress and that nitric oxide diminishes the vasoconstriction promoted by sympathetic nerve endings in an *in vivo* preparation.

METHODS: Fasted cats of either sex were anesthetized with sodium pentobarbital (32.5 mg/kg ip). Anesthesia was maintained by i.v. infusion (6 ml pentobarbital/500 ml dextran) adjusted as required. Body temperature was maintained at 37.5°C with a thermal control unit regulating a heated surgical table. Arterial pressure and central venous pressure were monitored via catheters in a carotid artery and a femoral vein. Laparotomy and occlusion of the inferior mesenteric artery were performed. The superior mesenteric artery was isolated, its nerve plexus separated and placed in a stimulating electrode. Both femoral arteries were isolated, cannulated, and connected to the superior mesenteric artery through a pump controlled long circuit as previously described [8]. This circuit allowed control and measurements of blood flow by the use of an electromagnetic flow probe (Carolina Medical Electronics EP408). Change in shear stress status was achieved by controlling the circuit blood flow. Increase in shear stress was achieved by maintaining the flow into the superior mesenteric artery constant during vasoconstriction induced by stimulating the sympathetic nerves. Shear stress was held constant by decreasing the blood flow to achieve constant mesenteric arterial pressure during sympathetic nerve stimulation.

PROTOCOL: Sympathetic nerve stimulation (2 Hz, 15 V square pulse, 1 msec duration) was performed under conditions of constant flow or constant pressure achieved by adjusting the perfusing circuit. Responses to nerve stimulation were determined in a control condition, following intravenous injection of NG-nitro-L-arginine methyl ester (L-NAME) (2.5 mg/kg), and after intravenous injection of L-arginine (75 mg/kg) to reverse the action of L-NAME.

DATA ANALYSIS: Results are expressed as $\pm$ SE. Statistical analysis was made using analysis of variance (ANOVA) for multiple comparasions followed by Tukey's (Honestly Significant Difference). A probability of less than or equal to 0.05 was considered significant.

RESULTS: Under constant flow conditions (shear stress increased), 2 Hz nerve stimulation promoted an increase in perfusion pressure of 24.8 $\pm$ 4.7 mmHg in the control state which represents an increase of 30.8 $\pm$ 4.9% and 74 $\pm$ 21.6 mmHg representing an increase of 88.7 $\pm$ 28.5% in the presence of L-NAME ($n=6$, $p<0.006$). When shear stress remained constant, under constant pressure perfusion, the constriction promoted by sympathetic nerve stimulation resulted in reduction of blood flow. In this condition the reduction in blood flow in response to 2 Hz nerve stimulation was not different in the control and in the presence of L-NAME (5.5 $\pm$ 2.1 and 5.2 $\pm$ 1.7 ml/Kg/min respectively).

DISCUSSION: Our studies provide evidence that nitric oxide may play an important role in modulating the sympathetic nerve plexus of the superior mesenteric artery. Previous studies have also suggested that inhibition of nitric oxide increases sympathetic neural activity in the kidney [9]. Others showed that with perfused pulmonary vessels, isolated segments of both rabbit carotid artery and rat tail artery, the responses to nerve stimulation were enhanced in the absence of the endothelium [4-6]. Modulation of nerve-induced constriction by a shear-dependent release of NO may serve a protective function in that a purely regional constriction will result in reduced flow (arterial pressure remaining constant) with no rise in shear stress at the site of constriction. A generalized constriction at multiple organs will, in contrast, result in a rise in perfusion pressure and elevated shear stress at the site of constriction. The shear stress leads to release of NO and suppression of the vasoconstriction thus affording protection to the endothelium.

ACKNOWLEDGEMENTS: We are grateful for the superb technical and analytical assistance of Dallas J. Legare. The authors thank Dr. Clive Greenway for his critical comments and suggestions. This work was funded by the Medical Research Council of Canada and the Manitoba Heart & Stroke Foundation. Manuscript preparation was by Karen Sanders.

REFERENCES:

1. Buga, G.M., Gold, M.E., Fukuto, J.M., and Ignarro, L.J. (1991) Shear stress-induced release of nitric oxide from endothelial cells grown on beads. Hypertension 17, 187-193.
2. Lamontagne, D., Pohl, U., and Busse, R. (1992) Mechanical deformation of vessel wall and shear stress determine the basal release of endothelium-derived relaxing factor in the intact rabbit coronary vascular bed. Circ. Res. 70, 123-130.
3. Smiesko, V., Lang, D.J., and Johnson, P.C. (1989) Dilator response of rat mesenteric arcading arterioles to increased blood flow velocity. Am. J. Physiol. 257, H1958-H1965.
4. Cohen, R.A. and Weisbrod, R.M. (1988) Endothelium inhibits norepinephrine release from adrenergic nerves of rabbit carotid artery. Am. J. Physiol. 254, H871-H878.
5. Greenberg, S., Diecke, F.P.J., Peevy, K., and Tanaka, T.P. (1989) The endothelium modulates adrenergic neurotransmission to canine pulmonary arteries and veins. Eur. J. Pharmacol. 162, 67-80.
6. Thorin, E. and Atkinson, J. (1994) Modulation by the endothelium of sympathetic vasoconstriction in an in vitro preparation of the rat tail artery. Br. J. Pharmacol. 111, 351-357.
7. Yasuhiro, E., Matsumara, Y., Murata, S. et al. (1994) The effect of N-nitro-L-arginine, a nitric oxide synthetase inhibitor, on norepinephrine overflow and antidiuresis induced by stimulation of renal nerves in anesthetized dog. J. Pharmacol. Exp. Ther. 269, 529-535.
8. Lockhart, L.K. and Lautt, W.W. (1990) Hypoxia-induced vasodilation of the feline superior mesenteric artery is not adenosine mediated. Am. J. Physiol. 259, G605-G610.
9. Sakuma, I., Togashi, H., Yoshioka, M. et al. (1992) N-Methyl-L-arginine, an inhibitor of L-arginine-derived nitric oxide synthesis, stimulates sympathetic nerve activity in vivo. Circ. Res. 70, 607-611.

CAPILLARY microvasculature of the heart is highly sensitive for NO in vivo: Autonomic regulation of capillary diameter by stimulation of endogenous NO release and exogenously supplied NO.

WILHELM BLOCH, LUMIR KOPALEK, DIRK REITZE and KLAUS ADDICKS

Department of Anatomy, University of Cologne, 50931 Cologne, FRG

introduction

The uneven distribution of endothelial NOS along the coronary vascular tree[1] suggests a differential role for endothelium-derived NO in the regulation of conductance and resistance arteries, veins and capillaries. Recent studies propose an active regulation of the capillary diffusion area in the heart independent from flow regulation[2]. The aim of the present study was to investigate the influence of basal and stimulated NO-release and exogenously supplied NO on the NOS-free capillaries.

material and method

For isolated heart perfusion, male wistar rats (350-450g) were anaestetized with carbon dioxide (CO_2) and were then killed by cervical dislocation. Subsequently the heart was quickley excised and retrogradely perfused via the aorta according to the Langendorff technique. The perfusion pressure was kept constant at 60cm H_2O with a modified Krebs Henseleit solution prewarmed to 37°C and equilibrated with 95% O_2/5% CO_2. The hearts were allowed to beat spontaneously. Experimental groups with animals matched in age and weight were created and compared with corresponding controls receiving saline only. The experimental and control groups were equilibrated for 20 minutes. After the equilibration period coronary flow was registrated for 10min respectively 25min. The switch over time for substance was the 25th minute point with exception of the groups receiving double treatment. In double treated groups, the first substance was added at the 10th minute and the second at the 25th minute. In order to study the effect of endogenously produced and exogenously administered NO of coronary flow and capillary diameter L-NA (10^{-4}M), Bradykinin(10^{-7}M), GTN ($5*10^{-5}$M) and SNP(10^{-4}M-10^{-9}M) were administered.

Abbreviations used: L-nitroarginine (L-NA), sodium nitroprusside (SNP), glyceroltrinitrate (GTN)

Following the pressure constant Langendorff perfusion the hearts were perfusionfixed at 60cm H_2O with 0.1M cacodylate buffered 2% glutaraldehyde/2% paraformaldehyde. The left papillary muscle was removed and subsequently fixated in the same fixative followed by postfixation in 2% osmium tetroxide buffered at pH 7.3 with 0.1M sodium cacodylate for 2 hours at 4°C. The specimens were rinsed in cacodylate buffer for three times in 1% uranyl acetate in 70% ethanol for 8 hours, dehydrated in a series of graded ethanol and embedded in araldite. Semithin sections of plastic embedded specimens were cut with a glass knife on a Reichert ultramicrotome and stained with methylene blue.

Morphometric data was collected on randomly sampled semithin transverse sections from the left papillary muscle which consited exclusively of muscle fibres and exchange vessels with a maximal diameter of 26μm. The papillary muscles are particulary suitable for a stereological analysis, since an axis of anisotropy can be detected. Capillary diameters were automatically detected and measured on a Leitz Medilux microscope connected with a Leitz CBA8000 image analysing system. The capillary diameter of 400 capillaries, was measured by recording the smallest profile diameter as the closest approximation of the true diameter. Statistical analysis was performed using adjusted t test with p values corrected by the Bonferroni method.

results

Administration of L-NA reduced coronary flow up to 40% in comparison to control but had no effect on capillary diameter (control 4,5±0,20; L-NA 4,68±0,14μm). Stimulation of endogenous NO-release by bradykinin for 5min increased coronary flow and capillary diameter (5,80±0,25μm). The increase of capillary diameter against control (4,5±0,2μm) by Bradykinin was prevented by pretreatment with L-NA (3,52±0,21μm). Bradykinin treatment for 20min resulted in a coronary flow equally to control and a decrease of capillary diameter (3,98±0,4μm). The spontaneous NO-donor SNP increased significantly the coronary flow in concentrations from 10^{-4}M to 10^{-7}M but capillary diameter was significantly increased from 10^{-4}M (5,15±0,15μm) to 10^{-9}M (5,1±0,1μm) (fig.1). GTN, a NO-donor not releasing NO in the capillaries increased coronary flow in a concentration ranging from 5*10-5M to 45% equal to the flow increase of SNP in concentrations ranging from 10-5M to 10-6M but had no effect on capillary diameter (4,61±0,31μm).

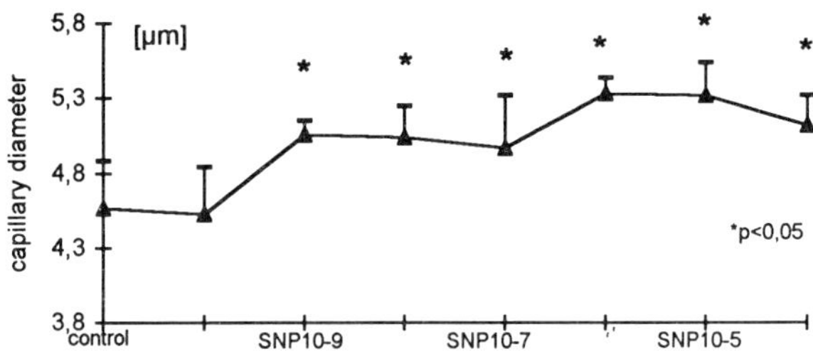

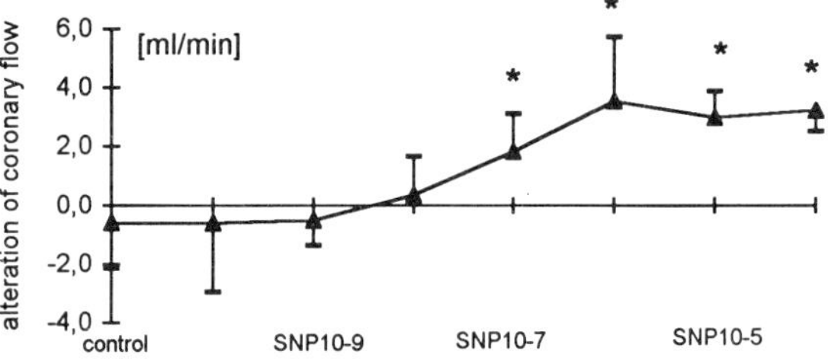

Figure 1 (A) Alteration of capillary diameter of isolatedly perfused hearts under NNP treatment in concentrations from 10^{-10}M to 10^{-4}M. (B) Difference of coronary flow before treatment (24.min) to coronary flow after treatment (44.min). Results are presented as the means ± SD of six to fourteen individual experiments per group.

discussion

A directly regulative effect of NO on intracardial capillaries independent from flow regulation has been demonstrated within this study using a combination of functional experiment and morphometrical measurement. Basal NO-release has no effect on capillaries. The results under basal release are consistent with the spacial distribution of NOS in the coronary vessels: No NOS can be found in the capillary endothelium [1]. In contrast an endogenous NO-release stimulated by bradykinin which exceeds the basal endogenous NO-release has a regulatory effect on the capillary diffusion area. Therefore a luminal overflow of NO in the upstream vessels, located high enough in order to influence the capillaries, will be the consequence. The distinct increase of capillary diameter as a result of short time treatment with bradykinin provides evidence of the existence of a highly sensitive autoregulation of the capillaries mediated by NO. NO-independent Bradykinin has a contractile effect on capillaries in accordance with the observed contractile effect which bradykinin has on endothelial cells in culture [3]. Our findings suggest a regulation for local tissue perfusion [4] and an autoregulation of blood flow in microvessels [3]. These findings are supported by the results obtained using exogenously supplied NO-donors which release NO through a different mechanism and at a different site [5,6] within the vessels. Local differences concerning the release of unstable NO by SNP and GTN are responsible for variations in capillary diameter regulation leading to a lack of capillary diameter increase in the GTN treated groups in contrast to the distinct increased coronary flow. The SNP results demonstrated that capillaries are 10^{-2}M times more sensitive to NO than the resistance vessels in the coronary bed. The high sensitivity of capillary microvasculature for NO and the local differences concerning the effect of NO-donors provides insights in the mechanisms of capillary autoregulation by NO and enables a search for new therapeutic approachs in microvasculare disease.

references

1. Addicks, K., Bloch, W. and Feelisch M. (1994) Microsc.Res.Tech. 29, 161-168
2. Bloch, W., Hoever, D., Reitze, D., Kopalek, L. and Addicks, K. (1995) Agents and Actions Suppl. 45, 151-156
3. Morel, N.M.L., Dodge, A.B., Patton, W.F., Herman, I.M., Hectman, H.B., and Shepro, D. (1989) J.Cell Physiol. 141, 653-659
4. Ragan, D.M.S., Schmidt, E.E., MacDonald, I.C. and Groom, A.C. (1988) Microvasc.Res. 36, 13-30
5. Feelisch, M. and Noack, E. (1987) Eur.J.Pharmacol. 142, 465-469
6. Sellke, F.W., Myers, P.R., Bates, J.N. and Harrison, D.G. (1990) Heart Circ.Physiol. 27, H515-H520

Endogenous and exogenous nitric oxide inhibits norepinephrine release from rat heart sympathetic nerves

PETRA SCHWARZ, RICARDA DIEM, NAE J. DUN* and ULRICH FÖRSTERMANN

Department of Pharmacology, Johannes Gutenberg University, 55101 Mainz, Germany, *Department of Anatomy and Neurobiology, Medical College of Ohio, Toledo, OH 43614

Nitric oxide (NO) is synthesized by three different isoforms of NO synthase (NOS) [for review see 1, 2]. NOS I was originally isolated from rat and porcine cerebellum [3], but immunohistochemical studies have also localized this isoform to sympathetic and parasympathetic neurons [4, 5].

NO is known to influence sympathetic neurotransmission in various vascular preparations, however the implications of prejunctional and postjunctional effects are controversial [6, 7].

In the rat heart, NO released by degranulating mast cells was regarded as a permissive mediator that enabled coreleased histamine to increase the evoked norepinephrine (NE) overflow via activation of prejunctional, facilitatory H_2 receptors on sympathetic nerve terminals [8].

In this study on the isolated perfused rat heart, we demonstrate a prejunctional inhibition of NE release by endogenous and exogenous NO. Persistence of this NO inhibition after functional damage of the endothelium with the non-denaturing detergent 3-[(3-cholamidopropyl)-dimethylammonio]-1-propanesulfonate (CHAPS) indicates that significant amounts of the NO are of non-endothelial origin. Immunohistochemistry localized NOS I and tyrosine hydroxylase (TH) to the same or adjacent neurons showing some overlap.

Effects of N^G-Nitro-L-arginine (L-NNA), N^G-Methyl-L-arginine (L-NMA) and [Arg⁸]-Vasopressin on Evoked NE Overflow

Hearts were perfused in the Langendorff mode with Tyrode's solution at a constant rate of 5 mL/min. The right sympathetic nerve was stimulated with trains of 3 Hz (270 pulses) and NE release was measured by HPLC.

The NOS inhibitor L-NNA, added in increasing concentrations (30, 100 and 300 µmol/L) 24.5 min before the sympathetic nerve stimulation periods SNS3, SNS4 and SNS5, enhanced significantly the evoked fractional NE release (S) in a concentration-dependent manner (30 µmol/L: 42.2±4.6%, 100 µmol/L: 87.2±8.6%, 300 µmol/L: 106.8±11.4%, n=4). The basal outflow of NE was not affected by any concentration of L-NNA. The effect of L-NNA (0.1 mmol/L) was stereospecifically prevented by L-arginine (1 mmol/L), but not by D-arginine (1 mmol/L) (Fig 1). Similar to L-NNA, the NOS inhibitor L-NMA (0.1 mmol/L) significantly enhanced stimulation-evoked NE overflow to 117.9±11.1% (n=4, p<.05; for control see Fig 1).

To mimic the vasoconstrictor effects of the two NOS inhibitors, 0.1 U/L [Arg⁸]-vasopressin was added 24.5 min before SNS3. This substance increased coronary perfusion pressure (CPP) by 113±9 mm Hg (mean±SEM, n=4) over the experimental period of 40 min. However, evoked NE overflow remained unchanged (67.9±5.5%; for control see Fig 1).

These findings suggest that the observed facilitation of evoked NE release by the NOS inhibitors is *not* due to non-specific effects such as ischemia.

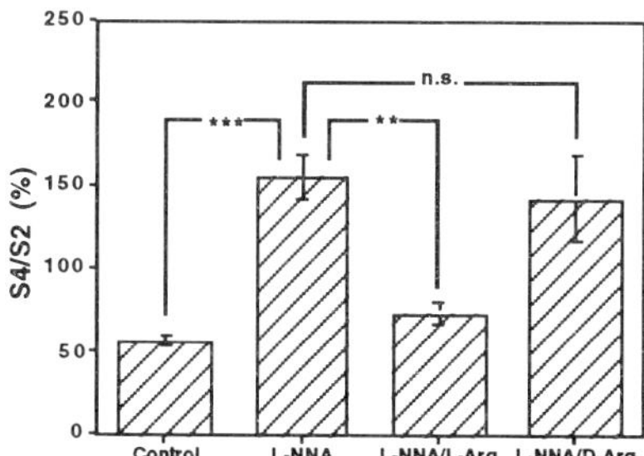

FIG 1. The effect of L-NNA (0.1 mmol/L) on the NE overflow evoked by 3 Hz (270 pulses). L-NNA was tested alone and in the presence of L-arginine (L-Arg, 1 mmol/L) or D-arginine (D-Arg, 1 mmol/L). Data represent mean±SEM (n=4). Significant differences (as calculated by factorial test) are indicated (***p<.001, **p<.01, n.s.=not significant).

Effects of S-Nitroso-N-acetyl-D,L-penicillamine (SNAP) and Linsidomine (SIN-1) on Evoked NE Overflow

The NO-donor compound SNAP (1 nmol/L to 10 µmol/L), when infused into the coronary system starting 5 min before SNS3, SNS4 and SNS5, inhibited the evoked NE overflow in a concentration-dependent manner (in the presence of 0.1 mmol/L L-NNA). The effect of SNAP (1 µmol/L) is shown in Fig 2. SIN-1, another NO-donor compound, produced a similar inhibition (Fig 2).

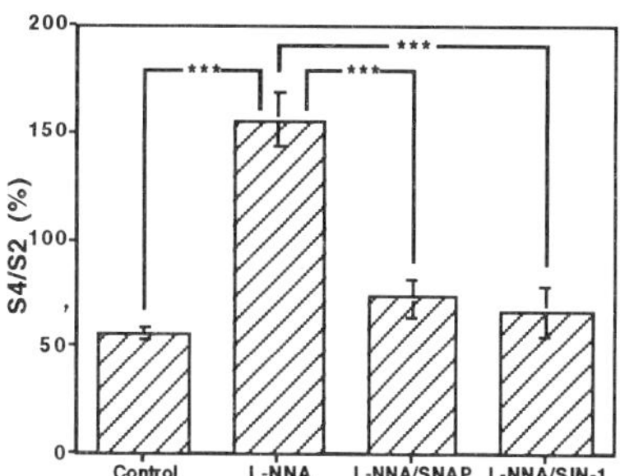

FIG 2. The effect of SNAP (1 µmol/L) and SIN-1 (1 µmol/L) on the NE overflow evoked by 3 Hz (270 pulses) in the presence of L-NNA (0.1 mmol/L). Data represent mean±SEM (n=4). Significant differences (as calculated by factorial ANOVA followed by Fisher's protected-least-significant difference test) are indicated (***p<.001).

Impairment of Endothelial NO Release Does not Alter NE Overflow

Perfusion of the hearts with the non-denaturing zwitterionic detergent CHAPS for 10 min increased CPP by about 30 mm Hg, which is compatible with a functional impairment of endothelial NO release. This treatment converted an endothelium-dependent vasodilation to ACh (0.1 µmol/L) into a constriction. At the same time, cardiac contractility and rhythm were not significantly changed.

Treatment of the hearts with CHAPS (1 mmol/L) did not alter basal outflow of NE compared with untreated hearts (63.8±8.3 pg/min versus 59.5±9.0 pg/min; mean±SEM, n=8). Also, the release of NE in response to electrical stimulation did not vary from that of control experiments. Furthermore, in CHAPS-treated hearts, L-NNA (0.1 mmol/L) still facilitated significantly the fractional overflow of NE, pointing to *non*-endothelial NO as being responsible for the repression of NE release.

Immunochemical Characterization of NOS Isoform

Western blotting of proteins from the whole rat heart (partially purified on 2',5'-ADP-Sepharose) demonstrated weak bands for neuronal NOS I at 160 kDa and endothelial NOS III at 135 kDa. No immunoreactivity for the inducible NOS II was detected in these normal hearts.

Double-labeling immunofluorescence using antisera to NOS I and TH demonstrated NOS I- and TH-immunoreactivities in the same or adjacent postganglionic neurons of the rat stellate ganglion and in nerve fibers innervating the right atrium of the heart.

We conclude that NO of predominantly neuronal origin exerts a negative control over the evoked NE release from sympathetic nerves in the rat heart. This may occur in a paracrine fashion (with NO being generated in nitrergic nerves adjacent to the sympathetic nerves) or in an autocrine fashion (with NO being an atypical cotransmitter in sympathetic neurons themselves). A potential functional significance of this control mechanism for cardiac function *in vivo* remains to be determined.

References

1. Förstermann U, Closs EI, Pollock JS, Nakane M, Schwarz P, Gath I, Kleinert H. *Hypertension* 1994; 23: 1121-1131
2. Nathan C, Xie QW. *Cell* 1994; 78: 915-918
3. Bredt DS, Snyder SH. *Proc Natl Acad Sci USA* 1990; 87: 682-685
4. Dun NJ, Dun SL, Wu SY, Förstermann U. *Neurosci Lett* 1993; 158: 51-54
5. Dun NJ, Dun SL, Wu SY, Förstermann U, Schmidt HHHW, Tseng LF. *Neuroscience* 1993; 54: 845-857
6. Bucher B, Ouedraogo S, Tschopl M, Paya D, Stoclet JC. *Br J Pharmacol* 1992; 107: 976-982
7. Greenberg S, Diecke FPJ, Peevy K, Tanaka TP. *Eur J Pharmacol* 1989; 162: 67-80
8. Fuder H, Ries P, Schwarz P. *Fundam Clin Pharmacol* 1994; 8: 477-490

Nitric Oxide Attenuates the Lusitropic Response to β-Adrenergic Receptor Stimulation in Humans

JOSHUA M. HARE, MICHAEL M. GIVERTZ, MARK A. CREAGER, and WILSON S. COLUCCI

Brigham and Women's Hospital, and Harvard Medical School, Boston, MA 02115

Introduction: NOS inhibitors augment the positive inotropic response to βAR stimulation in myocytes [1], dogs [2], and humans with heart failure [3]. Myocardial NOS activity appears to be increased in HF and may contribute to diminished βAR responses associated with this condition [4]. Since βAR modulation of both systole and the active phase of diastole is regulated by the production of cAMP, we hypothesized that NOS inhibition would also accentuate the lusitropic response to βAR agonists.

Methods: Left ventricular pressures and τ, the time constant of LV relaxation, were measured in 17 patients during atrial pacing using a high-fidelity micromanometer (Millar). τ was calculated from the digitized LV pressure signal using the method of Weiss [5]. Eleven patients had HF (EF, 23 ± 3 %) and 6 had normal LV function (EF, 69 ± 4%). All subjects were free of coronary artery disease and in normal sinus rhythm. The βAR agonist dob was infused via a peripheral vein before and during intracoronary infusion of the NOS inhibitor L-NMMA (20 μmol/min for 5 min). L-NMMA was infused directly into the left main coronary artery to avoid systemic pressor responses. Data are expressed as mean ± SEM.

Results: In HF and normals combined, dob decreased τ from 57.0 ± 4.0 to 45.4 ± 3.8 ms (-21 ± 2%, $p<0.0001$). Intracoronary L-NMMA further decreased τ to 41.7 ± 3.1 ms, a 38 ± 14% ($p<0.005$) accentuation of the dob-stimulated lusitropic effect. This enhancement was present in HF (26 ± 11%), and was not statistically different from the effect in normals (59 ± 35%) (Figure 1).

Discussion: These data show that inhibition of NOS in humans accentuates the positive lusitropic effect of βAR stimulation. L-NMMA when administered by the intracoronary route enhanced the ability of dob, a βAR agonist, to accelerate τ, a measure of the rate of LV relaxation. We have previously observed that in HF the βAR lusitropic response is preserved [6]. This is in contrast to a diminished βAR inotropic response associated with HF, and suggests a differential sensitivity of these two phases of the cardiac cycle to cAMP-mediated pathways [6]. As with the maintained βAR lusitropic response in HF, the NO-mediated influence on βAR lusitropic response was also preserved, and was similar to normals.

We have previously shown that L-NMMA enhances the hyporesponsive positive inotropic response to dob in

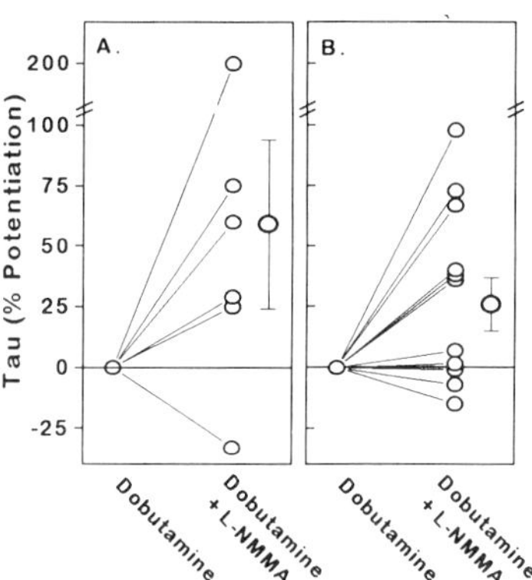

Figure1: Plots show the potentiation of the positive lusitropic response to dobutamine by an intracoronary infusion of L-NMMA in patients with normal LV function (A.) and with heart failure (B.). Depicted are the individual responses for the combination of dobutamine and L-NMMA relative to dobutamine alone.

patients with HF [3], possibly reflecting increased NO in HF. Taken together, these observations are consistent with an influence of NO on divergent βAR signaling pathways that regulate systolic and diastolic function. It is also possible that the accentuation of the rate of diastolic relaxation is due to ventricular mechanical properties (chamber geometry, elasticity, or increased systolic recoil), and not related to biochemical regulation.

The effect of NO on diastole in humans has previously been explored by Paulus et al., who demonstrated that intracoronary administration of nitroprusside (without βAR stimulation) hastens the onset of diastole and improves diastolic distensibility without affecting τ in humans [7]. Thus, it appears that NO may have differential effects in early vs. late diastole, decreasing the rate of relaxation while increasing its extent. The difference in direction of these effects may also reflect the significantly different experimental conditions, in which endogenous NO production is inhibited as opposed to when exogenous NO-donors are administered. In this regard, Shah and colleagues have shown in isolated rat myocytes that the NO effector, 8-bromo-cGMP, directly reduced the sensitivity of the contractile apparatus to Ca^{2+} and increased diastolic cell length [8].

Acknowledgments: This work was supported in part by grants K08 HL03238-01 and P50 HL52320-01 from the N.I.H. and a Physician Investigator Fellowship (JMH) from the American Heart Association, Massachusetts Affiliate.

References:

1. Balligand, J.-L., Kelly, R.A., Marsden, P.A., Smith, T.W., and Michel, T. (1993) Proc. Natl. Acad. Sci. USA **90**, 347-351
2. Hare, J.M., Keaney, J.F. Jr., Kelly R.A., Loscalzo, J., Smith, T.W., and Colucci, W.S. (1993) Circulation **88**:I-1905
3. Hare, J.M., Loh, E., Creager, M.A., Colucci, W.S. (1995) Circulation **92**, 2198-2203
4. Hare, J.M., and Colucci, W.S. (1995) Prog. Cardiovasc. Dis. **38**,155-166
5. Weiss, J.L., Frederiksen, J.W., and Weisfeldt, M.L. (1976) J. Clin. Invest. **58**,751-760
6. Parker, J.D., Landzberg, J.S., Bittl, J.A., Mirsky, I., and Colucci, W.S. (1991) Circulation **84**,1040-1048
7. Paulus, W.J., Vantrimpont, P.J., and Shah, A.M. (1994) Circulation **89**,2070-2078
8. Shah, A.M., Spurgeon, H.A., Sollott, S.J., Talo A., and Lakatta, E.G. (1994) Circ. Res. **74**,970-978

Abbreviations used: βAR, β-adrenergic receptor; cAMP, 3',5'-cyclic adenosine monophosphate; dob, dobutamine; EF, ejection fraction; HF, heart failure; L-NMMA, N^G-monomethyl-L-arginine; LV, left ventricle; NO, nitric oxide; NOS, nitric oxide synthase; τ, tau.

Present Addresses: J. Hare, The Johns Hopkins Hospital, 600 N. Wolfe St., Baltimore, MD 21287; W. Colucci, Boston University School of Medicine, 80 E. Concord St., Boston, MA 02118

ROLE OF NITRIC OXIDE IN CORONARY AUTOREGULATION: INTERPLAY WITH ADENOSINE AND HISTAMINE

M.M.Kostić, M.R.Petronijević and V.Lj.Jakovljević

Institute of Physiology, Faculty of Medicine, 34000 Kragujevac, P.O.B. 124, Yugoslavia

Coronary autoregulation (CA), the capacity of coronary vascular bed to maintain blood flow constant in the face of changes in perfusion pressure, reflects the interaction of myogenic tone with the opposing effects of one or more endogenous vasodilators /1/. The metabolites that are involved in the regulation of CA have not been fully defined. Although adenosine plays an important role in the adjustment of coronary flow /2/, there is no direct link between the myocardial release of adenosine and CA /3/. Recently it has been shown that NO is important regulator of coronary reactive hyperemia (RH) /4/ and modulator of CA /5/. Myocardial histamine also regulates RH and influences CA /6,7/.

Recent data indicate multitude of interactions between these three autacoids. For example, there is an inverse relationship between NO and adenosine in regulation of RH /4/. Changes in release of cardiac NO and histamine indicate their positive feed-back relationship in the regulation of RH /7,8/. Since these interactions are not clearly defined, neither in RH nor in CA, the aim of this work was interplay of NO, adenosine and histamine in the CA.

MATERIALS AND METHODS

Hearts, isolated from male Wistar albino rats with a body mass of about 200 g, were perfused according to Langendorff technique at constant pressure conditions /6,7/. After the isolated heart perfusion was set up, 30 minutes at 60 cm H_2O was allowed for the stabilization of preparation. Thereafter, the coronary perfusion pressure (CPP) was gradually lowered until 20 cm H_2O, then increased backward gradually to 60 cm H_2O, and further to 80, 100 and 120 cm H_2O. When flow was estimated as stable at each value of CPP samples of coronary effluent for determination of histamine /9/ and NO /10/ were collected. At the end of this series of CPP changes (basic protocol), the same sequence of CPP changes was performed but in the presence of different agents as indicated bellow.

Effects of NO on CA were assessed by determination of CF, release of NO and release of histamine from the hearts perfused with 30 uM L-NAME /4/. Effects of histamine were estimated on the basis of changes obtained with hearts perfused with specific antagonists of all three types of histamine receptors. Here, pyrilamine (H1-antagonist), cimetidine (H2-antagonist) and thioperamide (H3-antagonist) were infused at rate giving final concentrations of 10 nM, 1 μM and 10 nM, respectively /11/. Theophylline, which blocks the coronary vascular action of adenosine (and ATP) /12/, was infused in final concentration of 30 μM.

RESULTS AND DISCUSSION

Results of this study show CA in isolated hearts from rats 50 and 80 cm H_2O of CPP, which is slightly different from autoregulatory range in isolated guinea pig hearts /5,7/. Basal release (at 60 cm H_2O of CPP) of NO (as nitrite) and histamine amounts to 2.1±0.6 nmol/min/g wt and and 91±12 pmol/min/g wt, respectively (Table 1). These values are for about 50 % lower than those reported for guenea pig hearts /7/ and may explain why CF is lower in rats than in guinea pigs (3.7±0.6 vs 5.3±0.5 ml/min/g wt).

During autoregulation, as well as out of autoregulatory range, release of NO paralells strictly changes of CF. In contrast release of histamine shows a reciprocial relationship with CF, as well as with release of NO, which contrasts positive feed back of these two autacoids during RH /8/. However, there are no significant changes of histamine release above autoregulatory range (Table 1). Obtained changes of histamine release resemble very much those of adenosine /13/.

Table 1. Release of nitrite and histamine during coronary perfusion pressure changes (CPP): Effects of theophylline (30 μM, Theo).
Asterisks denote autoregulatory range. Values are mean ± S.E.M. for 6 paired experiments.

CPP (cm H_2O)	NO_2 (nmol/min/g wt)		Hist (pmol/min/g wt)	
	Control	Theo	Control	Theo
20	0.78±0.11	0.31±0.11	135±18	48±7
30	1.04±0.19	0.54±0.18	130±19	50±6
40	1.49±0.27	0.84±0.24	126±8	52±6
50*	1.75±0.30	1.20±0.22	115±17	60±8
60*	2.10±0.57	1.34±0.19	91±12	77±9
80*	2.81±0.70	2.22±0.36	78±14	76±14
100	4.17±0.60	2.70±0.50	71±9	96±6
120	5.70±1.42	3.86±0.69	70±16	123±11

Perfusion with L-NAME, but not D-NAME, significantly reduces CF over the entire range of CPP, and widens CA to the range 40-100 cm H_2O. Employment of histamine receptors antagonists decreases CA and shifts autoregulatory range to 30-60 cm H_2O. Inhibition of NO synthesis was accompanied by lower release of histamine, while blockade of histamine receptors decreases release of NO. Theophylline shifts CPP-CF curve downward leaving autoregulatory range unchanged. Release of NO was decreased in the same proportion as CF. However, in the presence of theophylline release of histamine shows completely different pattern as compared to that obtained under basal conditions (Table 1).

REFERENCES

1. Dole, W.P.(1987) Prog.Cardiovasc.Dis.29, 293-323
2. Berne, R.M.(1980) Circ.Res.47, 808-813
3. Hanley, F.L., Grattan, M.T., Stevens, M.B. and Hoffman,J.I.E.(1986) Am.J.Physiol.250, H558-H566
4. Kostić, M.M. and Schrader, J. (1992) Circ.Res.70, 208-212
5. Ueeda, M., Silvia, S.K. and Olsson, R.A.(1992) Circ.Res.70, 1296-1303
6. Rosić, G.L., Stojadinović, N.D., Petronijević, M.R. and Kostić, M.M.(1993) Iugoslav. Physiol. Pharmacol. Acta 29, 147-154
7. Kostić, M.M. and Petronijevic, M.R. (1995) in Mediators in the Cardiovascular System: Regional Ischemia (Schror, K. and Pace-Asciak, C.R.,eds.), pp.145-150, Birkhauser Verlag, Basel
8. Kostić, M.M. and Petronijević, M.R.(1994) J.Mol.Cel. Cardiol.26(6),CXXVIII
9. Hill, S.J.(1990) Pharmacol.Rev.42, 45-83
10. Green, L.C., Wagner, D.A., Glogowski, J., Skipper, P.L., Wishnok, J.S. and Tannenbaum, S.R.(1982) Anal. Biochem.126, 131-138
11. Bergendorff, A. and Unvas, B.(1972) Acta Physiol.Scand. 84, 320-331
12. Bunger, R., Haddy, F.J. and Gerlach, E.(1975) Pflugers Arch.358, H213-H224
13. Schrader, J., Haddy, F.J. and Gerlach, E.(1977) Pflugers Arch. 369, 1-6

Protective effect of relaxin from myocardial damage induced by ischemia-reperfusion: the role of nitric oxide.

Emanuela Masini, *Daniela Salvemini, Laura Mugnai, Maria Grazia Bello, #Daniele Bani, #Tatiana Bani-Sacchi and Pier Francesco Mannaioni.
Departments of Pharmacology and #Histology, Florence University, 50134 Florence, Italy and *Searle Research and Development, Monsanto Co, St. Louis, Mo 63167, USA.

Relaxin is a peptide hormone produced predominantly by the ovary and the decidua in females during the fertile life [1]. There is increasing evidence that relaxin is a multifunctional hormone which, besides its well known effects on the female reproductive system, is also able to influence the cardiovascular system. Previous studies of our group have shown that relaxin increases coronary flow in isolated, perfused guinea pig and rat heart [2], depresses platelet aggregation in isolated human and rabbit platelets [3] and modulates mast cell-histamine release. These effects of relaxin are mediated by a stimulation of the production of nitric oxide [4]. Reperfusion of the ischaemic myocardium, which is known to result in tissue damage, is associated with the formation of oxygen radicals which facilitate lipid peroxidation, calcium overload [5] and alteration in nitric oxide production. Moreover, the release of histamine by resident mast cells in response to the oxy-radicals generated during ischaemia-reperfusion [6] further increases myocardial injury. The present study was designed to evaluate whether relaxin influences nitrite and histamine release and attenuates lipid peroxidation and calcium overload in isolated perfused guinea pig hearts submitted to ischaemia-reperfusion.

Methods

Male guinea pig were anaesthetised, intubated and ventilated with room air using a Palmer pump. After thoracotomy, the pericardium was opened and two loose silk sutures were placed around the left coronary artery. The hearts were quickly excised animals and set up as a Langendorff preparation. Upon stabilization of the hearts, ischaemia was induced by tigthening the sutures and maintained for 20 min. Reperfusion was obtained by cutting the sutures and maintained for further 20 min. Some hearts were treated with pure porcine relaxin (a generous gift of Dr. O.D. Sherwood) dissolved in the Tyrode solution used for perfusion at a concentration of 30 ng/ml. The perfusion with relaxin was started at the beginning of ischaemia and maintained until the end of reperfusion. In both untreated and relaxin-treated hearts, coronary perfusates were collected at 10 min intervals for determination of coronary flow, nitrites, histamine, and LDH. At the end of the experiments, specimens of the left ventricle were taken for determination of malonyldialdehyde production and calcium content. Nitrites were measured by the Griess reaction; histamine and LDH were determined by fluorimetrically and spectrophotometrically; malonyldialdehyde was determined spectro-photometrically by the reaction with 2-thiobarbituric acid; calcium content was determined by atomic absorption spectrometry.

Results and discussion

In the untreated hearts, ischaemia-reperfusion caused a linear increase in the release of LDH, whereas histamine release occurred preferentially during reperfusion. The amount of nitrite released in the perfusates was significantly lower than before induction of ischaemia-reperfusion. This was accompanied by an increase in malonyldialehyde and calcium content in the tissue homogenates. In the relaxin-treated hearts, the coronary flow and the amount of nitrite in the perfusates increased significantly and histamine and LDH release were reduced. The peptide was also effective in inhibiting malonyldialdehyde production and in preventing the cardiac overload of calcium. The results of this study show that relaxin increases the coronary flow through an NO-mediated mechanism and protects the heart from tissue damage induced by ischaemia-reperfusion. These properties of relaxin, together with its antiaggregatory activity [3], raise the possibility for a future use of relaxin or relaxin-derived drugs in the prevention and treatment of coronary ischaemic disease.

References

1. Bryant-Greenwood,G.D. (1992) Endocr.Rev. **3**, 62-90.
2. Bani-Sacchi,T.,Bigazzi,M.,Bani,D.,Mannaioni,P.F. and Masini,E (1995) Br. J. Pharmacol. **116**, 1589-1594.
3. Bani,D.,Bigazzi,M.,Masini,E.,Bani,G. and Bani Sacchi,T. (1995) Lab.Invest.,in press.
4. Masini,E.,Bani,D.,Bigazzi,M.,Mannaioni,P.F. and Bani-Sacchi,T. (1994) J.Clin.Invest. **94**, 1974-1980.
5 Masini,E.,Bianchi,S.,Mugnai,L.,Gambassi,F., Lupini,M.,Pistelli,A. and Mannaioni,P.F. (1991) Agents Actions **33**,53-56.
6. Mannaioni,P.F. and Masini,E. (1988) Free Radical Biol. Med. **5**,177-197.

Diminished nitric oxide production during left ventricular bypass with nonpulsatile perfusion

MAHENDER MACHA, HIROAKI KONISHI, KENJI YAMAZAKI, PHILIP LITWAK, TIMOTHY R. BILLIAR, and BRACK G. HATTLER

Artificial Heart and Lung Program, Department of Surgery, University of Pittsburgh, 300 Technology Drive, Pittsburgh, PA, USA 15219

The physiologic significance of pulsatile flow has remained unclear. Nitric oxide, a highly reactive product of endothelial cells, has been shown to stimulate increased cyclic guanosine monophosphate levels in smooth muscle cells, leading to vasodilation[1,2]. Pulsatile flow and shear stress are considered to stimulate endothelial nitric oxide release[3]. Although the basic mechanism is unclear, mechanical stimulation of the endothelial cell is thought to be critical to the regulation of vascular tone. The loss of pulsatile perfusion may be associated with a reduction in nitric oxide mediated vasodilation, resulting in increased systemic arterial pressure.

Early studies with an extracorporeal centrifugal pump have been performed in an anesthetized sheep model. Animals were maintained on partial left ventricular (LV) bypass for three days, then nonpulsatile flow was created by increasing the pump flow to capture total LV outflow for sixty minutes. Pump outflow pressure rose markedly after pump speed was increased by 35% (1.6 liters/minute) to creast nonpulsatile perfusion (Figure 1). Mean carotid arterial flow remained between 0.9 to 1.4 liters/minute during both pulsatile and nonpulsatile perfusion.

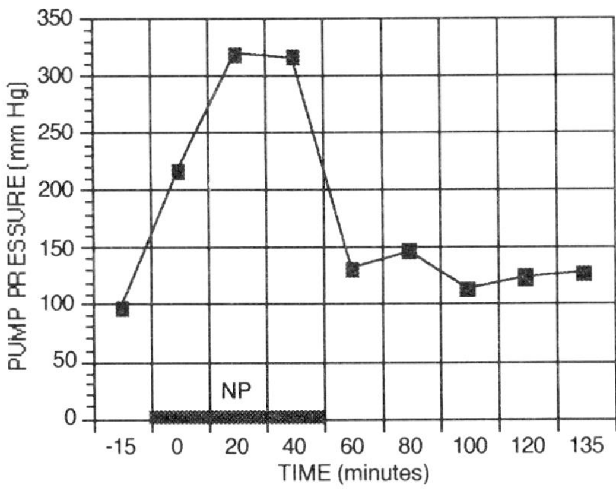

Figure 1. Mean pump outflow pressure during and after nonpulsatile (NP) stage.

Carotid mean arterial pressure was slightly elevated (95 mm Hg, vs. 89 mm Hg) during the nonpulsatile state as compared with the prior pulsatile state. Carotid mean arterial pressure returned to baseline (92 mm Hg) during the subsequent period of pulsatile perfusion (Figure 2).

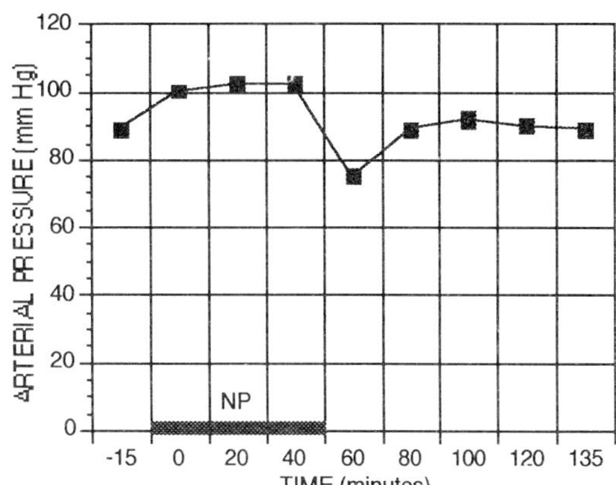

Figure 2. Mean arterial pressure during and after nonpulsatile (NP) stage.

Plasma nitrite and nitrate, as measured by the Greiss reaction, were diminished by 51% (13.8 micromolar, vs. 28.4 micromolar) after sixty minutes of non-pulsatile perfusion (Figure 3). Levels were returned to baseline within 75 minutes after resumption of pulsatile flow.

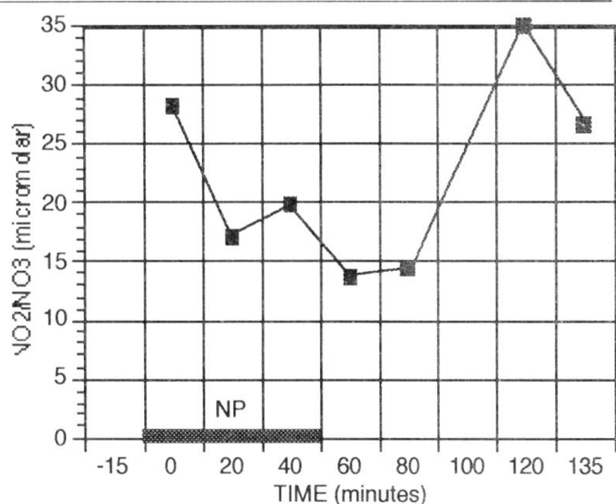

Figure 3. Mean plasma nitrite/nitrate concentration during and after nonpulsatile (NP) stage.

These results suggest that nitric oxide may be critical for maintaining a vasodilator tone during the basal physiologic state of pulsatile perfusion. This may have implications in the use of nonpulsatile cardiopulmonary bypass and cardiac assist devices.

References:

1. Ignarro LJ. Biosynthesis and metabolism of endothelium-derived nitric oxide. Annu Rev Pharmacol Toxicol. 1990;30:535-560.
2. Waldman SA, Murad F. Cyclic GMP synthesis and function. Pharmacol Rev. 1987;39:163-196.
3. Rubanyi GM, Romero JC, Vanhoutte PM. Flow-induced release of endothelium-derived relaxing factor. Am J Physiol 1991;261:H257-H262.

Impairment of nitric oxide mediated ß-adrenergic relaxation in the mesenteric vein from portal hypertensive rats.

Mª ANGELES MARTINEZ-CUESTA, LUCRECIA MORENO, JOSÉ Mª PIQUÉ[#], JAUME BOSCH[*] and JUAN V ESPLUGUES.

Department of Pharmacology, University of Valencia, 46021 Valencia, Spain; [*]Liver and [#]Gastroenterology Unit, Hospital Clínic i Provincial, University of Barcelona, 08036 Spain.

Nitric oxide (NO) has been associated with the changes in vascular reactivity induced by portal hypertension (PH) [1,2]. It has been recently suggested that the relaxation of arteries induced by stimulation of ß-adrenoceptors could be mediated by nitric oxide (NO) [3,4]. Furthermore, non-selective ß-adrenergic blockers are widely used in the pharmacological treatment of portal hypertension. The aim of the present study was to characterize the role of NO in the response to the specific β_2-adrenergic agonist salbutamol in the isolated mesenteric vein from portal hypertensive rats, territory that undergoes the most pronounced functional and morphological changes in this clinical syndrome.

Methods. PH was induced in male Sprague-Dawley rats by a calibratd stenosis of the portal vein with a partial ligation. Two weeks later, animals were sacrificed and rings from mesenteric vein were mounted in an organ bath (Krebs' solution at 37 °C and gassed with 95% O_2 - 5% CO_2) for isometric tension recording. After a 1 h stabilization period, tissue responsiveness was assessed by a contraction induced by KCl (60 mM). Following a washout period, concentration-responses curves to salbutamol (10^{-8} - 10^{-5} M) were evaluated after precontraction of mesenteric vein rings with KCl (30 mM) in the absence and presence of L-NAME (10^{-4}M). Likewise, responses to the non-specific smooth muscle relaxant papaverine (10^{-6} - 10^{-4} M) were also analyzed.

Results/Discussion.

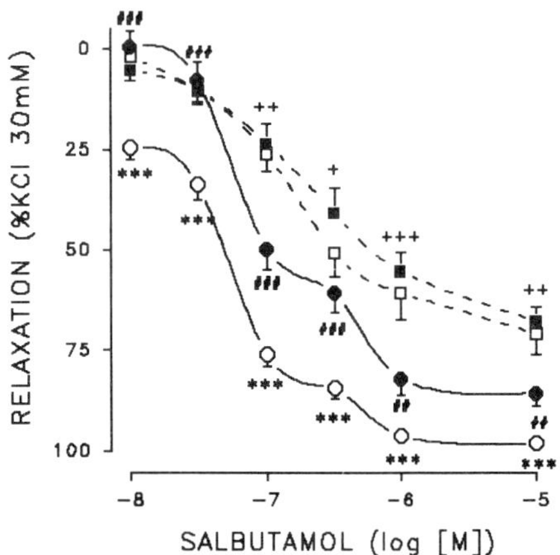

Figure 1. Concentration-response curves to salbutamol in the absence (circles) or the presence of L-NAME 10^{-4} M (squares) of mesenteric vein rings from sham operated (open symbols, n>9) and PH rats (filled symbols, n>10).
##p<0.01, ###p<0.001 vs sham-operated;
*** p<0.001 vs sham-operated + L-NAME;
+ p<0.05, ++p<0.01 and +++p<0.001 vs PH.

As shown in figure 1, the relaxation induced by salbutamol was significantly ($p<0.001$) reduced in the mesenteric vein from portal hypertensive rats. By contrast, the non-specific drug papaverine induced a similar relaxation in both tissues with 100% of maximal response (fig. 2). This finding supports that a non-specific failure of the smooth muscle relaxation can not explain the attenuated relaxation to salbutamol in the mesenteric vein from portal hypertensive rats.

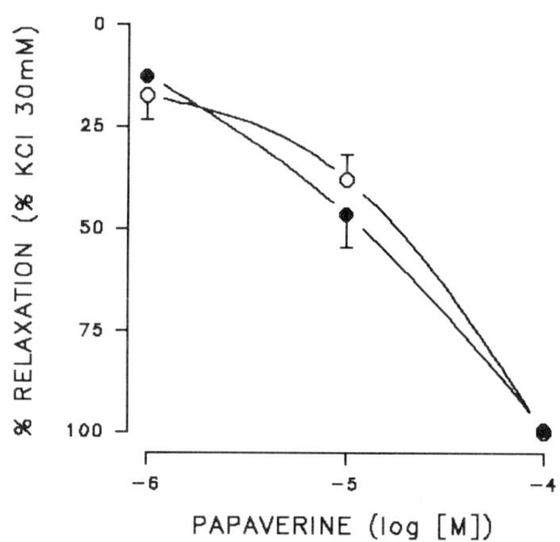

Figure 2. Concentration-response curves to papaverine of mesenteric vein rings from sham operated (open symbols, n>5) and PH rats (filled symbols, n>5).

Incubation of tissues with L-NAME markedly blocked the relaxation induced by this ß-adrenergic agonist (fig. 1). Thus, there is a role for NO in the modulation of ß-adrenergic relaxation in the mesenteric vein. However, the effect of L-NAME was significantly ($p<0.05$) greater in rings from sham-operated than in rings from portal hypertensive rats, suggesting a deficit of NO in such response of tissues from PH rats.
In summary, PH induces an impairment in the release of NO specifically after stimulation of β_2-adrenoceptors in the mesenteric vein.

References

1. Sieber, C.C. and Croszmann, R.J. (1992). Nitric oxide mediates the hyporeactivity to vasopressors in mesenteric vessels of portal hypertensive rats. Gastroenterology, **103**, 235-239.

2. Pizcueta, M.D., Piqué, J.M., Bosch, J., Whittle, B.J.R., Moncada, S and Rodés, J. (1991). Hemodynamic effects of endogenous nitric oxide inhibition in cirrhotic rats. Hepatology, **14**, 301.

3. Gray, D.W. and Marshall, I. (1992). Novel signal transduction pathway mediating endothelium-dependent ß-adrenoceptor vasorelaxation in rat thoracic aorta. Br. J. Pharmacol., **107**, 689-690.

4. Graves, J. and Poston, L. (1993). ß-adrenoceptor agonist mediated relaxation of isolated resistance arteries: a role for the endothelium and nitric oxide. Br. J. Pharmacol., **108**, 631-637.

Pathologic changes in rat heart after long-term nitric oxide synthase inhibition.

OĽGA PECHÁŇOVÁ, PAVEL BABÁL* and IVETA BERNÁTOVÁ

Institute of Normal and Pathological Physiology Slovak Academy of Sciences, Sienkiewiczova 1, *Department of Pathology, Medical Faculty, Comenius University, 813 71 Bratislava, Slovak Republic

The combined effect of hypertension and NO-deficiency was investigated in rat hearts after long-term administration of L-NAME.

15 weeks old, male, wistar rats were devided into 2 groups. First group (n=8) served as the control. Second group (n=8) was given L-NAME in the dose 40mg/kg/day in their drinking water for 4 weeks. Systolic blood pressure and heart rate were measured by tail-cuf plethysmography. After 4 weeks of L-NAME treatment heart/body weight ratio was calculated. Total RNA content was analysed by single step method [1]. For DNA analysis non phenol method was used [2]. The proteosynthesis was determined by [^{14}C]leucine incorporation into protein [3]. The heart samples for light microscopy were fixed in formalin and routinely processed in paraffin. Histological slides were stained with hematoxylin and eosin, Mallory's phosphotungstic acid and Van Gieson's staining. Morphometric evaluation was performed on Telemet II (Tesla, Piešťany, Slovakia). To determine lactate dehydrogenase activity in animals serum a comercial available kit was used (Sigma).

L-NAME in the dose 40mg/kg/day led to systolic pressure increase by 30%, while the heart rate decreased by 20%. Heart/body weight ratio was without significant change. However, total RNA and DNA content as well as [^{14}C]leucine incorporation into protein of myocardium significantly increased by 10%, 238%, and 98% respectively. Light microscopy of rat hearts after 4 weeks of L-NAME administration showed large areas with pathologic changes characterized by necrosis and different stages of scar development. Maximum changes were in the subendocardial location of the left ventricle, the papillary muscles, and in the interventricular septum. Similar changes were occassionally seen also in the right ventricle. Areas of acute necrosis with inflammatory cells accumulation right next to foci of fibrosis documented the dynamic of ischemic changes in the myocardium. Morphometric evaluation showed significant increase in myocardial fibrosis. Fibrous tissue/muscle ratio was 0.055 0.037 in control group and 1.492 0.14 in experimental group, $p<0.05$. Fibroblast proliferation in the wall of coronary microvessels resulted in remarkable reduction of their lumen. These alterations were not observed in control animals. In addition, lactate dehydrogenase activity, as one of the ischemic injury marker increased by 51% ($p<0.05$) after L-NAME administration.

Abbreviations used: L-NAME - N^G-nitro-L--arginine methyl ester

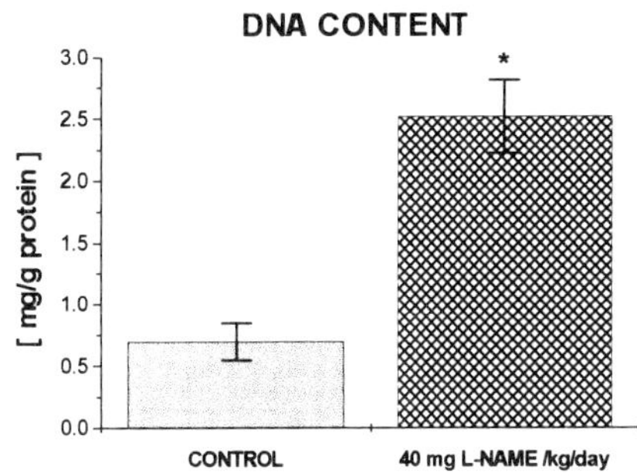

Fig. 1. Total DNA content. Results are presented as mean ± SEM, $p < 0.05$.

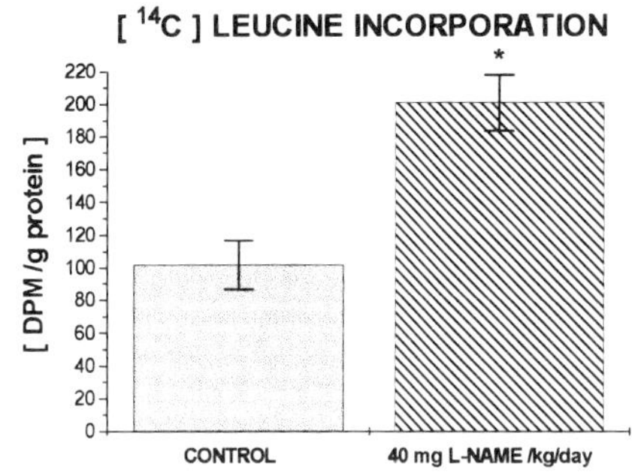

Fig. 2. [^{14}C] leucine incorporation into the protein. Results are presented as mean ± SEM, $p < 0.05$.

We hypothesize that the inhibition of NO-synthase leads to hypertension connected with the ischemic injury of the myocardium in response to decreased vasorelaxation of coronary vessels. As a result fibrotic reparative process occurs.

The study was supported in part by PECO grant BMH1-CT-92-1893 and Slovak National Bank.

1. Chomczynski, P. and Sacchi, N.(1987)Anal. Biochem. 162,156-159

2. Sambrook,J., Fritsch,E.F. and Maniatis,T. (1989)Molecular cloning. A laboratory manual second edition (Ford.N.,ed.), pp. 9.16-9.19, Gold Spring Harbor Laboratory Press, New York

3. Gerová.M., Pecháňová,O.,Stoev,V.,Kittová, M., Bernátová, I. and Bárta, E. (1995) Am.J. Physiol. In press

Nitric oxide synthesis is decreased in essential hypertension

MHAIRI COPLAND, LORNA M SMITH, JOHN SUTHERLAND AND NIGEL BENJAMIN

Department of Medicine and Therapeutics
University of Aberdeen Medical School, Aberdeen UK

Introduction

Nitric oxide (NO) synthesis by the vascular endothelium is important in the regulation of vasodilator tone and the control of blood pressure in man [1]. Previous studies have provided indirect evidence that impairment of NO production accounts for the abnormalities in vascular function that characterise many vascular diseases, including experimental hypertension in animals and human essential hypertension [2,3,4]. These studies in general have relied on disparate responses to either substances which release NO from vascular endothelium, such as muscarinic agonists, or inhibitors of NO synthesis from L-arginine, such as N^G-monomethyl-L-arginine (L-NMMA). Studies on the response to muscarinic agonists have shown conflicting results [5] as these agents may cause dilatation by mechanisms other than generation of NO.

The aim of the present study was to determine NO production more directly in patients with untreated essential hypertension by measurement of 24 hour urinary nitrate excretion. Urinary nitrate is endogenously produced in man and is thought to be the stable end-product of NO oxidation. Particular importance was paid to reducing the dietary intake of nitrate in those who participated, as this is generally greater in amount than endogenous nitrate synthesis, and accounts for the greatest proportion of nitrate excretion when on a normal diet.

Methods

Patients referred to the Aberdeen Hypertension Clinic for management of hypertension were invited to participate if the clinic blood pressure readings were above 160mmHg systolic or 90mmHg diastolic, measured by standard sphygnomanometry.

The patients had never been treated with antihypertensive drugs (8 patients) or had been off treatment for at least two weeks (atenolol in both cases). Secondary causes of hypertension were excluded in all patients.

Patients were studied as hospital in-patients during the time of the study whereas age and sex-matched control subjects (recruited from hospital and laboratory staff) were treated in an identical way but were not hospitalised.

A low (less than 90µmoles/day) nitrate diet was administered on both study days; the first 24 hours was a washout period, allowing previously ingested nitrate to be metabolised or excreted in the urine (the half life of nitrate is approximately 5 hours). Endogenous nitrate synthesis was estimated during the second 24 hours by measurement of total urinary nitrate excretion.

All subjects, patients and controls underwent 24 hour ambulatory blood pressure monitoring using SpaceLab model no. 90207, with hourly readings during the second study day. Only those with an average daytime (07.00-21.00hrs) mean arterial pressure (diastolic +1/3 pulse pressure) of >107mmHg were considered to be hypertensive. (Patients with a MAP < 107mmHg were included in the control group.)

Urinary and plasma nitrate were measured using a Greiss assay following reduction with a copper/cadmium column [7]. Statistical analysis was performed using unpaired Student's t-tests on log-transformed measurements of urinary nitrate excretion and plasma nitrate in the two study populations; $P < 0.05$ was taken to be statistically significant. Simple regression was then used to assess the correlation between mean daytime (7.00am-9.00pm) arterial blood pressure and 24 hour urinary nitrate excretion, urinary nitrate:creatinine ratio and urinary nitrate clearance.

Results

With the exception of blood pressure there were no significant differences in the variables considered apart from a slightly higher mean (±sd) plasma creatinine concentration in the

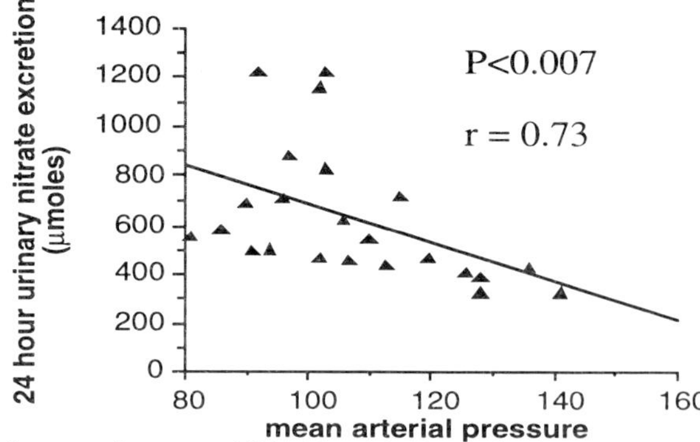

hypertensive group (100±16 vs 84±12 µmoles/l, P<0.01). Mean 24 hour urinary nitrate excretion in the hypertensive group was 450 ±37µmol/day (mean ± SEM) compared with a mean of 760 ±77µmol/day in the control group (P < 0.001, Student's t-testing on log-transformed data). In addition, there was an inverse correlation between mean daytime ambulatory blood pressure and nitrate excretion; (P<0.007; coefficient of correlation 0.73). This is illustrated in Figure 1. Within the hypertensive group nitrate excretion was also negatively correlated with mean daytime ambulatory blood pressure (P<0.01; coefficient of correlation 0.6).

There was no correlation between plasma nitrate and plasma creatinine concentrations. The urinary nitrate:creatinine ratio in the hypertensive group was 0.038 ±0.015 compared with 0.079 ±0.036 in the control group (P<0.005, Student's t-test). The relationship between blood pressure and urinary nitrate:creatinine ratio was very similar to that of blood pressure versus nitrate excretion; i.e. as blood pressure increased, the urinary nitrate:creatinine ratio decreased (P<0.009; coefficient of correlation 0.627). Plasma nitrate in the hypertensive group had a mean value of 25.8 ±2.6µmol/L compared with 26.6 ±3.1µmol/L in the control group (non-significant). There was no relationship between plasma nitrate and blood pressure.

Discussion

The results of this study provide further and more direct evidence that NO production is decreased in patients with essential hypertension. Furthermore, endogenous nitrate synthesis correlates inversely with the average mean arterial daytime ambulatory blood pressure over the whole group of subjects studied and the hypertensive group when analysed separately. The source of the nitrate excreted in this study cannot be determined. The careful limitation of dietary sources of nitrate considerably reduces this potentially important confounding factor.

References

1.Vallance P, Collier J, Moncada S. Effects of endothelium-derived nitric oxide on peripheral arteriolar tone in man. *Lancet* 1989; ii: 997-1000.

2.Linder L, Kiowski W, Buhler FR, Luscher TF. Indirect evidence for release of endothelium-derived relaxing factor in human forearm vasculature in vivo: blunted response in essential hypertension. *Circulation* 1990; 81: 1762-1767.

3.Panza JA, Quyyumi AA, Brush JE, Epstein SE. Abnormal endothelium- dependent vascular relaxation in patients with essential hypertension. *NEJM,* 1990; 323: 22-27.

4.Calver A, Collier J, Moncada S, Vallance P. Effect of local intra-arterial N^G-monomethyl-L-arginine in patients with hypertension: the nitric oxide dilator mechanism appears abnormal. *J. Hypertens*1992; 10: 1025-1031.

5.Cockcroft JR, Chowienczyk PJ, Benjamin N, Ritter JM. Preserved endothelium-dependent vasodilatation in patients with essential hypertension.*NEJM* 1994; 330: 1036-1040.

6.Green LC, Ruiz de Luzuriaga K, Wagner DA, Rand W, Istfan N, Young VR, Tannenbaum SR. Nitrate biosynthesis in man. *Proceedings of the National Academy of Sciences, USA* 1981; 78: 7764-7768.

7.Green LC, Wagner DA, Glogowski J, Skipper PL, Wishnok JS, Tannenbaum SR. Analysis of nitrate, nitrite and [^{15}N]nitrate in biological fluids. *Analytical Chemistry* 1982; 126: 131-138.

Responsiveness of high- and low- pressure vessels in nitric oxide -deficient hypertensive rats.

ANNA HOLÉCYOVÁ, JOZEF TÖRÖK, IVETA BERNÁTOVÁ and OĽGA PECHÁŇOVÁ

Institute of Normal and Pathological Physiology, Slovak Academy of Sciences, 813 71 Bratislava, Slovak Republic.

In the experimental model of hypertension induced by long-term inhibition of NO synthesis, except of nonspecific pressure load on vascular wall, a decrease in NO levels occurs. This interferes specifically with the modulatory capacity of the vessel.

The systemic (high-pressure) and pulmonary (low-pressure) vessels adapt differently to these influences. Moreover, controversies still exist as to the contribution of NO in control of pulmonary vascular tone [1,2].

The objective of this study was to determine the role of NO and/or hypertension in modulation of responsiveness of high- and low- pressure vessels during hypertension induced by 4 week inhibition of NO synthesis in rats.

Male Wistar rats (30) were offered to drink water containing L-NAME at a concentration 20 or 40 mg/kg/day to inhibit NO synthesis. Control group of animals (12) was studied in parallel. After 4 weeks, systolic blood pressure was higher by 34% in the group receiving 20 mg/kg/day and by 30% in that receiving 40 mg/kg/day of L-NAME as compared with control group. At that time, animals were sacrificed and vessels were dissected free. Ring preparations of high-(aorta, renal artery) and low-pressure vessels (pulmonary artery) were mounted for isometric tension recording. Cumulative concentration-response curves for individual agonists were compared by ANOVA with P values adjusted for multiple comparisons by Bonferroni method.

NA (10^{-9} - 10^{-5}M) elicited concentration-dependent contraction in both high- and low-pressure vessels. As demonstrates Fig.1, chronic L-NAME treatment in 20 mg/kg/day resulted in significant enhancement of maximum contraction by 23% in aorta, by 27% in renal artery and by 26% in pulmonary artery as compared with individual control responses. After 40 mg/kg/day, the enhancement was even more pronounced in renal and pulmonary artery, however, the difference was not significant.

In both types of vessels, precontracted with NA (3×10^{-7}M), ACh (10^{-8} - 3×10^{-6}M) induced concentration- dependent relaxation averaged in aorta 89%, in renal artery 86% and in pulmonary artery 79%. Treatment with L-NAME in both concentrations significantly reduced the relaxation in all 3 vessels tested (Fig.2).

In contrast to ACh, treatment with L-NAME had no effect on relaxation of the vessels to sodium nitroprusside (10^{-9} - 10^{-6}M). All the vessels relaxed completely, indicating unchanged ability of vascular smooth muscle to relax.

When indomethacin (10^{-5}M) was added to the bath to block cyclooxygenase activity, neither control, nor residual relaxation to ACh in L-NAME treated groups were influenced, suggesting no role of vasoactive prostaglandins in ACh-induced relaxation in the vessels.

Additional administration of L-NAME (10^{-4}M) to the bath, however, reduced the maximum relaxation to ACh in the control aortas by 68%, while reversed the reduced relaxation to constriction in the aortas from animals chronically treated with L-NAME. Similar results were obtained with pulmonary artery.

Acute pretreatment of the vessels from L-NAME treated animals with L-arginine (10^{-4}M) improved the reduced relaxation to ACh.

The results suggest that NO play similar role in modulation of vascular tone in both high- and low-pressure vessels. Relaxation of aorta and pulmonary artery to ACh was depressed independently on blood pressure in the respective vessel in vivo, indicating rather the decisive role of NO than of blood pressure.

NO- nitric oxide, L-NAME- NG- nitro- L- arginine methyl ester, NA- noradrenaline, ACh- acetylcholine, AO- aorta, RA- renal artery, PA- pulmonary artery

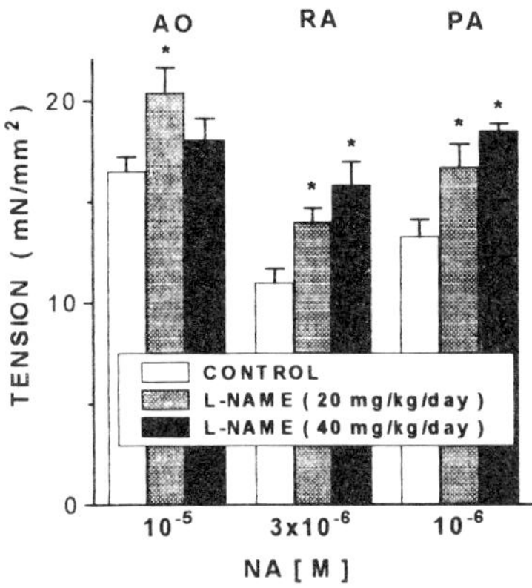

Fig. 1. Effect of 4 week L-NAME treatment on maximum contraction to NA in rat high- and low - pressure vessels. Results are expressed as the mean ± SEM. *$p < 0.05$.

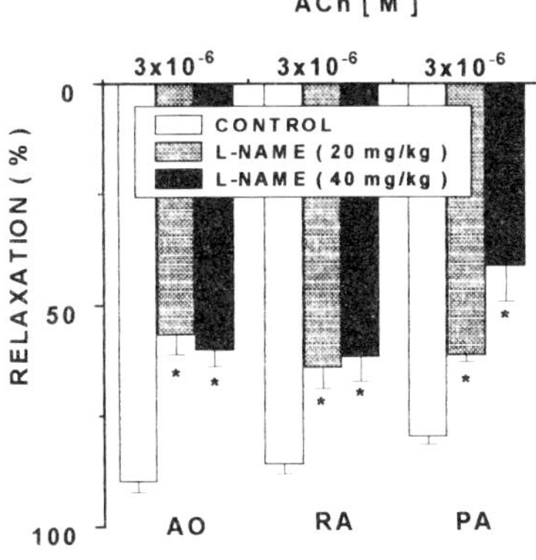

Fig. 2. Effect of 4 week L- NAME treatment on maximum endothelium - dependent relaxation in rat high- and low-pressure vessels. Results are expressed as the mean ± SEM. *$p < 0.05$.

The study was supported by PECO grant BMH1-CT-92-1893

1. Fineman, J.R., Heymann, M.A. and Soifer, S.J. (1991) Am. J, Physiol. 260, H1299-H1306

2. Ogata, M., Ohe, M., Katayose, D., and Takashima T. (1992) Am. J. Physiol. 262, H691-H697.

Cardiac hypertrophy and vascular remodeling in NO-deficient hypertension

František KRISTEK, Mária GEROVÁ, Lubomír DEVÁT[*], and Ivan VARGA[=]

Institute Normal and Pathological Physiology, Sienkiewiczova 1, 813 71 Bratislava, Slovakia, [*]Drug Research Institute,900 02 Modra, Slovakia [=]Slovakofarma Joint Stock Company, 920 01 Hlohovec, Slovakia.

Controversial data concerning the structure of the heart and vessel wall in NO deficient hypertension were published. The aim of the study was to follow the growth of the heart and the structure of the coronary (septal branch), and carotid artery during hypertension induced by long term inhibition of NO-synthase.

Control rats were given the tap drinking water, experimental rats L-NAME 50 mg/kg bw/day in the drinking water for a period of 8 weeks. Systolic blood pressure was measured weekly by a tail plethysmographic method. After this period the animals were sacrificed. In 7 control and 7 experimental animals the chest was opened, heart excised and the weight was estimated. In the other 7 controls and 7 experimental animals the cardiovascular system was perfused under the pressure of 120 mmHg by a glutaraldehyde fixative. After fixation middle part of carotid artery and upper half of septal branch was excised, postfixed with OsO_4, dehydrated and embedded in Durcupan ACM. Both inner circumference and thickness of arterial wall (tunica intima + tunica media) were measured on semithin sections in ligh microscopy. From this data inner diameter and the area of vessel wall were calculated.

At the end of experiment the blood pressure increased significantly. Significant decrease in heart rate was found. The heart weight and heart/body weight increased (Table 1), proving reliably the cardiac hypertrophy.

Table 1

	BP(mmHg)	HR (b/min)	HW (g)	HW/BW
Contr.	131.4±1.9	352.6±4.1	1.10±0.03	2.1±0.04
Exp.	187.2±4.2	334.4±7.0	1.32±0.08	3.0±0.15
Sign.	$p<0.01$	$p<0.05$	$p<0.05$	$p<0.01$

As far as geometry of the vessel wall concerns the inner circumference and/or inner diameter of septal branch decreased and the thickness of arterial wall in experimental animals increased (Fig.1,2). The wall lumen ratio was 1:10, remarkably altered in comparison to controls 1:21 (Table 2).

A similar decrease of inner circumference and/or inner diameter, and increase of wall thickness was observed in carotid artery of experimental animals (Fig. 3,4). The wall lumen ratio changed: 1:17 in experimental and 1:33 in control animals (Table 3).

L-NAME - N -nitro-L-arginine methyl ester, BP-blood pressure, HR-heart rate, HW-heart weight, HW/BW-heart/body weight ratio. TM-tunica media, Ci-circumference inner, Wall-area of arterial wall (tunica intima +tunica media), W/D - wall/diameter ratio. Each value represents the mean + S.E.M.

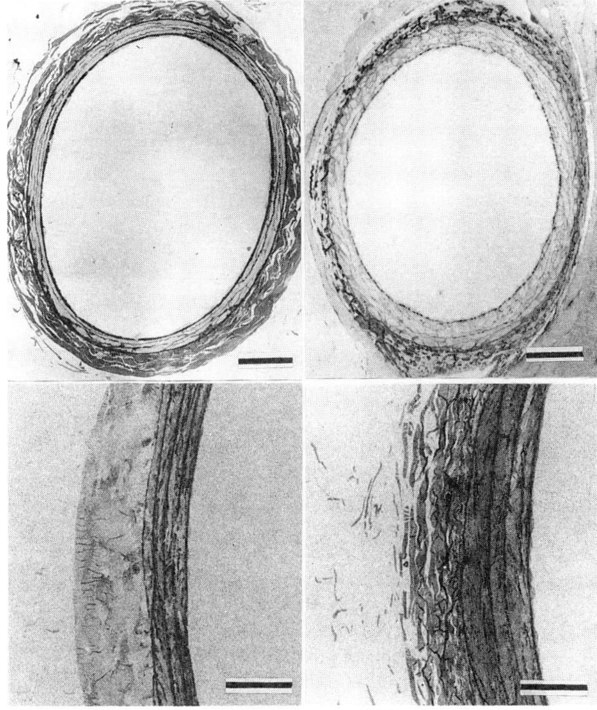

Fig. 1 - Septal branch - control rats.
Fig. 2 - Septal branch - L-NAME treated rats.
Fig. 3 - Carotid artery - control rats.
Fig. 4 - Carotid artery - L-NAME treated rats.
Bar 50 μm.

Table 2 - Septal branch

	Wall (μm)	Ci (μm)	Wall (μm²)	(W/D)x10³
Contr.	12.5±0.6	832±30	11.2±9.2x10³	48.0±1.9
Exp.	21.2±0.8	664±25	15.4±8.5x10³	106.5±3.7
Sign.	$p<0.01$	$p<0.01$	$p<0.01$	$p<0.01$

Table 3 - Arteria carotis

	Wall (μm)	Ci (μm)	Wall (μm²)	(W/D)x10³
Contr.	26.1±1.2	2733±66	73.3±3.6x10³	30.2±1.7
Exp.	45.1±1.4	2456±39	117.6±4.6x10³	57.9±1.7
Sign.	$p<0.01$	$p<0.01$	$p<0.01$	$p<0.01$

In conclusion eight weeks lasting inhibition of No-synthase induces the cardiac hypertrophy and remodeling of the wall of the above arteries. The mechanism and namely question whether this changes are a primary consequence of NO deficiency, or NO deficiency triggers a neurohumoral process which secondary induces the above changes remains open. Increase of arterial wall thickness together with decrease of the inner radius becomes a real jeopardy for blood supply of the respective areas.

The study was supported by the Slovakofarma Joint Stock Company, Hlohovec, Slovakia.

Assessment of endothelial function by pulse wave analysis in conscious hypercholesterolaemic rabbits

JØRGEN MATZ*#, DAVID W LAIGHT*, BEN CEASAR*, MARTIN J CARRIER* and ERIK E ÄNGGÅRD*.

*The William Harvey Research Institute, St. Bartholomew's Hospital Medical College, Charterhouse Square, London EC1M 6BQ, UK, and #The Royal Danish School of Pharmacy, 2 Universitetsparken, DK-2100 Copenhagen Ø, Denmark.

Assessment of NO-mediated vasodilation as a measure of endothelial function may be important in predicting the risk of atherosclerosis in asymptomatic individuals, and in evaluation of anti-atherogenic treatment of hypercholesterolaemia.

Computerised PPG provides a non-invasive method whereby endothelium-dependent vasodilation can be assessed via analysis of arterial pulse waves. The method measures the relative height of the DN; a parameter that relates to the functional and elastic properties of the larger arteries. The DN is influenced by NO synthesis [1], and is used clinically to assess effects of organic nitrates [2].

The present study uses PPG to assess the endothelial function in rabbits fed a control or 1 % CH-enriched diet for 8 weeks. Pulse wave curves were recorded following application of a PPG-sensor on the central ear artery after slight sedation. The pulse wave contour was analysed as a %-change in the relative height of the DN (Fig.1). The height of the DN was determined via its first derivative (dPPG/dt), and was averaged over 20 cycles at a representative interval. Ach was administrated intravenously for 5 min (0.4, 1.0 and 2.0 $\mu g{\cdot}min^{-1}{\cdot}kg^{-1}$) with 5 min interval between each dose. HR and MAP was measured via cannulation of the opposite ear artery.

PPG-recordings from control rabbits

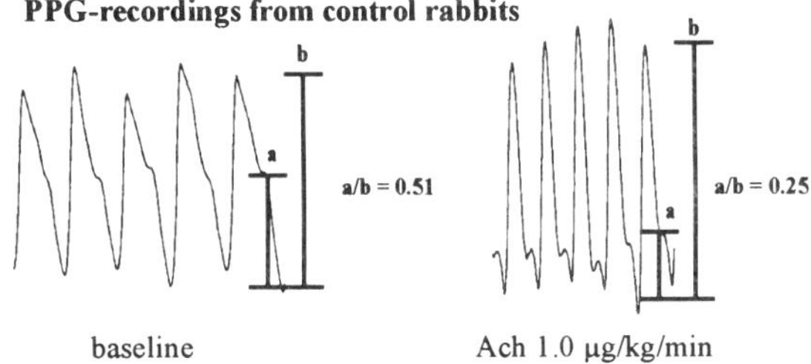

PPG-recordings from hypercholesterolaemic rabbits

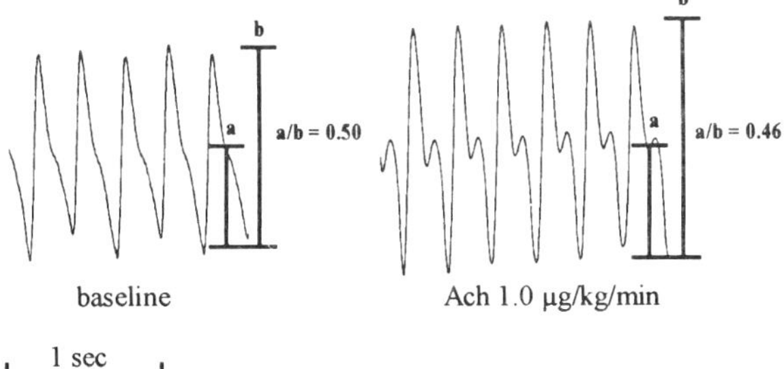

Fig.1. Representative trace recordings of arterial pulse waves from control (n=7) and hypercholesterolaemic rabbits (n=8). Ach caused a decrease in the relative height of the DN as evaluated by the a/b ratio. This response was significantly impaired in hypercholesterolaemic rabbits (P<0.05).

Abbreviations used: Cholesterol (CH), Nitric oxide (NO), Photoplethysmography (PPG), Dicrotic notch (DN), Acetylcholine (Ach), Mean arterial blood pressure (MAP), Heart rate (HR).

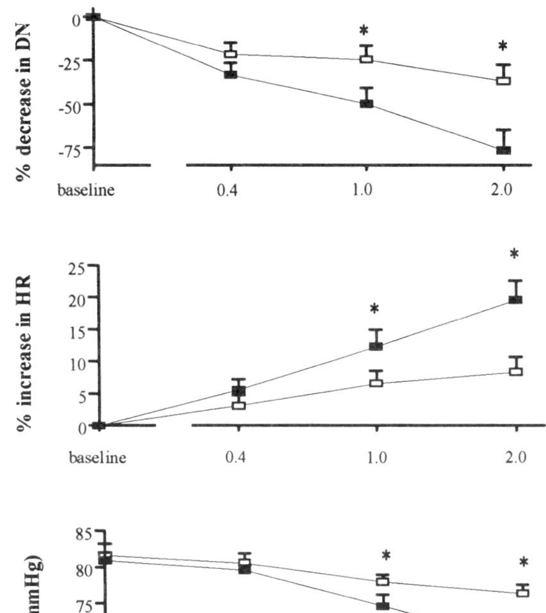

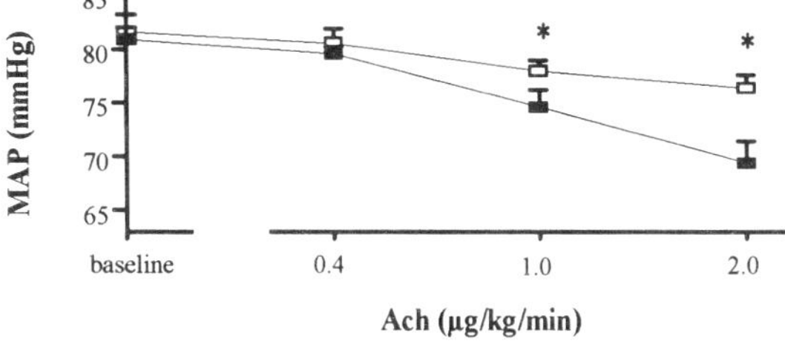

Fig.2. Effects of intravenous infusions of Ach on the relative height of the DN, HR and MAP in conscious rabbits after 8 weeks on control (■, n=7) or cholesterol (□,n=8) diet. Significance is denoted by * as compared to control (P<0.05) with error bars denoting mean ± SEM.

The current study using conscious rabbits and a photoplethysmographic method confirms that endothelium-dependent vasodilation to Ach is impaired in hypercholesterolaemic animals. We demonstrate that loss of NO-mediated vasodilation is evident on systemic haemodynamic parameters: The DN, HR and MAP (Fig.2).

There is a need for simple methods for detection of endothelial dysfunction. PPG measures blood volume fluctuations non-invasively via a reflection of infra-red light from movements of erythrocytes and the vessel wall. The height of the DN on arterial pulse waves, originating from a closure of the aortic valve, is influenced by tone and bioimpedance in the arterial tree [3]. Acetylcholine and organic nitrates reduce the height of the DN. This vasodilation is predominantly due to changes in larger arteries (unpublished observations). CH-feeding impaired the Ach-induced lowering of the DN. This impairment may reflect the loss of NO-mediated vasodilation and enhancement of arterial bioimpedance in hypercholesterolaemic rabbits confirming a previous study from our laboratory [1].

In conclusion, we demonstrated an endothelial dysfunction in CH-fed rabbits via non-invasive analysis of arterial pulse waves.

1. Klemsdal, T.O., Andersson, T.L.G., Matz, J., Ferns, G.A.A., Gjesdal, K. and Änggård, E.E. (1994) Cardiovasc. Res. **28**, 1397-1402

2. Klemsdal, T.O., Mundal, H.H., Rudberg, N. and Gjesdal K. (1994) Eur. J. Clin. Pharmacol. **47**, 351-54

3. O'Rourke, M.F., Safar, M.E. and Dzau, V.J. (1993) Arterial vasodilation: Mechanisms and Therapy, Edward Arnold, Great Britain.

The authors wish to thank the Danish Medical Research Council and ONO Pharmaceutical Co. Ltd. for financial support.

Inducible nitric oxide synthase expressed in an atheroma-like neo-intima in rabbit carotid artery

JANE F ARTHUR, HEATHER M YOUNG* and GREGORY J DUSTING

Departments of Physiology and *Anatomy & Cell Biology, The University of Melbourne, Parkville, Victoria 3052, Australia.

A silastic collar placed around the common carotid artery of rabbits causes the formation, within 7 days, of an atheroma-like neo-intima containing cells with the appearance of synthetic-state smooth muscle. The intimal thickening occurs while the endothelium remains morphologically normal. Preceding the development of the neo-intima, collared sections of artery, in common with the pre-atherosclerotic stages of human artery disease, shows impairment of endothelium-dependent vasodilatation [1,2]. This defect in endothelium-dependent vasodilatation even extends to the microcirculation downstream of the collar [3]. We examined by immunohistochemistry the isoforms of nitric oxide (NO) synthase present in the artery wall during development of these lesions.

Methods

Silastic, non-occlusive collars were applied to the carotid arteries of rabbits, as in previous studies [1]. Two and seven days after surgery, control (outside the collar) and collared artery segments were removed from the rabbit, fixed in 4% paraformaldehyde (for endothelial NO synthase, eNOS) or Zamboni's fixative (for neuronal or inducible NO synthase, nNOS or iNOS, respectively) for 24 hours at $4^{o}C$, washed and embedded in 50% OCT embedding medium (Tissue-Tek, Miles Inc., USA), then 10μm sections were cut onto gelatin-coated slides and air-dried for subsequent fluorescence immunohistochemistry. The primary antibodies used were: a soluble rabbit antiserum to NO synthase II from murine RAW 264.7 macrophages (iNOS, provided by Dr. R. Tracey) diluted at 1:400, a polyclonal antibody raised against eNOS (Transduction Laboratories, Lexington, Kentucky, USA) at 1 μg/ml, and an antibody raised against nNOS (N74)[4] diluted at 1:200. Antiserum to substance P diluted at 1:2000, and normal rabbit serum diluted at 1:400 were also used on the sections as negative controls. The following day, the sections were washed and incubated in the secondary antibody, a sheep anti-rabbit immunoglobulin conjugated with fluorescein isothiocyanate isomer (FITC). Chicago Blue was also applied to reduce autofluorescence.

Results and Discussion

iNOS fluorescence was clearly evident in the neo-intima of artery segments that had the collar in place for 7 days (Fig 1A). There was no evidence of iNOS expression in the arterial wall before the neo-intima was apparent (at 2 days), nor in control arterial segments at either time point. eNOS was detected in the endothelial layer of both control and collared sections at both 2 and 7 days (Fig. 1C and E). No specific fluorescence for nNOS could be detected in any sections. Some non-specific fluorescence was evident in the adventitia of all sections.

The demonstration of iNOS in the neo-intima is consistent with data previously obtained in organ bath studies. In arterial segments from 7 day-collared rabbits, the addition of N-nitro-l-arginine, an inhibitor of NO synthase, produced large increases in resting tone in collared segments, but had little effect in controls. This suggests that there is a higher basal release of NO in collared arterial segments than in controls.

These results suggest that NO is produced by the inducible isoform of NO synthase in modified smooth muscle cells of the developing neo-intima. Endothelial NOS is expressed normally in control and collared arteries at all stages. Activity of iNOS might deprive the endothelium of substrate for NO production, and might explain the compromised endothelium-dependent vasodilatation observed both in this model and in human coronary artery disease. Whether iNOS has other roles in development of these neo-intimal lesions needs further study.

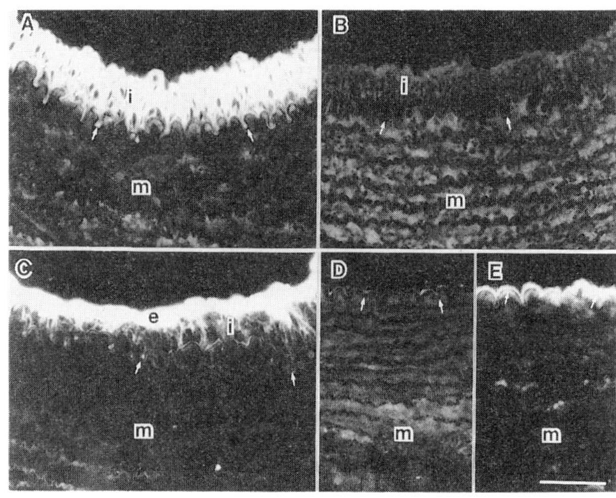

Figure 1. Fluorescence micrographs demonstrating the localisation of different isoforms of NO synthase in collar and control artery sections, taken 7 days after collar placement. Arrows indicate the internal elastic lamina. i, neo-intima; m, media; e, endothelium. Scale bar: 50 μm.
A) iNOS immunoreactivity is present in the collar-induced neo-intima (i) but not in the media (m).
B) nNOS immunoreactivity in collared section is minimal
C) eNOS immunoreactivity shows in endothelial cells (e) in collared arterial segments, but none in the neo-intima (i) or media (m).
D) iNOS immunoreactivity in control: none apparent
E) eNOS immunoreactivity in control segment: restricted to endothelial cells.

Supported by the National Health & Medical Research Council of Australia.

1. Dusting G.J., Curcio A., Harris P.J., Lima B., Zambetis M. and Martin J.F. (1990) Supersensitivity to vasoconstrictor action of serotonin precedes the development of atheroma-like lesions in the rabbit. J Cardiovasc Pharmacol **16**, 667-674.

2. Dusting, G.J. (1995) Nitric oxide in cardiovascular disorders. J. Vascular Res **32**, 143-161.

3. Arthur, J.F., Dusting, G.J. and Woodman O.L. (1994) Impaired vasodilator function of nitric oxide associated with developing neo-intima in conscious rabbits. J Vascular Res **31**, 187-194.

4. Anderson C.R., Furness J.B., Woodman H.L., Edwards S.L., Crack P.J. and Smith A.I. (1995) Characterisation of neurons with nitric oxide synthase immunoreactivity that project to prevertebral ganglia. Journal of the Autonomic Nervous System **52**, 107-116.

TNFα and inducible NO synthase in the myocardium of patients with dilated cardiomyopathy

FARIDA M HABIB, GRAHAM J DAVIES, CELIA M OAKLEY, MAGDY H YACOUB* , JULIA M POLAK and DAVID R SPRINGALL

Cardiology and Histochemistry Departments, Hammersmith Hospital (RPMS) Ducane Road London W12 ONN and *Harefield Hospital, Hill End Road, Harefield, Middlesex UB9 6JH.

INTRODUCTION

The aetiology and pathogenetic mechanism of dilated cardiomyopathy is unknown. Elevated plasma levels of tumour necrosis factor alpha (TNFα) [1] and evidence of increased basal production of nitric oxide [2,3] has been obtained in patients with heart failure, some of whom had dilated cardiomyopathy.
Recent experimental studies have shown that induced nitric oxide production has a negative inotropic effect on cardiac myocytes [4,5]. Furthermore,TNFα has been shown to stimulate the synthesis of iNOS and increased iNOS expression been found in cardiac tissue from patients with dilated cardiomyopathy [6]. These reports raise the possibility that nitric oxide production, induced by a cytokine such as tumour necrosis factor α, exerts a chronic negative inotropic effect on myocardial cells leading to the development of dilated cardiomyopathy and its complications.

MATERIAL AND METHODS

Tissues were obtained from patients undergoing heart transplantation (slices of right and left ventricle, 5mm thick) or endomyocardial biopsy (dilated cardiomyopathy, n=21; ischaemic heart disease, n=10) or from donor hearts (controls, n=9) and fixed in 1% paraformaldehyde. Frozen sections (7μm) were immunostained by the ABC-peroxidase method with antisera to a peptide from human iNOS and to TNFα. Immunoreactivity was measured by computer-assisted image analysis (Seescan Symphony system) For each case, 5 fields that included myocardium were selected randomly in each of 5 non-serial sections and the average optical density measured.

Differences in antigen expression between the groups were analysed by Student's t test for paired observations. A P value of less than 0.05 was considered to be statistically significant.

RESULTS

The cardiac myocytes of patients with dilated cardiomyopathy showed intense iNOS immunoreactivity (Figure 1.a) compared to the much weaker staining of myocytes of patients with ischaemic heart disease. Regional differences were seen in the number of immunoreactive myocytes in patients with dilated cardiomyopathy. They were abundant in the subendocardium but sparsely scattered in other areas.The mean intensity of iNOS staining was greater in patients with dilated cardiomyopathy (mean ±SD; 0.71±0.35) than in patients with ischaemic heart disease (0.18±0.06) (P =0.005). It was greater in both groups than in the hearts of normal controls (Figure 1 c) (0.009±0.007) (P =0.005).
In patients with dilated cardiomyopathy, intense TNFα immunostaining was found in endothelial and vascular smooth muscle cells of intramyocardial blood vessels (Figure 1b). Immunostaining of lesser intensity was found in the cardiac myocytes. There was with no staining in cardiac tissue of ischaemic heart disease or the controls.

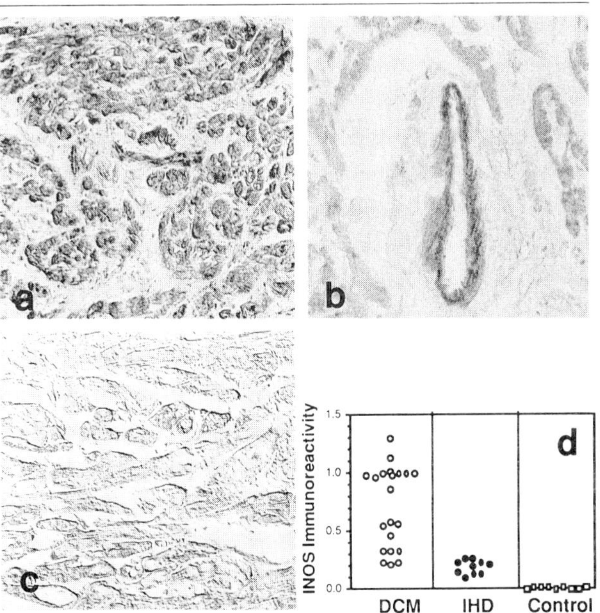

Figure 1. a-c) Cardiac tissue from a patient with dilated cardiomyopathy (a,b) and a control (c) showing staining for iNOS (a,c) and TNFα (b). d) Scattergram showing intensity of iNOS immunostaining i

DISCUSSION

The finding in this study of intense immunoreactivity for TNFα and its co-localisation with inducible nitric oxide synthase in cardiac tissue of patients with dilated cardiomyopathy, may be relevant to the pathogenesis of this disease. Elevated circulating levels of TNFα have been reported in patients with congestive heart failure [1]. TNFα mRNA can be expressed in the heart and biologically active TNFα has been found in cat heart [7]. It is of interest that TNFα has been shown to exert a direct negative inotropic effect in the intact heart [4]. It has also recently been shown that nitric oxide production in cardiac myocytes reduces their contractility [5].
Increased iNOS expression has been found in myocardial biopsy specimens from patients with dilated cardiomyopathy [6]. Furthermore, an indirect negative inotropic effect of TNFα has been found which is mediated by nitric oxide [8]. Heart failure patients have a significant risk of thromboembolism and this risk is particularly high in patients with dilated cardiomyopathy [9]. TNFα can induce tissue factor expression via a 55kD TNFα receptor and is therefore potentially procoagulant [10].
The findings of this study lead to the hypothesis that cytokines produced by an inflammatory response to the causative agent, reduce myocardial contractility by a direct mechanism and by the induction of nitric oxide synthesis in cardiac myocytes leading to abnormal production of nitric oxide. This negative inotropic effect of cytokines, exerted over long periods of time, may lead to a permanent depression of myocardial contractility and the features of dilated cardiomyopathy.

REFERENCES

1. Levine B, Kalman J, Mayer L, Fillit H.M, Packer M. Elevated circulating levels of tumour necrosis factor in severe chronic heart failure. N. Eng J.Med. 1990; 223: 236-241.
2. Drexler H, Hayoz D, Munzel T, Just H, Zelis R, Brunner HR. Endothelial function in congestive heart failure. Am Heart J 1993; 126:761-4.
3. Habib F, Dutka D, Crossman D, Oakley CM, Cleland JGF. Enhanced basal nitric oxide production in heart failure: another failed counter-regulatory vasodilator mechanism? Lancet 1994;344:371-373.
4. Matsumori A, Shioi T, Yamada T, Matsui S, Sasayama S. Vesnarinon, a new inotropic agent, inhibits cytokine production by stimulated human blood from patients with heart failure. Circulation 1994;89: 955-8
5. Brady AJB, Poole-Wilson PA, Harting SE, Warren JB. Nitric oxide production within cardiac myocytes reduces their contractility in endotoxemia. Am. J. Physiol. 1992; 263:H1963-6.
6. De belder A, Radmoski M, Why HJF, Richardson PJ, Bucknall C, Salas E, Martin J, Moncada S. Nitric oxide synthase activities in human myocardium. Lancet 1993;341:84-85.
7. Kapudia S, Lee J, Torre-Amione G, Ma TS, Mann DL. Functional effects of TNFα gene expression in the heart. J Am Coll Cardiol 1995; 91A (Abstract)
8. Finkel MS, Oddis CV, Jacob TD, Watkins SC, Hattler BG, Simmons RL. Negative inotropic effects of cytokines on the heart mediated by nitric oxide. Science 1992;257:387-389.
9. Diaz RA, Obsohan A, Oakley CM. Prediction of outcome in dilated cardiomyopathy. Br. heart J.1987; 58: 393-9.
10. Paleolog EM, Delasalle SA, Buurman WA, Feldmann M. Functional activities of receptors for tumour necrosis factor-alpha on human vascular endothelium. Blood 1994;84:2578-90.

Endotoxin reduces the inotropic action of isoprenaline via nitric oxide in rat isolated hearts

SHAN WEI, XIAOLU SUN and GREGORY J. DUSTING

Department of Physiology, The University of Melbourne, Parkville, Victoria, 3052, Australia

During septic shock the heart exhibits diminished contractility that is characterized by decreased left ventricular ejection fraction and increased end-systolic and diastolic volume indexes [1]. Endotoxin and cytokines induce the expression of a Ca^{2+}-independent nitric oxide (NO) synthase in cardiac myocytes [2]. Induction of NO synthase in ventricular myocytes also attenuates the positive inotropic effects elicited by isoprenaline [3]. We set out to examine the effects of endotoxin in working, perfused rat hearts, to determine the contribution of NO to these effects.

Methods

Sprague-Dawley rats (250-350g) were anesthetized with ether, the hearts rapidly excised and the left ventricle cannulated via the aorta. The hearts were then subjected to a modified Langendorff perfusion at a constant pressure of 60 mmHg, using Krebs' solution warmed to 37°C containing 2.5 mM Ca^{2+}. The perfusate was collected to determine coronary flow rate. Drugs were injected through a side-arm located upstream of the aortic cannula, and ventricular pressure was measured via a second side-arm. Hemodynamic measurements included the peak left ventricular developed pressure (LVP max), the peak rate of rise of LVP (dP/dt max), heart rate (HR), and the coronary flow rate (CF).

Rats were treated with: 1) Vehicle (controls); 2) endotoxin (lipopolysaccharide, 4mg/kg, i.p. 3 hr before perfusion); 3) endotoxin, preceded 30 min by dexamethasone (Dexam, 1mg/kg, i.p.); 4) endotoxin (heart perfused by L-nitroarginine, L-NA 10^{-5}M, 60 min before the second dose of isoprenaline); 5) endotoxin (heart similarly perfused by D-NA 10^{-5}M).

Results and Discussion

After an equilibration period of 30 min, baseline parameters immediately preceding each injection (at 0 min and 90 min perfusion) of isoprenaline were monitored for a period of 10 min. Treatment with endotoxin, or endotoxin and Dexam, L-NA or D-NA had no significant effects on baseline parameters at 0 min of LVPmax (103±3 mmHg), dP/dt max (640±52 mmHg/s) HR (265±15 beats/min) or CF (11.7±1.3 ml/min) (values are means±s.e.m. for 5 control hearts), but after a further 90 min baseline values of dP/dt max had declined significantly in all hearts pretreated with endotoxin. *In vivo* pretreatment with endotoxin also significantly attenuated the inotropic response to isoprenaline (0.15 μg) at both 0 and 90 min, as assessed by reduced percentage change in LVPmax and dP/dt max induced by isoprenaline (Fig.1).

Pretreatment with dexamethasone, 30 min before endotoxin, significantly prevented attenuation of the isoprenaline-induced increase of LVP max and dP/dt max, both at 0 and 90 min, the value of each parameter being similar to the control group (Fig. 1). Moreover, perfusion with the NO synthesis inhibitor L-NA for 60 min in endotoxin-pretreated hearts resulted in a significant restoration of the inotropic response to isoprenaline by 90 min, the percent change in LVPmax and dP/dt max again being similar to control levels. In contrast, the D-arginine isomer (D-NA), which does not block NO synthase, had no effect on the endotoxin-induced depression of inotropy in these hearts. These results suggest that the inhibitory effects of endotoxin on the inotropic response to stimulation of ß-adrenoceptors is mediated by an NO synthase, whose induction is suppressed by dexamethasone. On the other hand, the depressant actions of endotoxin on baseline contractile function in these hearts does not appear to involve NO, and dexamethasone does not protect the hearts from this depressant effect.

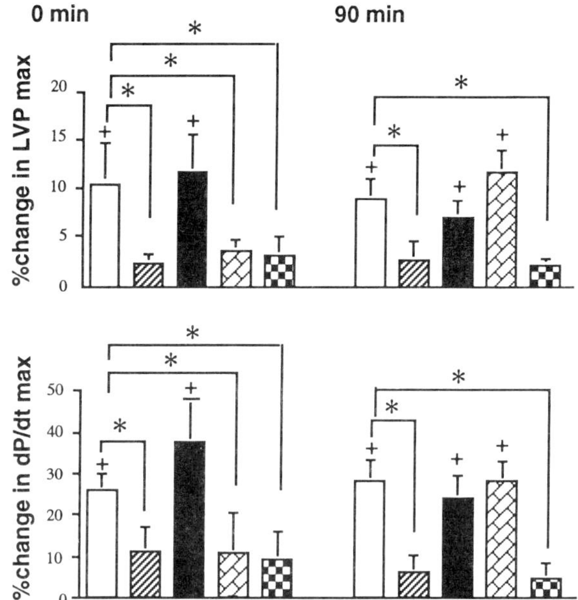

Figure1. Effects of different treatments on the increase of LVPmax and dP/d*t*max produced by each injection of isoprenaline (0.15μg) at 0 min and 90 min in perfused hearts. Control (□ , n=5), endotoxin (▨ , n=4), Dexamethasone (■ , n=4), L-NA (▧, n=4), D-NA (▩ , n=5). Values are mean ±SE. Results are given as the percentage change from baseline value. +significantly different from baseline value within group (Student's paired *t* test). * significantly different from control group at each time point (Student's unpaired *t* test).

In the present study, isoprenaline also caused increases in CF of 5.6±0.6 % and 12.4±4.2 % at 0 min and 90 min, respectively. The increase in CF induced by isoprenaline was abolished in endotoxin-pretreated hearts, but was restored by dexamethasone pretreatment (to 5.1±1.3 and 11.7±4.1 % at 0 and 90 min, respectively), and reversed by L-NA (to 16.3±2.8 % at 90 min), but not by D-NA (1.8±1.2 % at 90 min). It is likely that a substantial portion of the increase in coronary flow was metabolic in nature, resulting from the increased cardiac work. Thus, the decrease of cardiac work performed after endotoxin could account for the weaker flow response to isoprenaline, and for its restoration when NO production was blocked.

Administration of isoprenaline induced 19.5±5.9 % and 17.1±5.8 % increases in HR at 0 and 90 min. None of endotoxin or dexamethasone pretreatment, L-NA or D-NA perfusion affected this chronotropic response. It is possible that endotoxin-induced NO synthase alters those receptor mechanisms mediating inotropy but not chronotropy.

In conclusion, our experiments indicate that the effect of endotoxin on contractility and cardiac output is to reduce ß-adrenergic responsiveness, and that this depressant effect is mediated, at least in part, by products of an endogenous NO synthase.

Supported by the National Heart Foundation of Australia

1. Parker, MM, Shelhamer, JH, Bacharach, SL, et al.(1984) Annals Int. Med. 100, 483
2. Schulz, R., Nava, E. and Moncada, S. (1992) Br.J. Pharmacol. 105, 575-580
3. Baligand, JL., Kelly, RA., Marsden, PA, Smith, TW. and Michel, T. (1993) Proc. Nat. Acad. Sci. 90, 347-351.

Bacterial lipopolysaccharide induces rapid coronary vasodilatation in rats but vasoconstriction in rabbits.

Toby R. CANNON, Giovanni E. MANN and Anwar R. BAYDOUN.

Vascular Biology Research Centre, Biomedical Sciences Division, King's College London, Campden Hill Road, London W8 7AH

INTRODUCTION

In our previous studies we reported that bacterial lipopolysaccharides (LPS) caused a rapid vasodilatation in the rat isolated perfused heart [1]. This response is endothelium-dependent and blocked by inhibitors of nitric oxide synthase (NOS) or pre-treatment with the glucocorticoid dexamethasone. We have now extended these studies and conclusively demonstrated acute release of NO from isolated rat hearts in response to LPS. We have also investigated the concentration-dependency and time-course of inhibition by dexamethasone. Furthermore, we have conducted preliminary experiments on the effects of LPS in the isolated perfused *rabbit* heart, highlighting critical species differences in the response to acute LPS treatment.

EXPERIMENTAL

Materials

Bradykinin, *Eschericia coli* lipopolysaccharide (serotype 0111:B4), L-N^{G}-nitro arginine methyl ester (L-NAME), dexamethasone and 9,11-dideoxy-11α, 9α-epoxymethano-prostaglandin $F_{2\alpha}$ (U46619) were from Sigma Chemical Co. (Poole, U.K.). All other reagents were of analytical grade and purchased from British Drug House (Poole, U.K.).

Experimental Protocol

Hearts obtained from male Sprague-Dawley rats (200 - 350 g) or New Zealand White rabbits (2.5 - 3.5 kg) were perfused by a modified Langendorff technique [1] at constant flows of 10 and 20 ml min^{-1} respectively with Kreb's buffer maintained at 37°C and gassed with 95% O_2/5% CO_2. Coronary perfusion pressure (CPP) was measured as an index of coronary vascular tone using a pressure transducer attached to the aortic inflow via a side arm. Hearts were allowed to equilibrate for 20 minutes and then, where appropriate, the coronary vasculature was constricted with the thromboxane mimetic U46619 (5 nM). Bradykinin (0.1 nmol) was applied as a 10 μl bolus into the perfusion line 3 cm proximal to the aortic cannula. LPS and all other compounds were added to the buffer reservoir and perfused through the hearts. Release of NO by the isolated perfused heart was measured by chemiluminescent detection in samples of coronary effluent collected before, during and after perfusion of LPS (5μg ml^{-1}).

RESULTS

In agreement with our previous findings [1], perfusion of LPS (5μg ml^{-1}) through rat isolated hearts, preconstricted with U46619, resulted in coronary vasodilatation causing a fall in CPP of 34 $\pm$ 4 mmHg. In contrast, administration of LPS (1 μg ml^{-1}) to rabbit isolated perfused heart caused marked coronary vasoconstrtiction, increasing CPP by 80 $\pm$ 24 mmHg. Responses to LPS in both rat an rabbit hearts were rapid in onset, maximal within 10 minutes and sustained over prolonged periods (>20 min) during continued LPS infusion.

Analysis of coronary effluent from rat hearts perfused with LPS showed an increase in NO release up to 170 $\pm$ 6 % of basal levels within 10 min of perfusion of LPS (5μg ml^{-1}). This increase in NO production paralleled the fall in CPP with both effects abolished in the presence of the NOS inhibitor L-NAME (50 μM). These observations confirm our hypothesis that LPS causes acute release of NO within the vasculature of the rat heart thereby initiating coronary vasodilatation. The precise mechanisms underlying the vasoconstrictor effect of LPS in rabbit hearts is being investigated.

The rapid vasodilator action of LPS in the rat isolated perfused heart was blocked (87 $\pm$ 1% inhibition) by a low concentration (1 nM) of the glucocorticoid dexamethasone, indicating a highly selective effect. Furthermore, the inhibitory action of dexamethasone was prevented by pre-treatment of isolated perfused hearts with the type I glucocorticoid receptor antagonist RU-486 showing that Dexamethasone inhibits the response to LPS via an action on the steroid receptor

DISCUSSION

Previous reports have demonstrated that the acute response to LPS *in vivo* is a transient vasodilatation [2,3]. However, the underlying mechanism(s) mediating this effect appears to vary between species with NO implicated in the response observed in rats [2] but not in rabbits [3]. In this study we have observed diametrically opposite acute responses of rat and rabbit hearts, with LPS causing vasodilatation in the rat and marked vasoconstriction in the rabbit coronary vasculature. These findings provide further evidence of significant species differences in the rapid responses to LPS.

LPS-induced vasodilatation in the rat perfused heart may be analogous to the acute phase of the depressor response observed in the whole animal. Both these responses are markedly attenuated by inhibitors of NOS [1,2] and are accompanied by increased production of NO [4]. Interestingly, the response in rat hearts was susceptible to inhibition by dexamethasone which had no effect on the acute *in vivo* response [5].

The precise mechanisms by which the effects of dexamethasone are mediated in the rat heart is unclear. Most of the known actions of this compound involve gene regulation at the transcriptional level, and require relatively prolonged incubation periods [5] compared to those employed in this study. The fact that dexamethasone was effective in rat hearts at 1 nM and after just 15 min treatment would suggest an involvement of a rapidly turned-over intermediate in the response to LPS. Identification of this mediator and elucidation of the remainder of the pathway acutely activated by LPS in both rat and rabbit hearts is currently being investigated.

Acknowledgements

This work was funded by the Medical Research Council U.K. We gartefully acknowledge the support of the British Heart foundation and the Wellcome Trust for the award of a travel grant to ARB.

REFERENCES

[1] Baydoun, A.R., Foale, R.Dand Mann, G.E. (1993). Br. J. Pharmacol., **109**, 987-991

[2] Thiemmerman, C. & Vane, J.R. (1990). Eur. J. Pharmacol., **182**, 591-595.

[3] Wright, C.E., Rees, D.D., and Moncada, S. (1992). Cardiovasc. Res., **26**, 48-57.

[4] Kosaka, H., Watanabe, M., Yoshihara, H., Hurad, N. and Shiga, T. (1992). Biochem. Biophys. Res. Commun., **184**, 1119-1124.

[5]Szabo, C., Thiemerman, C. and Vane, J.R. (1993). Proc. Royal Soc. London, **253**, 233-238.

INFLUENCE of the complement- and contactsystem on the expression of nitric oxide-synthase in endotoxin treated rats.

WILHELM BLOCH, FRANK DOBERS, ANDREAS KRAHWINKEL, GERHARD DICKNEITE* and KLAUS ADDICKS

*Department of Pharmacology, Behring Werke AG Marburg, Department of Anatomy, University of Cologne, Joseph-Stelzmann-Str.9, 50931 Cologne, FRG

introduction

This study focuses on NO in its role as an important regulator in septic shock. NO is released by constitutive and inducible NOS isoforms [1]. There is evidence for an expression of the iNOS in a variety of cells including vascular smooth muscle cells [2], macrophages [3], neutrophils [4] and endothelial cells [5]. It has been established that endotoxin and cytokines induce the expression of iNOS in septic shock [6].It is wellknown that the activation of the complement and contact system via the classical pathway plays an important role in the pathological symptoms characteristic of septic shock [7]. The object of this study is to investigate the influence ot the complement and contact system on the endotoxin-induced expression of iNOS in different organs and cell typs at an early stage of septic shock.

material and methods

Female CD rats (Charles River-Wiga, Sulzfeld, Germany) were anesthetized i.p. initially with a mixture of 80mg/kg Ketamin and 4mg/kg Xylazin followed by i.v. infusion of 20mg/kg/h Ketamin and 1.6mg/kg/h Xylazin as a maintenance dose. Body temperature was maintained at 37°C with a heated underblanket. The rats were either treated with only LPS (500µg/kg/h) or with LPS (same concentration) in combination with the C1-Inh (200U/kg/h) via the tail vene. During the treatment blood pressure, heart rate and ventilation rate were registrated. The administration of endotoxin elicited a servere shock reaction. When the shock phases switched from hyper- to hypodynamic (between 180min and 240min after endotoxin) or latest after 240min the rats were perfused with 4% paraformaldehyde as a fixative. Samples of the spleen, heart, lung, kidney, gut and brain of the animals served as material for the study. NADPH-d staining for light- and electronmicroscopy (described in detail elsewhere) [8] and lightmicroscopical-immunostaining with iNOS-antibody (Biomol, Hamburg) was used in order to detect NOS within the respective tissues. After pretreatment with 3% H_2O_2, 0.2% Triton and 5% bovine serum albumine (BSA) paraffin slices of 7µm were incubated with the primary antibody in a solution of 0.8% BSA and 20 mM NaN_3 in PBS containing the iNOS (Biomol, Hamburg) specific rabbit antibody (1:500) . After rinsing, the sections were incubated with the secondary biotinylated goat-anti-rabbit antibody (1:100 dilution) (Vector Laboratories) for 1 hour at room temperature. Then a streptavidin-horseradish-peroxidase complex was utilized as a detection system (1:200 dilution) for another hour and finally developed for 15 minutes with 3,3 diaminobenzidine tetrahydrochloride in 0,05 M Tris HCl-buffer and 0,1% H_2O_2.

Abbreviations used: nitric oxide (NO), inducible nitric oxide synthase (iNOS), NADPH-diaphorase (NADPH-d), lipopolysaccharides (LPS), C1-esterase-inhibitor (C1-Inh)

results

In comparison with the control group, the LPS and LPS/C1-Inh treated groups expressed iNOS in different organs and cell types. There were distinct differences in the induction pattern between the investigated organs at this early time of septic shock. No correlation could be registrated between the degree of iNOS expression and the hemodynamic changes at the point in time at recording. A distinctly increased expression of iNOS was observed particularly in spleen, lung and surprisingly also in the heart. In the spleen the red pulp showed a distinct induction contrarily to the iNOS negative white pulp (fig.1). Contralily to the iNOS negative control in the lung bronchial epithelium, vasculare endothel cells, smooth muscle and alveolar makrophages were iNOS positive. The smooth muscle of the bronchi revealed a distinct induction of iNOS while in vasculare smooth muscle only a weak expression was observed. In the heart, endothelium and smooth muscle of coronary vessels were iNOS positive. Particularly the endothelium of the cardial capillaries, which under basal conditions are only weakly NADPH-d stained and show no iNOS immunoreaction, displayed strong NADPH-d staining and iNOS immunohistochemistry in both treated groups. Cardiomyocytes also showed weak NADPH-d upon light- and electronmicroscopical staining in LPS and LPS/C1-Inh group and a positive immunoreaction with the iNOS-antibody. A distinct expression of NADPH-d only was investigated in the lympathetic tissue of the gut. The other investigated organs as brain (only in vessels) and kidney only contained iNOS positive leukocytes. It looks as if the expression of iNOS in LPS/C1-Inh treated rats was more intensive and extensive compared to animals treated solely with LPS.

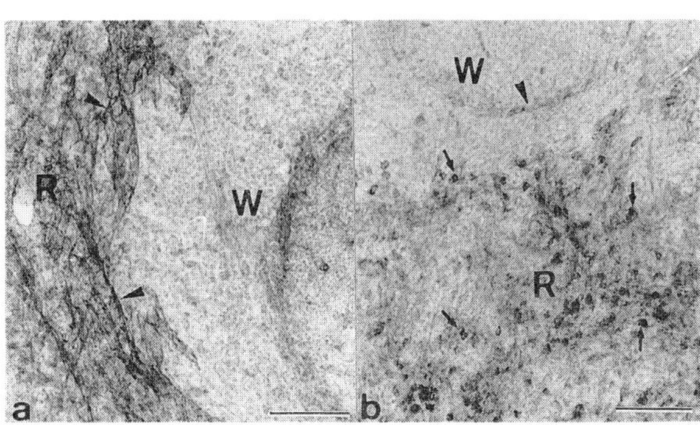

figure 1 (a) The spleen of an untreated only shows NADPH-d staining in the endothelium (arrowheads) of red (R) and white pulp (W) vessels. (b) Additionally in LPS/c1-Inh treated rat NADPH-d positive leucocytes(arrows) appear in the red pulp (R). bar 100µm

discussion

The inhibition of the classical pathway of complement activation does not prevent the induction of NOS on the contrary, it seems that there is a slight increase of iNOS expression. Although complement system activation enhances the expression of cytokines [10,11]. It can be suggested that factors directly or indirectly involved in the classical complement pathway negatively influence the induction of iNOS as descript for thrombin [9]. But further studies are needed to quantitatively evaluate the iNOS expression after LPS and LPS/C1-Inh treatment.

The variation of iNOS expression between the organs and cell types suggests the existence of equally varying, organ and cell specific induction mechanisms in vivo. Our results supported the findings of other groups which demonstrated different induction mechanism for the lung, liver, spleen and heart in endotoxin induced shock [2,12]. Moreover the observed difference of iNOS expression between organs and in different cell types in one organ is of major importance for the understanding of the heterogenous NO influence elicited in septic shock.

The lack of correlation between iNOS-expression and hemodynamic changes supports the theory of NO-independent regulation of the hemodynamic situation [13,14] at least at this early time of septic shock.

references

1. Förstermann, U., Schmidt, H.H.H.W., Pollock, J.S., Sheng, H., Mitchell, J.A., Warner, T.D., Nakane, M.M. and Murad, F. (1991) Biochem. Pharmacol., 42: 1849, 1991.
2. Fleming, I., Gray, G.A., Schott, C., & Stoclet, J.. (1991) Eur. J. Pharmacol. 200, 375-376.
3. Green, S.J., Mellouk, S., Hoffman, S.L., Meltzer, M.S., & Nacy, C.A.. (1990) Imm. Lett. 25, 15-20.
4. McCall, T.B., Boughton-Smith, N.K., Palmer, R.M.J., Whittle, B.J.R., & Moncada, S.. (1989) Biochem. J. 261,293-296.
5. Ohshima, H., Brouet, I.M., Bandaletova, T., Adachi, H., Oguchi, S., Iida, S., Kurashima, Y., Morishita, Y., Sugimura, T. & Esumi, H.. (1992) Biochem. Biophys. Res. Commun. 187,1291-1297.
6. Cunha, F.Q., Assreuly, J., Moss, D.W., Rees, D., Leal, L.M.C., Moncada, S., Carrier, M., Donnell, C.A. and Liew, F.Y. (1994) Immunology 81,211-215
7. Hack, C.E., Nuijens, J.J., Felt-Bersma, R.J.F., Schreuder, W.O., Eerenberg-Belmer, A.J.M., Paardekooper, J., Bronsveld, W. and Thijs, L.G. (1989) Amer.J.Med. 86,20-26
8. Addicks, K., Bloch, W. and Feelisch, M. (1994) Microsc.Res.Tech. 29,161-168
9. Schini, V.B., Catovsky, S., Scott-Burden, T. and Vanhoutte, P.M. (1992) J.Cardiol.Pharmacol. 20 Suppl.12,S142-S144
10. Haeffner-Cavaillon, N., Cavaillon, J.M., Laude, M. and Kazatchkine, M.D. (1987) J.Immunol. 139,794-799
11. Cavaillon, J.M., Fitting, C. and Haeffner-Cavaillon, N. (1990) Eur.J.Immunol. 20,253-257
12. Evans, T., Carpenter, A., Silva, A. and Cohen, J. (1992) Infect.Immun. 60,4133-9
13. van den Berg, C., van Amsterdam, J.G.C., Bisschop, A., Piet, J.J., Wemer, J. and de Widt, D.J. (1994) Eur.J.Pharmacol. 270,379-382
14. Preiser, J.C., Zhang, H., Wachel, D., Boeynaems, J.M., Buurman, W. and Vincent, J.L. (1994) Eur.Surg.Res. 26,10-18

IL-1β and TNFα mediate the circulatory failure elicited by lipoteichoic acid from *Staphylococcus aureus*

MURALITHARAN KENGATHARAN, SJEF J. DE KIMPE, DAMON SMITH* and CHRISTOPH THIEMERMANN

The William Harvey Research Institute and *Therapeutic Antibodies Ltd, St. Bartholomew's Hospital Medical College, Charterhouse Square, London EC1M 6BQ, United Kingdom

Introduction

Cytokines such as tumor necrosis factor-α (TNFα) and interleukin-1β (IL-1β) are important mediators of circulatory shock. Administration of TNFα or IL-1β to animals mimics several cardiovascular features of circulatory shock [1,2]. Agents that neutralise the effects of TNFα or IL-1β provide protection in animal models of endotoxin shock suggesting that these cytokines play an important role in sepsis [1,2]. In rats, lipoteichoic acid (LTA), a cell wall component of *Staphylococcus aureus* (an organism without endotoxin) causes circulatory failure characterised by hypotension and vascular hyporeactivity to norepinephrine (NE) [3]. This effect of LTA is associated with an increase in the plasma levels of TNFα and is attributed to an enhanced formation of the potent vasodilator nitric oxide (NO) following the induction of the NO synthase (iNOS) [3]. TNFα and IL-1β alone or in concert with bacterial cell wall components or cytokines such as interferon-γ (IFNγ) cause the induction of iNOS [1,2]. Here, we investigate the role of TNFα and IL-1β in the circulatory failure elicited by LTA in anesthetized rats.

Methods

Male Wistar rats (250-350g) were anesthetized with thiopentobarbitone sodium (120mg/kg, ip). The trachea was cannulated to facilitate spontaneous respiration, the carotid artery for the measurement of blood pressure (MAP) and heart rate; and the jugular vein for administration of compounds. Sheep polyclonal antibodies (PAb) against human TNFα (3mg/kg, TNFα-PAb, iv), IL-1β (3mg/kg, IL-1β-PAb, iv) or a mixture of PAb against human TNFα and IL-1β (both 3mg/kg, TNFα+IL-1β-PAb, iv) were administered 30 min prior to injection of LTA (10mg/kg, iv, time 0). At 90 min, a plasma sample was collected to measure TNFα by ELISA. The pressor response to NE (1μg/kg, iv) was assessed prior to and every 60 min after the injection of LTA. At 360 min, rats were killed and lungs removed to determine iNOS activity by measuring the conversion of [^{3}H]L-arginine to [^{3}H]L-citrulline [3] and a plasma sample was collected to measure IFNγ by ELISA. Statistical evaluation was performed using ANOVA followed by Bonferroni's *t* test.

Results

Administration of LTA (10mg/kg) resulted in an initial fall in MAP from 120±2 mmHg (time 0, control, n=4-6) to 95±5 mmHg at 60 min (p<0.05) which was followed by a further fall in MAP to 78±3 mmHg (by 360 min). The degree of hypotension was significantly greater than that observed in sham-operated animals (Table 1).

Table 1. Effect of PAb to TNF-α and IL-1β on the hypotension and vascular hyporeactivity to NE induced by LTA (at 360 min)
Each value represents the mean ± S.E.M. for 4-6 rats.
# *P*<0.05 vs sham operated control and † *P*<0.05 vs LTA control.

	MAP (mmHg)	NE pressor (mmHg.min)
Sham	113±3	48±4
LTA (10 mg/kg)	78±3#	23±4#
+ TNFα-PAb	87±7	32±5
+ IL-1β-PAb	79±3	35±5
+ TNFα+IL-1β-PAb	100±5†	48±8†

The pressor response elicited by NE (1μg/kg) was significantly attenuated, compared to sham-operated animals at 60-360 min after injection of LTA (Table 1). LTA also caused, a significant increase in the plasma level of TNFα (at 90 min) and IFNγ (at 360 min).

Pretreatment of rats with TNFα-PAb or TNFα+IL-1β-PAb, but not IL-1β-PAb, prevented the rise in the plasma levels of TNFα induced by LTA (Table 2). In contrast, the increase in the plasma levels of IFNγ was significantly reduced by the treatment of rats with IL-1β-PAb or TNFα+IL-1β-PAb, but not by TNFα-PAb. Furthermore, the delayed hypotension and vascular hyporeactivity to NE was significantly attenuated by TNFα+IL-1β-PAb. However, the effect of TNFα-PAb or IL-1β-PAb on either MAP or vascular hyporeactivity to NE was not significant.

Table 2. Effect of PAb to TNFα and IL-1β on the plasma level of TNFa, plasma level of IFN-γ and the induction of iNOS (at 360 min) in lungs caused by LTA
Each value represents the mean ± S.E.M. for 4-6 rats.
# *P*<0.05 vs sham operated control and † *P*<0.05 vs LTA control.

	TNFα at 90 min (ng/ml)	IFNγ at 360 min (ng/ml)	iNOS activity (% control)
sham	0.2±0.1	0.1±0.1	17±8
LTA (10 mg/kg)	3.5±0.6#	1.7±0.3#	100
+ TNFα-PAb	0.4±0.1†	1.0±0.4	70±12
+ IL-1β-PAb	4.1±1.0	0.4±0.3†	82±11
+ TNFα+IL-1β-PAb	1.0±0.2†	0.2±0.2†	44±9†

At 360min after injection of LTA, there was a significant increase in the calcium-independent iNOS activity in lung homogenates compared to lung homogenates from sham-operated animals (13±2 nmol L-citrulline/30min/g tissue for LTA treated animals vs 2±1 nmol L-citrulline/30min/g tissue for sham-operated animals). This increased iNOS activity was significantly attenuated by TNFα+IL-1β-PAb, but the reduction in iNOS activity by either TNFα-PAb or IL-1β-PAb was not statistically significant (Table 2).

There was no significant change in heart rate in any of the experimental groups (results not shown).

Discussion

The present study demonstrates that both TNFα and IL-1β play an important role in the circulatory failure (hypotension and vascular hyporeactivity to NE) induced by LTA.

Administration of LTA to anesthetized rats results in delayed hypotension and vascular hyporeactivity to NE due to an enhanced formation of NO by iNOS [3]. Release of TNFα by LTA is postulated to be one of the mechanisms by which LTA induces the expression of iNOS *in vivo* [3]. Our results show that LTA induces the release of TNFα and IFNγ both of which are known to induce iNOS. Interestingly, removal of IL-1β rather than TNFα with polyclonal antibodies leads to prevention of the rise in IFNγ indicating that IL-1β contributes to the release of IFNγ. Neutralisation of the effects of TNFα and IL-1β resulted in significant amelioration of the circulatory failure induced by LTA, and this effect is likely to be due to prevention of iNOS induction. Furthermore, neutralising TNFα and IL-1β is more effective in preventing circulatory failure than neutralising either TNFα or IL-1β (this study), as (i) IL-1β or TNFα alone can induce a shock-like state in animals [1,2], and (ii) these cytokines can be released into the circulation upon stimulation by bacterial wall components independent of each other [4]. In conclusion, both TNFα and IL-1β contribute to the induction of iNOS and circulatory failure caused by LTA in anesthetized rats suggesting that these cytokines are important mediators in the pathogenesis of Gram-positive sepsis.

References

1. Van der Poll, T. and Lowry, S.F. (1995) Shock. **3(1)**, 1-12.
2. Pruitt, J.H., Copeland, E.M. and Moldawer, L.L. (1995) Shock. **3(4)**, 235-251.
3. De Kimpe, S.J., Hunter, M.L., Bryant, C.E., Thiemermann, C. and Vane, J.R. (1995) Br. J. Pharmacol. **114**, 1317-1323.
4. Perretti, M., Duncan, G.S., Flower, R.J. and Peers, S.H. (1993) Br. J. Pharmacol. **110**, 868-874.

Abbreviations used: interluekin-1β, IL-1β; interferon-γ, IFNγ; lipoteichoic acid, LTA; mean blood pressure, MAP; nitric oxide, NO; norepinephrine, NE; polyclonal antibodies to TNFα, TNFα-PAb; poylclonal antibodies to IL-1β, IL-1β-PAb; tumor necrosis factor α, TNFα.

Inhibitors of nitric oxide synthase attenuate the liver dysfunction caused by endotoxin *in vivo*.

C. THIEMERMANN, H. RÜTTEN, C.C. WU and J.R. VANE

The William Harvey Research Institute, St. Bartholomew's Hospital Medical College, Charterhouse Square, London, EC1M 6BQ, U.K.

An enhanced formation of nitric oxide (NO) due to the induction of NO synthase (iNOS) contributes to the severe hypotension and vascular hyporeactivity to catecholamines in shock. There is, however, little information regarding the effects of NOS inhibitors on organ function in endotoxemia (1). The progression of shock to a multiple organ dysfunction syndrome (MODS) is associated with a substantial increase in mortality. Non-selective NOS inhibitors, such as N^G-methyl-L-arginine (L-NMMA) may cause excessive vasoconstriction as well as an increase in organ ischaemia and mortality by inhibiting the activity of endothelial NOS (eNOS) activity (1). Thus, the beneficial haemodynamic effects of non-selective NOS inhibitors may be due to inhibition of iNOS activity, while the adverse effects may be due to inhibition of eNOS activity. Here we have compared the effects of aminoethyl-isothiourea (AE-ITU), a relatively selective inhibitor of iNOS activity (2, 3) and L-NMMA, a non-selective inhibitor of iNOS and eNOS activity, on hemodynamics, MODS and iNOS activity in rats with endotoxic shock.

Male Wistar rats (260-340g) were anesthetised with thiopentome sodium (120 mg/kg i.p.) and instrumented for the measurement of mean arterial blood pressure (MAP) and heart rate as well as injection of drugs (femoral vein). Rats received vehicle or *E. coli* lipopolisaccharide (LPS, 10 mg/kg i.v. at time=0) and hemodynamic parameters were recorded for 6 h. At 2 h after injection of LPS or vehicle, animals received vehicle (saline, 0.6 ml/kg/h, n=8), L-NMMA (3 mg/kg i.v. bolus + 3 mg/kg/h) or AE-ITU (1 mg/kg i.v. bolus + 1 mg/kg/h). At 2 or 6 h after LPS, blood was collected and centrifuged (6000 rpm for 3 min) to prepare serum. Liver dysfunction was assessed by measuring the rises in the serum levels of alanine aminotransferase (GPT), aspartate aminotransferase (GOT), bilirubin and γ-glutamyl transferase (γ-GT). Renal dysfunction was assessed by measuring the rises in the serum levels of creatinine and urea. The activity of iNOS was determined by measuring the conversion of [^{3}H] L-arginine to [^{3}H] L-citrulline in lung or liver homogenates. A two-way analysis of variance (ANOVA) followed by a Dunnett's t-test was used to compare means between groups.

In the anesthetised rat, LPS caused a fall in MAP from 117±3 mmHg (time=0) to 97±4 mmHg at 2 h ($p<0.05$, n=15) and 84±4 mmHg at 6 h ($p<0.05$, n=15). The pressor effect of norepinephrine (NE, 1 µg/kg, i.v.) was also significantly reduced at 1-6 h after LPS (vascular hyporeactivity). Treatment of LPS-rats with AE-ITU only caused a transient rise in MAP, but significantly attenuated the delayed vascular hyporeactivity seen in LPS-rats. Infusion of L-NMMA caused a rapid and sustained rise in MAP and attenuated the delayed vascular hyporeactivity to NE. Neither AE-ITU nor L-NMMA had any effect on either MAP or the pressor effect elicited by NE in rats infused with saline rather than LPS. Endotoxemia for 6 h was associated with a significant rise in the serum levels of GOT, GPT, γ-GT and bilirubin and, hence, liver dysfunction. Treatment of LPS rats with AE-ITU significantly attenuated the observed rises in GOT, GPT, γ-GT and bilirubin ($p<0.05$, n=10). In contrast, L-NMMA reduced the increase in the serum levels of γ-GT and bilirubin, but not in GOT and GPT (n=5). Injection of LPS also caused a time-dependent, but rapid (almost maximal at 2 h), increase in the serum levels of urea and creatinine and, hence, renal dysfunction. This renal dysfunction was not affected by either AE-ITU (n=10) or L-NMMA (n=5). In rats infused with saline rather than LPS, neither AE-ITU (n=4) nor L-NMMA (n=4) had any effect on the serum levels of GOT, GPT, γ-GT, bilirubin, creatinine or urea.

Endotoxemia for 6 h resulted in a 4.5 fold rise in the serum levels of nitrite (9.1±0.8mM, $p<0.01$, n=15), which was significantly reduced by treatment with AE-ITU (6.3±0.5mM, $p<0.05$, n=10) or L-NMMA (5.1±0.4mM, $p<0.05$, n=5). In addition, endotoxemia was also associated with a significant increase in iNOS activity in lung (from 0.2±0.1 to 8.4±0.6 pmol/mg/min) and liver homogenates (from 0.3±0.2 to 3.3±0.6 pmol/mg/min), which was significantly reduced in lung or liver homogenates obtained from LPS-rats treated with either AE-ITU (lung: 4.9±0.2; liver: 1.6±0.4 pmol/mg/min) or L-NMMA. (lung: 5.5±0.6; liver: 1.4±0.4 pmol/mg/min).

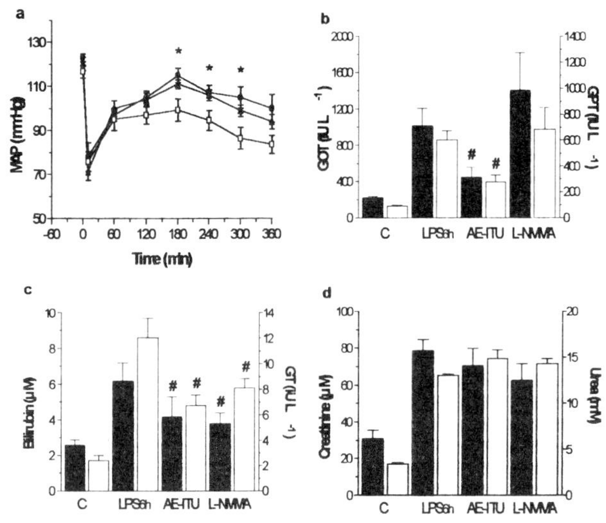

Figure 1: Effects of the NOS inhibitors aminoethyl-isothiourea (AE-ITU) or L-NMMA on (a) mean arterial blood pressure (MAP), (b) serum activities of alanine aminotransferase (GPT, open columns) and aspartate aminotransferase (GOT, solid columns), (c) bilirubin (solid columns) and γ-glutamyl transferase (GT, open columns) and (d) creatinine (solid columns) and urea (open columns) in rats with endotoxin shock. All animals received endotoxin (lipopolysaccaride, LPS, 10 mg/kg i.v. at time=0). Different groups of LPS-animals received continuous infusions of vehicle, AE-ITU (1mg/kg+1mg/kg/h, n=10) or L-NMMA (3 mg/kg i.v. +3 mg/kg/h, n=5) starting 2 h after LPS. Data are mean±s.e. mean of n observations. $P<0.05$ when compared to rats treated with LPS alone (LPS-controls); ANOVA followed by t-test.

Here we demonstrate that the NOS inhibitors AE-ITU or L-NMMA attenuate the delayed circulatory failure caused by endotoxemia in the rat. In addition, the iNOS selective NOS inhibitor AE-ITU abolished the rises in the serum levels of GOT and GPT (marker enzymes for hepatic parenchymal injury), bilirubin and γ-GT, (marker enzyme for cholestasis) and, therefore, the severe liver dysfunction caused by endotoxaemia. Interestingly, the non-isoenzyme selective NOS inhibitor L-NMMA also attenuated the rise in the serum levels of bilirubin and γ-GT, but not the rise in GPT or GOT caused by endotoxemia. Thus, L-NMMA attenuated, but did not abolish, the liver dysfunction caused by endotoxic shock.

Expression of iNOS activity in hepatocytes and Kupffer cells results in a reduction in the synthesis of proteins (4), prostaglandins and interleukin-6 (5); and inhibition of mitochondrial respiration possibly due to formation of peroxynitrite (6). We propose that the beneficial effects on liver dysfunction by the NOS inhibitor used are due to prevention of the above-mentioned cytotoxic effects of NO. The finding that the degree of inhibition of iNOS activity in the liver caused by L-NMMA and AE-ITU are similar, while AE-ITU, but not L-NMMA prevented the rise in the serum levels of GOT and GPT, indicates that the beneficial effects of AE-ITU (on GOT and GPT) are independent of the inhibition of iNOS activity.

1. Thiemermann, C. (1994). Adv. Pharmacol. **28**, 45-79.
2. Garvey, P.E., Oplinger, J.A., Tanoury, G.J. et al. (1994). J. Biol. Chem. **269**, 26669-26676.
3. Southan, G.J., Szabo, C. and Thiemermann, C. (1995). Br. J. Pharmacol. **114**, 510-516.
4. Billiar, T.R., Curran, R.D., Stuher, D.J., West, M.A., Bentz, B.G. and Simmons, R.L. (1989). **169**, 1467-1472.
5. Stadler, J., Harbrecht, B.G., Di Silvio, M., Curran, R.D., Jordan, M.L., Simons, R.L. and Billiar, T.R (1993). J. Leukoc. Biol. **53**, 165-172.
6. Szabo, C. and Salzman, A.L. (1995). Biochem. Biophys. Res. Comm. **209**, 739-743.

Serum and 24 hour urine excretion of nitrate in severe human septic shock.

Øystein A. Strand*, Tom Granli#, Roger Dahl#, Knut A. Kirkebøen+and Oluf C. Bøckman#.

*Dep. Inf. Dis., Ullevål Hospital, N-0407 Oslo, Norway #Norsk Hydro Research Center, N-3900 Porsgrunn, Norway, +Inst. Exp. Med. Res., Ullevål Hospital, N-0407 Oslo, Norway.

Inflammatory stimuli have been shown to induce an isoenzyme of nitric oxide (NO) synthase in experimental models of septic shock. NO produced in pathophysiological amounts by this enzyme is proposed to be of importance for the hypotension and cardiovascular hyporeactivity in septic shock [1].

Several studies have shown increased levels of nitrate ($NO3^-$), the stable bioreaction-product of NO, in plasma from septic patients [2-6]. The normal plasma half life of $NO3^-$ is between 5 and 8 hours [7,8]. $NO3^-$ is mainly excreted in the urine and the distribution volume is equal to or larger than the extracellular volume [7,8], but unknown during sepsis. In this study $NO3^-$ was measured in serum and urine by a new method of ion chromatography. This method was developed as the Griess reaction in our hands gave unspecific reactions in plasma, serum and especially in urine, as described by others [9].

The aim of the present study was to determine serum $NO3^-$ levels in patients with severe septic shock. All patients (Group A, n=7) were in septic shock on admission to hospital, had kidney failure and died of septic shock. Daily serum $NO3^-$ levels and 24 hour urine excretion of $NO3^-$ (Nex) was measured. All septic patients had severe impairment of renal function. Predialysis sera from chronic ambulatory hemodialysis (CADH) patients with comparable creatinine clearance values were included as controls (Group B, n=6). In group C, (n=7) serum $NO3^-$ levels were measured in healthy volunteers. Basal Nex was measured in six healthy volunteers (24 samples) on low $NO3^-$ diet ($NO3^-$ intake <6 mg /24 hours).

In group A serum $NO3^-$ on admission was 80±26 μM (Mean± SEM) and 75±10 μM in preterminal samples. Nex (11 samples from 2 patients) was 0,5±0,1 mmol. In group B serum $NO3^-$ was 85±8 μM. In group C serum $NO3^-$ was 46±10 μM and Nex 0,5±0,2 mmol. When compared to healthy volunteers, preterminal $NO3^-$ levels were elevated in patients with septic shock ($p<0.05$). No difference was, however, found between $NO3^-$ levels in septic patients and CAHD patients. Nex in septic shock (11 samples) was not elevated above control levels.

The findings of the present study is in accordance to studies having shown increased plasma NOx levels during sepsis [2-6]. In this study of severe sepsis the patients were matched to a control group of CAHD patients using creatinine clearance which are independent of sex and muscle mass. Other studies have either used serum creatinine values, a rough estimate of renal function, [2,5], or have not included a control group with renal impairment [3,4,6]. Due to the severety of disease in all our patients, the result of our study is not directly comparable to others. A possible downregulation of NO-synthesis in our more severely ill patients cannot be excluded, as downregulation of NO synthase has been shown to occur in vitro [10], and possibly also in vivo during sepsis in trauma patients [2,11].

The main finding of the present study is that serum $NO3^-$ levels are not consistently elevated in patients with irreversible septic shock when compared to $NO3^-$levels in CAHD patients with the same degree of renal failure. The high $NO3^-$ levels measured in our septic patients may be caused by $NO3^-$ accumulation due to loss of renal excretion. Septic patients cannot, however, be directly compared to patients with CAHD. Septic shock leads to changes in distribution volumes due to fluid resuscitation and capillary damage, and the distribution volume of $NO3^-$ during sepsis is probably increased. Total body $NO3^-$ may be therefore be elevated in sepsis even with normal serum levels. The unknown half-life of $NO3^-$ in serum during sepsis , the influence of renal dysfunction on $NO3^-$ levels, and the unknown distribution volume of $NO3^-$ during sepsis, suggest that a better parameter than $NO3^-$ is needed to further clarify the possible role of NO in the pathophysiology of human septic shock.

References:

1. Moncada S, Palmer RMJ, Higgs EA. Nitric oxide: physiology, pathophysiology and pharmacology. (1991) Pharmacol. Rev. **43**, 109-142.

2. Ochoa JB, Udekwu AO, Billiar TR et al. Nitrogen oxide levels in patients after trauma and during sepsis. (1991) Ann. Surg. **214**, 621-626.

3. Wong HR, Carcillo JA, Burckart G, Shah N, Janosky JE. Increased serum nitrite and nitrate concentrations in children with the sepsis syndrome. (1995) Crit. Care Med. **23**, 835-842.

4. Shi Y, Li H, Shen C, Wang J, Qin S, Pan J. Plasma nitric oxide levels in newborn infants with sepsis. (1993) J. Pediatrics. **123**, 435-438.

5. Gomes-Jimenez J, Selgado A, Mourelle M, Martin MC, Segura RM, Peracaula R, Moncada S. L-arginine: nitric oxide pathway in endotoxemia and human septic shock. (1995) Crit. Care Med. **23**, 253-258.

6. Evans T, Carpenter A, Kinderman H, Cohen J. Evidence of increased nitric oxide production in patients with the sepsis syndrome. (1993) Circ. Shock. **41**, 77-81.

7. Wagner DA, Schultz DS, Deen WM, Young VR, Tannenbaum SR. Metabolic fate of an oral dose of ^{15}N-labeled nitrate in humans: Effects of diet supplementation with ascorbic acid. (1983) Cancer Res. **43**, 1921-1925.

8. Wennmalm Å, Benthin G, Jungersten L, Edlund A, Peterson A-S. Nitric oxide formation in man as reflected by plasma levels of nitrate, with special focus on kinetics, confounding factors, and response to immunological challenge. (1993) Endothelium,**1(Suppl)**, 84.

9. Lee K, Greger JL, Consaud JR, Graham KL, Chinn BL. Nitrate, nitrite balance, and de novo synthesis of nitrate in humans consuming cured meats. (1986) Am. J. Clin. Nutr. **44**, 188-194.

10. Assruy J, Cunha FQ, Liew FY, Moncada S. Feedback inhibition of nitric oxide synthase activity by nitric oxide. (1993) Br. J. Pharmacol. **108**, 833-837.

11. Jacob T, Ochoa JB, Udekwu AO et al. Nitric oxide production is inhibited in trauma patients. (1993) J. Trauma **35**, 590-597.

Endotoxin-induced venous hyporesponsiveness in humans

K Bhagat, J. Collier, P. Vallance; Clinical Pharmacology Unit, St George's Hospital Medical School, London, SW17 0RE, UK.

Injection of endotoxin (ETX) into animals causes a biphasic vascular response. An initial and transient hypotensive effect occurs within 10-15min; a longer lasting hyporesponsiveness occurs 4-5h later. The delayed effects are inhibited by nitric oxide synthase inhibitors and can be prevented by glucocorticoids, but the mechanisms of the acute effects are not clear. Studies in which ETX has been infused systemically into healthy volunteers suggest that the hypotensive response to ETX starts within 1-2 h of administration and reaches a maximum by 8 h. However, these studies do not allow assessment of direct vascular effects of ETX or allow the analysis of mechanisms of changes seen. In this study the dorsal hand veins of healthy volunteers were used in order to study the direct vascular effects of ETX.

Subjects lay in a temperature-controlled laboratory(28-30°C) with one arm placed on an angled support with the hand above the level of the heart. The diameter of a single vein was recorded by measuring the linear displacement of a light weight probe placed on the skin overlying the summit of the vein when the pressure in a congesting cuff placed around the upper arm was deflated from 40-0mmHg. For the studies of reactivity, drugs or physiological saline were infused continuously (0.25ml/min) into the vein through a 23SWG needle placed 5-10mm downstream from the tip of probe.

A dose response curve to NA (5-1280pmol/min; doubling doses and each dose given for 5min) was constructed. The vein was then isolated from the circulation by means of two wedges and ETX (LOT EC-5, E.Coli, 100 ETX units) was instilled into the isolated segment. After 1h the contents of the veins were aspirated, the wedges removed, and further dose response curves to NA constructed at 1, 2, 3 and 4h using doses of NA that produced 0, 20-40, 40-70 and 70-100% venoconstriction prior to ETX.

ETX caused a significant rightward shift in the dose response to NA at 1, 2, 3h. The effect was maximal at 1h (maximum constriction pre-ETX: 87±4% and post ETX 52±8%; n=4 for each time point; p<0.05) and returned to baseline values by 4h (pre-ETX: 74±4% and post ETX 74±16%; Figure 1) In the control (saline-treated) vein there was no change in response to NA (Figure 2). To determine whether the effect was a local response, in 3 subjects dose response curves to NA were constructed in two separate veins on the same hand. One vein was isolated and received ETX while the other was left unoccluded. Dose responses to NA were established as before (in both veins simultaneously). In ETX-treated veins there was a rightward shift of the NA dose resonse curve, whereas no such change was seen in the control vein (Figure 3).

Locally infused L-NMMA (100nmol/min; n=5), aspirin (1g taken orally, 2h before the study; n=5) or a combination of L-NMMA and aspirin had no effects on hyporesponsiveness induced by ETX (Figure 4). However, hydrocortisone (100mg; n=5) taken 2 h before the study completely abolished effects of ETX (preETX: 76±4% and post ETX:70±5%; p<0.05; Figure 5).

Figure 1 Time course of the effects of endotoxin

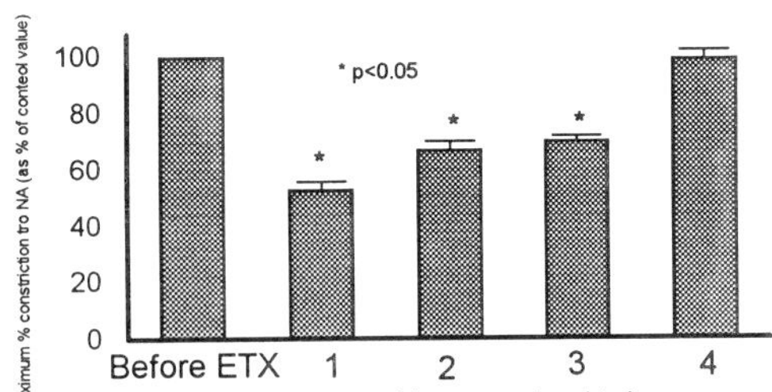

In 12 subjects the time course of the response to endotoxin was explored. Dose-response curves to NA were constructed to NA before and at 1, 2, ,3 and 4h after exposure to endotoxin. results are expressed as maximum percentage constriction to NA relative to the initial control value for that study.

Figure 2 Effects of endotoxin

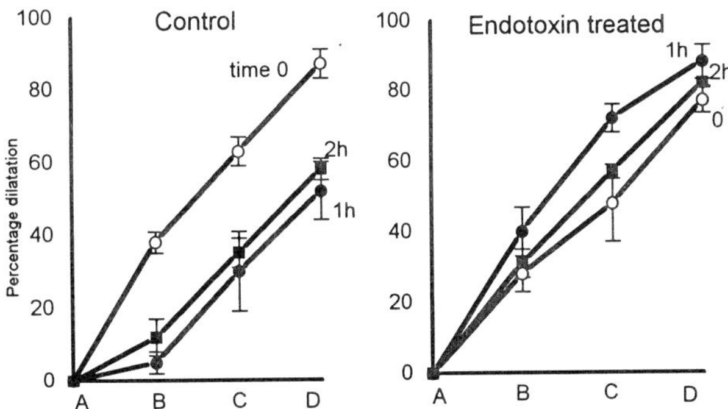

In 4 subjects dose response curves to NA were constructed before and 1h and 2h after the vein was exposed to endotoxin or saline. For repeat dose response curves to NA at 1 and 2h after saline or endotoxin 4 doses of NA were selected that produced 0, 20-40, 40-70, 70-100% constriction on the first occasion (designated A, B, C,D).

Figure 3 Effect of endotoxin: Local or systemic effect?

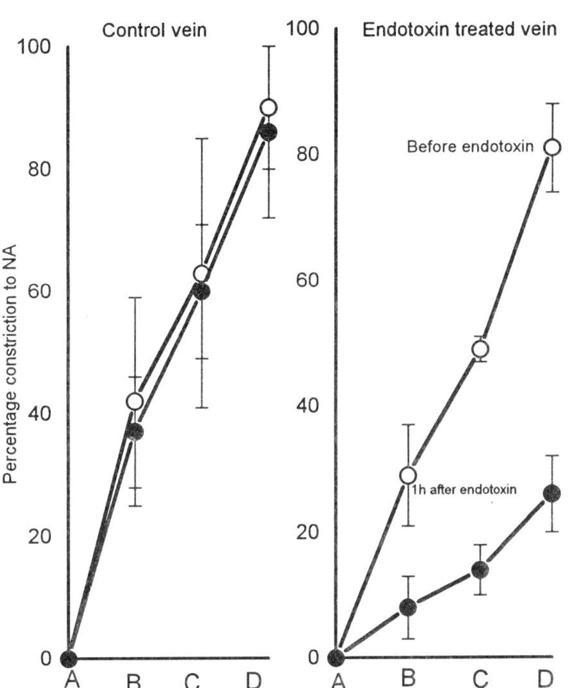

In 3 subjects dose response curves were constructed simultaneously in 2 adjacent veins. The control vein was left unoccluded while the other vein was isolated and received endotoxin. Dose response curves to noradrenaline were constructed in both veins before and 1h after one vein was exposed to endotoxin.

Figure 4 Effects of NO synthase and/or cyclo-oxygenase inhibitors on the response to endotoxin

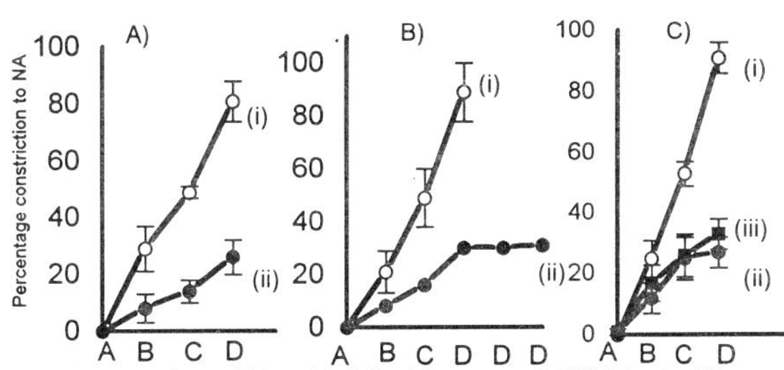

A) 5 Subjects were given soluble aspirin (1g) and the response to ETX determined 2h later. Dose response curves established to NA before (i) and 1h after exposure to endotoxin (ii).

B) In 5 subjects a dose response curve to NA was constructed before (i) and 1h after ETX (ii). At the end of the 2nd dose response curve L-NMMA was co-infused with the same dose of NA that produced the maximum constriction prior to ETX (ii).

C) The combined effects of L-NMMA and aspirin were studied in 5 subjects. All subjects received oral aspirin 2h before the study and the response to NA was determined before (i) and 1h after exposure to ETX (ii). In addition, immediately after the 2nd dose response curve to NA (ii) had been constructed, LNNMMA was co-infused with NA and the dose response curve repeated for a 3rd time (iii).

Figure 5 Effects of hydrocortisone on the response to endotoxin

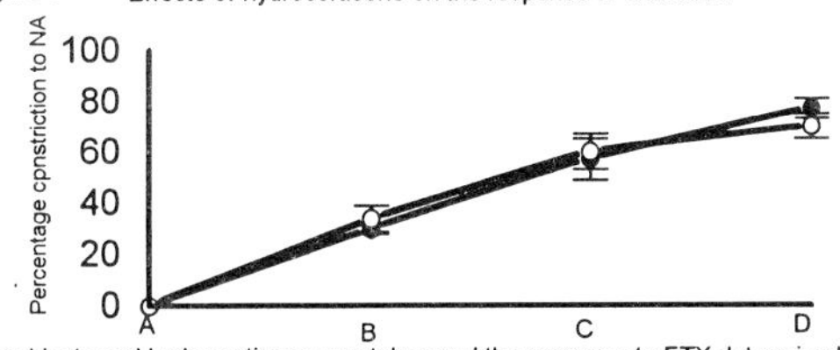

In 5 subjects oral hydrocortisone was taken and the response to ETX determined 2h later. Dose response curves were established to NA before (i) and 1h after exposure to ETX (ii).

These results provide direct evidence for a local vascular action of ETX in humans and demonstrate that this effect is supressed by a glucocorticoid. The effect was greatest at 1h and persisted for at least 3h. This acute vascular response does not appear to be mediated by nitric oxide or prostanoids.

Mesenteric vasodilator responses in cirrhotic rats: a role for nitric oxide?

Robert T. Mathie, Vera Ralevic*, Kevin P. Moore, and Geoffrey Burnstock*

Royal Postgraduate Medical School, London, W12 0NN, and *University College, London, WC1E 6BT, United Kingdom

Introduction. Induction of NO synthase by endotoxaemia or cytokines and the consequent increase in NO synthesis may be partly responsible for the mesenteric vasodilatation and hyperdynamic circulation associated with cirrhosis [1]. Experimental and clinical data, however, do not universally support this hypothesis. *In vivo* and *in vitro* studies in portal hypertensive animals have demonstrated an enhanced vascular sensitivity to the pressor effect of inhibitors of NO [2, 3], but NO synthase induction itself has been both confirmed [4] and denied [5]. Increased plasma nitrite and nitrate has been observed in cirrhotic patients with endotoxaemia [6], but not all investigations in human cirrhosis have revealed raised endotoxin or cytokine levels [7]. To date, no studies have directly examined mesenteric vasodilator responses in cirrhotic animals. The purpose of the present study was to investigate the contribution of NO to mesenteric arterial vasodilator responses elicited by the endothelium-dependent vasodilator agents acetylcholine (ACh) and adenosine 5'-triphosphate (ATP) and the endothelium-independent vasodilator sodium nitroprusside (SNP) in an experimental rat model of cirrhosis.

Methods. Experiments were carried out on 6 cirrhotic rats (induced by CCl_4/phenobarbitone); age-matched (n=9) and phenobarbitone-treated rats (n=9) served as controls. Responses to bolus injections of ACh, ATP and SNP were investigated in the isolated perfused mesenteric arterial bed after tone was raised by continuous infusion of methoxamine, before and during infusion of the NO synthesis inhibitor N^G-nitro-L-arginine methyl ester (L-NAME; 30μM) ± L-arginine (1 mM). The isolated mesentery was mounted in a humid chamber and perfused, without recirculation, at a constant rate of 5 $ml.min^{-1}$ with oxygenated Krebs solution at 37°C.

Results. Under basal conditions of raised tone (ie. in the absence of L-NAME), there was no difference in the vasodilator responses to ACh or ATP in the cirrhotic group compared with controls. NO inhibition by L-NAME caused a significant attenuation of vasodilator responses to most doses of ACh and ATP in all groups ($P<0.05$); under these conditions, however, ATP produced *vasoconstriction* only in cirrhotic animals (43.0 ± 10.6 mmHg for 5 x 10^{-8} mol, mean ± s.e.m. for n=4; $P<0.05$ compared to basal, Figure 1). L-arginine restored normal vasodilator responses in all groups. Under basal conditions, vasodilatation due to the highest dose of SNP was significantly attenuated in cirrhotic rats compared with phenobarbitone-treated animals ($P<0.05$). In the presence of L-NAME, responses to SNP were significantly potentiated in all groups, but cirrhotic preparations remained hyporesponsive compared with controls.

Discussion. The observation of normal responses to ACh and ATP under basal conditions does not support an increase in endothelial NO in cirrhosis. Moreover, during L-NAME infusion, ATP dilatation was not only attenuated, but was significantly *reversed* such that constriction was produced. This effect is unlikely to be mediated by up- or down-regulation of purinoceptors on vascular smooth muscle or endothelium, since it was not revealed during basal conditions of raised tone. The fact that it occurred only in the presence of L-NAME must implicate NO, suggesting that there may be *reduced endothelial* NO in cirrhosis. It has been postulated that the potentiation of SNP responses during NO inhibition is due to increased available smooth muscle guanylate cyclase [8, 9]. The attenuated vasodilator responses to SNP we have noted in cirrhotic preparations may, therefore, be a result of NO synthase induction in vascular smooth muscle. Recent work would support this view: attenuated responses to SNP in endotoxin-treated rats were found to be associated with a decrease in the activation of guanylate cyclase as a consequence of prolonged exposure to muscle-derived NO [10].

Conclusions. The normal responsiveness to vasodilatation by ACh and the vasoconstriction to ATP observed during NO blockade in cirrhotic animals indicate that mesenteric *endothelial* NO is unchanged or even diminished in experimental cirrhosis. The attenuated dilator responses to SNP may be a reflection of increased mesenteric smooth muscle NO, arising from NO synthase induction.

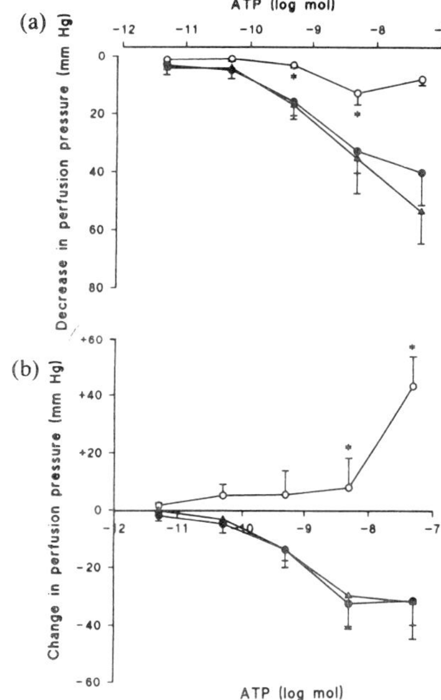

Figure 1. Effect of increasing bolus doses of ATP on perfusion pressure at raised tone in the mesenteric bed of (a) phenobarbitone-treated, and (b) cirrhotic rats, showing responses in the presence of methoxamine alone (●), during perfusion with methoxamine + L-NAME (○), and with methoxamine + L-NAME + L-arginine (▲). * Significant difference compared to response with methoxamine alone ($P < 0.05$).

References

1. Vallance, P. and Moncada, S. (1991). Hyperdynamic circulation in cirrhosis: A role for nitric oxide? Lancet **337**, 776-778.
2. Clària, J., Jiménez, W., Ros, J., *et al.* (1992). Pathogenesis of arterial hypotension in cirrhotic rats with ascites: Role of endogenous nitric oxide. Hepatology **15**, 343-349.
3. Sieber, C.C., Lopez-Talavera, J.C. and Groszmann, R.J. (1993). Role of nitric oxide in the in vitro splanchnic hyporeactivity in ascitic cirrhotic rats. Gastroenterology **104**, 1750-1754.
4. Michielsen, P.P, Boeckxstaens, G.E., Pelckmans, P.A., Herman, A.G. and Van Maercke, Y.M. (1993). Role of nitric oxide in vascular reactivity of isolated conductance vessels in portal hypertensive rats. Gastroenterology **104**, A954.
5. Fernández, M, García-Pagán, J.C., Casadevall, M., *et al.* (1995). Evidence against a role for inducible nitric oxide synthase in the hyperdynamic circulation of portal hypertensive rats. Gastroenterology **108**, 1487-1495.
6. Guarner, C., Soriano, G., Tomas, A., *et al.* (1993). Increased serum nitrite and nitrate levels in patients with cirrhosis: relationship to endotoxaemia. Hepatology **18**, 1139-1143.
7. Le Moine, O., Soupison, T, Sogni, P., *et al.* (1993). The role of endotoxins and TNF_α in the hyperkinetic circulation of cirrhosis. J. Hepatol. **18** (Suppl 1), S40.
8. Moncada, S., Rees, D.D, Schulz, R. and Palmer, R.M.J. (1991). Development and mechanisms of a specific supersensitivity to nitrovasodilators after inhibition of vascular nitric oxide synthesis in-vivo. Proc. Natl. Acad. Sci. U.S.A. **88**, 2166-2170.
9. Ralevic, V., Mathie, R.T., Alexander, B. and Burnstock, G. (1991). N^G-nitro-L-arginine methyl ester attenuates vasodilator responses to acetylcholine but enhances those to sodium nitroprusside. J. Pharm. Pharmacol. **43**, 871-874.
10. Tsuchida, S, Hitaoka, M, Sudo, M, Kigoshi, S. and Muramatsu, I. (1994). Attenuation of sodium nitroprusside responses after prolonged incubation of rat aorta with endotoxin. Am. J. Physiol. **267**, H2305-H2310.

Dynamic aspects and role of nitric oxide in endotoxin-induced liver injury

S. TAKEMURA†*, Y. MINAMIYAMA†, H. KINOSHITA* and M. INOUE†

Departments of †Biochemistry and *Surgery
Osaka City University Medical School, 1-5-7 Asahimachi, Abeno, Osaka *545*, Japan

Nitric oxide occurs in various cells and serves as a cytotoxic effector, neurotransmitter, immunoregulator, and endothelium-derived relaxing factor[1]. Although nitric oxide (NO) has been postulated to play important roles in the pathogenesis of endotoxin shock[2] and underlie the pathogenesis of tissue injury in endotoxemia-induced multiple organ failure, direct evidence supporting this hypothesis is lacking. Furthermore, quantitative aspects of the generation and *in vivo* fate of NO remain to be elucidated. The present study reports the dynamic aspects of the induction of NO synthase (NOS) in various tissue, formation of NO-Hemoglobin adduct (NO-Hb) in the circulating RBC and nitrosyl heme-iron complexes in the liver, and plasma nitrite + nitrate (NOx) in endotoxemic rats. The critical role of NO in the pathogenesis of endotoxemia was discussed.

Material and Method

Male Wistar rats (SLC, Shizuoka), weighing 220-240 g, fed laboratory chow and water *ad libitum*, were used. Endotoxemia was induced with *E. coli* lipopolysaccharides (LPS, type 055:B5, Difco Laboratories, Detroit, MI), dissolved in isotonic NaCl just before the experiments. Under light ether anesthesia, rats were given LPS (10 mg/kg) intravenously via the tail vein. Several hours (~24 hr) later, the animals were used for experiments. In some experiments, 5 mg/kg of N^G-iminoethyl-L-ornithine (L-NIO, Wako pure chemical, Osaka), a potent inhibitor of NOS was administrated intraperitoneally every 3 hr ESR spectra of blood and liver were obtained under liquid nitrogen at 110 K using a JES-RE1X spectrometer (JOEL, Tokyo) with 100kHz field modulation. Analytical conditions were as previously reported[3]. Levels of NOx in plasma and urine were analyzed by using a cadmium column and Griess regent. NOS activity was measured by the generation of L-citrulline from ^{3}H-L-arginine. Hepatic injury was evaluated by the change in plasma enzymes, such as aspartate aminotransferase (AST) and alanine aminotransferase (ALT). Survival rate of these animals was scored until 24 hr after treatment.

Results

All LPS treated rats showed typical symptoms of endotoxin shock, such as chills, tachypnea, and severe diarrhea. ESR analysis revealed that the ESR spectra of venous blood with a g value of 1.99 with distinct three-line hyperfine splitting became apparent after LPS injection. This spectra accounted for NO-Hb peaked at 8 hr (30 μM) after treatment and decreased thereafter to normal levels (<3 μM at 24 hr). Plasma levels of NOx increased 3 hr after LPS treatment, peaked at 8 hr and remained elevated after 24 hr. The maximal level of plasma NOx was about 50 times higher than the level of NO-Hb. Urinary levels of NOx changed similarly to those in plasma. LPS increased iNOS activities in liver, spleen, and lung, which peaked at 5 hr. The maximum activities of the enzyme in liver, spleen and lung were 93±30, 67±15, and 47±2 nmol/min/g protein, respectively.

Figure shows the ESR spectra of normal and LPS-treated rat liver. The ESR spectra includes cytochrome P450 which is characterized by a g value of 2.24. In LPS-treated rat, the distinctive spectra with a g value of 2.04 and triplet hyperfine structure appeared. This spectrum might seem to reflect the presence of nitrosyl P420 spectra. Double integrated analysis revealed that the amount of P450 slightly diminished after LPS treatment. The specimen included a small amount of hemoglobin, whose spectrum differed from that of NO-Hb with a g value of 1.99.

Plasma AST and ALT levels increased significantly and peaked at 12 hr after LPS treatment. Histological examination showed a marked infiltration of mononuclear and polymorphonuclear cells. The time course of NO formation, hepatic injury and histological alteration showed good correlation.

L-NIO completely inhibited the generation of NO-Hb and NOx. L-NIO also inhibited the generation of the heme-iron nitrosyl complexes of the LPS-treated rat. However, inhibition of NOS activity deteriorated the liver injury and increased the mortality of animals.

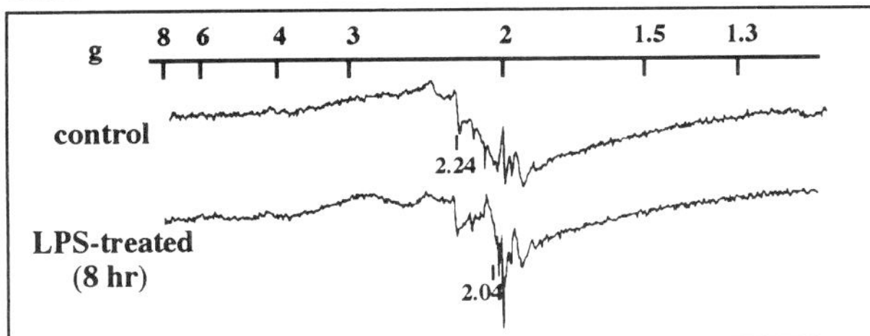

Figure ESR spectra of the liver from control and LPs-treated rats. The control spectra reflects the presence of cytochrome P450 which is characterized by a g value of 2.24. After LPS treatment the distinctive spectra with g value of 2.04 appeared. This signal seems to reflect the presence of NO-P420 derived from degenerated nitrosyl cytochrome P450.

Discussion

LPS markedly increased NOS in liver, spleen and lung. Significant fractions of NO formed NO-Hb adduct in the circulating RBC, which was metabolized to NOx in plasma, and slowly underwent urinary excretion. Since the formation of NO and hepatic injury occurred in parallel, NO may play a role in endotoxin-induced liver injury. Hepatic cytochrome P450 are heme proteins that function as terminal oxidases in the microsomal-mixed-function oxidase system. This system detoxifies a wide variety of hydrophobic xenobiotics and endogenous compounds including fatty acids and steroids. The P420-NO adduct lost the catalytic activity of cytochrome P450. Consequently, LPS seemed to inactivate the liver oxygenase. Thus, NO may have remarkable effect on the metabolism related to metal containing proteins, such as ferric enzymes. Stadler et al.[4] reported that cytochrome P450 families were inactivated by NO derived from NO-donor, such as nitroprusside, in a concentration-dependent manner. They also showed that cytokine induced NO inhibited cytochrome P450 and the transcription of NOS in cultured hepatocytes. The study describing the formation of nitrosyl heme complexes in endotoxemia has not been reported yet. Previous studies revealed that LPS aggravated the activity of cytochrome P450. The enzyme activity *in vivo* in pathological state was swayed by hemodynamics and enzyme induction. In fact, in endotoxemia blood flow decreased but there is a few decrease in the low spin signals of cytochrome P450 by ESR spectroscopy after LPS treatment, indicating cytochrome P450 induction. In endotoxin-induced inactivation of cytochrome P450, the inactivated protein rapidly turn over by *de novo* synthesis.

In contrast, inhibition of NO generation markedly deteriorated the liver injury. ESR analysis revealed that this inhibitor suppressed the occurrence of nitrosyl heme-iron complexes, and suggested that NO may have cytoprotective effect in endotoxemia. Previous report suggested that NO induced relaxations of resistant arteries and played a role as an anticoagulant and scavenged superoxide. We reported the regulatory role of NO in hepatic circulatory system[3,5]. The portal perfusion pressure and the tension of isolated portal vein rings were decreased by NO donor, such as nitroprusside[6]. Fat-storing cells which have been postulated to regulate of sinusoidal blood flow also underwent relaxation[7]. These observations suggested that NO may play a cytoprotective role through improving hepatic hemodynamics.

References

[1] Moncada, S., Palmer, R.M.J. and Higgs, E.A.(1991) Pharmacol. Rev., 43, 109-142

[2] Freeswick, P.D., Gellar, D.A., Lancaster, J.R. and Billiar, T.R. (1994) in The Liver; Biology and Pathobiology (Arias, I.M., Boyer, J., Fausto, N., Jacoby, W.B., Schachter, D.A. and Shafritz, D.A., ed.) pp1031-1045, Raven Press, New York

[3] Inoue, M., Minamiyama, Y. and Takemura, S. (1995) in Methods in Enzymology (Packer, L., ed.), in press Academic Press Inc., New York

[4] Stadler, J., Trockfeld, J., Schmalix, W.A., Brill, T., Siewert, J.R., Greim, H. and Doehmer, J. (1994) Proc. Natl. Acad. Sci., 91, 3559-3563

[5] Takemura, S., Kawada, N., Minamiyama, Y., Inoue, M. and Kinoshita, H. (1994) in Frontiers of Reactive Oxygen Species in Biology and Medicine (Asada, K. and Yoshikawa, T., ed.), pp233-234, Elsevier, Amsterdam

[6] Minamiyama, Y., Takemura, S., Kawada, N. and Inoue, M. (1994) in microcirculation annual (Tsuchia, M., Asano, M. and Ohashi, H., ed.) pp113-114, Nihon-Igakukan, Tokyo

[7] Kawada, N., Tran-Thi, T.A., Klein, H. and Decker, K. (1993) Eur. J. Biochem. 213, 815-823

Inducible nitric oxide synthase (iNOS) like immunoreactivity in the hepatocytes from various liver diseases

Shin-ichi Kasai, Toru Kono, Michio Mito and Jun Iwamoto[1]

Departments of Surgery and Physiology[1], Asahikawa Medical College, Asahikawa, Hokkaido 078 Japan.

It has been shown that, in rat hepatocytes (1) and Kupffer cells (2), iNOS is induced by the stimulation of LPS and NO can reduce mitochondrial respiration and protein synthesis (3). In the isolated human hepatocytes, stimulation with LPS plus cytokines can induce iNOS (4). However, there has been no report for the existence of iNOS in the liver diseases of human although the hepatotoxic effect of NO was suggested by an observation that cirrhotic patients exhibited the increased nitrite and nitrate (5). Hence we explored the existence of iNOS in liver tissues obtained from the various human patients with liver diseases by means of immunohistological technique for iNOS.

The tissues samples were obtained by either a regular liver biopsy or the wedge resection during surgery for pathological examination. After biopsy or resection, small tissue blocks are rapidly immersed in a chilled 4% paraformaldehyde adjusted at pH7.4 with 0.1 M phosphate buffer followed by ordinary paraffin embedding and sectioning. An iNOS-like immunoreativity (iNOS-LIR) was detected by incubation with the iNOS antibody (Transduction Laboratory). Secondary staining was carried out with DAB on the secondary antibody having peroxidase.

The iNOS-LIR was detected in the diseased liver tissues at various magnitudes whereas none or little iNOS-LIR was detected in the normal liver tissues. In the B-cirrhosis accompanied with hepatocellular carcinoma, iNOS-LIR was most prominent, while chronic hepatitis (B and C) exhibited less iNOS-LIR which was still considerably high. Higher magnification revealed that reaction product was confined to the cytoplasmic region of hepatocytes leaving the nucleus unstained. The iNOS-LIR was detected in the hepatocytes as well as Kupffer cells. Thus, the iNOS-LIR may be widely detected in the chronic viral hepatitis at various stages. In some cases, the localized iNOS-LIR was found in the lining hepatocytes around central veins of the lobules in the chronic hepatitis liver. In the fatty liver tissue, however, iNOS-LIR was least detected in the cirrhotic nodules. This may suggest that the viral infection could be essential for the occurrence of iNOS.

In general, a stronger iNOS-LIR was detected in the cirrhotic liver tissues (not in the fatty liver) than in the chronic hepatitis without severe fibrotic changes. This suggests that iNOS-LIR may be related to both inflammatory process and remodeling of the liver tissue. Pathophysiological role of NO in the human liver diseases is unknown, but presumed to be related to a hepatoprotective mechanism (6) as well as hepatotoxicity (2). The iNOS-positive hepatocytes from the cirrhotic liver didn't differ morphologically from other iNOS-negative cells. There was no tendency of massive necrosis of hepatocytes in the heavily stained iNOS-positive areas. Therefore the present result doesn't support one of the hypotheses that the high amount of NO released from the iNOS-occured cells would play a role for killing of adjacent cells. In some cases, iNOS-LIR was detected in a series of hepatocytes lining around central veins and connecting sinusoids, suggesting that the moderate endotoxemia that often accompanies with liver cirrhosis may contribute to the occurrence of iNOS via portal vein system. However, this was not always the case and the pattern of the positivity for iNOS was basically sporadic. Pathophysiological significance of NO in the human liver diseases is obscure and needs further investigation.

Acknowledgement : The authors thank to Ms. Yoko Ikoma for her expert immunohistological technique. This work is supported by research grants 07670077 and 07771421 from the Japanese Ministry of Art and Education.

References

1. Evans T, Carpenter A, Cohen J. Purification of a distinctive form of endotoxin-induced nitric oxide synthase from rat liver. (1992) Proc. Natl. Acad. Sci. USA 89, 5361-5365.

2. Billiar TR, Curran RD, Stuehr DJ, et al. An L-arginine-dependent mechanism mediates kupffer cell inhibition of hepatocyte protein synthesis *in vitro*. (1989) J. Exp. Med. 169, 1467-1472.

3. Stadler J, Billiar TR, Curran RD, et al. Effect of exogenous and endogenous nitric oxide on mitochondrial respiration of rat hepatocytes. (1991) Am. J. Physiol. 200, C910-C916.

4. Nussler AK, Di Silvio M, Billiar TR, et al. Stimulation of the nitric oxide synthase pathway in human hepatocytes by cytokines and endotoxin. (1992) J. Exp. Med. 176, 261-264.

5. Guarner C, Soriana G, Thomas A, et al. Increased serum nitrite and nitrate in patients with cirrhosis of the liver: relationships to endotoxemia. (1993) Hepatology 118, 1139-1143.

6. Billiar TR, Curran RD, Harbrecht BG, et al. Modulation of nitrogen oxide synthesis in vivo: N^G-monomethyl-L-arginine inhibits endotoxin-induced nitrite/nitrate biosynthesis while promoting hepatic damage. (1990) J. Leuk. Biol. 48, 565-569.

Innervation of nitric oxide-mediated vasodilator nerves in dog and monkey renal arteries

TOMIO OKAMURA and NOBORU TODA

Department of Pharmacology, Shiga University of Medical Science, Seta, Ohtsu 520-21, Japan

Kidney plays physiological roles in the regulation of blood volume and electrolyte balance. Retension of Na^+ and water by an impairment of kidney function is one of the important factors responsible for the genesis of hypertension. Inhibition of endogenous nitric oxide (NO) synthesis from L-arginine by NO synthase inhibitors decreases the urinary volume and Na^+ excretion, possibly by preferential actions on tubuli and also by glomerular blood flow [1], suggesting that NO derived from the kidney is acting as an endogenous diuretic substance.

Endothelium-derived relaxing factor is identified to be NO or NO-related molecule. It has been reported that renal arterial endothelium liberates NO in response to chemical stimuli [2]. Detection of mRNA of cNOS in the renal tubuli by polymerase chain reaction implys the presence of tubulus-derived NO [3]. Our recent studies provide an evidence that NO is derived also from perivascular nerves [4,5]. However, NO-mediated vasodilator innervation in the renal artery has not been reported.

The present study was aimed to determine the nature of the vasodilator nerves innervating the renal arterial wall in dogs and monkeys and to clarify mechanisms underlying neurogenic vasodilatation with special reference to NO.

Functional study

Isometric mechanical responses of the arterial strips denuded of the endothelium to nerve stimulation by nicotine were measured. Nicotine has been used as a chemical stimulant of perivascular nerves, since the agent produces consistent responses and shares pharmacological actions with transmural electrical stimulation, except for the fact that the response to electrical stimulation is abolished by tetrodotoxin but not by hexamethonium, and the opposite is the case in the response to nicotine [6]. Denudation of the endothelium of the strips were verified by the abolishment of relaxations caused by 1 μmol/L acetylcholine.

a) Dog

In dog renal arterial strips, nicotine (0.01 to 1 mmol/L) caused a concentration-related contraction which was abolished by treatment with 10 μmol/L hexamethonium. For the analysis of the nicotine-induced responses, single concentration of nicotine (0.1 mmol/L) was used because higher concentration of the drug sometimes develops tachyphylaxis. Nicotine-induced contractions were dose-dependently potentiated by N^G-nitro-L-arginine (1 and 10 μmol/L), a NO synthase inhibitor. The potentiating effect was reversed by L- but not by D-arginine. N^G-nitro-D-arginine was without effect. N^G-nitro-L-arginine did not potentiate the contraction caused by norepinephrine applied exogenously.

The contraction induced by nicotine was abolished by treatment with prazosin (10 μmol/L) under resting conditions and was reversed to a relaxation when the strips were partially contracted with prostaglandin $F_{2\alpha}$. The relaxation was not influenced by indomethacin (1 μmol/L), timolol (0.1 μmol/L) or atropine (0.1 μmol/L), but was abolished by hexamethonium (10 μmol/L), methylene blue (10 μmol/L), a soluble guanylate cyclase inhibitor, and oxyhemoglobin (16 μmol/L), a NO scavenger. Nicotine-induced relaxation was also abolished by L-NA (1 μmol/L) and reversed by L-arginine (0.3 mmol/L) but not by D-arginine (0.3 mmol/L). Relaxations caused by NO (0.1 μmol/L, $NaNO_2$ solution at pH2) were not influenced by N^G-nitro-L-arginine, but were abolished by methylene blue and oxyhemoglobin.

b) Monkey

Nicotine (0.1 mmol/L) elicited a contraction in monkey renal arterial strips under resting conditions. The contraction was potentiated by N^G-nitro-L-arginine (1 μmol/L) but not by N^G-nitro-D-arginine in the same concentration. The potentiation was reversed by L-arginine. Hexamethonium abolished the relaxant responses to nicotine.

The contraction by nicotine was reversed to a relaxation by treatment with prazosin in the precontracted strips. The relaxation was not altered by timolol, indomethacin or atropine but was abolished by methylene blue. The relaxant response was not affected by N^G-nitro-D-arginine, but was abolished by N^G-nitro-L-arginine; the response was restored by L- but not by D-arginine. NO-induced relaxations were abolished by oxyhemoglobin and methylene blue but not by N^G-nitro-L-arginine.

These functional data suggest that NO or NO-related molecule synthesized from L-arginine by NO synthase plays an important role in transmitting information from perivascular nerves to smooth muscle of dog and monkey renal arteries. NO liberated from the vasodilator nerve activates soluble guanylate cyclase in smooth muscle cells and increases the production of cyclic GMP, resulting in the vasodilatation.

Histochemical study

The dog and monkey renal arteries were fixed in ice-cold 0.1 mol/L phosphate-buffered saline containing a mixure of glutaraldehyde and paraformaldehyde, postfixed, and followed by a cryoprotection. The fixed blocks were cut into sections (20 μm thick) in a cryostat and used for NADPH diaphorase staining [7] or immunohistochemical staining of nerve-derived NO synthase [8].

In the renal, interlobar, arcuate and interlobular arteries, perivascular nerve fibers containing NADPH diaphorase and NO synthase were observed in the adventitia, and some fine fibers were also seen in the media.

Conclusion

NO is regarded as a diuretic substance that acts directly on renal tubules and also increases renal blood flow by dilating arteries and arterioles [9]. Many investigators suggest a role of NO derived from the arterial endothelium [10], however, the present study suggests that NO derived from the perivascular nerves also influences kidney function. Dog and monkey renal arteries appear to be innervated reciprocally by NO-mediated vasodilator and noradrenergic vasoconstrictor nerves as well as other peripheral arteries [11,12], and impairment of the vasodilator nerve function by NO synthase inhibitors may cause potentiation of neurogenic vasoconstriction and decrease renal blood flow.

References

1. Alberrola, A., Pinilla, J.M., Quesada, T., Romero, J.C., Salom, M.G. and Salazar, F.J. Hypertension **19**, 780-784
2. Cairns, H.S., Rogerson, M.E., Westwick, J. and Neild, G.H. (1991) J. Physiol (Lond) **436**, 421-429
3. Terada, Y., Tomita, K., Nonoguchi, H. and Marumo, F. (1992) J. Clin. Invest. **90**, 659-665
4. Toda, N. and Okamura, T. (1990) Am. J. Physiol. **259**, H1511- H1517
5. Toda, N. and Okamura, T. (1991) Am. J. Physiol. **261**, H1740-H1745
6. Toda, N. (1995) Hypertens. Res. **18**, 19-26
7. Toda, N., Ayajiki, K., Yoshida, K., Kimura, H. and Okamura, T. (1993) Circ. Res. **72**, 206-213
8. Yoshida, K., Okamura, T., Kimura, H., Bredt, D.S., Snyder, S.H. and Toda, N. (1993) Brain Res. **629**, 67-72
9. Lahera, V., Salom, M.G., Miranda-Guardiola, F., Moncada, S. and Romera, J.C. (1991) Am. J. Physiol. **261**, F1033-F1037
10. Moncada, S., Palmer, R.M.J. and Higgs, E.A. (1991) Pharmacol. Rev. **43**, 109-142
11. Toda, N. and Okamura, T. (1992) Hypertension **19**, 161-166
12. Toda, N., Kitamura, Y. and Okamura, T. (1991) J. Vasc. Biol. Med. **3**, 235-241

The role of nitric oxide in the control of renal blood flow. Hypoxia/resuscitation model of newborn piglets.

Ikuko Morikawa, Jyunzo Hyodo, Toshihiro Suzuki and Hajime Togari.
467 Japan, Nagoya City University Medical School Department of Pediatrics, 1- Kawasumi, Mizuhoku, Nagoya.

In newborn infants, vasodilator such as NO is supposed to be an important factor for renal blood flow development [1][2] and protection from posthypoxic renal damage [3]. However, the role of NO in immature renal hemodynamics is not fully understood. We attempted to study, in newborn piglets, the role of NO in the immature renal blood flow after hypoxia followed by resuscitation, and urinary cGMP as possible index of renal function. The role of NO was also investigated by comparing with and without inhibition of NO release when administering a NO synthesis analogue, Nω-nitro-L-arginine (L-NNA).
<Method and materials> Ten newborn piglets within 24 hr after birth (BW 1.62±0.6 kg) were anesthetized with Diethyl ether, intubated and controlled by mechanical ventilation. Renal blood flow was measured by Colored microspheres [4] (EZ-Trac, USA); diameter 15 µm, 5 colors, 2 million of each color. Catheterization was performed via umbilical artery for monitoring blood pressure, heart rate, and acid base balance, and for drawing reference blood sample, umbilical vein, left ventricle and left ureter catheterized for drip infusion and blood transfusion, infusion of microspheres and collecting urine sample.
Control group (n=5): After stabilization, red microspheres injected into the left ventricle for measurement of blood flow in prior to hypoxia. Hypoxia was induced by $O_2$5%$N_2$95% mixing gas for 20 minutes. The animals were then resuscitated by 100%O_2 for 15 minutes, followed by room air for 60 minutes. The blood flow of each episode was measured by injection of yellow, green, and black microspheres, in that order. In each injection of microspheres, reference blood sample was drawn at 2 ml/min via umbilical artery, and the equivalent volume of fresh blood from newborn piglet donors were simultaneously infused at 2 ml/min via umbilical vein. L-NNA group (n=5): After injection of red microspheres for measurement prior to hypoxia, L-NNA was administrated at 3 mg/bolus followed by 2 mg/20 min via umbilical vein. Blood pressure gradually increased after administration of L-NNA, followed by stability of blood pressure, blue microspheres was injected for measurement of blood flow change in thereafter. Hypoxia and resuscitation were performed by same protocol as control group. Urine and plasma samples were collected in each episode for measurement of Ccr, NAG/Cr., FENa and urinary cGMP as the marker of renal function. After removing the kidneys from animals at the end the experiment the cortex and medulla were separated. That tissue and reference blood sample at each episode were treated with processing reagent solution of cell segment and extracted microspheres, the each color of microspheres of tissue and reference blood sample were counted and calculated on Fuchs-Rosenthal counter. These sample were computed with formula for the value of blood flow. Results were presented as mean±SD. Wilcoxon test paired sample were used for comparison between two groups. P<0.05 was considered significant.
<Result> The renal blood change of pre and post L-NNA administration in physiological condition was shown in fig 1. The cortical blood flow decreased to 58%, while the medullary blood flow decreased to 67%. These data were significantly decreased in blood flow after L-NNA administration, suggesting NO is an important factor for maintaining renal circulation in physiologically based condition. The changes in blood flow of cortex and medulla in hypoxia and resuscitation were shown in fig 2 and 3. In hypoxia, renal blood flow remarkably decreased in both groups. While in resuscitation period for 60 minutes, the blood flow of control group recovered to prior to hypoxia, but L-NNA group was depressed to recover from blood flow comparing with control group. These data of resuscitation period were significantly different between the controls and L-NNA group, and NO was recognized to be possibly one of the important factors for recovery of renal blood flow after hypoxic insult. In the change of renal function during prior to hypoxia and recovery phase (after resuscitation), Ccr, NAG/Cr and FENa were not significantly different between control and L-NNA group, but urinary cGMP significantly decreased after administration of L-NNA, and control group showed an increase up to several times. L-NNA group, however, showed depression in the recovery phase, suggesting the possibility of urinary cGMP excretion as the biological marker for NO release. (fig 4)

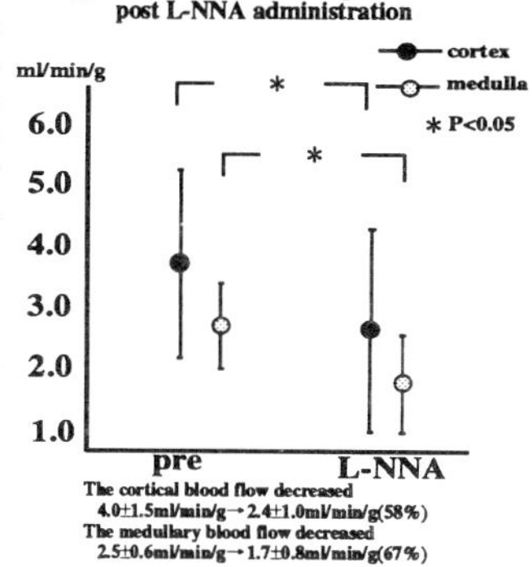

Fig 1 The renal blood flow pre and post L-NNA administration

The cortical blood flow decreased 4.0±1.5ml/min/g→2.4±1.0ml/min/g(58%)
The medullary blood flow decreased 2.5±0.6ml/min/g→1.7±0.8ml/min/g(67%)

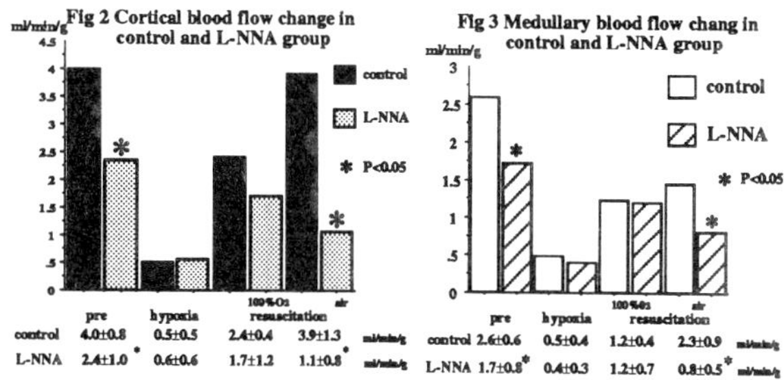

Fig 2 Cortical blood flow change in control and L-NNA group

Fig 3 Medullary blood flow chang in control and L-NNA group

<Conclusion> We conclude that NO is an important factor in the recovery phase of immature renal blood flow after hypoxia, and urinary cGMP excretion can be a biological marker for NO release.

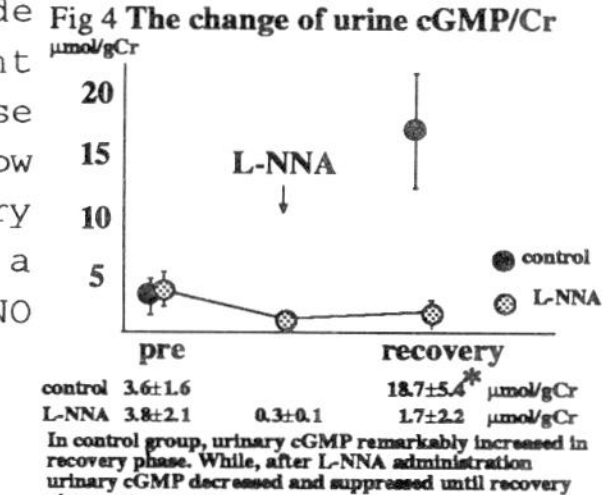

Fig 4 The change of urine cGMP/Cr

In control group, urinary cGMP remarkably increased in recovery phase. While, after L-NNA administration urinary cGMP decreased and suppressed until recovery phase. *:P<0.05

<Preference>
1. Solhaug, M.J., Wallace, M.R., Granger. J P, Endothelium-derived nitric oxide modulates renal hemodynamics in developing piglets. Pediatr Res 1993;34:750-754.
2. Bogaert,G.A., Kogan, B. A., Mevorach, A.R., Effects of endothelium-derived nitric oxide on renal hemodynamics and function in the sheep fetus. Pediatr Ress 1993;34:755-761.
3. Perrella,M. A., Edell,E., Krowka,M. J., Cortese, D. A., Burnett, J. C.Jr., Endothelium-derived relaxing factor in pulmonary and circulation during hypoxia. Am. J. Physiol 1992;263:R45-R50.
4.Hale,S. L., Kevin, J. A., Robert, A .K., Evaluation of nonradioactive, colored microspheres for measurement of regional myocardial blood flow in dogs. Circulation 1988;78:428-434.

The evidence of hemodynamic action of arginase infusion

FRIEDRICH LÄNGLE, ERICH ROTH, RUDOLF STEININGER, SUSANNE WINKLER and FERDINAND MÜHLBACHER

University Clinic of Surgery, Department of Transplantation, Währinger Gürtel 18-20, A-1090 Vienna, Austria.

Introduction The L-arginine-NOsystem is one of the most powerful systems regulating blood pressure which can be affected by the administration of either L-arginine, L-arginine analogs, or nitroso-containing compounds [1]. However, only little is known about the impact of arginase administration on hemodynamic parameters [2]. In a recent paper we reported the release of high amounts of arginase following graft-reperfusion in human liver transplantation, resulting in a plasma deficiency of L-arginine and in a reduction of plasma levels of nitrite hypothetizing that this L-arginine deficiency should evoke hemodynamic alterations [3,4]. In this study we monitored the hemodynamic and metabolic alterations following the administration of liver-arginase in catheterized anaesthetized pigs.

Materials and methods All experiments were performed on anaesthetized and catheterized pigs weighing 20 - 25 kg. Group 1: in 5 pigs a primed (8000 IU) continuous infusion (17000 IU) of bovine liver-arginase over a period of 20 minutes was applied Group 2: in 4 pigs 100 ml of bovine albumin was applied in the same manner. Hemodynamic measurements were continuously performed over a period of 1 hour. Blood samples for the determination of amino acids (separated by HPLC and determined with OPA [5] and dansyl hydrochloride for ornithine [6]), arginase (measured with an amino acid analyzer [7]), NO2- and NO3- (quantitated colorimetrically after the reaction with the Griess reagent), GOT, GPT, LDH and BUN were obtained every 10 minutes. Haemodynamic monitoring included MAP (mm Hg), MPAP (mm Hg), RAP (mm Hg), PCWP (mm Hg), CO (l/min), SVR (dynes s cm-5), PVR (dynes s cm-5), DPAP to PCWP gradient, blood flow of the thoracic aorta, portal vein, and the hepatic artery.

Results Blood levels of arginase increased from 58±42 IU/L to a maximum of 3690±962 IU/L after 20 minutes ($p<0.01$) and decreased to 2286±963 IU/L at 60 minutes. L-arginine dropped from 90±21 mmol/L to 21±17 mmol/L after 20 seconds and to a minimum of 17±22 mmol/L after 20 minutes ($p<0.01$) and increased to 66±29 mmol/L at 60 minutes. L-ornithine increased from 70±49 mmol/L to 163±71 mmol/L after 20 minutes ($p<0.05$) and decreased to 130±25 mmol/L after 60 minutes. BUN increased about 22% after 30 minutes. Nitrite as well as nitrate levels showed an increase at 60 minutes from 0.6±0.2 µmol/L to 1.4±0.4 µmol/L and from 38±26 µmol/L to 65±42 µmol/L respectively. There was a marked increase of GOT from 32±14 IU/L to a maximum of 369±57 IU/L at 20 minutes ($p<0.01$). No alteration of GPT and only a slight increase of LDH could be noticed. No metabolic alterations could be found in the albumin group. There was no difference of the time point 0 minute between the arginase group and the albumin group. 15 minutes after starting the infusion a remarkable increase of MPAP could be observed which reached a maximum level of 48±5 mmHg at 30 minutes with a following decrease to 27±7 mmHg at 60 minutes. PVR as well as DPAP to PCWP gradient increased and decreased in the same manner. PCWP increased from 11±2.3 mmHg to 14.5±0.7 mmHg after 20 minutes and decreased to the starting values after 60 minutes. RAP showed a slight increase from 9±1.5 mmHg to 11±2 mmHg after 30 minutes and a decrease to 8±1.6 mmHg after 60 minutes. The flow of the hepatic artery decreased about 30% within 20 minutes ($p<0.01$) and increased to 366±121 ml/min after 60. The flow of the portal vein decreased about 8% after 20 minutes, and about 17% after 60 minutes. These organ-specific hemodynamic changes could not be observed in the control group. No significant hemodynamic alterations could be found for cardiac output, mean arterial pressure, heart rate and total peripheral resistance for both groups.

Conclusions In this study we investigated the effect of the administration of arginase in catheterized anaesthetized pigs to see whether arginase can evoke hemodynamic effects. The rise of the plasma arginase concentrations to levels similar to those found after hepatic-reperfusion, evoked a significant increase of vascular pulmonary blood pressure and a decrease of arterial hepatic blood flow. These results indicate for the first time, that supraphysiological levels of arginase can evoke profound alterations on organ-specific hemodynamic behaviours, especially of the liver and the lung. The administration of an arginase-inhibitor as well as of L-arginine may be helpful in pathophysiological situations of an increased occurence of arginase [8].

References

1. Moncada, S., Palmer, R.M.J. and Higgs, E.A. (1991) Pharmacol Rev. 43, 109-116
2. Griffith, O.W., Parks, K.H., Aisaka, K., Levi, R. and Gross, S.S. (1992) in The biology of nitric oxide (Moncada, S., ed.), pp.6-9, Portland Press, London and Chapel Hill
3. Roth, E., Steininger, R., Winkler, S., Längle, F. and Mühlbacher, F. (1994) Transplantation. 57, 665-669
4. Längle, F., Roth, E., Steininger, R., Winkler, S. and Mühlbacher, F. (1995) Transplantation. 59, 1542-1549
5. Turnell, D.C., Cooper, J.D.H. (1991) Clin Chem. 28, 527-531
6. Kaneda, N., Sato, M. and Yagi, K. (1982) Anal Biochem. 127, 49-53
7. Bastone, A., Diomede, L., Parini, R., Carnevale, F. and Salmoma, M. (1990) Anal Biochem. 191, 384-389
8. Minor, T., Yamaguchi, T. and Isselhard, W. (1993) Workshop: Experimental Liver Transplantation, Frankfurt / Germany

Abbrevations used: BUN, blood urea nitrogen, CO, cardiac output; DPAP, diastolic pulmonary arterial pressure; GOT, glutamic oxaletic transaminase; GPT, glutamic pyruvic transaminase; LDH, lactate dehydrogenase; MAP, mean arterial pressure, MPAP, mean pulmonary arterial pressure; NO, nitric oxide; PCWP, pulmonary capillary wedge pressure; PVR, pulmonary vascular resistance; RAP, central venous pressure; SVR, systemic vascular resistance.

Role of nitric oxide synthesized during pancreas ischemia/reperfusion in rats

SATONORI TANAKA, WATARU KAMIIKE, TOSHINORI ITO, HIKARU MATSUDA, SHIGEOMI SHIMIZU*, EIJI KUMURA* and HIROAKI KOSAKA*

First Department of Surgery, *First Department of Physiology, Osaka University Medical School, 2-2 Yamadaoka, Suita, Osaka 565, Japan

Introduction: Although it has been reported that reactive oxygen-derived free radicals are involved in pancreatic ischemia / reperfusion injury, the pathogenesis of the acute pancreatitis has been yet unclear. The purpose of the present study is to elucidate whether NO is generated during pancreatic ischemia / reperfusion in rats.

Materials and Methods: Male Wistar rats, weighing 250-300 g, were anesthetized by intraperitoneal injection of pentobarbital sodium. Pancreas ischemia was produced by applying a small clamp to the celiac axis. To exclude any influence from gastric ischemia by occlusion of the celiac artery, total gastrectomy was performed prior to occlusion. The tissue blood flow in the pancreas was measured by fixing a laser doppler flowmeter. NO was detected by measuring NO end products, i.e., plasma nitrite plus nitrate and NO bound to hemoglobin (HbNO) by electron spin resonance (ESR). The plasma lipase level was measured as a marker of damage to pancreatic exocrine tissue.

Results: *Tissue blood flow in pancreas* During occlusion of the celiac axis, tissue blood flows in pancreas body and tail were reduced to 16.8 ± 7.6% and 12.6 ± 2.7%, respectively, whereas that in pancreas head was reduced to 82.7 ± 13.2% (Table. 1). After 1 hr of reperfusion, the tissue blood flow was restored to approximately 90%. But in the presence of N^G-nitro-L-arginine methyl ester (L-NAME), the blood flow was restored to approximately 50% after 1 hr of reperfusion. *Plasma levels of nitrite plus nitrate* Plasma levels of nitrite plus nitrate in veins after total gastrectomy, sham operation (exfoliation of the celiac axis) and 1 hr occlusion of the celiac axis were significantly increased than those with anesthesia alone (Table 2). The plasma levels of nitrite plus nitrate in systemic and portal veins after 1 hr of reperfusion were significantly elevated compared with those only after 1 hr occlusion of the celiac axis. The elevated levels of plasma NO end products in both systemic and portal veins were significantly decreased by the administration of L-NAME. L-Arginine counteracted the L-NAME-induced decrease in the NO end products. On the other hand, plasma levels of nitrite plus nitrate in both veins after 1 hr occlusion of splenic artery and 1 hr reperfusion were not increased than those after 1 hr occlusion of the celiac axis. *NbNO* An ESR signal at g=2.0, which was presumed to be of unidentified free radical origin, was observed in the venous blood of rats with total gastrectomy, sham operation and 1 hr of occlusion of the celiac axis. After 1 hr of reperfusion, the signals of HbNO with a three-line hyperfine structure appeared clearly in the vein [1]. *Plasma level of lipase activity* The plasma levels of lipase after 1 hr of ischemia or after 1 hr of reperfusion were significantly higher than those after the sham operation. The elevated plasma level of lipase after 1 hr of reperfusion was significantly augmented by administration of L-NAME.

Discussion The present study first demonstrated that NO was synthesized during reperfusion of rat pancreas after 1 hr of partial ischemia. After 1 hr reperfusion and 1 hr occlusion of the celiac axis, the level of plasma NO end products was significantly increased than that after 1 hr occlusion. The celiac artery supplies also the spleen with blood. But by 1 hr reperfusion after 1 hr occlusion of the splenic artery, the plasma level of NO end products were not elevated. Therefore, the pancreatic ischemia/reperfusion was responsible for the elevation of NO. On the pancreatic tissue, role of NO during pancreas ischemia / reperfusion is un clear. In the study, the plasma level of lipase was significantly increased after pancreas ischemia / reperfusion, and it was further increased by administration of L-NAME. This result suggested that NO generated during ischemia/reperfusion of rat pancreas might have a protective role for pancreas exocrine tissue.

Reference

[1] Kosaka, H., Sawai, Y., Sakaguchi, H., Kumura, E., Hrada, N., Watanabe, M. and Shiga, T. (1994) Am. J. Physiol. 266, C1400-1405.

Table 1.Tissue blood flow in rat pancreas

	Tissue Blood Flow (% Control)		
	Head	Body	Tail
Occlusion of celiac axis	82.7 ± 13.2	16.8 ± 7.6	12.6 ± 2.7
1 hr after reperfusion of celiac axis	90.9 ± 3.2	87.1 ± 10.8	92.3 ± 4.2
1 hr after reperfusion of the celiac axis + L-NAME	52.1 ± 13.6	51.4 ± 16.0	64.7 ± 15.3

The dose of L-NAME was 10 mg/kg body weight administered twice. Values are the mean ± SD of 4 rats per group.

Table 2. The Plasma level of nitrite plus nitrate during occlusion and reperfusion of the celiac axis in rats

Experimental group	Nitrite plus nitrate (μM)	
	Systemic vein	Portal vein
Anesthesia alone	10 ± 2	10 ± 2
Total gastrectomy	21 ± 5^a	20 ± 4^a
Sham operatiom	21 ± 5^a	21 ± 5.^{a}
1 hr reperfusion after 1h occlusion of the splenic artery	22 ± 3^a	23 ± 4^a
1 hr occlusion of the celiac axis	20 ± 2^a	20.± 4^a
1 hr reperfusion after 1 hr occlusion of the celiac axis	37 ± 12^{a,b}	37 ± 8^{a,b}
1 hr reperfusion after 1 hr occlusion of the celiac axis + L-NAME (20 mg/kg)	17 ± 2^c	22 ± 4^c
1 hr reperfusion after 1 hr occlusion of the celiac axis + L-NAME (20 mg/kg) + L-arginine (400 mg/kg)	27 ± 2^d	34± 7

$^a p < 0.05$ versus anesthesia alone, $^b p < 0.05$ versus 1 hr occlusion of the celiac axis, $^c p < 0.05$ versus 1 hr reperfusion, $^d p < 0.05$ versus1 hr reperfusion + L-NAME.

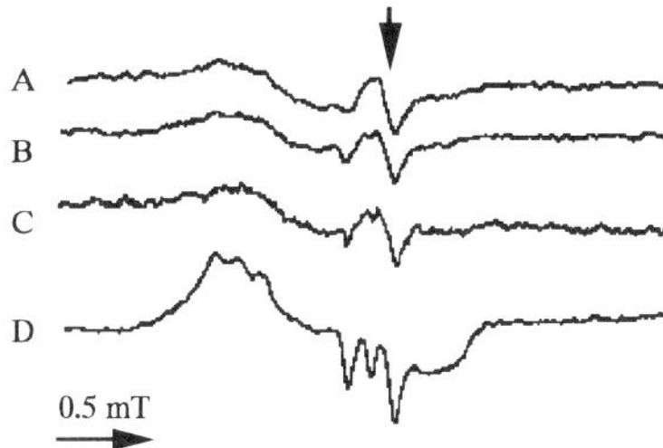

Fig. 1 ESR spectra of venous blood during occlusion and reperfusion of the celiac axis in rats.
Spectrum A is from the rat with total gastrectomy . Spectrum B is from the rat with sham operation. Spectrum C is from the rat with 1 hr occlusion. Spectrum D is from the rat with 1 hr reperfusion after 1 hr occlusion. The spectra were recorded at 110 k with an incident microwave power of 5 mW, a modulation frequency of 100 Khz and modulation amplitude of 0.5 mT.
The arrow denotes g=2.0.

Table 3. Changes in the plasma lipase level

Experimental group	Lipase activity (IU/L)
Anesthesia alone	20 ± 4
Total gastrectomy	27 ± 10
Sham operation	29 ± 4
1 hr occlusion of the celiac axis	113 ± 46^a
1 hr reperfusion after1hr occlusion of the celiac axis	140 ± 15^a
1 hr reperfusion after 1 hr occlusion of the celiac axis + L-NAME	259 ± 61^b
Total gastrectomy + L-NAME	84 ± 21^c

$^a p < 0.01$ versus anesthesia alone , total gastrectomy and sham operation, $^b p < 0.01$ versus 1 hr reperfusion after 1 hr occlusion of the celiac axis, $^c p < 0.01$ versus total gastrectomy

Immunohistochemical study of nitric oxide synthases in transgenic sickle cell mouse kidneys.

SUZETTE Y. OSEI, REXFORD S. AHIMA, JUDY QIU, MARY E. FABRY, RONALD L. NAGEL and NORMAN BANK

Montefiore Medical Center, Bronx, NY 10467

Young sickle cell patients often manifest glomerular hyperfiltration, which is postulated to be causally related to glomerulopathy later in life. In a previous study, we found in a transgenic mouse model of sickle cell disease (β^S TG) that the mice have renal hyperfiltration, elevated urinary nitrite excretion, and increased quantities of nitric oxide synthases III and II in renal protein extracts. Inhibition of NO corrected the hyperfiltration [1 and submitted]. In the present study, we carried out immunohistochemical localization of NOS III and II in the kidneys of β^S TG and control mice. In addition, we examined the hypothesis that hypoxia may be a mechanism of increased renal NOS abundance.

METHODS: Control mice used in this study were the C57BL/6J strain. Animals were maintained in individual cages under room air conditions, or were housed in glass environmental chambers for 4-5 days filled with constantly flowing 10% O_2/0.5% CO_2/89.5% N_2 gas. After anesthesia with ether, the mice were perfused transcardially with cold phosphate-buffered saline (PBS) followed by perfusion fixation with paraformaldehyde-lysine-periodate. Kidneys were removed, sections were post-fixed in PLO and then embedded. Serial 4 micron-thick sections were cut, dewaxed, rehydrated in ethanol and washed. Endogenous peroxidase was blocked by H_2O_2. Blocking of non-specific binding and permeabilization was carried out with normal goat serum and Triton x-100. Sections were then incubated with either anti-human endothelial NOS (NOS III) or anti-mouse activated macrophage NOS (NOS II) antibodies (Transduction Labs, Lexington, KY). Control sections were incubated with normal rabbit serum, or the primary antiserum was omitted. Sections were incubated with biotinylated goat antirabbit IgG and avidin-biotin-peroxidase and counterstained with hematoxylin.

RESULTS: NOS III was seen diffusely in almost all tubules in the renal cortex but not in the renal medulla of control mice. In the β^S TG mice NOS III was seen in the same locations but stained much more intensely. In addition, convoluted tubules in the outer and inner medulla also manifested NOS III reactivity. NOS II staining was not observed in the cortex and only weakly in the medulla of control mice under room air conditions. In contrast, NOS II was found in cortical collecting tubules and distal tubules of the β^S TG mice, as well as in medullary collecting ducts. Weakly positive staining for NOS II was seen in the glomeruli of the β^S TG mice but not in the control mice. Chronic exposure of the mice to 10% O_2 caused an intense increase in staining of NOS III in the convoluted tubules in the outer cortex of control mice. In the case of NOS II, hypoxia resulted in *de novo* appearance of this enzyme in cortical tubules of control mice, located in collecting and distal convoluted tubules, as well as medullary collecting tubules. In the β^S TG mice, chronic hypoxia caused a further increase in NOS II in the inner medulla and glomeruli. In addition, NOS III was seen in the glomerular mesangium. A semiquantitative analysis of the results is in Table 1.

Table 1. Semiquantitation of NOS III and II in normal and TG mouse kidneys

NOS III	Normal	Hypox	TG	Hypox
Cortex	2-3+	3-4+	3-4+	3-4+
Outer medulla	1-2+	3+	3-4+	3-4+
Inner medulla	1-2+	2+	3+	3+
Glomeruli	0	0	0	2-3+
NOS II				
Cortex	0	2+	2+	3+
Outer medulla	1+	2-3+	3+	3+
Inner medulla	1+	2-3+	1-2+	3+
Glomeruli	0	0	1+	3+

DISCUSSION: The finding of extensive NOS III constitutively in cortical tubules of normal mice has not previously been demonstrated, to our knowledge. Several *in vitro* studies have found that proximal tubule cells can generate NO, and that an EDRF-like substance is present in LLC-PK kidney epithelial cells [2]. Assays of NOS III mRNA in microdissected renal structures by RT/PCR, found mRNA expression in blood vessels but inconsistently in proximal tubules, thick ascending limbs, and cortical and inner medullary collecting ducts [3,4]. NOS II can be induced in certain segments of the nephron by TNF-α, IFN-γ and LPS, including the S_3 portion of the proximal tubule, cortical and medullary thick ascending limbs of Henle's loop, distal convoluted tubules and cortical and medullary collecting ducts [5,7]. The present observations demonstrate that NOS III is greatly increased in β^S sickle cell mice. In addition, the β^S sickle cell mice manifest markedly increased NOS II staining throughout the cortical and medullary tubules, as well as in the glomerular mesangial region even without exposure to hypoxia. Thus, both NOS III and II are increased in the kidneys of sickle cell mice under room air conditions, and increase further with exposure to hypoxia.

Sickle cells adhere to vascular endothelium in narrow blood vessels, and this is thought to result in intermittent occlusion and variations in blood flow [8]. Changes in flow activate NO production in endothelial cells [9]. However, cytokines are presumably responsible for increased NOS II in the tubular epithelial cells of the β^S TG mice. Based upon our finding that chronic hypoxia increased renal epithelial cell NOS in normal mice, we postulate that local tissue hypoxia in the β^S TG mice leads to induction of NOS III and II. Increased NO production may occur locally as a result, and contribute to glomerular hyperfiltration and perhaps other functional disturbances in sickle cell disease.

ACKNOWLEDGMENT

This study was supported by an NIH Sickle Cell Center Grant 2 P60 HL 38655.

REFERENCES

1. Fabry, M.E., Costantini, F.D., Pachnis, A., Suzuka, S.M., et al (1992) Proc. Natl. Acad. Sci. USA 89:12155-12159
2. Tracey, W.R., Pollock, J.S., Murad, F., Nakane, N., and Forstermann, U. (1994) Am. J. Physiol. 266:C22-C28
3. Ujiie, K., Yuen, J., Hogarth, L., Danziger, R., and Star, R.A. (1994) Am. J. Physiol. 267:F296-F302
4. Terada, Y., Tomita, K., Nonoguchi, H., Marumo, F. (1992) J. Clin. Invest. 90:659-665
5. Markewitz, B.A., Michael, J.R., and Kohan, D.E. (1993) J. Clin. Invest. 91:2138-2143
6. Ahn, K.Y., Mohaupt, M.G., Madsen, K.M., and Kone, B.C. (1994) Am. J. Physiol. 267:F748-F757
7. Mohaupt, M.G., Schwobel, J., Elzie, J.L., Kannan, G.S., and Kone, B.C. (1995) Am. J. Physiol. 268: F770-F777
8. Kaul, D.K., Fabry, M.E., and Nagel, R.L. (1989) Proc. Natl. Acad. Sci. USA 86:3356-3360
9. Kuchan, M.J., and Frangos, J.A. (1994) Am. J. Physiol. 266:C628-C636

Ontogeny of nitric oxide-mediated relaxation of fetal, neonatal and adult guinea pig renal arteries.

LOREN P. THOMPSON and CARL P. WEINER.

Department of Obstetrics/Gynecology and Perinatal Research Laboratory. The University of Iowa, Iowa City, IA 52242 USA.

Renal blood flow increases with fetal and postnatal kidney maturation by mechanisms not fully understood. The fetal kidney receives 2-3% of cardiac output which increases to 15-18% in the newborn and 20-25% in the adult kidney [1]. The increase in postnatal kidney perfusion relative to the fetal kidney is associated with decreased renal vascular resistance [2]. Endothelium-derived nitric oxide (NO) has been implicated as a modulator of fetal renal blood flow [3] and a contributor to the regulation of renal hemodynamics in the postnatal and adult kidney [4]. We hypothesize that NO release increases with maturation of the kidney and contributes to the progressive decrease in vascular resistance. To test this hypothesis we measured relaxation to acetylcholine (ACh), the prototypic endothelium-dependent relaxing agent, and sodium nitroprusside (SNP), an NO donor, of isolated renal arteries from fetal, neonatal and adult guinea pigs.

Methods. Kidneys were removed through an abdominal incision from anesthetized newborn (<50 days old) and reproductively-mature (600-800g) adult guinea pigs and near term fetuses from anesthetized time-mated pregnant guinea pigs. Renal arteries were excised and cut into rings under a dissecting microscope, mounted onto tungsten wires and placed in a horizontal myograph apparatus for measurement of isometric force. Tissues were maintained in temperature-regulated baths containing Krebs bicarbonate buffer and aerated with 95% oxygen and 5% carbon dioxide. Optimal passive force was determined using submaximal KCl (60mM) for each tissue.

Vascular Responses. Relaxation to cumulative addition of ACh and SNP (10^{-9}M - 10^{-5}M) was measured in renal arteries contracted with $PGF_{2\alpha}$ ($5x10^{-6}$M). Comparisons were made between renal arteries from fetal, neonatal, and adult guinea pigs. To determine the role of NO on ACh relaxation, rings were treated with the NO synthase inhibitor, nitro-L-arginine (10^{-4}M). *NO Synthase Activity.* Renal cortex was excised from the left kidneys of a separate group of fetal, neonatal and adult guinea pigs. Tissues were frozen in liquid nitrogen and homogenized at 0°C in 3 volumes of buffer brought to pH 7.0 at 20°C with HCl. NO synthase activity was measured by conversion of ^{14}C-arginine to ^{14}C-citrulline. Total activity was measured as the difference between ^{14}C-citrulline produced (pmol/min/g tissue) from control samples and samples containing 1mM EGTA and 2mM N^G-monomethyl-L-arginine (LNMMA). Ca^{2+}-independent activity was determined as the difference between samples containing EGTA only and both EGTA and LNMMA. Ca^{2+}-dependent NO synthase activity was calculated by subtracting Ca^{2+}-independent activity from total activity.

Results. Maximal relaxation to ACh was greater in renal arteries of both the neonatal and adult guinea pigs compared to the fetal guinea pigs. Vascular sensitivity (measured as the -log ED_{50} value) of arteries from fetal and neonatal guinea pigs was less than arteries from adult guinea pigs. LNA significantly inhibited the sensitivity to ACh in arteries of all age groups. Maximal relaxation to SNP did not differ among the three age groups. However, the vascular sensitivity to SNP of fetal arteries was significantly less than that of both neonatal and adult guinea pigs. Thus, a maturational increase in the vascular smooth muscle sensitivity to NO was measured.

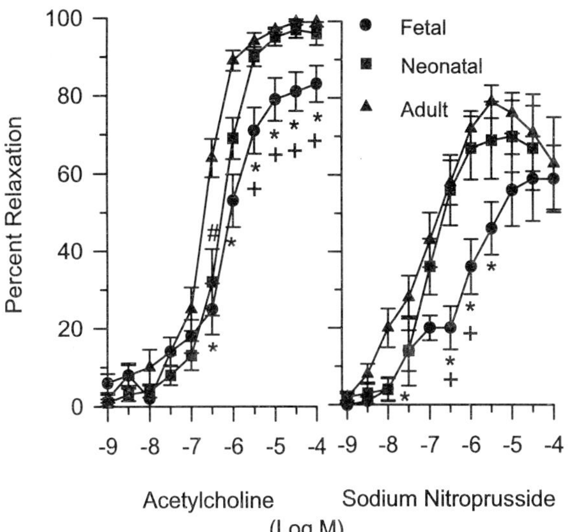

Figure 1. Relaxation of renal arteries of near term fetal, neonatal and adult guinea pig kidneys to acetylcholine (left) and sodium nitroprusside (right). (*, + = $P<0.05$ vs. fetal; # = $P<0.05$ vs. newborn.

Ca^{2+}-dependent NO synthase activity of the fetal renal cortex was near undetectable at 0.5 gestation and progressively increased to 577±161 pmol/min/g with 0.87 gestational age. After birth, activity of neonatal and adult kidneys was similar to that measured at 0.87 gestation.

Conclusions. This study demonstrates a maturational increase in endothelium-dependent relaxation to ACh of the guinea pig renal artery and of Ca^{2+}-dependent NO synthase activity of the renal cortex. This study suggests that stimulated NO release increases with maturation and occurs in both the main renal artery and also the microvasculature. As a result, NO may contribute to the changes in renal vascular resistance before and after birth. In conclusion, functional adaptations in both the endothelium and vascular smooth muscle contribute to the maturational changes in mechanisms regulating renal hemodynamics before and after birth.

References

1. Iwamoto, H.S., Oh, W., and Rudolph A.M. (1985) In Physiological Development of the Fetus and Newborn (Jones, CT and Nathanielz, PW., ed), pp.37-40. Academic Press, London.

2. Robillard, J..E., Nakamura, K.T., Matherne, G.P., and Jose, P.A., (1986) Renal hemodynamics and functional adjustments to postnatal life. Sem. Perinatal. **12**(2), 143-150.

3. Bogaert, G.A., Kogan, B.A., and Mevorach, R.A. (1993) Effects of endothelium-derived nitric oxide on renal hemodynamics and function in the sheep fetus. Pediatr. Res. **34**(6), 755-761.

4. Solhaug, M.J., Wallace, M.R., and Granger, J.P. (1993) Endothelium-derived nitric oxide modulates renal hemodynamics in the developing piglet. Pediatr. Res. **34**(5), 750-754.

Bradykinin-induced vasodilatation in perfused rat kidney is dependent on nitric oxide and potassium channel opening

Kênia S. POMPERMAYER*, Jamil ASSREUY# and Maria A.R. VIEIRA*

*Department of Physiology and Biophysics, Universidade Federal de Minas Gerais, Belo Horizonte, MG, 31270-901, Brazil and #Department of Pharmacology, Universidade Federal do Rio de Janeiro, RJ, PO 68013, 21944-970, Brazil.

Bradykinin (BK) is a potent vasodilatory peptide released from several tissues and organs [1]. In the isolated perfused rat kidney, BK causes a transient decrease in the relative renal vascular resistance (RVR) detected at 2 min after kinin addition [2]. This effect may be resultant from the release of one or more vasorelaxant factors such as prostaglandins [3], nitric oxide (NO) [4] and hyperpolarizing factor [5]. In this work, we have studied the involvement of NO and the potassium channel activation in the renal vasodilatory effect of BK.

All experiments were performed on the isolated rat kidney perfused in a closed circuit system as previously described [6]. Briefly, male Wistar rats (250-380 g) anesthetized with sodium thionembutal (40 mg/Kg) were used as donors for the isolated kidneys. The right kidney was exposed through an incision in peritoneal cavity. Nonfiltering kidneys were sistematically used. The urinary flow was stopped out by ureter ligation. The right renal artery was cannulated avoiding flow interruption. The kidney was isolated and transferred to the closed system. The perfusate consisting of a Krebs-Henseleit-bicarbonate buffer containing bovine serum albumin (7.5%), glucose (1 mg/ml), creatinine (0.5 mg/ml) and an aminoacid mixture (158 mg/ml) was kept at 37° C and gassed with a carbogen mixture (O_2:CO_2::95%:5%). The renal perfusion pressure and perfusate flow were continuously monitored . A 40 min period of stabilization was allowed and depending on the purpose of the experiments, the kidneys were preconstricted with phenylephrine (PHE, 5 µM). The compounds to be tested were first diluted in 1.0 ml of perfusate and then added to the system.

BK (0.5 µM) produced a transient decrease in RVR (1.00 to 0.78 ± 0.03) 2 min after addition (n = 20) (Tab. 1). The NO synthesis inhibitor, N^G-nitro-L-arginine methyl ester (L-NAME, 75 µM) increased the RVR by 100% reaching a plateau 30-40 min after addition to the system. When BK was added 20 min after inhibitor there was a reduction of only 40% (0.78 ± 0.03 versus 0.87 ± 0.02) in its vasodilatory effect since the renal NO synthesis was not completely inhibited at this time (n = 11). On the other hand, a second dose of BK added 50 min after L-NAME addition did not have any effect on RVR (n = 5) (Tab.1). The RVR was 0.83 ± 0.02 and 0.99 ± 0.02 in the kidneys perfused with BK alone and BK plus L-NAME, respectively. At this time, the increase in the RVR induced by L-NAME had already reached a plateau. Clotrimazole (CTZ, 150 µM), a potassium channel blocker, added 10 min before BK, produced a reduction of about 70% (0.78 ± 0.03 versus 0.94 ± 0.01) in its vasodilatory effect (n = 11) (Tab. 1). In these experiments, at 40 min after CTZ addition, a second dose of BK produced a decrease in the RVR similar to that observed with the first dose (n = 4). The association of L-NAME and CTZ although did not abolish the decrease in the RVR induced by the first dose of BK (n = 4), completely blocked the RVR decrease induced by the second dose of BK (n = 4) (Tab. 1).

Table 1 Effect of BK (0.5 µM) on the RVR in the presence of L-NAME (75 µM) and/or CTZ (150 µM)
The kidneys perfused with BK alone or BK plus CTZ were preconstricted with PHE (0.35 µM).

Perfusion with	BK effect on RVR	
	First dose	Second dose
None	1.00	1.00
BK alone	0.78 ± 0.03 *	0.83 ± 0.02*
BK + L-NAME	0.87 ± 0.02 **	0.99 ± 0.02**
BK + CTZ	0.94 ± 0.01 **	0.97 ± 0.01**
BK + CTZ + L-NAME	0.97 ± 0.01 **	1.02 ± 0.01**

*$p < 0.05$ None x BK alone
**$p < 0.05$ BK alone x BK + Compounds

Abreviations used: BK, bradykinin; CTZ, clotrimazole; L-NAME, N^G-nitro-L-arginine methyl ester; NO, nitric oxide; PHE, phenylephrine; RVR, relative renal vascular resistance.

The data indicate that normal vascular tone in isolated rat kidney depends on NO production and that both NO production and potassium channel opening mediate BK renal vasodilatory effect. Whether there a relationship between NO production and potassium channel opening is now being investigated in our laboratory.

We are grateful to Mr. César N. Oliveira for his tecnical assistance. This work was supported by grants from Conselho Nacional de Desenvolvimento Científico e Tecnológico (CNPq), Fundação de Amparo à Pesquisa do Estado de Minas Gerais (FAPEMIG) and Pró-reitoria de Pesquisa/UFMG (PRPq/UFMG).

REFERENCES:

1 Mills, I.H. (1979). Nephron **23**, 61-67
2 Guimarães, J.A., Vieira, M.A.R., Camargo, M.J.F. and Maack, T. (1986). Eur. J. Pharmacol. **130**, 177-185
3 McGiff, J.C., Terragno, N.A., Malic, K.U. and Lonigro, A.J. (1972). Circ. Res. **31**, 36-43
4 Moncada, S. and Higgs, A. (1993). N. Eng. J. Med. **32** (27), 2002-2012
5 Vanhoutte, P.M. (1987). Nature, **327**, 459-460
6 de Mello, G. and Maack, T. (1976). Am. J. Physiol. **231**, 1699-1707

Nitric oxide synthase activity in human kidney and renal cell carcinoma

OLOF JANSSON, EDWARD MARCOS, LOU BRUNDIN*, JAN ADOLFSSON and N. PETER WIKLUND

Department of Urology and *Department of Neurology, Karolinska Hospital, S-171 76 Stockholm, Sweden

Introduction: The role of NO in solid tumors is still unclear. In the tumor, NO can be generated by several different cells such as macrophages, endothelial cells in the neovasculature, fibroblasts and tumor cells. It has been described that human monocytes requires an L-arginine/nitric oxide synthase dependent effector mechanism for the generation of angiogenic activity [1]. Neovascularity has also been reported to be a prerequisite for tumor progression [2]. In human gynecological cancers and breast cancer, the NO synthase activity has recently been described to be positively correlated to tumor grade [3,4]. In contrary, human colon cancers have lower NO synthase expression than normal colonic mucosa [5]. The aim of this study was to investigate the NO synthase activity in human kidney and renal cell carcinoma.

Materials and methods: Tissue was collected from patients undergoing nefrectomy due to cancer. Pieces of tumor and normal parenchyma were collected and immediately frozen in liquid nitrogen. The study was approved by the local ethics committee. NO synthase activity was measured by the conversion of L-[U-^{14}C]arginine to L-[U-^{14}C]citrulline as described previously [6] or by the conversion of L-[2,3,4,5-^{3}H]arginine to L-[2,3,4,5-^{3}H]citrulline. Immunohistochemistry was performed with polyclonal antibodies from rabbit (Transduction Laboratories) raised against endothelial-, brain- and macrophage NO synthase respectively. Frozen tissue was cut at 10 μm in a cryostat and thawed on to object slides, fixed in 4% paraformaldehyd for 5 min. and incubated with antiserum and then washed. The tissue was then incubated with fluorescein isothiocyanat-conjugated anti-antibodies, rinsed and examined in a Nikon fluorescence microscope.

Results and discussion: We found significantly ($P < 0.0001$) decreased NO synthase activity in human renal cell carcinoma compared to nonmalignant renal tissue from the same kidney in patients undergoing nefrectomy. Calcium-dependent NO synthase activity was found in all nonmalignant tissues (range, 102-884 pmol/min/g tissue) but was significantly decreased in renal cell carcinoma (range, 0-75 pmol/min/g tissue). In contrary to human breast and gynecological cancers the NO synthase activity correlated negatively with tumor grade. Calcium-independent activity was found in some nonmalignant tissues (≥ 28 pmol/min/g tissue) and tumors (≥ 23 pmol/min/g tissue). When the NO synthase activity was measured in samples containing a mixture of equal volumes of nonmalignant and malignant supernatants the NO synthase activity was equal to the average between the two samples. The low activity is therefor not likely due to an endogenously produced NO synthase inhibitor in the tumor tissue. Immunohistochemical investigation using antibodies raised against endothelial, nerve and inducible NO synthase was performed. In the normal kidney a dense staining of the endothelium in vessels and glomeruli was evident, whereas in the tumors, endothelial NO synthase could not be demonstrated . We could also show nerve NO synthase-like immunoreactivity in nerves in the normal kidney but not in the tumors. Inducible NO synthase immunoreactivity could not be detected in the normal kidney or tumor tissue. In conclusion our data show that NO synthase activity is markedly decreased in renal cell carcinoma as compared to nonmalignant kidney tissue. High grade tumors have less NO synthase activity than low grade tumors. It may represent a decreased transcription and/or translation of the NO synthase gene in the tumor neovasculature.

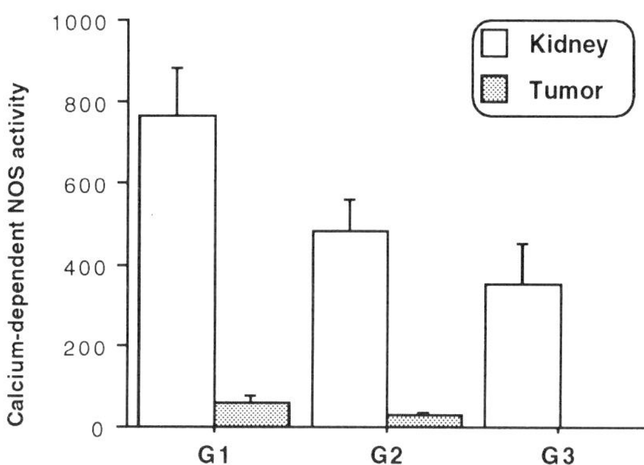

Fig. 1. Calcium-dependent NO synthase activity in normal kidney tissue and tumors of different grade from the same kidney. The activity is expressed by citrulline formation (pmol x min^{-1} x g^{-1}) and shows the mean +SEM (n = 2, 9 and 6 in G1, G2 and G3 respectively).

References

1. Leibovich, S.J., Polverini, P.J., Fong, T.W., Harlow, L.A. and Koch, A.E. Production of angiogenic activity by human monocytes requires an L-arginine/nitric oxide-synthase-dependent effector mechanism. (1994) Proc. Natl. Acad. Sci. USA **91**, 4190-4194

2. Brawner, M.K., Deering, R.E., Brown, M., Preston, S.D. and Bigler, S.A. Predictors of pathological stage in prostatic carcinoma: The role of neovascularity. (1994) Cancer **73**, 678-687

3. Thomsen, L.L., Lawton, F.G., Knowles, R.G., Beesley, J.E., Riveros-Moreno, V. and Moncada, S. Nitric oxide synthase in human gynecological cancer. (1994) Cancer Res. **54**, 1352-1354

4. Thomsen, L.L., Miles, D.W., Happerfield, L., Bobrow, L.G., Knowles, R.G. and Moncada, S. Nitric oxide synthase activity in human breast cancer. (1995) Br. J. Cancer **72**, 41-44

5. Chhatwal, V.J., Ngoi, S.S., Chan, S.T.F., Chia, Y.W. and Moochhala, S.M. Aberrant expression of nitric oxide in human polyps, neoplastic colonic mucosa and surrounding peritumoral normal mucosa. (1994) Carcinogenesis **15**, 2081-2085

6. Salter, M., Knowles, R.G. and Moncada, S. Widespread tissue distribution, species distribution and changes in activity of Ca^{2+}-dependent and Ca^{2+}-independent nitric oxide synthases. (1991) Fed. Eur. Biochem. Soc. **291**, 145-149

THE ROLE OF NITRIC OXIDE AS A SYNCHRONIZING CHEMICAL MESSENGER IN THE HYPERPERFUSION OF THE CANCEROUS BREAST

Michael Anbar
Department of Biophysical Sciences, School of Medicine and Biomedical Sciences, University at Buffalo, Buffalo, NY 14214.

The mechanism of the hyperthermia of the breast and the important role of nitric oxide (NO) in it have been previously described, as have experimental findings consistent with it.[1-3] In brief, our hypothesis is that excessive NO level associated with the interaction between macrophages and neoplastic cells and enhanced by the presence of ferritin in the neoplastic tissue,[4] interferes with the autonomic vasoconstrictive process. This results in regional hyperperfusion that is not reversed by vasoconstrictive stimuli, i.e, these regions lose their autonomic thermoregulatory control. These effects can be demonstrated by the use of dynamic area telethermometry (DAT),[5] where hundreds of thermal images of the breasts taken over a few minutes are used the analyze the dynamics of thermoregulation. This effect can be used in a novel computerized screening test for the early detection of breast cancer.[6] While DAT generally uses fast Fourier transformation (FFT) to extract the autonomic thermoregulatory frequencies,[5] it has been recently shown that DAT measurement of the rate of cooling of different subareas of the breast under a vasoconstrictive challenge, can be sufficient to demonstrate regions with impaired autonomic control. Instead of going into further details about breast cancer detection, which is not in the purview of this NO meeting, I wish to discuss the role of NO in the mechanism of breast hyperthermia in a broader biological perspective.

A basic premise in biology is that no biological function can survive the thermal, photochemical and radiolytic randomization processes in nature unless it has a survival value for the living organisms involved. It is my hypothesis that the biological functions of NO have survived billions of years because of its _unique_ synergic synchronous intracellular, intercellular and inter-organ functions, _all_ of which have high survival values.

It has been pointed out that NO is very inefficient as a chemical messenger.[7] NO must be produced *in situ* on demand, because it cannot be stored in any biological compartment and, therefore, the risetime of its concentration is relatively slow. Because of its non specific-binding to most iron-carrying proteins, its biological activity is less specific than that of most other chemical messengers. Because there is no effective enzymatic process to remove it irreversibly, its concentration at the target site is poorly controlled, which limits the rate of repeated chemical messages. Because of its rapid isotropic diffusion, NO must be produced in large quantities and only a very minute fraction of it reaches its target. In brief, unless it had a unique survival value, its functions, e.g., its role in vasodilation, would have been replaced by evolution by a more specific and, therefore, more effective chemical messenger.

While its most important role may be as an intracellular synergic synchronizing regulatory agent,[7,8] it is suggested that the survival of NO as an intercellular chemical messenger in mammalian systems implies a similar role in highly evolved metazoan systems. The production of NO in different organ systems may proceed by different pathways, involving different enzymes. If NO is generated in one organ system, e.g., in the vascular system, and it diffuses into another organ system, e.g., neurons behind the blood-brain barrier, it would exert a biological effect "reserved" for a genuine interneuron transmitter by NO produced inside the latter organ system. The same would be true if it was originally produced inside brain tissue and diffused into the surrounding arterioles and venules. If this crosstalk between organ systems induced by NO would be detrimental, evolution would have developed "better" chemical messengers for both systems, messengers that do not interfere with each other. To have survived evolution, this crosstalk must be beneficial. Taking again this example of inter-organ crosstalk, it is likely that when NO is used inside brain tissue as a neurotransmitter (e.g., for long-term potentiation)[9], there is a higher local demand for oxygen or glucose, which is met by local vasodilation. Another mechanism might involve the need for oxygen to quench the neurotransmitter signal by oxidation of NO via HO_2, thus increasing the rate of signal transmission. Furthermore, since NO can be neurotoxic when produced in excessive amounts,[10] the simultaneously induced vasodilation supplies oxygen to minimize this potentially detrimental effect. Alternatively, a vasodilatory signal might induce long-term potentiation. It has been actually claimed that endothelial (vascular) NO is responsible for long-term potentiation.[11] One could advance additional mechanisms to explain this synergic synchronous behavior in the brain.

As a second example of NO-mediated synergic synchronous inter-organ communication, let us consider the role of NO in hyperthermia associated with inflammation[3] or malignancy. NO may be produced by macrophages by a calcium independent enzyme in response to alien cells.[12] This NO can diffuse into the regional vasculature and induce non-neuronal cGMP production and subsequent smooth muscle relaxation, i.e., vasodilation. Let us examine the potential synergic synchronization associated with this mechanism. The vasodilation increases the supply of oxygen and nutrients to the affected area. It also increases the influx of additional lymphocytes to the affected area (positive feedback). The simultaneous vasodilation of the regional venules by the macrophage-produced NO enhances the removal of CO_2 and other metabolites, preventing local acidosis. This implies a dual mechanism of local hyperperfusion - an autonomic neuronal thermoregulatory mechanism and a regional non-neuronal mechanism associated with immune response. This dual mechanism is sufficient to explain the evolutional maintenance of the endothelial NO vasodilatory mechanism, because immune response induced localized vasodilation that enhances the effectiveness of macrophage activity, has undoubtedly a survival value. Next, it has been shown that NO sensitizes the response of nociceptors.[13] The association of pain with a local inflammation has again a survival value, as it might reduce the use of an infected limb or of an inflamed joint, helping their recovery. This might explain the positive response of certain nociceptors to NO. The interesting feature of this synergism is that it is synchronized by NO very much like in the case of brain function discussed above.

Another example of inter-organ system crosstalk involving NO that has a survival value, is the well known binding of NO to hemoglobin (Hb) inside erythrocytes, which may not be merely coincidental with the intravascular function of NO. The vasodilatory effect of endothelial NO will be reduced at high levels of Hb and enhanced under anemic conditions, as expected from a system that is regulating the oxygen supply to tissues. Another inter-organ system crosstalk related to hemodynamics is the well known effect of NO as an inhibitor of platelet aggregation, which enhances blood flow in vasodilated arterioles and venules. Guanylate cyclase activity and subsequent smooth muscle relaxation can be induced by chemical messengers other than NO, including 5-hydroxytryptamine[14] and brain natriuretic peptide,[15] thus NO is not absolutely necessary as mediator in vasodilation. Survival of the vasodilatory function of NO might thus have been thanks to the different beneficial interactions with other organ systems.

It is not clear whether the NO-induced nonadrenergic-noncholinergic (NANC) penile erection invokes a synchronous involvement of another proximal organ system, but it involves the synchronous action of NO as a neurotransmitting mediator at the CNS and penile levels.[16] It seems to be a unique manifestation of the nervous system when its "objective" is to cause substantial local vasodilation without involving other parts of the vasculature. Other examples of local smooth muscle relaxation under NANC conditions include the lower esophageal sphincter, esophageal peristalsis,[17] and pyloric motility.[18] In these cases, NO may diffuses into the microvasculature of the esophagus or the pylorus, enhancing local blood supply needed for the muscular activity. Similar synergism may exist in the NANC NO-mediated effect on the trachea and bronchi.[19] In the NANC NO-mediated action on the intestines it has been suggested that NO has a synergic synchronous role of protecting the mucosa from blood-borne toxins.[20]

There are rare situations where NO-mediated inter-organ system cross-talk may lead to catastrophic results. The NO-induced hypotension in sepsis and the subsequent multiple-organ failure,[21] caused by the autocatalytic overproduction of NO by macrophages[22] and neutrophils[23] in reaction to pervasive microorganisms is an example where the biological function of NO-mediated inter-organ system cross-talk has a negative survival value. Like lethal malignant hyperthermia or hypothermia[24] the occurrence of which does not degrade the survival value of physiological thermoregulation, sepsis is an example of an extreme catastrophic situation where NO-induced synergism runs out of control.

What can be concluded from this teleological discussion is that when the involvement of NO as a chemical messenger is observed, one must question the reason for this role of NO to have survived evolution in spite of the significant limitations of NO as a chemical messenger. From the examples discussed here it can be concluded that the inter-organ synergism, which is a rather unique feature of nitric oxide, has played a major role in preserving this premordial chemical messenger in most life forms up to date.

1. Anbar, M. (1994) *Cancer Lett.* 84, 23-29.
2. Anbar, M. (1995) *Biomed. Thermology* 15,135-139.
3. Anbar, M. (1995) *Thermologie Osterreich* 5,15-27.
4. Elliott, R.L., Elliott, M.C., Wang, F. and Head J.F. (1993) *Ann. N Y Acad. Sci.* 698,159-166.
5. Anbar, M. (1994) *Medical Electronics*, 146, 62-73; 147, 73-85.
6. Anbar, M. (1995) *SPIE Proc.* 2473, 312-322.
7. Anbar, M. (1995) *Experientia* 51, 545-550.
8. Anbar, M. (1995) These proceedings, paper #259 (poster presentation).
9. Chiang, L.W., Schweizer, F.E., Tsien R.W. and Schulman H. (1994) *Brain Res. Mol. Brain Res.* 27, 183-188.
10. Schulz, J.B., Matthews, R.T., Muqit, M.M. *et al.* (1995) *J. Neurochem.* 64,936-939.
11. O'Dell, T.J., Huang, P.L., Dawson, T.M. *et al.* (1994) *Science* 265, 542-546.
12. Bukrinsky, M.I., Nottet, H.S., Schmidtmayerova, H. *et al.* (1995) *J. Experim. Med.* 181,735-745.
13. Holthusen, H. and Arndt, J.O. (1994) *Neurosci. Lett.* 165,71-74.
14. Obi, T., Kabeyama, A. and Nishio A.(1994) *J. Vet. Pharmac. Therap.* 17, 218-225.
15. Akiho, H., Chijiiwa, Y., Okabe, H. *et al.* (1994) *Life Sci.* 55,1293-1299.
16. Argiolas, A. (1994) *Neuropharm.* 33,1339-1344.
17. Chakder, S., Rosenthal, G.J. and Rattan S. (1995) *Am. J. Physiol.* 268, G443-G450.
18. Allescher, H.D. and Daniel, E.E. (1994) *Digest. Diseas. & Sci.* 39, 73S-75S.
19. Ellis, J.L. and Conanan N. (1994) *J. Pharm. & Expl. Ther.* 269, 1073-1078.
20. Dilorenzo, M., Bass, J. and Krantis, A. (1995) *J. Pediat. Surg.* 30, 235-241.
21. Goode, H.F., Howdle, P.D., Walker, B.E. and Webster, N.R. (1995) *Clin. Sci.* 88, 131-133.
22. Harbrecht, B.G., Wang, S.C., Simmons, R.L. and Billiar T.R. (1995) *J. Leukocyte Biol.* 57, 297-302.
23. Neilly, I.J., Copland, M., Haj, M. *et al.* (1995) *Brit. J. Haemat.* 89, 199-202.
24. Collins, K.J. (1993) in *Autonomic Failure* 3rd Ed, (Bannister R. & Mathias C.J. Eds), pp. 212-230, Oxford Univ. Pr.

Nitric oxide: Role in human cervical cancer

LINDY L. THOMSEN*, ROBIN FARIAS-EISNER# and GAUTUM CHAUDHURI#≠

*WELLCOME RESEARCH LABORATORIES, LANGLEY COURT, BECKENHAM, KENT BR3 3BS, UNITED KINGDOM and #DEPARTMENT OF OBSTETRICS, GYNECOLOGY AND PHARMACOLOGY≠, UCLA SCHOOL OF MEDICINE, LOS ANGELES, CALIFORNIA 90024-1740, USA

Introduction

Recent studies suggest a dual role for nitric oxide (NO) in tumor biology. Farias-Eisner *et al.* [1] and Thomsen *et al.* [2] suggest that high NO synthase activity and the putative high concentrations of NO generated mediate tumoricidal activity *in vivo*. In contrast, at lower concentrations, NO has been shown to promote tumor growth [3]. Further, Thomsen *et al.* [4,5] recently demonstrated a positive correlation between low level NO synthase activity and grade of human breast and ovarian malignancies, thus supporting the hypothesis that low concentrations of NO provide a positive signal for tumor growth and progression. We therefore wanted to elucidate whether there was a relationship between NO production and tumor growth in other gynecological malignancies such as cervical cancer.

Methods

Cervical biopsy specimens measuring 3 x 3 mm were obtained from 23 patients. Of these, 14 specimens were from women diagnosed as having grade III malignant squamous cell cervical lesions, 2 had adenocarcinoma, and 7 specimens (controls) were obtained from women without cancer at the time of elective hysterectomy. The specimens were snap frozen in liquid nitrogen and stored at -70°C for later analysis of NO synthase activity.

Assay for NO synthase was determined by the conversion of L-[U-^{14}C]arginine to [U-^{14}C]citrulline [4]. The limit of detection in this assay was ≥0.7 pmol/min per mg protein.

Results

Nitric oxide synthase activity was detectable in only 4 squamous cell carcinoma specimens. The NO synthase activity was exclusively calcium-independent and ranged from 0.7 to 2.2 pmol/min per mg protein. Nitric oxide synthase activity was not detected in control specimens.

Of the 4 patients whose cervical specimens showed NO synthase activity, 1 had persistent disease in spite of standard therapy with curative intent (5040 cGy, external beam radiation and 3500 cGy intracavity implants). This patient was subsequently treated with adjuvant hysterectomy.

Of the remaining 10 patients with squamous cell malignancies without detectable NO synthase activity, the tumor extent was variable. Some patients had less, some similar and some greater extent of disease compared with those which demonstrated NO synthase activity.

Only 1 out of 12 patients with out NO synthase activity had recurrent disease. By contrast, 1 out of 4 patients with NO synthase had persistent disease, and 1 was lost to follow up.

Discussion

We have shown that calcium-independent NO synthase activity, indicative of the inducible NO synthase iso-form, is present in poorly differentiated cervical cancers. The range of activity found was similar to that previously reported for human breast cancer [4]. However, in contrast to breast cancer in which expression of the enzyme was found in all high grade cancers, enzyme activity was detected in only 25 % of the cervical cancer cases.

Our data suggests that the role of NO in promotion of tumor growth is not ubiquitous for all human cancers. Follow up studies are needed to elucidate whether tumors with NO synthase activity are predictive of recurrent or persistent disease.

References

1. Farias-Eisner, R., Sherman, M.P., Aeberhard, E. and Chaudhuri, G. (1994) Proc. Natl. Acad. Sci. USA. **71**, 9407-9411
2. Thomsen, L.L., Baguley, B.C. and Wilson, W.R. (1992) Cancer Chemother. Pharmacol. **31**, 151-155
3. Jenkins, D.C., Charles, I.G., Thomsen, L.L., Moss, D.W., Holmes, L.S., Baylis, S.A., Rhodes, P., Westmore, K., Emson, P.C. and Moncada, S. (1995) Proc. Natl. Acad. Sci. USA. **92**, 4392-4396
4. Thomsen, L.L., Miles, D.W., Happerfield, L., Bobrow, L.G., Knowles, R.G. and Moncada, S. (1995) Brit. J. Cancer **72**, 41-44
5. Thomsen, L.L., Lawton, F.G., Knowles, R.G., Beesley, J.E., Riveros-Moreno, V. and Moncada, S. (1994) Cancer Res. **54**, 1352-1354

OXYGEN-DEPENDENT REGULATION OF ENERGY METABOLISM IN PERITONEAL TUMOR CELLS BY NO

M. Inoue, M. Nishikawa, E. Sato, Y. Takehara, K. Utsumi, Dept. Biochem., Osaka City Univ. Med. School, Osaka and Inst. Med. Sci., Center for Adult Diseases, Kurashiki, Japan

INTRODUCTION

Because of gaseous nature of NO, it easily penetrates through plasma membranes and modulates cell function through formation of its adducts with thiols and hemeproteins. Although the lifetime of NO is extremely short under air atmosphere, it is fairly stable under low oxygen tensions. Since most of work was performed under high O_2, *in vivo* effect of NO would be quite different from those observed *in vitro*. To know the critical roles of NO under physiologi-cally low oxygen tensions, its effect on the energy metabolism of ascites tumor cell AH-130 was studied under varying oxygen tensions.

RESULTS AND DISCUSSION

AH130 cells revealed a marked respiration which was inhibited reversibly by NO. The inhibitory effect of NO was more marked at low O_2 tensions than at high concentrations (Fig. 1). Similar inhibition was also observed with digitonin-treated cells. Analysis using antimycin A and rotenone revealed that cytochrome c oxidase might be the primary site of inhibition by NO. Since hemoglobin has high affinity for NO, both HbO_2 and RBC suppressed the inhibitory effect of NO. NO reacts with O_2^- to form $ONOO^-$. Since $ONOO^-$ inactivated mitochondrial electron transport and ATPase irreversibly without affecting cytochrome c oxidase, NO *per se* might account for the inhibition. Although O_2^- reacts with NO 3-times faster than with SOD, high levels of SOD would have effectively scavenged O_2^- and minimized the occurrence of $ONOO^-$.

The inhibitory effect of NO was more marked with digitonin-permealized cells than with untreated cells. Since cytosol also contains compounds which react with NO, they might affect the fate of NO metabolism. In fact, the inhibitory effect was markedly reduced by adding GSH to the permealized cells. Since the lifetime of NO is significantly longer at physiologically low O_2 than previously expected from experiments performed under air atmosphere, effects of NO would be more significant *in vivo*. Because the peritoneal cavity is fairly anerobic, energy metabolism of peritoneal cells would be inhibited by NO. When activated by ligands, peritoneal macrophages express high level of NO synthase and generate NO. Thus, NO released from these cells might inhibit the synthesis of ATP and DNA thereby suppressing the proliferation of tumor cells. Since oxyhemoglobin rapidly reacts with NO, antitumor effect of NO would be suppressed by bleeding into peritoneal cavity. Thus, reaction of NO with Hb, O_2 and mitochondria is of critical importance for the host-defense against tumor invasion.

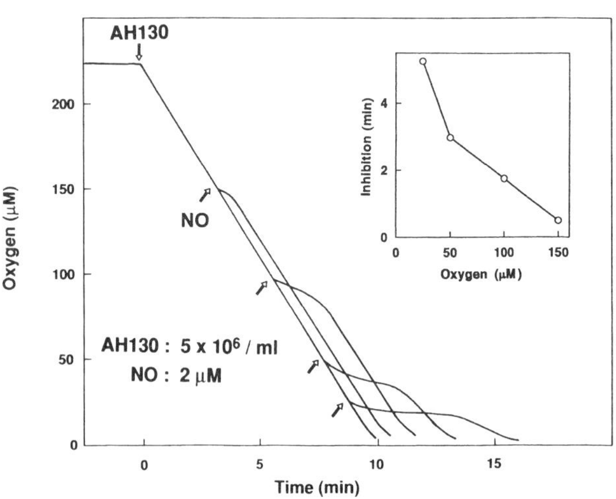

Fig. 1 O_2-dependent inhibition of tumor cell respiration by NO

Effect of 2 μM NO on the respiration of normal and digitonin-treated AH130 cells (5 x 10^6/ml) was observed at different oxygen tensions. Insert shows the inhibition at high (o) and low oxygen tensions (o).

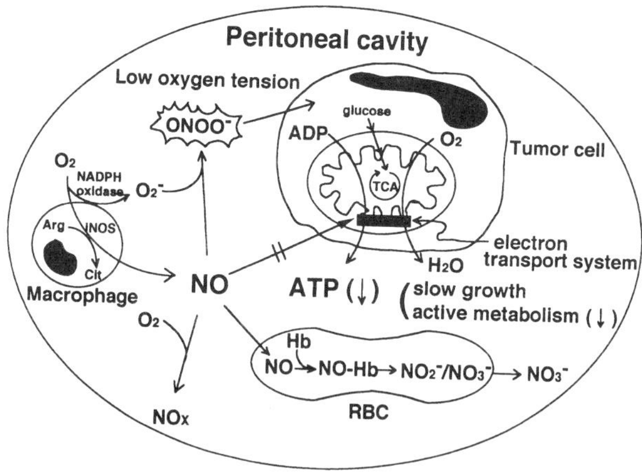

Fig. 2 NO-dependent regulation of energy metabolism of tumor cells

Effects of long-term inhibition of nitric oxide synthase on the development of cholangiocellular preneoplastic lesions in hamsters infected with liver fluke (*Opisthorchis viverrini*)

Hiroshi OHSHIMA,* Paiboon SITHITHAWORN,# Satoru TAKAHASHI,*§ Vladimir YERMILOV* and Dominique GALENDO*

*International Agency for Research on Cancer, 150 cours Albert-Thomas, 69372, Lyon, Cedex 08, France, and #Faculty of Medicine, Khon Kaen University, Khon Kaen, 40002, Thailand

INTRODUCTION

Infection with the liver flukes *Opisthorchis viverrini* or *Clonorchis sinensis* has been associated with an increased risk of cholangiocarcinoma in south-east Asia, especially in northeast Thailand [1]. Human subjects infected with liver flukes excrete elevated levels of nitrate and *N*-nitrosamino acids in their urine, suggesting that nitric oxide (NO) synthase activity is increased by infection [2,3]. We recently demonstrated that an inducible type of NO synthase is expressed in the inflammation zone surrounding the parasite-containing bile ducts in the liver of Syrian golden hamsters experimentally infected with *O. viverrini* liver fluke [4]. Since high concentrations of NO exert cytotoxic and mutagenic effects *per se*, excess NO production in chronically infected or inflamed tissues may play a role in the development of cholangiocarcinoma [5]. In the present study we investigated effects of long-term administration of an NO synthase inhibitor (nitroarginine methyl ester (NA)) on the induction of cholangiocellular preneoplastic lesions in *O. viverrini*-infected hamsters.

MATERIALS AND METHODS

We used a Syrian golden hamster model for induction of cholangiocarcinoma. *O. viverrini* infection has been reported to significantly increase yields of cholangiocellular preneoplastic and neoplastic lesions induced by carcinogenic *N*-nitrosodimethylamine (NDMA) in hamsters [6,7].

Cyprinoid fish harbouring metacercarial cysts were bought in a high-risk area for cholangiocarcinoma in northern Thailand. Cysts of *O. viverrini* were isolated and identified under a dissecting microscope as previously reported [4].

A total of 110 male hamsters (60 infected by single intragastric intubation of *O. viverrini* metacercariae (50/hamster) and 50 non-infected) were divided into eight groups as follows: Group 1, untreated (n=15); Group 2, NDMA (n=10); Group 3, NA (n=15); Group 4, *O. viverrini* (n=10); Group 5, NDMA + NA (n=10), Group 6, *O. viverrini* + NA (n=10); Group 7, *O. viverrini* + NDMA (n=20), Group 8, *O. viverrini* + NDMA + NA (n=20). The hamsters in Groups 2, 5, 7 and 8 received NDMA by gavage (0.2 mg/hamster/day, 3 times/week) for 4 weeks starting at age 8 weeks. Those in Groups 3, 5, 6 and 8 were given NA in drinking water (1 g/l) between 13 and 31 weeks. Samples of 24-h urine and faeces were collected during the weeks 30 and 31 from each hamster. The animals were placed in individual metabolic cages and were given only distilled water *ad libitum* to avoid the effect of dietary intake on urinary nitrate. Nitrate was measured using an automated analyser which was based on the reaction of nitrite with a Griess reagent after reduction of nitrate to nitrite by a cadmium column [4]. The hamsters were sacrificed at week 34.

RESULTS AND DISCUSSION

All animals, except one in Group 1, survived until killing at week 34. As shown in Figure 1, infected hamsters excreted significantly elevated levels of nitrate (an oxidized product of NO) in the urine, compared with untreated hamsters (means ± SD: 2.92 ± 0.33 for Groups 4 + 7 combined versus 2.34 ± 0.60 µmol/hamster/day for Groups 1 + 2, $P < 0.001$). Administration of NA significantly reduced the urinary levels of nitrate by 55-70% (1.08 ± 0.90 for Groups 3 + 5 and 0.78 ± 0.30 for Groups 6 + 8). These results imply that (i) the hamsters infected with *O. viverrini* have increased production of NO due to elevated expression or activation of NO synthase(s) and (ii) NA administered in drinking water effectively inhibited NO synthase(s). Because NA reduced the urinary nitrate levels in both infected and non-infected hamsters, NA appeared to inhibit effectively both constitutive and inducible isoforms of NO synthase.

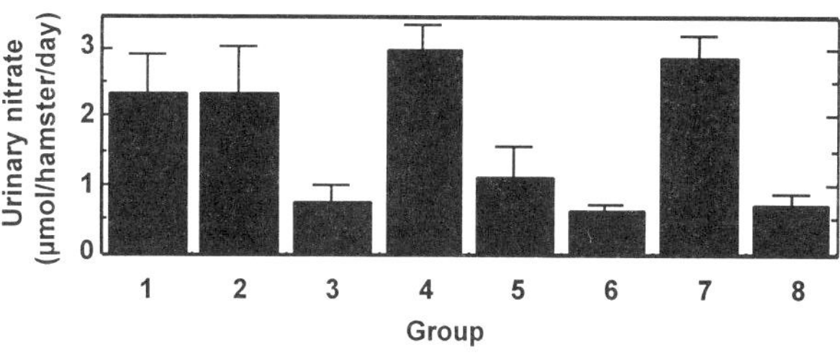

Figure 1 Urinary levels of nitrate
(Means ± SD excreted by different groups of hamsters treated as described in MATERIALS AND METHODS)

The administration of NA in drinking water for 19 weeks resulted in ~10% reduction in the rate of body weight increase, compared with animals which received no NA. However, there were no significant gross lesions or histological changes in the livers of the non-infected hamsters which were given NA.

All hamsters administered *O. viverrini* metacercariae and given normal water developed adult liver flukes in the liver and excreted significant numbers of parasite eggs in the faeces. Adult parasites, however, were not detected in the liver of over half of the infected hamsters which received NA. These hamsters also excreted much fewer parasite eggs in the faeces than the infected hamsters given normal water (means of number of eggs excreted for 24 h; 315 for Group 4 versus 171 for Group 6, $P < 0.05$; 381 for Group 7 versus 87 for Group 8, $P < 0.01$).

The liver of infected and NDMA-treated hamsters (Group 7) showed histological lesions such as cholangiofibrosis and increased bile duct proliferation. On the other hand, the administration of NA to similarly treated hamsters (Group 8) diminished the incidence and severity of such lesions. Since the occurrence of these lesions was closely associated with the presence of adult flukes in the liver, this diminished effect of NA was mainly due to elimination of the parasite by NA. Possible explanations for these findings would be that (1) NA is toxic to *O. viverrini*, (2) the parasite requires NO to live in bile ducts or (3) NA possibly induces intrahepatic bile duct contraction, which results in parasite elimination.

We thank Dr. J. Cheney for editing the manuscript.

REFERENCES

1 Parkin, D.M., Ohshima, H., Srivatanakul, P. and Vatanasapt, V. (1993) Cancer Epidemiol. Biomarkers. Prev. **2**, 537-544
2 Srivatanakul, P., Ohshima, H., Khlat, M., Parkin, M., Sukarayodhin, S., Brouet, I. and Bartsch, H. (1991) Int. J. Cancer **48**, 821-825
3 Haswell Elkins, M.R., Satarug, S., Tsuda, M., Mairiang, E., Esumi, H., Sithithaworn, P., Mairiang, P., Saitoh, M., Yongvanit, P. and Elkins, D.B. (1994) Mutat. Res. **305**, 241-252
4 Ohshima, H., Bandaletova, T.Y., Brouet, I., Bartsch, H., Kirby, G., Ogunbiyi, F., Vatanasapt, V. and Pipitgool, V. (1994) Carcinogenesis **15**, 271-275
5 Ohshima, H. and Bartsch, H. (1994) Mutat. Res. **305**, 253-264
6 Thamavit, W., Bhamarapravati, N., Sahaphong, S., Vajrasthira, S. and Angsubhakorn, S. (1978) Cancer Res. **38**, 4634-4639
7 Thamavit, W., Kongkanuntn, R., Tiwawech, D. and Moore, M.A. (1987) Virchows Arch. B. Cell Pathol. **54**, 52-58

§ Present address: Nagoya City University Medical School, Mizuho-ku, Nagoya 467, Japan

Chemical synthesis of nitric oxide in the human stomach

R.S. DRUMMOND, C.W. DUNCAN, L.M. SMITH, G.M. McKNIGHT and N. BENJAMIN

Department of Medicine and Therapeutics,
University of Aberdeen Medical School,
Aberdeen, U.K. AB9 2ZD

Introduction

It is now well established that mammalian cells produce nitric oxide (NO) from the amino acid L-arginine by a group of NO synthase enzymes. We have recently proposed that NO may be synthesised by an alternative mechanism which relies on sequential reduction of nitrate [1]. Dietary nitrate (mainly derived from green vegetables) absorbed from the stomach and proximal small intestine is subsequently concentrated in saliva [2]. Nitrate-reducing facultative anaerobic bacteria on the dorsal surface of the tongue rapidly reduce nitrate to nitrite under hypoxic conditions, therefore concentration of nitrite in saliva increases subsequent to oral nitrate intake [3]. When swallowed this nitrite is readily protonated under acidic conditions of the stomach (acid dissociation constant ~ 3.2) to form nitrous acid which in turn decomposes to form oxides of nitrogen.

It has been shown that NO formation from L-arginine is important in host defence. [4] We have suggested that acidification of salivary nitrite is important in augmenting the antimicrobial effects of stomach acid by the subsequent generation of NO by the destruction or inhibition of swallowed pathogens [5]. Several studies suggest that a variety of organisms are sensitive to this substance [6]. The generation of NO in the stomach has been demonstrated by measurement of NO in expelled air [7]. The purpose of the present study was to more clearly quantify NO synthesis in the human stomach and determine the temporal relationship of chemical NO production following nitrate ingestion in healthy volunteers.

Methods

Six healthy fully informed male volunteers (aged 21-39 years) were recruited with ethical approval. Subjects were studied on two separate days at least one day apart after overnight fasting.

Experimental protocol on each day was identical apart from the intake of a solution of potassium nitrate BP (2mmoles) in 100 ml of water or distilled water alone as control. The order of experiment was allocated on a blind randomised basis. Following insertion of a nasogastric feeding tube, gastric headspace gas (2ml in a glass syringe), gastric juice (5ml) and saliva samples (0.5ml) were taken at 15 minute intervals for 2 hours. Following aspiration of gastric contents air (50ml) was injected through the nasogastic tube to achieve adequate insufflation for sampling. The pH of gastric juice samples was measured before alkalinisation with 50μl 5M NaOH to prevent further reduction of NO_2^-. Nitrate and nitrite assays were as previously described [8,9]. NO was analysed using a chemiluminescence meter calibrated with standard NO/nitrogen mixtures (BOC Special Gases) following dilution with laboratory air (NO concentration <10ppb) to achieve final concentration less than 20ppm.

Statistical Analysis

The results were compared using 2 way analysis of variance with repeated measures and post-hoc analysis using Wilcoxon's signed rank test. $P<0.05$ was considered significant.

Results

On the control day fasting concentrations of gastric nitrate (105.3±23.1μM), nitrite (19.2±6.1μM) and headspace gas NO (6.1±1.8 ppm) changed little over the time course of the experiment (Fig 1). Similarly there was little variation in basal salivary nitrate (30.5±17.8μM) and nitrite (37.6±17.5μM).

In contrast, following oral administration of 2mmoles of potassium nitrate solution there were marked increases in gastric nitrate, salivary nitrate, salivary nitrite, and gastric headspace NO concentration which remained considerably elevated until the end of the study at 2 hours (Fig 1). Gastric nitrite concentration was non-significantly elevated at 75 and 120 minutes.

Gastric nitrate concentration peaked at 15 mins after intake and salivary nitrate and nitrite at 90 and 75 minutes respectively. Gastric headspace NO concentration peaked at 45 minutes and remained significantly elevated for the duration of the study ($P= 0.0015$; ANOVA). The highest measured gastric headspace gas NO value was 482 parts per million (ppm).

Gastric acidity decreased over the study period on both days but was not significantly affected by nitrate intake. There was a negative relationship between gastric juice pH and stomach headspace gas NO concentration but no significant relationship between stomach NO and salivary nitrite concentrations.

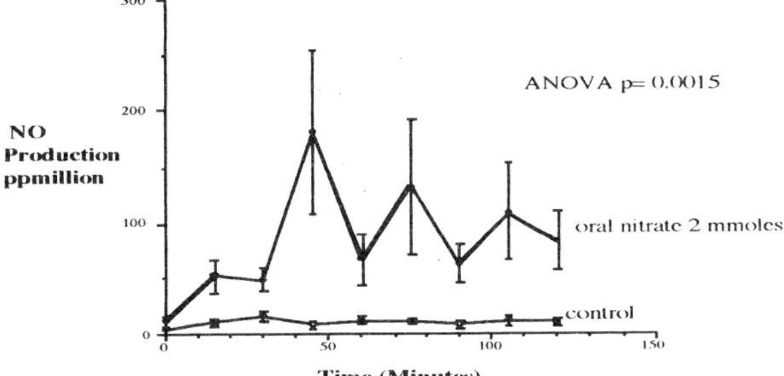

Gastric NO production after ingestion of 200mg potassium nitrate or distilled water
Fig 1

Discussion

This study demonstrates large concentrations of NO are generated in the stomach following a dose of nitrate which represents 1-2 day's average adult intake in the UK [10]. The maximum concentration of NO was seen approximately 45 minutes and averaged nearly 200ppm, about 10,000 times that found in exhaled breath [7]. At 2 hours the concentration remained considerably higher than baseline, suggesting a prolonged effect of dietary nitrate on gastric NO synthesis. The concentrations of gastric headspace gas NO were considerably higher in this study than those reported in expelled gastric air (basal NO 600-800ppb) perhaps due to dilution of gastric gas in the previous study with carbon dioxide generated from the carbonated drink. The higher values measured in this study may be an underestimate of true NO concentrations in the stomach as NO at this concentration will readily combine with oxygen to form NO_2, which is not measured by the chemiluminescence analyser we used. Furthermore, as we had to inject air into the stomach 5 minutes before sampling the concentration in headspace gas may not have reached equilibrium.

The timecourse of increased NO generation and rise in salivary nitrate and nitrite suggests that stomach NO synthesis may derive from acidification of salivary nitrite, even though there was no direct correlation between stomach NO and salivary nitrite. Gastric headspace NO however positively correlated with gastric acidity.

NO generation in the stomach greatly exceeds that expected from *in vitro* studies with acidified nitrite, suggesting the presence of another reducing agent (e.g. ascorbate) in saliva or gastric juice.

In summary, this study shows that dietary nitrate, by bacterial and chemical reduction and enterosalivary circulation, may generate very large concentrations of NO in the stomach which may be beneficial by augmenting the antiseptic properties of gastric acid and modifying physiological gastric function.

References

1.Benjamin, N.,Driscoll, F., Dougall, H., Duncan, C., Smith,L.M and Golden, M.(1994) Nature,**368**, 502

2.Tannenbaum, S.R., Weissman, M. and Fett, B. (1976)Food Cosmet. Toxicol.,**14**, 549-552

3.Sasaki, K. and Matano, K. (1979) J. Food. Hyg.Soc. Jap.,**20**, 363-369

4.Hibbs, J.B.Jnr.(1991) Respiratory Immunology, **142**, 565-569

5.Duncan,C., Dougall, H., Johnston, P.et al (1995)Nature Medicine, **1**, 546-551

6.Klebanoff, S.J.(1993) Free Radical Biol. Med.,**14**, 351-360

7.Lundberg, J.O.N., Weitzberg, E., Lundberg, J.M. and Alving, K.(1994) Gut, **35**,1543-1546

8.Green, L.C., Wagner, D.A., Glogowski, J, Skipper,P.L., Wishnok, J.S. and Tannenbaum, S.R. (1982) Anal. Biochem., **126**, 131-138

9.Phizackerly, P.J.R. and Al-Dabbagh, S.A.(1983) Anal. Biochem.,**131**, 242-245

10.Knight, T.M., Forman, D., Al-Dabbagh, S.A.and Doll, R.(1987) Food Chem. Toxicol., **25**, 277-285

Nitric oxide synthase in the rat gastric mucosa: isoforms, localization and regulation.

PETER J. HANSON, KENNETH J. PRICE, CLARE R. BYRNE and BRENDAN J. R. WHITTLE*

Pharmaceutical Sciences Institute, Aston University, Aston Triangle, Birmingham B4 7ET, UK, *Wellcome Research Laboratories, Langley Court, Beckenham, Kent BR3 3BS, UK.

Gastric nitric oxide synthase (NOS) activity is high relative to many other rat tissues [1]. NOS is present in the mucosa [2], isolated mucosal cells [3], and the smooth muscle [2]. Nitric oxide (NO) regulates blood flow and secretory activity [2] in the mucosa, and sensory neuropeptides, NO and prostanoids interact to maintain mucosal integrity [4]. The aims of the present work were to identify the isoforms of NOS present in gastric mucosa, to localize them and to initiate an investigation of the regulation of mucosal NO production.

An antibody to the carboxyl-terminal hexadecapeptide of rat brain NOS (nNOS) was produced in rabbits. Monoclonal antibodies raised against a 22.3 kDa fragment corresponding to amino acids 1095-1289 of human nNOS, and to a 20.4 kDa fragment corresponding to amino acids 1030-1209 of human endothelial NO synthase (eNOS) were obtained from Transduction Laboratories (Lexington, Kentucky, USA). Immunoblotting, with enhanced chemiluminescence detection, was performed with nNOS polyclonal antiserum at a dilution of 1/1250 or with nNOS or eNOS monoclonal antibodies at concentrations of 1 μg/ml. Cryostat sections (10 μm), fixed by immersion in acetone at 4°C for 5 minutes, were exposed to antibodies to NOS for 30-60 min at room temperature. Rabbit antiserum to nNOS was used at a dilution of 1/320, with similarly diluted pre-immune serum, or antiserum plus 100 μg/ml of peptide, as controls. Monoclonal antibodies to nNOS and eNOS were applied at concentrations of 5 and 10 μg/ml respectively, with mouse monoclonal antibody of irrelevant specificity as the control. Sections were developed using a Vectastain Elite ABC kit (Vector Laboratories, Peterborough, UK) with diaminobenzidine as the peroxidase substrate. Mucosa was homogenized (Ultra-Turrax) in buffer with protease inhibitors [5] and centrifuged at 100,000 x g for 30 min at 4°C. Isolated gastric mucosal cells [6] were incubated with 3 μCi/ml of L-[2,3,4,5-^{3}H] arginine for 15 min at 37°C. Cell pellets were extracted with 65% ethanol and products were separated from arginine by thin-layer or ion-exchange chromatography. Arginase activity was assayed according to Cook et al. [7].

Immunoblots of gastric mucosal extracts exhibited bands of 160 kDa and 140 kDa with antibodies to nNOS and eNOS respectively. Addition of the peptide to the antiserum virtually abolished the 160 kDa band. Image analysis of blots indicated that 88 ± 2% (n=3) of eNOS was in the pellet and 96 ± 3% (n=3) of nNOS was in the supernatant.

Abbreviations used: NO, nitric oxide; NOS, nitric oxide synthase; nNOS, neuronal nitric oxide synthase; eNOS, endothelial nitric oxide synthase; BAPTA-AM, 1,2-bis-(o-aminophenoxy)-ethane-N,N,N',N'-tetraacetic acid tetra-(acetoxymethyl)-ester.

Cells on the mucosal surface of cryostat sections were stained by antibodies directed to nNOS. The antibody to eNOS stained blood vessels in the submucosa, but did not stain surface cells. The antibodies to nNOS also stained myenteric plexi in the duodenum. Staining of mucosal cells by the antipeptide antiserum was reduced by the addition of the peptide.

After 15 min incubation with L-[2,3,4,5-^{3}H]arginine, 6.2 ± 1.7% of the ^{3}H associated with the gastric cells was found in ornithine. The label in citrulline was 11.2 ± 2.7% (by thin-layer chromatography) or 12.4 ± 2.2% (by ion-exchange analysis) of the total (means ± S. E. M. from five cell preparations). The above results were not significantly affected by the preincubation of the cells for 15 min with N^{G}-nitro-L-arginine (300 μM), by the chelation of extracellular Ca^{2+} with 2 mM EGTA or by chelation of both intracellular and extracellular Ca^{2+} with 2 mM EGTA and 25μM BAPTA-AM. Arginase was detected in the gastric mucosa at an activity of 34 ± 6 nmol min^{-1} mg protein^{-1}.

The antibodies were selective for NOS isoforms. Thus, although both eNOS and nNOS immunoreactivity was present in gastric mucosa, no single antibody produced bands at both 160 and 140 kDa and recognized both isoforms. The subcellular distribution of eNOS in gastric mucosa was compatible with recent findings on endothelial cells [8], but the negligible particulate nNOS contrasts with findings in rat cerebellum [9]. The pathway forming citrulline from arginine, apparently via arginase, could potentially compete with NOS for substrate, and might therefore have regulatory significance.

In conclusion, gastric mucosa contains NOS isoforms which are immunologically related to both nNOS and eNOS. The nNOS-like isoform is present in surface mucosal cells, where it may be involved with regulation of mucus secretion [10]. The presence of eNOS in submucosal blood vessels is compatible with a role for NO in regulation of mucosal blood flow [2].

Acknowledgments. Supported by an MRC project grant and by a Collaborative Studentship to Clare Byrne from MRC and the Wellcome Research Laboratories.

References

1. Salter, M., Knowles, R. G. and Moncada, S. (1991) FEBS Lett. **291**, 145-149
2. Whittle B. J. R. (1994) in The Physiology of the Gastrointestinal Tract (Johnson, L. R. , ed.), pp. 267-294, Raven Press, New York
3. Brown J. F. , Tepperman B. L., Hanson P. J., Whittle B. J. R. and Moncada S. (1992) Biochem. Biophys. Res. Commun. **184**, 680-685
4. Whittle B. J. R., Lopez-Belmonte, J. and Moncada S. (1990) Br. J. Pharmacol. **99**, 607-611
5. Nakane, M., Mitchell, J., Forstermann, U. and Murad, F. (1991) Biochem. Biophys. Res. Commun. **180**, 1396-1402
6. Hatt, J. F. and Hanson, P. J. (1989) Am. J. Physiol. **256**, G129-G138
7. Cook, T. H., Jansen, A., Lewis, S., Largen, P., O'Donnel, M., Reaveley, D. and Cattel, V. (1994) Am . J. Physiol. **267**, F646-F653
8. Hecker, M., Mulsch, A., Bassenge, E., Forstermann, U. and Busse, R. (1994) Biochem. J. **299**, 247-252
9. Hecker, M., Mulsch, A. and Busse, R. (1994) J. Neurochemistry **62**, 1524-1529
10. Brown J. F., Keates A. C., Hanson P. J. and Whittle B. J. R. (1993) Am. J. Physiol. **265**, G418-G422

Cerebral nitric oxide mediates the inhibition by stress of gastric acid secretion

M. DOLORES BARRACHINA, SARA CALATAYUD, BRENDAN J.R. WHITTLE*, SALVADOR MONCADA* and JUAN V. ESPLUGUES

Department of Pharmacology, University of Valencia, 46021 Valencia, Spain and *Wellcome Research Laboratories, Beckenham, Kent, BR3 3BS UK

Introduction. Changes in gastrointestinal functions are one of the main features appearing during stress. Previous investigations have demonstrated a reduction in rat gastric acid output following exposure to stress induced by a great diversity of stimuli and, furthermore, there is frequent clinical evidence that stress erosions and ulcers are usually combined with acid hyposecretion. Studies with the NO synthase inhibitor, N^G-nitro-L-arginine methyl ester (L-NAME) have shown that acute acid inhibition induced by the stress represented by bolus i.v. administration of endotoxin requires the synthesis, or release, of NO [1] and the integrity of the peripheral nervous system [2]. These observations suggested that endotoxin triggers a nervous reflex which leads to the reduction of acid production by the stomach and involves NO, possibly acting as a non-adrenergic/ non-cholinergic (NANC) neurotransmitter. In the present study we have analyzed further the role, and the location of the synthesis, of the NO implicated in the inhibitory actions of stress on acid production induced by a stimulus known to be neuronally mediated such as gastric distension.

Methods. The stomach of anaesthetized Wistar rats (urethane 1.5 g/kg, i.p.) was continuously perfused with saline (0.9 ml/min). Once acid secretion had remained constant for 60 min stress was induced by the administration of E. coli lipopolysaccharide (800 ng/kg i.c.v. or 5 μg/kg i.v.) or by mild levels of hyperthermia (38-39° C maintained for the duration of the experimental period). Thereafter the stomach was distended for 90 min with an intragastric pressure of 20 cm H_2O. Animals were treated (i.c.v.), 15 min before stress, with L-NAME (0.8 mg/kg), L-arginine (12 mg/kg) or D-arginine (12 mg/kg).

Results. *a) Effects of endotoxin on gastric acid secretion.* Distension of the stomach with an intragastric pressure of 20 cm water produced a progressive increase of acid secretion which, after 90 min, reached a submaximal value of H^+ production with 23±5 μEq/100g/30 min. The total acid output during the 90 min experimental period was 36±7 μEq H^+/100g/90 min (n=9). The i.c.v. injection of endotoxin (800 ng/kg, n=8) inhibited distension-induced acid production by 83±6% ($p<0.01$), with significant levels of acid inhibition already present within the first 30 min following administration of i.c.v. endotoxin. H^+ output stimulated by distension was not modified by intravenous administration of this low dose of endotoxin (800 ng/kg, n=4), and higher doses of i.v. endotoxin (5 μg/kg, n=15) were needed to achieve a significant ($p<0.001$) reduction of acid production (76±4% reduction). The i.c.v. administration of L-NAME (800 μg/kg) restored acid production to control distension levels in rats receiving endotoxin either i.c.v (n=6) or i.v. (n=7). During the experimental period blood pressure was not significantly modified by the injection of endotoxin (i.c.v. or i.v.) or L-NAME (i.c.v.) at the doses used in the present study. *b) Effects of hyperthermia on gastric acid secretion.* In further experiments the increase in the rectal temperature from 36-36,5° C to 38,5-39° C for the 90 min period of acid stimulation by distension, induced a significant ($p<0.001$) reduction in the total production of acid (74±5% diminution, n=9). Prior i.c.v. administration of L-NAME (800 μg/kg) prevented the inhibitory effects of hyperthermia with values of acid production of 37±8 μEq H^+/100g/90 min (n=8). This dose of L-NAME (800 μg/kg) if administered i.v. did not influence the inhibitory effects of either endotoxin or hyperthermia on distension-stimulated acid production (n=3 in each). *c) Reversal of effects of L-NAME by L-arginine.* The reversal by L-NAME (800 μg/kg, i.c.v.) of the inhibition of distension-stimulated acid secretion by both methods of stress was prevented by the i.c.v. administration of L-arginine (12 mg/kg, n=6 in each). Pretreatment with D-arginine (12 mg/kg, i.c.v., n=3 in each) did not modify the acid recovery induced by i.c.v. L-NAME in both groups of experiments.

Discussion. In the present study induction of stress by i.c.v. administration of endotoxin abolished distension-stimulated acid responses, at doses that had not such effect when administered i.v. and which did not affect systemic blood pressure. The nature of the inhibitory actions of endotoxin appear to involve the local generation of NO in brain tissue since i.c.v. injection of L-NAME reversed the actions of endotoxin, administered either centrally or peripherally. Furthermore, i.c.v. administration of low doses of L-arginine restored the acid inhibitory effects of endotoxin in animals treated with L-NAME. Such findings extend previous results [1,2,3] and confirm that the inhibition of acid production by endotoxin results from activation of the L-arginine:NO pathway in the central nervous system. A similar cerebral pathway involving NO is triggered by hyperthermia, thus implying that it is not characteristic for endotoxin and could represent a common mechanism mediating the acid inhibitory responses to all types of stress. The inhibitory reflex so evoked may also involve other neurotransmitters and the release of inhibitory mediators in the vicinity of the parietal cell.

Furthermore, it seems likely that this neuronal pathway may be associated with other gastric responses to stress, as suggested by the recent implication [4] of NO in the protective effects of low-dose endotoxin on gastric mucosal damage.

Acknowledgements. The present study has been supported by grants SAF95-0472 and FIS95/1343. M.D.B and S.C are the recipients of a "Schering-Plough Fellowship for Gastrointestinal Research".

References

1. Martinez-Cuesta, M.A., Barrachina, M.D., Piqué, J.M., Whittle, B.J.R. and Esplugues, J.V. (1992) The role of nitric oxide and platelet-activating factor in the inhibition by endotoxin of pentagastrin-stimulated gastric acid secretion. Eur. J. Pharmacol., **218**, 351-354.

2. Martinez-Cuesta, M.A., Barrachina, M.D., Whittle, B.J.R., Piqué, J.M., and Esplugues, J.V. (1994) Involvement of neuronal processes and nitric oxide in the inhibition by endotoxin of pentagastrin-stimulated gastric acid secretion. Naunyn-Schmied. Arch. Pharmacol., **349**, 523-527.

3. Barrachina, M.D., Whittle, B.J.R., Moncada, S. and Esplugues, J.V. (1995) Endotoxin inhibition of distension-stimulated gastric acid secretion in rat: mediation by NO in the central nervous system. Br. J. Pharmacol., **114**, 8-12

4. Barrachina, M.D., Calatayud, S., Moreno, L., Martinez-Cuesta, M.A., Whittle, B.J.R. and Esplugues, J.V. (1995) Nitric oxide and sensory afferent neurones modulate the protective effects of low-dose endotoxin on rat gastric mucosal damage. Eur. J. Pharmacol., **280**, 339-342.

Visualisation of nitric oxide release from rabbit hypogastric nerve trunk

Selim CELLEK, Anna M. LEONE, Vanessa W. FURST, N. Peter Wiklund* and Salvador MONCADA;

Wellcome Research Laboratories, Langley Court, Beckenham, Kent, BR3 3BS, UK; * Department of Physiology and Pharmacology, Institute of Environmental Medicine, Karolinska Institute, Stockholm, Sweden.

Nitric oxide (NO) is a neurotransmitter, synthesised and released from non-adrenergic non-cholinergic nerves or so called "nitrergic nerves" [1] as well as some neurones in the CNS [2]. Being a gaseous molecule and having a relatively short half-life, NO can diffuse through biological membranes and exerts its function at nearby target organs. So far, the release pattern and diffusion profile of NO have been demonstrated only on a theoretical basis [3]. In order to understand the nature of its release and diffusion in biological systems, visualisation of NO is required. In this study, we have visualised NO released from the rabbit hypogastric nerve trunk (RHNT) by electrical stimulation. This preparation is known to be involved in nitrergic innervation of the pelvic organs and has no anatomical connections with any smooth muscle.

The hypogastric nerve trunk (~3-4 cm) from male New Zealand rabbits was isolated between the level of the inferior mesenteric artery and the bifurcation of the abdominal aorta [4]. After cleaning of adjacent tissues, RHNT was incubated in modified Krebs' medium oxygenated with 95% O_2 and 5% CO_2 for 45 minutes at 37°C. RHNT was then carefully transferred to a plastic transparent culture well and covered by a glass cover-slip (22 mm diameter) so that each end of the tissue projected out of the cover-slip. The ends of the tissue were hooked on platinum electrodes which were connected to a Grass S88 stimulator. The tissue was immersed in 200 µl of modified Krebs' solution so that electrodes were not touching the medium. Before adding the imaging medium (IM) containing 5 mM luminol and 15 µM H_2O_2 in Krebs, a light image of the tissue was obtained (10x magnification). In the presence of IM, photon counting was performed for 1 minute periods until the counts stabilised. After stabilisation, electrical stimulation was applied for 1 min with trains of rectangular pulses of 30 V, 0.3 ms pulse duration and frequencies ranging from 5 to 30 Hz. Before, during and after 1 min stimulations, sequential photon countings for 5 sec were performed in order to visualise the sequential pattern of release. In order to inhibit NO release, the tissue was incubated with N^G-nitro-L-arginine methylester (L-NAME) (1mM) or tetrodotoxin (TTX) (2µM) for 15 min. After incubation with the inhibitor, electrical stimulation and photon counting were performed using the same parameters.

Images were obtained using an Argus 50 photon counting system (Hamamatsu Photonics UK Ltd.) coupled to a Zeiss Axiavert 135 TV inverted microscope (Carl Zeiss, Oberkerken Ltd). The camera was mounted directly under the microscope objective, allowing maximum light transmission. The unit was housed in a light-tight enclosure during acquisitions. Bright field images were overlaid with pseudo-colourised photon counted images in slice mode allowing intensity and colours to be adjusted. Photon counts were also counted in gravity mode for quantification and acquired/integrated over 5-60 seconds.

Although the background photon count without any electrical stimulation was variable, electrical stimulation for 1 minute at 5-30 Hz caused a significant increase in the photon count ($p<0.05$, for paired observations). This increment was frequency-dependent with a threshold frequency of 10-20 Hz. L-NAME (1 mM) and TTX (2 µM) inhibited the increase in the photon count by 64.5±8.7% and 90% respectively ($p<0.02$, for paired observations). The effect of the inhibitors on background photon count was not significant ($p=0.0993$, for paired observations). Analysis of variance for these results showed that the effect of electrical stimulation or treatment with inhibitors was significant in spite of the high variability among the tissues ($F=9.572$ and $p=0.0009$).

Sequential photon counting for 6 seconds before, during and after 1 min stimulation revealed the pattern of NO release from the nerve tissue. In the first 18 seconds the central part of the nerve tissue became brighter, then all of the tissue became illuminated and finally diffusion of the light to the dark field near the tissue was observed. The speed of diffusion of the light was calculated as 300 µ/ 6 s by measuring the change in light intensity at a known distance from the nerve. After termination of the stimulation the light intensity returned to background levels within approximately 5-10 s.

The hypogastric nerve trunk is a plexiform nerve between the superior hypogastric plexus and inferior hypogastric plexus (pelvic plexus) and contains mainly ganglion cells, preganglionic axons and postganglionic fibres from prevertebral ganglion cells [4,5]. It conveys sympathetic and parasympathetic transmission as well as inhibitory non-adrenergic non-cholinergic transmission to the pelvic organs, including the penis [5]. It is now established that the main inhibitory non-adrenergic non-cholinergic transmission in the penis which produces erection is mediated by the L-arginine:NO pathway [6]. Electrical stimulation of the hypogastric nerve trunk causes erectile responses in the cat and rabbit [7,8]. Histological studies showing the presence of NO synthase immunoreactivity and NADPH-diaphorase activity in ganglion cells, preganglionic efferent pathways to the pelvic viscera [9] and postganglionic efferent pathways to the urethra and penis [10,11] further suggest the presence of nitrergic neurotransmission in the hypogastric nerve trunk. For this reason, we used RHNT as a neuronal source of NO in our experiments, and it gave a significant light signal when electrically stimulated.

NO does not react with luminol directly [12]; however NO can react with hydrogen peroxide to form peroxynitrite or singlet oxygen, each of which produces light when it reacts with luminol [12,13]. Thus the image seen in our experiments may involve either peroxynitrite or singlet oxygen or, more likely, both mechanisms. Nonetheless, whatever the actual mechanism, NO release initiates the process of light generation. Thus we conclude that the light generated in our experiments is due to NO release from RHNT.

The light signal elicited by electrical stimulation of the tissues was significantly inhibited by L-NAME and TTX, suggesting that the signal was nitrergic in nature and neurogenic in origin. NO differs from known neurotransmitters in that it is not stored in vesicles and can diffuse through biological membranes. The diffusion characteristics of NO have been widely debated. Models of diffusional spread of NO have been suggested [3], but so far the diffusion of NO in a biological system has not been demonstrated. In this study, we present the first data showing the diffusion pattern of NO from a biological source.

1. Rand, M.J. & Li, C.G. (1995) in Nitric Oxide In the Nervous System (Vincent, S., ed.), pp. 227-280, Academic Press, San Diego.
2. Garthwaite, J. & Boulton, C.L. (1995) Annu. Rev. Physiol. **57**, 683-706.
3. Wood, J. & Garthwaite, J. (1994) Neuropharmacol. **33**, 1235-1244.
4. Langley, J.N & Anderson, H.K. (1895-1896) J.Physiol. **19**, 384.
5. Andersson, K.E. & Wagner,G. (1995) J.Physiol. **75**, 191-235.
6. Burnett, A.L. (1995) Biol. Reprod. **52**, 485-489.
7. Root, W.S & Bard, P. (1947) Am. J. Physiol. **151**, 80-90.
8. Sjostrand, N.O. & Klinge, E. (1979) Acta Physiol. Scand. **106**, 199-214.
9. Vizzard, M.A., Erdman, S.L. & de Groat, W.C. (1993) Neurosci. Lett. **152**, 72-76.
10. Vizzard, M.A., Erdman, S.L., Forstermann, S.L. & de Groat, W.C. (1994) Brain Res. **646**, 279-291.
11. Schirar, A., Giuliano, F., Rampin, O. & Rousseau, J.P.(1994) Cell Tiss. Res. **278**: 517-525.
12. Radi, R., Cosgrove, T.P., Beckman, J.S. & Freeman, B.A. (1993) Biochem. J. **290**, 51-57.
13. Kikuchi, K., Nagano, T., Hayakawa, H., Hirata, Y. & Hirobe, M. (1993) J. Biol. Chem. **268**, 3106-23110.

Characterization of calcium-independent NO synthase expressed in the ileum of healthy rats

MALGORZATA J. ZEMBOWICZ, ARTUR ZEMBOWICZ, A.RIZWAN KHAN, SANDRA C. HIGHAM, NORMAN W. WEISBRODT, FRANK G. MOODY, THOMAS A. PRESSLEY and ROBERT F. LODATO

Departments of Surgery, Physiology and Medicine, University of Texas Medical School, Houston, 6431 Fannin, TX 77030

Nitric oxide synthase type II (NOS-II) is believed to be an inducible enzyme, expression of which requires stimulation of cells with endotoxin or other cell-specific signals such as cytokines tumor necrosis factor-α (TNF-α) or interleukin-1β (IL-1β) [1]. Consistently, NOS-II is absent from quiescent cultured cells and most tissues in healthy animals. Recently, however, expression of NOS-II has been reported in the epithelial cells in the bronchi [2] and the kidney [3]. Whether the expression of NOS-II at these locations is constitutive or is due to induction of NOS-II by local stimuli has not been established.

Studying the role of NO in ileal physiology and in multiple organ failure, we made an original observation of continuous expression of Ca^{++}-independent NOS activity in the ileum but not other intestinal segments in the healthy rats [4]. Results described below provide biochemical and molecular evidence that this Ca^{++}-independent NOS activity is due to expression of NOS-II in the ileal mucosa. We also show that inhibitors of the cytokines, IL-1β and TNF-α, decrease the levels of NOS-II in the ileum. These findings argue against a constitutive nature of NOS-II expression in the ileum and suggest that it is a result of a basal stimulation of mucosal cells by cytokines, perhaps locally produced.

Methods

Animals. Healthy untreated male Sprague-Dawley rats, 250 - 350 g, were used as controls. In some experiments, animals were treated for 6 hours with the combination of IL-1β receptor antagonist (IL-1ra, 100 mg/kg s.c.) and TNF-α binding protein (TNFbp, 1.5 mg/kg i.v.), or respective vehicles (1ml/kg). Animals were sacrificed and ileal segments were rapidly isolated, washed in phosphate-buffered saline, snap-frozen in liquid nitrogen and stored at -70 C until assayed.

Nitric oxide synthase activity, RNA isolation and Northern blotting analysis, and Immunoblotting. NOS activity was assayed as nitro-L-arginine methyl ester-inhibitable conversion of ^{3}H-L-arginine to ^{3}H-L-citrulline [5]. RNA was isolated using RNAzol B, electrophoresed on 1% formaldehyde agarose, transferred to nitrocellulose membrane and hybridized to NOS-II cDNA probe radiolabeled to high specific activity (~10^{8} cpm/μg) with deoxycytosine 5'-[α-^{32}P]triphosphate obtained by random oligonucleotide priming method. Prior to immunoblotting analysis, NO synthases present in tissue homogenates were concentrated by adsorption to 2'-5'-ADP sepharose bead (Pharmacia Biotech AB, Uppsala, Sweden) at 4°C for 2 hours. Bound protein was released to the solution by boiling in 100 μl of Tris buffer containing 1% SDS for 5 min. Equal amounts of total protein present in the supernatants were electrophoresed on 7% SDS/PAGE and transferred to nitrocellulose paper by electroblotting. NOS-II was visualized using affinity-purified polyclonal anti-NOS-II antibodies. Antibody-antigen interactions were visualized using ECL reagents (Amersham).

Results and Discussion

To determine the contribution of Ca^{++}-independent NOS activity to the total NO synthase activity present in the ileal homogenates we assayed NOS activity in the presence and absence of Ca^{++}. Chelation of Ca^{++} with EDTA (1 mM) decreased NOS activity by approximately 40 % . To verify that this activity is, indeed, due to expression of NOS-II we performed molecular and biochemical characterization of NO synthase expression. Northern blotting analysis of RNA isolated from the segments of ileum demonstrated the presence of a single mRNA band that hybridized to NOS-II probe. Molecular weight of this mRNA was ca 4.5 kb, consistent with the molecular weight of NOS-II mRNA. Western blotting analysis of the ileal homogenates revealed that they contain NOS-II protein. Immunoblotting analysis of homogenates prepared from the ileal mucosa, submucosa and longitudinal smooth muscle revealed that NOS-II is expressed only in the mucosal layer. NOS-II has generally been considered to be an inducible enzyme and constitutive expression of NOS-II in the ileum has not been reported. To differentiate between constitutive and cytokine-induced nature of NOS-II expression in the ileum we compared the levels of NOS-II protein in the lysates isolated from control animals to those isolated from rats treated with IL-1ra and TNFbp. As shown on the figure below inhibition of IL-1β and TNF-α significantly decreased NOS-II protein levels. These findings strongly argue against the constitutive

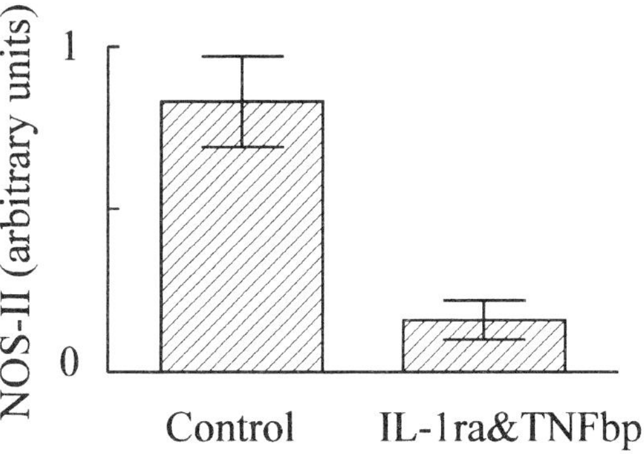

nature of NOS-II expression in the ileum and suggest that NOS-II is induced by IL-1β and/or TNF-α, probably produced locally by mucosal cells. We have previously demonstrated that ileal epithelium is the primary site of NOS-II expression in the rat model of endotoxic shock [6]. The cellular source of cytokines remains to be established. Thus, similarly to NOS-II expression observed in epithelial cells in the lung [2], which is suppressed by glucocorticoids, inhibition of NOS-II expression in the ileum was reduced by cytokine inhibitors, suggesting that NOS-II is locally induced. Whether NOS-II expression observed in the kidney [3] is constitutive or locally-induced has not been established.

The physiological role of NO derived from continuously expressed NOS-II in the ileum remains to be established. However, the biological actions of NOS-II-derived NO suggest that NOS-II expression may be a part of mucosal immunity directed against pathogen invasion. Whether it can also play a role in other ileum-specific processes such as absorption, regulation of mucosal blood flow or ileal motility is still an open question.

References

1. Nathan,C., Xie, Q.W. (1994) **J. Biol. Chem.** 269, 13725-13728
2. Guo, F.H., De Raeve, H.R., Rice, T.W., Stuehr, D.J., Thunnissen, F.B., Erzurum, S.C. (1995) **Proc. Natl. Acad. Sci. U S A.**, 92, 7809-7813
3. Ahn, K.Y., Mohaupt, M,G., Madsen, K.M., Kone, B.C. (1994) **Am. J. Physiol.** 267, F748-7572
4. Zembowicz, M.J., Higham, S.C., Fallaw, C.H., Zembowicz, A., Moody, F.G., and Pressley T.A. (1995) **FASEB J.** 9(4), A704
5. Zembowicz, A., Hatchett, R.J., Radziszewski W., Gryglewski R. J. (1993) **J. Pharmacol. Exp. Ther.** 267, 1112-1118
6. Zembowicz, M.J., Higham S.C., Fallaw C.W., Pressley T.A., Weisbrodt N.W. and Moody F.G. (1995) **Gastroenterology** 108(4), A949

Nitric oxide production in infective gastroenteritis and inflammatory bowel disease.

Roelf S. Dykhuizen*, John Masson#, Ashley N.G. Mowat#, Christopher C. Smith*, Graham J. Douglas*, Lorna Smith+, Mhairi Copland+ and Nigel Benjamin+.

*Infection Unit Aberdeen Royal Infirmary, Foresterhill, Aberdeen AB9 2ZD, Scotland.
#Department of Gastroenterology, Aberdeen Royal Infirmary.
+Department of Medicine and Therapeutics, Medical School, Forresterhill, Aberdeen AB9 2ZD, Scotland.

Introduction

Endogenous production of nitric oxide (NO) via the L-arginine-NO pathway [1] is thought to occur in response to bacterial infection and may cause hypotension in patients with septic shock [2]. Nitrate is the stable end product of NO oxidation and plasma and urinary concentrations reflect endogenous NO production [3]. Even in the abscence of septic shock, plasma nitrate concentration and urinary nitrate excretion are increased in patients with infective gastroenteritis and levels correlate with disease activity [4].

Here we report on plasma nitrate concentrations in healthy controls, patients with irritable bowel syndrome, exacerbations of inflammatory bowel disease and infective gastroenteritis to estimate the extent of nitric oxide production in these conditions.

Patients and methods

Controls: healthy volunteer clerical and laboratory staff, 21-57 years of age, on a normal diet (n=20). Irritable bowel syndrome: patients presenting to the Department of Gastroenterology outpatients clinic consecutively with the characteristic symptomatology of irritable bowel syndrome, with a stool frequency of more than three times per day, negative stoolcultures and no abnormalities on rectal biopsy or barium studies (n=12). Inflammatory bowel disease: patients admitted to the Gastro-intestinal Unit consecutively because of an exacerbation of inflammatory bowel disease, with a stoolfrequency of more than three times per day, negative stoolcultures, and histological and/or radiological evidence of inflammatory bowel disease (n=18, 11 ulcerative colitis and seven Crohn' disease; six patients were on corticosteroid therapy on admission to hospital [four with ulcerative colitis and two with Crohn's disease]). Infective gastroenteritis: patients admitted to the Infection Unit consecutively because of infective gastroenteritis with a stool frequency of more than three times per day (n=20; *Campylobacter jejuni* x 10, *Shigella sonnei* x 5, *Clostridium difficile* [toxin+] x 3, *Salmonella enteriditis* x 1 and *Escherichia coli 0157* x1.

Venous blood was obtained from patients and controls and centrifuged for 10 minutes at 2000 rpm. Plasma urea and creatinine were measured to ascertain that renal function was unimpaired. The plasma nitrate was determined using a copper/cadmium reduction column and spectophotometry on a HPLC system as described by Green et al. [5], modified by replacing the carrier fluid with 1.5% glycine at PH 9.4.

Results

There was no difference in the median plasma nitrate concentration of controls, patients with irritable bowel syndrome, or patients with inflammatory bowel disease, whereas patients with infective gastroenteritis showed significantly raised median plasma nitrate levels (Table 1).

Table 1. Plasma nitrate concentration (µmol/L) in patients with irritable bowel syndrome, inflammatory bowel disease and infective gastroenteritis

Controls		Irritable Bowel Syndrome	
mean	35.88 µmol/L	mean	33.43 µmol/L
median	32.75 µmol/L	median	35.53 µmol/L
S.D.	9.44 µmol/L	S.D.	12.92 µmol/L
Inflammatory Bowel Disease		**Infective Gastroenteritis**	
mean	48.28 µmol/L	mean	228.47 µmol/L
median	33.73 µmol/L	median*	117.92 µmol/L
S.D.	35.75 µmol/L	S.D.	322.16 µmol/L

*Mann-Whitney distribution free rank testing P<0.0001

No significant difference between median plasma nitrate concentrations in ulcerative colitis and Crohn's disease was observed, and no influence of corticosteroid medication (n=6, data not shown). Only four patients with inflammatory bowel disease (2 ulcerative colitis and 2 Crohn's) showed a significantly raised plasma concentration (>2S.D. from the mean) compared to controls.

Conclusions

Exacerbations of inflammatory bowel disease do not switch on a production of nitric oxide large enough to give elevation of nitrate concentration in the blood.

Measurement of plasma nitrate concentration in patients presenting with diarrhoea could serve as a diagnostic tool in the differentiation of infective pathology from functional or inflammatory disease.

References

1. Moncada S., Palmer R.M.J. and Higgs E.A. (1991) Nitric oxide: physiology, pathophysiology and pharmacology. Pharmacol. Rev. **43**,109-142
2. Petros A., Lamb G., Leone A., Moncada S., Bennett D. and Vallance P.(1994) Effects of nitric oxide synthase inhibitor in humans with septic shock. Cardiovasc. Res. **28**, 34-39
3. Hibbs J.B. Jr., Westenfelder C., Taintor R. et al. (1992) Evidence for cytokine-inducible nitric oxide synthesis from L-arginine in patients receiving interleukin-2 therapy. J. Clin. Invest. **89**, 867-877
4. Dykhuizen R.S., Copland M., Smith C.C., Douglas J.G. and Benjamin N. (1995) Plasma Nitrate concentration and Urinary Nitrate excretion in patients with Gastroenteritis. J. Inf. **31**, 73-75
5. Green L.C., Wagner D.A., Glogowski J. (1982) Analysis of nitrate, nitrite and ^{15}N nitrate in biological fluids. Anal. Biochem **126**, 131

Inducible nitric oxide synthase and nitrotyrosine are localized in damaged intestinal epithelium during human inflammatory bowel disease (IBD)

Irwin I. Singer*, Douglas W. Kawka*, Sol Scott*, Jeffrey R. Weldner*, Richard A. Mumford*, Terrence E. Riehl# and William F. Stenson#

*Merck Research Laboratories, Rahway, NJ 07065, USA
#Washington Univ. School of Medicine, St. Louis, MO 63110, USA

Introduction

Inflammatory bowel disease includes ulcerative colitis (UC) and Crohn's disease (CD): chronic diseases marked by mucosal inflammation, ulceration, and diarrhea (1). Several IBD sequelae including mucosal vasodilation, and enhanced vascular and epithelial permeability are consistent with the effects of increased nitric oxide (NO) synthesis (1,2). Inducible nitric oxide synthase (iNOS) dependent NO production has been demonstrated in UC, but its cellular source has not been identified (3-6). Also, the role of iNOS-mediated NO in causing tissue damage via the formation of nitrotyrosine (7,8) has not been explored in IBD. Here we report the cellular distribution of iNOS and nitrotyrosine (NT) in normal colonic mucosa, and in colonic resections for UC, CD, and diverticulitis (another acute inflammatory disease of the colon).

Results

NO-53, our anti-human iNOS antibody, specifically detected a 130 kDa species in Western blots of epithelial cells isolated from the inflamed human UC colon, and in Sf9.10 insect cells transfected with human iNOS cDNA, but did not react with epithelial extracts of uninflamed colon. Likewise, using immunoperoxidase microscopy, no iNOS staining was detected in histologically normal ileum and colon resected from cancer patients (Fig. 1A), but intense iNOS labeling was localized in the colonic epithelium of inflamed mucosal foci of patients with ulcerative colitis (Fig. 1B). iNOS staining was localized in both crypt and surface epithelia and abruptly diminished subbasal to the basement membrane (Fig. 1B). These iNOS-positive epithelial cells were immediately juxtaposed to ulcers with intense inflammatory cell infiltration of the lamina propria and crypt abscesses. Although iNOS staining was also localized in lamina propria macrophages and neutrophils (PMNs) within crypt abscesses, epithelial iNOS staining was always more intense and widespread than in the adjacent inflammatory cells (Fig. 1B). All iNOS immunolabeling was inhibited by pre-absorption of NO-53 with a peptide whose seven C-terminal amino acids are equivalent to the C-terminus of human iNOS (YRASLEMSAL), but not by a control peptide lacking the C-terminal leucine (YRASLEMSA). When these UC sections were stained for nitrotyrosine, NT labeling was also concentrated focally in the epithelium, and always co-localized with high concentrations of iNOS.

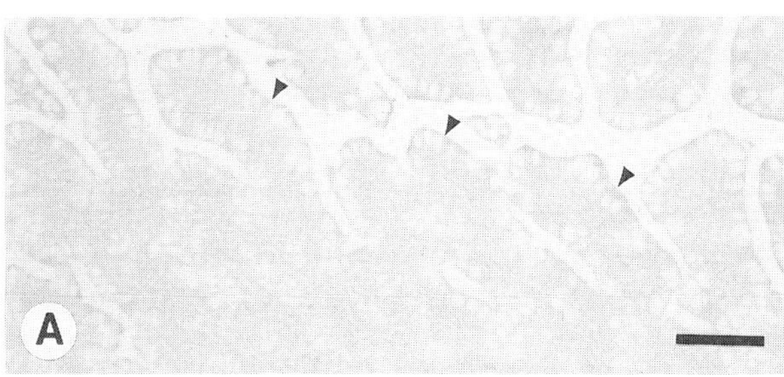

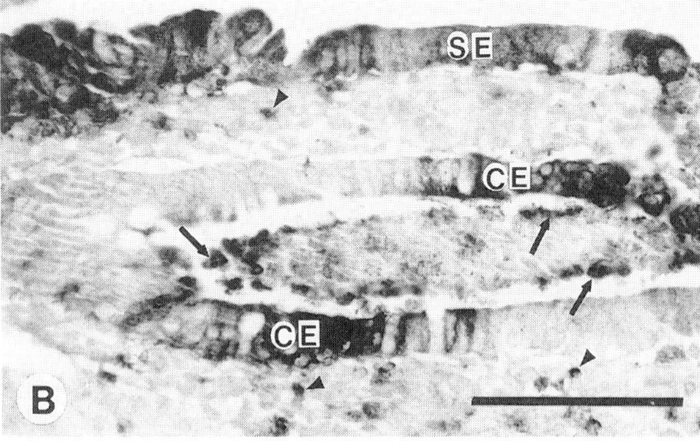

Figure 1. Immunoperoxidase localization of iNOS in normal and UC human bowel. (A) No iNOS expression is detected in the epithelium (arrowheads) or lamina propria of the normal ileum. (B) Colonic section from a UC patient. Intense focal iNOS labeling is localized in the superficial (SE) and crypt epithelium (CE). Lamina propria macrophages (arrowheads), and PMNs within the crypt abscess (arrows), are also iNOS-positive. Bars = 100 µM.

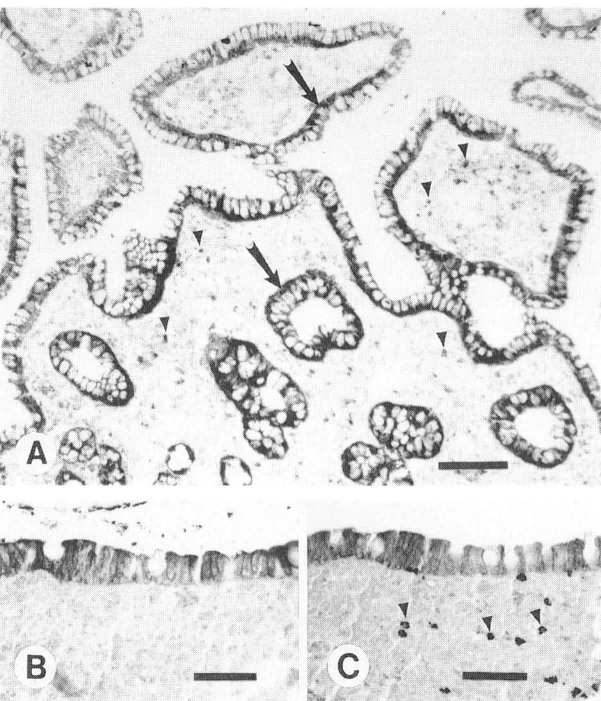

Figure 2. Co-localization of iNOS and NT in bowel mucosal epithelia of CD patients. (A) CD ileum: intense iNOS labeling is localized in villus and crypt epithelia (arrows), and in a few lamina propria macrophages (arrowheads). (B & C) Serial sections of a CD colon stained for iNOS and NT. (B) iNOS is conspicuously concentrated in the epithelium, but not in the lamina propria. (C) Adjacent section shows NT co-localized in the epithelium, and also present in lamina propria macrophages (arrowheads). Bars = 100 µM.

iNOS and NT labeling were also strikingly co-localized in the Crohn's colitis mucosa. Staining for iNOS using NO-53 IgG showed focally positive epithelial staining in both ileal and colonic lesions (Fig. 2). As with UC, the most intense iNOS labeling was localized along the epithelium, and expression appeared to be limited by the basement membrane, although some iNOS-positive mononuclear cells were observed in the lamina propria (Fig. 2A). Intense nitrotyrosine labeling was also co-localized with iNOS in serial sections of the CD epithelium, but NT staining did not co-distribute with iNOS in lamina propria macrophages (Figs. 2B and 2C). Further, expression of iNOS and NT were co-distributed in the inflamed mucosal epithelium of diverticulitis patients.

Conclusions

iNOS was expressed in mucosal epithelial cells of inflamed colonic foci from patients with ulcerative colitis, Crohn's disease and diverticulitis, but not by epithelial cells from uninflamed regions of the bowel, indicating that the inflamed epithelium produced nitric oxide.

Nitrotyrosine (NT) was co-localized with iNOS in epithelial cells of inflamed colonic ulcers, suggesting that nitric oxide contributed to epithelial damage in inflammatory bowel disease via the formation of peroxynitrite (7,8).

Although iNOS and NT were also induced in macrophages and PMNs of the inflamed lamina propria, the number of positive cells was a small fraction of the total inflammatory cell population. In contrast to the epithelium, iNOS and NT did not consistently co-localize in these inflammatory cells.

These findings suggest that the generation of iNOS in the colonic epithelium results in the formation of peroxynitrite and in the indiscriminate nitration of cellular proteins, leading to epithelial cell damage and pathology in inflammatory bowel disease and diverticulitis.

References

1. Stenson, W.F. (1995) Inflammatory bowel disease. in Textbook of Gastroenterology, 2nd edition. (Yamada, T., ed.) pp1748-1805 Lippincott, Philadelphia.
2. Kubes, P. (1992) Am. J. Physiol 262 (Gastrointest. Liver Physiol. 25): G1138-G1142
3. Roediger, W.E.W., Lawson, M.J., Nance, S.H., and Radcliffe, B.C. (1986) Digestion. 35:199-204
4. Boughton-Smith, N.K., Evans, S.M., Hawkey, C.J., Cole, A.T., Balsitis, M., Whittle, B.J.R., and Moncada, S. (1993) Lancet 342:336-340
5. Middleton, S.J., Shorthouse, M., and Hunter, J.O. (1993) Lancet 341:465-466
6. Lundberg, J.O.N., Hellstrom, P.M., Lundberg, J.M., and Alving, K. (1994) Lancet. 344: 1673-1674
7. Beckman, J.S., Ye, Y.Z., Anderson, P., Chen, J., Accavitti, M., Tarpey, M.M., and White, C.R. (1994) Biol. Chem. Hoppe-Seyler. 375:81-88
8. Haddad, Y.I., Pataki, G. , Hu, P., Galliani, C. , Beckman, J.S., and Matalon, S. (1994) J. Clin. Invest. 94:2407-2413

Nitric oxide synthase inhibition attenuates chronic granulomatous colitis without affecting colonic blood flow.

Satoshi AIKO, Elaine CONNER, Jonathan DAVIS, Matthew B. GRISHAM.

Department of Physiology and Biophysics, Louisiana State University Medical Center, Shreveport, LA 71130.

Introduction

Recent studies from our laboratory, as well as others, have demonstrated that chronic gut inflammation is associated with the overproduction of nitric oxide (NO) and that oral administration of certain NOS inhibitors attenuates tissue injury and inflammation in different models of gut pathobiology [1-4]. Although these data have been interpreted to suggest that NO may directly or indirectly play an important role in promoting gut inflammation and dysfunction, it is possible that inhibition of vascular NOS may protect the gut by promoting vasoconstriction which would decrease blood flow thereby limiting the delivery of inflammatory cells and mediators to the tissue. Therefore, the objective of this study was to quantify blood flow to the splanchnic organs before and after acute or chronic administration of two different NOS inhibitors known to be protective in different models of chronic colitis.

Results and Discussion

A total of 46 male Sprague-Dawley rats (300-325g) were randomized into 5 groups consisting of one untreated control group (n=15), two N^G-nitro-L-arginine methyl ester (L-NAME) groups which were treated for 1 or 21 days with 15 μmoles/kg/day (p.o.; n=8 for each group) and two aminoguanidine (AG) groups which were treated for 1 or 21 days with 15 μmoles/kg/day (p.o.; n=7 for 1 day and n=8 for 21 day group). L-NAME and AG were chosen as the NOS inhibitors because of their relative selectivities for the constitutive and inducible NOS, respectively. The radiolabeled microsphere/reference organ method was used to determine tissue blood flows in the splanchnic organs and plasma nitrate and nitrite levels were quantified using the Griess reaction following reduction of all nitrate to nitrite. We found that acute and chronic administration of L-NAME but not AG enhanced mean arterial pressure (134 vs 163 and 170 mmHg for 1 and 21 day, respectively, $p<0.005$). Plasma levels of nitrate and nitrite were significantly inhibited only in the 21 day L-NAME group (14 ± 1 vs 20 ± 1 μM; $p<0.05$). Acute (1 day) administration L-NAME produced a significant decrease in hepatic arterial blood flow (28 ± 3 vs 54 ± 5 ml/min/100g wet wt; $p<0.05$) whereas acute administration of AG reduced blood flow in the stomach (39 ± 5 vs 64 ± 7 ml/min/100 g wet wt; $p<0.05$), pancreas (66 ± 8 vs 105 ± 10 ml/min/100 g wet wt; $p<0.05$) and mesentery (43 ± 8 vs 80 ± 10 ml/min/100 g wet wt; $p<0.05$). Blood flow to all other splanchnic organs (including the small intestine and colon) was unaffected by acute administration of either L-NAME or AG. Interestingly, chronic (21 day) administration of L-NAME reduced hepatic arterial blood flow by approximately 70% (15 ± 3 vs 54 ± 5 ml/min/100 g wet wt; $p<0.05$) whereas chronic administration of AG did not affect blood flow to any splanchnic organ. We conclude that the protective effects of L-NAME or AG in models of intestinal or colonic injury and inflammation are not due to reductions in blood flow to these organs. However, L-NAME does reduce hepatic arterial BF and thus this effect may be important in models of hepatic pathophysiology.

Acknowledgements: This work was supported by a grant from the National Institute of Health (DK47663).

References

1. Yamada, T., Sartor, R.B., Marshall, S., Specian, R.D., and Grisham, M.B. (1993) Gastroenterology **104**:759-777.
2. Miller, M.J.S., Sadowska-Krowicka, H. Chorinaruemol, S., Kakkis, J.L., and Clark, D.A. (1992) J. Pharm. Exp. Ther. **264**:11-16.
3. Grisham, M.B., Specian, R.D. and Zimmerman, T.E. (1994) J. Pharm. Exp. Ther. **271**:1114-1121.
4. Aiko, S. and Grisham, M.B. (1995) Gastroenterology **109**:142-150, 1995.

L(+)-amino-4-phosphonobutyric acid prevents nitric oxide-induced functional inhibition in the rabbit retina

KENNETH I. MAYNARD, PABLO M. ARANGO and CHRISTOPHER S. OGILVY

Neurosurgical Service, Massachusetts General Hospital and Harvard Medical School, Boston, MA 02114.

We previously reported that exogenous nitric oxide (NO) inhibits the light-evoked electroretinogram and compound action potentials (CAPs) in the rabbit retina [1]. Since L(+)-amino-4-phosphonobutyric acid (L-AP4), a metabotrophic glutamate receptor (mGluR) agonist, protects neurons against NO exposure [2], and all subclasses of the L-AP4-sensitive mGluR (i.e. mGluR4, mGluR6, and mGluR7) are present in the mammalian retina [3], we examined whether their activation could block the NO-induced inhibition of the light-evoked responses.

Retinas with optic nerve attached were isolated from anesthetized, dark-adapted New Zealand White rabbits and maintained in Ames' medium (Sigma) at 36-37 °C. Preparations were stimulated by dim light flashes (1 s), which evoked an electroretinogram (i.e. the PIII which is the vitreous negative response which reflects the light-evoked interruption of the dark current in the photoreceptors) recorded transretinally, and ON and OFF CAPs, recorded from the optic nerve [1]. All agents (Sigma) were added to the Ames' medium and left for 20 - 60 min to record the maximum effect (an average of 2 or 3 recordings) with washing between the application of drugs. Sodium nitroprusside (SNP) was used as an NO donor, and potassium ferricyanide (PFC) as a control agent, which has a similar chemical structure to SNP, but which does not release NO.

Table I. L-AP4 pretreatment prevents SNP-induced attenuation of light-evoked responses, whereas PFC and L-AP4, alone or in combination are not inhibitory.

All data mean ± S.E.M. are % of control light-evoked responses, n = number of retinas and * = P<0.05.

	n	PIII	CAPs ON	CAPs OFF
SNP (1 mM)	3	16±8*	26±11*	35±9*
L-AP4 (100 μM)	3	123±28	0*	124±26
L-AP4 (100 μM) + SNP (1 mM)	3	64 ±10	0*	67±11
PFC (1 mM)	3	155±23	77±24	72±26
L-AP4 (100 μM) + PFC (1 mM)	3	135±11*	0*	80±10

The PIII, and the optic nerve responses (CAPs) to the "ON" and "OFF" of a dim light stimulus presented every 5 - 15 min were measured under control conditions, and following the addition of the test agents at the concentrations shown.

Abbreviations used: L(+)-amino-4-phosphonobutyric acid (L-AP4), nitric oxide (NO), PIII of the electroretinogram (PIII), compound action potentials (CAPs), sodium nitroprusside (SNP), potassium ferricyanide (PFC), metabotrophic glutamate receptor (mGluR)

As we previously reported [1], SNP inhibited the PIII and the CAPs. Although L-AP4 blocked the ON CAPs (as expected, since mGluR6 receptors are located on the ON bipolar cells in the retina [3]), the ERG and the OFF CAPs were not reduced. L-AP4 pretreatment (20 min) prevented the SNP-induced inhibition of the light-evoked PIII and the OFF CAPs. Although there was a tendency towards an increase in the PIII following exogenous PFC, it did not significantly affect either the PIII or the CAPs. Following pretreatment with L-AP4, PFC did cause a significant increase on the PIII, but still had no effect on the OFF CAPs.

We conclude that the activation of L-AP4-sensitive mGluR in the rabbit retina blocks the NO-induced inhibition of the light-evoked PIII (representative of phototransduction) and the OFF CAPs (i.e. neurotransmission), since SNP, but neither L-AP4 nor PFC alone or combined have any significant inhibitory effect. Whether this functional neuroprotection is mediated by a specific subclass of L-AP4-sensitive mGluR (i.e. mGluR4, mGluR6, mGluR7) found on a particular cell type(s) (e.g., ON-bipolar, ganglion, amacrine or horizontal cells) in the retina, and what the mechanism of action of this effect is, remain to be investigated. Since both cerebral ischemia and neuroexcitotoxicity are influenced by mGluR and NO, further clarification of their roles in this model may be useful for developing strategies against stroke and neurodegenerative diseases.

Acknowledgements: The authors thank Dr. Adelbert Ames III for helpful discussion and Yasaman Vafai for technical assistance.

References

1. Maynard, K.I., Yanez, P., and Ogilvy, C.S. (1995) Nitric oxide modulates light-evoked compound action potentials in the intact rabbit retina. NeuroReport **6**, 850-852.

2. Maiese, K., Greenberg, R., Boccone, L. and Swiriduk, M. (1995) Activation of the metabotrophic receptor is neuroprotective during nitric oxide toxicity in primary hippocampal neurons of rats. Neurosci. Letts. **194**, 173-176.

3. Akazawa, C., Ohishi, H., Nakajima, Y. *et al.*. (1994) Expression of mRNAs of L-AP4-sensitive metabotrophic glutamate receptors (mGluR4, mGluR6, mGluR7) in the rat retina. Neurosci. Letts. **171**, 52-54.

Vasodilatation by nitric oxide derived from nerve and endothelium in canine retinal arteries and arterioles

NOBORU TODA and TOMIO OKAMURA

Department of Pharmacology, Shiga University of Medical Sciences, Seta, Ohtsu 520-21, Japan

Nitric oxide (NO) in vasculature is derived not only from the endothelium but also from perivascular nerve. We first demonstrated the NO-mediated neural vasodilatation in canine cerebral arteries [1,2]. Although the role of NO in the control of vascular tone has been determined in a variety of arteries and veins, there is no information concerning NO-mediated responses of the retinal artery and arteriole. This presentation includes evidence showing that the tone of isolated retinal central artery just before entering into the eyeball and of intraocular retinal arteriole in vivo is regulated by NO liberated from the endothelium and perivascular nerve.

I) Endothelium-derived NO

In isolated canine retinal arteries, substance P and Ca^{++} ionophore A23187 produced an endothelium-dependent relaxation [3], which was abolished by treatment with NO synthase inhibitors, oxyhemoglobin and methylene blue. The inhibitory effect of N^G-nitro-L-arginine (L-NA), but not the other blockers, was reversed by L-arginine.

In the anesthetized dog, retinal vasculature in the ocular fundus was monitored by a camera and videotaped, and the arteriolar size was measured [3]. Intraarterial injections of substance P dilated the arterioles. Treatment with intravenous L-NA depressed the effect of the peptide but did not alter the response to nitroglycerin. Substance P-induced hypotension was seen after the retinal vasodilatation was completed. L-arginine reversed the vasodilator effect of substance P in L-NA-treated dogs. These findings indicate that NO derived from the endothelium mediates substance P-induced relaxation of retinal arteries, and the peptide dilates the arterioles in vivo by a mediation of NO possibly from the endothelium.

II) Nerve-derived NO

In isolated retinal arteries denuded of the endothelium, transmural electrical stimulation and nicotine produced a transient contraction followed by a relaxation [4]. The contraction was abolished by prazosin. The relaxant response under -receptor blockade was depressed by L-NA and methylene blue but not by D-NA [4]. L-arginine restored the response suppressed by L-NA. The typical tracing with 10^{-4} M nicotine (N) is shown in Fig. 1 (10^{-6} M L-NA, 3 x 10^{-4} M L-arginine and 10^{-4} M papaverine, PA). The effect of nicotine was abolished by hexamethonium. Histochemical study demonstrated the presence of perivascular nerves containing NO synthase immunoreactivity or NADPH diaphorase in the retinal artery and arteriole.

In anesthetized dogs, intraarterial injections of nicotine produced an arteriolar dilatation which was abolished by treatment with L-NA [5] and hexamethonium. L-arginine reversed the effect of L-NA. Intravenous injections of L-NA constricted the retinal arterioles, suggesting that the release of NO from the nerve and endothelium contributes physiologically to the vasodilatation.

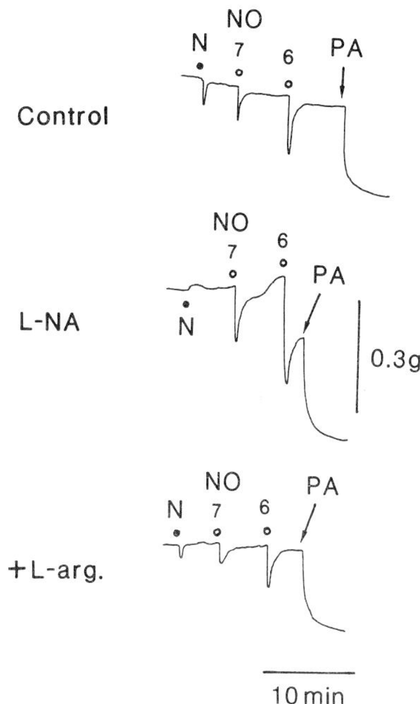

Fig. 1

This is the first demonstration of the effect of substances liberating NO from the endothelium and perivascular nerve in retinal arteries in vitro and arterioles in vivo. The experimental conditions used for the in vivo study are quite useful for determining the role of NO in resistance vessels with minimal surgical interventions, devoid of non-physiological maneuvers, including the perfusion with artificial solutions, the exposure to air, etc. It is concluded that NO derived from the endothelium and vasodilator nerve plays an important role in the regulation of retinal arterial and arteriolar tone in the dog.

References

1. Toda, N. and Okamura, T. (1990) Biochem. Biophys. Res. Commun. 170, 308-313
2. Toda, N. and Okamura, T. (1990) Am. J. Physiol. 259, H1511-H1517
3. Kitamura, Y., Okamura,T., Kani K. and Toda, N. (1993) Invest. Ophthal. Vis. Sci. 34, 2859-2865
4. Toda, N., Ayajiki, K., Kimura, H. and Okamura, T (1993) Circ. Res. 72, 206-213
5. Toda, N., Kitamura, Y. and Okamura, T. (1994) Am. J. Physiol. 266, H1985-H1992

Role of nitric oxide in the generation of a motor signal in physiological conditions: a study in the oculomotor system

CARMEN ESTRADA*, BERNARDO MORENO-LOPEZ*# and MIGUEL ESCUDERO#.

*Departamento de Fisiología, Facultad de Medicina, Universidad Autónoma de Madrid, Arzobispo Morcillo 1, 28029 Madrid, and #Laboratorio de Neurociencia, Facultad de Biología, Universidad de Sevilla, Av. Reina Mercedes 6, 41006, Sevilla, Spain.

Neurons containing nitric oxide synthase (NOS) are widely distributed in the central nervous system; however the participation of nitric oxide (NO) in specific cerebral processes remains largely unknown. The role of nitric oxide in motor behavior was investigated by using the cat horizontal oculomotor system as a model. This system offers several advantages since eye movements can be accurately measured in alert animals and are generated by motor and premotor nuclei well localized in the brain stem and accessible to local injections and recordings. Eye movements in the horizontal plane are conducted by only two antagonist muscles driven by neurons in the abducens (ABD) nucleus of the brain stem. This nucleus is controlled by burst neurons in the paramedian pontine reticular formation providing velocity signals, tonic medial vestibular nucleus (MVN) neurons whose activity is mainly related with head velocity, and prepositus hipoglossi (PH) neurons carrying eye position signals (1).

NADPH-diaphorase staining and NOS immunohistochemistry of cat brain stem sections (Figure 1) revealed the presence of a large number of positive neurons in the PH nucleus. Some scattered labeled neurons were also observed in ABD nucleus, reticular formation, and MVN.

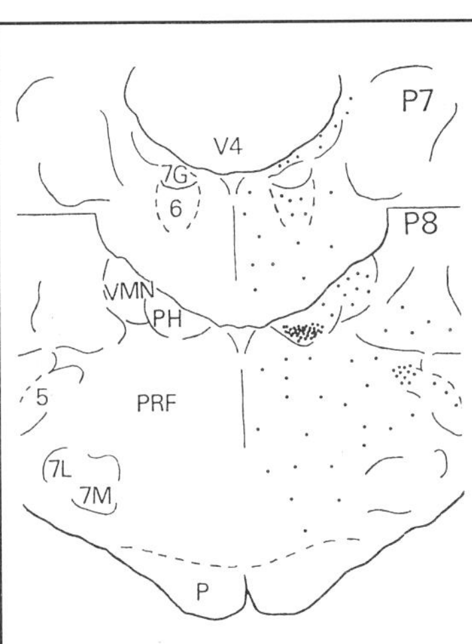

Figure 1.- Diagram showing the location of NOS containing neurons through two brain stem coronal sections at stereotaxic planes P7 and P8. 5, trigeminal nucleus; 6, abducens nucleus; 7G, facial genu; 7L and 7M, lateral and medial facial nucleus, P, pyramidal tract; PH, prepositus hipoglossi nucleus; PRF, pontine reticular formation; V4, fourth ventricle; VMN, medial vestibular nucleus.

In order to analyze the possible functional role of NO in the oculomotor system, 4 adult cats were prepared for chronic recording of eye movements with the magnetic search coil technique and for local injection and neuronal recording in the brain stem (2). Unilateral injections of the NOS inhibitors L-nitroarginine methyl ester (L-NAME; 25-90 nmol) and L-N-monomethylarginine (L-NMMA; 30-100 nmol) in the anterior third of the PH nucleus produced conjugated nystagmic eye movements with slow phases directed to the contralateral side (Figure 2A). The nistagmus was strongest in complete darkness. The velocity of the slow phases, which is indicative of the intensity of the nistagmus, was dose-dependent, appeared within 1-2 minutes after injection, and lasted for 30-60 minutes (Figure 2B). The effect of NOS inhibitors was stereospecific, because no alteration was observed upon D-NAME (90 nmol) injections, and was abolished by simultaneous administration of L-arginine.

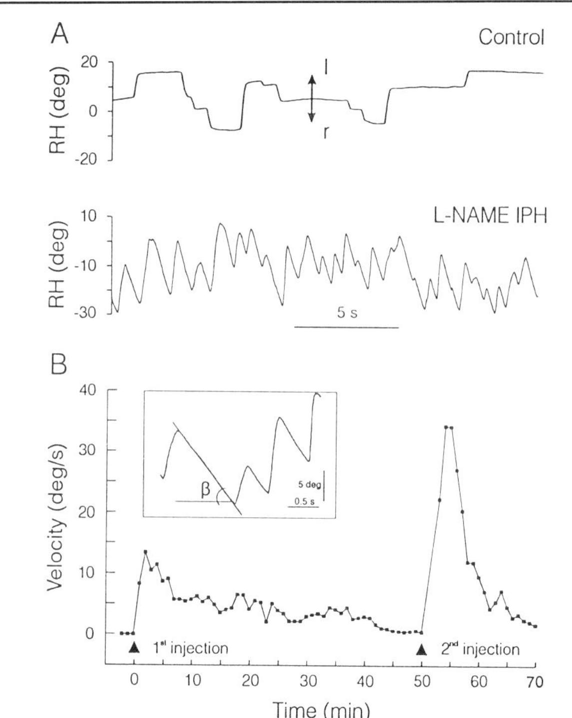

Figure 2.- A. Horizontal component of right eye position to the left (l) and to the right (r) in control conditions and after injection of L-NAME in the left prepositus hipoglossi (lPH) nucleus. B.- Velocity of the eye movement, as assessed by measurement of the slow phase slope (inset), after two consecutive injections with increasing doses of L-NAME in the PH nucleus.

Unilateral injections of the NO donor sodium nitroprusside (SNP) in the PH nucleus produced abnormal eye movements with slow phases in a direction opposite to that observed upon NOS inhibition. Changes induced by the NO donor were immediate, as may be expected from the diffusibility of the released NO. The permeant analog of cGMP, 8-Br-cGMP, produced an effect similar to that of SNP, thus suggesting that guanylate cyclase was the intracellular target for NO.

These results indicate that NO produced by PH neurons modulates the generation of the motor signals that control horizontal eye movements. Since it is a highly diffusible gas, an anterograde effect of NO should not be expected, but rather a change in the neuron microenvironment that may affect neighboring structures such as afferent nerve endings. The results are compatible with a modulator effect of NO on tonic signals reaching PH neurons from the MVN. This is the first time that a physiological effect of neuronal NO is demonstrated in an alert animal.

Acknowledgments

This work was supported by grant 94/0388 from Fondo de Investigaciones Sanitarias.

References

1.- Escudero, M., de la Cruz, R.R. and Delgado-García, J.M. (1992) A physiological study of vestibular and prepositus hipoglossi neurons projecting to the abducens nucleus in the alert cat. J. Physiol. **458**, 539-560

2.-Delgado-García, J.M., del Pozo, F. and Baker, R. (1986) Behavior of neurons in the abducens nucleus of the alert cat. I. Motoneurons. Neuroscience **17**, 929-952.

Heterogeneity of nitric oxide production sites in striated muscle

REINHART GOSSRAU and ZARKO GROZDANOVIC

Department of Anatomy, Free University of Berlin, D-14195 Berlin, Germany

Introduction

Recently, Balon and Nadler [1] demonstrated the release of NO from rat isolated skeletal muscle preparations at rest. NO efflux was enhanced by addition of either arginine or sodium nitroprusside, or in response to electrical stimulation. L-NMMA reduced both NO release and basal 2-deoxyglucose transport. These data raise the question about the source(s) of NO in the skeletal muscle. Previously, Nakane et al. [2] reported the expression of human brain NOS mRNA in human, but not rat skeletal muscle. Kobzik et al. [3] detected NOS immunoreactivity at the sarcolemma of type II skeletal muscle fibers in rats. Our group localized NOS immunostain and catalytic activity to the sarcolemmal region of type I and II somatic and visceral striated muscle fibers in several mammalian species [4]. Since, however, NO could be derived from extraparenchymal NOS-containing sources in the striated musculature, such as microvascular endothelial cells [5] or nerves, we have analysed the occurrence of NOS protein and activity in a wide range of somatic and visceral striated muscles of mammals.

Material and methods

The indirect immunofluorescent technique was performed with a polyclonal antibody raised against the pig brain NOS [6]. The histochemical NADPHd activity was visualized on fresh, formaldehyde-postfixed cryosections, either according to Scherer-Singler et al. [7] with or without 0.5 mM potassium permanganate to eliminate non-NOS NADPHds, or according to Nakos and Gossrau [8] in the presence of 1% formaldehyde in the incubation medium.

Results

NOS immunoreactivity and NADPHd staining always labeled identical structures. In the visceral striated muscles (tongue, pharynx, esophagus), NOS was present in the region of the fiber surface, in nerve cell bodies and processes as well as in arterial and arteriolar endothelial cells. NOS was predominantly localized in the extrasynaptic portion of the myofiber membrane. In longitudinal sections of the tongue, some apparantly tangentially cut muscle fibers were seen to be invested, at least in part, by undulating, transversely running (tigroid), intensely NADPHd-positive membranes. Rarely, there was a co-existence between AChE and NADPHd in certain MPEs (especially in the tongue). NOS-containing nerve fascicles were observed in the vicinity of blood vessels and muscle fibers. In the somatic striated muscles, NOS was found in the region of sarcolemma of most of the extrafusal muscle fibers, including the myotendinous junction. In addition, NOS was detected at the membrane surface of intrafusal fibers in muscle spindles. Nerves, either motor or autonomic, did not react for NOS activity. NOS immunoreactivity and NADPHd staining were readily observed in arterial and venous endothelial cells, but not in mast cells.

Abbreviations used: AChE, acetylcholinesterase; NADPHd, NADPH diaphorase; L-NMMA, N^G-monomethyl-L-arginine; MEP, motor end-plate; NO, nitric oxide; NOS, nitric oxide synthase; sGC, soluble guanylate cyclase.

Discussion

According to our findings, there are multiple sources of NO in the striated musculature: 1) muscle fibers (extra- and intrafusal), 2) endothelial cells and, in the case of visceral muscles, 3) intrinsic neuronal elements. Light microscopic evidence suggests that NOS protein and activity are located at the myofiber surface. Indeed, immunolocalization at the ultrastructural level has revealed labeling directly in the plasma membrane [Langer et al., this issue].

The functions of NO in skeletal muscle are probably determined by the spatial distribution of sGC, the presumably most important "receptor" molecule for NO, in target cells. Kobzik and co-workers [3] have demonstrated cyclic GMP immunostaining at the muscle fiber membrane. However, our immunohistochemical studies with antisera raised against ß1- and ß2-subunits of sGC (kindly provided by Dr. B. Koesling, Berlin) have yielded only weak or no labeling [9]. This is in accordance with biochemical analyses (G. Schulz, personal communication). Yet, physiological data suggest an involvement of NO in the regulation of glucose transport [1] and force development [3]. Based on the results presented herein, NO produced either within muscle fibers or blood vessel endothelial cells may be responsible for these effects. Thus, NO released from endothelial cells in exercising muscle [10] could promote glucose uptake, providing a long suspected link between blood flow and metabolism in skeletal muscle. On the other hand, NO generated by the sarcolemmal NOS, which is regulated by Ca^{2+}, could diffuse to nearby microvessels and stimulate smooth muscle relaxation, leading to augmented blood flow (i.e., oxygen supply) to the contracting muscle fibers. Further experiments are needed to define the targets and functions of NO in skeletal muscle.

Summing up, our data demonstrate the expression of a plasma membrane-anchored neuronal-type NOS in extrafusal, both type I and II, and intrafusal fibers of somatic and visceral striated muscles, which certainly constitute the richest source of NO in mammals. In addition, we have identified further NO production sites, such as blood vessel (i.e., arterial, venous, and arteriolar) endothelial cells and intrinsic ganglion cells (tongue, pharynx, esophagus).

Acknowledgements

We are indebted to Dr. B. Mayer for donating the NOS antiserum and thank Ms H. Richter for technical assistence.

References

1. Balon, T.W. and Nadler, J.L. (1994) J. Appl. Physiol. **77**, 2519-2521
2. Nakane, M., Schmidt, H.H.H.W., Pollock, J.S., Förstermann, U. and Murad, F. (1993) FEBS Lett. **316**, 175-180
3. Kobzik, L., Reid, M.B., Bredt, D.S. and Stamler, J.S. (1994) Nature **372**, 546-548
4. Grozdanovic, Z., Nakos, G., Dahrmann, G., Mayer, B. and Gossrau, R. (1995) Cell Tissue Res. **281**, 493-499
5. Segal, S.S. (1994) J. Appl. Physiol. **77**, 2517-2518
6. Mayer, B., John, M. and Böhme, E. (1990) FEBS Lett. **277**, 215-219
7. Scherer-Singler, U., Vincent, S.R., Kimura, H. and McGeer, E.G. (1983) J. Neurosci. Meth. **9**, 229-234
8. Nakos, G. and Gossrau, R. (1994) Acta histochem. **96**, 335-343
9. Grozdanovic, Z. and Gossrau, R. (1996) Ann. Anat. In press
10. Sun, D., Huang, A., Koller, A. and Kaley, G. (1994) J. Appl. Physiol. **76**, 2241-2247

Selective visualization of the NADPH diaphorase activity of nitric oxide synthase in mammalian striated muscle fibers

REINHART GOSSRAU+, GEORGIOS NAKOS+, TATJANA CHRISTOVA* and ZARKO GROZDANOVIC+

+Department of Anatomy, Free University of Berlin, D-14195 Berlin, Germany and *Department of Anatomy and Histology, Medical University, Sofia 1431, Bulgaria

Introduction

Recent evidence indicates that the skeletal musculature constitutes the richest source of nitric oxide synthase (NOS) in mammals [1,2,3]. Immunolabeling discerned neuronal NOS in the myofiber sarcolemma [3]. NADPH diaphorase (NADPHd) histochemistry with NADPH as the substrate and nitro BT as the electron acceptor, however, revealed formazan formation in both, the sarcolemmal region and sarcoplasm and/or mitochondria. Since NADPH is primarily not a specific substrate for the NADPHd activity of NOS but can be oxidized by several other NADPH dehydrogenating enzymes [4], we set out to improve the specificity of the NADPHd reaction for NOS detection in skeletal muscles.

Material and methods

Samples of the tongue and diaphragm were removed from Wistar rats, sacrificed under nembutal anesthesia, frozen in liquid nitrogen, and cut into 10 µm cryosections. Sections were either fixed in formaldehyde or glutaraldehyde, or were used unfixed. NADPHd was demonstrated according to Scherer-Singler et al. [5]. Fresh cryosections were incubated in the presence of 0.5-1 % formaldehyde. Fixed sections were reacted in the absence (controls) or presence of chemicals listed in Table 1 or other compounds (see Results).

Results

Table 1 summarizes the data obtained with compounds which can be used to differentiate between NOS-associated NADPH diaphorase (NOSad) and NADPHds not associated with NOS (nNOSad).

Table 1. Response of NOSad and nNOSad to different chemicals in the tongue or diaphragm of rats

	NOSad	nNOSad
Aldehydes		
Formaldehyde	+	0/(+)
Glutaraldehyde	+	0/(+)
Acrolein	+	0/(+)
Alcohols		
Ethanol	+	0/(+)
Methanol	+	0/(+)
Oxidizers		
Permanganate	+	0/(+)
H_2O_2	+	0/(+)
Diamide	+	0/(+)
Thiol inhibitors		
pCMB	+	0/(+)
Ethyl maleimide	+	0/(+)
NADPH analogs		
α-NADPH	+	0/(+)
CPR inhibitors		
Cytochrome C	+	0/(+)
Oxidizing enzymes		
Peroxidase	0/(+)	+

\+ still present 0/(+) absent or significantly reduced, depending on the concentration used

All other reagents tested, i.e., the L-Arg analog L-nitroarginine, the NADPH cytochrome P450 reductase (CPR) inhibitors β-NADP and miconazole, the iron chelator cyanide, anoxia, the Ca^{2+} chelator EDTA, the Ca^{2+} agonist EGTA, the calmodulin agonist trifluoperazine, the flavoprotein inhibitor diphenyl iodonium chloride, the DT diaphorase inhibitor dicoumarol and the oxidizing enzyme catalase produced identical data for NOSad and nNOSad, i.e., they could not be used to differentiate between NOSad and nNOSad.

Discussion

Our results show that reagents which can influence the oxygenase (N-terminal) domain (i.e., L-Arg analogs, iron chelators, anoxia) or the "hinge" region (i.e., Ca^{2+} chelators, Ca^{2+} agonists) of the NOS molecule were without effect, suggesting that these parts of the molecule do not participate in NADPH oxidation and formazan production. Thus, the reductase (C-terminal part), i.e., the diaphorase segment, is likely to be responsible for formazan formation.

These findings are paralleled by the observation that there is a one-to-one colocalization of NOS immunostaining and NADPHd activity in the sarcolemma of muscle fibers [3]. In support of this view, we have recently noted the loss of NOS protein and catalytic activity in skeletal muscle sarcolemma of patients with Duchenne muscular dystrophy [6]

Compared with nNOSad, which are primarily localized in the sarcoplasm and/or mitochondria, NOSad is obviously an aldehyde- and oxidation-resistant enzyme. This is possibly due to a low concentration of thiol-containing cystein residues in the NOSad molecule. The residual activity in the sarcoplasm in the presence of some of the chemicals listed in Table 1 may be represented by the NADPHd activity of inducible NOS (Gath et al., this issue) or the endothelial type enzyme [7].

In summary, especially in the presence of fixation aldehydes or oxidizers in the incubation medium, there is a close correlation between NADPHd activity and anti-NOS labeling in striated muscle fibers.

Acknowledgement

The authors are thankful to Ms H. Richter for technical assistence and Ms U. Saykam for the preparation of the manuscript.

References

1. Nakos, G. and Gossrau, R. (1994) Acta histochem. **96**, 335-343
2. Kobzik, L., Reid, M.B., Bredt, D.S. and Stamler, J.S. (1994) Nature **372**, 546-548
3. Grozdanovic, Z., Nakos, G., Dahrmann, G., Mayer, B. and Gossrau, R. (1995) Cell Tissue Res. **281**, 493-499
4. Stoward, P.J., Meijer, A.E.F.H., Seidler, E. and Wohlrab, F. (1991) In Enzyme Histochemistry (Stoward, P.J. and Pearse, A.G.E., eds.), pp. 27-71, Churchill Livingstone, Edinburgh
5. Scherer-Singler, U., Vincent, S.R., Kimura, H. and McGeer, E.G. (1983) J. Neurosci. Meth. **9**, 229-234
6. Grozdanovic, Z., Gosztony, G. and Gossrau, R. Manuscript submitted
7. Kobzik, L., Stringer, B., Balligand, J.-L., Reid, M.B. and Stamler, J.S. (1995) Biochem. Biophys. Res. Commun. **211**, 375-381

NADPH sources for nitric oxide synthase in mammalian striated muscle fibers

GEORGIOS NAKOS, REINHART GOSSRAU and ZARKO GROZDANOVIC

Department of Anatomy, Free University of Berlin, D-14195 Berlin, Germany

Introduction

In visceral and somatic striated muscles nitric oxide synthase (NOS) is associated with the sarcolemma [1,2,3]. As at other expression sites the enzyme needs NADPH as the co-substrate (so-called NADPH diaphorase [NADPHd] activity of NOS) but the sources of NADPH (which may serve as a regulator molecule of NOS activity) are not yet known. Therefore, in this communication we report the localization and estimated activity of certain NADPH-generating dehydrogenases in several rat striated muscles.

Material and methods

Male adult Wistar rats, kept under standardized conditions with tap water and diet ad libitum, were sacrificed in deep ether anesthesia, and visceral (tongue, pharynx, larynx, esophagus) and somatic (diaphragm, gastrocnemius, soleus) muscles were removed quickly, mounted on cork plates, wrapped with plastic foil, frozen in liquid nitrogen-cooled propane and cut in 10 μm cryostat sections. NADPH-dependent isocitrate dehydrogenase (ICDH; 1.1.1.42), glucose-6-phosphate dehydrogenase (G6PDH; 1.1.1.49) and malate dehydrogenase ("malic enzyme", MDHd, 1.1.1.40) were visualized according to Stoward and Pearse [4] using 10% polyvinyl alcohol (PVA, M_r 10.000, cold-water soluble) as diffusion protectant and nitro BT (NBT) as the final electron acceptor.

In parallel, the NADPHd activity of NOS was demonstrated according to Scherer-Singler et al. [5] but using formaldehyde-fixed cryostat sections [1], since in perfusion-fixed skeletal muscle NOS cannot be reliably visualized, and permanganate in the incubation medium to inhibit NADPHds not associated to NOS [3].

The controls, which were performed without substrate, always yielded negative results.

Results

ICDH (Fig. 1) produced the highest amount of the final reaction product NBT-formazan (NBTF) in visceral (tongue, esophagus, pharynx, larynx) and somatic (diaphragm, gastrocnemius, soleus) muscles. NBTF was seen in the sarcoplasm and sarcolemma region of type I and in lower amounts in type II fibers. Comparatively less NBTF was produced at the same sites in type I and II fibers of somatic muscles by G6PDH. MDHd generated the lowest amounts of NBTF, which was also found in somatic muscle fibers. Compared with the intensity of the NADPHd reaction in the sarcolemma region of visceral (Fig. 2) and somatic fibers, the NADPH quantities generated by one of the three NADP-dependent dehydrogenases were relatively low.

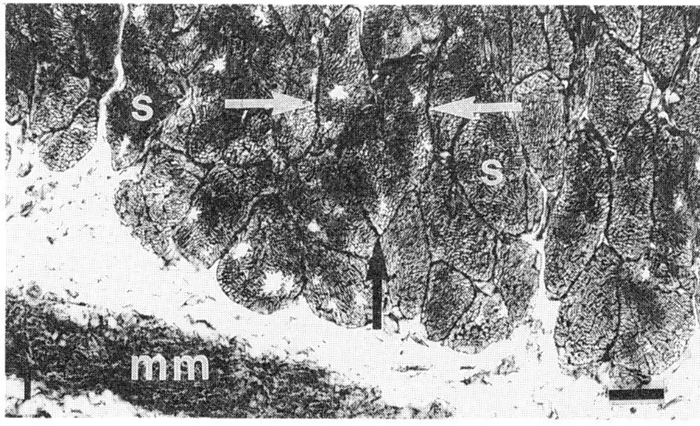

Fig.1. NBT-formazan produced by ICDH is present in the sarcolemma region (arrows) and sarcoplasm (s) of type I fibers in the esophagus. mm = muscularis mucosae. Bar = 20 μm

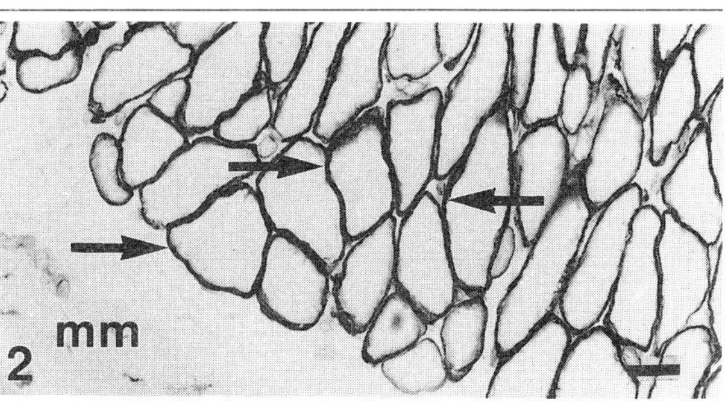

Fig. 2. The NADPHd activity of NOS produces high amounts of NBT formazan in the region of the sarcolemma (arrows) of type I fibers. mm = muscularis mucosae. Bar = 20 μm.

Discussion

Our study shows that in somatic striated muscle fibers ICDH, G6PDH and MDHd can serve as NADPH sources for NOS [6], while visceral muscle fibers can only use ICDH as an NADPH generator. Since, on the other hand, the NADPHd reaction for NOS is more active than the dehydrogenases are, i.e., consumes more NADPH than these enzymes appear to generate, there may be further NADPH sources especially in the visceral muscle fibers of the tongue, esophagus, pharynx and larynx, where the NADPH activity of NOS is especially high. Possibly, the generation of NADPH from NAD by the NAD kinase [7] followed by NADP reduction represents a further pathway, or NADPH from extracellular sources may be used by the enzyme.

Acknowledgement

We are thankful to Ms H. Richter for technical assistence, Ms U. Sauerbier for photographic help and Ms U. Saykam for the preparation of the manuscript.

References

1. Nakos, G. and Gossrau, R. (1994) Acta histochem. **96**, 335-343
2. Kobzik, L., Reid, M.B., Bredt, D.S. and Stamler, J.S. (1994 Nature **372**, 546-548
3. Grozdanovic, Z., Nakos, G., Dahrmann, G., Mayer, B. and Gossrau, R. (1995) Cell Tissue Res. **281**, 493-499
4. Stoward, P.J., Meijer, A.E.F.H., Seidler, E. and Wohlrab, F. (1991) In Enzyme Histochemistry (Stoward, P.J. and Pearse, A.G.E., eds.), pp. 27-71, Churchill Livingstone, Edinburgh
5. Scherer-Singler, U., Vincent, S.R., Kimura, H. and McGeer, E.G. (1983) J. Neurosci. Meth. **9**, 229-234
6. Meijer, A.E.F.H. (1991) Prog. Histochem. Cytochem. **22**, 1-118
7. Barman, Th.E. (1969) Enzyme Handbook, Vol. I, Springer, Berlin, Heidelberg, New York

Immunoelectron microscopic localization of nitric oxide synthase in striated muscle fibers

JENS-UWE LANGER+, REINHART GOSSRAU+, BERND MAYER* and ZARKO GROZDANOVIC+

+Department of Anatomy, Free University of Berlin, D-14195 Berlin, Germany and *Department of Pharmacology and Toxicology, Karl Franzens University, A-8010 Graz, Austria

Introduction

Recently, a new source of neuronal nitric oxide synthase (NOS) has been discovered [1,2,3] in mammalian skeletal muscle. Using light microscopic NOS immunohistochemistry with an antibody against constitutive neuronal NOS and a NOS-specific histochemical NADPH diaphorase (NADPHd) reaction, NOS has been identified in the sarcolemma region of type I and, especially, type II fibers [3] in visceral (tongue, pharynx, larynx, esophagus) and somatic muscles (all skeletal muscles) of several laboratory animals (rat, mouse, gerbil, hamster, guinea-pigs, monkeys) and man. Since, however, light microscopy does not allow information on actual expression sites, NOS electron microscopic immunohistochemistry was performed. Rat tongue was used as the test tissue because of its comparatively high NOS activity [3].

Material and methods

Small samples of tongue tissue of male Wistar rats were processed according to Loesch and Burnstock [4].

Incubation without the primary antibody served as control.

Results

The controls yielded negative data.

After pre-embedding the DAB reaction product osmium black (Fig. 1,2) and gold particles were predominantly found in the longitudinal sarcolemma with local differences. Occasionally, far less osmium black was associated with membranes of some subsarcolemmal structures and the basal lamina. The plasma membrane of transversal tubules, membranes of the sarcoplasmic reticulum and most motor end-plates were free of DAB-reaction product or gold labeling. With the postembedding procedure labeling was exclusively present in mitochondria using the gold method, while osmium black was found in the sarcolemma region.

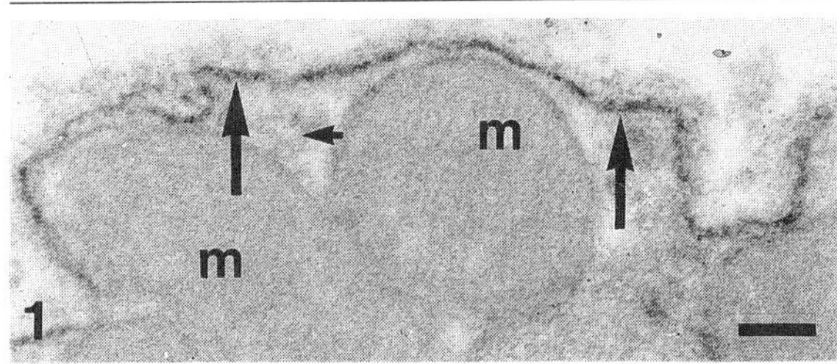

Fig. 1. After pre-embedding osmium black is found in the sarcolemma (arrows) of a striated muscle fiber of the rat tongue. m = mitochondria, short arrow = sarcoplasmic reticulum. Bar = 0.08 μm.

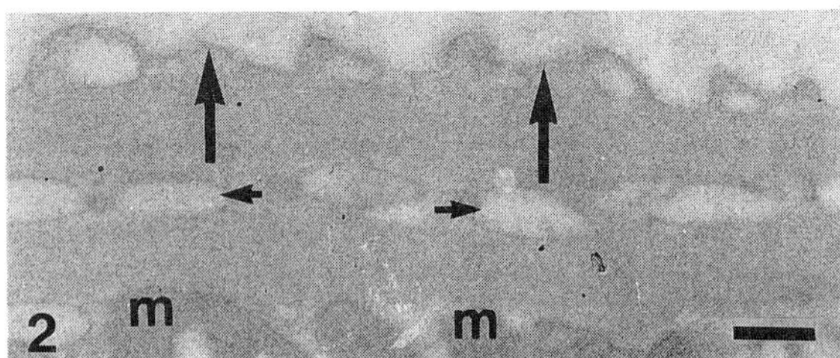

Fig. 2. In the absence of the primary antibody the sarcolemma (arrows) is free of osmium black. m = mitochondria, short arrows = sarcoplasmic reticulum. Bar = 0.2 μm.

Discussion

Our data suggest that in the rat tongue the longitudinal sarcolemma appears to be the main expression site of NOS and therefore source of NO, while the plasma membrane of the transversal tubules is not or less concerned. Since mitochondrial labeling is only obtained with one procedure, i.e., after post-labeling with gold, this is likely to represent an artifact the reasons of which are unclear at present.

Additionally, we cannot rule out the presence of some NOS immunoreactivity above the sarcolemma in the basal membrane and below the sarcolemma in relation to vesicles, or in the cytoplasm.

The sarcolemma is the main location of NOS. Whether the enzyme represents an integral or associated membrane protein has to be cleared. Already clear appear to be differences to the other membrane-type NOS, i.e., NOS in vascular endothelial cells, where the enzyme is said to be translocated from the cytoplasm to the plasma membrane via a myristylated anchor. However, different from NOS in the sarcolemma, using immunelectronmicroscopy NOS in vascular endothelial cells has never been shown in the plasma membrane but in the cytoplasm and various organelles [4,5,6].

The sarcolemmal localization of NOS raises the question about its regulation. It can be assumed that it is stimulated by Ca^{2+} ions, the source of which has to be determined.

The absence of NOS in the sarcolemma of the transversal tubules shows the molecular heterogeneity of the plasma membrane and points to different roles of the longitudinal and transversal sarcolemma in striated muscle fiber function.

Acknowledgement

We are thankful to Ms H. Richter for technical assistence, Ms U. Sauerbier for photographic help and Ms U. Saykam for the preparation of the manuscript.

References

1. Nakos, G. and Gossrau, R. (1994) Acta histochem. **96**, 335-343
2. Kobzik, L., Reid, M.B., Bredt, D.S. and Stamler, J.S. (1994) Nature **372**, 546-548
3. Grozdanovic, Z., Nakos, G., Dahrmann, G., Mayer, B. and Gossrau, R. (1995) Cell Tissue Res. **281**, 493-499
4. Loesch, A. and Burnstock, G. (1993) Endothelium **1**, 23-29
5. Loesch, A. and Burnstock, G. (1995) Cell Tissue Res. **279**, 475-483
6. O'Brien, A.J., Young, H.M., Povey, J.M. and Furness, J.B. (1995) Histochemistry **103**, 221-225

The role of nitric oxide in proteoglycan turnover by bovine articular cartilage organ cultures

MAJA STEFANOVIC-RACIC[*], TERESA I. MORALES[#], DILEK TASKIRAN[+], LORI A. MCINTYRE[*] and CHRISTOPHER H. EVANS[*]
[*]Ferguson Laboratory, 986 Scaife Hall, University of Pittsburgh School of Medicine, Pittsburgh, PA 15261 USA
[#]National Institute of Dental Research, N.I.H., Bethesda, MD 20092 USA
[+]Department of Biochemistry, Ege University School of Medicine, Bornova, Izmir, Turkey

INTRODUCTION - Articular chondrocytes produce large amounts of nitric oxide (NO) in response to interleukin-1 (IL-1) and a limited number of other cytokines [1-3]. However, little is known of the influence, if any, of NO on the turnover of the extracellular matrix of cartilage. Any such effect would be of major importance to the physiology of joints, as disturbances in the turnover of matrix components are associated with the erosion of cartilage that occurs in arthritis.

The main structural components of cartilage are collagen and proteoglycan. The former macromolecules appear to turn over very slowly in normal adult tissues, whereas cartilage proteoglycans are actively remodeled with half-lives ranging from days to weeks. Alterations in the rates at which matrix macromolecules are synthesized and degraded have been implicated in the loss of cartilage in disease. IL-1 is particularly damaging as it both enhances the catabolism and inhibits the synthesis of matrix components.

NO inhibits the synthesis of proteoglycans by articular chondrocytes obtained from rabbits [4], humans [5] and rats [6]. However, there is no information on the possible effects of NO on the catabolism of matrix macromolecules. In the present study, we have measured the effects of NO on proteoglycan turnover using bovine articular cartilage, a well characterized system for this type of study.

METHODS - Proteoglycan turnover was measured by previously described methods [7]. Briefly, hooves were obtained from local slaughterhouses, and slices of articular cartilage were shaved aseptically from the metacarpophalangeal joints. Cartilage fragments were cultured in 24-well plates (150-200mg wet weight in 1.5ml medium). Newly synthesized proteoglycans were labeled with $Na_2{}^{35}SO_4$ (30μCi/well) for 24h. Labeling medium was then replaced with fresh medium in the presence or absence of 20U/ml human recombinant IL-1β (hrIL-1β) 1mM L-N^G-monomethyl arginine (L-NMA) or L-arginine, as appropriate. Cultures were maintained for an additional 14 days, with a daily change of medium and re-addition of the appropriate chemicals. At the end of the experiment, the glycosaminoglycan (GAG) chains of the residual proteoglycans were extracted with 0.5M NaOH at 4° for 48h. Proteoglycan breakdown was monitored both as the release from the cartilage fragments of ^{35}S and of GAG measured by the dimethylmethylene blue assay [8]. Proteoglycan synthesis was measured as the uptake of $^{35}SO_4{}^{2-}$ into macromolecular material.

Production of NO was measured as the concentration of NO_2^- in the medium, using the Griess method. Neutral proteinase activity was measured by the degradation of ^3H-casein, using 1mM aminophenylmercuric acetate to activate latent enzyme.

RESULTS - Bovine cartilage produced little NO spontaneously, but hrIL-1β (20U/ml) induced a marked synthesis of NO which peaked during days 1-2 and then declined to a lower steady rate. High levels of NO production could be regained by increasing the concentration of IL-1 to 200U/ml. NO production was inhibited by 1mM L-NMA and substantially restored by L-arginine.

Cartilage responded to hrIL-1β by increasing the rate of proteoglycan loss L-NMA (1mM) further increased the rate of proteoglycan loss, whether this was measured as the release of ^{35}S or by the dimethylmethylene blue method. This effect was partially reversed by L-arginine. Neutral proteinase activity increased in response to hrIL-1β, increased further in the additional presence of 1mM L-NMA and was lowered by the co-addition of L-arginine.

Synthesis of proteoglycans was strongly suppressed by IL-1 and addition of L-NMA was unable to affect this process.

DISCUSSION - Contrary to expectations, NO appeared to protect cartilage matrix proteoglycans from degradation in response to IL-1. This may be coupled to an inhibitory effect of NO on the matrix metalloproteinases that mediate proteoglycan breakdown. It was also surprising to find that L-NMA did not affect the suppression of proteoglycan synthesis that occurs in response to hrIL-1β. In this respect bovine articular cartilage differs from rabbit [4], human [5] and rat [6]. The reason for this species difference is not obvious.

Although NO is clearly an important endogenous modulator of cartilage proteoglycan turnover, present data do not indicate unequivocally whether it should be viewed as a chondrodestructive or chondroprotective molecule.

ACKNOWLEDGEMENTS - Supported, in part, by NIH grant number RO1 AR42025. We thank Liz Arner of DuPont-Merck for kindly providing the hrIL-1β used in these experiments.

REFERENCES

1. Stadler, J., Stefanovic-Racic, M., Billiar, R.D., et al (1991) J. Immunol. **147**, 3915-3920.
2. Palmer, R.J., Hickery, M.S., Charles, I.G., Moncada, S. and Bayliss, M.T. (1993) Biochem. Biophys. Res. Commun. **193**, 398-405.
3. Rediske, J.J., Koehne, C.F., Zhang, B. and Lotz, M. (1994) Osteoarthritic Cart. **2**, 199-206.
4. Taskiran, D., Stefanovic-Racic, M., Georgescu, H.I. and Evans, C.H. (1994) Biochem. Biophys. Res. Commun. **200**, 142-148.
5. Hauselmann, H.J., Oppliger, L., Michel, B.A., Stefanovic-Racic, M. And Evans, C.H. (1994) FEBS Lett. **352**, 361-364.
6. Jarvinen, T.A.H., Moilanen, T., Jarvinen, T.L.N. and Moilanen, E. (1995) Mediat. Inflam. **4**, 107-111.
7. Morales, T.I. and Hascall, V.C. (1988) J Biol. **263**, 3632-3638.
8. Farndale, D.W., Buttle, D.J. and Barrett, A.J. (1986) Biochim. Biophys. Acta **883**, 173-177.

ABBREVIATIONS used: GAG - Glycosaminoglycan; hrIL-1β - human, recombinant interleukin-1β; NO - nitric oxide; L-NMA - L-N^G-monomethyl-arginine

Inducible nitric oxide synthase and cyclooxygenase-2 in human failed total hip arthroplasties

MIKA HUKKANEN, JEREMY BATTEN*, IAN D McCARTHY*, SEAN PF HUGHES* and JULIA M POLAK

Departments of Histochemistry and *Orthopaedic Surgery, Royal Postgraduate Medical School, London W12 ONN, UK.

Introduction

Aseptic loosening of the total hip prosthesis is a major clinical problem which limits the long-term use of joint prostheses to alleviate the painful symptoms of osteoarthritis and rheumatoid arthritis. The pathophysiological mechanisms of prosthesis loosening are not well understood and so the only treatment is revision surgery. There is growing evidence to show that the adverse tissue response to prosthesis wear particles is an important contributor to both aggressive local osteolysis and linear bone resorption around the implants. Both forms of cortical bone loss are caused by activation of a cascade of mediators, such as cytokines and prostanoids (PGs), as a result of pseudosynovial membrane formation between the bone and implant [1].

Inflammatory and resident cell activation can result in production of inflammatory mediators, such as IL-1, IL-6, TNF-α and prostaglandin E_2 (PGE_2), known to stimulate bone resorption. It is now well established that nitric oxide (NO) is an important secondary messenger molecule for many of the actions of the cytokines in bone [2,3] and can induce the synthesis of the PGE_2 converting enzyme cyclooxygenase-2 (COX-2). As iNOS and COX-2 seem to be two of the key elements mediating many of the cytokine actions, we have investigated the possibility that the enzymes are synthesised in pseudosynovial membranes in patients with failed total hip arthroplasties.

Materials and Methods

Patients and tissues

Pseudosynovial membranes were obtained from patients undergoing revision surgery for failed total hip arthroplasties (n=10). Preoperative synovial samples were taken from osteoarthritic patients undergoing total hip replacement surgery (n=5). Tissues were either directly frozen for protein extraction or were fixed in 1% PFA for immunocytochemistry.

Measurement of NOS and COX enzyme activities in tissue homogenates

Tissues were homogenized in the presence of protease inhibitors, centrifuged and analysed using a citrulline assay in the presence of necessary co-factors, both with and without the NOS inhibitor L-NMA. After incubation, the samples were centrifuged and aliquots of supernatants were applied to TLC plates. After chromatographic separation, the locations of specific amino acids were determined by ninhydrin reaction; spots corresponding to arginine, citrulline and ornithine were scraped off for scintillation counting. NOS activity was defined as the amount of ^{3}H-L-citrulline (nM) produced per hour and the results were corrected for total protein concentrations. PGE_2, as an indicator of COX-2 enzyme activity, was measured using a commercial enzymeimmunoassay system (Boehringer; Biotrak) after solid phase extraction procedure with Amprep C-18 columns.

Immunocytochemical localisation of iNOS and COX-2 proteins

Cellular localizations of iNOS and COX-2 were studied by immunocytochemistry. Polyclonal antibodies used in the study included antibodies raised against a synthetic peptide corresponding to amino acid residues 53-77 of human hepatocyte iNOS, or against a synthetic peptide corresponding to amino acid residues 47-71 of murine macrophage iNOS sequence. COX-2 antibodies were raised against a peptide sequence of murine COX-2. Co-localization of iNOS and COX-2 immunoreactivities with CD 68 positive macrophages was performed in serial sections using a modified avidin-biotin-peroxidase technique.

Results

In both pseudosynovial and synovial tissue there were relatively low calcium independent iNOS enzyme activity in failed hip arthroplasties and pre-operative osteoarthritic samples (0.945 $\pm$ 0.072 vs 1.121 $\pm$ 0.174 pmol/mg/min). L-NAME reduced the activity by more than 95% ($p<0.01$) supporting the view that authentic NOS enzyme activity was measured in homogenates. PGE_2 concentration was 27.8 $\pm$ 2.93 pg/mg of pseudosynovial membrane and 27.4 $\pm$ 3.14 pg/mg of pre-operative osteoarthritic hip synovium.

Antibodies against human hepatocyte and murine macrophage iNOS sequences gave similar profiles for iNOS-immunoreactive cells in pseudosynovial tissue. Macrophage-like cells, containing prosthesis wear debri, were the most prominent cell type showing immunoreactivity for both antisera. Vascular smooth muscle and endothelial cells were occassionally found to be immunoreactive for iNOS. Antibodies raised against murine COX-2 sequence showed immunoreactivity in many macrophage-like cells. Strong immunoreactivity was found frequently in both vascular smooth muscle and endothelial cells. Antibodies for the macrophage cell surface molecule CD68 were used to confirm the nature of iNOS and COX-2 immunoreactive cells. Clusters of CD68 positive macrophages were found around the vasculature, and in serial sections these cells were characterised by their immunoreactivity for iNOS and COX-2 (Figure 1).

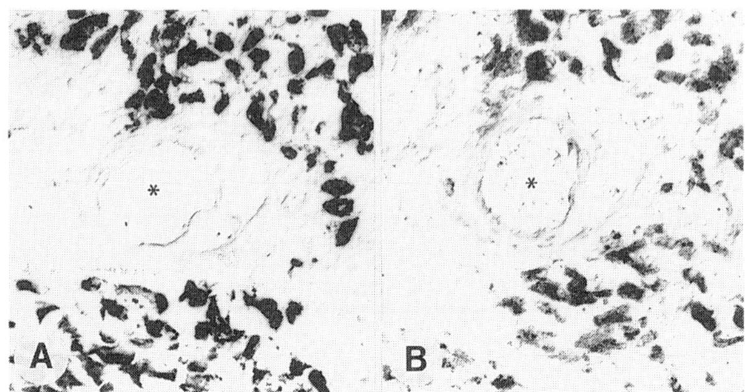

Figure. 1. Immunolocalisation of CD68 positive macrophages (A) and iNOS (B) in pseudosynovial membrane tissue from a patient with aseptic loosening of total hip prosthesis. iNOS and COX-2 were both found to localise in prosthesis wear particle laden CD68 positive macrophages scattered around a small artery (asterisks).

Discussion

In this study, we have investigated the possibility that iNOS and COX-2 are synthesised in the pseudosynovial membrane between the prosthesis and cancellous bone in patients with aseptic loosening of total hip prosthesis. The results of the present experiments suggest that in both pseudosynovial membranes and osteoarthritic synovium the iNOS activity is rather low and at the same levels in both groups. Similarly, PGE_2 concentrations were found to be almost identical in the two groups studied. Nonetheless, immunocytochemical detection of iNOS and COX-2 enzymes provided clear evidence their expression in vascular smooth muscle cells, endothelial cells and, perhaps most prominently, in CD68 positive macrophages. These data indicate that induction mechanisms for iNOS and COX-2 are present and functioning in both failed total hip arthroplasties and in pre-operative osteoarthritic synovial tissue. As both NO and PGs have been shown to induce bone loss *in vivo* in chronic inflammatory conditions, these molecules may account for the aseptic loosening and failure of human total hip and other arthroplasties in addition to aggravation of local inflammation and stimulation of prostanoid-induced pain production.

Acknowledgements: This work was in part supported by the Medical Research Council, UK. Antibodies were kindly provided by Dr S Moncada, Wellcome Research Laboratories, UK, Dr T Evans, RPMS, and Prof J Maclouf, INSERM, Hopital Lariboisiere, France.

References

1. Chiba J, Rubash HE, Kim KJ, Iwaki Y (1994) The characterization of cytokines in the interface tissue obtained from failed cementless total hip arthroplasty with and without femoral osteolysis. *Clin Orthop* 300:304-12.

2. Hukkanen M, Hughes F, Buttery LDK, et al. (1995) Cytokine-stimulated expression of inducible nitric oxide synthase by mouse, rat and human osteoblast-like cells and its functional role in osteoblast metabolism. *Endocrinology*, in press (12/1/95 issue).

3. Brandi ML, Hukkanen M, Umeda T, et al. (1995) Bidirectional regulation of osteoclast function by nitric oxide synthase isoforms. *Proc Natl Acad Sci USA* 92:2954-2958.

Investigations into nitric oxide modulation of methamphetamine-stimulated dopamine release.

RUSSELL A. GAZZARA*, R. ROBERT HOLSON*, PETER CLAUSING[+], BOBBY GOUGH*, W. SLIKKER JR.[+], and JOHN F. BOWYER[+].

Divisions of Reproductive & Developmental Toxicology* and Neurotoxicology[+], National Center for Toxicological Research, Jefferson, AR 72079, USA.

These studies were designed to investigate the possible role of NO in METH-induced DA release in the NS of the rat. The mechanisms for production and action of NO have been found in the NS [1-2]. Therefore we postulated that NO could potentially influence DA release through its interaction with N-methyl-D-aspartate receptors located on DA terminals in the NS that have been shown to regulate DA release [3]. The experiments presented here used both inhibitors of NOS, the enzyme that synthesizes NO in vivo, as well as NO_x generators.

Subjects were 4- to 6-month-old adult male Sprague-Dawley rats from the NCTR colony. The microdialysis method used in these studies to measure DA release has been described elsewhere [4]. Briefly, rats were prepared for microdialysis of DA in the NS of freely-moving conscious rats. The microdialysis probe was perfused with aCSF and samples collected at 20-min intervals. In those experiments in which NOS inhibitors (NOARG or L-NAME) or NO_x generators (SNP or ISON) were used, these compounds were added to the aCSF and perfused directly into the NS. Two hours after the start of perfusion, a series of four i.p. injections of 5 mg/kg METH was given at two hours intervals.

Figure 1 shows that METH greatly increased extracellular levels of DA in the NS. The addition of the NOS inhibitor NOARG clearly reduced the effectiveness of METH, suggesting that NO may be involved in METH-induced DA release. Addition of the NOS inhibitor L-NAME produced a similar effect (data not shown). The addition of L-arginine (substrate for NOS) did not significantly alter the response to METH.

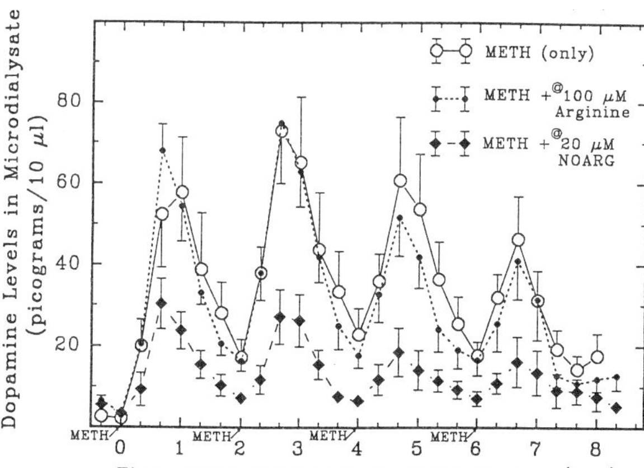

Figure 1. NOARG inhibition of METH-induced DA levels in NS microdialysate samples. The METH injections occurred just prior to sample collection (arrows on the x-axis). The DA levels for the NOARG group were significantly lower (p<0.05) than controls starting 20-40 min after the first METH dose, with the exception only of the 80-100 min sample following the third METH dose. N = 9, 8, and 12, in legend sequence.

Figure 2 shows the effect of blocking the inhibition of NOS by NOARG on METH-induced DA release in the NS. The addition of an excess of either the substrate for NOS (L-arginine) or a precursor of L-arginine (L-citrulline) to the microdialysate blocked the inhibitory effect of NOARG on METH-induced DA release. The levels of METH-induced DA release in the NS after the addition of L-arginine or L-citrulline were not significantly different from levels found after injection of METH alone. Thus, these data suggest that the NOARG inhibition of METH-induced DA release in the NS is mediated through its inhibition of NOS.

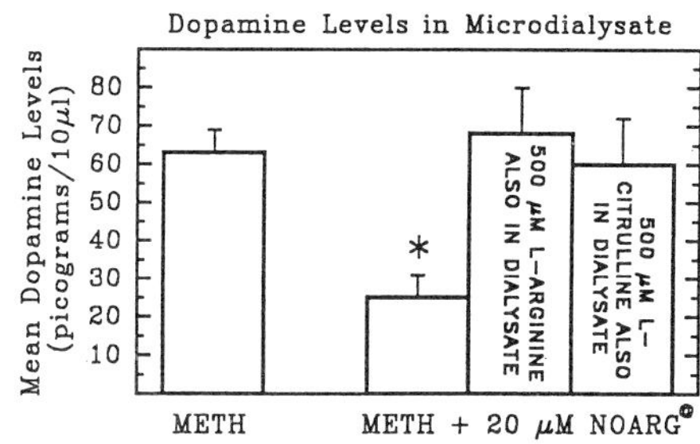

Figure 2. NOARG inhibition of METH-induced DA release is reversed by addition of either L-arginine or L-citrulline to the microdialysate. The average levels of DA in the samples over the entire four METH injections were calculated for each rat. These values when then used to create a group average (± SEM). *Release was significantly inhibited by NOARG (p<0.025). N = 9, 12, 4, and 4, in bar sequence.

Figure 3 shows the effect of the addition of ISON (an NO_x generator) on METH-induced DA release in the NS. The addition of either 2 μM or 20 μM ISON produced a significant decrease in METH-induced DA release in the NS. The addition of an NO generator (SNP) produced a similar effect (data not shown). These results suggest that an increase in NO (or NO_x) also results in an inhibition of METH-induced DA release in the NS.

There is a complex and poorly understood interrelationship between amino acid neurotransmitters, monoamine neurotransmitters and NO. The results of this study suggest that these systems interact to modulate METH-induced DA release in the NS. The portion of these experiments dealing with NOS inhibition strongly suggests that the physiological effect of NO is to stimulate METH-induced DA release. We can not currently explain why NO releasers also showed this effect, but we suspect that it may be due to a non-specific flooding of these mechanisms by NO.

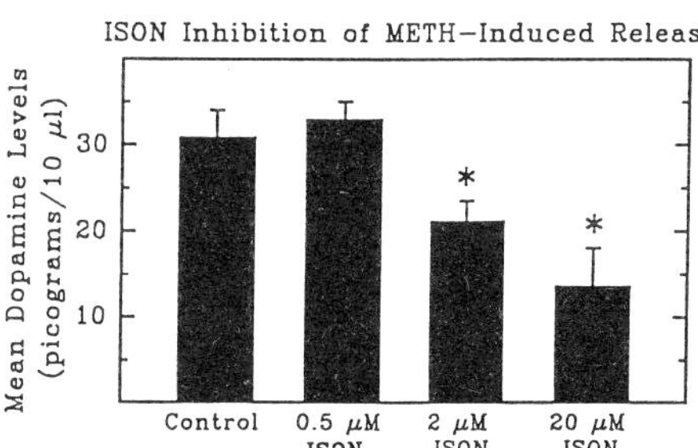

Figure. 3. The inhibition of METH-induced DA release by ISON. The DA levels were averaged over the 4 METH injections for each rat. These values were then used to generate a group average (± SEM). ISON was added to the microdialysate from 2 hrs before METH administration to the end of the experiment. *Release was significantly inhibited by ISON (p<0.05). N = 8, 2, 2, and 4, in bar sequence.

Abbreviations used: aCSF, artificial cerebrospinal fluid; DA, dopamine; ISON, isosorbide dinitrate; L-NAME, N^G-nitro-L-arginine methyl ester; METH, methamphetamine; NO, nitric oxide; NOARG, N^G-nitro-L-arginine; NOS, nitric oxide synthase; NS, neostriatum; SNP, sodium nitroprusside

1. Roberts, P.J. and Anderson, S.D. (1979) J. Neurochem. **32**, 1539-1545.
2. Vincent, S.R. (1994) Prog. Neurobiol. **42**, 129-160.
3. Krebs, M.O., Desce, J.M., Kemel, M.L., et al. (1991) J. Neurochem. **56**, 81-85.
4. Bowyer, J.F., Gough, B., Slikker, W. Jr., Lipe, G.W., Newport, G.D. and Holson, R.R. (1993) Pharmacol. Biochem. Behavior. **44**, 87-98.

Ischemia promotes formation of nitric oxide in different regions of rat brain *in vivo*

SØREN-P. OLESEN*, ARNE MØLLER*, PETER I. MORDVINTCEV#, RUDI BUSSE# and ALEXANDER MÜLSCH#
*NeuroSearch A/S, 26B Smedeland, DK-2600 Glostrup, Denmark;
#Zentrum der Physiologie, J.W.G. Universität, 60590 Frankfurt, Germany.

Objective

Ischemia elicits N^G-nitro-L-arginine methylester-inhibitable NO formation in the rat brain [1, 2]. By using *in vivo* NO spin trapping and *ex vivo* cryogenic EPR spectroscopy [3, 4] we assessed the influence of a neuronal NO synthase inhibitor (7-nitroindazole, 7-NI), a non-NMDA glutamate receptor antagonist (NBQX), and a blocker of L-type Ca^{2+} channels (NS 638) on NO formation and cell damage in rat and gerbil brains exposed to 2 and 7 min of global ischemia.

Materials and methods

Rat global brain ischemia. At day 1 the vertebral arteries were occluded by microcoagulation. Next day, the rats were anaesthetized with halothane, the carotid arteries located and occluded in 2 or 7 minutes, respectively.

Gerbil global brain ischemia. The gerbils were exposed to 4 min of global ischemia, and the damage was scored on day 4 as described [5].

Determination of NO formation in vivo by EPR spectroscopy. NO was spintrapped in brain tissues *in vivo* by dithiocarbamato iron-complex ($Fe(DETC)_2$), thus generating a stable paramagnetic mononitrosyl-iron complex ($NOFe(DETC)_2$) [3, 4]. Fifteen min before ischemia the animals obtained a s.c. injection of $FeSO_4$ (50 mg/kg) and sodium citrate (250 mg/kg). Sodium-DETC (250 mg/kg i.p.) was given 10 min before and another 250 mg/kg i.v. immediately before ischemia. Following ischemia (lasting 2 or 7 min) and 1 min recirculation the brains were rapidly removed, dissected and frozen in liquid nitrogen for cryogenic EPR spectroscopy. The amount of NO trapped was calculated by double integration of the triplet signal at $g\perp = 2.035$ of the $NOFe(DETC)_2$ complex after subtraction of the $Cu(DETC)_2$ signal from the original recording [3, 4].

Results

The typical EPR spectrum of $NOFe(DETC)_2$ in frozen tissues was characterised by an anisotropy signal with $g\perp$ 2.035 and $g\|$ 2.02 and a triplet hyperfine splitting at $g\perp$. This signal appeared in the ischemic brain tissues in addition to the signal of $Cu(DETC)_2$ which dominated the EPR spectrum of the control tissues. In the hippocampus 0.18 ± 0.04 (mean ± S.E.M.), 0.42 ± 0.14 and 1.11 ± 0.15 nmole/g NO were trapped in controls and in rats subjected to 2 and 7 min of ischemia, respectively (Fig. 1). The rate of NO formation, which is the amount of NO trapped divided by period of DETC exposure, was 0.01 nmole/g/min in control animals, and it increased by 13- and 12-fold at 2 and 7 min of ischemia ($p < 0.05$).

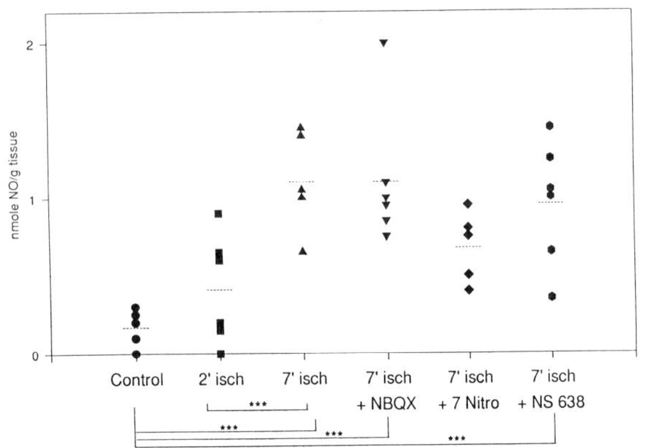

Fig. 1. Influence of NBQX, 7-NI, and NS 638 on formation of NO (nmole/g) in rat hippocampus after global ischemia. ***: p<0.05.

In the rat cerebellum NO formation increased significantly after 7 min of ischemia, but not after 2 min. NO also increased moderately in the striatum and the neocortex (data not shown). Pretreatment 15 min before ischemia with NBQX (30 mg/kg i.p.), 7-nitroindazole (40 mg/kg i.p.), or NS 638 (30 mg/kg i.p.) did not significantly reduce NO formation elicited in different brain regions by 7 min ischemia, albeit there was a tendency in this direction.

In the gerbil whole brain control NO formation was 0.25 ± 0.04 nmole/g, which was increased by ischemia (2 min: 0.87 ± 0.03 nmole/g; 7 min: 0.5 ± 0.06 nmole/g). The cytotoxic effect of ischemia and the possible protective effects of treatment with NBQX, 7-nitroindazole and NS 638 were evaluated in gerbils. The degree of cell death in hippocampus was scored in animals sacrificed 4 days after the ischemic insult. In non-ischemic animals all cells were intact (score 2), and in non-treated ischemic animals the cells were significantly damaged (score 6). NBQX protected the cells partly against degeneration ($p < 0.01$), whereas 7-nitroindazole and NS 638 were devoid of effect. Each group contained at least 10 animals.

Discussion

By spintrapping of NO *in vivo* and cryogenic EPR spectrometry *ex vivo* we detected a significant increase in NO formation (nmole/g) elicited by short ischemic episodes in different regions of the rat and gerbil brain. Since NO was continuously trapped during exposure of the tissue to the trapping agent (maximally 18 min), and the resulting $NOFe(DETC)_2$ complex was stable during this time [3], the amount of NO detected represents NO formation integrated over the trapping period. Therefore, despite different accumulation periods the rates (nmole/min) of NO formation could be calculated and taken for comparison. Thus, the NO formation in the rat hippocampus increased rapidly by more than 10-fold to a constant rate during 2 min and 7 min of ischemia compared to non-ischemic conditions, while cerebellar NO increased only after 7 min. Previously, Tominaga et al. [1, 2] reported qualitatively similar findings using the same NO trapping technique but a slightly different protocol to elicit cerebral ischemia in rats. However, in their studies the amount of NO detected depended on the ischemia-eliciting protocol and the amount of spintrap added. In the present study the neuronal NO synthase inhibitor 7-NI, applied in a dose previously shown to be effective against kainate-elicited NO formation in the rat brain [4], was ineffective. This suggests that NO formation during ischemia does not primarily depend on enzymatic conversion of L-arginine to NO via the neuronal NO synthase. Indeed, nitrite was identified as a major source of NO in isolated ischemic rat hearts [6]. NO formation from nitrite was correlated to acidification of the myocardial tissue. Since ischemic acidification develops within a minute in the brain [7], NO formed in ischemic brains might have originated from nitrite or some other NO_x pool activatable by tissue hypoxia and acidosis. The very weak effect of the Ca^{2+}-antagonist NS 638 and the non-NMDA-glutamate receptor antagonist NBQX on NO formation were thus intelligable, since the effector cascade interrupted by these agents linking brain ischemia to depolarization of neurones, Ca^{2+} influx and activation of NO synthase [8] would not account for the major part of NO formation in this case.

NBQX afforded protection against ischemia-elicited neuronal damage, which suggest, that either ischemic damage in our model occurred independently of NO formation, or that NBQX prevented damage at a step distal to NO production.

References

1. Tominaga, T. et al. (1993) Brain Res. **614**, 342-346.
2. Tominaga, T. et al. (1994) J. Cereb. Blood Flow Metab. **14**, 715-722.
3. Mülsch, A., Mordvintcev, P. and Vanin, A. (1992) Neuroprotocols **1**, 165-73.
4. Mülsch, A. et al. (1994) NeuroReport **5**, 2325-28.
5. Jensen, L.H. and Møller, A. (1992) Acta Neurol. Scand. **85**, 187-90.
6. Zweier, J.L. et al. (1995) Nature Medicine **1**, 804-809.
7. Mutsch, W.A.C. and Hansen, A.J. (1984) J. Cereb. Blood Flow Metab. **4**, 17-27.
8. Dawson, V.L. (1995) Clin. Exp. Pharmacol. Physiol. **22**, 305-308.

Increase in intrajugular nitric oxide during reperfusion after rat focal cerebral ischemia

Eiji Kumura, Satonori Tanaka, Toshiki Yoshimine[#], Shigeki Kubo[#], Toru Hayakawa[#], Takeshi Shiga and Hiroaki Kosaka.

Departments of Physiology and Neurosurgery[#],
Osaka University Medical School, 2-2 Yamadaoka, Suita,
Osaka 565, Japan.

Nitric oxide (NO) binds strongly to erythrocyte hemoglobin to yield nitrosyl hemoglobin (HbNO). Because HbNO is paramagnetic and relatively stable, which can be detected with electron spin resonance (ESR) spectroscopy [1]. In the present study, we determined jugular levels of HbNO by ESR during and after middle cerebral artery (MCA) occlusion in rats. The levels of plasma nitrite+ nitrate in the same blood were compared with the ESR results [2, 3] .

[Material and Methods]

Twenty four male Sprague-Dawley rats (300-320g) were anesthetized with intraperitoneal injection of ethyl carbamate (1g/kg). The proximal portion of the left MCA was occluded by a miniature clip and recirculated by removing the clip. Throughout the experiments, the rectal temperature of each animal was kept between 36.5 and 37°C by servo-controlled heating pads. Arterial blood pressure was monitored continuously. Arterial blood gases and hematocrit were analyzed at 5 min before the arterial occlusion and just before sacrifice.

The left jugular blood was collected at 2 h MCA occlusion (n=4), and 2 h and 30 min after sham operation (n=4). The same procedure was done at 30 min reperfusion after 2 h occlusion (n=4). In the other 4 animals, N^{G}-nitro-L-arginine methyl ester (L-NAME, 10mg/kg) was administered intraperitoneally twice at 30 min before MCA occlusion and 1 h after MCA occlusion. In the other 4 animals, L-NAME and L-arginine (100 mg/kg) were administered simultaneously in the same manner. In the remaining animals, human Cu, Zn-superoxide dismutase (SOD, 5 mg/kg) was injected intravenously at 10 min before reperfusion (n=4). The jugular blood was collected at 30 min reperfusion after 2 h occlusion.

The jugular blood, 0.4 ml, was immediately transferred to a ESR tube and frozen in liquid nitrogen. ESR spectra were recorded at 110 K and the concentration of HbNO was determined based on double integration of ESR spectrum [3]. As a standard, authentic HbNO prepared anaerobically with NO gas was used [1, 3]. Residual blood sample was centrifuged immediately, and the plasma fraction was diluted with nitrite/nitrate free-distilled water. Then, diluted plasma was centrifuged with micropore filter to remove protein. The filtrates were analyzed with automated procedure based on the Griess reaction [4]. Briefly, after reducing nitrate to nitrite through copper-plated cadmium column, absorbance at 540nm was measured after the reaction with the Griess reagent. The value was expressed as plasma nitrite+nitrate [2].

[Results]

In animals with sham operation and with 2 h occlusion, only small amount of HbNO was detected. The plasma nitrite+nitrate was significantly increased only in the latter condition ($p<0.01$). At 30 min reperfusion after 2 h occlusion, both levels of HbNO and nitrite+nitrate were further increased than those with 2 h occlusion alone ($p<0.01$). The increase in HbNO and plasma nitrite+nitrate at 30 min reperfusion were suppressed by the intraperitoneal administration of L-NAME ($p<0.01$). The effects of L-NAME were reversed by the simultaneous administration of L-arginine ($p<0.01$). When SOD was administered, the levels of HbNO were increased significantly ($p<0.05$), but not plasma nitrite+nitrate.

[Discussion]

The major finding of the present study is the increased concentrations of HbNO and plasma nitrite+nitrate in the jugular blood during reperfusion after focal cerebral ischemia. The increase was suppressed by the administration of L-NAME. But the suppression was counteracted by simultaneous administration of L-arginine. Those findings indicate that L-arginine-NO pathway is responsible for the increase in both jugular levels of HbNO and nitrite+nitrate during reperfusion.

In the circulating blood, NO is mainly decomposed to nitrate by a rapid reaction of NO with oxyhemoglobin in the erythrocytes. Reaction of NO and deoxyhemoglobin produces HbNO. Although NO also reacts with dissolved oxygen molecule in the plasma to yielding nitrite, nitrite is converted to nitrate autocatalytically by the reaction with oxyhemoglobin [5]. When SOD was administered before reperfusion, only the levels of HbNO were increased without affecting the levels of plasma nitrite+nitrate. This finding suggested that generation of superoxide upon reperfusion decreases intravascular NO. Because NO reacts with superoxide to produce peroxynitrite which may be converted to nitrite+nitrate finally.

Recent studies demonstrated increased NO production by using porphyrinic microelectrode [6] or by measuring brain concentration of nitrite [7] in the early ischemic brain. In contract to those findings, the present findings suggested that the increased NO generation in the reperfused brain which resulted in the ESR detectable HbNO formation in the cerebral circulating blood.

The role of NO in the cerebral ischemia-reperfusion is controversial. The administration of either NO donors or L-arginine improved cerebral circulation and reduced ischemic damage in rat focal ischemia models [8]. At least in the early stage of cerebral ischemia, the increase in NO seems beneficial. Because NO ameliorates microcirculation in the ischemic brain by inhibiting platelet aggregation and suppressing neutrophil adhesion to the endothelium. On the other hand, NO may also participate in neurotoxicity. One of which is glutamate mediated, and the other is peroxynitrite formation. As expected upon reperfusion, when the increase in NO associates with the increase in superoxide, a cytotoxic peroxynitrite may elaborate. Peroxynitrite is a strong oxidant which initiates lipid peroxidation, oxidizes with sulfhydryls and inactivates aconitase of the mitochondria [9]. Therefore, the *in situ* balance of superoxide and NO generations can be an important factor which determines the net biological effect of NO. The balance between superoxide and NO may alter dynamically during the time course of cerebral ischemia and reperfusion. Those issues, however, require further investigations.

[Conclusion]

The present study demonstrated that intrajugular NO increased during reperfusion after focal cerebral ischemia via NO synthase activation. The results of SOD administration implied that intravascular net NO may decrease by the reaction with superoxide to form peroxynitrite during reperfusion which may be finally converted to nitrate.

[References]

1. Shiga, T., Hwang, K. J. and Tyuma, I. (1969) Biochemistry 8, 378-383
2. Kumura, E., Kosaka, H., Shiga, T., Yoshimine, T. and Hayakawa, T. (1994) J. Cereb. Blood. Flow. Metab. 14, 487-491
3. Kosaka, H., Sawai, Y., Sakaguchi, H., et al. (1994) Am. J. Physiol. 266, C1400-C1405.
4. Green, L. C., Wagner, D.A., Glogowski, J., Skipper, P. L., Wishnok, J. S. and Tannenbaum, S. R. (1982) Anal. Biochem. 126, 131-138
5. Kosaka, H., Imaizumi, K. and Tyuma, I. (1982) Biochim. Biophys. Acta. 702, 237-241
6. Malinski, T., Bailey, F., Zhang, Z. G. and Chopp, M. (1993) J. Cereb. Blood. Flow. Metab. 13, 355-358
7. Kader, A., Franzzini, V. I., Solomon, R. A. and Trifilletti, R. R. (1993) Stroke 24, 1709-1716.
8. Zhang, F., White, J. G. and Iadecola, C. (1994) J. Cereb. Blood. Flow. Metab. 14, 217-226
9. Radi, R., Beckman, J. S., Bush, K.M. and Freeman, B.A. (1991) Arch. Biochem. Biophys. 288, 481-487.

Nitric oxide synthase inhibition augments postischemic leukocyte adhesion in pial cerebral venules in vivo

ANTAL G. HUDETZ*#, JAMES D. WOOD*, PETER J. NEWMAN+, and JOHN P. KAMPINE*#

Departments of *Anesthesiology, #Physiology, +Cell Biology and +Pharmacology, Medical College of Wisconsin, and +Blood Center of Southeastern Wisconsin, Milwaukee, WI 53226, U.S.A.

Introduction

Leukocytes are critically involved in ischemia/reperfusion injury in many tissues but their exact role in cerebral ischemia is unclear. Ischemia facilitates rolling and adhesion of leukocytes in mesenteric venules [7], however, this has not been demonstrated in the cerebral microcirculation. Nitric oxide has antiadhesion properties for leukocytes in vivo [6] and may modulate leukocyte adhesion during cerebral ischemia/reperfusion.

In this work we examined if decreased production of endogenous antiadhesion molecule nitric oxide (NO) promoted leukocyte firm adhesion to the endothelium of pial cerebral venules during cerebral ischemia and reperfusion.

Methods

The cerebrocortical microcirculation was studied using intravital video-microscopy with techniques established in our laboratory [4]. Adult male Sprague-Dawley rats were anesthetized with sodium pentobarbital (65mg/kg) and underwent implantation of a closed cranial window with ports for perfusion and measurement of intracranial pressure (ICP) in the right parietal area. Circulating nucleated cells in microvessels were visualized using fluorescence video-microscopy after infusion of Rhodamine 6-G (20mg/ml, 0.1ml/h iv.). Forebrain ischemia was induced by bilateral common carotid artery occlusion (BCO) followed by elevation of intracranial pressure (ICP) to 20mmHg for 60 minutes. These maneuvers decreased the velocity of flow in pial venules to about 30% of control. Occlusions were then released and ICP was restored to 5mmHg. Three to five vascular areas were video-recorded under control conditions, after BCO, BCO + ICP elevation (5 and 60 minutes), and reperfusion (5 and 60 minutes), for 1 minute each. In four experiments, N^{ω}-nitro-L-arginine methyl ester (L-NAME) was infused at 20mg/kg iv. over a period of 10 minutes, beginning at 30 minutes before reperfusion. The control group (N=4) was treated with D-NAME. Leukocytes firmly adhering to the endothelium of 15 to 80 μm diameter venules for longer than 3 seconds were counted from the video recordings.

Results and Discussion

Before ischemia the number of adhering leukocytes was $5.4 \pm 2.3 \times 10^{-4} mm^{-2}$ (SD). During ischemia, leukocyte adhesion increased about 2-fold at 5 minutes (see Figure, 5'I), and about 3-fold at 60 minutes of ischemia (60'I). There was no difference in leukocyte firm adhesion during ischemia between the L-NAME and D-NAME treated groups. In the D-NAME group, the number of adhering leukocytes returned to control level by 60 minutes of reperfusion (60'R). However, in the L-NAME group, leukocyte adhesion remained elevated during the 60 minutes of reperfusion.

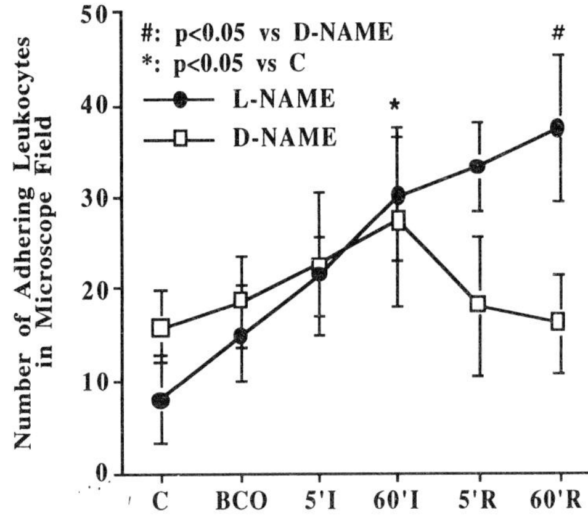

Our finding is similar to that of Dirnagl et al [3] who demonstrated only a slight increase in the number of leukocytes rolling along or sticking to the venular endothelium after ischemia in the cerebral cortex. However, the significant increase in leukocyte adhesion during cerebral ischemia has not been demonstrated before. In addition, the finding that after NOS inhibition leukocyte adhesion remains elevated during reperfusion is new. L-NAME decreases cerebral blood flow and flow shear rate in the microcirculation in both the normal and the ischemic brain. While L-NAME administration caused no difference in leukocyte firm adhesion during ischemia, it is possible that postischemic cerebral blood flow was lower in the L-NAME than in the D-NAME treated animals. Therefore, the enhancement of postischemic leukocyte adhesion by NOS inhibition may be due in part to decreased wall shear rate in addition to the elimination of NO as antiadhesion molecule.

The cellular mechanism of the antiadhesion effect of NO is yet unclear. In the rat mesentery, superfusion with L-NAME increases leukocyte rolling and P-selectin expression [1], however, the antiadhesion effect of NO donors may not be mediated by a direct effect on P-selectin or integrin CD18 [5]. In human saphenous vein endothelial cells, NO represses and L-NAME induces the expression of VCAM-1 and NO decreases the endothelial expression of E-selectin and ICAM-1 [2]. Whether these mechanisms are responsible for the L-NAME induced augmentation of postischemic leukocyte adhesion in the cerebral microcirculation remain to be investigated.

In summary, the present results suggest that NO synthase inhibition promotes postischemic leukocyte adhesion in pial venules. Therefore, endothelial dysfunction with diminished NO production may contribute to leukocyte-mediated cerebral injury.

This work supported in part by the NSF Grant BES-9411631 and NIH Training Grant HL-07902.

References

1. Davenpeck, K.L., Gauthier, T.W. and Lefer, A.M. (1994) Gastroenterol. 107:1050-1058
2. De Caterina, R., Libby, P., Peng, H-B,m Thannickal, V.J., Rajavashisth, T.P., Gimbrone, M.A., Jr., Soo Shin, W., and Liao J.K. (1995) J. Clin. Invest. 96:60-68
3. Dirnagl, U., Niwa, K, Sixt, G. and Villringer, A. (1994) Stroke 25:1028-1038
4. Hudetz, A.G., Fehér, G., Weigle C.G.M., Knuese, D.E. and Kampine J.P. (1995) Am. J. Physiol. 268:H2202-H2210.
5. Kubes, P., Kurose, I. and Granger, D.N. (1994) Am. J. Physiol. 267: H931-H937
6. Kurose, I., Wolf, R., Grisham, M.B. and Granger, D.N. (1994) Circ. Res. 74:376-382
7. Perry, M.A. and Granger D.N. (1992) Am. J. Physiol. 263:H810-815

Histamine induces both immediate and delayed headache in migraineurs due to H1 receptor activation. Support for the NO hypothesis of migraine.

LISBETH H LASSEN, LARS L THOMSEN and JES OLESEN.

Department of Neurology, Glostrup Hospital, University of Copenhagen, Denmark.

Introduction: We have recently suggested that nitric oxide (NO) is crucially involved in the initiation of migraine attacks (1). This was based on experimental studies of vascular headache using the exogenous NO-donor glyceryl trinitrate (GTN) (2,3). During GTN infusion migraineurs developed a significantly stronger headache than controls (2). In normal controls the GTN induced headache disappeared rapidly after the infusion, but in migraineurs it lasted considerably longer. At an average of 5.5 hours after GTN, migraineurs developed a migraine attack similar to their spontaneous attacks and fulfilling the International Headache Society criteria (IHS criteria) for migraine without aura (3,4).
Histamine is also known to induce headache (5). In human and non-human primates histamine dilates cerebral arteries due to interaction with a H_1 receptor in the endothelium which activates constitutive nitric oxide synthase (cNOS) with subsequent formation of NO (6). Thus, histamine is a trigger of endogenous NO formation in cerebral arteries. The purpose of the present study was to study whether histamine induces migraine attacks and whether this could be prevented by pretreatment with the histamine H_1-blocker, mepyramine.
Material and methods: Twenty patients suffering from migraine without aura according to the classification of IHS (4) received pretreatment with placebo or the histamine- H_1-receptor antagonist, mepyramine, in a randomized, double blind fashion, followed in both groups by intravenous histamine 0.5 μg/kg/min for 20 minutes. Headache intensity and characteristics were - recorded before the infusions and then every 10 minutes the first three hours in hospital and then every hour in the subsequent 9 hours at home. In both groups headache intensity was scored on a verbal scale form 0 to 10. Other headache characteristics necessary for precise classification of the induced headaches according to the IHS criteria were recorded, and at every recording it was evaluated whether the headache fulfilled the IHS criteria for migraine before the code was broken (4). Blood pressure, and heart rate were measured every 5 minutes during the first 3 hours in hospital. The study was conducted according to the Helsinki II declaration of 1964. All participants gave their informed consent.
Results: Immediate headache was defined as headache occurring during the histamine infusion or during the subsequent 20 minutes i.e. a period of 40 minutes. Number of patients developing immediate -headache and/or -migraine, see table 1. In the same period the median peak headache score was 5 (range 3-7) in the placebo pretreated group and 1 (range 0-3) in the mepyramine pretreated group ($p<0.001$ Mann Whitney).
Delayed headache was defined as headache occurring in the period from 40 minutes until 12 hours after start of infusion. Number of patients developing delayed - headache and/or -migraine see table 1. The median peak headache score in the this time period was 3.5 (range 0-8) in the placebo pretreated group and 0 (range 0-3) in the mepyramine pretreated group (-$p<0.001$ Mann Whitney). The peak headache intensity was reached 5.1 ± 1.0 hours (mean$\pm$SEM) after start of the histamine infusion.
During the hole 12 hours observation period histamine infusion caused a headache that fulfilled the IHS criteria for migraine without aura in 7 of 10 patients pretreated with placebo but in none of the 10 mepyramine pretreated patients ($p<0.01$, χ^2 test).
Haemodynamics: Mepyramine did not change the sum of percent changes in systolic nor diastolic blood pressure ($p>0.05$, unpaired t-test). Neither did the sum of percent changes in heart rate differ significantly between the placebo and the mepyramine pretreated group ($p>0.05$, unpaired t-test).

Table 1. Number of subjects developing histamine induced headache after pretreatment with placebo or with mepyramine.

	0-40 min		40min-12hours	
	headache	migraine (IHS)	headache	migraine (IHS)
placebo	10	4	9	5
mepyramine	7	0	3	0
χ^2-test	n.s.	n.s	$p<0.05$	$p<0.05$

Discussion: Previous findings have shown that the exogenous NO -donor, GTN, provokes migraine in migraineurs (2,3). Migraine sufferers are particularly sensitive to GTN as compared to tension-type headache sufferers and healthy controls. Thus, they develop more severe headaches during a GTN infusion and most migraineurs develop a typical migraine attack on average 5 hours after infusion
(2,3). Histamine in primates and certain other species causes endogenous formation of NO in intracranial arteries, due to activation of endothelial H_1 receptors, with subsequent activation of the constitutive NO-synthase (cNOS) and formation of NO from L-arginine (6). The development of histamine induced migraine, as demonstrated for the first time in the present study shows exactly the same time profile as after GTN infusion (2,3). Furthermore, migraineurs are more sensitive to both substances than tension-type headache patients and healthy controls (2,5). A common molecular mechanism is therefore highly likely. NO as the common mediator of GTN and histamine induced headache has been proposed (1). Controversially, it has been suggested that GTN induces liberation of histamine from mast cells and basophils (7). In this way histamine could be the common mediator. However, mepyramine in doses as used in the present study had no effect on GTN induced headache, (8). Thus GTN induced headache is not mediated via H_1-receptor stimulation. The present study supports the theory that activation of the cNOS-NO-cGMP pathway plays a pivotal role in the triggering of migraine attacks (1).
It remains to be shown whether migraine pain is strictly related to the NO-cGMP induced dilatation of the cerebral arteries or whether it is caused by a noxious effect of NO on perivascular sensory nerveendings.

References:

1. Olesen J, Thomsen LL, Iversen HK. TIPS **15**,149-153 (1994).
2. Olesen J, Iversen HK, Thomsen LL. Neuroreport **4**,1027-1030 (1993).
3. Thomsen LL, Kruuse C, Iversen HK et al. Eur J Neurol.**1**,73-80 (1994).
4. Classification and diagnostic criteria for headache disorders, cranial neuralgias and facial pain. Cephalalgia **8**(suppl 7),19-28 (1988).
5. Krabbe AE, Olesen J. Pain.**8**,253-259 (1980).
6. Toda N. Am J Physiol. **258**,H311-H317 (-1990).
7. Rozniecki JJ, Kuzminska B, Prusinski A. Cephalalgia **9**(Suppl 10),80-81 (1989).
8. Lassen LH, Thomsen LL, Kruuse C, Iversen HK, Olesen J. Eur J Clin Pharmacol. In print

Local production of nitric oxide in viral and autoimmune diseases of the central nervous system

D. Craig Hooper*, S. Tsuyoshi Ohnishi†, Rhonda Kean*, Jean Champion*, Yoshihiro Numagami†, Bernhard Dietzschold*, and Hilary Koprowski*

*Center for Neurovirology, Department of Microbiology and Immunology, and Jefferson Cancer Institute, Thomas Jefferson University, Philadelphia, PA 19107-6799, USA; †Philadelphia Biomedical Research Institute, King of Prussia, PA, USA.

Cells of the macrophage/monocyte lineage, activated by immune or inflammatory stimuli, mediate their well known toxic effects through the production of free radicals including nitric oxide (NO). The nature of the stimulus leading to macrophage activation, invading pathogen or self-antigen, is clearly important in determining whether the destructive activities of these cells have a beneficial or deleterious effect on the host. The site of activation of macrophages and related cells is also of enormous significance to the outcome of a disease process. In the central nervous system (CNS), for example, the multiple biological roles of NO in the control of vasodilation, in neurotransmission, and in immunity [1] may all converge during infection or an autoimmune disease. Thus the high levels of NO potentially produced in an immune or inflammatory response in the CNS may perturb both the local circulation and neural function. Previous investigations from our laboratory have provided evidence that levels of mRNA specific for inducible nitric oxide synthase (iNOS), the enzyme responsible for the production of NO by macrophages, are enhanced in the brains of rats infected with rabies virus and Borna disease virus (BDV) or suffering from experimentally-induced allergic encephalomyelitis (EAE) [2,3]. To determine, by direct measurement, whether this is associated with similar increases in NO levels in the CNS during infection, we have employed spin trapping coupled with electron paramagnetic resonance (EPR) spectroscopy [4] to semi-quantitate NO produced in rabies and Borna disease as well as two distinct models of EAE and lipopolysaccharide (LPS)-induced shock.

NO analysis of brain and spinal cord tissue was performed, as previously described [5], using diethyldithiocarbamate (DETC) as a spin trap and Fe to stabilize the NO-DETC complex. EPR spectra were examined in a Varian E-109 spectrometer under conditions described previously [5] and NO concentrations estimated from EPR spectra as described elsewhere [6]. The average levels of NO detected in at least two similarly treated animals were scored as (-) < 0.5 μM, (+/-) 0.5 to 1 μM, (+) 1 to 2.5 μM, (++) 2.5 to 5 μM, (+++) 5 to 10 μM, and (++++) > 10 μM.

EAE was induced in rats by the i.v. administration of activated cells of an MBP-specific T cell line as previously described [8] and in SJL mice by i.m. immunization with 10 μg guinea pig MBP in complete Freund's adjuvant. In the rat model a progressive paralysis starting in the tail appears 4 days after the cells are introduced and culminates in a hind-quarters or general paralysis by day 5 after which the rats often recover. In this model, high levels of NO are generated predominantly in the spinal cord [5]. In contrast, SJL mice immunized with MBP undergo an acute onset, general paralysis approximately 18 days after immunization. Heightened NO levels can be detected in both brain and spinal cord in these animals around the time, several days before, that initial symptoms appear. Rats infected with the neurotropic viruses, rabies CVS-24 or BDV, as previously described [5], generally develop signs of neurological disease approximately 5 and 19 days, respectively, after infection and die within 3 to 5 days afterwards. High levels of NO are detectable in the brains of these animals in the last few days of the infections. The i.v. administration of LPS to SJL mice induces shock, through the activation of inflammatory cells, within 48 hours, which is associated with high levels of NO in both brain and spinal cord. For the various disease processes, clinical signs were scored as mild (+) through increasing signs of disease (++,+++) to moribund (++++).

As summarized in Table 1, our results indicate that these diverse disease processes all share the development of high levels of NO production in the appropriate sites. For the neurotropic rabies and borna disease virus infections, substantial quantities of NO were detected in infected rat brain. While spinal cord was the predominant target of NO production for EAE induced in rats by adoptive transfer of myelin basic protein (MBP)-specific T cells, SJL mice immunized with MBP in CFA or treated with 20 μg LPS i.v. elaborate NO in both spinal cord and brain.

Table 1. NO levels in brain and spinal cord

Disease process	Stage of disease	NO level brain	NO level spinal cord
none	-	+/-	-
Rabies	-	+/-	N.D.*
	+	+	N.D.
	+++	++	N.D.
Borna disease	++	+++	N.D.
EAE (T cell transfer)	++	++	++++
EAE (MBP immunization)	+/-	+	+
Endotoxin shock	++++	+++	++++

* not done

These latter results confirm observations that NO is produced in the spinal cord and brain of mice suffering from MBP-induced EAE [7] or LPS-induced shock [8] using related methodology.

In addition to being involved in toxicity, it is conceivable that high levels of NO produced by activated inflammatory cells may contribute to the pathogenesis of neurological diseases by directly perturbing neurotransmission and local circulation. Treatments designed to interfere with the production or availability of NO have met with some success in the treatment of MBP-elicited EAE [9] and LPS-induced shock [8].

Table 2. Treatment of adoptive transfer EAE

Treatment	Clinical symptoms	NO level (cord)
none	++	++++
L-NMMA*	+++	++++
PTIO†	+	N.D.

* 66 mg/kg/day
† 100 mg/kg/day

As summarized in Table 2, we have examined treatments that may alleviate an ongoing disease process, such as would be seen in multiple sclerosis and myasthenia gravis in humans, using the T-cell adoptive transfer model of EAE, which bypasses the T-cell priming and activation steps required in the MBP-induced model. Administration of N^{ω}-monomethyl-L-arginine (L-NMMA), a competitive inhibitor of INOS, neither interfered with the development of EAE, as described elsewhere [10], nor decreased the level of NO released in the spinal cord. On the other hand, our results indicate that clinical EAE can be limited by treatment of rats that have received MBP-specific T-cells with an NO scavenger, 2-phenyl-4, 4, 5, 5-tetramethyl-imidazoline-1-oxyl-3-oxide (PTIO). A derivative of PTIO has previously been successfully used for the treatment of LPS-induced shock in rats [8]. Experiments to quantify the effects of PTIO on brain and spinal cord NO levels during EAE are underway.

1. Snyder, S.H. and Bredt, D.S. (1992) Sci.Amer. **5**, 68-77.
2. Koprowski, H., Zheng, Y.M., Heber-Katz, E., Fraser, N., Rorke, L., Fu, Z.F., Hanlon, C. and Dietzschold, B. (1993) Proc.Natl.Acad.Sci. USA. **90**, 3024-3027.
3. Zheng, Y.M., Schafer, M.K.-H., Weihe, E., Sheng, H., Corisdeo, S., Fu, Z.F., Koprowski, H. and Dietzschold, B. (1993) J.Virol. **67**, 5786-5791.
4. Kubrina, L.N., Caldwell, W.S., Mordvintcev, P.I., Malenkova, I.V. and Vanin, A.F. (1992) Biochem.Biophys Acta **1099**, 233-237.
5. Hooper, D.C., Ohnishi, S.T., Kean, R., Numagami, Y., Dietzschold, B. and Koprowski, H. (1995) Proc.Natl. Acad.Sci. USA. **92**, 5312-5316.
6. Sato, S., Tominaga, T., Ohnishi, T. and Ohnishi, S.T. (1994) Brain Research **647**, 91-96.
7. Lin, R.F., Lin, T.-S., Tilton, R.G. and Cross, A.H. (1993) J.Exp.Med. **178**, 643-648.
8. Yoshida, M., Akaike, T., Wada, Y., Sato, K., Ikeda, K., Ueda, S. and Maeda, H. (1994). Biochem. and Biophys.Res. Commun. **202**, 923-930.
9. Cross, A.H., Misko, T.P., Lin, R.F., Hickey, W.F., Trotter, J.L., and Tilton, R.G. (1994) J.Clin.Invest. **93**, 2684-2690.
10 Zielasek, J., Jung, S., Gold, R., Liew, F.Y., Toyka, K.V. and Hartung, H.P. (1995) J.Neuroimm. **58**, 81-88.

Nitric oxide modulates cerebrovascular permeability in a rodent model of meningitis

Kathleen M. K. Boje
Department of Pharmaceutics
School of Pharmacy
University of Buffalo
Buffalo, New York, USA 14260

Nature has marvelously utilized nitric oxide (NO) as a versatile biological emissary, commissioned to serve either as an signal transduction agent or immunological response mediator. Pivotal physiological roles for NO have been identified in the cardiovascular, endocrinological, gastrointestinal, immune, pulmonary, renal and central nervous systems [1,2]. Conversely, excessive NO synthesis has been found to initiate or exacerbate various pathological conditions [1].

The seminal discovery of the presence of NO in the central nervous system [3] virtually ushered in "The Decade of the Brain" (a.k.a., the 1990's). Extensive research has implicated excessive NO synthase activity in a variety of neurological diseases, e.g., multiple sclerosis [4,5], experimental autoimmune encephalomyelitis [6,7], cerebral ischemia [8], lymphocytic choriomeningitis [9], acute hypertension [10] and meningitis [11-15].

Our laboratory is especially interested in NO modulation of cerebrovascular permeability. Physiologic synthesis of NO by brain endothelial cells is vitally important in the regulation of cerebral blood flow and vasodilatation, and in the maintenance of the ionic membrane characteristics of the blood-brain barrier (BBB) [16]. Pathological production of NO during meningitis [14, 15] is intimately involved in disturbances of cerebral blood flow [14, 17] pial arteriolar dilatation [14] and disruption of the integrity of the blood-cerebrospinal fluid (B-CSF) [15] and BBB barriers [KMK Boje, unpublished data].

To test our hypothesis that disruption of the BBB and B-CSF barriers during experimental meningitis may be mediated by pathological production of NO, we needed to demonstrate that during meningeal inflammation (a) excessive quantities of NO were synthesized; (b) the permeabilities of the B-CSF and BBB barriers were increased; and (c) pharmacologic inhibition of NO synthesis reversed barrier permeability alterations. Using a sensitive chemiluminescent NO detector, we observed NO production from meningeal tissues from rats previously dosed with intracisternal LPS (to induce experimental meningitis), whereas NO was not detected from a control group of rats [15,18]. It was also necessary to pharmacokinetically quantify disruption of the B-CSF and BBB barriers during meningitis using Evans Blue dye [15] or ^{14}C-sucrose [19], respectively. Compared to control rats, statistically significant increases in CSF Evans Blue dye and brain tissue ^{14}C-sucrose penetration were observed in rats dosed intracisternally 8 hours earlier with 25 or 200 μg LPS [15,19]. These observations merely suggested, but did not prove, that there was an association between elevated NO synthesis during meningitis and increased permeability of the B-CSF and BBB barriers.

A rigorous test of our hypothesis required the selection of an appropriate NO synthase inhibitor and subsequent design of a dosing regimen which would block meningeal NO synthesis. Aminoguanidine was the NO synthase inhibitor of choice, since we and others [15, 18, 20-22] established that it was relatively specific for the induced isoform of NO synthase without eliciting excessive systemic hypertension. Moreover, the data of Beaven et al. [23] permitted estimation of key pharmacokinetic parameters, thereby enabling the semi-rational design of an appropriate aminoguanidine infusion regimen [15]. Rats were dosed intracisternally with LPS (0, 25 or 200 μg) to induce meningeal inflammation, followed by an 6 - 8 hour intravenous infusion of saline or aminoguanidine. In rats with meningeal inflammation, not only did aminoguanidine remarkably attenuate permeability alterations in the B-CSF and BBB barriers, but also blocked meningeal NO synthesis [15, KMK Boje, unpublished data]. Compared to control rats, aminoguanidine treatment itself did not alter the integrity of the cerebrovascular barriers.

Thus, we were able to experimentally demonstrate critical elements of our hypothesis: (a) NO is produced in our rat model of meningitis; (b) disruption of the cerebrovascular barriers occurs during meningeal inflammation; and (c) aminoguanidine therapy blocks both NO production and disruption of the cerebrovascular barriers. These exciting experiments argue convincingly that NO is an important inflammatory mediator of BBB and B-CSF barrier integrity during meningitis.

Abbreviations used: AG, aminoguanidine; BBB, blood-brain barrier; B-CSF, blood-cerebrospinal fluid; LPS, lipopolysaccharides; NO, nitric oxide

Acknowledgements

The author thanks Mr. Steven Jodish and Ms. Patricia Neubauer for their excellent technical assistance. Supported by NIH (NS 31939).

References

1. Dawson TM and Dawson, VL (1995) The Neuroscientist **1**, 7-18.
2. Schmidt, HHHW and Walter,U. (1994) Cell **78**, 919-925.
3. Garthwaite, J., Charles, S.J. and Chess-Williams, R. (1988) Nature **336**, 385-388.
4. Johnson, A.W.,Land, J.M., Thompson, E.J. et al., (1995) J. Neurol. **58**, 107.
5. Bö, L., Dawson, T.M., Wesselingh, S. et al., (1994) Ann. Neurol. **36**, 778-86.
6. MacMicking, J.D., Willenborg, D.O., Wedemann, M.J., Rockett, K.A. and Cowden, W.B. (1992) J. Exp. Med. **176**, 303-7.
7. Cross, A.H., Misko, T.P., Lin, R.F., Hickey, W.F. , Trotter, J.L., and Tilton, R.G. (1994) J. Clin. Invest. **93**, 2684-90.
8. Nowicki, J.P., Duval, D., Poignet, H. and Scatton, B. (1991) Eur. J. Pharmacol. **204**, 339-40.
9. Campbell, I.L., Samimi, A. and Chiang, C.-S. (1994) J. Immunol. **153**, 3622-9.
10. Mayhan, W.G. (1995) Brain Res. **686**, 99-103.
11. Visser, J.J., Scholten, R.J.P.M. and Hoekman, K. (1994) Ann. Internal Med. **120**, 345-6.
12. Milstien, S., Sakai, N., Brew, B.J. et al., (1994) J. Neurochem. **63**, 1178-80.
13. Pfister, H.-W., Bernatowicz, A., Ködel, U. and Wick, M. (1995) Neurol. Neurosurg. Psychiat. **58**, 384.
14. Koedel, U., Bernatowicz, A., Paul, R., Frei, K., Fontana, A. and Pfister, H.W (1995) Ann. Neurol. **37**, 313-23.
15. K.M.K. Boje (1995) Eur. J. Pharmacol. **272**, 297-300.
16. Jangro, D., West, G.A., Nguygen, T.-S. and Winn, H.R. (1994) Circ. Res. **75**, 528-38.
17. Tureen, J. (1995) J. Clin. Invest. **95**, 1086-91.
18. K.M.K. Boje, Neuropharmacol., submitted and under review.
19. K.M.K. Boje, (1995) J. Pharmacol Exp. Ther. **274**, In Press.
20. Corbet, J.A., Tilton, R.G., Chang, K. et al., (1992) Diabetes **41**, 552-6.
21. Misko, T.P., Moore, W.M., Kasten, T.P. et al., (1993) Eur. J. Pharmacol. **233**, 119-25.
22. Griffiths, M.J.D., Messent, M., MacAllister, R.J. and Evans, T.W. (1993) Br. J. Pharmacol. **110**, 963-8.
23. Beaven, M.A., Gordon, J.W., Jacobsen, S. and Severs, W.B., (1969) J. Pharmacol. Exp. Ther. **165**, 14-22.

Nitric oxide production during tidal breathing: an effort-independent assay.

Jon Woods M.D., Steven Sears M.D., John Hunt M.D., Sarah Walton M.D., Con Yee Ling M.D., Michael Strunc M.D. and Benjamin Gaston M.D.

Department of Pediatrics, Pulmonary Division, Naval Medical Center, San Diego CA 92134.

High vital capacity (VC) expirate NO· levels in asthmatic subjects [1-3] appear to reflect increased inflammatory nitric oxide synthase expression in asthmatic airway biopsies [4]. Moreover, condensate nitrite and nitrate concentrations from asthmatic subjects are high during tidal breathing [5]. We hypothesized that measurement of net tidal NO· production ($V_{NO\cdot}$) might be useful as a marker for disease severity in patients with lung disease too young or incapacitated to cooperate with conventional pulmonary function testing.

Materials and Methods. Control (no history of respiratory disease or evidence of airflow obstruction) and asthmatic subjects (acute exacerbations of chronic airway reactivity) wearing noseclips breathed quietly for five minutes in a 760 liter sealed plethysmograph while NO· was measured by chemiluminescence (Thermo Envirnmental, Franklin, MA) through a gas sampling port with continuous in-line capnography. Nitric oxide in exhaust gas from ventilated (Servo 900C, Siemens, Sweden) children with pneumonia was assayed by chemiluminescence at the 0 amplitude CO_2 waveform point of a 700 cc open-ended ventilator exhaust reservoir, and $V_{NO\cdot}$ calculated on the basis of expired minute ventilation determined by pneumotachometry. Control experiments (n = 8) were performed on the exhaust limb of this cicuit using known quantities of NO· gas (40 - 800 ppb) demonstrating a small but consistent loss (10%) during normal operation, an effect which was accounted for in final $V_{NO\cdot}$ calculations. Means were compared using the Mann-Whitney Rank Sum test, with $p < 0.05$ considered to be significant. Results are presented as mean, ± standard error.

Results. Plethysmographic $V_{NO\cdot}$ was highly reproducible (intrasubject coefficient of variation, 2.7%, n = 10) and linearly related to body surface area (BSA) (r = 0.88). Moreover, it was readily measured in young children. Specific $V_{NO\cdot}$ ($V_{NO\cdot}$/BSA, $SpV_{NO\cdot}$), 0.17 ±0.07 μl/min/m^2 in control subjects (n = 10), was elevated in asthmatic subjects (0.35 ±0.09 μl/min/m^2, n = 6, $p < 0.005$). Levels of CO_2 in the plethysmograph never rose to > 1%, nor did any test require termination. Values did not differ when measured sequentially with 10% and 90% BSA exposed (p = NS, n = 5) suggesting that differences did not arise from dermal artifact. Endotracheally intubated patients with pneumonitis had $V_{NO\cdot}$ over ten-fold higher than normal (2.7 ± 1.3 μl/min/m^2, n = 6, $p < 0.001$) consistent with previous reports that the lower airway is the principal source of expired NO· in subjects with inflammatory lung disease [6].

Discussion. As is the case with CO_2, we have shown that net metabolic NO· production in expired air can be quantitated during tidal breathing with good reproducibility. Moreover, $V_{NO\cdot}$ indexed to BSA is significantly abnormal in patients with airway inflammation, consistent with several previous observations on gas and aqueous phase nitrogen oxides collected by invasive or effort-dependent techniques [1-3,5-7]. We conclude that $V_{NO\cdot}$ measurements are 1) reproducible; 2) effort-independent; 3) applicable to incapacitated and uncooperative patients; and 4) abnormal in the presence of pulmonary inflammation. We speculate that these measurements may be useful for non-invasive monitoring of lung inflammation in the critical care setting and the pediatric pulmonary function laboratory.

Supported by US Navy grant NMCSD S-93-LH-113.

References.

1. Gaston B, Drazen J, Chee CBE, Wohl MEB, Stamler JS. (1993) Endothelium 1:87.
2. Alving K, Weitzber E, Lundberg JM. (1993) Eur Respir J 6:1368-1370.
3. Persson MG, Gustafsson LE. (1993) Acta Physiol Scand 149:461-466.
4. Hamid Q, Springall DR, Riveros-Moreno V, et al. (1993) Lancet 342:1510-1513.
5. Hunt J, Byrns R, Igarro LJ, Gaston B. (1995) Lancet, in press.
6. Massaro A, Mehta S, Lilly C, Kobzik L, Reilly J, Drazen JM. (1995) Am J Respir Crit Care Med, in press.
7. Gaston B, Reilly J, Drazen JM, et al. (1993) Proc Natl Acad Sci USA 90:110957-10961.

Home Measurement of Airway Condensate Nitrite as a Marker for Asthmatic Airway Inflammation.

John Hunt, M.D.[+], Russell E. Byrns[*], Louis J Ignarro[*], Ph.D. and Benjamin Gaston, M.D.[+¶]

[+]Department of Pediatrics, Naval Medical Center, San Diego, CA 92134; [*]Department of Pharmacology, University of California, Los Angeles, School of Medicine, Los Angeles, CA 90024. [¶] Department of Pediatrics, University of California, San Diego, School of Medicine, San Diego, CA 92134

Treatments for inflammatory lung diseases are initiated and followed on the basis of tests which measure airflow obstruction but which do not measure the severity of inflammation [1]. Evidence that expression of inducible nitric oxide synthase is increased in the airway inflammatory and epithelial cells of patients with asthma [2] is supported by studies revealing high concentrations of nitric oxide (NO·) in asthmatic exhalates which decrease with anti-inflammatory treatment [3-7]. However, the potential for widespread use of NO· assays in asthma management is limited by cumbersome and expensive technology. We hypothesized that aqueous phase nitrogen oxides (NO_x) may be elevated and conveniently measured in the condensed exhalate of asthmatic subjects.

<u>Materials and Methods.</u> Control subjects had cardiopulmonary disease or evidence of airflow obstruction as assessed by history, exam and spirometry (PK Morgan, Andover, MA) [1]. Asthmatic patients had acute, β_2 agonist-reversible (> 10% improvement in airflow) bronchospasm at the time of study entry. Subjects wearing noseclips breathed quietly through a mouthpiece. Minute ventilation was determined in series (Ohmeda RM311, Madison, WI). Expired air was filtered (0.3 μ; Marquest Medical Products, Englewood, CO) and channelled through a one-way valve to a 60 cm Tygon © (Norton, Worcester, MA) condensing chamber. Condensate nitrite (NO_2^-) was measured colorimetrically after diazotization [9], and nitrate (NO_3^-) was measured as NO· by chemiluminescence after reduction in vanadium chloride [10]. Amylase was measured colorimetrically (Boehringer-Manheim 911, Indianapolis, IN). Reduced thiol and S-nitrosothiol (RS-NO) concentrations were determined after reaction with Ellman's reagent and mercuric chloride (Saville reaction), respectively [9]. Expirate was also collected at home (350 μl) as described above and reacted 1:2 volume/volume with Greiss reagent in 0.5 N HCl. Reactivity was estimated by the subject at home using an analogue visual scale, and later re-analyzed spectrophotometrically. Reagents were purchased from Sigma (St. Louis, MO). Concentrations were compared using the Mann-Whitney rank sum test. Chemical, spirometric and demographic data were compared using standard regression analysis. Results are presented as mean, ± standard error. P values of < 0.05 were considered significant. This investigation was approved by the institutional Committee for the Protection of Human Subjects.

<u>Results.</u> Asthmatic subjects (n = 10) had condensed exhalate NO_2^- concentrations (2.63 μM ± 0.54) over three-fold higher than those of controls (0.78 μM ± 0.072; n = 15; $p < 0.001$). These groups did not differ with regard to age (20 ± 3.5 vs 24 ± 2.3 years) or sex (50% vs 47 % female). An inverse linear relationship existed between NO_2^- concentration and both forced expiratory volume at 1 second ($r = 0.94$, $p < 0.001$) and forced expiratory flow between 25% and 75% of vital capacity ($r = 0.87$, $p < 0.005$). Likewise, NO_3^- concentrations were elevated in asthmatic subjects (9.48 ± 1.25 μM) relative to controls (5.27 ± 0.51 μM, $p < 0.01$). Absence of salivary contamination was confirmed by failure to detect amylase in six randomly selected samples. Failure of turbulent flow-induced particle aerosolization to account for asthmatic differences was suggested by reduced thiol concentrations in asthmatic and control subjects (n = 9) of less than 1 μM, over two log orders lower than that ordinarily present in airway lining fluid (ALF), and RS-NO concentrations less than 500 nM in all samples tested (n = 6) including two additional patients with cystic fibrosis (CF) [9]. Intrasubject minute ventilation during the collection procedure was reproducibly autoregulated with a coefficient of variation over 3 consecutive 5 minute intervals of 3.1% (n = 6). Condensate assays also were highly reproducible with a coefficient of variation of 6.2% (n = 11). Results when breathing ambient and NO·-free air did not differ (n = 5).

Condensate collection for NO_2^- assay was readily performed at home, distinguished asthmatic (>1 μM) from non-asthmatic (<1 μM) subjects and was accurate on spectrophotometric reanalysis (n = 7). The total cost of equipment and reagents for visual-scale analysis was under $50.00.

<u>Discussion.</u> Consistent with previous reports demonstrating high concentrations of gas-phase NO· in asthmatic expirates [2-8], we have shown using two unrelated chemical methods that concentrations of the stable NO· metabolites NO_2^- and NO_3^- are two- to three-fold higher than normal in condensed water collected from expired air of subjects with acute asthma. Moreover, NO_2^- can be assayed simply, reliably and inexpensively at home as a marker for the severity of asthmatic airway inflammation.

Measured NO_x concentrations appear to have arisen primarily from reactions of endogenously produced NO· with oxygen and, in turn, with nitrogen dioxide dissolved in condensed water. The possibility that NO_x arose from nebulization of ALF resulting from turbulent airflow is less likely, given 1) the absence of reduced thiol in condensate samples [9]; 2) the absence of RS-NO, particularly in patients with CF who have both high ALF RS-NO concentrations and markedly turbulent airflow [9]; and 3) size filtration of particles less than 0.3μ in diameter. Inhaled NO_x did not appear to contribute significantly to measured values as, consistent with studies on expired NO·, values were unaffected by breathing medical air [6,7]. An effect of nasal artifact cannot be ruled out completely, though normal and asthmatic differences in NO· production are preserved after elective awake intubation [7], and subjects with allergic rhinitis and asthma have vital capacity expired NO· concentrations no greater than subjects with asthma alone [8].

In summary, we report the first colorimetric test for acute asthma, a test which can be performed reproducibly and inexpensively at home. We speculate that home NO_2^- assays may have a role in the longitudinal assessment of asthma severity.

Supported by United States Navy grant NMCSD S-93-LH113.

REFERENCES.

1. Crapo R. (1994) N Engl J Med. 331:25-30.
2. Hamid Q, Springall DR, Riveros-Moreno V, et al. (1993) Lancet. 342:1510-1513.
3. Alving K, Weitzberg E, Lundberg JM. (1993) Eur Respir J. 6:368-1370.
4. Persson MG, Gustafsson LE. (1993) Acta Physiol Scand. 149:461-466.
5. Gaston B, Drazen J, Chee C, et al. (1993) Endothelium. 1:S87.
6. Massaro A, Gaston B, Kita D, Fanta C, Stamler JS, Drazen JM. (1995) Am J Respir Crit Care Med 800-803.
7. Massaro A, Mehta S, Lilly C, Kobzik L, Reilly J, Drazen J. (1995) Am J Respir Crit Care Med (in press).
8. Alving K, Lundberg JON, Nordvall SL. (1995) Am J Respir Crit Care Med. 151:A129.
9. Gaston B, Reilly J, Drazen JM, et al. (1993) Proc Natl Acad Sci USA. 90:10957-10961.
10. Ignarro LJ, Fukuto JM, Griscavage JM, Rogers NE, Byrns RE. (1993) Proc Natl Acad Sci USA. 90:8103-8107.

®noXon - The NO_2-Filter Material for a Safe Application of NO-Gas and First Results under Clinical Conditions

T.Vaahs, A.Schleicher, G.Frank, W.Sixl
Hoechst AG, D-65926 Frankfurt/M. Germany

P.Germann, G.Urak, Ch.Krebs*, C.Leitner, G.Röder, M.Zimpfer
Department of Anesthesia and General Intensive Care, University of Vienna, Austria
***Messer-Griesheim Austria**

1. Introduction

One of the main problems in the clinical administration of NO gas in the treatment of ARF etc. is the harmful contaminant NO_2. Once deposited, NO_2 dissolves in lung fluids, and various chemical reactions occur that give rise to undesired products found in the blood and other body fluids. To avoid this the ®noXon filter material removes completely and in a selective way all NO_2 from the inspiration gas stream. Because of the performed chemical reaction of the ®noXon filter material there is no possibility of desorbing NO_2 during pressure peaks or mishandling of the equipment. Applying the ®noXon filter gives way to pure NO for the treatment of lung diseases.

2. Problem

NO reacts with oxygen to form a thermodynamic equilibrium with NO_2.

$$2\ NO + O_2 \Leftrightarrow 2\ NO_2$$

The reaction rate for the formation of NO_2 is depending on the concentration of NO to the second.

$$[NO_2] = k \times [NO]^2 \times [O_2]$$

Applying concentrations of NO from 20 to 100 ppm means that there will not be an equal amount of NO_2 but some NO_2 will be formed. Depending on country regulations the maximum working place concentrations allowed are in the range of 1 to 5 ppm NO_2.

One can expect NO_2-concentrations during a normal every day use of NO gas in the range between 0.5 ppm (30 ppm NO) and 3.5 ppm (80 ppm NO) and NO_2-concentrations in worst case scenarios variing between 4 and 32 ppm. These investigations were made in the laboratory by down scaling medical conditions giving qualitatively comparable results.

3. Solution

®noXon as a polymeric material reacts with the undesired NO_2 by converting it into the desired NO.

$$[-C_6H_4-S-] + NO_2 \longrightarrow [-C_6H_4-S(=O)-] + NO$$

XPS data of virgin ®noXon and ®noXon that has filtered NO_2 show the transition of ®noXon from the unoxidized into the oxidized state by the filtering of NO_2.

To satisfy needs like pressure drop as low as possible and compatibility with close-to-patient filter systems we developed different morphologies for the use as filter material, i.e. granules, fibers, vleeces etc..

4. Filter Ability

To test the filter characteristics of ®noXon we applied a continuous flow of only NO_2 within oxygen or air and measured NO and NO_2 behind the filter element. The fact that NO comes out of the filter shows the conversion of NO_2 into NO. Because the medical application of ®noXon takes place in a close-to-patient filter element there is always a 100% humidity in the inspiration gas mixture. Therefore all tests were performed not only under dry laboratory conditions but also under 100% humidity. The filtering data show that ®noXon works even better under humid conditions.

5. Feasibility Study of ®noXon Filter Material Using an Artificial Thorax System

An artificial thorax connected to a respirator system (servoventilator 300) was used to test ®noXon under pulsating inspiration gas flow. Granules were put into a filter element and brought into the inspiration gas stream near the mouth. No optimisation of the filter element was made prior to the measurement.
We mixed the NO gas with the humid inspiration air and measured the NO_2 concentrations with and without ®noXon. In the case of applied 90 ppm NO ®noXon filtered 75% of the formed NO_2, whereas with 50 ppm NO ®noXon filtered NO_2 to the detection limit, i.e. ca. 100%.

6. Possible Points of Use for ®noXon Filter Elements

There are three possible points of use for ®noXon filters. For every solution different characteristics of the ®noXon filters are necessary. For instance if you use it at the NO cylinder or in the NO-dosing device to avoid high NO_2 concentrations either by mishandling or by flushing the remaining NO-air mixture there is less concern about pressure drop compared with a filter element in the inspiration gas flow close to the patient.

7. Methods in the Clinical Use

®noXon was tested in a close-to-patient filter integrated into the inspiratory part of the respiratory cycle in five patients using NO concentrations of 10, 15 and 18 ppm in a time course of 1 min, 10 min, 1 h, 2 h, 4 h and 10 h.

8. Results

NO/NO_2 analysis (CLD) with and without ®noXon filter at 18 ppm inspiratory NO:

	conc. NO	conc. NO_2	conc. NO	conc. NO_2
	without ®noXon		with ®noXon	
	(ppm)	(ppm)	(ppm)	(ppm)
1 min	17.88+-0.26	0.24+-0.05	18.52+-0.35	0+-0.00
10 min	17.98+-0.38	0.22+-0.06	18.50+-0.40	0+-0.00
1 h	18.08+-0.30	0.20+-0.06	18.48+-0.42	0+-0.00
2 h	18.14+-0.29	0.18+-0.03	18.54+-0.48	0+-0.00
4 h	17.48+-0.46	0.24+-0.09	18.62+-0.46	0+-0.00
10 h	17.70+-0.48	0.22+-0.09	18.84+-0.47	0+-0.00

n=5, mean + SEM $FiO_2=0.55$

The results using 10 and 15 ppm NO are similar. With ®noXon no NO_2 could be detected, which means that in every case the concentrations of NO_2 were less than the detection limit (around 0.01 ppm).

9. Summary

®noXon is a polymeric material, which filters NO_2 in a selective way and converts it into NO. This ability makes ®noXon a unique material to be used within the NO-therapy to get rid of the inherent NO_2-problem. The filter properties are in humid environments even better than in dry air. There are several possible points of use for the prospected ®noXon filter element in the clinical administration of NO. As shown in our results ®noXon abolished the detection of NO_2 regardless of the different NO concentrations used.

Systemic effect of inhaled nitric oxide: modification of renal hemodynamics in pigs.

ERIC TRONCY, MARTIN FRANCOEUR, IGOR SALAZKIN, PATRICK VINAY* and GILBERT A BLAISE.

Anesthesia Laboratory, *Medicine Department, Hôpital Notre-Dame, Université de Montréal, CP1560, Montreal (Qc) H2L 4K8, CANADA.

Intravenous nitrovasodilators such as sodium nitroprusside and nitroglycerine (ivNTG) act as direct nitric oxide (NO) donors and are very efficient in reducing pulmonary pressure but their use is limited by their side effects[1,2]. Opposed to the iv medication, inhaled nitric oxide (inhNO) is commonly accepted as a selective pulmonary vasodilator during experimental[3,4] or clinical[5,6] conditions. However, we have observed occasional cardiac and systemic effects of inhNO[7]. In order to evaluate extrapulmonary effects of inhNO, we used renal hemodynamics as a probe and compared the effects of inhNO and ivNTG on urinary flow (UF), glomerular filtration rate (GFR) and renal blood flow corrected for PAH extraction (RBFc) in pigs with / without systemic or pulmonary hypertension.

Female pigs (22.2 ± 0.67 Kg; n = 30) were anesthetised, paralysed and mechanically ventilated. The cardiovascular (ECG, SBP, PAP, CVP, PWP, HR, CO, Hte) and respiratory (arterial and mixed venous blood gases and pH) parameters were monitored. The left ovarian vein was ligated and catheters were installed in both ureters, carotid artery and left renal vein. A continuous infusion of PAH and inulin was maintained following suitable primes of each marker and blood samples from the two last-cited vessels allowed to measure PAH and inulin clearances and calculate respectively RBFc and GFR. The experimental protocol (Table 1) was divided into four stages of 20 minutes each. Data analyses were performed, within groups with the Super ANOVA procedures and Fisher's protected LSD *post-hoc* tests ($P_1 < 0.05$), and between groups with unpaired t-tests on the differences between two consecutive stages ($P_2 < 0.05$).

Table 1. Experimental design

inhNO: 40 PPM controlled by chemiluminescence;
Hypoxia: $FiO_2 = 15\%$;
ivNTG: progressive infusion with a mean dosage of 94.43 µg/Kg/min;
PE: phenylephrine infusion at 14.7 ± 1.22 µg/Kg/min.

Stages	I 30-50'	II 50-70'	III 70-90'	IV 90-110'	Sample size
Effector					
A	Nil	Nil	Nil	Nil	n = 5
B	Nil	inhNO	inhNO	Nil	n = 5
C	Nil	Hypoxia	Hypoxia	Nil	n = 4
D	Nil	Hypoxia	Hypoxia-inhNO	Nil	n = 4
E	Nil	Hypoxia	Hypoxia-ivNTG	Nil	n = 4
F	Nil	PE	PE -inhNO	Nil	n = 4
G	Nil	PE	PE -ivNTG	Nil	n = 4

The renal hemodynamics and UF were influenced by time (stage IV different from the three precedent in *Effector A*). InhNO (*B*) (Table 2) increased UF (+70%), RBFc (+119%) and GFR (+92%; Fig 1), besides during hypoxia (*C*), UF and GFR remained constant while RBFc decreased of 31% ($P_1 = 0.0009$). In animals with hypoxia and phenylephrine-induced hypertensions, inhNO decreased the pulmonary artery pressure ((*D*)-30% and (*F*)-15%) and increased the UF ((*D*)+41%, $P_1 =$ 0.001 and (*F*)+128%, $P_1 = 0.0004$) almost to the same extend as ivNTG ((*E*)+11%, $P_1 = 0.01$ and (*G*)+153%, $P_1 = 0.0005$). In presence of hypoxia-induced hypertension, the two treatments reversed the hypoxic decrease of RBFc and increased significantly GFR ($P_2 = 0.04$). Only ivNTG decreased the systemic arterial pressure ((*E*)-16% and (*G*)-40%).

Abbreviations used: ivNTG = intravenous nitroglycerine; NO = nitric oxide; inhNO = inhaled nitric oxide; UF = urinary flow; GFR = glomerular filtration rate; RBFc = renal blood flow corrected for PAH extraction; ECG = electrocardiogram; SBP = systemic blood pressure; PAP = pulmonary artery pressure; CVP = central venous pressure; PWP = pulmonary wedge pressure; HR = heart rate; CO = cardiac output; Hte = hematocrite; PAH = para-amino hippuric acid.

Table 2. Stage III compared to stage I (in percentage)

The statistical significative differences are noticed in bold.

	control	inhNO	ivNTG	Hyp-inhNO	Hyp-ivNTG	PE-inhNO	PE-ivNTG
RBFc	96%	**215%**	**65%**	78%	54%	**48%**	**75%**
GFR	99%	**191%**	96%	**115%**	**115%**	110%	109%
UF	120%	**190%**	112%	**153%**	**123%**	**240%**	**265%**

Renal hemodynamics and especially GFR were influenced by inhNO indicating the presence of extrapulmonary effects. NO is known to influence GFR through vasodilation of afferent and efferent glomerular arteries[8]. Hemoglobin, nitrosothiols[9,10], dinitrosyl iron complexes[9,11,12] are potent NO carriers: inhNO is not likely to raise free plasma NO. Renal hemodynamics may be influenced during inhNO administration by NO metabolites (*e.g.* NO_2 and NO_3) or by local NO delivery through reversing of nitrosothiols formation (albumin[13]). In relation with the instantaneous response (application and recovery) to inhNO, we can postulate that the effector inhNO on pulmonary circulation may be accompanied by non selective extrapulmonary effects due to local delivery of free NO on peripheral vascular territories such as the renal bed.

1. Vlahakes, G.J., Turley, K. and Hoffman, J.I.E. (1981) Circulation. **63**, 87-95.
2. Pearl, R.G. (1995) W.J.M. **162**, 52-53.
3. Channick, R.N., Newhart, J.W., Jonhson, FW., Moser, K.M. (1994) Chest. **105**, 1842-1847

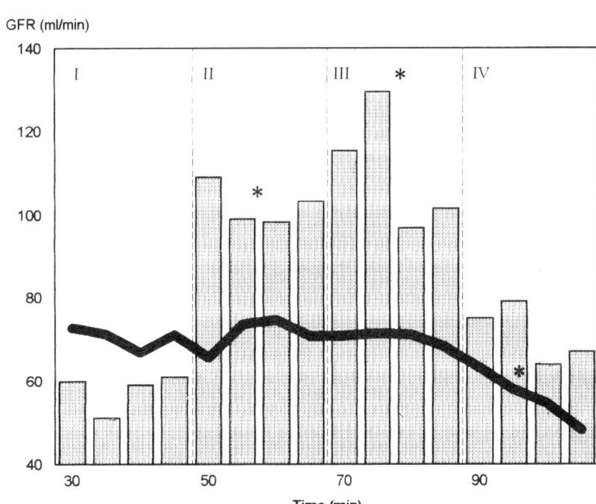

Fig 1. inhNO influences GFR
Dark ligne: control group (*A*); Clear columns: inhNO group (*B*); *stage II, III or IV statistically different from stage I.

4. Zappol, W.M., Falke, H.J., Hurford, W.E. and Roberts, J.D.Jr. (1994) Chest Suppl. **105**, 87S-91S.
5. Pepke-Zaba, J., Higenbottam, T.W., Dinh-Xuan, A.T., Stone, D. and Wallwork, J. (1991) Lancet. **338**, 1173-1174.
6. Rich, G.F., Murphy, G.D.Jr., Roos, C.M. and Johns, R.A. (1993) Anesthesiology. **78**, 1028-1035.
7. Troncy, E., Francoeur, M., Salazkin, I. and Blaise, G. (1995) Can.J.Anæsth. Suppl. **42**(5, II), A65B.
8. Ito, S. (1995) Cur. Op. Nephr. Hyp. **4**, 23-30.
9. Kuo, P.C. and Shroeder, R.A. (1995) Ann.Surg. **221**, 220-235.
10. Stamler, J.S., Jaraki, O., Osborne, J.A., et al. (1992) Proc.Natl.Acad.Sci.USA. **89**, 7674-7677.
11. Vědérnikov, Y.P., Mordvintcev, P.I., Malenkova, I.V. and Vanin, A.F. (1992) Eur.J.Pharmacol. **211**, 313-317.
12. Mulsch, A., Mordvintcev, P.I., Vanin, A.F. and Busse, R. (1993) Biochem.Biophys.Res. Commun. **196**, 1303-1313.
13. Kaufmann, M.A., Castelli, I., Pargger, H. and Drop, L.J. (1995) J.Pharmacol.Exp.Ther. **273**, 855-862.

Nitric oxide (NO) prevents platelet consumption and platelet activation during Extra Corporeal Life Support (ECLS)

KARIN MELLGREN, LARS-GÖRAN FRIBERG, HANS WADENVIK* GÖSTA MELLGREN, ANN-SOFIE PETERSON# and ÅKE WENNMALM#.

From the Department of Pediatric Surgery, Östra Hospital, S-416 85, Göteborg, and Departments of Medicin* and Clinical Physiology#, Sahlgrenska Hospital, S-413 45, Göteborg, Sweden.

INTRODUCTION

Extracorporeal circulation activates various cascade reactions in the blood, resulting in an increased bleeding tendency. Hemorrhagic complications are experienced in 10-35% of neonates treated with extra corporeal life support (ECLS) [1]. Different drugs have been used to prevent the ECLS-induced platelet lesions, e.g. dipyramidole [2], prostacyclin [3] or combinations of various drugs [4].

Nitric oxide (NO) donors, e.g. nitroglycerin, are known to inhibit platelet aggregation [5] as well as platelet adhesion to endothelial cells [6,7]. Treatment with organic nitrates and inhalation of nitric oxide prolongs the bleeding time in animals [8] and humans [9], indicating that inhaled NO has a significant effect on the circulating platelets.

The aim of the present study was to evaluate the effect of NO, added to the oxygenator sweep gases, on the ECLS-induced platelet consumption and activation.

METHODS

Experimental design

Two identical in vitro ECLS circuits were primed with fresh, heparinized human blood (one liter in each circuit) and circulated for 24 hours. NO was added as a gas (15, 40 and 75 ppm, respectively), directly to the oxygenator (Dideco D-901 hollow fiber oxygenator, Dideco, Italy) in one of the circuits; the other served as a control. The perfusion flow was maintained at 0.6 L/min employing roller pumps (Sarn Inc., USA). The oxygenators were supplied with air supplemented with 5% carbon dioxide. Blood gases, electrolytes and glucose were followed and kept within physiological limits. In total eight paired experiments were performed. Thirty ml blood samples were withdrawn from each circuit before start and at 0.5, 1, 3, 12 and 24 hours of perfusion.

Platelet membrane glycoproteins

5 ml blood was collected into Diatube H collecting tubes (Diagnostica Stago, Belgium) containing a platelet inhibitor cocktail provided by the manufacturer. The anticoagulated blood was centrifuged and a platelet-rich plasma (PRP) prepared. Platelet membrane GPIb was analysed using flow cytometry as described elsewhere [10 .

Plasma betathromboglobulin (BTG)

5 ml blood was collected into Diatube H collecting tubes (Diagnostica Stago, Belgium) and incubated on ice for 15 minutes. Platelet-poor plasma (PPP) was prepared and plasma concentration of BTG was analysed using a commercially available ELISA (Asserachrom, Diagnostica Stago, Belgium).

Platelet serotonin content

Platelet serotonin content was analysed by using high-performance liquid chromatography (HPLC).

Plasma cGMP and cAMP

Determination of plasma concentrations of cGMP and cAMP was made by radioimmunoassay, as described elsewhere [10].

Plasma nitrate

Plasma nitrate was determined with a stable isotope dilution assay, utilising gas chromatography/mass spectrometry, as described elsewhere [11.

RESULTS

A marked increase in plasma nitrate concentration was seen in the NO circuits, while the concentration remained stable in the control circuits during the 24 hours experiment.

Plasma cGMP concentration decreased over time in the control circuits. Conversely, in the NO circuits an increase in plasma cGMP was observed during the first three hours of perfusion. The difference in plasma cGMP between the NO and the control circuits was statistically significant. As regards the cAMP concentration no statistically significant difference was seen between the NO and the control circuits.

A marked decline in mean platelet count was observed in the control circuits. A similar decline, but less pronounced, was also seen in the NO circuits. However, statistically significant lower plasma BTG levels were observed in the NO circuits compared to the control circuits.

No statistically significant difference in the platelet membrane GPIb expression was observed between the two groups.

DISCUSSION

During extra corporeal circulation platelets adhere to the artificial surfaces, become activated and release their granula content. Previous studies have shown that even at almost normal platelet counts, a large percentage of the circulating platelets have lost important membrane receptors [9,11]. As a consequence their hemostatic function is impaired. This is believed to account for an important part of the incresed bleeding tendency seen in patients treated with extra corporeal circulation [12].

The present study shows that NO, administered as a gas to the oxygenator in an in vitro ECLS system, preserves platelet count and decreases platelet activation. NO can be added as a gas to the oxygenator and, due to its short half-life, it is conceivable to assume that it only acts locally in the extra corporeal device.

To our knowledge this is the first study in which NO, administered as a gas, has been used to inhibit platelet activation during extra corporeal circulation. We conclude that NO might be useful to reduce the platelet exhaustion, and thereby to reduce the frequency of bleeding complications encountered after extra corporeal circulation. Clinical studies are, however, requested to evaluate the possible benificial effect of such a therapy.

REFERENCES

1. Zwichenberger JB, Nguyen TT, Upp JR et al. (1994) J Thorac Cardiovasc Surg;**107**,838-49
2. Toeh KH, Christasis GT, Weisel RD et al. (1988) J Thorac Cardiovasc Surg;**96**,332-41
3. Feddersen K. (1985) Prostacyclin infusion during cardiopulmonary bypass, Minab/Gotab, Kungälv.
4. Bernabei A, Gikakis N, Kowolska MA, Niewiarowski S, Edmunds LH Jr. (1995) Ann Thorac Surg;**59**:149-53
5. Hampton JR, Harrison AJ, Honour AJ, Mitchell JR. (1967) Cardiovasc Res;**1**:101-6
6. Radomski MW, Palmer RMJ, Moncada S. (1987) Lancet;**ii**:1057-8
8. Sneddon JM, Vane RR. (1988) Proc Natl Acad Sci USA;**85**:2800-4
9. Högman M, Frostell C, Arnberg H, Sandhagen B, Hedenstierna G. (1994) Acta Physiol Scand;**151**:125-9
10. Högman M, Frostell C, Arnberg H, Hedenstierna G. (1993) Lancet;**341**:1664-5
11. Mellgren K, Friberg LG, Hedner T, Mellgren G, Wadenvik H. (1995) Int J Artif Organs; in press
12. Wennmalm Å, Benthin G, Edlund A et al. (1993) Circ Res;**73**:1121-7
13. George JN, Pickett EB, Saucerman S, McEver RP, Kunicki TJ, Kieffer N, Newman PJ. (1986) J Clin Invest;**78**:340-8
14. Plötz FB, van Oeveren W, Bartlett RH, Wildevuur CRH. (1993) J Cardiovasc Surg;**105**:823-32

Effects of inhaled nitric oxide on right ventricular function after mitral valve replacement.

PATRICK M. DUPUY*, MARIE-CARMEN GOMEZ*, MICHEL WILKENING*, and MICHEL DAVID#

Departments of *CardioVascular Intensive Care, and #Surgery
Bocage University Hospital - F-21034 Dijon - France

Pulmonary hypertension is common in patients with mitral valve disease and is especially important after surgical repair of the valve where it appears to be exacerbated by cardiopulmonary bypass [1], and may cause right ventricular (RV) dysfunction. RV failure has been shown to be a factor limiting survival in the post-operative period of mitral valve replacement [2]. In humans, endogenous levels of nitric oxide (NO) maintain both systemic and pulmonary vascular beds in an active dilated state [3]. Cardiopulmonary bypass damages pulmonary endothelium and causes a failure of endothelium-dependent vasodilation [4]. Inhaled NO selectively induces pulmonary vasodilation without causing systemic hypotension in cardiac surgical patients [5]. The aim of the study was to examine whether inhaled NO could improve RV function after mitral valve replacement.

An open-labeled, prospective trial was undertaken enrolling 12 consecutive patients meeting the entry criteria: mean pulmonary artery pressure >25 mmHg, sinus rythm (ECG), and no significant tricuspid regurgitation (2D echo and pulsed Doppler examination). The sudy was approved by the University Ethics Committee, and informed consent was obtained. Five hours after completion of the surgical procedure, all patients were sedated and mechanically ventilated. Following parameters were measured: heart rate (HR), mean systemic arterial pressure (MAP) (radial artery catheter), central venous (CVP), pulmonary artery occluded (PAPo) and mean pulmonary arterial (MPAP) pressures (mmHg)(Swan-Ganz catheter, RVEF/Ox TD, Baxter-Edwards), and arterial oxygen blood tension (PaO_2)(mmHg), or calculated: sytemic (SVRI) and pulmonary (PVRI) vascular resistance index. ($dyn.sec.cm^{-5}.m^{-2}$). Cardiac index (CI)($l.min^{-1}.m^{-2}$) and RV ejection fraction (RVEF)(%) were estimated by thermodilution using 4 to 5 injections of 10 ml iced (<4°C) 5% dextrose solution in water, administered at random to the ventilatory cycle. RVEF was evaluated from the exponential analysis of the thermal curve using a computer (Explorer, Baxter-Edwards), and end-diastolic (EDVI) and end-systolic (ESVI)($ml.m^{-2}$) volume index were calculated. All curves were inspected for normal thermal decay profile. RVEF were calculated from runs in which HR was constant with no arrythmia. Baseline measurements were made (C1), and a cumulative dose-response was assessed with a stepwise increase in the rate of inhaled NO at 10, 20, 40 and 80 ppm (AGA medical) and at 20 min intervals. Twenty min after completion of the NO inhalation study, a second set of baseline measurements were made (C2). Infusion rate of vasoactive agents were not modified during the protocol. All data are reported as mean±SD. ANOVA (* $p<0.05$ versus C1).

Inhaled NO induced a dose dependent decrease of MPAP and PVRI (table 1). The onset was rapid, beginning within sec after starting NO inhalation, and was rapidly reversible. At the lowest NO inhaled NO concentration (10 ppm), MPAP significantly decreased (figure 1).

Table 1:

	C1	NO10	NO20	NO40	NO80	C2
CI	2.8±0.2	2.8±0.2	2.9±0.1	3.0±0.1	3.0±0.1	2.8±0.1
MAP	85±4	82±4	82±4	81±4	76±4	78±5
CVP	9±1	9±1	8±1	8±1	7±1	9±1
SVRI	691±63	664±52	633±43	630±58	590±57	597±58
MPAP	31±2	26±1*	24±1*	22±1*	21±1*	28±2
PAPo	15±1	15±1	14±1	13±1	12±1*	14±1
PVRI	141±18	108±13*	98±11*	89±12*	85±12*	129±13
EDVI	111±8	107±7*	101±5*	101±6*	100±6*	116±8
ESVI	81±7	76±6*	68±5*	71±4*	70±4*	85±6
RVEF	30±1	32±2*	33±1*	34±1*	34±2*	30±1
PaO_2	136±15	138±13	137±12	129±11	121±11	131±14

RV function parameters are given in table 1. RV volumes significantly increased at all doses of inhaled NO, in a ratio leading to an increased RVEF (figure 2).

Figure 1:

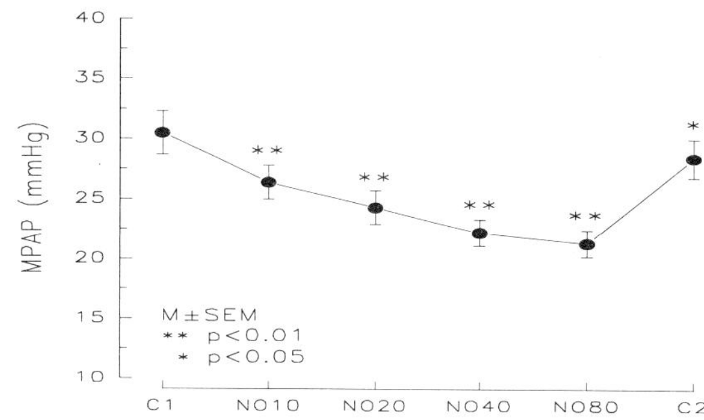

Among all patients, all had satisfactory echocardiograms, with tricuspid regurgitation below grade 2. Despite a trend toward decreased PaO_2 at the highest inhaled NO concentrations (40 and 80 ppm), no significant change in PaO_2 was observed.

Figure 2:

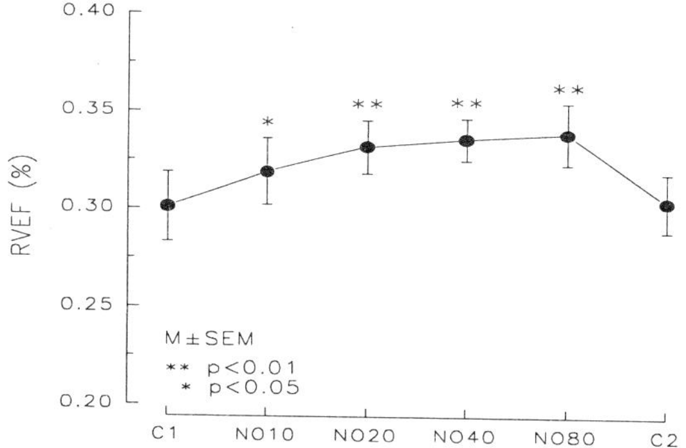

Inhaled NO improves the loading conditions of the RV without changing RV output and blood gas exchange in the postoperative period of mitral valve replacement. This effect is dose-dependent. As a consequence of decreased RV dilation, a decrease in PAPo suggests that NO inhalation improves left ventricular filling. The selective pulmonary vasodilator effect of inhaled NO is consistent with previous reports in cardiac surgical patients.

References:

1- Heath, D., Edwards, J.E. (1958) Circulation 18, 533-544. **2-** Wheller, J., George, B.L., Mudler, D.G., Jarmakani, J.M. (1979) Circulation 70, 1640-1644. **3-** Stamler, J.S., Loh, E., Roddy, M.A., Currie, K.E., Creager, M.A. (1994) Circulation 89, 2035-2040. **4-** Wessel, D.L., Adatia, I., Giglia, T.M., Thompson, J.E., Kulik, T.J. (1993) Circulation 88, 2128-2138. **5-** Rich, G.F., Murphy, G.D., Roos, C.M., Johns, R.A. (1993) Anesthesiology 78, 1028-1035.

Inhaled Nitric Oxide in Postoperative Congenital Heart Disease

I. Schulze-Neick, F. Uhlemann, M. Bültmann, B. Stiller, N. Haas, I. Dähnert, H. Werner, R. Rossaint*, R. Hetzer, V. Alexi-Meskishvili, and P.E. Lange

German Heart Institute Berlin, and *Department of Anaesthesiology and Operative Intensive Care Medicine, Virchow Clinic, Humboldt Universtity of Berlin, 13353 Berlin, Germany

Introduction:

Pioneering studies from Frostell in animals [1] and Rossaint in patients with ARDS [2] have demonstrated that inhaled nitric oxide (iNO) effectively lowers pulmonary hypertension (PHT) and improves ventilation-perfusion mismatch by vasodilating ventilated areas. We used this effect of vasodilation to treat PHT in patients after cardiac operation or to decrease right ventricular afterload in patients with right ventricular dysfunction.

Methods and Patients:

After cardiac operation, complete hemodynamic and pulmonary gas exchange monitoring was installed. Measurements of cardiac performance included evaluation of right and left ventricular function by transoesophageal echocardiography, measurements of transthoracic echocardiographic flow velocity integral as an indirect measurement of cardiac output, and determination of cardiac output by conventional thermodilution technique.
Indications for NO-application included >33% systemic pressure in the pulmonary artery, right ventricular function as demonstrated by echocardiography or elevated right atrial filling pressures (>10mmHg), and an increased oxygenation index (FiO2*100*mean airway pressure/pO2[mmHg]) >10 .
NO was applied using a specially designed equipment for pediatric and adult patients (Messer-Griesheim, Germany/Austria). It included continuous measurement of concentrations of nitric oxide and nitric dioxide. Concentrations of 2-50 ppm were used. NO was given synchronous with the respiratory cycle by means of a built-in mass flow controller, thus enabling a low variability of the NO concentration applied. Methhemoglobin levels were determined 1-3times daily. Assessment of NO efficacy consisted of on-off-on measurements according to the clinical stability of the patient including all the above mentioned monitored parameters of hemodynamic and pulmonary function and transesophageal echocardiographic ventricular function.

Results:

Application of NO occured in 23 patients and its efficacy was completely examined in 32 situations. It was applied 0-628h (26.1 days) postoperatively. There was improvement of oxygenation (135±34 to 155±42 pO2 (mmHg), $p<0.02$), decrease of pulmonary artery pressure (41±7 to 33±8 mmHg PA pressure, $p<0.0001$) and improvement in cardiac output (3.2±0.25 to 3.6±0.32 L/min*BSA, $p<0.025$) was shown. In 8 patients (n=4 immediately postoperative, and n=4 with sepsis and multi-organ failure), no NO response could be elicited. In 7 other patients, no reduction of pulmonary artery pressure occured, but echocardiographic examination of right ventricular function or cardiac output measurements showed an improvement (increase in stroke volume 42±13%, increase in ejection fraction 16±7%), while heart rate remained constant.

Conclusions:

For a special group of patients after cardiac operation, NO has become an important part in the clinical postoperative care. The reduction of right ventricular afterload by NO's selective effect on the pulmonary vascular bed can improve depressed right ventricular function in the postoperative period. Thus, NO-application can be justified in right ventricular failure even when pulmonary artery pressure seems to be unresponsive.

1. Frostell C, Fratacci M-D, Wain JC, Jones R, Zapol WM. Inhaled Nitric Oxide. A Selective Pulmonary Vasodilator Reversing Hypoxic Pulmonary Vasoconstriction. Circulation 1991;83:2038-2047.

2. Rossaint R, Falke KJ, Lopez F, Slama K, Pison U, Zapol WM. Inhaled nitric oxide for the adult respiratory distress syndrome. N Engl J Med 1993;238:399-405.

Glyceryl trinitrate in the treatment of preterm labor - another placebo?

JANE E NORMAN, LINDA M WARD, *WILLIAM MARTIN, NICHOLAS S MACKLON, ALAN D CAMERON, *MARGARET R MACLEAN, *JOHN C MCGRATH, IAN A GREER, IAIN T CAMERON.

Department of Obstetrics and Gynaecology and *Institute of Biological and Life Sciences, University of Glasgow, G12 8QQ, UK.

Introduction and aims

Prematurity is responsible for a significant proportion of neonatal handicap and neonatal death in developed countries. Thirty percent of preterm births are due to idiopathic preterm labor, yet there are no effective therapeutic agents to inhibit uterine contractions and thus prevent preterm delivery in this condition. Nitric oxide (NO) is a myometrial relaxant, and a recent uncontrolled study has suggested that glyceryl trinitrate (GTN) might be useful in the treatment of preterm labor[1]. The aim of the present study was to quantify the effect of GTN on human myometrial contractility *in vivo* and *in vitro*.

Methods

In vivo, twelve women undergoing medical termination of pregnancy between 12 and 16 weeks gestation were recruited into the study and randomised to a "treatment" or "placebo" group. Intrauterine pressure recordings were made 48hrs after administration of 200mg mifepristone orally. In the "treatment" group, a 15 minute infusion of glyceryl trinitrate (80μg/ml) was administered into the antecubital vein at a rate of 0.25ml/minute, giving a dose of 20μg GTN/minute. Women in the "placebo" group, received an infusion of normal saline at the same rate.

In vitro, studies were performed on myometrial strips obtained from women undergoing lower uterine caesarean section prior to the onset of labor at term. After an equilibration period of 2 hours the contractile response of the tissue was recorded in response to increasing doses of GTN. Both studies had local ethics committee approval and informed consent was obtained from all women.

Results

In vivo, an infusion of 20μg GTN / minute in patients undergoing termination of pregnancy at 12-16 weeks gestation had no significant inhibitory effect on mifepristone induced myometrial contractions, compared with placebo infusion of normal saline (figure 1 and 2).

In vitro, increasing concentrations of GTN caused a progressive inhibition in the amplitude of spontaneous myometrial contractions (figure 3). A 40% inhibition in the amplitude of myometrial contractions was effected by 10^{-4}M GTN.

Conclusions

- we have failed to find a consistent effect of GTN on myometrial contractility *in vivo* in the second trimester
- *in vitro*, GTN was applied in concentrations of 10^{-4}M in order to suppress the amplitude of spontaneous contractions in isolated strips of term myometrium by 40%
- although the administration of nitric oxide might provide a novel mechanism for the treatment of preterm labor, the currently available nitric oxide donors are unlikely to be clinically useful in this setting unless the sensitivity of myometrium to nitric oxide can be increased

Figure 1
The effect of a placebo infusion (normal saline) and glyceryl trinitrate (GTN) infusion (20μg / minute) on myometrial activity in Montevideo units expressed as a mean (SE) percentage of the baseline activity

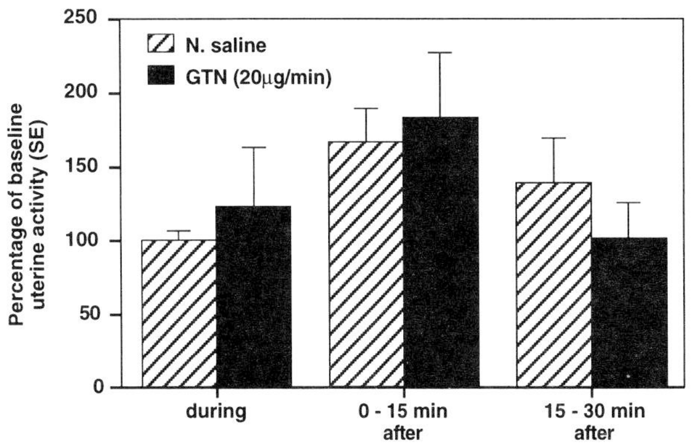

Figure 2
The effect of glyceryl trinitrate on the frequency and amplitude of myometrial contractions *in vitro*. Increasing concentrations of glyceryl trinitrate (GTN) (10^{-7}M to 10^{-4}M) were administered as shown. Contractions in an untreated strip are shown as control.

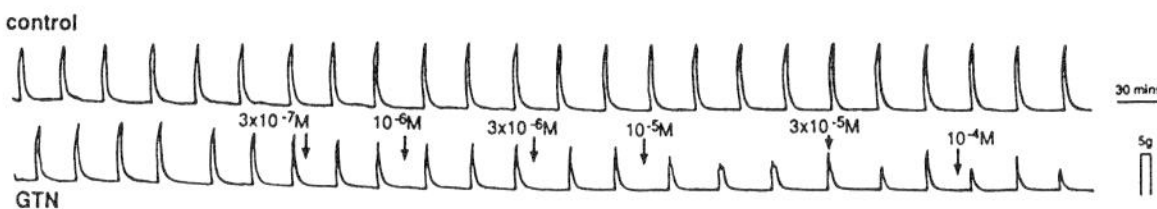

Figure 3
Dose response curve for the inhibition of amplitude of myometrial contractions in response to glyceryl trinitrate (GTN).

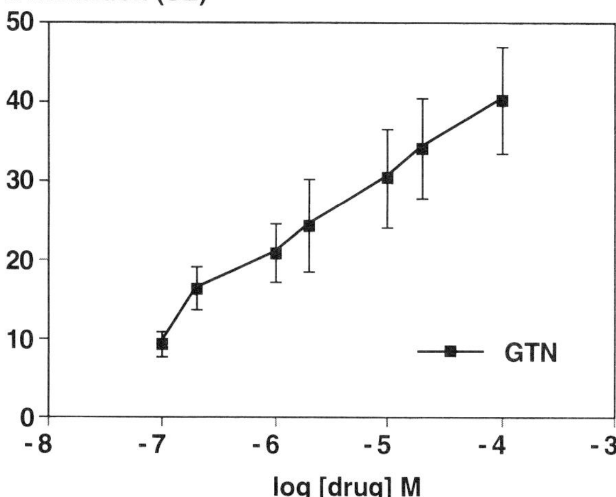

Acknowledgement
This work was supported by grants from Tenovus and from the Yorkhill NHS Trust for which we are very grateful.

Improved survival in a pig model of abdominal sepsis by treatment with N^G-monomethyl-L-arginine.

Øystein A. Strand*, Anna E. Leone#, Karl-Erik Giercksky+ and Knut A. Kirkebøen.•

*Dept. Inf. Dis., Ullevål Hospital, N-0407 Oslo, Norway. #Glaxo-Wellcome Research, Beckenham, Kent BR 3 3BS, England +Dept. Surg. Oncol., National Cancer Hospital, N-0310, Oslo, Norway. •Inst. Exp. Med. Res., Ullevål Hospital, N-0407 Oslo, Norway.

During normal physiological conditions nitric oxide (NO) is produced continuously in the endothelium by the constitutive isoform of the NO-synthase. Immunological and inflammatory stimuli have been shown to induce an isoenzyme of nitric oxide synthase (INOS) in numerous tissues. NO produced in pathophysiological amounts by INOS is proposed to be of importance for myocardial depression, hypotension and vascular hyporeactivity during septic shock [1,2]. Treatment with inhibitors of the NO synthase has been shown to reverse hypotension and vascular hyporesponsiveness induced by LPS-infusion in several species [3,4]. Although inhibitors of the NO synthase are able to reverse some of the hemodynamic effects of LPS-infusion, studies using LPS have serious limitations as analogues to human sepsis.

In this study we tested if a continuous infusion of the NO synthase inhibitor N^G-monomethyl-L-arginine (LNMMA) improves hemodynamics and survival in a pig model of severe endogenous peritoneal sepsis. This model was chosen as it closely mimicks severe abdominal sepsis seen in patients after intestinal leaks. We specifically investigated if treatment with LNMMA reduces coronary blood flow and promotes myocardial ischemia, as suggested by Avontuur et al. [5].

ANIMAL PREPARATION: Open-chest pentobarbital anesthetized pigs. Aortic flow was measured by an electromagnetic flowmeter. Coronary blood flow (CBF) was measured by a Transonic transit time flow probe on the mid-LAD. Pulmonary and right atrial pressures were measured by a catheter introduced via the right jugular vein. A femoral artery was cannulated for pressure recordings and blood sampling. A femoral vein was cannulated for infusions and blood sampling. Urine was sampled through a cystostoma. Cecal content was obtained through an enterotomy. Sepsis was induced after 45 minutes of postsurgical stabilisation. Half a gram of cecal content pr kg body weight (BW) was diluted in 150 ml of saline and installed intraperitoneally at time zero. Intravenous fluid was given as a continuous infusion of Ringer lactate in a dose of 20 ml/kg BW/hour during surgery and 15 ml/kg BW/hour for the rest of the experiment. Animals were observed for 9 hours after surgery or until death.

Three groups of pigs were used (fig. 1). Group A (n=6): Time control. After completion of surgery animals were given 15 ml of Ringer lactate/kg BW/hour, and observed for 9 hours. Group B (n=9): Peritoneal sepsis. Sepsis was induced by intraperitoneal fecal administration (IPFA). Fluid treatment was given as in group A. Animals were observed for 9 hours or until death. Group C (n=6): Peritoneal sepsis + LNMMA. Sepsis was induced by IPFA and fluids were given as in group A and B. Three hours after IPFA, a continuous infusion of LNMMA (10 mg/kg BW/hour) was given for six hours or until death. Hemodynamic variables and specimens for blood and urine analysis were sampled hourly. Plasma nitrate was measured by capillary electrophoreses [6].

RESULTS: During the first hour of LNMMA infusion MAP rose from 80 to 93 mmHg ($p<0.01$) and pulmonary artery pressure (PAP) rose from 14,5 to 19 mmHg ($p<0.05$). During the observation period of 9 hours the LNMMA treated septic animals showed significantly increased survival compared to non-treated septic animals. Only 1 of 9 animals in the non-treated septic group survived the observation period of 9 hours, while 5 of 6 LNMMA treated septic animals survived for 9 hours. Sepsis survival increased from 9 % in untreated animals to 83 % in animals treated with LNMMA ($p<0,01$). PAP rose gradually during the LNMMA infusion and was significantly elevated at the end of the experiment in the LNMMA treated animals, 27±7 vs. controls 17±4 mmHg ($p=0,016$). In the LNMMA treated animals MAP, CBF and urinary production were preserved during the experiment and values did not differ significantly from time controls. Normalised plasma nitrate at T0 and T9/ time of death were for group A: 100 / 57±8 ($p<0,05$ vs. T0), Group B: 100 / 92±15 (NS v.s. T0), and for Group C 100 / 55±7 ($p<0.05$ vs.T0).

CONCLUSION: A continuous infusion of the NO synthase inhibitor LNMMA at a dose of 10 mg/kg BW/ hour improves survival and hemodynamics in a pig model of severe endogenous peritoneal sepsis. Coronary blood flow was not reduced during infusion of LNMMA. Plasma nitrate fell to control levels in the LNMMA treated group during treatment.

REFERENCES

1. Moncada S., Palmer R.M.J. and Higgs E.A. Nitric Oxide: Physiology, pathophysiology and pharmacology. (1991) Pharmacol. Rev. **43:** 109-142.

2. Schulz R., Nava E. and Moncada S. Induction and potential biological relevance of a Ca^{2+}-independent nitric oxide synthase in the myocardium. (1992) Br. J. Pharmacol. **105:** 575-580.

3. Kilbourne R.G., Jubran A., Gross S.S., Griffith O.W., Levi R., Adams J. and Lodato R.F. Reversal of endotoxin mediated shock by N-g-monomethyl-arginine, an inhibitor of nitric oxide synthesis. (1990) Biochem. Biophys. Res. Commun. **172:** 1132-1138.

4. Thiemermann C. and Vane J. Inhibition of nitric oxide synthesis reduces the hypotension induced by bacterial lipopolysaccaride in the rat in vivo. (1990) Eur. J. Pharmacol. **182:** 591-595.

5. Avontuur J.A.M., Bruining H.A. and Ince C. Inhibition of nitric oxide synthesis causes myocardial ischemia in endotoxemic rats. (1995) Circ. Res. **75:** 418-425.

6. Leone A., Francis P.L., Rhodes P. and Moncada S. A rapid and simple method for the measurement of nitrite and nitrate in plasma by high performance capillary electrophoreses. (1994) Biochem. Biophys. Res. Comm. **200 (2):** 951-957.

Effects of aminoguanidine on multiple organ failure elicited by lipoteichoic acid and peptidoglycan in anaesthetised rats

SJEF J. DE KIMPE, MURALITHARAN KENGATHARAN, CHRISTOPH THIEMERMANN and JOHN R VANE

The William Harvey Research Institute, St Bartholomew's Hospital, Medical College, Charterhouse Square, London EC1M 6BQ, United Kingdom

Gram-positive as well as Gram-negative organisms can cause septic shock. However, Gram-positive organisms cause septic shock without the involvement of lipopolysaccharide (LPS) [1]. The cell walls of Gram-positive organisms contain various components, such as lipoteichoic acid (LTA) and peptidoglycan (PepG) which can produce an inflammatory response [1-4]. We have previously shown that a high dose of LTA elicits TNF-α release, induction of nitric oxide synthase (iNOS) and circulatory failure in anaesthetised rats [2]. However, in contrast to LPS, LTA or PepG does not cause lethality or organ toxicity at an immunostimulatory dose in mice or rats [3-4]. Here, we demonstrate that LTA and PepG (from *Staphylococcus aureus*) synergise to cause systemic release of TNF-α and IFN-γ, induction of iNOS, circulatory and respiratory failure.

Male Wistar rats (200-325g) were anaesthetised with thiopentobarbitone sodium (120mg/kg, ip). The trachea was cannulated to facilitate spontaneous respiration, the carotid artery to monitor mean arterial blood pressure (MAP) and heart rate, and the jugular vein to administer compounds. Respiratory function was assessed by the determination of partial oxygen pressure (P_AO_2) in carotid blood. Plasma levels of TNF-α and IFN-γ were measured by ELISA. Induction of iNOS activity was determined by the conversion of [^{3}H]L-arginine to [^{3}H]L-citrulline in lung homogenates. Statistical evaluation was performed using Bonferoni's *t* test for multiple comparison of single means.

Table 1. Inflammatory response to PepG & LTA
Each value represents the mean ± S.E.M. of 5-8 rats.

	TNF-α at 90 min (ng/ml)	IFN-γ at 360 min (ng/ml)	iNOS activity at 360 min (pmol/min/mg)
Vehicle control	<0.035	0.1±0.1	0.3 ± 0.1
LTA (3 mg/kg)	11.7±3.0	0.2±0.1	3.2 ± 0.9
PepG (10 mg/kg)	7.9±2.3	0.1±0.1	2.5 ± 0.4
LTA & PepG	39.1±4.3	8.9±2.2	22.0 ± 4.0

Intravenous injection of LTA (3 mg/kg) or PepG (10 mg/kg) alone resulted in an increase in (i) plasma concentration of TNF-α at 90 min (P<0.05) and (ii) iNOS activity in lungs obtained at 360 min (P<0.05) after the administration of bacterial components (Table 1). However, plasma levels of IFN-γ were not elevated. Combined administration of LTA (3 mg/kg) and PepG (10 mg/kg) caused a 4-6 fold increase in the plasma levels of TNF-α (P<0.05) and iNOS activity in lung (P<0.05) compared to administration of the cell wall components alone (Table 1). Furthermore, the combined treatment also caused the release of IFN-γ (P<0.05).

Injection of LTA (3 mg/kg) and PepG (10 mg/kg) caused a delayed hypotension at 360 min (Table 2). Although PepG or LTA alone decreased MAP, the hypotension was less pronounced than that caused by the combination (Table 2). Neither PepG nor LTA caused a significant alteration in heart rate, while the combination of PepG and LTA caused tachycardia (Table 2). Furthermore, co-administration of PepG and LTA decreased the pressor response to noradrenaline (1 μg/kg, iv) at 360 min by 84±2% from control (time 0), while the pressor response at 360 min was not significantly influenced after administration of LTA (-2±17%) or PepG (0±16%) alone. Injection of both PepG and LTA also resulted in a significant decrease in P_AO_2 at 360 min indicating a respiratory failure (Table 2).

Treatment of rats with the iNOS inhibitor aminoguanidine (AMG; 5mg/kg iv, 30 min prior to the injection of PepG and LTA, followed by an intravenous infusion of 10mg/kg/h until the end of the experiment) inhibited the hypotension, the hyporeactivity to noradrenaline (23±19% vs 84±2% for vehicle; P<0.05) and the decrease in P_AO_2 at 360 min after administration of the bacterial cell wall components (Table 2). The iNOS activity in lungs treated with aminoguanidine was decreased from 22.0±4.0 to 6.7±1.3 pmol/min/mg protein (P<0.05).

Table 2. PepG & LTA cause circulatory and respiratory failure
Each value represents the mean ± S.E.M. at 360 min of 5-8 rats.
† P<0.05 vs vehicle control and # P<0.05 vs LTA&PepG.

	MAP (mmHg)	heart rate (beats/min)	P_AO_2 (mmHg)
Vehicle control	113±4	405±10	78±2
LTA (3 mg/kg)	97±6	426±15	74±5
PepG (10 mg/kg)	86±5†	407±13	72±3
LTA & PepG	73±4†	496± 7†	54±4 †
+ AMG (10 mg/kg/h)	107±7#	414±19#	69±4 #

Thus, co-administration of two major cell wall components from *Staphylococcus aureus*, LTA and PepG, results in delayed hyperdynamic circulatory (hypotension, vascular hyporeactivity and tachycardia) and respiratory failure. How Gram- positive organisms initiate sepsis is not yet clear. Plasma levels of TNF-α, IL-1 and IL-6 are increased in human septic shock [5]. Indeed, administration of both LTA and PepG resulted in the release of TNF-α and IFN-γ. These cytokines cause the expression of iNOS in a wide variety of cells including macrophages and vascular smooth muscle cells. Interestingly, plasma levels of nitrite and nitrate are increased in patients with septic shock [6]. The present results show a pronounced increase in iNOS activity in lungs obtained from rats treated with the combination of LTA and PepG. The effect of combined administration of PepG and LTA on the release of cytokines and induction of iNOS is significantly greater than the expected additive effect of the two components alone. These results suggest that PepG and LTA synergise via different pathways in inducing an inflammatory response. Nitric oxide (NO) is a powerful vasodilator, binds to mitochondrial enzymes inhibiting cellular respiration, and reacts with superoxide anion to form peroxynitrite resulting in nitrosylation of proteins or formation of highly reactive hydroxyl radicals [7]. Inhibition of iNOS activity by aminoguanidine prevented the delayed circulatory failure and attenuated the respiratory failure elicited by LTA and PepG.

Thus, the joint activity of LTA and PepG, two cell wall components common to Gram-positive organisms, may explain the initiation of Gram-positive septic shock in general. Furthermore, the present results support the hypothesis that the vasodilator and cytotoxic actions of NO following the induction of iNOS is a general pathway in the pathophysiology of septic shock irrespective of the nature of the microorganism.

This study is partially supported by Cassella AG (Germany). SJDK is a recipient of a fellowship of the European Union.

1. Bone, R. C. (1994) Arch. Intern. Med. **154**, 26-34.
2. De Kimpe, S. J., Hunter, M. L., Bryant, C. E., Thiemermann, C. & Vane, J. R. (1995) Br. J. Pharmacol. **114**, 1317-1323.
3. Sverko, V., Marotti, T., Gavella, M., Lipovac, V. & Hrsak, I. (1994) Immunopharmacology **28**, 193-199.
4. Takada, H., Kawabata, Y., Arakaki, R., et al. (1995) Inf. Immun. **63**, 57-65.
5. Casey, L. C., Balk, R. A. & Bone, R. C. (1993) Ann. Intern. Med. **119**, 771-778.
6. Ochoa, J. B., Udeko, A. O., Billiar, T. R., et al. (1991) Ann. Sur. **214**, 621-626.
7. Wizemann, R. M., Gardner, C. R., Laskin, J. D., et al. (1994) J. Leukocyte Biol. **56**, 759-768.

Abbreviations used: iNOS: inducible nitric oxide synthase; IFN-γ: interferon-γ; LPS: lipopolysaccharide; LTA: lipoteichoic acid; MAP: mean arterial pressure; NO: nitric oxide; P_AO_2: partial arterial oxygen pressure; PepG: peptidoglycan; TNF-α: tumour necrosis factor-α

Endoscopic measurements of localized gaseous nitric oxide in the inflammatory bowel disease (IBD): A clinical application

Jun Iwamoto[1] Toru Kono, Kazunori Kamiya, Akitoshi Kakisaka and Shin-ichi Kasai

Department of Surgery and Department of Physiology[1] Asahikawa Medical College, Asahikawa, 078 Japan.

Nitric oxide (NO) has been detected in the luminal gas obtained from the colon of six ulcerative colitis (UC) patients (1). Although this was the first report, overall methodology and data were not entirely detailed. In addition, usage of the environmental NO analyzer in their study seems to be insufficient for a quantitative measurement of NO. Measurement of the luminal gas requires quantitative collection of sample gases, since ventilatory technique is inevitable for gas collection.

We attempted a quantitative measurement of luminal NO in the IBD patients (UC=4, Crohn's disease=2, control=12) during colonoscopy. Having started to establish the gas collection technique, we finally set a twenty-seconds stay-time for the infused air (100 ml, NO-free) after repeated washout of the colonal lumen with compressed air. A 10 ml of sample gas was drawn by an air-tight syringe after a 20 sec stay-time. This procedure was performed separately in each colon segment (ascending, transverse, descending colon, and rectum) followed by a quick measurement of NO by chemiluminescence (Model 270B, Sievers). This NO analyzer allows fine measurement for small aliquot of gas samples via fast transduction of photon signals and yields a real quantity of NO molecules. Thus it becomes possible that, by means of simple calculations, NO levels are expressed in either an original concentration in the air-tight syringe or a releasing rate of NO from the colon mucosa.

Figure 1. NO levels in the colon from healthy subjects

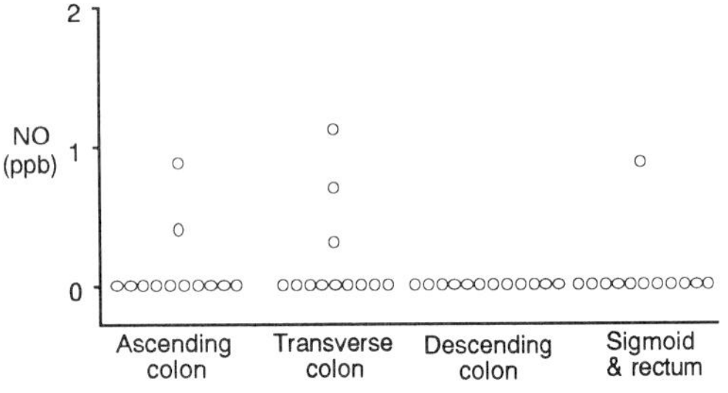

Figure 2. NO levels in the colon of UC and Crohn's disease

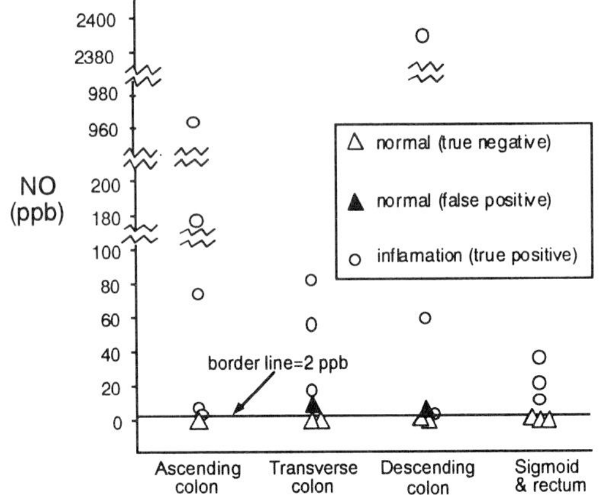

The NO levels in diseased segments varied from 12.4 to 2389.7 ppb (Fig. 2) whereas the NO levels in most of segments in the control subjects exhibited less than 1 ppb (Fig. 1). We took a 2 ppb of NO level as the critical value that could separate diseased segments from normal segments. Accordingly, the matching of NO levels vs endoscopic findings for each colon segment was highly paralleled among 24 sites (true positive=11, false positive=2, false negative=0, true negative=11). However, this critical level can be higher depending on the stay-time of NO-free gas in the colon.

Expression of NO levels in the GI tract needs to be considered, since the NO level as expressed in concentration (ppb) is often misleading unless the ventilation technique is fully described. The concentration of NO is totally dependent on the ventilated (diluted) air volume. The present study demonstrates that one of the colonal lumens of UC which was washed and deflated can acquire NO gas to become a space (100 cm^3) filled with a high concentration NO gas at 2.4 parts per million (2400 ppb) within 20 seconds. Here, the net amount of NO is 240 nl and the NO excretion rate (or VNO) is estimated 720nl/min. It is surprizing that such small area of colon can excrete NO gas at high rate. Indeed, this value exceeds the total NO output from human lung which ranges from100 to 300 nl/min (2)(3).

Our technique confirms application of the quantitative measurement of the luminal NO in localized colonal segments in IBD patients. The technique simply consists of " flush and fill" of NO-free air into the segment that is to be investigated. Probably the amount of filling air needs to be considered depending on the size and structure of the lumen. If the colonal NO level is directly related the magnitude of inflammation, the quantitative measurement of luminal NO in the diseased colon may be a good tool for assessment of IBD.

Acknowledgement : We thank to Drs. Y. Saito and T. Ayabe for superb skill of endoscopy. This work is supported by research grants 07557185 and 07670077 from the Japanese Ministry of Art and Education.

References

1. Lundberg JON, Hellstrom PM, Lundberg JM, Alving K. Greatly increased luminal nitric oxide in ulcerative colitis. (1994) Lancet 344, 1673-1674.

2. Borland C, Cox Y, Higenbottam T. Measurement of exhaled nitric oxide in man. (1993) Thorax 48,1160-1162.

3. Iwamoto J, Pendergast DR, Suzuki H, Krasney JA. Effect of graded exercise on nitric oxide in expired air in humans. (1994) Respir. Physiol. 97, 333-345.

Growth stimulation and increase in nitric oxide synthase activity in oestrogen receptor-negative MDA 231 human breast tumour xenograft by oestrogen implants.

EDWIN C. CHINJE, SHIRLEY COLE, DEBBIE A. POCOCK, TERRY HACKER, PAULINE J. WOOD and IAN J. STRATFORD

Medical Research Council, Radiobiology Unit, Chilton, Didcot, Oxon, OX11 ORD, United Kingdom.

INTRODUCTION

Many *in vitro* studies suggest that oestrogen has a direct effect on proliferation of human breast cancer cell lines which contain oestrogen receptors (ER). The human breast cancer cell line MDA231 has been well characterised as ER-negative and often serves as a prototype for hormone-independent breast cancer that demonstrates oestrogen- independence. It has also been shown that these cells do not transcribe the ER gene[1]. Recent studies by other workers have demonstrated that oestrogen stimulates the growth of MDA231 tumours in immunodeficient mice but does not alter cell proliferation [2]. Oestrogen is also important in modulating NO synthase levels in certain tissues. We have therefore investigated the effect of oestrogen implants on the growth of ER-negative MDA231 cells in nude mice and on the expression of NO synthase activity in these tumour xenografts. Thus NO generation may be implicated in angiogenesis and hence tumour growth and progression.

MATERIALS AND METHODS

Animals and growth experiments *in vivo*

Female nu/nu mice (12-15 weeks old) bred in-house were used in the experiments. MDA231 tumours were maintained by continuous *in vivo* passage of 1-2mm tumour pieces. For experiment they were implanted subcutaneously into the mid dorsal region of the mice. Subcutaneous pellets of either placebo or 1.7mg β-oestradiol (60 day release, Innovative Research of America, U.S.A.) were implanted in the scruff of the neck two days before tumours were implanted. Tumours were measured in three orthogonal diameters at regular intervals once they became visible and continued until they were about 600 mm^3 in volume. Tumours were then excised and a piece snap frozen in liquid nitrogen for subsequent enzyme assays and the other portion processed for immunohistological analysis using a rabbit polyclonal antibody to inducible NO synthase (TCS Biologicals Ltd, Buckingham, UK).

Assay of NO synthase

NO synthase activity in the samples was determined by measuring the conversion of L-[*U*-^{14}C]arginine to [*U*-^{14}C]citrulline at 37°C[3]. The activity of the calcium-dependent enzyme was calculated as the difference between the L-[*U*-^{14}C]citrulline generated by control samples and by those containing 2mM EGTA. The activity of the calcium-independent enzyme was calculated as the difference between activity in samples containing 2mM EGTA and samples containing both EGTA and 1 mM of the NO synthase inhibitor L-NMMA.

Abbreviations: ER, Oestrogen receptor; NO, Nitric oxide; L-NMMA, L-N^G-monomethyl-arginine; EGTA, [(ethyleneglycol bis(oxy-ethylenenitrilo)] tetracetic acid.

RESULTS

Table 1. Effect of oestrogen implants on the growth of MDA 231 human breast tumour xenografts.

Data are means ± S.E.M. for at least 9 animals per group and represents the time taken for the tumours to reach a volume of 300 mm^3. *Significantly different from control and placebo groups ($p < 0.05$)

Treatment Group	Days taken to reach 300 mm^3
Oestrogen-implant	29.3 ± 6.8*
Control	52.3 ± 5.6
Placebo	58.3 ± 9.8

Table 2. NO synthase activity in MDA 231 human breast tumour xenografts.

Data are means ± S.E.M. from triplicate determinations for at least 4 animals per group and assayed by the conversion of [^{14}C]-L-arginine to [^{14}C]-citrulline. *Significantly different from control and placebo groups ($p < 0.05$)

Treatment Group	NO synthase activity (pmol/min per mg protein)	
	Total	Ca^{2+}-independent
Oestrogen-implant	2.92 ± 0.68*	2.18 ± 0.45*
Control	0.69 ± 0.10	0.42 ± 0.10
Placebo	0.56 ± 0.11	0.32 ± 0.10

DISCUSSION AND CONCLUSIONS

Our results show that oestrogen implants lead to the stimulation of ER-negative MDA 231 human breast tumour suggesting a sustained growth promoting effect by the hormone. Oestrogen treatment also lead to a 4-fold increase in total NO synthase activity measured and was mainly the calcium-independent form of the enzyme. Immunohistochemistry using a rabbit polyclonal antibody to inducible-NO synthase (data not shown), indicated positive staining for NO synthase across all treatment groups particularly in the endothelium of blood vessels, the perineurium of nerve fibres and some connective tissue cells. A weaker staining of tumour cells was observed. However, in the oestrogen-treated group, there was consistent localisation of NO synthase expression (intense staining) in a region between 'viable' tumour cells and true necrosis, consistent with cells in a state of hypoxia or undergoing apoptosis. Potentially, oestrogen treatment could enhance angiogenesis, mediated by NO production thus resulting in the increased growth observed and this is currently under investigation.

REFERENCES

1. Weigel, R.J. and deConnick, E.C. (1994) Transcriptional control of estrogen receptor in estrogen receptor-negative breast carcinoma. Cancer Res. **53**, 3472-3474.

2. Friedl, A. and Jordan, V.C. (1994) Oestradiol stimulates growth of oestrogen receptor-negative MDA-MB-231 breast cancer cells in immunodeficient mice by reducing cell loss. Eur. J. Cancer. **10A**, 1559-1564.

3. Salter, M., Knowles, R.G. and Moncada, S. (1991) Widespread tissue distribution, species distribution and changes in activity of Ca^{2+}-dependent and Ca^{2+}-independent nitric oxide synthases. Fed. Eur. Biochem. Soc. **291,** 145-149.

Alterations in energy metabolism of murine transplantable tumours *in vivo* by the inhibition of nitric oxide synthase.

PAULINE J WOOD*, JANET M SANSOM*, IAN J STRATFORD*, GERALD E ADAMS*, LINDY L THOMSEN#, D CONWIL JENKINS# and SALVADOR MONCADA#.

*MRC RADIOBIOLOGY UNIT, CHILTON, DIDCOT, OXON, OX11 0RD, UK. and #WELLCOME RESEARCH LABORATORIES, LANGLEY COURT, BECKENHAM, KENT, BR3 3BS, UK.

A recent approach to improving the efficacy of some anti-cancer therapies has been the use of vasoactive agents to manipulate tumour blood flow and hence oxygenation. Increased tumour oxygenation would improve sensitivity to X-rays, while increased tumour hypoxia would enhance the toxicity of bioreductive agents.

The involvement of nitric oxide (NO) in the maintenance of vascular homeostasis, together with the finding of NO synthase (NOS) enzyme activity in solid tumours [1,2], has led to the examination of NO and NOS as potential target molecules for the manipulation of tumour blood flow and hence oxygenation. The administration of NO donors, such as sodium nitroprusside or SIN-1 has been shown to increase tumour oxygenation and sensitivity to X-rays [3,4]. Conversely, tumour hypoxia is increased by the administration of the NOS inhibitor nitro-L-arginine [4].

The aim of the present study was to compare the changes in metabolism of the murine transplantable tumour SCCVII/Ha *in vivo*, induced by a range of NOS inhibitors, which are known to have differing selectivities for the NO synthase isoforms in normal tissues.

SCCVII/Ha murine tumours were implanted intradermally on the back of C_3H mice, 2 cm from the tail base, and were used at volumes of 250-400 mm^3. Mice were unanaesthetised for experiments but gently restrained in jigs, which exposed the tumour on the mouse back. Changes in tumour energy metabolism were monitored using *in vivo* ^{31}P magnetic resonance spectroscopy (MRS) with a 4.7 T, 30 cm horizontal bore magnet, and a 7 mm diameter surface coil placed over the tumour. Each spectrum comprised 256 scans with a 2 sec delay, giving a total acquisition time of 8 min. A tail vein catheter was used to allow injection of the agents without moving the mouse from its position in the magnet. Spectra were analysed using a baseline and Lorentzian curve fitting routine, and data were expressed as the ratio of the inorganic phosphate peak area to the sum of all peak areas, or Pi/total. Previous studies have demonstrated that an increase in tumour hypoxia results in increased Pi/total [5].

The time courses of the metabolic response *in vivo* of SCCVII/Ha tumours to 1-20 mg/kg i.v. nitro-L-arginine (NOARG) or methylthiocitrulline (MTC), and to 10-200 mg/kg i.v. monomethyl-L-arginine (L-NMMA) or iminoethyl-L-ornithine (L-NIO) were investigated, and a summary of the results is shown in table 1.

Table 1. Increases in Pi/total in SCCVII/Ha tumours *in vivo* after injection of NOS inhibitors

NOS inhibitor	Dose at maximal response	Increase in Pi/total		
		30 min	60 min	120 min
NOARG	10 mg/kg	2.6	2.6	2.5
MTC	20 mg/kg	2.5	3.1	3.3
L-NMMA	200 mg/kg	1.3	1.8	1.9
L-NIO	200 mg/kg	2.1	1.8	2.4

All four NOS inhibitors examined were capable of increasing Pi/total in this tumour model, which is consistent with increased tumour hypoxia. The size of the increase in this ratio induced by these agents is not as great as that produced when the blood supply to the tumours is fully occluded by application of a clamp, which gives a 4-5 fold increase in Pi/total [5]. However, an increase in Pi/total of a factor of 2-3, as demonstrated for NOARG in the SCCVII/Ha and other tumours [4,6] is sufficient to produce maximal increase in radiation resistance, and to significantly increase the toxicity of bioreductive agents [6]. The results presented here indicate that MTC is as effective as NOARG in increasing Pi/total and hence hypoxia in this tumour. However, the other two agents, L-NMMA and L-NIO do not appear to be as effective as NOARG or MTC at increasing Pi/total. This is reflected in the size of the increase, which is less than that for NOARG or MTC, and also the dose of L-NMMA or L-NIO required to produce this increase, which is at least 10 times greater than that for NOARG or MTC.

NOARG and MTC are NOS inhibitors considered to have greater selectivity towards the calcium dependent NOS, whereas L-NMMA and L-NIO are considered to be more selective towards the calcium independent NOS. The greater ability of NOARG and MTC over L-NMMA and L-NIO, to induce hypoxia in the SCCVII/Ha tumour may reflect the relative activities of the NOS isoforms within the solid tumour.

1. Thomsen, L.L., Lawton, F.G., Knowles, R.G. *et al.* (1994). Cancer Res. **54**, 1352-1354.

2. Cobbs, C.S., Brenman, J.E., Aldape, K.D. Bredt, D.S. and Israel, M.A. (1995). Cancer Res. **55**, 727-730.

3. Teicher, B.A., Holden, S.A., Northey, D., Dewhirst, M.W. and Herman, T.S. (1993). Int. J. Radiat. Oncol. Biol. Phys. **26**, 103-109.

4. Wood, P.J., Stratford, I.J., Adams, G.E., Szabo, C., Thiemermann, C and Vane, J.R. (1993). Biochem. Biophys. Res. Commun. **192**, 505-510.

5. Bremner, J.C.M., Counsell, C.J.R., Adams, G.E *et al.* (1991). Br. J. Cancer **64**, 862-866.

6. Wood, P.J., Sansom, J. M., Butler, S.A. *et al.* (1994). Cancer Res. **54**, 6458-6463.

A GluRs--NOSs--RNI/ROIs cascade mechanism may be involved in experimental complex partial seizures

Chang-Kai Sun[1],# Yuan-Gui Huang[1], You-Sheng Jia[2], Gong Ju[2], Duo-Ning Wang[3], Jian Mo[3] and Cheng-Ji Wang[4].

[1]Department of Neurology in Xijing Hospital, [2]Institute of Neuroscience, [3]Free Radical Research Laboratory and [4]Laborotary of Biochemistry and Molecular Biology, The Fourth Military Medical University, Xi'an 710032, P.R. China

Epilepsy is a common but intractable disease, which is still invaliding a quite large population of human beings. The rising interest on NOS-NO* pathway for biology and medicine gives a new clue to the limited research of molecular mechanisms for epilepsy, but is rather unclear up to now. To clarify the involvement of NOS-NO pathway in epilepsy, we carried out some indirect investigations concerned with a reliable model of complex partial seizure induced by kainic acid (KA; Sigma; 20mg/kg bw i.p.) in young adult male BALB/c mice.

A. Colorimetric detection of the production of nitrite (NO_2^-) and thiobarbituric--acid--reactive substances (TBARSs) in the mouse brain homogenate during the seizure

A total of 66 mice (weighing 21-35g) were used in this study (n=6 for each group of after KA administration 5, 10, 15, 30, 60min, 3, 6h, L-Arg-$\xrightarrow{30'}$KA 3h and L-NNA-$\xrightarrow{30'}$KA 30min, 3h), in which we took our choice of NO_2^- in the brain homogenate as a colorimetric detective target at 546nm for the production of NO during the seizure, and made choice of TBARSs as the colorimetric examination index at 532nm for the lipid peroxidation damage caused by some reactive nitrogen/oxygen intermediates (RNI/ROIs). Statistical significances were tested by Student *t*-test. The results show that the production of both NO_2^- and TBARSs is related to different seizure stages. NO_2^- highly increases but TBARSs keep low at the early stage ($p<0.01$), and soon NO_2^- decreases again but TBARSs go high along with the seizure lasting ($p<0.01$). The variation of the both NO_2^- and TBARSs may be influenced ($p<0.01$) by L-Arg (Zhangjiagang, China; 40mg/kg bw/12h i.p. for 9 times, the last just 30 min prior to KA injection) and L-NNA (Sigma; 50mg/kg bw/12h i.p., 9 times, the last 30 min prior to KA).

B. Pharmalogical interventions of NOS-NO pathway

30 mice (weighing 20-34g) were used (n=6 for each group of different drugs and the controls). All mice were pretreated with some intervention drug or vehicle intraperitoneally 30 min before KA). The behaviour of each mouse was continuously evaluated during 3h (9 20-min periods) after KA administration by comparison with the number of wet shakes (ws) tested by Student *t*-test and the KA-seizure scores in every 20-min period tested by Mann-Whitmey *U*-test[1]. The results suggest that inhibitions of NOS and superoxide dismutase (SOD), blocking calcium channel, and scavenging ROI/RNIs may give rise to different effects on the KA-seizure as follows: Nimodipine injection (Tianjin, China; 3mg/kg bw i.p.), an antagonist of Ca^{2+} channel, shows an apparent antiseizure action whenever early ($p<0.01$) or late ($p<0.001$); the effect of L-NNA seems complex, at the begining of seizure, it shows an action of proconvulsion ($p<0.05$), but at the late stage it can attenuate the seizure severity ($p<0.01$); diethyldithiocarbamate (Shanghai, China; 400mg/kg bw i.p.), an inhibitor of Cu-Zn SOD, may augment the seizure intensity ($p<0.05$), but this effect is only limited at the late stage; as a scaverger of ROI/RNIs, Vitamin E injection (Shanghai, China; 50mg/kg bw i.p.) also shows a complex action on the seizures: proconvulsion at the early (slightly) and antiseizure at the late ($p<0.05$), but is not as potent as L-NNA.

To whom correspondence should be addressed, at: Room 401, Postgraduate Dorimitory, The Fourth Military Medical University, Xi'an 710032, P.R. China

* Abbreviations used: NOSs, nitric oxide synthases; NO, nitric oxide; GluRs, glutamate receptors; RNI/ROIs, Reactive nitrogen / oxygen intermediates; KA, kainic acid; L-Arg, L-arginine; L-NNA, N^G-nitro-L-arginine; $O_2^{\cdot-}$, superoxide anion; $ONOO^-$, peroxynitrite; $HO^\cdot$, hydroxyl radical

From our data we like to propose that[2-4]: (1) NOS-NO pathway is definitely involved in the KA-seizure; (2) the implications of NOS-NO in the seizure may be complex. During the same course of seizure NO may firstly act as a homeostasis agent then insult brain. These paradoxical actions of antiseizure/anticovulsion, or proconvulsion even neurotoxic insult may be related to the different seizure stages (early, middle and late) and the different output even different states ($NO^\cdot/NO^+/NO^-$) of NO during the seizure; (3) during the KA-seizure, NOS-NO pathway may play a central role in a cascade mechanism which may be elucidated as a GluRs-NOSs-RNI/ROIs cascade mechanism. In this cascade mechanism, NOSs (nNOS, eNOS even iNOS) may act as the downstream of the activation of GluRs (NMDAR, KAR etc.) and other environmental or endogenous pathway which can caused some seizure, and the upstream of many pathophysiologic events in epilepsy such as co-excitation of neurons, cerebral vasodilation and damage of neurons/glial cells etc. directly and/or indirectly caused or influenced by some RNI/ROSs such as NO, $O_2^{\cdot-}$, $ONOO^-$, $HO^\cdot$ etc.; (4) carrying out further studies for this cascade mechanism may be benificial to the perfect solution of epilepsy problems.

References

1. Baran, H., Löscher, W. and Mevissen, M. (1994) The glucine/NMDA receptor partial agonist D-cycloserine blocks kainate-induced seizures in rats. Brain Res. 625, 195-200
2. Moncada, S. and Higgs, A. (1993) The L-arginine-nitric oxide pathway. N. Engl. J. Med. 329, 2002-2012
3. Snyder, S.H. (1993) Janus faces of nitric oxide. Nature (Lond.) 364: 577
4. Stamler, J.S., Singel D.J. and Loscalzo J. (1992) Biochemistry of nitric oxide and its redox-activated forms. Science (Wash. D.C.) 258: 1898-1902

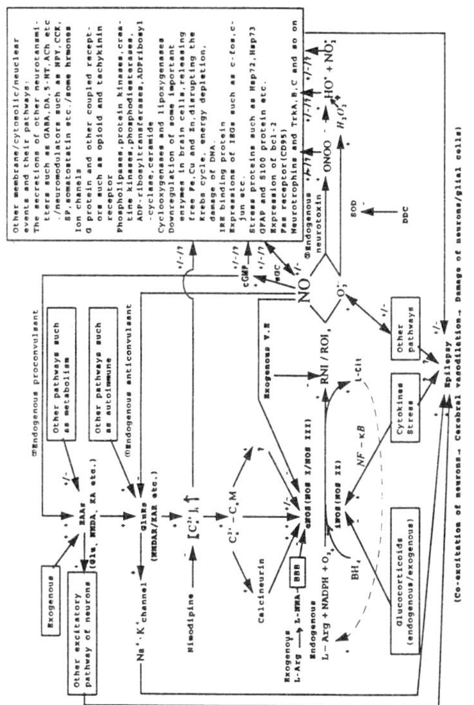

A sketch of hypothesis: GluRs-NOSs-RNI/ROIs CASCADE MECHANISM in seizure
+denotes upregulate, augment, facilitate, produce, or cause
-repesents downregulate, attenuate, suppress, scavenge, or terminate
?indicates some pathways possible or unknown

Ruthenium complexes as nitric oxide scavengers: a new therapeutic approach to nitric oxide mediated disease.

SIMON P. FRICKER, ELIZABETH SLADE, NIGEL A. POWELL, BARRY A. MURRER, IAN L. MEGSON*, GORDON D. KENNOVIN*, STUART K. BISLAND*, MARK LOVELAND* and FREDERICK W. FLITNEY*.

Johnson Matthey Technology Centre, Blount's Court, Sonning Common, Reading RG4 9NH, UK and *School of Biological & Medical Sciences, University of St Andrews, St Andrews, Fife KY16 9TS, UK.

Introduction. Induction of the inducible isoform of nitric oxide synthase (iNOS) by cytokines and bacterial endotoxin and the concomitant production of excess nitric oxide (NO) are strongly implicated in the pathogenesis of septic or endotoxic shock. NO is a key mediator of the hypotension, peripheral vasodilation and diminished sensitivity to vasoconstrictor agents in endotoxaemia [1]. The selective inhibition of iNOS is one therapeutic approach currently being pursued for treatment of NO mediated diseases. The design of drugs able to scavenge and remove NO is an alternative approach.

Ruthenium(III) complexes react rapidly with NO to form stable, inert, six coordinate Ru(II) mononitrosyls. We have adopted a strategy whereby the metal ion is chelated with a suitable ligand set, with the aim of conferring water solubility, rapid *in vivo* clearance and low toxicity, whilst providing an available binding site for NO with an overall high affinity and rapid rate of NO binding. Encouraging results have been obtained with JM1226, K[Ru(Hedta)Cl]. JM1226 reacts with NO in aqueous solution to form a nitrosyl complex. The chloride ion is replaced by water in solution to yield the aqua species [Ru(Hedta)(H_2O)], JM6245 [2]. Formation of the JM1226/NO product has been confirmed by infra-red spectroscopy, which reveals a peak at 1897 cm^{-1} characteristic of a Ru-NO adduct. A similar result was obtained after reacting JM1226 with *S*-nitroso-*N*-acetyl penicillamine (SNAP). Kinetic studies indicate that NO binding is rapid, with a rate constant of $2.2 \times 10^7.M^{-1}.s^{-1}$ (M.T. Wilson, personal communication).

Methods. The ability of JM1226 and JM6245 to inhibit several NO mediated processes has been studied. First, RAW264 macrophages were cultured on 24 well plates and activated with 10μg/ml LPS+100units/ml interferon-γ (IFN-γ). NO production was asssayed 18hr later by measuring NO_2^- in the culture medium using the Griess reagant [3]. Second, tumour cell killing by activated RAW264 cells was assessed in co-culture using non-adherent P815 murine mastocytoma 'target' cells on 96 well plates. Viability of P815 cells was determined 18hr post-stimulation by the MTT assay [3]. Third, the ability of JM1226 and JM6245 to inhibit the vasodilator effect of an NO donor drug (SNAP) was studied using precontracted, internally-perfused, rat tail arteries (RTAs) [4]. Fourth, we investigated the effects of JM1226 and JM6245 on hyporesponsive RTAs taken from rats previously treated with LPS to induce endotoxic shock. Finally, the potential of JM1226 to reverse the hypotension associated with endotoxic shock was studied *in vivo* using LPS-injected (4mg/kg, i.p) Wistar rats.

Results and Discussion JM1226 (100μM) reduced NO_2^- levels in culture medium of RAW264 cells stimulated with LPS+IFN-γ, from 20.9+/-0.4μM to 7.6+/-0.2μM. Similar results were obtained using JM6245 and LNMMA (250μM). JM1226 also afforded substantial protection against NO-mediated tumour cell killing, increasing P815 cell viability from 50.2+/-4.5% to 74.8+/-4.6%.

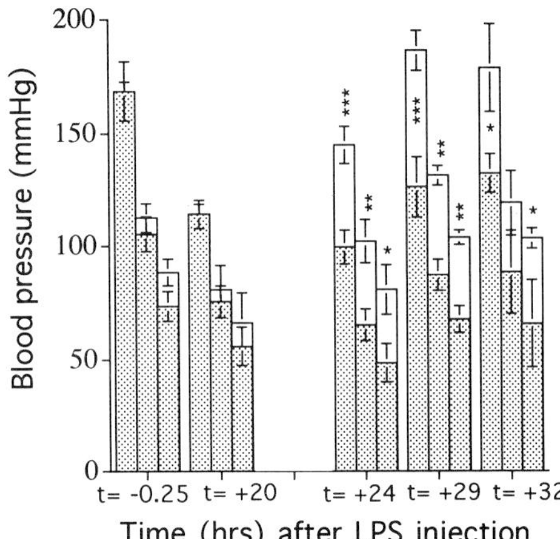

Figure 1. Blood pressures (systolic, mean and diastolic) of LPS-treated (4mg/kg) rats after a single i.p. injection of JM1226 (100mg/kg; empty columns) or 0.9% saline (stippled columns), measured by the rat tail cuff method. (Statistics: Student's unpaired t test; *** = $p<0.001$; ** = $p<0.01$; * $p< 0.05$; JM1226 group compared with saline group). Times relative to time of LPS injection (t=0hr).

JM1226 and JM6245 attenuated vasodilator responses of precontracted RTAs to bolus injections (10μl) of SNAP delivered into the tinternal perfusate: the control ED_{50} value for SNAP was 6.0μM and in the presence of JM6245 (10^{-4}M) 1.8mM. The ability of JM1226 and JM6245 (both 100μM) to reverse the hyporesponsiveness of isolated RTAs taken from LPS-injected rats was not significantly different from that of LNMMA (100μM). Finally, the hypotension induced in rats injected with LPS was fully reversed by JM1226 (100mg/kg i.p) after 9-12hr (Figure 1).

In conclusion, JM1226 and its aqua derivative JM6245, the postulated active species in solution, are both able to inhibit NO mediated processes in a variety of biological systems. Preliminary studies indicate that JM1226 has low toxicity and a promising pharmacokinetic profile. These data indicate that this class of compound has therapeutic potential against NO-mediated diseases such as septic shock.

References

1 Moncada, S., Palmer, R.M.J., and Higgs, E.A. (1991). Pharmacol. Rev. **43**, 109-142.

2. Matsubara, T. and Creutz, C. (1979). Inorg. Chem. **18**, 1956-1966.

3. Fricker, S.P., Slade, E. and Powell, N.A. (1995). Biochem. Soc. Trans. **23**, 231S.

4. Flitney, F.W., Megson, I.L., Flitney, D.E. and Butler, A.R. (1992). Brit. J. Pharmacol. **107**, 842-848.

Interferon (IFN)-γ -induced upregulation of Type 1 nitric oxide synthase (NOS) activity in neurons inhibits viral replication

Zhengbiao BI* and Carol S. REISS*,#

*Biology Dept., #Center for Neural Science and Kaplan Comprehensive Cancer Center, New York University, New York, NY 10003 USA.

Intranasal infection of mice with Vesicular stomatitis virus (VSV) results in infection of nasal neuroepithelial cells and retrograde transport via the olfactory nerve to the olfactory bulb of the central nervous system, by 24h post application [1]. At that time, virus is limited to the neurons of the olfactory bulb and accessory olfactory nucleus [2]. Immunohistochemical staining of frozen tissue sections reveals expression of IFN-γR (receptor) and induction of Type 1 NOS in olfactory neurons and Type 3 NOS in adjoining astrocytes [3]. We examined this further *in vitro*.

NB41A3 neuroblastoma cells, like the olfactory bulb neurons in immunohistochemical staining of frozen sections, expressed IFN-γR, neurofilaments, and Type 1 NOS, but not express GFAP, Types 2 or 3 NOS. Treatment of the NB41A3 cells with NMDA or with IFN-γ resulted in increased NO_2- production (Table 1) which could be inhibited by L-NMA and reconstituted with L-Arg but not D-Arg (not shown).

Biosynthetic labelling of IFN-γ treated cells followed by immunoprecipitation revealed increased incorporation of ^{35}S into Type 1 NOS. When quantitated using a BioRad Phosphoimager, this IFN-γ-induced protein

Table 1. Type 1 NOS Activity is induced by IFN-γ and NMDA

	NB41A3 cells			murine macrophages		
	med	IFN-γ	NMDA#	med	IFN-γ	NMDA
med.	10*	23	31	15	63	14
L-NMA+	9	11	10	14	21	16

*μM in Greiss assay; #500μM NMDA; +500μM L-NMA.

synthesis was determined to be 2.2x the incorporation into NOS-1 than in unstimulated cell group (not shown).

We had previously seen that VSV replication was inhibi-ted in NB41A3 cells when activated through their glutamate receptors or provided with nitrate donors (eg SNAP [4]). Evidence from other laboratories had shown that *iNOS in macrophages* was effective at limiting Herpes, vaccinia virus replication [5,6] and HIV infection [7]. We tested whether this inhibition of viral replication by *Type 1 NOS in neurons* were limited to VSV (negative-stranded unsegmented enveloped virus) or whether it was a general phenomenon. VSV is inhibited by IFN-γ-induced antiviral activity, all of which is inhibitable by the arginine analog L-NMA, suggesting that all the IFN-γ-inducible antiviral activity is attributable to NO. In contrast, influenza A/NWS virus (segmented negative stranded enveloped virus) was inhibited by IFN-γ treatment of NB41A3 cells, but not via NO as the ultimate mediator (Table 2). Thus, other interferon-inducible proteins [reviewed in 8] are likely to be the active antiviral effector molecules for influenza.

This work has been extended to show that Herpes simplex-1 (dsDNA enveloped virus) and Type 1 Polio (positive-stranded encapsidated virus) viruses are inhibited in neurons in culture by NO (Table 3). This

Table 2. Inhibitory Effect of IFN-γ on VSV and Influenza A/NWS Virus Replication (pfu) in NB41A3 cells.

Conditions	Virus	Medium	L-NMA
medium	VSV	2x10^5	4x10^5
IFN-γ	"	3x10^3	2x10^5
IFN-γ + *a*-IFN-γR	"	1x10^5	5x10^5
NMDA (500μM)	"	5x10^2	4x10^5
medium	A/NWS	5x10^4	7x10^4
IFN-γ	"	8x10^2	5x10^2

Table 3. NO-induced Inhibition Replication of HSV-1 and Polio Type 1 Viruses in Neuroblastomas.

Conditions	HSV-1(NB41A3)	Polio (N2-A-1 cells)
medium	1.5x10^5	5x10^3
100 μM SNAP	2.0x10^3	6x10^2
100 μM NAP	1.6x10^5	4x10^3
500 μM NMDA	4.0x10^3	nd
NMDA+L-NMA	1.3x10^5	nd

suggests strongly that upregulation of Type 1 NOS activity directly in neurons or via IFN-γ-induction of protein synthesis may promote viral clearance and recovery from neuronal viral infection. We have indirect evidence of this effect, *in vivo* [9], and are testing this hypothesis. Furthermore, we are investigating the mechanism (whether nitrosylation of nucleic acids, viral proteins, host proteins essential for replication, or viral morphogenesis) by which NO inhibits viral replication in neurons.

Acknowledgements: This work was supported by AI18083 and a Bridge grant from NYU to CSR. Polyclonal antibody to Type 1 NOS was generously provided by Ted Dawson (Johns Hopkins School of Medicine), Influenza A/NWS virus was the gift of Dr. P. Palese (Mt. Sinai School of Medicine, New York), Herpes simplex virus was from Dr. P. Schaffer (Dana-Farber Cancer Institute, Boston, MA), Polio virus type 1 and human poliovirus-receptor transfected murine neuroblastomas and laboratory space for that BL-2 work was generously provided by Dr. V. Racaniello (College of Physicians and Surgeons, New York). This data has been submitted for publication in a peer reviewed journal.

1. Plakhov, I.V., Arlund, E.E., Aoki, C. and Reiss, C.S. (1995) Virol, **209**, 257-262.
2. Huneycutt, B.S., Plakhov, I.V. Schusterman, Z.,et al. (1994) Brain Res. 635, 81-95.
3. Barna, M., Komatsu, T., Bi, Z. and Reiss, C.S. (1995) in preparation.
4. Bi, Z. and Reiss, C.S. (1995) J. Virol, **69**, 2208-2213.
5. Karupiah, G., Xie, Q., Buller, R.M.L., et al. (1993) Science **261**, 1445-1448.
6. Harris, N., Buller, R.M.L. and Karupiah, G. (1995) J. Virol., **69**, 910-915.
7. Mannick, J., Stamler, J.S., Tang, E. and Finberg, R. (1995) this volume.
8. Staeheli, P. (1990) Adv. Virus Res. **38**, 147-175.
9. Bi, Z., Barna, M., Komatsu, T. and Reiss, C.S. (1995) J. Immunol., in press.

Role of inducible NOS in renal ischemia: antisense oligodeoxynucleotide-based approach

Eisei NOIRI, Tatyana Y. PERESLENI, Frederick MILLER, and Michael S. GOLIGORSKY

State University of New York, Stony Brook, NY 11794-8152, USA

Determining cellular sequelae of NO production in a uniform population of cells expressing a single isoform of NOS represents a relatively straightforward problem. In contrast, the heterologous cell population expressing different isoforms of NOS imposes challenging obstacles for study of NO effects. A paradigm of such a mosaic distribution of NOSs is represented by the kidney where different isozymes are expressed in different segments of the nephron [1,2]. This complex topography of NOSs, probably, underlies a well established paradox: L-arginine-based inhibitors of NOS are cytoprotective in cultured epithelial cells exposed to various noxious agents [3], whereas the same inhibitors aggravate renal dysfunction in the whole animal [4]. We have reasoned that such a discrepancy may be related to ❶ the functional diversity of individual NOS isoforms and ❷ poor selectivity of the existing inhibitors.

In an attempt to resolve the problem, we elected to utilize an antisense oligodeoxynucleotide (ODN) strategy. Antisense ODN to iNOS targeted a sequence within the open reading frame of murine iNOS shown to be conserved in the rat kidney [1,5,6]. Antisense ODN: (5'-3') CTT CAG AGT CTG CCC ATT GCT, was prepared using a solid-phase DNA synthesizer. Sense and scrambled sequences were used as control. All ODN constructs were phosphorothioated and used at the concentration of 1 mg/kg for systemic injection into rats ca. 8 h prior to the induction of ischemic acute renal failure (45 min bilateral renal ischemia). After 24 h reperfusion, the blood was drawn for serum creatinine (Cr) determination and animals were sacrificed for morphologic evaluation of the kidneys.

Renal ischemia resulted in a 3-fold elevation of serum Cr (Fig.1) in untreated rats. Animals pretreated with antisense ODN showed a remarkable preservation of renal function which was not reproduced with sense or scrambled ODNs. In sharp contrast with these results, rats pretreated with L-NAME exhibited a more severe renal dysfunction, than control rats (** $p<0.001$).

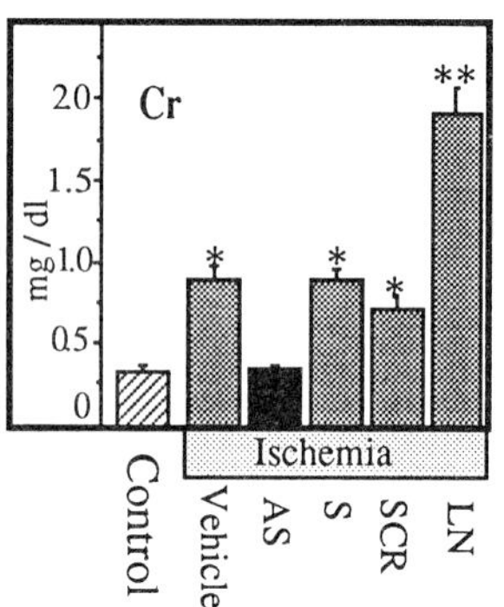

Fig. 1
Serum Creatinine in rats subjected to 45 min renal ischemia after pretreatment with ODNs or L-NAME (LN).

Abbreviations: AS-antisense, S-sense, SCR-scrambled.
* denotes $p<0.05$ compared to the sham operated control.

To test whether the observed effect of antisense ODN was due to the knock-down of iNOS, three sets of experiments were performed. Renal tubules isolated from ischemic kidneys showed a 4-5-fold increase in nitrite production compared to those obtained from ODN-pretreated animals (not shown). In addition, Western blot analysis revealed iNOS expression in homogenates from ischemic, but not from pretreated, kidneys (Fig. 2).

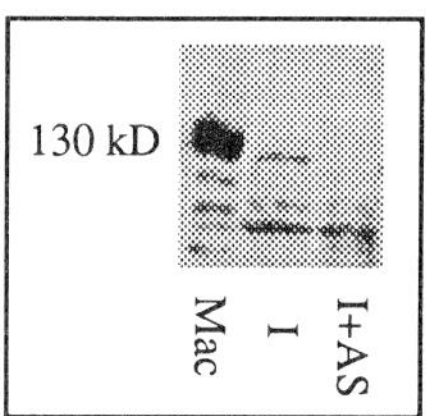

Fig. 2
Western analysis of iNOS expression in kidney homogenates obtained from control ischemic (I) and antisense (AS) pretreated ischemic group.
Mac- macNOS (positive control).

Finally, kidney sections from ischemic animals displayed immunoreactive staining for iNOS, which was spectacularly reduced in the pretreated group (not shown). These findings confirm the adequacy of iNOS knock-down. Morphologic analysis of kidney sections, performed in a blind fashion, was based on a well-established scoring criteria [7]. As shown in Fig.3, rats pretreated with antisense ODN had significantly lesser pathological score. The most prominent element of morphologic preservation in this group was a dramatic decrease in the number of necrotic tubular epithelial cells.

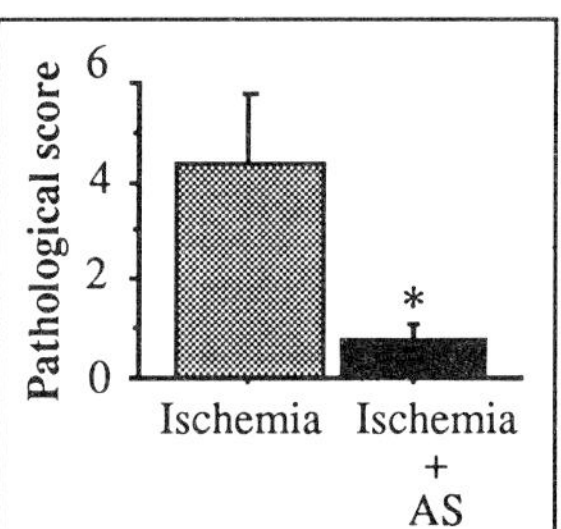

Fig. 3
Pathological scores of renal damage in ischemic kidneys treated with antisense (AS).

* denotes $p<0.05$ vs. control ischemia.

The above data provide evidence for ❶ the cytotoxic role of NO produced via iNOS in proximal tubular epithelial cells, ❷ the prevalent organoprotective effect of NO generated via eNOS, and ❸ establish *in vivo* use of antisense ODN to iNOS for prevention of ischemic acute renal failure.

References

1. Mohaupt M.G., Elzie J.L., Ahn K.Y., Clapp W.L., Wilcox C.S., and Kone B.C. (1994) Kidney Int. 46, 653-665
2. Bachmann S., Bosse H., and Mundel P. (1995) Am.J.Physiol. 268, F885-F898
3. Yu L., Gengaro P., Niederberger M., Burke T., Schrier R.W. (1994) Proc.Natl.Acad.Sci.USA 91, 1691-1695
4. Conger J., Robinette J., Villar A., Raij L, Shultz P. (1995) J.Clin.Invest. 96, 631-638
5. Lyons C.R., Orloff G., and Cunningham J.M. (1992) J.Biol.Chem. 267, 6370-6374
6. Nunokawa Y., Ishida N., and Tanaka S. (1993) Biochem.Biophys.Res.Comm. 191, 89-94
7. Conger J., Shultz M., Miller F., and Robinette J. (1994) Kidney Int. 46, 318-323

Heat shock protein 70 reduces expression of astroglial inducible nitric oxide synthase.

Douglas L. Feinstein, Elena Galea, Hui Xu, and Donald. J. Reis.

Division of Neurobiology, Cornell University Medical College, 411 East 69th Street, New York, N.Y., 10021, U.S.A.

In brain, expression of Ca^{2+}-independent nitric oxide synthase (iNOS) is associated with numerous pathologies, including ischemia, demyelinating disease, Alzheimer's disease, and viral infection. Development of methods to reduce or prevent iNOS expression are desirable as possible therapeutic means. One endogenous mechanism to prevent cellular damage due to inflammation is the heat shock response (HSR). The HSR includes down-regulation of general cellular activity, e.g. transcription and translation, and rapid induction of specific HS proteins (HSPs). The HSR can be elicited by means other than heat, including hypoxia, chemical stimulation, heavy metals, and hypoglycemia. In brain, HSPs are induced during ischemia, excitoxicity, and hypoxia [4].

The mechanisms of cellular protection by HSR are not completely understood. However, in some cases HSP expression suppresses inflammatory response, such as TNF-α production [5] and IL-6 release [6]. Incubation of rat astrocytes or C6 glioma cells with bacterial endotoxin (LPS) and/or pro-inflammatory cytokines leads to *de novo* induction of iNOS mRNA and protein expression [1-3]. We therefore examined if HSR and/or HSP expression could modulate iNOS induction in cultured rat glial cells.

Astrocytes were heat shocked at 43°C for 40 minutes, placed back at 37°C to recover, and then LPS added after different times (Fig. 1). Subsequent iNOS induction, assessed by measurement of accumulated nitrites in the cell culture media, was significantly reduced (32 ± 5% of non-heated cells) in heat shocked cells compared to control cells. The suppression due to HS was dependent upon recovery time at 37°C, and by 4 hr iNOS induction returned to control levels. Similar HS-dependent effects were observed if iNOS was induced using cytokines rather than LPS, or C6 cells rather than astrocytes.

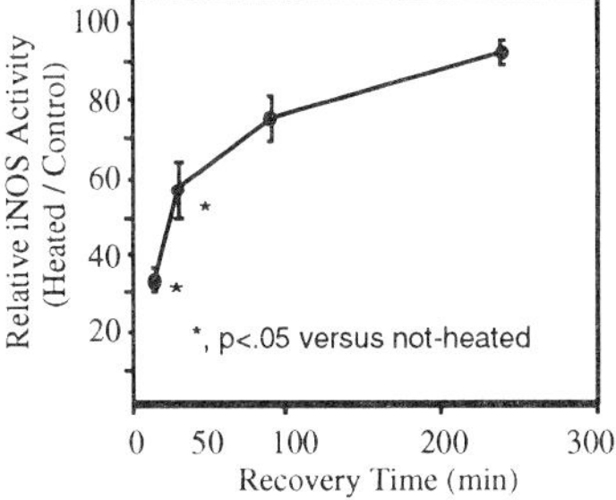

Figure 1. Heat shock reduces LPS-induced iNOS expression. Astrocytes were heat shocked (43°C, 40 min), placed at 37°C, and LPS added at the indicated times. Accumulated NO_2 was determined 20 hr later using Griess reagent . Data is mean ± s.e..m. of 3 expts.

To rule out possible effects of HS on cofactor or substrate availability in intact cells, we directly measured iNOS activity in cell lysates 20 hr after LPS addition. HS reduced the rate of L-arginine to L-citrulline conversion when assayed in saturating concentrations of all cofactors and 20 μM L-arginine, indicating that HS reduced final iNOS protein levels.

To determine if HS influenced iNOS mRNA levels, we used a semi-quantitative RT-PCR assay [3] to measure iNOS mRNA levels in samples prepared 4 hr following LPS addition. Prior HS treatment decreased steady state iNOS mRNA levels approximately 3-fold, from 1.6 to 0.6 fg per 50 ng of total RNA. At the same time, levels of glyceraldehyde-3-phosphate dehydrogenase mRNA were not modified by HS.

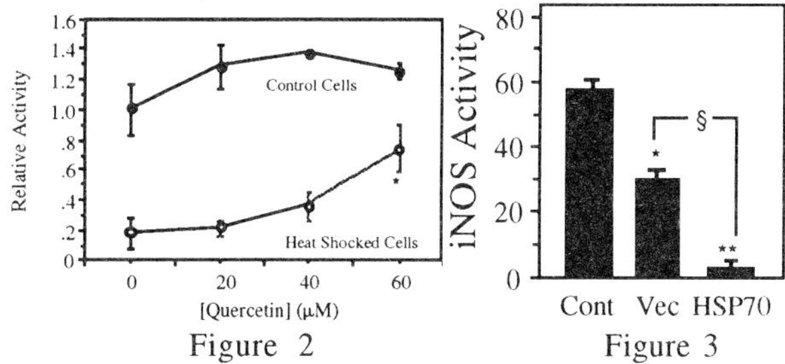

Figure 2. Quercetin reverses heat shock effects. Astrocytes were incubated with the indicated concentration of quercetin 1 hr, heated at 43°C for 0 or 40 min, then LPS added. Data is mean ± s.em. of 3 expts. *, p<.05 versus 0 quercetin.

Figure 3. HSP70 expression blocks iNOS induction. Rat-1 fibroblasts, stablely transfected with vector only or human iHSP70, were incubated with LPS / TNF-α / IFN-γ / IL1-β, and iNOS activity assessed by Griess assay at 24 hr. *, p<.05; **, p<.0005 vs non-transfected (control) cells. §, p<.05.

To determine if HSP expression was necessary to observe suppression of iNOS induction, we treated cells with the bioflavenoid quercetin (fig. 2), which inhibits HSP expression. Quercetin (60 μ M) had no significant effect on iNOS activity of control cells, but almost completely reversed the suppression due to prior HS. To determine if HSP70 was responsible for HS effects, we examined iNOS induction in Rat-1 fibroblasts stablely transfected with human iHSP70 [7], (Fig. 3). Cells transfected with vector had a reduced level of iNOS induction, however transfection with HSP70 almost completely abolished iNOS induction.

These results demonstrate that HS reduces iNOS induction in glial cells, and that this effect can be replicated by HSP70 expression. How HSP70 inhibits iNOS expression is unknown, but preliminary data suggest that HSP70 prevents nuclear translocation of NfkB, necessary for iNOS induction to occur. The regulation of HSP expression may provide a novel means of preventing pathological iNOS expression.

We thank Dr. Dennis Aquino for advice, Dr. Gloria Li for Rat-1 cells, and Liubov Lyandvert for technical support. This work was supported in part by grants from the National Multiple Sclerosis Society (D.L.F. and E.G.)

1. Galea, E., Feinstein, D.L., and Reis, D.J. (1992) PNAS 89, 10945-10949.
2. Feinstein, D.L., Galea, E., Roberts, S., Berquist, H., Wang, H., and Reis, D.J. (1994) J. Neurochem. 62, 315-321.
3. Galea, E., Reis, D.J., and Feinstein, D.L. (1994) J. Neurosci. Res. 37, 406-414.
4. Koroshetz, W.J. and Bonventre, J.V. (1994) Experientia 50, 1085-1091.
5. Synder, Y. M., Guthrie, L, Evans, G.F., and Zuckerman, S.H. (1992) J. Leuko. Biol. 51, 181-187.
6. Simon, M.M., Reikerstorfer, A., Schwarz, A., et al. (1995) J. Clin. Invest. 95, 926-933.
7. Li, G., Li, L., Liu, R., Rehman, M., and Lee, W.M.F. (1992) PNAS 89, 2036-2040.

Stretch as a major stimulus of pulmonary nitric oxide formation

LARS E GUSTAFSSON#*, GERARD BANNENBERG# and STEFAN STRÖMBERG#

#Department of physiology and pharmacology and *Institute of environmental medicine, Karolinska Institute, S-17177 Stockholm, Sweden

Nitric oxide is a regulator of the pulmonary circulation and is necessary for normal oxygenation of the blood, at least in animals [1]. Nitric oxide is present in the exhaled air of animals and man [2], and in normal animals the exhaled nitric oxide is formed in the lungs in a mainly calcium-dependent way [3]. A considerable part of exhaled nitric oxide is formed in the airways [4], but little is known about stimuli for the nitric oxide formation. We have observed that expansion of the lungs by positive end-expiratory pressure can increase exhaled nitric oxide in rabbits[5]. We therefore investigated whether exhaled nitric oxide could be related to changes in lung volume, and whether gadolinium, a blocker of stretch activated calcium channels [6], could affect exhaled nitric oxide.

Rabbits and guinea pigs were anaesthetised with pentobarbital and put on mechanical ventilation with air, using a constant volume ventilator set for 250 ml/kg/min at 36 breaths/min. Positive end-expiratory pressure was applied by immersion of the ventilator exhaust tube into a graded water cylinder. In the rabbits, negative or positive extrathoracic pressure could be applied by means of a plastic box surrounding the animal. The box could also be used for ventilation of the animal by intermittent negative extrathoracic pressure. Changes in pulmonary volume was registered as change in functional residual capacity by means of a pneumotachygraph. Exhaled nitric oxide was determined by chemiluminescence [2-5].

In both rabbits and guinea pigs application of positive end-expiratory pressure caused marked increments in exhaled nitric oxide. Expansion or compression of the thorax by positive or negative extrathoracic pressure caused graded changes in FRC in rabbits. Ventilation with intermittent negative extrathoracic pressure exhibited normal exhaled nitric oxide concentrations, and expansion of the lungs by further negative extrathoracic pressure caused increments in exhaled nitric oxide comparable with those seen during positive end-expiratory pressure during intermittent positive pressure ventilation. A positive correlation was obtained when comparing changes in lung volume (as measured by change in functional residual capacity) with exhaled nitric oxide (r=0.88 from 29 measurements in 8 animals).

Infusion of gadolinium chloride (50 mg/kg) intravenously in guinea pigs inhibited basal exhaled nitric oxide by 75% and abolished the increase in exhaled nitric oxide during positive end-expiratory pressure (Fig 1).

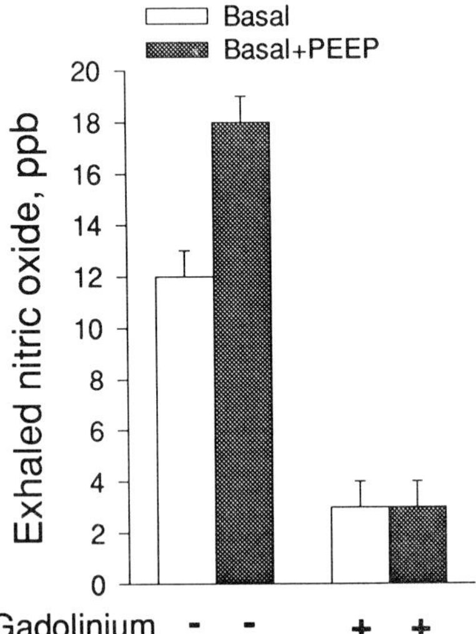

Figure 1. Inhibition of exhaled nitric oxide in anaesthetised guinea pigs by gadolinium ($GdCl_3$, 50 mg /kg iv). Open bars denote basal exhaled nitric oxide, hatched bars denote exhaled nitric oxide concentration during lung expansion by application of 7 cm H_2O positive end-expiratory pressure (PEEP). Note that gadolinium totally abolished the effect of PEEP, and that basal exhaled nitric oxide was inhibited by 75%. n=5.

Preliminary experiments in rabbits showed a marked increase in pulmonary artery pressure (>100% increase) during infusion of gadolinium, without any simultaneous increase in left atrial pressure.

The present observations show that expansion of the lungs caused an increase in exhaled nitric oxide which was directly correlated to the degree of expansion of the lungs. Furthermore, gadolinium fully blocked the stretch-induced effect on exhaled nitric oxide. Since gadolinium is a blocker of stretch-activated calcium channels [6], it seems reasonable to suggest that the stretch effect on nitric oxide in the lungs is dependent on calcium entry. If so, this would imply that also basal nitric oxide production might be dependent on activation of stretch-activated calcium channels.This indeed seems to be the case since gadolinium strongly inhibited basal nitric oxide production. This is also in agreement with a significant part of exhaled nitric oxide being dependent on the presence of calcium in isolated perfused lungs [3].

Gadolinium infusion in rabbits caused a pulmonary artery hypertension without raising left atrial pressure. This strongly suggests that increased pulmonary vascular resistance, rather than change in left ventricular function, was the cause for the gadolinium-induced pulmonary hypertension.

Nitric oxide has been suggested to be involved in regulation of the pulmonary circulation by evoking vasodilatation in well ventilated areas of the lungs [1] and to be responsible for the adaptation of the pulmonary circulation at birth [7], a situation where a marked expansion of the lungs occurs. We would like to suggest that both during normal ventilation and during adaptation at birth a prominent stimulus for nitric oxide formation will be stretch of the parenchyma or airways, leading to calcium entry into some type of pulmonary cell. This nitric oxide formation is of importance for regulation of the pulmonary circulation as indicated by our finding of increased pulmonary artery pressure during application of gadolinium.

Supported by the Swedish National Environment Protection Board, the Swedish MRC (07919), the Swedish Heart-Lung Foundation, and the Karolinska Institute

1. Persson, M.G., Gustafsson, L.E., Wiklund, N.P., Moncada, S. and Hedqvist, P. (1990) Acta Physiol. Scand., **140**, 449-457
2. Gustafsson, L.E., Leone, A.M., Persson, M.G., Wiklund, N.P. and Moncada, S (1991) Biochem. Biophys. Res. Communic. **181**, 852-857
3. Persson, M.G., Midtvedt, T., Leone, A.M. and Gustafsson, L.E. (1994) Eur. J. Pharmacol. **264**, 13-20
4. Persson, M.G., Wiklund, N.P. and Gustafsson, L.E. (1993) Am. Rev. Respir. Dis. **148**, 1210-1214
5. Persson, M.G., Lönnqvist, P.A. and Gustafsson, L.E. (1995) Anaesthesiology **82**, 969-974
6. Swerup, C., Purali, N. and Rydqvist, B. (1991) Acta Physiol. Scand. **143**, 21-26
7. Abman, S.H., Chatfield, B.A., Hall, S.L. and McMurtry, I.F. (1990) Am. J. Physiol. **259**, H1921-H1927

Tissue-selective delivery of nitric oxide using diazeniumdiolates (formerly "NONOates")

LARRY K. KEEFER[*], DANIEL J. SMITH[#], SHARON PULFER[#], STEPHEN R. HANSON[+], JOSEPH E. SAAVEDRA[‡], TIMOTHY R. BILLIAR[@], and PETER P. ROLLER[**]

[*]Chemistry Section, Laboratory of Comparative Carcinogenesis, National Cancer Institute, Frederick, MD 21702, USA; [#]Department of Chemistry, University of Akron, Akron, OH 44325, USA; [+]Division of Hematology and Yerkes Regional Primate Research Center, Emory University, Atlanta, GA 30322, USA; [‡]Biological Carcinogenesis and Development Program, SAIC Frederick, NCI-FCRDC, Frederick, MD 21702, USA; [@]University of Pittsburgh Medical Center, Pittsburgh, PA 15261, USA; and [**]Laboratory of Medicinal Chemistry, Division of Basic Sciences, National Cancer Institute, Bethesda, MD 20892, USA.

A frequent problem in nitric oxide research is how to deliver NO to the specific organ or cell type where it is needed without adversely affecting other NO-sensitive parts of the body. We are working to accomplish this by exploiting the extraordinary chemical versatility of the diazeniumdiolates (compounds of structure **1**). We have synthesized a variety that, in their anionic form, release NO without activation at physiological pH [1, 2]. In accord with the spontaneity of this process (equation 1), systemic administration normally results in system-wide effects. Nevertheless, there are at least three different strategies by which tissue-selective NO delivery can be arranged based on this chemistry.

Incorporation into a stationary solid. One approach is to allow the diazeniumdiolate function to dissociate to NO spontaneously but anchor the nucleophile residue to which it is attached (X, equation 1) in an insoluble polymer. In this way, exposure to NO can be limited to the tissue with which the solid is in physical contact. As one example, we have coated vascular grafts with crosslinked polyethyleneimine that was exposed to NO and shown them to be markedly less thrombogenic when installed in the baboon circulatory system than control grafts that were releasing no NO [2].

$$X - [N(O)NO]^- \xrightarrow[37\,^\circ C]{pH\ 7.4} X^- + 2\,NO \qquad (1)$$

1 (a diazeniumdiolate) (X = nucleophile residue)

Prodrug design. The second approach involves the opposite strategy, i.e., allowing the diazeniumdiolate to move freely through the circulatory system but prevent it from spontaneously dissociating. This can be effected by covalently attaching a protecting group to its terminal oxygen [3]. If the protecting group is selected such that it cannot be removed until it is metabolically cleaved by enzymes found only in the desired target organ, generation of NO from the resulting anion (**1**) should be concentrated in that organ.

To be effective, this "prodrug" strategy of NO delivery must produce an intermediate diazeniumdiolate anion with a half-life short enough that it dissociates essentially completely to NO within the metabolizing organ, because escape of free **1** into the general circulation could produce significant systemic effects. Thus **2**, the diazeniumdiolate derivative of pyrrolidine, has been chosen as a starting point for the study described elsewhere in this volume by Billiar *et al.* because its half-life at physiological pH and temperature is 3 seconds. Attachment of a vinyl group at the O^2-position as in equation 2 produced **3**, a stable product. As an enol ether, we predicted that **3** might undergo epoxidation by enzymes of the cytochrome P450 family, which are most abundant in the liver. If the resulting oxirane is hydrolyzed sufficiently rapidly, possibly with catalysis by epoxide hydrolases also present in that organ, exposure to NO should be liver-selective. Evidence that this hypothetical metabolic route might actually proceed with hepatocyte-specific consequences will be summarized in the chapter by Billiar *et al.* elsewhere in this proceedings.

Abbreviations used: NO, nitric oxide; BSA, bovine serum albumin.

(pyrrolidine)N-N(→O)=N-O⁻ Na⁺ (**2**) ⟶ (pyrrolidine)N-N(→O)=N-O-CH=CH₂ (**3**) (2)

Attachment to a protein. The third strategy we are exploring as a means of effecting tissue-selective NO delivery involves binding the diazeniumdiolate function to peptides and proteins. In this way, we postulated that specific cellular interactions unique to the proteins to be adducted might allow exquisite targeting even though the adduct is systemically administered and NO release is spontaneous.

To begin this study, we needed a relatively long-lived diazeniumdiolate with at least one other functional group that could be used for binding to the protein. For this purpose, we chose **4**, whose half-life for NO release at 37°C and pH 7.4 is approximately two weeks. This was converted to its nucleophilic N-4 mercaptoethyl derivative, **5**. In our initial studies, we have activated the model macromolecular carrier, BSA, toward coupling reactions by reacting it with γ-maleimidobutyric acid *N*-hydroxysuccinimide ester. Compound **5** was then covalently attached to the activated BSA through its maleimide functionality. The adduct, **6**, was found to generate NO steadily over several days in pH 7.4 phosphate buffer at 37°C. It should be emphasized that the procedure has only been attempted twice (both times successfully), and problems with purification and stability during storage remain to be solved. Additionally, NO release must be quantified, the yield needs to be optimized, and our assumption that similar procedures can be successfully extended to other proteins must be confirmed. Nevertheless, these preliminary results suggest that peptide hormones, antibodies, and other proteins might be similarly adducted, allowing use of their receptor-mediated interactions to target NO delivery to specific points in the body.

RN(piperazine)N-N(→O)=N-OCH_2OCH_3

4: R = H-
5: R = $HSCH_2CH_2$-
6: R = BSA-C(=O)$CH_2CH_2CH_2$-N(succinimide)-SCH_2CH_2-

REFERENCES

1. Keefer, L.K., Christodoulou, D., Dunams, T.M, *et al.* (1994) Am. Chem. Soc. Symposium Ser. **553**, 136-146
2. Hanson, S.R., Hutsell, T.C., Keefer, L.K., Mooradian, D.L., and Smith, D.J. (1995) Adv. Pharmacol. **34**, in press
3. Saavedra, J.E., Dunams, T.M., Flippen-Anderson, J.L., and Keefer, L.K. (1992) J. Org. Chem. **57**, 6134-6138

Subject index